Advances in Intellig

Volume 1363

The series "Advances in Intelligent Systems and Computing" contains publications on theory, applications, and design methods of Intelligent Systems and Intelligent Computing. Virtually all disciplines such as engineering, natural sciences, computer and information science, ICT, economics, business, e-commerce, environment, healthcare, life science are covered. The list of topics spans all the areas of modern intelligent systems and computing such as: computational intelligence, soft computing including neural networks, fuzzy systems, evolutionary computing and the fusion of these paradigms, social intelligence, ambient intelligence, computational neuroscience, artificial life, virtual worlds and society, cognitive science and systems, Perception and Vision, DNA and immune based systems, self-organizing and adaptive systems, e-Learning and teaching, human-centered and human-centric computing, recommender systems, intelligent control, robotics and mechatronics including human-machine teaming, knowledge-based paradigms, learning paradigms, machine ethics, intelligent data analysis, knowledge management, intelligent agents, intelligent decision making and support, intelligent network security, trust management, interactive entertainment, Web intelligence and multimedia.

The publications within "Advances in Intelligent Systems and Computing" are primarily proceedings of important conferences, symposia and congresses. They cover significant recent developments in the field, both of a foundational and applicable character. An important characteristic feature of the series is the short publication time and world-wide distribution. This permits a rapid and broad dissemination of research results.

Indexed by Scopus, DBLP, EI Compendex, INSPEC, WTI Frankfurt eG, zbMATH, Japanese Science and Technology Agency (JST), SCImago.

All books published in the series are submitted for consideration in Web of Science.

More information about this series at http://www.springer.com/series/11156

Kohei Arai
Editor

Advances in Information and Communication

Proceedings of the 2021 Future of Information and Communication Conference (FICC), Volume 1

Springer

Editor
Kohei Arai
Faculty of Science and Engineering
Saga University
Saga, Japan

ISSN 2194-5357 ISSN 2194-5365 (electronic)
Advances in Intelligent Systems and Computing
ISBN 978-3-030-73099-4 ISBN 978-3-030-73100-7 (eBook)
https://doi.org/10.1007/978-3-030-73100-7

This Springer imprint is published by the registered company Springer Nature Switzerland AG
The registered company address is: Gewerbestrasse 11, 6330 Cham, Switzerland

Preface

In these extraordinary and challenging times, it is our great pleasure to present you the proceedings of the Future of Information and Communication Conference (FICC) 2021, which was held virtually during 29 and 30 April 2021.

Over the past several decades, information and communication technologies have become more pervasive in the lives of people around the world and now they are being used in various facets of daily life. The future of these technologies will be shaped with multidisciplinary collaboration and applying new technologies. With this aim in mind, FICC provides a platform to discuss information and communication technologies with participants from around the globe, from both academia and industry. The success of this conference is reflected in the papers received, both topical and strategic, allowing a real exchange of experiences and ideas. The main topics of the conference include communication, security and privacy, networking, ambient intelligence, data science, and computing.

A total of 152 full papers (from initial 464 submitted papers) were presented at FICC 2021. Each submission has been double-blind peer-reviewed by two to four reviewers in the right area. The book chapters in this edition showcase new research, demos of new technologies, and poster presentations of late-breaking research results.

The conference also witnessed inspiring keynote speakers and moderated networking sessions for participants to explore and respond to big challenging questions. The conference success was due to the collective efforts of many people. Therefore, we would like to express our sincere gratitude to the technical program committee and the reviewers who helped ensure the quality of the papers as well as their invaluable input and advice and to the keynote speakers for their active and fruitful contribution. Also, special thanks go to Springer Publication for making the publication possible.

We hope that this publication can be insightful and learned by broader society and eventually giving benefits for today's development. We look forward to welcoming you all next year at the conference.

Regards,

Kohei Arai

Contents

Body Area Networks: A Data Sharing and Use Model Based on the Blackboard Architecture and Boundary Node Discovery

Jeremy Straub(✉)

Department of Computer Science, North Dakota State University, Fargo, ND 58102, USA
jeremy.straub@ndsu.edu

Abstract. Body area networks are growing in use and complexity, with many individuals carrying around several devices that have interconnection capabilities. Devices may offer internet connectivity, storage, media recording and playback, image or audio capture, and numerous other capabilities. Some may house recently captured data; others may house users' older data or have direct or indirect access to data from subscription providers. Users may enter into range of other providers who may provide a myriad of services ranging from information to environmental interaction. While limited interoperability standards (such as Bluetooth) exist, these are not full stack connection solutions. In many cases, users are left unable to access (or unsure as to how they access) services available from their devices or infrastructure providers. This paper presents a paradigm for data sharing and use as part of a Blackboard Architecture boundary node discovery (BABND) protocol for defining available data and other services. The basics of the BABND protocol are described and its use for data sharing is detailed. The efficacy of this protocol is evaluated through the consideration of case studies that are presented.

Keywords: Body area networks · Blackboard architecture · Boundary nodes · Service discovery · Data sharing

1 Introduction

The devices carried by individuals' everyday are growing in number and the functionality that they provide. A given person may regularly carry a collection of devices that includes a smart phone, smart watch, fitness tracker, headphones, a medical sensor, payment cards, a WiFi portable hotspot and one or more security devices. Individuals may temporarily carry a smart camera, hotel room keys, conference or event badges, sports or event tracking devices and any one of a plethora of Bluetooth-enabled peripherals. Further, they may interact with numerous infrastructure devices that are designed to connect with one of their devices to capture information or provide information. Some infrastructure devices may provide services that individuals interact with directly, potentially using data or other capabilities from a device or cloud service. Other infrastructure devices may provide the opportunity to interact with the user's environment; however, they may need to be connected to and commanded by one of the user's devices.

K. Arai (Ed.): FICC 2021, AISC 1363, pp. 1–14, 2021.
https://doi.org/10.1007/978-3-030-73100-7_1

Adding to the complexity of this ocean of data, services and devices is the fact that users may not know how or bother to properly configure or associate their devices and they may lose, break, replace or augment their devices. Malicious individuals may seek to use device interconnectivity as a mechanism to obtain access to data or capabilities that they should not have access to, either on a user's devices or on a service that the user subscribes to.

This paper uses the Blackboard Architecture, proposed by Hayes-Roth [1] for autonomous decision making, augmented by the concept of boundary nodes which were proposed, in [2], to act as an intermediary between different segments of a Blackboard Architecture network. It proposes to expand upon these as part of a solution to the device-service-capability discovery and interaction challenge, described above. A Blackboard Architecture boundary node discovery (BABND) protocol is presented and demonstrated for use in data discovery, sharing and use. It is evaluated through the use of several case studies that demonstrate its capabilities.

This paper continues with a discussion of prior work in two areas that provides a foundation for the work presented herein. Next, the Blackboard Architecture Boundary Node Discovery Protocol is presented and its data types, implementation and use for data searches are discussed. Then, two case studies are presented demonstrating the use of this protocol. Finally, the paper ends with a review of key conclusions and a discussion of future work.

2 Background

In this section, prior work in two areas, which provides a foundation for the work presented herein, is reviewed. First, prior work on the Blackboard Architecture is presented. Next, the concept of boundary nodes is discussed.

2.1 Blackboard Architecture

The Blackboard Architecture was proposed by Hayes-Roth in 1985 [1] and was based off of prior work on a system called Hearsay-II [3, 4]. It extends the concept of expert systems, which are commonly used for providing recommendations and decision support [5]. While expert systems have rule-fact networks, the Blackboard Architecture adds the concept of actions [6] which allow the system to effect its surrounding environment. For example, in [6], actions were used to task robots to perform additional data collection required for decision making, by tasking robots to move around the operating environment and collect data.

The Blackboard Architecture has been demonstrated for numerous uses which include the development of mathematical proofs [7] and software testing [8], in addition to the aforementioned use in robotics [6] and numerous uses in other areas.

A variety of distributed command concepts have been proposed, based on the Blackboard Architecture. These have included approaches based on using shared memory [9], centralized storage [10], a client-server model [11] and hierarchy-based systems [12].

2.2 Boundary Nodes

In [13], van Liere, et al. minimized the amount of data that needed to be moved between Blackboard Architecture systems through electing to only synchronize data that is used on target systems. In [2], the concept of using boundary nodes in Blackboard Architecture networks was introduced, based on the concept of boundary objects [14].

The approach proposed in [2] dictated that certain fact objects would be identified that serve as the interface between different areas of a Blackboard Architecture network. Aside from these interface nodes, the components of the network could operate separately as their own independent Blackboards, in most regards. An example of the use of boundary nodes, in this way, is presented in Fig. 1.

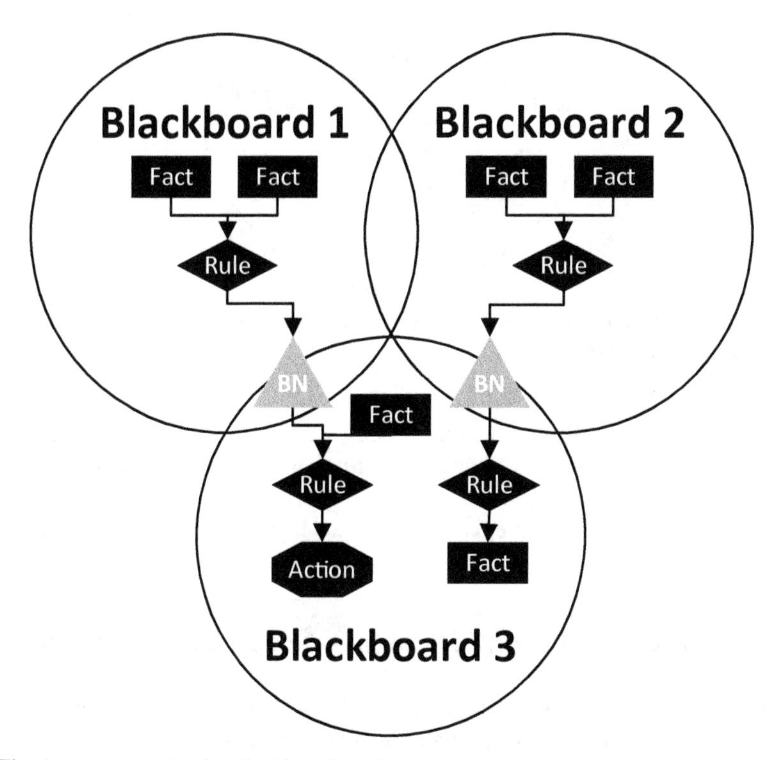

Fig. 1. An example of a Blackboard Architecture network utilizing boundary nodes.

Rules or actions could prospectively change these facts (rules through logical operations and actions through data collection) and the change in the fact's status would be replicated to other blackboards that contained the node. The remote blackboards could then use the facts as relevant data when processing their own rules and actions. While [2] did not demonstrate this concept, a boundary node fact could potentially span more than two networks. It would simply need to be replicated between all of the blackboards sharing it. Figure 2 shows how a boundary node fact could potentially connect three or more networks.

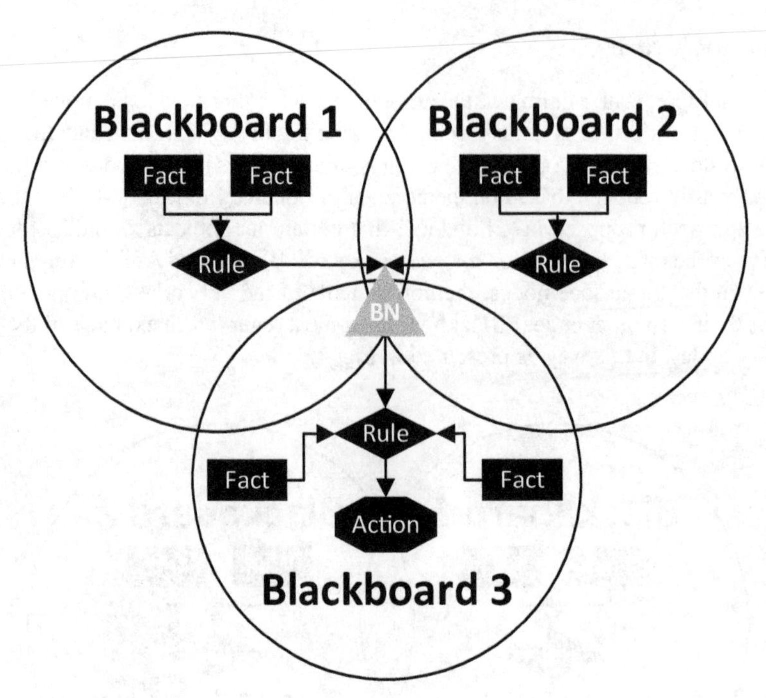

Fig. 2. An example of a Blackboard Architecture network utilizing a single boundary node to connect multiple blackboards.

In [2], several features of boundary nodes were described. These included non-directionality (i.e., they could be changed by either/any blackboard that they are associated with) and uniqueness within the system (each was assigned a GUID). Potentially, multiple boundary nodes could exist between two blackboards that might be changed by different rule-fact-action paths within the blackboard.

3 Blackboard Architecture Boundary Node Discovery Protocol

This section discusses the implementation of the BABND protocol, which utilizes boundary nodes as an interface between personal area network devices. A conceptual example of this is presented in Fig. 3, which shows three devices connected by boundary nodes.

3.1 BABND Protocol

The BABND protocol is based on a Blackboard Architecture network being implemented on each device (conceivably, boundary nodes could be utilized to connect Blackboard Architecture devices/nodes to non-Blackboard Architecture ones; however, this is left as a subject for future work). Each service that could be provided or consumed by a device is exposed as boundary node.

These nodes could, thus, take two forms. In the simpler case, they could be a data element. Examples of data could be transferred using this approach include text, pictures,

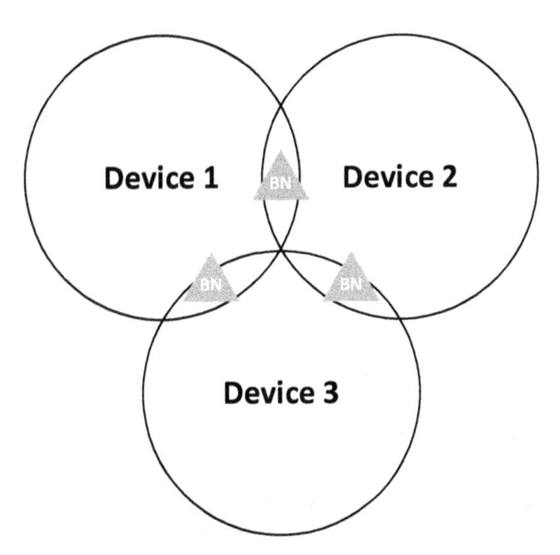

Fig. 3. Boundary nodes employed between devices.

pre-recorded videos and similar. In this case, the boundary node would have a data element that includes meta data and a loosely typed object data storage area. At a minimum, the node would need to include its GUID, the data type of the object that it stores and the object data. To facilitate replication, the date and time that it was last changed and the system that changed it should also be stored. These required fields are summarized in Table 1.

Table 1. Logical model for basic boundary node data object contents.

Field	Data contents
ID	GUID that uniquely identifies the node
DataType	Identifier for the type of data contained in the node. Standard MIME data types [15] (e.g., "text/plain") are used
DataObject	Data storage area, non-typed
LastEdit	Date/Time of last edit
LastEditor	GUID of system that last edited the object

The second form is a data stream. This would be used for content that is being produced or distributed in real time that cannot be stored and replicated. The primary differences between a data stream object and a basic object are the lack of the data object storage, the use of a streaming type for the DataType field (the basic octet-stream type [16] or similar could be used for content types that have not been registered with IANA [17]) and the incorporation of an IP address or DNS name and port to access the stream via. As the stream is being sent in real time, the LastEdit and LastEditor fields are not needed. Table 2 summarizes these required fields.

Table 2. Logical model for streaming type boundary node data object contents.

Field	Data contents
ID	GUID that uniquely identifies the node
DataType	Identifier for the type of data contained in the node. Standard MIME data types [17] (e.g., "application/octet-stream") are used
StreamLocation	IP address or DNS name and port

As a practical matter, two different storage mechanisms are not needed for these two logical data types. The DataType field indicates when a stream is present and, in this case, the DataObject field can simply be used to store the StreamLocation data.

A device, such as the smart phone (or internet connected tablet shown in Fig. 4) might have numerous services that it provides or consumes. In this case, it would create a separate boundary node for each. On the other hand, some devices may provide or consume a single service. Figure 5 depicts two examples of this: a treadmill that only

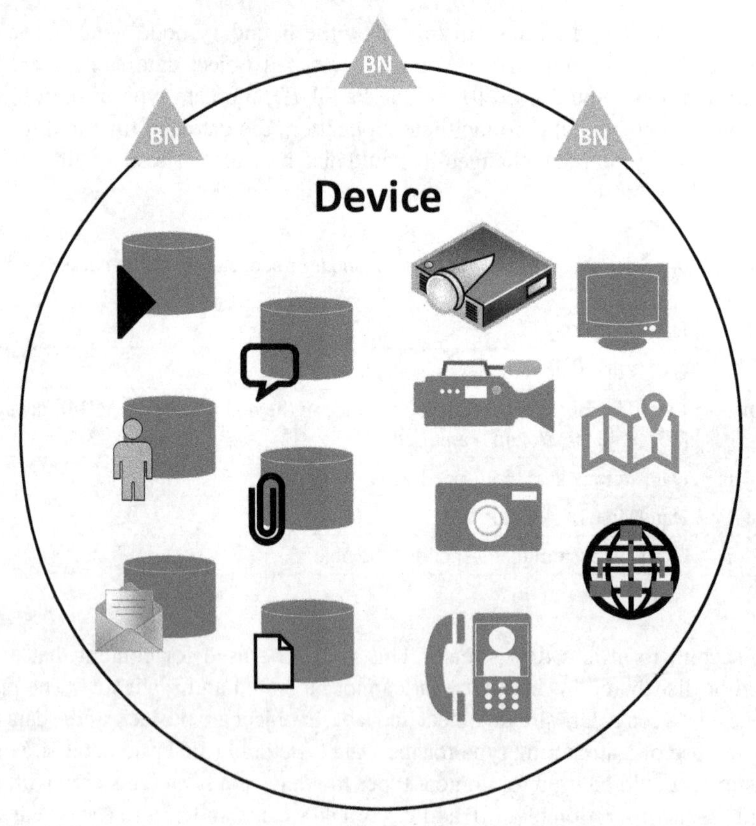

Fig. 4. Multi-function device showing the services potentially available via a smart phone or internet-connected tablet.

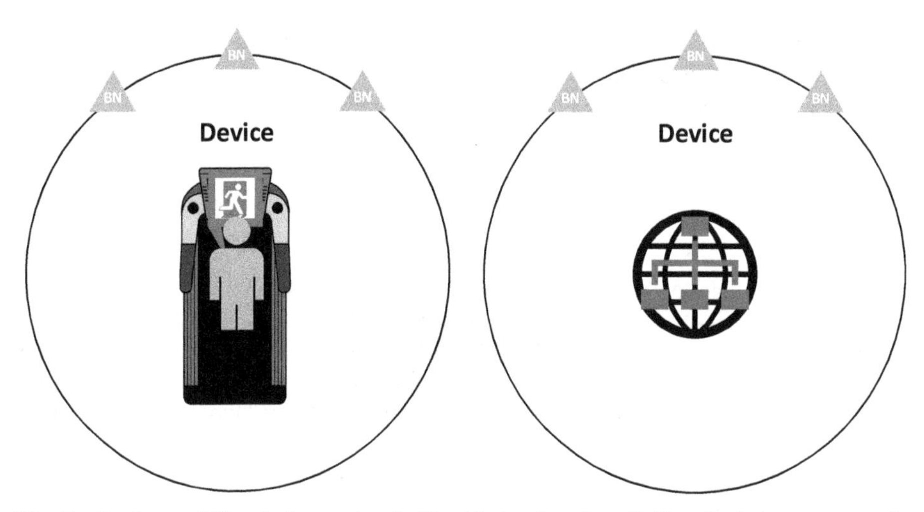

Fig. 5. Single capability devices: a treadmill with data interface (left) and wireless access point (right).

provides an exercise data service and a wireless access point that only provides that service.

3.2 BABND Protocol and Data Types

The proposed BABND protocol presents an additional challenge in that it uses a number of data types that are not conventionally used with expert systems or Blackboard Architecture implementations. A mechanism is needed to deal with these data types that is extensible. Four basic object types are used for this purpose.

The first two are facts and rules that are data type agnostic. These ObjFacts (object facts) have an identifier and are externally referenceable just as other facts are. However, they store a variety of different types of data (as described for the boundary facts, previously) and cannot be processed by normal rules. Notably, while boundary facts can contain stream data (to facilitate communications) normal facts (which are designed to store data) cannot and are limited to storable data types. ObjRules are used to process ObjFacts (and can also work with normal facts as well). ObjRules utilize custom code to process ObjFacts and can update normal facts and ObjFacts when they trigger. They can also trigger any type of action.

In addition, to support ObjFacts, two new types of actions are included. ObjConvertActions are used to convert between ObjFact data types. Again, these are custom code and can be triggered by normal rules or ObjRules. ObjConvertActions can convert between different data types of ObjFacts or between normal facts and ObjFacts. Notably ObjConvertActions can be used to convert a boundary fact's streaming data to the storable data required for normal facts.

A second type of action, the ObjAction, is used to perform actions that return data that is stored in an ObjFact (including data that will be streamed by a boundary fact).

Aside from their capability to work with non-standard data types, ObjFacts, ObjActions and ObjRules function in the Blackboard network like other facts, actions and rules and can be used to develop application logic that includes these non-standard (as well as conventional) data types.

3.3 BABND Implementation Example

A simple example illustrates how BABND could be used to consume a data storage service, which is the focus of this paper. In this case, a single boundary node is located between the consuming device (it will be assumed that this is a smart phone however, the exact details of this are not significant for the purposes of this example) and the data provider (which will be assumed to be an exercise tracking device). Only the relevant aspects of the two devices are illustrated, for purposes of understandability, in Fig. 6.

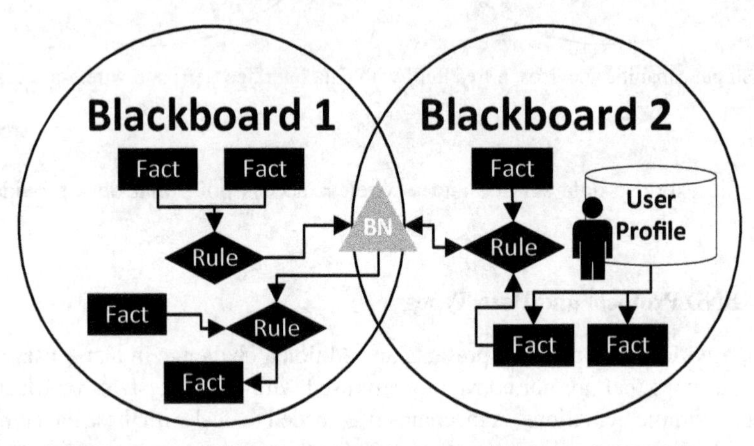

Fig. 6. Consumption of a user profile data service from a PAN device.

Because software on the smart phone is designed to connect to this device, it has a boundary node that encapsulates the data that it seeks to consumer from or update in the data service on the fitness tracker device. Notably, BABND doesn't handle the physical (wireless) network connection between the two devices. However, once the devices have this connection, it operates over it to locate, transmit and apply business logic to the data. In the example shown in Fig. 6, the profile data can be updated by the phone device (i.e., as part of a configuration protocol). If new data is received, a rule is triggered to apply business logic to the data change request before it is stored in the user profile (the storage aspect is not depicted in the figure). Alternately, if the profile data is changed in another way (i.e., via a desktop computer or another mobile device) the device will update the boundary node with this information, notifying the phone, which applies its own business logic rule to this.

3.4 Searching for Data with BABND

Up until this point, the connections discussed and the approach being used for connecting has been based on identifying the boundary nodes via their GUID. While this works well for an established network, it doesn't solve the problem of service and data discovery and making the initial connection. To support this, one additional change is needed to the boundary node data object: a service/data descriptor is added. The final form of the boundary node data object is presented in Table 3.

Table 3. Boundary node data object contents.

Field	Data contents
ID	GUID that uniquely identifies the node
DataType	Identifier for the type of data contained in the node. Standard MIME data types [15] (e.g., "text/plain") are used. Streaming MIME types are used to signal that the DataObject field contains connection details
DataObject	Data storage area, non-typed. For streaming DataTypes, this field is used to store the IP address or DNS name and port number of the stream
ServiceDataID	GUID that uniquely identifies a particular data type or service or type of data/service
LastEdit	Date/Time of last edit
LastEditor	GUID of system that last edited the object

The ServiceDataID field contains a GUID that either uniquely identifies a specific format of data (for example, the exact data sent by a particular model of fitness monitoring equipment or a data format shared between several similar models) or a type of data that can be met by multiple specific formats. The ServiceDataIDs inherit from their parent

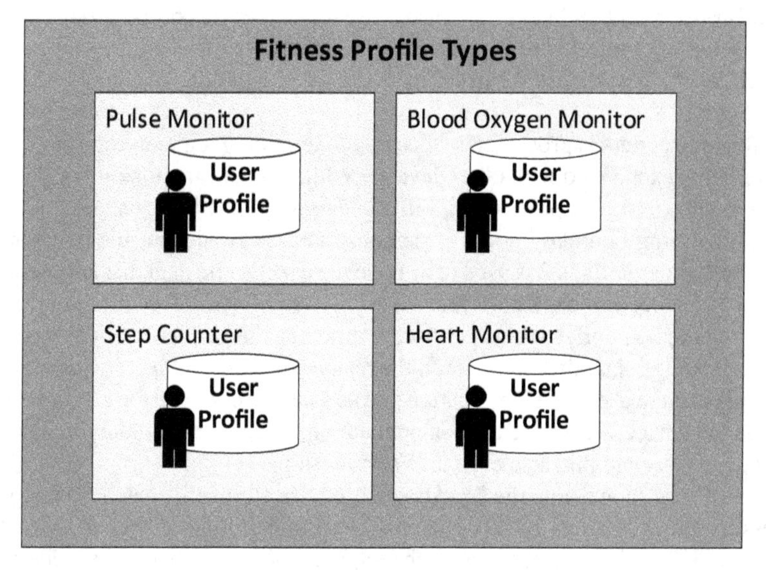

Fig. 7. Inheritance of ServiceDataIDs.

ID types, as shown in Fig. 7, thus a high level type of data (fitness profile types) might have multiple subtypes (e.g., for heart monitors or blood oxygen meters) and these subtypes may have multiple device or device collection-specific formats. The more specific formats must incorporate the fields required by their parent formats, but can add additional fields to their specific format to suite device or device type-specific needs.

When searching for data, the system can look for boundary nodes (on all connected devices) that match a particular ServiceDataID. If the ServiceDataID is a specific type for a specific device (or group of closely related devices), only these nodes would be located. If looking for data of a particular type (e.g., heart monitor data), the more general ServiceDataID for this category could be used in the search instead.

It is important to note that each device and consuming piece of software would have their own security policies and business logic. Thus, for services or data types requiring authentication beyond that needed to connect the devices, the boundary node would need to be defined (using an appropriate ServiceData format associated with a ServiceDataID) to provide these credentials or other information.

4 Case Studies and BABND

Focus now turns to two case studies that illustrate the efficacy of the BABND protocol. The first builds on the previously described fitness data collectors and adds a storage location and wireless transmission to this model. The second shows how BABND could be used to capture streaming video data and use it as part of a decision support system.

4.1 Fitness Data

In the first case study, the fitness data example is expanded somewhat. For this case study, it is presumed that blackboard 1 is located on a general purpose device (such as a tablet) which is connected to the fitness device (blackboard 2), an internet connection (provided by the blackboard 4 device) and a storage location (blackboard 3). This is depicted in Fig. 8.

As illustrated, data is processed by business/operational logic on each device before and after being exchanged with other devices. While not all functionality of this type of a system is shown (for purposes of legibility and brevity), the basic encapsulation of the relevant data using boundary nodes is demonstrated. In particular, the boundary node between blackboards 1 and 2 is used to provide the user profile data that is generated and updated by a collection capability present within the system. This profile information includes data that is collected in real-time (while that functionality is enabled and the user is exercising), data that is based on the analysis of this and historic data and basic user details that are used for identification purposes. If access to historical (or other) data stored on the device was needed for an application, a separate boundary node could be created to encapsulate this data exchange relationship.

Figure 8 also shows how the blackboard 1 device stores this data (after processing business/operational rules) on a connected persistent storage device (blackboard 3). The connection to the internet is also shown, embodied as a boundary node between blackboards 1 and 4. In this case, this connection is used for the storage of basic fact

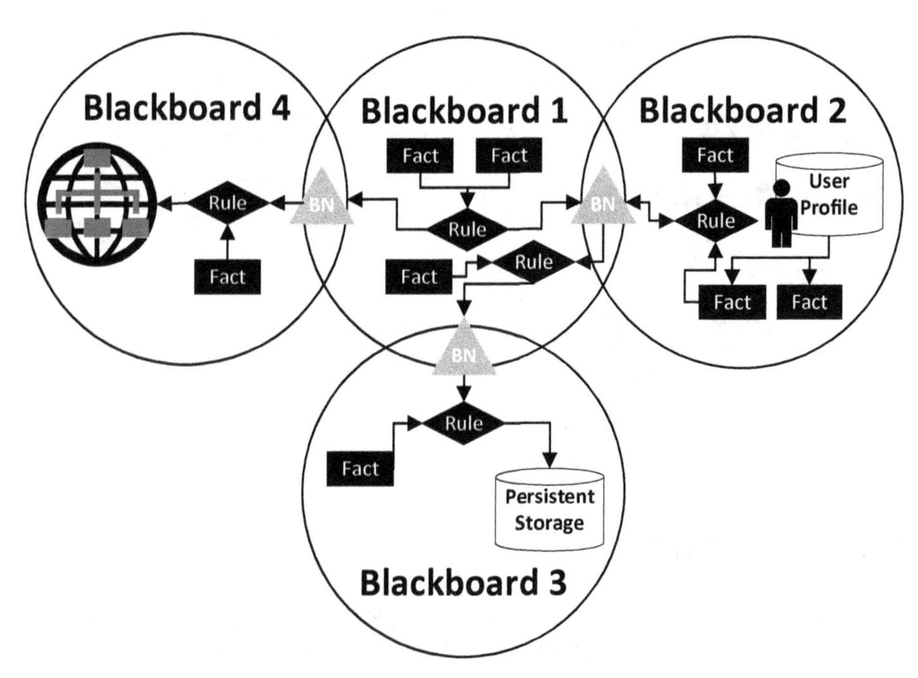

Fig. 8. Fitness data case study diagram.

data using a well-defined interface between the two devices (such as might be used for storing the tablet's configuration data or other similar data). If a greater level of data transfer was needed, a stream connection could be used instead. The use of such a connection is described in the subsequent section.

4.2 Streaming Data and Processing

The second case study involves the use of an ObjConvertAction to make streamed data (in this example, video) available to the more general Blackboard Architecture member rules and facts for use. This is depicted in Fig. 9.

In this case, the streaming data is provided by Blackboard 2 after performing its business/operational logic (in this case, to determine that the blackboard 1 device is authorized to access the stream). Note that, as discussed previously, the stream data does not flow through the rule shown in blackboard 2, it just passes the stream details through.

The stream connection details pass through the boundary node between blackboard 2 and blackboard 1 and are received by the ObjConvertAction, which opens the stream and uses its custom code to process the incoming stream data. In this case, it the custom code provides a transcription of the audio included with the video. This data is stored in several facts, based on business logic embedded in the ObjConvertAction's custom code (which separates the program into segments in the transcription). This data is then in a form that can be processed by the remainder of the rule/fact/action members of the Blackboard Architecture (as is shown by its use in another rule within blackboard 1).

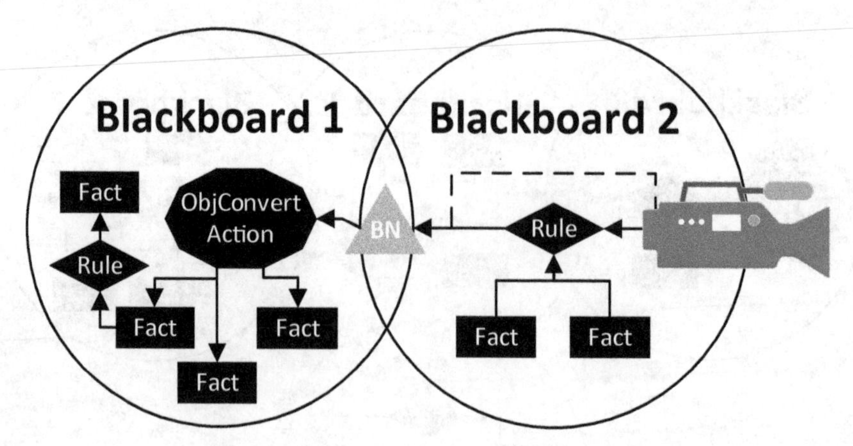

Fig. 9. Streaming data case study diagram.

5 Efficacy of BABND

The proposed BABND protocol adds some complexity to the provisioning of basic services. If all that two devices are interconnecting for is making a video stream available (or similar), as in the most basic concept behind the second case study, there may be little point for using the BABND protocol for this type of interconnection. Where BABND adds value is when the data must be discovered or must be stored or used amongst several devices. Its value is particularly pronounced when the data product from processing on one device is used as an input to processing, decision making or other activity (e.g., display to the user) on another device. In these cases, the BABND protocol implements business logic making only the appropriate data products available for each use.

To this end, many peripheral devices could implement BABND and other more limited protocols as well. BABND would be used for complex connections as part of processing workflows and in situations where multiple data producer and consumer nodes exist as part of a process. More basic protocols could be used when all that is required is the movement of data (either of a streaming or non-streaming type) between devices.

The concept of having multiple protocols supported by devices is not new. In fact, many devices support multiple protocols. Most Wi-Fi devices, for example, support more than one of the 802.11 sub-standards (e.g., supporting both '802.11b' and '802.11n'). It may seem like BABND would be needed and could, thus, only be implemented on higher-end devices that participate in complex workflows. However, this would potentially limit the participation of highly capable 'consumer grade' devices in consumers' everyday activities. Consumers may desire to listen to, watch and read (or some subset of these) incoming information while also having it concurrently processed by their device to enable other related functionality.

Thus, while device manufacturers must identify features to add to more advanced devices to differentiate them from lower-cost, lower-end devices, BABND could be implemented across all levels of device, providing interoperability. The capability limitation, for differentiation purposes, could instead be the services that are exposed by the

BABND protocol and the underlying data collection, storage and processing capabilities of the device.

6 Conclusions and Future Work

This paper has shown how a Blackboard Architecture network can be used to support interconnection between devices on a personal area network. The use of boundary nodes to encapsulate data that is shared and transferred between devices has been demonstrated. The use of boundary nodes to facilitate the transfer of stream data has also been discussed. Additionally, new action types have been developed that are designed to specifically process this streaming data and make it available (in a non-streaming form) for use by other nodes (with conventional capabilities) with a Blackboard Architecture network.

This paper has focused on the description of the BABND protocol and the consideration of its potential uses and utility. To this end, several case studies have been presented and considered. However, this work does not address all of the practical and logistical implications of the actual implementation of the BABND protocol. This is left as a topic for future work.

The work described herein also did not deal with circumstances where some connected devices do not implement the BABND protocol and how data exchange and service provision/consumption could occur with a combination of BABND and non-BABND devices. This also remains a key topic for future work. The exploration and resolution of this is integral to the potential utility of BABND as part of a body area network implementation that does not necessitate all connected devices to be part of a single proprietary or non-interoperable standard ecosystem. Bridging mechanisms between BABND and capability-specialized and general-purpose protocols would enable interoperability, though some capabilities may not be available outside the BABND network. Additionally, the characterization of the performance of the BABND protocol for the multitude of its prospective applications is planned as future work.

References

1. Hayes-Roth, B.: A blackboard architecture for control. Artif. Intell. **26**, 251–321 (1985)
2. Straub, J.: A distributed blackboard approach based upon a boundary node concept. J. Intell. Robot. Syst. **82**, 467–478 (2016). https://doi.org/10.1007/s10846-015-0275-2
3. Lesser, V., Fennell, R., Erman, L., Reddy, D.: Organization of the HEARSAY II speech understanding system. Acoust. Speech Sig. Process. IEEE Trans. **23**, 11–24 (1975)
4. Erman, L.D., Hayes-Roth, F., Lesser, V.R., Reddy, D.R.: The Hearsay-II speech-understanding system: Integrating knowledge to resolve uncertainty. ACM Comput. Surv. **12**, 213–253 (1980)
5. Turban, E., Watkins, P.R.: Integrating expert systems and decision support systems. MIS Q. Manag. Inf. Syst. **10**, 121–136 (1986). https://doi.org/10.2307/249031
6. Straub, J., Reza, H.: A Blackboard-style decision-making system for multi-tier craft control and its evaluation. J. Exp. Theor. Artif. Intell. **27** (2015) https://doi.org/10.1080/0952813X.2015.1020569
7. Benzmüller, C., Sorge, V.: A blackboard architecture for guiding interactive proofs. In: Artificial Intelligence: Methodology, Systems, and Applications, pp. 102–114. Springer (1998)

8. Chu, H.-D.: A blackboard-based decision support framework for testing client/server applications. In: 2012 Third World Congress on Software Engineering (WCSE), pp 131–135. IEEE (2012)
9. Compatangelo, E., Vasconcelos, W., Scharlau, B.: The ontology versioning manifold at its genesis: a distributed blackboard architecture for reasoning with and about ontology versions. Technical report (2004)
10. Kerminen, A., Jokinen, K.: Distributed dialogue management in a blackboard architecture. In: Proceedings of the Workshop on Dialogue Systems: Interaction, Adaptation and Styles of Management. 10th Conference of the EACL (2003)
11. Tait, R.J., Schaefer, G., Hopgood, A.A., Nolle, L.: Automated visual inspection using a distributed blackboard architecture. Int. J. Simul. Man Cybern. **7**, 12–20 (2006)
12. Weiss, M., Stetter, F.: A hierarchical blackboard architecture for distributed AI systems. In: Proceedings Fourth International Conference on Software Engineering and Knowledge Engineering, pp 349–355. IEEE (1992)
13. van Liere, R., Harkes, J., De Leeuw, W.: A distributed blackboard architecture for interactive data visualization. In: Proceedings of the Conference on Visualization'98, pp 225–231. IEEE Computer Society Press (1998)
14. Star, S.L.: This is not a boundary object: reflections on the origin of a concept. Sci. Technol. Hum. Values **35**, 601–617 (2010)
15. Common MIME types. In: Mozilla Developer Network (2020). https://developer.mozilla.org/en-US/docs/Web/HTTP/Basics_of_HTTP/MIME_types/Common_types. Accessed 15 Jul 2020
16. Octet-Stream Media Type. In: IANA Website (1996). https://www.iana.org/assignments/media-types/application/octet-stream. Accessed 15 Jul 2020
17. Media Types. In: IANA Website (2020). https://www.iana.org/assignments/media-types/media-types.xhtml. Accessed 15 Jul 2020

In the Footsteps of Pierre Duhem: How a Modern Theory of Value Relates to XIX Century Physics

William M. Saade(✉)

Formerly with Stanford Research Institute, 333, Ravenswood Ave, Menlo Park, CA 94305, USA

Abstract. Having in the last 40 years focused on a function generated from a theory of morphogenesis called Elementary Catastrophe Theory (ECT) [19], the Swallowtail, to show its fitting the most respected landmark criteria of Utility or Value in Economics I felt the need to explain the resolution of the controversy surrounding the shape of this function via a polynomial solution against the trend of mainstream literature [10]. A little noticed theorem dating back to 1963 by Rene Thom, the author of ECT [22], set the ground for the development of ECT while linking it to the thermodynamic Potentials (Entropy and Internal Energy) as the XIX century ended up characterizing them with an exclusive feature: Differentiability [2]. The importance of the synthesis attempted by P. Duhem comes from not assuming any atomism without precluding it while seeking a unified field theory between gravitational, electrical, magnetic and chemical theories based on Newton law as reinforced with Coulomb law for electrical charges [26]. One of my basic finding was the negativity of the Schwarz' derivative the highest order (the third) differential invariant coming from the heart of Pratt's paper on risk in Economics [31] as a characteristic displacing concavity. My result so far has been that only for a polynomial representation we have a theory of self-observation fitting uniquely with the procedure of Stimulus-Response known since 150 years under the title of Psycho-Physics and opening up vistas to the world of Linguistics [32].

Keywords: Entropy · Potential thermodynamics · Differentiability

1 Introduction

Having followed the advice of H. Poincare I could indeed grasp the gist of such a solution, polynomial, in my naive indifference to the surrounding controversy in a field so loaded with taboos, pre-judgments and emotions: the human preferences. This came from my attraction to the beauty of a symmetrical form before making any rational

Since it's a survey and review paper I am following Sir Peter Medawar suggestion to present the interactions of people and ideas in producing results, a kind of extended mind storming. The synthesis achieved brings a new paradigm completely substituting to the expected utility hypothesis already outliving its usefulness for several generations.

K. Arai (Ed.): FICC 2021, AISC 1363, pp. 15–27, 2021.
https://doi.org/10.1007/978-3-030-73100-7_2

sense of it except to its intimate relationship to the three dimensionality of the physical space. My only guiding light was that only on the catastrophic set that function becomes non-decreasing and that feature was intimately linked to the basic assumption about the underlying physiology [1], a dynamic of gradient of a potential. Together nevertheless with my mastering of the desiderata of Control Theory: Observability, Reachability, and Controllability. The next step was to test its fitting the famous Arrow Impossibility Theorem for the aggregation of preferences [33]. My task was both simple and complicated: simple because I was dealing with a specific cardinal functional where the criterion for minimum impossibility was clear: One had to disaggregate the representative functional in a unique fashion into 3 components one real and 2 complex conjugates. Complicated because so far the literature dealt solely, until today, with ordinal axioms with which I was not familiar to say the least. Next came under the surprising result of S. Ross [25] with a new classification of risk aversion from a topological perspective fitting the approach of ECT and generalizing the concept of uncertainty as first introduced in Pratt's paper.

As soon as I verified that my functional "behaved" nicely following S. Ross criterion I sent my first paper "the essential tension " to Pr M. Allais in Paris whom I thought to be a brilliant psychologist having set up a double experiment in 1952 colloquium putting all the participants in contradiction with their own axioms for choices under uncertainty and promoting a polynomial representation contrary to the wishes of the majority. Little I knew that Pr Allais was graduated on top of his class from one of the most prestigious mathematical institutes in France, Ecole Polytechnique, recipient of the golden medal after 50 years of achievements combining theoretical and empirical findings, a rarity in such a field. The same Pr Allais was the only experimenter of a XIX century relic called the Conical Pendulum and the NASA set up an experiment with its satellite observing the Allais Effect during the 1999 solar eclipse. Indeed, he had voluminous set of data proving the non-uniformity of the speed of light in all directions as assumed by the theory of Relativity. He immediately gave me his unreserved support and was the first to put me on the tracks of Psycho-Physics [30]. Thus Pr Arrow became my thesis supervisor and confirmed experimentally via Pr Ward Edwards the precise fitness of my quintic polynomial. Later on, at one of those Conference meetings I met Pr Edwards and asked him point blank about his verification of the quintic polynomial. He answered with another question: "Why would Pr Arrow tell YOU that?"

Little did he know that I was with Pr Arrow in his room when he called him to tell about his findings! Then it hit me like a rock remembering what Pr Aumann comment had been: "You have destroyed everything they built since WWII". I was facing a cover-up at the hand of people who had sworn that such a function does not exist because it could not while all the same continuing at teaching, publishing and building careers as if it did exist "not even in Plato's heaven". One could understand better the situation if one had lived through the American educational system in Economics where literally speaking one had, before starting an essay putting his(her) name, the title, the date, one had to state the act of faith in concavity. This reminded me when in religious high school one had to state "God is watching me" to ascertain that one is being truthful and not cheating before starting any essay. To make matters surreal you could find tee-shirts with slogans like" at Harvard they found (such and such), at Stanford we prove the impossible". In the hard sciences like physics, one such instance happened with Einstein when he built on the phenomenon of non-additivity of speed to imagine a new framework made by Minkowski with a new algebra and a new rule of addition other than the arithmetic one. Otherwise there would not have been a theory of Special relativity hence no Einstein. So impossibility was only the beginning of a process of creativity and not the conclusion. Digging further one could find the twist that such an impossibility could serve an ideological purpose. Indeed, it was used to prove that the whole system of representation in Parliament based on one man one vote in western style democracies was baseless and bound to be replaced by a "benevolent" dictator. The shortest proof I found in the annual calendar of the year 1982 produced annually by Springer-Verlag, which titled "The dictator Theorem" for that year. Every page you flipped there was a series of titles of theorems dealing with the impossibility of aggregation. Perhaps even until today you could still find publications of that sort.

2 The Bifurcation Year 1963

I went to the conference of Oslo in 1982 to present my results with the sponsorship of Pr Allais to the dislike of Pr Arrow who was sponsoring the Prospect Theory of Pr Tversky who never showed up. Indeed there was nothing to show, no function, no new arguments, just a repetition of a set of beliefs already dealt with in my first paper" The essential tension" together with a comparison with G.L. Schackle [3]: to be specific, Prospect Theory confuse reference point and origin of the coordinates 'axis, more importantly it assume the existence of super-additive weights nobody knows where they are coming from. I could still remember that a year earlier, just after verifying my quintic with Pr Ward Edwards, Pr Arrow seemed shaken then introduced me to Pr Tversky while qualifying him with words I will not repeat here by sheer respect for the dead. Stunned by the turn around I remembered that the famous Arrow-Pratt risk premium were coming from a 1963 paper by Pr Pratt then at Harvard with Pr Arrow which became a seminal work dealing with the shape of the utility function in presence of uncertainty. The singular feature was that Pr Allais had already acknowledged the contribution of the paper because its approach, driving the probability function to zero, within the framework of expected utility, would leave the characteristics of utility untainted by probability. It is a common procedure in applied physics where you can only gather data on two inseparable ingredients taken together. Then you perform a "gedanken experiment" where you imagine a situation where you can drive one of the ingredient to zero in order to collect the features of the remaining one. The curious thing, though, was that Pr Allais was adamantly opposed to the very conclusion of the Pratt's paper proving apparently the concavity of the utility function. This position was coherent with Pr Allais promoting a polynomial representation since World War II as made clear with the Allais so called paradox of 1952 where his purpose was to prove that for large sums and small probability the utility cannot be concave. During my presentation at the same conference at Oslo Pr Gunther Menges [4] interrupted me to claim, when I was explaining the minimax phenomenon giving rise to the non-decreasing feature of the 5th degree polynomial, that this should resolve the quantum riddle. It made sense to me since this showed that the abscissa where the horizontal inflexion point appeared was coming from the control space and was parametrizing it almost completely. Pr H. Skala, at the same conference, whispered that I was dealing with a super filter. Could this be the "hidden" variable that D. Bohm was talking about in his formulation? Coming from the editor of Statistische Hefte, promoter of the concept of Etiality, a view of a thermodynamical equilibrium of the human body, this constituted indeed an endorsement. Sure enough, Pr Menges proposed to publish my work in his review together with the suggestion of a new title" A synthetic approach to Value as a standard of reference" [28, 29]. Unfortunately, he died that same year before we finalized the paper.

When I went back to Pratt's paper my mathematics from the level of high school led directly and rigorously to a polynomial solution, the only stable one instead of a concave one. When one goes back to the logic leading to concavity one is struck with the irrationality of questions begging the answer. More importantly with time passing and the same teaching going on one cannot escape the conclusion that such an outcome was crucial to sustain the very foundation of the Arrow-Pratt general equilibrium model since Pr M. Allais who was supervising Pr G. Debreu thesis approved it with his utmost reservation about the concavity hypothesis, just to acknowledge the beauty of the mathematical construction. But the Emperor had no clothes…

Only recently, 6 months ago, I received the vol. II of R. Thom mathematical works where I discovered the 1963 paper proving the equivalence between differentiable functions and polynomials up to a diffeomorphism. Diffeomorphism is a smooth and reversible transformation. Smooth is a function whose successive derivatives up to a certain point are continuous. Reversible is a characteristic of stability of equilibrium which bring back to me remembrances of Irreversible Physics as in I. Prigogine who acknowledged Pierre Duhem credit in his inspiration about the subject. In a famous debate about determinism in 1990 [18] R. Thom with a distinguished set of famous scholars among them I. Prigogine, Fields Medalists or Nobel Prize recipients, debated at length about the subject. Mr Jean Petitot [6–8] who gave me the hint about the vol. II of Thom works, last X'mas, was present and active. His contributions in the field of linguistics stemming from ECT are numerous and continuous for more than 40 years. This subject will not be dealt with here for short space. Interestingly 1963 was the year when R. Thom reached the age of 40, meaning all the achievements which led to his Fields medal were behind him, i.e. the equivalence between Differentiability and Polynomials was the crowning of his innovation in biology and the stepping stone towards further progress.

Going back to differentiable functions I learned that toward the end of XIX century, Physics had reached a turning point where the innovation was not coming from a basic assumption about matter like at the time of Newton, instead all new entities called State Functionals were derived from purely logical and rigorous mathematical developments assumed to exist at a certain SCALE of matter and only justified by empirical observations after the fact, like in changes of phases or states of matter [2].

So now we understand why R. Thom dealing with biology, was to clash with the mainstream approach to explain things in biology through chemical reactions and was calling from all his strength for an experiment where his ECT would reach a result impossible for chemical reactions to dream of. He was even proposing that one should use ECT as a given in a certain circumstance and draw the consequences. It was to imagine a heroic assumption which I adopted in going back to the heart of Psycho-Physics and its famous logarithmic shape coinciding with Entropy, the dear symbol of theoretical Economists to ground their field in the prestigious aura of Theoretical Physics. This paradigm was not only going back to the XIX century. It was actually and unashamedly "borrowed" from D. Bernouilli 3 hundred years ago as his reaction to the St Petersbourg gambler behavior and written in Latin [27], when the concept of polynomial didn't exist yet and had to wait another century before Galois showed up on the scene. Curiously no scholar had any reservations about this choice since entropy was and is still considered a measure of disorder when every beginner knows that human beings are an island of "Negentropy" a symbolic denial of entropy as measuring decay and death. But of course the main interest of Entropy was its concave shape to suit the purpose of its proponents. More on this in the annex II under the heading of "The missing link"

In any case I was not asking for that much when I first simply took the gradient of such a potential as the basis of my investigation. My attraction was that I was dealing with a world of FORMS long neglected since Aristotle because it needed a special kind of mathematics and approach only made available in the sixties of last century and being misused, to my sense, by the 90% of the practitioners adopting the quartic polynomial i.e. the 2 dimensional space of Control in ECT.

The picture I drew was that a new body, like a planet, the control variable, was introduced from the outset putting order in the trajectories of surrounding planets left unexplained without it. The incredible conclusion was that my quintic could have been guessed at a century earlier with the available logic without any theory of observation [32] but the handicap was a lack of empirical data due to the limitations of the stimuli-response procedure to the positive quadrant. More telling was the attempt at Stanford in 1961 to match the data with the cubic form which obviously failed. Is this what triggered the distortion of the Taylor series expansion of the Pratt's approach in order to force the concavity assumption 2 years later? In any case, it showed the tatonnement process at its best.

3 Negative Schwarz' Derivative (or Schwarzian)

My conclusion coming from the Pratt's approach displaces concavity with a much more sophisticated entity the "Schwarz' Derivative Negative" [26, 31]. It's well known from the literature that this is a differential invariant of the highest order, the third, and that it defines the limit of what's known as "chaos" in dynamical systems. Actually, as showed I proved a stronger differential inequality which lead directly to the Negative Schwarzian. The invariance of the Schwarzian is up to substitution of the main variable with a hyperbolic transformation. Indeed, the positivity of Schwarz' Derivative is a sure sign of bizarre behavior. I was, later on, comforted to learn that only at that level did the concept of control start to take hold. Moreover, it's so intimately linked to the polynomial forms that I developed an algorithm finding the real roots of polynomials of any degree based solely on the necessary condition of negative Schwarzian for convergence of the algorithm [27]. The intuitive basis of substituting Negative Schwarzian to concavity goes back to R. May who first did it in his studies on population dynamics. The whole concepts go back to the theoretical findings of Julia and Fatou around a century ago. In order to understand the upheaval one has to remember that concavity was the basis of the whole model of General Equilibrium in Economics as elaborated by G. Debreu and K. Arrow in the 2 nd half of last century. Of course there was a previous such model from Walras but that was in the XIX century. And there was another model made by Y. Balasko in 1978 [34] based on the same mathematics than ECT the one I was using for a unique individual as obviously required by its author.

4 Resonance

What kept me going during the last decades was of course the experimental verification done by both prestigious scholars like Pr Allais and Pr Arrow, although separately and for opposing purposes, but also the continuous development of related fields and the physical features that ECT unfolded like the structure of the synapse where this swallowtail was taking place, alternatively called a self-intersecting curve. The same structure, the swallowtail, was predicted to occur at the level of the human vision by the same theory (ECT) the whole transmitted through the brain by a network of 10^{14} synapses via a resonance phenomenon forecast by Poincare-Dulac theorem for the shape of forms around a critical point. With the added quality of being the locus of the interaction subject/object, where else could anyone hope to find a better locus to embody the identification of the Observer with the Observed? Just a reminder that at the beginning of the decade of the brain, "Brain project" at this side of the Atlantic had its twin brother on the other side "The Synapse project". Remembering that the deepest efforts to reconcile quantum theory and Relativity the last 4 decades led to a deadlock due to the dimensionality problem, it seems fit to propose that at the human level at least a new physics could emerge between the scale of the infinitely small and the one of the infinitely large filling a gap by being coherent with the double slit experiment at the synapse level and the three dimension feature of relativity theory. With a square integrable functional away from statistics, as the dearest wishes of Einstein and others would dream of, but also a shape where only the prime variables appear without additives except their exponents and distances, we do understand now the vitalist bias that such a theory would be inclined to, if we were to follow the ambition of P. Duhem [16–18] by replacing Newton law with our functional. This would not seem too ambitious though if one notice a unique feature of ECT that its structure, although coming from biology, is actually independent from its underlying substratum. What else could one had hoped in order to generalize one theory to every other field of matter? Finally staying in the area of Economics, one should mention the shape of the total cost function familiar with the logic uncovered in the preference function (Axiom of Non-satiation) to give a quantitative twist to a consultancy field dear to a maverick engineer by the name of Ralph Keeney advising governments and big corporations about allocation of resources [11].

5 Extensions and Conclusions

"The closest resemblance in this field to the concepts we have been putting together comes from Richard Jeffrey's "The Logic of Decision" [13]. Although he dismisses the St Petersburg game as a big lie, he clearly sees the need to define a scale with three points instead of two. His utilities, when referring to the lottery are unbounded. More importantly, he is dealing with a world of propositions giving a linguistic flavor to his endeavor. The method he adopts is not causal but constitute a variant from Ramsey's procedure to elicit probabilities from preferences' profiles:" [26].

"Here, the elementary logical operations on propositions (denial, conjunction, disjunction) do the work which is done by the operations of forming gambles in the "classical" theory of Ramsey and Savage…but here (see Sect. 6), the preference ranking of propositions determines the utility function only up to a fractional linear transformation with positive determinant…The classical case is obtained here if the preference ranking is of the sort that can only be represented by a utility function that is unbounded both above and below; and it is shown (Sect. 10) that the present theory is immune to the St Petersburg paradox, so that one can reasonably be a Bayesian in the present sense and still have an unbounded utility function."(Richard C. Jeffery in the preface of The Logic of Decision'1983).

One last remark in Economics should mention the extension of ECT to the impulse with two parameters: it produces what is called the Umbilics, in two different shapes. I chose one of them the one symmetric in the two variables on esthetic grounds but also because it was more fitting the empirical data, a bell shaped curve inclined −45 degrees asymptotically, that mainstream economics tried to hide because it didn't fit their well-known Cobb-Douglas curves, all based on speculative reasoning,

To finish off with the field of biology one should recall the active investigations of Rupert Sheldrake with his "Presence of the Past" and his latest version (2020) of his "Science Delusion" giving grounds to ECT anti atomistic conclusion and explaining phenomena impossible for materialistic physics to even handle [23].

Finally R. Thom had some drastic news for Neo-Darwinism all stemming from his choice of stability (structural) among half a dozen available at the time and putting in question the finality of the DNA structure in inheritance as other studies recently tended to show [21].

This will lead to the interest of R. Thom in a new reading of Aristotle as did almost all epistemologists of his tradition like P. Duhem [20]. A quick remark here that some famous economists like Pr Gerard Debreu and Yves Balasko are graduate from the same prestigious school combining Mathematics and Physics in Paris:

Ecole Normale Superieure.

For myself I am still holding the late Pr Arrow in great respect despite the tribulations surrounding our relationship. I still remember with fondness one instance when he was in good mood and pointed with his unique sense of humor to Sherlock Holmes: It is Elementary, my dear Watson" to catch the spirit & the letter of the moment. Indeed after reading my answers to all his questions, in a 42 pages document, he had left half of the blackboard filled with my handwriting for six months while promising me" You deserve an answer" I did not understand at the time that he was taking things personally nor that he was planning to ask Pr Ward Edwards for verification without my knowledge. Even after he told me in our last correspondence that Schwarz' Derivative was a complete unknown to him he managed to hold at Stanford an international conference about the subject in 1995. I can say that I held my promise at his insistence to not divulgate these souvenirs until he was gone. More importantly I am grateful that I was given the opportunity to rub shoulders with giants and that I could contribute with one theory to unify a whole field of such an importance to human endeavors in such a way reminding us with Occam's razor. Sufficient to say that the eminent scholar Pr Pierre Buser in his last book [15] dedicated a whole chapter on Psycho-Physics calling the XIX century logarithmic representation as the central problem of Neuro-Philosophy and yet still qualifying it as "subjective" as did many academicians of the same caliber. I took it to myself to prove it eminently "objective" by providing an exact unit of measurement for the human brain, taking into account our subjective factor, the control variable, that we don't need the schizophrenic approach any longer and that we could restore our self-respect by going back to the Greeks: not only Heraclitus and Aristotle but finally to Socrates "know thyself". The more gratifying was the statement of one specialist of ECT to the extent of qualifying my quintic of no use!

I wouldn't like to end this paragraph without a quote from Pierre Duhem summarizing his philosophy: "Logic can afford to be patient, because it is eternal".

Acknowledgments. I owe a great debt to the distinguished Pr Jean Petitot for informing me of the edition of Vol II of R. Thom Mathematical Works, this past X'mas. I wish also to address my warmest thanks to my dear friend Freddy Hasbani for his relentless efforts in obtaining and delivering the needed material on time in spite of Corona and in complete dedication of his person and time. The alumnus from my French undergraduate school, Mr Nagib Yahchouchi deserves my greatest thanks for his availability and his unexpected help with his masterly technical assistance with my laptop.

In the intellectual arena I wish to dedicate my work, among the living, to the unforgettable Pr Robert Aumann and our laughters at Stanford coffee store, and for among the departed to the living memory of Pr Maurice Allais with his encyclopedic knowledge and his dedicated search for the truth. I extend my thanks to his daughter Ms Christine Allais for her courtesy and hospitality, encouraging her to handle the M. Allais Foundation in the spirit of her departed extraordinary father and urging her to use part of my donation to repeat the experiment of Pr M. Allais 40 years ago, to either confirm my quintic or infirm it for the sake of progress of Science like Pr Allais would have done it.

Appendix

Annex I

The double slit experiment. In the small, the big surprise is that the very same function than for our visual system, the Swallowtail, is predicted to occur at the level of the synapse by the same theory (E.C.T.). But if one considers that resonance is really a two way street one could admit that it works in both directions up and down. We quote R. Thom directly: [19].

"I would not have carried these very hypothetical considerations so far if they did not give a good representation of the behavior of nervous activity in the nerve centers, where an excitation (called a stimulus in physiology) remains relatively canalized until it results in a well-defined motor reflex; here the role of diffusion seems to be strictly controlled, if not absent. It is known (the Tonusthal theorem of Uexkull) that, when the associated first reflex of a stimulus is inhibited by artificially preventing the movement, there is a second reflex which, if inhibited, leads to a third reflex, and so forth. This seems to suggest that diffusion of the excitation is, in fact, present, but that, as soon as the excitation find an exit in an effective reflex, all the excitation will be absorbed in the execution of this reflex. This gives a curious analogy with the mysterious phenomenon of the reduction of a wave packet in wave mechanics." pp. 149–150 [5].

Briefly, considering the general 5th degree polynomial, we see on its graph in the positive quadrant that a necessary condition for self-duality is that the local Max and Min, respectively Min and Max to each other, are confounded at the same point. It is an analog to von Neumann's Mini Max concept. With symmetry toward the origin this condition becomes a sufficient one (Fig. 1).

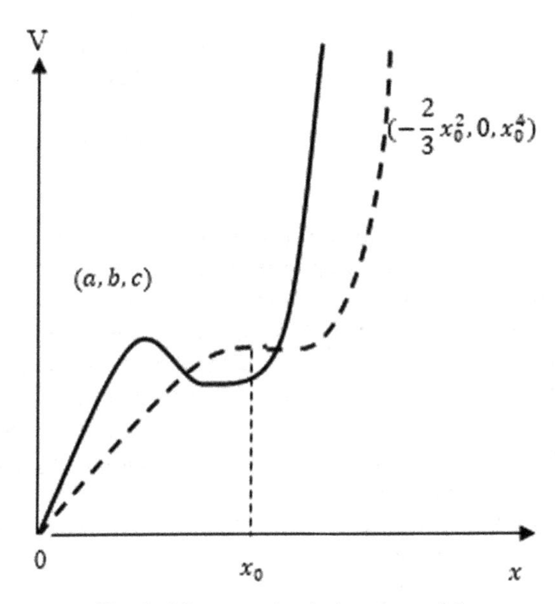

Fig. 1. The general quintic polynomial

$$V = \frac{1}{5}x^5 - \frac{2}{3}x_0^2 x^3 + x_0^4 x$$

Notice the start with the highest exponent stressing the fact of the Polynomial representation.

This was written over 50 years ago. By now many Quantum Scientists like Henry Stapp [5] firmly believe the following empirical facts related in Jose R. Dos Santos latest book" La clé de Salomon" pp. 452:

"In general, these quantum jumps are only possible in spaces whose width is equivalent to seven atoms. But in rare cases they could happen in widths reaching 180 atoms at a maximum. Well it so happen, by sheer coincidence or not, that the width of the synaptic slit is precisely 180 atoms. Since electrons are in constant movement, they could try billions of times to cross the synaptic tissue during the one thousandth of a second that the electrically polarized synapse takes to get activated. This will bring their rate of success in the quantum tunnel for that width to 50%. By studying with scrutiny the structure of a synapse one realized, again with a strange coincidence, that its architecture is perfectly suited to exploit an effect of quantum tunnel when an impulse arrive at a synapse, the slit becomes electrically polarized and this powerful electric field allows the quantum tunnel phenomenon. That is why one can assume that the wave function collapses in the synapses when a thought occurs, and that from this phenomenon consciousness emerges."

Annex II

The missing link to fulfill R. Thom last wish. "Going back to the last attempt by Pr Maurice Allais to axiomatize the utility function, I noticed 2 new axioms that Pr Allais added, after reviewing his 1952 experiments and analyzing the diverse reactions for over a generation, more precisely in "The expected utility hypotheses and the Allais paradox", 1979:

Axiom (VI): Axiom of invariance and homogeneity of the index of psychological value and

Axiom (VII): Axiom of cardinal isovariation." [14]

After analyzing two cases, the log linear and the non-log linear approximation one, he finds an excellent fit with a behavior verifying his axioms, up to approximation to the errors due to psychological introspection. In the meantime I sent my first paper, "the essential tension" seeking his feedback, thinking I was dealing with a prominent psychologist who set up the "Allais Paradox" at the Paris colloquium of 1952. He sent me, a few weeks later, the first English version, I was to learn later, of any of his works, the 1979 book with O. Hagen summarizing many centuries of dealings with the subject since before the Physiocrats. I learned later that Pr Allais, considering his disappointment from not having won the Nobel freshly set up in Economics in 1969, was considering emigrating to the USA! Indeed his own daughter told me later that he and his wife spent few months considering seriously the pros and cons of such a drastic move at the age of 60! Putting this in perspective one could understand the determination of such a character. Since he was finally awarded the Prize in 1988 for work done in 1943, he acknowledged his own handicaps. Indeed he was publishing only in the French language during and after WWII. Secondly all his memoirs were no less than 100 pages long. The conclusion was that all the intelligent people knowing French could plagiarize him at length during decades without being caught and without giving him any credit! Add to that the fact that he decided to redo his second preparatory year for entrance at the prestigious Ecole Polytechnique because although admitted from the first attempt he was not the "major", or the first of his class, which he ended up with again at the exit exam. When one considers the risk of not being accepted that the competitors face with the odds of 10 to one, one measure the caliber of such straight bullets!

So when he sent me the book he also wrote a short note: "To William M. Saade with my cordial homage" I was flabbergasted because I had not forgotten yet my French language. I asked myself "what did I do to deserve such an honor?" I shared the news with the guru of Decision Analysis at my department, E.E.S. The next thing I remember was Pr Arrow, from the Economics department, approaching me with the statement: "If I don't put my nose in it nobody will"!

Looking back at that time, I understand why ECT seemed like a glove fitting perfectly the hand:

"First: the new conceptualization of ECT brings an essential feature under the name of critical point or Catastrophe point, fitting the bill for a reference point, but the novelty being that it shows on the numerator side, instead of the denominator one, retained since the time of Weber-Fechner.

Second: the basic problematic, also from the time of Weber-Fechner, as to the search for an origin and scale of the logarithmic shape inherited from the time of Bernouilli, precisely represented by his two new axioms, was addressed with success by ECT, by the hypothesis of "diffeomorphism". Now, the final punch line to relate it to our problem resides in the methodology brought about by ECT. Indeed, one recalls from classical Analysis, two basic results:

1. Any function could be approximated to any degree (of approximation) by a polynomial.
2. The Taylor series expansion cannot be sure to converge. And even when it does, it's not sure the convergence will be to the original function which led to the local Taylor expansion [26].

ECT formalism had, a decade earlier, solved this problem in an original fashion.

Inspired by his thorough correspondence with Waddington and Zeeman about physiology, (chreods…) R. Thom embarked himself and others on a research work leading to the emergence of potential functions in finite number, the famous seven catastrophes, corresponding to a combination of the dimensions of two spaces, the control and state spaces. It was the first time the control space was introduced from the outset in conjunction with the state space and I identified it with the physical space corresponding to the brain wiring. Hence I focused on the dimension 3 in order to have a chance to have a measurement by the human brain corresponding to an observable result.

The basic procedure of ECT states that, starting from the "germ", highest term of the Taylor expansion at which one wishes to stop the approximation, a diffeomorphic change of coordinates leads to a UNIQUE representation with "determinacy", i.e. exact polynomial. And this is possible only by the introduction of the catastrophe point which insures both the "structural stability" and the "analytical continuation" i.e. the passage from the local to the global. It was the striking answer to the problem dormant since the XIX century as if Weber-Fechner had commandeered R. Thom to do that job a century earlier!

How could this connect with the logarithmic shape? Taking the Taylor expansion of the neural quantal function:

$$\mathrm{Log}(1+\mathrm{x}) = \mathrm{x} - x^2\big/2 + x^3\big/3 - x^4\big/4 + x^5\big/5 - \ldots \quad (1)$$

Retaining the "germ" $x^5/5$, corresponding to the dimension 3 by definition, since the germ has to have two degrees higher than the dimension of the control space, ECT leads to a function, called "self-intersection curve", defined by the cancellation of its first and second derivatives at the critical(catastrophic)point. This was the perfect topological translation of a non-decreasing function with its derivative reaching its strict minimum (zero) at the reference point, parameterizing the control space, by definition. Symmetry towards the origin completes the representation from – to + infinity. The most striking aspect which hit me was the remark of Wigner about "The incredible precision of mathematics in describing the real world". But still what we have here is the incredible continuity in human reasoning across the centuries finally leading to a conclusion of the primordial place of the human in his world [24]."

References

1. Bruter, C.P.: "Topologie et perception" Tome.II, Aspects neurophysiologique. Maloine et Doin editeurs (1976)
2. Verdet, C.: La Physique du Potentiel. Etude d'une lignee de Lagrange a Duhem. CNRS editions. Paris (2018)
3. Schackle, G.L.: Decision, Order and Time/Epistemics and Economics, Cambridge (1972)
4. Menges, G.: Information, Inference and Decision. Theory and Decision Library, vol. 1. Dordrecht, Boston (1974)
5. Stapp, H.P.: Mindful Universe, Quantum Mechanics and the Participating Observer, Springer, New York (2014)
6. Petitot, J.: Logos et Theorie des Catastrophes, Colloque de Cerisy sur l'œuvre de R.Thom, ed. Patino (1988)
7. Petitot, J.: Issues in Contemporary Phenomenology and Cognitive Science. Stanford University Press (1999)
8. Petitot, J.: In Mathématiques et Sciences Humaines Numéro Spécial linguistique et Mathématiques – III (1982)
9. Delboeuf, J.: Etude critique de la loi psychophysique de Fechner, l'Harmattan (2006)
10. Steve, K., Edward, F.: Improbable, incorrect or impossible? The persuasive but flawed mathematics of microeconomics. In: A Guide to What's Wrong with Economics, pp. 209–222. Anthem Press, London (2004)
11. Ralph, K.: Value-Focused Thinking. Harvard U.Press, Cambridge (1992)
12. Le débat: La querelle du déterminisme. Editions Gallimard (1990)
13. Jeffrey, R.: The Logic of Decision, 2nd edn. University of Chicago Press, Chicago (1983)
14. Allais, M., Hagen, O.: Expected Utility Hypothesis and Allais Paradox. Theory and Decision Library, vol. 21. Dordrecht, Reidel (1979)
15. Buser, P.: Neurophilosophie de l'Esprit, Odile Jacob (2013)
16. Duhem, P.: traité énergetique ou de thermodynamique générale: Editions Jaques Gallay (1911)
17. Duhem, P.: l'évolution de la mécanique mathesis. Paris VRIN (1992)
18. Duhem, P.: Sauver les apparences. Paris VRIN (1992)
19. Thom, R.: Structural Stability and Morphogenesis. Addison-Wesley (1989)
20. Thom, R.: Esquisse d'une Semiophysique. InterEditions (1991)
21. Thom, R.: Apologie du Logos. Hachette (1990)
22. Thom, R.: "L'Equivalence d'une fonction differentiable et d'un Polynome" received 1 November 1963 reprinted in Documents Mathematiques: Oeuvres Mathematiques Vol.II par Societe Mathematique de France in 2019, pp. 247–257
23. Sheldrake, R.: The Science Delusion Coronet (2013)
24. Morris, S.C.: Life's Solution. Cambridge University Press, Cambridge (2013)
25. Ross, S.: Some Stronger Measures of Risk Aversion in the Small and the Large with Applications, Series n°39. Yale University (1979)
26. Saade, W.M.: Arrow and Pratt Revisited. IARAS Publication (2017). www.iaras.org/iaras/home/caijmcm/arrow-and-pratt=revisited

27. Saade, W.M.: From D. Bernouilli to M. Allais: a new paradigm in the making? In: WSEAS Conference MACISE 2017–128, London, 27–29 Oct 2017
28. Saade, W.M.: A synthetic approach to value as a standard of reference. In: Proceedings of FTC Conference, London, 18–20 July 2017
29. Saade, W.M.: Toward a convergence between utility and value through a polynomial representation. In: Proceedings of FTC conference in San Francisco, December 2016, http://www.ieeexplore.ieee.org/xpl/most Recentissue.jsp? punumber7802033
30. Saade, W.M.: An alternative vision to Weber-Fechner Psycho-Physics, WSEAS, NAUM joint conference ICANCM, Malta. SLIEMA, August 2015
31. Saade, W.M.: The identification of Negative Schwarz' derivative with systematic risk aversion, same publication as following reference below
32. Saade, W.M.: A phenomenological approach to value. In: Post Conference Proceedings of the 18th World Multi-Conference on Systemics, Cybernetics and Informatics, Orlando, Florida (2014)
33. Saade, W.M.: An Engineering-Economy, presented at Oslo First international Conference on the Foundation of Utility and Risk Theory, July 1982
34. Balasko, Y.: Economie et Théorie des Catastrophes. In: Cahiers du Séminaire d'Economie (1978)

Application of DSM and Supervised Image Classification Method for Sun-Exposed Rooftops Extraction

Balakrishnan Mullachery(✉)

Claremont Graduate University, Claremont, CA 91711, USA
balakrishnan.mullachery@cgu.edu

Abstract. In urban areas, building rooftops are used for PV system installation for commercial or individual household purposes. An accurate solar map is essential to determine the amount of solar radiation falling on the rooftop of a building for PV installation. This study focuses on developing a methodology on how to configure a Drone to capture imageries for extracting features that are useful in assessing sun-exposed building rooftops. The focus is also to develop a geospatial model using GeoAI for processing Drone images and DSM to delineate actual solar exposed rooftops. Then, the model output was evaluated and compared with the existing models. The study was designed using DSR principles to develop software artifacts, a workflow model, and evaluation methods. This study provides a novel approach by combining image classifications using GeoAI for information extraction from Drone images for practical utility purposes. The study fulfilled the three research objectives. First, a flying configuration of a Drone to capture imagery for DSM creation. Second, the study fulfilled the development of a geospatial model using the DSM and GeoAI algorithm for delineating sun-exposed building rooftop. Third, this study evaluated the output of the model efficiency and reliability to determine whether this model can be used for enhancing the existing solar maps available from the government or other agencies. The utility of this model is a relevant topic in the shared economy research community, and no studies so far are focused designing a process that can create or augment inexpensive solar maps from Drone images.

Keywords: Drone · Digital surface model · DSM · GeoAI · Supervised classification · Green energy · Solar radiation · Rooftop

1 Introduction

The most abundant form of energy on earth is solar energy. Photovoltaic (PV) systems convert solar energy into electricity and are a good source of alternative energy by reducing fossil fuel-based energy generation systems. The amount of solar radiation falling on the earth's surface depends on many determinates such as weather, location on the earth, shadows of trees or structures, the slope of the surface, and open fields versus urban areas and others. In urban areas, building rooftops are used for PV system installation

K. Arai (Ed.): FICC 2021, AISC 1363, pp. 28–46, 2021.
https://doi.org/10.1007/978-3-030-73100-7_3

for commercial and individual household purposes. The PV system installation agencies use the existing solar maps available from local government or private agencies. These maps are not created for individual buildings or houses by considering shadows of trees or building structures but generalized using conventionally available information. Therefore, an accurate solar map is essential to quantify the solar radiation availability per building rooftops. Being that this problem is related to geographic context, geospatial technology can provide an innovative solution to create or enhance the existing solar map. The Drones make the technology affordable to perform data collection and information generation on-demand for practical applications in one's world.

The main goal of this research is to develop a methodology using Drone and geospatial technology for delineating solar exposed building rooftop locations for PV system installation and to generate electric energy efficiently. In order to meet this goal, the three main objectives of the research are the following. First, develop a methodology for configuring a Drone to capture overlapping imageries for extracting features that are useful for assessing sun-exposed building rooftops. The flight path, flying height, images overlap percent, and image shot locations are important in configuring Drones for capturing the required imagery data. Second, a geospatial model was developed using Geospatial Artificial Intelligence (GeoAI) for processing Drone images and Digital Surface Model (DSM) to delineate solar exposed building rooftop features in the urbanized area. Identifying noises such as tree and building shadows, rooftop slope, contiguous surface area in the same elevation, and sun angles are important parameters to consider during the sun-exposed building rooftops delineation. Finally, the model output was evaluated mathematically, and comparing with the existing solar maps available from a local government agency.

Drones are increasingly becoming popular in scientific, industrial, and recreation industries. Drones can carry various information capturing detectors such as different types of sensors, different spectrum length cameras, and video cameras. The Drones can provide a huge amount of data and this data can be used to generate valuable information for science and technology applications. In recent years the government has been promoting the importance of green energy and is providing financial subsidiaries for interested parties to install the hardware to generate green energy. However, it is important to ensure that the public money reserved for green energy promotion is used appropriately and have methodologies for auditing. Also, the PV system users for green energy generation want to make certain that their money is invested for a better cause. The local government and private agencies develop solar radiation maps for green energy generation awareness and the benefit of utilizing the unused rooftop of buildings and homes for solar energy generation. The development of an accurate and reliable solar radiation map is essential for a cost-effective solar energy generation. Using conventional technologies, it would be very expensive to develop such an accurate solar radiation map.

Spatial technology is widely used to solve problems related to geography. The suitable PV installation location identification is a spatial problem. Hence, GIS technology can use for suitability analysis for PV installation. Since 1995, Geographic Information Systems (GIS) technology has undergone several transitions with the emergence of Information Technology (IT) infrastructure and improved its data storage mechanism and application diversity. In recent years GIS applications have become an essential part of human life and are growing fast due to the emergence of Cloud Computing technology. The application of GIS and hosting of data from various scientific and engineering disciplines has increased. GIS has undergone many theoretical and technological challenges to keep up with the advancement of computing hardware and software technologies. It has transitioned from a conventional digital mapping and spatial analysis system to a fully supported Information Technology (IT). However, unlike other IT systems, the advancement of GIS is influenced by other technologies like Global Positioning System (GPS), Drone, aerial and satellite technologies, advancement in Remote Sensing, mobile data capturing technologies, and crowdsourcing [1]. According to previous studies, these additional technologies further changed the design philosophy of GIS. Moreover, the advancement of web technologies facilitated the creation of a web GIS platform where remote computers or mobile devices (that are connected to a wired or wireless internet) access and perform GIS analysis. The industry that provides spatial data should be reliable and ameliorate the spatial accuracy of the data deluge. This is a long technical and managerial process and requires a huge monetary investment. The emergence of Drones and its application in spatial science has enhanced the integration between the traditional data capture technologies and ad-hoc spatial data requirements. Also, Drones made the technology available for the users to perform data collection and information generation rather than approaching the data collection corporate agencies.

Refining existing IT methodologies is mandatory to fit for the information deluge from various remote sensors and other devices for application enrichment. In this situation, an idiosyncratic approach is required as a framework with a model and software artifacts. Based on my studies of previous researchers, the spatial representation of geographic objects received from remote sensors can be derived and used for solving many spatial problems by adopting technology workflows and available algorithms [3, 10, 19].

Identifying the rooftops from a Drone image can be better analyzed using Artificial Intelligence (AI) algorithms. Currently, AI is one of the most discussed buzzwords in the technology industry. It refers to a self-learning system based on data. It augments the human decision-making process using the experiences or observations of a human brain with machine-learned patterns from data. The two main components of AI are data and mathematical algorithms. The data can be structured or unstructured. AI algorithms are classified into supervised, unsupervised, and reinforced learning processes based on how the algorithm learns patterns from data. The supervised learning algorithm uses human-labeled data for training. In the unsupervised learning process, the algorithm creates labels from data and trains itself. The third class, reinforced learning, uses iterative interactions of an agent and the environment for actions to change the status either as a reward or penalty. During the interaction of the agent and environment, the agent learns the environment through the iteration of action and the reward/penalty process. For solving problems related to geographic context a new interdisciplinary field has emerged for knowledge recovery using spatial analysis with the integration of AI, geospatial artificial intelligence (GeoAI), which learns and predicts or forecasts an event in a geographic frame of reference.

In this study, I demonstrate how Drone images assess and appraise the potential sun-exposed building rooftops for solar energy harvesting, with the help of advanced image processing and spatial science techniques. Also, the information generated in the process can be useful for enhancing or validating the reliability of the existing solar maps. Flightpath configurations of Drone are very necessary for capturing the suitable stereo or overlapped images and are also included in this study.

There are two main contributions of this paper to the scientific community and practice. First, it discusses the current and potential applications of Drone to improve solar mapping capabilities using geospatial technology. Second, it proposes a comprehensive geospatial model for an efficient solution for sun-exposed rooftops feature extraction using the combination of DSM created from Drone images and supervised learning computing processes. The model considers noises, shadows, the elevation of the building, slope of the roof, and sun angle. The solution providers in the solar energy sector can use this model to facilitate their business development.

The novelty of this study is the approach of combining image classifications using GeoAI along with the DSM for information extraction from Drone images for a practical utility like mapping sun-exposed unused rooftops for assessing the PV system installation. Hence, this promotes renewable green energy.

The remainder of this paper is structured as follows. The next section discusses the related work done.in this direction followed by the goal and objectives of the study. Then, the data acquisition method, which details a Drone configuration for capturing the images for the study area, flight planning such as the height of the drone, image shot locations, and final Drone deployment. The logistics of the Drone survey such as the clearance process from the local airport agencies are also included in this section. Next, a methodology and model were presented to demonstrate the proposed solution. This section includes the philosophical approach, research methods, and artifacts design. After that, the Results and Discussion section which provides the details of the model output details. Next is evaluation, which incorporates the experimental aspects and validation criteria and discussed the evaluation of the model and usefulness are summarized. The contribution section summarizes the research contributions and potential implications of the study.

2 Related Work

There have been interesting Drone-based mapping studies conducted in many parts of the world. Drone based technology is evolved around many industries [20]. Infrastructure management, smart city initiatives, digital twin, constructions, package delivery, goods transportation, pipeline, and energy sectors are a few. Shared drone infrastructure services like "Platform as a Service" (PaaS) created the trust between stakeholders which helped Drone usage to become widely accepted among the users [15]. The DSM was used for solar radiation analysis in the past studied. Yokozawa et al. [10] studied using high-resolution DSM to reconstruct the shape of partial shadows that covers PV systems [10]. In this study, a pyranometer was used for solar radiation analysis considering weather conditions. A Drone was used by Lim et al. (2020) for creating a 3D Urban digital twin model and was used for infrastructure monitoring and analysis [16]. Developing an intelligent environment for connectivity and efficiency using technology are the key principles behind smart cities and the technologies integration for constructions and infrastructure. This helps the policymakers to assess the impact of infrastructure development before its commencement [17]. Structure damages in urban areas after a disaster were studied by Norman Kerle et al. [2] using UAV with oblique images. In this study, they created a 3D point cloud and used object-based image analysis to extract damage indicators for both the façade and roof of buildings. The study by Kerle et al. developed a methodology to assess the different damage levels. Another study was conducted by Eisenbeiss [3], for photogrammetric documenting of cultural heritage in Thailand using UAV by producing DEM at the scale of 1:7000 [3].

Xu et al. [4] developed an automated rooftop extraction process using color aerial imagery. The process is based on Multi-level image segmentation and integrating four RGB-D priors such as depth cue, uniqueness prior, the shape prior, and transition surface prior [4]. Some of the limitations identified in their study were the process mislabeled the shadows, vegetations, and dark rooftops that matched the shadows. Drones are widely using in surveillance monitoring. The capabilities of Drone based PV system surveillance was studied for automating Drone for monitoring PV power plant and data logging [18]. The installation location of photovoltaic (PV) systems for green energy generation depends on many determinates such as weather conditions, shadows of trees or other structures, and sun angle. The high-resolution DSM model was used for mapping the detailed shadow location for PV installation and the pyranometer used to correct the result of solar radiation analysis considering weather conditions [10]. Rakha [11] discussed Drone flightpath configuration procedures and 3D imaging methodology for building inspections [11]. Dhamankar et al. [21] in a study highlighted the benefits of large-scale mapping of rooftops using Unmanned Aerial Vehicle Survey (UAV) for solar harvesting in a built environment [21]. In this workflow, the aspect ratio of the rooftops is used to determine the solar energy potential. In most developed urban areas, the shadows of the trees and buildings obstruct the sun. Removing the effect of trees and buildings that obstruct the sun is essential to calculate the net rooftops area for solar harvesting. There is a gap in the past literature for a comprehensive model for building rooftops solar harvesting considering the different characteristics of rooftops such as shadows, trees, the elevation of the roofs, and sun angle. There was no study or model found that can delineate the sun-exposed rooftops of building from Drone images considering the noise such as trees and shadows.

3 Goal and Objectives

This study aimed to develop a methodology using Drone and geospatial technology for delineating solar exposed building rooftop locations for PV system installation and to generate electric energy efficiently. The study attempts to solve the aforementioned gap identified from the previous studies to provide a solution to the industry and contributes to geospatial science. Hence, this study showcases how Drone imagery can be used to assess and appraise solar harvesting from unused space of building rooftops.

Thus, the research objectives of this paper are:

- Assess and appraise how a Drone can be configured to capture images for deriving the sun-exposed building rooftops.
- Design and develop a geospatial model and software artifacts for delineating solar energy falling on unused space of a building rooftop from Drone images

- Evaluate the model output, discuss efficiency and reliability, and determine whether this model can enhance the existing solar maps available from the local government and other agencies.

4 Data Acquisition

The data collection was one of the initial tasks in this study. The pilot site selection was done with the help of an expert from the field. The pilot area covered was approximately 9 acres of land area. This area and the building belong to a university campus. Also, the ground knowledge of this area was a promising factor in its selection, as there was no external permission needed for the study. Permission was availed from the adjacent local airport authority for flying Drones in this area.

The flying was conducted using the Drone, DJI Phantom 4 [5]. This Drone had a built-in RGB camera, GPS, and IMU. It can operate remotely and can be controlled using a remote. The Drone was configured with MAPS Made Easy [6] application installed on a smartphone. This software was used for flight path planning purposes. The flight plan was created, and the Drone was configured and charged to takeoff in an autopilot mode. The elevation of the flying was set to 40 m and the overlap between images was 60%. This was determined by multiple experiments to get suitable imageries for DSM creation. The flight was set to fly and take pictures in an auto mode. It took 20–30 min to complete the survey. The local wind caused the Drone to fly slower than expected. Figure 1 below illustrates the flight path and flying parameters.

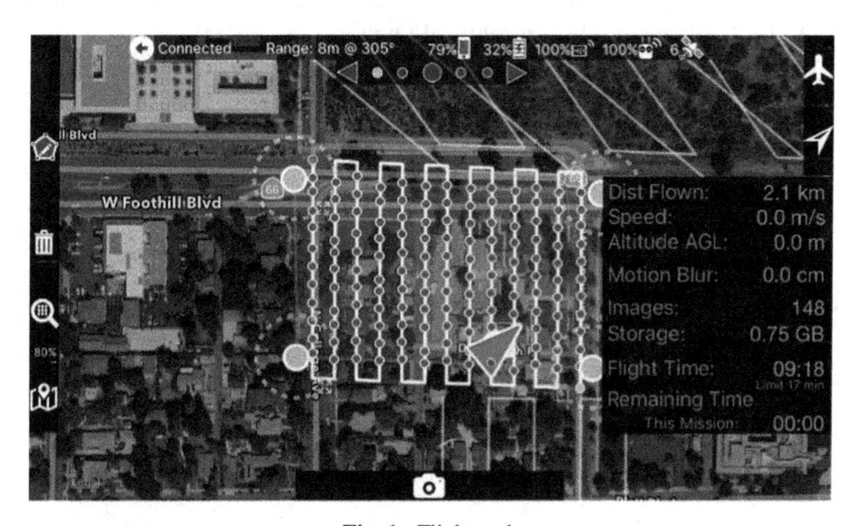

Fig. 1. Flight path

The Drone captured 144 images for this pilot area. The image quality was inspected from the field and transferred to the laptop.

The raw images were processed using ArcGIS Drone 2 Map image processing software and created orthomosaic and DSM. The spatial reference system used was WGS84/UTM Zone 11 N. Figure 2 illustrates the actual imagery shoots during the flight along the pre-loaded flight path.

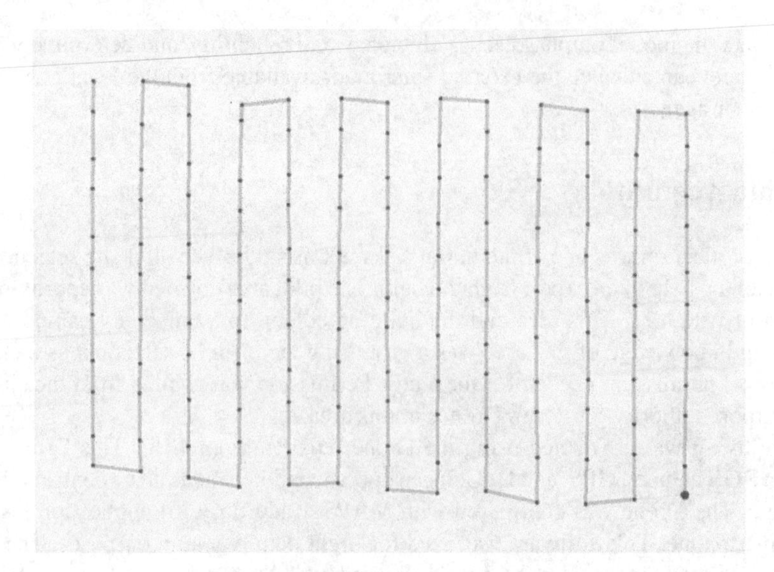

Fig. 2. Image capture locations along the flight path

Figure 3 illustrates the calibration performed within Drone2Map software using automatically collected Ground Control Points (GCP) by the software. Then the images were rectified and registered to the true ground location on the earth. The blue dots represent the original location of the shots and the green dots represent the calculated location. After orthorectification, the ground sample distance (GSD) was 1.64 cm.

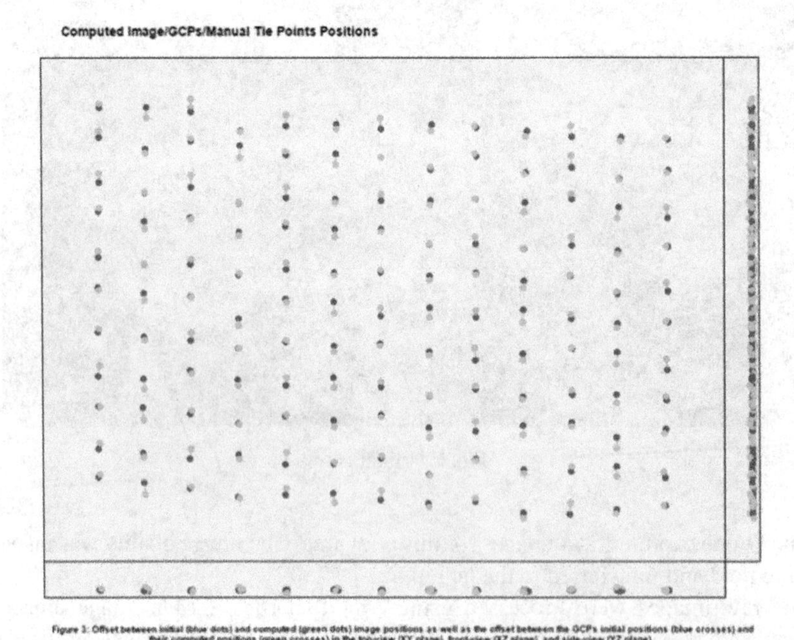

Fig. 3. Actual versus the calibrated locations of image

Figure 4 represents the matching between the images used for creating orthomosaic. The software performs a photogrammetry techniques based on Structure from Motion Multi-View Stereo (SfM-MVS) matching for sparse cloud densification, and then, the inverse distance weighting algorithm (IDW) interpolation is used for DSM generation [9].

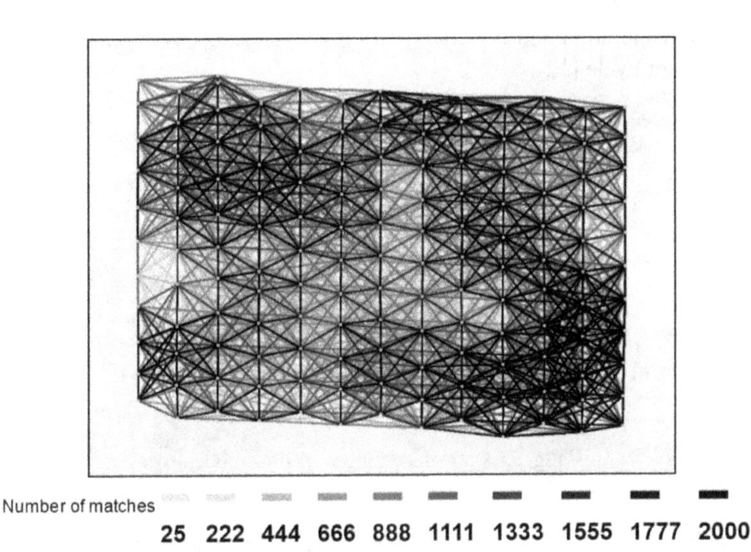

Fig. 4. Image matching for orthomosaic creation

4.1 System

The computer with 16 GB of RAM and nearly 1 TB of disk space running on Windows 8.1 Enterprise, 64 bits was used for data processing. The software used for this study was ArcGIS, ArcGIS Spatial Analyst, and ArcGIS Drone2Map.

5 Methodology

The study was designed using design science research (DSR) principles to develop software artifacts, workflow, and evaluation methods that help to map the solar potential of unused spaces of building rooftops.

Figure 5 illustrates the different steps as processes and their relationships within the model. Each step in the model is an artifact which is a group of data preparation, processing, and analyzing geoprocessing algorithms (tools). The order of the processes was experimented during the design process to meet the relevance of the study. Hevner et al. [7] had stated that the discovery of truth may lag the application of its utility and design science research paradigm seeks to create "what is effective?" [7]. So, the efficacy of the model and the individual process were experimented with and evaluated for its performance and scalability measures before finalizing a generic and complete working model. If any discoveries or knowledge are identified later, then those can be added as an additional process and incorporated into the model. This design approach will help the model to be very agile and can be enhanced or expanded as needed. Innovation in design is one of the key design principles of DSR. Herbert A Simon, in his book *Science of the Artificial*, describes how an artificial system interfaces the inner and outer environments of the problem concerning the natural sciences (Simon Third Edition). The DSR follows this design consideration and is iterative in three cycles: The relevance cycle, Design Cycle, and Rigor cycle [12].

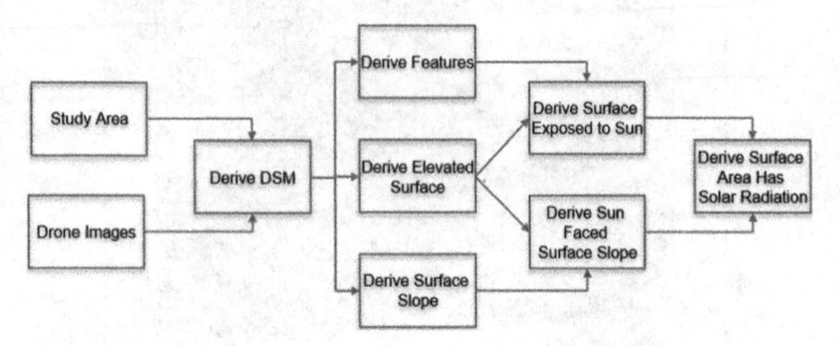

Fig. 5. Data processing workflow model

The DSR starts with an application environment in which an effective and innovative design should have the ability to improve the characteristics of the environment. This must be an iterative process adopting the field study and further evaluation methods. In this study iterative method of refining the design with the help of a field study to meet the study goal. A Constant interplay between proposing and checking the requirements and design was implemented during the design process.

The utility and efficacy of the design and its artifacts are evaluated for their performance and scalability. A pilot testing was conducted to study the utility and efficacy of the design of the artifacts. In the design cycle, there was a constant interplay between the relevance of the design, the rigor of the design by searching the body of knowledge, and evaluation of utility and usability using objective and/or subjective approaches.

DSR is concerned with finding a solution for solving practical problems of interest [13] using a product, in this case, IT artifacts, and a process, in this case, workflow, with activities. According to the recommended DSR guidelines [14], the existing kernel theory from the body of knowledge sets the foundation for specifying the requirements for DSR and designing the IT artifacts.

The design evolved after repeated experimentation of rooftops feature extractions and filtering the noises from the images in various conditions. The main challenges were applying technology tools for extracting the elevated surface, surface slope, shadows, and delineating the surface without any sun obstructions. The finalized tools are bound together in the ArcGIS Model builder for repeated operational needs. The abstract model of the workflow is illustrated in Fig. 5 – Data Processing Workflow Model. The novelty of the design is the integration of DSM and image processing algorithms to eliminate sun obstruction noises and extraction of elevation with sloped surfaces from the ground.

Once the Drone images were captured, the workflow starts with the creation of the orthomosaic and Digital Surface Model (DSM) for the study area. Figure 6 represents the DSM created from the Drone images. Drone2Map software is used for this purpose. Then using the study area, the images are clipped for further data processing. The feature extraction process was implemented to extract building boundaries, elevated rooftops surface, and surface slope of the roof from the DSM. Different surface levels were extracted from the DSM and vector output polygons were created.

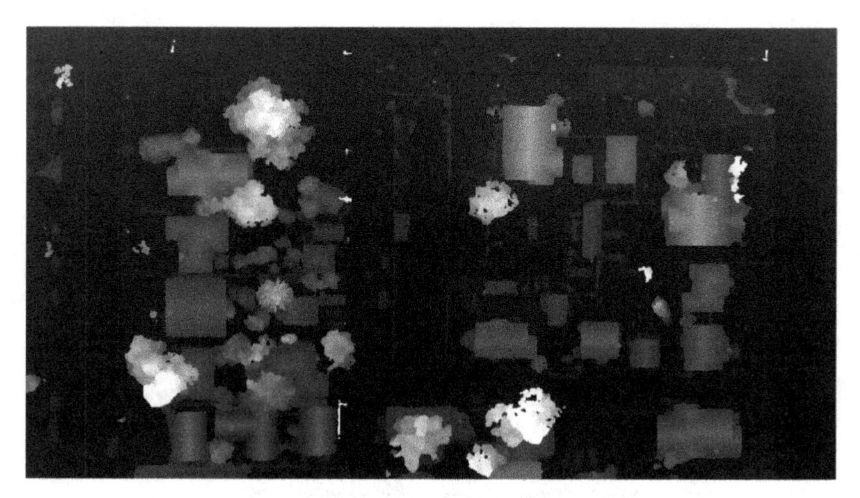

Fig. 6. DSM created from the Drone images

Next, supervised image classifications were performed using the orthomosaic (RGB) images and extracted vegetation and shadowed features to isolate rooftops with minimum anomalies. Most of the anomalies are either due to the tree canopies or shadows or shadows of the adjacent buildings. The portion of the rooftops that limits the solar radiation by the trees or structures are eliminated using the image classification process. The rooftop slope and direction of the slope is calculated from the DSM. The slope and the directions are used to extract the rooftop surface slope in the South West, South, and South-East side of the buildings. These extracted polygons were proposed as a suitable area on the rooftop where solar radiation was available for energy generation.

6 Results and Discussions

There haven't been many studies steered using a Drone for detailed mapping. Hence many unknown facts exist to prove. One of the main requirements for Drone mapping is the computing infrastructure requirements for storing and processing a large number of images and extracting the information needed for the study. A small area mapping requires thousands of images. In this study, for 8.68 acres of the area, there were 144 images. The study requires maximum overlap to generate a DSM dataset which is a key source of information for further data analysis. So, a 60 percent overlap was used between consecutive images.

The vegetation including tall trees and shadows (trees or buildings or other infrastructure objects) were classified and extracted using the Maximum Likelihood (MHL) supervised classification algorithm. This process helped to eliminate or automatically detect the influence of external objects that reduce solar radiation in the immediate neighboring area. Figure 7 demonstrates the extracted vegetation and shadows from the RGB imagery.

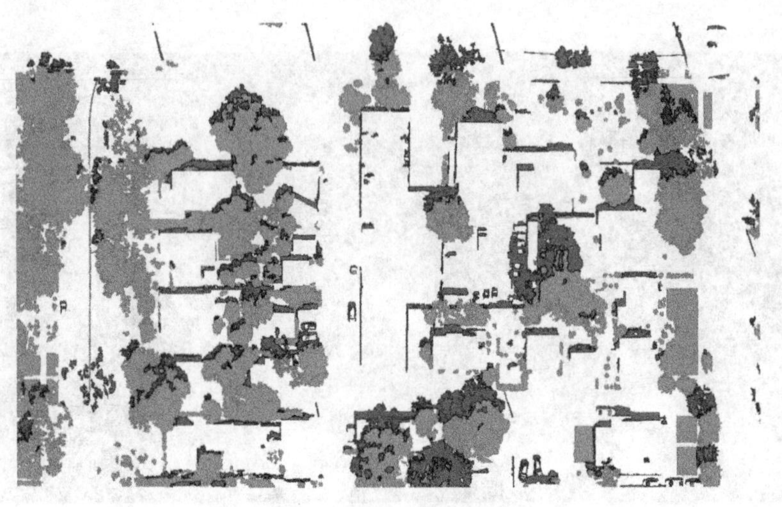

Fig. 7. Trees (green) versus Shadows (grey)

The efficacy of the workflow configured for the extraction of rooftops surface from Drone images is encouraging. Comparatively the generated rooftop polygons' metes and bounds are closely matched with the existing building footprints available from the local government agency. There is future research to be done for filtering out the anomalies from the generated building footprints, which requires additional algorithm development. At the same time, the spatial accuracy of the building footprint derived from the Drone images was better than the existing building footprints from the local agencies. This workflow was able to identify all the buildings belong to the study area without any omission errors. Figure 8 illustrates the generated rooftop polygons overlaid on Drone imagery.

The polygons with red dot represent anomalies in the process. These building rooftops area was not entirely included in the output surface polygons. This is because of the actual slope of the roof versus the interpreted surface didn't come close to reality. The algorithm requires enhancement which will be included in the future study.

Figure 9 illustrates the area identified from the rooftops boundary that was free from solar radiation obstructions and suitable for maximum energy generation. The star symbol in Fig. 9 represents the rooftop exposed to solar radiation. The red polygons represent the area proposed for solar radiation available during the daytime to generate energy. This surface area can be used for energy output calculation using the existing scientific formula. The building polygons adjacent to the red dots indicate the rooftops

Fig. 8. Extracted rooftop polygons

that are not useful for solar energy generation. These buildings are either adjacent to tall trees or to the other objects which block the sun during the daytime.

Fig. 9. The structures identified for solar radiation

Figure 10 illustrates the solar radiation available during the day from 6 AM to 5 PM for the 24 identified locations. This is generated using the solar radiation tool available in the ArcGIS software platform.

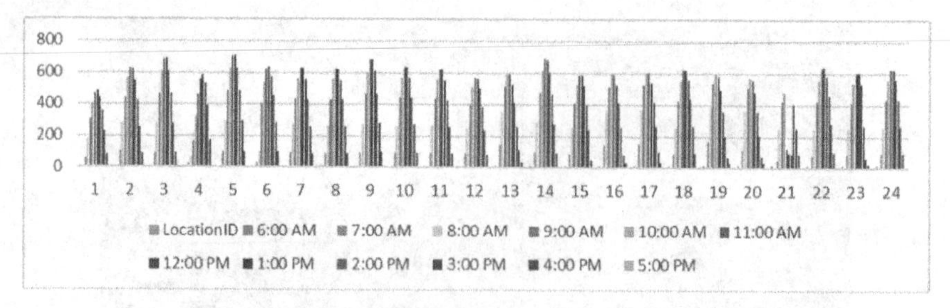

Fig. 10. Solar radiation graph of a day

7 Evaluation

The entire workflow artifacts were tested during the study development phase using Drone images. The model consists of many geoprocessing tools to derive the solar exposed rooftops. So, the output of the model was evaluated mathematically and compared with the existing solar map models. The individual geoprocessing algorithms (tools) output parameters from the model were not evaluated, because the tools used were from an industry well-known ArcGIS software platform. Also, the study's purpose was how to create or argument solar maps for PV installation. Hence the output results were evaluated for their validity and reliability. First, the output dataset is evaluated manually by comparing it to the input data. Aerial imagery data from Google and ArcGIS Online were used in the evaluation process to understand the geographic pattern of the study area and the generated result. This evaluation process has helped to refine the processes, tools, and parameters used in the model. The model ran repeatedly a few times to get the refined result and to meet the required output accuracy level. Then the output result is visually evaluated (ground truth) from the field. The scientific methodology is not used in the evaluation process which will be an enhancement in the future study. The manually inspected building structure for solar radiation will be binary data and will be classified as either "Yes" or "No." In the current sample, visual inspection provided 100% accuracy. However, between 80% and 85% accuracy is considered an adequate result in an image analysis science [8].

The output of the workflow is a binary data classified as either "Yes" or "No." The entire workflow model is evaluated by running the workflow using Drone images collected from the study area where the buildings and neighboring environmental objects are familiar for verification.

After processing the images, the results were evaluated using the confusion matrix as illustrated in Table 1. A confusion matrix is a table used to describe the performance of a binary classification on a set of test data with known true values. In this study, the result can be in two values, either "Yes" or "No". "Yes" indicates that the model identified rooftops with solar radiation. On the other hand, "No" signifies that the model didn't identify the rooftops with solar radiation.

The model correctly identified the availability of solar radiation on the rooftops. The information was inserted into a confusion matrix as illustrated in Table 1. Out of 30 rooftop locations, 21 of them interpreted as solar radiation "Yes", 6 of them interpreted as solar radiation "No" which correct from the ground truth validation, One

Table 1. Confusion matrix – classification table

Total = 30	Failure of observation	Success of observation	Total	
Failure of prediction	TN = 6	FN = 2	8	Predicted negative
Success of prediction	FP = 1	TP = 21	22	Predicted positive
Total	7	23	30	
	Observed negative	Observed positive	Total	

interpreted as "No" and it supposes to be "Yes" and two of them omitted. Based on these numbers, accuracy, precision, specificity, and statistical sensitivity values were calculated. The confusion matrix and further statistics generated from the confusion matrix were validated for the correctness and effectiveness of this model.

- True Positives (TP): The tool predicted "Yes" (availability of solar radiation on twenty-one rooftops) and does have the availability of solar radiation on 21 rooftops. Figure 9 illustrates the outcome.
- True Negatives (TN): The tool predicted "No," but doesn't have the availability of solar radiation on six rooftops. Figure 8 illustrates the outcome.
- False Positives (FP): The tool predicted "Yes," but doesn't have the availability of solar radiation on one rooftop. ("Type I error.") Fig. 8 illustrates the outcome
- False Negatives (FN): The tool predicted "No," but do have the availability of solar radiation on two rooftops. ("Type II error.") Fig. 7 illustrates the outcome

The Confusion matrix parameters calculated are provided in Table 2 below. Accuracy is a measure of the fit of the model which means that the model gives an accurate prediction of 76.67% of the time or 76.67% of the rooftops identified for PV installation correctly. Sensitivity (recall) measures the ability of the model to correctly classify rooftops into sun-exposed versus not sun-exposed. Higher the value more predictive power. High specificity means that there are few false positive, means the rooftops will rarely select for PV system when it is not sun-exposed.

Typical binary classification ROC curve measurement is used for evaluation. ROC curve use specificity and sensitivity to provide maximum Area under Curve (AUC) value, which can be used for the evaluation of the output. In the future with more study locations and images, the ROC curve will be plotted and the AOC value will be included in the evaluation process. Evaluating the solution using more variety of images will provide the efficacy of the model, which will be the immediate future goal of this study. The following evaluation methods will be implemented in the future:

- DSM - Potentially using LiDAR data or manual measurements of the samples
- Rooftop elevation from the ground – Potentially using LiDAR data

Table 2. Classification table statistics

Sensitivity	91.30%
Specificity	85.71%
Accuracy	76.67%
Error rate	26.67%
False positive rate	14.29%
Positive predictive value	95.45%
Negative predictive value	25.00%

- Rooftop slope - Potentially using LiDAR data
- Shadows – Multiple data collections in a day and further interpolation

The feature extraction accuracy of the model is compared with google and ArcGIS online maps. Figure 11 illustrates visually the accuracy of the three different maps: 1) Google, 2) ArcGIS Online, 3) The model output. The model output shows improvement in building edge detection compared to the other two. This proves that the DSM created from Drone images can be used for rooftop feature extraction more accurately and for detailed mapping purposes such as building footprint extraction.

Fig. 11. Accuracy between Google, ArcGIS Online, and Model output

Figure 12 illustrates the DSM derived from the Drone images and the model delineated sun-exposed buildings marked in pink stars. The building within tall trees shadow was not detected by the model as sun-exposed for PV installation.

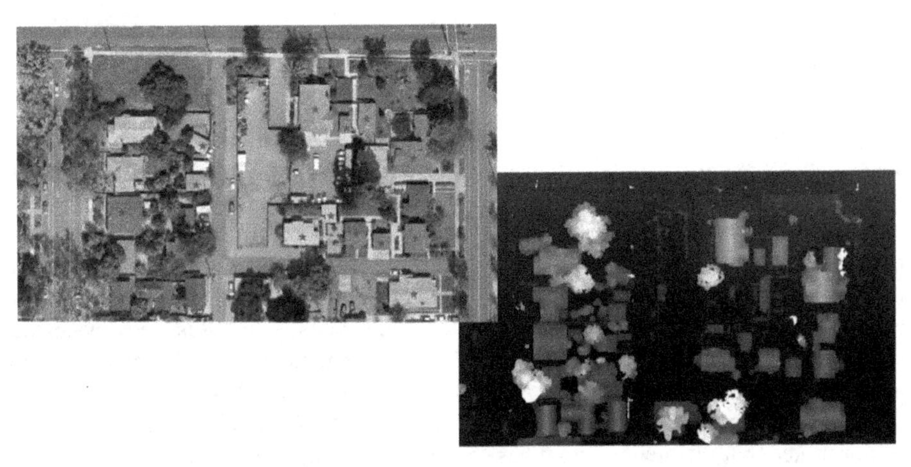

Fig. 12. DSM and Sun-exposed buildings detected by the model

The sun-exposed rooftops identified by this model were compared with Los Angeles (LA) county solar map. In Fig. 13, comparison of the model output with the LA County solar map, the red box shows the area of comparison used for evaluating with the LA county solar map that corresponds to the study area. In the LA County solar map, the sun-exposed rooftops are color-coded in orange. In the area of comparison, none of the rooftops are coded with orange and it reveals there are no building rooftops available

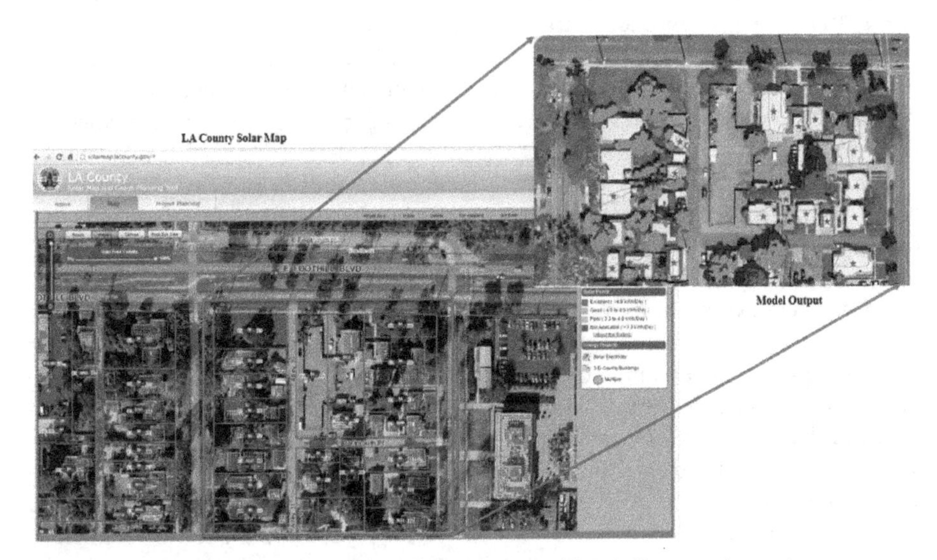

Fig. 13. Comparison of the model output with LA County solar map

for PV installation. However, the model output identified most of the building rooftops either completely or partially sun-exposed (marked with a red star), which are available for PV system installation. This evaluation proves the accuracy of the model compared to the existing solar map.

8 Conclusion

Drones are becoming popular in the geospatial science for ad-hoc data collection. This study showcases a use case of Drone and geospatial technology to create or enhance or validate existing solar maps of building rooftops by considering the actual sun exposure. This study fulfilled the three research questions and main objectives. First, a flying configuration of Drone captured aerial imagery for DSM creation. Second, the study fulfilled the development of a geospatial model using the DSM and GeoAI algorithm for delineating sun-exposed building rooftop from Drone imageries. Third, this study evaluated the output of the model efficiency and reliability to determine whether this model can be used for enhancing the existing solar maps available from the local government and other agencies. Comparing the model output with Los Angeles County's existing solar map demonstrated that the model output has shown better results. Also, the study helped to effectively assess the rooftop locations exposed to solar radiation. The evaluation of the study proved that the methodology implemented is helpful to improve the existing solar maps without omission errors. The utility of this model is very relevant. The awareness of global warming and green energy promotions are very eminent globally. The fossil fuel-based energy generation processes are degrading slowly and transitioning to green energy generation. Also, non-fossil fuel-based automobiles are becoming more popular. So, the efficient way to convert maximum solar energy falling on a building rooftop to supplement energy requirements for energizing all electric apparatus in that building, cooking energy, and operating electric automobiles will hugely promote the green energy transition process and contribute to reducing the global warming process. This study provides a model that helps to create accurate and up to date solar maps which help to identify locations for PV system installation for maximum electric energy generation. No studies are so far focused to design a process that can create or augment inexpensive solar maps from Drone images.

The main contributions of this paper to the scientific body of knowledge and practice are twofold. First, it discusses the current and potential applications of Drone and interconnected technology solutions to improve solar mapping capabilities using geospatial technology. Second, it proposes a comprehensive geospatial model for an efficient solution for feature extraction using the combination of DSM created from the overlapped Drone images and series of supervised learning computing processes (considering three different dimensions: elevation of the building, slope of the roof, and sun angle). The proposed model can also be used by solution providers in the solar energy sector to facilitate their business development. The Geospatial model developed as part of this study will be the main contribution to the academic community and the industries. The study demonstrated how the Drone is used for feature extraction considering the elevation of an object. In the future, such technology can be deployed on-demand for commercial map development and/or enhancement of the existing solar maps. This use case will also beneficial to the data acquisition and processing for smart city initiatives and drone-based 3D urban digital twin models. Furthermore, it proves that a camera-mounted Drone can be used to appraise the site location for solar energy estimation and calculation for return of investment (ROI).

The novelty of this study is the approach of combining supervised image classifications using GeoAI along with DSM for feature extraction from Drone images for a practical utility such as mapping sun-exposed unused rooftops to assess the PV system installation and promote renewable green energy. The study provided adequate results, which are supported by the ground truth verification. When compared with the local government solar map, there are no omission errors in this study. In addition, the study identified the supplementary rooftops where solar radiation is available for energy generation. The current solar maps are not created using a detailed survey. They use land classifications from satellite imageries and which cannot provide a better small-scale map with more details. Drones can provide high-resolution images for localized areas. So, all buildings in the study area can be analyzed in detail. As a result, more sun-exposed building rooftops are identified. More samples from different data conditions will be beneficial to generalize the workflow. The identification of the main artifacts, their functions, and the framework model are part of this study. However, an extension of this study should include enhancing the feature extraction techniques using advanced AI algorithms, such as deep learning and other image classification machine learning algorithms, within the geospatial context. This will eliminate the current limitations of this study: different and complex data conditions, different implementation scenarios, and multiple users' business requirements. The following future studies can enhance the model:

- Integration of various analytics such as deep learning and image classification machine learning algorithms can be developed for faster inferential and/or predictive analysis.
- By examining different datasets, the spatial semantic representation of different objects can be derived.
- The evaluation step can be improved using the additional methodology and analytic tools.

Acknowledgment. I thank Prof. Andrew Max, Claremont Graduate University for his initial guidance on this study and for helping me get permission to fly a Drone.

References

1. Song, W., Keller, J.M., Haithcoat, T.L., Davis, C.H.: Relaxation-based point feature matching for vector map conflation. Trans. GIS **15**(1), 43–60 (2011)
2. Kerle, N., Galarreta, J.F., Gerke, M.: Urban Structural Damage Assessment With Oblique UAV Imagery, Object-Based Image Analysis and Sematic Reasoning, p. 7 (2014)
3. Eisenbeiss, H.: A Mini Unmanned Aerial Vehicle (UAV): System Overview and Image Acquisition, p. 7 (2004)
4. Xu, S., et al.: Automatic building rooftop extraction from aerial images via hierarchical RGB-D priors. IEEE Trans. Geosci. Remote Sens. **56**(12), 7369–7387 (2018)

5. DJI Phantom 4 – Specs, FAQ, Tutorials and Downloads, DJI Official (2020). https://www.dji.com/phantom-4/info. Accessed 07 Jan 2020
6. Easy, M.M.: 3D Olympic Training Center (2020) https://www.mapsmadeeasy.com/maps/public_3D/fd8859995dae42e59668cfedc0da9611/. Accessed 07 Jan 07
7. Hevner, A.R., March, S.T., Park, J., Ram, S.: Design science in information systems research. MIS Q. **28**(1), 75–105 (2004)
8. Congalton, R.G., Green, K.: Assessing the Accuracy of Remotely Sensed Data: Principles and Practices, 3rd edn. CRC Press, Boca Raton (2019)
9. Jairo, E.V.R., et al.: DEM generation from fixed-wing UAV imaging and LiDAR-derived ground control points for flood estimations. Sensors **19**(14), 3205 (2019). https://doi.org/10.3390/s19143205 (Basel, Switzerland)
10. Yokozawa, K., Fuchino, G., Shiot, A., Mitani, Y.: Solar radiation analysis on GIS considering influence of weather condition and partial shadow analyzed by high-resolution digital surface model. IJSGCE, pp. 48–53 (2018)
11. Rakha, T., Gorodetsky, A.: Review of Unmanned Aerial System (UAS) applications in the built environment: Towards automated building inspection procedures using drones. Auto. Constr. **93**, 252–264 (2018)
12. Alan, H., Chatterjee, S.: Design Research in Information Systems: Theory and Practice. Integrated Series in Information Systems. Springer, USA (2010). https://www.springer.com/de/book/9781441956521
13. William, L.K., Vaishnavi, V.K.: A framework for theory development in design science research: multiple perspectives. J. AIS **13**, 3 (2012). https://doi.org/10.17705/1jais.00300
14. Shirley, G., Hevner, A.R.: Positioning and presenting design science research for maximum impact. MIS Q. **37**, 337–55 (2013). https://doi.org/10.25300/MISQ/2013/37.2.01
15. Grigoropoulos, N., Lalis, S.: Simulation and Digital Twin Support for Managed Drone Applications. IEEE (2020). https://doi.org/10.1109/DS-RT50469.2020.9213676
16. Seong-Ha, L., Kyu-Myeong, C., Gi-Sung, C.: A study on 3D model building of drones-based urban digital twin. J. Cadastre Land **50**(1), 163–180 (2020). pISSN 2508-3384. eISSN 2508-3392
17. Shirowzhan, S., Tan, W., Sepasgozar, S.M.E.: Digital twin and cybergis for improving connectivity and measuring the impact of infrastructure construction planning in smart cities. ISPRS Int. J. Geo-Inf. **9**(4), 240 (2020). https://doi.org/10.3390/ijgi9040240
18. Kumara, N.M., Sudhakara, K., Samykanoa, B.M., Jayaseelan, V.: On the technologies empowering drones for intelligent monitoring of solar photovoltaic power plants. Proc. Comput. Sci. **133**(2018), 585–593 (2018)
19. Grubesic, T.H., Nelson, J.R.: Unmanned aerial systems for energy infrastructure assessment. In: UAVs and Urban Spatial Analysis. Springer, Cham (2020). https://doi-org.ccl.idm.oclc.org/10.1007/978-3-030-35865-5_8
20. Jenkins, N.A.L.: An Application Of Aerial Drones In Zoning And Urban Land Use Planning In Canada (2015)
21. Dhamankar, N., Kulkarni, P., Dixit, N.: Estimating potential solar energy on rooftops using unmanned aerial vehicle. In: 2019 International Conference on Computational Intelligence and Knowledge Economy (ICCIKE), pp. 275–278. Dubai, United Arab Emirates (2019). https://doi.org/10.1109/iccike47802.2019.9004418

Tensor Multi-linear MMSE Estimation Using the Einstein Product

Divyanshu Pandey(✉) and Harry Leib

Department of Electrical and Computer Engineering, McGill University, Montreal, QC H3A 0E9, Canada
divyanshu.pandey@mail.mcgill.ca, harry.leib@mcgill.ca

Abstract. Tensors are multi-way arrays which are natural generalization of vectors and matrices to higher orders, and can be used to represent data and signals depending on a multitude of indices. Hence they are heavily used in various disciplines including signal processing, multi-domain communications, machine learning and Big Data. This paper introduces a tensor framework for multi-linear minimum mean square error (MMSE) estimation using the Einstein Product. The classical notion of linear MMSE estimator for vectors has been extended to tensor case leading to the notion of multi-linear MMSE estimation. A numerical approach for approximating tensor inversion based on Newton Method is also presented, and its use for computing the proposed multi-linear MMSE estimator is considered. A comparison with the existing multi-dimensional filtering using the Tucker product which relies on n-mode unfolding of tensors is also presented. An application of the tensor based estimation in a multiple antenna Orthogonal Frequency Division Multiplexing (MIMO OFDM) system is considered where the tensor formulation allows a convenient treatment of inter-carrier interference, leading to a tensor equalizer which outperforms the per sub-carrier equalizer more commonly used for MIMO OFDM.

Keywords: Einstein product · Multi-linear MMSE · Tensors · Tensor inversion

1 Introduction

Estimation based on minimizing the mean square error is one of the most fundamental methods used in Statistical Signal Processing. The general problem at hand is to estimate a signal based on a measurement which contains information about the quantity to be estimated albeit corrupted with some noise. Traditionally, this problem has been extensively studied, and well established techniques exist which have found many applications in various disciplines. However, most of these solutions apply to vector based signals, and not much has been explored for the cases where the quantity to be estimated is a tensor. Tensors are multi-way arrays whose elements are indexed by more than two indices, also known as

K. Arai (Ed.): FICC 2021, AISC 1363, pp. 47–64, 2021.
https://doi.org/10.1007/978-3-030-73100-7_4

modes. The number of indices is called the *order* of the tensor. Hence, a matrix can be seen as an order 2 tensor while a vector as an order 1 tensor [1]. Tensors have an inherent ability to characterize processes with dependence on more than two variables. Hence they have found widespread applications in various engineering disciplines including signal processing [2,3], big data and machine learning [4–6], communications [7–9], and multi-linear system theory [10]. Our work considers the MMSE estimation problem in context of tensors, and hence it extends this basic subject beyond the common vectors and matrices settings.

So far, tensor based estimation techniques have been addressed in the literature for specific applications. For instance, use of tensors for channel estimation in relay based MIMO systems has been considered in [11]. Blind receivers based on Parallel Factors (PARAFAC) decomposition of tensors have been considered for Direct Sequence-Code Division Multiple Access (DS-CDMA) systems in [2] and for MIMO systems in [7,8] and references within. Recently, a tensor based MMSE channel estimation technique using compressive sensing was proposed in [12] for massive MIMO OFDM systems. In Image Processing, tensor based estimation has been used for de-noising by treating coloured images as third order tensors and applying filtering on different mode unfoldings of the tensor into a matrix by using the Tucker operator [13,14]. All these cases use either PARAFAC or Tucker operators, thereby employing n-mode product or outer product of tensors with factor matrices and vectors depending on specific system requirements. However, Tucker and PARAFAC operations can be represented by the Einstein product as explained in [15]. Hence, Einstein product can be used to develop a more generic framework for MMSE estimation in many applications. The Einstein product can be seen as a natural generalization of the matrix product to higher order arrays, thereby providing an easy mechanism to extend linear algebra tools to tensors [15,16].

In this paper, a multi-linear framework for tensor estimation using Einstein product is proposed. The Einstein product of tensors is a form of tensor contraction initially used in Physics and Continuum Mechanics. The Einstein product has recently garnered attention in various engineering disciplines because of its ability to generalize well-known linear algebra notions to a multi-linear setting without compromising on the tensor structure of the associated quantities. Since different modes of a tensor can attribute different physical meanings depending on the system model, retaining the distinction between modes is of paramount importance in order to leverage the information encapsulated in the structure of a tensor. For instance, in a communication system different modes can represent different domains such as space, time, frequency, codes, channel taps and users. Such a multi-domain approach towards communication systems can be observed in many multi-carrier and multi-antenna schemes such as MIMO OFDM, GFDM (Generalized Frequency Division Multiplexing). The use of tensors for representing the system model in such communication systems has been considered in [9]. The tensor MMSE estimation framework presented in this paper can be used in many such multi-domain settings. Our contribution builds on the Einstein Product properties used for solving multi-linear systems of equations initially suggested in [15] to develop a multi-linear framework for MMSE estimation. We also present an extension of Newton method for finding the inverse of a tensor.

This paper is organised as follows: we first present a brief review of tensor algebra in Sect. 2 needed for developing the framework. We also present a numerical method for solving tensor inversion problems. Sect. 3 introduces the tensor multi-linear MMSE estimation problem. Sect. 4 considers applications of the tensor framework for multi-linear estimation in MIMO OFDM systems. The paper is concluded in Sect. 5.

2 Tools from Tensor Algebra

2.1 Notations

Throughout this paper, deterministic vectors, matrices and tensors will be represented using lower case underlined fonts, upper case fonts, and upper case calligraphic fonts respectively e.g. $\underline{\mathrm{x}}, \mathrm{X}$, and $\mathcal{X}$. Their corresponding random quantities will be denoted by bold fonts, e.g. $\underline{\mathbf{x}}, \mathbf{X}$ and $\boldsymbol{\mathcal{X}}$. Individual entries of a tensor are denoted with the indices in subscript, e.g. the $(i, j, k)^{th}$ element of a third order tensor $\mathcal{X}$ is denoted by $\mathcal{X}_{i,j,k}$. A colon in subscript is used to indicate all elements of a mode, e.g. $\mathcal{X}_{:,j,k}$ represents all the elements of first mode corresponding to j^{th} second and k^{th} third mode. The nth element in a sequence is denoted by a superscript in parentheses, e.g. $\mathcal{A}^{(n)}$ denotes the nth tensor in a sequence of tensors. Furthermore, $()^*$ represents the complex conjugate, $\mathbb{E}[.]$ represents expectation, and $0_{\mathcal{T}}$ represents an all zero tensor.

2.2 Basic Definitions

Definition 1. Tensor Linear Space: *The set of all tensors of size* $I_1 \times \ldots \times I_K$ *over* $\mathbb{C}$ *forms a linear space, denoted as* $\mathbb{T}_{I_1,\ldots,I_K}(\mathbb{C})$. *For* $\mathcal{A}, \mathcal{B} \in \mathbb{T}_{I_1,\ldots,I_K}(\mathbb{C})$ *and* $\alpha \in \mathbb{C}$, *the sum* $\mathcal{A} + \mathcal{B} = \mathcal{C} \in \mathbb{T}_{I_1,\ldots,I_K}(\mathbb{C})$ *where* $\mathcal{C}_{i_1,\ldots,i_k} = \mathcal{A}_{i_1,\ldots,i_k} + \mathcal{B}_{i_1,\ldots,i_k}$, *and scalar multiplication* $\alpha \cdot \mathcal{A} = \mathcal{D} \in \mathbb{T}_{I_1,\ldots,I_K}(\mathbb{C})$ *where* $\mathcal{D}_{i_1,\ldots,i_k} = \alpha \mathcal{A}_{i_1,\ldots,i_k}$.

A tensor valued function $g(\mathcal{X})$ can be defined as a function whose domain and range are both tensors. Notationally, $g : \mathbb{C}^{I_1 \times \ldots \times I_N} \rightarrow \mathbb{C}^{J_1 \times \ldots \times J_M}$ represents a function where $\mathbb{C}^{I_1 \times \ldots \times I_N}$ and $\mathbb{C}^{J_1 \times \ldots \times J_M}$ are tensor linear spaces spanned by the domain and range of g respectively.

Definition 2. Einstein product [15]: *For any* N, *the Einstein product is defined using the operation* $*_N$, *where:*

$$(\mathcal{A} *_N \mathcal{B})_{i_1,\ldots,i_N,k_{N+1},\ldots,k_M} = \sum_{k_1,\ldots,k_N} \mathcal{A}_{i_1,i_2,\ldots,i_N,k_1,\ldots,k_N} \mathcal{B}_{k_1,\ldots k_N,k_{N+1},k_{N+2},\ldots,k_M} \tag{1}$$

$\mathcal{A} \in \mathbb{C}^{I_1 \times \ldots \times I_N \times K_1 \ldots \times K_N}$, $\mathcal{B} \in \mathbb{C}^{K_1 \times \ldots \times K_N \times K_{N+1} \ldots \times K_M}$, *and* $\mathcal{A} *_N \mathcal{B} \in \mathbb{C}^{I_1 \times \ldots \times I_N \times K_{N+1} \ldots \times K_M}$.

The Einstein product is a special case of tensor contracted product defined in [1] where contraction is over N consecutive modes. Furthermore, for tensors $\mathcal{X}, \mathcal{Y} \in \mathbb{C}^{I_1 \times I_2 \times \ldots \times I_N}$ and $\mathcal{Z} \in \mathbb{C}^{J_1 \times J_2 \times \ldots \times J_M}$, using Einstein product we can defined the following:

$$\textit{Inner Product}: \langle \mathcal{X}, \mathcal{Y} \rangle = \sum_{i_1=1}^{I_1} \sum_{i_2=1}^{I_2} \cdots \sum_{i_N=1}^{I_N} \mathcal{X}_{i_1,i_2,\ldots,i_N} \mathcal{Y}^*_{i_1,i_2,\ldots,i_N} = \mathcal{X} *_N \mathcal{Y}^* \quad (2)$$

$$\textit{Outer Product}: (\mathcal{X} \circ \mathcal{Z})_{i_1,i_2,\ldots,i_N,j_1,j_2,\ldots,j_M} = \mathcal{X}_{i_1,i_2,\ldots,i_N} \mathcal{Z}_{j_1,j_2,\ldots,j_M} = \mathcal{X} *_0 \mathcal{Z} \quad (3)$$

$$\textit{Norm}: ||\mathcal{X}|| = \sqrt{(\mathcal{X} *_N \mathcal{X}^*)} \quad (4)$$

A bijective matrix transformation is defined in [15] where a tensor $\mathcal{A} \in \mathbb{C}^{I_1 \times \ldots \times I_N \times J_1 \times \ldots \times J_M}$ is mapped to a matrix $\mathrm{A} \in \mathbb{C}^{I_1 \cdots I_N \times J_1 \cdots J_M}$ through transformation $f_{N|M}(\mathcal{A}) = \mathrm{A}$, such that

$$\mathcal{A}_{i_1,i_2,\ldots,i_N,j_1,j_2,\ldots,j_M} = \mathrm{A}_{i_1 + \sum_{k=2}^{N} (i_k - 1) \prod_{l=1}^{k-1} I_l, j_1 + \sum_{k=2}^{M} (j_k - 1) \prod_{l=1}^{k-1} J_l}. \quad (5)$$

The bar in subscript of $f_{N|M}$ represents the partitioning after N modes of an $N + M$ order tensor where first N modes correspond to the rows of the representing matrix, and the last M modes correspond to the columns of the representing matrix. The following lemma is shown in [15,17]:

Lemma 1. *For $\mathcal{A} \in \mathbb{C}^{I_1 \times \ldots \times I_N \times J_1 \times \ldots \times J_M}$ and $\mathcal{B} \in \mathbb{C}^{J_1 \times \ldots \times J_M \times K_1 \times \ldots \times K_P}$, we have $f_{N|P}(\mathcal{A} *_M \mathcal{B}) = f_{N|M}(\mathcal{A}) \cdot f_{M|P}(\mathcal{B})$ where $\cdot$ refers to usual matrix multiplication.*

Using the Einstein product and Lemma 1, we can extend many linear algebra concepts to tensor algebra as follows:

1. A tensor $\mathcal{A} \in \mathbb{C}^{I_1 \times \ldots \times I_N \times J_1 \times \ldots \times J_M}$ is called a *square* tensor if $N = M$ and $I_k = J_k$ for $k = 1, \ldots, N$ [16].
2. A square tensor, say $\mathcal{D} \in \mathbb{C}^{I_1 \times \ldots \times I_N \times I_1 \times \ldots \times I_N}$ is pseudo-diagonal if all its entries $\mathcal{D}_{i_1,\ldots,i_N,j_1,\ldots,j_N}$ are zero except when $i_1 = j_1, i_2 = j_2, \ldots, i_N = j_N$. Such a tensor is referred as diagonal tensor in [15], but we call it pseudo-diagonal for our purpose of discussion, so as to make a clear distinction from another diagonal tensor definition found in literature which states that a diagonal tensor is one where entries $\mathcal{D}_{i_1,\ldots,i_N}$ are zero except when $i_1 = i_2 \cdots = i_N$ [1]. A pictorial illustration of pseudo-diagonal structure of tensors is provided in [9].
3. An *identity tensor*, $\mathcal{I} \in \mathbb{C}^{I_1 \times \ldots \times I_N \times I_1 \times \ldots \times I_N}$ is a square pseudo-diagonal tensor where $\mathcal{I}_{i_1,\ldots,i_n,i_1\ldots,i_N} = 1$ for all $i_1, \ldots, i_N$.
4. The tensor $\mathcal{A}^{-1} \in \mathbb{C}^{I_1 \times \ldots \times I_N \times I_1 \times \ldots \times I_N}$ is an *inverse* of a square tensor of same size, $\mathcal{A} \in \mathbb{C}^{I_1 \times \ldots \times I_N \times I_1 \times \ldots \times I_N}$ if $\mathcal{A} *_N \mathcal{A}^{-1} = \mathcal{A}^{-1} *_N \mathcal{A} = \mathcal{I}$.
5. The *Hermitian* of a tensor $\mathcal{A} \in \mathbb{C}^{I_1 \times \ldots \times I_N \times J_1 \times \ldots \times J_M}$ is given by a tensor $\mathcal{B} \in \mathbb{C}^{J_1 \times \ldots \times J_M \times I_1 \times \ldots \times I_N}$ with entries $\mathcal{B}^*_{j_1,j_2,\ldots,j_M,i_1,i_2,\ldots,i_N} = \mathcal{A}_{i_1,i_2,\ldots,i_N,j_1,j_2,\ldots,j_M}$ and is denoted as $\mathcal{A}^H$. Similarly the transpose of a tensor is denoted as $\mathcal{A}^T$.

6. A square tensor $\mathcal{X} \in \mathbb{C}^{I_1\times\ldots\times I_N\times I_1\times\ldots\times I_N}$ is called *Hermitian tensor* if $\mathcal{X} = \mathcal{X}^H$.
7. A square tensor, $\mathcal{U} \in \mathbb{C}^{I_1\times\ldots\times I_N\times I_1\times\ldots\times I_N}$ is *unitary* if $\mathcal{U}^H *_N \mathcal{U} = \mathcal{U} *_N \mathcal{U}^H = \mathcal{I}$.
8. Einstein product is not commutative in general. However for the specific case where the contraction is taken over all the N modes of one of the tensors, say for tensors $\mathcal{A} \in \mathbb{C}^{I_1\times\ldots\times I_P\times J_1\times\ldots\times J_N}$ and $\mathcal{B} \in \mathbb{C}^{J_1\times\ldots\times J_N}$, the following can be established:

$$\mathcal{A} *_N \mathcal{B} = \mathcal{B} *_N \mathcal{A}^T \Rightarrow (\mathcal{A} *_N \mathcal{B})^* = \mathcal{B}^* *_N \mathcal{A}^H \tag{6}$$

9. Let $\mathcal{A} \in \mathbb{C}^{I_1\times\ldots\times I_N\times I_1\times\ldots\times I_N}, \mathcal{X} \in \mathbb{C}^{I_1\times\ldots\times I_N}$, $\lambda \in \mathbb{C}$, where $\mathcal{X}$ and λ satisfy $\mathcal{A} *_N \mathcal{X} = \lambda\mathcal{X}$, then we call $\mathcal{X}$ and λ as *eigentensor* and *eigenvalue* of $\mathcal{A}$ respectively [18].
10. The *tensor Eigenvalue Decomposition (EVD)* of a Hermitian tensor $\mathcal{A} \in \mathbb{C}^{I_1\times\ldots\times I_N\times I_1\times\ldots\times I_N}$ can be written as $\mathcal{A} = \mathcal{U} *_N \mathcal{D} *_N \mathcal{U}^H$ where $\mathcal{U} \in \mathbb{C}^{I_1\times\ldots\times I_N\times I_1\times\ldots\times I_N}$ is a unitary tensor and $\mathcal{D} \in \mathbb{C}^{I_1\times\ldots\times I_N\times I_1\times\ldots\times I_N}$ is a square pseudo-diagonal tensor with its non-zero values being the eigenvalues of $\mathcal{A}$ and $\mathcal{U}$ containing the eigentensors of $\mathcal{A}$. Note that the eigenvalues of the tensor $\mathcal{A}$ are same as the eigenvalues of $f_{N|M}(\mathcal{A})$.
11. *Trace* of a tensor, denoted as $\text{tr}(\mathcal{A})$ is defined as the sum of its pseudo-diagonal entries, which is same as the sum of its eigenvalues. For two tensors $\mathcal{A}, \mathcal{B} \in \mathbb{C}^{I_1\times\ldots\times I_N}$ of same size and order N, we have

$$\mathcal{A} *_N \mathcal{B} = \mathcal{B} *_N \mathcal{A} = \text{tr}(\mathcal{A} \circ \mathcal{B}) = \text{tr}(\mathcal{B} \circ \mathcal{A}) \tag{7}$$

and in general for any two tensors $\mathcal{A} \in \mathbb{C}^{I_1\times\ldots\times I_N\times J_1\times\ldots\times J_M}$ and $\mathcal{B} \in \mathbb{C}^{J_1\times\ldots\times J_M\times I_1\times\ldots\times I_N}$, we have:

$$\text{tr}(\mathcal{A} *_M \mathcal{B}) = \text{tr}(\mathcal{B} *_N \mathcal{A}) \tag{8}$$

In view of all the definitions and properties presented above, we can generalize most of the linear algebraic relations of matrix and vector cases to their corresponding tensor relations using the Einstein product and Lemma 1.

Second Order Characteristics of Complex Random Tensors: The covariance of a tensor $\boldsymbol{\mathcal{X}} \in \mathbb{C}^{I_1\times I_2\times\ldots\times I_N}$ can be defined as a tensor of size $I_1 \times I_2 \times \ldots \times I_N \times I_1 \times I_2 \times \ldots \times I_N$ represented by $\mathcal{Q} = \mathbb{E}[(\boldsymbol{\mathcal{X}} - \mathcal{M}) \circ (\boldsymbol{\mathcal{X}} - \mathcal{M})^*]$ where $\mathcal{M} = \mathbb{E}[\boldsymbol{\mathcal{X}}]$ is the mean tensor. However, a complete second-order characterization also requires defining the pseudo-covariance which is given as $\tilde{\mathcal{Q}} = \mathbb{E}[(\boldsymbol{\mathcal{X}} - \mathcal{M}) \circ (\boldsymbol{\mathcal{X}} - \mathcal{M})]$. Similarly, cross covariance and cross pseudo-covariance between random tensors $\boldsymbol{\mathcal{X}}$ and $\boldsymbol{\mathcal{Y}}$ can be defined as $\mathcal{Q}_{\boldsymbol{\mathcal{X}}\boldsymbol{\mathcal{Y}}} = \mathbb{E}[(\boldsymbol{\mathcal{X}} - \mathbb{E}[\boldsymbol{\mathcal{X}}]) \circ (\boldsymbol{\mathcal{Y}} - \mathbb{E}[\boldsymbol{\mathcal{Y}}])^*]$ and $\tilde{\mathcal{Q}}_{\boldsymbol{\mathcal{X}}\boldsymbol{\mathcal{Y}}} = \mathbb{E}[(\boldsymbol{\mathcal{X}} - \mathbb{E}[\boldsymbol{\mathcal{X}}]) \circ (\boldsymbol{\mathcal{Y}} - \mathbb{E}[\boldsymbol{\mathcal{Y}}])]$ respectively. We define a complex random tensor to be *proper* if its pseudo-covariance vanishes, i.e. $\tilde{\mathcal{Q}} = 0_{\mathcal{T}}$.

2.3 Approximating Tensor Inversion

Solving tensor equations without resorting to matrix transformation has become an active area of research in the past few years [15–19]. The reasons for avoiding a matrix transformation of the tensors are many folds. In many applications, tensors arise naturally as part of the problem. Hence there is no point in unfolding the tensors for computations, only to revert back to the original structure afterwards, since it adds additional steps going to and from tensor space to matrix space. Also, in several applications particularly in Big Data, the tensors are stored using Tensor Train (TT) decomposition format for reducing storage complexity [20,21]. The TT decomposition breaks a higher order tensor into a set of sparsely connected lower order tensors called cores or components. Different cores are connected with each other using tensor contracted product over one or more modes. Hence any mathematical operation to be performed on the tensor should not require any rearranging of the data but should be able to act on the tensors in TT format. Algorithms to compute the Einstein product between tensors stored in TT format are presented in [22].

Among other tensor operations, one of the more frequently used one is tensor inversion [15]. Higher Order Bi-conjugate Gradient (HOBG) and Jacobi methods are presented in [15,23] for tensor inversion. In this paper, we present another numerical method of approximating tensor inversion, without using any matrix transformation, by extending the Newton Method (NM) used for matrix inversion of [24]. To find the inverse of tensor $\mathcal{A} \in \mathbb{C}^{I_1 \times \ldots \times I_N \times I_1 \times \ldots \times I_N}$, we use the following iterations:

$$\mathcal{B}^{(k+1)} = (2\mathcal{I}_N - \mathcal{B}^{(k)} *_N \mathcal{A}) *_N \mathcal{B}^{(k)} \tag{9}$$

where the initial $\mathcal{B}^{(0)}$ can be set to $a \cdot \mathcal{A}^H$. Using similar line of arguments as in Theorem 2 in [24], it can be shown that this method converges if $0 < a < 2/\sigma_{max}^2$ where σ_{max}^2 is the largest eigenvalue of $\mathcal{C} = \mathcal{A}^H *_N \mathcal{A}$. To avoid the calculation of σ_{max}^2, we use the bound suggested for matrix inversion in [25]. We define:

$$\lambda = m + s(\prod_{n=1}^{N} I_n - 1)^{1/2} \tag{10}$$

where $m = \text{tr}(\mathcal{C})/(\prod_{n=1}^{N} I_n)$ and $s^2 = \text{tr}(\mathcal{C} *_N \mathcal{C}^H)/(\prod_{n=1}^{N} I_n) - m^2$. It can be shown that $\lambda \geq \sigma_{max}^2$, hence we use $a = 2/\lambda$ to ensure convergence.

The time complexity of each iteration in the Newton method for finding tensor inversion using (9) mainly depends on the complexity of the Einstein product between tensors of order $2N$. Directly using the formula for Einstein product $\mathcal{A} *_N \mathcal{B}$ for tensors $\mathcal{A}, \mathcal{B} \in \mathbb{C}^{I_1 \times \ldots \times I_N \times I_1 \times \ldots \times I_N}$ requires $\mathcal{O}((I_1 \cdots I_N)^3)$ operations. More efficient algorithms to compute the Einstein product or tensor contraction using parallel processing are discussed in [22,26]. An advantage of Newton method is its fast convergence as compared to HOBG method. A comparison of number of required iterations by HOBG method and Jacobi method is presented in [15] which shows that HOBG outperforms Jacobi method. Here, we present a comparison between HOBG and NM by comparing the performance of both the algorithms to find the inverse of tensors, having i.i.d. zero mean unit variance circular complex Gaussian entries of various size and the number of iterations required to find the inverse were averaged over 100 different realizations of tensors of each size. Tolerance was kept as 10^{-6}. Figure 1 shows number of iterations against order of the tensor when each dimension has size 3. Clearly, as the order increases, iterations required by NM are significantly lower than HOBG. Similarly, Fig. 2 shows number of iterations against dimensions of individual domains for order 4 tensor, where similar observation can be made. The number of iterations required by NM do not increase drastically with the size of the tensor, as opposed to HOBG.

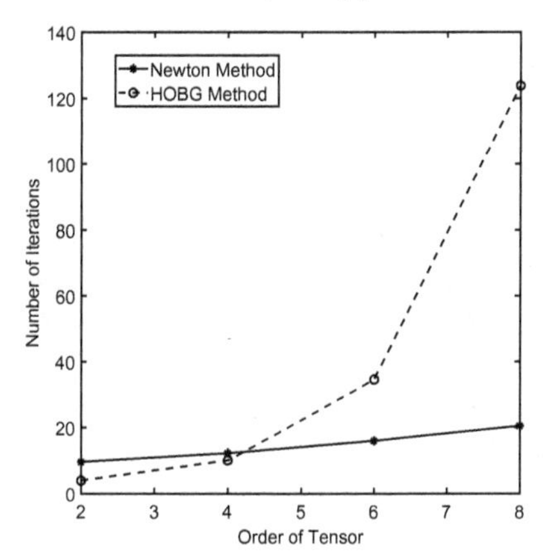

Fig. 1. Number of iterations vs Order of tensor with dimension 3 for Tensor Inversion Algorithms

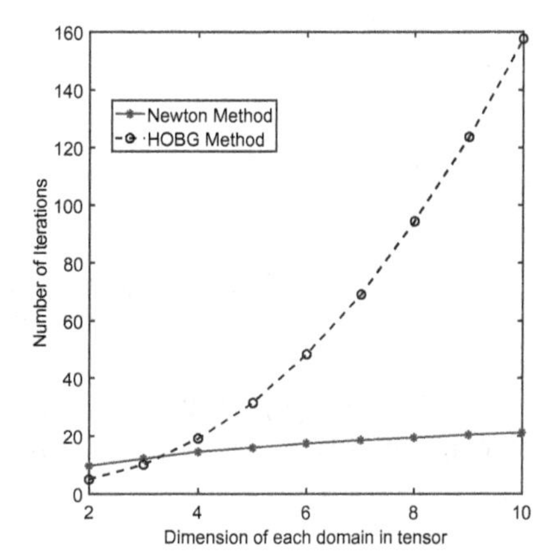

Fig. 2. Number of iterations vs Dimension of order 4 tensor for Tensor Inversion Algorithms

3 MMSE Estimation for Tensors

The general problem at hand is the estimation of a complex tensor $\mathcal{X} \in \mathbb{C}^{I_1 \times \ldots \times I_N}$ from an observed complex tensor $\mathcal{Y} \in \mathbb{C}^{J_1 \times \ldots \times J_M}$. Throughout this section, we will assume the observed tensor and the tensor to be estimated are proper and have zero mean.

3.1 Multi-linear MMSE Estimation Using Einstein Product

We are interested in the class of estimators which depend linearly on all the elements of the received tensor $\mathcal{Y}$. Hence the desired estimator takes the form:

$$\hat{\mathcal{X}} = g(\mathcal{Y}) = \mathcal{A} *_M \mathcal{Y} \tag{11}$$

where $\mathcal{A} \in \mathbb{C}^{I_1 \times \ldots \times I_N \times J_1 \times \ldots \times J_M}$. We will refer to the form of function $g(\mathcal{Y})$ defined by (11) as a multi-linear function of $\mathcal{Y}$ [15]. We first establish an Orthogonality principle for tensors through the following theorem:

Theorem 1. *Let* $g, h : \mathbb{C}^{J_1 \times \ldots \times J_M} \rightarrow \mathbb{C}^{I_1 \times \ldots \times I_N}$ *be multi-linear tensor valued functions of tensors such that* $g(\mathcal{Y}) = \mathcal{A} *_M \mathcal{Y}$ *and* $h(\mathcal{Y}) = \mathcal{B} *_M \mathcal{Y}$ *where* $\mathcal{A}, \mathcal{B} \in \mathbb{C}^{I_1 \times \ldots I_N \times J_1 \times \ldots \times J_M}$*. Let* $g(\mathcal{Y})$ *be an estimator of tensor* $\mathcal{X} \in \mathbb{C}^{I_1 \times \ldots \times I_N}$ *based on the observation tensor* $\mathcal{Y} \in \mathbb{C}^{J_1 \times \ldots \times J_M}$*. Then for the error tensor* $\mathcal{E} = \mathcal{X} - g(\mathcal{Y})$*, if*

$$\mathbb{E}[\langle \mathcal{E}, h(\mathcal{Y}) \rangle] = 0 \quad \textit{for all } \mathcal{B} \in \mathbb{C}^{I_1 \times \ldots I_N \times J_1 \times \ldots \times J_M} \tag{12}$$

then,

$$\mathbb{E}[||\mathcal{E}||^2] \leq \mathbb{E}[||\mathcal{X} - h(\mathcal{Y})||^2] \quad \textit{for all } \mathcal{B} \in \mathbb{C}^{I_1 \times \ldots I_N \times J_1 \times \ldots \times J_M} \tag{13}$$

Proof.

$$\mathbb{E}[||\mathcal{X} - h(\mathcal{Y})||^2] = \mathbb{E}[||\underbrace{\mathcal{X} - g(\mathcal{Y})}_{\mathcal{E}} + \underbrace{g(\mathcal{Y}) - h(\mathcal{Y})}_{\bar{h}(\mathcal{Y})}||^2] \tag{14}$$

$$= \mathbb{E}[(\mathcal{E} + \bar{h}(\mathcal{Y})) *_N (\mathcal{E} + \bar{h}(\mathcal{Y}))^*] \tag{15}$$

$$= \mathbb{E}[\mathcal{E} *_N \mathcal{E}^* + \mathcal{E} *_N \bar{h}(\mathcal{Y})^* + \bar{h}(\mathcal{Y}) *_N \mathcal{E}^* + \bar{h}(\mathcal{Y}) *_N \bar{h}(\mathcal{Y})^*] \tag{16}$$

$$= \mathbb{E}[||\mathcal{E}||^2] + \underbrace{\mathbb{E}[\mathcal{E} *_N \bar{h}(\mathcal{Y})^*] + \mathbb{E}[\bar{h}(\mathcal{Y}) *_N \mathcal{E}^*]}_{\text{cross-terms}} + \mathbb{E}[||\bar{h}(\mathcal{Y})||^2] \tag{17}$$

Since $\bar{h}(\mathcal{Y}) = g(\mathcal{Y}) - h(\mathcal{Y}) = (\mathcal{A} - \mathcal{B}) *_M \mathcal{Y}$ is just another multi-linear function of $\mathcal{Y}$, hence if (12) holds, then the cross terms above would be zero, which results into:

$$\mathbb{E}[||\mathcal{X} - h(\mathcal{Y})||^2] = \mathbb{E}[||\mathcal{E}||^2] + \underbrace{\mathbb{E}[||\bar{h}(\mathcal{Y})||^2]}_{\geq 0} \tag{18}$$

$$\geq \mathbb{E}[||\mathcal{E}||^2] \tag{19}$$

In order to estimate the tensor $\mathcal{X} \in \mathbb{C}^{I_1 \times \ldots \times I_N}$ from an observed complex tensor $\mathcal{Y} \in \mathbb{C}^{J_1 \times \ldots \times J_M}$, our objective is to find the tensor $\mathcal{A} \in \mathbb{C}^{I_1 \times \ldots I_N \times J_1 \times \ldots \times J_M}$ such that the estimator:

$$\hat{\mathcal{X}}_L = \mathcal{A} *_M \mathcal{Y} \tag{20}$$

satisfies

$$\mathbb{E}[||\mathcal{X} - (\mathcal{A} *_M \mathcal{Y})||^2] \leq \mathbb{E}[||\mathcal{X} - (\mathcal{B} *_M \mathcal{Y})||^2] \tag{21}$$

for any other tensor $\mathcal{B} \in \mathbb{C}^{I_1 \times \ldots \times I_N \times J_1 \times \ldots \times J_M}$. From Theorem 1, we know that the optimal $\mathcal{A}$ will be such that:

$$\mathbb{E}[\langle (\mathcal{X} - \hat{\mathcal{X}}_L), (\mathcal{B} *_M \mathcal{Y}) \rangle] = 0 \tag{22}$$

for any choice of $\mathcal{B} \in \mathbb{C}^{I_1 \times \ldots \times I_N \times J_1 \times \ldots \times J_M}$. From (2) and (20), we can write (22) as:

$$\mathbb{E}[(\mathcal{X} - \mathcal{A} *_M \mathcal{Y}) *_N (\mathcal{B} *_M \mathcal{Y})^*] = 0 \tag{23}$$

From (6), we can write $(\mathcal{B} *_M \mathcal{Y})^* = (\mathcal{Y}^* *_M \mathcal{B}^H)$, and hence the left hand side of (23) can be written as:

$$\mathbb{E}[(\mathcal{X} - \mathcal{A} *_M \mathcal{Y}) *_N (\mathcal{Y}^* *_M \mathcal{B}^H)] \tag{24}$$
$$= \mathbb{E}[\operatorname{tr}\{(\mathcal{X} - \mathcal{A} *_M \mathcal{Y}) \circ (\mathcal{Y}^* *_M \mathcal{B}^H)\}] \quad \text{(from (7))} \tag{25}$$
$$= \mathbb{E}[\operatorname{tr}\{\mathcal{X} \circ \mathcal{Y}^* *_M \mathcal{B}^H - \mathcal{A} *_M \mathcal{Y} \circ \mathcal{Y}^* *_M \mathcal{B}^H\}] \tag{26}$$
$$= \operatorname{tr}\{\underbrace{\mathbb{E}[\mathcal{X} \circ \mathcal{Y}^*]}_{\mathcal{C}_{\mathcal{X}\mathcal{Y}}} *_M \mathcal{B}^H - \mathcal{A} *_M \underbrace{\mathbb{E}[\mathcal{Y} \circ \mathcal{Y}^*]}_{\mathcal{C}_{\mathcal{Y}}} *_M \mathcal{B}^H\} \tag{27}$$
$$= \operatorname{tr}\{\underbrace{(\mathcal{C}_{\mathcal{X}\mathcal{Y}} - \mathcal{A} *_M \mathcal{C}_{\mathcal{Y}})}_{\bar{\mathcal{B}}} *_M \mathcal{B}^H\} \tag{28}$$

For (28) to be 0 for any $\mathcal{B}$, we need $\bar{\mathcal{B}}$ to be all zero tensor, which gives us the conditions for optimal $\mathcal{A}$:

$$\mathcal{C}_{\mathcal{X}\mathcal{Y}} = \mathcal{A} *_M \mathcal{C}_{\mathcal{Y}} \tag{29}$$

Equation (29) represents a system of multi-linear equations which can be solved for $\mathcal{A}$ using methods described in [15]. If the inverse of $\mathcal{C}_{\mathcal{Y}}$ does not exist, a minimum norm least square solution is suggested in [19] leading to Moore-Penrose inverse of tensors. If the inverse of $\mathcal{C}_{\mathcal{Y}}$ exists, then we have

$$\mathcal{A} = \mathcal{C}_{\mathcal{X}\mathcal{Y}} *_M \mathcal{C}_{\mathcal{Y}}^{-1} \tag{30}$$

and the multi-linear MMSE estimate of $\mathcal{X}$ is given by:

$$\hat{\mathcal{X}}_L = (\mathcal{C}_{\mathcal{X}\mathcal{Y}} *_M \mathcal{C}_{\mathcal{Y}}^{-1}) *_M \mathcal{Y} \tag{31}$$

With error tensor defined as $\mathcal{E} = (\mathcal{X} - \hat{\mathcal{X}}_L)$, the covariance of the corresponding error tensor is:

$$\mathcal{Q}_{\mathcal{E}} = \mathbb{E}[(\mathcal{X} - \hat{\mathcal{X}}_L) \circ (\mathcal{X} - \hat{\mathcal{X}}_L)^*] \tag{32}$$
$$= \mathbb{E}[\mathcal{X} \circ \mathcal{X}^*] - \mathbb{E}[\mathcal{X} \circ \hat{\mathcal{X}}_L^*] - \mathbb{E}[\hat{\mathcal{X}}_L \circ \mathcal{X}^*] + \mathbb{E}[\hat{\mathcal{X}}_L \circ \hat{\mathcal{X}}_L^*] \tag{33}$$

Using (6), we know that $(\mathcal{A} *_M \mathcal{Y})^* = (\mathcal{Y}^* *_M \mathcal{A}^H)$. So individual terms in (33) can be simplified as:

$$\mathbb{E}[\mathcal{X} \circ \hat{\mathcal{X}}_L^*] = \mathbb{E}[\mathcal{X} \circ (\mathcal{A}^* *_M \mathcal{Y}^*)] = \mathbb{E}[(\mathcal{X} \circ \mathcal{Y}^*) *_M \mathcal{A}^H] = \mathcal{C}_{\mathcal{X}\mathcal{Y}} *_M \mathcal{A}^H \tag{34}$$

$$\mathbb{E}[\hat{\mathcal{X}}_L \circ \mathcal{X}^*] = \mathbb{E}[(\mathcal{A} *_M \mathcal{Y}) \circ \mathcal{X}^*] = \mathcal{A} *_M \mathbb{E}[\mathcal{Y} \circ \mathcal{X}^*] = \mathcal{A} *_M \mathcal{C}_{\mathcal{X}\mathcal{Y}}^H \tag{35}$$

$$\mathbb{E}[\hat{\mathcal{X}}_L \circ \hat{\mathcal{X}}_L^*] = \mathbb{E}[(\mathcal{A} *_M \mathcal{Y}) \circ (\mathcal{A}^* *_M \mathcal{Y}^*)] = \mathbb{E}[(\mathcal{A} *_M \mathcal{Y}) \circ (\mathcal{Y}^* *_M \mathcal{A}^H)] \tag{36}$$
$$= \mathcal{A} *_M \mathcal{C}_{\mathcal{Y}} *_M \mathcal{A}^H = \mathcal{C}_{\mathcal{X}\mathcal{Y}} *_M \mathcal{A}^H \quad \text{(using (29))} \tag{37}$$

Substituting $\mathbb{E}[\mathcal{X} \circ \mathcal{X}^*] = \mathcal{C}_{\mathcal{X}}$ along with (34), (35) and (37) into (33), we get:

$$\mathcal{Q}_{\mathcal{E}} = \mathcal{C}_{\mathcal{X}} - \mathcal{A} *_M \mathcal{C}_{\mathcal{X}\mathcal{Y}}^H \tag{38}$$

The corresponding mean square error is given by

$$MSE = \mathbb{E}[||\mathcal{E}||^2] = \mathbb{E}\Big[\sum_{i_1,\ldots,i_N} |\mathcal{E}_{i_1,\ldots,i_N}|^2\Big] = \sum_{i_1,\ldots,i_N} \mathbb{E}[|\mathcal{E}_{i_1,\ldots,i_N}|^2] = \operatorname{tr}(\mathcal{Q}_{\mathcal{E}})$$
$$= \operatorname{tr}(\mathcal{C}_{\mathcal{X}} - \mathcal{A} *_M \mathcal{C}_{\mathcal{X}\mathcal{Y}}^H) = \operatorname{tr}(\mathcal{C}_{\mathcal{X}} - \mathcal{C}_{\mathcal{X}\mathcal{Y}} *_M \mathcal{C}_{\mathcal{Y}}^{-1} *_M \mathcal{C}_{\mathcal{X}\mathcal{Y}}^H) \tag{39}$$

3.2 Comparison with Tucker Operator Based MMSE Filter

A commonly used approach for multi-dimensional MMSE filtering is the n-mode Wiener filtering which makes use of the Tucker operator [13,27,28]. The objective is to estimate $\mathcal{X} \in \mathbb{C}^{I_1 \times \ldots \times I_N}$ based on observation $\mathcal{Y} \in \mathbb{C}^{I_1 \times \ldots \times I_N}$. Note that this class of estimators assume that the observation and the signal to be estimated have same order and dimensions. The estimate of the signal tensor $\mathcal{X}$ is achieved by N n-mode filters $\mathrm{A}^{(n)} \in \mathbb{C}^{I_n \times I_n}$ using Tucker operator as follows [28]:

$$\hat{\mathcal{X}}_T = \mathcal{Y} \times_1 \mathrm{A}^{(1)} \times_2 \mathrm{A}^{(2)} \times_3 \ldots \times_N \mathrm{A}^{(N)} \tag{40}$$

where $\times_n$ denotes the n-mode product [1]. The criteria to obtain the optimal n-mode filters $\mathrm{A}^{(n)}$ is the minimization of the mean square error between $\mathcal{X}$ and $\hat{\mathcal{X}}_T$, defined as:

$$e(\mathrm{A}^{(1)}, \mathrm{A}^{(2)}, \ldots, \mathrm{A}^{(N)}) = \mathbb{E}[||\mathcal{X} - \hat{\mathcal{X}}_T||^2] \tag{41}$$

$$= \mathbb{E}[||\mathcal{X} - \mathcal{Y} \times_1 \mathrm{A}^{(1)} \times_2 \mathrm{A}^{(2)} \times_3 \ldots \times_N \mathrm{A}^{(N)}||^2] \tag{42}$$

The optimal choice of the n-mode filters $\mathrm{A}^{(n)}$ which ensures minimum mean-square error between $\mathcal{X}$ and $\hat{\mathcal{X}}_T$ is calculated by finding different n-mode matrix unfolding of the tensor $\mathcal{Y}$, and using Wiener filtering on each matrix unfolding [27]. For finding $\mathrm{A}^{(n)}$ for each n, it is assumed that $\mathrm{A}^{(m)}$ for $m \neq n$ are known. Hence an alternative least squares approach is required to calculate all the optimal $\mathrm{A}^{(n)}$ where $\mathrm{A}^{(m)}$ for $m \neq n$ is fixed to find $\mathrm{A}^{(n)}$ for all n, and then we repeat for all n until a convergence criteria is met. A detailed derivation of the solution and the algorithm to calculate the optimal n-mode matrix filters are presented in [13,27,28].

Note that such a Tucker based estimator can be seen as a specific case of the multi-linear MMSE estimator presented in this paper. On writing (40) element-wise, we get:

$$\hat{\mathcal{X}}_{Ti_1,\ldots,i_N} = \sum_{j_N=1}^{I_N} \cdots \sum_{j_1=1}^{I_1} \mathcal{Y}_{j_1,\ldots,j_N} \cdot \mathrm{A}^{(1)}_{j_1,i_1} \cdot \mathrm{A}^{(2)}_{j_2,i_2} \cdots \mathrm{A}^{(N)}_{j_N,i_N} \tag{43}$$

We define a tensor $\mathcal{A} \in \mathbb{C}^{I_1 \times \ldots \times I_N \times I_1 \times \ldots \times I_N}$ such that

$$\mathcal{A}_{i_1,\ldots,i_N,j_1,\ldots,j_N} = \mathrm{A}^{(1)}_{j_1,i_1} \cdot \mathrm{A}^{(2)}_{j_2,i_2} \cdots \mathrm{A}^{(N)}_{j_N,i_N} \tag{44}$$

In this case we can re-write (43) as:

$$\hat{\mathcal{X}}_{Ti_1,\ldots,i_N} = \sum_{j_N=1}^{I_N} \cdots \sum_{j_1=1}^{I_1} \mathcal{Y}_{j_1,\ldots,j_N} \cdot \mathcal{A}_{i_1,\ldots,i_N,j_1,\ldots,j_N} \tag{45}$$

$$\Rightarrow \hat{\mathcal{X}}_T = \mathcal{A} *_N \mathcal{Y} \tag{46}$$

The solution for the optimal tensor $\mathcal{A}$ which minimizes the mean square-error between $\mathcal{X}$ and $\hat{\mathcal{X}}_T$ in (46) is the multi-linear MMSE estimator as given by (30). Hence the Tucker based estimator can be seen as a special case of the Einstein product based estimator with an additional constraint that assumes the elements of tensor $\mathcal{A}$ have a form as in (43).

4 Application of Tensor Multi-linear MMSE Estimation

In this section, we explore such an application in multi-domain communication systems. One of the standard tasks of a receiver in a MIMO communication system is to estimate the transmitted vector $\underline{\mathbf{x}}$ based on a noisy observation $\underline{\mathbf{y}} = \mathrm{H}\underline{\mathbf{x}} + \underline{\mathbf{n}}$ where $\underline{\mathbf{n}}$ is the additive noise vector and H is the channel matrix. If H is known, a common approach is to apply a linear MMSE filter. However, in most modern communication systems, the transmitted and received signals have an inherent multi-linear structure and they can be considered as tensors rather than vectors. A communication system can span multiple domains such as space, time, frequency, users, spreading sequence, to name a few. Hence, tensors can be used to model such multi-domain communication systems [9]. For a generic case, let us consider an order N input $\boldsymbol{\mathcal{X}} \in \mathbb{C}^{I_1 \times \ldots \times I_N}$ and order M output $\boldsymbol{\mathcal{Y}} \in \mathbb{C}^{J_1 \times \ldots \times J_M}$, and hence the channel $\mathcal{H} \in \mathbb{C}^{J_1 \times \ldots \times J_M \times I_1 \times \ldots \times I_N}$ as an order $N+M$ tensor. The system model can be written as:

$$\boldsymbol{\mathcal{Y}} = \mathcal{H} *_N \boldsymbol{\mathcal{X}} + \boldsymbol{\mathcal{N}} \tag{47}$$

Such a system model can be used to represent multi-domain communication systems, such as MIMO OFDM and MIMO GFDM, where the tensor based MMSE estimation techniques developed in this paper can be employed at the receiver. For a known channel $\mathcal{H}$, assuming $\boldsymbol{\mathcal{X}}$ and $\boldsymbol{\mathcal{N}}$ to be independent and zero mean, using (47) we can write the received covariance and cross covariance tensors as:

$$\mathcal{C}_{\boldsymbol{\mathcal{Y}}} = \mathcal{H} *_N \mathcal{C}_{\boldsymbol{\mathcal{X}}} *_N \mathcal{H}^H + \mathcal{C}_{\boldsymbol{\mathcal{N}}} \tag{48}$$

$$\mathcal{C}_{\boldsymbol{\mathcal{X}}\boldsymbol{\mathcal{Y}}} = \mathcal{C}_{\boldsymbol{\mathcal{X}}} *_N \mathcal{H}^H \tag{49}$$

Substituting (48) and (49) into (31) and (39) gives the receiver structure based on multi-linear estimation and the associated mean square error, respectively. The multi-linear MMSE estimate of $\boldsymbol{\mathcal{X}}$ is given as:

$$\hat{\boldsymbol{\mathcal{X}}}_L = \mathcal{C}_{\boldsymbol{\mathcal{X}}} *_N \mathcal{H}^H *_M (\mathcal{H} *_N \mathcal{C}_{\boldsymbol{\mathcal{X}}} *_N \mathcal{H}^H + \mathcal{C}_{\boldsymbol{\mathcal{N}}})^{-1} *_M \boldsymbol{\mathcal{Y}} \tag{50}$$

Let's assume that the transmitted tensor has symbols normalised to unit power with transmit covariance $\mathcal{C}_{\boldsymbol{\mathcal{X}}} = \mathcal{I}_N$ which is an identity tensor of size $I_1 \times \ldots \times I_N \times I_1 \times \ldots \times I_N$ such that $\mathrm{tr}(\mathcal{C}_{\boldsymbol{\mathcal{X}}}) = I_1 \cdot I_2 \cdots I_N$. Let the noise be additive circular Gaussian with mean zero and variance σ^2, such that its covariance is $\mathcal{C}_{\boldsymbol{\mathcal{N}}} = \sigma^2 \mathcal{I}_M$. Hence, the mean square error from multi-linear estimation (based on (39)) can be written as:

$$MSE = \mathrm{tr}\left(\mathcal{C}_{\boldsymbol{\mathcal{X}}} - \mathcal{C}_{\boldsymbol{\mathcal{X}}} *_N \mathcal{H}^H *_M (\mathcal{H} *_N \mathcal{C}_{\boldsymbol{\mathcal{X}}} *_N \mathcal{H}^H + \mathcal{C}_{\boldsymbol{\mathcal{N}}})^{-1} *_M \mathcal{H} *_N \mathcal{C}_{\boldsymbol{\mathcal{X}}}\right) \tag{51}$$

$$\begin{aligned} &= \mathrm{tr}\left(\mathcal{I}_N - \mathcal{H}^H *_M (\mathcal{H} *_N \mathcal{H}^H + \sigma^2 \mathcal{I}_M)^{-1} *_M \mathcal{H}\right) \\ &= \mathrm{tr}(\mathcal{I}_N) - \mathrm{tr}(\mathcal{H} *_N \mathcal{H}^H *_M (\mathcal{H} *_N \mathcal{H}^H + \sigma^2 \mathcal{I}_M)^{-1}) \quad \text{(from (8))} \\ &= \prod_{n=1}^{N} I_n - \sum_{j_1,\ldots,j_M} \left(\frac{d_{j_1,\ldots,j_M}}{d_{j_1,\ldots,j_M} + \sigma^2}\right) \end{aligned} \tag{52}$$

where $d_{j_1,\ldots,j_M}$ represents the eigenvalues of the tensor $\mathcal{H} *_N \mathcal{H}^H$. Since $\mathcal{H} *_N \mathcal{H}^H$ and $\mathcal{H}^H *_M \mathcal{H}$ have same non-zero eigenvalues, we can also write the mean square error as:

$$MSE = \prod_{n=1}^{N} I_n - \sum_{\substack{i_1,\ldots,i_N \\ d_{i_1,\ldots,i_N} \neq 0}} \frac{1}{1+\sigma^2/d_{i_1,\ldots,i_N}} \tag{53}$$

where $d_{i_1,\ldots,i_N}$ represents the eigenvalues of $\mathcal{H}^H *_M \mathcal{H}$. The above expression shows how the mean square error scales with noise power and the channel strength which is reflected in the eigenvalues of $\mathcal{H}^H *_M \mathcal{H}$. At very low noise power, the mean square error can be written as:

$$\lim_{\sigma^2 \to 0} MSE = \prod_{n=1}^{N} I_n - \sum_{\substack{i_1,\ldots,i_N \\ d_{i_1,\ldots,i_N} \neq 0}} 1 = \prod_{n=1}^{N} I_n - \Psi \tag{54}$$

where Ψ denotes the number of non-zero eigenvalues of $\mathcal{H}^H *_M \mathcal{H}$. If all the eigenvalues of $\mathcal{H}^H *_M \mathcal{H}$ are non zero, then $\Psi = \prod_{n=1}^{N} I_n$ and hence as $\sigma^2 \to 0$, mean square error tends to zero. This is also intuitive, as when noise power goes to zero and the channel is known, we can recover the transmitted signal perfectly. For non-zero noise power and a given size of transmit symbol, we can see that as the eigenvalues of $\mathcal{H}^H *_M \mathcal{H}$ increase, the mean square error decreases.

4.1 MIMO OFDM System

A conventional system model for MIMO OFDM in the frequency domain is given by [29]:

$$\underline{\mathbf{y}}^{(p)} = \mathrm{H}^{(p,p)}\underline{\mathbf{x}}^{(p)} + \sum_{q=1,q\neq p}^{N_{sc}} \mathrm{H}^{(p,q)}\underline{\mathbf{x}}^{(q)} + \underline{\mathbf{n}}^{(p)}, \quad p = 1,\ldots,N_{sc} \tag{55}$$

where $\underline{\mathbf{y}}^{(p)} \in \mathbb{C}^{N_R}, \underline{\mathbf{x}}^{(p)} \in \mathbb{C}^{N_T}$ and $\underline{\mathbf{n}}^{(p)} \in \mathbb{C}^{N_R}$ are the frequency domain received, transmitted and noise symbol vectors at sub-carrier p, and N_R, N_T and N_{sc} denote the number of receive antennas, transmit antennas and sub-carriers respectively. The frequency domain channel matrix of size $N_R \times N_T$ between transmit sub-carrier q and receive sub-carrier p is $\mathrm{H}^{(p,q)}$.

One common approach used in MIMO OFDM receiver design is to assume that there is no inter-carrier interference, i.e. $\mathrm{H}^{(p,q)} = 0$ if $p \neq q$, and perform linear MMSE estimation on a per sub-carrier basis [30–32]. Assuming that the input $\underline{\mathbf{x}}^{(p)}$ have independent zero mean unit variance entries such that input covariance is an identity matrix, $\mathrm{C}_{\underline{\mathbf{x}}^{(p)}} = \mathrm{I}$ for each p, and is independent of noise $\underline{\mathbf{n}}^{(p)}$, the LMMSE receiver structure with per sub-carrier estimation is given as [30]:

$$\hat{\underline{\mathbf{x}}}^{(p)} = \mathrm{H}^{(p,p)H} \cdot (\mathrm{H}^{(p,p)} \cdot \mathrm{H}^{(p,p)H} + \mathrm{C}_N)^{-1} \cdot \underline{\mathbf{y}}^{(p)} \qquad \text{for } p = 1, \ldots, N_{sc} \tag{56}$$

where $\mathrm{C}_N \in \mathbb{C}^{N_R \times N_R}$ is the noise covariance matrix. The per-sub-carrier estimation in (56) is based on the standard linear MMSE filter used in matrix based systems. This can also be seen as a special case of the multi-linear MMSE estimator presented in (50) with vector (order 1) input, output and matrix (order 2) channel where the Einstein product reduces to simple matrix multiplication.

Notice that a receiver based on (56) does not make use of the inter-carrier interference (ICI) terms to extract any signal information. In many cases such as where terminals have high mobility, channel will be doubly selective leading to a significant inter-carrier interference in which case ignoring the interference terms or treating them as noise would lead to sub-optimal performance. In literature, a doubly selective channel for MIMO OFDM is handled by concatenating the transmit and receive vectors for each sub-carrier into a long vector and thereby representing the channel by a large matrix [33,34]. However, the tensor framework is more intuitive as it retains the distinction between the domains. Besides, for future communication systems where more than 1024 sub-carriers or hundreds of antennas in case of massive MIMO are envisioned, the number of sub-carriers and antennas can be very large. Hence concatenating the input and output as vectors will result into a large matrix channel which is difficult to store and perform operations such as inversion. The system model suggested in (47) can be used in such a scenario where the input, output and noise can be treated as order two tensors with the two domains corresponding to antenna and sub-carriers and the channel as an order four tensor accounting for interference between antennas and between sub-carriers as well. The system model would become

$$\mathbf{Y} = \mathcal{H} *_2 \mathbf{X} + \mathbf{N} \tag{57}$$

where each vector $\underline{\mathbf{y}}^{(p)}$ and $\underline{\mathbf{n}}^{(p)}$ for $p = 1, \ldots, N_{sc}$ form the columns of matrices $\mathbf{Y}$ and $\mathbf{N}$ of size $N_R \times N_{sc}$, each vector $\underline{\mathbf{x}}^{(p)}$ form the columns of matrix $\mathbf{X} \in \mathbb{C}^{N_T \times N_{sc}}$. The channel connecting input and output is a fourth order tensor $\mathcal{H} \in \mathbb{C}^{N_R \times N_{sc} \times N_T \times N_{sc}}$ such that $\mathcal{H}_{:,p,:,q} = \mathrm{H}^{(p,q)}$ from (56). With such a system model in place, one can use the tensor based receiver structure from (31) as the tensor formulation gives an easy mechanism to take into account the information provided by the interfering terms as well. With input covariance as identity tensor and noise covariance as $\sigma^2\mathcal{I}$, the estimate of the transmitted signal can be given using the multi-linear estimator as (using (50)):

$$\hat{\mathbf{X}} = \mathcal{H}^H *_2 (\mathcal{H} *_2 \mathcal{H}^H + \sigma^2\mathcal{I})^{-1} *_2 \mathbf{Y} \tag{58}$$

We present simulation results using the multi-linear MMSE estimation for a MIMO OFDM system with $N_{sc} = 64$ sub-carriers, 2 transmit and 2 receive antennas. The channel between each transmit and receive pair of antennas was generated as in [35,36]. The channel impulse response matrix between n_t^{th} transmit and n_r^{th} receive antenna denoted as $\bar{\mathrm{H}}^{(n_r,n_t)} \in \mathbb{C}^{N_{sc}\times N_{sc}}$ is generated by employing a two tap multipath fading channel following Jakes' model [37] using exponential power profile. This matrix is further converted to frequency domain using the DFT matrix $\mathrm{W} \in \mathbb{C}^{N_{sc}\times N_{sc}}$ with elements $\mathrm{W}_{m,n} = 1/\sqrt{N_{sc}}\exp(-j2\pi mn/N_{sc})$, which then forms the sub-tensor of the frequency domain channel tensor as $\mathcal{H}_{n_r,:,n_t,:} = W\bar{\mathrm{H}}^{(n_r,n_t)}W^H$. The channels are generated for different values of Doppler d, normalised to the symbol rate, to induce inter-carrier interference.

We consider three different cases based on the estimation technique that the receiver employs:
Case 1: Per sub-carrier estimation from (56) is used where interference term is completely ignored, such that $\mathrm{C}_N = \sigma^2 \cdot \mathrm{I}$.
Case 2: Per sub-carrier estimation from (56) is used while treating interference as noise, such that $\mathrm{C}_N = (\sigma^2 \cdot \mathrm{I} + \sum_{q\neq p} \mathrm{H}^{(p,q)} \cdot \mathrm{H}^{(p,q)H})$.
Case 3: Multi-linear MMSE estimation from (58) is used where the interference terms are also captured in the framework and aids in estimation.

We compare estimation in MIMO OFDM system with 4QAM constellation using all the three receivers. It is assumed that the channel is known at the receiver. The output of the estimator in all the cases is passed through a QAM demodulator to determine the transmitted symbol. The plots presented are calculated using Monte Carlo simulations with averaging over 500 channel realisations, and 100 bit errors were collected for each channel. The BER and MSE results are plotted against the SNR per bit defined as E_b/N_0. Figure 3 and 4 present the MSE and BER for all the three cases for different values of the Doppler parameter d respectively. The mean square error is normalised with the respect to the number of elements in transmit tensor. It can be clearly seen that as d increases, there is a significant performance degradation for case 1 and case 2 as compared to case 3. Case 2 slightly performs better than case 1 as it accounts for interference albeit as noise. Case 3, which is the tensor based estimation, exploits the information present in the interfering terms as well for estimating the transmitted symbols. Hence, it also shows robustness to a change in d, as the MSE or BER performance does not change significantly between $d = 0$ to $d = 2$. It can be observed in Fig. 3 that for case 1 at $d = 1$, the mean square error increases at very high SNR. This is because at high SNR, the receiver of case 1 tends to a zero forcing receiver which tries to invert the channel while ignoring the interference terms totally. Hence at high value of d when the interfering terms are dominant in the received signal, the channel inversion further amplifies the interference part of the received signal leading to higher mean square error. This is easily remedied by making simultaneous use of information from all the domains with the tensor multi-linear MMSE estimator, as done in case 3. Note that one limitation of the scheme is that it assumes perfect channel state information is available at the receiver.

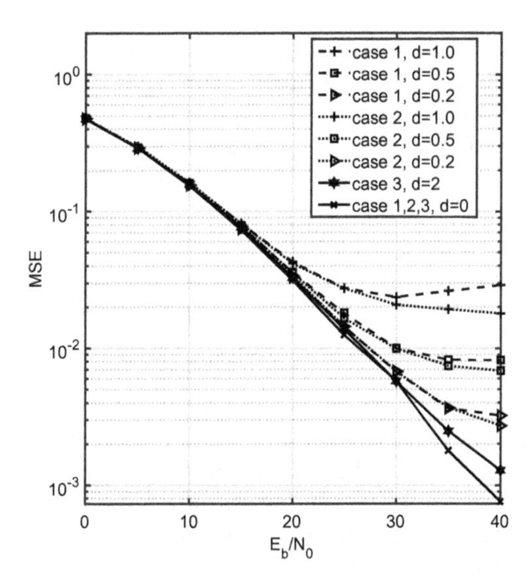

Fig. 3. MSE vs E_b/N_0 for 2×2 MIMO OFDM system with 4QAM modulation for different Doppler values, d using tensor and per sub-carrier estimation

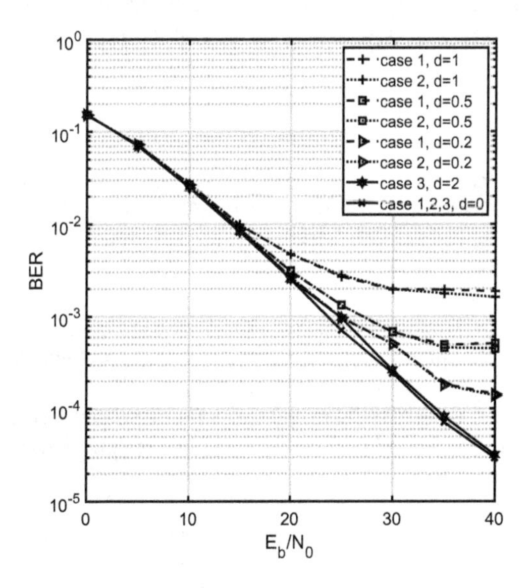

Fig. 4. BER vs E_b/N_0 for 2×2 MIMO OFDM system with 4QAM modulation for different Doppler values, d using tensor and per sub-carrier estimation

5 Conclusions

This paper presents a tensor framework for multi-linear estimation. Multi-linear MMSE estimation techniques have been formulated while keeping the tensor structure intact. The framework is developed using the Einstein product of tensors, whose properties enable a natural extension of basic linear algebra tools to a multi-linear setting. Furthermore, a numerical approach based on Newton method for approximating the inverse of a tensor without resorting to any matrix transformation was proposed. The tensor framework allows for a joint processing of the signal across all the domains, thereby capturing and harnessing mutual effects. As an example, we presented a tensor based system model for a MIMO OFDM communication system in a doubly selective channel, where the input and output can be seen as order two tensors and channel as an order four tensor, accounting for inter-carrier and inter-antenna interference in a single framework. It was shown that the tensor based MMSE estimation technique outperforms per sub-carrier estimation by a significant margin. For any such multi-domain communication scheme, the tensor framework can be used to represent the system model and the multi-linear MMSE estimation technique proposed in this paper can be employed at the receiver. The work presented here can be extended to develop estimation techniques for improper complex tensors also, which forms the basis of our future work.

Acknowledgment. This work was supported by grant RGPIN-2016-03647 entitled "Tensor modulation for space-time-frequency communication systems", awarded by the Natural Sciences and Engineering Research Council of Canada (NSERC).

References

1. Kolda, T.G., Bader, B.W.: Tensor decompositions and applications. SIAM Rev. **51**(3), 455–500 (2009). https://doi.org/10.1137/07070111X
2. Sidiropoulos, N.D., Giannakis, G.B., Bro, R.: Blind PARAFAC receivers for DS-CDMA systems. IEEE Trans. Signal Process. **48**(3), 810–823 (2000)
3. Cichocki, A., Mandic, D.P., Phan, A.H., Caiafa, C.F., Zhou, G., Zhao, Q., Lathauwer, L.D.: Tensor decompositions for signal processing applications from two-way to multiway component analysis, *CoRR*, vol. abs/1403.4462 (2014). http://arxiv.org/abs/1403.4462
4. Kuang, L., Hao, F., Yang, L.T., Lin, M., Luo, C., Min, G.: A tensor-based approach for big data representation and dimensionality reduction. IEEE Trans. Emerg. Top. Comput. **2**(3), 280–291 (2014)
5. Vervliet, N., Debals, O., Sorber, L., De Lathauwer, L.: Breaking the curse of dimensionality using decompositions of incomplete tensors: tensor-based scientific computing in big data analysis. IEEE Signal Process. Mag. **31**(5), 71–79 (2014)
6. Sidiropoulos, N.D., Lathauwer, L.D., Fu, X., Huang, K., Papalexakis, E.E., Faloutsos, C.: Tensor decomposition for signal processing and machine learning. IEEE Trans. Signal Process. **65**(13), 3551–3582 (2017)
7. Favier, G., de Almeida, A.L.F.: Tensor space-time-frequency coding with semi-blind receivers for mimo wireless communication systems. IEEE Trans. Signal Process. **62**(22), 5987–6002 (2014)
8. Fernandes, C.A., Favier, G., Mota, J.C.: PARAFAC-based channel estimation and data recovery in nonlinear MIMO spread spectrum communication systems. Signal Process. **91**(2), 311–322 (2011)

9. Venugopal, A., Leib, H.: A tensor based framework for multi-domain communication systems. IEEE Open J. Commun. Soc. **1**, 606–633 (2020)
10. Chen, C., Surana, A., Bloch, A., Rajapakse, I.: Multilinear time invariant system theory. In: 2019 Proceedings of the Conference on Control and its Applications, pp. 118–125. SIAM (2019). https://epubs.siam.org/doi/abs/10.1137/1.9781611975758.18
11. Han, X., de Almeida, A.L., Yang, Z.: Channel estimation for MIMO multi-relay systems using a tensor approach. EURASIP J. Adv. Signal Process. **2014**(1), 163 (2014)
12. Araújo, D.C., De Almeida, A.L., Da Costa, J.P., de Sousa, R.T.: Tensor-based channel estimation for massive MIMO-OFDM systems. IEEE Access **7**, 42 133–42 147 (2019)
13. Muti, D., Bourennane, S.: Multidimensional filtering based on a tensor approach. Signal Process. **85**(12), 2338–2353 (2005)
14. Marot, J., Fossati, C., Bourennane, S.: About advances in tensor data denoising methods. EURASIP J. Adv. Signal Process. **2008**(1), 1–12 (2008)
15. Brazell, M., Li, N., Navasca, C., Tamon, C.: Solving multilinear systems via tensor inversion. SIAM J. Matrix Anal. Appl. **34**(2), 542–570 (2013). https://doi.org/10.1137/100804577
16. Liang, M.-L., Zheng, B., Zhao, R.-J.: Tensor inversion and its application to the tensor equations with Einstein product. Linear Multilinear Algebra **67**(4), 843–870 (2019). https://doi.org/10.1080/03081087.2018.1500993
17. Wang, Q.-W., Xu, X.: Iterative algorithms for solving some tensor equations. Linear Multilinear Algebra **67**(7), 1325–1349 (2019). https://doi.org/10.1080/03081087.2018.1452889
18. Cui, L.-B., Chen, C., Li, W., Ng, M.K.: An Eigenvalue problem for even order tensors with its applications. Linear Multilinear Algebra **64**(4), 602–621 (2016). https://doi.org/10.1080/03081087.2015.1071311
19. Sun, L., Zheng, B., Bu, C., Wei, Y.: Moore-Penrose inverse of tensors via Einstein product. Linear Multilinear Algebra **64**(4), 686–698 (2016). https://doi.org/10.1080/03081087.2015.1083933
20. Oseledets, I.V.: Tensor-train decomposition. SIAM J. Sci. Comput. **33**(5), 2295–2317 (2011)
21. Cichocki, A.: Era of big data processing: a new approach via tensor networks and tensor decompositions, arXiv preprint arXiv:1403.2048, (2014)
22. Liu, H., Yang, L.T., Ding, J., Guo, Y., Yau, S.S.: Tensor-train-based high-order dominant eigen decomposition for multimodal prediction services. IEEE Trans. Eng. Manage. **68**(1), 1–15 (2019)
23. Can, C., Surana, A., Bloch, A., Rajapakse, I.: Multilinear Time Invariant System Theory (Longer version), arXiv preprint arXiv:1905.08783v1, (2019)
24. Pan, V., Schreiber, R.: An improved Newton iteration for the generalized inverse of a matrix, with applications. SIAM J. Sci. Stat. Comput. **12**(5), 1109–1130 (1991)
25. Wang, Y., Leib, H.: Sphere decoding for MIMO systems with Newton iterative matrix inversion. IEEE Commun. Lett. **17**(2), 389–392 (2013)
26. Matthews, D.A.: High-performance tensor contraction without transposition. SIAM J. Sci. Comput. **40**(1), C1–C24 (2018). https://doi.org/10.1137/16M108968X
27. Muti, D., Bourennane, S., Marot, J.: Lower-rank tensor approximation and multiway filtering. SIAM J. Matrix Anal. Appl. **30**(3), 1172–1204 (2008). https://doi.org/10.1137/060653263

28. Muti, D., Bourennane, S.: Multidimensional estimation based on a tensor decomposition. In: IEEE Workshop on Statistical Signal Processing, vol. 2003, pp. 98–101. IEEE (2003)
29. Stamoulis, A., Diggavi, S.N., Al-Dhahir, N.: Intercarrier interference in MIMO OFDM. IEEE Tran. Signal Process. **50**(10), 2451–2464 (2002)
30. Burg, A., Haene, S., Perels, D., Luethi, P., Felber, N., Fichtner, W.: Algorithm and VLSI architecture for linear MMSE detection in MIMO-OFDM systems. In: IEEE International Symposium on Circuits and Systems. IEEE (2006)
31. Liu, D.N., Fitz, M.P.: Low complexity affine MMSE detector for iterative detection-decoding MIMO OFDM systems. IEEE Trans. Commun. **56**(1), 150–158 (2008)
32. Wu, M., Yin, B., Wang, G., Dick, C., Cavallaro, J.R., Studer, C.: Large-scale MIMO detection for 3GPP LTE: algorithms and FPGA implementations. IEEE J. Sel. Top. Signal Process. **8**(5), 916–929 (2014)
33. Kim, K., Park, H.: Modified successive interference cancellation for MIMO OFDM on doubly selective channels. In: VTC Spring 2009-IEEE 69th Vehicular Technology Conference, pp. 1–5. IEEE (2009)
34. Rugini, L., Banelli, P., Fang, K., Leus, G.: Enhanced turbo MMSE equalization for MIMO-OFDM over rapidly time-varying frequency-selective channels. In: IEEE 10th Workshop on Signal Processing Advances in Wireless Communications, vol. 2009, pp. 36–40. IEEE (2009)
35. Yi, W., Leib, H.: OFDM symbol detection integrated with channel multipath gains estimation for doubly-selective fading channels. Phys. Commun. **22**, 19–31 (2017)
36. Teo, K.A.D., Ohno, S.: Pilot-aided channel estimation and viterbi equalization for OFDM over doubly-selective channel. In: IEEE Globecom 2006, pp. 1–5, November 2006
37. Alimohammad, A., Fard, S.F., Cockburn, B.F., Schlegel, C.: An improved SOS-based fading channel emulator. In: IEEE 66th Vehicular Technology Conference, vol. 2007, pp. 931–935. IEEE (2007)

Capacity Enhancement of High Throughput Low Earth Orbit Satellites in a Constellation (HTS-LEO) in a 5G Network

Arooj Mubashara Siddiqui[1(✉)], Barry Evans[1], Yingnan Zhang[2], and Pei Xiao[1]

[1] Home of 5G Innovation Centre, Institute for Communication Systems, University of Surrey, Guildford, Surrey GU2 7XH, UK
{a.siddiqui,b.evans,p.xiao}@surrey.ac.uk
[2] CAST, Xi'an, China
https://www.surrey.ac.uk/institute-communication-systems/5g-innovation-centre

Abstract. The global telecommunication market aims to fulfil future ubiquitous coverage and rate requirements by integrating terrestrial communications with multiple spot beam high throughput satellites (HTS). In this paper a new scheme is proposed to connect multiple Low Earth Orbit (LEO) satellites in a constellation to a single gateway to support integrated-satellite-terrestrial networks. A single gateway with multiple steerable antenna arrays is proposed for reduction in gateway numbers and cost. Using a power allocation strategy, the target is to maximize the gateway link capacity of the HTS-LEO satellites for operation including feasible used cases studies of 3GPP for necessary adaptations in a 5G system. Firstly, an objective function is established to find the optimal power levels required. Secondly the interference from neighbouring satellite beams is considered to achieve maximum capacity. Mathematical formulations are developed for this non-convex problem. Simulation results show that the proposed system architecture improves capacity and meets the dynamic demand better than traditional methods.

Keywords: Capacity · Leo satellite · Optimization · Power allocation · 5G network

1 Introduction

In the past few years, advancements in satellite technology have promised a massive increase in throughput delivered from narrow focused beam high throughput satellites (HTS) [6]. These high throughput satellites ensure the requirements of high capacity, and large coverage can be met. According to the 3rd Generation Partnership Project (3GPP) recent release 16, integrated satellite and terrestrial networks in a 5G infrastructure has attracted increasing attention from both

K. Arai (Ed.): FICC 2021, AISC 1363, pp. 65–76, 2021.
https://doi.org/10.1007/978-3-030-73100-7_5

academia and industry targeting high speed mobile users [2,14]. Thus extended 5G coverage could be achieved by such integrated systems.

Low Earth Orbit (LEO) satellites emerged as a solution to the large delays in Geo stationary satellites (GEO), while providing global coverage and enhanced capacity [7]. The integration of terrestrial systems with non- Geo stationary orbit (NGSO) satellite constellations aim to meet the 5G requirements by using multiple spot beams HTS-LEO's and has aroused significant commercial interests [16]. Major LEO satellite companies such as SpaceX, and OneWeb have started to launch satellites aim at providing low-cost internet access to remote locations [8,10]. Such systems are targeting service operation from 2021 [5].

In spite of the extensive available literature on LEO satellites, the problem of optimally allocating bandwidth to beams and optimally operating the payload from a power perspective according to the amount of requested traffic has not yet been addressed. One of the main difficulties is the large number of variables in the capacity and power allocation problem [8]. Traditionally, this problem has often been addressed from a ground segment viewpoint, the feeder-link capacity will be taken into account in the optimization framework for propagation impairments (e.g., rain) and interference from other beams. In [15], authors applied the beam-hopping technique with fixed signal-to-noise ratio for rate allocation on the basis of a fairness function, but in this strategy the remaining unused slots causes resource under utilization.

In multi-beam satellite systems, users from co-frequency interference beams especially those at the edges provide limiting performance. Fixed power allocation was adopted in [12] to each beam, leading to rate mis-match. More over, the golden section theory was used in [9] for a joint power and bandwidth resource allocation scheme, in which the total system capacity was improved to minimize the gap between bandwidth requirement and resource allocation. No existing works in the literature have considered how to maximize capacity by integrating multiple antenna gateways with multiple satellites especially at lower altitudes. Several dimensions in the HTS-LEO system design can be exploited to minimise the impact of the interference.

Recently the beam-forming technique has been introduced to the terrestrial-satellite architecture [11]. Using multiple-input multiple-output (MIMO) in the non-terrestrial domain reduces the operating cost and hardware complexity [17]. Although using many antennas at transmitter/receiver nodes is not feasible, a system model was proposed in [17] where multiple LEO satellites with spot beams are controlled by a single gateway.

In addition, the key challenge in LEO constellation is the reduction in gateway numbers and cost. Major satellite companies, e.g., SpaceX need to deploy a very large number of ground stations and gateways to operate at full power [1]. The challenge is to reduce the number of gateways which are costly in the case of LEO constellation. Therefore, in the proposed system model, a single gateway with multiple steerable antenna arrays is proposed for reduction in gateway numbers and cost. Using multiple beams from the same gateway cause interference and this needs to be taken account of in the capacity evaluation.

This paper focuses on a LEO constellation satellite in which at any designated time, multiple satellites of the constellation could communicate with a single fixed earth ground gateway station as long as the gateway is within the satellites' coverage. This scheme is similar to the concept of beam-forming where multiple transmitting antennas are connected to the multiple receiving antennas [3]. In existing studies on the feeder link capacity, the calculation of the received signal-to-noise plus interference ratio (SINR) at the satellite from gateway is generally based on standard power link budgets without considering the inter-feeder-beam interference.

The main contribution of this paper comes from the fact that we incorporate this type of interference into the feeder link capacity analysis. Specifically, we investigate and discuss the impact of capacity on the SINR from each LEO satellite. This paper presents a novel system model to calculate the received SINR at the satellite from a gateway on standard link budget with the inter-feeder-beam interference. To this end, we start by calculating the individual rates of each satellite and studying the impact of inter-beam interference, which is further used to measure the interference in the system from adjacent satellite beams. Next we utilize the non-convex Lagrangian duality method to optimize the power allocation. Also we propose cost reduction solution in LEO satellites by enabling a single gateway to simultaneously control more that one satellite. Finally we demonstrate that the proposed method can maximize the capacity by using different number of transmitting/receiving antennas.

The rest of the paper is organized as follows: Sect. 2 describes system model for the HTS-LEO system under study, Sect. 3 emphasis on the optimal power allocation technique. Finally, numerical results and discussion are given in Sect. 4, followed by the conclusion.

2 System Model

2.1 Model of HTS-LEO Satellite Communication System

This paper focuses on a scenario in which multiple LEO satellites in the same constellation could simultaneously communicate with one gateway. For the sake of simplicity, we assume at any time, a gateway could only view M spot beams from satellites with reasonable elevation angles, and the gateway has multiple steerable beams pointing at those multiple satellites.[1] The multiple-beam satellite communication system is shown in Fig. 1.[2] Also the terrestrial links to the gateway are traffic flows shown as dynamically changing users in Fig. 1.

[1] It should be noted that this analysis could be straightforwardly extended to any number of satellites within the gateway view field.

[2] In reality these satellite user beams are dynamic and change with respect to mobile users.

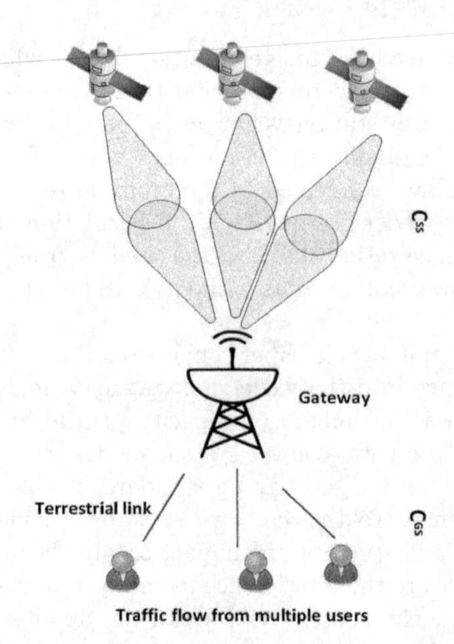

Fig. 1. System model.

2.2 Capable Allocation Model with Inter-beam Interference

We assume multiple co-interference paths in this system. The essential property of beams decides the inter-beam interference in-terms of main and side lobes. These side lobe beams will interfere with the adjacent satellite beams to create inter-beam interference. In this system model adjusting the power between the beams leads to a trade off between the reduction of capacity due to the link budget on the direct link and the interference between the beams. The latter would depend on the spacing between the satellites in the constellation.

As in the LEO satellites the distance from the earth is relatively short but even then the useful signal will have the same path length as the interference signal. This distance is expressed as d_e. The distance between the satellite and earth ground gateway station is calculated by the cosine law given as $d_e = \sqrt{(h_s + \mathrm{R})(h_s + \mathrm{R} - 2Rcos(\beta) + \mathrm{R}^2}$. Here h_s represents satellite altitude, R is spot beam radius and α represents the central angle. Assuming that locations of satellite and user are known, the length of arc, l_a between user and the satellite point is used to calculate the central angle, i.e. $\beta = \dfrac{180 l_a}{\pi \mathrm{R}}$. Also, the free path-loss $\left(\dfrac{4\pi d_e}{\lambda}\right)^2$ and channel fading parameter γ are explicit. Therefore, the required power P_R can be written as

$$P_R = \gamma P_{\mathrm{t}} G_{max} G_{rd} \left(\frac{\lambda}{4\pi d_e}\right)^2 \tag{1}$$

where G_{max} is the maximum antenna gain[3] and G_{rd} is the received antenna gain of satellite. For each satellite the received power will adapt to its individual powers from satellites, e.g. $P_t = P_1, P_2, P_3$. Now, we introduce the interference[4] I_{in} as

$$\mathrm{I}_{in} = \sum_{i=1}^{M} \gamma P_{\mathrm{t}} G_{max} G_{rd} (\frac{\lambda}{4\pi d_e})^2 \tag{2}$$

2.3 Capacity Maximization

The system model presented in Fig. 1 has two capacity segments: a) Space segment; from gateway to satellite, C_{ss}, and b) ground segment, C_{gs}, from user (terrestrial links to the gateway as traffic flows) to gateway. The gateway performs a signal regeneration operation, hence the capacity is decided by the minimum of C_{ss} and C_{gs}, given by

$$C = \min\left(C_{ss}, C_{gs}\right) \tag{3}$$

C_{ss} which is the bottle neck in the system will be calculated individually to see how it can improve total system capacity.

It is important to note that if there were no inter-beam interference, the optimal power allocation for each beam is straight forward since the model mathematically is equivalent to an independent parallel sub-channel model, and the classic Water-filling algorithm could be applied directly. When taking the interference into account, the problem is non-convex and is difficult to solve. Therefore, in-order to solve this non-convex optimization problem, we start by maximizing the rate from the summation of individual satellite, given as

$$C = \max(R_1 + R_2 + R_3) \tag{4}$$

where

$$R_1 = f_1(\mathrm{SINR}_1), \quad \mathrm{SINR}_1 = g_1(P_1.P_2.P_3) \tag{5}$$

$$R_2 = f_2(\mathrm{SINR}_2), \quad \mathrm{SINR}_2 = g_2(P_1.P_2.P_3) \tag{6}$$

$$R_3 = f_3(\mathrm{SINR}_3), \quad \mathrm{SINR}_3 = g_3(P_1.P_2.P_3) \tag{7}$$

where R_1, R_2, R_3, represent rate[5] and $\mathrm{P}_1, \mathrm{P}_2, \mathrm{P}_3$, define power from individual HTS-LEO satellites. Also f_1, f_2, f_3 shows the function of SINR which is dependent on individual rates and g_1, g_2, g_3 represents function of individual

[3] Generally side lobes contribute to the inter-beam interference. G_{max} is the antenna gain for the main lobe. Lets denote G_{sl} as the power level of the side lobe, interference is a function of G_{sl}. Here for simplicity we assume $G_{sl} = G_{max}$, which is only true for antennas with grating lobes.

[4] Since this is a novel idea to introduce co-channel interference; therefore to start with we are considering all the interferes are equidistant.

[5] Rates, R_1, R_2, R_3, represents the throughput of system which are the rate at which some information is processed.

powers from the satellites. The signal-to-interference-plus-noise ratio for individual satellites can be expressed as:

$$\mathrm{SINR} = \frac{P_R(\gamma)}{\mathrm{I}_t + N_o} \tag{8}$$

where $\mathrm{I}_t = \mathrm{I}_{in} + \mathrm{I}_{out}$. Since I_{out} is the interference occurred outside the system.[6] The Shannon capacity for this system can then be defined as

$$C = B \log_2 (1 + \mathrm{SINR}) \tag{9}$$

where B represents the system bandwidth. Therefore, the Shannon bounded capacity C can be expressed by incorporating (8), (1), and (2) as:

$$C = B \log_2 \left(1 + \frac{\gamma P_{\mathrm{t}} G_{max} G_{rd} \left(\frac{\lambda}{4\pi d} \right)^2}{N_o + \sum_{i=1}^{M} \gamma P_{\mathrm{t}} G_{max} G_{rd} \left(\frac{\lambda}{4\pi d_e} \right)^2} \right) \tag{10}$$

Normalizing the whole function given in (10) with factor $V = \frac{1}{\gamma G_{max} G_{rd}} \left(\frac{4\pi d_e}{\lambda} \right)^2$, yields

$$C = B \log_2 \left(1 + \frac{P_{\mathrm{t}}}{N_o V + \sum_{i=1}^{M} P_{\mathrm{t}} H} \right) \tag{11}$$

Here channel coefficient H can be generated by using the number of transmitting/receiving antennas, nT_x and nR_x written as, $H = rand\left(\frac{1}{\mathrm{nT}_x}, \frac{1}{\mathrm{nR}_x} \right)$.

3 Optimal Power Allocation

To evaluate the system performance, we first need to select the objective function that optimizes the capacity and power allocation. We first formulate the capacity maximization with average input power constraint:

$$\max_{P_{\mathrm{t}}(\gamma) \geq 0} \quad C \tag{12}$$

$$\text{subject to:} \quad 0 \leq P_{\mathrm{t}}(\gamma) \leq 0.25 P_{\max}, \tag{13}$$

$$\text{subject to:} \quad 0 \leq \sum_{i=0}^{M} P_{\mathrm{t}}(\gamma) \leq P_{\max}, \tag{14}$$

$$\text{subject to:} \quad P_{\mathrm{t}}(\gamma) \geq 0, \tag{15}$$

[6] The interference outside the system refers to physical factors that can include multipath fading, rain attenuation, clean sky and scintillation. Since this is the initial findings for the simplicity we have considered it 30 dB.

Table 1. Table of parameters and values

Parameters	Value
Spot beams, M	3
Satellite altitude, h_s	500 km
Spot beam radius, R	100 km
Maximum antenna gain, G_{max}	30 dB
Received antenna gain, G_{rd}	30 dB
Noise power spectral density, VN_o	−10 dB
Number of iteration,	10,000
Interference outside system, I_{out}	30 dB

where $P_{\max}$ is the average transmission power limit. (13) refers to a constraint which means a gateway is capable of adjusting a maximum of 25% total transmission power to one beam, and which is only an assumption dependent on the gateway actual transmission power adjustment capability. The maximum value of 100% means that the gateway could use all its transmission power for only one satellite.

3.1 Optimal Approximation

Let us introduce the non negative Lagrangian multipliers μ, ϕ and Φ of the optimization problem mentioned in (12) with respect to required power P_R. The problem can be formulated as,

$$L(P,\gamma) = C + \mu\big(P_R(\gamma) - 0.25P_{\max}\big) + \phi\Big(\sum_{i=0}^{M} P_R(\gamma) - P_{\max}\Big) + \Phi P_R(\gamma) \tag{16}$$

Since the optimization problem presented in (16) is a non-convex problem therefore we need to approximate with the help of dual problem. The optimal solution of the problem and dual variable have a relationship which can be first expressed as a primary problem and represented as

$$\begin{aligned} f(\mu,\phi,\Phi) &= \min L(P,\gamma) \\ &= \min \Big[C + \mu\big(P_R(\gamma) - 0.25P_{\max}\big) \\ &\quad + \phi\Big(\sum_{i=0}^{3} P_R(\gamma) - P_{\max}\Big) + \Phi P_R(\gamma)\Big] \le f(P^*,\gamma^*) \\ &\quad \Big(C^* + \mu\big(P_R^*(\gamma) - 0.25P_{\max}\big) \\ &\quad + \phi\Big(\sum_{i=0}^{M} P_R^*(\gamma) - P_{\max}\Big) + \Phi P_R^*(\gamma)\Big) \end{aligned} \tag{17}$$

$$\leq f(P^*, \gamma^*) = p^* \tag{18}$$

Here p^* is the primary problem and p_d^* is the dual problem which can be obtained by Lagrangian duality theory [13]. Further golden section method is used to calculate the update $P_R^*(\gamma)$.

4 Numerical Results

In this section, simulation results are presented to evaluate the capacity of HTS-LEO system with antennas beams to be aligned as beam-forming networks. Simulation results are conducted in MATLAB and the experimental design is being verified through beamplot software. The results shed light on how the transmission rate of the HTS can be improved with respect to Shannon capacity, in bits/s/Hz, while fulfilling the constraints. Simulation parameters are summarized in Table 1. In particular, we consider the number of transmitting antennas nT_x equal to number of receiving antennas nR_x, $M = 3$, $h_s = 500\,\mathrm{km}$, and $G_{max} = G_{rd} = 30\,\mathrm{dB}$, unless otherwise indicated.

Figure 2 shows the capacity of the HTS-LEO system versus SNR in dB for different values of nT_x, nR_x. We can see from the figure that the larger the number of transmitting and receiving antennas, the higher the capacity that can be achieved (the result of the HTS-LEO system model result is indicated by ***). When interference is introduced from adjacent beams, the capacity degrades but our optimized scheme performs better than the state of art MIMO system with equal power allocation (shown in coloured curves).

Figure 3 illustrates the performance degradation when considering the inter-beam (or inter-path) for systems with $\mathrm{nT}_x = \mathrm{nR}_x = 3$. From the figure we can observe that without considering interference greater capacity can be achieved. Interference here depends on the adjacent beams from other satellites. Therefore for each satellite, the interference from the rest of the side-lobe is considered as it can not cause interference to itself.

SINR versus Power, P, with varying transmitting and receiving antennas is presented in Fig. 4. The results show that even with inter-feeder beam interference and increasing the number of antennas at the transmitter and receiver, better SNR can be achieved in the proposed system. Although in real systems, keeping the cost of multiple gateways is not a practical solution.

Figure 5 illustrates a special case in which capacity versus SNR with equal transmit/receive antenna beams and varying RF chains is being considered. This figure depicts hybrid beam-forming is used for the first two cases, where full digital array is used for RF = 3. Beam-forming design using limited radio frequency (RF) chains[7] has recently received significant attention [4]. These RF chains are introduced to improve system gain and reduce noise levels. As we can see from

[7] RF chain is a cascaded electronic component which includes low-noise amplifier, down-converter, digital to analog converter (DAC) and analog to digital converter (ADC). In digital beam-forming network, each antenna element at the base station and user equipment requires one dedicated RF chain.

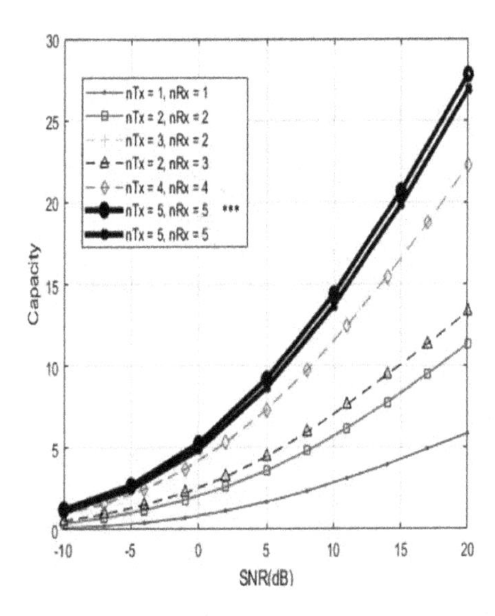

Fig. 2. Capacity versus SNR with different values of transmitting/receiving antennas

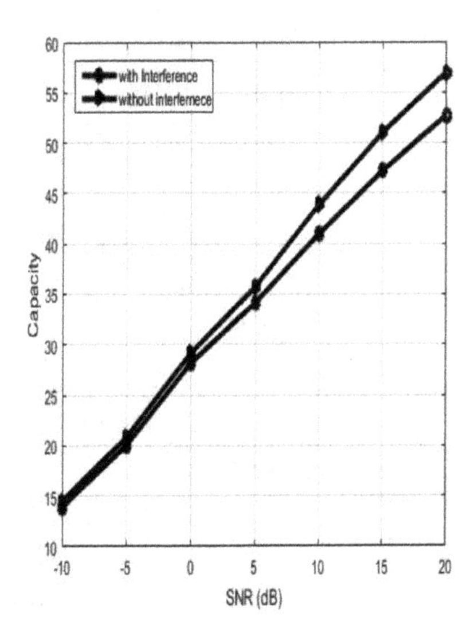

Fig. 3. Capacity versus SNR for **nT_x = nR_x = 3**, with and without interference

the figure that when SNR ¡ 0 dB, RF = 2, 3 coincide with each other and after SNR ¿ 0 dB, more RF chains, i.e., RF = 3 yield higher capacity. Although the idea is to reduce RF chains in order to reduce cost and power consumption. For a feeder link (gateway to satellite) it is not practical to have so many antennas. We have to consider the trade-off and this setup is more suitable for the systems where we can increase capacity by limiting the transmission power.

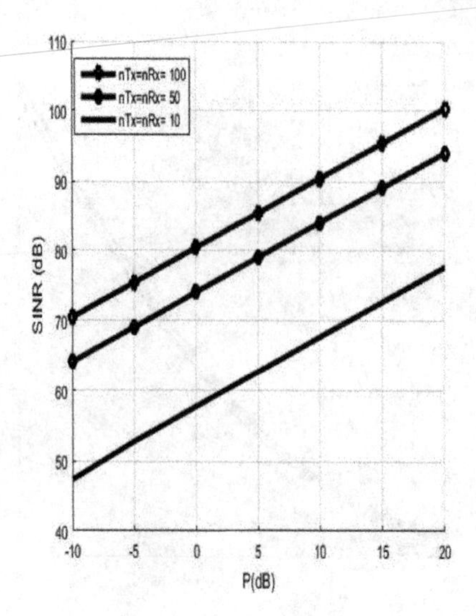

Fig. 4. SNR versus Power, P(dB), with multiple transmitting/receiving antennas including interference

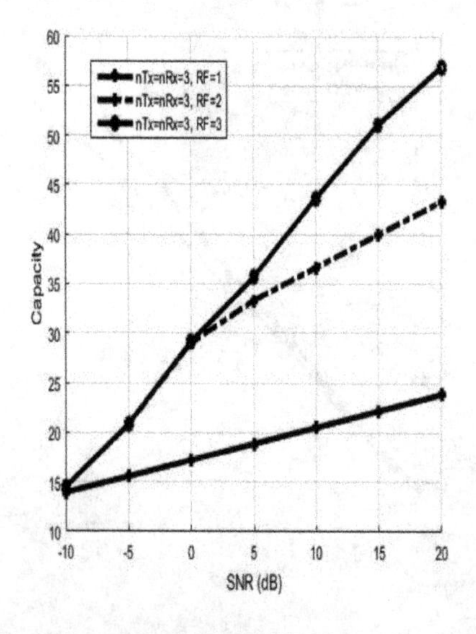

Fig. 5. Capacity versus SNR with $\mathbf{nT\_x} = \mathbf{nR\_x} = \mathbf{3}$ and different values of RF chains RF $= 1, 2, 3$.

5 Conclusion

In this paper, an optimal power allocation scheme for the HTS-LEO satellite was proposed in which multiple LEO satellites in the same constellation could simultaneously communicate with one gateway. We assumed at any time, a gateway could only view multiple satellites with reasonable elevation angles, and the gateway has multiple steerable beams pointing at those multiple satellites. Our power allocation strategy is used to maximize capacity of the HTS-LEO satellites for a system that might be used for 5G services. Interference from neighbouring satellite beams was taken into account and its impact on capacity evaluated. We calculated the optimal power level by using Lagrangian duality to solve this non-convex problem. Simulation results showed that the proposed scheme improves capacity of the HTS system and meet the dynamic demand better than traditional methods. Using single gateway with multiple steerable antenna arrays showed reduction in gateway numbers and cost. In addition the multiple beams from the same gateway caused interference and that was considered in improving capacity evaluation. This study could be straightforwardly extended to any number of satellites within the multiple gateway field of view for future work.

Acknowledgment. This work was sponsored by China Academy of Space Technology (CAST)-Xian. The authors also would like to acknowledge the support of the University of Surrey 5GIC (http://www.surrey.ac.uk/5gic).

References

1. Adriaensen, M., Antoni, N., Giannopapa, C.G., Papadimitriou, A.: Comparative analysis of ESA member states space and security governance and strategy in the frame of European integration. In: 69th International Astronautical Congress, p. 47624, October 2018
2. Anttonen, A., Ruuska, P., Kiviranta, M.: 3GPP nonterrestrial networks: a concise review and look ahead (2019)
3. Bandi, A., Joroughi, V., Grotz, J., Ottersten, B., et al.: Sparsity-aided low-implementation cost based on-board beamforming design for high throughput satellite systems. In: 9th Advanced Satellite Multimedia Systems Conference and the 15th Signal Processing for Space Communications Workshop (ASMS/SPSC) (2018)
4. Bogale, T.E., Le, L.B., Haghighat, A., Vandendorpe, L.: On the number of RF chains and phase shifters, and scheduling design with hybrid analog-digital beamforming. IEEE Trans. Wirel. Commun. **15**(5), 3311–3326 (2016)
5. Del Portillo, I., Cameron, B.G., Crawley, E.F.: A technical comparison of three low earth orbit satellite constellation systems to provide global broadband. Acta Astronautica **159**, 123–135 (2019)
6. Dimitrov, S., et al.: Capacity enhancing techniques for high throughput satellite communications. In: International Conference on Wireless and Satellite Systems, pp. 77–91, September 2015
7. Guidotti, A., Vanelli-Coralli, A., Conti, M., Andrenacci, S., Chatzinotas, S., Maturo, N., Evans, B., Awoseyila, A., Ugolini, A., Foggi, T., et al.: Architectures and key technical challenges for 5G systems incorporating satellites. IEEE Trans. Veh. Technol. **68**(3), 2624–2639 (2019)
8. Huang, J., Cao, J.: Recent development of commercial satellite communications systems. In: Artificial Intelligence in China, pp. 531–536 (2020)
9. Jia, M., Zhang, X., Xuemai, G., Guo, Q., Li, Y., Lin, P.: Interbeam interference constrained resource allocation for shared spectrum multibeam satellite communication systems. IEEE Internet Things J. **6**(4), 6052–6059 (2018)

10. McDowell, J.C.: The low earth orbit satellite population and impacts of the spaceX starlink constellation (2020). arXiv preprint arXiv:2003.07446
11. Mosquera, C., López-Valcarce, R., Ramírez, T., Joroughi, V.: Distributed precoding systems in multi-gateway multibeam satellites: regularization and coarse beamforming. IEEE Trans. Wirel. Commun. **17**(10), 6389–6403 (2018)
12. Park, U., Kim, H.W., Oh, D.S., Chang, D.I.: Performance analysis of dynamic resource allocation for interference mitigation in integrated satellite and terrestrial systems. In: 2015 9th International Conference on Next Generation Mobile Applications, Services and Technologies, pp. 217–221 (2015)
13. Radio Regulations: Satellite antenna radiation pattern for use as a design objective in the fixed-satellite service employing geostationary satellite. In: Recommendation ITU-R S, pp. 672–4 (1997)
14. Siddiqui, A.M., Musavian, L., Aissa, S., Ni, Q.: Performance analysis of relaying systems with fixed and energy harvesting batteries. IEEE Trans. Commun. **66**(4), 1386–1398 (2017)
15. Wang, L., Zhang, C., Qu, D., Zhang, G.: Resource allocation for beam-hopping user downlinks in multi-beam satellite system. In: 15th International Wireless Communications & Mobile Computing Conference (IWCMC), pp. 925–929 (2019)
16. Wang, P., Zhang, J., Zhang, X., Yan, Z., Evans, B.G., Wang, W.: Convergence of satellite and terrestrial networks: a comprehensive survey. IEEE Access **8**, 5550–5588 (2019)
17. Zhang, X., Wang, J., Jiang, C., Yan, C., Ren, Y., Hanzo, L.: Robust beamforming for multibeam satellite communication in the face of phase perturbations. IEEE Trans. Veh. Technol. **68**(3), 3043–3047 (2019)

Measurement of Work as a Basis for Improving Processes and Simulation of Standards: A Scoping Literature Review

Paul V. Ronquillo-Freire[1] and Marcelo V. Garcia[1,2(✉)]

[1] Universidad Técnica de Ambato, UTA, 180103 Ambato, Ecuador
{pronquillo4211,mv.garcia}@uta.edu.ec
[2] University of Basque Country, UPV/EHU, 48013 Bilbao, Spain
mgarcia294@ehu.eus

Abstract. All companies, regardless of size, all over the world to cope with COVID-19 pandemia must innovate to overcome the difficult economic situation. This research paper shows a scoping review based on PRISMA guidelines to present the most used techniques for calculating the standard work times. At the beginning were selected 400 papers from IEEEexplore, Scopus, Web of Sciences, MDPI, and ACM databases. 33 papers finally were selected to answer the research questions which are based on this study. This study concluded that the use of these time systems allows the job quantification before implementing its production. Besides, the high cost when a new product is developed cab be reduced, and all the administrative process are optimized.

Keywords: Work measurement · Time studies · Maynard Operation Sequence Technique (MOST) · Modular Arrangement of Predetermined Time Standards (MODAPTS) · Methods Time Measurement (MTM)

1 Introduction

These hard times have forced the companies to implement processes and systems to know the environment's needs of each business sector and allows to establish strategies in order to control materials and humans resources. These strategies are developed to reduce costs, increase profits and have a long term presence in the market. [40]. This is an augmented version of previous papers published by the authors [5] with the explanation of the PRISMA methodology and the new challenges of this measurements work techniques.

Actually, the Covid 19 pandemia changed the economic and productive field of enterprises, a lot of workers lost their jobs and the companies must face all these obstacles if they want to survive in this hard economic moment [30]. The bio-security measures taken by enterprises to mitigate the virus consequences caused difficulties in productive activities and the decrease in product demand and specialized services. This huge problem caused by pandemia affects the fixed capital formation, losses in the scale's economies, and the manufacturing of fewer products [28].

K. Arai (Ed.): FICC 2021, AISC 1363, pp. 77–92, 2021.
https://doi.org/10.1007/978-3-030-73100-7_6

The economic crisis made by the pandemia accelerated the changes that companies should be coped a decade ago. During these months the changes implemented by enterprises are about e-commerce, virtual negotiations, and home office [8]. This new way to cope with clients and product developments made that the enterprises think about what happens if they need to store information from shopfloor production areas or how the enterprises can develop or release new products in the market. To solve these problems companies make researches to find new approaches to base their decisions on up-to-date information taken by the production processes.

The enterprises must be implemented new methodologies that joint, improve or adapt the old standards with new requirements of actual enterprises. This scoping review tries to explain the methodologies to establish standard times and how these evolved and adapted to each company environment. These new approaches allowed the issuance of information without timed the shopfloor process [1]. In consideration of the traditional methodologies do not face the new reality in enterprises, the new approaches and techniques must expand the previous methodologies focusing on progressive digitalization information to achieve productive results [14].

The aim of this research is to present a scoping literature review to show updated methodologies on work measurement techniques. These techniques, although were developed a long time ago, could be adapted to the present times, allowing the development of standards either by a system of predetermined times or by direct observation of activities.

This article consists of 5 sections: Sect. 2 describes the methodology used to select the articles and how other articles were discarded. Section ?? presents the literature review. Section 3 discusses and analyzes the results. And finally, in Sect. 4 the conclusions and future works are evidenced.

2 Article Selection Methodology

As the prisma guidlines explain the seection and review of papers follows a three phase methodology: (i) The proposition of research questions which coduce the guide the investigation, (ii) a list of keywords for searching information and (iii) data extraction using a exclusion criteria. The research was focus on main databases like: MDPI, Scopus; IEEE Explore; ACM, and Web of Science and others.

2.1 Scoping Review Research Questions

For this article three research questions were formuled (see Table. 1). This questions fulfill the aim of this research and are a guide in the search process. These questions persue the explanation of how the measurement of the work time to understand its importance, as well as exploring the new techniques which are giving results.

Table 1. Research questions

Number	Questions	Motivation
RQ1	What is the author's outlook about the work measurement?	Establish the work measurement importance
RQ2	What are the work measurement techniques used in theses times?	Identify new methodologies used in industrial process
RQ3	There are a methodology that measure and analyzed the time in the workers activities?	Identify if the joint of different work techniques improve the analysis

2.2 Article Search

Specific keywords were used to stablish the search the literature. For the viewpoint 1 (VP1) corresponding to the measurement of work the keywords are: "work measurement" OR "time studies" AND "predetermined time system". The second point-view (VP2) face the techniques most used to measure the work were the keywords: "methods-time measurement" OR "Maynard operation sequence technique AND "the modular arrangement of predetermined time standards". Finally, the viewpoint three (VP3) show the new standards that enterprises measure the work : "preset timing standards" OR "time formulas", AND "production" OR "manufacturing". In the search engines, a variation or combination of the keywords was used in order to get articles about the use of work measurement techniques.

The literature search was achieved in eleven databases (see Fig. 1), with a result of 64 papers in English and Spanish corresponding to articles in journals or conferences, books, master's thesis and reports; The documents search spanning correspond from 2000 to 2019.

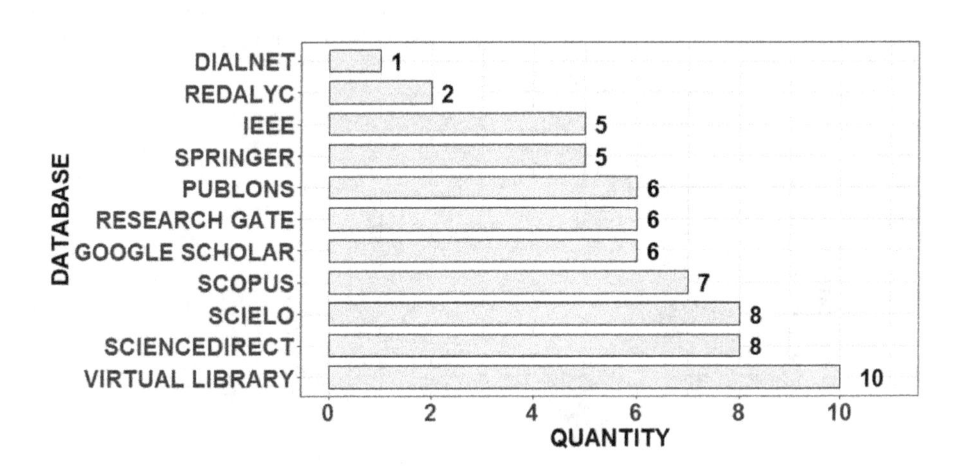

Fig. 1. Number of documents per database

Table 2. Inclusion and exclusion criteria.

Number	Inclusion	Exclusion
C1	Articles related to the aim of using measurement work time in industrial process	Papers duplicates from all databases included in this scoping review
C2	Articles published from 2015 to 2020	Articles not related to work time measurement
C3	English written articles	Low citation and low quartil of the journal or congress
C4	Articles related to aim of using new methodologies of time work measurement in industrial process	Low h-index of the authors and the journals

2.3 Selection of Information

The first phase, exclusion and inclusion criteria were applied. Several aspects were applied, for example the language of the papers, the date of publication, h-index of the authors, the citations among others (see Table 2). In the second phase, documents were organized by title, relevance, abstract, perspective, and keywords. It granted a faster and efficient review of each research investigation. In the third step, if the information shown in the introduction and conclusions section provided is the evidence permit to answer the research questions. The summary of the obtained documents in each phase is presented according to the PRISMA guidelines.

The exclusion criteria applied in the classification of documents are detailed below. (i) The articles which are not published in the last five years were discarded. About the books, this criteria was not used because this methodology was discovered years ago. (ii) The second exclusion criteria was based on citation and the quartil of the journal or congress were the paper were published. (iii) Finally, the h-index of the authors and the journals are used to order the articles by title, abstract and perspective.

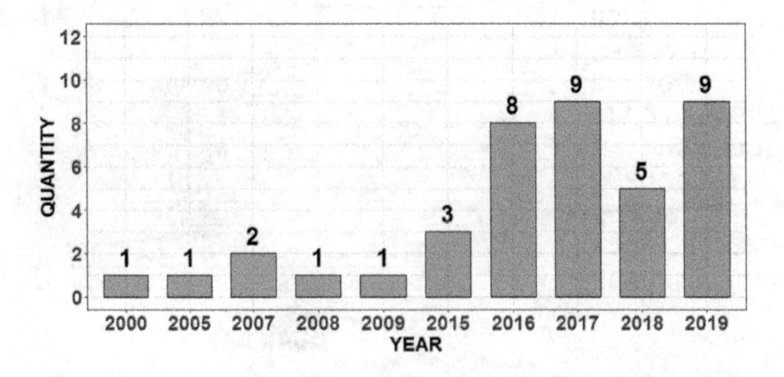

Fig. 2. Select articles by year

Based on the exclusion criteria, 30 articles were selected (see Fig. 2), 24 are in English and 6 in Spanish. Furthermore, from a total of 10 books, 7 books were selected. Figure 3 shows that most of the articles were found in the databases: Scopus, Web of Science and Scielo.

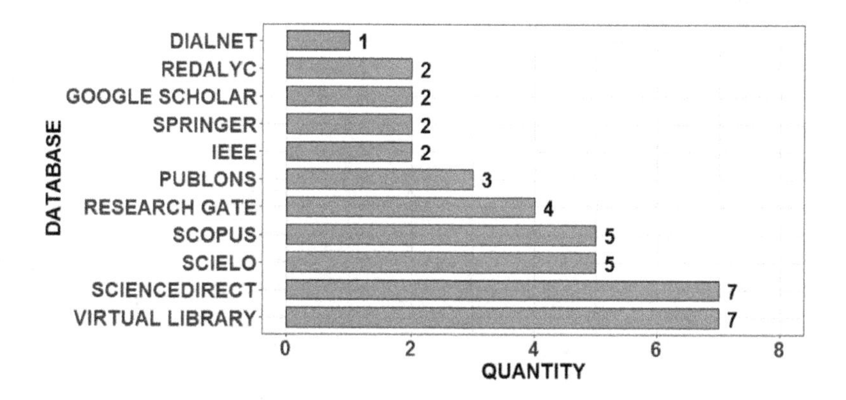

Fig. 3. Articles chosen by each database

2.4 Data Extraction

Table 3, presents the 30 articles arranged by relevance and high citations. The information drawn from each document is based on answering the research questions. Various issues such as the fields in which work time measurements is developed for industrial process, technology, new techniques and efficiency, advantages, disadvantages, and the gaps of this standards have been considered.

At a final step, all the references were verified. It was aimed to confirm each one is correctly placed and with all the items detailing important information and that it aligns with the context of each research questions. The summary of the reviewed documents in each phase is showed according to the PRISMA guidelines [39].

Table 3. Selected papers for scoping review.

Code	Title	Sample size	Year	Viewpoint	Authors	Goals
P1	Estudio del trabajo [11]	N = 10	2005	VP1	Garcia Criollo, R	The measurement of work can be defined in different ways according to the perspective of execution time, trained worker and established procedures
P2	Métodos y tiempos con manufactura ágil [9]	N = 23	2016	VP1	Escalante Lago, A	Considers work measurement as one of the necessary supports for improve productivity
P3	Estudios de tiempos y movimientos para la manufactura ágil [26]	N = 8	2000	VP3	Meyers, F.E	Is a tool for managers to make important decisions intelligently even before production starts
P4	Manual de tiempos y movimientos [16]	N = 40	2008	VP2	Janania, A.C	Work measurement continually improving and recognizes it as a instrument to get high performance from machines and equipment
P5	Ingeniería de métodos. Globalizacion: Técnicas para el manejo eficiente de recursos en organizaciones fabriles, de servicios y hospitalarias. [7].	N = 100	2007	VP2	Durán, Freddy	Reduce or mitigate the activities that cause downtime and study methods to Reduce the content of work by eliminating unnecessary movement
P6	Introducción al estudio del trabajo [17]	N = 1	2007	VP1	Kanawaty, G	The aim of work measurement is the application of techniques focused on determining the time that a qualified worker executes a defined task
P7	Ingenieria industrial: Métodos, estándares y diseño del trabajo [29]	N = 20	2009	VP3	Niebel, Benjamin Willard and Freivalds, Andris	Present new techniques to measure the work made by a qualified worker
P8	Ingenieria industrial: Methodological analysis for the performance of studies of methods and times [32]	N = 30	2007	VP1	Polanco Vides, Evis Ximena and Lauren Andrea, Díaz Jiménez and Jorge Junior, Gutiérrez Rodríguez	The author review how companies find out the most appropriate methodologies to integrate or improve direct observation tools
P9	Estudio de Tiempos y Movimientos para Incrementar la Eficiencia en una Empresa de Producción de Calzado [2]	N = 9	2019	VP1	Andrade, Adrián M. and A. Del Río, César and Alvear, Daissy L.	The work in the industrial process was balanced using standard times per worker. As a result, an increase in productivity of 5.5% was got
P10	Tiempos estándar para balanceo de línea en área soldadura del automóvil modelo cuatro [6]	N = 35	2019	VP1	Cascante, Gloria Miño and Alulema, Julio Moyano and Mariño, Carlos Santillán	The authors organized the workers to implement the activities evenly on both sides of the assembly of the vehicle

(*continued*)

Table 3. (*continued*)

Code	Title	Sample size	Year	Viewpoint	Authors	Goals
P11	Estudio de tiempos y movimientos de producción para Fratello Vegan Restaurant [25]	N = 17	2019	VP2	Mendoza Novillo, Paulina Alejandra and Erazo Álvarez, Juan Carlos and Narváez Zurita, Cecilia Ivonne	Aims to increase resources and normalize operations in a restaurant company. The goal was to minimize the complaint of the clients about the delay on the service due to a long time in the service of the dishes
P12	Medición de Tiempos en un Sistema de Distribución bajo un Estudio de Métodos y Tiempos [12]	N = 61	2018	VP1	Henríquez-Fuentes, Gustavo R. and Cardona, Diego A. and Rada-Llanos, Jesús A. and Robles, Nilka R.	Study case of time measurement application and the authors presents high performance on distribution system of a marketer
P13	Productivity improvement in logistical work systems of the genuine parts supply chain [21]	N = 10	2015	VP3	Kuhlang, P. and Sunk, A	Use the Methods Time Measurement (MTM) methodology to determine worker times and increase profits in logistics systems
P14	Work Standards in Selected Third Party Logistics Operations: MTM-LOGISTICS Case Study [20]	N=85	2017	VP3	Koptak, Michal and Džubáková, Martina and Vasilienė-Vasiliauskienė, Virgilija and Vasiliauskas, Aidas Vasilis	Use Methods Time Measurement (MTM) methodology to reduce 0.5% in the cycle time
P15	Automatic Proposal of Assembly Work Plans with a Controlled Natural Language [24]	N = 237	2015	VP3	Manns, M. and Wallis, R. and Deuse, J.	Used time measurement techniques in joint with data mining to estimate the time and sequence of operations in assembly line process
P16	An automated time and hand motion analysis based on planar motion capture extended to a virtual environment [38]	N = 115	2015	VP3	Tinoco, Hector A. and Ovalle, Alex M. and Vargas, Carlos A. and Cardona, María J.	Introduce the MTM methodology automation platform using a motion capture system embedded into a virtual reality environment
P17	Towards Method Time Measurement Identification Using Virtual Reality and Gesture Recognition [4]	N = 50	2015	VP1	Bellarbi, Abdelkader and Jessel, Jean-Pierre and Da Dalto, Laurent	Design a system capable of automatically generating a MTM code using the movements capture of the head and both hands in a virtual environment. This platform is used to train technicians
P18	Empirical validation of the time accuracy of the novel process language Human Work Design (MTM-HWD®) [10]	N = 62	2019	VP2	Faber, Marco and Przybysz, Philipp and Latos, Benedikt A. and Mertens, Alexander and Brandl, Christopher and Finsterbusch, Thomas and Härtel, Jörg and Kuhlang, Peter and Nitsch, Verena	Shows a revision on the accuracy of the MTM human work design technique (MTM-HWD) with MTM-1. The case study presents a strong relationship between these two techniques
P19	Teaching Methods-Time Measurement (MTM) for Workplace Design in Learning Factories [27]	N = 215	2017	VP3	Morlock, Friedrich and Kreggenfeld, Niklas and Louw, Louis and Kreimeier, Dieter and Kuhlenkötter, Bernd	Show the use MTM in the design of workplaces. The aim of this paper emphasizing the usefulness to manage production times

(*continued*)

Table 3. (*continued*)

Code	Title	Sample size	Year	Viewpoint	Authors	Goals
P20	A Software Tool for the Calculation of Time Standards by Means of Predetermined Motion Time Systems and Motion Sensing Technology [22]	N = 17	2019	VP3	León-Duarte, Jaime and Aguilar-Yocupicio, Luis and Romero-Dessens, Luis	The authors developed a software which analyzes movements and set a time using the Modular Arrangement of Predetermined Time Standards (MODAPTS) methodology improving times of the assembly in electrical harness process
P21	A Model for Complexity Assessment in Manual Assembly Operations Through Predetermined Motion Time Systems [1]	N = 200	2016	VP1	Alkan, Bugra and Vera, Daniel and Ahmad, Mussawar and Ahmad, Bilal and Harrison, Robert	Proposes a methodology to evaluate the complexity of work process in order to give an optima time approach, uses the MODAPTS with a virtual manufacturing tool called vueOne
P22	Incorporating motion analysis technology into modular arrangement of predetermined time standard (MODAPTS) [41]	N = 60	2016	VP3	Wu, Shuang and Wang, Yao and BolaBola, Joëlle Zita and Qin, Hua and Ding, Wenying and Wen, Wei and Niu, Jianwei	Presents a joint methodology between motion analysis technology (PCA) and MODAPTS to oimprove the preset time system. Using the PCA methodology a precision of 80% was got in the calculation of the work times
P23	Application of The MODAPTS Method with Innovative Solutions in The Cement Packing Process [13]	N = 20	2018	VP1	Ita Erliana, Cut and Abdullah, Dahlan	Uses MODAPTS methodology to reduce working time in cement industry process. Compare the industrial processes with the standard worker time determined by applying the technique into the system. The results present the decrease in cycle time to the fulfillment the production process
P24	Quality Engineering with Taguchi Loss Function Method and Improvement of Work Method in Anode Changing [37]	N = 1	2019	VP3	Siregar, Ikhsan	Uses the MODAPTS methodology to improve the working methods and calculate the processing time. Reduce in 126.65 secs, between the actual and MODAPTS method
P25	Implementation of Maynard Operation Sequence Technique (Most) To Improve Productivity and Workflow – a Case Study [35]	N = 1	2018	VP1	Rahman, M.S., Karim, R., Mollah, J., Miah, S.	The authors establish non-value-added activities and decrease bottlenecks to improve productivity and reduce cycle time. Reduce 30 s in the production time, increasing the production from 600 pieces to 1600 pieces

(*continued*)

Table 3. (*continued*)

Code	Title	Sample size	Year	Viewpoint	Authors	Goals
P26	A review on optimization in total operation time through Maynard Operation Sequence Technique [31]	N = 60	2017	VP1	Patel, Divyang and Tomar, Prashantsingh	Shows a work measurement using Maynard Operation SequenceTechnique (MOST) reducing 65% of time process increasing the production. The production time apliying this technique can be got before manufacturing begins
P27	Assembly line productivity improvement as re-engineered by MOST [19]	N = 1	2016	VP1	Karim, A.N. Mustafizul and Tuan, Saravanan Tanjong and Emrul Kays, H.M.	Achieved a reduction of 2 min in cycle time on the industrial process and increase a 29.7% of the workers performance
P28	Productivity Improvement By Maynard Operation Sequence Technique [18]	N = 5	2016	VP1	Karad, Prof A A and Waychale, Nikhil K and Tidke, Nitesh G	Use MOST technique in an automotive company allowed to redesign the processes and designing the correct flow reducing the workforce from 17 to 11 workers and the reduction of 2 min in worker cycle time
P29	Productivity Improvement through Maynard Operation Sequence Technique [15]	N = 3	2017	VP3	Jadhav, Mangesh and Mungase, Samadhan and Karad, Prof A A	Analyze the data got from automotive industry in comparing with the estimates, accomplish the reduction of a production shift because the demand was fulfilled in one shift
P30	Implementation of Maynard operation sequence technique in dry pack operation-a case study [34]	N = 10	2019	VP2	Puvanasvaran, A. P. and Yap, Y. Y. and Yoong, S. S.	Highlights the implementation MOST technique joint with the direct measurement technique by timing to reveal hidden wasting time in the dry packaging process. The study of times presents the real cycle time and the equation of the worker's performance is calculated monitoring the daily activities of the workers

3 Discussion of the Results

The scoping review shows a positive perspective in the application of predetermined time systems to simulate standards without beginning the manufacturing. All of these methodologies can be recognized as useful instruments to control the increase of production cost in companies when develop new products, to improve the operations already implemented and to bring information for the cost analysis. The methodologies discussed in this article could become a starting point for developing researches that meet the needs of each organization.

Seeing the results on the second research question, the authors of this paper verify the widely used techniques at present are mostly the use of the classification of synthetic methods (see Fig. 4): MOST (8), MODAPTS (9), MTM (8), GSD (1). The last technique are the traditional method of direct observation ie: the study of times (5). This means that enterprises are focusing on optimizing their processes using pre-established methods.

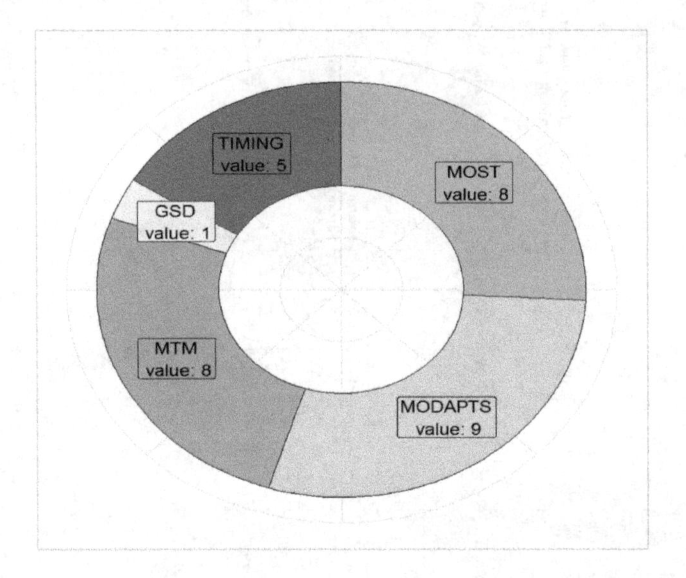

Fig. 4. Number of methodologies used currently in industry

This research could established the year in which the different techniques were most researched. The synthetic method were increased their analysis throughout the period from 2019 with because 4 selected research articles (see Fig. 5). The study of times as an individual application was checked in articles for most of the period between 2017 and 2019. However, as Fig. 5 shows the combination of methodologies has a wide field of research, because having the support of direct observation methodology in alliance with any of the predetermined time techniques can design new systems.

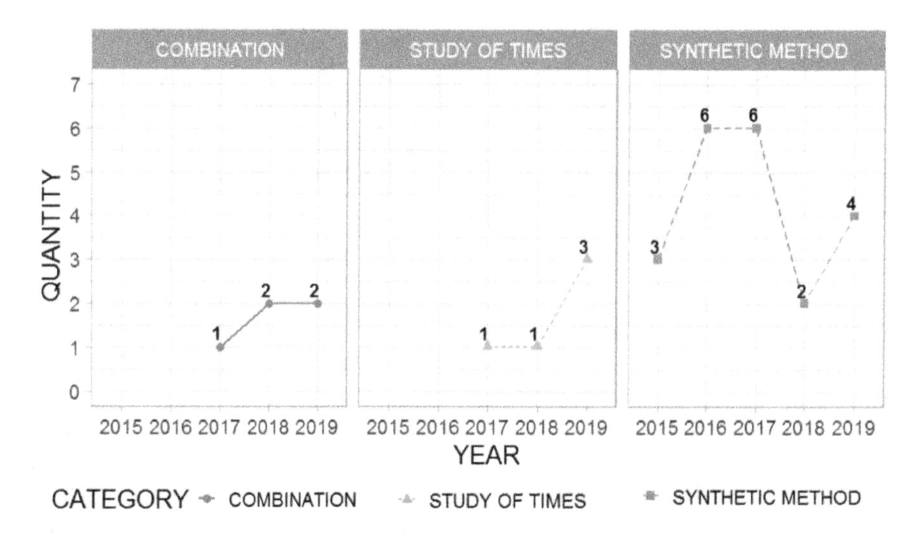

Fig. 5. Work measurement times methodologies

3.1 Challenges in the Use of Work Time Measurement

All the 30 selected papers in this article shows that theses methodologies become a fundamental tool to improve the production of any type of organization. Some of the articles are study cases were authors decrease the production time improving the profits of the enterprises using the combination of new techniques and traditional ones. The goal of this subsection is answer the research questions presented above.

RQ1: What is the Author's Outlook about the Work Measurement?

For the analysis method of work by direct observation, five articles were reviewed from which one of them [32] was used to expose the research about the times using a chronometer. Unfortunately, this research does not have an real use at the industrial field. Articles like [2,6,25], presents the sequence of steps for the correct implementation of the methodology in real industry shop-floor cases. Summarizing, theses papers uses the standard formulas for calculating the optimal number of samples in the workers shifts, as the selection of a qualified operator, the standardization of the process and the operations are deeply studied. The application in real industry cases was showed in the research article [12], in which the methodology was correctly used, focusing on improving the activities in a logistics process.

RQ2: What are the Work Measurement Techniques used in Theses Times?

The article [27] do not use MTM in a real industry cases but this research focus on the necessary factors that should be considered when teaching the methodology for engineering students. Any other way, articles [10,24,36], used this methodology in a correct way in real industry cases of companies and the results were compared with productive activities and the use of new methodologies of predetermined times. Likewise, the use of MTM methodology is showed in articles like [20, 21], were each MTM phases is running correctly and systematically. The difference with the other papers is that the case studies were carried out in logistics operations.

Oppositely, the investigations carried out by [4, 38], validate their goals with the MTM methodology using simulated study cases in controlled environments, in which using digital applications studied manual activities. The two articles use the methodology correctly , but in the last one, the activities do not add value to the process causing the estimated time not approximately to the real industry process.

Currently, the technologies will provide a huge support for the study of work times. However, at the moment articles [13, 23, 33, 37] presents the steps to estimate time work using MODAPTS methodology without using digital platforms. In these researches, the methodology is applied in real industrial case studies, looking for the aim of continuous improvement. Another example of implementation in indutrial processes can seen in the article [3], this study analyzes the estimated work times using MODAPTS and direct observation. The authors of this research apply the methodologies following the standards and the results are observed in each case. If any organization wants to implements any type of system must have a huge knowledge of the process to be studied because theses techniques were affectes if researchers estimate or consider operations into the process that do not add value to improve the process.

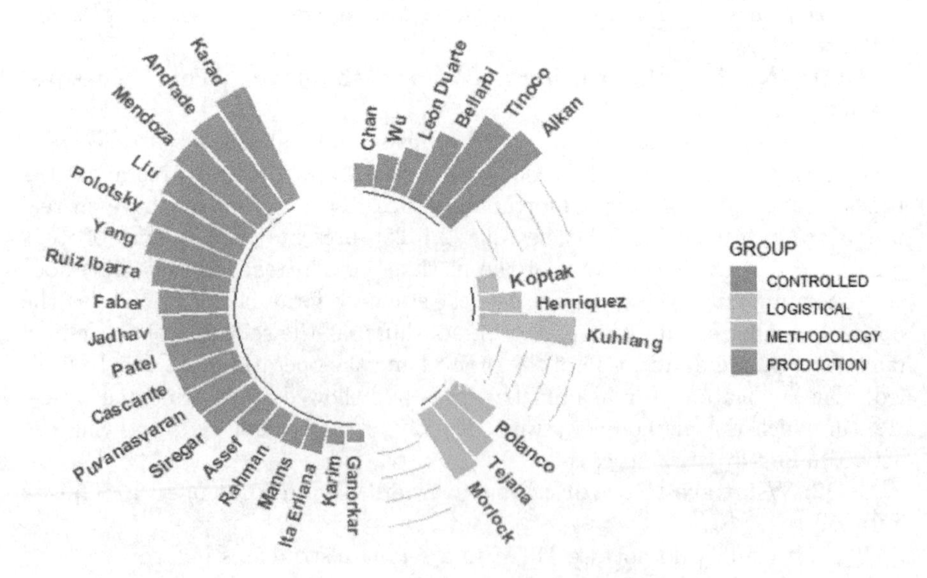

Fig. 6. Classification of cases

With the MOST methodology, the authors reviewed in this article implements in industry cases or logistic studies this methodology, getting improvements on reducing the waste of time and the cost of the production. Summarizing, from the total of thirty-one articles reviewed, nineteen of them are investigations applied to industry cases into production processes, three articles applied theses techniques in logistical processes, six focus their investigations in controlled environments, and three papers the authors did not apply the methodology in productive activities (See Fig. 6).

RQ3: There are a Methodology that Measure and Analyzed the Time in the Workers Activities?

Analyzing the feasibility to link methodologies with other tools. The literature review shows in twelve articles there is a combination of work-study methodologies with statistical tools, virtual analysis, meta-mechanisms, movement analysis, among others. These new techniques are used to estimate production times or compare current times with pre-established standards. Besides, if the industry wants to estimate times, improve processes or operations, innovating or adapting existing methodologies to a specific process, a combination of techniques with statistical tools must be carried out. As a result, it will improve research and generate new knowledge and applications in the industrial field.

4 Conclusions and Ongoing Works

This paper has presented a scoping review that summarizes information from the last six years on how different measurements work times methodologies has been implemented in the field of industrial process. Different combinations of techniques have been analyzed which allowed establishing a way of using assets, reducing costs to face the new production horizon after the pandemic and keeping the integrity of the value chain.

If theses methodologies are combined with digital tools, will become popular in the industry field due to its versatility, ease of implementation and constantly evolving technology. In fact, what is persued by this tools is to optimize working times, reduce costs of production, and dabble in the fourth industrial revolution.

As can be seen, the majority articles are aimed at combining the capabilities of different technologies of time measurement with direct observation in industrial environment. Unfortunately the techniques was design to analyze manual operations this is were the future researches are facing to develop tools that encompass machine conditions in theses work measurements standards. In the case of the designing and implementation the MOST methodology in the assembly lines the ongoing works in this technique is to explore how the system behaves in the mixed assembly models.

Acknowledgment. This work was financed by Universidad Técnica de Ambato (UTA) and their Research and Development Department (DIDE) under project CONIN-P-256–2019.

References

1. Alkan, B., Vera, D., Ahmad, M., Ahmad, B., Harrison, R.: A model for complexity assessment in manual assembly operations through predetermined motion time systems. Procedia CIRP **44**, 429–434 (2016)
2. Adrián, M., César A. Del Río, A., Alvear, D.L.: Estudio de Tiempos y Movimientos para Incrementar la Eficiencia en una Empresa de Producción de Calzado. Información tecnológica, **30**(3), 83–94 (2019)
3. Assef, F., Scarpin, C.T., Steiner, M.T.: Confrontation between techniques of time measurement. J. Manufact. Technol. Manage. **29**(5), 789–810 (2018)
4. Bellarbi, A., Jessel, J.P., Da Dalto, L.: Towards method time measurement identification using virtual reality and gesture recognition. In: 2019 IEEE International Conference on Artificial Intelligence and Virtual Reality (AIVR), pp. 191–1913 (2019)
5. Caiza, G., Ronquillo-Freire, P.V., Garcia, C.A., Garcia, M.V.: Scoping review of the work measurement for improving processes and simulation of standards. In: Advances in Intelligent Systems and Computing, pp. 543–560. Springer International Publishing (2021)
6. Miño Cascante, G., Alulema, J.M., Santillán Mariño, C.: Tiempos estándar para balanceo de línea en área soldadura del automóvil modelo cuatro (2019)
7. Durán, F.: Ingeniería de métodos. Globalizacion: Técnicas para el manejo eficiente de recursos en organizaciones fabriles, de servicios y hospitalarias (2007)
8. Hernández, J., Galante, E.: Edición especial Impacto económico. Technical report (2020)
9. Lago, A.E.: Métodos y tiempos con manufactura ágil. Alfaomega Grupo Editor (2016)
10. Faber, M., Przybysz, P., Latos, B.A., Mertens, A., Brandl, C., Finsterbusch, T., Härtel, J., Kuhlang, P., Nitsch, V.: Empirical validation of the time accuracy of the novel process language human work design (MTM-HWD®). Product. Manufact. Res. **7**(1), 350–363 (2019)
11. Criollo, R.G.: Estudio del trabajo. McGraw Hill (2005)
12. Henríquez-Fuentes, G.R., Cardona, D.A., Rada-Llanos, J.A., Robles, N.R.: Medición de Tiempos en un Sistema de Distribución bajo un Estudio de Métodos y Tiempos. Información tecnológica **29**(6), 277–286 (2018)
13. Erliana, C.I., Abdullah, D.: Application of the MODAPTS method with innovative solutions in the cement packing process. Int. J. Eng. Technol. **7**(2.14), 470 (2018)
14. Izquierdo, R., Novillo, L., Mocha, J.: El impacto de la cuarta revolución industrial en las relaciones sociales y productivas. Universidad y Sociedad **9**(2), 313–318 (2017)
15. Jadhav, M., Mungase, S., Karad, A.A.: productivity improvement through Maynard operation sequence technique, **3**(1), 565–568 (2017)
16. Jананía, A.C.: Manual de tiempos y movimientos. Limusa, México (2008)
17. Kanawaty, G.: Introducción al estudio del trabajo. Textos académicos. Oficina Internacional del Trabajo, cuarta edition (2007)
18. Karad, A.A., Waychale, N.K., Tidke, N.G.: Productivity improvement by Maynard operation sequence technique. Int. J. Eng. Gen. Sci. **4**(2), 657–662 (2016)
19. Karim, A.N.M., Tuan, S.T., Kays, M.H.E.: Assembly line productivity improvement as re-engineered by MOST. Int. J. Product. Perform. Manage. **65**(7), 977–994 (2016)

20. MTM-LOGISTICS Case Study: Koptak, M., Džubáková, M., Vasilienė-Vasiliauskienė, V., Vasiliauskas, A.V.: Work standards in selected third party logistics operations. Procedia Eng. **187**, 160–166 (2017)
21. Kuhlang, P., Sunk, A.: Productivity improvement in logistical work systems of the genuine parts supply chain. In: 2015 IEEE International Conference on Industrial Engineering and Engineering Management (IEEM), pp. 280–284. IEEE, December 2015
22. León-Duarte, J., Aguilar-Yocupicio, L., Romero-Dessens, L.: A software tool for the calculation of time standards by means of predetermined motion time systems and motion sensing technology. In: Ahram, T., Karwowski, W., Taiar, R., (eds.), Human Systems Engineering and Design, pp. 1088–1093. Cham, Springer International Publishing (2019)
23. Liu, L., Jiang, Z., Song, B., Zhu, H., Li, X.: A novel method for acquiring engineering-oriented operational empirical knowledge. Math. Prob. Engi. **1–19**, 2016 (2016)
24. Manns, M., Wallis, R., Deuse, J.: Automatic proposal of assembly work plans with a controlled natural language. Procedia CIRP **33**, 345–350 (2015)
25. Novillo, P.A.M., Álvarez, J.C.E., Zurita, C.I.N.: Estudio de tiempos y movimientos de producción para Fratello Vegan Restaurant. CIENCIAMATRIA **5**(1), 271–297 (2019)
26. Meyers, F.E.: Estudios de tiempos y movimientos para la manufactura ágil. México, segunda edition (2000)
27. Morlock, F., Kreggenfeld, N., Louw, L., Kreimeier, D., Kuhlenkötter, B.: Teaching methods-time measurement (MTM) for workplace design in learning factories. Procedia Manufact. **9**, 369–375 (2017)
28. Naciones Unidas. América Latina y el Caribe ante la pandemia del COVID-19Efectos económicos y sociales. Technical Report, 1 (2020)
29. Niebel, B.W., Freivalds, A.: Ingenieria industrial: Métodos, estándares y diseño del trabajo. McGraw Hill, duodécima edition (2009)
30. Organización Internacional del Trabajo (OIT). El COVID-19 y el mundo del trabajo: Repercusiones y respuestas. Technical report (2020)
31. Patel, D., Tomar, P.: A review on optimization in total operation time through Maynard operation sequence technique. Int. J. Sci. Technol. Eng. **3**(09), 13–16 (2017)
32. Vides, E.X.V., Andrea, D.J.L., Junior, G.R.J.: Análisis metodológico para la realización de estudios de métodos y tiempos Methodological analysis for the performance of studies of methods and times. (1), 3–10 (2017)
33. Polotski, V., Beauregard, Y.: Work-time identification and effort assessment: application to fenestration industry and case study. IFAC-PapersOnLine **51**(15), 569–574 (2018)
34. Puvanasvaran, A.P., Yap, Y.Y., Yoong, S.S.: Implementation of Maynard operation sequence technique in dry pack operation-a case study. **14**, 21 (2019)
35. Rahman, S., Karim, M.S., Mollah, R., Miah, J.: Implementation of Maynard operation sequence technique (most) to improve productivity and workflow - a case study. Int. J. Emerg. Technol. Innov. Res. **5**(6), 270–278 (2018)
36. Ruíz-Ibarra, J., Ramírez-Leyva, A., Luna-Soto, K., Estrada-Beltran, J.A., Soto-Rivera, O.J.: Optimización de tiempos de proceso en desestibadora y en llenadora. Ra Ximhai, **13**(3), 291–298 (2017)
37. Siregar, I.: Quality engineering with Taguchi loss function method and improvement of work method in anode changing. In: MATEC Web of Conferences, vol. 296, p. 02008, October 2019

38. Tinoco, H.A., Ovalle, A.M., Vargas, C.A., Cardona, M.J.: An automated time and hand motion analysis based on planar motion capture extended to a virtual environment. J. Ind. Eng. Int. **11**(3), 391–402 (2015)
39. Tricco, A.C., Lillie, E., Zarin, W., O'Brien, K.K., Colquhoun, H., Levac, D., Moher, D., Peters, M.D.J., Horsley, T., Weeks, L., et al.: Prisma extension for scoping reviews (prisma-scr): checklist and explanation. Ann. Intern. Med. **169**(7), 467-473 (2018)
40. Valderrama, Y., de Carmona, L.C., Colmenares, K., Jaimes, R.: Costo de la gestión laboral en el proceso productivo de una empresa manufacturera trujillana. Caso: Industrias Kel, C.A. Actualidad Contable FACES, **2**(33), 96–111 (2016)
41. Wu, S., Wang, Y., BolaBola, J.Z., Qin, H., Ding, W., Wen, W., Niu, J.: Incorporating motion analysis technology into modular arrangement of predetermined time standard (MODAPTS). Int. J. Ind. Ergon. **53**, 291–298 (2016)

Booting a Linux Kernel Under Bhyve on ARMv7

Darius Mihai, Maria-Elena Mihailescu(✉), Mihai Carabas, and Nicolae Țăpuș

University Politehnica of Bucharest, Bucharest, Romania
{darius.mihai,maria.mihailescu,mihai.carabas,
nicolae.tapus}@upb.ro

Abstract. ARM processors are more energy efficient when compared to their older and more powerful x86 counterparts. As such, more complex systems (e.g., servers) would greatly benefit from using them should they become powerful enough to be able to handle complex tasks. One such task, that is an essential tool for system administrators, is the ability to run virtual machines in order to provide secure and isolated environments for certain applications. With ARM-powered servers being under development for years already, anticipating the needs of system administrators and adding relevant features to the operating system may prove critical to increase the user base. Linux is by far the most successful free operating system, so any virtualization mechanism will need to be able to run a virtual machine with Linux before it may be considered viable for use in large-scale deployments. Consequently, bhyve, FreeBSD's virtual machine manager requires a proof of concept that runs a Linux-based operating system.

Keywords: FreeBSD · Bhyve · ARMv7 · Linux guest · VirtIO devices

1 Introduction

To create a functional virtualization environment that is able to run any guest operating system under FreeBSD, *bhyve* [1], the virtual machine manager, must support the VirtIO devices [2]. VirtIO devices are a paravirtualization mechanism for efficient communication between guest and host, created by the OASIS Committee. The virtualized devices are supported by all major operating systems (i.e., specialized device drivers are implemented for each operating system) for low overhead data transfers. The devices operate by creating an emulated data bus on the host and adding the VirtIO devices to this bus. From the guest's point of view, the devices should be accessible using regular bus access mechanisms (e.g., the x86 implementation uses an emulated PCI bus, where device iteration and polling can be used), and can be identified by the drivers using certain magic values.

The development of bhvye on ARMv7 started in summer 2015 as a Google Summer of Code project [3]. Two years later, in summer 2017 a working proof of concept was finished. The proof of concept was able to virtualize a one core CPU, system memory, generic interrupt controller and the device timer for a FreeBSD guest.

Bhyve is a type 2 hypervisor and relies on the host operating system's kernel to manage part of the functionality. Without using VirtIO devices, communication between

K. Arai (Ed.): FICC 2021, AISC 1363, pp. 93–101, 2021.
https://doi.org/10.1007/978-3-030-73100-7_7

guest and host is done by reserving a specific I/O port or memory region on the host. When the guest attempts to access that port or memory region, an exception is automatically generated since the guest does not have control over the resource. In order to obtain a simple console for debugging purposes on the ARM platform, the *bvmconsole* device is used by FreeBSD guests. The console uses a specific I/O port for x86 systems, or memory address on ARMv7, to write data one byte at a time and relies on polling to check if a data transfer is pending from the host. Since console devices usually have a small throughput requirement, the overhead generated by the many context switches (i.e., each transferred byte will require a context switch between the guest's and host's user spaces, going through each of their kernel spaces, and through the hypervisor) can be considered acceptable in most cases despite incurring a large overhead relative to the amount of data transferred. However, block or network devices must have a much higher data transfer expectancies, reducing the system performance drastically. VirtIO devices provide a mechanism to transfer blocks of data by allocating buffers that can be accessed by both guest and host. The number of context switches required can be controlled by intentionally generating notifications when sending data from guest to host or injecting an interrupt when transferring information from host to guest.

According to Popek and Goldberg's formal requirements of virtualization [4], VirtIO devices fall into the paravirtualization category, but are required for a virtualization mechanism that performs well. Additionally, specific, non-standardized device implementations, such as the bvmconsole that is used by bhyve cannot be used when working with other operating systems.

In order to extend support for virtualization on ARMv7 systems and include the new features into the upstream, bhyve on ARMv7 (i.e., the bhyve implementation for ARM processors) requires support for VirtIO device virtualization, code refactoring and a proof-of-concept run of a Linux guest. The implementation of the VirtIO devices or their extended functionality for ARM-based systems will not be covered by this paper.

The following sections will cover the following topics:

- **Section 2:** An overview of the VirtIO devices
- **Section 3:** Work on VirtIO devices on ARM and code refactoring
- **Section 4:** Booting a Linux kernel
- **Section 5:** Results using FreeBSD and Linux
- **Section 6:** Conclusions and plans for future work

2 State of the Art and Related Work

VirtIO Devices [2] are a mechanism used by virtual machines to provide a generic mechanism for communication between guests and host with reduced overhead. The host creates an emulated device that is presented to the guest as a regular device connected to a data bus (i.e., the specification [2] describes how devices communicate over the PCI bus, memory-mapped I/O and Channel I/O in Sect. 4 of the specifications document). The guest must implement special drivers for the devices that are tasked with allocating the shared resources and handling data transfers between the host and the guest.

In Sect. 5 of the VirtIO specifications document [2], six device types are described. The implementation for bhyve on ARM implements the following four VirtIO devices:

- Network Device
- Block Device
- Console Device
- Entropy Device

The ARM device emulation is heavily based on the PCI-specific implementation. The devices for bhyve on ARM were originally a carbon copy of the devices created for bhyve on x86, with the names of relevant data structures and functions changed to MMIO, instead of PCI (ARMv7 devices are unable to use the PCI bus; instead, memory-mapped I/O is used to transfer data to devices). Additionally, since MMIO devices cannot be automatically discovered because the memory location that is used may be randomly selected (as opposed to PCI devices that use a hierarchical bus structure where free device slots can be identified), the device allocation mechanism had to be reworked. Furthermore, the hierarchical structure of the PCI bus can be used to identify available interrupt pins for a device in order to send interrupts; since the MMIO devices are non-hierarchical in nature, interrupts must statically be associated with each device (implying another mechanism for interrupt handling was required).

The FreeBSD virtual machine monitor *bhyve* [5] was originally implemented for systems with x86 family processors. bhyve supports virtualization of multi-core processors, memory and is able to emulate PCI devices (e.g., UART, Intel e82545 network interface) and a subset of the VirtIO devices (entropy, console, network, block and SCSI).

The performance of a system is defined by the power of its CPU and memory, but also how well it can transfer data to peripheral devices (e.g., network and disk). A naive implementation of a console device that was used for early debugging required reading/writing using a specific resource (e.g., I/O port). Since context switches to/from the virtual machine were handled by the host kernel, but exception handling was done in userspace, every read or write of one data byte required multiple context switches between the host and guest.

VirtIO devices [2] were added to improve the throughput of data transfers between the guest and host. The devices are emulated by the host (i.e., have no hardware backing) and use shared resources between the host and the guest for improved performance. The mechanism implies that guests must know that they run in a virtualized environment in order to use the proper devices, and thus are a form of paravirtualization [4]. The VirtIO devices were written for x86 based systems, so the device emulation heavily relies on the PCI specifications (a hierarchical distribution of devices, with MSI or MSI-X capabilities).

3 VirtIO on ARMv7 and Code Refactoring

In order to run a Linux operating system guest under bhyve on ARM and prepare the implementation for upstreaming, finishing the VirtIO device [2] implementation using the MMIO interconnect mechanism and refactoring the emulation code base were required.

3.1 VirtIO Device Implementation

The original implementation of bhyve on ARMv7 used a trivial console device that polled a specific memory location in order to send information between the guest and the host. Transferring information one byte at a time by polling a memory location is highly ineffective, because it requires multiple context switches between guest and host.

As opposed to the x86 implementation that relies on PCI device emulation, ARMv7 devices can only use MMIO addressing and implement a different interrupt mechanism. Each device must have an individual memory location and interrupt number assigned for them. Operating systems for ARMv7 systems use a Flattened Device Tree [8] to specify the board's device structure in a unique pseudo-hierarchical manner. The hierarchical structure enables device drivers to iterate over the installed devices and identify compatible devices.

Some initialization parameters (e.g., compatibility, memory mappings and interrupts) can be specified for each device. As an example, the parameters from Listing 1 are used for the VirtIO network device over MMIO (the *compatible* parameter specifies a MMIO VirtIO device) in the guest's device tree.

```
virtio_net@6000 {
compatible = "virtio,mmio";
reg = <0x6000 0x200>;
interrupt-parent=<&gic>;
interrupts = <23>;
};
```

Listing 1. VirtIO Network FDT Entry

The device emulation reserves a memory region of size 0x200 starting at address 0x6000; memory access in this region will have to be handled by the host that will simulate the expected device functionality. The *interrupt-parent* option is used to link the device to the Generic Interrupt Controller (GIC) to receive interrupts; interrupt 23 (as specified by option *interrupt*) will be routed to the device when received by the interrupt controller. To start the device emulation, a symmetrical device specification has to be passed to the virtualization manager. Listing 2 presents the command can be used to start a basic virtual machine with VirtIO network and bvmconsole support.

The virtual machine will attach a VirtIO network device emulation to the virtual interface tap0 and pass it to the virtual machine. Using the tap interface network packets can be exchanged between the guest and host.

```
bhyve -m 0x200@0x6000#23:virtio-net,tap0 -b vm_name
```

Listing 2. Run bhyve with VirtIO network

The syntax specifies that the created VirtIO device must have *0x200* bytes of memory allocated starting with address *0x6000* and will use interrupt number *23*.

The other widely used device bhyve on ARM emulates (i.e., fourth, including the entropy and console devices) is the block device. Using the VirtIO block device on the guest will cause the guest to add a *dev* entry that can be mounted as any other block device.

Assuming the block device uses a device tree entry similar to that of the VirtIO network device (Listing 1), the command from Listing 3 will start the virtual machine with an emulated block device backed by the file called *disk*.

```
bhyve -m 0x200@0x7000#24:virtio-blk,disk -b vm_name
```

Listing 3. Run bhyve with VirtIO disk

The two devices have an implementation that is almost identical to their x86 counterparts, with the exception that the device identifiers differ to show that they are meant for MMIO, instead of PCI.

3.2 Refactoring Bhyve

Bhyve was originally written for x86 devices only; consequently, device emulation including VirtIO devices was highly dependent on some architecture specifics (e.g., all emulated device were connected to a virtual PCI bus with MSI/MSI-X capabilities). Because of this, part of the application had to be refactored since creating a separate implementation of the code for ARMv7 device emulation over MMIO generated a large amount of duplicate code - almost the entire device-specific implementations and a large part of the infrastructure are the same.

The major differences are related to how device resources are managed: PCI devices use the PCI bus for data transmissions and the MSI/MSI-X interrupt type, while MMIO devices create a memory mapping to transfer data and use a virtualized Generic Interrupt Controller for interrupts.

Emulated devices running over PCI were using a *pci_** naming convention (e.g., the generic component of the emulation were in files named *pci_emul.c*, *pci_emul.h*, *pci_irq.c*, etc.; device-specific implementation files were *pci_virtio_rnd.c*, *pci_virtio_console.c*", etc.; furthermore, all data structures and files had the *pci_* prefix). For MMIO devices this naming convention would have been confusing; the initial emulation used a similar *mmio_** naming convention but was not maintainable.

To accommodate a more generic device implementation, files and code have been renamed to use a *devemu_** naming convention. Files used to emulate the VirtIO devices were renamed to fit this new naming convention (e.g., the *pci_emul.c* and *pci_irq.c* files became *devemu.c* and *devemu_irq.c* respectively, and the *pci_devemu* data structure was renamed to *devemu_dev*). All files that contained architecture-specific code (PCI for x86 and MMIO for ARM) were placed in specific subdirectories, depending on the underlying data bus (namely, *mmio/* and *pci/*). The VirtIO device emulation files were adapted to use the refactored infrastructure; the identifier initialization has been moved to the bus specific files to eliminate all specific code. The files of these devices and some other files that are identical for both PCI and MMIO are kept in the root of the *bhyve/* directory and are used by either architecture.

It is worth mentioning that PCI-specific code and device implementations that had no MMIO equivalent were left unchanged at the time of writing this paper (e.g., the *pci_e82545.c* was only changed in order to use the refactored infrastructure; the *pci_populate_msicap* and a number of other functions that dealt with PCI -specific capabilities such as MSI/MSI-X).

4 Booting Linux on ARMv7

The previous tests were done with a FreeBSD guest, since it was the only one that allows using the *bvmconsole* for debugging and communication. This meant that running any other operating system (e.g., Linux) was not possible until a more generic communication mechanism was implemented. This is the reason why having working support for VirtIO devices was critical before being able to start testing with a Linux guest.

The work on starting the Linux guest was initially heavily influenced by the work written by Virtual Open Systems about porting KVM to ARM Cortex-A15 Fast Models [7, 6], since this was a work implemented on the same platform that I was using. Their approach relied on cross-compiling a Linux kernel and using a certain configuration file that they had provided to boot the guest system, but the configuration they provided did not work on my system.

The devices used by the guest operating system must be provided through a compiled flattened device tree [8] file, called a flattened device blob. The device blob should be integrated with the file that also contains the kernel, in a manner that allows the kernel to recognize its location. This is where the first issue with booting the Linux kernel arose - while the FreeBSD kernel allows integrating the device tree blob file into the kernel image at compile time, and being able to identify it at runtime automatically, the Linux kernel does not. Instead, the Linux kernel is compiled separately, and the device tree blob is appended to the binary file at an offset that guarantees that the blob does not overwrite part of the kernel (e.g., using *dd*). The Linux binary interface used the first three ARM registers to identify where the kernel was loaded into memory (*r0*), the platform ID (*r1*) and the location where the device tree blob was loaded (*r2*). While the FreeBSD kernel should have used the same registers for the boot process, the *r2* register was not used since it could identify the location of the device tree blob without reading it from *r2*.

Once the device tree was successfully identified by the guest Linux guest, we managed to create a working device tree based on the one used for the FreeBSD guest through trial and error. There were two major issues with the original configuration file.

The first issue was that the configuration file specified that the guest's memory should start at address *0xc0000000* (for ease of use, the guest's memory was passed through, beginning with 3 GB). The Linux guest refused to run, however, since it considered that the memory region starting with 3 GB was reserved for the system highmem, and consequently refused to allocate memory in that region. This caused the guest to crash, because it could not allocate memory to run the operating system. Moving the memory start location to *0x80000000* (the 2 GB boundary) solved this issue.

The second issue was caused by the lack of a timer. The configuration file did not properly pass the device timer through to the guest, making it unable to read the time properly. The Linux kernel operates based on an internal counter called *jiffies* to measure the passing of time. Jiffies are periodically updated to allow kernel modules to measure how much time had passed between two events. When the kernel attempts to measure that a functional timer is present, it enters a *while (true)* loop that will exit only if the value read from the timer is greater than the value it initially read (i.e., the timer counter reflects time passing). However, because the timer was not properly passed to the guest, the value stayed the same, resulting in a silent system freeze. This issue was, however, fixed when the timer was properly passed to the guest.

Once the issues described above were solved, the kernel could finally boot. However, no data was displayed to the terminal where the virtual machine was started. This meant that the mechanism used to print information through the *early_printk* mechanism for debugging that was described by the article from Virtual Open Systems [7] was not working for my setup. After doing some research into the internals of the Linux kernel code, I had discovered that the *printascii* function that was called when they were testing was not called in our case. To solve this issue, we changed the code for the *printk_emit* function to call the *printascii* function directly, which restored the expected functionality.

The boot process would then fail when trying to load the device driver for the VirtIO devices. The devices implemented by bhyve implemented the v0.9.4 standard, but failed to present the proper device version for the legacy standard. While testing using a FreeBSD guest, the device drivers did not throw any errors, while the Linux counterpart generated an assertion error. After fixing this issue, the boot process reached the userspace, where no more output was shown.

The first attempt at testing the userspace, relied on building a ramdisk file that would be included with the kernel, similar to how the FreeBSD guests were tested. The FreeBSD guest configuration used a rule adding a *ROOTDEVNAME* ramdisk to the guest's image file that relied on the *rescue* binary for most basic functionality. With this idea in mind, we tried creating a ramdisk that relied on the *busybox* binary to get a similar functionality. However, this approach did not work, as the ramdisk did not seem to be reached. The next approach was building a root filesystem using the *buildroot* automated toolchain for Linux. This proved much more useful, since the *init* script from the filesystem was ran, but failed since it could not find the */dev/console* device.

By using the *ttyprintk* option in the kernel's build configuration, the */dev/ttyprintk* could be created manually (the node was not created automatically), and a link from */dev/console* to */dev/ttyprintk* could be added through the userspace initialization script. This provided the Linux guest with a working console, even though it did not allow interactive access.

By adding commands to the init script, to explore the filesystem, mount a VirtIO block device or ping the host, the functionality of both the VirtIO block and network devices could be tested.

5 Results

Four VirtIO devices that were implemented have been tested and are functional on the ARM FastModels [9], an ARMv7 and ARMv8 system emulator created by ARM Ltd. Testing on an ARMv7 development board is impossible because there are some issues with the Generic Interrupt Controller virtualization that impede guest interrupt injections.

The Linux kernel is compatible with the implementation of the VirtIO devices from bhyve. Despite not having an interactive command line interface, the guest could be set to run commands by adding them to the init script when building the root filesystem. Since the console's output was functional, the tests could be performed without an interactive mode. As such, we could run tests to check that the root filesystem was created and was visible to the host, and that the VirtIO devices were indeed functional.

For the networking device, a tap virtual interface is created on the host and attached to a virtual bridge; the virtual device binds to the tap interface, to allow packet exchanges between the guest and the host. To test the functionality of the device, a bridge device on the host has an IP address set to allow pings from inside the guest to be sent. Listing 4 shows the output of a ping command sent from inside the guest operating system (the IP address of the bridge interface on the host is 10.0.4.86).

```
# ping 10.0.4.86
PING 10.0.4.86 (10.0.4.86): 56 data bytes
10.0.4.86: icmp_seq=0 ttl=64
time=235.110 ms
64 bytes from 10.0.4.86: icmp_seq=1
ttl=64 time=9.093 ms
```

Listing 4. Ping from guest to host

The block device uses a virtual disk image that can be loaded into the device memory. The ramdisk image file, called *disk*, was created outside the of the virtualization environment since it is a portable format (i.e., was created in a FreeBSD environment, but does not have hardware dependencies, and the partition format could be read by both FreeBSD and Linux), and contains a plain text file that contains the *"Hello VirtIO"* text. When the VirtIO device driver connects to the device, it creates the */dev/vda1* device in the guest operating system. Mounting this device will allow exploring its contents. Listing 5 shows the mount command and a basic exploration of the virtual block device.

```
guest$ mount /dev/vda1 /mnt
guest$ ls /mnt
test.txt
guest$ cat /mnt/test.txt
Hello VirtIO
guest$ umount /mnt
```

Listing 5. Mount and explore block device

6 Conclusions and Future Work

Currently, the FreeBSD virtual machine monitor, *bhyve*, is able to virtualize basic hardware components (e.g., CPU, memory, Generic Interrupt Controller), but also provide more advanced virtualization mechanisms, such as VirtIO devices. The Linux guest operating system can make use of the VirtIO devices such as block and network, but does not have a working interactive console sas of yet.

The implementation of bhyve on ARM must move on to the ARMv8 platform to ensure compatibility with ARM processors for workstations and servers that will be used to run virtual machines. The implementation for ARMv7 has so far been a good enough proof of concept, but its drawbacks, such as the inability to run FreeBSD virtual machines on ARMv7 development boards makes it not acceptable. The hypervisor on ARMv8 was developed by Alexandru Elisei, and tested on the Fast Models emulator for ARMv8; his work is currently being continued by Andrei Martin who is working to bring the project into the upstream, with a functional bringup on the EspressoBin development board. Once everything is set, we will work with him to begin testing with a Linux guest on that platform as well.

Acknowledgments. This work was supported by CONDEGRID project (no. 07/10.03.2020): National contribution to the development of the LCG computing grid for elementary particle physics.

References

1. FreeBSD Project. Freebsd as a Host with bhyve. https://www.freebsd.org/doc/handbook/virtualization-host-bhyve.html. Accessed 14 Feb 2020
2. OASIS Committee Specification Draft 01/ Public Review Draft 01. Virtual I/O Device (VIRTIO) Version 1.1. https://docs.oasis-open.org/virtio/virtio/v1.1/csprd01/virtio-v1.1-csprd01.html. Accessed 15 Feb 2020
3. Carabas, M., Grehan, P.: Porting bhyve to ARM. https://www.bsdcan.org/2016/schedule/attachments/370bsdcan2016.pdf. Accessed 12 Feb 2020
4. Popek, G.J., Goldberg, R.P.: Formal requirements for Virtualizable Third Generation Architectures, July 1974
5. FreeBSD Project. Bhyve. [Source code], https://github.com/FreeBSD-UPB/freebsd/tree/master/usr.sbin/bhyve. Accessed 12 Feb 2020
6. Dall, C., Nieh, J.: KVM on ARM. http://systems.cs.columbia.edu/les/wpid-asplos2014-kvm.pdf. Accessed 12 Feb 2020
7. KVM port on ARM Cortex-A15 Fast Models. http://www.virtualopensystems.com/en/solutions/guides/kvm-on-arm/. Accessed 15 Feb 2020
8. FreeBSD Project wiki. Flattened Device Tree. https://wiki.freebsd.org/FlattenedDeviceTree. Accessed 15 Feb 2020
9. ARM Ltd. Fast Models. https://developer.arm.com/tools-and-software/simulation-models/fast-models. Accessed 15 Feb 2020

S-Band Patch Antenna System for CubeSat Applications

Ravneet Kaur(✉), Mayank, Tushar Tandon, Johannes Schumacher, Guhan Sundaramoorthy, and Udit Kumar Sahoo

Department of Engineering, Celestial Space Technologies GmbH, Oedenberger Street 159, 90491 Nüremberg, Germany

Abstract. A novel design of S-band based circularly polarized small satellite patch antenna is presented. It comprises three slots and an overall dimensions of 98 mm x 98 mm. The antenna has been designed for an impedance of 50 ohm and shows optimum performance for frequency range of 2.3 GHz to 2.5 GHz. Absolute gain observed is 7.8 dBi with main lobe at 20 normal to the antenna plane, making it close to unidirectional antenna. A diamond shaped patch with suitable location of coaxial feed allows the antenna to be circular polarized. Additionally, three identical slots were embedded in the radiating patch to obtain the high bandwidth compact antenna with right circular polarization. The variation of feed position allows switching between Left and Right Hand Circular Polarization (LHCP, RHCP) with 5.48 dBi gain. The S11 parameter was measured using Network Analyzer and similar results were obtained as software simulation output. The designed antenna is compatible with CubeSat configuration and hence can be used for various small satellite applications.

Keywords: Impedance · S-band · Circularly polarized · Patch antenna

1 Introduction

Small satellites are revolutionizing the new space industry. The advancement in technology has resulted in miniaturization of hardware and consequently, the size of satellites has also been miniaturized without any significant compromise on performance and capabilities. The advancement of technologies has in fact improved the capabilities of small satellites so much that NASA has recently demonstrated the capabilities of CubeSat (MarCO) in a Mars mission [1]. The advantages of Cube Sat are immense and because of smaller size and less weight which leads to overall mission cost. Companies like SpaceX (Starlin k) [2], OneWeb [3] and Amazon (Kuiper) [4] are planning and have already started developing small satellite based mega constellations with the vision to provide connectivity to everyone and every part of earth.

One of the main systems of the satellite is the communication system for uplink and downlink of the telemetry and commands from ground station to satellite and vice versa. In communication system, antenna system is a critical component which ensures the compliance with small size without compromising on the system performance. Antenna

K. Arai (Ed.): FICC 2021, AISC 1363, pp. 102–112, 2021.
https://doi.org/10.1007/978-3-030-73100-7_8

system has two functions, firstly, it provides the strength to the signal transmitted by the satellite and secondly, it directs the signal towards the receiving end by providing directivity to the transmitted signal.

Typical CubeSats rely on monopole or dipole antenna for VHF/UHF frequencies. Frequencies ranging from 2 GHz to 4 GHz is reserved for amateur radio applications by International Telecommunication Union [5]. The radiation pattern of antenna is designed based on satellite orbit and operation modes. Indeed, depending upon different modes of satellite operations, various antenna radiation patterns are required. One of the most desired property of antenna radiation pattern is circular polarization. This is where a patch antenna offers advantage over monopole or dipole antenna. Furthermore, the satellite system downlink budget also plays an important role in selecting the antenna system. The highly directional antenna will offer more gain due to reduced beam width angle. Hence, directional antenna will be preferred for a low downlink power budget system.

The paper is structured as follow: In Sect. 2 the geometrical constraints and system design is presented. In Sect. 3, Simulation boundary conditions and material for antenna explained. Various simulation and test results are explained in Sect. 4 followed by the result discussions in Sect. 5.

2 Antenna System

Figure 1 shows the SDR (Software Defined Radio) based communication system with the Small satellite patch antenna which is described in this work. The system is designed to be available as a plug play system off the shelf for CubeSat and larger missions. The system shall be usable by a 1U (100 mm x 100 mm) CubeSat mission. The System is based on SDR and thus, is configurable by software and commands. Such system can be upgraded based on different modes of mission without updating any hardware. Considering CubeSat size and mass constraints, the best option for a communication system is printed patch antenna. However, to manufacture a circularly polarized antenna the patch is rotated 45° and power supply is supplied close to the diagonal of the patch using a coaxial connection as shown in Fig. 2.

There are two configurations possible for the shown patch antenna. One is Right Hand Circularly Polarized (RHCP) and nother is Left Hand Circularly Polarized (LHCP). In this work, only RHCP configuration is explained. The antenna shown in Fig. 2 is the second prototype of the patch antenna. The first one was with a micro strip feed and was linearly polarized.

The material chosen for the antenna is Copper (Cu) as a radiative material for patch and ground station. The substrate material is RO4003 PCB printing material.

Fig. 1. SDR based communication system integrated with Small satellite slotted patch antenna

Fig. 2. Small satellite slotted patch antenna

3 Design Approach

There are many design techniques to enhance the performance parameters of patch antenna like bandwidth, gain, directivity. The techniques adopted by Celestial targets bandwidth and gain for low profile antennas. The resonant frequency has been considered as the main element for designing antenna in order to acquire the specific performance characteristics mentioned above. The model in Fig. 3 is designed at a resonant frequency of 2.48 GHz.

The main radiating element is a square patch antenna with 45° rotation. 45° rotated square excites two orthogonal modes with 90° phase difference resulting in change to current path to obtain circular polarization. The physical dimensions of the patch are function of the dielectric constant, effective dielectric constant, height of the substrate

Fig. 3. Small satellite slotted patch antenna (Dimensions in mm)

and resonant frequency as mentioned in the formulas [6] below:

$$L = (0.47 - 0.49)\lambda_0 \sqrt{\varepsilon r} \tag{1}$$

$$W = c \div (2fr \times \sqrt{\varepsilon r} + 1) \tag{2}$$

The major impact of length and width is observed on the resonant frequency and radiation efficiency respectively of the antenna. By varying the effective length of antenna, the target resonant frequency can be achieved for desirable performance whereas the width of antenna only has effect on the S11 parameter and bandwidth of antenna. The effective length is more for lower frequencies and vice versa. Generally, width and length of patch effect the performance of antenna independently, but, in his case, both the parameters are kept equal to enable circular polarization.

3.1 Slotted Patch

Symmetrical and rectangular embedded slots were incorporated in the design to achieve a high bandwidth compact antenna. The addition of slots in the radiating patch results in lowering of the resonant frequency. Therefore, antenna size can be reduced for the desired operating frequency. Additionally, spacing and orientation of slots affects the performance of antenna. The slots were orientated at 45° to x-axis to obtain right hand circular polarization. Furthermore, the parallel orientation of slots results in reducing the perturbations due to probe inductance. The parallel slots introduce the capacitance and thus compensates for probe inductance. The slot dimensions and spacing between the slots has major impact in the reduction of resonant frequency results from the spacing between slots [7]. In the present design i.e. Fig. 5, the three identical narrow slots of 15 x 1 mm are placed at equal spacing of ≈ L/4.

3.2 Substrate Height

The substrate height is a major design decision to achieve the desirable bandwidth without compromising on efficiency of antenna. Thicker substrate h > 4 mm leads to enhanced impedance bandwidth whereas thinner substrate h < 1 mm reduces the fringing effect. Along with high impedance bandwidth for height more than 4 mm, reduction in antenna efficiency was observed due to surface wave propagation and spurious coupling [8]. Therefore, the height was varied between 2 to 3 mm as per Eq. (3) by keeping dielectric constant fixed to obtain higher bandwidth with minimum fringing affect. A substrate material with low dielectric constant is chosen to allow the use of thin substrate i.e. between 2 and 3 mm which results in desired performance without compromising the efficiency.

$$0.003\,\lambda_o \leq h \leq 0.05\,\lambda_o \tag{3}$$

3.3 Feed Position

In Rf applications, matching the impedance of interconnected blocks is very essential including antenna to LNA or Power amplifier to antenna to avoid any signal loss. The feed location plays an important role in obtaining the good impedance matching [10]. For this case, the feed location chosen at (16, 4.065) i.e. point 1 in the first quadrant [Fig. 3] for 50-ohm impedance matching, further discussed in smith chart results. The feed location in figure is chosen more towards x-axis to achieve right hand circular polarization. If the feed location is shifted on y-axis, left hand polarization can be obtained. Therefore, the antenna can be switched between LHCP and RHCP just by changing the feed position.

4 Simulation Boundary Conditions and Materials

The simulations for the antenna are performed in CST software. The antenna is designed in the design toolbox of the above-mentioned software and consists of four sections, namely, Substrate, Ground plane, Radiating patch and Coaxial connector. The antenna substrate and ground plane are a square plates of dimensions 98 mm by 98 mm. The radiating patch is 30 mm by 30 mm with a standard dimension coaxial connector. The design of the antenna is shown in Fig. 4 below:

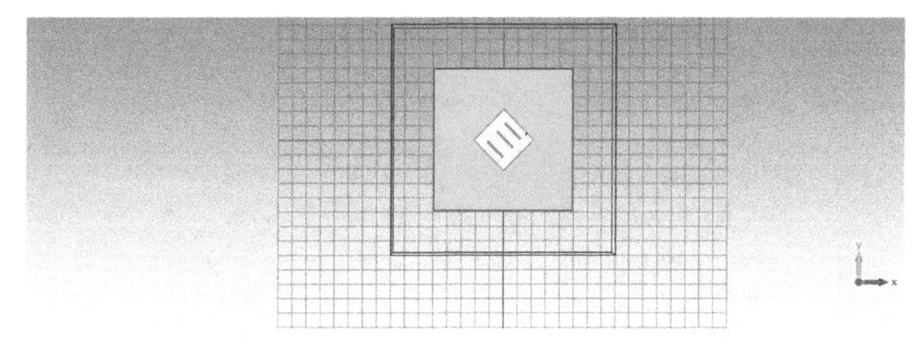

Fig. 4. Design of patch antenna in CST microwave

4.1 Material Selection

The material selection plays an important role for antenna performance in different environments. The main properties affecting antenna performance are dielectric constant of substrate material, thermal properties of the substrate and patch material. The effects of these material properties are mentioned below:

Dielectric of Substrate Material
There are various microwave materials which are available in the software libraries with dielectric constant varying between $2 \leq \varepsilon r \leq 12$. Choosing the higher value of dielectric constant for thin substrates results in poor antenna efficiency and lower bandwidth, however, also leads to size reduction of antenna. Using materials with lower dielectric constant comes at a cost of increased antenna size. A tradeoff has been performed between thickness and dielectric constant value for better efficiency without affecting the antenna size. A value between $2 \leq \varepsilon r \leq 3.5$ was targeted for the final selection.

Thermal Properties
Material with high thermal conductivity and lower thermal expansion shall be targeted to avoid any heat shocks on the radiating material. Additionally, Thermal expansion could lead to variation in the dielectric thickness further resulting the change of resonant frequency [8].

The properties of both the materials for patch and substrate are shown in Table 1 below:

4.2 Defining the Port

The S-matrix describes the transmission of electromagnetic field energy between different ports of a structure. These ports need to be defined in CST MICROWAVE STUDIO. Three different kinds of ports exist in software are waveguide port and discrete port and plane wave. In present design the excitation is applied by the means of discrete port. It consists of a perfect conducting wire connecting start and end points with a lumped element in the center of the wire. The typical way to define a discrete port is to pick its two end points from the structure using the common pick tools and then to enter the discrete port dialog box [11].

Table 1. Selected material properties

Materials	Copper [9]	RO4003 [10]	Unit
Material properties			
1. Epsilon	NA	3.38	NA
2. Insertion loss	5	0.017	dB
3. Thermal conductivity	394	0.64	W/m/°K
4. Density	8.89	1.79	gm/cm^3
5. Electrical conductivity	101	NA	%IACS
6. Relative permeability	0.99	NA	μ/μ_0

4.3 Mesh Property Definition

For mesh based on hexahedral grid structure, the Finite Integration Method (FI-Method) in addition with the Perfect Boundary Approximation (PBA) is used. The simulated structure and the electromagnetic fields are mapped to hexagonal mesh [12].

The mesh is generated in time domain by the automatic mesh generator feature in the software [12]. The number of mesh cells obtained my mesh generator were 188160. This generator allows a good trade-off between the need of an accurate structure and simulation time.

5 Simulation and Results

The results of simulation of patch antenna made by CST software are discussed below:

5.1 Radiation Pattern/Directivity

Figure 5 shows that the distribution of power radiation around the antenna as a function of direction represented by the π angle at 2.483 GHz. RF field pattern is observed at operating frequency 2.483 GHz. Radiation pattern obtained is almost unidirectional with main lobe directed at an angle of 2 degree, having angular beamwidth of 79.2 degree. The magnitude of the main lobe is 5.85 dBi.

The Directivity plot (3D view) in Fig. 6 represents amount of radiation intensity i.e. is equal to 5.842 dBi. The simulated antenna radiates more energy in a specific direction by an amount of 5.842 dBi.

5.2 VSWR

VSWR is the ratio of maximum value of standing wave voltage to its minimum value. The ideal VSWR is 1 i.e. no loss occurs. VSWR more than 2 means that the antenna is poorly matched. In simulation results as shown in Fig. 7. The VSWR obtained is less than 2 for frequency range 2 GHz–3 GHz. From this we can deduce that the antenna can perform from range of 2 GHz to 3 GHz without damaging the RF front end circuit to which antenna might be connected.

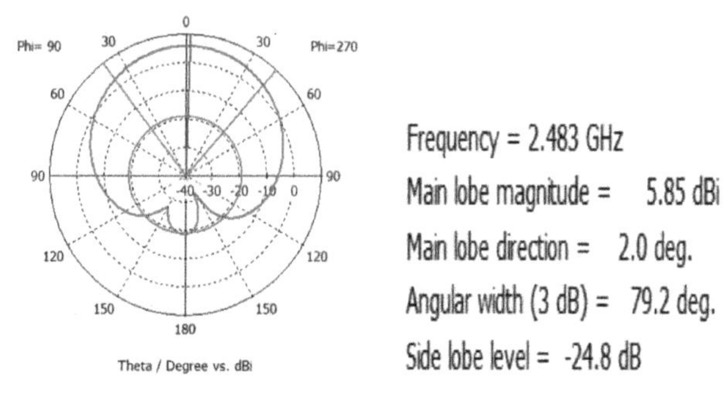

Fig. 5. Polar graph of radiation pattern

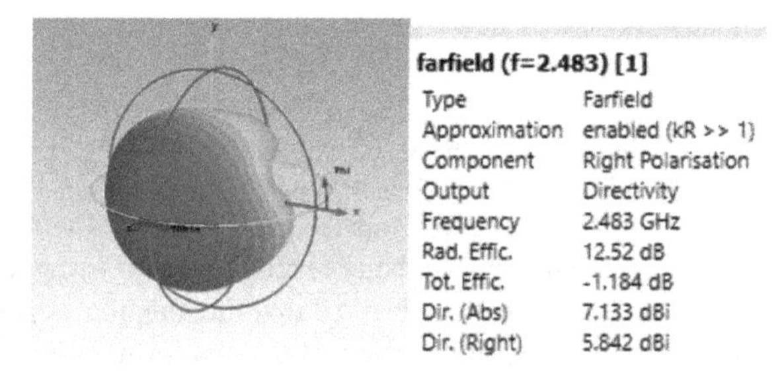

Fig. 6. 3D view of radiation pattern

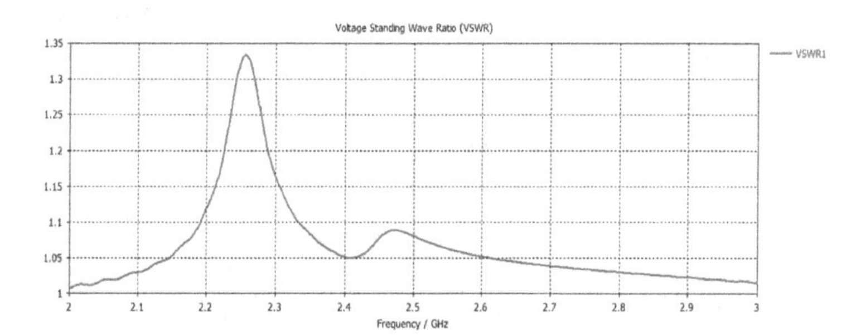

Fig. 7. VSWR between 2–3 GHz

5.3 Efficiencies Plot

The radiation efficiency is defined as the ratio of the total power radiated by the antenna to the antenna efficiency that considers the reflection, conduction, and dielectric losses [13]. In case of simulated antenna, the maximum efficiency as shown in Fig. 8 is ranging from 2.4 to 2.6 with the value of –0.1 to –2 dBi i.e. 90 to 70%

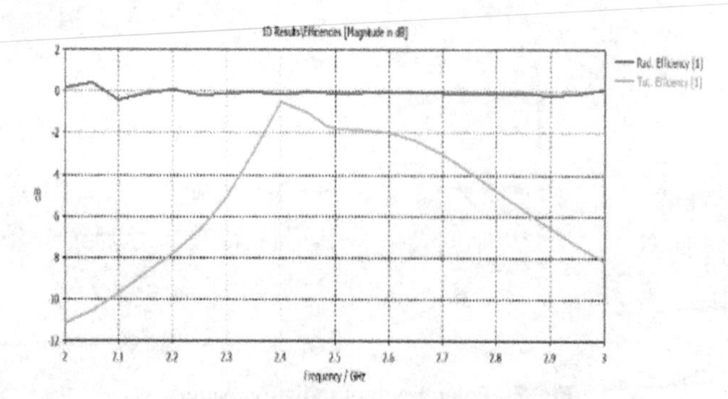

Fig. 8. Antenna efficiency

5.4 S11 Parameters

This parameter measures the power reflected by antenna due to mismatch. As per graph shown in Fig. 9, the return loss is low in the frequency range of 2.4 GHz to 2.5 GHz. Therefore, less power reflectance occurs in this frequency range. The results from simulation graph were compared with experimental results obtained using Vector Network Analyzer (VNA). It can be observed from Fig. 10 that the antenna works in range of 2 to 3 GHz with a greater peak value in comparison to simulated antenna. Additionally the variation of S parameters with frequency shows similar pattern as simulated antenna thus verifying the simulation results. From here we can speculate that the gain and other performance parameters estimated in simulation are close to actual performance of the antenna.

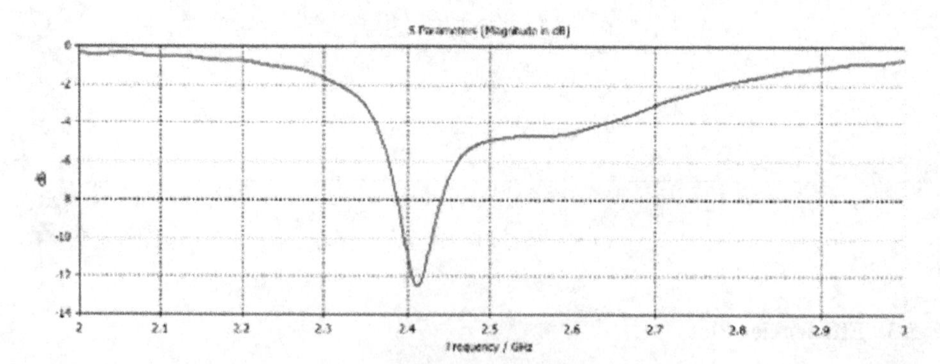

Fig. 9. S11 parameters

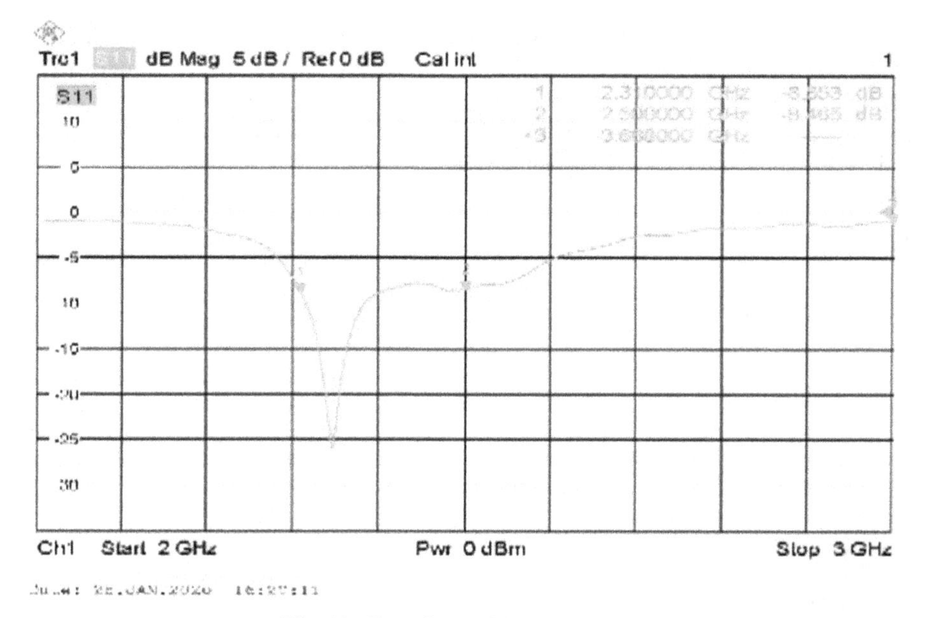

Fig. 10. Experimental S11 parameters

6 Conclusion

The simulated result provides that, lower value of return loss, VSWR < 2 and better efficiencies leads to an optimum feasibility of designed patch antenna. The added slots, in scope of gain improvement, increases gain by 20% in comparison to antenna with no slots. The current PCB material-based design can be optimized by adding space grade materials like aluminum alloys (AlSi10Mg) for patch and polymers for substrate (Arlon), making it possible to integrate them with 1U CubeSat and demonstrate space applications. The satellite communication systems need very light weight antenna which can be easily attached with the main systems and which does not make the system bulky, making it compatible with CubeSat configurations. Furthermore, making a gridded substrate the weight of the antenna can be further reduced. Gridded substrate will improve the antenna performance, because of the trapped air/vacuum. Since the dielectric constant of air/vacuum is 1 (ideal), this will result in improved antenna performance. Another configuration of the same antenna could be by reducing the size of antenna substrate and ground plane to 70 mm by 70 mm. This will have advantage of smaller antenna size.

7 Future Scope

The tested antenna will be further developed to operate in space environment by replacing the existing substrate material with the space grade material. Additionally, an adaptable antenna will be redesigned by implementing parasitic patch elements to cover the allocated S-band frequencies for commercial space activities. Patch array antennas are popularly being researched for the interplanetary space communication. The future missions to moon and mars will require high gain directional antennas for establishing required link margin. The designed patch element can be further used in the array network of multiple patch elements by implementing a feed network technique. There will be a continuous scope of patch array antennas in the future missions due to its low profile and ease of integration with the space structures.

References

1. MarCO Homepage. https://www.jpl.nasa.gov/cubesat/missions/marco.php
2. https://www.starlink.com/
3. https://www.oneweb.world/
4. https://spacenews.com/amazon-moving-project-kuiper-team-to-new-rd-headquarters/
5. Panel on frequency allocation spectrum protection for scientific users, committee on radio frequencies, national research council. Handbook of frequency Allocation and Spectrum Protection for Scientific Uses. The National Academies Press (2007)
6. Wong, K.-L.: Compact and Broadband Microstrip Antennas. John Wiley & Sons Inc., New York (2002)
7. Sebastian, M., Ubic, R., Jantunen, H.: Microwave Materials and Applications. Wiley, Hoboken (2017)
8. Balanis, C.A.: Microwave Material and Applications, 4th edn. John Wiley & Sons Inc., Hoboken (2016)
9. https://copperalliance.eu/about-copper/copper-and-its-alloys/properties/
10. https://www.pcbexpress.eu/assets/documents/Specification%20RO4003C%20,%20RO43 50B%20.pdf
11. http://www.mweda.com/cst/cst2013/mergedProjects/CST_CABLE_STUDIO/special_solv opt/special_solvopt_discreteports.htm
12. http://www.mweda.com/cst/cst2013/mergedProjects/CST_EM_STUDIO/common_ove rview/common_overview_mesh.htm
13. http://www.antenna-theory.com/basics/efficiency.php

Cardiovascular Signal Processing: State of the Art and Algorithms

Hiwot Birhanu(✉) and Amare Kassaw

Bahir Dar Institute of Technology, Bahir Dar University, Bahir Dar, Ethiopia
https://www.bdu.edu.et/

Abstract. The emergence of Artificial Intelligence (AI) has brought many advancements in biomedical signal processing and analysis. It has opened the way for having efficient systems in the diagnosis and treatment of diseases such as Cardiovascular (CV) disorder. CV disorder is one of the critical health problems causing death to lots of peoples globally. Electrocardiogram (ECG) signal is the signal taken from the human body to diagnosis the status of CV and heart conditions. Earlier to the introduction of computers, the diagnosis of heart conditions was made by experts manually and that caused various mistakes. Currently, the usage of advancing signal processing devices help to reduce those errors and enables to develop effective signal detection and parameter estimation algorithms that are useful to analyze the parameters of ECG signals. Which intern supports to decide if the person is in critical condition and take an appropriate action. In this work, we analyze the performances of classical techniques and machine learning algorithms for ECG based CV parameters estimation. For this, first an in-depth review is done for both classical techniques and machine learning algorithms. Specifically, the benefits and challenges of machine learning and deep-learning algorithms for CV signal processing and parameter estimation is discussed. Then, we evaluate the performances of both classical (Kalman Filtering) and machine learning algorithms. The machine learning based algorithms are modeled with Butterworth low pass filter, wavelet transform and linear regression for parameter estimation. Besides, we propose an algorithm that combines adaptive Kalman filter (AKF) and discrete wavelet transform (DWT). In this algorithm, the ECG signal is filtered using AKF. Then, segmentation is performed and features are extracted by using DWT. Numerical simulation is done to validate the performances of these algorithms. The results show that at 20% false positive rate, the detection performance of Kalman filtering, the proposed algorithm and machine learning algorithm are 83%, 94% and 97%, respectively. That shows the proposed algorithm gives better performance than classical Kalman filtering and has nearly the same performance with machine learning algorithms.

Keywords: ECG signal · CV parameter estimation · Kalman filter · Machine learning · DWT

K. Arai (Ed.): FICC 2021, AISC 1363, pp. 113–127, 2021.
https://doi.org/10.1007/978-3-030-73100-7_9

1 Introduction

Cardiovascular disease (CVD) is the most common chronic disease that cause massive amount of economic and social problems. Various researches done on this area shows that, CVD is the main cause for the death of lots of people of all age range. So having a system with better diagnosis and prediction of heart rate is necessary so as to combat the effects of this disease. ECG is one of the devices that capture the electrical activity of the heart, which in turn is used for the diagnosis of cardiac problems [1,2]. The study of cardiovascular parameters from ECG signal is mostly dependent on a cardiac cycle due to the cardiac muscle depolarization and repolarization (the PQSRT pattern). This pattern varies from person to person as well as from time to time. As shown in Fig. 1, one cycle has atrial contraction (P-wave), contraction of the ventricles (QRS-complex) and relaxation of the ventricles (T-wave). The amplitude of these waves and the RR, PR, QT, and QRS complex intervals indicate the condition of the heart. The variation of these waves may or may not be visible to the physicians. Even if it is visible, the variation might occur within a fraction of seconds and the time the physicians takes to make the decision might be very long. Due to that, the life of the patient at risk will be endangered. Hence, due to such variation in addition to the noise and other motion artifacts, having a signal processing system with efficient algorithms is vital to study the condition of the patient, and to make effective and wise decisions [3].

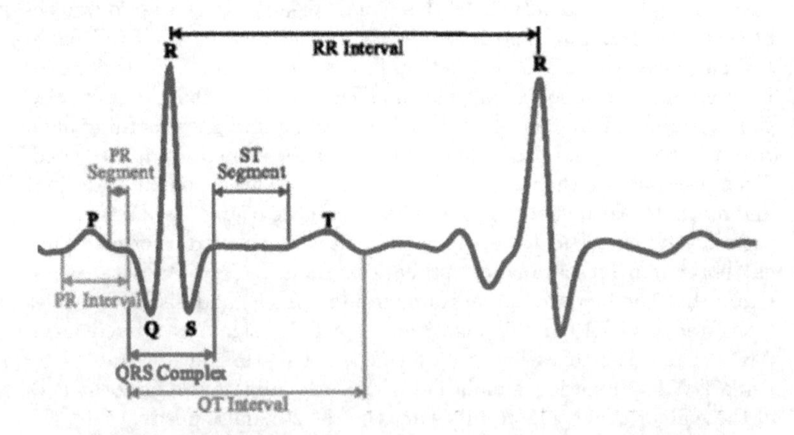

Fig. 1. Heart beat cycle in the ECG waveform [3].

In this work, an in-depth overview of classical techniques and machine learning algorithms is done with the analysis of their benefits, challenges and future directions. Then, the performance of Kalman filter and machine learning based algorithms are evaluated. Based on that, an algorithm that uses the combination of Adaptive Kalman Filter (AKF) and Discrete Wavelet Transform (DWT) is proposed. And numerical simulation is done to show the performances of the algorithms.

The rest of the paper is organized as follows. State of the art and review of related works for cardiovascular signal processing is presented in Sect. 2. In Sect. 3 the system and signal model is provided. Section 4 comprises of the performance metrics for cardiovascular parameter estimation. In Sect. 5 the simulation results are discussed and finally conclusions are drawn in Sect. 6.

Notations: Vectors and matrices are expressed in lower and upper case boldface letters, respectively. $\mathbf{A}^{\mathrm{T}}$ denotes conjugate transpose of matrix and $\mathbb{E}\{.\}$ is the expectation operator.

2 State of the Art and Review of Related Works

In the past decades, various algorithms with different level of accuracy has been developed and tested to estimate cardiovascular parameters. Stroke volume (SV), cardiac output (CO), and total peripheral resistance (TPR) are some of the parameters that determine the output blood per minute from the heart and give information about the status of patient's heart condition [4].

One of the main problems in ECG signals is noise which needs noise filtering or denoising to study the characteristics of those parameters. Wavelet is one of the commonly used denoising methods for ECG signal filtering. In this method, the signal is decomposed in to various bands and each coefficient is processed. Besides, threshold setting and wavelet reconstruction is done to attain the filtered ECG signal [5,6]. This needs an effective thresholding to set the right threshold value that give a better performance. If the ECG signal is too noisy, the wavelet method might not be reliable. In addition to ECG signals, arterial blood pressure (ABP) pulse is used in [7] to study the health condition of the patient. But this result in additional complexity to align the R-peak of ECG signal with ABP pulse, since both are not aligned.

Recursive denoising method is done by using Kalman filtering. The noise in ECG signals makes it difficult to get a good estimation of time varying cardiovascular state. But, Kalman denoising, which uses a continual parameter update, enables to attain an estimation of parameters with lower minimum mean square error (MMSE). The extended Kalman filers and adaptive Kalman filters were used for non linear models [8,9]. Empirical mode decomposition (EMD) based adaptive signal processing method has been very essential for non stationary and non linear signal analysis. With the existence of high frequency component in ECG signal, EMD have high probability of filtering the noise but it causes some sort of distortion on the amplitude of the peaks [10,11]. Parameter estimation methods such as Gaussian derivative model [12], autoregressive moving average, Wesseling correlated impedance and other methods were analyzed in [13].

With the classical algorithms, the parameter estimation is mainly based on the shapes of QRS complex, and P and T waves that provide good accuracy for normal ECG signals. However, with the existence of variation in the ECG signal shapes having abnormal rhythms, their performance becomes unreliable. Due to that, data driven approaches including machine-learning models and neural networks have been introduced. Currently, that is more expanded into researches using deep learning algorithms for parameter estimation and classification of ECG signals. The performance of those algorithms depends on the availability of data besides other factors. That is, with the existence of high amount of training and test data deep learning algorithms are providing good results while other machine learning based algorithms works with better accuracy even with the existence of small amount of data [14,15].

Over the past decade, several machine-learning techniques have been used for cardiovascular disease diagnosis and prediction. These techniques are mainly categorized as Supervised learning, Unsupervised learning and Reinforcement learning. Supervised learning uses a data for a serious of observations and uses it to form a predictive model to classify the outcomes. Supervised learning problems include; classification (category prediction), regression (value prediction) and anomaly detection (unusual pattern prediction). It is mainly used in prediction diagnosis and treatment of CVD. The prediction accuracy mainly depends on algorithms, dataset and hypothesis. So, knowing which algorithm for which hypothesis is key for efficiency and accuracy. Algorithms used includes; Linear regression, random forest, Artificial Neural Network (ANN), Support Vector Machine (SVM), K-Nearest Neighbor and others. Its limitations are mainly the need of higher training datasets and labeling, so as not to lead to inaccurate decision. Unsupervised learning is the other machine learning algorithm used in CV medicine that works in finding hidden pattern from the data without feedback from humans (train and learn by itself to make the decision). It has clustering algorithms and association rule learning algorithms to uncover relationships between unrelated data items. But, this algorithm may face difficulty in identifying the initial pattern which will lead to inaccuracy as final pattern is dependent on initial pattern. Reinforcement learning is the other machine learning algorithm that enables the system to learn in an interactive environment by trial and error using feedback from its own actions and experiences [16–20]. Each problem requires some degree of understanding of the problem, in terms of cardiovascular medicine and statistics, to apply the optimal machine-learning algorithm.

Recently, Deep Learning has been emerging as of great use for CVD treatment and prediction. Deep learning algorithm uses multiple layers of ANN that can generate automated prediction from input training datasets. It works well even with noisy data, having higher amount of dataset. However, due to the existence of multiple layers and many parameters, this algorithm is highly prone to over-fitting. Over-fitting is when the model learned patterns for the training data to give a good fit and not works well on the other unseen data. So, it needs careful problem formulation, appropriate algorithm selection, appropriate data selection and usage in addition to appropriate interpretation of the outcomes. Deep Learning in CVD mainly focuses on exploration of hidden risk factors, classification, risk prediction, complex decision making and better treatment pathway selection in complex CVD [16,18]. All in all, although there are several challenges in the clinical use of data driven based applications, it has still a massive potential in CVD treatment and prediction application with proper problem formulation and algorithm usage.

3 The System and Signal Model

We consider an ECG signal which is contaminated by nose and modeled as [21, 22]

$$\mathbf{y}(n) = \mathbf{x}(n) + \mathbf{v}(n) \tag{1}$$

where the vector $\mathbf{y}(n)$ represents the measured ECG signal, $\mathbf{x}(n)$ is the desired ECG signal, $\mathbf{v}(n)$ is a vector of additive Gaussian noise with $\mathcal{N}(0, \sigma^2)$ due to the sensor and amplifier noise and $n = 0, 1, 2, \cdots, N-1$ when N is the length of the signal. Muscle contractions or electromyographic spikes, movements that affects the skin-electrode interface, respiration, electromagnetic interference and noise from other electronic devices that couple into the device are some of the sources of noise in ECG signal. With the existence of such noise, studying and characterizing the health condition of the patient is very difficult. Due to that, having an algorithm that is capable of filtering out the noise while extracting the ECG signal is essential. With that, the estimation and classification of cardiovascular parameters will be effective. Those algorithms have their different techniques and performances as discussed in the subsequent sections.

3.1 Kalman Filtering Algorithm for Cardiovascular Parameter Estimation

Kalman filter is an estimator of optimal state for systems with random noises. It can be used for filtering, smoothing and state prediction of a linear dynamic system depending on the quantity of information available. For that the Kalman filter has two phases: the prediction and correction of the system which allows to estimate the state of a dynamics system of one information, a priori and the actual measurements using the current state of the system, the estimate of the previous state and current measures. Thus the linear representation of the received ECG signal in Eq. (1) can be reformulated as [21,23,24].

$$y_k = \mathbf{H}x_k + v_k \tag{2}$$

where $\mathbf{H}$ is the dynamic constant matrix and v_k is the noise with a covariance matrix Q_k.

Let $\bar{\bar{x}}_k$ be a priori state estimation at step k given the information of the process prior to step k, and $\bar{x}_k$ be a posteriori state estimate at the step k with given measurement y_k. Then, the priori and a posteriori estimation errors can be respectively defined as [21,23].

$$\begin{aligned} \bar{\bar{e}}_k &\triangleq x_k - \bar{\bar{x}}_k \\ \bar{e}_k &\triangleq x_k - \bar{x}_k. \end{aligned} \tag{3}$$

And a priori estimate error and posteriori estimate error covariance is given by [21,23,24].

$$\begin{aligned} \bar{\bar{P}}_k &= \mathbb{E}\{\bar{\bar{e}}_k \bar{\bar{e}}_k^T\} \\ \bar{P}_k &= \mathbb{E}\{\bar{e}_k \bar{e}_k^T\}. \end{aligned} \tag{4}$$

Then, the posteriori state estimate is given by [21,23,24]

$$\bar{x}_k = \bar{\bar{x}}_k + K(y_k - \mathbf{H}\bar{\bar{x}}_k). \tag{5}$$

The difference in the above equation is termed as the measurement residual which shows the discrepancy between actual and predicted state where K is the gain matrix given by

$$K_k = \bar{P}_k \mathbf{H}^T (\mathbf{H}\bar{P}_k \mathbf{H}^T + \mathbf{R})^{-1}. \tag{6}$$

where $\mathbf{R}$ is the covariance matrix. Hence, with Eq. (5) the filtered ECG signal can be achieved [21,23,24].

3.2 Machine Learning Algorithm for Cardiovascular Parameter Estimation

Machine learning algorithm is capable to make decisions autonomously without any external support by learning from the data and understanding the patterns that are contained within it. Using that pattern, the machine can perform tasks such as classification and prediction based on the type of algorithm. The three main machine learning algorithms at present includes supervised learning for classification including diagnosis and regression tasks for risk assessment, unsupervised learning for dimensionality reduction and clustering and reinforcement learning for gaming, robot interaction and other applications [21,25]. The basic building blocks of machine learning algorithms is shown in Fig. 2. The description of each block is outline below.

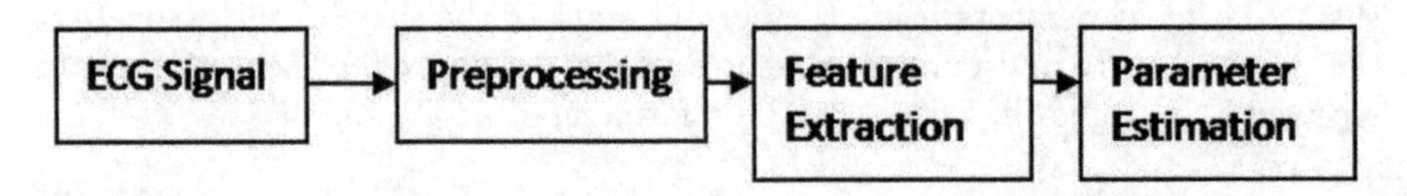

Fig. 2. Basic steps for ECG signal parameter estimation with machine learning algorithm.

- **Preprocessing:** Due to the existence of noise in the ECG signal, first the signal has to be preprocessed in order to extract its features and perform the estimation task. Filtering is one of the techniques used to preprocess the ECG signal. In this work, we use a second order Butterworth low pass filter to filter the unwanted signal [21,26].

- **Feature Extraction:** The task of feature extraction for ECG signal starts from feature selection which includes PQRST wave detection using various algorithms. R-peak detection is essential task while other PQST components can be detected by taking the location of R-peak as a reference and tracking to/from R-peak relative position. The features might be time, amplitude, distance, angle and others that can be useful to characterize the ECG signal and estimation. Wavelet transform approach and time domain analysis is used to detect R-peak and others, respectively as in [21,26].
- **Parameter Estimation:** At this stage the cardiovascular parameters can be estimated based on the extracted features. Linear regression is one of the most known techniques to estimate cardiovascular parameters. The new sequence of the signal to be estimated with linear regression is given by [21,27].

$$y^{'}(x) = ax + v. \tag{7}$$

The value of the coefficient a and the value of v that describe the best estimated signal, $y^{'}(j)$ from the real $y(j)$ is given by [27]

$$a = \frac{m\sum_{j=l_1}^{l_2} x(j)y(j) - \sum_{j=l_1}^{l_2} x(j) \sum_{j=l_1}^{l_2} y(j)}{m\sum_{j=l_1}^{l_2} x^2(j) - (\sum_{j=l_1}^{l_2} x(j))^2}$$

$$v = \frac{\sum_{j=l_1}^{l_2} x^2(j) \sum_{j=l_1}^{l_2} y(j) - \sum_{j=l_1}^{l_2} x(j)y(j) \sum_{j=l_1}^{l_2} x(j)}{m\sum_{j=l_1}^{l_2} x^2(j) - (\sum_{j=l_1}^{l_2} x(j))^2} \tag{8}$$

with Butterworth filter $H(f) = \frac{b_1(jf)^n + b_2(jf)^{n-1} + b_{n+1}}{a_1(jf)^n + a_2(jf)^{n-1} + a_{n+1}}$ having filter order n (0,1,..., N − 1) and cut-off frequency f = 100 Hz. And $i \in \{w, w+1, \ldots N-w-1\}$ where $l_1 = i - \omega$ and $l_2 = i + \omega$, $m = 2\omega + 1$ with window $\omega = 0.03 f_s$ when f_s being the sampling frequency and $j \in \{i-w, i-w+1, \ldots, i+w\}$ [27].

3.3 Proposed Algorithm for Cardiovascular Parameter Estimation

In this section, we propose a new algorithm for cardiovascular parameter estimation by combining adaptive Kalman filtering and discreet Wavelet transform as shown in Fig. 3. In this algorithm, first the noise in the ECG signal is filtered using an adaptive Kalman filter (AKF). AKF predicts the state and error covariance ahead, computes the gain, used the estimate with measurement and then update the error until it gets a minimum defined acceptable error that is the difference of desired signal and the filtered noisy signal.

After the noise filtering, the signal is decomposed by Wavelet transform. We use the discrete wavelet transform (DWT) which incorporates low pass filters and high pass filters to reduce the noise that couldn't be filtered with Kalman filter. The DWT signal is given by

$$X[a,b] = \sum_{n=-\alpha}^{\alpha} y(n)\phi_{a,b}(n) \tag{9}$$

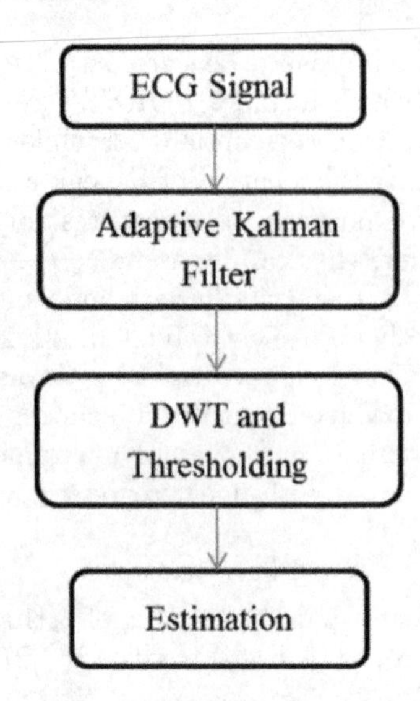

Fig. 3. Block diagram of proposed algorithm for cardiovascular parameter estimation.

where $y(n)$ is the output signal of the Kalman filter to be transformed with n length. And $\phi(n)$ is the window of finite length having a window translation parameter b and dilation/ contraction parameter a. The next step is to set the threshold by using hard thresholding approach proposed in [28] to estimate the wavelet coefficients as

$$cD_j^{'} = \begin{cases} cD_j, & |cD_j| \geq t \\ 0, & |cD_j| \leq t \end{cases} \tag{10}$$

where t is obtained by universal threshold selection with a value of $t = \sigma\sqrt{2\log N}$ with the standard deviation of the noise $\sigma = (\text{median}\{|cD_j| /0.6457\})$ where N is the data length, cD_j and $cD_j^{'}$ are wavelet coefficients before and after thresholding [28].

4 Performance Metrics for Cardiovascular Parameter Estimation

To evaluate the performances of cardiovascular parameters estimation algorithms, different parameters are used. Parameters such as True positive Rate (TPR), False Positive Rate (FPR), True positive (TP), False Positive (FP), False negative (FN), True negative (TN), accuracy, sensitivity and others can be used to check the performance of parameter estimation algorithms. Additionally, parameters such as mean absolute deviation (MAD) and Percentage root-mean-square difference (PRD) can be used to evaluate the performance of the algorithms. The mathematical expression for FPR, TPR and average positive predictive value (PPV) for N-data length is given as follows [7,29,30]:

$$\mathrm{FPR} = \frac{1}{N}\left(\sum_{i=1}^{N}\frac{\mathrm{FP}_i}{\mathrm{FP}_i + \mathrm{TN}_i} \times 100\%\right) \tag{11}$$

$$\mathrm{TPR} = \frac{1}{N}\left(\sum_{i=1}^{N}\frac{\mathrm{TP}_i}{\mathrm{TP}_i + \mathrm{FN}_i} \times 100\%\right) \tag{12}$$

$$\mathrm{PPV} = \frac{1}{N}\left(\sum_{i=1}^{N}\frac{\mathrm{TP}_i}{\mathrm{TP}_i + \mathrm{FP}_i} \times 100\%\right). \tag{13}$$

5 Simulation Results and Discussion

5.1 Simulation Setup and Parameters

To analyze the comparative performances of the proposed algorithms, we use a data from MIT-BIH AR dataset [31,32] which contains the ECG signal that is affected by noise. We use the standard ECG signal data from a publicly available MIT-BIH AR dataset. Each ECG signal is sampled at 360 samples per second. The data are used to test the performances of Kalman filter, machine learning and proposed algorithms for cardiovascular parameter estimation using the parameters in Table 1.

Table 1. Simulation parameters used

Type	Parameter	Methods/value
	fs	500 Hz [21,26]
Machine learning	Preprocessing	Butterworth with order 2 [21,26]
	Feature extraction	Wavelet with hard thresholding [21,28]
	Parameter estimation	Linear regression [21,27] w = 0.03 fs [21,27]
	Proposed	AKF with DWT [21,27,28]

5.2 Simulation Results

The ECG signal corrupted by noise in Fig. 4 is a sample data from the MIT-BIH AR dataset used to analyze the performances of the algorithms stated in Sect. 3.

Figure 5 shows the estimation of the ECG signal after Kalman filtering of the original ECG signal corrupted by noise. The first plot shows the noisy ECG signal and the second plot is the result of Kalman filtered ECG signal. The result shows that Kalman filter reduces the noise and makes the signal more smoother. This helps to study and evaluate the shape of P wave, QRS pattern and T wave to determine the state of the heart. But Kalman filtering affects the amplitude of the wave and this reduces the performance for R-peak detection.

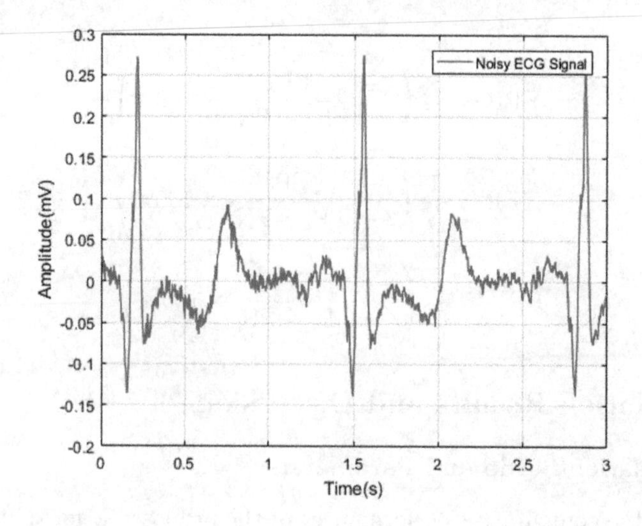

Fig. 4. Sample noisy ECG signal from MIT-BIH AR dataset.

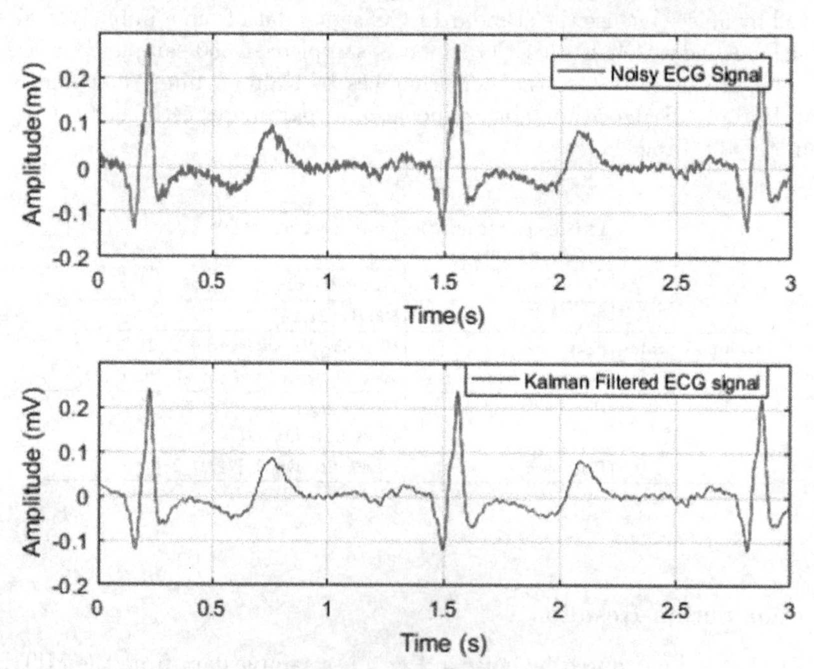

Fig. 5. ECG signal filtered by Kalman filter.

In the case of machine learning algorithm, a Butterworth low pass filter is used for preprocessing and then DWT with hard thresholding followed by linear regression is used to detect the R-peak and QRS complex as shown in Fig. 6. As can be seen from the result, a second order Butterworth low pass

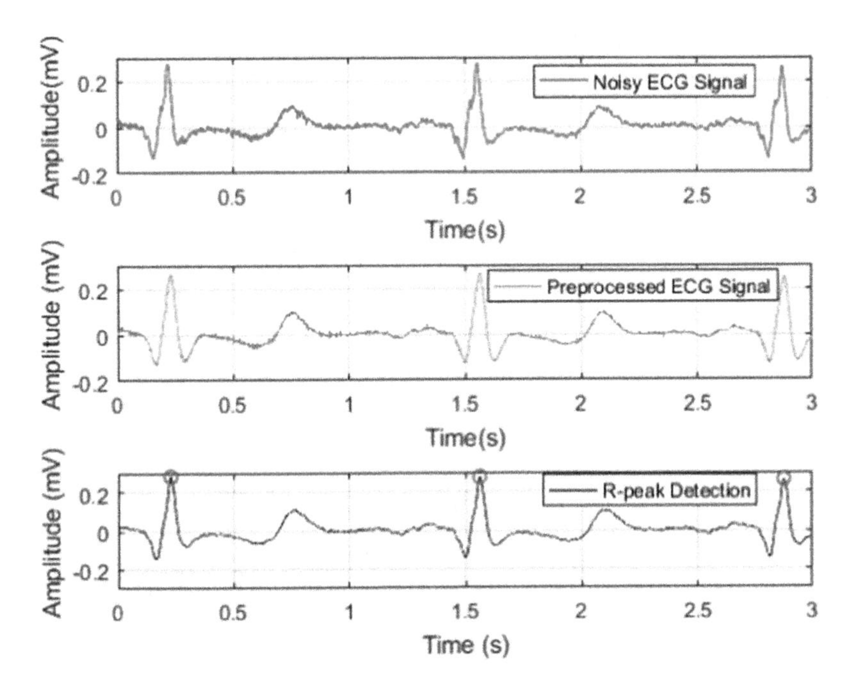

Fig. 6. ECG signal preprocessed by Butterworth LPF and R-peak detected by DWT.

filter smooths the noisy ECG signal and then, after the processing with DWT and linear regression R-peak is detected correctly. Hence, accurate detection of R-peak helps for better prediction of the state of the heart.

Then the performance of the proposed combined AKF with DWT algorithm is simulated and gives a result filtered ECG signal shown in Fig. 7. The noise is filtered with AKF and then the use of DWT enables to have a better capability in providing less noisy signal. With this algorithm studying P-wave, QRS pattern and T-wave of the ECG signal gives better accuracy than the simple Kalman filtering technique. Finally, to evaluate the detection performances of the stated algorithms, we plot the receiver operating characteristics (ROC) curve as shown in Fig. 8. The result depicts that by using machine learning algorithm, it is possible to achieve a better estimation of ECG signal parameter even with a reduced false positive rate. This shows that although it has high complexity, machine learning algorithms give better performance than classical techniques. However, it needs high amount of data and also has high computational complexity. In this case, employing the proposed classical technique with lower complexity can provide a good result with low computational cost. The result in Fig. 8 also shows that the proposed classical algorithm provides better estimation accuracy than Kalman filtering and a comparative performance with machine learning algorithm. This is achieved due to implementation DWT block in the system.

At the end, Table 2 shows the true positive rate results from Kalman filtering, machine learning algorithm and proposed AKF with DWT algorithms. From the

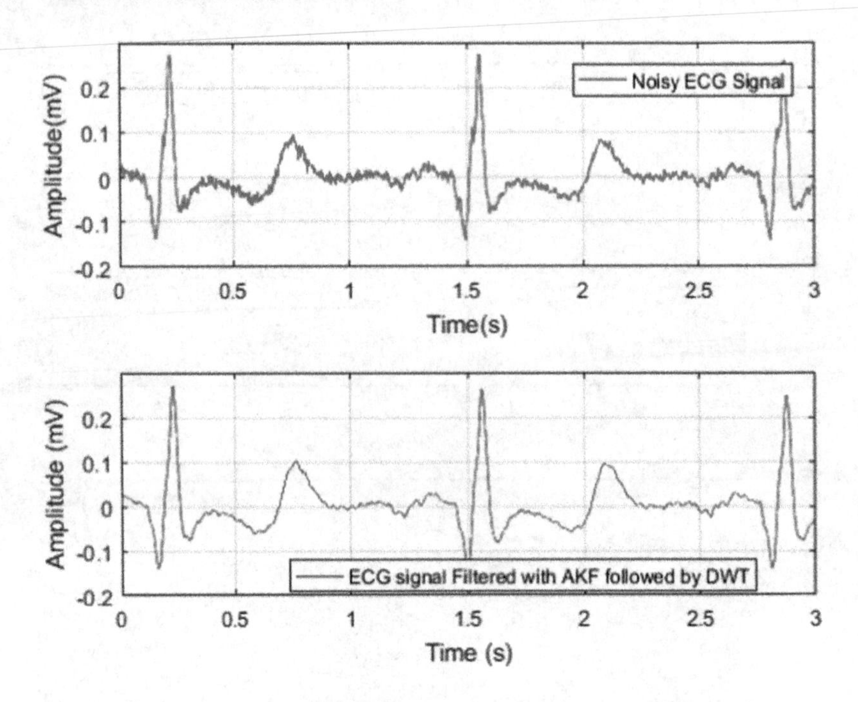

Fig. 7. ECG signal filtered by AKF combined with DWT.

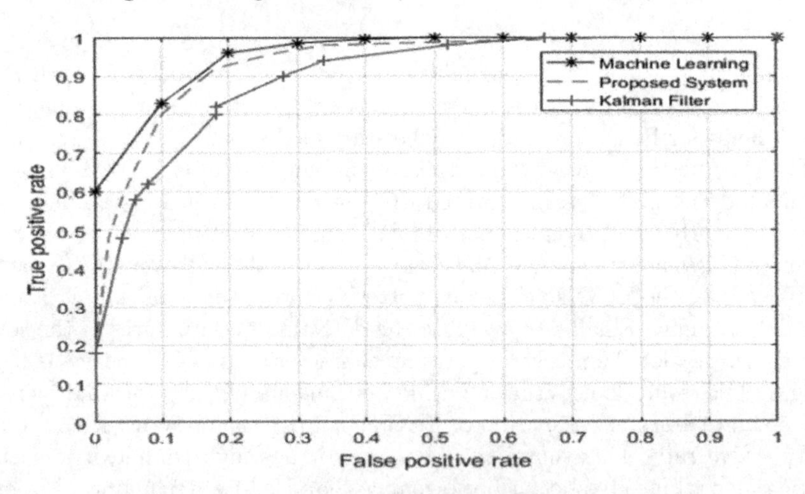

Fig. 8. The ROC for the proposed algorithms.

results, we can see that when the false positive rate is above 30%, the Kalman filter and the proposed AKF with DWT technique provides a true positive rate of more than 92% and 97% whereas the machine learning algorithm provides a true positive rate above 98%.

Table 2. Summary of ROC for the three algorithms.

FPR	0	0.1	0.2	0.3	0.4	0.5	0.6	0.7	0.8	0.9	1
TPR (Kalman Filter)	0.19	0.65	0.83	0.92	0.96	0.97	0.99	1	1	1	1
TPR (Machine Learning)	0.6	0.82	0.97	0.98	1	1	1	1	1	1	1
TPR (AKF with DWT)	0.2	0.8	0.94	0.97	0.98	0.98	0.99	1	1	1	1

6 Conclusion

In this work, we study the state of the art of signal processing algorithms for cardiovascular signal processing. Both classical and machine learning algorithms are reviewed and analyzed for ECG signal parameter estimation. We also propose an algorithm which uses Kalman filtering and DWT and analyze its performance with numerical simulation. The simulation result shows that, machine learning algorithm gives much better detection performance than classical Kalman filtering techniques. The results also show that the proposed algorithm (AKF followed by DWT) has a better performance than classical Kalman filtering techniques. Besides, classical techniques can give nearly the same performance with machine learning algorithms with less computational complexity. Hence, this algorithm can be employed when there is no availability of enough training datasets. The study of deep learning algorithms for effective CV parameter estimation and feature extraction is the future direction of this work.

References

1. Lyon, A., Mincholé, A., Martínez, J.P., Laguna, P., Rodriguez, B.: Computational techniques for ECG analysis and interpretation in light of their contribution to medical advances. J. Roy. Soc. Interface **15**(138), 20170821 (2018)
2. Ghasemi, Z., et al.: Estimation of cardiovascular risk predictors from non-invasively measured diametric pulse volume waveforms via multiple measurement information fusion. Sci. Rep. **8**(1), 1–11 (2018)
3. Arumugam, M., Sangaiah, A.K.: Arrhythmia identification and classification using wavelet centered methodology in ECG signals. Concurr. Comput.: Pract. Experience **32**(17), e5553 (2019)
4. Yao, Y., Shin, S., Mousavi, A., Kim, C.S., Xu, L., Mukkamala, R., Hahn, J.O.: Unobtrusive estimation of cardiovascular parameters with limb ballistocardiography. Sensors **19**(13), 2922 (2019)
5. Smital, L., Vítek, M., Kozumplík, J., Provazník, I.: Adaptive wavelet wiener filtering of ECG signals. IEEE Trans. Biomed. Eng. **60**(2), 437–445 (2013)
6. Wang, Z., Zhu, J., Yan, T., Yang, L.: A new modified wavelet-based ECG denoising. Comput. Assist. Surg. **24**(sup1), 174–183 (2019)
7. Singh, O., Sunkaria, R.K.: A new approach for identification of heartbeats in multimodal physiological signals. J. Med. Eng. Technol. **42**(3), 182–186 (2018)
8. Vullings, R., De Vries, B., Bergmans, J.W.: An adaptive Kalman filter for ECG signal enhancement. IEEE Trans. Biomed. Eng. **58**(4), 1094–1103 (2011)

9. Kostoglou, K., Robertson, A.D., MacIntosh, B.J., Mitsis, G.D.: A novel framework for estimating time-varying multivariate autoregressive models and application to cardiovascular responses to acute exercise. IEEE Trans. Biomed. Eng. **66**(11), 3257–3266 (2019)
10. Rakshit, M., Das, S.: An efficient ECG denoising methodology using empirical mode decomposition and adaptive switching mean filter. Biomed. Signal Process. Control **40**, 140–148 (2018)
11. Han, G., Lin, B., Xu, Z.: Electrocardiogram signal denoising based on empirical mode decomposition technique: an overview. J. Instrum. **12**(03), P03010 (2017)
12. Spicher, N., Kukuk, M.: ECG delineation using a piecewise Gaussian derivative model with parameters estimated from scale-dependent algebraic expressions. In: 41st Annual International Conference of the IEEE Engineering in Medicine and Biology Society (EMBC), October 2019
13. Arai, T., Lee, K., Cohen, R.J.: Comparison of cardiovascular parameter estimation methods using swine data. J. Clin. Monit. Comput. **34**(2), 261–270 (2019)
14. Mykoliuk, I., Jancarczyk, D., Karpinski, M.,Kifer, V.: Machine learning methods in Electrocardiography classification. J. Adv. Comput. Inf. Technol. **2300**, 102–105 (2018)
15. Al Rahhal, M.M., Bazi, Y., AlHichri, H., Alajlan, N., Melgani, F., Yager, R.R.: Deep learning approach for active classification of electrocardiogram signals. Inf. Sci. **345**(2016), 340–354 (2016)
16. Krittanawong, C., Zhang, H., Wang, Z., Aydar, M., Kitai, T.: Artificial intelligence in precision cardiovascular medicine. J. Am. College Cardiol. **69**(21), 2657–2664 (2017)
17. Mathur, P., Srivastava, S., Xu, X., Mehta, J.L.: Artificial Intelligence, Machine Learning, and Cardiovascular Disease. SAGE, September 2020
18. Krittanawong, C., Johnson, K.W., Rosenson, R.S., Wang, Z., Aydar, M., Baber, U., Min, J.K., Tang, W.W., Halperin, J.L., Narayan, S.M.: Deep learning for cardiovascular medicine: a practical primer. Eur. Heart J. **40**(25), 2058–2073 (2019)
19. Princy, R.J.P., Parthasarathy, S., Jose, P.S.H., Lakshminarayanan, A.R., Jeganathan, S.: Prediction of Cardiac Disease using Supervised Machine Learning Algorithms. IEEE, June 2020
20. Singh, A., Kumar, R.: Heart Disease Prediction Using Machine Learning Algorithms. IEEE, June 2020
21. Birhanu, H., Kassaw, A.: Comparative analysis of Kalman filtering and machine learning based cardiovascular signal processing algorithm. In: EAI-ICAST. Springer, Accepted (2020)
22. Lastre-Dominguez, C., et al.: ECG Signal denoising and features extraction using unbiased FIR smoothing. BioMed. Res. Int. **2019**, 1–16 (2019)
23. Reddy, D.V.R., Rahim, B.A., Fahimuddin, S.: Gaussian noise filtering from ECG signal using improved Kalman filter. Int. J. Eng. Res. Rev. **3**(2), 118–126 (2015)
24. Sharma, B., Suji, R.J., Basu, A.: Adaptive Kalman filter approach and Butterworth filter technique for ECG signal enhancement. In: Information and Communication Technology for Sustainable Development. Lecture Notes in Networks and Systems, vol. 10. Springer, Singapore, November 2017
25. Schmidt, J., Marques, M.R., Botti, S., Marques, M.A.: Recent advances and applications of machine learning in solid-state materials science. Nat. Partner J. Comput. Mater. **5**(1), 1–36 (2019)
26. Patro, K.K., Kumar, P.R.: Effective feature extraction of ECG for biometric application. In: 7th International Conference on Advances in Computing and Communications (ICACC), Cochin, India (2017)

27. Aspuru, J., et al.: Segmentation of the ECG signal by means of a linear regression algorithm. Sensors **19**(4), 775 (2019)
28. Lin, H.Y., Liang, S.Y., Ho, Y.L., Lin, Y.H., Ma, H.P.: Discrete-wavelet-transform-based noise removal and feature extraction for ECG signals. Irbm **35**(6), 351–361 (2014)
29. Plawiak, P.: Novel generic ensembles of classifiers applied to myocardium dysfunction recognition based on ECG signals. Swarm Evol. Comput. **39**, 192–208 (2017)
30. Yadav, O.P., Ray, S.: ECG signal characterization using Lagrange-Chebyshev polynomials. Radioelectron. Commun. Syst. **62**(2), 72–85 (2019)
31. Goldberger, A.L., Amaral, L.A., Glass, L., Hausdorff, J.M., Ivanov, P.C., Mark, R.G., Mietus, J.E., Moody, G.B., Peng, C.K., Stanley, H.E.: PhysioBank, PhysioToolkit and PhysioNet: components of a new research resource for complex physiologic signals. Circ. Electron. Page **101**(23), e215–e220 (2003)
32. Moody, G.B., Muldrow, W.E.: A noise stress test for arrhythmia detectors. Comput. Cardiol., 381–384 (1984)

Towards an Efficient and Secure Virtual Wireless Sensor Networks Framework

Mahmoud Osama Ouf[1](✉), Raafat A. EL Kammar[2], and Abdelwahab AlSammak[2]

[1] Computer Engineering Department, Al Safwa High Institute of Engineering Ministry of High Education, Cairo, Egypt
mahmoud.ouf@alsafwa.edu.eg

[2] Computer Engineering Department, Faculty of Engineering (Shoubra), Benha University, Cairo, Egypt
asammak1@feng.bu.edu.eg

Abstract. Wireless Sensor Networks (WSNs) are vastly used for a huge range of applications such as object tracking, observation surrounding environment, health care, manufacturing automation, and smart homes. A WSN application would have central controlling modules that control the deployment and management process for the underlying WSN. Virtualization in WSN is a recent trend that aims to create a framework that helps in separating the traditional structure of the WSN. This paper purposes a novel framework for virtual sensor networks. The proposed framework uses a virtual middleware layer that separates the application layer and the WSN sensor node infrastructure layer. This layer offers mechanisms to handle different issues, including security, routing, and aggregating resources from multiple Physical WSN infrastructures to be used by a single application. In addition, Virtual sensor networks will impose a number of unique and new conceptual and technical challenges to be considered in the traditional wireless sensor network. There are a few reasons that motivate us to work towards providing a novel virtual sensor network architecture: (i) Provide flexibility by Allowing WSN to perform different tasks even when deployed in the same geographical region, (ii) Provide cost effective solutions, (iii) Promote variety, and (iv) Increase manageability and ensure security. An evaluation will be performed to benchmark the proposed framework against a traditional baseline. The evaluation results will demonstrate the capabilities of the proposed framework against other benchmarks in terms of packet delivery rate, network lifetime, and stability period.

Keywords: Virtualization in WSN · Wireless sensor network · Virtualization · System architecture

1 Introduction

A wireless sensor network (WSN) consists of a large-scale wireless event-based network system. A WSN typically consists of low cost distributed tiny heterogeneous sensor nodes devices. The WSN nodes main function is to monitor the physical and environmental conditions at different locations and collect readings such as vibration, Pollution motion,

K. Arai (Ed.): FICC 2021, AISC 1363, pp. 128–137, 2021.
https://doi.org/10.1007/978-3-030-73100-7_10

sound, pressure, and temperature [2, 3]. The concept of a Virtual Sensor Network (VSN) can be seen as a similar technique to the one used in the case of deploying Virtual Machines (VM) over a cluster of compute servers. Usually, compute clusters have a large number of resources such as CPU elements or storage. In this case, a VM represents a group of these resources dedicated to one application [4, 5].

In WSN, the same concept can be applied so that more than one application can use the same WSN infrastructure capabilities. Virtualization in WSN will facilitate the ability to host multiple applications or services on the same physical network such as environmental data collection, security and data monitoring, and sensor node tracking. Each application will have its own characteristics that dictate the choice of system components such as routing or security mechanisms. The proper choice of such components will affect the efficiency of the whole system.

In this work, we aim to design a novel framework to better handle the virtualization of WSN. Throughout the framework, we are going to propose and implement various techniques to mitigate the security and the resources aggregation issues. The proposed technique is efficient in terms of reduced communication bandwidth and reduced sensor nodes power consumption. It is a challenging task to introduce a VSN framework that can provide good power management, cost effective, flexible, and security at the same time.

The remainder of this paper is organized as follows: In Sect. 2, we present an overview of the related work. In Sect. 3, details about the proposed framework are given. In Sect. 4, we present the implemented model and the simulation details. Evaluation and results are discussed in Sect. 5. Finally, the paper is concluded in Sect. 6.

2 Related Work

Virtualization has gained more attention recently due to its efficient resource utilization, energy efficient networked computing, as well as reduced cost of operation and management for the network. In addition to that, virtualization offers more flexibility, manageability, and interoperability within different computing sensor node devices. In recent years, research into sensor networks virtualization has many areas of interest including Sensor virtualization, OS virtualization, Network virtualization, Middleware layers, and Virtual machines.

The virtualization of a sensor node could be implemented by adding an abstraction layer between the functions of the application and the driver of the sensor. The purpose of this layer is to address issues such incomplete prior knowledge of the operation area, re-calibration of the sensors' hardware, conditions change, etc. Due to constraints in the sensor hardware, running multiple Operating Systems on the same node level is not an efficient choice. Majority of the Operating Systems are constructed using the event-driven framework (e.g. TinyOS [13], Contiki [14]).

The main areas that WSNs virtualization pays the most attention to are protocols and algorithms usability, adapting to changes, information and maintenance of the subset of sensor nodes collaborating together to perform specific tasks and being organized as a VSN using shared physical infrastructure resources [6, 10]. To make VSNs, we need the following mechanisms to maintain the VSN network such as deleting nodes, adding nodes, merging or splitting of VSN [6, 9].

In the last few years, researchers have been providing many middleware-based solutions to address the issues in the WSN. Each proposed solution creates its own framework design which is affected by the change of the application model and the underlying assumptions that are largely based on the applications of WSN. As a result, each solution supports virtualization features in a different way. Common features for existing framework solutions increase sensor data query level by separating services into two parts: (i) sensor infrastructure provider (SInP); and (ii) sensor virtualization network service provider (SVNSP), SInP is responsible for administering the physical infrastructure, and, SVNSP offers end to end services in the VSN by improving resource aggregated from different SInPs. Also, the solutions try to mask the distribution and network heterogeneity and address resource restrictions by providing a different routing and diversion query protocols, that are aware of the energy usage [1].

3 The Proposed Framework

In this paper, we introduce a new framework to create a Virtual Sensor Network. This framework includes: (i) application layer, (ii) virtualization layer, and (iii) WSN physical infrastructure layer. The work in this paper is mainly related to the Virtualization middleware layer. As shown in Fig. 1, the components of this layer will be the WSN routing algorithm, the cryptography (Encryption/decryption) algorithm and the publish/subscribe paradigm. The first two components are used in the traditional WSN sensor, while the publish/subscribe model is mainly used to enable using a group of resources to certain applications.

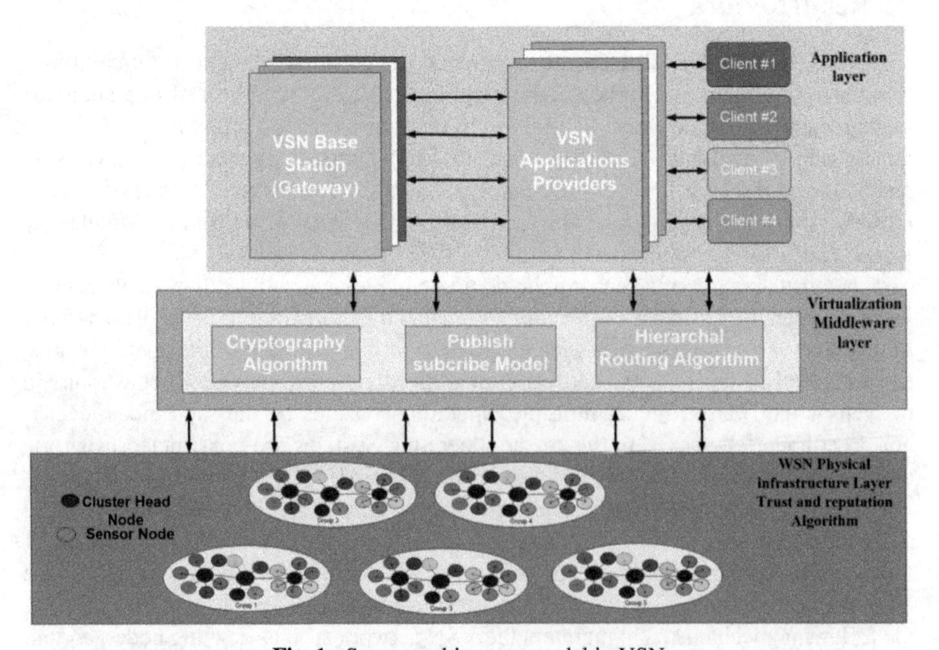

Fig. 1. Secure architecture model in VSN

The publish/subscribe model can be described as two sets of nodes. Patterns of events are produced by nodes are called publishers and the other nodes called subscribers are interested in. The subscribers will be notified by the information or notification when the events from the publishers are available. Until the notification service delivers notification, subscribers can continue their tasks.

The publish-subscribe based models introduce three types of decoupling between the publishers and the subscribers. The first is called space decoupling where publishers and subscribers don't need to know each other location. The second is called time decoupling in which publishers and subscribers do not need to run at the same time. The third is synchronization decoupling where publishers and subscribers' operations and tasks are not stopped during publishing and receiving notifications. The subscription processes can be represented by one of two models. The first is called Topic or subject based subscription model where a subscriber registers for a certain topic and receives filtered notifications according to the registered topic. It is similar to joining a Group, but it is a little bit more dynamic.

The other model is called content-based subscription model. In this model, the content itself is the factor that initiates subscription and notification process. Any subscription language can be used for filtering such as <temperature> <30>. Publish and subscribe communication model is based on several group communication primitives such as broadcast, multicast or uni-cast data delivery. In the virtual sensor network, the sink nodes are represented by the subscriber while the normal sensor nodes are considered as a publisher. There are several Advantages of the Publish Subscribe model such as enhanced response time, publishers-subscribers loosely coupled relationship, scalability, and finally, low delay (latency).

To create an efficient VSN, sensor nodes are divided into clusters. Each cluster will have one head node and several cluster member sensor nodes. The head node responsible for data aggregation from all member nodes within this cluster. In order to enable different applications to share the same resource infrastructure sensor nodes, different routing protocols would be used depending on the nature of each application. The proper choice of the routing protocol will have a massive impact on the network life span, load balance and network stability [15].

Virtualization for WSN can be classified into three categories: network level virtualization, node level virtualization, and hybrid solutions. As shown in Fig. 2 each of these categories are classified into two subcategories as will be discussed below.

In network level virtualization, a subset of WSN's nodes is formed. This subset is dedicated to one application at a given time. Network level virtualization can be subcategorized into virtual network overlay based solutions and cluster-based solutions. Virtual network overlay based solutions can be viewed as node level virtualization in which a logical network is created on top of the physical network, then, application overlays are utilized to achieve network level virtualization. In cluster based solution, the nodes in the physical network are clustered in order to work together in connected groups. This can be viewed as a physical partitioning of the network where one part of the network is used to one application and another part is used by a different application.

Node level virtualization in WSN allows multiple applications to run their tasks concurrently on a single sensor node, so that a sensor node can essentially become a

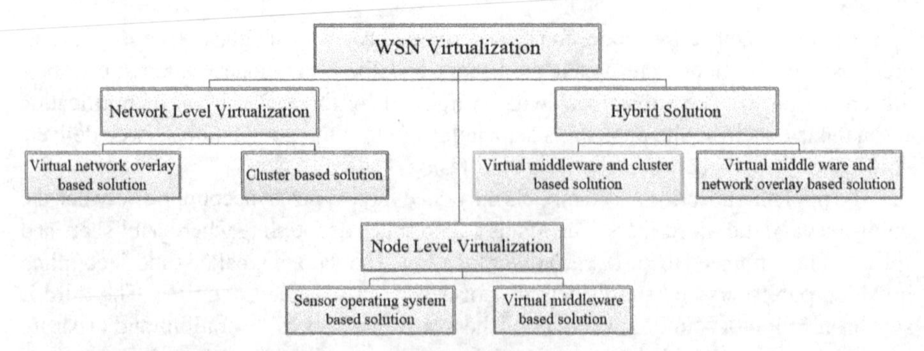

Fig. 2. WSN virtualization categories. The proposed framework belongs to the highlighted sub-category

multi-purpose device. Node level virtualization can be sub-categorized as sensor operating system-based solution and virtual middleware-based solution. In sensor operating system-based solutions, node level virtualization is part of the sensor Operating system. While in virtual middleware-based solutions, node level virtualization is performed by a component running on top of the sensor operating system. The programming model used in node level virtualization can be either event driven or thread-based programming model.

Hybrid solutions in WSN combine both network and node level visualization mechanisms which can be either of the following types: virtual middleware, cluster-based solution, virtual middleware, and network overlays-based solution. In virtual middleware and cluster-based solution, the middleware handles node level virtualization while network level virtualization is achieved by grouping sensor nodes into clusters. In virtual middleware and network overlays-based solution, the middleware handles node level virtualization while network level virtualization is achieved using virtual network overlays. The proposed framework belongs to the category hybrid solutions and the subcategory virtual middleware and cluster based solutions.

We are going to use the OAEMRP [11] algorithm which is an Optimal, adaptive, and energy driven mesh routing protocol that can be used in VSN. Relay nodes selection process function in this scheme is done by using balance between node energy, distance between nodes, and value of angle between CH, BS, and each node. Also, new scheme limits the choice of relay nodes for reducing broadcasting messages during selection process of relay path. This leads to the lowering of entire network energy consumption, also, achieves a better delivery rate and a better load balance compared to the state of the art in network virtualization routing techniques.

Trust and reputation management systems for VSN could be very useful for detecting operational issues in nodes. These issues can happen as a result to a faulty or as a result of a malicious action. In addition, the trust and reputation system helps the decision-making process to achieve a secure infrastructure. Special attention should be given to the collection of information method as well as the relevancy of the collected information. As a result of that, reasons of mistrust might be dropping packets, joining, or leaving the network without any clear cause. And due to constraints related to the nature and properties of those kinds of networks such as being dynamic or static, energy constraint also depending on the root issue that the trust and reputation management systems is trying to solve. The main challenges are accuracy, scalability and power consumption [8].

A lot of existing models for trust and reputation (LFTM, Eign Trust, power trust, BTRM-WSN, PEER trust) can be used as a part of the proposed framework. The choice of the best model depends on the nature of the underlying application or the type of the network. LFTM is used with static network on the other hand Power Trust used with dynamic network. The reason for these choices is that each of them give the best energy consumption, the highest accuracy, and the best average path length corresponding to each type of these networks. So, based on application and network types the trust and reputation model will be chosen [8].

Through our work, the following assumptions are going to be considered: (i) Sensor nodes are either distributed randomly or according to certain pattern in the target area, (ii) The sensor nodes have identical transmission range, (iii) Each sensor node can locate its neighbor within the transmission range, and (iv) Reliable communication exists between different sensor nodes.

Due to the weakness and highly vulnerable nature of wireless sensor networks against attacks, it was necessary to choose some techniques to safeguard the network from all kinds of known attacks. Such mechanisms would guarantee that the system is protected all the time as well as after any incident of attack. Public key cryptography provides protection from the participation of unauthorized external entities and also eliminates the problem of having a malicious insider trying to use more than one identity. The three most common cryptography methods to solve the problem of key management through a pre-distribution stage, asymmetric cryptography, symmetric cryptography, and hybrid cryptography systems. Encryption keys are stored in nodes memory in the WSN which is the pre-distribution stage prior to nodes deployment.

The classification of key management and distribution models can be divided into two groups. The first group contains the asymmetrical cryptography schemes based on Public key infrastructure and nodes identity. The second group contains symmetrical cryptography schemes based on encryption key pre-distribution either master or random [7]. The key management and distribution will be chosen depending on nature of the application.

4 System Simulation Tools

WSNs simulation is an essential methodology in evaluating proposed ideas. Protocols, algorithms, and new techniques can be tested in a large-scale using simulation tools. WSNs simulators enable researchers to test different scenarios by changing configuration parameters. WSNs simulators main aspects includes the simulation models being correct as well as choosing the appropriate tools to implement that model.

There are seven widely used simulation tools used in WSNs: the first simulator group which don't support GUI interface: NS-2, Avrora. The second group is divided to two groups. First, simulator tools group which support GUI interface such as OMNeT++, J-Sim. Second, emulator tools group which support GUI interface such as TOSSIM, EMStar and ATEMU. All this simulator and emulator provide open source and documentation online.

We used OMNeT++ to simulate the proposed framework. This simulator tool supports MAC protocols and some routing protocols in WSN. Also, channel control and network power consumption problems in WSNs can be simulated by using OMNET++. The limitations of this study are tied up with the simulation tool accuracy as well as the processing power of the machine used to perform the simulation. Better simulation tools and more powerful machines would enable better and larger scale simulations yielding more accurate results [12].

5 Evaluation

To evaluate the performance of the proposed framework, we conducted three different experiments using the OMNeT++ simulator. In these experiments, we measure different metrics for the proposed framework comparing it to a traditional framework as a baseline. The metrics used are the packet delivery rate, network lifetime, and stability period. The packet delivery rate is the rate of packets delivered from non-cluster nodes to the cluster head CH. The second evaluation metric is the network lifetime which can be defined as the time from the starting of operation until the moment when no transmission is possible. In this period of time the network remains functional for sending data between non-cluster nodes, CH, and BS. The third evaluation experiment measured the stability period. The stability period is measured as the ratio between the time until First Node Dies (FND) to the number of events created during the simulation. The time until First Node Dies (FND) can be defined as the time since the staring of operation of the network until the death of the first node. The node is considered dead whenever its energy drops to 0.005 J or below.

The metrics has been measured in different setups that controls different hyperparameters. A summary of the simulation parameters are shown in Table 1. As shown in the table, some of the parameters are set to a fixed value. These hyperparameters includes simulation area with two different sizes of the monitoring area (150 m* 150 m) and (300 m* 300 m). Also, number of nodes which is in the range of 50 to 200. Another important hyperparameter is the locations of the BS node that varies from (75,75) to (150,150).

We found that the packet delivery rate of the proposed framework is 2.4% better than the baseline framework as shown in Fig. 3. This boost in performance came as a result of using of a weighted sum of balanced factors function. This function balances distance, angle, and energy to the number of broadcast messages. The CH receives responses only from nodes that has the value of Cos αj greater than 0 and less than 90, Where αj is the angle between BS, Node j, and CH. This will led to a decrease in the number of broadcast messages the CH and non-cluster nodes as well as inter base stations messages. With regard to the second metric, network lifetime, we found that the proposed framework is 25.04% better than the baseline framework.

This substantial difference came as a result of the enhanced power consumption profile of the proposed framework. The proposed framework consumes less energy compared to the baseline framework because of the limitations imposed on the participation of non-cluster nodes in communicating events. With regard to the third metric, the stability period of the proposed framework was 37.9% better when compared to the baseline

Table 1. Simulation parameters

Parameters	Values
Percentage of cluster head (popt)	0.1
Initial energy of nodes	0.5 joules
Data packet size	6400 bit
Transmission & receiving energy (Eelec)	50 nJ/bit
Free space transmitter amplifier energy (E fs)	10 pJ/bit/m2
Multipath fading transmitter amplifier energy (E mp)	0.0013 pJ/bit/m4
Data aggregation energy (EDA)	5 nJ
Type of distribution	Random

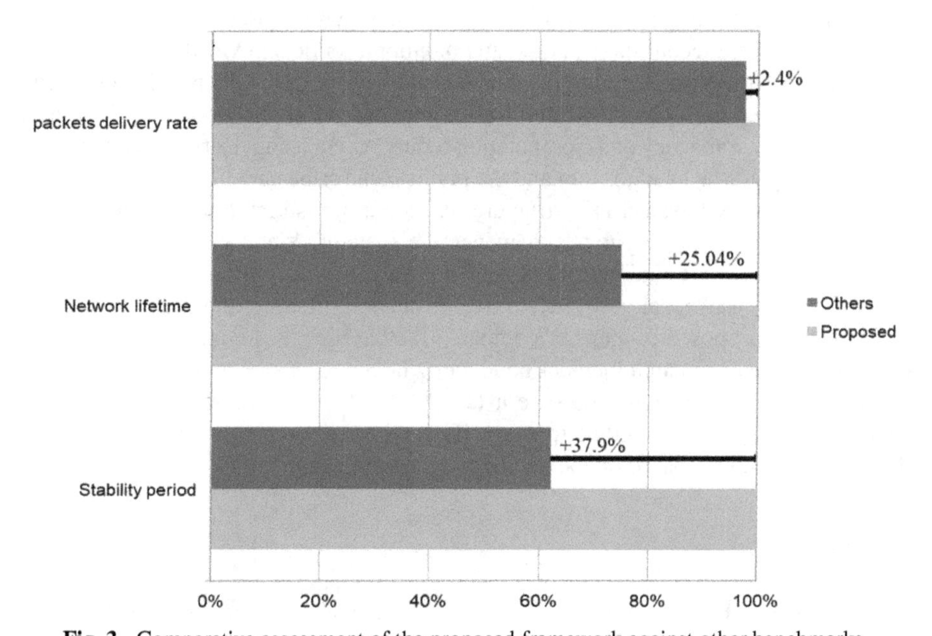

Fig. 3. Comparative assessment of the proposed framework against other benchmarks

framework. The big positive jump in the stability period came as a result of the enhanced power consumption profile of the proposed framework. In my opinion, the middleware layer did not just enable the virtualization of WSN but also introduced a flexible substrate where functionality can adapt to the nature of the application. This adaptability came naturally as the middleware has the ability to host different components and different algorithms separating the application layer and physical layer.

6 Conclusions

In this paper we presented an efficient and secure architecture for virtual sensor networks to enable the secure sharing of the same sensor node resources by multiple applications. This sharing is enabled by authorizing multiple heterogeneous wireless sensor nodes to be divided into clusters. Each cluster head is responsible for aggregating the sensed data from cluster member nodes using the OAEMRP algorithm. Forming different sensor network clusters hides the system details form the running applications while communication is done via a middleware layer between the different applications and a shared physical substrate.

We discussed and presented a novel framework of VSN that allows secure management of the network resources. That VSN architecture pushes the virtualization concept all the way down to the sensor nodes meanwhile maintaining acceptable levels of performance and service discovery. The VSN architecture is designed based on three main pillars. The first pillar is the application Layer that allows the user applications to subscribe to services, execute specific task, and monitor a service provided by sensors. The second pillar is the Virtualization Layer or the middleware layer which is the intermediate layer between the WSN physical infrastructure and the application layer. This layer provides energy-efficiency and Computation reduction via using the right WSN routing algorithm, cryptography algorithm and the publish/subscribe paradigm. The third pillar is the secure WSN Physical infrastructure. A secure infrastructure has been achieved by using the Trust and Reputation management systems. We evaluated the proposed framework against a traditional baseline. The evaluation results showed an enhancement of 2.4% in packet delivery rate, 25.04% in network lifetime, and 37.9% in stability period for the proposed framework compared the baseline. A future work that can be done in this line of research includes conducting larger-scale simulations as better tools and more powerful machines become available. Another line could be implementing a dynamic cryptography module that would use different cryptography algorithms for different applications based on the nature of each application.

References

1. Islam, Md.M., Huh, E.: Virtualization in wireless sensor network: challenges and opportunities. J. Netw. **7**(3), 412 (2012)
2. Singh, K.U., Chandra Phuleriya, K., Bunkar, K., Bhumarkar, S.: Exploration of wireless sensor networks technology and development. IJETTCS **1**(1) (2012). ISSN 2278-6856
3. Sohraby, K., Minoli, D., Znati, T.: Wireless Sensor Networks Technology, Protocols and Applications. John Wiley & Sons Inc., Hoboken (2007)
4. Islam, Md.M., Huh, E.: Virtualization in wireless sensor network challenges and opportunities. J. Netw. **7**(3) (2012)
5. Islam, Md.M., Mehedi, M., Lee, G., Huh, E.: A survey on virtualization of wireless sensor networks. Sensors **12**(2), 2175–2207 (2012)
6. Islam, Md.M., Mehedi, M., Lee, G., Huh, E.: A Survey on Virtualization of Wireless Sensor Networks (2012). ISSN 1424-8220
7. Maleh, Y., Ezzati, A.: An advanced study on cryptography mechanisms for wireless sensor networks. Mediterranean Telecommun. J. **6**(2), June 2016, ISSN: 2458-6765

8. Alkalbani, A.S., Tap, A.O. Md., Mantoro, T.: Energy consumption evaluation in trust and reputation models for wireless sensor networks. In: 5th International Conference on Information and Communication Technology for the Muslim World (2013). 978-1-4799-0136-4/13
9. Jayasumana, A., Han, Q., Illangasekare, T.H.: Virtual sensor networks – a resource efficient approach for concurrent applications. In: Proceedings of ITNG 2007, Las Vegas, April 2007
10. Sarakis, L., Zahariadis, T., Leligou, H., Dohler, M.: A framework for service provisioning in virtual sensor networks. EURASIP J. Wirel. Commun. Netw. **2012**, 135 (2012). https://doi.org/10.1186/1687-1499-2012-135
11. Ouf, M.O., EL Kammar, R.A., AlSammak, A.: Improving energy efficiency for EMRP routing protocol for virtualization in wireless sensor network. Int. Res. J. Eng. Technol. (IRJET) **7**(4), April 2020, e- (2008). ISSN: 2395-0056
12. Varga, A., Hornig, R.: An Overview of the OMNeT++ Simulation Environment (2008). ISBN 978-963-9799-20-2
13. Levis, P., et al.: TinyOS: An Operating System for Sensor Networks. In: Weber, W., Rabaey, J.M., Aarts, E. (eds.) Ambient Intelligence, pp. 115–148. Springer, Berlin, Heidelberg (2005). https://doi.org/10.1007/3-540-27139-2_7
14. Dunkels, A., Gronvall, B., Voigt, T.: Contiki–a lightweight and flexible operating system for tiny networked sensors. In: Proceedings of the 29th Annual IEEE International Conference on Local Computer Networks (LCN 2004), Washington, DC, USA, pp. 455–462 (2004)
15. Ouf, M.O., Ahmed, A.E.S., EL Kammar, R.A., AlSammak, A.: Analytical study of hierarchical routing protocols for virtual wireless sensor network. Int. Res. J. Eng. Technol. (IRJET), **7**(2) (2020). e-ISSN: 2395-0056

Interface Design of Transport Mobile Commerce APP: A Case Study of Taiwan Railway

Yi-Heng Lin(✉) and Li-Hao Chen

Fu Jen Catholic University, No. 510, Zhongzheng Road, Xinzhuang District, New Taipei City 242, Taiwan (R.O.C.)

Abstract. Mobile Apps have emerged in the past ten years. Transportation Apps have emerged accordingly, and are no longer restricted to purchase tickets on-site. The purpose of this study is to explore the user's needs for the Taiwan Railway App, and based on the user's needs, revise and design the Taiwan Railway App interface to increase intuition. This study analyzes and compares the three existing Taiwan Railway Apps, using observation methods, semi-structured interviews, and questionnaires to summarize the needs of the participants: 1) The overall satisfaction of the Taiwan Railway Reservation (80%) is the highest, followed by Taiwan Railways e-booking (20%), Taiwan Railway Assistant is the lowest (0%); 2) refer to the participant's suggestions to modify the new APP, the items are as follows: 1) Enlarge the booking block; 2) change the small icon Simplify re-composition; 3) add one-way and round-trip ticket interface; 4) set the booking button after selecting the booking column; and 5) separate the date and time menu. After completing the pre-testing, using the Taiwan Railway Reservation as the prototype, the App interface was revised and the new App Prototype was made. After the test, it can be found that users have become more intuitive in using the interface and the average score has improved. Based on the above analysis results, a traffic mobile business App was designed and produced, and user tests were conducted. The results are consistent with the original expectations and more intuitive to use.

Keywords: Mobile e-commerce · App · Taiwan Railway · Interface design

1 Introduction

Smart phones are developing rapidly. Since the first smart phone came out in 2007, everyone has a smart phone. Mobile apps have emerged in the past decade. Now we are in an era of fast-paced life, everything is convenient and fast. Transportation apps are also born accordingly. Such as: train tickets, air tickets, etc. No longer limited to on-site ticket purchases. According to the Ministry of Transport and communications R.O.C Taiwan Railway Passenger Traffic Statistics Data [4], the number of tourists has been increasing year by year since 2010. By 2019, it has grown to approximately 236,151 thousand. App Annie's 2019 Global Online App Usage Survey Report [8] shows that global App downloads have exceeded 194 billion times and $101 billion dollars has been spent in

K. Arai (Ed.): FICC 2021, AISC 1363, pp. 138–149, 2021.
https://doi.org/10.1007/978-3-030-73100-7_11

App stores. This shows the importance of mobile e-commerce. Five development trends are sorted out in the existing apps, including augmented reality experience, health and fitness, entertainment, e-learning, and mobile e-commerce.

In real experience, according to the statistics of Sensor Tower, a mobile analytics company in 2018 [3, 6], the number of AR App downloads from September 2017 to March 2018 has exceeded 13 million, and nearly half (47%) are games; Health and Fitness: Medical applications make health data tracking easier, users are more comfortable to operate. They encourage people to engage in health life. With the increase of wearable devices, maintaining exercise habits has been valued by more and more people in recent years. Fitness Apps are expected to develop steadily; Entertainment: According to Sensor Tower 2018 statistics [5], audiovisual platforms and music streaming media occupy Ranked among the top 10 downloadable apps. Although some are only free services, audio-visual platforms and music streaming media have commercial potential in the future; E-learning: In addition to personal learning and e-mobile learning apps are also used in corporate education and training, so that employees' learning ability can be improved. Finally, electronic mobile commerce: Mobile Commerce (M-Commerce) is electronic commerce (Electronic Commerce). In the past, e-commerce mainly used desktop devices as the main interface. Lin & Li [7] indicate mobile commerce has upgraded the media to handheld mobile terminal devices, including mobile phones, laptops, tablets, etc. Li, Jin, Zhang, Zhang [1] indicate the rapid development of mobile Internet technology has laid a solid foundation for the development of mobile e-commerce. Mobile e-commerce is creating a brand-new consumption model, which has had a huge impact on the public's consumption behavior.

In recent years, everyone has at least one mobile phone, which makes life more convenient. In addition to increasing real-life experience, sports, entertainment and learning. Shopping and ticket purchase on mobile phones are also a brand-new consumption model, which is more convenient than physical stores. Because I am accustomed to buying transportation tickets with mobile phones, I choose mobile e-commerce as my research direction. The new ticketing system of Taiwan Railways was officially launched on April 9, 2019. The system includes two forms: website and App. Because the new system has just been launched, it is not stable and its functions are complex. Gu [2] indicates that although commuters have used the Taiwan Railway official Apps, the App developed by the private sector are more intuitive and can find information more quickly. Therefore, the interface of Taiwan Railways App has been redesigned and simplified. Allowing users to spend less time operating and more directly use the Taiwan Railway App to query and purchase tickets. This research hopes that through the analysis of the existing Taiwan Railway App and the interviews with users, we can revise and produce an app that provides ticket purchase needs for Taiwan Railway and is used by people aged 19–64. Therefore, the main purpose of this research is: 1) To explore the needs of users for existing Taiwan Railway App. 2) Based on user needs, revise and design the Taiwan Railway App interface to increase intuition.

2 Method

This chapter is divided into three parts. The first part is the research process. This research is mainly divided into pre-test and post-test. Both phases use observation method, interview method, and questionnaire method. The second part is the information and task flow of the participants in the pre-test. It analyzes the three existing Taiwan Railway Apps, operates the same operating process. Find the differences from them. Finally, the interface design and usability test for the post-test are completed. Prototype Production to subsequent improvements.

2.1 Study Process

This study focuses on the ethnic groups (participants) aged 19–64 who have the need to purchase tickets from Taiwan Railway. In the first step, the users' experience of using the three existing Taiwan Railway Apps was learned through testing, and the mistakes in the operation process were interviewed, and questionnaires on usability and satisfaction were designed. Then, based on the collected data, Adobe Xd was used to create the new Taiwan Railway Reservation. After the prototype was completed, the second step of usability testing was carried out. Including the operation of the new app, interviews with its process and suggestions. And the same design for usability and satisfaction. After that, we will further explore whether the shortcomings of the existing apps have been improved in the new apps. The three Taiwan Railway Ticketing Apps analyzed in the pre-test are the official Taiwan Railway e-booking, the Taiwan Railways Reservation and the Taiwan Railway Assistant. The basic information is shown in the following Table 1:

Table 1. Basic information of sample app.

App	Taiwan Railway e-booking	Taiwan railway Reservation	Taiwan Railway Assistant
Added Time	2019/04	2015/12	2016/10
Application Platform	iOS9.0/Android	iOS9.0/Android	iOS8.0/Android
Version	iOS:1.0.10	iOS:7.2	iOS:1.8.52
Memory Size	61.2 M	26.2 M	78.3 M

2.2 Pre-test Subject Information and Task

In the pre-test, a total of 10 participants aged 19–64 were found, 6 female and 4 male. With a college degree to a master's degree, and occupations such as student, public office, designer, and trade. It is divided into two groups:5 people who have never used the Taiwan Railways App and online booking at all, and the other is 5 people who have used the Taiwan Railways booking website but did not use the App to book tickets. The two groups have the same number. During the test, the test subject will be asked to perform uniform actions for subsequent differences and error comparisons. The process is to check the time according to the time period after opening the App and click the station. The departure station is Yilan Station, Yilan County, and arrival station is Banqiao Station, New Taipei City. Choose the departure date (2019/5/11) and the Puyuma Express closest to 9:30 in the morning. After confirming the schedule, select the ticket and enter the personal information and the number of tickets (adult 2 tickets) and purchase tickets with a credit card. Finally confirm whether the order is established and return to the main screen. From query time, selection of shifts, booking, entering information for credit card purchase, to the final confirmation of the order, all must be completed.

2.3 Interface Design and Usability Test

In the post-test, a total of 5 participants aged 19–64 were found, 2 female and 3 male. With a college degree to a master's degree, and occupations such as students, public jobs, designers, etc. And it is divided into two groups: 3 people who have never used the Taiwan Railways App and online booking at all, and the other is 2 people who have used the Taiwan Railways booking website but did not use the App to book tickets. The information of the 5 subjects is shown in the table below Table 2:

Table 2. Information of five post-test subject.

Item\No.	N1	N2	N3	N4	N5
Gender	Male	Female	Male	Male	Female
Age	19–30	31–40	51–64	31–40	19–30
Education	Master	College	College	College	Master
Occupation	Student	Public officer	Retirement	Designer	Student
Usage	A	B	A	B	A

A. Those who have never used the Taiwan Railways APP and website to book tickets.
B. Those who have used the Taiwan Railways ticket booking website but have not used the APP to book tickets.

During the test, the test subject will be asked to perform a unified action for subsequent differences and error comparisons. The process is to check the time according to the time period after opening the App, and click the station. The departure station is Taipei Station, Taipei City, and arrival station is Kaohsiung Station, Kaohsiung City. Select the departure date (2019/6/11) and the Tze-Chiang Express closest to 11:00 in the morning. After confirming the schedule, select booking and enter personal information, the number of tickets (adult 2 tickets) and purchase a ticket with a credit card. Finally confirm whether the order is established and return to the main screen. From query time, selection of shifts, booking, entering information for credit card purchase, to the final confirmation of the order, all must be completed.

3 Results

This chapter mainly explains the results of the pre-test and post-test subject observation method, interview method, and questionnaire method. Using comprehensive usability analysis, and improving the existing design in order to repair the better product.

3.1 Pre-test Results

The following will explain the results of the observation method, interview method and questionnaire method of 10 participants. Record the 10 participants of the test operation process, the time it takes for the three sample apps to complete the task. The number of seconds to one decimal place. To use the smart phone stopwatch to time the process, where errors occurred in the process the operation is also recorded. Test results:The operating error rate of the participants of the Taiwan Railway e-booking: 20%, the number of errors is accumulated, the starting and ending station selection function has 2 people, and the time selection has 1; the operating error rate of the participants of the Taiwan Railway Reservation: 80%. Accumulated number of errors, 3 for date button, 1 for search button, 1 for time button, 7 for booking button; operation error rate of the participants of the Taiwan Railway assistant: 80%, cumulative error times:starting and ending stations are 3 for selection, 2 for the query record button, 1 for the time button, and 6 for the booking button. Operation error rate of participants:Taiwan Railway e-booking(20%) < Taiwan Railway Reservation(80%) = Taiwan Railway assistant(80%); Taiwan Railways e-booking pass: the most errors are in the starting and ending stations With the typing function 2 times, there is a choice of time. Taiwan Railway Ticketing Pass: The error is concentrated on the selection of the ticket button up to 7 times, and the date button, time button, and search button. Taiwan Railway Assistant: The error is concentrated on the selection of the ticket button up to 6 times, the selection of the start and end stations 3 times, and the selection of the record query and time button. There is no problem in checking the order after booking for the three apps.

According to the results of the subject interview method, after implementing the observation method on ten participants. A semi-structured interview was conducted, and each subject was asked three questions. They are the ranking of the overall usefulness of the three sample Taiwan Railway Apps. Why there will be hesitations or errors in using Apps. The three sample Taiwan Railway Apps are easy to use, difficult to use and suggestions. For the first question, please rank the overall case of use:the highest overall ease of use is the Taiwan Railway Reservation (80%), the Taiwan Railway e-booking (20%), and the Taiwan Railway Assistant (0%). The ones with the lowest overall ease of use are the Taiwan Railway Assistant (70%), the Taiwan Railway e-booking (30%), and the Taiwan Railway Reservation (0%). The second question is why there are hesitations or errors? In the Taiwan Railway e-booking (official) there are errors or hesitations. 1) Typing and dragging when selecting stations causes confusion, and the morning and afternoon options are too close. It will be wrong. 2) Typing and drag-and-drop functions appear at the same time, which will make people hesitate. 3) The booking button and the fare button are too close, and the booking button is not easy to find at a glance. There are errors or hesitating items in the Taiwan Railway Reservation:1) The time and date block partition is not obvious. It takes time to search. The color of the search key is similar to the background, so it can't be found. In addition, I clicked into the booking column intuitively and didn't see the booking button. 2) The date button is not separated by blocks and it is difficult to find. After querying the vehicle type, you will directly click the booking column. So the booking button on the right is not found at all. 3) The input time block is not obvious, so it took a little time to look for it. 4) The booking button is too close to the fare button, and the booking button is not easy to spot from the edge. 5) The booking button is too close to the price color, so I didn't find it. 6) The booking button is too similar to the advertisement, making the icon inconspicuous. 7) The ticket booking button is too unobvious, and people will directly click on the ticket booking column and then run to the station. The mistakes or hesitations of the Taiwan Railway assistant are: 1) The ticket booking button is too unobvious, only a small one, which is not intuitive at first glance. 2) I don't know what the key is, I just press the query record directly. In addition, the booking key is small and I didn't find it at all. Then I clicked on the arrival time. 3) The icon is not obvious. I accidentally clicked on the historical record. In addition, the booking button was too narrow and I didn't find it at all. Then I clicked on each arrival time. 4) The time menu text is too close, it is easy to choose the wrong column in the morning and afternoon. And the ticket button block is too small for people to notice. 5) The booking button is too unobvious, and the intuition is low at first glance. 6) The booking button is too small. 7) The booking button is too unobvious. It's just a small strip, and it's also in English. 8) The booking button is really not obvious and will lead to neglect. 9) The ticket booking button is too inconspicuous, and the starting and ending station names are too closely spaced, which is visually confusing. The third question is where the three apps are easy to use and difficult to use or suggestions? The items that can be modified in the Taiwan Railway e-booking (official) are: 1) The station selection arrow icon for the start and end stations can be changed from right to down. 2) Change the information function block too flat to a wider range. 3) Separate the time menu text to enlarge the booking button block. 4. Re-modify the App Icon. The items that can be modified by the Taiwan Railway Reservation are: 1) Change the start and end block menu and date and time query to a drag menu. 2) Reform the small Icon and simplify the color. 3) It can increase the round-trip and one-way interface. 4) Set the booking button after selecting the booking column. The items that the Taiwan Railway Assistant can modify are: 1) Delete or change the difficult-to-identify Icon to be displayed in text to avoid clicking errors, and to distinguish Taipei from the New Taipei area. 2) Separate the time menu text. 3) The BOOK NOW booking button is enlarged, the interface is re-formatted, and the notes in the booking column are enlarged in word level, and the color of the vehicle type in the timetable is simplified.

Pre-test questionnaire method results. After completing the observation method and the interview method, let the participants fill in the questionnaire. This questionnaire uses the Richter five-point scale, strongly agree (5 points), agree (4 points), none Opinion (3 points), disagree (2 points), and strongly disagree (1 point). According to the usability and interface, there are a total of ten questions. After the test, sort out the average, maximum, and minimum of each question Value, and standard deviation, and compare the data. From the three sample App questionnaires, it can be found that the average number of the Taiwan Railway e-booking and the Taiwan Railway Reservation has four questions higher than or equal to 4.0, and none of the Taiwan Railway assistant has an average number higher than or equal to 4.0. In terms of standard deviation, the Taiwan Railway Reservation has three questions less than 1.0. The Taiwan Railway e-booking has one question less than 1.0, and the Taiwan Railway Assistant has no question less than 1.0. It can be seen that the Taiwan Railway Reservation Pass is a sample of three types. The highest score in the App.

In terms of usability, the Taiwan Railway Reservation is the best among the three sample apps. Therefore, the original Taiwan Railway Reservation is used as a reference prototype, and modifications are suggested in accordance with the interview method in the previous test. The modified items are as follows in Table 3:

Table 3. New Taiwan Railway Reservation modification project.

App	Editable items
Taiwan Railway e-booking(official)	1. Separate the time menu text to enlarge the booking button block
Taiwan Railway Reservation	1. Reform the small Icon and simplify the color 2. It can increase the round-trip and one-way interface 3. Set the booking button after selecting the booking column
Taiwan Railway Assistant	1. Delete or change the Icon that is difficult to identify and present it in text to avoid clicking mistakes, and to distinguish Taipei from the New Taipei area 2. Separate the time menu text

3.2 Post-test Results

The following will explain the results of the observation method, interview method and questionnaire method of 5 post-test participants. The subjects observed the task results and recorded the 5 participants of the test operation process. The time taken by the New Taiwan Railway Reservation to complete the task, the number of seconds is recorded to one decimal place. And the smart phone is used for timing. Where errors occurred during the process also recorded. After the 5 participants used the New Taiwan Railways Reservation, the interface was simplified and easier to operate. There was only one error, which was that the orange triangle (Fig. 1) was not intuitive for the selection of the starting and ending stations. And there is no problem with the query order after booking.

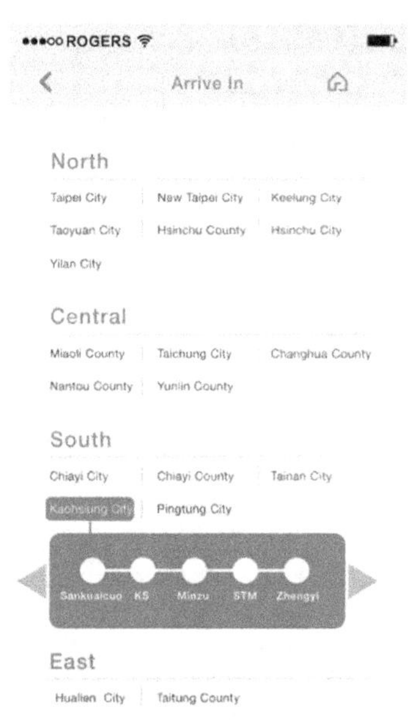

Fig. 1. User error Interface

The results of the interview with the participants. After the observation method was implemented on the 5 participants, a semi-structured interview was conducted, and each participant was asked three questions. Please describe in one sentence how you feel after the operation and why you are using the App. If there are hesitations or errors, the New Taiwan Railway Reservation App is easy to use, difficult to use and suggestions. Focus on collation of interview data, organize the interview information into Table 4, Table 5 and Table 6.

Table 4. Post-sensing items list.

App	One sentence thought
New Taiwan Railway Reservation	1. The overall use is very smooth, better than the official one 2. The picture is uniformly composed of blue and orange, which makes me feel that the picture looks very comfortable 3. The fonts of some explanatory texts are too thin and too small, which will affect reading

Data Source:Created by author.

Q1. Please tell me your feelings after the operation in one sentence.

Q2. Why are there hesitations or errors when using the App?

Q3. What are the easy and difficult places or suggestions for the New Taiwan Railway Reservation?

Table 5. List of post-test operation delays.

App	Wrong or hesitating item
New Taiwan Railway Reservation	1. The orange triangle area of the selected site is too small, and at first glance you will not think it is a button 2. After querying the page that displays the vehicle type booking, the vehicle type label on the left is a bit too small for people to see clearly

Data Source:Created by author.

Table 6. List of Items that can be Modified after Testing.

App	Editable items
New Taiwan Railway Reservation	1. In the site selection section, you can change the orange triangle that selects other sites to a blue block that is all sites 2. On the page where you are going to book a ticket after checking the car type, the car type label on the left is a bit too small, making it difficult to see whether it is a Tze-Chiang, Chu-Kuang or a shuttle bus. I think I can enlarge the entire column. The five columns are arranged in one, you can try to adjust the screen to four columns, the font can be enlarged, and the image of the train remains the original size 3. In terms of fonts, the characters can be enlarged again, and the characters can be bolded, and the layout distribution block is larger

Data Source:Created by author.

The post-test results of the questionnaire method. After completing the observation method and the interview method, let the participants fill in the questionnaire. This questionnaire uses the Richter five-point scale. They strongly agree (5 points), agree (4 points), and have no comments (3 points), disagree (2 points), and strongly disagree (1 point). According to the usability and interface, there are ten questions. After the test, sort out the average, minimum, standard deviation, and compare the data.

Compared with the three sample apps in the previous test, it can be found that the average number of five questions in the new Taiwan Railway Reservation app is higher than or equal to 4.0, which is higher than the three apps in the previous test, and the standard deviation is also less than 1.0.

4 Conclusions and Suggestions

The purpose of this research is to use the analysis of existing Taiwan Railway apps, interviews with users to revise and create apps for people aged 19–64 who need to buy tickets for Taiwan Railways. The purpose is mainly to explore the user's needs for the use of the existing Taiwan Railway App, and to re-modify and design the interface of the Taiwan Railway App based on user needs to increase intuition. By providing test subjects to test three sample Taiwan Railway App, using observation method to operate screen recording, semi-structured interviews, and questionnaire information collation, the use of the Taiwan Railway Reservation App is the highest among the three. Use this App As a modified prototype, it focused on the function of ordering tickets, made a prototype, and performed post-testing. From the values, it can be found that it is easier to use than the first three sample Apps, and modification suggestions are also obtained to make more complete modifications for subsequent Apps.

4.1 Summary of Three Apps of Taiwan Railway

The observation method of the previous test found that when the subjects used the three apps, most of them had no problems in selecting the site, date and time, etc. The only reason was the hesitation and error when selecting the ticket button. The reason was the vehicle type information page. The color is messy, the blocks are not obvious, and it does not make people intuitively want to choose the button. The error rate of the participants in operating the App is: Taiwan Railway e-booking (20%) < Taiwan Railways Reservation (80%) = Taiwan Railways Assistant (80%). All three apps have no problem in checking orders after booking. In the previous test interviews, it was found that the most usable was the Taiwan Railway Assistant (80%), and the least usable was the Taiwan Railway Assistant (0%). They asked the participants why they made mistakes or operations. Most of them think that the ticket booking button is not obvious, which delays their operation steps. Color clutter is also a problem. There are more than four colors in an interface, and the vision will be confused. In addition, refer to the participant's suggestions to modify the new App. The items are as follows: (1) Enlarge the booking area. (2) Re-format and simplify the small icon. (3) Add one-way and round-trip ticket interface. (4) Set the booking button after selecting the booking column. (5) Date and time menu separate. The previous test questionnaire found that the highest average number of questions for the Taiwan Railway Reservation has four questions higher than or equal to 4.0. The participants are most satisfied, the standard deviation has three questions lower than 1.0. Taiwan Railway Reservation is the lowest of the three and the opinions are more consistent.

4.2 Summary of App Design Process

Many results of this research are presented in the App interface. The results obtained during the design process are summarized as follows: 1) The App design of this research is based on the research of existing App interfaces and functions. Through comparative analysis of similar products and its operation methods. Conducive to own product design. 2) The final results of the research are presented in Prototype and used as a sample of usability testing and interviewees, effectively saving development costs.

4.3 Post-test Usability Conclusion

From the post-test observation method, it is found that the interface is simplified and easier to operate. There is only 1 error. The orange triangle (Fig. 1) is not intuitive for people to click on the station. The query after booking there is no problem with the order. From the post-test interviews, it was found that the test subjects felt that the overall use was smooth. The screen was uniformly composed of blue and orange, which looked very comfortable. Some explanatory texts were too thin and too small to affect reading, and they felt that they chose the triangle block of the site. And ordered the vehicle type indications on the left side of the ticket page are too small to be intuitively selected. In terms of fonts, the fonts can also be enlarged. From the post-test questionnaire method, it is found that compared with the three sample Apps in the previous test. It can be found that the average number of five questions in the new Taiwan Railway Reservation App is higher than or equal to 4.0, which is higher than the three apps in the previous test. The standard deviation is also three questions are less than 1.0.

4.4 Suggestions

There are still shortcomings in the research process of this research. After sorting it out, research recommendations are written for follow-up researchers to refer to. This research proposes the following recommendations based on the above study process and conclusions: 1) The prototype software for this research is Adobe Xd. In the future, it is recommended to use Axture to make prototype software. Although Adobe Xd is fast and convenient, it is compatible with Adobe AI (Illustrator) and Adobe PS (Photoshop), but the prototype produced cannot be completely realistic and online publishing simulation, only images can be used Replacement, unable to type and other special effects. 2) The standard iOS design guidelines referred to in this research have been updated many times. The three sample Apps have different levels of version updates and their active numbers have also changed significantly. It shows that this kind of App market has great room for development.

This research was supported by the grant from the Ministry of Science and Technology, Taiwan, Grant MOST 109-2410-H-030-014.

References

1. Li, Q., Jin, D., Zhang, W., Zhang, T.: Research on the development trend of M-commerce. Telecommun. Sci. **27**(6), 6–13 (2011)
2. Gu, T.-H.: An exploration of usage of taiwan railway app on the commuters based on user experiences. Zhongyuan University Department of Business Design Master's thesis, Taiwan (2019)
3. Dreaming News, Top 5 App Trends in (2019). https://ek21.com/news/tech/35526/. Accessed on 15 Aug 2020
4. The ministry of transportation and communication R.O.C Statistics query, Taiwan Railway Passenger Traffic Statistic Data. https://stat.motc.gov.tw/mocdb/stmain.jsp?sys=220&ym=10300&ymt=10906&kind=21&type=1&funid=b210101&cycle=41&outmode=0&compmode=0&outkind=1&fld0=1&rdm=deUr3eaq. Accessed 15 Aug 2020
5. SensorTower, The Top Mobile Apps, Games, and Publishers of 2018: Sensor Tower's Data Digest. https://sensortower.com/blog/top-apps-games-publishers-2018. Accessed 15 Aug 2020
6. SensorTower, ARKit-only Apps Surpass 13 Million Downloads in First Six Months, Nearly Half from Games. https://sensortower.com/blog/arkit-six-months. Accessed 15 Aug 2020

7. Lin, J.T., Li, Y.S.: Mobile commerce fundamentals, practices & application: ubiquitous cloud computing, mobile devices, RFID, and the Internet of Things. Taipei City (ROC) (2012)
8. App Annie Releases Annual State of Mobile 2019 Report, https://www.appannie.com/en/about/press/releases/app-annie-releases-annual-state-of-mobile-2019-report/, last accessed 2020/8/15

A Survey of 6G Wireless Communications: Emerging Technologies

Yang Zhao[1(✉)], Jun Zhao[1], Wenchao Zhai[2], Sumei Sun[3,4], Dusit Niyato[1], and Kwok-Yan Lam[1]

[1] School of Computer Science and Engineering, Nanyang Technological University, Nanyang, Singapore
s180049@e.ntu.edu.sg, {junzhao,dniyato,kwokyan.lam}@ntu.edu.sg
[2] College of Information Engineering, Jiliang University, Hangzhou, China
zhaiwenchao@cjlu.edu.cn
[3] Institute for Infocomm Research (I2R), Agency for Science, Technology and Research (A*STAR), Singapore, Singapore
sunsm@i2r.a-star.edu.sg
[4] Infocomm Technology Cluster, Singapore Institute of Technology, Singapore, Singapore

Abstract. While fifth-generation (5G) communications are being rolled out around the world, sixth-generation (6G) communications have attracted much attention from both the industry and the academia. Compared with 5G, 6G will have a wider frequency band, higher transmission rate, spectrum efficiency, greater connection capacity, shorter delay, wider coverage and stronger anti-interference capability, so as to meet the various network requirements for industries. In this paper, we present a survey of potential essential technologies in 6G. In particular, we will introduce index modulation, artificial intelligence, intelligent surfaces, and terahertz communications technologies in detail, while giving a brief introduction to other potential technologies, including visible light communications, blockchain-enabled wireless network, advanced duplex, holographic radio and network in box.

Keywords: 6G · Wireless communications · Index Modulation (IM) · Artificial Intelligence (AI) · Intelligent reflecting surfaces · IRS · Terahertz (THz)

1 Introduction

Fifth-generation (5G) networks are being deployed commercially [76]. However, the rapid growth of data-centric intelligent systems has brought significant challenges to the capabilities of 5G wireless systems. For example, the 5G air interface delay standard of less than 1 ms (ms) is challenging to meet the air interface delay of less than 0.1 ms required by haptic Internet-based telemedicine [51]. To overcome the performance limitations of 5G to deal with new challenges, countries are researching the sixth-Generation (6G) mobile communication system.

K. Arai (Ed.): FICC 2021, AISC 1363, pp. 150–170, 2021.
https://doi.org/10.1007/978-3-030-73100-7_12

Upcoming technologies such as artificial intelligence (AI), virtual reality, and the Internet of Everything (IoE) require low latency, ultra-high data rates and reliability. The various applications as shown in Fig. 1 cannot be satisfied by existing 5G's ubiquitous mobile ultra-broadband, ultrahigh data density, and ultrahigh-speed-with-low-latency communications [12,34,53,63]. Performance limitations of 5G and the emerging revolutionary technologies drive the development of 6G networks [89].

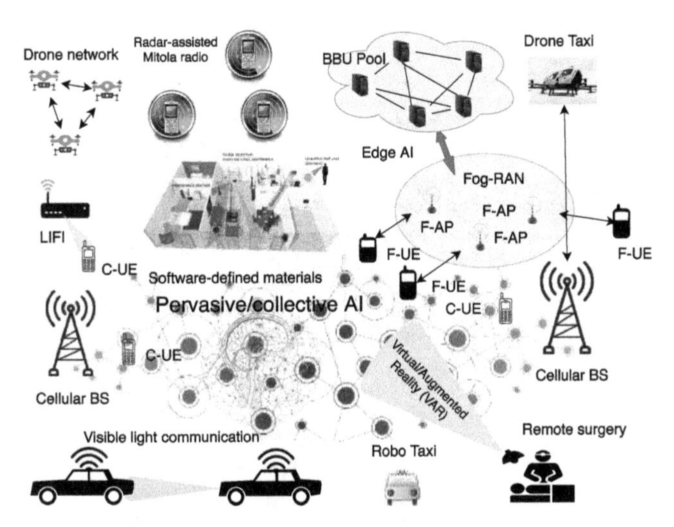

Fig. 1. The vision of 6G.

Past generations of wireless networks utilize micro-wave communications over the sub-6 GHz band, whose resources are almost used up [88]. Hence, the Terahertz (THz) bands will be the major candidate technology for the 6G wireless communications [3,26,63,73,80,88]. Due to the propagation loss, the THz will be used for high bit-rate short-range communications [57]. Besides, the 90-200 GHz spectrum is often not used in the past generations of wireless networks. The sub-THz radio spectrum above 90 GHz has not been exploited for radio wireless communications yet; thus, it is envisioned to support the increased wireless network capacity [14]. 6G will undergo the transition from radio to sub-terahertz (sub-THz), visible light communication and terahertz to support explosive 6G applications [18].

Furthermore, the 6G system is envisioned to support new services such as smart wearable, computing reality devices, autonomous vehicles, implants, sensing, and 3D mapping [12]. 6G's architecture is expected to be a paradigm-shift

Table 1. Requirements and features of 6G [17,34,58,60,73].

Requirements	6G
Service types	MBRLLC/mURLLC/HCS/MPS
Service level	Tactile
Device types	Sensors and DLT devices/CRAS/ XR and BCI equipment/Smart implants
Jitters	1 μs
Individual data rate	100 Gbps
Peak DL data rate	≥1 Tbps
Latency	0.1 ms
Mobility	up to 1000 km/h
Reliability	up to 99.99999%
Frequency bands	- sub-THz band - Non-RF, e.g., optical, VLC, laser ···
Power consumption	Ultra low
Security and privacy	Very high
Network orientation	Service-centric
Wireless power transfer /Wireless charging	Support (BS to devices power transfer)
Smart city components	Integrated
Autonomous V2X	Fully
Localization Precision	1 cm on 3D
Architecture	Intelligent Surface
Core network	Internet of Everything
Satellite integration	Full
Operating frequency	1 THz
Highlight	Security, secrecy, privacy
Multiplexing	Smart OFDMA plus IM

and carries higher data rates with low latency [63]. Ho *et al.* [34] and Piran *et al.* [60] present their predictions for 6G's requirements and features, which are summarized in Table 1. With these advanced features, 6G wireless communication networks will integrate space-air-ground-sea networks to achieve the global coverage as shown in Fig. 2 [27].

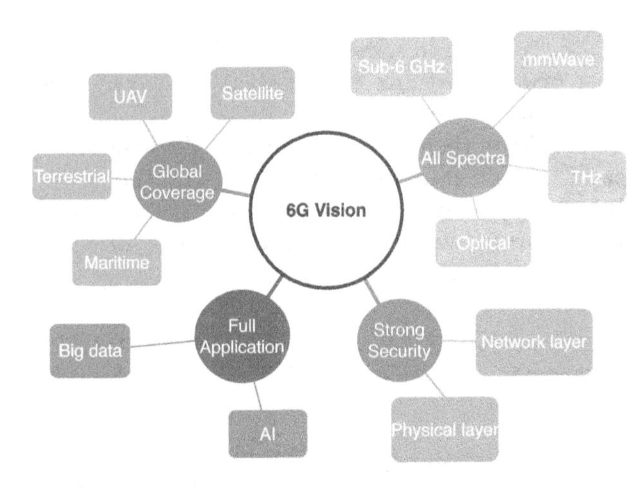

Fig. 2. 6G wireless communication networks.

Contributions of this Survey: In this survey, we focus on the emerging technologies that will be used in the 6G. The contributions of this survey are summarized below.

- Current papers on 6G pay more attention to predicting technologies that may be used in the future, and none of them gives a summary. Our paper surveys existing visions of 6G and summarize them.
- We highlight four technologies that are envisioned to play a significant role in 6G, including index modulation, AI, intelligent reflecting surface, and THz communication.
- We briefly explain several potential technologies that have been discussed recently, such as visible light communication, blockchain-based network, satellite communication, holographic radio, and network in box.

Organization. The rest of paper is organized as follows. Section 2 introduces emerging technologies that enable the paradigm shift in 6G wireless networks. Section 3 concludes this paper.

2 6G Architecture: A Paradigm Shift

In this section, we will explain the most eye-catching ideas pertaining to 6G in detail, including index modulation, artificial intelligence, active/passive intelligent reflecting surfaces and THz communications [10,17,21,29,40,67,72,82]. Also, we will present some promising technologies such as blockchain, satellite communication, full-duplex, and holographic radio in short.

2.1 Index Modulation

Index Modulation (IM) has high spectral efficiency and high power efficiency due to its idea of sending extra information through the indexed resource entities such as the time slots, the transmit/receive antennas, the subcarriers and the channel states. Its low deployment cost and high throughput attract much attention in the upcoming 6G communications.

Based on the entities of the indexed resources, IM can be classified into time-domain IM, spatial-domain IM, frequency-domain IM, and channel-domain IM. Each kind of IM technique divides the transmitted bits into two parts, as illustrated in Fig. 3: one part for classical modulation, for example, phase shift keying and quadrature amplitude modulation, etc. the other part for the activation of the indexed resources used for transmitting the additional bits. Because the additional information requires neither spectrum resources nor power resources, IM can improve the throughput while consuming low power compared with its non-IM-aided counterpart.

TD-IM Technique. Time Division Duplex (TDD) is one of the generally used kind of wireless techniques, and will be inherited in 6G communications without doubt. In TDD transmission, a data frame consists of several time slots, and each time slot can be used to transmit the source information. However, for TD-IM technique, only a fraction of time slots activated by the transmitting bits are used for signal transmission. To further improve the demodulation performance, space-time shift keying (STSK) technique is proposed by Sugiura *et al.* in [70] to combine TD-IM and space-time block code (STBC) together where a dispersion matrix can be activated for index selection. Subsequently, the same authors extend STSK to a general format in [71], where indices of multiple activated

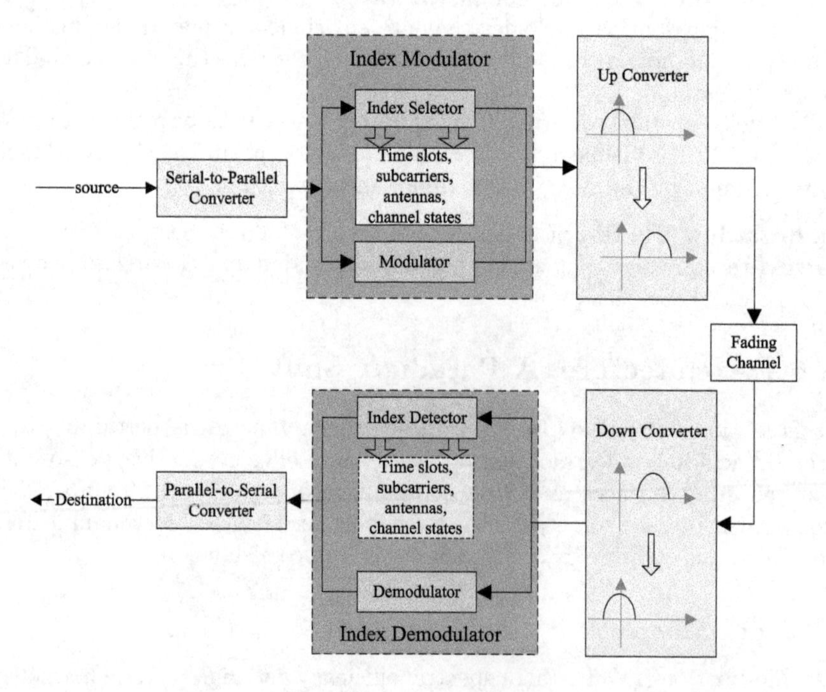

Fig. 3. The structure of IM-aided systems.

dispersion matrices are chosen based on the transmission information bits, further improving the capacity.

FD-IM Technique. The orthogonal frequency division multiplexing (OFDM) technique is widely used due to its high spectral efficiency in 4G and 5G communications and will still be used in the 6G network. The bandwidth is divided into several subcarriers orthogonally, and each subcarrier transmits its own data bits individually. For the FD-IM technique, additional bits are used to choose the indices of the activated subcarriers. Therefore, FD-IM is also called subcarrier-indexed OFDM (SIM-OFDM). Different from TD-IM, the FD-IM technique has much more resource entities. Thus, FD-IM often divides the subcarriers into several blocks, and activates one subcarrier (SIM-OFDM) [1] or more than one subcarriers (generalized SIM-OFDM, GSIM-OFDM) [24] in each subblock, reducing the number of index patterns at the transmitter as well as the demodulation complexity at the receiver. When combing with MIMO systems, SIM-OFDM can also be extended to MIMO-OFDM-IM to achieve considerable performance gain [4,5].

SD-IM Technique. SD-IM is also called spatial modulation (SM), where the spare information bits are used to activate the transmit antennas [81]. Compared with MIMO systems, SM needs no inter-antenna synchronization and is free of inter-antenna interference, which leads to low receiver complexity. The SM can also be extended to general SM (GSM) to improve the data rate when multiple antennas are activated to transmit the modulated signals. On the other hand, precoding SM (PSM) or general precoding SM (GPSM) can be applied by activating the receive antennas to exploit the advantage of beamforming [48, 86]. In PSM/GPSM, a precoding scheme is implemented at the transmitter to identify the desired antennas/antennas, thus providing transmit diversity and improving the detect performance at the receiver.

CD-IM Technique. Unlike the aforementioned IM schemes, CD-IM can change the property of radio frequency (RF) environment by employing RF mirrors or electronic switches [6,8,56,65]. Therefore, CD-IM has also been named media based modulation (MBM). MBM uses several RF mirrors/electronic switches around the transmit antennas. It allows the signal to transmit to the receiver through distinct channel paths according to the on/off status of the RF mirrors/electronic switches. Compared with the MIMO system, where the channel matrix is generally nonorthogonal, MBM can further randomize the channel by perturbing the wireless environment, enhancing the achievable rate. MBM can be combined with MIMO systems (MIMO-MBM) [8,65], Alamouti STBC systems (STCM) [6], etc. to improve the capacity further or detect performance.

In addition, there are some other complex IM schemes. In [68], Schamasundar *et al.* propose a scheme by combining the TD-IM and MBM, namely, time-indexed MBM. Such hybrid IM technique can increase the data rate considerably at the cost of detect complexity. In [2], Ertugrul *et al.* give a novel sparse code multiple access (SM-SCMA) scheme operating in uplink transmission, which is used for organizing the accessing of multiple users. SM-SCMA is an example of non-orthogonal multiple access aided SM (NOMA-SM), which is remarkable in reducing the inter-user interference in 6G multi-user communications [75]. IM-aided systems can also be designed with bit-interleaved coded modulation (BICM) where soft information between the channel decoder and index pattern can be exchanged iteratively to obtain near-capacity performance [74]. In [79], a reduced complexity detector is presented by introducing the compressing theory into the IM-aided system, where the sparsity of the activated resource entities is exploited to detect the information.

2.2 Artificial Intelligence

Artificial Intelligence (AI) provides intelligence for wireless networks by simulating some human thought processes and intelligent behaviors. By leveraging AI, 6G will enable more applications illustrated in Fig. 4 to be intelligent such as smart city, cellular network, connected autonomous electric vehicles, and unlicensed spectrum access [9,18,22,25,44,60,69,84,85,87,89]. In particular, AI, residing in new local "clouds" and "fog" environments, will help to create many novel applications using sensors that will embedded into every corner of our life [26]. In the following, we present some essential AI techniques and corresponding applications for 6G.

Deep learning, considered as the vital ingredient of AI technologies, has been widely used in the wireless networks [50]. It will play an essential role in various areas, including, semantic communications, holistic management of communication, control resources areas, caching, and computation, etc. which push the paradigm-shift of 6G.

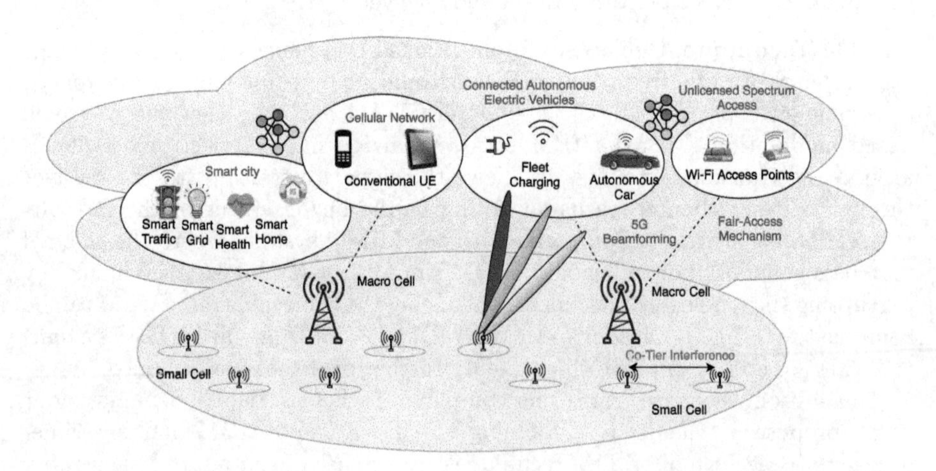

Fig. 4. An AI-Enabled 6G wireless network and related applications.

Artificial Intelligence Algorithms. In this section, we summarize some potential AI techniques: supervised and unsupervised learning, model-driven deep learning, deep reinforcement learning, federated learning, and explainable artificial intelligence as follows.

Supervised Learning. The supervised Learning trains the machine model using labelled training data [60]. There are some well developed algorithms that can be used in the 6G network, such as support vector machines, linear regression, logistic regression, linear discriminant analysis, naive Bayes, k-nearest neighbors and decision tree, etc. Supervised learning techniques can be used in both physical layer and network layer. In physical layer, we can utilize supervised learning for channel states estimation, channel decoding, etc. Supervised learning techniques can be deployed for caching, traffic classification, and delay mitigation and so on in the network layer.

Unsupervised Learning. Unsupervised learning is leveraged to find undefined patterns in the dataset without using labels. Commonly used unsupervised learning techniques include clustering, anomaly detection, autoencoders, deep belief nets, generative adversarial networks, and expectation–maximization algorithm. At the physical layer, unsupervised learning techniques are applicable to optimal modulation, channel-aware feature-extraction, etc. In addition, unsupervised learning technologies can be used for routing, traffic control and parameter prediction, etc., in network layer.

Model-Driven Deep Learning. The model-driven approach is to train an artificial neural network (ANN) with prior information based on professional knowledge [25,33,84]. The model-driven approach is more suitable for most communication devices than the pure data-driven deep learning approach, because it does not require tremendous computing resources and considerable time to train what the data-driven method needs [33]. The approach to apply model-driven deep learning proposed by Zappone *et al.* [33] includes two steps: first, we can use theoretical models derived from wireless communication problems as prior expert information. Secondly, we can subsequently tune ANN with small sets of live data even though initial theoretical models are inaccurate.

Deep Reinforcement Learning. Deep reinforcement learning (DRL) leverages Markov decision models to select the next "action" based on the state transition models [50]. DRL technique is considered as one of the promising solutions to maximize some notion of cumulative reward by sequential decision-making [60]. It is an approach to solve resource allocation problems in 6G [22,85]. As 6G wireless networks serve a wider variety of users in the future, the radio-resource will become extremely scarce. Hence, efficient radio-resource allocation is urgent and challenging [22].

Federated Learning. Federated Learning (FL) aims to train a machine learning model with training data remaining distributed at clients in order to protect data owners' privacy [42]. As 6G heads towards a distributed architecture, FL technologies can contribute to enabling the shift of AI moving from a centralized cloud-based model to the decentralized devices based [44,66,73]. In addition, since the edge computing and edge devices are gaining popularity, AI computing tasks can be distributed from a central node to multiple decentralized edge nodes. Thus, FL is one of the essential machine learning methods to enable the deployment of accurately generalized models across multiple devices [15].

Explainable Artificial Intelligence. Since there will emerge a large scale of applications such as remote surgery and self-driving in the 6G era, it is necessary to make artificial intelligence explainable for building trust between humans and machines. Currently, most AI approaches in PHY and MAC layers of 5G wireless networks are inexplicable [30]. AI applications such as self-driving and remote surgery are considered to be widely used in 6G, which requires explainability to enable trust. AI decisions should be explainable and understood by human experts to be considered as trustworthy. Existing methods, including visualization with case studies, hypothesis testing, and didactic statements, can improve deep learning explainability.

Artificial Intelligence Applications in 6G. In this section, we present some potential use cases of AI in 6G such as AI in network management and AI in autonomy.

AI in Network Management. As 6G network becomes complex, it may utilize deep learning instead of human operators to improve the flexibility and efficiency in the network management [60]. AI technologies are applicable to both the physical layer and network layers. In physical layer, AI techniques have involved in design and resource allocation in wireless communications [34]. For example, unsupervised learning are applicable to interference cancellation, optimal modulation, channel-aware feature-extraction, and channel estimation, etc. [60]. Deep reinforcement learning is possible to be employed for link preservation, scheduling, transmission optimization, on-demand beamforming, and energy harvesting, etc. [60,66]. In addition, AI technologies can be used to the network layer as well. Supervised learning techniques can tackle problems such as resource allocation, fault prediction, etc. [60]. Besides, unsupervised learning algorithms can help in routing, traffic control, parameter prediction, resource allocations, etc. [60]. Reinforcement learning can be important for traffic prediction, packet scheduling, multi-objective routing, security, and classification, etc. [60,66].

AI in Autonomy. AI technologies are potential to enable 6G wireless systems to be autonomous [25,49,87]. Agents with intelligence can detect and resolve network issues actively and autonomously. AI-based network management contributes to monitoring network status in real-time and keep network health. Also, AI techniques can provide intelligence at the edge devices and edge computing, which enables edge devices and edge computing to learn to solve security problems autonomously [49,52,61]. In addition, autonomous applications such as autonomous aerial vehicles and autonomous robots are envisioned to be available in 6G [34].

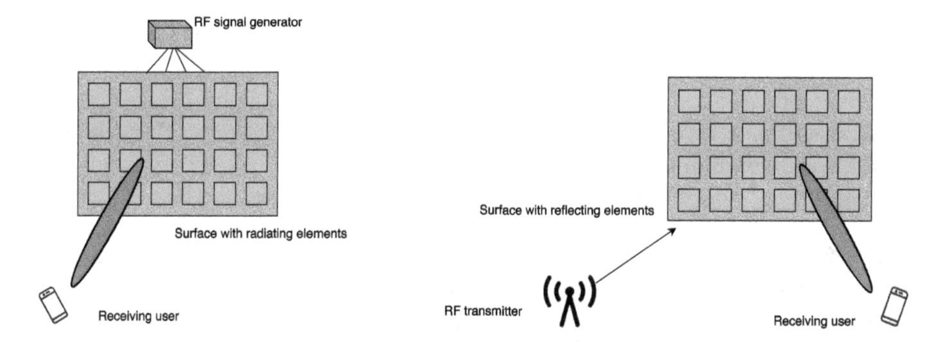

Fig. 5. Left: Large Intelligent Surfaces (an RF signal generator locates at the backside). Right: Intelligent Reflecting Surfaces (an RF signal generator locates at another location).

2.3 Intelligent Surfaces

Currently, two types of intelligent surfaces shown in Fig. 5 attract researchers' attention - Large Intelligent Surfaces (LISs) and Intelligent Reflecting Surfaces (IRSs). LISs are useful for constructing an intelligent and active environment with integrated electronics and wireless communications [23,38]. Renzo *et al.* [20] believe that IRSs will be utilized in 6G, because they predict that future's wireless networks will serve as an intelligent platform connecting the physical world and the digital world seamlessly. They foresee that wireless networks will be smart radio environments which are potential to realize uninterrupted wireless connectivity and use existing radio waves to transmit data without generating new signals.

Large Intelligent Surfaces (LISs). The concept of deploying antenna arrays as LISs in massive MIMO systems was originally proposed by Hu *et al.* [37]. LISs are electromagnetically active in the physical environment, where each part of an LIS can send and receive electromagnetic fields. Buildings, streets, and walls are expected to be electronically active after decorating with LISs [23]. LISs have the following main favorable features [23]: (i) Generate perfect LoS indoor and outdoor propagation environments. (ii) They put little restriction on the spread of antenna elements. Therefore, antenna correlations and effects of mutual coupling can be avoided more easily, such that sub-arrays are large and the channel is well-conditioned for propagation. Thus, LISs can be realized via THz Ultra-Massive MIMO (UM-MIMO). LISs are very useful for applications with low-latency, because channel estimation techniques and feedback mechanisms that LISs support are simple.

Intelligent Reflecting Surfaces (IRSs). The IRS is considered as a promising candidate in improving the quality of the signal at the receiver by modifying the phase of incident waves [19,28,35–39,55]. IRSs are made of electromagnetic

(EM) material that are electronically controlled with integrated low-cost passive reflecting elements, so that they contribute to forming the smart radio environment [7]. IRS can change the wireless signal propagation environment by adjusting the phase shift of the reflecting elements. Besides, IRSs help to enhance the communication between a sender and a receiver by reflecting the incident wave [7,45,59]. By adjusting the reflection coefficients, IRSs enable the reflected signals being coherently added to the receiver without adding additional noise [45]. Besides, IRSs can increase signal power and modify signal phase [23]. In particular, by utilizing local tuning, graphene-based plasmonic reconfigurable metasurfaces can obtain some benefits, including beam focusing, beam steering, and control on wave vorticity [46]. Unlike LISs, IRSs use passive array architecture for reflecting purpose [78]. Distinguishable features of IRSs summarized by Basar *et al.* [7] and Wu *et al.* [78] include:

- They comprise low-cost passive elements which are controlled by the software programming.
- They do not require specific energy source to support during transmission.
- They do not need any backhaul connections to exchange traffics.
- The IRS is a configurable surface, so that points on its surface can shape the wave impinging upon it.
- They are fabricated with low profile, lightweight, and conformal geometry such that they can be easily deployed.
- They work in the full-duplex mode.
- No self-interference.
- The noise level does not increase.

Different from existing technologies such as backscatter communication, active relay, and active surface based massive MIMO, the IRS-aided network includes both active components (BS, AP, user terminal) and passive component (IRS). We highlight some differences between IRS and well-known technologies as follows.

Massive MIMO. IRSs and massive MIMO consist of different array architectures (passive versus active) and operating mechanisms (reflecting versus transmitting) [78]. Benefit from the passive elements, IRSs achieve much more gains compared to massive MIMO while consuming low energy [83].

Amplify-and-Forward (AF) Relay. Relay uses active transmit elements to assist the source-destination communication, but the IRS serves as a passive surface, which reflects the received signal [78,83]. Relays help to reduce the rate of the available link if they are in half-duplex mode. When they operate in full-duplex mode, they suffer the severe self-interference. Active relay usually works in half-duplex mode for reduced self-interference. While IRS can work in full-duplex mode, improving the spectrum efficiency compared to the former. Active relay usually works in the half-duplex mode, which wastes spectrum compare to IRS, which works under full-duplex mode. If AF implements full-duplex mode, it will require costly self-interference cancellation techniques to support. But IRS overcomes above outstanding shortcomings of AF relays.

Backscatter. Backscatter requires the reader to realize self-interference cancellation at the receiver to decode the radio frequency identification (RFID) tag's message. RFID communicates with the reader by modulating its reflected signal sent from the reader [78]. However, IRS only reflects received signals without modifying information; thus, the receiver can add both the direct-path and reflect-path signals to improve the decoding's signal strength.

2.4 Terahertz Communications

Currently, wireless communication systems are unable to catch up with the ever-increasing number of applications in 6G. Terahertz (THz) frequency band, which ranges from 0.1 to 10 THz, is the unexplored span of radio spectrum [64,83]. THz communications provide new communication paradigms with ultra-high bandwidth and ultra-low latency [64]. It is envisioned the data rate should be as high as Tbps to satisfy 6G applications' requirements of high throughput and low latency [83]. A novel approach to generate the THz frequency is discovered by Chevalier *et al.* [11]. They build a compact device that can use the nitrous oxide or laughing gas to produce a THz laser. The frequency of the laser can be tuned over a wide range at room temperature. Traditionally, the THz frequency band limits the widespread use of THz. THz transceiver design is regarded as the most critical factor in facilitating THz communications [83].

Terahertz Source Technique. Recent technology advancements in THz transceivers, such as photonics-based devices and electronics-based devices, overcome the THz gap, and enable some potential use cases in 6G [64]. The electronic technologies such as silicon-germanium BiCMOS, III-V semiconductor, and standard silicon CMOS related technologies (III and V represents the old numbering of the periodic system groups), have been vastly advanced, such that mixers and amplifiers can operate at around 1 THz frequency [32,83]. The photonic technologies are possiblely be used in the practical THz communication systems [32,83]. In addition, the combination of electronic-based transmitter and photonics-based receiver is feasible. Recent nanomaterials may help to develop novel devices that can used for THz communications [32].

Applications of Terahertz. Due to the high transmission RF, signals transmitted through THz frequency band suffer from a high pass loss. According to the Friis' law, the pass loss in free space increases quadratically with the operating frequency [31]. This feature limits the use of THz to short-distance transmission such as indoor communications [41]. Meanwhile, THz band can satisfy the requirement of ultra-high data rate; therefore, ultra-broadband applications, for example, virtual reality (VR) and wireless personal area networks can also exploit THz band to transmit signals [31]. THz technique can also be used in secure wireless communications. Since THz signals possess a narrow beam, it's very hard to wiretap the information for the eavesdropper when locating outside the transmitter beam [41].

2.5 Other Potential Technologies

In this section, we introduce some potential technologies that will be used in 6G, for example, visible light communications, blockchain-enabled network, satellite communication, full-duplex technology, holographic radio, and network in box.

Visible Light Communications. Visible light communication (VLC) is considered as one of the techniques that will be used in 6G communications. VLC contains transmitters and receivers. For short-range communication, either data-modulated white laser diodes or light-emitting diodes are used as transmitters, while photodetectors are utilized as receivers. Besides, VLC is considered as a complementary technology of RF communications since it can utilize an unlicensed spectrum for communication [26].

The laser diode (LD)-phosphor conversion lighting technology can provide better performance in efficiency and brightness, and larger illumination range compared with traditional lighting techniques [83]. Thus, it is considered as the most promising technology for 6G. The speed LD-based VLC system is possible to reach 100 Gbps, which meets the requirements of ultra-high data density (uHDD) services in 6G. Besides, the upcoming new light sources based on microLED will overcome the limitation of low speed in short range communication [9]. As massive parallelization of microLED arrays, spatial multiplexing techniques, CMOS driver arrays, and THz communications develop, VLC's data rate is expected to reach Tbps in the short range indoor scenario by the year of 2027 [9,64].

VLC can be used in indoor scenarios because it has limited coverage range, and it needs an illumination source and suffers from interference from other sources of light (e.g. the sun) [26]. Also, the space-air-terrestrial-sea integrated network can use VLC to provide better coverage [57]. In addition, traditional electromagnetic-wave signals cannot achieve high data transmission speed using laser beams in the free space and underwater, but VLC has ultra-high bandwidth and high data transmission speed [13]. Therefore, VLC are useful in cases where traditional RF communication is less active, for example, in-cabin internet service [73]. Furthermore, VLC is envisioned to be widely used in vehicle-to-vehicle communications, which depend on the head and tail lights of cars for communications [13,73,83]. Besides, VLC serves as a potential solution to build gigabit wireless networks underwater.

Blockchain-Based Network. Blockchain is a chain of blocks which constitute a distributed database. It is designed for cryptocurrencies (e.g. bitcoin) initially. However, nowadays, blockchain can do more than just in cryptocurrencies but run Turing-complete programs such as smart contracts in a distributed way (e.g. Ethereum) [77]. Blockchain provides a distributed and secure database for storing records of transactions, and each node includes the previous block's cryptographic hash, transaction data, and a time stamp [54,87]. Besides, blockchain-like mechanisms are expected to provide distributed authentication, control by leverging digital actions provided by the smart contracts [9]. Combining with federated learning, blockchain-based AI architectures are shifting AI processing to the edge [61]. Recently, a blockchain radio access network (B-RAN) has been proposed with prototype [43,47]. Thus, blockchain can help to form a secure and decentralized environment in 6G. Blockchain can provide a secure architecture for 6G wireless networks as shown in Fig. 6 [16].

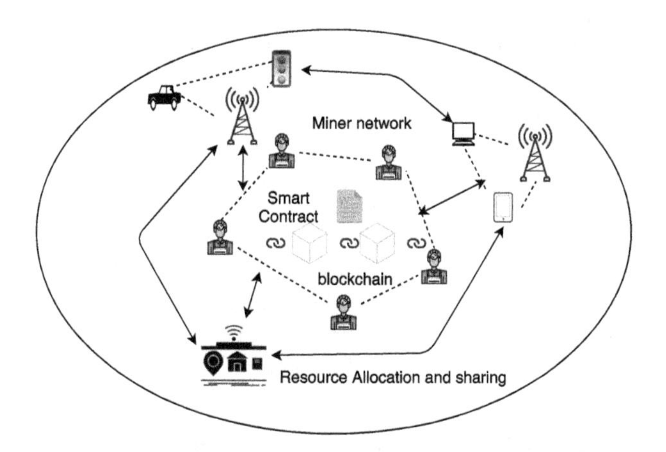

Fig. 6. Blockchain-based network.

Satellite Communication. Satellite communication means that earth stations communicate with each other via satellites. Satellite communication is a promising solution to the global coverage (i.e. space-air-ground-sea integrated network) in 6G era as shown in Fig. 7. By integrating with satellite communication, 6G can provide localization services, broadcast, Internet connectivity, and weather information to cellular users [60].

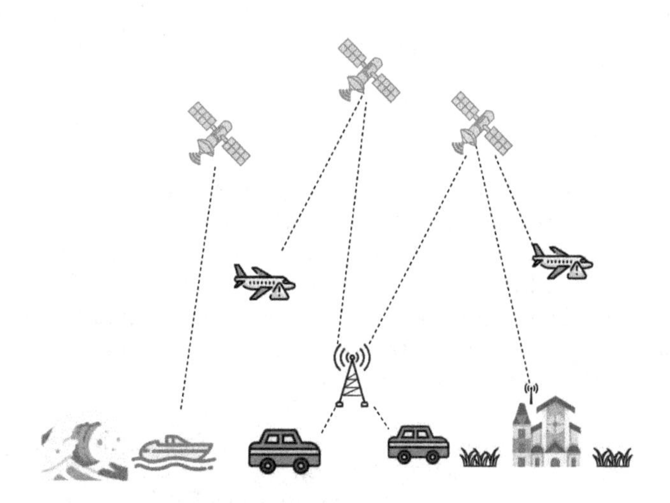

Fig. 7. Satellite communication.

Full-Duplex Technology. Full-duplex and in-band full-duplex (IBFD) technologies improve the communication efficiency by allowing devices to transmit and receive a signal in the same frequency band [62]. Full-duplex technologies are able to make the current efficiency of sharing spectrum double and increase the networks and communication systems' throughput. In the 4G/5G wireless systems, transmission and reception have to be done at the different frequency bands because half-duplex such as frequency division duplex (FDD) or time division duplex (TDD) do not support performing the transmission and reception at the same time slot. In addition, compared with half-duplex, full-duplex technology leverages self-interference cancellation technology to increase the utility of spectral resources, reduce the transmission delay, and improve the throughput between transceiver and receiver links. The hard part is to eliminate the self-interference, and now transmit signal generates over 100 dB higher noise than the receiver noise floor. Thus, the new scheduling algorithms should be designed. Currently, three types of self-interference cancellation techniques are proposed, including digital cancellation, analog cancellation, and passive suppression.

Holographic Radio. Unwanted signals are treated as harmful interference in traditional wireless networks, but they are considered as useful resources to develop holographic communication systems [89]. Computational holographic radio is one of the most promising interference-exploiting technologies [83].

Network in Box (NIB). More and more kinds of technologies will be embedded in the 6G wireless network, such as autonomous vehicles, factory automation, etc. To satisfy the real-time and reliable features of the network, the NIB technique is paid much attention to industrial automation because NIB offers a device that can provide seamless connectivity between different services. On the other hand, NIB covers the wireless environments, including terrestrial, air, and marine well agreed to the vision of Internet of Everything in 6G network.

3 Conclusion

In this paper, we highlight some promising technologies in 6G networks. We present a detailed explanation of artificial intelligence, intelligent reflecting surfaces, and THz communications. Furthermore, we briefly introduce some promising technologies, including blockchain, satellite communication, full-duplex, holographic radio, and THz communication. We envision that industry and academia will pay more attention to these technologies in 6G. The aforementioned technologies will contribute to 6G in the future.

Acknowledgments. The research is supported by 1) Nanyang Technological University (NTU) Startup Grant, 2) Alibaba-NTU Singapore Joint Research Institute (JRI), 3) Singapore Ministry of Education Academic Research Fund Tier 1 RG128/18, Tier 1 RG115/19, Tier 1 RG24/20, Tier 1 RT07/19, Tier 1 RT01/19, and Tier 2 MOE2019-T2-1-176, 4) NTU-WASP Joint Project, 5) Energy Research Institute @NTU (ERIAN),

6) Singapore NRF National Satellite of Excellence, Design Science and Technology for Secure Critical Infrastructure NSoE DeST-SCI2019-0012, 7) AI Singapore (AISG) 100 Experiments (100E) programme, and 8) NTU Project for Large Vertical Take-Off & Landing (VTOL) Research Platform.

References

1. Abu-Alhiga, R., Haas, H.: Subcarrier-index modulation OFDM. In: 2009 IEEE 20th International Symposium on Personal, Indoor and Mobile Radio Communications, pp. 177–181. IEEE (2009)
2. Al-Nahhal, I., Dobre, O.A., Basar, E., Ikki, S.: Low-cost uplink sparse code multiple access for spatial modulation. IEEE Trans. Veh. Technol. **68**(9), 9313–9317 (2019)
3. Andrews, J.G., Bai, T., Kulkarni, M.N., Alkhateeb, A., Gupta, A.K., Heath, R.W.: Modeling and analyzing millimeter wave cellular systems. IEEE Trans. Commun. **65**(1), 403–430 (2016)
4. Başar, E.: Multiple-input multiple-output OFDM with index modulation. IEEE Sig. Process. Lett. **22**(12), 2259–2263 (2015)
5. Basar, E.: On multiple-input multiple-output OFDM with index modulation for next generation wireless networks. IEEE Trans. Sig. Process. **64**(15), 3868–3878 (2016)
6. Basar, E., Altunbas, I.: Space-time channel modulation. IEEE Trans. Veh. Technol. **66**(8), 7609–7614 (2017)
7. Basar, E., Di Renzo, M., de Rosny, J., Debbah, M., Alouini, M.S., Zhang, R.: Wireless communications through reconfigurable intelligent surfaces. arXiv preprint arXiv:1906.09490 (2019)
8. Bouida, Z., El-Sallabi, H., Ghrayeb, A., Qaraqe, K.A.: Reconfigurable antenna-based space-shift keying (SSK) for MIMO rician channels. IEEE Trans. Wirel. Commun. **15**(1), 446–457 (2015)
9. Strinati, E.C., Barbarossa, S., Gonzalez-Jimenez, J., Cassiau, D.N., Dehos, C.: 6G: The next frontier. arXiv preprint arXiv:1901.03239 (2019)
10. Chen, S., Liang, Y.C., Sun, S., Kang, S., Cheng, W., Peng, M.: Vision, requirements, and technology trend of 6G: how to tackle the challenges of system coverage, capacity, user data-rate and movement speed. IEEE Wirel Commun. **27**(2), 218–228 (2020)
11. Chevalier, P., Armizhan, A., Wang, F., Piccardo, M., Johnson, S.G., Capasso, F., Everitt, H.O.: Widely tunable compact terahertz gas lasers. Science **366**(6467), 856–860 (2019)
12. Chowdhury, M.Z., Shahjalal, M., Ahmed, S., Jang, Y.M.: 6G wireless communication systems: Applications, requirements, technologies, challenges, and research directions. arXiv preprint arXiv:1909.11315 (2019)
13. Chowdhury, M.Z., Shahjalal, M., Hasan, M., Jang, Y.M.: The role of optical wireless communication technologies in 5G/6G and IoT solutions: prospects, directions, and challenges. Appl. Sci. **9**(20), 4367 (2019)
14. Corre, Y., Gougeon, G., Doré, J.B., Bicaïs, S., Miscopein, B., Faussurier, E., Saad, M., Palicot, J., Bader, F.: Sub-thz spectrum as enabler for 6G wireless communications up to 1 tbit/s (2019)
15. Cousik, T., Shafin, R., Zhou, Z., Kleine, K., Reed, J., Liu, L.: CogRF: A new frontier for machine learning and artificial intelligence for 6G RF systems. arXiv preprint arXiv:1909.06862 (2019)

16. Dai, Y., Du, X., Maharjan, S., Chen, Z., He, Q., Zhang, Y.: Blockchain and deep reinforcement learning empowered intelligent 5G beyond. IEEE Netw. **33**(3), 10–17 (2019)
17. Dang, S., Amin, O., Shihada, B., Alouini, M.S.: What should 6G be? Nat. Electron. **3**(1), 20–29 (2020)
18. David, K., Elmirghani, J., Haas, H., You, X.H.: Defining 6G: challenges and opportunities [from the guest editors]. IEEE Veh. Technol. Mag. **14**(3), 14–16 (2019)
19. De Carvalho, E., Ali, A., Amiri, A., Angjelichinoski, M., Heath Jr, R.W.: Nonstationarities in extra-large scale massive mimo. arXiv preprint arXiv:1903.03085 (2019)
20. Di Renzo, M., Debbah, M., Phan-Huy, D.T., Zappone, A., Alouini, M.S., Yuen, C., Sciancalepore, V., Alexandropoulos, G.C., Hoydis, J., Gacanin, H., et al.: Smart radio environments empowered by AI reconfigurable meta-surfaces: An idea whose time has come. arXiv preprint arXiv:1903.08925 (2019)
21. Elmeadawy, S., Shubair, R.M.: Enabling technologies for 6G future wireless communications: Opportunities and challenges. arXiv preprint arXiv:2002.06068 (2020)
22. Elsayed, M., Erol-Kantarci, M.: AI-enabled future wireless networks: challenges, opportunities, and open issues. IEEE Veh. Technol. Mag. **14**(3), 70–77 (2019)
23. Faisal, A., Sarieddeen, H., Dahrouj, H., Al-Naffouri, T.Y., Alouini, M.S.: Ultra-massive mimo systems at terahertz bands: Prospects and challenges. arXiv preprint arXiv:1902.11090 (2019)
24. Fan, R., Yu, Y.J., Guan, Y.L.: Generalization of orthogonal frequency division multiplexing with index modulation. IEEE Trans. Wirel. Commun. **14**(10), 5350–5359 (2015)
25. Gacanin, H.: Autonomous wireless systems with artificial intelligence: a knowledge management perspective. IEEE Veh. Technol. Mag. **14**(3), 51–59 (2019)
26. Giordani, M., Polese, M., Mezzavilla, M., Rangan, S., Zorzi, M.: Towards 6G networks: Use cases and technologies. arXiv preprint arXiv:1903.12216 (2019)
27. Giordani, M., Zorzi, M.: Satellite communication at millimeter waves: a key enabler of the 6G era. In: 2020 International Conference on Computing, Networking and Communications (ICNC), pp. 383–388. IEEE (2020)
28. Gong, S., Lu, X., Hoang, D.T., Niyato, D., Shu, L., Kim, D.I., Liang, Y.C.: Towards smart radio environment for wireless communications via intelligent reflecting surfaces: a comprehensive survey. arXiv preprint arXiv:1912.07794 (2019)
29. Gui, G., Liu, M., Tang, F., Kato, N., Adachi, F.: 6G: Opening new horizons for integration of comfort, security and intelligence. IEEE Wireless Communications (2020)
30. Guo, W.: Explainable artificial intelligence (XAI) for 6G: Improving trust between human and machine. arXiv preprint arXiv:1911.04542 (2019)
31. Han, C., Chen, Y.I.: Propagation modeling for wireless communications in the terahertz band. IEEE Commun. Mag. **56**(6), 96–101 (2018)
32. Han, C., Wu, Y., Chen, Z., Wang, X.: Terahertz communications (teracom): Challenges and impact on 6G wireless systems. arXiv preprint arXiv:1912.06040 (2019)
33. He, H., Jin, S., Wen, C.K., Gao, F., Li, G.Y., Xu, Z.: Model-driven deep learning for physical layer communications. IEEE Wirel. Commun. (2019)
34. Ho, T.M., Tran, T.D., Nguyen, T.T., Kazmi, S.M., Le, L.B., Hong, C.S., Hanzo, L.: Next-generation wireless solutions for the smart factory, smart vehicles, the smart grid and smart cities. arXiv preprint arXiv:1907.10102 (2019)

35. Hu, S., Chitti, K., Rusek, F., Edfors, O.: User assignment with distributed large intelligent surface (lis) systems. In: 2018 IEEE 29th Annual International Symposium on Personal, Indoor and Mobile Radio Communications (PIMRC), pp. 1–6. IEEE (2018)
36. Hu, S., Rusek, F., Edfors, O.: Cramér-rao lower bounds for positioning with large intelligent surfaces. In: 2017 IEEE 86th Vehicular Technology Conference (VTC-Fall), pp. 1–6. IEEE (2017)
37. Hu, S., Rusek, F., Edfors, O.: The potential of using large antenna arrays on intelligent surfaces. In: 2017 IEEE 85th Vehicular Technology Conference (VTC Spring), pp. 1–6. IEEE (2017)
38. Hu, S., Rusek, F., Edfors, O.: Beyond massive mimo: the potential of data transmission with large intelligent surfaces. IEEE Trans. Sig. Process. **66**(10), 2746–2758 (2018)
39. Jung, M., Saad, W., Kong, G.: Performance analysis of large intelligent surfaces (liss): Uplink spectral efficiency and pilot training. arXiv preprint arXiv:1904.00453 (2019)
40. Kato, N., Mao, B., Tang, F., Kawamoto, Y., Liu, J.: Ten challenges in advancing machine learning technologies toward 6G. IEEE Wirel. Commun. (2020)
41. Khalid, N., Akan, O.B.: Wideband THz communication channel measurements for 5G indoor wireless networks. In: 2016 IEEE International Conference on Communications (ICC), pp. 1–6. IEEE (2016)
42. Konečný, J., McMahan, H.B., Yu, F.X., Richtárik, P., Suresh, A.T., Bacon, D.: Federated learning: Strategies for improving communication efficiency. arXiv preprint arXiv:1610.05492 (2016)
43. Le, Y., Ling, X., Wang, J., Ding, Z.: Prototype design and test of blockchain radio access network. In: 2019 IEEE International Conference on Communications Workshops (ICC Workshops), pp. 1–6. IEEE (2019)
44. Letaief, K.B., Chen, W., Shi, Y., Zhang, J., Zhang, Y.J.A.: The roadmap to 6G-AI empowered wireless networks. arXiv preprint arXiv:1904.11686 (2019)
45. Liang, Y.C., Long, R., Zhang, Q., Chen, J., Cheng, H.V., Guo, H.: Large intelligent surface/antennas (lisa): Making reflective radios smart. arXiv preprint arXiv:1906.06578 (2019)
46. Liaskos, C., Nie, S., Tsioliaridou, A., Pitsillides, A., Ioannidis, S., Akyildiz, I.: A new wireless communication paradigm through software-controlled metasurfaces. IEEE Commun. Mag **56**(9), 162–169 (2018)
47. Ling, X., Wang, J., Bouchoucha, T., Levy, B.C., Ding, Z.: Blockchain radio access network (B-RAN): Towards decentralized secure radio access paradigm. IEEE Access, **7**, 9714–9723 (2019)
48. Liu, C., Yang, L.L., Wang, W.: Transmitter-precoding-aided spatial modulation achieving both transmit and receive diversity. IEEE Trans. Veh. Technol. **67**(2), 1375–1388 (2017)
49. Lovén, L., Leppänen, T., Peltonen, E., Partala, J., Harjula, E., Porambage, P., Ylianttila, M., Riekki. J.: EdgeAI: a vision for distributed, edge-native artificial intelligence in future 6G networks. In: The 1st 6G Wireless Summit, pp. 1–2 (2019)
50. Mao, Q., Hu, F., Hao, Q.: Deep learning for intelligent wireless networks: a comprehensive survey. IEEE Commun. Surv. Tutorials **20**(4), 2595–2621 (2018)
51. Matti, L., Kari, L.: Key drivers and research challenges for 6G ubiquitous wireless intelligence. 6G Flagship, Oulu, Finland, White Paper (2019)
52. Mollah, M.B., Azad, M.A.K., Vasilakos, A.: Secure data sharing and searching at the edge of cloud-assisted Internet of Things. IEEE Cloud Comput. **4**(1), 34–42 (2017)

53. Mollah, M.B., Zeadally, S., Azad, M.A.K.: Emerging wireless technologies for Internet of Things applications: opportunities and challenges. In: Encyclopedia of Wireless Networks, pp. 1–11. Springer International Publishing Cham (2019)
54. Mollah, M.B., Zhao, J., Niyato, D., Lam, K.Y., Zhang, X., Ghias, A.M.Y.M., Koh, L.H., Yang, L.: Blockchain for future smart grid: a comprehensive survey. IEEE Internet Things J. **8**(1), 18–43 (2020)
55. Nadeem, Q.U.A., Kammoun, A., Chaaban, A., Debbah, M., Alouini, M.S.: Large intelligent surface assisted mimo communications. arXiv preprint arXiv:1903.08127 (2019)
56. Naresh, Y., Chockalingam, A.: On media-based modulation using RF mirrors. IEEE Trans. Veh. Technol. **66**(6), 4967–4983 (2016)
57. Nawaz, S.J., Sharma, S.K., Wyne, S., Patwary, M.N., Asaduzzaman, M.: Quantum machine learning for 6G communication networks: state-of-the-art and vision for the future. IEEE Access **7**, 46317–46350 (2019)
58. Nayak, S., Patgiri, R.: 6G: Envisioning the key issues and challenges. arXiv preprint arXiv:2004.04024 (2020)
59. Özdogan, O., Björnson, E., Larsson, E.G.: Intelligent reflecting surfaces: Physics, propagation, and pathloss modeling. arXiv preprint arXiv:1911.03359 (2019)
60. Piran, J., Suh, D.Y.: Learning-driven wireless communications, towards 6G. In: 2019 International Conference on Computing, Electronics and Communications Engineering (ICCECE), pp. 219–224. IEEE (2019)
61. Porambage, P., Kumar, T., Liyanage, M., Lauri Lovén, J.P., Ylianttila, M., Seppänen, T.: Sec-EdgeAI: AI for edge security vs security for edge AI (2019)
62. Rajatheva, N., Atzeni, I., Bjornson, E., Bourdoux, A., Buzzi, S., Dore, J.B., Erkucuk, S., Fuentes, M., Guan, K., Hu, Y., Huang, X., Hulkkonen, J., Jornet, J.M., Katz, M., Nilsson, R., Panayirci, E., Rabie, K., Rajapaksha, N., Salehi, M.J., Sarieddeen, H., Svensson, T., Tervo, O., Tolli, A., Wu, Q., Xu, W.: White paper on broadband connectivity in 6G. arXiv preprint arXiv:2004.14247 (2020)
63. Saad, W., Bennis, M., Chen, M.: A vision of 6G wireless systems: Applications, trends, technologies, and open research problems. arXiv preprint arXiv:1902.10265 (2019)
64. Sarieddeen, H., Saeed, N., Al-Naffouri, T.Y., Alouini, M.S.: Next generation terahertz communications: A rendezvous of sensing, imaging and localization. arXiv preprint arXiv:1909.10462 (2019)
65. Seifi, E., Atamanesh, M., Khandani, A.K.: Media-based mimo: Outperforming known limits in wireless. In: 2016 IEEE International Conference on Communications (ICC), pp. 1–7. IEEE (2016)
66. Shafin, R., Liu, L., Chandrasekhar, V., Chen, H., Reed, J., Zhang, J.: Artificial intelligence-enabled cellular networks: A critical path to beyond-5G and 6G. arXiv preprint arXiv:1907.07862 (2019)
67. Shafin, R., Liu, L., Chandrasekhar, V., Chen, H., Reed, J., Zhang, J.C.: Artificial intelligence-enabled cellular networks: a critical path to beyond-5G and 6G. IEEE Wirel. Commun. **27**(2), 212–217 (2020)
68. Shamasundar, B., Jacob, S., Chockalingam, A.: Time-indexed media-based modulation. In: 2017 IEEE 85th Vehicular Technology Conference (VTC Spring), pp. 1–5. IEEE (2017)
69. Stoica, R.A., de Abreu, G.T.F.: 6G: the wireless communications network for collaborative and AI applications. arXiv preprint arXiv:1904.03413 (2019)
70. Sugiura, S., Chen, S., Hanzo, L.: Coherent and differential space-time shift keying: a dispersion matrix approach. IEEE Trans. Commun. **58**(11), 3219–3230 (2010)

71. Sugiura, S., Chen, S., Hanzo, L.: Generalized space-time shift keying designed for flexible diversity-, multiplexing-and complexity-tradeoffs. IEEE Trans. Wirel. Commun. **10**(4), 1144–1153 (2011)
72. Tan, J., Dai, L.: THz precoding for 6G: Applications, challenges, solutions, and opportunities. arXiv preprint arXiv:2005.10752 (2020)
73. Tariq, F., Khandaker, M., Wong, K.K., Imran, M., Bennis, M., Debbah, M.: A speculative study on 6G. arXiv preprint arXiv:1902.06700 (2019)
74. Wang, Q., Wang, Z., Chen, S., Hanzo, L.: Enhancing the decoding performance of optical wireless communication systems using receiver-side predistortion. Opt. Express **21**(25), 30295–30305 (2013)
75. Wang, X., Wang, J., He, L., Tang, Z., Song, J.: On the achievable spectral efficiency of spatial modulation aided downlink non-orthogonal multiple access. IEEE Commun. Lett. **21**(9), 1937–1940 (2017)
76. Wills, J.: 5G technology: Which country will be the first to adapt? 23 April 2020. https://www.investopedia.com/articles/markets-economy/090916/5g-technology-which-country-will-be-first-adapt.asp
77. Wood, G.: Ethereum: a secure decentralised generalised transaction ledger. Ethereum Proj. Yellow Pap. **151**, 1–32 (2014)
78. Wu, Q., Zhang, R.: Towards smart and reconfigurable environment: intelligent reflecting surface aided wireless network. IEEE Commun. Mag. **58**(1), 106–112 (2019)
79. Xiao, L., Yang, P., Xiao, Y., Fan, S., Di Renzo, M., Xiang, W., Li, S.: Efficient compressive sensing detectors for generalized spatial modulation systems. IEEE Trans. Veh. Technol. **66**(2), 1284–1298 (2016)
80. Xiao, M., Mumtaz, S., Huang, Y., Dai, L., Li, Y., Matthaiou, M., Karagiannidis, G.K., Björnson, E., Yang, K., Chih-Lin, I., Ghosh, A.: Millimeter wave communications for future mobile networks. IEEE J. Select. Areas Commun. **35**(9), 1909–1935 (2017)
81. Yang, Y.: Spatial modulation exploited in non-reciprocal two-way relay channels: efficient protocols and capacity analysis. IEEE Trans. Wirel. Commun. **64**(7), 2821–2834 (2016)
82. You, X., Wang, C., Huang, J., Gao, X., Zhang, Z., Wang, M., Huang, Y., Zhang, C., Jiang, Wang, Y.J., Zhu, M., Sheng, B., Wang, D., Pan, Z., Zhu, P., Yang, Y., Liu, Z., Zhang, P., Tao, X., Li, S., Chen, Z., Ma, X., Chihlin, I., Han, S., Li, K., Pan, C., Zheng, Z., Hanzo, L., Shen, X., Guo, Y.J., Ding, Z., Haas, H., Tong, W., Zhu, P., Yang, G., Wang, J., Larsson, E.G., Ngo, H., Hong, W., Wang, H., Hou, D., Chen, J., Zhangcheng Hao, C., Li, G., Tafazolli, R., Gao, Y., Poor, V., Fettweis, G., Liang, Y.: Towards 6G wireless communication networks: Vision, enabling technologies, and new paradigm shifts. SCIENCE CHINA Information Sciences
83. Yuan, Y., Zhao, Y., Zong, B., Parolari, S.: Potential key technologies for 6G mobile communications. arXiv preprint arXiv:1910.00730 (2019)
84. Zappone, A., Di Renzo, M., Debbah, M., Lam, T.T., Qian, X.: Model-aided wireless artificial intelligence: embedding expert knowledge in deep neural networks for wireless system optimization. IEEE Veh. Technol. Mag. **14**(3), 60–69 (2019)
85. Zhang, L., Liang, Y.C., Niyato, D.: 6G visions: mobile ultra-broadband, super Internet-of-Things, and artificial intelligence. China Commun. **16**(8), 1–14 (2019)
86. Zhang, R., Yang, L.L., Hanzo, L.: Performance analysis of non-linear generalized pre-coding aided spatial modulation. IEEE Trans. Wirel. Commun. **15**(10), 6731–6741 (2016)

87. Zhang, Z., Xiao, Y., Ma, Z., Xiao, M., Ding, Z., Lei, X., Karagiannidis, G.K., Fan, P.: 6G wireless networks: vision, requirements, architecture, and key technologies. IEEE Veh. Technol. Mag. **14**(3), 28–41 (2019)
88. Zhu, L., Xiao, Z., Xia, X.G., Wu, D.O.: Millimeter-wave communications with non-orthogonal multiple access for B5G/6G. IEEE Access **7**, 116123–116132 (2019)
89. Zong, B., Fan, C., Wang, X., Duan, X., Wang, B., Wang, J.: 6G technologies: Key drivers, core requirements, system architectures, and enabling technologies. IEEE Veh. Technol. Mag. **14**(3), 18–27 (2019)

New Paradigm of Hydrometeorological Support for Consumers

Evgenii D. Viazilov(✉)

RIHMI-WDC, 249035 Obninsk, Kaluga Region, Russia
vjaz@meteo.ru

Abstract. Shown are the existing barriers to increase the level of adaptation of the population and enterprises to climate change and related disasters. The perspective directions of development of hydrometeorological support of consumers presented, they connected with the introduction of a new paradigm for processing, communicating and using information about the state of the environment. New ways of processing data to serve consumers briefly reviewed. All processing of observed, forecast and climate data should be basing on the integration of heterogeneous and distributed data. To increase the speed of communicating information about disasters and clarifying the areas of their manifestation, it is proposing to organize automatic detection of disasters based on local threshold values of indicators of disasters. Communication of information about disasters to consumers should also be automating, including the delivery of this information to mobile Internet devices, to information systems of enterprises and even to their business processes. The obtained information on disasters should be using to predict the possible impacts of disasters on enterprises and the population. Based on the information about the impacts, it is proposing to issue recommendations to the heads of enterprises and the public for decision-making. The necessity of assessing the possible damage from the impacts of disasters and calculating the cost of preventive actions before the onset of disasters indicated.

Keywords: Hydrometeorological data · Disasters · Integration · Applied processing · Communication · Decision making · Data exchange

1 Introduction

The main barriers for adaptation of the population and enterprises to natural disasters are the inadequacy of assessing the situation using digital values of hydrometeorological parameters, the lack of systematized and formalized information on the possible impacts of the disasters, and recommendations for decision-making. There are cases of untimely communication of information about a possible manifestation of disasters to government bodies, heads of enterprises and the population, which are in the danger zone. Governments bodies are busy with their own business and do not always respond in time to the predictions of the disasters. Therefore, here are necessary to organize a compulsory notification to familiarize managers with the current situation using modern means - using mobile Internet devices, to provide a forecast of disasters impacts on enterprises

K. Arai (Ed.): FICC 2021, AISC 1363, pp. 171–181, 2021.
https://doi.org/10.1007/978-3-030-73100-7_13

and the population, recommendations for decision-making. Enterprise managers need to assess the possible damage before disasters beginning; otherwise, they will not take preventive action.

That is, in fact, it is necessary to introduce a new paradigm of the hydrometeorological support of the population and heads of enterprises. New paradigm is creating a radical change in the hydrometeorological support (HMS) of enterprises, development of new business models for the use of hydrometeorological information in the business processes of enterprises. New business models are including:

- fast search for the required data through the use of integrated, distributed and heterogeneous data from various sources;
- increasing the level of automation of the processes for identifying disasters;
- transferring the necessary data to consumers in accordance with their requirements;
- loading data into external information systems of enterprises;
- visualization of warnings about disasters, operational and prognostic data using information panels for presenting of hydrometeorological situations.

All these directions are developing to one degree in Russia and other countries. Adaptation to climate changes is writing about for many years. Hundreds of books have been publishing in this field, for example [1]. Unfortunately, this is mostly scattered and unformalized information. The idea of creating a version of the information system "MeteoMonitor" to warn about disasters for St. Petersburg expressed back in 2011 [2]. An interesting implementations of a decision support system (DSS) are available for aviation [3], marine operations [4]. Abroad and in the Hydrometeorological Center of Russia [5], the display of the danger level, using the "Traffic light" has used for a long time. Here the general criteria of disasters indicators are mainly used. To reduce the risk, the World Meteorological Organization proposes to use a climate services platform [6].

The objectives of the preparation of the article are following:

- comprehensive presentation of the results of the author's previous researches, which allows us to present the end-to-end automated technology of consumers "from observation to decision-making";
- development of a new paradigm of HMS.

2 Data Integration

In order for distributed, heterogeneous data to be available for use from one or more sources, through one interface, so that data can be easily delivering to any consumer automatically, they must be integrating. The technological scheme for the integration, processing and use of integrated data shown in Fig. 1. This scheme based on the experience of the Unified State System of Information about the World Ocean (ESIMO) operation (http://esimo.ru). The main decisions on the creation of ESIMO are next [7]:

- use of ISO 19115 standards for metadata, Open Geospatial Consortium for spatial data, ISO 19100 series standards for metadata and data;

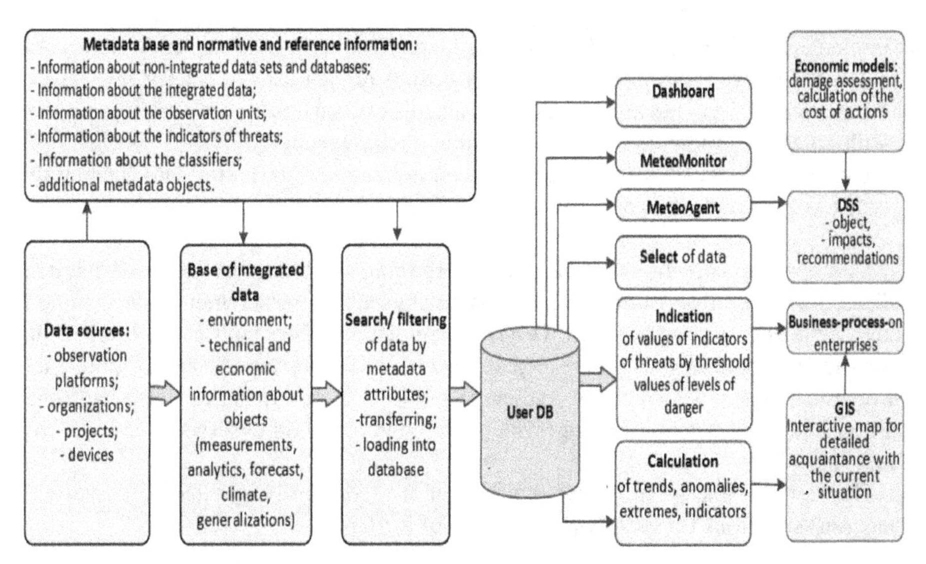

Fig. 1. Technological scheme of integration, applied processing and use of hydrometeorological data.

- remote input of metadata;
- maintaining a single dictionary of parameters;
- mapping of classifies, which allows using any coding notation;
- use of cross-platform tools - J2EE, XML, PostgreSQL database server, JBoss Portal application server, GIS SERVER, GIS visualizer OpenLayers;
- data classification by the level of data processing (observation, diagnostic, forecast, aggregation), presentation form (point, profile, grid, object files);
- unification of attributes names using of parameters vocabulary.

These approaches provide a solution to the problems of managing heterogeneous information resources through uniform access to all resources and the use of search attributes for different forms of information presentation (text, digital, graphic, and spatial), the use of several access methods depending from of the data source type.

Integration of heterogeneous data makes it possible to organize a unified technological process of automatic identification of disasters, bringing information about them to the consumer or another information system.

3 Processing of Integrated Data

3.1 Automatic Detection of Disasters

Enterprises need hydrometeorological information when the values of disasters indicators exceed the threshold values for specific facilities, types of activities, indicating the dangerous level. This requires:

- implement technology for integrated and targeted information services through automatic delivery of information about disasters to the heads of enterprises;
- develop means of assessing the state of indicators of disasters for enterprises, activities with access to a more detailed consideration of situations;
- create a database of threshold values of disasters indicators, taking into account the type of object and types of activity using al"traffic light".

The input information for exceeding determining the threshold values of disasters indicators are the values of the measured parameters at the measurement points (station, buoy), or nodes of a regular grid. Disasters are distinguished both at the level of the observation point (the excess of the indicator was registered in one observation point) and the region (disasters were registered at several points in space). In this case, Roshydromet warnings about disasters are using, which identified based on observed and predictive data.

The result of the work of this application is a constantly updated database with dangerous situations for each object and type of activity.

3.2 Applications for Transferring of Information About Disasters

When providing HMS to consumers, it is necessary to use the official "Storm warnings and messages" issued by the Hydrometeorological Center of Russia, the Arctic and Antarctic Research Institute, regional offices of Roshydromet. These warnings compiled by observers and forecasters, transmitted over the Global Telecommunication System (GTS) and uploaded to an integrated database. In the first variant, information about disasters detected by observers at the hydrometeorological station and transmitted in the WAREP code. These messages can be using to identify the objects and impacts on them. The second variant is when forecasters, based on the analysis of the current situation on forecasts maps and other materials, give forecast disasters and transmit it to interested enterprises.

There should be possible to reconfigure the composition of monitored disasters and threshold values of indicators for a particular enterprise. When disasters are identifying, the cause of its occurrence is established. Knowledge of the causes of disasters allows us to predict its development and the emergence of new phenomena, the causes of which may be the first phenomenon.

The organization of automatic communication to enterprises heads of information about disasters is designing to increase the awareness of enterprise managers, and quick acquaintance with the current situation. It are using:

- automatic selection of indicators values in the form of "traffic light" for different levels of dangerous;
- selection and transmission of received storm alerts;
- transmission of information about disasters;
- visualization the state of indicators of the situation for a specific object;
- spatial and temporal presentation of hydrometeorological data using an interactive map).

Consumers do need no data, but information about disasters or abnormal deviations of current and forecast values of indicators from climatic values for specific enterprises and activities. Information deliver not at the initiative of the consumer, but by automatic to an Internet device of consumers.

The generated information about disasters must meet the following requirements: the minimum amount of information given to the head; visual presentation of data in the form of an information panel, the presence of a link to a more detailed presentation of information; displaying in a form that is directly usable without the need for additional calculations or data transformations.

3.3 "Dashboard" and "MeteoMonitor" Applications

In addition to an SMS message, the enterprise manager should see the status of indicators about the current situation. For this, an application for visualizing indicators in the form of a dashboard or information panel should be using.

The information panel is configuring by the consumer for a specific object, hydrometeorological station or geographic area, threshold values of indicators. Each of the types of economic activity has its own restrictions on activities, for example, those associated with strong winds, which may not coincide with the general gradations of strong winds established in the existing regulatory documents Roshydromet [8, 9].

The application shows changes in the values of the observed and prognostic parameters marked on the icons of the devices, indicating the level of their danger. Next to each parameter, there are graphs of their changes - the last 5–10 observations, as well as anomalies and trends. When the data in the sources is updating, the data on the dashboard is automatically keep up to date too. In such an application, information is displaying in a more compact form on the screen. Moreover, one glance at this form of display is enough to understand the current situation.

Scheme reflection hydrometeorological conditions with an indication of threshold values on the information panel is given in Fig. 2.

If necessary, the manager can obtain detailed information on an interactive map using the "MeteoMonitor" application on the ESIMO portal the by region, displaying the dynamics of changes in indicators in the form of a map with layers for various parameters [7]. The program interface of application should provide the following forms of information presentation:

- lists of messages about disasters;
- maps of the distribution of disasters in spatial;
- graphs of changes in indicators over time;
- tables of values of environmental parameters at specific observation points or nodes of the regular grid closest to the settlement;
- results of identification of values showing the state of indicators for individual objects and types of activity;
- means of signaling about danger by sound and color.

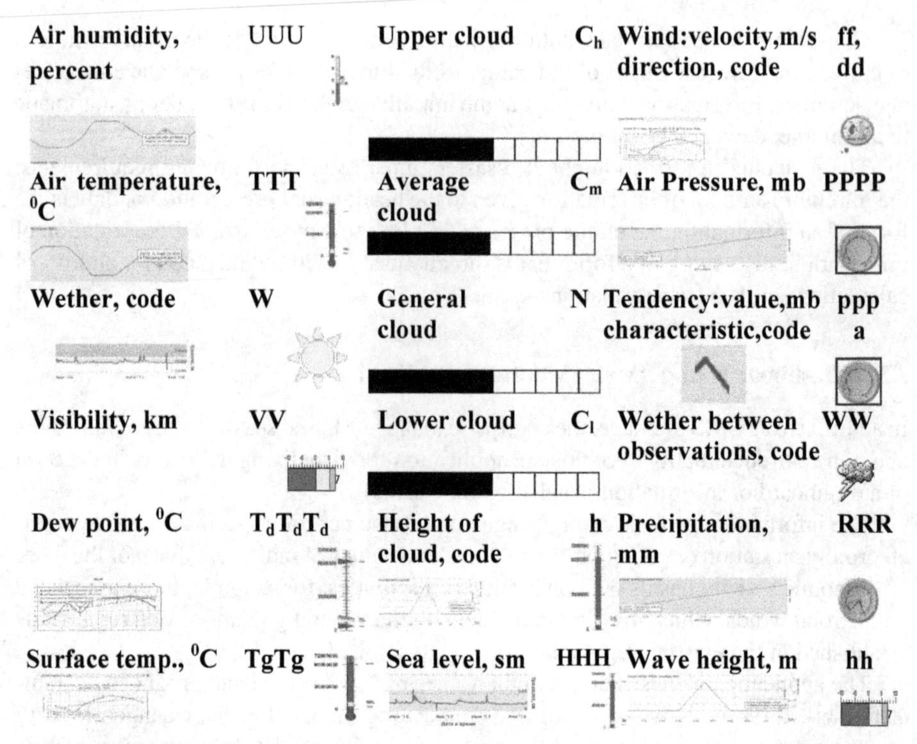

Fig. 2. Scheme reflection hydrometeorological conditions with an indication of threshold values on the dashboard.

3.4 Transferring of Data to External Information Systems of Enterprises

Currently, it is necessary that data on disasters be used in automated business processes of enterprises for their constant accounting, for example, wind speed and direction must be taken into account when coal unloading in marine port. Data selected in point of observations or in the grid point for forecast information. At the same time, depending on the business process, data can be suppling for a point, area or trajectory. If the enterprise is located at a certain point, then the observed data is suppling by closest station to the object. If the enterprise is a dynamic object (vessel), then observations are selecting on trajectory of the vessel.

Means of transferring information to consumers include subscribing to necessary data and posting it on remote ftp servers or sending data by e-mail in form news feeds for documents, news, messages, telegrams received on the GTS.

To develop business processes that are solving using hydrometeorological data, it is necessary:

- to clarify the indicators that affect the performance of business processes;
- to detect the dangerous situations on threshold values of indicators on a regular basis;
- to communication information about the disasters to the object and load the values of the indicators into the database;

- for the most efficient data delivery, the following capabilities should be implementing;
- transferring data upon update;
- instant receipt of notifications for any disruptions in delivery for their immediate elimination, avoiding the impact of the consequences on the business processes of consumers;
- monitoring all data deliveries to assess their compliance with the settings.

3.5 Decision Support

DSS designed to visualize information about disasters based on the received SMS message with the application address to obtain a list of possible impacts and recommendations [10–12], as well as links to applications for damage assessment and calculation of the cost of preventive actions.

The main idea of creating a DSS is as follows. Knowing hydrometeorological conditions it is possible to determine in advance the list of environmental impacts on enterprises and the population. Knowing the impacts, you can determine the recommendations for decision-making.

Decision support is most easily implemented based on a knowledge base in the form of rules "if.., that.., else…". With a large number of rules, it takes a lot of time to test them.

For hydrometeorological situations, in which decision support is required, information contained in the regulations can be dividing into information about disasters and information about possible impacts and recommendations for making decisions at various objects. The detection of disasters based on forecasted data or in the stream of operational data is a data processing process that relies on the threshold values of disasters indicators (precipitation, wave height, water level, etc.). It is proposed to organize the knowledge in the form of a database of threshold values of the indicators of disasters by dangerous levels, based on which they are identified. The knowledge represented as two databases:

- information about the threshold values of the indicators of the disasters;
- information on impacts and recommendations for various objects.

Information on the threshold values of the indicators of the disasters include the name, threshold values for dangerous levels, enterprises name and types of activity, geographical area, season of the year, name of the climatic zone. In different regions of the country, enterprises and the population are prepared in different ways for the same disasters, for example, in areas of constant exposure to strong winds, frosts, the population has already adapted to survive in such conditions. The information on impacts and recommendations include the name of the disaster, the dangerous level, the type of data (observed, predictive, climatic), information about the impacts for various objects and activities, recommendations for different management levels.

An example of issuing information on impacts and recommendations (fragment for disaster "Waves in the sea") is presented below.

Impacts and Recommendation for Disaster "Waves in the Sea"

Waves in the sea are vibrational movements of a fluid in a certain layer. Waves are wind, swell, tidal, seiche.

Causes: Waves generated by natural causes in the atmosphere or ocean floor. Wind is one of the main reasons for generating short sea waves during storms, tropical cyclones, hurricanes, typhoons. Swell waves are wind waves escaping from the area of the wind. Tidal waves are causing by the mutual attraction of the Earth, Moon and Sun. A sharp change in atmospheric pressure leads to the formation of so-called standing waves - seiches. Huge waves up to 40–50 m in height arise suddenly on the sea surface and are the result of a random addition of the amplitudes of ordinary waves.

Object: ship.
Information type: observed data.
Indicators: wave height in the open sea> = 6 m, ocean> = 8 m.
Danger level: red.
Impacts:
Operation of the vessel:

There is a threat of a significant loss of stability during the waves, which can cause capsizing and death of the vessel.
A roll appears at heading angles of wind of 30 ° -120 °.
Speed of movement decreases.
Controllability of the vessel worsens.
Especially dangerous is the case of resonant rolling, i.e. coincidence of the rolling period with the period of natural oscillations of the vessel.
The cargo is shifting in the hold, on the deck, as a result, the list of the vessel increases.
There is a threat of collision of all vehicles at sea.

Recommendations:
Captain of a vessel on the open seas:

Remove water regularly from wells and bilges, using a drainage system.
Reduce speed, or change course, or both, to prevent the boat from being hit by violent wave shocks.
Choose the optimal course in unfavorable meteorological conditions, which depends on the structure of the vessel, its size, cargo and the nature of the load.

Passengers:

Do not panic.
Follow all instructions of the captain.
Put on life jackets on yourself and your loved ones.
Take with you documents, matches, medicines, money, wrapping them in a plastic bag and putting them under your underwear.

To develop a complete DSS, it is necessary to solve the following tasks:

- collect, formalize, enter into the database and store information on the impact of various disasters on enterprises and the population;
- formalize recommendations for decision-making, taking into account possible impacts for various objects, types of activities, danger levels, decision-making levels;
- develop software tools for automatic detection, communication and visualization of information about disasters; and
- implement means of input, editing and visualization of information about impacts and recommendations, depending on the current situation.

3.6 Damage Assessments and Calculation of the Cost of Preventive Actions

Material damage is assessing in natural and monetary units. Material damage is reported in physical units, for example, the number of destroyed bridges, houses, industrial buildings, the number of damaged buildings, the number of deaths; the number of sick people; number of wounded, and etc.

Potential damage from disasters may contain material damage associated with the loss of property of enterprises, cultural values, and public health, of complete or partial destruction of various properties of objects. In addition, damage to the environment is possible. There may be costs associated with claims by injured parties, with fines for the consequences of impacts (chemical spill), with the cost of restoration work and the elimination of the consequences of disasters. In addition, losses from disasters are determining by the termination of certain types of production work, by loss of cargo or deterioration in the quality of transported products. It also occurs a decrease in the book value of fixed assets of objects (recreation, tourism, production, and housing stock), a decrease in the output of products (materials), the downtime of an object.

After the apparition need the expenses of salvation, emergency recovery actions; treatment of the wounded; on salaries of employees during emergency recovery actions. These costs too include the cost of renovating buildings; payments to the population from the social insurance fund for the period of temporary and permanent disability; lost profits.

The assessment of economic consequences (E) is presented in (1):

$$E = \sum\nolimits_{A}^{N} P(A) * E_{A,} \tag{1}$$

Where A - a disaster, P (A) - the probability of a disaster, E_A - economic consequences.

To make an optimal decision, in addition to the possible damage, it is necessary to know the cost of preventive actions. Calculations of the costs of preventive measures consist of the salaries of workers, rent of equipment for the removal of goods from the disaster zone, training of personnel, the acquisition of additional equipment, the construction of temporary enclosing dams, embankments in case of floods, evacuation of people, forced movement of material and technical means, shelter of materials, consumables materials.

4 Conclusions

The scheme of using integrated, distributed, heterogeneous data for automatic detection of disasters, delivery of information about them to the population and heads of enterprises, their use for predicting impacts, assessing the current hydrometeorological situation is presenting.

For the first time, a new paradigm of hydrometeorological support of consumers is proposing, which includes not only forecasting disasters and their impacts, but also visualizing the situation in the form of an information panel, issuing recommendations for decision-making, assessing damage and calculating the cost of preventive actions before the onset of the phenomenon.

Modern technologies and tools proposed that are necessary for the transition from traditional automation to end-to-end customer service technology from observation to decision making. As a result, new ways of providing services should change the very nature of the use of hydrometeorological information in the business processes of enterprises.

The proposed ideas and decision can be using in National Hydrometeorological Services in various countries to improve the efficiency of HMS for consumers.

Acknowledgments. The work financially supported by the Ministry of Science –and Higher Education of the Russian Federation, a unique project identifier RFMEFI61618X0103.

References

1. Kobysheva, N.V., Akentieva, L.M., Galyuk, L.P.: Climate Risks and Adaptation to Climate Change and Variability in the Technical Field, vol. 256. Cyrillic Publishing House, St. Petersburg (2015)
2. Grachev, N.R., Dikinis, A.V., Ivanov, M.E., Kuzmin, V.A., Smyshlyaev, S.P., Surkov, A.G.: Automated information system "MeteoMonitor" for early warning of dangerous weather phenomena in St. Petersburg and the Leningrad region, using assimilation of heterogeneous meteorological information. In: Proceedings of the International Supercomputer Conference on Scientific Service on the Internet: Exascale Future, pp. 432–438. Publishing house of Moscow State University, Novorossiysk (2011)
3. Michael, D.E., et al.: The Aviation Weather Decision Support System: Data Integration and Technologies in Support of Aviation Operations, Weather Decision Technologies Inc. https://www.researchgate.net/publication/228455668. Accessed on 09 July 2020
4. Grasso, R., Cococcioni, M., Rixen, M., Baldacci, A.: Decision support architecture for maritime operations exploiting multiple METOC centres and uncertainty. Int. J. Strateg. Dec. Sci. **2**(1), 1–27 (2011)
5. Arutyunyan, R.V., et al.: "Traffic Light": an early warning system for meteorological threats. IBRAE RAS, FGBU "Hydrometeorological Center of Russia" **36** (2015)
6. Disaster Risk Reduction for the Global Framework for Climate Services user interface platform. WMO **58** (2014)
7. Vyazilov, E.D., Mikhailov, N.N.: Data integration for marine environment and marine activities Infrastructure of satellite geoinformation resources and their integration. Popov, M.A., Kudashev, E.B. (eds.) Collect. Sci. Articles. Carbon-Service, Kyiv, 174–181 (2013)
8. RD 52.27.881-2019. Guide to the hydrometeorological support of marine activities. M., FSBI Hydrometeorological Center of Russia, 132 (2019)
9. RD 52.88.699. Regulation on the procedure for actions of institutions and organizations in case of threat of occurrence and occurrence of natural hazards. M., Roshydromet, 31 (2008)
10. Vyazilov, E.D.: New approaches for bringing information about dangerous hydrometeorological phenomena and raise awareness of persons taking decisions. In: XVI All-Russian Scientific Conference: "Problems of Forecasting Emergency Situations", Collection of Materials, EMERCOM of Russia. FKU "Antistichia", M. (27–28 Sep. 2017), pp. 40–44. https://static.mchs.ru/upload/site1/document_file/w5PgTGcnn5.pdf

11. Viazilov, E.D.: Development of hydrometeorological services to support decisions of enterprise leaders: examples from the Russian Federation. United Nations Office for Disaster Risk Reduction. The UN Global assessment report on disaster risk reduction 2019 (GAR19), Contributing Paper to GAR 2019, p. 26 (2019). https://www.preventionweb.net/publications/view/66441/
12. Viazilov, E.D., Mikheev, A.S.: Formalization of disasters impacts on enterprises and population, and recommendations for decision-making. In: Selected Papers of the XX International Conference on Data Analytics and Management in Data Intensive Domains (DAMDID/RCDL-2019). Kazan, Russia, 15–18 Oct 2019, pp. 95–106 (2019). http://ceur-ws.org/Vol-2523/paper11.pdf

Optimization of Logistic Flows in the Mail Processing System in the Republic of Kazakhstan

Askar Boranbayev[1](✉), Seilkhan Boranbayev[2], Malik Baimukhamedov[3], and Askar Nurbekov[2]

[1] Nazarbayev University, Nur-Sultan, Kazakhstan
aboranbayev@nu.edu.kz
[2] L.N.Gumilyov Eurasian National University, Nur-Sultan, Kazakhstan
sboranba@yandex.kz
[3] Kostanay Socio-Technical University named after Academician Z. Aldamzhar, Kostanay, Kazakhstan

Abstract. The process of processing and sending mail is quite complex and requires automation. Sometimes several automated and information systems can be involved in this work. And then the question arises of integrating such systems both within the country and with international postal systems. The article analyzes the ways of integrating the postal system of the Republic of Kazakhstan with international postal systems. Also, an important issue is the optimization of logistics flows in mail processing systems. A software system for deadlines has been developed for automation, operational processing and optimization of logistics flows when sending mail.

Keywords: Information system · Automated system · Mail processing · Logistics flow · Optimization · Management · Program

1 Introduction

Many enterprises use various systems to automate work, optimize workflow, and increase labor productivity. At the same time, sometimes one information system may not perform all possible functions of the production process. The creation of an information system that takes into account all the details of business processes and quickly fulfills them is rather complicated. Also, enterprises are developing, expanding and in this regard, they change old business processes and new business processes appear. Enterprises have to create new information systems while preserving the old ones. This leads to the fact that enterprises serve several information systems. These systems perform various tasks, however, they must work in concert, exchange data, etc. At the same time, for some organizations, it is necessary to link internal information systems with external IS. In this regard, information systems of enterprises should be integrated both with each other and with external IS. The article analyzes the ways of integrating the postal system of the Republic of Kazakhstan with international postal systems. The questions of optimization of logistics flows in mail processing systems are investigated.

K. Arai (Ed.): FICC 2021, AISC 1363, pp. 182–191, 2021.
https://doi.org/10.1007/978-3-030-73100-7_14

2 Integration of the Postal System of the Republic of Kazakhstan and the World Postal Union

Integration in the Form of File Exchange. This type of integration is often used. It is easy to implement and uses a data warehouse, which makes it highly resilient. Files exchanged by information systems must be of the same format and structure in both systems, which is a disadvantage of this type of integration (Fig. 1).

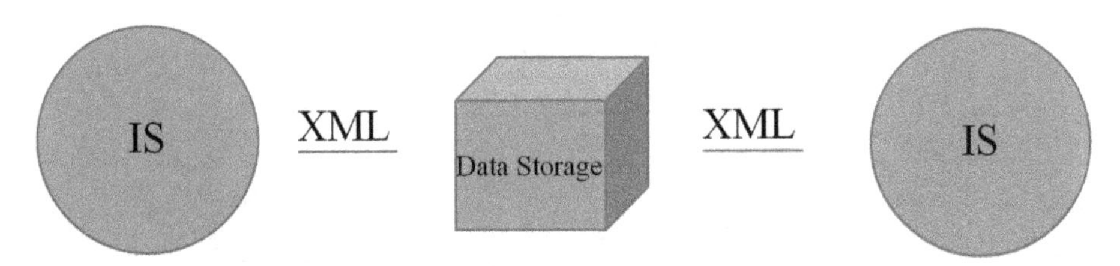

Fig. 1. Data exchange [1].

Integration in the Form of Data Replication Between Databases. With this type of integration, information from the main database is copied to the subscription servers. This type of integration is often used to store and use data when the main server is down. With this type of integration, there is a heavy load on the system, which slows down its work (Fig. 2).

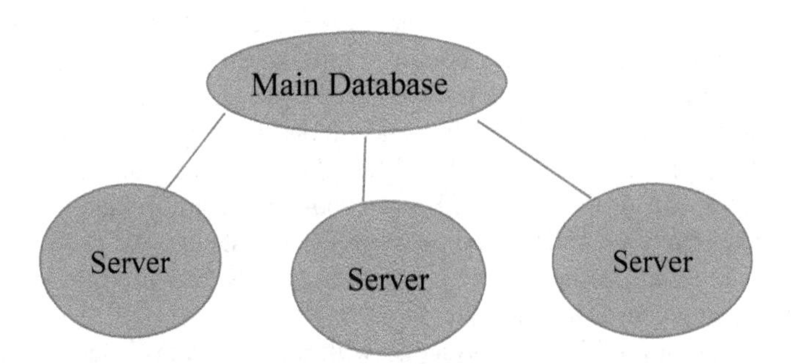

Fig. 2. Data replication [1].

Integration in the Form of Web Services Technologies. With this type of integration, services have their own inherent functionality, where files are exchanged without transactions and only in real time. With this type of integration, the exchange of a large amount of information is impossible, because processing a large data stream loads the system slows down its work (Fig. 3).

Integration in the Form of Messaging Through a Message Broker. With this kind of integration, the software sends the information to the queue as a message. The information is collected in a queue and then sent to the recipient, where it is processed in the form necessary for the second system (Fig. 4).

Fig. 3. The work of Web service technologies [1]

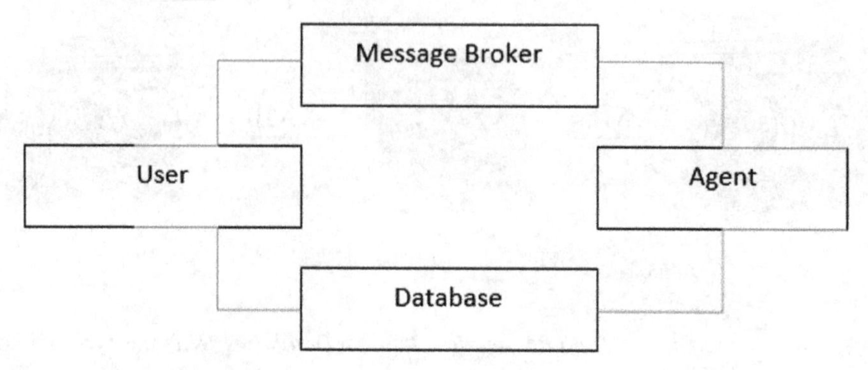

Fig. 4. Messaging through a message broker [2]

The postal service of the Republic of Kazakhstan is represented by the Kazpost organization, it is the national postal operator. Currently, e-commerce is rapidly developing in Kazakhstan and a huge number of international parcels pass through Kazpost, which arrive daily in large volumes. The e-commerce market in Kazakhstan is not as strong as in Europe or the United States, but the competition is quite high. Service is turning into a competitive advantage, where fast and timely delivery of goods is of no small importance. For the period from 2010 to 2019, there has been an increase in postage from 15% to 25% [3]. At the moment, Kazpost has several information and automated systems. Various links are established between these systems. The process of integration of information and automated systems of Kazpost with various systems of other countries and companies is underway.

For the integration of Kazpost systems with other systems, message queues are mainly used (Fig. 5). If one of the systems does not work, messages will accumulate in the queue, and when the receiving system starts functioning, the messages will be processed. Also, the advantage of the queue is that the interacting systems can have different interfaces and work with different data. Ultimately, each system processes the data in the appropriate format.

The main recipient of messages from Kazpost is the Universal Postal Union (UPU). There are a number of problems that require system integration to solve. These problems are mainly of an economic nature. First, there is the problem of violation of the delivery time of postal items. Secondly, there are problems with logistics, it is necessary to resolve issues with its optimization, taking into account the existing routes.

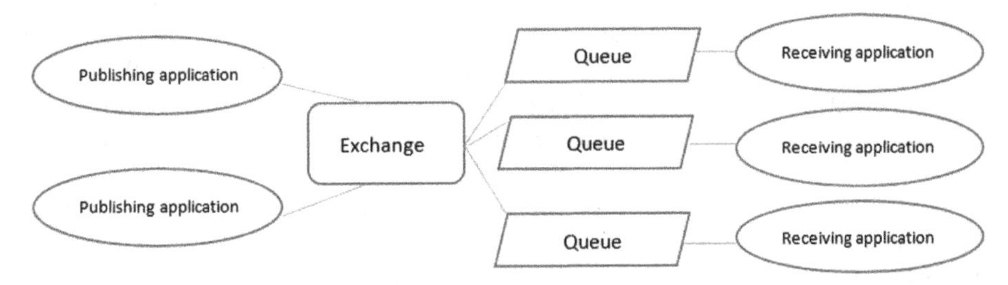

Fig. 5. Principle of how queues work [4].

At the moment, messages of two types go from the postal accounting system of Kazpost to the Universal Postal Union. Both types of messages go to the queue, but they have a different structure, and are formed in different places.

The first type of messages is electronic data interchange (EDI) (Fig. 6).

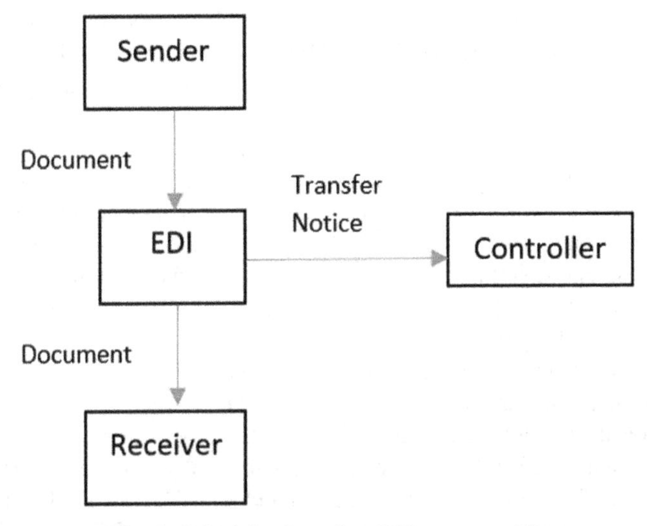

Fig. 6. Principle of sending EDI messages [4].

EDI is a generic abbreviation that encompasses electronic data interchange, usually between different parties. Many organizations exchange data over EDI networks. It uses standards such as EDIFACT. Technically, EDI transmission can be thought of as a set of delimited tags. The purpose of the tag is to tell you how to handle the following data. Most EDI mailings are based on the EDIFACT standard.

Integration of the existing at the enterprise accounting system with EDI, significantly increases the efficiency of work on the exchange of documents. Due to the fact that the manual labor of transferring data from documents to the accounting system is reduced, the number of errors is reduced [5].

The second type of messages is XML (Extended Markup Language), this type is gaining popularity lately. It is a structured presentation of information, each tag of which is tagged. Tags are descriptive. PostalAddress, EventTime, Identifier, DespatchDate are all examples of data descriptors that serve as tags. XML is a W3C (World Wide Web Consortium) standard and is very widely used in Internet applications. An important feature of XML is its flexibility. XML schemas define rules for the structure and valid content of XML documents.

The EDIFACT standard was developed long before XML was introduced. EDI postal exchanges have always been originally in EDIFACT format. Thus, today there are two exchange options. Should a postal exchange be based on EDIFACT or XML? Each approach has its advantages:

- EDIFACT is compact, reliable and widespread.
- XML is flexible, easy to implement, and can work with any alphabet or language (the technical capability that allows this is called Unicode, which is an encoding mechanism extensive enough to cover all existing alphabets).

With regard to the flexibility of the alphabet and languages, it should be noted that while EDIFACT is not as flexible as XML, it still provides the required level of flexibility. It should also be noted that the flexibility of XML can be limited if needed. The flexibility to handle any alphabet or language may not always be seen as an advantage in exchange. It is only important for the field trip that the party receiving the message understands the alphabet and language used by the party sending the message. With the same content, the XML message can be five times larger than the equivalent EDIFACT representation. Since transmission costs are based on message size, this is in favor of EDIFACT. On the other hand, solutions for compressing XML messages are currently being developed. XML messaging is just getting started in the postal industry. Therefore, Kazpost needs to wait for the postal industry to gain more experience with XML before deciding which format is best in the long run.

Some EDI messaging businesses use IT solutions called EDI translators. EDI translator is a tool that converts data between formats. These translators help interact with systems by converting formats between EDIFACT and other system formats and storage architectures. Using such a tool makes it easier to implement EDI messaging, especially in the case of EDIFACT. EDI translator includes all EDIFACT rules and grammar. When configuring such a tool to handle EDI mailing based on EDIFACT, it is important to inform the tool about the EDIFACT message and the version on which the EDI mailing is based. This information is presented in the UPU messaging standards (Fig. 7).

3 Optimization of Logistic Flows in the Mail Processing System

At present, postal services are actively developing, and Internet commerce is growing. Especially during a pandemic. Parcels are sent from different parts of the world to different countries using mail. In this case, the speed of delivery of postal items is important. Delivery speed is an important element of any logistics process. In order to timely and quickly deliver a postal item, it is necessary to control its timely delivery on certain sections of the route, automate and optimize the management of logistics flows to ensure timely delivery.

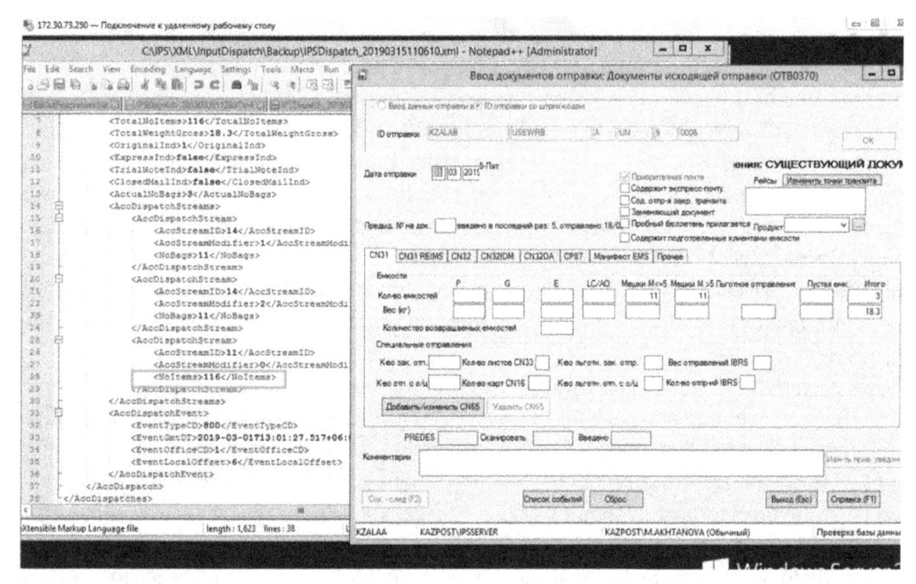

Fig. 7. Data transmitted by the UPU in the form of an XML message and processed by the UPU information system.

The model of the problem under study can be represented as a graph. The vertices of the graph indicate the settlements from where the postal items depart and arrive. The arcs of the graph indicate movements between settlements. The trajectory of movement of a postal item from the place of origin to the place of destination will be called a forwarding path. A shipment path consists of one or more segments.

Each postal item has its own delivery deadline. The package can be delivered earlier than this date, but not later. The postal item can be transported using various types of transport (train, plane, car). Each of the types of transportation on each of the sections of the route has its own duration in time and cost. The aim is to ensure the minimum shipping cost of the mail item within the target time frame. The developed software system of deadlines calculates the optimal route and method of transportation of postal items, taking into account the transportation parameters. The software system of deadlines allows you to make management decisions to optimize the delivery of postal items, provides timely information about the time spent on a particular section of the route, as well as other information. The deadlines for the carriage of postal items are the time from the moment of sending to the delivery to the addressee.

In the development of software systems, the level of its reliability and safety plays an important role. It is impossible to achieve uninterrupted and trouble-free operation of the control system without ensuring the reliability and safety of automated information systems [6–32]. When developing a software system for calculating the optimal route and method of transportation of postal items, technologies and methods were used to increase the level of its reliability and safety.

The deadline software system calculates the optimal route and method of transportation for postal items.

The deadline for transportation is influenced by intra-milestone dates. Internally, the milestones represent their time of processing postal items at the post production facilities at each stage of transportation. Also, the deadline for transportation depends on the path along which the postal item will pass from the sender to the addressee, and the transport schedule. For the software system of target dates to work, reference books are needed: within milestone schedules; referral plans; schedules of mail routes, etc. For each type of mail (parcel, Express Mail Service, etc.), its own target dates are calculated. At each stage, for each post office, its own milestone control dates are calculated. Each postal item has its own categories (incoming, outgoing, transit). The software system of control deadlines makes the calculation for each internal stage and along the entire route of the postal item.

The system calculates the optimal route and method of transportation of the mail item. In this case, transportation parameters and reference books are used. To calculate the optimal route and method of cargo movement, the system consults the reference books. To obtain information about the end time of processing at the internal stages, it refers to the directory of schedules at the internal stages of departments. To determine on which postal route it is necessary to load the postal item and when this item gets to an intermediate point, it turns to the directory of the postal route schedules of post offices. To obtain information about when the processing at the internal stages in the intermediate point should end, one refers to the directory of schedules at the internal stages of the intermediate point. To determine to which next intermediate point the postal item should be sent, refer to the directory of plans for the directions of the intermediate point. To determine on which postal route it is necessary to load the shipment and when this shipment gets to the next intermediate point, it refers to the directory of the timetables of the postal route of the current intermediate point. To calculate when a shipment should reach its destination, it refers to the directory of timetables at the internal stages of the destination. The directories are updated every six months, tables are filled with routes and methods of transportation of all types of shipments from all divisions to all other divisions. These guides are used to calculate the optimal shipping times for postal items. Forwarding steps directory data is correlated with statuses from tracking and tracking of mailings. A record is kept of the time spent on processing postal items within the post office.

A handbook of deadline schedules is shown in Fig. 8.

Fig. 8. Key date scheduling reference.

4 Conclusion

To create a single space for managing the logistics flows of postal items, the integration of the information systems of Kazpost and the Universal Postal Union is needed. This will increase the speed of delivery and improve the quality of service, and reduce costs. Messaging follows certain standards. EDI and XML are used, which differ in composition and method of sending. Messages are sent at night when the mail system is less congested. A queuing system is used to send messages. Special applications on the respective servers read the data sent to the queue and put them in folders, from which the corresponding mail systems read the data they need.

The software system of deadlines calculates the optimal route and method of transportation of postal items, taking into account the transportation parameters, allows making management decisions to optimize the delivery of postal items, provides timely information about the time spent on a particular section of the route, as well as other information.

References

1. Levina, T.M., Barvin, S.K, Pokalo, S.D., Budnikov, V.A.: Modern methods of integration of multilevel information systems. Innov. Sci. **11**(60), 16–20 (2016)
2. Samokhin, N.Y., Khoruzhnikov, S.E., Trubnikova, V.M., Akhmedzyanova, R.R., Bulykina, A.B.: Information on utilization of data center resources with message broker implementation. Sci. Tech. J. Inf. Technol. Mech. Opt. **18**(5), 858–862 (2018). (in Russian)

3. Muhadieva, K.S.: The impact of the growth of the e-commerce market on the postal and logistics infrastructure in the Republic of Kazakhstan. In: Collection of Scientific Articles XIV International Correspondence Scientific and Practical Conference, pp. 42–45 (2018). (in Russian)
4. Glushkova, Y.O.: The influence of integration on logistics processes between countries. Logistic Syst. Glob. Econ. **7**, 104–107 (2017). (in Russian)
5. Boranbayev, S.N., Tasmagambetov, O.K.: Forms and methods of integration of information systems of law enforcement agencies in Kazakhstan. ENU Bull. **2**, 26–32 (2016). (in Russian)
6. Boranbayev, A., Boranbayev, S., Nurusheva, A., Yersakhanov, K.: Development of a software system to ensure the reliability and fault tolerance in information systems. J. Eng. Appl. Sci. **13**(23), 10080–10085 (2018)
7. Boranbayev, S., Goranin, N., Nurusheva, A.: The methods and technologies of reliability and security of information systems and information and communication infrastructures. J. Theor. Appl. Inf. Technol. **96**(18), 6172–6188 (2018)
8. Boranbayev, A., Boranbayev, S., Nurusheva, A.: Development of a software system to ensure the reliability and fault tolerance in information systems based on expert estimates. Adv. Intell. Syst. Comput. **869**, 924–935 (2018)
9. Boranbayev, A., Boranbayev, S., Yersakhanov, K., Nurusheva, A., Taberkhan, R.: Methods of ensuring the reliability and fault tolerance of information systems. Adv. Intell. Syst. Comput. **738**, 729–730 (2018)
10. Boranbayev, S., Altayev, S., Boranbayev, A.: Applying the method of diverse redundancy in cloud based systems for increasing reliability. In: The 12th International Conference on Information Technology: New Generations (ITNG 2015). April 13–15, 2015, Las Vegas, Nevada, USA, pp. 796–799
11. Turskis, Z., Goranin, N., Nurusheva, A., Boranbayev, S.: A fuzzy WASPAS-based approach to determine critical information infrastructures of EU sustainable development. Sustainability (Switzerland) **11**(2), 424 (2019)
12. Turskis, Z., Goranin, N., Nurusheva, A., Boranbayev, S.: Information security risk assessment in critical infrastructure: a hybrid MCDM approach. Informatica (Netherlands) **30**(1), 187–211 (2019)
13. Boranbayev, A.S., Boranbayev, S.N., Nurusheva, A.M., Yersakhanov, K.B., Seitkulov, Y.N.: Development of web application for detection and mitigation of risks of information and automated systems. Eurasian J. Math. Comput. Appl. **7**(1), 4–22 (2019)
14. Boranbayev, A.S., Boranbayev, S.N., Nurusheva, A.M., Seitkulov, Y.N., Sissenov, N.M.: A method to determine the level of the information system fault-tolerance. Eurasian J. Math. Comput. Appl. **7**(3), 13–32 (2019)
15. Askar, B., Seilkhan, B., Askar, N., Roman, T.: The development of a software system for solving the problem of data classification and data processing. In: 16th International Conference on Information Technology - New Generations (ITNG 2019), vol. 800, pp. 621–623 (2019)
16. Askar, B., Seilkhan, B., Assel, N., Kuanysh, Y., Yerzhan, S.: A software system for risk management of information systems. In: Proceedings of thc 2018 IEEE 12th International Conference on Application of Information and Communication Technologies (AICT 2018), 17–19 Oct. 2018, Almaty, Kazakhstan, pp. 284–289 (2018)
17. Seilkhan, B., Askar, B., Sanzhar, A., Askar, N.: Mathematical model for optimal designing of reliable information systems. In: Proceedings of the 2014 IEEE 8th International Conference on Application of Information and Communication Technologies (AICT 2014), Astana, Kazakhstan, October 15–17, 2014, pp. 123–127 (2014)

18. Seilkhan, B., Sanzhar, A., Askar, B., Yerzhan, S.: Application of diversity method for reliability of cloud computing. In: Proceedings of the 2014 IEEE 8th International Conference on Application of Information and Communication Technologies (AICT 2014), Astana, Kazakhstan, October 15–17, 2014, pp. 244–248 (2014)
19. Seilkhan, B.: Mathematical model for the development and performance of sustainable economic programs. Int. J. Ecol. Dev. **6**(1), 15–20 (2007)
20. Boranbayev, A., Boranbayev, S., Nurusheva, A.: Analyzing methods of recognition, classification and development of a software system. In: Advances in Intelligent Systems and Computing, vol. 869, pp. 690–702 (2018)
21. Boranbayev, A.S., Boranbayev, S.N.: Development and optimization of information systems for health insurance billing. In: ITNG2010 – 7th International Conference on Information Technology: New Generations, pp. 1282–1284 (2010)
22. Akhmetova, Z., Zhuzbayev, S., Boranbayev, S., Sarsenov, B.: Development of the system with component for the numerical calculation and visualization of non-stationary waves propagation in solids. Front. Artif. Intell. Appl. **293**, 353–359 (2016)
23. Boranbayev, S.N., Nurbekov, A.B.: Development of the methods and technologies for the information system designing and implementation. J. Theor. Appl. Inf. Technol. **82**(2), 212–220 (2015)
24. Boranbayev, A., Shuitenov, G., Boranbayev, S.: The method of data analysis from social networks using apache hadoop. In: Advances in Intelligent Systems and Computing, vol. 558, pp. 281–288 (2018)
25. Boranbayev, S., Nurkas, A., Tulebayev, Y., Tashtai, B.: Method of processing big data. In: Advances in Intelligent Systems and Computing, vol. 738, pp. 757–758 (2018)
26. Boranbayev, A., Boranbayev, S., Nurbekov, A.: Estimation of the degree of reliability and safety of software systems. In: Advances in Intelligent Systems and Computing (AISC), vol. 1129, pp. 743–755 (2020)
27. Boranbayev, A., Boranbayev, S., Nurbekov, A.: Development of the technique for the identification, assessment and neutralization of risks in information systems. In: Advances in Intelligent Systems and Computing (AISC), vol. 1129, pp. 733–742 (2020)
28. Boranbayev, A., Boranbayev, S., Nurusheva, A., Seitkulov, Y., Nurbekov, A.: Multi criteria method for determining the failure resistance of information system components. In: Advances in Intelligent Systems and Computing, vol. 1070, pp. 324–337 (2020)
29. Boranbayev, A., Boranbayev, S., Nurbekov, A., Taberkhan, R.: The software system for solving the problem of recognition and classification. In: Advances in Intelligent Systems and Computing, vol. 997, pp. 1063–1074 (2019)
30. Akhmetova, Z., Boranbayev, S., Zhuzbayev, S.: The visual representation of numerical solution for a non-stationary deformation in a solid body. In: Advances in Intelligent Systems and Computing, vol. 448, pp. 473–482 (2016)
31. Boranbayev, A., Boranbayev, S., Nurbekov, A.: Evaluating and applying risk remission strategy approaches to prevent prospective failures in information systems. In: Advances in Intelligent Systems and Computing, vol. 1134, pp. 647–651 (2020)
32. Boranbayev, A., Boranbayev, S., Nurbekov, A.: A proposed software developed for identifying and reducing risks at early stages of implementation to improve the dependability of information systems. In: Advances in Intelligent Systems and Computing, vol. 1134, pp. 163–168 (2020)

Layered Interoperability for Collaborative IoT Applications

Daniel Flores-Martin[1](✉), Niko Mäkitalo[2], Javier Berrocal[1], José García-Alonso[1], Tommi Mikkonen[2], and Juan M. Murillo[1]

[1] University of Extremadura, Badajoz, Spain
{dfloresm,jberolm,jgaralo,juanmamu}@unex.es
[2] University of Helsinki, Helsinki, Finland
{niko.makitalo,tommi.mikkonen}@helsinki.fi

Abstract. Today, smart environments require attention from the network level for connecting devices to the application level for adapting their behavior to continuously changing environmental conditions. Just as a smartphone can choose between using mobile internet or a WLAN connection, IoT applications that manage smart environments must know how to establish relationships among heterogeneous devices. However, establishing connections and relationships between devices may require many manual actions from both the application developer and the user. In this paper, we propose two new layers that can make programming IoT applications more intuitive and convenient for the software developers and enable them to leverage the various connectivity types. These new layers, Social Layer and Situational Context Layer, are crosscutting with IoT-A layers, complement many of the existing IoT architectures and, in addition, to take advance of the ever-growing set of devices and their resources around the user in their applications easily.

Keywords: Internet of Things · IoT · Interoperability · Collaboration · Layered architectures · Situational context · Social computing.

1 Introduction

The term Internet of Things (IoT) was first used in 1999 by Kevin Ashton [2]. For him, people's lack of time creates the need to connect to the Internet in new and autonomous ways. This fostered the creation of devices that execute tasks automatically or with as little human interaction as possible. IoT applications are being developed for many application domains, such as smart home, automotive, smart cities, healthcare, and so on. These systems integrate the Internet-connected devices to achieve a higher collaboration and automatize different activities.

This approach ties IoT systems to the integrated hardware elements, increasing the complexity of these systems as the integrated infrastructure grows. In

K. Arai (Ed.): FICC 2021, AISC 1363, pp. 192–211, 2021.
https://doi.org/10.1007/978-3-030-73100-7_15

particular, interoperability between devices is one of the greatest challenges of the IoT due to the large number of protocols, manufacturers and devices that exist in the market [33]. This interoperability is even more difficult to achieve among devices from different domains. This leads to the development of vertical silos, hindering the development of cross-domain application and reducing the benefit that fully-interoperable IoT systems would provide.

Current works develop architectural models and standards to address these interoperability problems, such as layered reference models [19,21] or the IoT-A model [5]. For instance, the IoT-A model creates a five layer generic reference IoT architecture to cover different aspect of the interaction among devices, from the physical connection to the data exchange. The application of these reference architecture greatly improves the interoperability and reduces the required effort to develop IoT systems. Nevertheless, these reference architectural models do not focus on using the users contextual information to dynamically modify what devices should be orchestrated and what communication technologies should be used. This directly impact on social and dynamic environment in which the interaction among devices highly depend on their context.

In social environments, the collaboration among devices is crucial to meet people's needs and preferences [36]. This communication should be achieved using different communication protocols and architectural styles, such peer-to-peer or client-server, depending on the specific devices and context. In addition, these social environments cause many different situations that devices must know how to detect correctly to adapt their behavior. Currently, there are some alternatives that allow smart devices to learn and identify the environment around them. Thus, works such as [18] and [10], are focused on reference models and middlewares adding some context layer to adapt the system behaviour and the information provided/exchanged to the context. Nevertheless, the whole system should be adapted to the context including the behaviour, from the specific communication protocol used to how and when smart devices should interact.

In this paper, we propose an extension of the IoT-A reference model adding two new layers. The *Social Layer*, in charge of achieving a better integration between intelligent devices and people in social environments; and the *Situational Context Layer*, which detects the context and the different situations that may occur in a smart environment. This model will allow the development of context-aware IoT applications in an easier way for developers, by considering the social integration between devices and the detection of situations.

The rest of this paper is structured as follows. Section 2 relates our work to other recent researches in the field while Sect. 3 provides the background and motivation of this paper. Section 4 presents our goals for this work, and Sect. 5 introduces our concrete design. Section 6 discusses some implementation details of the design, while Sect. 7 provides two example systems, representing two alternate IoT domains. Section 8 highlights our lessons learned and key observations. Finally, Sect. 9 draws some final conclusions.

2 Related Work

Nowadays, there are works that allow a higher level of interoperability to be achieved, but in many cases, are restricted to specific application domains or do not contemplate people's contextual information. For instance, Bandyopadhyay and Sen's work [3] shows that, without a standardized technological approach, the proliferation of architectures, identification schemes and protocols are likely to occur in parallel, each of them with a specific goal and for a particular use but with a lack of collaboration among devices. Different approaches have been proposed for solving specific problems such as for improving the interoperability among devices using different communication protocols, or for orchestrating smart devices for solving specific problems. In addition, some reference architectural models have also been defined to provide some guidelines to developers and reduce the effort of developing these applications.

To reduce problems caused by the heterogeneity of smart devices and the variety of communication protocols, different approaches have been developed that abstract the applications of the specific communication protocol, such as oneM2M [35] or OpenIoT [34], or middleware such as the one proposed by Dubois et al. [13]. Nevertheless, IoT systems requires to dynamically select the communication protocol depending on the specific devices connected. To that end, some works propose the creation of opportunistic social networks depending on the concrete devices to interact with. In [1], the authors propose a smartphone-based mobile gateway acting as a flexible and transparent interface between different IoT devices and the Internet. However, its main limitation is an excessive energy consumption. Social Internet of Things [29] includes the social networking aspects to build social relationships and, thus, improve the support to new applications, provide reliable networks and enhance the information sharing. Nevertheless, these relationships and the opportunistic networks should be based on previous interactions and created at run time depending on the concrete devices interacting and their history.

In addition, the IoT devices connected to an opportunistic network or just working together to solve some specific problem should be adapted and orchestrated to that specific situation. Different frameworks, [9,37], propose devices to expose their capabilities using web services and, through different techniques, to compose those capabilities depending on the users' needs. To that end, the relationships among services are identified at design time and, at run-time, the present relationships are activated. Focusing on an specific domain, in [18], the authors propose a semantic framework for IoT-based smart farming applications, which integrate multiple cross-domain data streams ensuring seamless interoperability among sensors, services and actors. In this framework, different cross-domain elements are integrated but, in the end, it should be the domain expert and the software developer the ones that have to relate at design time how those elements should collaborate. Therefore, if one relationship is not correctly identified, it will be almost impossible to integrate that device in the system. Desai et al. [11] present a semantic gateway to be used in IoT systems to provide interoperability between systems, which utilizes established communication and data standards in order to support semantic reasoning to obtain higher-level actionable knowledge from low-level sensor data. Thus, providing interoperability of vertical silos of applications. However, this solution can lead to greater ambiguity because instead of having a broad scope, they focus on domain-specific solutions. In [24], Maarala et al. study how to use the data obtained from the IoT to reason processable knowledge through the application of state-of-the-art semantic technologies. To this end, they have developed a semantic reasoning system that works in a realistic IoT environment. Although these works solve some of the previous aspects, unfortunately they do not cover all the different drawbacks detected. In our proposal, the relationships and the coordination strategies emerge from past behaviors without them being defined at design time.

Finally, other works also focus on defining reference architectures and middlewares for improving the interoperability of IoT systems. For instance, in [10], the authors present a reference architecture model for an IoT middleware detailing the best operation method for each module and the security features that should be implemented to have a safe IoT system. They indicate that these middleware help developer during the integration of different devices from different manufacturers. Even, they propose a context-aware layer for adapting the devices to the users' context. In the work presented in this paper the social aspects and the context are mainly considered for creating more secure application.

3 Motivations

One of the fundamental approaches to systematically develop interoperable systems is an architectural style called layering. In particular, the approach is a fundamental element in many protocol stacks. Examples include the OSI model [6], where Briscoe defines a 7-layer model to facilitate the exchange of information from traditional models, such as TCP/IP, at the network level and allow better compatibility between different network environments; the 3 Layer architecture model [21], which allows the adaptation and reconfiguration of the runtime user interface for a smooth interaction of users with applications; or the IoT-A Model [5], where it was presented as an architectural reference model (IoT ARM) as a common basis for IoT, where the central idea is that IoT ARM provides a common structure and guidelines to address the central aspects of the development, use and analysis of IoT systems.

For instance, IoT-A (IoT Architecture Reference Model [4]) was designed as a reference for the generation of compliant IoT concrete architectures that are tailored to one's specific needs. Depending from the viewpoint, several layered structures are present in the model. Furthermore, the model is comprehensive in the sense that, in addition to plain communication, it also takes into account things like service composition, orchestration, and choreography in a coherent fashion. This architectural model is composed by the following layers:

- A **Physical** layer focused on the physical characteristics of the communication technologies, allowing the communication among devices without forcing the adoption of specific technologies.

- **Link** layer. It abstracts from the physical characteristics for linking different devices and enabling direct communication. At the same time, it provides upper layers with standardised capabilities and interfaces.
- The **IP/ID** layer provides networking with identifiers to provide a common communication paradigm and ensure that the devices' identifiers are supported or can be resolved via appropriate methods.
- **End-to-end** layer. This layer provides interoperability aspects on top of the IP/ID layer to improve the reliability when the communication crosses different networks and to achieve a global M2M communication model.
- The **Data** layer models data exchange between two devices or IoT entities providing meta-information for supporting data transfer ranging from raw data to complex structures or data conversion.
- The **Application** layer is the closest to the end user and interacts with software applications providing the capabilities and abstraction obtained by the other layers.

Also, to improve and facilitate the development of IoT systems, the Institute of Electrical and Electronics Engineers (IEEE), is working on a number of standards, such as [15,16], that include descriptions of several application domains, definitions of domain abstractions and identification of common elements between different domains. In this sense, the World Wide Web Consortium (W3C) has developed an abstract architecture for the Web of Things (WoT) [17]. The goal of this work is that WoT is intended to enable interoperability across IoT platforms and different application domains. To achieve this, the W3C describes a thing interface (WoT Interface) so that other things can interact with other services and things. Although these proposals for standards contribute in many ways to achieving a greater interoperability, however, they do not consider the social aspect that IoT applications must take into account. So, some work is still needed to adapt these standard to social and collaborative environment where the context of involved entities is key to adapt the behaviour and relationships among devices. In particular, the social aspect has received considerable attention that we believe should be taken into account [7]. While often overlooked in exchange for more technical considerations, the social part should also be taken into account when designing IoT applications.

Due to the lack of standards that meet the needs detected and the potential offered by the IoT-A model, the IoT-A project is suitable for us to incorporate social and situational aspects into the development of IoT applications. Since this model can not fully address the problems of interoperability between heterogeneous objects, much less to interpret the social and situational elements, in the following sections we specify two additional layers addressing these problems.

4 Towards Human-Centric IoT Applications

Next, we outline a number of goals for building Human-Centric, collaborative IoT applications that could even be combined with a Human-Data model [26]. These goals are complemented by the characteristics that are commonly associated with enabling levels of IoT applications, including technology, the domain at hand, semantics of the interactions between devices, and the context-aware at hand.

According to the approach proposed in [28], the current IoT technology needs improvements focused on people and, therefore, on the social part. These characteristics are:

1. **Social:** The goals to be met by IoT applications will be provided by the users in order to be solved by the different devices around them, who decides what information to share and with whom.
2. **Personalized:** This customization allows devices to be adapted more optimally to users. In addition, this must be done in a transparent way to users allowing a higher level of comfort to be achieved.
3. **Proactive:** The interactions should be detected by intelligent applications and devices. This proactivity can involve security risks, so users are the only ones who can decide what information to share and with whom.
4. **Predictable:** Applications must be activated according to a predictable context and be able to identify the information coming from the context as well as to specify the expected behaviour of the devices or people involved.

These features give IoT applications greater autonomy and social awareness in performing their functions. In this way, interactions will encourage the development of collaborative environments between IoT applications, devices and users, and allow IoT applications, whose are based on the information that users choose to share, to achieve a social and proactive environment.

In addition, the concern to define certain enabling levels between smart devices is present within the scientific community [30,32]. In accordance with [14], the collaboration among devices is achieved at different levels that progressively increase the user experience. The levels of interoperability generally match, although with some nuances or different nomenclature. These levels are:

1. **Technology:** The diversity of manufacturers makes communication between devices from different manufacturers difficult, because each one develops its own communication mechanisms, protocols and technologies.
2. **Domains:** Devices designed for a specific domain should be reused to perform other complementary tasks and to interact with devices from different IoT domains.
3. **Semantic:** The semantic distance enriching them with specifying services and parameters must be reduced. Smart devices can have similar characteristics or provide similar services, and we still have to get all their details in order to know how to interact with them.
4. **Situation:** Detecting the situation and its characteristics is key for detailing how the different devices should act on it and agreeing to establish a communication. This is why each situation requires that the services work in a specific way.

Currently, these enabling levels of IoT are only achieved by using devices from the same manufacturer, or relying on a simpler version of a dominant platform, in which case the feature set is often limited. New technologies and layers for the reference architectures are required in order to achieve fully interoperable and human-centric IoT applications.

5 Cross-Cutting Layered IoT-A Model

To improve the IoT collaboration mismatches, we propose a layered approach consisting of two different layers: *Cross-cutting Social Layer* and *Situational Context Layer*. Below we describe the main ideas behind these new layers.

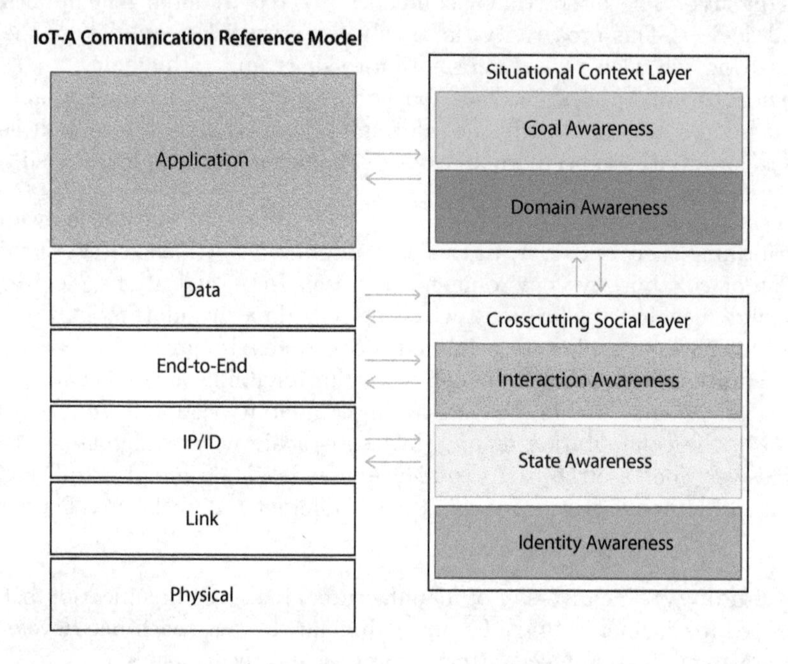

Fig. 1. Crosscutting social layer and situational context layer depicted.

5.1 Cross-Cutting Social Layer and Trusted Connection

The goal of Cross-cutting Social Layer is that the various IoT devices could self-organize by opportunistically forming social mobile networks. The layer utilizes and complements the existing layers of IoT-A Communication Model, as has been depicted in Fig. 1.

The social relations between entity owners[1] are reflected to the connections between devices. These new socially-aware connections are called as *Trusted*

[1] An owner in this context can be a single user, a group of users, or an organization, for example.

Connections as the connections will ensure that personal information used in IoT applications is shared only with entities known by the user. Additionally, by using the Trusted Connections, the layer makes sure that the interactions can take place only between trusted entities.

For establishing the Trusted Connections, the layer needs information about the social relationships and status of the owners. For this purpose, we leverage the concept of *Self-Awareness*. In computer systems, Self-Awareness can be considered as an umbrella term, enclosing a variety of different but partially overlapping aspects [20,22]. Key aspects of Self-Awareness for implementing the Cross-cutting Social Layer are described in the following.

Identity Awareness: Current Internet Protocol and BLE technology provide identifiers that can often dynamically change. Such changing identities are not very well suited in the scope of IoT. The reason is that when two IoT devices connect the first time, they establish a trusted relationship by exchanging configurations. The purpose of that exchange is to ensure more accessible communications and trust in future connections. Such exchanges can even create trusted circles, including other entities. For example, a voice assistant communicating with a smart-home hub makes it accessible and configurable for every other connected smart device. Hence, when two devices communicate, it is expected that they are the same, they were in the past.

For this reason, it is essential that the devices can recognize each other by the social relationships instead of a changing IP address. According to Lewis et al. identity awareness aims modeling experiences that contribute towards a conceptualization of a coherent identity over time. The Cross-cutting Social Layer provides unique and understandable identifiers that are formed over time, based on the devices and their owners social relationships.

State Awareness: We argue, that the plain networking (and possible service information) is not yet providing enough information for intuitive communication infrastructure of the modern IoT applications. We propose complementing the present approaches with state awareness aspects that can directly influence how the connections are established opportunistically in a self-organizing manner between the entities. This will allow the entities to select the most optimal network topology based on their computing capabilities, battery status, or the role of the device in the intended activity.

According to Lewis et al. state awareness refers to the knowledge about the entity itself, the world, and other entities in the world. State awareness aims to form a coherent model of these elements. On the networking side, such information traditionally has only included information about the networking itself (although some protocols have introduced service discovery approaches).

Interaction Awareness: The current service discovery protocols enable querying for services that are being advertised in the device's current networking environment. The process of detecting these services, however, is tied to specific connectivity types and protocols, and many are vendor-specific. Although, there are improved approaches such as UPnP these yet do not consider the social aspect, and how the social relationships could improve the process of detecting the services provided by trusted entities.

In self-aware computing systems, the idea of interaction awareness is to model the patterns of interactions between entities. As an example, Lewis et al. mention how on the protocol level the system could detect whether a message is a response to another message and thus an isolated action, or a message. On the other hand, interaction awareness can also build upon state awareness and hence considering the causality between different actions that typically follow each other. Cross-cutting Social Layer aims learning how the entities interact with each other, and proactively then encapsulate various communication approaches to Skills that the software developers can then use for implementing IoT applications.

5.2 Situational Context Layer

IoT applications drive automated actuation, such as in intelligent environments, distributed microgrids, or user-centric automation. These applications operate on dynamically changing and ontologically defined data called context, data whose type, range and sources are specified in an interface. The current state of the art adapts the end-to-end implementations of each application to its initial infrastructure and platforms, making it difficult to adapt to the number and types of devices. That is why we need to somehow model the goals and entities in context. In this way, devices are aware of what needs are present to try to solve them. This is where the Situational Context layer takes on special interest through its two main components: Goal-Awareness and Domain-Awareness.

Goals Awareness: In accordance with Lewis et al. [22], Goal Awareness includes the ability to conceptualize the internal factors that drive the behavior, such as a system's goals, objectives, and constraints. In relation to this, smart devices aim at satisfying people's and applications goals by adapting their behaviour to them. These goals can be of many different types and in each situation they can vary. To allow an application to meet one of their goals, it can use a specific device that has the capability to solve it, such as to control the traffic lights to avoid accidents. This process takes a considerable amount of time and requires having minimal technical knowledge. This is why the *Situational Context Layer* aims to model the particular situation in which one or more entities are involved, as well as the smart devices around them. Thus, the devices are aware of the specific goals that other entities have in the situation in which they find themselves, thanks to contextual information and personal information that can be shared between the entities in the environment. For example, smart light application can increase the luminosity of a room because it is getting night and there are some people inside. But we can go further. In simple situations it is easy to interpret how the devices should behave, but in complex situations this becomes more complicated. A complex situation can be determined by the relationship of two or more entities for the resolution of a goal. Continuing with the previous example, if the smart light does not offer enough luminosity, several lights could be turned on in a collaborative way.

In addition to this, the complexity of the situation has to do with the number of entities present in the environment (same goal for many different entities). For example, if instead of one person entering the room enter several, it is necessary to determine what level of brightness to establish, because the goals can be completely different. This adaptability requires that certain information about entities is available, such as their goals and associated information (place, desired value, date and time, etc.), as well as the application domain where they are involved and even the role they play within the situation to establish a priority.

Domain Awareness: In a computer system, a domain refers to all areas, things from, above, below, related, adjacent or bordering a specific space, which could affect activities, infrastructure, people, load, moving or operating within space. A domain can be physical, i.e., a room or a house, or conceptual, i.e., the domain where a person works. The Domain Awareness is understanding how entities operate and interact within their environment and how they might affect the adaptability of computer systems. In this sense, the different entities must be aware of the domain in which they find themselves to be able to adapt to it.

Domain Awareness improves the adaptability and security of the entities involved in a given situation. In this sense, there are several aspects to consider, including: quickly identifying actual or potential events; making informed decisions; taking effective actions; and sharing knowledge with appropriate entities.

This definition may lead one to think that the Domain Awareness and the situations in which the entities are involved are the same concept. However, from our point of view, they are different. Within a domain there can be different situations, that is to say, all the elements that belong to the same domain can be involved in different situations. Imagine a person that is at work and has different smart devices around him/her and other people as well. In this case, the domain that affects this person would be everything related to his work, but there may be situations created from interactions with those entities. If we also pay attention to another person who is within the same domain as the previous one, the situations may not be the same.

Therefore, the purpose of the *Situational Context Layer* is to analyze the behavior and information of entities within the environment to promote interactions with other devices or people. In this way, devices are able to adapt without the need for human intervention, and perform tasks more automatically. For this, the parameters associated with the goals of people and the available capabilities of the devices in the particular situation must be taken into account. That is, to favor proactivity when it comes to solving the detected goals and to take maximum advantage of the capabilities of smart devices in each situation that may arise in an environment.

6 Implementation

In the following we discuss aspects related to a practical implementation of the Cross-cutting two-layer model. We start with the Cross-cutting Social Layer, and then advance to Situational Context layer.

6.1 Cross-Cutting Social Layer

Based on the above, relying on the layered networking approach relies on three essential features. These are unified entity identifiers, trusted connections between the different parties, and skills these parties possess. In the following, each of these features are discussed separately.

Unified Entity Identities: Cross-cutting Social Layer homogenizes the identities despite of the used connectivity type or protocol. The idea of the layer is that social relationships are built in the communication stack, and the identities are meaningful for the end-users to support their understanding and predictability aspect of the IoT applications. Single and meaningful identity referring to any entity (user or device), and further to their resources will help the user to understand what resources are included in the interaction.

A central element for implementing unified identities is Identity Awareness component. The component learns over the time with which entities the user interacts. The learning process is fostered by feeding information about the social relationships. Possible sources for such information include, for example, the social media platforms and contact applications of the mobile devices.

For enabling this self-adaptivity, the Interaction Awareness component needs to work seamlessly together with the Identity Awareness component to make it intuitive for the programmers to define with which kind of entities the interactions are targeted to take place. Similarly, if there is a need for the human in the loop, the end-user would be able to understand with which entity the interaction is taking place. We propose implementing a cloud-based repository where such identities could be maintained and then disseminated to the devices. An identity file could be combining multiple different types of identities from the physical world entities as well as identities from the cyber world, like MAC addresses, current IP addresses, BLE service UUIDs, social media usernames and so on. Naturally, revealing such information to be available for various devices opens possibilities for security threats. However, most of such information is already available online or can be detected for example with service discovery protocols. Hence the actual threat lies in the combination of such information as this may threat the users privacy. An example identity file is depicted in Listing 1.1.

IoT-A or other similar solutions do not yet consider the social relationships between the entity owners already on the identity management level. Naturally, there is a need to communicate with previously unknown entities. However, on their daily activities, the users need to communicate with entities they already know and trust. For this reason, we argue that the social relationships between

Listing 1.1. Example identity file.

```
{
    username: "bob",
    identity: "bob@hdm",
    companion:
        "FB694B90-F49E-4597-8306-171BBA78F844",
    facebookID: "102684690214746",
    devices: {
        "bob@iphoneSE":
            "5BF2E050-4730-46DE-B6A7-2C8BE4D9FA36",
        "bob@ibeacon":
            "8B034F7B-FA9B-540F-ACF3-88C0CA70C84F"
    }
};
```

the entities must be built in already on the low levels of the communication stack so that these can be used for referring to certain users' devices and resources.

Trusted Connection: Presently, communication in IoT systems is mainly based on the Internet Protocol. The protocol has proved to be one of the greatest engineering feats due to its robustness and ability to adapt. The Internet based communication has, however, its deficits. For example, the lag in the communication may become an obstacle in the most mission critical systems. Moreover, trusting only global communication can be risky from functional-safety perspective. Hence we propose using peer-to-peer communication approaches along with global Internet-based communication.

Establishing and maintaining P2P-connections between entities is yet cumbersome process. The developer is often forced to implement own protocols for the task. Alternatively, the developer may leverage solutions provided by the current devices. The problem with both approaches is that these make the applications dependent on certain platforms and hence often rule out any other devices. Moreover, from end-user perspective, understanding and remembering multitude of pairing and connection protocols requiring too much manual efforts.

Skills: Implementing the Interaction Awareness, we propose combining the awareness about the current communication protocols, technologies, and connections with state awareness to understand what kind of input and output modalities or capabilities are presently available to be leveraged in the interactions. With the Interaction Awareness, the Cross-cutting Social Layer then abstracts the complexity of various synchronous and asynchronous communication protocols and provide more meaningful programming concepts for the IoT application developers despite of the used connectivity type. Such concepts provide access to information resources as well as be used for programming the actuation with the world.

Modelling the interaction awareness requires run-time approach since the environment or situations are in continuous change (e.g., changes are produced by the entry or exit of entities to the environment). At this point, the system must be able to identify the interactions between devices and people (entities), so that the relationships between them can be established and recognized. To achieve this, the identity awareness must be achieved so that each entity can be correctly identified to detect its relationships with other entities and the established, trusted connections.

The interactions between entities are determined at the time of resolving the different situations that can occur in an environment. These situations refer to the different skills and goals that each entity possesses. Each situation is different, and the devices and people who may be involved in them will be constantly changing. Therefore, interactions will be given by the direct relationship between the skills of one device and the goal of another, when these can be related and are available in the environment. In relation to this, it is also important to detect what type of relationship is being established or which entities are being involved as well as the state in which they are.

6.2 Situational Context Layer

The features or skills of smart devices are often provided by manufacturers. For example, when one buys a smart bulb, the manufacturer indicates that it is possible to regulate the light intensity, color or even program it to turn on or off at a certain time or when an event occurs. In the case of IoT applications, these features are provided by the developers. However, the needs or goals that are present in a context belong to people or entities and they are the only ones who know them. Also, only people or applications know what would be necessary to meet a specific need. For example, the need to light a room with gentle intensity because someone is inside to reading a book. These needs can be detected to a greater extent by analyzing entity's behavior patterns.

Based on the above, to implement the Goal-Awareness component, we propose that the different IoT applications have an associated file where their goals are described, in the same way as the personal information seen in the previous section. In the first place, the information of these goals would be introduced in a manual way, thanks to the developers and experts in the application domain. Then, through the use of the application, the goals information and the actions carried out to solve them would be analyzed with the objective of detecting patterns and behavior that help to detect new goals.

In addition to the goals, a context is also influenced by the domain to which it belongs. Within IoT applications we have several domains of action such as healthcare, smart city, agriculture or industry. That is why the information associated with the goals of an entity must be complemented with the domain in which it is. That is, because the same goal can be solved in different ways depending on the domain in which it is. For each domain, the expert developer must provide the necessary knowledge so that the detected needs are resolved in the most effective way possible. For example, if the need to watch television is detected, the volume of television will be different if the person is at home, in the workplace, or in a hospital. This allows IoT applications to be placed within the correct state and action domain when it comes to solving present needs. Therefore, to implement the Domain-Awareness component to establish the domain of an IoT application it is necessary to have the figure of the expert developer of the domain in question, which allows the state of the applications to be adequate and to be able to interpret the information in the best way. In this way, domains also help to abstract us from the different situations that may occur in a given place, as expert knowledge is able to identify some of the most common situations, facilitating interactions between entities.

6.3 Developing Human-Centric IoT Applications

We describe how software developers can use the approach for implementing the new Human-Centric IoT applications. We also introduce two example use cases that leverage the approach. Different roles have been depicted in Fig. 2.

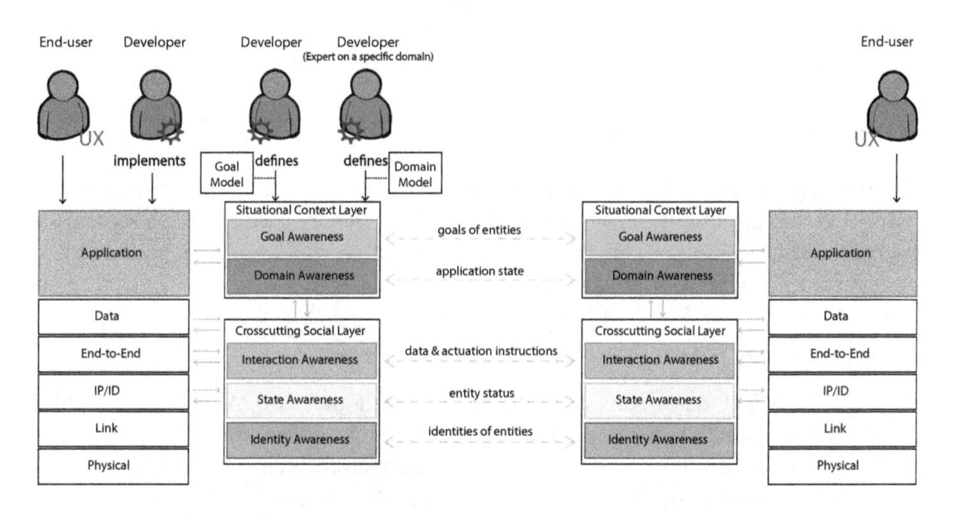

Fig. 2. Initial sketch about the use of our approach.

IoT Application Developer: IoT application developer is the main person who implements the actual end-user applications. Hence it is up to this person to actually leverage the benefits that our approach provides for implementing more Human-Centric IoT applications.

The developer imports a framework library that provides the Situational Context for the application. By using the provided Situational Context, the developer can easily send data for example to the user's own devices, the user's friends' devices, or provide the data for public devices. The protocol is hence based on the social relationships instead of complex physical addressed (e.g., MAC) or dynamic addresses (e.g., IP address).

Domain Expert (Developer): can be a software developer or otherwise technically oriented person who is an expert on a specific domain. These include for example traffic planners, logistics experts, or health care professionals that understand what kind of information is available and how this information can be used in the name and spirit of the law.

The domains should be carefully designed and audited, and in many cases even regulated. The domains define the rules for what kind of information the IoT applications are allowed to handle. The organizations and developers implementing IoT applications should approve a contract with the domain providers to assure that they are using the information responsibly.

Goal Model Developer: is a role in between the domain model expert and the IoT application developer. The role can be taken by a public organization (e.g., government office, institute, etc.), or the role can be taken by a company providing a certain kind of Human-Centric IoT solutions). For instance, an organization may have a set of goal models for different applications.

The goal model defines how the Situational Context is formed based on the data, and hence what information the Situational Context is providing for the applications to be leveraged by the IoT application developer.

7 Sample Use Cases

There are numerous use cases where the layered interoperability model for IoT introduces benefits to application design. In the following, we outline two sample use cases, first a Human-Centric healthcare based on IoT applications, and then a smart traffic related application.

7.1 Use Case I: Human-Centric IoT Healthcare

This use case is focused on an application for monitoring the health of elder people living alone in a smart home environment, called SmartElderly. This environment should have different Internet-connected devices for making elders live easier (smart bulbs, air conditioners, blinds, coffee makers, etc.). In addition, it should have deployed other devices for monitoring their health, such as air quality monitor, oximeter, tensiometer, and so on. Elders' smartphones act as a gateway for getting and storing the sensed information and controlling them. All the connection among devices are P2P. This use case mainly focuses on the advantages provided by the previous defined layers.

The social layer allows developers to better and more securely control the interactions among devices. Thus, the Identity Awareness layer automates the connection among the individual devices independently of the specific communication interface or stack used. For instance, for the air quality monitor a WiFi technology is usually used, whilst oximeters are usually connected through Bluetooth. The proposed layer will store the identity of each smart device and previous connections to determine how to interact with them. The developer of this application will not have to manually include those operations.

The State Awareness layer will organize ad-hoc networks depending on the role of the devices. Thus, the devices shared among the elders living in the smart house (the smart bulbs, the coffee maker or the air quality monitor) will be in one network. Instead, for personal devices, opportunistic ad-hoc networks will be created between the user and the its smartphone. For instance, an ad-hoc network is created when one user is using the tensiometer to store the sensed data in his/her smartphone. New ad-hoc networks are also created when the healthcare professional is in the connection range of the elders' phone and (s)he needs that information. The State Awareness layer creates the network and the Interactions Awareness layer encapsulates the interaction among devices. Instead of developing this behaviour, these rules are defined by the Domain Expert and reused by the Developer.

Finally, the Goal modeler developer specifies some desirable values and goals for improving the elders' health. For instance, the recommended air quality (the CO2 should be between 350 ppm and 1.000 ppm) indoor or the appropriate temperature (between 20 and 24°). That information is automatically used by the application from the Domain Awareness to automate the interactions among device, i.e., to control the air conditioning and the window opening. The recommended range is narrowed by the application developer and the user specifying in the application the desired values. The elder can specify that (s)he wants 560 ppm of CO2. So that, if that level is higher, the app identifies nearby devices with the skill of lowering that level and, then, all the presented layers will be used to start an interaction with the window opener through the ad-hoc network. For those values not detailed by the user or the developer, the values detailed by the Goal Model Developer are used.

7.2 Use Case II: Smart Traffic

This use case is focused on a smart traffic application for reducing the traffic jam, alerting about accidents and obtaining a better traffic flow in smart cities. The presented approach greatly facilitates the development of this application. This use case mainly focuses on the work of the different roles.

First, the Goal Models Developer has to define the goals of the systems. In this sense, two different kinds of goals could be defined: general goals and user specific goals. The general goals are those generic for the whole application for improving the traffic flow in the city such as improve the traffic flow, reduce the number of accidents or pile-ups accidents. The user goals are those specific for each user such as reducing the driving time to get to the destination and the number of traffic jams in his/her route.

Second, the Domain Expert defines what information can be exchanged. In this case, (s)he defines that the vehicles can send information about their routes to the structural elements of the traffic network (traffic lights, traffic signals, etc.), so that they can detect traffic jams. In addition, the structural elements can send to the vehicles anonymous information about possible traffic jams or congested routes. Thus, vehicles can automatically suggest alternative routes to the driver. Similarly, vehicle to vehicle communications are allowed to transmit information about a possible accident, so that it alert the driver about that accident before (s)he arrives to the area.

Finally, the developer makes use of the defined goals and domain rules to implement IoT systems with less effort and without having to face the integration of devices from different domains for the data exchange. The social layer automates the integration of different devices and the secure control. For instance, it controls that no personal information is exchanged among vehicles (such as the destination, the owner, etc.) and that the information about traffic accidents can be exchanged independently of the manufacturer by creating ad-hoc networks. The Situational Context layer allows developer to control when the information is exchanged. For instance, only when two vehicles are in the same area and the target vehicle is heading to the accident the information is exchanged.

8 Discussion

The presented layered approach is targeted for improving human-centric IoT applications. In such applications, plenty of personal and intimate data is typically involved. Additionally, since the IoT devices act on behalf of the user, it becomes essential the users trust that their devices. In the following, we discuss about security, privacy, and trust aspects that the proposed layered model provides.

The transferring of data is based on trusted connection which is established on top of the typical connectivity types. The communication is based on trusted device coalition framework that we studied before in [12,25,27]. However, there are plenty of communication approaches that can be used similarly [8,23,31]. Hence, in this paper the focus is not on the communication details.

Therefore, the work required to validate this research is based on two main tasks: First, the proposed approach yet requires several experiments with real developers that are dedicated to these specific roles. Only then can we gain information how different developers and technically oriented people take the proposed approach. Second, after the real developers from different organizations have implemented the IoT applications, we also need to run user studies with real end-users to collect their feedback how the proposed approach should be improved. This will help us to better understand the approach as a whole.

9 Conclusions

The Internet of Things has meant a revolution in people's daily lives, allowing them to develop tasks more simply or automatically. To obtain the maximum benefit from this technology, the devices must be able to interact with the environment, obtain complex information and even adapt their behavior. Thanks to IoT applications, a multitude of tools can be developed to manipulate the behaviour of these devices, and ensure that their behaviour adapts to people's needs. However, to achieve a total adaptation to the environment, the social and situational aspect must be considered of the devices and the people around them, which means having to develop increasingly complex IoT applications that can identify the needs and capabilities available in the environment.

Nowadays, some proposals manage to interpret the environment of the devices in an increasingly precise way to adapt their behavior or make decisions. However, most of these solutions are aimed at achieving the adaptation of devices at specific situations and often previously defined in design time, and without taking into account the conditions of the devices and people available in the environment. That is why this document has proposed an extension of the IoT-A model, a reference model in communication within the IoT, to provide IoT applications the ability to model the social and situational character of devices and people who are in a particular situation. This extension has been achieved through the introduction of two layers (social and situational), and is intended to be a tool that facilitates the development of IoT applications to developers.

As a future work, we are currently working on the validation of the proposed model by creating cross-domain applications to detect vulnerabilities and aspects to improve from developers perspective. In addition, efforts are also being made to improve the communication and the distribution of social information to ensure users privacy.

Acknowledgments. This work was supported by the project RTI2018-094591-B-I00 (MCI/AEI/FEDER, UE) and FPU17/02251 grant, by 4IE project (0045-4IE-4-P) funded by the Interreg V-A España-Portugal (POCTEP) 2014–2020 program, by the Department of Economy, Science and Digital Agenda of the Government of Extremadura (GR18112, IB18030), and by the European Regional Development Fund. The work of N. Mäkitalo was supported by the Academy of Finland (projects 313973 and 328729).

References

1. Aloi, G., Caliciuri, G., Fortino, G., Gravina, R., Pace, P., Russo, W., Savaglio, C.: Enabling IoT interoperability through opportunistic smartphone-based mobile gateways. J. Netw. Comput. Appl. **81**, 74–84 (2017)
2. Ashton, K., et al.: That 'Internet of Things' thing. RFID J. **22**(7), 97–114 (2009)
3. Bandyopadhyay, D., Sen, J.: Internet of Things: applications and challenges in technology and standardization. Wirel. Pers. Commun. **58**(1), 49–69 (2011)
4. Bassi, A., Bauer, M., Fiedler, M., van Kranenburg, R.: Enabling Things to Talk. Springer-Verlag GmbH, Heidelberg (2013)
5. Bauer, M., Walewski, J.W.: The IoT architectural reference model as enabler. In: Enabling Things to Talk, pp. 17–25. Springer, Berlin, Heidelberg (2013)
6. Neil, B.: Understanding the OSI 7-layer model. PC Netw. Advisor **120**(2), 13–15 (2000)
7. Bródka, P., Filipowski, T., Kazienko, P.: An introduction to community detection in multi-layered social network. In: World Summit on Knowledge Society, pp. 185–190. Springer (2011)
8. Bródka, P., Kazienko, P., Musiał, K., Skibicki, K.: Analysis of neighbourhoods in multi-layered dynamic social networks. Int. J. Comput. Intell. Syst. **5**(3), 582–596 (2012)

9. Cabitza, F., Fogli, D., Lanzilotti, R., Piccinno, A.: Rule-based tools for the configuration of ambient intelligence systems: a comparative user study. Multimedia Tools Appl. **76**(4), 5221–5241 (2017)
10. da Cruz, M.A.A., Rodrigues, J.J.P.C., Al-Muhtadi, J., Korotaev, V.V., de Albuquerque, V.H.C.: A reference model for internet of things middleware. IEEE Internet Things J. **5**(2), 871–883 (2018)
11. Desai, P., Sheth, A., Anantharam, P.: Semantic gateway as a service architecture for IoT interoperability. In: 2015 IEEE International Conference on Mobile Services, pp. 313–319. IEEE (2015)
12. Devos, M., Ometov, A., Aaltonen, M.T., Andreev, S., Koucheryavy, Y.: D2D communications for mobile devices: technology overview and prototype implementation. In: 8th International Congress on Ultra Modern Telecommunications and Control Systems and Workshops (ICUMT), pp. 124–129. IEEE (2016)
13. Dubois, D.J., Bando, Y., Watanabe, K., Miyamoto, A., Sato, M., Papper, W., Bove, V.M.: Supporting heterogeneous networks and pervasive storage in mobile content-sharing middleware. In: 2015 12th Annual IEEE Consumer Communications and Networking Conference (CCNC), pp. 841–847, January 2015
14. Flores-Martin, D., Velasco, J.C.C., Berrocal, J., Murillo, J., et al.: Abordando los distintos niveles de colaboración entre dispositivos en entornos IoT (2019)
15. IEEE. IEEE standards enabling products with real-world applications. https://standards.ieee.org/initiatives/iot/stds.html. Accessed 05 Nov 2019
16. IEEE. IEEE standards activities in the internet of things (IoT). https://standards.ieee.org/content/dam/ieee-standards/standards/web/documents/other/iot.pdf/2018. Accessed 05 Nov 2019
17. Kaebisch, S., Kamiya, T., McCool, M., Charpenay, V.: Web of things (WoT) thing description. candidate recommendation, W3C (2019)
18. Kamilaris, A., Gao, F., Prenafeta-Boldú, F.X., Ali, M.I.: Agri-IoT: a semantic framework for internet of things-enabled smart farming applications. In: 2016 IEEE 3rd World Forum on Internet of Things (WF-IoT), pp. 442–447. IEEE (2016)
19. Kearns, A.: 5 layer architecture. White paper, version v0.2, 22 November 2010. https://morphological.files.wordpress.com/2011/08/5-layer-architecture-draft.pdf
20. Kounev, S., Kephart, J.O., Milenkoski, A., Zhu, X.: Self-Aware Computing Systems, 1st edn. Springer Publishing Company, Incorporated (2017)
21. Lehmann, G., Rieger, A., Blumendorf, M., Albayrak, S.: A 3-layer architecture for smart environment models. In: 2010 8th IEEE International Conference on Pervasive Computing and Communications Workshops (PERCOM Workshops) (2010)
22. Lewis, P., Bellman, K.L., Landauer, C., Esterle, L., Glette, K., Diaconescu, A., Giese, H.: Towards a framework for the levels and aspects of self-aware computing systems. In: Self-Aware Computing Systems, pp. 51–85. Springer, Heidelberg (2017)
23. Li, N., Chen, G.: Multi-layered friendship modeling for location-based mobile social networks. In: 2009 6th Annual International Mobile and Ubiquitous Systems: Networking & Services, MobiQuitous, pp. 1–10. IEEE (2009)
24. Maarala, A.I., Su, X., Riekki, J.: Semantic reasoning for context-aware Internet of Things applications. IEEE Internet Things J. **4**(2), 461–473 (2017)
25. Mäkitalo, N., Aaltonen, T., Raatikainen, M., Ometov, A.: Sergey action-oriented programming model: collective executions and interactions in the fog. J. Syst. Softw. **157**, 110391 (2019)
26. Makitalo, N., Flores-Martin, D., Berrocal, J., Garcia-Alonso, J.M., Ihantola, P., Ometov, A., Murillo, J.M., Mikkonen, T.: The Internet of bodies needs a human data model. IEEE Internet Comput. **24**(5), 28–37 (2020)

27. Niko, M., Aleksandr, O., Kannisto, J.: Yevgeni Safe, secure executions at the network edge. IEEE Softw. Sergey Andreev (2018)
28. Miranda, J., Mäkitalo, N., Garcia-Alonso, J., Berrocal, J.: Tommi from the Internet of Things to the Internet of people. Internet Comput. IEEE **19**(2), 40–47 (2015)
29. Roopa, M.S., Pattar, S., Buyya, R., Venugopal K.R., Iyengar, S.S., Patnaik, L.M.: Social internet of things (SIoT): oundations, thrust areas, systematic review and future directions. Comput. Commun. **139**, 32–57 (2019)
30. Noura, M., Atiquzzaman, M., Gaedke, M.: Interoperability in internet of things: taxonomies and open challenges. Mob. Netw. Appl. **24**(3), 796–809 (2019)
31. Ometov, A., Zhidanov, K., Bezzateev, S., Florea, R., Andreev, S., Koucheryavy, Y.: Securing network-assisted direct communication: the case of unreliable cellular connectivity. In: Proceedings of IEEE Trustcom/BigDataSE/ISPA (2015)
32. Patel, K.K., Patel, S.M., et al.: Internet of things-IoT: definition, characteristics, architecture, enabling technologies, application & future challenges. Int. J. Eng. Sci. Comput. **6**(5) (2016)
33. Razzaque, M.A., Milojevic-Jevric, M., Palade, A., Clarke, S.: Middleware for Internet of Things: a survey. IEEE Internet Things J. **3**(1), 70–95 (2015)
34. Soldatos, J., Kefalakis, N., Hauswirth, M., Serrano, M., Calbimonte, J.P., Riahi, M., Aberer, K., Jayaraman, P.P., Zaslavsky, A., Žarko, I.P., Skorin-Kapov, L., Herzog, R.: Openiot: Open source internet-of-things in the cloud. In: Podnar Žarko, I., Pripužić, K., Serrano, M., (eds.), Interoperability and Open-Source Solutions for the Internet of Things, pp. 13–25. Cham, Springer International Publishing (2015)
35. Swetina, J., Guang, L., Jacobs, P., Ennesser, F., Song, J.S.: Toward a standardized common M2M service layer platform: Introduction to one M2M. IEEE Wirel. Commun. **21**(3), 20–26 (2014)
36. Thangavel, G., Memedi, M., Hedström, K.: A systematic review of social Internet of Things: concepts and application areas (2019)
37. Yachir, A., Amirat, Y., Chibani, A., Badache, N.: Event-aware framework for dynamic services discovery and selection in the context of ambient intelligence and Internet of Things. IEEE Trans. Autom. Sci. Eng. **13**(1), 85–102 (2015)

Smart Soil Monitoring Application (Case Study: Rwanda)

Eric Byiringiro(✉), Emmanuel Ndashimye, and Innocent Kabandana

College of Science and Technology, University of Rwanda, Kigali, Rwanda

Abstract. Internet of things has changed the way of working and monitoring things in different domain; through this paper; the internet of things is applied in agriculture for monitor the composition of soil parameters such as soil potential Hydrogen (pH), moisture, humidity and temperature. The proposed Smart Soil Monitoring Application (SSMA) collects accurate, reliable, timely data of soil status. Collected data play a big role in decision making for selecting the appropriate type of crops and the type and the quantity of required fertilizer for a particular soil in different areas in Rwanda. In this paper, data for soil compositions were collected by sensors and collected data are sent to microcontroller unit (nodeMCU) to be analyzed. The Smart Soil Monitoring Application system consists of a nodeMCU, Soil pH sensor for collecting pH data, moisture sensor to sense presense of water in soil and DHT11 for collecting data about humidity and temperature. Also, a ESP8266 WiFi module enables the system to send the sensed data to the cloud for storage and processing. A Processed data are displayed on dashboard (web based application) and accessed by farmers via an android application installed in their smart phone. The obtained preliminary results confirm the efficiency of the proposed SSMA for real-time monitoring soil parameters and thus contributing to the growth of farm production in Rwanda.

Keywords: Internet of things · Mobile application · Soil composition

1 Introduction

Rwanda is a land locked country, its economy is based on knowledge and agriculture sector. A large number of population depends on farming. As shown by statistics; a big number of working population (72%) are farmers [1].

However, the poor performance of the agricultural sector has been a major threat to Rwandan economic at national level. As precaution the Government of Rwanda has proposed policies aimed at enhancing the farming productivity such as reinforcing farmers' capacity to use fertilizers and identifying crops that showed more profitable than others [2].

Through the strategic plan for the transformation of agriculture in Rwanda, Ministry of Agriculture and Animal Resources (MINAGRI) indicated that the agriculture sector contributes to 4% with 31% contribution to the national GDP in 2018 [3]. Despite continuous advancement in contribution to national GDP, Agriculture in Rwanda faced

K. Arai (Ed.): FICC 2021, AISC 1363, pp. 212–224, 2021.
https://doi.org/10.1007/978-3-030-73100-7_16

with challenges like high cost of investment or finance, poor farming technologies and techniques, selection of crops based on agricultural zones and limited market access [1]. Thus not only these issues in agriculture also a luck of soil parameters monitoring techniques is a barrier to the agriculture in Rwanda. Generally, digitizing agriculture has positive impact. With the use of ICTs, farmers can access to information on markets and weather situational changes. In addition, the rapid emergence of the Internet of Things remodeled almost every industry including agriculture [4]. Using sensor network can quickly track soil composition parameters and overcome some barriers in agriculture.

Tracking agricultural environment for various soil parameters such as pH, moisture, humidity and temperatures are of great significance [5]. It is time consuming and tedious task to manually measure these parameters and control the moisture or pH levels. Moreover, luck of real time information on the above parameters results to many problems such as decision on which type of fertilizer to apply, crop appropriate to soil and failure to plan for policy makers. These problems result to poor production and costly utilization of chemicals in soil. As of now, farmers in Rwanda decide on crop to cultivate either based on information about the latest crop production or use the fertilizer without knowing the presence of nutrients in soil which is both less reliable and inefficient [6]. These problems could be addressed by applying IoT technology for monitoring the soil composition for deciding the appropriate crops and required quantity and type of fertilizer.

On the other hand, IoT is a revolutionary technology that leads the future of computing and communications [7]. Most of Rwandan depend on agriculture [1]. Because of this reason Internet of Things is needed to migrate with traditional agriculture methods. With the advent and growth of IoT, monitoring techniques are also becoming very modernized. The main objectives of soil parameters monitoring are to determine the status of soil composition through the sensors. The real time data collected by sensors allows farmers to plan for the coming seasons and automating multiple process including fertilization and crop selection. Temperature and humidity reports by sensors can utilized to choose the right crop for a particular season. With right amount of fertilizer and water can help in controlling the quality of crop. As result of using Internet of Things farmers can get an improvement in crop production and benefits financially.

Smart Soil Monitoring Application (SSMA) proposed in this paper provides an easy and quick method of measuring soil composition to provide accurate soil information to farmers and policy makers for deciding on the type of crops, as well as the type and the quantity of fertilizer to apply according to the soil parameters.

The remainder of this paper is organized as follows. Section 2 discusses literature review by IoT researchers on previously proposed solutions for smart agriculture and soil parameters monitoring. Further, Sect. 3 discusses identified gap and justification, Sect. 4 proposed Smart Soil Monitoring application description. Section 5 discusses the evaluation and results of the SSMA. Section 6 concludes the paper restating the achievements.

2 Literature Review

The existing literature includes different IoT based approach to monitor soil parameters for irrigation purpose, nutrients detection and modernize agriculture.

Researchers motivated by new technology (Internet of things) were developed a monitoring and controlling of greenhouse that uses sensors to measure humidity and temperature and send data to microcontroller to be processed. The data collected used to monitor and control the greenhouse parameters to achieve maximum plant growth in a greenhouse [8]. IoT Based Smart Farming System assisting farmer in getting live data (temperature and humidity) for efficient environment monitoring which will enable them to increase their overall yield and quality of products [9]. YASIR FAHIM, in his research recommends future researchers to increase sensors to fetch more data to full-fledged the agriculture monitoring [9]. LabView software was used to view humidity, temperature and conductivity of soil to monitor land and data were collected using sensor nodes [10].

To modernize agriculture a smart GPS based remote controlled robot to perform tasks like weeding, spraying, moisture sensing, bird and animal scaring, keeping vigilance, was developed in IoT based smart agriculture. It includes smart irrigation with smart control and intelligent decision making based on accurate real time data [11]. A system for precision agriculture which provides farmers with useful data about the soil, the water supply, and the general condition of their fields in a user friendly, easily accessible manner was discussed in literature and the system aims to make cultivation and irrigation more efficient as the farmer is able to make better informed decisions and thus save time and resources [12].

Internet of Things used to automate irrigation. A real time soil monitoring system for the application of agriculture designed to collect data sets for use in the analysis of selection of crops and their vulnerabilities for regulating the irrigation parameters which will help in the agricultural practices. The system automates water pump if the water level in the soil decreases below the threshold level [13]. Some other authors design and construct a microcontroller based automatic irrigation system to manage irrigation for a small area of land based on real time values of soil moisture and temperature [14]. In addition to the automatic irrigation, system use GSM to alert users to the measurements taken by sensors. In smart agriculture monitoring system using IoT, researcher recommend future work to remote data to ease monitoring [15, 16]. An advanced irrigation system was discussed and the system provided adequate amount of water to the crops [17, 18].

Based on the color sensor TCS3200, soil analysis using IoT was developed and measures soil nutrients (N, P, K) for rice crop. The system utilizes a device that can use the calorimetry to check the nutrients of the soil by generating color reactions. In a variety of literature, soil nutrients were detected using optical transducer and color sensor [19, 20]. Monitoring system in agriculture evolved and remotely control moisture, temperature, intruders, leaf wetness and proper irrigation facilities. This system was simulated in Proteus simulator [21].

3 Identified Gap and Justification

In different literatures, soil parameters were studied and analyzed for different usage. In [8] soil parameters were correctly locally and the users were not able to access on the data remotely. As in many literature, the authors developed system to collect data for irrigation purpose they focused only on temperature and humidity [14, 15, 22]. Nutrients were analyzed through the use of pH sensor or optical transducer [16, 19, 20]. GSM is a communication technology which is costly and consume a lot of power. To monitor soil parameters most of literature were used GSM as a communication technology as discussed in [15, 16, 18].

To lower system cost, human interaction and monitor a variety of soil composition such as soil nutrients, acidity or alkalinity of soil, moisture, humidity and temperature a new system was developed. In this paper, Smart Soil Monitoring Application measure the value of pH in soil, soil nutrients, soil moisture, soil humidity and temperature. The sensed values sent to cloud and accessed by users either on his mobile through SSMA android application or through web application.

4 Proposed System Description

This paper discusses a Smart Soil Monitoring Application proposed for monitoring soil composition to facilitate farmers and policy makers in deciding the type of crops and fertilizer needed in different regions in Rwanda. Proposed Smart Soil Monitoring Application consists of a pH sensor, moisture sensor and DHT11 controlled by a nodeMCU. pH sensor for collecting pH data, moisture sensor to sense presence of water in soiland and DHT11 for collecting data about humidity and temperature. The ESP8266 WiFi module enables the system to send the sensed data to the cloud through the protocols such as HTTP and MQTT for storage and processing. Processed data are displayed on dashboard (web based application) and accessed by farmers via an android application installed in their smart phone. The proposed system is presented in Fig. 1.

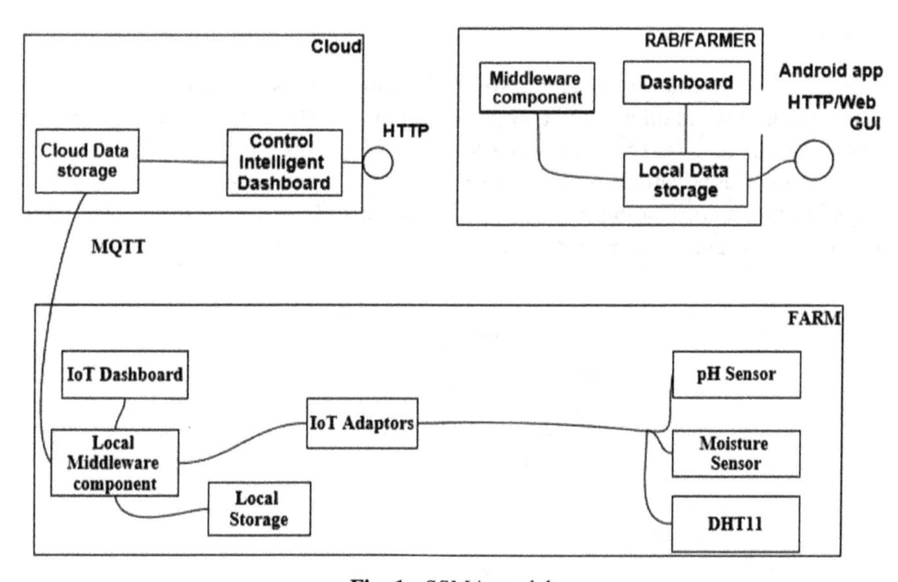

Fig. 1. SSMA model

4.1 Soil pH Sensor

The pH is a measure of acidity and alkalinity, or the caustic and base present in a given solution. It is generally expressed with a numeric scale ranging from 0–14. The value

7 represents neutrality. The numbers on the scale increase with increasing alkalinity while the numbers on scale decrease with increasing acidity. In general, Nitrogen(N) and Potassium available within pH 6.5 to 8 while Phosphorus is most available within soil pH 5.5 to 7.5 [23]. Figure 2 shows pH sensor.

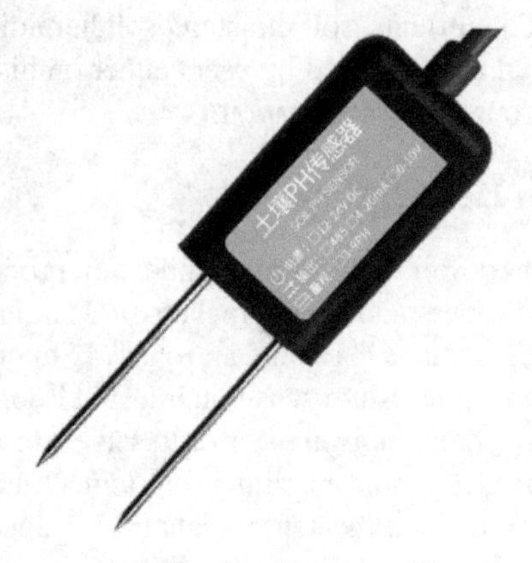

Fig. 2. pH sensor

4.2 Moisture Sensor

Soil moisture is the content of water in soil. Soil moisture sensor measure the volumetric water content in soil. Soil water content is a soil physical state variable which is defined as the water contained in the unsaturated soil zone. Measuring soil moisture is important for agricultural applications to help farmers manage moisture to increase yields and the quality of the crop by improving management of soil moisture during critical plant growth stages. Figure 3 shows a moisture sensor.

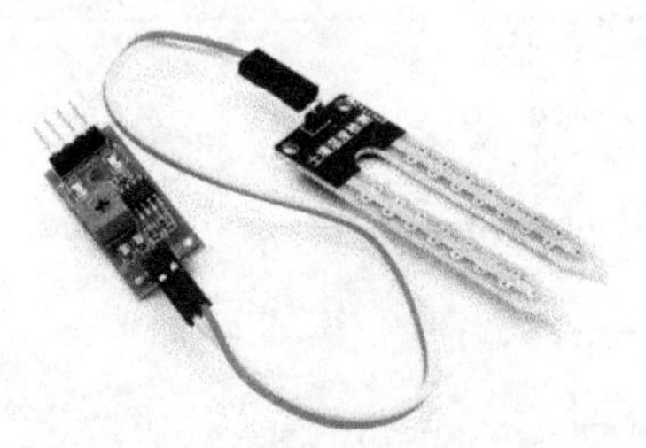

Fig. 3. Moisture sensor

4.3 DHT11

DHT11 is a basic, ultra-low-cost digital temperature and humidity sensor. It uses a capacitive humidity sensor and a thermistor for measuring the surrounding air and spits out a digital signal on the data pin. It is simple to use and get data every 2 s. Figure 4 shows a Dht11 sensor.

Fig. 4. DHT11 for humidity and temperature

4.4 NodeMCU

NodeMCU is an open source firmware and development kit that helps you to prototype your IoT product. It is based on ESP8266. Figure 5 exhibits a nodeMCU.

Fig. 5. nodeMCU for coordinating sensors

5 Evaluation and Results

Developed system consists of a sensor node to sense data and send data to cloud and accessed by users via an android application. Temperature, humidity, moisture and pH values are displayed on a dashboard.

5.1 Evaluation of Temperature and Humidity

To evaluate temperature and humidity a DHT11 sensor was used. Figure 6 presents a daily average of temperature in line graph. Figure 7 presents daily average of humidity in line graph. Figure 8 shows numerical values for both temperature and humidity and these values are the ones displayed on android application for end users, farmers. Displayed data are updated every 15 s. With real time temperature and humidity data provided within SSMA system, farmers can adopt the given situation and take additional farm management practices to minimize crop losses. Therefore, accurate information regarding the weather is important so that farm activities can be planned.

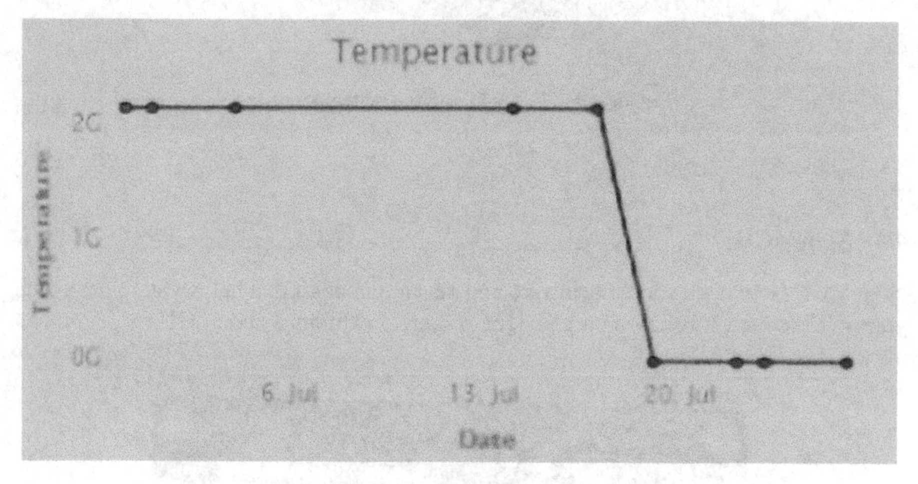

Fig. 6. Temperature graph

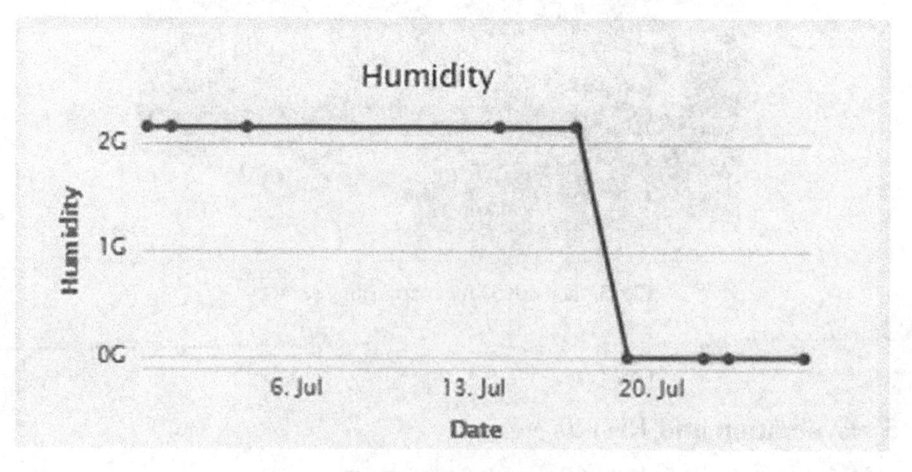

Fig. 7. Humidity graph

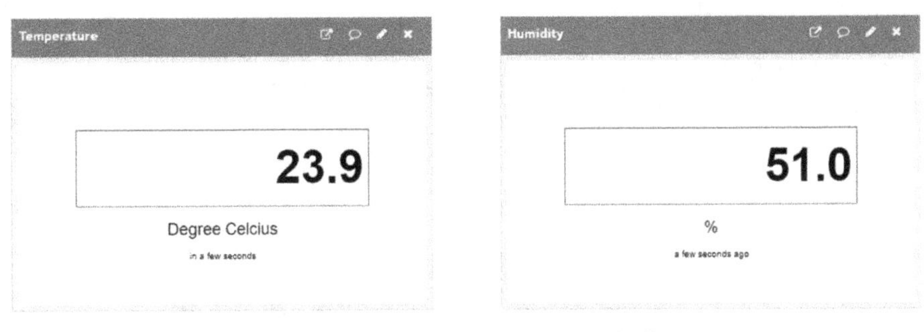

Fig. 8. dht11 data displayed numerically

5.2 Evaluating Soil Moisture

This section evaluates soil moisture of the proposed system. Data sensed by moisture sensor are presented in Fig. 9. Collected data are displayed numerically and updated every 15 s as shown in Fig. 10. In this paper, monitoring soil moisture help farmers understanding the actual soil water condition for deciding the correct measures. In addition, having soil moisture data of a particular region help farmers to select a crop to cultivate based on estimate quantity of moisture.

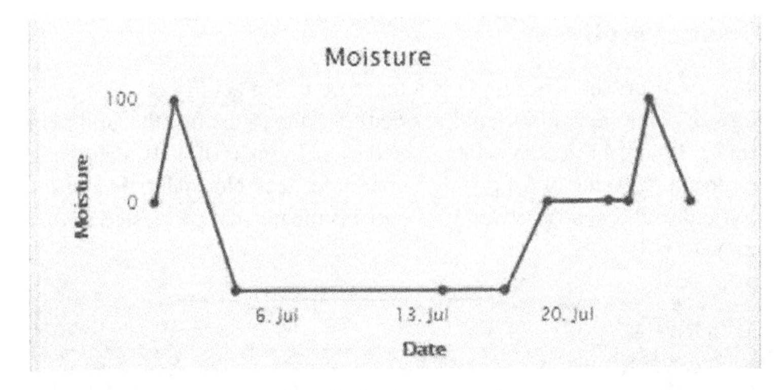

Fig. 9. Moisture graph

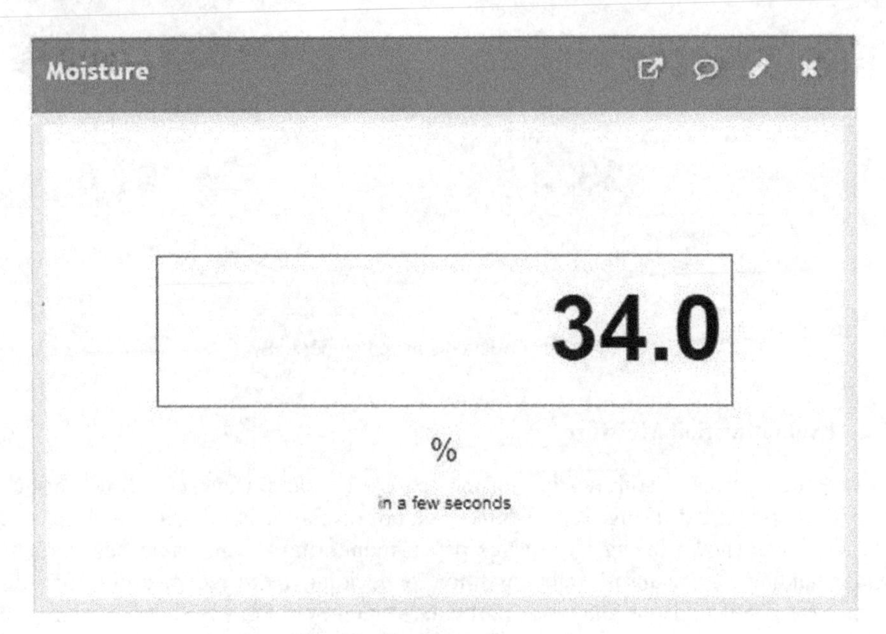

Fig. 10. Current moisture value

5.3 Evaluating Soil pH

Furthermore, as discussed before, pH is a measure of the acidity or alkalinity of a soil. Monitoring soil pH is very important in agriculture due to the fact that soil pH regulates plant nutrients. Figure 11 presents data sensed by pH sensor of daily average. Figure 12 shows the current pH value of 6.1 during a conducted test. Normally, the optimum range for most agricultural crops is between 5.5 and 7.5 means that soil tested in this paper is in good range of pH.

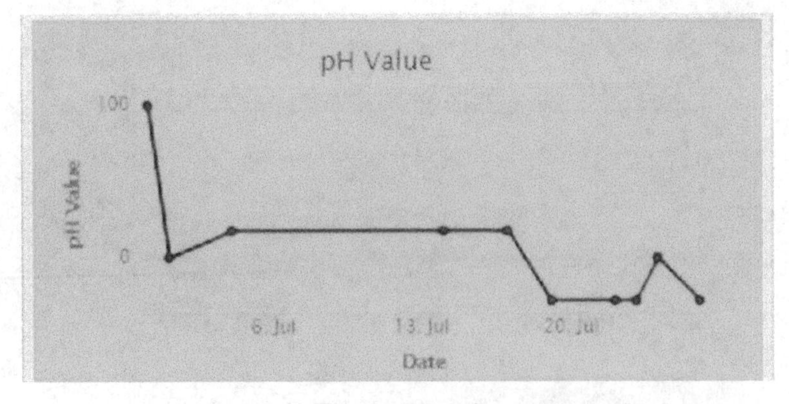

Fig. 11. pH graph

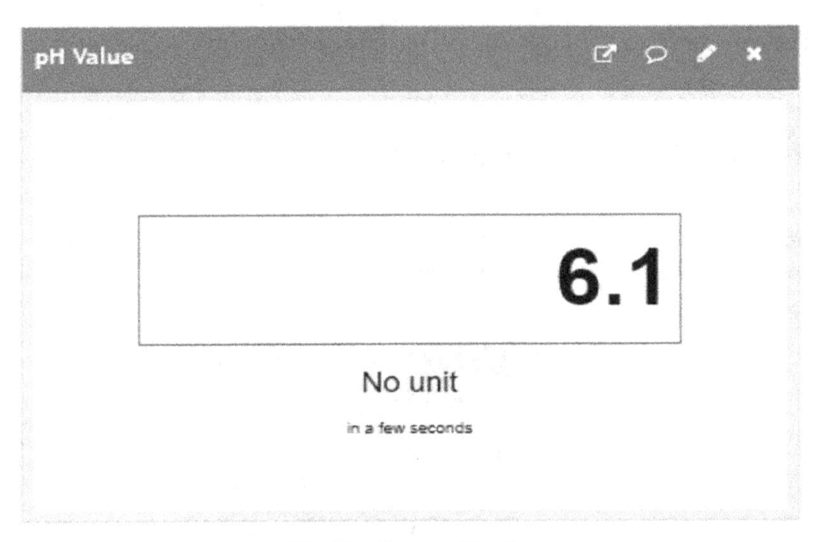

Fig. 12. Current pH value

5.4 Sensor Node Location

Farmers will be able to track the geographical location of our developed embedded device once deployed in a farm. During conducted test, one sensor node was used and deployed in a farm. Figure 13 indicate the real geographical location (latitude and longitude) at which our node was deployed. Tracking node locations facilitates policy makers (such as Rwanda Agriculture Board) with an exact map of farms with specific needs for a

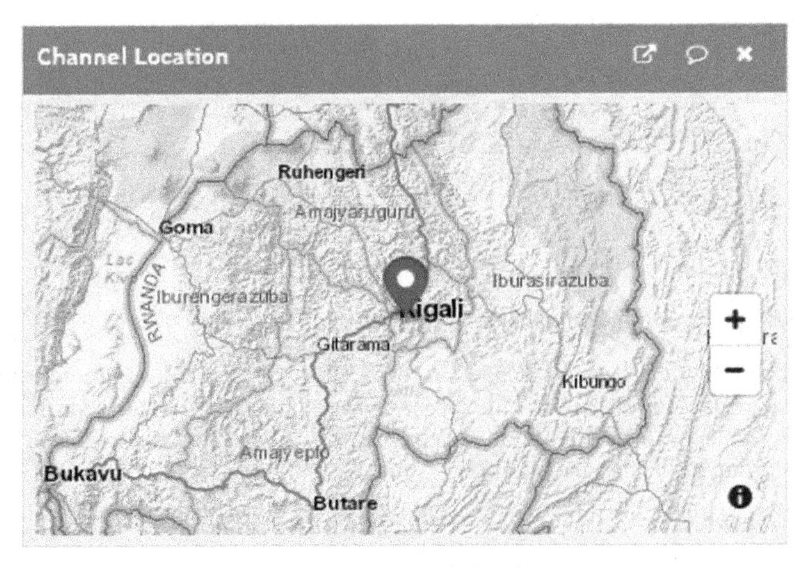

Fig. 13. Sensor node location

direct action. Moreover, fertilizer vendors make use of this map to locate zones needing fertilizer.

5.5 Evaluating SSMA Mobile Application

An android application was developed for farmers to easily communicate with the SSMA system. In the system, farmers monitor their soil status with an android application installed in their smart phone. Figure 14 shows SSMA apk running in android phone.

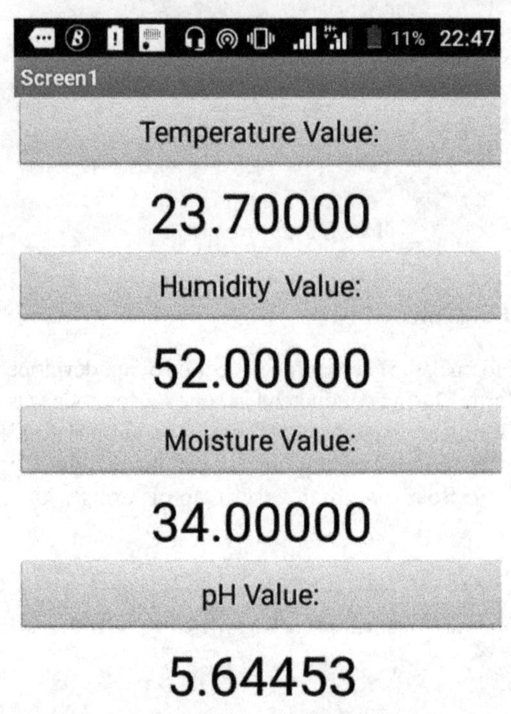

Fig. 14. SSMA android application

5.6 Seven Days Data Comparison

The proposed system stores data collected for analysis and reporting. Figure 15 exhibits temperature data collected over a week period. The highest temperature is identified on sixth day while the lowest temperature is on fifth day. This illustration helps the system users to compare the data in a week.

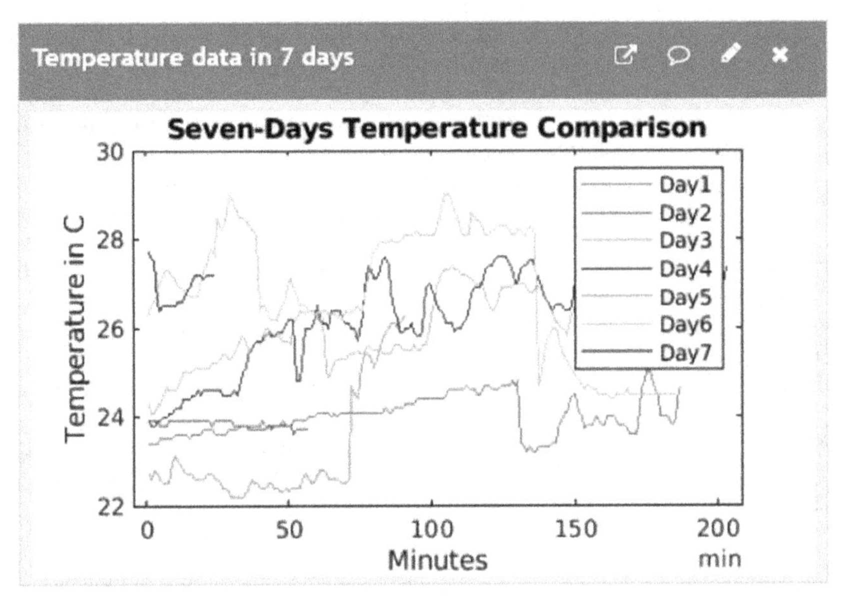

Fig. 15. Data comparison

6 Conclusion and Future Scope

A Smart Soil Monitoring Application is presented, demonstrating its functionality for retrieving data from sensors, pushing these data through a gateway, storing the data on cloud. Subsequently, the results are presented to users via a web application and an android application for end users. The system collects accurate, reliable and timely data of soil status. This helps farmers to select a typical type of crops and fertilization activities for a specify region in Rwanda based on the soil composition. Moreover, the monitored data assist decision makers to take action and plan for different farming seasons, hence an increase in revenue and food production and thus contributing to the production growth in Rwanda. A comparison of our results with data from Rwanda Agriculture Board (RAB) collected and processed in RAB soil laboratory, is suggested as an extension of the work presented here.

References

1. IPAR Rwanda: Rwandan agriculture sector situational analysis. MINAGRI, Kigali (2018)
2. Murekezi, A., Kelly, V.: Fertilizer response and profitability in Rwanda. FSRP and FAO, Kigali (2000)
3. MINISTRY OF AGRICULTURE AND ANIMAL RESOURCES: Strategic plan for agriculture transformation 2018–24, MINAGRI, Kigali (2018)
4. Konstantinos, P.B., Liakos, G.: Machine learning in agriculture: a review. Sensors **18**(10), 29 (2018)
5. Ayaz, M.: Internet-of-things (IoT)-based smart agriculture: toward making the fields talk. Research Challenges and Opportunities, p. 33 (2019)

6. Rwirahira, J.: Rwandan agriculture sector situational analysis, IPAR Rwanda, Kigali (2011)
7. Mekala, M.S.: Smart agriculture with IoT and cloud computing. In: International Conference on Microelectronic Device, Circuits and System (2017)
8. Mathew, S.: Monitoring and controlling of greenhouse using microcontroller," International Journal of Pure and Applied Mathematics, vol. 118, no. 20, p. 6, (2018)
9. Fahim, Y.: IoT based Smart Farming System. CIT, KOKRAJHAR (2018)
10. Georgiera, T.: Design of sensor network for monitoring soil parameters. Agric. Life, Life Agric. **10**(10), 7 (2016)
11. Kawitkar, R.S., Gondchawar, N.: IoT based smart agriculture. Int. J. Adv. Res. Comput. Commun. Eng. **5**(6), 5 (2016)
12. Zografos, A.: Wireless Sensor-Based Agricultural Monitoring System. KTH Royal Institute of Technology, Stockholm (2014)
13. Balakrishna, K.: Real-Time Soil Monitoring System for the Application of Agriculture, vol. 6, no. 5, p. 7. Maharaja Research Foundation (2018)
14. Aderemi, A.A.: Design and Construction of a Microcontroller Based Automatic Irrigation System. In: World Congress on Engineering and Computer Science, San Francisco, USA (2015)
15. Patil, S.N., Jadhav, M.B.: Smart agriculture monitoring system using IOT. Int. J. Adv. Res. Comput. Commun. Eng. **8**(4), 5 (2019)
16. Marwah, G.K.: Acquisition of soil pH parameter and data logging using PIC microcontroller. Int. J. Eng. Res. Technol. **3**(9), 4 (2014)
17. Mali, P.M.: Microcontroller based soil parameter monitoring for automatic irrigation system. Int. J. Fut. Revol. Comput. Sci. Commun. Eng. **4**(4), 3 (2018)
18. Aswathy, S., Manimegalai, S., Maria Rashmi Fernando, M., Frank Vijay, J.: Smart soil testing. Int. J. Innov. Res. Sci. Eng. Technol. **7**(4), 4 (2018)
19. Vivekanandreddy, P., Reshmi, S.: IOT based soil nutrients detection system. Int. J. Res. Appl. Sci. Eng. Technol. **7**(6), 5 (2019)
20. Abhishek, A.M., et al.: Soil Analysis Using IoT. K.J.Somaiya Institute Of Engineering I.T, Mumbai (2018)
21. Suma, N.: IOT based smart agriculture monitoring system. Int. J. Recent Innov. Trends. Comput. Commun. **5**(2), 5 (2017)
22. Fenila Naomi, J.: A soil quality analysis and an efficient irrigation system using agro-sensors **8**(5), 4 (2019)
23. McCauley, A.: Soil pH and organic matter. In Nutrient Management, p. 16. Montana State University, Bozeman (2017)

IoT Beehives and Open Data to Gauge Urban Biodiversity

Gerard Schouten[1,2(✉)], Mirella Sangiovanni[2], and Willem-Jan van den Heuvel[2]

[1] Fontys University of Applied Sciences, Eindhoven, The Netherlands
g.schouten@fontys.nl
[2] Tilburg University - JADS, Tilburg, The Netherlands

Abstract. Environmental sustainability is a key element of modern society. It has received global attention in recent years, both at scientific and administrative levels. Despite the scrupulous studies addressing this theme, many issues remain largely unresolved. A (big) data and AI approach is a promising alternative for tackling societal or environmental problems that are hard to grasp. We apply this approach to assess urban biodiversity. More specifically, the concept of *intelligent beehives* is introduced. This concept encapsulates and leverages biotic elements, such as harnessing bees as biomonitoring agents, with technologies like IoT instrumentation and AI. Together they comprise the data-driven services that shape the backbone of a real-time environmental dashboard. In this vision paper, our solution architecture and prototypization for such service-enabled beehives are sketched and discussed. We focus on the role of the IoT beehive network and open data for predictive modelling of biodiversity and argue how MLOps practices support a transformative process for creating awareness and maintaining or even increasing urban biodiversity.

Keywords: Urban biodiversity · IoT Beehive · Open data · MLOps and Stratified Architecture

1 Introduction

Urban areas are expanding rapidly [15]. It is estimated that by 2050 about two-thirds of the global population lives in cities. On the one hand, the consequence of urbanization is a large-scale loss of natural habitats, affecting people's health and well-being [20]. As stated in the alarming IPBES 2019 report of the UN[1], unsustainable 'grey' cities have pervasive and irremediable effects on the quality of ecosystems [9] and accelerate biodiversity decline [10]. On the other hand, if planned well, 'green and blue' cities can maintain nature-friendly habitats and provide ample opportunities for a diverse flora and fauna ecology (e.g., [2,23]), with various micro-climates and places to breed and shelter. Supporting

[1] https://ipbes.net/global-assessment-report-biodiversity-ecosystem-services.

K. Arai (Ed.): FICC 2021, AISC 1363, pp. 225–235, 2021.
https://doi.org/10.1007/978-3-030-73100-7_17

environmental sustainability[2] is a global challenge and a necessity for the survival of the organisms that are part of it (including humans). It is an indispensable prerogative to guarantee vital ecosystem services, like water purification, nutrient cycling, waste management, pollination, supply of medicines and other natural products. Protection as well as careful design and usage of man-made habitats is of crucial importance for this [7,14].

In this paper, a hybrid solution to materialize this urgency is proposed, as an alternative to costly surveys performed by humans. With a simple yet realistic concept – a network of intelligent beehives that function as a proxy for monitoring biodiversity in urban areas – we show how biotic elements, along with Internet-of-Things (IoT) instrumentation, state-of-the-art Artificial Intelligence (AI), and software engineering [4,27], can create a breakthrough for the preservation of urban ecosystems. AI algorithms are able to spot and expose biological dependencies in large and varied multi-sensory datasets. This approach allows us to grasp nature's intricate relationships [25]. Informative and expressive dashboards will increase awareness for biodiversity. Digital city twin platforms can be build to run simulations, evaluate scenarios, and to do preemptive contingency planning for biodiversity. These technologies have the potential to raise the quality of complex decisions beyond state-of-the-art [21].

The basic idea of the IoT beehive network is to commission honeybees (*Apis mellifera*) as biomonitoring agents and mine their 'data stories' to gauge local biodiversity. Honeybees are highly sensitive to the quality of their environment [12,19]. They live in colonies of up to 50,000 individuals. In order to feed the colony, including the queen bee and her larvae, the bees collect pollen and nectar from flowers in the environment and bring them back to the hive. Forager and scout bees travel up to three kilometres from the hive to do this.[3] More specifically, a network of glass-walled beehives with multi-sensory equipment is envisioned, intermixing such IoT sensing with open data, like: (1) satellite images, (2) public databases of planted greenery, (3) city layout, and (4) chemical analyzes of air, soil and water.

The IoT beehive network is an ideal vehicle to reach out to citizens, from school children to urban planners. To increase awareness and support transformative practices on environmental issues [24], these stakeholders will be involved in both the design and operational phase of our proposed platform. An agile way of working is adopted, where activities like: (1) experimenting with data, (2) development of the hardware and software, and (3) deployment of the IoT beehives in the field, are interwoven, and continuously improved.

Section 2 of this paper[4] outlines how AI – fed with data collected 24/7 by instrumented hives and open data – can be used to predict ecological quality measures. Section 3 describes the transformative process in more detail. The layered architecture of our proposed solution is explained in Sect. 4. This architecture is able to process large amounts of data and supports urban biodiversity

[2] https://www.un.org/sustainabledevelopment/news/communications-material.
[3] Note that pollinators directly affect 35% of the world's food crop production [13].
[4] This is an extended version of an earlier concept presented at SummerSoC 2020 [22].

services, like an environmental dashboard. Future outlook and technical as well as non-technical challenges are discussed in Sect. 5. The conclusions are summarized in Sect. 6.

2 AI for Urban Biodiversity

The ambitious, long-term objective is to plan cities taking into account biodiversity as a major factor. In order to do this, it is indispensable to quantify the concept of urban biodiversity cost-effectively. The current way of dealing with biodiversity is based on costly field monitoring programs and is largely an afterthought. The lack of AI models, that combine multiple aspects to predict urban biodiversity, makes it difficult to set goals for policies and targets for an actionable approach [16]. Our objective is to build a first-of-a-kind supervised machine learning model, that is able to predict urban biodiversity, based on data that can be collected real-time. As *feature set* (i.e., input) for this AI model we propose to start with: (1) IoT beehive data, and (2) open environmental data. To digest the huge amount of data in inference mode, a data collection and storage platform is conceptualized as shown in Fig. 1. In addition, a flat glass-walled beehive is proposed to preserve visual access to bee behavior for the general public. Note that this hive design mimics the natural vertical nesting of honey bees in tree cavities.

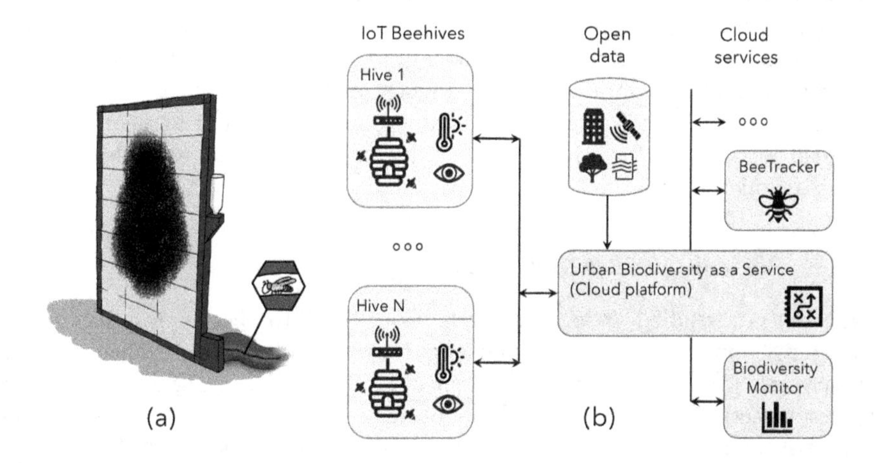

Fig. 1. (a) Schematic prototype of the intelligent beehive. (b) Conceptual setup with data collection at IoT nodes, use of open data, and computational cloud services.

2.1 Data from the IoT Beehive Network

Honeybees are excellent bio-indicators [5]. They have an intriguing colony behavior [26], and strongly interact with their natural environment. Some illustrative examples, how 'data stories' of bees can be used for assessing biodiversity of the area surrounding the IoT beehive, are listed below:

- The so-called waggle dance, [3,18] performed by worker bees, communicates the location of food resources to unemployed foragers. They simply check the dancer's movements, decodes them into the direction and distance from the hive, and start to search for food. This amazing behavior can be detected and captured by high-speed camera's mounted to the transparent beehive. Validated computer vision algorithms decode the waggle dance and convert it to 'whereabouts' information [30].
- A constant and diverse offering of pollen and nectar is beneficial for colony health. There are a number of indicators for colony well-being, e.g. (1) the number of individual worker bees, (2) the amount of offspring produced, (3) the amount and composition of stored honey [19], and (4) the number of holes (empty cells) in the honeycomb. During the experimentation phase, data about these factors will be collected with various IoT devices.
- Bees bring back pollen to the hive. These small particles are a direct indicator of the flower richness in the environment. To identify the pollen we start with offline eDNA pollen analysis 'in the lab', and finally want to evolve towards real-time 'in vivo' analysis based on infrared spectroscopy [31].

2.2 Open Environmental Data

In addition, open environmental data complements and enriches the feature set that is needed to reliably assess biodiversity. For instance, classification of tree species in urban areas can be derived from satellite images and LiDAR with a data fusion and deep learning approach [11]. Other useful information related to biodiversity might be obtained either directly from public databases of planted greenery, or indirectly from air pollution measurements, water and soil samples, etc. Also weather, traffic, city layout, and even social media are relevant data sources to take into account.

2.3 Ecological Quality Measures

Biodiversity is studied at various levels, from DNA to species and even entire ecosystems. Our envisioned AI aims for a system producing ecological quality measures [6], rather than detailed species counts. These aggregate measures express aspects of ecosystem health, focusing on species groups (insects, flowers, birds, etc.). These indicative measures (environmental index) are designed for easy use and interpretation by decision makers and citizens, without having in-depth biological knowledge. So we propose to use ecological quality measures as *labels* (i.e. output) to train the AI. The indicators are derived from systematic field monitoring surveys with a well-defined mapping. The protocolized ground-truth counts will be collected, during the next few years, in several (at least three) Dutch cities. The set of established AI algorithms – random forest, support vectors machines, neural networks, etc. – will be probed and tuned in order to achieve optimal performance.

2.4 Biodiversity as a Service

As explained above, the AI model for urban biodiversity and its underlying data engineering pipeline enable an *Urban Biodiversity Monitor* as a real-time service. An *Interactive Bee Tracker*, that shows the flight patterns as indicated by the waggle dances projected on an actual city map, and a *Bee Health Dashboard* might be considered as informative visualisations. By adding more data sources to this concept, the 'app store' could be extended with for instance a *Bio-hazard Reporter* or a digital twin-like *Nature-inclusive Urban Planning Module*, to mention a few.

3 MLOps Approach for Systemic Change

MLOps has emerged from the DevOps philosophy and associated practices that streamline the software development workflow and delivery processes. In particular, DevOps adopts the continuous integration and testing cycle to produce and deploy production-ready new micro-releases of applications at fast pace [8]. As shown in Fig. 2, MLOps is an extension of this iterative approach, targeted at data-driven (AI-like) applications. In addition to traditional software development, it includes a third experimental 'data' cycle that is composed of the following activities: (1) business understanding, (2) data acquisition, (3) exploratory data analysis, and (4) building AI models. To put this into practice, MLOps requires policies based on metrics and telemetry such as algorithmic performance indicators for accuracy (like precision and recall), as well as software quality metrics [17]. It also demands improved dependency management (and thus transparency) between model development, training, validation and deployment.

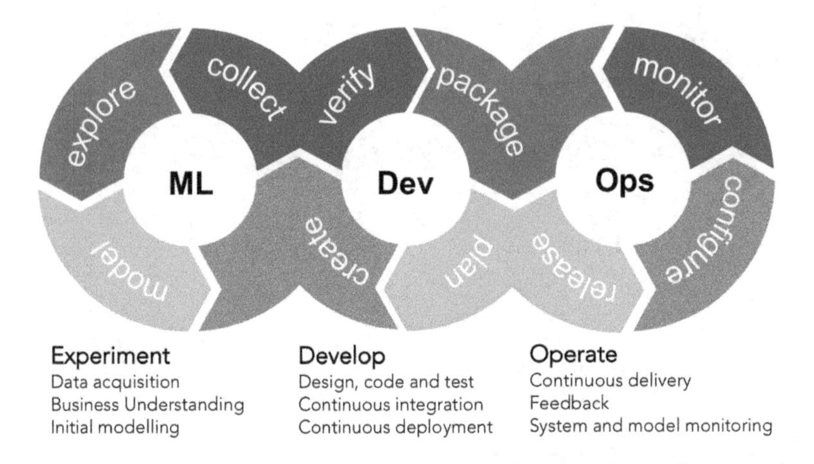

Fig. 2. MLOps lifecycle steps.

Adopting MLOps implies a culture shift between data scientists, data engineers, software engineers, and domain experts. This project engages not only domain experts like biologists and urban planners (decision makers with respect to city layout and planning), but also school children, the future citizens of our planet. All these stakeholders are involved in the development and operations processes of the IoT beehive network, for requirements elicitation but also to experiment with early designs and prototypization. An MLOps approach is paramount for projects that aim at creating impact with smart services and contributing to SDG goals [27].

4 Beehive Network Architecture

This project embraces a stratified data architecture, with layers that are separated in terms of the type of data and logic they combine, and built on top of each other. Each layer is self-contained, and addresses a specific concern, such as data collection, preprocessing, storage and analytics, visualization, and is inhabited by service-enabled components that can be accessed, queried and managed through a well-defined interface. In this way, a smart service taxonomy for IoT beehives is systematically developed, ranging from coarse-grained components, to fine-grained micro-services.

In the following, the technical solution architecture is explained, starting from the physical layers and working up in the stack to the logical (software) layers. Graphically, the IoT beehive data architecture is depicted in Fig. 3.

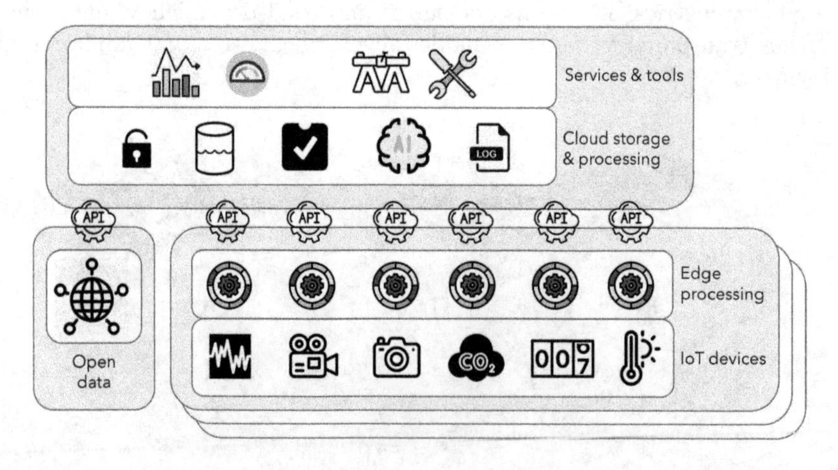

Fig. 3. The IoT beehive stratified data architecture.

4.1 IoT Beehive Device Layer

Each beehive is to be instrumented with a series of IoT devices to measure several factors, including, but not restricted to: (1) the number of bees entering and leaving the hive, (2) climate and weather conditions inside and outside the hive, (3) the type of pollen collected by forager bees upon entering the hive, (4) the composition of the honeycomb, and (5) the registration of waggle dances. The *Device Layer* consists of a mixture of existing off-the-shelf sensor systems (such as the Melixa[5] or BEEP[6] scale hive monitoring system), a weather station with a 24/7 datalink, vision with high-speed camera's, audio, and even completely new experimental sensors (such as miniaturized infrared spectroscopy for pollen classification).

4.2 IoT Beehive Edge Layer

It is evident that the above IoT devices generate huge volumes of data, requiring vast amounts of bandwidth to be transported from its source to the target computing environment. The *Edge Layer* consists of a set of intermediate data storage and preprocessing modules close to the source (the edge of the cloud). Edge modules serve as decentral proxies of the central processing unit. This can be achieved either through data preprocessing in microprocessors that are embedded in the IoT device (e.g., in the weather station), or by intelligent data gateways (e.g. that fuse images from multiple cameras). A special edge processing decoder will be developed to decipher bee waggle dance video into polar coordinates pointing to food supply in the field. Processing at the edge of an IoT device that is attached to each beehive, minimizes communication demands and allows for real-time and local status monitoring at each beehive in the array network.

4.3 Cloud Data Layer

This logical layer sits on top of the physical data architecture stratosphere. Data from each geographically distributed beehive is transmitted from the IoT microprocessors or edge-gateways through RESTful APIs to the central cloud datalake that is the beating heart-and-soul of the *Cloud Data Layer*. In this way, the datalake collects and stores all data generated by the IoT beehives as well as mined open data sources (public databases of planted greenery, satellite images, LiDAR, air pollution data, traffic, social media, etc.). Next to the datalake, this layer also offers dedicated data repositories for test and validation purposes. A data quality monitor continually measures the level of data quality along the axes of consistency, accuracy and timeliness, and employs advanced methods, such as machine learning, to predict and automatically resolve potential data omissions or anomalies. A special facility is in place to guarantee data safety and security, and allows for example citizens (including primary school kids) and professionals (e.g., policy makers) to access, manipulate and query data according to their own access/control schemes, and organizational/legal rules, norms, and policies. But above all, it contains an AI-engine that is able to gauge urban biodiversity based on the data that is present in the datalake.

[5] http://melixa.eu/en/.
[6] https://beep.nl/home-english.

4.4 Cloud Service Layer

This layer provides the central access point for all users of the urban biodiversity platform, offering: (1) an AI-workbench for modifying (e.g., re-training) AI-models, (2) configuration and testing tools for building new models promoted by the MLOps process [28], and (3) an interactive user-friendly and easy-to-understand environmental dashboard. Lastly, this layer will be the habitat of a digital city twin constituting a virtual system that accommodates through the layers below the real-time confluence of data across beehive IoT devices and other repositories and uses the AI-engine to facilitate urban planning with respect to biodiversity. Through mechanisms such as the ability to run 'what-if' scenarios the digital city twin is a powerful tool to better predict and appreciate the impact of novel policies on the environment in which we live.

5 Discussion

Recent developments in (big) data and AI enable a new scientific approach to so-called 'wicked' problems, such as global challenges [27]. Big data facilitates a high level of automation in mining, storing, processing, analyzing, and visualizing data. AI algorithms can uncover hidden correlations in large datasets that are difficult (if not impossible) to find using traditional hypothesis-based empirical inquiry.

We argue that 'data stories told by bees' enriched with open data can be exploited to gauge urban biodiversity. Note that our proposed system taps-in on insects and flowers, key players for ecosystems, and hence accounts already directly for a portion of biodiversity. However, in the 'web of life', long and complex food chains may produce unanticipated relationships. A simplified example: voles feed on certain herbaceous plants, kestrels eat voles, so detecting these plants' pollen in the hive may indicate the presence of this bird of prey. Traditional approaches [29] might miss such a pattern, whereas our approach will find it, if present in the data. Harnessing this power for biological research is one of the breakthroughs of this research.

The proposed architecture serves several user scenarios. The interactive visualization tools create for citizens awareness and a sense of urgency for the biodiversity crash. The city digital twin combined with predictive AI tools enable policy makers to come up with city layouts that are nature friendly. And finally, biologists might use the wealth of data that will be collected (and published as open data) to study bee behavior.

Some technical and non-technical challenges that we foresee are: (1) before decoding the waggle dance we should be able to detect real-time where exactly this happens on the honeycomb dance floor. This requires a special-purpose visual attention software module, (2) it is known that the development of AI services can easily create unwanted data dependencies making the solution not maintainable [1]. This is counteracted by defining a componetized setup of models (instead of one monolithic end-to-end machine learning model) supported by well-defined interfaces, (3) bee colonies might die, interrupting the data collection process (this is inherent to working with living creatures), and finally, (4) is not clear whether our AI model – based on a limited number of hive locations in a few cities – can be generalized to other cities and/or other years. In order to mitigate this and reinforce the concept we are currently experimenting with large glass-walled beehive housings (size: 1m80 × 1m20 × 0m20) that can host two colonies. Two independent colonies, not only provide an extended sampling of the area surrounding the hive, they also double the number of probing IoT devices.

In the near future, we foresee novel research routes in various directions in the field of hi-tech and environmental sustainability. Firstly, we aim to develop reusable models, based on observations and abstractions, for predicting biodiversity in rural and other areas. Second, we intend to explore the possibilities of other hybrid biotic hi-tech systems and combine it with open data sources. Note that our architecture is generic and open. It can easily accommodate for new IoT devices. Lastly, accumulated future research will further flesh out a smart service ontology to accommodate semantic reasoning, connecting discrete and instantaneous events in the real world with it's impact on biodiversity.

6 Conclusions

In this vision paper, a novel architecture is introduced to facilitate IoT instrumented beehives. This stratified architecture is dedicated towards assessing urban biodiversity and supports sustainable urban ecosystems. We have illustrated and preliminary explored the proposed architecture in a case study and identified major data sources and types of analysis that should be pursued. We also emphasized, that in order to increase impact, this data engineering effort will adopt an MLOps approach with a diverse, multi-disciplinary group of stakeholders. In the coming years, the idea of monitoring urban biodiversity with intelligent beehives and open data will be probed, tested, and elaborated further in three Dutch cities.

Acknowledgment. The authors thank Michiel Groenemeijer for his valuable and creative beehive design contributions and his dedicated beekeeping activities for this project.

References

1. Amershi, S., Begel, A., Bird, C., DeLine, R., Gall, H., Kamar, E., Nagappan, N., Nushi, B., Zimmermann, T.: Software engineering for machine learning: a case study. In: Proceedings of the 41st International Conference on Software Engineering: Software Engineering in Practice, pp. 291–300. IEEE Press (2019)
2. Baldock, K.C.R., Goddard, M.A., Hicks, D.M., Kunin, W.E., Mitschunas, N., Osgathorpe, L.M., Potts, S.G., Robertson, K.M., Scott, A.V., Stone, G.N., Vaughan, I.P., Memmott, J.: Where is the UK's pollinator biodiversity? The importance of urban areas for flower-visiting insects. Proc. Roy. Soc. B: Biol. Sci. **282**(1803), 20142849 (2015)

3. Biesmeijer, J.C., Seeley, T.D.: The use of waggle dance information by honeybees throughout their foraging careers. Behav. Ecol. Sociobiol. **59**(1), 133–142 (2005)
4. Bublitz, F.M., Oetomo, A., Sahu, K.S., Kuang, A., Fadrique, L.X., Velmovitsky, P.E., Nobrega, R.M., Morita, P.P.: Disruptive technologies for environment and health research: an overview of artificial intelligence, blockchain, and internet of things. Int. J. Environ. Res. Public Health **16**(20), 3847 (2019)
5. Celli, G., Maccagnani, B.: Honey bees as bioindicators of environmental pollution. Bull. Insectol. **56**(1), 137–139 (2003)
6. Dale, V.H., Beyeler, S.C.: Challenges in the development and use of ecological indicators. Ecol. Ind. **1**(1), 3–10 (2001)
7. de Palma, A., Kuhlmann, M., Roberts, S.P.M., Potts, S.G., Borger, L., Hudson, L.N., Lysenko, I., Newbold, T., Purvis, A.: Ecological traits affect the sensitivity of bees to land-use pressures in European agricultural landscapes. J. Appl. Ecol. **52**(6), 1567–1577 (2015)
8. Ebert, C., Gallardo, G., Hernantes, J., Serrano, N.: DevOps. IEEE Softw. **33**(3), 94–100 (2016)
9. Guetté, A., Gaüzère, P., Devictor, V., Jiguet, F., Godet, L.: Measuring the synanthropy of species and communities to monitor the effects of urbanization on biodiversity. Ecol. Ind. **79**, 139–154 (2017)
10. Hallmann, C.A., Sorg, M., Jongejans, E., Siepel, H., Hofland, N., Schwan, H., Stenmans, W., Müller, A., Sumser, H., Hörren, T., Goulson, D., de Kroon, H.: More than 75 percent decline over 27 years in total flying insect biomass in protected areas. PloS One **12**(10), e0185809 (2017)
11. Hartling, S., Sagan, V., Sidike, P., Maimaitijiang, M., Carron, J.: Urban tree species classification using a worldview-2/3 and LiDAR data fusion approach and deep learning. Sensors **19**(6), 1284 (2019)
12. Henry, M., Béguin, M., Requier, F., Rollin, O., Odoux, J.F., Aupinel, P., Aptel, J., Tchamitchian, S., Decourtye, A.: A common pesticide decreases foraging success and survival in honeybees. Science **336**, 348–350 (2012)
13. Klein, A.-M., Vaissiere, B.E., Cane, J.H., Steffan-Dewenter, I., Cunningham, S.A., Kremen, C., Tscharntke, T.: Importance of pollinators in changing landscapes for world crops. Proc. Roy. Soc. B: Biol. Sci. **274**(1608), 303–313 (2007)
14. Lepczyk, C.A., Aronson, M.F.J., Evans, K.L., Goddard, M.A., Lerman, S.B., MacIvor, J.S.: Biodiversity in the city: fundamental questions for understanding the ecology of urban green spaces for biodiversity conservation. BioScience **67**(9), 799–807 (2017)
15. Maksimović, Č., Kurian, M., Ardakanian, R.: Rethinking Infrastructure Design for Multi-use Water Services. Springer International Publishing (2015)
16. Nilon, C.H., Aronson, M.F., Cilliers, S.S., Dobbs, C., Frazee, L.J., Goddard, M.A., O'Neill, K.M., Roberts, D., Stander, E.K., Werner, P., Winter, M., Yocom, K.P.: Planning for the future of urban biodiversity: a global review of city-scale initiatives. BioScience **67**(4), 332–342 (2017)
17. Nogueira, A.F., Ribeiro, J.C., Zenha-Rela, M.A., Craske, A.: Improving La Redoute's CI/CD pipeline and DevOps processes by applying machine learning techniques. In: 11th International Conference on the Quality of Information and Communications Technology, pp. 282–286 (2018)
18. Nürnberger, F., Keller, A., Härtel, S., Steffan-Dewenter, I.: Honey bee waggle dance communication increases diversity of pollen diets in intensively managed agricultural landscapes. Mol. Ecol. **28**(15), 3602–3611 (2019)

19. Porrini, C., Sabatini, A.G., Girotti, S., Ghini, S., Medrzycki, P., Grillenzoni, F., Bortolotti, L., Gattavecchia, E., Celli, G.: Honey bees and bee products as monitors of the environmental contamination. Apiacta **38**(1), 63–70 (2003)
20. Reeves, J.P., Knight, A.T., Strong, E.A., Heng, V., Neale, C., Cromie, R., Vercammen, A.: The application of wearable technology to quantify health and well-being co-benefits from urban wetlands. Front. Psychol. **10**, 1840 (2019)
21. Rolnick, D., Donti, P.L., Kaack, L.H., Kochanski, K., Lacoste, A., Sankaran, K., Ross, A.S., Milojevic-Dupont, N., Jaques, N., Waldman-Brown, A., Luccioni, A., Bengio, Y.: Tackling climate change with machine learning. arXiv preprint:1906.05433 (2019)
22. Sangiovanni, M., Schouten, G., van den Heuvel, W.-J.: An IoT beehive network for monitoring urban biodiversity: vision, method, and architecture. In: Communications in Computer and Information Science. Springer International Publishing (2020)
23. Savard, J.-P.L., Clergeau, P., Mennechez, G.: Biodiversity concepts and urban ecosystems. Landscape Urban Plan. **48**(3–4), 131–142 (2000)
24. Sebba, R.: The landscapes of childhood: the reflection of childhood's environment in adult memories and in children's attitudes. Environ. Behav. **23**(4), 395–422 (1991)
25. Sedjo, R.A.: Perspectives on biodiversity: valuing its role in an everchanging world. J. Forestry **98**(2), 45 (2000)
26. Seeley, T.D.: Honeybee Democracy. Princeton University Press, Princeton (2010)
27. Sinha, A., Sengupta, T., Alvarado, R.: Interplay between technological innovation and environmental quality: formulating the SDG policies for next 11 economies. J. Clean. Prod. **242**, 118549 (2020)
28. van den Heuvel, W.-J., Tamburri, D.A.: Model-driven ML-Ops for intelligent enterprise applications: vision, approaches and challenges. In: Business Modeling and Software Design, pp. 169–181. Springer International Publishing (2020)
29. Wainwright, J., Mulligan, M.: Environmental Modelling: Finding Simplicity in Complexity. Wiley, Hoboken (2013)
30. Wario, F., Wild, B., Rojas, R., Landgraf, T.: Automatic detection and decoding of honey bee waggle dances. PLoS ONE **12**(12), e0188626 (2017)
31. Zimmermann, B., Kohler, A.: Infrared spectroscopy of pollen identifies plant species and genus as well as environmental conditions. PLoS ONE **9**(4), (2014)

Gesture Recognition in an IoT Environment: A Machine Learning-Based Prototype

Mariagrazia Fugini and Jacopo Finocchi(✉)

Dipartimento di Elettronica, Informazione e Bioingegneria, Politecnico di Milano, Piazza L. da Vinci 32, Milano, Italy
mariagrazia.fugini@polimi.it, jacopo.finocchi@mail.polimi.it

Abstract. The spread of IoT and wearable devices is bringing out gesture interfaces as a solution for a more natural and immediate human-machine interaction. The "Seamless" project is an industrial research and development experience, aimed to build a virtual environment where data collected by IoT sensors can be navigated through a *gesture interface* and *virtual reality* tools. This paper presents the portion of the project concerning gesture analysis, focused on the problem of automatically understanding a set of hand gestures, in order to give commands through a wearable control device. The tackled issue is to build a *real time gesture recognition system* based on inertial data that can easily adapt to different users and to an extensible set of gestures. This *gesture data variability* is addressed by means of a supervised Machine Learning approach that allows adapting the system response to different gestures and to various ways of performing them by different people. A context-aware adapter allows interfacing the gesture recognition system to various applications.

Keywords: Human-Machine Interaction · Inertial Measurement Unit · Gesture recognition · Machine Learning · Neural networks · Context aware architecture

1 Introduction

This paper presents an industrial research and development experience dealing with gesture recognition as part of the "*Seamless*" Project, funded by Regione Lombardia (Italy) and partially by the H2020 WorkingAge EU Project[1].

The Project, conducted in cooperation between Politecnico di Milano - DEIB, Next Industries[2] and Digisoft System Engineering (DSE)[3], was aimed to build a *virtual environment* where a set of three-dimensional structural data can be navigated through virtual reality tools and a gesture interface. Structural data regarding constructions, e.g.,

Address questions and comments to Mariagrazia Fugini

[1] This work was supported by the European Union's Horizon 2020 research and innovation programme under grant agreement No.826232, project WorkingAge (Smart Working environments for all Ages).

[2] https://www.nextind.eu.

[3] https://www.gruppodse.it.

K. Arai (Ed.): FICC 2021, AISC 1363, pp. 236–248, 2021.
https://doi.org/10.1007/978-3-030-73100-7_18

bridges, buildings, risky frameworks, etc. were collected by IoT sensors, monitoring the vibrations and movements of the structures.

In particular, recognition of movements was performed using a wearable device to give commands by drawing gestures in the air. The spread of IoT and wearable devices is bringing out gesture interfaces as a solution for Human-Machine Interaction (HCI) in a more natural and immediate way [1, 2].

The work presented here is framed in *Seamless* as follows. DSE developed a *virtual environment* where multiple subjects can interact collaboratively by connecting remotely through the web to a cooperative space of IoT "objects". This environment can be set up quickly in any room, even with no pre-setting or knowledge of the physical layout, by using a simple mobile equipment consisting of a set of IoT devices. It is controlled by a software application based on the Unreal Engine, which is interfaced with the Tactigon-Skin™ provided by Next Industries, as a gesture control input device. The Tactigon-Skin is developed by Next Industries as an IoT device to be worn on the user's hand and is equipped with a 6-axis Inertial Measurement Unit (IMU).

When analyzing the inertial data collected through Tactigon-Skin, we observed a remarkable *data variability*, attributable to the difference between gestures and for hand gestures are performed differently depending on people and circumstances. For instance, the gesture has different intensity of speed and acceleration (e.g., rapid acceleration and slow braking or vice versa), different duration amplitude (e.g., in relation to the length of the user's arm), trajectory or slight hand vibrations. The need to handle such data variability led us to adopt a supervised Machine Learning (ML) approach, which is described in the paper.

The gesture recognition task focused on the problem of automatically recognizing a set of hand gestures, to enable users to send commands to applications through the Tactigon-Skin device. The analysis aims at building a real-time *hand gesture recognition system* that can adapt to different users, able to recognize a set of gestures that can be defined and extended in different application environments.

Under these premises, our goal was to develop a prototype software capable to perform gesture recognition under four constraints:

- based on *inertial data only*, with no additional sensors, whereas most of the previous works rely on other types of sensors, especially vision-based [3, 4];
- in *real-time*, so that the recognized gesture prediction occurs within a predetermined time interval from the hand gesture conclusion, to allow for a smooth man-machine interaction;
- *user independent*, so as not to require a specific user training activity for each new user, after the completion of the general training sessions;
- with *no explicit signals* marking the gesture "start" and "end" (like pressing or releasing a button), in order to achieve a better user experience.

During our experimentation, two different solutions were tested, that rely on different ML techniques: an approach based on the Dynamic Time Warping (DTW) technique and a classical neural network approach.

The paper is organized as follows. In Sect. 2 we present relevant related works. In Sect. 3, we introduce our gesture recognition solution, illustrating the inertial data and the gesture recognition process. Section 4 describes the inertial data pre-processing stage and Sects. 5 and 3 the neural network classifier. Section 6 focuses on the context aware adapter that connects the gesture recognition system to applications.

2 Related Work

In the field of gesture recognition, most systems are based on some additional sensor technology, merging the inertial data with another data source, like magnetometers, visual information from a camera, or infrared sensor data.

Various challenges, such as interposition of obstacles, variations in lighting or complex background, make an accurate detection of hand gestures difficult in vision-based approaches.

A vision-based hand gesture recognition system is presented in [5]. Several researchers developed solutions based on depth cameras, as in [6], where an authentication system is proposed. In [7] a binocular vision system provides the depth information needed to build a three-dimensional map that allows to detect the gestures on a complex background. In [8] a reinforcement learning solution is described, based on two neural networks, trained to analyze video images to segment and classify surgical gestures.

The use of neural networks is frequent among systems that use inertial data as well.

In [9] gestures capturing and recognition is based on inertial and magnetic measurements units (IMMU) assembled on a fully cabled glove. The paper describes recognition methods based on Extreme Learning Machine, a kind of feed-forward neural networks, both for static and dynamic gestures classification.

A user-independent hand gesture recognition system is described in [10], based on an accelerometer-based pen device. Basic gestures are classified by a feed-forward neural network, while a segmentation algorithm is proposed to split complex gestures into several basic gestures, through a similarity matching procedure. In [11] a Wiener-type recurrent neural network is used for handwriting trajectory reconstruction, based only on inertial data. Some classifiers are compared in [12] to recognize hand gestures from two different inertial data sets, using various ML methods, such as Extremely Randomized Trees, Gradient Boosting and Ridge Classifiers.

For the analysis of inertial data, other researchers used ML techniques different from neural networks, such as the DTW algorithm, a technique that evaluates the distance between two sequences [13] through a non-linear distortion with respect to an independent variable, such as time.

DTW is particularly suitable for the analysis of time series and actually is used in various fields of application, from Speech Recognition to the recognition of human activities.

It was applied with good results in several inertial data studies, such as in [14], that presents a continuous hand gestures recognition technique, using a three-axis accelerometer and gyroscope. The solution is based on a gesture-coding algorithm, where gestures are recognized by comparing the gesture code against a gesture database, using the DTW algorithm.

The previous stage of developing a gesture recognition solution within *Seamless* Project, under the constraints described in Sect. 1, was also based on DTW technique [15].

In the second stage of the project, explained in this paper, we tested a neural network architecture and we paid more attention to the application interfacing questions, which also involve aspects of *context awareness*. ML and context awareness are topics that often appear together in smart technologies applications [16] and context-aware solutions are emerging in different fields and areas, in particular in smart environments and IoT [17]. We are exploring the feature of people movements through gesture recognition, aimed to the development of innovative HCI methods, in the WorkingAge EU Project [18].

The concept that the same gesture may have different meanings depending on the context has been proposed in gesture recognition, starting from [19], that focuses on gesture recognition in the medical field. The aim is to reduce the gesture set and to improve recognition reliability and usability. In [20], context awareness is achieved through a finite state machine, applied to a gesture recognition system based on neural networks.

3 Gesture Analysis Process

3.1 The Gestures Set

The T-Skin device contains an IMU chip, based on Micro Electro-Mechanical Systems (MEMS) technology, which collects inertial data, measuring three-axis acceleration and three-angular velocities with a 50 Hz frequency.

Sensor data are collected on-line from the T-Skin device, trough Bluetooth transmission or through a serial interface. Inertial data can also be acquired off-line from data files recorded during the previous on-line sessions. Recorded sessions allow us to perform a fast system training without the need of a gestures live session and to compare different algorithms under the same conditions.

During the experimentation we tested the recognition process on a set of 15 gestures: the simple translations and rotations along each of the space axes (one-dimensional motions) plus a few more complex gestures (two-dimensional motions), and a short hand movements that we call "grab", that is performed without moving the wrist.

In detail, the system was trained on the following hand motions:

- Simple translations: Forward, Backward, Up, Down, Right, Left;
- Simple rotations: Pitch Down, Pitch Up, Roll Right, Roll Left;
- Complex gestures: Circle, Square, Slope Up, Slope Down.
- Small gestures: Grab.

These gestures were selected both for the clear inertial characterization and for their suitability to give intuitive commands in a three-dimensional spatial context. A gesture is typically performed with the T-skin device in about half a second for the shorter gestures, up to about one second for the more complex ones, with a short pause between gestures, of the order of a few tenths of a second.

3.2 The Gesture Recognition

Raw inertial data are first processed by a *filtering stage* with the purpose of reducing signal disturbances and perturbations. Then, data are analyzed to *detect and separate* the single gestures, using a segmentation algorithm based on thresholds, developed by the authors. Subsequently, the data set corresponding to each gesture is normalized in length and amplitude and forwarded to the *classification stage.*

The classification is performed by a *supervised ML* system, previously trained through a series of labeled gestures recognition sessions. The recognized gesture is finally made available to the adapter module that implements the interface with the surrounding *Seamless* virtual environment.

The different steps of the described data analysis process were implemented in the gesture recognition prototype software developed during the project. The software was developed in Octave environment and partially in Python language.

The logical structure of the implemented process is summarized in Fig. 1.

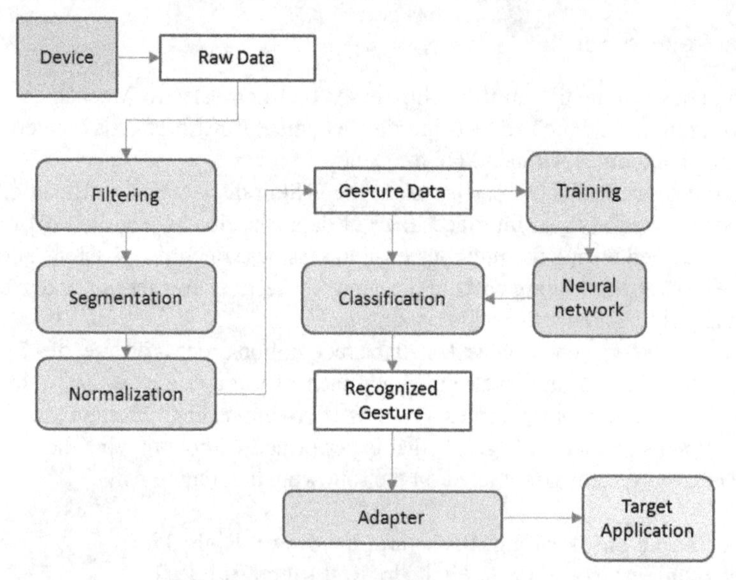

Fig. 1. The gesture recognition process

4 Gesture Pre-processing

4.1 Data Filtering

Raw data acquired by the sensors are subject to various perturbations, namely the gravity acceleration, the drift due to loss of calibration or measurement inaccuracy and high frequency noise, produced by vibrations and sensor thermal noise. For this reason, the acquired signals are firstly processed through a correction and filtering step.

The signal offsets due to gravity are removed by subtracting the projection of gravity along each axis from each acceleration sample.

Then, simple low-pass and high-pass filters are applied to remove noise and fixed components. A low-pass Butterworth filter is applied to remove high-frequency noise and a mild high-pass filter is applied to remove drift effects. The effects of high frequency noise are also mitigated by the normalization step performed after the gesture segmentation.

The drifting effects are limited by the short duration of a single gesture and by levelling the acceleration values to zero before computing each new gesture.

4.2 Gesture Segmentation

The goal of the segmentation stage is to cut in real-time the subset of data corresponding to a single gesture from a continuous inertial data flow.

The current segmentation algorithm is able to detect the short pause that divides consecutive gestures, based on a non-linear combination of angular velocities and accelerations that is evaluated against a value threshold and a duration threshold. The thresholds were obtained experimentally, by comparing the effectiveness of different combinations on the inertial data recorded during mixed gestures sessions and with the help of visual charts generated from the same data. The segmentation algorithm so far developed, works well with wide and fast gestures, separated by a short movement pause, and shows a lower success rate on softer movements.

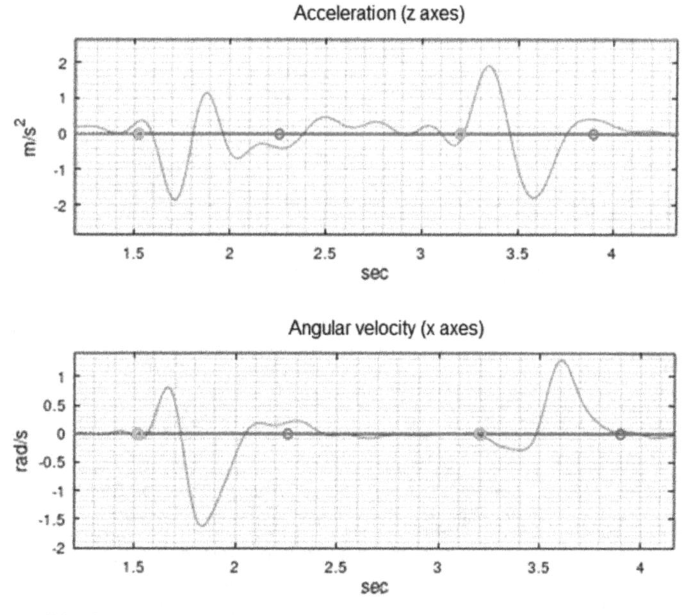

Fig. 2. Segmentation of two successive gestures: Up and Down.

An example of successful segmentation is shown in Fig. 2. The figure shows the inertial data for two of the six sensor axes, provided in particular by one of the accelerometers and one of the gyroscopes.

The data section shown as an example, includes two consecutive gestures (an upward translation followed by a downward translation) separated by a short pause. The inertial data shown are already filtered. Green dots mark the gesture start while red dots mark the gesture ending, as detected by our segmentation algorithm.

Only data between the beginning and the end of each segment are considered *significant as gestures* and hence passed to the subsequent processing phases. Other values are considered as not belonging to a gesture and are hence excluded from further processing.

4.3 Gesture Normalization

After the detection of a data segment corresponding to a single gesture, the six inertial values are normalized in length and in amplitude.

First, a length normalization is performed that uniforms the gesture data to a predefined standard number of samples, through an interpolation algorithm. After that, an amplitude normalization is performed by means of a custom algorithm that recalculates each value on all axes to ensure that the peak value is set to 100%.

Figure 3 shows an example of gesture data normalization.

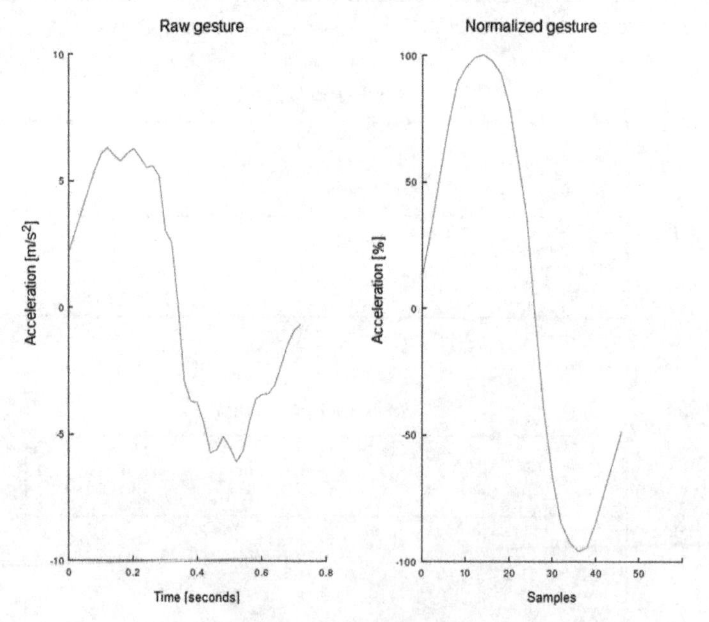

Fig. 3. Gesture raw data and normalized data

Figure 3 displays a simple translation gesture along one of the device axis, where first acceleration and then a deceleration occur. On the left hand side, the values returned by one of the three accelerometers are depicted, while on the right side the resulting values after the application of the interpolation and normalization algorithms are reported. The first step dampens the small variations due to vibrations or small differences in the path and also the abscissa axis takes on the meaning of the virtual sample number. The second step scales the values along the ordinates to an arbitrarily fixed scale between -100 and 100.

These processing steps allow bringing together data segments corresponding to short and long span gestures as well as faster and slower gestures, that otherwise show very different raw acceleration values.

It should be observed that the standardization of the samples number, by merging adjacent samples, also causes a value smoothing that reduces the effect of noise and outliers.

5 Gesture Classification

The gesture recognition task was carried out as *a problem of classification* of the segmented input data.

As described above, the gesture classification tasks are usually based on ML techniques. We also in our project make use of a ML approach, to manage the high variability in the input data.

During our experimentation, two different solutions were tested, that rely on different Machine Learning techniques. In the first project phase, we experimented a DTW based approach, as described in [15]. The second ML approach tested in the project is based on a classical neural network architecture. We successfully experimented a simple feed-forward neural network, designed as follows.

An output node is set for each gesture to be recognized, where the expected output value is converging to one for the actual gesture and converging towards zero for all other nodes.

As for the features to feed the network with, we found it effective to provide the network with the temporal succession of the values of each dimensional axis, in a single step. After the normalization stage, the detected gesture is in fact represented by a fixed number of samples for each axis of the gyroscope and for each accelerometer, regardless of the actual duration of the gesture. These values are provided directly to the neural network input nodes.

An input node is set for each subsequent data sample on each data dimension. The number of input nodes is therefore equal to six times the standardized gesture length: typically, we standardize the detected gestures to 50 samples and taking into account the three gyroscope axes and the three orthogonal accelerometers, we feed the network with 300 input nodes.

This solution is outlined in Fig. 4.

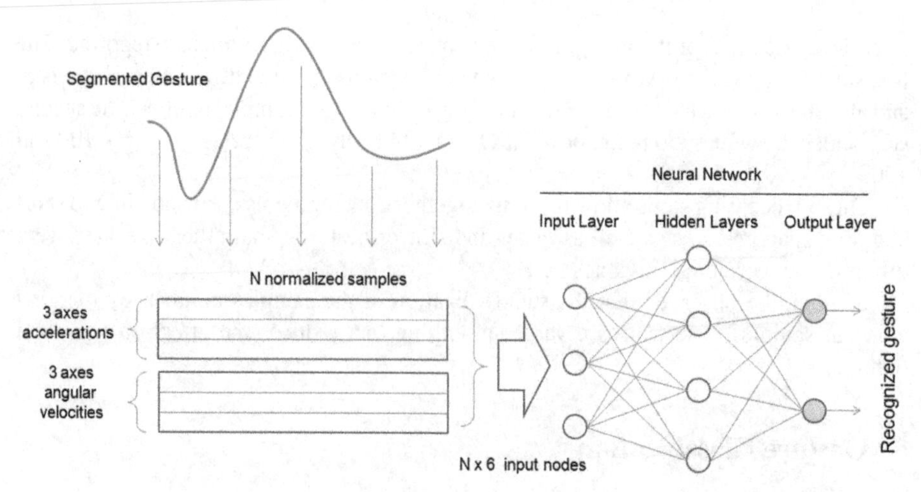

Fig. 4. The neural network classifier used in the gesture recognition process.

Each output node provides a value between 0 and 1 and the node with the highest value shows the gesture recognized by the classifier.

On the output values we considered two classification *confidence indexes*, to rate the result provided by the neural network.

The greatest value itself returned from the output nodes, provides a first confidence index. Then, by sorting the output values in descending order and taking into consideration the two greatest ones, we used the quotient between the second and the first value as a measure of uncertainty of the classification result proposed by the system. The second highest value represents the second most likely gesture according to the classification stage.

Values greater than 0.1 show a slight uncertainty, values greater than 0.5 denote a strong uncertainty. When the classifier should assign the same value to the best two output nodes, we have the same probability that the gesture belongs to one class or another and the uncertainty is equal to 1. For practical use, we choose to consider the results with uncertainty index lower than 0.1 as validly recognized gestures.

As for the management of the data variability depending on the type of gesture, in the current state of experimentation we find the gestures that resulted more difficult to recognize are the slow or small amplitude motions, like the "grab" gesture, which have a less clearly characterized pattern than others. Introducing this kind of motion in the gesture set, we observed that the classifier does not show a low successful recognition rate on that gesture, but rather the recognition rate gets worse on some other gestures, which are often confused with the new one.

Concerning the data variability in dependence on the different ways of performing the same gesture, the system behaviour was tested by setting up different training sessions, conducted by a different number of users.

We performed the training sessions starting from one system trainer up to five trainers. While adding more people to the classifier training sessions, the system reached a better performance rate with 3 trainers and showed a slight decay with 4 or 5 subjects. A possible explanation is that the different way of performing the same gestures reduces the separation between the ensembles of data trajectories that define each gesture, though on this aspect we believe that further experimentation is necessary.

The system was trained with a set of 15 gestures, as described in Sect. 3.1. Many real time test sessions were carried out in the company laboratory. The recognition software reached a performance ranging from 86% to 97% of correctly recognized gestures, depending from the single user and the gesture performing conditions.

Each test session included a sequence of 10 to 40 different gestures. In the longest test sequences, users showed a decreasing accuracy in the hand movement execution, leading to a slight decline in the recognition success rate.

The user independence of the recognition system was also successfully tested on random people at an exhibition fair, after a brief instruction about the gestures they could perform.

6 Context-Aware Adapter

After the recognition phase, an *adapter module* is placed which acts as a joint with the application interfaced, to enable a flexible association of learned gestures to an arbitrary target system.

This adapter is performing the function of a *semantic layer*, which gives the gestures a meaning that is relevant in the context of the specific interfaced application, translating gestures into actions that make sense for the application. This layer therefore allows giving the same gesture a different meaning according to the controlled application (e.g., the selection of an option in a graphic user interface or a command issued to a robot).

A sort of vocabulary, or correspondence map, translates each gesture into an action performed on the application. This vocabulary is used to define and adapt the gesture-based interface, based on the given usage context.

We however noticed that even dealing with a single application, the set of available gestures is often too limited to control all the desired functions. We have two possible solutions to this problem: i) to provide a wider gestures set (e.g., the letters of an alphabet) or ii) to translate the same gesture in different ways according to the application workflow context. This second solution seems to preserve the naturalness of the gesture, which is a characteristic of gestural interfaces, without requiring the user to learn the associations of various meanings with a series of abstract symbols.

The meaning generated by the semantic layer therefore depends both on the events generated by the user (gestures) and on the state of the application context, such as, for example, the presence or absence of an item, or the effects produced on the environment by previous gestures.

In our system, this solution is implemented by introducing the definition of different modes for each application and associating a different vocabulary to each mode. In this way, a gesture produces a certain effect depending on the current mode. For each mode, a separate configuration is set up. A mode can model an application workflow stage or a context situation, so that the same gestures set can be used for different purposes based on the context.

The solution operates like a finite state machine, where the transitions between modes can take place either through one action set in the vocabulary or based on the state of the interfaced application.

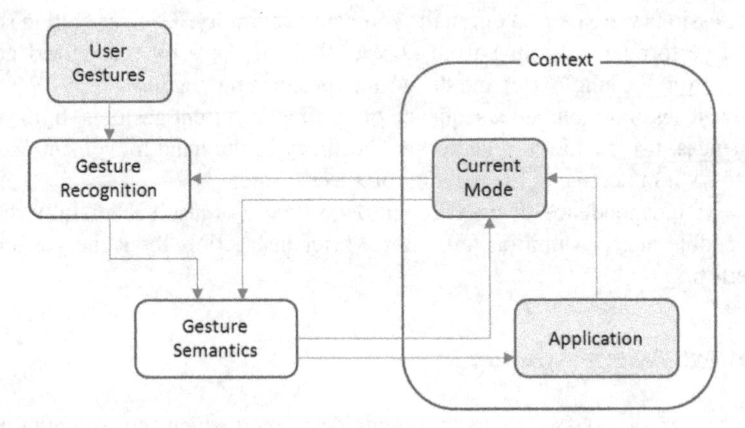

Fig. 5. Interactions of the context-aware gesture semantics module.

The semantic layer is implemented by an adapter module that controls the specific behaviour of a single application. As shown in Fig. 5, this module is triggered by user gestures, taking into account the current context, to generate the actions that drive the application and possibly change the mode state. The current mode may affect the recognition process too, limiting the valid gestures to a reduced set of allowed options.

The first version of our application adapter was configured to control a 3D modelling software, adding as actions to the vocabulary the mouse pointer movements, the mouse click and some keyboard shortcuts. For this purpose, the application adapter software directly controls the mouse and keyboard actions of a generic graphic user interface. In this scenario, the "Shift", "Move" and "Edit" modes were defined, which respectively allow to move the pointer and create or select an object, to rotate or move an object and to modify an object by extruding its size along one axis.

The system response can be adapted based on the application workflow state and on the current environment situation. The same gesture, for example a rightwards translation, can mean to "move the pointer", or to "move the object", or to "extrude the object", depending on the situational awareness about the objects placed in this virtual space the user can interact with, and on the previous actions performed on them (e.g., selecting the object or activating a menu option).

A reduction in the gestures set, by increasing the separation between the features space of each gesture, enhances the user independence and the recognition accuracy.

7 Future Works

Further planned developments include enhancing the ability to face gesture variability, in particular by improving the user independence and the recognition accuracy of *smaller or slower movements*.

The user independence, or more generally, the ability to deal with the differences within each gesture class, will be tackled by introducing an automated clusterization stage, where the different ways of performing the same gesture can be grouped as if they were different gestures. The recognition rate of small and slow movements will be tackled by enriching the features provided to the neural network with more information about each segmented gesture, such as the gesture duration and peak values, that are currently lost during the normalization stage.

Another aspect under development is the design of a *configurable adapter* module, to easy interface the gesture controller with different applications, without limiting the gesture recognition solution to the scope of the *Seamless* environment. A visual tool is now under study to allow customers to easily define the application adapter configuration, defining the different modes and associating for each mode the available gestures with the desired actions.

8 Concluding Remarks

The paper has presented a prototype about gesture recognition, relying on a ML approach, based on a neural network classifier. In particular, we have illustrated the processing steps and the obtained results.

The adoption of a supervised ML technique has proved effective in handling the variability of different types of gesture patterns, performing a continuous data flow analysis in real time. The prototype tested the feasibility of a user-independent gesture recognition that is based only on inertial data.

The design of a context-aware adapter allowed the gesture control system to adapt to different application contexts based on its configuration and provided the system the ability to switch between different operating modes, so as to issue a larger set of instructions with fewer gestures.

References

1. Vuletic, T., Duffy, A., Hay, L., McTeague, C.P., Campbell, G., Choo, P.L., Grealy, M.: Natural and intuitive gesture interaction for 3D object manipulation in conceptual design. In: DS 92, Proceedings of the DESIGN 2018 15th International Design Conference, pp. 103–114 (2018)
2. Roda-Sanchez, L., Olivares, T., Garrido-Hidalgo, C., Fernández-Caballero, A.: Gesture control wearables for human-machine interaction in Industry 4.0. In: International Work-Conference on the Interplay Between Natural and Artificial Computation, pp. 99–108. Springer, Cham (2019)
3. Al-Shamayleh, A.S., Ahmad, R., Abushariah, M.A., Alam, K.A., Jomhari, N.: A systematic literature review on vision based gesture recognition techniques. Multimedia Tools Appl. **77**(21), 28121–28184 (2018)
4. Rautaray, S.S., Agrawal, A.: Vision based hand gesture recognition for human computer interaction: a survey. Artif. Intell. Rev. **43**(1), 1–54 (2015)
5. Singha, J., Roy, A., Laskar, R.H.: Dynamic hand gesture recognition using vision-based approach for human–computer interaction. Neural Comput. Appl. **29**(4), 1129–1141 (2018)
6. Zhao, J., Tanaka, J.: Hand gesture authentication using depth camera. In: Future of Information and Communication Conference, pp. 641–654. Springer, Cham (2018)

7. Jiang, D., Zheng, Z., Li, G., Sun, Y., Kong, J., Jiang, G., Liu, H.: Gesture recognition based on binocular vision. Clust. Comput. **22**(6), 13261–13271 (2019)
8. Gao, X., Jin, Y., Dou, Q., Heng, P.A.: Automatic gesture recognition in robot-assisted surgery with reinforcement learning and tree search. arXiv preprint, arXiv:2002.08718 (2020)
9. Fang, B., Sun, F., Liu, H., Liu, C.: 3D human gesture capturing and recognition by the IMMU-based data glove. Neurocomputing **277**, 198–207 (2018)
10. Xie, R., Cao, J.: Accelerometer-based hand gesture recognition by neural network and similarity matching. IEEE Sens. J. **16**(11), 4537–4545 (2016)
11. Hsu, Y.L., Chou, P.H., Kuo, Y.C.: Drift modeling and compensation for MEMS-based gyroscope using a Wiener-type recurrent neural network. In: 2017 IEEE International Symposium on Inertial Sensors and Systems (INERTIAL), pp. 39–42. IEEE (2017)
12. Krishna, G.G., Nathan, K.S., Kumar, B.Y., Prabhu, A.A., Kannan, A., Vijayaraghavan, V.: A generic multi-modal dynamic gesture recognition system using machine learning. In: Future of Information and Communication Conference, pp. 603–615. Springer, Cham (2018)
13. Berndt, D.J., Clifford, J.: Using dynamic time warping to find patterns in time series. In: KDD Workshop, vol. 10, no. 16, pp. 359–370 (1994)
14. Gupta, H.P., Chudgar, H.S., Mukherjee, S., Dutta, T., Sharma, K.: A continuous hand gestures recognition technique for human-machine interaction using accelerometer and gyroscope sensors. IEEE Sens. J. **16**(16), 6425–6432 (2016)
15. Fugini, M., Finocchi, J., Trasa, G.: Gesture recognition using dynamic time warping. In: Proceedings of the 29th International Conference on Enabling Technologies: Infrastructure for Collaborative Enterprises (WETICE). IEEE (2020)
16. Fugini, M., Bonacin, R., Martoglia, R., Nabuco, O., Sais, F.: Web2Touch 2019: semantic technologies for smart information sharing and web collaboration. In: 2019 IEEE 28th International Conference on Enabling Technologies: Infrastructure for Collaborative Enterprises (WETICE), pp. 255–258. IEEE (2019)
17. Raibulet, C., Drira, K., Fugini, M., Pelliccione, P., Bures, T.: Software architectures for context-aware smart sytems. Inform. Software. Tech. (2019). Elsevier Science Publication
18. Fugini, M., et al.: WorkingAge: providing occupational safety through pervasive sensing and data driven behavior modeling. In: Proceedings of the 30th European Safety and Reliability Conference (ESREL 2020), pp. 1–8. Research Publishings (2020)
19. Bigdelou, A., Schwarz, L., Benz, T., Navab, N.: A flexible platform for developing context-aware 3D gesture-based interfaces. In: Proceedings of the 2012 ACM international conference on Intelligent User Interfaces, pp. 335–336 (2012)
20. Hakim, N.L., Shih, T.K., Kasthuri Arachchi, S.P., Aditya, W., Chen, Y.C., Lin, C.Y.: Dynamic hand gesture recognition using 3DCNN and LSTM with FSM context-aware model. Sensors **19**(24), 5429 (2019)

Location Awareness in the Internet of Things

Swen Leugner[1(✉)] and Horst Hellbrück[1,2(✉)]

[1] Lübeck University of Applied Sciences, 23562 Lübeck, Germany
{swen.leugner,horst.hellbrueck}@th-luebeck.de
[2] University of Lübeck, 23562 Lübeck, Germany
http://cosa.th-luebeck.de

Abstract. Received signal strength (RSS) measurements are easily available and provide context information to IoT-devices. However, indoor localization via RSS is error prone. Due to walls and ceilings multipath propagation occurs and leads to constructive and destructive interference and therefore higher or lower RSS values in an unpredictable and therefore random fashion. This random variation decreases localization accuracy if not considered right. In our work, we develop a pathloss model via a ray tracing engine and rate several measured RSS values by statistical means. We find a good estimate of the RSS of the line of sight connection which improves the distance estimation significantly. Our algorithm is based on the assumption that several frequencies are used for localization, which is true for Bluetooth Low Energy (BLE). We demonstrate the improved RSS-based-localization by real experiments and BLE hardware.

Keywords: Bluetooth · BLE · Frequency hopping · Multipath · IoT · IIoT · Coherence bandwidth

1 Introduction

In the Internet of Things, physical and virtual objects are connected to each other in order to make them work together and to solve a variety of problems. Typical Internet of Things devices are embedded, battery-operated systems with communication interfaces. In recent years, these battery-operated devices became more powerful and thus enable new fields of application.

One of these application fields are hospitals, where a variety of devices are used to care for patients. Typical devices are for example ventilators, which are needed in the hospital ward. The devices are kept in stock and are needed quickly, for example, when emergency patients are admitted. Ideally, in such an emergency situation, staff would know immediately where the nearest available ventilator is located. Here, Internet of Things devices can help by localizing the ventilator in rooms and corridors. Localization in this case means determining the location of the ventilator and displaying it on a map. A further field of

K. Arai (Ed.): FICC 2021, AISC 1363, pp. 249–265, 2021.
https://doi.org/10.1007/978-3-030-73100-7_19

application is in the industrial environment. In Industry 4.0, the production becomes more transparent and work pieces more intelligent. The so-called smart work piece, an Internet of Things device knows where it is in production and how long its production will take. Thus, optimizing the production on demand is improved. For instance, if it is known where and in which production step the work piece is, the process controlling can compare the target and actual times of the production step. If the difference between the target and actual production step time is too large, errors can be reported and process optimization can take place at an early stage. If the workpiece does not move for a certain time interval, an error has occurred. Thus the localization goal is to fix the position reliably. One challenge is the localization of objects within buildings. The Global Positioning System (GPS) is not available for indoor localization due to walls and ceilings that block satellite signals. To locate devices or objects, at least a room-accurate localization is required. Therefore, a large number of technologies have been developed and investigated in the past to close this gap. One option is Bluetooth technology, in particular the Low Energy extension. Due to the low power consumption of the extension, the device lifetime increases and is ideal for battery operation. Since the Bluetooth Low Energy 4.0, which is frequently used today, was not primarily designed for the localization of objects and devices, i.e. no location awareness was provided and various approaches were developed to enable this functionality. Many approaches are based on received signal strength measurements and do not consider multipath. As we will show in this paper, this is a major source of error that prevents precise localization if not considered in an appropriate way.

In this paper, we focus on two elements: First, the sources of error which have a negative influence on the localization results. Second, we perform experiments, create models and provide a solution for a more accurate determination of the received signal strength value appropriate for localization.

The contributions of our work are as follows:

- Simulation-based investigation of relevant RSSI measurement errors with particular focus on indoor localization.
- Detailed investigation of radio propagation in the Industrial Scientific Medical Band with focus on Bluetooth Low Energy.
- A comparative evaluation of simulative ray tracing models with a focus on constructive and destructive interference.
- Development and evaluation of an approach to minimize localization errors based on multipath reception suited for Bluetooth Low Energy.

The rest of the paper is organized as follows. Section 2 discusses related work for localization with Bluetooth Low Energy, multipath and measurement errors. In Sect. 3 we provide a detailed overview of localization techniques for the Internet of Things including an overview how to localize with Bluetooth Low Energy. We introduce our findings for measurement errors in Sect. 4. Section 5 presents our approach. Section 6 provides a description of the experimental setup, metrics and measurement results. We draw final conclusions and suggest future work in Sect. 7.

2 Related Work

In this section we focus on related work in the field of radio-based localization technologies. In particular, we look at work that investigates or applies received signal strength based techniques. In contrast to the previous work, we consider multipath propagation as a function of frequency and present a suitable method that allows a more accurate estimation of the received signal strength.

Huang et al. describe an approach for robust position determination by utilizing multiple Bluetooth Low Energy channels [6]. For their robust position determination, they need the channel information on which the frame was received and 1000 measurements to start the algorithm and decide which channel is the best. Furthermore, Huang et al. use a smartphone instead of a dedicated locator in their test setup and perform a trilateration. The results reach absolute mean errors of 12.8 m. In contrast to Huang et al. we assume in our approach that we cannot identify the channel number on which the frame has been received. Besides that, an indoor location with average errors up to 12.8 m is too large for many applications even too much to identify the room where the devices are placed. In addition we show that the use of multiple channels is only useful if the transmission channel is flat or multipath reception is taken into account. Furthermore, we show by simulation and measurements that this information can be invalid already 5 cm further away from the measuring point.

Canton et al. show a broad selection of relevant studies that have already been carried out in [3]. Many of these attempts use a combination of Bluetooth low energy and fingerprinting. In fingerprinting, individual measuring points are characterized and stored in a database. The size of the testbeds ranged up to 100 m $\times$ 100 m and errors up to 50 m. We will see that the size of the testbed is not a crucial criterion to evaluate BLE based localization systems, because the RSSI characteristics change already within a few centimeters. More important is the environment, i.e. office space, open space and which material especially walls have.

Wang et al. utilized ray tracing for an Indoor Bayesian Algorithm [14]. They use WiFi access points that support Bluetooth 4.0 and use a Huawei P9 mobile phone as receiver. For localization the access point sends localization messages. These messages are received and processed by the mobile phone to perform a localization. This is of advantage if the location information is to be available on the mobile phone. We are investigating a centralized approach, which means that there is a central localization instance in the network which collects the data, analyzes and processes it and provides the location. In the experiment of Wang et al. a received signal strength distribution is estimated by ray tracing and compared with the real measurements at 19 points. Wang et al. use fingerprinting methods without using real measurements but replacing the measurement by the ray tracing model. In contrast to Wang et al. our ray tracing model takes into account reflections up to the second order. As we show in our evaluation, that these reflections change the received signal by superposition and cancellation significantly.

In [10], Nezhadasl et al. conclude that ray tracing can be used to evaluate reflections caused, for example, by walls. The results of this simulation are verifiable in reality. Krishna et al. investigated localization using WiFi in the approach Zee: Zero-Effort Crowdsourcing for Indoor Localization [11]. The zero effort is achieved by using metadata from additional sensors such as acceleration sensors, compass sensors and gyroscope sensors. Afterwards a motion estimation is performed, which estimates the position using a particle filter. For this purpose, the particle filter accesses a WiFi database, in which received signal strength measurements of individual access points are stored for each x and y position. The database is continuously updated by integrating the results of the motion data. In contrast to Krishna et al. we do not use WiFi data or fingerprinting. However, our findings are also adaptable to the methods of Krishna et al. and can therefore lead to more accurate results.

Another problem of WiFi-based sensor nodes in the battery powered Internet of Things is shown by Mesquita et al. in [9]. With an ESP8266, and a transmission rate 1 Hz with a 1000 mAh, they only achieved a battery life of 40 h. This battery life means an average current consumption of 25 mA. In comparison, Spachos et al. measured an average current consumption about 0.05 mA with a Cyalkit-E02 Bluetooth Low Energie Device [12].

Thaljaoui et al. utilize Bluetooth Low Energy Devices in [13] to localize elderly people in a smart home. They use RSSI measurements and the iRingLA algorithm to determine the position of the BLE device. For localization, the average of the received signal strength measurement is taken for three stationary devices. The measured values are converted into distances. In the next step, inner and outer rings are drawn around the measured distance, which are generated over a predetermined error term. The two smallest, overlapping rings are then used for an initial search to determine the position. For this purpose, a rectangle is deposited in the overlapping points, tested against the overlapping rings and reduced to possible positions. With the reduced possible positions, the third ring is used for the final position determination. The remaining possible positions are combined to a centroid and output as final position. For evaluation, Thaljaoui et al. use an empty room with a size of $4\,\text{m} \times 4\,\text{m}$ and 9 evaluation points. They achieve an accuracy of 0.4 m. As we show in this paper, averaging over several RSSI measurements in rooms that are not empty is no longer reliable because of multipath reception. Therefore we use averaging over RSSI values as a state-of-the-art reference to compare our method.

In summary, a large number of algorithms and systems have already been investigated in the field of radio-based localization. Some of the investigations show good results, others cannot be reproduced one-to-one. As stated already in the introduction, we focus on two elements: First, the sources of errors and second, models and experiments to provide a more accurate received signal strength value appropriate for localization. Finally, the solution is evaluated in an unknown radio environment.

In the next section we will briefly introduce Bluetooth Low Energy localization and put it into the context of the Internet of Things.

3 Localization with Bluetooth Low Energy

In the Internet of Things world there are a lot of possibilities to localize devices. Figure 1 shows the most common localization techniques in the Internet of Things area. The first and simplest approach is to take advantage of the received signal strength indicator (RSSI) at the receiver. If this is known, a conversion to pseudoranges is performed. Subsequently, a trilateration or multilateration is possible. The second approach is to measure the arrival time. This approach is particularly useful in the area of ultra-wideband transceivers. However, these require highly accurate clocks and precise synchronization, as we showed in [7,8]. If the Time of Arrival is available, pseudo distances are also determined from it. Subsequently, the localization is performed using trilateration and multilateration algorithms.

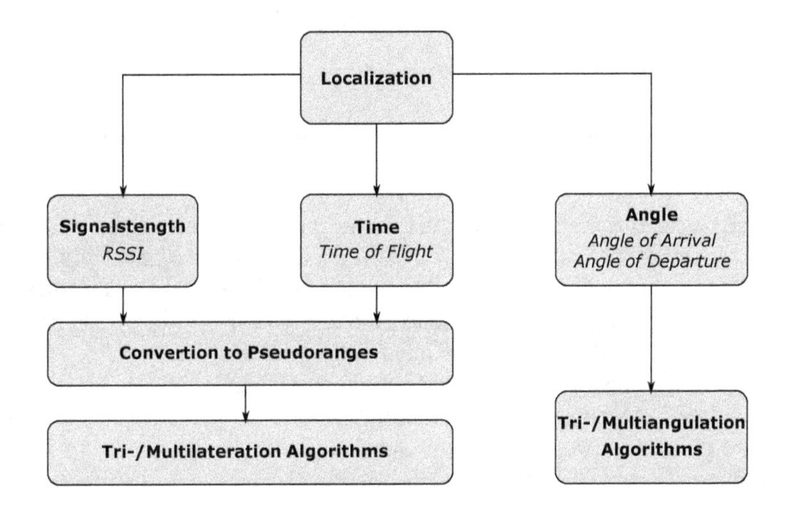

Fig. 1. Common localization techniques for the Internet of Things

The last approach is to use the angle of arrival or the angle of departure at the transmitter or receiver. If the angle is known, the position is determined from at least 2 angles by triangulation. Triangulation requires that the transmitter or receiver is equipped with an antenna array. Due to the required number of antennas and antenna sizes, these are often installed in a stationary receiver. However, small errors in angle measurements lead to rage depending localization errors. To reduce these errors several receivers are installed. This increases the bill of materials and thus the cost of such systems.

For Bluetooth Low Energy (BLE), Time Of Flight approaches are not implemented yet and Angle of Arrival systems are not cost effective. Hence, the more cost effective approach of RSSI is still popular for many applications. In BLE, the objects to be located are called beacons. A beacon transmits on channel 37, 38, 39 of the 40 available channels. These three channels are also called advertising channels. The transmitted frame corresponds to the so-called advertising frame. An advertising frame consists of a header, advertising address, flags, advertising data and a CRC and has a maximum size of 42 Bytes. There are a total of four Beacon Advertising Data Standards: iBeacon, AltBeacon, Eddystone and Geobeacon. Despite the different standards, all four Beacon Standards have the unique ID (UID) and transmit power within the frame.

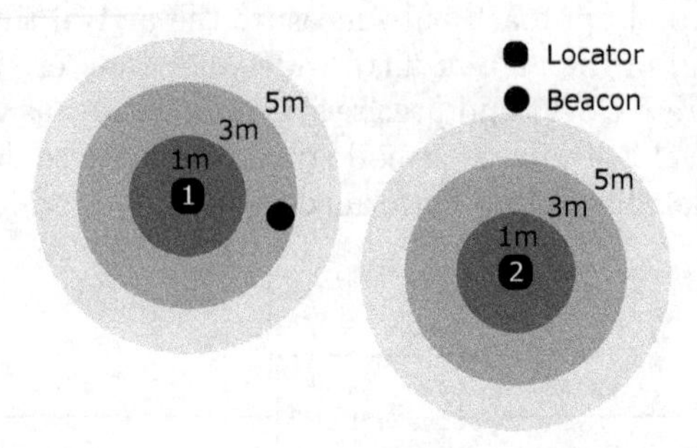

Fig. 2. Localization with BLE Beacon and two BLE Locators

While in localization with BLE the beacon represents the sender, the receiver is called the locator. The locator takes the advertising frames and logs the received signal strength to the UID of the beacon. Since the advertising frame is sent as a broadcast and is authenticationless, the frames are received by all locators and processed on a central localization server. The localization server calculates the beacon position. This is done as shown in Fig. 1 either by converting the RSSI measurement to distances or by using a predefined RSSI range to which a distance is assigned, as shown in Fig. 2. For instance, in Fig. 2 we illustrated two Locators. Locator 1 and Locator 2 have three distance ranges at 1 m, 2 m and 3 m. The beacon is placed near the 3 m distance range of Locator 1 and outside the distance range of Locator 2, so the localization server assigns the beacon to Locator 1. If Locator 1 and Locator 2 are placed in different rooms, the localisation server assigns the beacon according to the room. As mentioned before, path loss models are used for continuous distance determination. Although the approach is simple and straightforward, a challenge in RF system is multipath propagation. Path loss models often assume a line-of-sight between beacon and locator, which do not contain reflection paths. If the models consider reflection paths it is via a random variable. For instance, a widely used path loss model for BLE is the log distance model [6]. The log distance model defines the pathloss (PL) as

$$PL = P_{TX} - P_{RX} = PL_0 + 10\gamma log_{10}\frac{d}{d_0} + X_g, \tag{1}$$

where P_{TX} is the power of the transmitted signal in dBm, T_{RX} is the power of the received signal in dBm. PL_0 is the pathloss at a reference in dB, d is the distance between transmitter and receiver in m, d_0 is the distance of the reference and X_g is a Gaussian random variable to account for multipath. In the next section we measure RSSI values and model the multipath part of the signal strength values.

4 Investigation of Measurement Errors

As we noted in the previous section, the log distance path loss model takes into account the power loss over the distance from receiver to transmitter. A factor that is not considered in the model as such is the antenna radiation pattern of the transmitter or of the receiver. Therefore, we first evaluate the RSSI measurement error for the specific hardware caused by the antenna characteristic. Subsequently we present the RSSI measurement errors caused by multipath propagation in Section B.

4.1 Locator Receiving Angle

For the investigation of the RSSI measurement error caused by the antenna characteristics, we apply the setup according to Fig. 3.

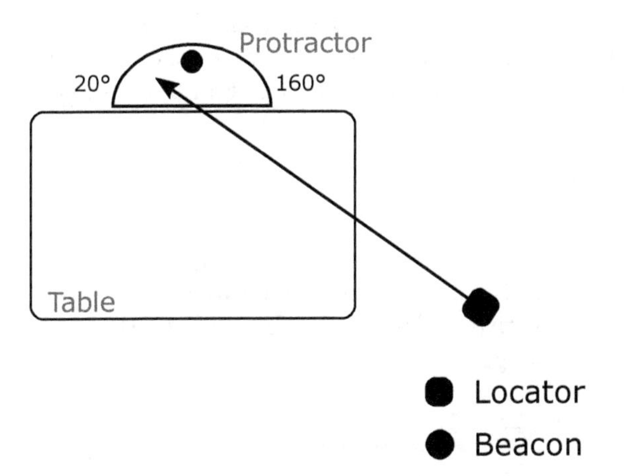

Fig. 3. Measurement setup for RSSI angle

On the edge of a table we attach a protractor which has a 1 m wooden batten on which the locator is fixed. In the experimental setup the locator points to the beacon. We rotate the locator from 20 to 160° around the beacon and record the RSSI values from the locator. The beacon sends out an advertising frame in Eddystone format once per second. The locator is from the company Hypros GmbH [2] and the Smart Beacon Go Mini from the company blukii [1].

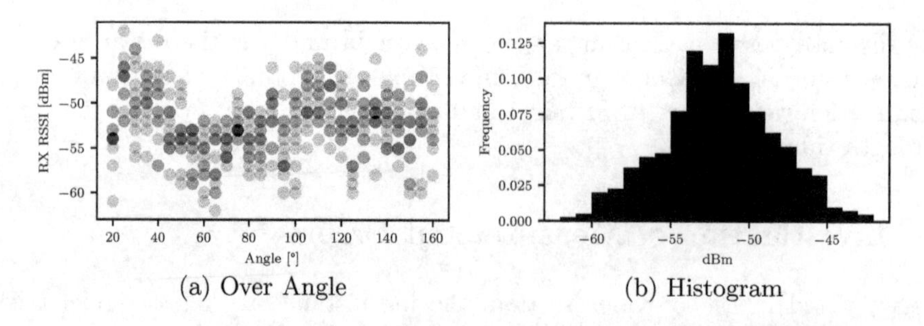

(a) Over Angle (b) Histogram

Fig. 4. Measured received signal strength

In Fig. 4a the RSSI of the beacon frames received by the locator are plotted over the rotation angle. Figure 4a shows a scattering of the received signal strength from −41 dBm to −63 dBm. The transparency of data points indicates the frequency of occurrence. If several data points are on top of each other, they are displayed in a darker color. A small reduction in RSSI values occurs at an angle around 60°. Here, we assume that the antenna placement within the devices or the antenna pattern causes this effect. The rest of the values show a comparable random behavior. As the effect is not large, we summarize all data set in a histogram in Fig. 4b.

As Fig. 4b shows, the majority of the measured RSSI data is in the range around −53 dB with a deviation of ±5 dBm. In the following we assume this deviation as measurement error due to the antenna radiation behaviour.

4.2 Multipath

Multipath propagation is one of the most challenging aspects of wireless communication systems. Signals are received from multiple directions by reflections on surfaces such as walls. Similar to light, the locator then receives a superposition that is either constructive or destructive. If we model the narrowband signal as a continuous wave, we get

$$\sum_{n=0}^{N-1} Ae^{i\varphi_n} = 0 \quad \text{with} \quad \varphi_n - \varphi_{n-1} = \frac{2\pi}{N} \tag{2}$$

where A is the amplitude, φ is phase difference and N is the number of waves. The phase shift is caused by the longer path of the reflective signal and therefore occurs mainly inside buildings.

From the locator's point of view, the superposition causes fluctuations in the received signal. The fluctuations lead to wrong estimations of the distance, e.g. using the path loss model in Eq. 1. We simulate this behavior using the building plan section shown in Fig. 5. For this purpose, we take into account reflections from walls, ceilings, and floors and allow for a maximum of 2 reflections. Walls, ceiling and floor are assumed to be of concrete. The height of the ceiling is 2.5 m. In the ray tracing simulation [5] we placed a locator in the building and simulated

10 beacon positions. The beacons are placed into two distinct areas. Each area has five beacon positions that are equally spaced five centimeters apart, as the legend of Fig. 5 shows. Area 1 (A1) is located 8.16 m from the locator, Area 2 (A2) at 1.96 m. Within the areas shown in Fig. 5 we denoted the position as 00 to 20 according to the spacing. Both areas are 0.75 m away from the walls and the beacon height is 1.75 m. The height of the locator is 2 m.

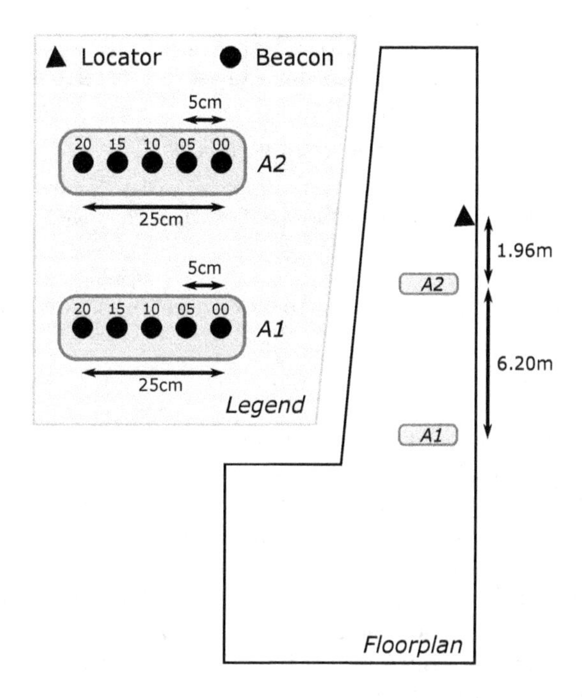

Fig. 5. Floorplan of the multipath ray tracing simulation

For each beacon position from the floorplan in Fig. 5, we have simulated the received signal strength at the locator, which results from superposition of the beacon signals. To simplify the simulation we assumed an isotopically radiating antenna. In addition, we varied the center frequency of the beacon signals and set it according to the frequencies of the channels of BLE. The resulting signal strength from two of these positions is shown in Fig. 6. In position A1, beacon position 10, shown as A110, we see that the RSSI increases by about 10 dB over the entire BLE frequency band. At this position, it is considered as a flat channel, meaning that the received signal strength over the frequency band has only a small variation. The behavior is different at position A1 (10 cm further), beacon position 20, shown as A120, where RSSI shows strong variation of more than 10 dB at position labeled A120.

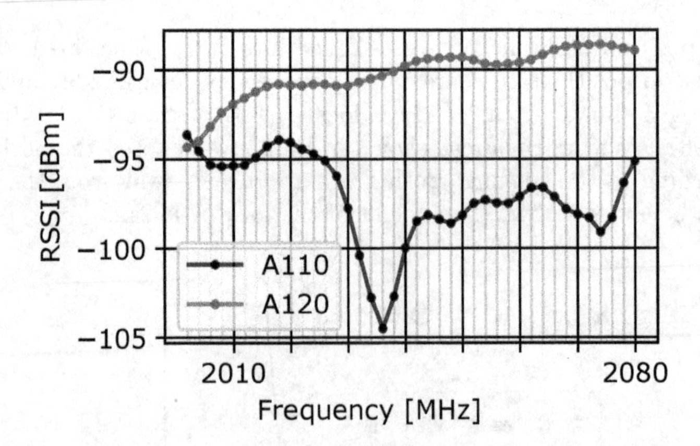

Fig. 6. Simulation results of all BLE channels

This circumstance is often referred to as Coherence Bandwidth

$$B_c \approx \frac{1}{2\pi \times D_s}, \tag{3}$$

where B_c is the Coherence Bandwidth and D_s is the delay spread [4]. In this example, the Coherence Bandwidth for A110 is 78 MHz, for A120 it is 28 MHz. However, the textbook approach is not valid because only a small part of the part needed within the Coherence bandwidth is needed and used. Typically channel 37, 38 and 39 are used for advertisement, as we discussed in Sect. 3 already. Although the channel numbering indicates adjacent channels, the center frequencies of the channels are 2402 MHz, 2426 MHz and 2480 MHz. Hence, we utilize our new metric of a selective coherence bandwidth to define a channel as flat. A flat channel satisfies

$$\begin{aligned} max(R) - min(R) < 10\text{ dB} \quad \text{with} \\ R = \{rssi_{37}, rssi_{38}, rssi_{39}\} \end{aligned} \tag{4}$$

where R is a set of RSSI measurements on the channels 37, 38, 39 as the indices denote. We derive the 10 dB threshold from empirical research from. The threshold corresponds to the deviation caused by the antenna. In practice, however, the RSSI values are measured without providing information of the transmission channels by the Bluetooth protocol stack BlueZ. BlueZ does not forward the information on which channel the RSSI value was recorded. We therefore consider R without information on which channel the RSSI was measured.

4.3 Investigation Results

In conclusion, there are a number of measurement errors that occur so that a localization via RSSI is not accurate. First the angle between beacon and locator results in a variance of ±5 dB in RSSI measurements. In addition, multi-path propagation occurs within buildings, which varies with location and frequency. The frequency dependence occurs in the 80 MHz range of 2.4 GHz ISM band, which BLE uses, and must be taken into account. In the following section we will present our approach to improve localization.

5 Approach

As we have seen in the previous section, constructive or destructive interference occurs at the receiver by multipath propagation. The interference depends on the location and frequency and results in a random behavior of RSSI measurements. The superimposition of two signals with the same amplitude and very different phases degrades the signal strength. As shown in Eq. 2, this attenuation can be very strong. In a worst case complete cancellations of the signal occurs, as long as the amplitude of the signal of the reflecting wave is equal and the phase difference is 180°. We illustrated this in Fig. 7a, where N is the wave number.

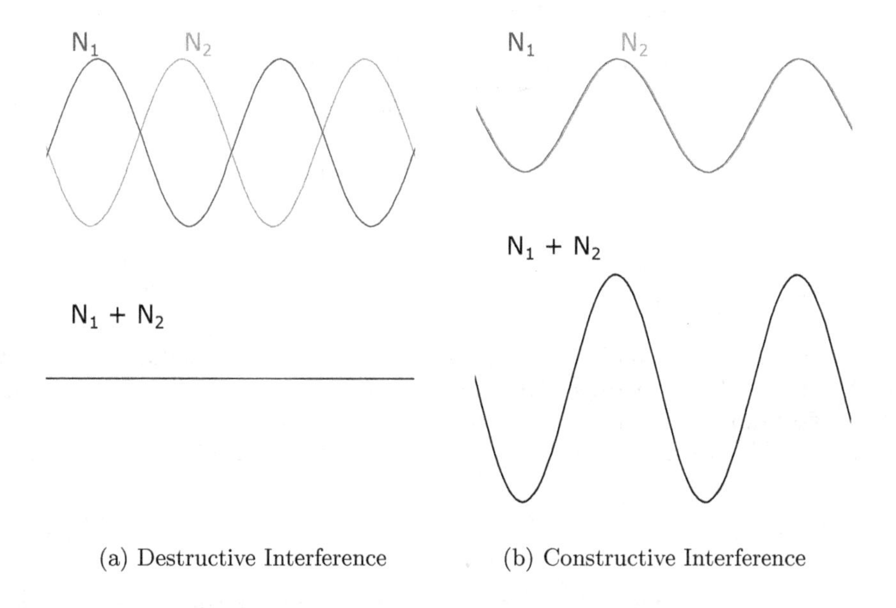

(a) Destructive Interference (b) Constructive Interference

Fig. 7. Interferences caused by reflected waves

On the other hand, a wave superposition that leads to an increase of the RSSI value is limited. For instance, if amplitude and phase are equal, a gain of 3 dB per phase and amplitude correct reflection occurs when the phase is the same. Figure 7b illustrates this effect. We focus our approach on the constructive wave superposition by proceeding it as illustrated in Fig. 8.

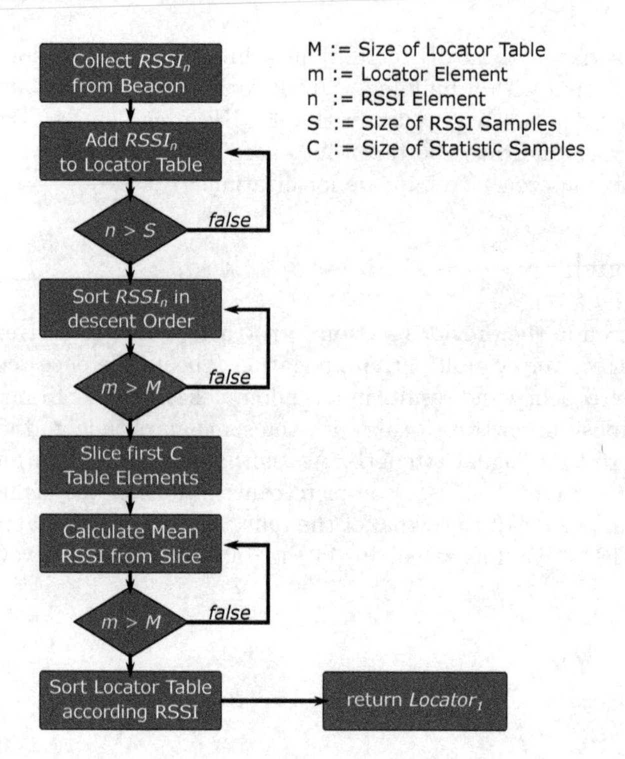

Fig. 8. Algorithmic approach for multipath compensation

First we collect the RSSI values of the Eddystone Beacon frame. We then assign the received RSSI of the frames to the locators of the locator elements m. When we have reached a sample size S for each locator, we sort the RSSI measurements of the individual locator elements m in descending order. We take the C highest measured RSSI values and average them. After averaging, the locator table M is sorted in descending order. The locator with the S highest average RSSI is then determined as the closest locator named Locator_1 in Fig. 8.

6 Evaluation

In this section we will first evaluate the constructive and destructive interference of the simulation in real measurements. Second, we refine the algorithm from the last section and provide parameters from the real environment.

6.1 Multipath

To evaluate the radio environment simulation from the previous section, we use the setup as shown in Fig. 5. The dimensions and wall, floor and ceiling materials in the figure correspond to the real corridor. However, we had to clear the hallway to reduce obstacles as much as possible. Nevertheless, we could not remove ceiling panels or lamps. The measurement itself was carried out over several hours, whereby the data set used was recorded at night in order to avoid interference from people. The devices for the test were the locators from Hypros and Smart Beacon Go Mini from blukii. We fixed the beacon at the position of Fig. 5. The beacons were attached to the ceiling with cord and hooks, so the beacons hang freely in the air without using additional, interfering support structure. By this mounting method, we achieved a direct line of sight to the locator. The locators were plugged into wall sockets at 2.4 m height. The beacon sends out Eddystone frames, which are received by the locator and sent to a PC for recording. The data was recorded over 4 h, resulting in a data set of about 8000 measurements per position.

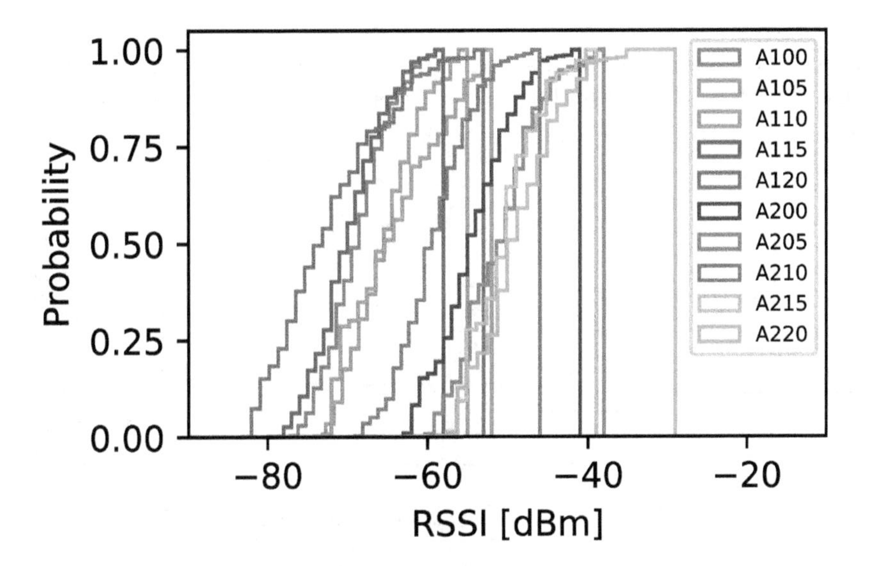

Fig. 9. Measurement result of real Environment

Figure 9 shows the result. We use an empirical cumulative distribution function (ECDF) to represent several distributions in the same way and thus simplify a comparison. The labeling was chosen so that A1xx corresponds to Area 1 at position xx. We start counting of positions at 00. The position on Area 1, 5 cm from the first measuring position corresponds to the notation A105. As in the simulation from Sect. 4, the areas are located 0.75 m from the wall. First we see that the dispersion of the measured values within Area 1 is 20 dB and in Area 2 40 dB. Whereby A120 has the lowest and A110 the highest RSSI values in Area 1, if the beginning of the ECDF plots is considered. In Area 2, A220 and A210 have the lowest RSSI values. The dashes at the end of the respective ECDF plots mark the largest RSSI value of the respective ECDF plot. For position A120, there is a variance of about 30 dB from the minimum to maximum measured RSSI value. The scattering is larger than expected from our simulations.

As we mentioned at the beginning, we have assumed a simplified corridor in the simulation e.g. without lamps. Furthermore we limited simulation results to second order reflections. In a large corridor like the one here, reflections of higher order occur frequently which additionally explains the deviations from the simulations. However, the results show the expected large variation of RSSI values due to constructive and destructive interferences. We confirm that the constructive interference and thus the maximum RSSI values are a good metric for the distance between beacon and locator and allows to distinguish the two areas from each other. Except for the outliers at A220 all positions of the areas are close together at maximum. The maximum RSSI are reached in the top 5% of the measurements, thus we define S of our algorithm as 100 and C as 5.

6.2 Algorithm

We evaluated the algorithm in a different setup in another building to prove the transferability of the approach presented in Sect. 5. Figure 10 shows the building plan.

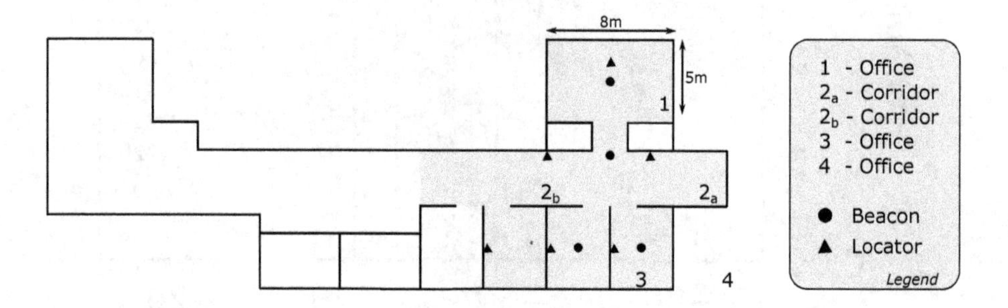

Fig. 10. Floorplan of the evaluation building

The circles represent the beacon positions. The beacons were placed at a height of 1.75 m. The locators are shown as rounded squares. For the evaluation we consider the grey-shaded area of the building that we divide into five zones. Four of these zones are in offices and one zone is on the corridor. To get an idea of the size of the room, we have noted the dimensions for the room of zone 5. The office rooms 3, 4 and 5 have a smaller area and are about half the size. The offices are in full use and therefore not empty. This means that they have computers, tables and other objects. A special feature of the rooms are the walls between rooms 3, 4 and 5, as they are made of plasterboard and are easily penetrated by radio signals. Towards the corridor the walls of rooms 3, 4 and 5 are made of glass. The walls to the extreme north and south are solid. For simplification, we have not marked any doors in the building plan. The doors are located on the open wall areas facing the corridor and are made of glass. The setup is the same than for the previous measurements. The only difference is that the beacons send with a higher frequency 5 Hz. The locators again log the unique ID of the beacons and their RSSI value. With the setup we evaluate if we manage to detect the correct room of the beacon with our algorithm. To provide a fast decision, we processed data for 20 s. During these 20 s we get 100 measurements from each locator if in range of a beacon. Then we determine the most likely location using the algorithm from Sect. 5.

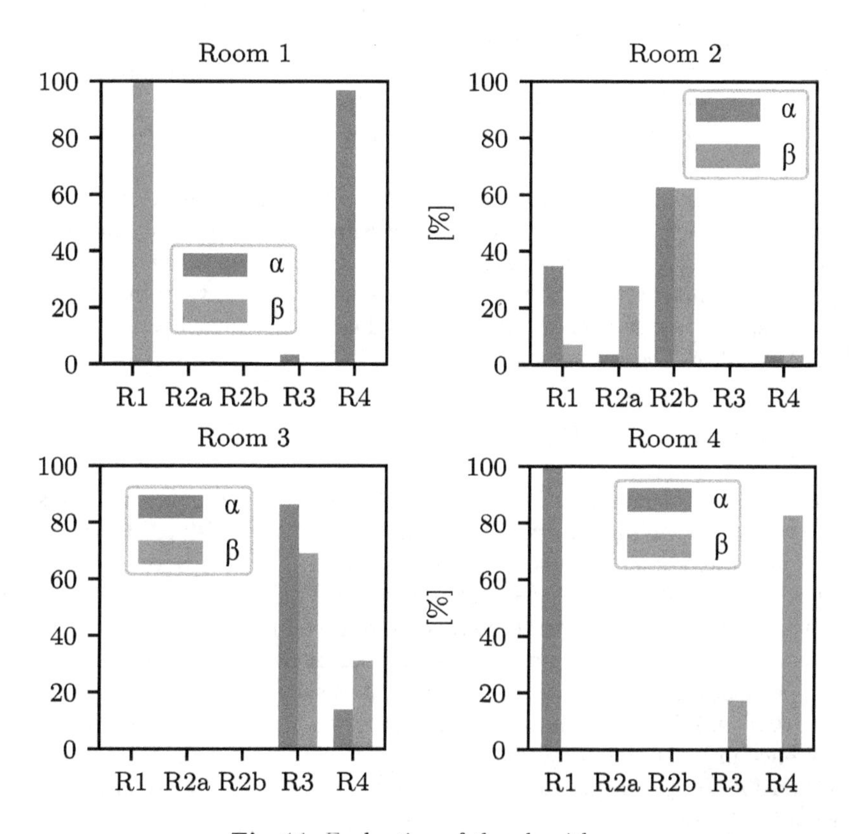

Fig. 11. Evaluation of the algorithms

In the evaluation we present the algorithm of Thaljaoui et al. [13] against our algorithm. We call the algorithm of Thaljaoui et al. α, and name our algorithm β. The title of each plot in Fig. 11 gives the room where the BLE beacon was placed. On the x-axis, we have plotted all rooms to show which room is detected by the algorithm and provide the ratio in percentage. The individual rooms on the x-axis were noted in short form to save space. The short form for Room 1 is $R1$, for Room 2 $R2$ and so on. Since Room 2 is a corridor, we have marked it as $R2a$ and $R2b$. $R2a$ is the front part of the corridor and $R2b$ the back part. The best algorithm would show 100% $R1$ for the plot entitled Room 1. The other rooms like $R2a$, $R2b$ and so on should have values of 0%.

For the true Room 1 in Fig. 11, we see that the algorithm α by Thaljaoui et al. recognizes it as Room 4 in 95% of the cases. In comparison, our algorithm recognizes Room 1 in 100% of the cases. Room 2 is recognized by algorithm α as Room 2b (61%) and as Room 1(38%). In contrast, our algorithm β recognizes Room 2b with probability of 60% and Room 2a with 30%. Room 3 was recognized by algorithm α with probability of 85%. Unlike algorithm α, our algorithm β recognizes Room 3 (70%) and Room 4 (30%). Room 4 is recognized as Room 1 in 100% of the cases by the algorithm α. Our algorithm β recognizes Room 4 in 81% and Room 3 in 19% of the cases.

We see that the algorithm α does not recognize rooms reliably. Only for Room 3 the algorithm shows a higher recognition rate than our proposed algorithm β. However, that Room 3 and Room 4 are recognized in combination by both algorithms is significant. We explain this by the fact that the partition walls between Room 3 and Room 4 are dry walls. Dry walls, in turn, can be penetrated very well by radio waves and therefore represent almost no attenuation in the received signal strength. Therefore the difference in the strength of the received signal between the two rooms is not significant enough to allow a reliable distinction between the two rooms. The algorithms therefore recognize either Room 3 or Room 4.

7 Conclusion and Future Work

In this paper, we have studied the multipath effects in communication systems for the Internet of Things to provide location awareness. Using Bluetooth Low Energy as an example, we first investigated sources of error that lead to signal variation. We simulated the effects of constructive and destructive interference, with the result that these have to be taken into account, as a small variation of a distance e.g. 5 cm results in an increased attenuation of 10 dB in the worst case. In the experimental evaluation we have proven that such attenuation occurs in real scenarios. To minimize the negative effects due to constructive and destructive interference in localization systems based on RSSI measurements, we proposed a method for compensation. The method improves the spatial accuracy for the location in almost all cases.

Acknowledgments. This publication is a result of the research work of the Center of Excellence CoSA. Horst Hellbrück is adjunct professor at the Institute of Telematics of University of Lübeck.

References

1. Blukii Smart Beacon Go Mini
2. Hypros GmbH
3. Cantón Paterna, V., Calveras Auge, A., Paradells Aspas, J., Perez Bullones, M.A.: A Bluetooth low energy indoor positioning system with channel diversity, weighted trilateration and Kalman filtering. Sensors **17**(12), 2927 (2017)
4. Farahani, S.: RF propagation, antennas, and regulatory requirements (chap. 5). In: Farahani, S. (ed.) ZigBee Wireless Networks and Transceivers, pp. 171–206. Newnes, Burlington (2008)

5. Hosseinzadeh, S.: 3D ray tracing for indoor radio propagation (2019)
6. Huang, B., Liu, J., Sun, W., Yang, F.: A robust indoor positioning method based on Bluetooth low energy with separate channel information. Sensors **19**(16), 3487 (2019)
7. Leugner, S., Constapel, M., Hellbrueck, H.: Triclock - clock synchronization compensating drift, offset and propagation delay. In: 2018 IEEE International Conference on Communications (ICC), pp. 1–6 (2018)
8. Leugner, S., Pelka, M., Hellbrück, H.: Comparison of wired and wireless synchronization with clock drift compensation suited for U-TDoA localization. In: 2016 13th Workshop on Positioning, Navigation and Communications (WPNC), pp. 1–4, October 2016
9. Mesquita, J., Guimarães, D., Pereira, C., Santos, F., Almeida, L.: Assessing the ESP8266 WiFi module for the internet of things. In: 2018 IEEE 23rd International Conference on Emerging Technologies and Factory Automation (ETFA), vol. 1, pp. 784–791. IEEE (2018)
10. Nezhadasl, M., Howard, I.: Localization of Bluetooth smart equipped assets based on building information models. In: Asset Intelligence through Integration and Interoperability and Contemporary Vibration Engineering Technologies, pp. 423–431. Springer (2019)
11. Rai, A., Chintalapudi, K.K., Padmanabhan, V.N., Sen, R.: Zee: zero-effort crowdsourcing for indoor localization. In: Proceedings of the 18th Annual International Conference on Mobile Computing and Networking, pp. 293–304 (2012)
12. Spachos, P., Mackey, A.: Energy efficiency and accuracy of solar powered BLE beacons. Comput. Commun. **119**, 94–100 (2018)
13. Thaljaoui, A., Val, T., Nasri, N., Brulin, D.: BLE localization using RSSI measurements and iRingLA. In: 2015 IEEE International Conference on Industrial Technology (ICIT), pp. 2178–2183. IEEE (2015)
14. Wang, H., Deng, Z., Fu, X., Li, J.: Ray-tracing aided indoor Bayesian positioning algorithm. In: Sun, J., Yang, C., Guo, S. (eds.) China Satellite Navigation Conference (CSNC) 2018 Proceedings, Singapore, pp. 589–599. Springer, Singapore (2018)

Towards Designing a Li-Fi-Based Indoor Positioning and Navigation System in an IoT Context

Ivan Madjarov[1,2](✉), Jean Luc Damoiseaux[1,2], and Rabah Iguernaissi[1,2]

[1] Aix-Marseille University, Marseille, France
{ivan.madjarov,jean-luc.damoiseaux,rabah.iguernaissi}@lis-lab.fr
[2] Université de Toulon, CNRS, LIS, DIAMS, I&M, Toulon, France
https://www.lis-lab.fr

Abstract. People generally find themselves lost while visiting a new indoor location because they are unaware of the building's architecture, especially when it is a large one or a shopping mall. The Global Position System (GPS) does not work properly in the indoor environment because of the satellite signal attenuation. In this paper, to assist people in finding their path, a Li-Fi (Light-Fidelity) based Indoor Positioning System (IPS) is proposed. A framework is developed based on a Li-Fi LED lamp transmitter and a dongle receiver connected to an Android smartphone to decode the received sequence. The pathfinding graph-based algorithm is proceeded in a REST architecture by a Web service consulting the graph-path database both installed on a Raspberry pi 4. The proposed solution is a low cost and does not require any additional infrastructure. It is easy to implement in most indoor environments like hospitals, buildings, and campuses. A short survey of techniques and algorithms for indoor positioning and navigation with the help of smartphones is also presented.

Keywords: Indoor positioning · Indoor navigation · Li-Fi localization · RESTful web service · Data integration

1 Introduction

The main objective of a positioning system is to provide location-based service with the computation of location coordinates supplied by a receiver. As an illustration, the Global Positioning System (GPS) makes outdoor localization and navigation continuously reliable with a smartphone. The same service is not as efficient for indoor localization since satellite signals are significantly degraded due to many obstacles and signal interferences as maintained in [4]. Thus, indoor navigation with smartphones became a relevant subject of research in recent years.

The authors in [5] review several technologies and techniques that have been proposed and implemented to improve indoor positioning and navigation. Among

K. Arai (Ed.): FICC 2021, AISC 1363, pp. 266–277, 2021.
https://doi.org/10.1007/978-3-030-73100-7_20

them are Infrared, Bluetooth, Wi-Fi, Li-Fi, RFID, NFC, BLE while the positioning algorithms are Triangulation, Trilateration, Scene Analysis, and Fingerprinting. The IPS localization methods are classified in [11] into two groups: (1) distance estimation process based on the signal strength and/or on the elapsed time between two signals; and (2) mapping-based methods where the localization works with pre-stored signals (tags) values in a database associated to a floorplan or a building's map.

In this work, we consider the Li-Fi technology as part of an IoT (Internet of Things) context that is used for local data transmission. Instead, Li-Fi is a wireless data transmission technology that uses the infrared and visible light spectrum. Our research is focused on the use of Li-Fi in the field of indoor location services. Authors in [3] present Li-Fi as a technology that uses visible light from a Light Emitting Diode (LED) to transmit data to a photodetector connected to a smartphone. Founded on this functional scheme, we developed a Li-Fi-based framework for indoor localization and navigation based on a mapping localization approach computed in a RESTful architecture as a data integration platform.

The remainder of this paper is organized as follows: Sect. 2 presents some motivation notes and a summary of works related to indoor positioning systems. Section 3 describes the design of the developed location-aware RESTful-based service. Section 4 shows the results of the test implementation. Finally, some conclusion notes and future work are presented in Sect. 5.

2 Motivation and Related Work

As specified in [8], there exists a remarkable number of proposals for IPS involving a multitude of technologies, techniques, and approaches. In this paper, we focus on three components that we argue to be essential for an IPS, namely Wi-Fi, Camera, and Li-Fi.

2.1 Wi-Fi-Based Localization Mechanism

In [5], an overview of the indoor Wi-Fi-based localization mechanism is presented. For a multitude of research projects, denoted in the overview, it exists similar approaches between the indoor Wi-Fi-based positioning and the outdoor GPS localization. In an indoor Wi-Fi-based positioning system, the received signal strength is used to calculate the distance between the Wi-Fi receiver and transmitter. The trilateration technique is the key computation approach to determine the location of a user. To achieve this, three or more Wi-Fi transmitters are required. So, to cover a building or even a single corridor, the number of Wi-Fi transmitters increases exponentially, and the downside of their multiplication is the possible interference with other sensible devices (i.e. in hospital). To improve the localization accuracy of the trilateration technique, in some research projects, a fingerprinting based Wi-Fi localization approach is proposed besides, which consists of signal phase analyses. Hybrid solutions also emerged that combine Wi-Fi with other techniques (e.g. Li-Fi, Bluetooth, optical camera) as presented in [9].

2.2 Camera-Based Positioning Systems

The optical and vision-based positioning system is discussed in [5] as a system where the position of a person is determined in a building by identifying an image captured by a camera and the processing power of a related device (e.g., a distant server with dedicated software for image processing). Besides, these systems significantly reduce installation and maintenance costs when compared with some other positioning systems (e.g., Wi-Fi-based). However, the camera-based positioning systems suffer, in general, from low accuracy, interference from multiple effects such as bright light and motion blur as pointed in [10]. On the other hand, in some cases, privacy concerns may be an issue since server stores image information for tracking and position processing.

2.3 Li-Fi-Based Approach

Li-Fi (Light Fidelity) was introduced in 2011 to provide pervasive data communication especially in areas susceptible to electromagnetic interference, e.g. aircraft cabins, hospitals. Li-Fi works based on VLC (Visible Light Communication) principle. This communication system can transmit data at high speeds over the visible light, ultraviolet, and infrared spectrums. It consists of two main components that communicate, i.e., a LED-lamp, namely, *Transmitter* and a Li-Fi-dongle, namely *Receiver*.

Li-Fi is optical communication technology, whereas Wi-Fi is a radio communication technology, and for this undergoes reduced interference, limited range, and finally increased security. A Li-Fi positioning system uses transmitters and receivers to determine the position of an object or a person. Li-Fi positioning systems are reliable, accurate, scalable, and real-time, requiring low maintenance [5], and fosters the reuse of already available artificial light infrastructure, so the cost of implementation could be considerably reduced.

As detailed in [10], the principle for Li-Fi positioning system is that each of the fixed LED-lamps has different flicker encoding, so the sensor, carried by the user, receives the light and compares the modulation against the known encoding schemes and thus associating the sensor location with the corresponding lamp. Trilateration and triangulation techniques are used for an IPS deployment in a larger environment, that it is in continuously lighted areas like shopping malls, airports, and museums. Note that, in areas where lights are not switched on continuously in a day, a hybrid method with Li-Fi is still an alternative as Wi-Fi, Camera, Bluetooth, BLE.

While Li-Fi IPS has a lot of advantages over other positioning systems, it nonetheless has limitations depending on coverage area, infrastructure, and computational costs, as pointed in [4].

To conclude this section, we note that the application of various techniques results in varying degrees of accuracy, complexity, and performance. Though, Li-Fi-based IPS are more simple, reliable, accurate, low-cost, and scalable.

3 Retrieving Indoor Paths

This section presents: (1) the system model with the source of data dissemination, and (2) the graph-based approach to retrieve all indoor paths stored in a relational database.

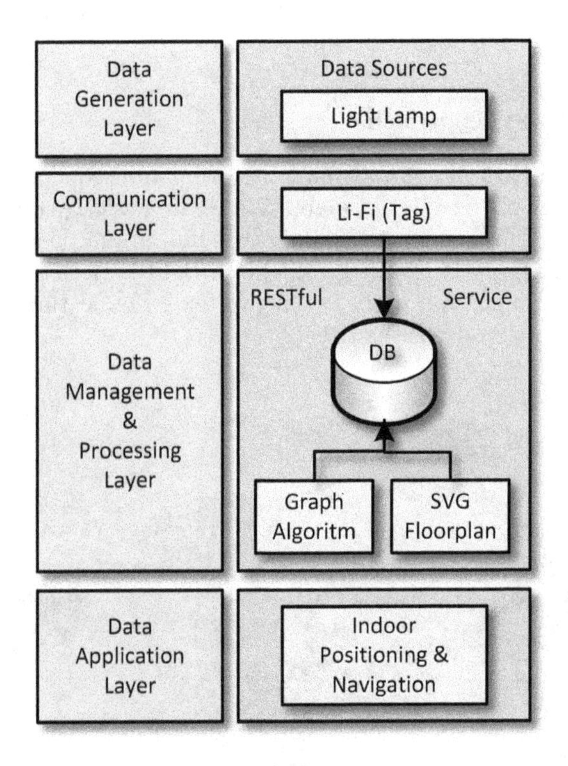

Fig. 1. System model for Li-Fi indoor positioning and navigation

3.1 System Model and Data Integration

In the proposed system model, the source of data dissemination is a Li-Fi lamp, whereas the source data collection is a user device with a photoreceptor. The collected data are processed, and the localization is performed via a Web service. In our work, we opt for a mapping-based localization method with pre-stored

Li-Fi tags values in a relational database. This localization method has an advantage in terms of the low cost of implementation, low power consumption, and processing accessibility.

In Fig. 1, the four layers system architecture is presented: (1) data generation, (2) communication technology, (3) data management and processing, and (4) application for data interpretation.

The data source is a lamp emitting a Li-Fi signal. The Li-Fi Tag is retrieved by a dongle connected to the user's smartphone. The guidance application transmits via a Wi-Fi the tag value to a server running a Web service. The user's position is confirmed on the building's plan. The next step to follow is displayed on the user's screen. The path points, to confirmed by the user, are extracted from the database.

3.2 Graph-Based Approach

In graph theory, a path in a graph is a finite or infinite sequence of edges which joins a sequence of vertices [1]. Two approaches are basically put forward when on trying to find a path between two vertices in a graph: (1) Either Breadth-First Search (BFS) or (2) Depth First Search (DFS). Both methods take the first vertex as a source, and if the second vertex is found through the finding process, then the method returns true [2].

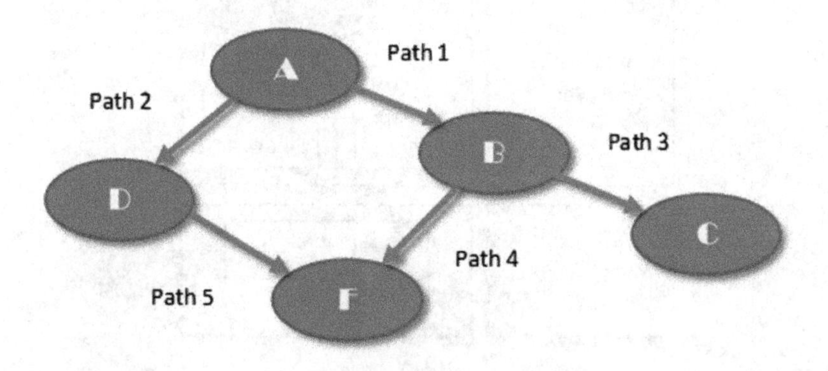

Fig. 2. Indoor unidirectional graph

To retrieve indoor paths, we consider a hierarchical, unidirectional graph schematically presented in Fig. 2. The graph-path is stored in a database with the following structure: (1) the graph consists of semantic nodes that can be combined hierarchically to create indoor paths; (2) the paths can be of arbitrary length.

In our case of use, we consider the nodes: *node A*, *node B*, *node C*, *node D*, and *node F*. Each node is represented by a Li-Fi LED lamp. Together, they form a limited number of paths and sub-paths as follows:

1. Path 1 consists of *lamp A* > *lamp B*
2. Path 3 consists of *lamp A* > *lamp B* > *lamp C*
3. Path 2 consists of *lamp A* > *lamp D*
4. Path 4 consists of *lamp A* > *lamp B* > *lamp F*
5. Path 5 consists of *lamp A* > *lamp D* > *lamp F*

3.3 Indoor Navigation System

For proof of concept (POC), in testing the indoor unidirectional graph navigation, we opt for a relational database instead of a graph-oriented database. For instance, the use of MySQL[1] (RDBMS) is less restrictive and more commonly used than Neo4j[2] (Graph Database Platform) within a RESTful Web application framework (i.e., Web services based on REST architecture) with JavaScript-based server utilities like Node.js[3] and Express.js[4].

The indoor navigation database schema consists of three tables, i.e. (1) *nodes* for storing the graph-vertices, (2) *paths* for storing the graph-edges, and (3) *results* for storing dynamically all valid paths in the indoor-graph. The resulting relational schema is presented in Fig. 3.

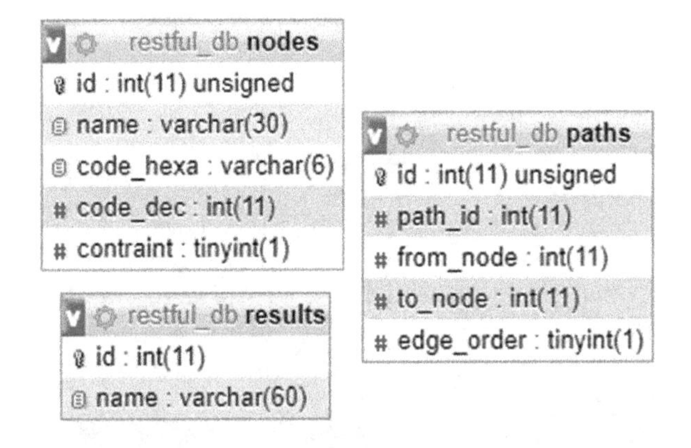

Fig. 3. Indoor graph database-relational schema

As presented in Fig. 3, the *nodes* table contains: (1) the identifier key of the record, (2) the name of the vertices, (3) the hexadecimal code of the IoT, (4) the decimal code of the same IoT, and (5) a Boolean value to test the vertices accessibility.

Nevertheless, for storing edges in a relational database table (e.g. *paths* table) is more complicated. To manipulate data about the indoor structure and associated resources per-row in a table is idiomatic and queryable in an RDBMS.

[1] https://www.mysql.com.
[2] https://neo4j.com.
[3] https://nodejs.org.
[4] https://expressjs.com.

id	path_id	from_node	to_node	edge_order
1	1	1	2	0
2	3	1	4	0
3	4	1	2	0
4	4	2	3	1
5	5	1	2	0
6	5	2	5	1
7	6	1	4	0
8	6	4	5	1

Fig. 4. The paths description for the graph presented in Fig. 2

Because we are using MySQL instead of a graph database system, it is not easy or useful to store a full path in one row of the table. Instead, we store each edge between the nodes of a path as a separate row using the identifier of the node FROM and the identifier of the node TO referring to the identifier key of the *nodes* table (see Fig. 3, *paths* table). For grouping edges, a common path-identifier key is used. In a unidirectional graph, we need to define which part of the path the edge represents, i.e. the sequences of paths in the graph. Therefore, a column with the edge order is added to the table structure. Finally, the resulting content for the *paths* table is presented in Fig. 4, with the possibility to query every sub-path.

The next step in our testing approach is to retrieve all possible paths for indoor navigation by querying the database. The query-results are saved as records in the *results* table presented in Fig. 5 in accordance with the graph presented in Fig. 2.

<table>
<tr><th>id</th><th>name</th></tr>
<tr><td>1</td><td>Lamp A|0xab67>Lamp B|0xcd45</td></tr>
<tr><td>3</td><td>Lamp A|0xab67>Lamp D|0x25a7</td></tr>
<tr><td>4</td><td>Lamp A|0xab67>Lamp B|0xcd45>Lamp C|0x24a9</td></tr>
<tr><td>5</td><td>Lamp A|0xab67>Lamp B|0xcd45>Lamp F|0x26c4</td></tr>
<tr><td>6</td><td>Lamp A|0xab67>Lamp D|0x25a7>Lamp F|0x26c4</td></tr>
</table>

Fig. 5. All paths are traced with the node name and the associated hexadecimal code

Figure 5 represents the querying results to retrieve all valid paths from the *paths* table. The *name* column contents a string “*node* x > *node* y > ...”. This means, beginning from “*node* x” on going to “*node* y” then to “*node* z”, and so on through the endpoint of the path. To achieve this task, the presented bellow query procedure is applied.

```
foreach (id in paths) {
    if (paths.edge_order > 0) then
      results=>name += nodes.name=>paths.to_node
    elseif (nodes.name=>paths.to_node ==
       nodes.name=>paths.from_node) then
          results=>name = nodes.name=>paths.to_node
        else
          results=>name = nodes.name=>paths.from_node +
         nodes.name=>paths.to_node
}
```

Instead of creating a view based on this query, we preferred to fully integrate the results-path in the database structure as a separate table (*results*). This is an easier way to manipulate multiple users' requests, their related data, and to consider constraints depending on the accessibility of a particular node in the path. For instance, if the node *lamp B* is not accessible the only way to navigate to the node *lamp F* is through the node *lamp D*. This means that the field constraint in the *nodes* table was marked as *false* for the node *lamp B*.

4 Technical Contribution and Test

In this section, we follow a REST-based indoor positioning system architecture and we discuss our test system configuration presented in Fig. 6.

4.1 Technical Support

The REST (REpresentational State Transfer) Web Services-based technology has certain advantages in an IoT (connected object) context over SOAP-based Web Services technology such as reduced parsing complexity and more efficient integration with the HTTP protocol as argued in [12]. The RESTful software architecture uses several operations for access and process data. These operations, specific to the HTTP protocol, represent an essential CRUD functionality (*Create, Read, Update, Delete*). Based on these operations and an address-resource name (URL), we build a REST API with an endpoint (route) for each one.

A lot of programming platforms can be used to build a REST API. In this work, our choice is focusing on Node.js framework and associated utilities for multiple grounds.

Node.js is a server-side and JavaScript-based runtime environment that we use to build not only REST APIs operations but also to invoke external services, e.g., interactions with the MySQL database. This is an asset for our project using REST system architecture with the Node.js scalability designed for building powerful network applications with a few lines of code and setting it up even on a micromachine, like Raspberry pi 4.

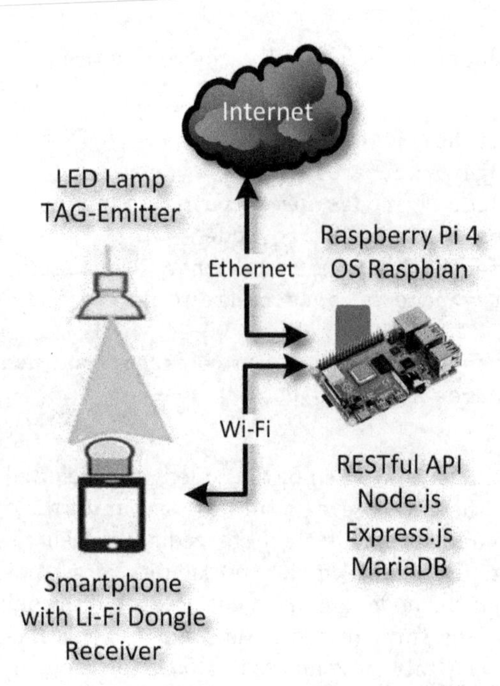

Fig. 6. RESTful API-based framework for Li-Fi indoor navigation

4.2 Use Case Scenario

The presented indoor navigation approach is based on a smartphone's interaction with a Web service. The main purpose of this Web service is to provide information on the path to be taken to arrive at the desired destination.

In the beginning, the smartphone (Android activity) receives at the point A (building entry point) the plan from the REST-website by scanning a QR code. Then, the application interrogates the user on the desired destination point. The REST service returns to the smartphone's screen the route to be followed from the point A to the final point X. This process was detailed in Sect. 3 above and showed in Fig. 7. The Li-Fi based application, a smartphone with a light sensor (dongle), is within the range of a Li-Fi lamplight (TAG emitter) trigger events confirming the right position throughout the path of destination or not. A notification is sent to the user's smartphone when the user accidentally left the path.

4.3 Test Scenario

The tested scenario is presented in Fig. 6.

We develop a location-aware Li-Fi application in which when an Android phone with a light sensor, is within the range of a Li-Fi lamplight, it will compare the emitted from the lamp tag with the tag listed in the path. The developed Android activity is based on the Oledcomm GEOLiFi Kit [6], with the GEOLiFi LED lamp, GEOLiFi Dongle, and GEOLiFi SDK Library. This is an intermediate technical solution while waiting for Li-Fi integration as IEEE standard for a mobile OS to directly use the camera or an integrated light sensor as a receiver, as suggested in [7].

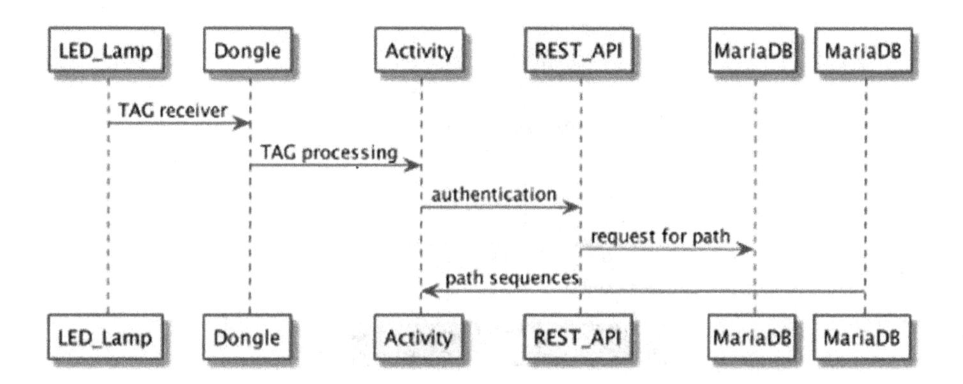

Fig. 7. Indoor navigation sequence diagram

As presented in Fig. 6, each time under a LED-Lamp the smartphone receives the lamp hexadecimal tag via the dongle. The Android activity transmits the processed tag to the REST service which queries the database on the veracity of the path, and if necessary, updates are sent to the activity, as presented in Fig. 8.

Our development approach is essentially server-side based processing. So, the developed REST service can handle multiple requests at once. In this schema, the system's client interaction is reduced to minimal. Thus, the client's mobile device is not overloaded with information that is not specific to it and that is likely to be used sporadically. On the other hand, this centralization approach for indoor navigation process management allows server-side service to track different paths simultaneously without interferences between users.

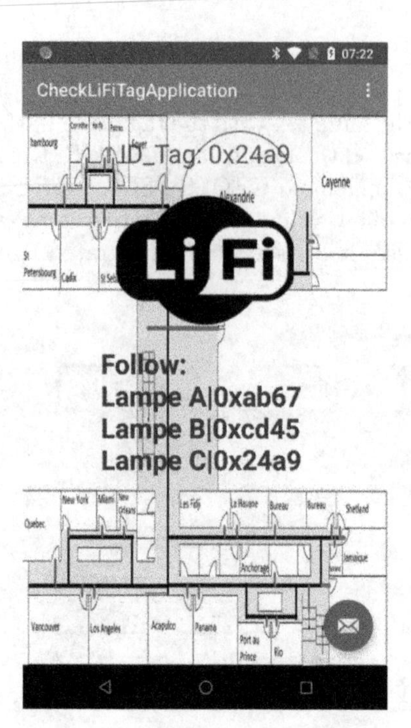

Fig. 8. Android pathfinder application for user's indoor navigation

5 Conclusion

In this paper, we present our Li-Fi-based IPS project: the LED lights emit visible light with location data, and a smartphone with a visible light receiver receives the data. Data are transferred to a REST-based Web service that calculates the optimal path to a destination point, considering the possible constraints. Nowadays, indoor navigation is a convenient phenomenon for everyone. For instance, the Li-Fi indoor navigation may be especially indispensable for the visually impaired. Smartphone presents the optimal path to a designation and speaks to the visually impaired through a headphone. As mentioned in [8], the principal motivation for most of the Li-Fi-based IPS proposals is the opportunity to reuse existing LED lighting infrastructure (i.e. to replace only bulbs) in opposition to the Wi-Fi-based solutions that require significant investment with proprietary hardware and software for the localization process and/or navigation management. In the continuity of this project, we will opt for a hybrid system to compensate for each other's shortcomings and take advantage of each other's strengths. We consider an IPS based on Li-Fi technology with path positioning from cameras placed in shadow zones. A systemic analysis of the QoS will be studied.

References

1. Robertson, N., Seymour, P.D.: Graph structure theory. In: Proceedings of the AMS-IMS-SIAM Joint Summer Research Conference on Graph Minors, 22 June–5 July 1991
2. https://www.geeksforgeeks.org/print-paths-given-source-destination-using-bfs/
3. Reyes, S., Segura, V.: Android app of location awareness using Li-Fi. Res. Comput. Sci. **118**, 107–114 (2016)
4. Naveed, U.L.H., et al.: Indoor positioning using visible LED lights: a survey. ACM Comput. Surv. **48**(2), 1–32 (2015)
5. Sakpere, W., Adeyeye-Oshin, M., Mlitwa, N.B.W.: A state-of-the-art survey of indoor positioning and navigation systems and technologies. S. Afr. Comput. J. **29**(3), 145–197 (2017). https://doi.org/10.18489/sacj.v29i3.452
6. Oledcomm, GEOLiFi kit (2020). https://www.oledcomm.net/lifimax-discovery-kit/
7. IEEE, VipPress.net. https://vipress.net/lieee-a-officialise-la-creation-du-futur-standard-lifi-802-11bb/
8. Mendoza-Silva, G.M., Torres-Sospedra, J., Huerta, J.: A meta-review of indoor positioning systems. Sensors **19**, 4507 (2019). https://doi.org/10.3390/s19204507
9. Wu, X., Mohammad Dehghani, S., et al.: Hybrid LiFi and WiFi Networks: a Survey, Published in ArXiv 2020. Mathematics, Computer Science (2020)
10. Brena, R., García-Vázquez, J.P., et al.: Evolution of Indoor positioning technologies: a survey. J. Sensors **2017** (2017). https://doi.org/10.1155/2017/2630413, Article ID 2630413, 21 pages
11. Rahman, A.B.M., Li, T., Wang, Y.: Recent advances in indoor localization via visible lights: a survey. Sensors **20**(5), 1382 (2020)
12. Madjarov, I., Slaimi, F.: A multigraph for restful services discovery in IoT ecosystem. In: Proceedings of the IEEE SERVICES 2019, IEEE World Congress on Services, 8–13 July 2019, Milan, Italy, pp. 366–367 (2019). https://doi.org/10.1109/SERVICES.2019.00107

Internet of Things (IoT) Based Plant Monitoring Using Machine Learning

Juancho D. Espineli[1](✉) and Khenilyn P. Lewis[2](✉)

[1] AMA University, Quezon City, Philippines
[2] Cavite State University, Cavite, Philippines

Abstract. The concept of Internet of Things (IoT) is one of the emerging technologies nowadays. In agriculture, there are seen utilization of IoT specifically in monitoring purposes. The IoT concept together with implementation of Machine Learning can lead to the development of a more reliable and accurate developed technologies to agriculture. Since plant monitoring is indeed a necessity for identifying possible preventive measures in plant's health. This study utilized IoT and Machine Learning in developing a prototype for plant's monitoring. Raspberry pi, Raspberry pi camera, SQL database, Python and web server are used. The algorithm was trained and validated in Orange Visual Programming and Waikato Environment of Knowledge Analysis (WEKA) using images of plants with nutritional deficiency. The machine learning algorithm has AUC of 0.974, CA of 0.824, F1 of 0.822 and Recall of 0.824. Evaluation of the system using ISO/IEC 25010 and Kappa statistics resulted 4.39 and 0.90 respectively. The results of validation and evaluation implies that the used of IoT and Machine Learning is acceptable in monitoring the nutritional health condition of coffee plants in farm and nursery.

Keywords: IoT · Machine Learning · Plant monitoring

1 Introduction

Philippines is an agricultural country and contributes to the total labor force of 39.8% with 20% of GDP [1]. One major crop commodity in the country is coffee. Robusta variety was considered as the most produced type with 69.2% of production [2]. It is remarkable that there is an increase of consumption of crop commodity in the country. Coffee farm monitoring in the country is manually performed by the farmers and agriculturists. They used labor-intensive procedures of identifying the nutritional status of plants in the farms. Because of the manual process of monitoring the nutritional health condition of the plants, it is considered as time consuming and there is a need for additional manpower. Aside from that, agriculture office needs to have monitored reports of the farm's current situation. One of the reports needed is the list of nutritional deficiencies present in the farm. The monitored status of the farm will help the farmers and growers to provide immediate preventive measure in the plants. The sending and receiving of data are also a problem in the physical process because there are no available technologies that can send and fetch data real time.

K. Arai (Ed.): FICC 2021, AISC 1363, pp. 278–289, 2021.
https://doi.org/10.1007/978-3-030-73100-7_21

Several aspects of agriculture utilized the Internet of Things (IoT) concept. The IoT notion is the used of connected devices to collect and process data needed for retrieval [3]. Smart agriculture is known as a developing concept in this area. It is also expected that the global smart agriculture will be tripled by 2025 because of the application of IoT concept [4]. The purpose of IoT is to provide relevant information real time to its clients. It is also noted that the employment of IoT in agriculture can lessen time, energy and saves money in an organization [5]. Precision agriculture is the framework integrated in IoT to incorporate detecting, measuring and responding to the need of the environment [6]. On the other hand, Machine Learning in IoT ensures the development of applications that contains massive data. Thus, it helps in analyzing, observing and processing large amount of data in different fields most specifically in agriculture [7]. In addition, machine learning used algorithms and predictive models to interpret dataset [8]. The model is being trained and tested to validate the classification process; thus, it can appropriate for any embedded systems or technologies.

It is evidently identified that the integration of IoT and Machine Learning can provide a more accurate and real-time monitoring in agriculture. With the real-time process, the clients pertaining to coffee farm workers and agriculturists can monitor the nutritional health conditions of the plants and can provide necessary intervention as needed. This study is conducted to utilize these two major concepts; IoT and Machine Learning for monitoring coffee plants nutritional deficiencies and send data in the web server real-time. The result of the study is intended to propose a method of manual monitoring of nutritional deficiencies in leaf by utilizing an algorithm for classification and web-based sending and fetching of data for farm monitoring.

2 Related Works

2.1 Plant Monitoring

The current status of the Philippines in monitoring the nutritional deficiency in plants is through manual system. Ideally, a coffee farmer or agriculturist go to the farm/field to visit the area. The monitoring of the plants is done through visual inspection and took several hours or days to finish. One of the problems arise with this manual method is the time consumed in one-by-one visual inspection. Other than being time consuming, additional manpower is needed for this kind of process [9]. In order to be successful in monitoring the health status of the plants, one must be familiar with the factors that affects its development and yield produced [10]. Monitoring the health of the plants can addressed several problems in the field. The ability to identify the plant's nutritional deficiencies and monitored it can help in avoiding costly attempts in solving the problems in the farm [11].

Another way of monitoring the health status of the plants is by conducting laboratory procedures. The plant with suspected issue will be brought to laboratory and test will be done. However, this process is costly and the availability of experts to do the process is limited. In this case, immediate response to the monitored plant can be deemed necessary [12]. Another issue that encountered in monitoring is the availability of data real time. Since the limited resources and embedded technologies occurred, it will be hard to agriculturist and coffee growers to monitor the plant's status as needed.

2.2 Internet of Things (IoT) in Agriculture

Smart Agriculture is one of the emerging trends nowadays. The used of Arduino or raspberry pi, wireless technologies and IoT concept helped in producing a smart agriculture system. The main components of IoT includes embedded system, hardware and software technologies, internet connection and other components [6].

In an IoT implementation, a cloud or web server in needed as dataset or storage medium of data. This data stored in the database is connected as input or output to the users. Arduino or Raspberry is considered as IoT gateway and an android smart farming application serve as user's view in manipulating the data in the repository. The process is real time and data can be accessed anytime.

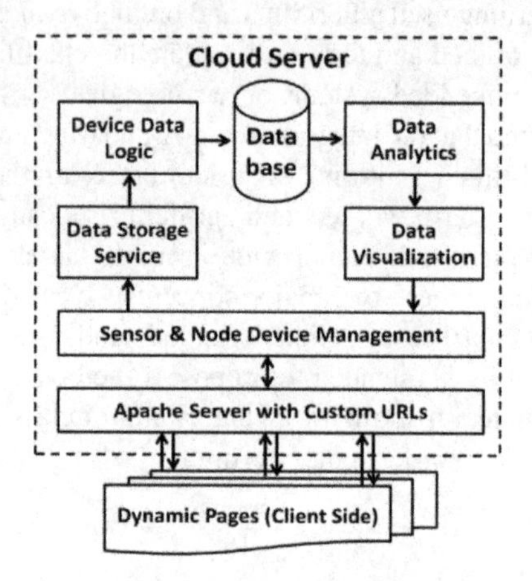

Fig. 1. Cloud server [13]

Cloud data supports the implementation of IoT in an application (see Fig. 1). Thus, it is necessary in a remote and real time environment. The used of monitoring system in agriculture is the collection of data and provide access to the processed data [14]. An embedded system is connected to cloud where data acquisition is performed and analyzed [15].

Three modules were created in the integration of Information Communication Technology (ICT) in smart agriculture. These modules are farm side, server side and client side [14]. In these three modules, six methods were implemented such as sensing local agricultural patterns, identification of location and data collection, transferring data from crop fields, data analysis, actuation and control, and crop monitoring. The remote monitoring approach provides access in agricultural facilities with the utilization of ICT. The study of [16] discussed the interactive system for future IoT based agriculture. The camera module is connected to the raspberry pi, cloud and big data is composed of the produced datasets and patterns for prediction and internet connection as gateway.

The main purposed of automating the monitoring in agriculture is to lessen human interaction and provide comfort. The monitoring process can be used as medium for recording and understanding the status of plants. The data can be accessed by the user anytime as needed specifically for generating reports [17].

2.3 Machine Learning

In Machine Learning, there are three general types that are known, namely: supervised learning, unsupervised learning and reinforcement learning. Supervised learning is giving labelled on data to get the desired output. In unsupervised learning, the data is unlabeled, hence, the algorithm is trained based on patterns. Lastly, reinforcement learning is in dynamic environment where the algorithm interacts [18]. Machine learning with IoT can create reliable source in data science [7].

MobileNet CNN is known for being simple and low-cost computational type of Convolutional Neural Network (CNN). It is widely used for image processing, object detection and real-world application [13]. It is a streamlined architecture that uses depthwise separable convolutions to construct lightweight deep convolutional neural networks and provides an efficient embedded vision applications [19].

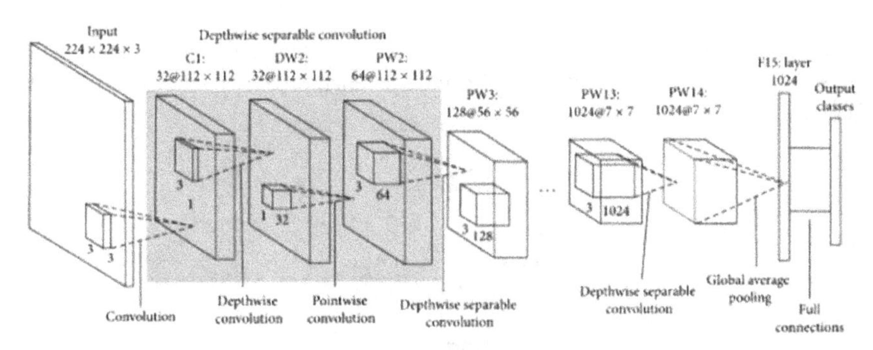

Fig. 2. MobileNet architecture [20]

The model processed the tasks within the dataset containing in the image. It is composed of convolutions, depth wise convolutions, pointwise convolutions, depth wise separable convolutions, pooling, fully connected and the output classes (see Fig. 2). The standard computation cost of the model is shown in Eq. 1 where $D^2{}_f$ is the dimension of input feature maps, M is the input channel, N is the output channel and $D^2{}_K$ is the kernel size.

$$Computation\,cost = D^2f * M * N * D^2K \tag{1}$$

The used of Machine Learning can be used in easily detection of plant behavior pertaining to its environment. The data can be easily access and stored in the database. The dataset is connected to the cloud server so that the retrieval can be made anytime. Thus it is indeed necessary to integrate the approach in any embedded technologies [21].

3 Methodology

3.1 Data Preparation

The dataset is composed of 1127 instances with 1000 features of images from coffee farm and nurseries in Cavite, Philippines. The images were captured using the raspberry pi camera with 3.68 × 2.76 mm (4.6 mm diagonal) image area, 1.12 μm × 1.12 μm pixel size and 3280 × 2464 pixels resolution.

3.2 Training the Model

The model was trained using Orange Visual Programming and Waikato Analysis of Knowledge Management (WEKA). Orange Visual Programming is used for data analysis by means of stacking components in the work flow [22]. The Waikato Analysis of Knowledge Management (WEKA) on the hand is a Machine Learning software for graphical user interface and standard terminations [23]. The process of 10-fold cross validation is to provide accurate result of the model and avoid over fitting.

3.3 Proposed Model

The proposed method of the study is the utilization of IoT components and Machine learning using MobileNet CNN in monitoring the nutritional deficiency detected in the coffee farms.

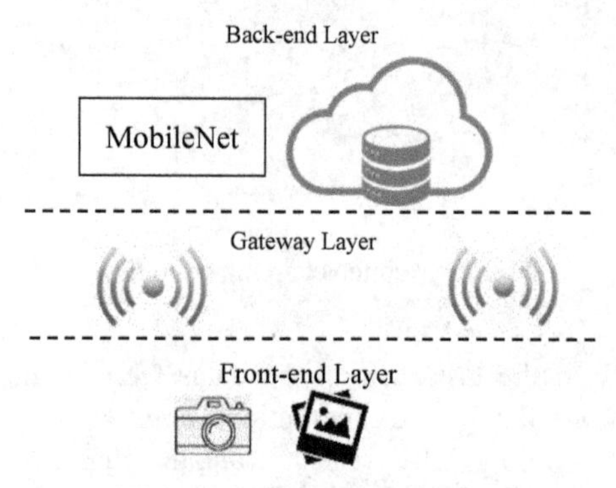

Fig. 3. Proposed architecture

The proposed architecture is composed of three major layers, these are front-end layer, gateway layer and back-end layer (see Fig. 3). In the front-end layer, a camera is needed to capture the images of coffee leaf that were identified to have a nutritional deficiency. The images are sent through the gateway layer in which internet connectivity is needed. The images are saved in the cloud database and the MobileNet algorithm will create patterns to identify which nutritional deficiencies it belongs.

Three main approaches are presented in the proposed model, the farm/nursery site, server and client (see Fig. 4). The camera is situated in the farm/nursey site where the data about images is sent in the server. The application of IoT components and algorithm is being implemented in the server. The last module is the client view, wherein the website can be seen containing the number of nutritional deficiencies identified in the coffee leaf. In the client view also, graphical view can be seen and used to generate reports regarding the real-time status of the coffee plants in the nursery. In the client view, the viewing of

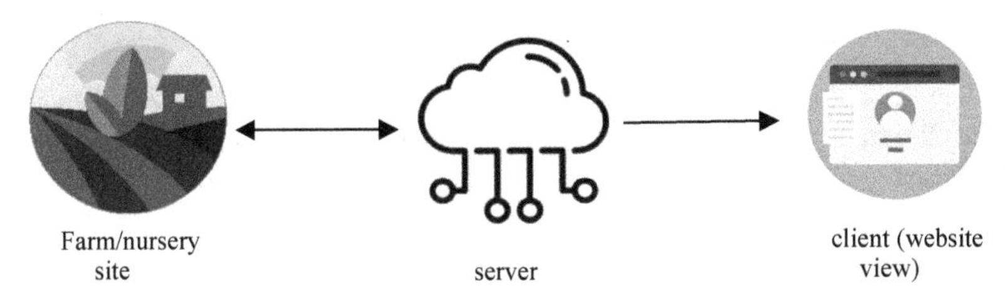

Fig. 4. Proposed approach

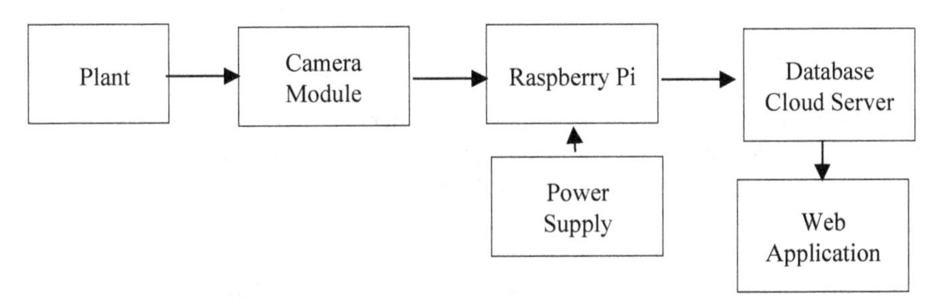

Fig. 5. Block diagram of the study

the web application pertaining to plant's health status can be done using mobile phone, laptop or desktop computers with internet connectivity as requirement.

The block diagram of the study is composed of plant, camera module, Raspberry pi, power supply, database cloud server and web application (see Fig. 5). The image of plants is captured using a Raspberry Pi camera module with 5 megapixels specifications. The camera is connected to the Raspberry Pi module connected in a power supply. The program is installed and run in the Raspberry Pi using Python programming language. The trained image serves as dataset and will undergone the MobileNet CNN algorithm. The classified nutritional deficiencies are sent in the cloud server and can be viewed in a web application using internet connectivity. SQL database serve as the data storage and repository.

3.4 Materials

The development of the prototype was made possible through the used to raspberry pi, raspberry pi camera, SQL database, web server and python as programming language. The Raspberry Pi 4 model B [24] include a high-performance 64-bit quad-core processor, dual-display support at resolutions up to 4K via a pair of micro-HDMI ports, hardware video decode at up to 4Kp60, up to 8 GB of RAM, dual-band 2.4/5.0 GHz wireless LAN, Bluetooth® 5.0, Gigabit Ethernet, USB 3.0, and PoE capability. The Raspberry pi camera is in 5 megapixels and 15 cm in length.

3.5 Validation and Evaluation

To validate the algorithm, Precision, Recall and F measure are used [25]. Precision is defined as the fraction of dataset relevant of the query (see Eq. 2). Recall is the fraction of dataset that are successfully retrieved (see Eq. 3). F measure is the weighted average of Precision and Recall (see Eq. 4).

$$Precision = tp/(\frac{tp}{fp}) \tag{2}$$

$$Recall = tp/(\frac{tp}{fn}) \tag{3}$$

$$F\ Score = 2 * (Recall * Precision)/(Recall + Precision) \tag{4}$$

To determine the Inter-reliability among classifiers, Cohen's Kappa statistics is used [26] (see Eq. 5). Po is the relative observed agreement among raters, Pe is the hypothetical probability of chance agreement and K is the Kappa value.

$$K = (Po - Pe)/1 - Pe \tag{5}$$

The system quality model is evaluated using ISO/IEC 25010 which identify the degree of quality in the implementation of the system. The criteria used are functionality suitability, performance efficiency, compatibility, usability, reliability, maintainability and portability [27]. The weighted mean is used to determine the acceptability of the proposed model. The prototype was evaluated by Information Technology experts, agriculturist and farm workers.

4 Results and Discussions

The study proposed the utilization of Machine Learning through MobileNet CNN with IoT concept. The algorithm was trained to identify the possible nutritional deficiency

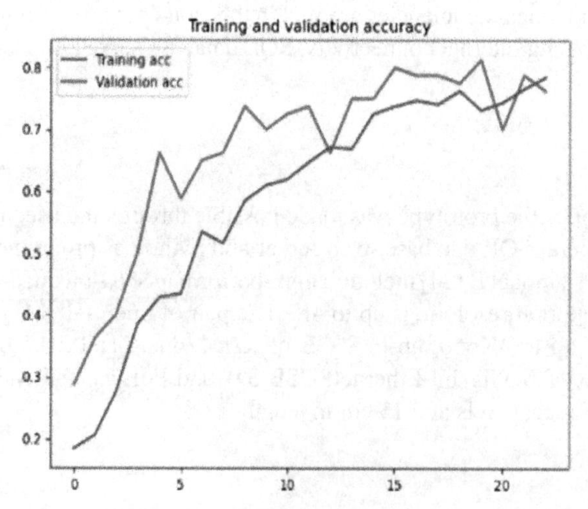

Fig. 6. Training and validation accuracy

in coffee leaf. The result of training is sent to the web server to provide clients view. The training and testing dataset were obtained from the images captures by raspberry pi camera, processed by the algorithm and sent in the web server.

The training and validation accuracy and loss of MobileNet CNN is shown in Fig. 6 and Fig. 7. For every step-match in the datasets, the computed percentage of accuracy and loss is presented.

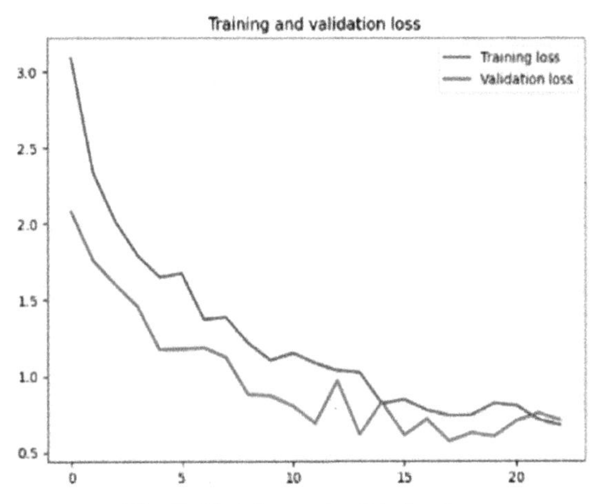

Fig. 7. Training and validation loss

Hence, once the highest percentage for accuracy and loss identified, this is the result submitted to the web server. Identifying the highest accuracy in the steps will end the iteration because it meets the highest point of accuracy in the model.

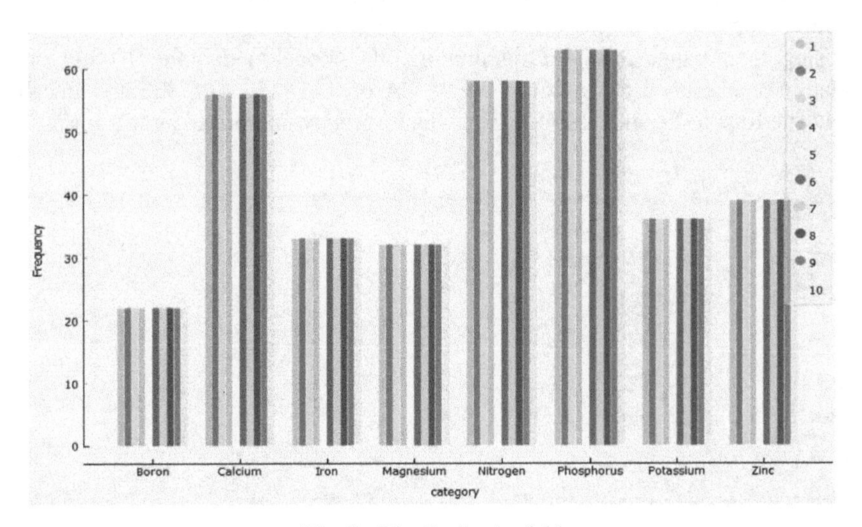

Fig. 8. Distribution by fold

The distribution of the nutritional deficiency and its frequency by category (see Fig. 8). The distribution was presented using the 10-fold cross validation of training the dataset of images. Eight nutritional deficiencies were captures during the testing and evaluation of the study.

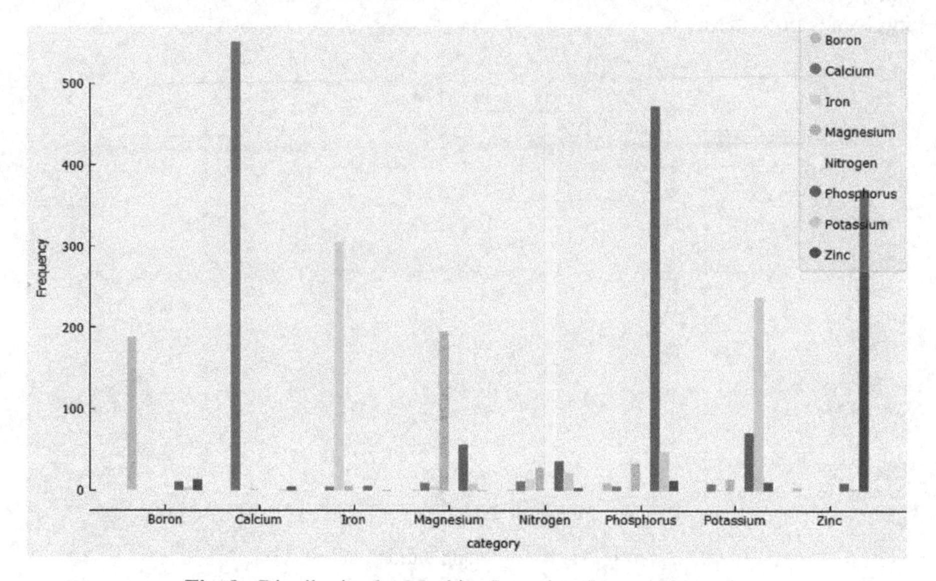

Fig. 9. Distribution by Machine Learning (Neural Network)

On the other hand, the distribution using the algorithm under the Neural Network and frequency was also illustrated (see Fig. 9). Aside from the 10-fold cross validation, the categories in view with neural network shows the frequency of each nutritional deficiency of plants.

Using the Orange Visual Programming, the model undergone 10-fold cross-validation and resulted the AUC of 0.974, CA of 0.824, F1 of 0.822 and Recall of 0.824. The Kappa statistics resulted 0.90 which verbally interpreted as “Strong”.

Robusta	Phosphorus	208	0	208
Robusta	Potassium	120	0	120
Robusta	Calcium	187	0	187
Robusta	Boron	74	0	74
Robusta	Iron	110	0	110
Robusta	Magnesium	107	0	107
Robusta	Nitrogen	199	0	199
Robusta	Zinc	129	0	129

Fig. 10. Webpage view

The fetched data from the farm/nursery site is represented in a web application (see Fig. 10). This includes the number of identified coffee leaf that can be used for monitoring the status of the plants.

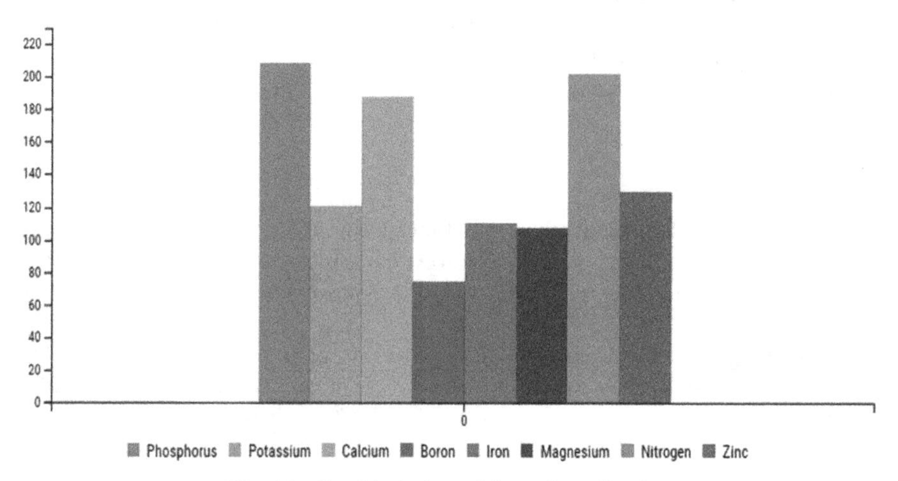

Fig. 11. Graphical view of the web application

The number of identified deficiency of plants can be seen in the graphical view of the web application (see Fig. 11). The user of the web application can immediately identify the up-to-date monitoring of the plants. The data is being sent from the farm/nursery site real time. Because of the real-time monitoring, the farmers or agriculturist can easily see the status of the plants in the farm and nursery.

In terms of the acceptability of the proposed model, the ISO/IEC 25010 is used. The result of evaluation shows a 4.39 rating with verbal interpretation of "Agree" which implies that the proposed model can be used for plant monitoring using IoT and machine learning.

5 Conclusions

The need for a real time monitoring of coffee farm is necessary to identify and see its condition. As the current status of farms are monitored manually of farmers, growers and agriculturists which exert too much effort of monitoring. Because of these encountered problems in farm monitoring, the study was conducted to propose a model that will monitor plant's nutritional deficiencies status using IoT and machine learning. Three layers were presented in the architecture; front-end layer, gateway layer and front-end layer. The MobileNet CNN model was used in training and testing the datasets. The training and validation accuracy and loss were presented to see the implementation of the machine learning algorithm. Orange Visual Programming and Waikato Environment for Knowledge Analysis (WEKA) were used for training the datasets with 70% and 30% in the total number of datasets. Kappa statistics, precision, recall and f measure were utilized to validate the classifiers. The images were trained so that the algorithm can identify the best classification possible within the datasets. The data is sent to the web server and database coming from the farm/nursery where the device is situated. Once sent, the graphical view of the monitored plant status can be viewed in a web application which can be access through mobile, laptop and computer views. The results of Kappa statistics and ISO evaluation is verbally interpreted as "Strong" and "Agree". By this monitoring approach, the farmer or agriculturist can identify easily the status of the plants real-time. Hence, preventive measures can be easily performed.

6 Future Works

The results of the study present a method of integrating IoT and Machine Learning for monitoring plants in farm and nurseries. However, it is recommended as future works to go deeper in these two concepts and integrate other embedded systems for enhancement.

Acknowledgment. The researchers would like to acknowledge the help and assistance of coffee farms in Cavite, Philippines for allowing them to gather the necessary data for the completion of the study.

References

1. Philippines Agriculture, Information about Agriculture in Philippines (2018). https://www.nationsencyclopedia.com/economies/Asia-and-the-Pacific/Philippines-AGRICULTURE.html
2. Philippine Statistics Authority: Philippine Statistics Authority|Republic of the Philippines, Philippine Statistics Authority (2018)
3. Chandra, A., Patil, S., Patil, A., Sakpal, S., Shanbhag, K.: Review on smart helmet using Internet of Things. Int. J. Adv. Eng. Res. **4**, 707–711 (2017)
4. Eastern Peak: IoT in Agriculture: 5 Technology Use Cases for Smart Farming (and 4 Challenges to Consider), Easternpeak (2019). https://easternpeak.com/blog/iot-in-agriculture-5-technology-use-cases-for-smart-farming-and-4-challenges-to-consider/
5. Mahdavinejad, M.S., Rezvan, M., Barekatain, M., Adibi, P., Barnaghi, P., Sheth, A.P.: Machine learning for internet of things data analysis: a survey. Digit. Commun. Netw. **4**(3), 161–175 (2018). https://doi.org/10.1016/j.dcan.2017.10.002
6. Pallavi, K., Mallapur, J.D., Bendigeri, K.Y.: Remote sensing and controlling of greenhouse agriculture parameters based on IoT. In: 2017 International Conference Big Data, IoT Data Science BID 2017, January 2018, pp. 44–48 (2018). https://doi.org/10.1109/BID.2017.8336571
7. Sharma, K., Nandal, R.: A literature study on machine learning fusion with IoT. In: Proceedings of International Conference Trends Electronics Informatics, ICOEI 2019, April 2019 (Icoei), pp. 1440–1445 (2019). https://doi.org/10.1109/icoei.2019.8862656.
8. Singh, B.K., Singh, R.P., Bisen, T., Kharayat, S.: Disease manifestation prediction from weather data using extreme learning machine. In: Proceedings 2018 3rd International Conference Internet Things Smart Innovation and Usages, IoT-SIU 2018, pp. 1–6 (2018). https://doi.org/10.1109/IoT-SIU.2018.8519908.
9. Home|Official Portal of the Department of Agriculture: Department of Agriculture, Republic of Philippines (2019). https://www.da.gov.ph/
10. Guy, S.: Visual Identification of Nutrient Deficiencies, pp. 4–7 (2018). https://www.smart-fertilizer.com/articles/nutrient-deficiencies

11. Nutrients, W., Require, P., Environment, U.Y., Step, F.: Diagnosing crop nutrient deficiencies in the, no. Figure 2. 1–7 (2017)
12. Beede, R.H., Brown, P.H., Kallsen, C., Weinbaum, S. A.: Diagnosing and Correcting Nutrient Deficiencies. Pist. Prod. Man. 147–157 (2005). https://ucce.ucdavis.edu/rics/fnric2/crops/papers/Chapter_16.pdf
13. Khattab, A., Abdelgawad, A., Yelmarthi, K.: Design and implementation of a cloud-based IoT scheme for precision agriculture. In: Proceedings International Conference Microelectron. ICM, pp. 201–204 (2016). https://doi.org/10.1109/ICM.2016.7847850
14. Anusha, A., Guptha, A., Sivanageswar Rao, G., Tenali, R.K.: A model for smart agriculture using IOT. Int. J. Innov. Technol. Explor. Eng. **8**(6), 1656–1659 (2019)
15. Mishra, D., Khan, A., Tiwari, R., Upadhay, S.: Automated irrigation system-IoT based approach. In: Proceedings 2018 3rd International Conference Internet Things Smart Innovation Usages, IoT-SIU 2018, pp. 1–4 (2018). https://doi.org/10.1109/IoT-SIU.2018.8519886
16. Veloo, K., Kojima, H., Takata, S., Nakamura, M., Nakajo, H.: Interactive cultivation system for the future IoT-based agriculture. In: Proceedings 2019 7th International Symposium Computing Networking Workshop, CANDARW 2019, pp. 298–304 (2019). https://doi.org/10.1109/CANDARW.2019.00059.
17. Elangovan, R., Santhanakrishnan, D.N., Rozario, R., Banu, A.: "Tomen: A Plant monitoring and smart gardening system using IoT. Int. J. Pure Appl. Math. **119**(March), 703–710 (2018). https://www.researchgate.net/publication/323811801_TomenA_Plant_monitoring_and_smart_gardening_system_using_IoT
18. Society, I.: Artificial Intelligence & Machine Learning: Policy Paper|Internet Society (2017). https://www.internetsociety.org/resources/doc/2017/artificial-intelligence-and-machine-learning-policy-paper/?gclid=Cj0KCQiAkePyBRCEARIsAMy5ScsN0mCd1mRbkGc7EFVqY1nxPryej-SNgVFcVlHLRHThV3ZyotNjIaEaAqo5EALw_wcB
19. Wang, W., Li, Y., Zou, T., Wang, X., You, J., Luo, Y.: A novel image classification approach via dense-mobilenet models. Mob. Inf. Syst. **2020** (2020). https://doi.org/10.1155/2020/7602384.
20. Tutorial, N.N.: "Transfer learning in TensorFlow 2 tutorial
21. Chauhan, G., Sharma, Y., College, K.P.G.: Cloud based Intelligent Plant Monitoring Device, pp. 231–237 (2019)
22. Orange programacion visual: , "Orange Data Mining - Visual Programming. https://orange.biolab.si/home/visual-_programming/
23. University of Waikato: Weka 3 - Data Mining with Open Source Machine Learning Software in Java, The University of Waikato (2016). https://www.cs.waikato.ac.nz/ml/weka/
24. Raspberry Pi: Raspberry Pi 4 Model B specifications – Raspberry Pi, no. June 2019. https://www.raspberrypi.org/products/raspberry-pi-4-model-b/specifications/
25. Shung, K.P.: Accuracy, precision, recall or F1?|by Koo Ping Shung|Towards Data Science (2018). https://towardsdatascience.com/accuracy-precision-recall-or-f1-331fb37c5cb9
26. Diamond, J.J.: Cohen's kappa. J. Clin. Epidemiol. **44**(6), 609 (1991). https://doi.org/10.1016/0895-4356(91)90224-W
27. ISO - ISO_IEC 25010_2011 - Systems and software engineering—Systems and software Quality Requirements and Evaluation (SQuaRE)—System and software quality models (2011)

Convergence of 5G with Internet of Things for Enhanced Privacy

Amreen Batool[1], Baoshan Sun[1], Ali Saleem[2], and Jawad Ali[3](✉)

[1] School of Computer Science and Software Engineering, Tiangong University, Tianjin, China
{Amreen,sunbaoshan}@tiangong.edu.cn

[2] Department of Computer Science, COMSATS University Islamabad, Lahore, Pakistan
Ddp-sp15-bse-115@ciitlahore.edu.pk

[3] Malaysian Institute of Information Technology, Universiti Kuala Lumpur, Kuala Lumpur, Malaysia
jawad.ali@s.unikl.edu.my

Abstract. In this paper, we address the issue of privacy in 5th generation (5G) driven Internet of Things (IoT) and related technologies while presenting a comparison with previous technologies for communication and unaddressed issues in 5G. Initially, an overview of 5G driven IoT is presented with details about both technologies and eventually leading to problems that 5th generation will face. Details about 5G are also presented while comparing them with previous technologies. The architecture of 5G is presented hence explaining layers of 5G and technologies like SDN, NFV and cloud computing that compose these layers. The architecture for 5g based IoT is also presented for providing visual understanding as well as explained based on how this addresses the issues present in 4G. Privacy is highlighted in 5G driven IoT while providing details about how SDN, NFV and cloud computing helps in elimination of this issue. The issues presented will be compared with 4G based IoT and solutions are provided about mitigation of these issues particularly bandwidth and security. Moreover, techniques used by 4G and 5G technologies for handling the issues of privacy in IoT are presented in a nutshell as a table. Paper also presents a detailed overview of technologies making 5G possible meanwhile giving an explanation about how these technologies resolve privacy issues in 5G.

Keywords: Privacy in 5G · SDN · Cloud computing · NFV · Internet of Things (IoT) · Privacy

1 Introduction

Internet of Things (IoT) can be considered as a network of objects, animals or machines that transfer data without involvement of human or computer interaction. The statement above defines it as a more open and independent system that benefits in terms of content transfer and interaction, but it also possesses challenges like the need of more bandwidth for transfer of massive content as well as low latency and security. IoT is an open network in terms of devices connected as any node is malicious to another node which rises trust

K. Arai (Ed.): FICC 2021, AISC 1363, pp. 290–300, 2021.
https://doi.org/10.1007/978-3-030-73100-7_22

and privacy issue in IoT [1–3]. Currently, IoT is deployed on current technology 4G that lacks the inability to address the challenges of IoT specifically bandwidth and privacy [4].

Scarcity of resources, as well as security and privacy issues, have raised the need for a more versatile and secure technology that make sure the integrity of data as it is transferred over the channel. 5G is on the verge of deployment making massive IoT and tactile internet technologies possible by addressing data rate and latency issues. Meanwhile providing enhanced security and privacy for these technologies.

Massive IoT implementation over 5G will be possible with software-defined networking that will not only enhance the speed but also will address many privacy issues faced by the current IoT network [5]. Hence in an environment where devices are connected wirelessly, it is of critical importance that devices communication is secure and nodes privacy is not compromised. 5G will provide a higher data rate of up to 10 Gbps and with more bandwidth and ability to choose among various service providers for QoS [6].

The ability to choose among several service provider will raise authentication and data integrity issues in the IoT environment. Moreover, data confidentiality, privacy availability and access control will also be an issue in open networks [7]. Furthermore, as 5G will be deployed over the current 4G IP base network it will by default inherit the problems specific to IP bases communication. Hence, to ensure the security and privacy in 5G for IoT and related technologies will certainly be a problem [8].

5G will not only benefit the IoT of things but also cloud, social computing, and cognitive radio technology as well. Moreover, it is designed by keeping in mind the needs for virtualization that will enable novice services that will also raise security and privacy issues. IoT has revolutionized the communication among nodes by the distribution of processing and storage power among mobile devices [9, 10]. IoT generic layered architecture constitutes of three layers that enable the IoT services provided to applications. This distribution of processing power and storage among multiple issues raise much security, privacy and trust issues in IoT enables devices [11]. Thus, a key step towards the resolution of these problems is IoT middleware that is a logical layer.

Meanwhile, Integration of 5G with IoT provides an SOA approach for IoT that will enable the decomposition of IoT system and provide a service-oriented approach for privacy threats in IoT architecture [12].

Moreover, our paper is portioned in 3 different sections. The first section is about the introduction of the topic as well a brief overview of what has already been done in the concerned field. The second section discusses the evolution of technologies from the first generation to 5G as well as a comparison table about what privacy techniques used in all these technologies to mitigate privacy and security concerns. The third section discusses the architecture of 5G along with its different layers and what measurements can be taken to enhance the technology that is on the verge of deployment. Forth section discusses the integration of 5G with IoT and various other domains like mobile computing and cloud computing and how the integration of these technologies will control the privacy risks while providing delay requirement of less than 10 s for real-time communication.

2 1G to 5G Transition

Wireless network communication architecture starting its journey from a 1G analog system to an era of 4.5G more flexible and secure IP-based network designed specifically for mobile networks has been successfully implemented. Although a couple of years back 4.5G was a secure enough system for IoT, cloud communication and related technologies [13, 14]. On the contrary the expectations proved to be wrong; with the arrival of the internet of everything (IOE) and tactile internet the network required should have ample bandwidth along with greater data rate to achieve the lesser delay requirement for real-time communication [15]. Although the hype created about the deployment of 5G is believed to not fulfil many expectations of the industry. Starting our comparison from a 1G analog system that was developed for only voice communication. Considering the security concerns in first-generation communication system 2G was introduced as a digital voice and SMS system that provided enhanced coverage, as well as the better data rate of 64kbps for faster communication and better security measures in comparison to 1G TDMA and CDMA, was used for multiplexing [16].

3G, on the other hand, introduced proved to fulfil promises of higher data rate and provided real broadband experience with a data rate of 2Mbps and used W-CDMA for multiplexing. Currently used 4G provide 20Mbps data rate and an IP based network [17].

4G followed by 4.5G were supposed to an efficient wireless communication system for real-time data transmission but with advents of technologies like cloud computing, IoT and mobile computing [18] the scarcity of resources raised problem, as well as these technologies, allow more open communication among nodes that raise privacy, authentication and trust issues. 5G is being developed with keeping in mind the requirement that previous technologies failed to fulfil [19]. Figure 1 illustrate the network management in 5G based IoT Architecture.

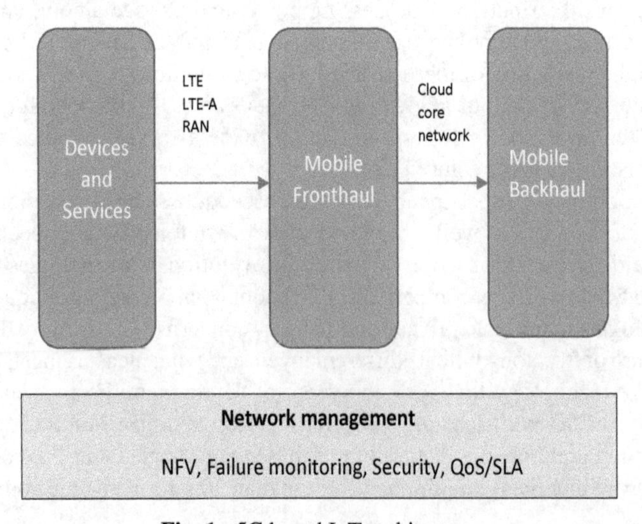

Fig. 1. 5G based IoT architecture

Table 1. 4G vs 5G

	4G	5G
Latency	10 ms	<1 ms
Data traffic	7.2 Exabytes/month	50 Exabytes/month
Peak data rate	1Gbps	20Gbps
Available spectrum	2–8 GHz	3–300 GHz
Connection density	1000/km^2	1 million/km^2
Battery consumption	High	10% more efficient

3 5G Architectural Vision

The next-generation wireless architecture will bring in a wide range of functions and services because of the diverse quality of service (QoS) requirements as well as the proliferation of connected devices [15]. 5G is designed to fulfil the needs for highly mobile and low latency networks with more resources as shown in Table 1. Hence the architecture of 5G depicted in Fig. 2 is sliced in 4 various layers by keeping in mind the diverse requirement of Human to Human and Human to Machine connections. Layers are named as the network layer, controller layer, management and orchestration layer, and service layer [20].

5G architecture will provide dedicated services to various sectors. This on-demand network slicing will enable the automated SLA allocation for sliced instances [21]. Hence the first layer of 5G architecture is a customer-oriented layer for personalized services including augmented reality as well as personalized internet services [22].

The second layer is the most important layer of 5G network as it enables cognitive procedures implementation throughout the slice lifecycle. This layer is further divided into two sub-layers network function virtualization (NFV) and software-defined network (SDN) [23]. The heterogeneity in the network will be enabled because of automatic instantiation and allocation of resources based on needs of service. Based on the function of the network services will be allocated to various service providers that will be connected to a centralized core network thus achieving low latency and lower number of handoffs. Moreover, the software-defined nature of the network will enable the orchestration of resources their efficient management and a trustworthy secure mechanism in a heterogeneous network. The software-defined network will implement various security measures to enable the privacy of users because of the open nature of the network [24, 25].

Next layer of the architecture is radio access network (RAN) infrastructure. It acts as a pipeline for recursive allocation of resources in a highly dense network where resources need to be allocated repeatedly to fulfil similar customer-oriented requests within a network [26]. This recursive allocation makes 5G network a scalable network by the allocation of a certain instance to multiple entities to handle the workload and bringing in elasticity as network services; as part of the network is allocated recursively based on customer needs [4, 27].

The last layer of this architecture will constitute end-user devices including IoT enabled sensor devices, mobile devices, and wearable devices. The recursive allocation of resources will allow a sliced instance to be allocated to multiple tenants as virtual infrastructure for the deployment of services [28].

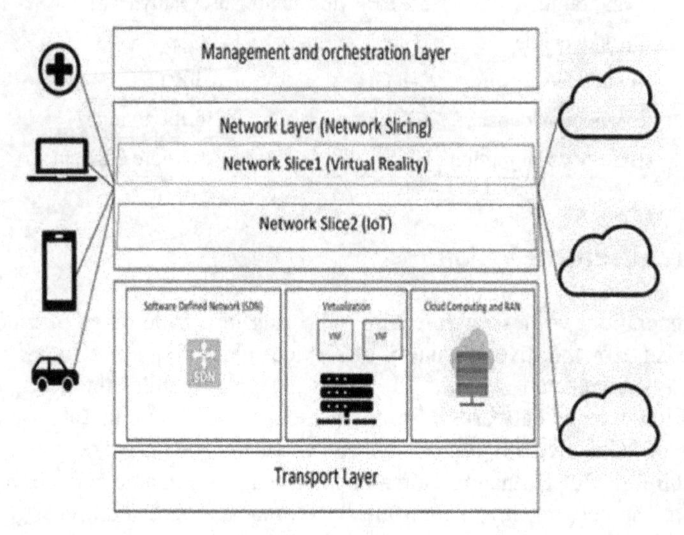

Fig. 2. 5G architecture

In a nutshell, 5G will be deployed on top of current 4G infrastructure but it will bring in multiple technologies like NFV and software-defined networking (SDN) [29] to handle the issues related to heterogeneity and security that openness of the 5G network will bring in. Moreover, network slicing will enable scalability and elasticity of heterogeneous network along with greater bandwidth, availability and low latency [30].

4 Integration of IoT and Related Technologies in Next-Generation 5G Network and Privacy

Next-generation wireless technology will be a heterogeneous network that constitutes of various services. Due to the heterogeneity and complexity 5G is divides into four various layers each providing different functionality hence possessing different security threat.

5G core architecture uses services like virtualization of resources and their assignment according to need to multiple users on a single domain. Privacy will be a major concern as third parties will be responsible for the provision of cloud services and data management. Moreover, as IoT network is a network of various types of devices [12], hence various virtualization services provider will be involved that will raise privacy and authentication issues for content and end-users might have trust issues for the service providers [22]. In a nutshell, IoT and cloud computing can be considered as two most prominent technologies in the deployment of next-generation wireless network hence possessing the privacy threats of both the technologies as well [31]. Integration of cloud into 5G is concentrated on three main cloud technologies SDN, NFV and RAN, Intelligent edge computing [32].

4.1 Integration of Cloud Computing in 5G

Trust, confidentiality, and privacy of content and user are most important and concerned areas of study. Moreover, privacy is a broader area as compared to the other two. Hence the integration of cloud services for virtualization in 5G will face two issues. One is in cloud user are not aware of where their data is kept or stored as user lose direct control on the content once it's in the cloud. Other is user is not aware of how their personal information is processed and in which region it is kept moreover as the users are not aware of content location, so they are also blank about what kind of privileges both public and private organizations have on the content [33].

A solution to this problem is to encrypt the content at the user end so no one except the service provider will be able to access the content. Moreover, various ciphertext decryption techniques are used some of which are symmetric, and some are based on the Bloom Filter method [34].

4.2 5G and SDN

5G will be a ubiquitous network of multiple technologies spread over a wide coverage area to provide a programmatic control of the whole network for the provision of services according to user needs. In SDN because of centralized control plan privacy of user and content will be at stake [35]. As there will be centralized control of the whole network pool of resources distributed denial of services attack and identity spoofing can be a major issue to mitigate resource provision to a node of a network of nodes and identity theft will be easy in SDN because of the centralized control plane. Privacy and trust are a major concern in SDN under discussion that needs to be resolved [36].

A major concern related to SDN security is Denial of service attack. SDN has a centralized controller. Controller control the entire network traffic. So, if controller is under attack then whole network faces the consequence.

Other potential threats in SDN architecture are illegal interceptor, privacy breach, fake base station, security policy conflicts. Moreover, deployment of 5G architecture over previous generation architecture so it also inherits threats from previous generations as well [37].

4.3 4G/5G Integration and Privacy Issues

5G will be deployed on top of existing IP-based network it will inherit privacy threats present in the 4G network along with its perks. Attacks present is 4G that 5G will inherit presented as man-in-the-middle attack, spoofing, eavesdropping, impersonation, privacy violation attack, chosen ciphertext attack, tracing attack, replay attack, collaborated attack, disclosure attack, parallel session attack and masquerade attack [38].

Although the 5G will be a ubiquitous network of various technologies but because of heterogeneity in the network it faces privacy theft issue because of increased number of handoffs with change of network that will not only affect the low latency condition of 5G but also present privacy problems like identity privacy, content privacy, location privacy, user anonymity as well as conditional and forward privacy [39].

5 Solutions for Privacy Preservation in 5G Enabled IoT

The need for 5G raised due to the increased number of IoT devices that need to be connected constantly for a reliable connection that 5G promises to provide in the future. State of the art of IoT considers data storage and processing in cloud for the extraction of useful information. Integration of various architecture exposes 5G to various security, privacy, and trust issues that will be highlighted at the deployment of 5G if not resolved. 5G based IoT devices are prone to user privacy threats that are subdivided into three concerns. Data privacy concerns, location privacy concerns, and identity privacy concerns.

Due to the inclusion of multiple stakeholders and heterogeneity (HetNet) of network into 5G architecture leakage of user information is not a problem. Data storage or analysis companies can access user information store it for a period longer than agreed and give it to third parties to earn an enormous amount of revenue. Table 2 show some major data privacy issues and their counter-measures.

Table 2. Data privacy threats

Threat	Solution
Leakage of user's personal information	Data protection and privacy techniques
Usage of data without user's consent	End-to-End encryption schemes
Due to Heterogeneity third party can derive user's data using data mining	End-to-End encryption schemes
Malicious nodes	Authentication and exclusion
Trusted network issue	Authorized access to nodes

Location-based privacy threats can be considered the most concerning attacks as service providers will be able to access user's location without them knowing. Location-based services (LBS) has the capability to turn on user location information without user knowledge also they are capable to access user's real-time location information. Table 3 presents location-based threats to the 5G-IoT network.

Hence 5G will be a network with an enhanced number of handovers raising the risk of identity theft, trust, and authentication issues. A solution proposed to preserve privacy theft in the next-generation wireless network is to software-defined networking by sharing user information among trusted access points [40]. Moreover, privacy-oriented encryption techniques are also a solution to ensure privacy in 5G enabled network [41]. Beside this using proxy with RFID tags to conceal users, location can also help in preserving user's location information when eavesdroppers have high computational resources

Table 3. Location privacy threats

Threat	Solution
IoT device position revealed	Anonymity-based solutions/outlier detection and database consistency monitoring
Tracking user movements	Data perturbation and obfuscation methods
Database corruption (wrong location reference)	Outlier detection and database consistency monitoring

[42] A solution proposed to this problem is to distribute cloud data files into specific services and files, in the end, replicating them into edge devices hence not uploading the content into the cloud for the training of the model [43]. IoT also covers the paradigm of mobile edge computing where privacy protection of edges is essential [44]. Moreover, ToRPEDO, PIERCER and IMSI cracking attack are new threats to 5G enabled IoT devices. Along with previous privacy attacks present in 5G due to the integration with past technologies like IoT 5G is vulnerable to 3 new type of identity theft attacks which are presented in Table 4 along with countermeasures Literature Proposes solutions like user identity verification for Cloud services, encryption of user location for privacy, subscriber identity encryption for preservation against IMSI attacks.

Table 4. Attacks and countermeasures

Attack	Comments	Potential target	Counter measures
ToRPEDO attack	Access user location; inject fabricated messages and enable DOS attack it is the enabler of the other two attacks	Centralized control elements	It adds noise as fake paging message [38]
PIERCER attack	Enable association of victims IMSI with attacker's phone it can also enable previous attacks on the device	Identity and information theft	A proposed solution is to never send IMSI in the paging message
IMSI cracking attack	Using IMSI brute force attack can be initiated onto a victim's device	Roaming and user equipment	To-add IMSI catcher in the path IMSI paging sniffer connected to device [33, 34]

6 Conclusion

5G is the technology of future hence possessing all the previous technologies and their flaws. Firstly, as 5G will be deployed on top of existing 4G network it will entail all the privacy issues including spoofing and impersonation threat.

Furthermore, technologies like SDN, IoT, Cloud Computing are what make 5G network so the deployment of 5G will encompass all the privacy issues of these technologies as well. In a nutshell, the paper is divided into 4 various section presenting the evolution of 5G, its architecture and privacy issues processed by each of the layers individually are also presented. Along with these privacy flaws inherited by 5G because of integration with previous technologies like SDN, IoT and Cloud Computing are also discussed in detail.

Discussion in this paper is limited to the privacy aspect of the 5G hence excluding security domain. Future work can include security of 5G and flaws presented in 5G because of integration with other technologies.

References

1. Ali, J., Ali, T., Musa, S., Zahrani, A.: Towards secure IoT communication with smart contracts in a blockchain infrastructure. Int. J. Adv. Comput. Sci. Appl. **9**(10), 584–591 (2018)
2. Bendale, S.P., Rajesh Prasad, J.: Security threats and challenges in future mobile wireless networks. In: Proceedings of the 2018 IEEE Global Conference on Wireless Computing and Networking, GCWCN 2018, pp. 146–150 (2019)
3. Ali, J., Ali, T., Alsaawy, Y., Khalid, A.S., Musa, S.: Blockchain-based smart-IoT trust zone measurement architecture. In: ACM International Conference on Proceeding Series, vol. Part F148162, pp. 152–157 (2019)
4. Panwar, N., Sharma, S., Singh, A.K.: A survey on 5G: the next generation of mobile communication. Phys. Commun. **18**, 64–84 (2016)
5. Li, S., Da Xu, L., Zhao, S.: 5G Internet of Things: A survey. J. Ind. Inf. Integr. **10**, 1–9 (2018)
6. Mitra, R.N., Agrawal, D.P.: 5G mobile technology: a survey. ICT Express **1**(3), 132–137 (2015)
7. Zhou, S., Zhang, Z., Luo, Z., Wong, E.C.: A lightweight anti-desynchronization RFID authentication protocol. Inf. Syst. Front. **12**(5), 521–528 (2010)
8. Ni, J., Lin, X., Shen, X.S.: Efficient and secure service-oriented authentication supporting network slicing for 5G-enabled IoT. IEEE J. Sel. Areas Commun. **36**(3), 644–657 (2018)
9. Ali, J., Khalid, A.S., Yafi, E., Musa, S., Ahmed, W.: Towards a secure behavior modeling for IoT networks using blockchain. In: CEUR Workshop Proceedings, vol. 2486, pp. 244–258 (2019)
10. Institute of Electrical and Electronics Engineers, IEEE Solid-State Circuits Society, and S. European Solid State Device Research Conference (46th: 2016 : Lausanne, ESSCIRC Conference 2016: 42nd European Solid-State Circuits Conference: 12–15 September 2016, June 2016, pp. 21–24 (2020)
11. Ferrag, M.A., Maglaras, L., Argyriou, A., Kosmanos, D., Janicke, H.: Security for 4G and 5G cellular networks: a survey of existing authentication and privacy-preserving schemes. J. Netw. Comput. Appl. **101**, 55–82 (2018)
12. Mavromoustakis, C.X., Mastorakis, G., Batalla, J.M.: Internet of Things (IoT) in 5G Mobile Technologies (2016)

13. Rost, P., et al.: Mobile network architecture evolution toward 5G. Infocommunications J. **9**(1), 24–31 (2017)
14. Rumney, M., et al.: Testing 5G: evolution or revolution? IET Semin. Dig. **2016**(5), 1–9 (2016)
15. Zhong, S., et al.: Networking Cyber-Physical Systems: Algorithm Fundamentals of Security and Privacy for Next-Generation Wireless Networks (2019)
16. Lee, J., et al.: LTE-advanced in 3GPP Rel -13/14: an evolution toward 5G. IEEE Commun. Mag. **54**(3), 36–42 (2016)
17. Al-Falahy, N., Alani, O.Y.: Technologies for 5G networks: challenges and opportunities. IT Prof. **19**(1), 12–20 (2017)
18. Nordrum, A.: 5 myths about 5G. IEEE Spectr., 2–3 (2016)
19. Alpaydın, E.: View on 5G Architecture. Version 3.0, June 2019, pp. 21–470 (2019)
20. Foukas, X., Patounas, G., Elmokashfi, A., Marina, M.K.: Network slicing in 5G: survey and challenges. IEEE Commun. Mag. **55**(5), 94–100 (2017)
21. Westphal, C.: Challenges in networking to support augmented reality and virtual reality. In: ICNC 2017 (2017)
22. Sayadi, B., et al.: SDN for 5G mobile networks: NORMA perspective. Lecture notes of the Institute for Computer Sciences, Social Informatics and Telecommunications Engineering, LNICST, vol. 172, pp. 741–753 (2016)
23. Ordonez-Lucena, J., Ameigeiras, P., Lopez, D., Ramos-Munoz, J.J., Lorca, J., Folgueira, J.: Network slicing for 5G with SDN/NFV: concepts, architectures, and challenges. IEEE Commun. Mag. **55**(5), 80–87 (2017)
24. Yousaf, F.Z., Bredel, M., Schaller, S., Schneider, F.: NFV and SDN-Key technology enablers for 5G networks. IEEE J. Sel. Areas Commun. **35**(11), 2468–2478 (2017)
25. Blanco, B., et al.: Technology pillars in the architecture of future 5G mobile networks: NFV, MEC and SDN. Comput. Stand. Interfaces **54**, 216–228 (2017)
26. Katsalis, K., Nikaein, N., Schiller, E., Favraud, R., Braun, T.I.: 5G architectural design patterns. In: 2016 IEEE International Conference on Communications Workshops, ICC 2016, 5GArch, pp. 32–37 (2016)
27. Ferrús, R., Sallent, O., Pérez-Romero, J., Agustí, R.: On 5G radio access network slicing: radio interface protocol features and configuration. IEEE Commun. Mag. **56**(5), 184–192 (2018)
28. Ejaz, W., Ibnkahla, M.: Multiband spectrum sensing and resource allocation for IoT in cognitive 5G networks. IEEE Internet Things J. **5**(1), 150–163 (2018)
29. Vukobratovic, D., et al.: CONDENSE: a reconfigurable knowledge acquisition architecture for future 5G IoT. IEEE Access **4**, 3360–3378 (2016)
30. Amaral, L.A., de Matos, E., Tiburski, R.T., Hessel, F., Lunardi, W.T., Marczak, S.: Middleware technology for IoT systems: challenges and perspectives toward 5G, vol. 8, October 2017 (2016)
31. Ahmad, I., Kumar, T., Liyanage, M., Okwuibe, J., Ylianttila, M., Gurtov, A.: Overview of 5G security challenges and solutions. IEEE Commun. Stand. Mag. **2**(1), 36–43 (2018)
32. ϒ. Η. Λ. Μ. Ηυανικων et al., "Internet of Things „ Γιαγικστο Σων Πραγμασων " □ Δ Τπηρδ □ Ιδ □Τγδια □ Μδ Δμφα □ Η □ Σην Κασ " Οικον Παρακολοτθη □ Η Α □ Θδνων Internet of Things „ Γιαγικστο Σων Πραγμασων " □ Δ Τπηρδ □ Ιδ □ Τγδια □ Μδ Δμφα □ Η □ Σην Κασ " Οικον Παρακολοτθη □ Η Α □ Θδνων," Cyber Resil. Syst. Networks, no. July 2016, pp. 1–150 (2009)
33. Schober, R., Wing, D., Ng, K., Wang, L.: Key Technologies for 5G Wireless Systems (2017)
34. Feinstein, L.: Zhou2017. IEEE Cloud Computing, pp. 26–33, January 2017 (2017)
35. Zhang, N., Yang, P., Ren, J., Chen, D., Yu, L., Shen, X.: Synergy of big data and 5G wireless networks: opportunities, approaches, and challenges. IEEE Wirel. Commun. **25**(1), 12–18 (2018)

36. Ahmad, I., Shahabuddin, S., Kumar, T., Okwuibe, J., Gurtov, A., Ylianttila, M.: Security for 5G and beyond. IEEE Commun. Surv. Tutor. **21**(4), 3682–3722 (2019)
37. Mehmood, Y., Haider, N., Imran, M., Timm-Giel, A., Guizani, M.: M2M communications in 5G: state-of-the-art architecture, recent advances, and research challenges. IEEE Commun. Mag. **55**(9), 194–201 (2017)
38. Ezhilarasan, E., Dinakaran, M.: A review on mobile technologies: 3G, 4G and 5G. In: Proceedings of the 2017 2nd International Conference on Recent Trends and Challenges in Computational Models, ICRTCCM 2017, pp. 369–373 (2017)
39. Duan, X., Wang, X.: Authentication handover and privacy protection in 5G hetnets using software-defined networking. IEEE Commun. Mag. **53**(4), 28–35 (2015)
40. Norrman, K., Näslund, M., Dubrova, E.: Protecting IMSI and User Privacy in 5G Networks, no. 1 (2016)
41. Porambage, P., Ylianttila, M., Schmitt, C., Kumar, P., Gurtov, A., Vasilakos, A.V.: The quest for privacy in the Internet of Things. IEEE Cloud Comput. **3**(2), 36–45 (2016)
42. Al Ridhawi, I., Aloqaily, M., Kotb, Y., Al Ridhawi, Y., Jararweh, Y.: A collaborative mobile edge computing and user solution for service composition in 5G systems. Trans. Emerg. Telecommun. Technol. **29**(11), 1–19 (2018)
43. Aigner, M.: Security in the Internet of Things. Cryptol. Inf. Secur. Ser. **4**, 109–124 (2010)
44. Ahmad, I., Kumar, T., Liyanage, M., Okwuibe, J., Ylianttila, M., Gurtov, A.: 5G security: analysis of threats and solutions. In: 2017 IEEE Conference on Standards for Communications and Networking, CSCN 2017, pp. 193–199 (2017)

Distributed Density Measurement with Mitigation of Edge Effects

Theodore Lee Ward[1(✉)], Ernst Bekkering[1], and Johnson Thomas[2]

[1] Northeastern State University, Tahlequah, OK, USA
{ward68,bekkerin}@nsuok.edu
[2] Oklahoma State University, Stillwater, OK, USA
jpt@cs.okstate.edu

Abstract. Hardware miniaturization and emerging technologies, such as Cyber Physical Systems and Nanotechnology make possible spatial areas that are densely populated with autonomous devices that are capable of peer to peer network communication. These devices may have the need to determine the density of their population, for instance to self-organize into factions for reasons, such as avoiding contention for network resources or to provide redundant services within overlapping areas where a minimum number of neighbors for each node is required, or a maximum is to be enforced. A number of challenges arise when attempting to determine population density in a peer to peer environment. A distributed algorithm is needed and then simply measuring the mean density of a spatial population is subject to a phenomenon known as "edge effect" which is a skewing of the mean when measuring the population density within a spatial model. This is because the count of members residing on the edges of a spatial area do not accurately represent the density of the inner populated area. This causes the measured density to be artificially sparse to a degree depending on the ratio of nodes residing along edges of the spatial area encapsulating the population. Approaches to solve this issue typically involve costly operations that require knowledge of where the edges of the population lie. However in an area populated by autonomous members, this knowledge is likely to be unknown and either impossible or impractical to calculate. This paper presents a distributed method whereby autonomous members of a population can agree on a density while reducing the impact of edge effect. Our method is statistics based and so also works well with complex spatial areas.

Keywords: Autonomous network · Population density · Edge effect · Cyber Physical Systems

1 Introduction

The future of computing hardware is predicted to be filled with large numbers of connected sensing devices albeit with limited power and limited capability with names such as Cyber Physical Systems (CPS) and the Internet Of Things (IOT)

K. Arai (Ed.): FICC 2021, AISC 1363, pp. 301–314, 2021.
https://doi.org/10.1007/978-3-030-73100-7_23

[3,9]. According to Akyildiz et al. [1] systems may be composed of hundreds or thousands or even millions of minuscule low power devices.

When dealing with dense populations of autonomous devices the need to determine the average degree of the nodes that make up this population may arise for a variety of reasons. Our motivation is discussed in Sect. 1.1, however the issue of density measurement and in particular the "edge effect" or "border effect" is a problem that is more commonly associated with biological studies of living populations but has also been noted when dealing with computer science models [2,4,8].

Edge effect refers to a phenomenon when measuring population density in which the nodes that lie near the perimeter of a population contribute lower degree counts to the mean causing the measured mean to be significantly lower than the direct density calculation. To get an accurate density estimate we must find a way to counteract this edge effect.

1.1 Motivation

A proliferation of inexpensive connected devices connotes the existence of very dense networks of sensing devices [6] which implies a high level of contention for wireless communication resources resulting in higher power consumption and reduced reliability and throughput of network communications. This also however offers the possibility of redundant services provided within overlapping areas of a given space. For instance redundant sensing or communications networks. By dividing nodes into logical partitions we can reduce contention for wireless resources if each partition is assigned to a different communication channel. Moreover the various partitions can be used as redundant information channels resulting in a more resilient systems.

Dividing dense populations into some number of overlapping partitions such that each partition falls within specified parameters requires knowledge of the overall density. Before we can effectively partition a single population of nodes, we must first decide how many partitions to create. This requires us to know the overall *density* of the population and be able to evaluate it with respect to an *ideal density* which we would like each partition to have.

Our need to estimate population density was originally motivated by the need to divide a dense population of sensing nodes into overlapping logical networks (LNETs) in our previous work [11]. The case for dividing a population is improving network reliability, mitigating heavy congestion and providing resiliency to failure and attack by creating overlapping ad-hoc LNETs that sense and then transmit data via redundant routes over multiple wireless channels to some data sink. In brief, an event E will probabilistically be sensed by nodes that are members of a subset of LNETs $S \in N$ where N is the set of all LNETs. Nodes in each LNET in N communicate exclusively on some channel $c \in C$ where C is a set of available communication channels allotted to LNETs. Hence data concerning event E will by transmitted via $|S|$ redundant systems where $|S| \leq |N| \leq |C|$.

1.2 Definitions

In this paper we will refer to a spatial area populated by computing nodes as a "nodescape". The term density refers to a measure of how many nodes exist per some unit of area. Density can be calculated δ, or measured μ, Fig. 1 illustrates nodescapes with different population densities.

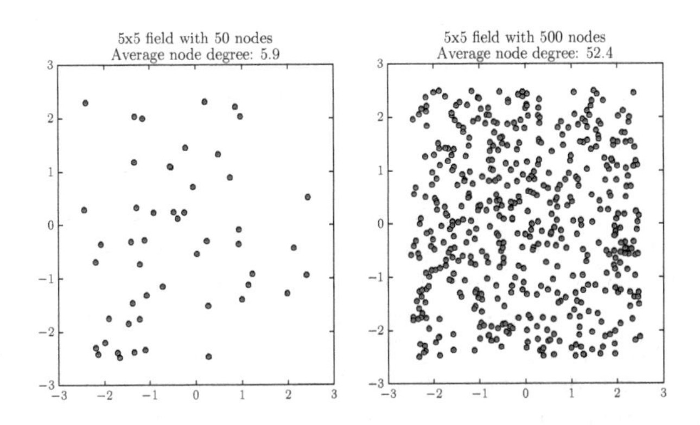

Fig. 1. Populated regions with varying densities.

Degree $deg(n)$ is a function that gives number of one hop neighbors for a node n and is analogous to the term degree as used in graph theory to denote the number of edges of a given vertex. Given that the number of neighbors a node has is directly related to its range of communication, a node's degree can be used as a sample of population size for unit of area U, which is typically the reachable area of n's radio. We want to determine the average number of nodes in an arbitrary area the size of U. By using a unit of area equal to 1, $U = U_1$ we get a population density with respect to a fundamental unit such as square meter.

Calculated Density: $\boldsymbol{\delta}$ Calculated density is directly computed as opposed to measuring populations or taking samples. Calculated density gives an exact representation with no significant effort, however it requires knowledge of the area of the nodescape which is often difficult or impossible to determine. Node density is calculated with (1) described by Bettstetter in [2], here the LNET population size $|M|$ is divided by the area A. This is the case where the unit of area is equal to 1 and gives the number of nodes we should expect to find on average within any arbitrary unit of area. By Eq. (2) we can multiply the physical density δ_1 by a unit of area U_x in order to obtain the density with respect to that unit of area, this gives the node population expected within an arbitrary unit of area the size of U.

$$\delta_1 = \delta(U_1) = \frac{|M|}{A} \tag{1}$$

$$\delta(U_x) = \delta_1 U_x = \frac{|M|}{A} U_x \tag{2}$$

Example. A $10 \times 10\,\text{m}$ area containing one node would yield $\delta_1 = \frac{1}{100\,\text{m}^2} = 0.01$, one hundredth of a node per square meter, or one node per $100\,\text{m}^2$. If the same area had 200 occupants, we would get $\delta_1 = \frac{200}{100\,\text{m}^2} = 2$, or two nodes per square meter. To apply a unit of area, assume $U = \pi$ resulting in $\delta = 2 \cdot \pi \approx 6.28$.

Measured Density: $\boldsymbol{\mu}$ Within a distributed network populated by autonomous nodes it is likely that density must be measured rather than calculated. Measured density μ is simply the *mean degree* observed within a population of nodes. Like computed density, measured density represents the population size we expect to find in an arbitrary unit of area, however there are key differences: *a)* the unit of area is inherent to the system being measured rather than being given; *b)* μ is the mean node degree, however degree is the number of neighbors observed by some node n and does not include n itself, hence μ will be one less than δ.

Whenever comparing μ to δ, μ is acting as an approximation to population size per unit of area, so we must either add 1 to μ or subtract 1 from δ.

The equation for measuring the density of a node population as given in [2] is in (3), this simply sums degrees for all nodes in N and divides the result by the population size. For the sake of efficiency sampling a subset of N would likely be used in a real world environment.

$$\mu(M) = \frac{1}{|M|} \sum_{m \in M} deg(m) \tag{3}$$

When compared to calculating δ as discussed in Sect. 2, the measured equivalent of μ requires a large expenditure of resources since, the degrees of a statistically significant number of nodes must be measured. Furthermore δ gives a precise value while μ will contain error. The error in μ is due to statistical error margins associated with sampling, and the fact that the needed measurements take some time to complete and so a nodes degree can change before the process completes resulting in inaccurate measurements, but a larger source of error is that introduced into measurements by the presence of "edge nodes".

1.3 Metrics

To determine whether our density estimation method is effective we identify an objective function in Sect. 2 which represents an ideal value we can use for comparison. The objective function is a calculation of the expected node density, since it is a calculated value as opposed to a measurement, the impact of edge nodes is eliminated and the *true* mean degree being sought is represented. The baseline value upon which we wish to improve is the measured mean degree of all nodes in the nodescape, so edge effect is not mitigated in any way.

2 Measuring Density

Simple mean calculation is a straightforward procedure whereby nodes determine their own degree and then share this information throughout the network in order to determine the mean. A node can trivially calculate its own degree which is simply the number of unique neighbors it is able to communicate with directly, that is to say nodes within range of radio transmission.

To measure our ability to calculate density we give Eq. (4) as our *objective function*. This is derived from the formula given by Bettstetter in [2] to represent the expected number of nodes for a unit of area. We scale Bettstetter's equation by the radio coverage area U_r which gives density with respect to the transmission coverage of the radio used for wireless communication resulting in a more intuitive value that is directly representative of node degree.

$$E(\delta) = \delta(U_r) = \frac{|M|}{A} \cdot U_r \tag{4}$$

2.1 Distributed Measurement

Measuring the mean degree of nodes within a distributed environment is an instance of the *distributed average consensus* problem [12]. Though this is a simple process using Eq. (3), it is costly in simulation since all nodes must locate and count their neighbors and in a distributed environment it requires information exchange between every pair of nodes in N which is costly and difficult to manage. This problem is exacerbated in a dynamic environment where nodes can be added, die or move to other locations as the populations and topography are constantly shifting. We avoid this by computing the average degree of all nodes using only direct communication between nodes (single hop distance).

The iterative equation in 5 is widely used for arriving at consensus in a distributed environment [12]. We can use this within each node in an LNET to iteratively hone in on a shared estimate of some value Δ. In order to calculate the standard deviation we need to know the distribution of degree values, or how many nodes have each given degree. This requires that the data being distributed include not only δ, but an array of "bins" in which each element contains two sub elements, element 0 is a degree d_i and element 1 is the number of nodes that have that degree $count(d_i)$. Given the node with the largest degree d_{max} and the node with the smallest degree d_{min}, the size of *bins* is at most $d_{max} - d_{min}$.

$$\Delta_n(t+1) = \Delta_n(t) + \frac{1}{|N_n|} \sum_{k \in N_n} (\Delta_k(t) - \Delta_n(t)) \tag{5}$$

$$\Delta = [\hat{\delta}, bins] \tag{6}$$

$$bins = \{(d_{min}, count(d_{min})), \cdots, (d_{max}, count(d_{max}))\} \tag{7}$$

The value Δ in Eq. (6) contains node n's estimate of network density $\hat{\delta}$ and its knowledge of how many nodes have a given degree *bins*, t is an iteration counter, N_n is the set of neighbors for node n, and $|N_n|$ is node n's degree. Each node $n \in N$ initializes Δ_1 using its own degree $\hat{\delta}_n(0) = |N_n|$, and the values in *bins* are calculated using its own neighbor set. Node n then broadcasts its values to, and collects estimates from, its neighbors. Once n has collected its neighbor's estimates, it can refine its own by calculating Δ_2 using 5, after which it will broadcast its improved estimate. This process repeats until all estimates are within a tolerance of one another at which point nodes will have agreed on a network density value $\delta \equiv \hat{\delta}_n \forall n \in N$ as well as the *bins* which can be used to calculate both the mean and the standard deviation σ. Equation (8) illustrates how we can use the contents of *bins* to determine the mean degree (our density estimate), we will use this in Sect. 3 to recalculate $\hat{\delta}$ by multiplying each degree ($b[0]$) by the number of nodes having that degree ($b[1]$), then dividing by the number of nodes, finally we add one since degree is one less than density due to the fact that the node observing the degree excludes itself.

$$\hat{\delta} = \frac{\sum\limits_{b \in bins} b[0] \times b[1]}{\sum\limits_{b \in bins} b[1]} + 1 \tag{8}$$

3 The Edge Effect

In [2] Bettstetter notes that if we were to *measure* the mean degree of a nodescape using Eq. (3) we would expect the measured value to closely resemble the objective function in (4). We would however find the resulting density to be lower than anticipated due to the *edge effect* which refers to the impact edge nodes have on density measurement. An edge node is a node within transmission range of the perimeter of the LNET, these nodes have fewer neighbors on their 'exterior' side and thus a lower average degree than interior nodes. The relative lack of neighbors edge nodes have does not impact their communication abilities since the neighbor density available to carry traffic inwards is not affected.

The root cause of the edge effect stems from the fact that Eq. (4) is an analytical derivation and assumes an infinite spatial area but an actual implementation, either physical or simulated, will have a perimeter.

If we measure average degree using only *interior nodes*, those not within transmission distance of the field edge, we will get an accurate representation of expected degree since the nodes that would skew our results are excluded from the calculation [2,5,7].

Figure 2 illustrates the edge effect, having four corners, four sides and a single center node with degrees of 3, 5 and 8, respectively. The measured average degree is $(4 \times 3 + 4 \times 5 + 1 \times 8) \div 9 \approx 4.4$, all nodes have the same *effective density* as the center node, but have differing degrees since they can only form connections towards the center [2].

Another illustration of edge effect is shown in Fig. 3 in which a central node positioned near the center, edge and corner choose a set of neighbors that are mutually exclusive from each other.

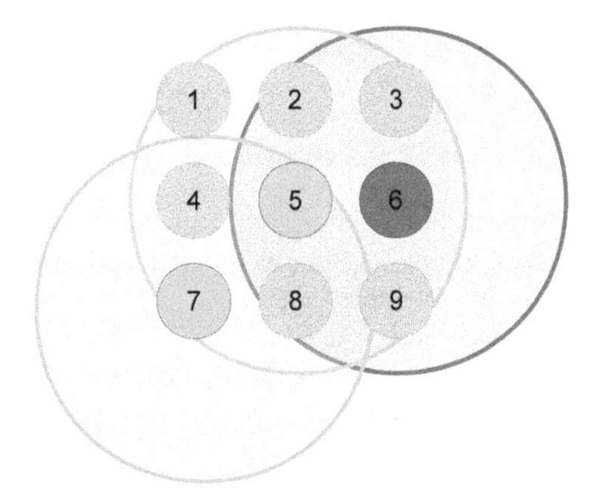

Fig. 2. The edge effect

3.1 Countering Edge Effect

Common strategies for counter edge effect in simulation include wrapping the edges around to form a virtual sphere or a torus ensures that no edges are ever encountered. Requiring that nodes positioned within r_0 of an edge be excluded from the mean calculation ensures that no edge nodes contribute to the mean. Both of these methods require knowledge of relative position of nodes to the perimeter, moreover if the nodescape has a complex or irregular shape or internal discontinuities the computational complexity involved in determining which

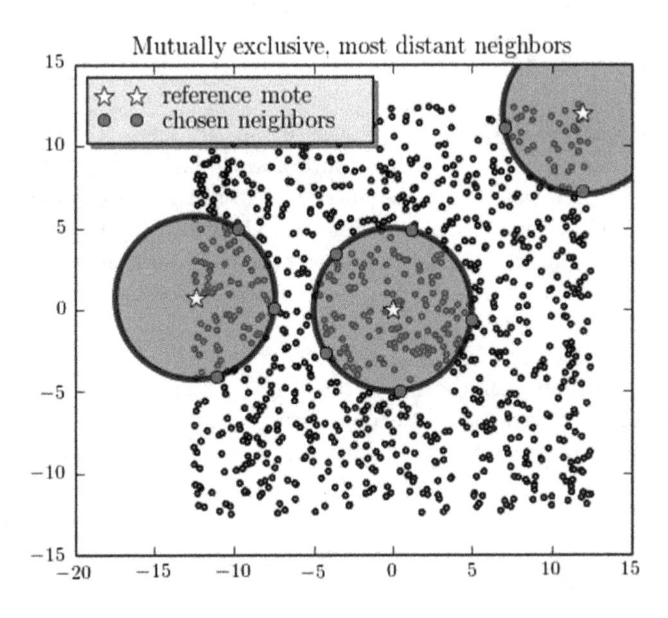

Fig. 3. Mutually exclusive neighbors of center, edge and corner nodes.

nodes lie on an edge outstrips the benefit of doing so [4]. To our knowledge compensating for edge effect hasn't been addressed for a physical realm.

$$\hat{\delta} = \mu(\{m \in M \mid deg(m) > \mu - \sigma\}) + 1 \tag{9}$$

The equation in (9) is the logical expression of our approximation. We begin by calculating the mean node degree μ and its standard deviation σ. This gives us an estimate of density but includes edge nodes and is thus skewed. To correct for this we then re-compute density as the mean of nodes having degree greater than one standard deviation below the mean, $deg(m) > \mu - \sigma$. We add one to the final estimate because when nodes measure degree they do not include themselves but we wish to approximate $E(\delta)$ from Eq. (4) which is a measure of *population* per unit area.

Describing Eq. (9) more explicitly, we first identify the degree values below our threshold of $mu - \sigma$ using Eq. (10), we then calculate our refined density estimate excluding these values as shown in Eq. (11).

$$bins\prime = \{b \in bins : b[0] > \mu - \sigma\} \tag{10}$$

$$\hat{\delta}\prime = \frac{\sum_{b \in bins\prime} b[0] \times b[1]}{\sum_{b \in bins\prime} b[1]} + 1 \tag{11}$$

Figure 4 provides an illustration of using $\hat{\delta}$ and includes the mean degree of all nodes μ, the minimum degree required for inclusion in the calculation of $\hat{\delta}$ $deg(m) \leq \mu - \sigma$, as well as the calculated density we use as our objective $E[\delta]$ and the resulting density calculated with our approximation $\hat{\delta}$. The approximated value is much closer to the objective than the mean, and the visual cue as to the nodes excluded from the density calculation make it clear that the majority of nodes being excluded are indeed edge nodes.

Not only does this provide an efficient means for density estimation but it has the advantage that it will work without additional complexity on fields that are odd shaped or contain impediments that would make calculation difficult or impossible to compute otherwise. Figure 5 shows a graph with several *holes* representative of natural barriers such as water or buildings. The histogram beneath this figure represents the degree bins used for calculation of the standard deviation.

The idea to use mean and standard deviation as a threshold for inclusion in our degree measurements is based on intuition. The exact threshold of one standard deviation below the mean comes from extensive simulation with various factors of the standard deviation for differing nodescape configurations.

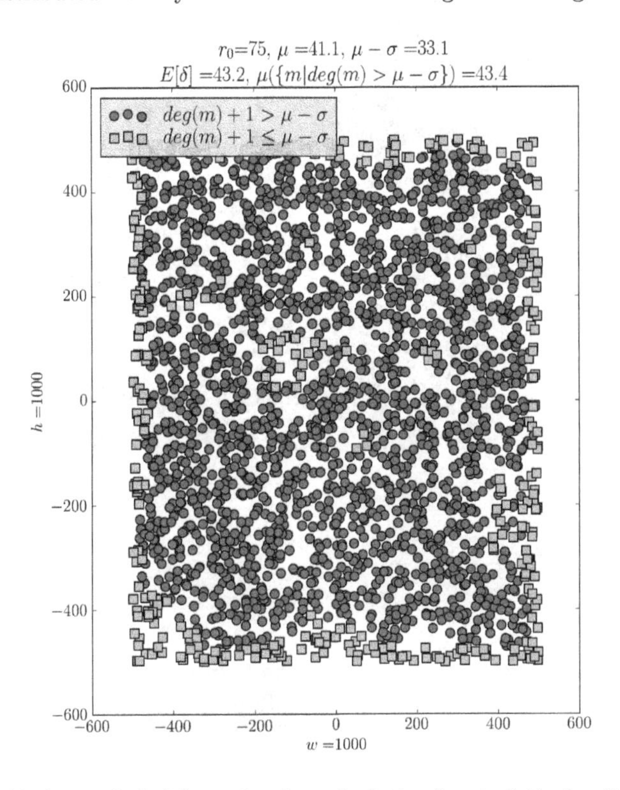

Fig. 4. Nodes excluded from density calculation by statistical estimation.

To measure the accuracy of our estimated density, we generate random rectangular nodescapes with sizes ranging from 300×300 to 2000×2000.

The size of the node population is calculated using Eq. (12) with the maximum area $A = 2000^2$, unit of area $U = \pi 75^2$, and $\delta = 6$ is the density we want the nodescape to have once populated.

By holding the expected density value $E(\delta)$ invariant, the only variable that affects our density measurements is the size and/or shape of the nodescape.

$$population = \delta \frac{A}{U} \tag{12}$$

With each experiment conducted we compile the expected density $E(\delta)$ using Eq. (4), the measured density $\mu(M)$ from Eq. (3), and the approximated density $\hat{\delta}$ measured in nodes selected by our approximation algorithm from (9). In Fig. 6, we graph the deviation of δ and $\hat{\delta}$ from the objective function. It is clear that as the field size grows the impact of edge nodes decreases. However our approximation provides a more accurate density value than the simple mean approach, even as field size grows very large.

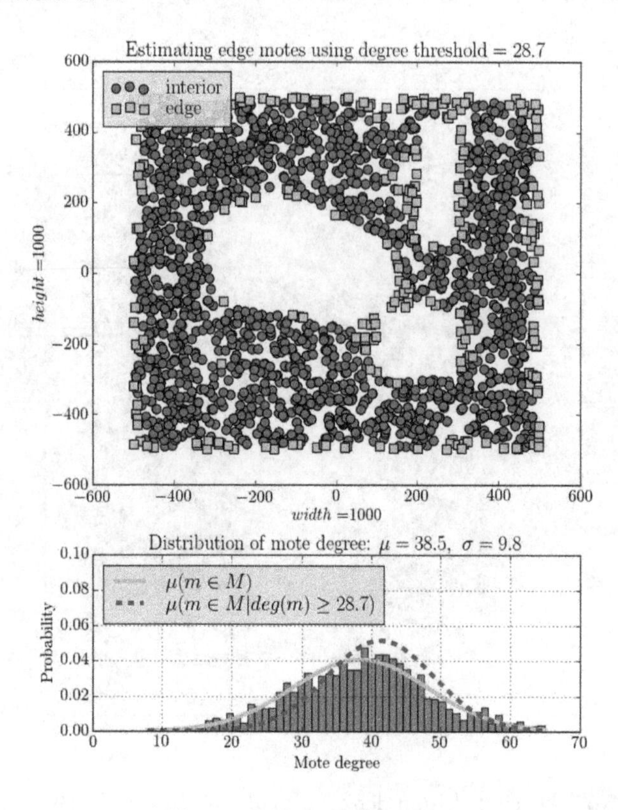

Fig. 5. Odd shaped network

Example: Assume we have a $10 \times 10\,\mathrm{m}$ field ($w = 10, h = 10$) containing 250 nodes ($|M| = 250$), and transmission range of a meter and a half ($r_0 = 1.5\,\mathrm{m}$). We can calculate our unit of area ($U_r = \pi r_0^2 \approx 7.07\,\mathrm{m}^2$) and total field area ($A = wh = 100\,\mathrm{m}^2$). Using our objective function in (4), we multiply our population size by the ratio of unit coverage area to total area to get the expected density: $E[\delta] = 250\frac{7.07}{100} = 17.675$ which should be the average degree of interior nodes.

This objective function depends on a random distribution and in fact will not be accurate for precise grid aligned nodes where changes in the transmission power will result in either no change to the number of neighbors or the sudden inclusion or exclusion or a large group of neighbors since neighbors on all sides are equidistant. For example if node 5 in Fig. 2 expands or contracts its radio range, even a small amount it can change the degree from 0 to 4, in a large grid this same effect would take place for all nodes in the grid simultaneously if they are all using the same radio radius.

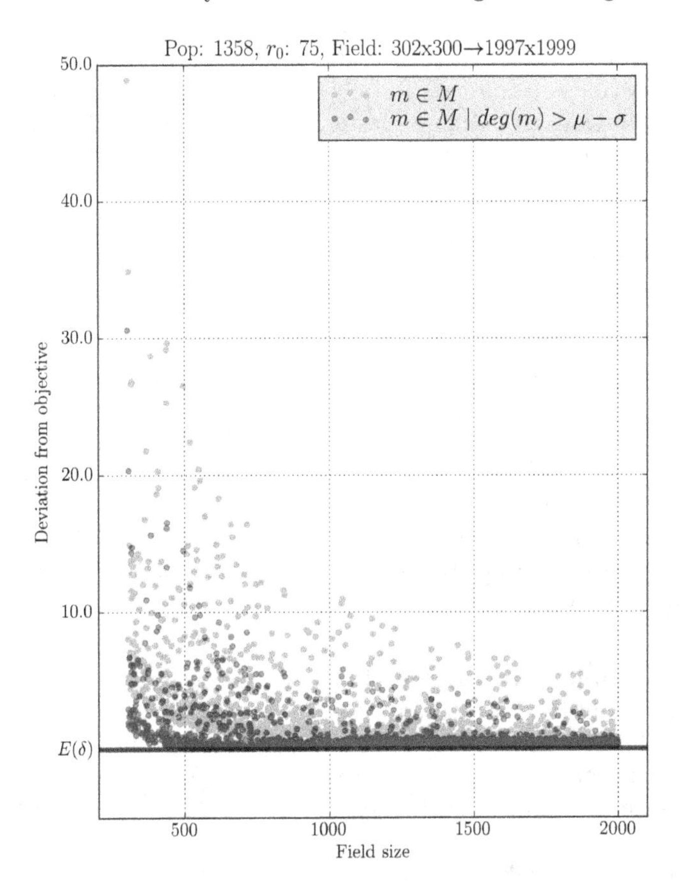

Fig. 6. Deviation from $E(\delta)$ when varying field size.

3.2 Effective Density

In his article on node degree and connectivity Bettstetter gives equations to show the effective area and node count of a two dimensional square field after taking the edge effect into account [2]. We expand Bettstetter's methods to work with any 2d rectangular nodescape in (13a, 13b and 13c) or 3d cuboid nodescape in Eqs. (14a, 14b and 14c).

Two Dimensional Nodescape. We can split the total area of a two dimensional rectangular field into edge area (13b) and interior area (13a) by calculating the area of a rectangle using a width and height that has been reduced such that edges are not included. The width of one edge is the radius of our radio range r_0, we must multiply it by two, once for each side of the nodescape to get the formula in (13a). We can then compute the amount of edge area as the remainder after subtracting the interior area from the total area using Eq. (13b). Finally with Eq. (13c) we take the ratio of edge to total area to quantify the intensity of the edge effect for a given rectangular field with the requirement that the range of radio transmission can be fully encapsulated within the perimeter of the field. The interior area A grows faster than A_{edge} so the edge effect becomes less significant as field size grows or r_0 shrinks.

$$A^2 = wh \qquad \text{Area of 2d rectangular nodescape}$$

$$w_{edge} = h_{edge} = 2r_0 \qquad \text{An edge is within } r_0 \text{ of perimeter}$$

$$A^2_{interior} = (w - 2r_0)(h - 2r_0) \tag{13a}$$

$$A^2_{edge} = A^2 - A^2_{interior} \tag{13b}$$

$$intensity = \frac{A^2_{edge}}{A^2} \tag{13c}$$

In these equations A^2 is the total area of a two dimensional field, w and h are its width and height, r_0 is the transmission radius of the radio used for communication which determines the width of the field's *edge*. These values allow us to state the ratio of edge area to total area in Eq. (13c) which represents the intensity of the edge effect.

Three Dimensional Nodescape. Both contention due to density as well as the edge effect will be exacerbated by cyber physical systems where the nodes are embedded within three dimensional space. This will result in a higher edge to area ratio, hence more nodes will negatively impact density calculation. Li, Pan and Fang study the subject of density in three dimensional networks in detail [10].

In determining the effect of edges within a cuboid area, we proceed in much the same way as with a rectangular one. We subtract twice the size of the edge from each dimension when calculating the area in Eq. (14a) giving the area of an inner cuboid which does not include the edges. We then use Eq. (14b) and subtract the inner area from the total area to get the area occupied by edge nodes. Finally we determine the intensity of edge effect for the given shape with Eq. (14c).

$$A^3 = whl \qquad \text{Area of cuboid nodescape}$$

$$A^3_{interior} = (w - 2r_0)(h - 2r_0)(l - 2r_0) \tag{14a}$$

$$A^3_{edge} = A^3 - A_{interior} \tag{14b}$$

$$intensity = \frac{A^3_{edge}}{A^3} \tag{14c}$$

In Fig. 7, we show the relative intensity of edge effect for a cube, square, sphere and circle as the area of the field increases.

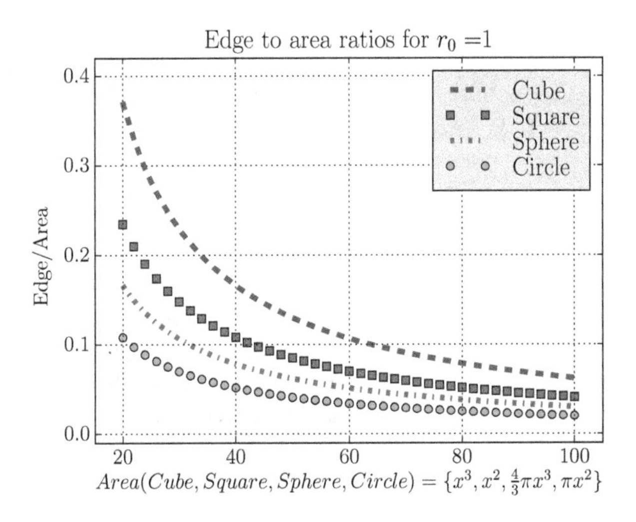

Fig. 7. Impact of field size on edge

4 Proactive Density Manipulation

With modern wireless communication, the radio transmission power is software configurable within a given range, we can utilize this to achieve a desired mean degree within a nodescape. Since μ represents the average node degree, then if we scale the radio coverage area by a factor of k, the expected degree of n will likewise be reduced by a factor of k.

In Eq. (15) we take the ratio of the desired density $\hat{\delta}$ to the measured mean degree μ, ensuring that we adjust by one since degree is from the perspective of some node and it does not include itself.

$$k = \frac{\hat{\delta}}{\mu + 1} \tag{15}$$

The equations in (16) illustrate how we can scale a unit of area U by a percentage k. If r_a is the radius used by the nodes in N, and we determine a desired percentage of change using Eq. (15) then we can calculate a new transmission radius r_b that will result in the desired mean degree as $r_b = r_a\sqrt{k}$.

$$\begin{aligned} U =& \pi r_a^2 \text{ rewritten gives } r_a = \sqrt{\frac{U}{\pi}} \\ kU =& \pi r_b^2 \text{ rewritten gives } r_b = \sqrt{\frac{kU}{\pi}} \\ r_b =& \sqrt{k} \cdot \sqrt{\frac{U}{\pi}} = \sqrt{k} \cdot r_a \end{aligned} \tag{16}$$

5 Conclusion

In this article, we described the meaning of density and the unit of area for which it is calculated. We explained the need to determine population density. We gave a formula for computing population density and reasons why it may not be possible to calculate in a distributed environment. We explained how mean degree is related to density, and gave a method by which distributed nodes can arrive at a consensus for mean degree. We then showed how measuring mean degree results in a flawed density estimate due to the inclusion of "edge nodes". We derived equations to quantify the effect of edge nodes within both a two dimensional and a three dimensional nodescape. And finally we gave a method for approximating density by using the measured mean degree and eliminating contributions that come from nodes that are statistically likely to lie within the edge of the nodescape.

References

1. Akyildiz, I., Su, W., Sankarasubramaniam, Y., Cayirci, E.: Wireless sensor networks: a survey. Comput. Netw. **38**(4), 393–422 (2002)
2. Bettstetter, C.: On the minimum node degree and connectivity of a wireless multihop network. In: Proceedings of the 3rd ACM International Symposium on Mobile Ad Hoc Networking & Computing, MobiHoc 2002, p. 80 (2002)
3. Ermolov, V., Heino, M., Karkkainen, A., Lehtiniemi, R., Nefedov, N., Pasanen, P., Radivojevic, Z., Rouvala, M., Ryhanen, T., Seppala, E., Uusitalo, M.A.: Significance of nanotechnology for future wireless devices and communications. In: 2007 IEEE 18th International Symposium on Personal, Indoor and Mobile Radio Communications, pp. 1–5. IEEE (2007)
4. Gupta, H.P., Rao, S.V., Venkatesh, T.: Critical sensor density for partial coverage under border effects in wireless sensor networks. IEEE Trans. Wirel. Commun. 1–9 (2014)
5. Kleinrock, L., Silvester, J.: Optimum transmission radii for packet radio networks or why six is a magic number. In: Proceedings of the IEEE National Telecommunications Conference (1978)
6. Krishnamachari, B., Wicker, S.B., Pearlman, M.: Critical density thresholds in distributed wireless networks. In: Communications, Information and Network Security, pp. 1–15. Kluwer Publishers (2002)
7. Kuo, J.C., Liao, W., Hou, T.C.: Impact of node density on throughput and delay scaling in multi-hop wireless networks. IEEE Wirel. Commun. **8**(10), 5103–5111 (2009)
8. Lazos, L., Poovendran, R.: Stochastic coverage in heterogeneous sensor networks. ACM Trans. Sens. Netw. **2**(3), 325–358 (2006)
9. Lee, E.A.: Cyber physical systems: design challenges. In: 2008 11th IEEE International Symposium on Object and Component-Oriented Real-Time Distributed Computing (ISORC), pp. 363–369, May 2008
10. Li, P., Pan, M., Fang, Y.: Capacity bounds of three-dimensional wireless ad hoc networks. IEEE/ACM Trans. Netw. **20**(4), 1304–1315 (2012)
11. Ward, T.L.: Robust multi-hop communication in high density wireless networks. Int. J. Wirel. Mob. Comput. **7**(5), 428–434 (2014)
12. Xiao, L., Boyd, S., Kim, S.J.: Distributed average consensus with least-mean-square deviation. J. Parallel Distrib. Comput. **67**, 33–46 (2007)

Techno-economic Assessment in Communications: Models for Access Network Technologies

Carlos Bendicho(✉)

Bilbao, Spain
carlos.bendicho@coit.es

Abstract. This article shows State of the Art of techno-economic modeling for access network technologies, presents the characteristics a universal techno-economic model should have, and shows a classification and analysis of techno-economic models in the literature based on such characteristics. In order to reduce the gap detected in the literature, the author created and developed a Universal Techno-Economic Model and the corresponding methodology for techno-economic assessment in multiple domains, currently available for industry stakeholders under specific licence of use.

Keywords: Techno-economic · Model · Access network · Technical · Economic · Viability · Feasibility · Access · Modeling · Assessment · SDN · SD-WAN · NFV

1 Introduction

The development and evolution of Software Defined Networking (SDN) and Network Function Virtualization (NFV) technologies have created a myriad of different SD-WAN vendor solutions. Besides, some telecom operators are offering SD-WAN as a service. SD-WAN adoption rate is still slow but current context requires organizations to evolve WAN networks towards SD-WAN leveraging enhanced monitoring capabilities, QoS and Security policies application and Virtual Network Functions (VNF) execution either from Customer Premises Equipment (CPE) or from any other point in network as for example a datacenter.

All this increases technical complexity and makes it difficult for organizations to take decision about choosing a SD-WAN solution as they need an effective techno-economic assessment of different SD-WAN solutions.

Based on the author's doctoral dissertation [82], this paper presents State of the Art of Techno-Economic Assessment in Communications, focusing on models for access network technologies.

Dr. Carlos Bendicho holds M.Sc. and Ph.D. degrees in Telecommunications Engineering from Bilbao School of Engineering, University of the Basque Country, Spain. He is also MBA from Technical University of Madrid and MIT Sloan Executive Program in Artificial Intelligence and Strategy, Massachusetts Institute of Technology.

He is ACM Member and IEEE Communications Society Member.

K. Arai (Ed.): FICC 2021, AISC 1363, pp. 315–341, 2021.
https://doi.org/10.1007/978-3-030-73100-7_24

Techno-economic models in the literature are based on the traditional definition of techno-economic model as "method for evaluating the economic feasibility of complex technical systems", according to the thesis of Smura [1].

Regarding the origin of the techno-economic modeling, Smura writes [1]:

"The nature of techno-economic modeling and analysis is usually future oriented and uses and combines a number of methods from the field of Future-Oriented Technology Analysis (FTA). Among these methods is the cost-benefit analysis, scenario analysis, trends, expert panels and quantitative modeling (for an exhaustive list of other families and FTA methods, see TFAMWG, 2004, and [Scapolo & Porter 2008, p. 152]). Although these methods and their combinations have been widely used by both academics and practitioners, academic work under the term "techno-economic" (e.g..: modeling, analysis, evaluation, assessment) has mainly been published related to energy (e.g. [Zoulias & Lymberopoulos 2006]), biotechnology (e.g. ..: [Hamelinck et al. 2005]) and telecommunications (ej.:[2]) especially by European research groups.

In the context of telecommunications, the term 'techno- economic' was introduced during the European research programme RACE (Research into Advanced Communicacions for Europe) in 1985-1995. Early techno-economic modeling work was done in the RACE 1014 ATMOSPHERIC project [7–9] and the RACE 1044 project [10] where alternative scenarios and strategies for evolution towards broadband systems were analysed. Later, the RACE 2087 TITAN (Tool for Introduction scenarios and Techno-economic studies for the Access Network) project developed a methodology and a tool for techno-economic evaluation of new narrowband and broadband services and access networks (see [2, 11]). Since the late 1990s, many European research projects have used and extended the methodologies and tools created in the early projects."

Note the following exception found in the literature to the traditional definition of a techno-economic model indicated by Smura in 2012. In 2010, [12] states: "Every business modeling should be accompanied by a technical and economic assessment so as to provide the reader with information on the financial perspective and the technical feasibility of a proposed investment project in telecommunications". The mentioned reference introduced the need for performance analysis of the access network, considering cost and reliability, but limited to relate both aspects in an indicator or specific figure of merit. Therefore, [12] suggests the assessment of technical feasibility, but eventually does not develop it.

Following a review and detailed analysis of the models in the literature, they are imbued with the traditional definition of techno-economic model indicated by Smura [1] and are eminently oriented to deployment of access technologies from the perspective of telecom operators, manufacturers and standards organizations. Only some models have capacity for evaluation of different access technologies, and a few of them include evaluation for combination of access technologies. Only one model has shown a slight hint of guidance to end users or agents other than those mentioned. Furthermore, all output parameters are economic, except for some very exceptional wink.

We proceed to show the review and analysis of the literature of techno-economic modeling of access technologies and the timing of projects related to public funding, given the interest in this research area shown by the institutions of the European Union (EU).

2 Review and Analysis of the Literature

We have reviewed and analyzed the literature of techno-economic assessment models for access technologies, finding that it is based on the aforementioned traditional concept of techno-economic model enunciated by Smura [1], with the outstanding suggestion already indicated, and not developed [12].

We have selected a representative sample of the most relevant articles in literature which allow to provide insight on the State of the Art of techno-economic models for access technologies. Those papers are listed below in Table 1.

Table 1. Papers that develop or use models of techno-economic evaluation

Author, year	Title	Techno-economic model	Access technologies
Reed & Sirbu (1989) [13]	'An Optimal Investment Strategy Model for Fiber to the Home'	Dynamic programming	FTTH
Lu et al. (1990) [14]	'System and Cost Analyzes of Broad-Band Fiber Loop Architectures'	Cost modeling	B-ISDN, four alternative architectures for fiber loop (loop fiber) (ADS, PPL HPPL, PON)
Graff et al. (1990) [7]	'Techno-Economic Evaluation of the Transition to Broadband Networks'	STEM	Evolution of STM to ATM
Ims et al. (1996) [11]	'Multiservice Access nework Upgrading in Europe: A Techno-Economic Analysis'	TITAN	xDSL, FTTx, HFC, FTTH (PON)
Olsen et al. (1996) [2]	'Techno-Economic Evaluation of Narrowband and Broadband Access Network Alternatives and Evolution Scenario Assessment'	TITAN	ADSL, PON, CATV, ISDN, FTTx, HFC
Ims et al. (1997) [15]	'Risk Analysis of Residential Broadband upgrade in a Changing Market and Competitive'	TITAN	xDSL, HFC, ATM PON

(continued)

Table 1. (*continued*)

Author, year	Title	Techno-economic model	Access technologies
Stordahl et al. (1998) [16]	'Risk Analysis of Residential Broadband upgrade based on Market Evolution and Competition'	OPTIMUM (based on TITAN)	FTTN, FTTB, HFC
Jankovic et al. (2000) [17]	'A Techno-Economic Study of Broadband Access Network Implementation Models'	P614	ISDN, xDSL, HFC, FTTx, WLL, Satellite
Katsianis et al. (2001) [18]	'The Financial Perspective of the Mobile Networks in Europe'	TERA	GPRS, UMTS
Welling et al. (2003) [19]	'Techno-Economic Evaluation of 3G & WLAN Business Case Under Varying Conditions Feasibility'	TONIC	UMTS, WLAN
Smura (2005) [20]	'Competitive Potential of WiMAX in the Broadband Access Market: A Techno-Economic Analysis'	based on ECOSYS/TONIC	WiMAX
Monath et al. (2005) [21]	'Muse Techno-economics for fixed access network evolution scenarios - DA3.2p'	MUSE	FTTx, ADSL, SHDSL, VDSL, xDSL over Optics
Sananes et al. (2005) [22]	'Techno-Economic Comparison of Optical Access Networks'	e-Photon/One	FTTH
Lahteenoja et al. (2006) [23]	'ECOSYS "techno-economics of integrated communication systems and services". Deliverable 16: "Report on techno-economic methology" '	ECOSYS	ISDN, B-ISDN (FITL), xDSL, HFC, FTTx, WLL, Satellite, WiMAX

(*continued*)

Table 1. (*continued*)

Author, year	Title	Techno-economic model	Access technologies
Olsen et al. (2006) [24]	'Technoeconomic Evaluation of the Major Telecommunication Investment Options for European Players'	ECOSYS/TONIC	HFC, ADSL, VDSL, LMDS, satellite, 3G, WLAN, FTTC, FTTH
Pereira (2007) [5]	'A Cost Model for Broadband Access Networks: FTTx versus WiMAX'	Proprietary (BATET)	FTTx, WiMAX
Chowdhury et al. (2008) [25]	'Comparative Cost Study of Broadband Access Technologies'	Proprietary	xDSL, cable modem, FTTx, WiFi, WiFi + Hybrid FTTx, Hybrid FTTx + WiMAX (WOBAN)
Pereira & Ferreira (2009) [3]	'Access Networks for Mobility: A Techno-Economic Model for Broadband Access Technologies'	Proprietary (BATET)	Static layer: FTTH (PON), xDSL, HFC, PLC; Nomadica layer (mobile users): WiMAX
Van der Merwe et al. (2009) [26]	'A Model-based Techno-Economic Comparison of Optical Access Technologies'	Proprietary	FTTH Optical Networks: GPON, AON/Active Ethernet (AE), P2P
Odling et al. (2009) [27]	'The Fourth Generation Broadband Concept'	ECOSYS	FTTdp (G.fast)
Ghazisaidi & Maier (2009) [28]	'Fiber-Wireless (Fiwi) Networks: A Comparative Analysis of Techno-Economic EPON and WiMAX'	Proprietary	FTTH + WiMAX
Verbrugge et al. (2009) [29]	'White Paper: Practical Steps in Techno-Economic Evaluation of Network Deployment Planning'	OASE	FTTH
Casier et al. (2010) [30]	'"Overview of Methods and Tools" Deliverable 5.1. OASE'	OASE	FTTH

(*continued*)

Table 1. (*continued*)

Author, year	Title	Techno-economic model	Access technologies
Zagar & Krizanovic (2010) [31]	'Analyzes and Comparisons of Technologies for Rural Broadband Implementation'	Proprietary (Rural Broadband in Croatia)	ADSL, WiMAX
Vergara et al. (2010) [32]	'COSTA: A Model to Analyze Next Generation Broadband Access Platform Competition'	COSTA (based on BREAD & TONIC & MUSE)	FTTH/GPON, FTTN/VDSL, FTTH/P2P, HFC/DOCSIS, WiMAX, LTE
Chatzi et al. (2010) [33]	'Techno-economic Comparison of Current and Next Generation Optical Access Networks Long Reach'	BONE	FTTH duplicated for reliability and FTTH ring WDM/TDM PON fibers (SARDANA architecture)
Rokkas et al. (2010) [34]	'Techno-Economic Evaluation of FTTC/VDSL and FTTH Roll-Out Scenarios: Discounted Cash Flows and Real Option Valuation'	ECOSYS	FTTC/VDSL and FTTH
Casier et al. (2011) [35]	'Techno-Economic Study of Optical Networks'	OASE	FTTH
Feijóo et al. (2011) [6]	'An Analysis of Next Generation Access Networks Deployment in Rural Areas'	Proprietary (model costs)	FTTH (GPON), FTTC/FTTB/VDSL, HFC DOCSIS 3.0, LTE (4G)
Martin et al. (2011) [36]	'Which Could be the Hybrid Fiber Coax role of Next Generation Access Networks in?'	Proprietary (model costs)	FTTH (GPON), HFC DOCSIS 3.0
Machuca et al. (2012) [37]	'Cost-based assessment of NGOA Architectures and Its Impact in the business model'	OASE	Wavelength-routed WDM PON, Ultra Dense WDM, PON, AON with WDM

(*continued*)

Table 1. *(continued)*

Author, year	Title	Techno-economic model	Access technologies
Van der Wee et al. (2012) [38]	'A modular and hierarchically structured techno-economic model for FTTH Deployments'	OASE	FTTH (PON) FTTH (AON)
Walcyk & Gravey (2012) [39]	'Techno-Economic Comparison of Next-Generation Access Networks for the French Market'	BONE	xDSL, FTTH (GPON), FTTH (LROA-SARDANA)
Pecur (2013) [4]	'Techno-Economic Analysis of Long-Tailed Hybrid Fixed Wireless Access'	Proprietary	FIWI (Fixed-Wireless); Fixed: xDSL, FTTx, FSO; Wireless: WiFi, WiMAX, LTE (4G)
Bock et al. (2014) [40]	'Techno-Economics and Performance of Convergent Radio and Fiber Architectures'	TITAN cost analysis	Active Remote Node PON FTTH combining + Radio Base Station (architecture Sodales)
Moreira & Zucchi (2014) [41]	'Techno-economic evaluation of wireless access technologies for network environments campi'	TONIC & ECOSYS	WiFi, WiMAX, LTE
Ruffini et al. (2014) [42]	'DISCUS: An End-to-End Solution for Ubiquitous Broadband Optical Access'	OASE	FTTP
Katsianis & Smura (2015) [43]	'A model cost data for radio access networks'	Proprietary	LTE
Forzati et al. (2015) [44]	'Next-Generation Optical Access Seamless Evolution: Results of the European Concluding FP7 Project OASE'	OASE	FTTH

(continued)

Table 1. (*continued*)

Author, year	Title	Techno-economic model	Access technologies
Van der Wee et al. (2015) [45]	'Techno-Economic Evaluation of Open Access on FTTH Networks'	OASE	FTTH
Shahid & Mas (2017) [80]	'Dimensioning and Assessment of Protected Converged Optical Access Networks'	Proprietary	Converged Access Networks (FTTB/FTTH//LTE)
Oughton et al. (2019) [81]	'An Open-Source Techno-Economic Assessment Framework for 5G Deployment'	Proprietary	5G

From the literature review, it is found that there is an American seed in the field of techno-economic modeling for access networks, in the late 80s and early 90s. Specifically, in 1989, predictions were published regarding the most appropriate moment to invest massively in FTTH (Fiber To The Home) access technology deployment using dynamic programming [13], identifying possible paths for investment from a pure copper access network to a FTTH network through hybrid networks, concluding that the optimum time to launch a massive deployment would not be before 2010, considering prediction of costs, income and interest rates. In the 90s and starting from [7, 14], studies begin to focus, as well as Smura commented in his doctoral dissertation [1], on the detailed cost analysis, out of the components with a 'bottom-up' approach but ignoring the end-user perspective, and always oriented to the deployment of access networks, in order to compare the economic feasibility of different technical alternatives and identify parts of the access network that have a greater contribution to costs, considering different scenarios of evolution of the access network, as well as changing patterns of demand. It is noteworthy that in the US they were contemplating FTTH scenarios [14], starting from ISDN Narrowband (integrated services digital network or N-ISDN) to the Broadband ISDN (B-ISDN: Broadband ISDN), with focus in the detailed cost analysis using learning curves for predicting component costs [46].

In the 90s, it also germinates techno-economic modeling for access networks in Europe, with the first European projects with public funding from the EU (European Union), also focused on the assessment of the costs and oriented towards evaluating technical alternatives for deployment and network evolution. We find highlighted in this European germination the STEM [7] model as the precursor of a more complete model TITAN [11, 15], including a costs prediction model based on the so-called extended learning curve [2], which provides greater accuracy to successive predictive models as OPTIMUM [16] and sets the basis to more complete models as TONIC [19, 47], all based on evolutions from TITAN and always oriented to choose the most appropriate alternative for access network deployment by telecom operators, with the aim also to promote standards and recommendations.

The stage of development for more complete techno-economic models, initiated and inspired by TITAN, begins its consolidation with ECOSYS model [20, 48], which enhances traditional techno-economic modeling based on the calculation of economic indicators such as NPV (Net Present Value) with the DCF analysis (Discounted Cash flows), ROA (Real options analysis) inspired in the so-called financial options or futures, in order to improve the accuracy of economic output parameters, and allows the techno-economic evaluation of wireline, wireless, and mixed or hybrid technologies in different scenarios and geographical areas to cover [23, 24, 49].

As a result of this aforementioned consolidation with the ECOSYS model, a dissemination effect occurs beyond the projects with public funding from the EU, which is detected by identifying new proprietary models as the model [50] for PLC technology (Power Line Communications), [51] for optical networks, [52] for 3G-LTE, the BATET model [5, 53], and its subsequent evolution distinguishing between fixed and nomadic layers, the latter for mobile users [3], identifying also general input parameters in which there is a slight orientation to end user incorporating bandwidth transmission and reception requirements. Outside the public funding of the EU, the COSTA proprietary model [32] arises modeling costs of access network and based on MUSE, an extension from TONIC for the whole access and aggregation network [21], which runs parallel to ECOSYS project. More proprietary models appear like [26] oriented to optical access technologies comparison, and [25, 28] for hybrid networks that combine FTTx and WiFi or WiMAX.

The consolidation and dissemination continues. Multiple papers targeting specific scenarios that rely on the ECOSYS model are published as [27] oriented to FTTdp (Fiber To The Distribution Point) within the framework of the 4GBB initiative leading to the current G.fast standard [34].

The BONE project emerges, aimed at a future European optical network incorporating cost modeling for optical networks in the field of access/metro networks, seeking long range optical networks architectures to provide high reliability [33, 39, 54, 55]. Under the BONE project, the article mentioned at the beginning of this section [12] is published, in which the assessment of the technical feasibility is suggested, after introducing the need for performance analysis of the access/metro network, considering the relationship between cost and reliability of the network. As stated before, [12] did not finally develop the evaluation of technical feasibility, but only limited to relate cost and reliability in a specific indicator or figure of merit, in order to evaluate various technical alternatives for long range optical networks with different mechanisms to increase reliability (e.g.: duplication of fibers, duplication of fibers and OLTs, duplication of fiber-OLT-ONU).

Fostered by public funding, emerges OASE project [29], which proposes a methodology based on the 'Plan-Do-Check-Act' (PDCA) by Shewhart/Deming, and adapts and reformulates it as 'Scope-Model-Evaluate-Refine', as well as a modular design of techno-economic modeling that integrates models and auxiliary methods around TONIC as 'framework tool' [30]. OASE extends view from ECOSYS, designing a modular framework for techno-economic modeling with the above methodology, enabling top-down and bottom-up approaches in techno-economic evaluation of optical access networks, becoming a model of relevance [35, 37, 38, 44, 45, 56–64]. The OASE model is used by other European projects in the field of optical networks, as DISCUS [42].

As a result of the mentioned spread effect, more proprietary models arise such as [65–67, 83, 84], for optical networks, [6, 31, 68–72] for deployment of broadband in rural areas, [36] comparing FTTH and HFC DOCSIS 3.0, [4] for hybrid networks FiWi (Fixed-Wireless) which particularly distinguishes between investors and lenders related to financial agents. Reference [43] to deploy LTE networks in Finland, models power consumption of radio access data networks as a function of data traffic to deploy wireless access points [73]. Studies are also published based on the TONIC and ECOSYS models for specific scenarios (e.g.: wireless networks on campus) [41]. Projects with public funding from the EU emerge, relying on these consolidated tools, in order to assess new technologies, such as the IST-Sodales project developing an Active Remote Node FTTH combining PON + Radio Base Station (SODALES architecture), which evaluates it techno-economically using the TITAN tool [40]. More recently, [80] aims at converged accessed networks combining FTTH/FTTB and LTE, and [81] for 5G deployment. As mentioned, models in literature are based on the traditional definition of techno-economic model indicated by Smura [1], and are mainly oriented towards deployment of access network technologies from the perspective of operators, manufacturers and standardization bodies.

Given the limited features that are detected in the literature regarding the ability of multiaccess comparison, combination of technologies, orientation to end user or other agents in the telecommunications sector, the market dynamics and the context already mentioned, we conclude that it is interesting to deepen into the characteristics that a theoretical, universal and generalizable techno-economic model should have, which will be discussed in Sect. 4.

3 Chronology of Publicly Funded Projects

At European level, public institutions of the European Union, have promoted and financed various projects in the past two decades, aimed at developing models for techno-economic evaluation of access technologies from the pioneers RACE 1014 ATMOSPHERIC, RACE 1028 REVOLVE, IBC 1044 RACE, RACE TITAN 2087, OPTIMUM AC226, AC364 TERA, IST-25172 TONIC, through EURESCOM, MUSE, BREAD, ECOSYS, OASE, etc. projects [1, 7, 8, 23, 74–76].

Publicly funded projects mentioned above give rise to much of the literature, as can be seen in Table 1, since many of the techno-economic models carry the name of the project that defines them. Other publicly funded projects also arise with different but related objectives, for example, related to backhaul or backbone networks that use and rely on techno-economic models developed by previous or parallel projects [40, 42, 77].

Techno-economic evaluation scenarios of above projects are closely related to the evolution of access technologies. In Fig. 1 the historical evolution of access technologies is shown, along with the timing of projects that develop or use techno-economic models, subject of literature in relevant publications and conferences.

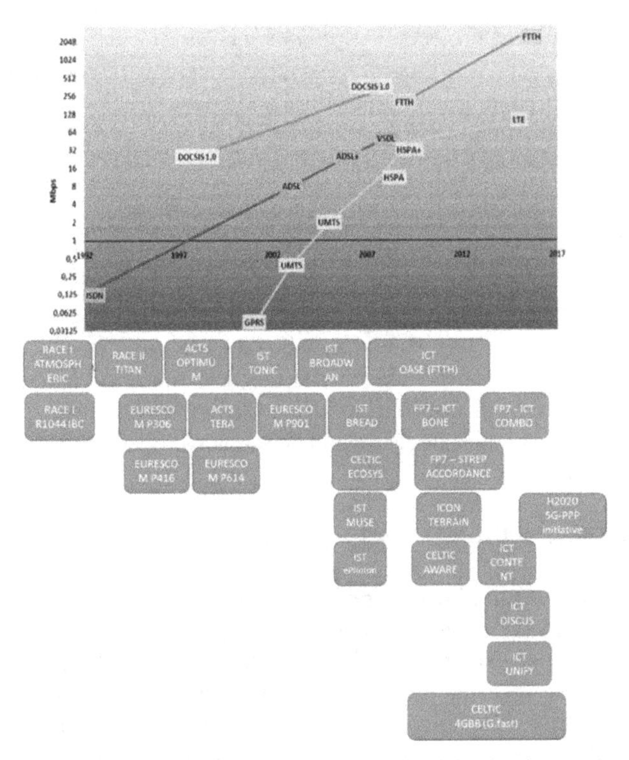

Fig. 1. Historical evolution of projects that develop or use models of techno-economic assessment for access networks, along with the historical evolution of access technologies. Source: ICP-ANACOM, European Commission (ec.europa.eu) and websites of each project.

All publicly funded projects identified are European. No other publicly funded projects have been found in other continents. The literature references from other continents, on the other hand minority, come from private companies and some universities, as can be seen in Table 1 [4, 13, 14, 28] and [79] related to 'Digital India' initiative that is based on a European model [32, 78].

According to the above, and in light of the number of research projects in this area, funded by institutions of the European Union, there is an economic interest and public funding, which justifies further deepening into the development of techno-economic evaluation models for access technologies.

It is observed that, despite having developed techno-economic models for access technologies by publicly funded projects, up to date it has not been identified a universal model for comparing any access technologies in any configuration or combination; which is oriented to any agent of the telecommunications sector; that makes it possible evaluation and comparison of technical feasibility and not only economic one, and that is flexible, extensible and integrable with other techno-economic models.

Therefore, the review and analysis of the literature, the existence of a wide variety of settings and access technologies, the high cost of investment and maintenance of the access network as well as the significant volume of scientific production supported by EU public funding for projects that promote and use techno-economic modeling, lead to the conclusion that it is interesting to deepen and lay the foundations regarding the characteristics required to have a universal and generalizable theoretical model for techno-economic evaluation of access technologies, in order to develop a specific classification and more accurately detect areas of improvement in techno-economic models of literature.

4 Characteristics of a Universal Techno-Economic Model

The features or characteristics a universal and generalizable theoretical model for techno-economic evaluation of access technologies should have are discussed below. The definition of these characteristics is based on a thorough study of the State of the Art, supplemented by author´s professional experience as a telecommunications engineer in designing innovative solutions in the field of access networks for different agents in the telecommunications sector:

- **Multiaccess Universality:** It must allow to compare different current and future access technologies.
- **Universality in Combination of Technologies:** It must allow technical and economic assessment of accesses made by combining different technologies.
- **Universality in User orientation of Technology:** It should be a model oriented to telecom operators and customers, telecommunications services end users, as well as to any other agent of the telecommunications market, such as regulators, Communication Service Providers (CSP) or technological consultancy firms.
- **Universality in Incorporating "Micro" and "Macro" Approaches:** It must incorporate "micro" ('bottom-up') approach (from the perspective of customer or end user) and "macro" approach ('top-down') (from the perspective of the deployment), when assessing techno-economically access technologies.
- **Orientation to User of the Model Requirements:** It must consider the requirements by the user of the model, be this customer, operator or any other agent to appraise access technologies. This feature is related to the model methodology of application.
- **Geographical Universality:** It must allow its application in any geographic area or geotype, whatever its population density, population segments (households, businesses) and distribution are.
- **Technical and Economic Universality:** It must provide technical and economic input parameters and technical and economic output, in order to allow both technical and economic assessment of different access technologies.
- **Extensibility and Flexibility:** the model must be extensible and flexible. It should provide easiness to add new input and output parameters contributing to its universality.
- **Technical and Economic Comparability:** It must make it easy to compare both technical and economic results with other models.
- **Predictive Ability:** It must allow to incorporate and make predictions for a certain period of time.
- **Ability to Integrate with Other Models:** It must allow integration with other techno-economic models to favor the most complete assessment possible and facilitate the evolution of the Art.

5 Classification and Analysis of Techno-Economic Models for Access Networks

Then we proceed to classify a selection of techno-economic models for access networks identified in the literature, based on the above characteristics for a universal and generalizable techno-economic model.

We select a sample of 16 articles out of the 42 listed in Table 1. The models for classification are selected by choosing 8 articles corresponding to models from public projects with EU funding, and 8 related to proprietary models (for which public funding has not been identified).

Every characteristic is composed by several items in order to identify whether the model under study is compliant with each item and evaluate its degree of compliance with the whole characteristic. We list items considered for each characteristic in Table 2.

5.1 Overall Rating and Ranking

In order to assess for each selected model the degree of compliance with the set of characteristics that is considered for a universal, scalable, flexible and generalizable techno-economic model, we use the following method, considering, for simplicity, that all items that compose a characteristic have the same weight:

- We assign '1' value for each item of a characteristic when the model under study is compliant with such item, and assign '0' value if the model is not compliant.
- We calculate the evaluation of each characteristic as the total sum of values for items that compose it.
- The total valuation for each model is the sum of the evaluations of all characteristics.

The result is shown in Table 3. Normalizing for each feature in base 100, we obtain the degree of compliance of each model of the literature regarding the maximum possible score.

The maximum score in 2016, when the author elaborated this study for first time, corresponded to [3] with a fulfillment of 56%, thereby identifying a gap of 44 points to 100%, showing the opportunity to deepen and research in the development of proposals that reach a higher degree of compliance. After the last update of the literature review in 2020, the máximum score corresponds to [80] with a fulfillment of 62% which shows an improvement but still leaving a gap of 38 points.

The ranking of models presented in Table 3 shows, taking as reference the models [3, 80] with greater compliance, that the path of improvement must focus on the following characteristics:

- Universality in Combination of access technologies
- Universality in User orientation
- Universality in incorporating "micro" and "macro" approaches
- Orientation to User of the Model Requirements ([80] reached compliance but not the rest)

Table 2. Items considered for each characteristic

Characteristic	Items
Multiaccess universality	– Fixed Access Technologies – Wireless Access Technologies – Mixed Access Technologies (Hybrid)
Universality in combination of technologies	– Series Combination of Fixed Technologies – Series Combination of Fixed and Wireless Technologies – Parallel Combination of Different Technologies
Universality in user orientation of technology	– Oriented to Telcos (deployment KPIs) – Oriented to Customers (Usage KPIs) – Oriented to Other Agents
Universality in incorporating "micro" and "macro" approaches	– Incorporates "micro" approach (end user perspective) – Incorporates "macro" approach (telco deployment perspective)
Orientation to user of the model requirements	– Economic Requirements – Technical Requirements
Geographical universality	– Geographical Area Description – Existing infrastructure – Population Mix Description
Technical and economic universality	– Input Technical Parameters – Input Economic Parameters – Output Technical Parameters – Output Economic Parameters
Extensibility and flexibility	– Easiness to add new input parameters – Easiness to add new output parameters
Technical and economic comparability	– Technical Output Comparability – Economic Output Comparability
Predictive ability	– Period of Study as input parameter – Allows input parameters with time prediction – Produces time prediction for output parameters
Ability to integrate with other models	– Allows to integrate as input other model output – Model logic allows to incorporate easily other models parameters

- Technical and Economic Universality ([80] reached compliance but not the rest)
- Flexibility and Extensibility
- Technical and Economic Comparability ([80] reached compliance but not the rest)
- Ability to integrate with other models.

Table 3. Overall rating and ranking of literature depending on the degree of compliance with the characteristics of a theoretical universal, generalizable, scalable and flexible techno-economic model (compliance normalized by feature in 100 basis).

	Multiaccess universality	Universality in combination of access technologies	Universality in user orientation	Universality in incorporating "micro" and "macro" approaches	Orientation to user of the model requirements	Geographic universality	Technical and economic Universality	Flexibility and extensibility	Technical and economic comparability	Predictive ability	Ability to integrate with other models	ASSESSMENT	%FULFILLMENT
Maximum possible score	100	100	100	100	100	100	100	100	100	100	100	1100	100%
Shahid & Machuca [80] (2017)	100	25	67	50	100	100	100	0	75	67	0	684	62%
Pereira & Ferreira [3] (2009)	100	25	67	50	50	100	75	0	50	100	0	617	56%
Pereira [5] (2007)	100	0	67	50	50	100	75	0	50	100	0	592	54%
Olsen et al. ECOSYS [24] (2006)	67	25	34	50	0	100	75	0	50	100	50	551	50%
Monath et al. [21]. MUSE (2005)	67	25	34	50	0	100	75	0	50	100	50	551	50%
Oughton et al.]81] (2019)	34	0	34	50	0	100	100	0	75	100	50	543	49%
Feijoo et al. [6]. RURAL (2011)	67	25	34	50	50	100	50	0	50	100	0	526	48%
Vergara et al. [32]. Model COSTA (2010)	67	25	67	50	0	100	50	0	50	100	0	509	46%

(continued)

Table 3. *(continued)*

	Multiaccess universality	Universality in combination of access technologies	Universality in user orientation	Universality in incorporating "micro" and "macro" approaches	Orientation to user of the model requirements	Geographic universality	Technical and economic Universality	Flexibility and extensibility	Technical and economic comparability	Predictive ability	Ability to integrate with other models	ASSESSMENT	%FULFILLMENT
Olsen et al. [2]. TITAN (1996)	34	50	34	50	0	100	50	0	50	100	0	468	43%
Jankovich et al. [17]. EURESCOM (2000)	67	25	34	50	0	100	50	0	50	100	0	476	43%
Smura [20]. WiMAX only. TONIC & ECOSYS (2005)	34	0	34	50	0	100	75	0	50	100	50	493	45%
Zagar et al. [31] (rural broadband in Croatia) (2010)	67	0	34	50	0	100	75	0	50	100	0	476	43%
Pecur [4] FIWI (2013)	100	25	0	50	0	100	75	0	50	67	0	467	42%
Martin et al. [36]. Only HFC (2011)	34	0	34	50	0	100	50	0	50	100	0	418	38%
Van der Wee et al. [38]. FTTH only. OASE (2012)	34	0	34	50	0	100	50	0	50	100	0	418	38%

(*continued*)

Table 3. (*continued*)

	Multiaccess universality	Universality in combination of access technologies	Universality in user orientation	Universality in incorporating "micro" and "macro" approaches	Orientation to user of the model requirements	Geographic universality	Technical and economic Universality	Flexibility and extensibility	Technical and economic comparability	Predictive ability	Ability to integrate with other models	ASSESSMENT	%FULFILLMENT
Van der Merwe et al. [26]. FTTH only (2009)	34	0	34	50	0	100	75	0	50	67	0	410	37%

6 Conclusions

In this paper, we have shown:

- In Sect. 2, a review and analysis of the literature
- In Sect. 3, a chronology of public projects with EU funding that develop and/or use models of techno-economic evaluation.
- In Sect. 4, the characteristics of a theoretical universal, scalable, flexible and generalizable techno-economic model for access technologies are discussed.
- In Sect. 5, a classification of techno-economic models of the literature is made, based on the characteristics of the theoretical universal and generalizable techno-economic model set out in Sect. 4. A ranking of techno-economic models of the literature is presented based on this classification.

After the review and classification of literature, it has been found that all models are oriented to deployment from the operator's perspective, and none is oriented to the end user, except for some exceptional wink in [3] which includes as input parameter the minimum transmission and reception bandwidth, and [5] that talks about QoS and a concurrency factor. All of them incorporate the "macro" approach from the perspective of deployment, but none incorporates the "micro" approach (end-user perspective).

Up to 2016, no model developed and provided technical output parameters, except BONE [12] suggesting an analysis of network performance, that eventually did not develop. However, in this last update after 4 years, two additional models have been identified with technical output parameters: [81] that includes capacity, coverage and energy efficiency as output parameters, and [80] which includes power consumption, availability and length of fiber related output parameters. Therefore, only these last two models of the literature allow the assessment of the technical performance of access technologies, while the rest lack technical comparability, in line with the traditional concept of techno-economic model stated by Smura [1].

Less than half of the models in the ranking, address a series combination of fixed access technologies. No model addresses the parallel combination of the same access technologies to improve technical performance. Eventually, in 2017, [80] addresses the parallel combination of different access technologies, in order to increase technical performance of the equivalent access, although there was a slight glimmer in [2] with HFC (CATV) in parallel with TPON by 1996. Only [3, 4] include the series combination of fixed + wireless technologies.

No model includes technical and economic requirements of the model user except [80], although oriented only to telecom operators. There is some slight hint of incorporating input information in [3] (minimum transmission and reception bandwidth), [5] (QoS and concurrency factor) and [6] which includes the Guaranteed Data Rate per User. However, no other model develops further this feature, incorporating, for example, a catalog or matrix of technical and economic requirements, as they all are oriented to the deployment of access technologies by telecom operators.

No model is identified that allows to add new input and output parameters in a flexible and simple way, so it is concluded that they are not flexible and extensible, probably motivated by the fact that all focus mainly on the assessment of economic viability.

No model includes the default logic for incorporation of other models parameters, thus limiting their ability to integrate with others. On the other hand, as stated before, all models but recent [80, 81], aim to evaluate only the economic feasibility, lacking the assessment of technical feasibility.

Therefore, the review, classification and analysis of the literature shows that there is currently room for improvement in order to develop models that meet the characteristics of a flexible, generalizable and scalable universal techno-economic model, which allow analysis and comparison of network access technologies, and probably their extension and applicability to other domains.

As explained, it makes sense to deepen and research in the development of models for techno-economic assessment of network access technologies to achieve a higher degree of overall compliance, and in each of the characteristics, thus approaching the theoretical universal and generalizable techno-economic model.

Such universal and generalizable models could also be used for assessment in other domains beyond network access technologies as, for example, to satisfy the aforementioned needs of SD-WAN solutions techno-economic assessment.

In order to reduce the gap detected in the literature, the author created and developed a Universal Techno-Economic Model (UTEM) and the corresponding methodology to use it for techno-economic analysis, assessment and decision-making in multiple domains. Table 4 shows author´s model reaches an overall compliance of 92% as validated in [82].

This model is currently available for industry stakeholders under specific license of use.

Table 4. Overall rating of author's model and ranking of literature depending on the degree of compliance with the characteristics of a theoretical universal, generalizable, scalable and flexible techno-economic model (compliance normalized by feature in 100 basis).

	Multiaccess universality	Universality in combination of access technologies	Universality in user orientation	Universality in incorporating "micro" and "macro" approaches	Orientation to user of the model requirements	Geographic universality	Technical and economic universality	Flexibility and extensibility	Technical and economic comparability	Predictive ability	Ability to integrate with other models	ASSESSMENT	%FULFILLMENT
Maximum possible score	100	100	100	100	100	100	100	100	100	100	100	1100	100%
Author's UTEM model – Bendicho (2016)	100	100	84	100	100	100	100	50	75	100	100	1009	92%
Shahid & Machuca [80] (2017)	100	25	67	50	100	100	100	0	75	67	0	684	62%
Pereira & Ferreira [3] (2009)	100	25	67	50	50	100	75	0	50	100	0	617	56%
Pereira [5] (2007)	100	0	67	50	50	100	75	0	50	100	0	592	54%
Olsen et al. ECOSYS [24] (2006)	67	25	34	50	0	100	75	0	50	100	50	551	50%
Monath et al. [21]. MUSE (2005)	67	25	34	50	0	100	75	0	50	100	50	551	50%
Oughton et al. [81] (2019)	34	0	34	50	0	100	100	0	75	100	50	543	49%
Feijoo et al. [6]. RURAL (2011)	67	25	34	50	50	100	50	0	50	100	0	526	48%

(continued)

Table 4. (*continued*)

	Multiaccess universality	Universality in combination of access technologies	Universality in user orientation	Universality in incorporating "micro" and "macro" approaches	Orientation to user of the model requirements	Geographic universality	Technical and economic universality	Flexibility and extensibility	Technical and economic comparability	Predictive ability	Ability to integrate with other models	ASSESS-MENT	%FULFILL-MENT
Vergara et al. [32]. Model COSTA (2010)	67	25	67	50	0	100	50	0	50	100	0	509	46%
Olsen et al. [2]. TITAN (1996)	34	50	34	50	0	100	50	0	50	100	0	468	43%
Jankovich et al. [17]. EURESCOM (2000)	67	25	34	50	0	100	50	0	50	100	0	476	43%
Smura [20]. WiMAX only. TONIC & ECOSYS (2005)	34	0	34	50	0	100	75	0	50	100	50	493	45%
Zagar et al. [31] (rural broadband in Croatia) (2010)	67	0	34	50	0	100	75	0	50	100	0	476	43%
Pecur [4] FIWI (2013)	100	25	0	50	0	100	75	0	50	67	0	467	42%
Martin et al. [36]. Only HFC (2011)	34	0	34	50	0	100	50	0	50	100	0	418	38%
Van der Wee et al. [38]. FTTH only. OASE (2012)	34	0	34	50	0	100	50	0	50	100	0	418	38%

(*continued*)

Table 4. *(continued)*

	Multiaccess universality	Universality in combination of access technologies	Universality in user orientation	Universality in incorporating "micro" and "macro" approaches	Orientation to user of the model requirements	Geographic universality	Technical and economic universality	Flexibility and extensibility	Technical and economic comparability	Predictive ability	Ability to integrate with other models	ASSESS-MENT	%FULFILL-MENT
Van der Merwe et al. [26]. FTTH only (2009)	34	0	34	50	0	100	75	0	50	67	0	410	37%

References

1. Smura, T.: Tecno-economic modelling of wireless network and industry architecture. Doctoral Dissertations, 23/2012. Aalto University Publication Series, Aalto University School of Science and Technology, Finland (2012). ISBN 978-952-60-4525-2
2. Olsen, B.T., et al.: Techno-economic evaluation of narrowband and broadband access network alternatives and evolution scenario assessment. J. Sel. Areas Commun. **14**(6), 1184–1203 (1996)
3. Pereira, J.P., Ferreira, R.: Access networks for mobility: a techno-economic model for broadband access technologies. In: Testbeds and Research Infrastructures for the Development of Networks & Communities and Workshops, TridentCom 2009 (2009)
4. Pecur, D.: Techno-economic analysis of long tailed hybrid fixed-wireless access. In: 2013 12th International Conference on Telecommunications (ConTEL), pp. 191–198. IEEE (2013)
5. Pereira, J.P.: A cost model for broadband access networks: FTTx versus WiMAX. In: Optics East 2007. International Society for Optics and Photonics (2007)
6. Feijóo, C., Gómez-Barroso, J.L., Ramos, S.: an analysis of next generation access networks development in rural areas. In: FITCE Congress (FITCE). IEEE (2011)
7. Graff, P., et al.: Techno-economic evaluation of the transition to broadband networks. In: International Conference on Integrated Broadband Services and Networks, pp. 35–40. IET (1990)
8. Fisher, G.D., Tat, N., Djuran, M.: An open network architecture for integrated broadband communications. In: International Conference on Integrated Broadband Services and Networks. IET (1990)
9. Fox, A.L., et al.: RACE BLNT: a technology solution for the broadband local network. In: International Conference on Integrated Broadband Services and Networks, pp. 47–57. IET (1990)
10. Maggi, W., Polese, P.: Integrated broadband communications development and implementation strategies. Electron. Commun. Eng. J. **5**(5), 315–320 (1993)
11. Ims, L., et al.: Multiservice access network upgrading in Europe: a techno-economic analysis. Commun. Mag. IEEE **34**(12), 124–134 (1996)
12. Kantor, M., et al.: General framework for techno-economic analysis of next generation access networks. In: 2010 12th International Conference on Transparent Optical Networks (ICTON). IEEE (2010)
13. Reed, D., Sirbu, M.A.: An optimal investment strategy model for fiber to the home. J. Lightwave Technol. **7**(11), 1868–1875 (1989)
14. Lu, K.W., Eiger, M.I., Lemberg, H.L.: System and cost analyses of broad-band fiber loop architectures. IEEE J. Sel. Areas Commun. **8**(6), 1058–1067 (1990)
15. Ims, L.A., Stordahl, K., Olsen, B.T.: Risk analysis of residential broadband upgrade in a competitive and changing market. Commun. Mag. IEEE **35**(6), 96–103 (1997)
16. Stordahl, K., Ims, L.A., Olsen, B.T.: Risk analysis of residential broadband upgrade based on market evolution and competition. In: International Conference on Communications (ICC98). Conference Record, vol. 2, pp. 937–941. IEEE (1998)
17. Jankovic, M., Petrovic, Z., Dukic, M.: A techno-economic study of broadband access network implementation models. In: 10th Mediterranean Electrotechnical Conference, vol. I, MEleCon 2000 (2000)
18. Katsianis, D., et al.: The financial perspective of the mobile networks in Europe. IEEE Pers. Commun. **8**(6), 58–64 (2001)
19. Welling, I., et al.: Techno-economic evaluation of 3G & WLAN business case feasibility under varying conditions. In: 10th International Conference on Telecommunications, ICT 2003, vol. 1, pp. 33–38. IEEE (2003)

20. Smura, T.: Competitive potential of WiMAX in the broadband access market: a techno-economic analysis. ITS (2005)
21. Monath, T., et al.: MUSE- techno-economics for fixed access network evolution scenarios. Project Deliverable MUSE, DA3.2p (2005)
22. Sananes, R., Bock, C., Prat, J.: Tecno-economic comparison of optical access networks. In: 7th International Conference on Transparent Optical Networks, vol. 2, pp. 201–2014. IEEE (2005)
23. Lähteenoja, M., et al.: ECOSYS: techno-ECOnomics of Integrated Communication SYStem and Services. Deliverable 16: Report on Techno-Economic Methodology. ECOSYS (2006)
24. Olsen, B.T., et al.: Technoeconomic evaluation of the major telecommunication investment options for European players. IEEE Netw. **20**(4), 6–15 (2006)
25. Chowdhury, P., Sarkar, S., Reaz, A.A.: Comparative cost study of broadband access technologies. In: International Symposium on Advanced Networks and Telecommunication Systems (ANTS) (2008)
26. Van der Merwe, S., et al.: A model-based techno-economic comparison of optical access technologies. In: IEEE Globecom Workshops (2009)
27. Ödling, P., et al.: The fourth generation broadband concept. IEEE Commun. Mag. **47**(1), 62–69 (2009)
28. Ghazisaidi, N., Maier, M.: Fiber-wireless (FiWi) networks: a comparative techno-economic analysis of EPON and WiMAX. In: Global Telecommunications Conference (GLOBECOM) 2009. IEEE (2009)
29. Verbrugge, S., et al.: White Paper: Practical Steps in Techno-Economic Evaluation of Network Deployment Planning. Ghent University, Belgium, IBCN (2009)
30. Casier, K., et al.: OASE: overview of methods and tools. deliverable 5.1 (2010). http://www.ict-oase.eu/public/files/OASE_D5.1_WP5_DTAG_rev2012.pdf
31. Zagar, D., Krianovic, V.: Analyses and comparisons of technologies for rural broadband implementation. In: 17th International Conference on Software, Telecommunications and Computer Networks (SoftCOM) 2009, pp. 292–296. IEEE (2009)
32. Vergara, A., Moral, A., Pérez, J.: COSTA: a model to analyze next generation broadband access platform competition. In: 2010 14th International Symposium on Telecommunications Network Strategy and Planning (NETWORKS). IEEE (2010)
33. Chatzi, S., Lazaro, J., Tomkos, I.: Techno-economic comparison of current and next generation long reach optical access networks. In: 2010 9th Conference on Telecommunications Internet and Media Techno Economics (CTTE). IEEE (2010)
34. Rokkas, T., Katsianis, D., Varoutas, D.: Techno-economic evaluation of FTTC/VDSL and FTTH roll-out scenarios: discounted cash flows and real option valuation. J. Opt. Commun. Netw. **2**(9), 760–772 (2010)
35. Casier, K., Verbrugge, S., Machuca, C.M.: Techno-economic study of optical networks. In: 24th Annual Meeting on IEEE Phonic Society (PHO) (2011)
36. Martín, A., Coomonte, R., Feijóo, C.: Which could be the role of hybrid fibre coax in next generation access networks? In: 10th Conference on Telecommunication, Media and Internet Techno-Economics (CTTE), pp. 1–12. IEEE (2011)
37. Machuca, C.M., et al.: Cost-based assessment of NGOA architectures and its impact in the business model. In: 11th Conference on Telecommunication, Media and Internet Techno-Economics (CTTE) (2012)
38. Van der Wee, M., et al.: A modular and hierarchically structured techno-economic model for FTTH deployments. Comparison of technology and equipment placement as function of population density and number of flexibility points. In: 16th International Conference on Optical Network Desing and Modeling (ONDM) 2012, pp. 1–6. IEEE (2012)

39. Walczyk, K., Gravey, A.: Techno-economic comparison of next-generation access networks for the french market. In: Information and Communication Technologies, pp. 136–147. Springer, Berlin, Heidelberg (2012)
40. Bock, C., et al.: Techno-economics and performance of convergent radio and fibre architectures. In: 2014 16th International Conference on Transparent Optical Networks (ICTON), pp. 1–4. IEEE (2014)
41. Moreira, L., Zucchi, W.L.: Techno-economic evaluation of wireless access technologies for campi network environments. In: 2014 International on Telecommunications Symposium (ITS). IEEE (2014)
42. Ruffini, M., et al.: DISCUS: an end-to-end solution for ubiquitous broadband optical access. IEEE Commun. Mag. **52**(2), S24–S32 (2014)
43. Katsigiannis, M., Smura, T.: A cost model for radio access data networks. Info **17**(1), 39–53 (2015)
44. Forzati, M., et al.: Next generation optical access seamless evolution: concluding results of the European FP7 project OASE. J. Opt. Commun. Netw. **7**(2), 109–123 (2015)
45. Van der Wee, M., et al.: Techno-economic evaluation of open access on FTTH networks. IEE/OSA J. Opt. Commun. Netw. **7**(5), 433–444 (2015)
46. Lu, K., Wolff, R., Gratzer, F.: Installed first cost economics of fiber/broadband access home. In: Global Telecommunications Conference and Exhibition. Communications for the Information Age, vol. 3, pp. 1584–1590. IEEE (1988)
47. Monath, T., et al.: Economics of fixed broadband access network strategies. IEEE Commun. Mag. **41**(9), 132–139 (2003)
48. Elnegaard, N.K., et al.: ECOSYS: techno-ECOnomics of integrated communications SYStems and services. deliverable 11: risk analysis and portfolio optimisation. In: ECOSYS (2005)
49. Autio, T.: Broadcast mobile television service in finland: a techno-economic analysis. Master's thesis, March 2007. Denmark-Technical and Economic Aspects (2007)
50. Tongia, R.: Can broadband over powerline carrier (PLC) Compete?. A techno-economic analysis. Telecommun. Policy **28**(7), 559–578 (2004)
51. Tran, A.V., Chae, C.-J., Tucker, R.S.: Ethernet PON or WDM PON: a comparison of cost and reliability. In: TENCON 2005. IEEE Region 10 Conference (2005)
52. Hoikkanen, A.: A techno-economic analysis of 3G long-term evolution for broadband access. In: CTTE 2007 6th Conference on Telecommunication Techno-Economics. IEEE (2007)
53. Ribeiro Pereira, J.P., Pires, J.A.: Broadband access technologies evaluation tool (BATET). Technol. Econ. Dev. Econ. **13**(4), 288–294 (2007)
54. Chatzi, S., et al.: A quantitative techno-economic comparison of current and next generation metro/access converged optical networks. In: 36th European Conference and Exhibition on Optical Communication (ECOC) (2010)
55. Tomkos, I.: Techno-economic evaluation of NGA architectures: how much does it cost to deploy FTTx per household passed? In: 13th International Conference on Transparent Optical Networks. IEEE (2011)
56. Casier, K.: Techno-economic evaluation of a next generation access network development in a competitive setting. Ph.D. thesis. Ghent University, Belgium (2009)
57. Havic, Z., Mikac, B.: Economic model for FTTH access network design. In: 2011, 10th Conference on Telecommunication, Media and Internet Techno-Economics (CTTE), pp. 1–5. IEEE (2011)
58. Katsigiannis, M., et al.: Quantitative modeling of public local area access value network configurations. 10th Conference on Telecommunication, Media and Internet Techno-Economics (CTTE), pp. 1–9. IEEE (2011)
59. Machuca, C.M., et al.: OASE: process modeling and first version of TCO evaluation tool. D5.2 (2011). http://www.ict-oase.eu/public/files/OASE_D5_2_WP5_IBBT_301211_V3_0.pdf

60. Battistella, C., et al.: Methodology of business ecosystems network analysis: a case study in telecom Italia future centre. Tecnol. Forecast. Soc. Change **80**(6), 1194–1210 (2003)
61. Katsigiannis, M., et al.: Techno-economic modeling of value network configurations for public wireless local area access. NETNOMICS Econ. Res. Electron. Netw. **14**(1-2), 27–46 (2013)
62. Romero, R.R., Zhao, R., Machuca, C.M.: Advanced dynamic migration planning toward FTTH. IEEE Commun. Mag. **52**(1), 77–83 (2014)
63. Zukowski, C., Payne, D.B., Ruffini, M.: Modelling accurate planning of PON networks to reduce initial investment in rural areas. In: 2014 International Conference on Optical Network Design and Modeling, pp. 138–143. IEEE (2014)
64. Van der Wee, M., et al.: Evaluation of the techno-economic viability of point-to-point dark fiber access infrastructure in Europe. J. Opt. Commun. Netw. IEE/OSA **6**(3), 238–249 (2014)
65. López Bonilla, M., Mosching, E., Rudge, F.: Techno-economical comparison between GPON and EPON networks. In: Innovations for Digital Inclusions (K-IDI) 2009. ITU-T Kaleidoscope. IEEE (2009)
66. Ricciardi, S.: GPON and EP2P: a techno-economic study. In: 2012 17th European Conference on Networks and Optical Communications (NOC). IEEE (2012)
67. Bozinović, Z., Dizdarević, H., Dizdarević, S.: Optimal techno-economic selection of the optical access network topologies. In: MIPRO, 2012 Proceedings of the 35th International Convention, pp. 562–567. IEEE (2012)
68. Vidmar, L., Peternel, B., Kos, A.: Broadband access network investment optimization in rural areas. In: MELECON 2010, 15th IEEE Mediterranean Electrotechnical Conference, pp. 482–486. IEEE (2010)
69. Krizanovic, V., Grgic, K., Zagar, D.: Analyses and comparisons of fixed access technologies for rural broadband implementation. In: 2010 32th International Conference on Information Technology Interfaces (ITI), pp. 483–488. IEEE (2010)
70. Krizanovic, V., Zagar, D., Grgic, K.: Techno-economic analysis of wireline and wireless broadband access networks deployment in croatian rural areas. In: 2011 11th International Conference on Telecommunications (conTEL), pp. 265–272. IEEE (2011)
71. Drago, Z., Visnja, K., Kresimir, G.: business case assessment of fixed and mobile broadband access networks deployments. In: 20th International Conference on Software, Telecommunications and Computer Networks (softCOM 2012), pp. 1–5. IEEE (2012)
72. Krizanović, V., Zagar, D., Martinović, G.: Mobile broadband access networks planning and evaluation using techno-economic criteria. In: ITI 2012, 34th International Conference on Information Technology Interfaces, pp. 281–286. IEEE (2012)
73. Kang, D.H., Sung, K.W., Zander, J.: High capacity indoor and hotspot wireless system in shared spectrum: a techno-economic analysis. IEEE Commun. Mag. **51**(12), 102–109 (2013)
74. Böcker, G.J., et al.: A techno-economic comparison of the RACE 2024 BAF and other broadband access system. In: 5th Conference on Optical/Hybrid Access Networks. IEEE (1993)
75. Verbrugge, S., et al.: Methodology and input availability parameters for calculating OpEx and CapEx costs for realistic network scenarios. J. Opt. Netw. **5**(6), 509–520 (2006)
76. Francis, J.C.: Techno-eonomic analysis of the open broadband access network wholesale business case. In: Mobile and Wireless Communications Summit, 2007 16th IST. IEEE (2007)
77. The ICT-DISCUS Project 2013. ICT-DISCUS. Project Website. http://www.discus-fp7.eu
78. Ovando, C., et al.: LTE techno-economic assessment: the case of rural areas in Spain. Telecommun. Policy **39**(3–4), 269–283 (2015)
79. Ashutosh, J., Debashis, S.: Techno-economic assessment of the potential for LTE based 4G mobile services in rural India. In: IEEE ANTS (2015)
80. Shahid, A., Mas Machuca, C.: Dimensioning and assessment of protected converged optical access networks. IEEE Commun. Mag. **55**(8), 179–187 (2017)

81. Oughton, E.J., et al.: An open-source techno-economic assessment framework for 5G deployment. IEEE Access **7**, 155930–155940 (2019)
82. Bendicho, C.: Model for techno-economic assessment of access network technologies. Doctoral dissertation, Bilbao School of Engineering (2016). https://arxiv.org/abs/2008.07286
83. Silveirinha, H., De Oliveira, A.M.: FTTH-GPON access networks: dimensioning and optimization. In: 21st Telecommunications Forum Telfor (TELFOR), pp. 164–167 (2013)
84. Tiantong, L., Saivichit, C.: A proposed network economic strategic model for FTTH access network. In: 2013 10th International Conference Electrical Engineering/Electronics, Computer, Telecommunications and Information Technology (ECTI-CON). IEEE (2013)

Application of Actor-Network Theory (ANT) to Depict the Use of NRENs at Higher Education Institutions

Zelalem Assefa[1(✉)] and Bankole Felix[2,3]

[1] General EthERNet, Addis Ababa, Ethiopia
zelalem@ethernet.edu.et
[2] University of South Africa, Pretoria, South Africa
[3] University of Technology Tshwane, Pretoria, South Africa

Abstract. Polarized views on the development of numerous satellite campuses in higher education have made it even harder to improve educational quality and research output in higher learning institutions. While the decision to use information and communication technology (ICT) related projects may seem pedagogically sound, short of resources such as libraries, laboratories, ICT, among others, may have reduced higher learning institutions' capacity to improve educational quality and research output. This paper explores the association between the various actors who have influenced user requirements in the National Research and Education Network (NREN) services and Ethiopian Education and Research Network (EthERNet). The results indicated that there is a strong association between the main actors and others. It was also noted that other actors had an indirect relationship that influenced other actors; they are not related. From the Actor Network Theory (ANT), three concepts that were examined, namely the obligatory passage point (OPPs) recognition, problematization, and the inscription process. The outcome of the research is an ANT model that can assist in the alignment of heterogeneous interests among various users.

Keywords: Actor-Network theory · NRENs implementation · Reliable networks

1 Introduction

The appeal for efficient network services is that it bears limitless prospects for improving the quality of learning and research output in most institutions of higher learning. In Ethiopia, NREN and the Ethiopian Education and Research Network (EthERNet) has been established to interconnect research organizations and educational institutions within the country.

Typically, the EthERNet is anchored in a national network in which campuses can interconnect. This has further been broadened at the regional level through the establishment of the Regional Research and Education Networks (RRENs). Unfortunately, most Ethiopian public universities are spread across a wide geographical region. Various challenges have influenced the effective use of EthERNet to achieve a high quality of

K. Arai (Ed.): FICC 2021, AISC 1363, pp. 342–354, 2021.
https://doi.org/10.1007/978-3-030-73100-7_25

education and research using the EthERNet. A study in developed countries shows that the socio-economic development of a country can improves significantly because of the presence of an efficient national research and education network [1].

Studies in the developed nations have demonstrated that there is a significant improvement in socio-economic development because of the development of a common NREN network [2]. Asabere, Togo, Acakpovi, Torgby, and Ampadu [3] discovered that most African universities pursue different interests, which often do no assist in improving the quality of education and research output.

From a global perspective, there are notable limitations regarding research that address the effects of reliable networks on the performance of research and education in higher education within the context of developing countries. Therefore, this study concentrates on the applicability of the Actor-Network Theory (ANT) to depict NREN service implementation requirements in Ethiopian institutions of higher education. Two research inquiries which are addressed in the study included the factors influencing the relationships amongst actor-networks in improving the quality of education and research output at Ethiopian higher education institutions and how ANT concepts can be used to understand NREN service requirements within Ethiopian institutions of higher learning? This will provide an opportunity to understand the relationships amongst the actors in the network, which is intended to improve the quality of education in the Ethiopian institution of higher education. Additionally, the study will also assess how the Actor-Network Theory (ANT) concept can elaborate on the NREN service requirements, trace, and describe actors influencing the required NREN service, their relative significance, and their relationships.

The actor-network is expected to promote the efficiency of Ethiopian institutions of higher learning as a way of promoting socio-economic development. The development of the EthERNet relies on different projects in which the specific steps for acknowledging the negotiations, roles, actors, and their interest alignment. Despite the challenges faced, the Actor-Network Theory (ANT) is relied upon to understand how the EthERNet could assist scientists, lecturers, and researchers in Ethiopian public institutions. The study is expected to contribute to the existing literature by developing an ANT model that can assist in the re-alignment of heterogeneous interests among various users.

2 Literature Review

2.1 Theoretical Framework

The Actor-Network theory was initially developed by Bruno Latour [4] during the 1970s and later improved by John Law and Michael Callon [5] in the 1980s. It was intended to simplify the association between scientific as well as emphasize more on how knowledge is constructed socially, the interaction among scientists, lecturers and researchers, their working environment, and how the interactions among researchers and scientists could lead to positive ideas accepted by all [1, 6].

However, more emphasis has been placed on the one main aspect of ANT that guides its application. This is based on the concept that the theory treats both human and non-human actors as equals and assumes they have the same potential to influence the outcome in a network.

The ANT approach to improved quality of education and research output in higher institutions of learning identifies the main factors which influence the use of NREN services, and the challenges faced and the expected services. The study does not commence with establishing the various categories, identifying the participants and objects (actors) with established roles. Additionally, the study employs the use of surveys to establish the association and the impact of reliable networks that could be significant. The impact of a reliable network can be examined by examining a certain group of people who have been involved in the management of networks as well as those who use it. This consists of, but not limited to, lecturers, researchers, students, with clear administrative duties such as collaboration among researchers, provision of networking services and materials for learning, video conferencing services, among others.

There is a need for linear sets of connections that create a network and are reshaped in a manner that propagates action in the selective elements [7], such that the users can know the nature of connections between actors in a single network and the nature of connections between actors in different networks. The creation of ANT involves the following four elements; actors that can exist separately or as a group of actors that symbolize a single unit in the network [8].

According to Rodger [7], the prominence of ANT has been based on its ability to restructure assorted networks into an identifiable pattern that has each actor prominently represented within the network. This study showed how networks could be created, the resources required to maintain given types of networks, and how different networks react in the presence of other networks. Figure 1 presents a visual representation of the requisite mechanics for the creation and maintenance of a network [2, 7].

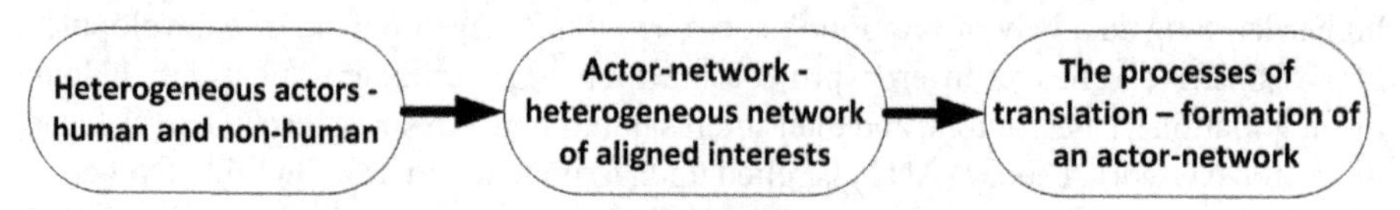

Fig. 1. The three parts of the ANT as postulated by Rodger [7].

In the three-step process, most scholars have discovered that the "black boxes" have contributed to the various challenges which have been faced. As indicated by Myhra [6, 9], to accomplish unwavering associations and an objective overview, the members must set an OPP (Obligatory Passage Point) to divert all benefits in a single route. This will make a black box with the goal in which interpretation procedures are executed naturally and are never discussed bit by bit. Martin claims that the OPP is a hub that fills in as a middle person between actors, systems, or system segments [1, 3, 10]. The OPP is robust when it practices control over actors' assets and can guarantee responsibility for the successful achievement of the network. According to Tatnall [11], the procedures of translation can be partitioned into segments. The aspect of black boxes remains unopened and, therefore, calling for the need for assumptions that can encapsulate the affected factors during analysis.

2.2 Challenges Affecting NREN Services

Numerous studies have been conducted to identify the barriers affecting the integration and use of ICTs for learning and research projects. Calis, Gulben, and Orhan [12] classified the obstacles that hinder the incorporation of ICT in the education sector, namely exterior and interior obstacles. Common barriers that were encountered outside the organization include the costly hardware and software facilities, inadequate time allocation, and lack of ICT experts, management disregard of appropriate locations to install ICT infrastructures [10, 13]. Some of the obstacles within an institution involve the negative perception of new technologies by the staff, as they perceive traditional approaches to be equally successful. Kunda indicated that inadequate allocation of time, absence of incentives, poor quality of ICT services, and lack of adequate personnel with ICT skills, are some of the main barriers faced when integrating ICT in education and research [14]. New inventions have been proposed to solve these problems, such as increasing the bandwidth. As such, the cost of investing in standard campus networks can offer challenges that are common in many educational institutions in developing nations.

3 Methodology

The section presents the research methodology employed to assess how reliable networks have impacted on the quality of education offered and the research output in Ethiopian institutions of higher learning. This study relied on the use of an unorthodox approach to evaluate NREN services within the precincts of institutions of higher learning through the theoretical lens of the Actor-Network Theory (ANT). The purpose of this paper is based on the identification of ANT's key actors, explains their recruitment, and how they contributed to the problem-solving phase in the construction. The identification of the actors to provide relevant feedback is drawn from four groups of participants, namely researchers, lecturers, EthERNet staff members, and ICT directors. This research design is deemed appropriate as there is a need for the study to explore, describe, and understand how reliable networks have impacted the network services and the challenges affected by the ANT perspective.

The study concentrates on the use of the actor-network approach to comprehend the construction and extension process of EthERNet as well as understand how each actor affects the others. This comprised determining project participation, the extent of their alignment in terms of the attributes, the required implementation measures for various institutions, and the need of the NREN to use service dissemination. Essentially, the stabilization and construction of the dynamic ANT actors aim to establish the use of NREN services in Ethiopia's institutions of higher education.

In the study, various challenges were faced ranging from geographical restrictions to budgetary constraints. To overcome most of these shortcomings, the study employed the use of structured questionnaires that were administered to researchers, instructors, ICT Directors from Ethiopian public higher education institutions and the Chief Technology Officer (CTO) of EthERNet via email from approximately 29 public universities in Ethiopia to understand the implications of ANT. The main aspects that the participants were requested to provide included 11 latent variables gathered from 54 different constructs. Additionally, the study considered outsourcing subject matter experts who could offer the best advice on the approach that needs to be used to understand the NREN services required by end-users, challenges faced by Ethiopian institutions of higher learning concerning the network services and future services that can be obtained from the EthERNet. Respondents were required to fill all sections. Most questions that were asked employed the use of the Likert-scale and ratings ranging from neither agree (1) to strongly agree (5). Specific questions were based on a careful analysis of the tasks involved in education and research. Each question was also complemented with a question on the impact of the end-user's network on that topic area.

The study employs the use of descriptive statistical methods to examine the NREN services required by end-users and develop the NREN service portfolio and roadmap for EthERNet. Structural equation modeling (SEM) is used for generating the values using the PLS regression calculations to establish whether reliable networks have a non-linear association with NREN services.

4 Results

This section presented the results of the analysis from the data collected through quantitative methods. The study concentrated on determining the impact of the Actor-Network Theory (ANT) on the NREN service requirement in Ethiopian institutions of higher learning. This was intended to trace and explain the main actors influencing the required NREN service, their relative importance, and associations. The study administered questionnaires to an aggregate of 253 participants; however, 172 responses were able to be complete while 81 responses were incomplete, representing a 68% response rate. From the completed responses, 157 were received through the online survey platform, while 15 survey forms were received using email.

In the study, there were three main groups of actors that were used to examine the impact of the ANT on the NREN service requirement. These include the educators or lecturers, researchers, and ICT directors. The Distribution of the 172 responses among the different groups of participants is as follows: 154 responses were from either educators or lecturers, and 129 responses were provided by both educators or lecturers and researchers who involved in teaching and research. The study also collected from the ICT Directors, which forms a significant part of this study.

The data revealed that the key actors formed a network that is aligned with the interest of enhancing the quality of education and research output. Furthermore, it was observed that a stronger NREN had a positive impact on supporting research and education activities. This implied that the NREN services positively influenced socio-economic development.

Regarding the challenges that negatively impacted on the research output, it was noted that there were quite a several challenges that were highlighted from the participants' feedback such as the absence of a reliable network service from EthERNet, no proper ICT policies and strategies in place, and poor campus network infrastructure which affected the network building process.

4.1 Categories of Actors Affecting NREN Services

The contribution of the ANT is examined from the perspective of the three groups of actors as follows.

Lecturers. In this case, the study intended to understand the association between lecturers and other actors within the network. Several actors were identified in the study, namely online materials, students, education-related networked technologies and services, reliable institutional network/campus network, the university actor, Massive open online courseware, and video or web conferencing services. The relationship between these actors is provided in Fig. 2 below.

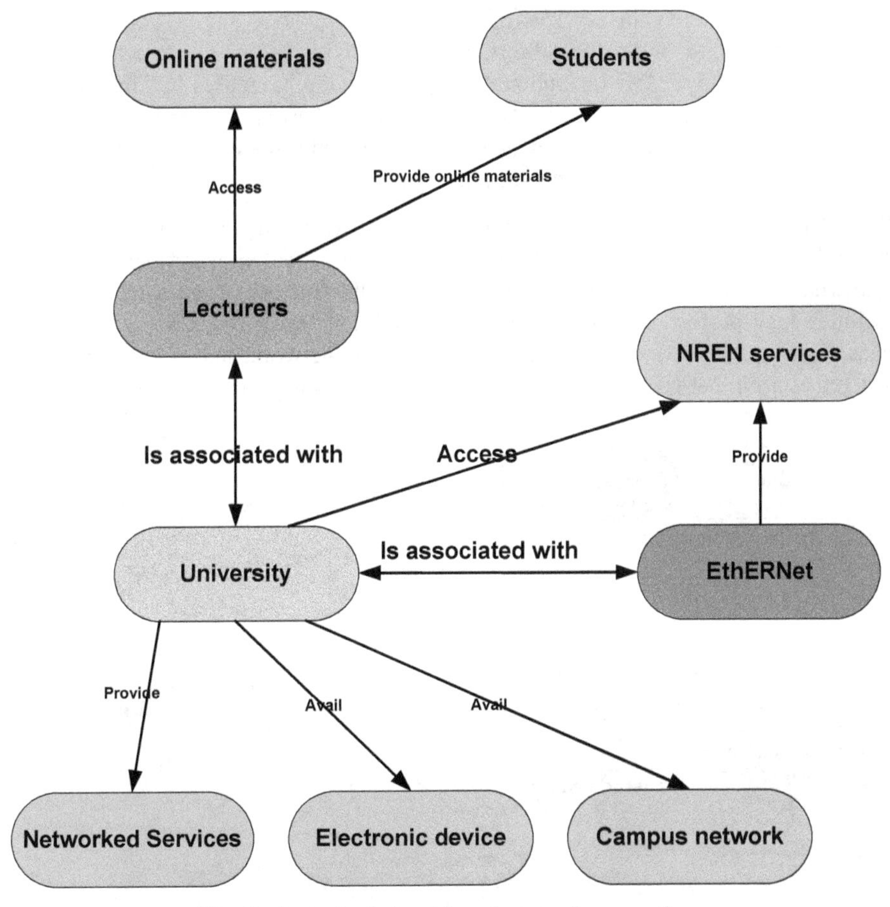

Fig. 2. Actors' relationships – lecturers' perspective

The data revealed that there was a strong association between the lecturers and other actors such as online materials (92.21%), education-related networked technologies (83%+), students (72%), reliable institutional net-work/campus network (60% and 80%). This is probably attributed to the lecturers' ability to utilize various resources to improve the quality of education, such as accessing online materials, supporting the teaching-learning process, providing materials online regularly to support the students, among others. This suggests that lecturers can enroll the students and other actors into the actor-network.

Additionally, the data revealed that there was an indirect association amongst the actors. On the one hand, the lecturers provided the needs and importance of an actor that were not directly linked. This suggests that actors could influence the associations of other actors to which they are indirectly related. For instance, the lecturers may account for issues that they observe about actors over which they do not have control. The EthERNet could be used to provide the required NREN service for member institutions in an organized and reliable manner such as credentials to log in at another institute, video, or web conferencing services, among others.

Accordingly, it is the responsibility of the university, as one of the actors, to provide a standard and reliable institutional/campus network that makes ICT infrastructure available to the lecturers and students. Actors require these and other requirements to access a higher quality of education as well as improved research output while using a reliable institutional network or campus network.

Role of Researchers. In this case, the study intended to identify the association between researchers and other actors within the network. Several actors were identified in the study, namely other researchers, online materials, the university and local computing facilities, high-performance computing (HPC) facilities, networked services, NREN service, Electronic Devices, institutional network/Campus network, and internet bandwidth. The relationship between these actors and the researcher is provided in Fig. 3 below.

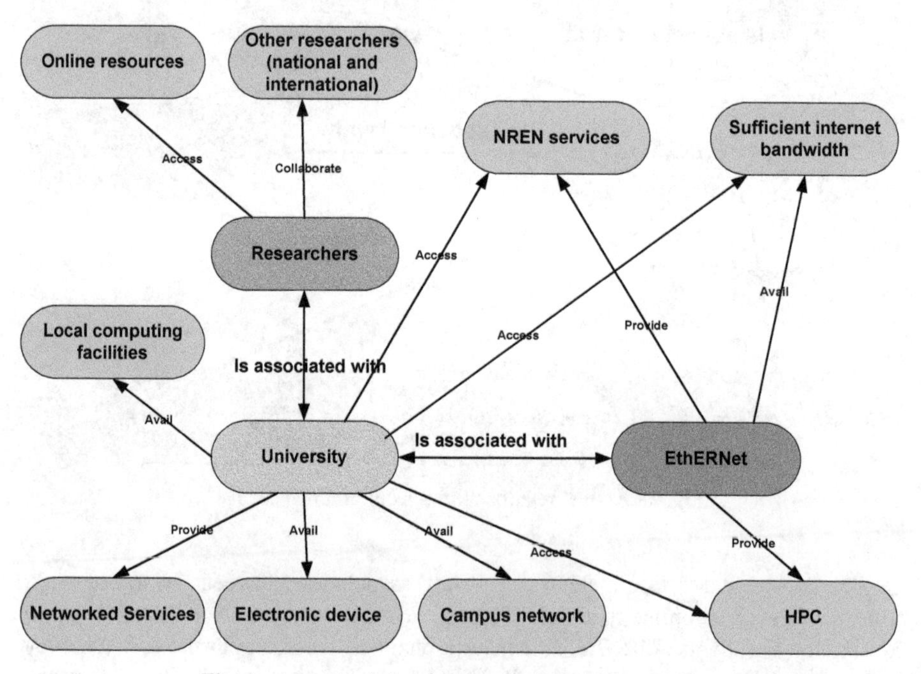

Fig. 3. Actors' relationships – researchers' perspective

The data revealed that there was a relationship between the researchers and the actors mentioned above with varying percentages, namely, such as other researchers (minimum

92.25%), NREN service (90%), institutional network/Campus network (74.42%). This suggests that the above-mentioned networked services and EthERNet can provide the required NREN service for member institutions in an organized and reliable manner. The results also indicated that there was a significant impact of a reliable network on research at both the national and international levels, which enabled the establishment of new partnerships and collaboration with their counterparts around the globe.

Additionally, there was some evidence that indicated that there was an indirect relationship among the actors. This is based on the comments which are made by researchers on the critical role of other actors, which were not directly associated. This suggests that there is some evidence which indicates that the actors can influence other actors which are not directly related.

ICT Directors and EthERNet Perspective. The study sought to understand how the ICT directors, as one of the key actors, contributed to the development of the network and how they influenced other actors. It relied on the ICT Directors analysis, members' input, the association of the actors, and the significant terms of ANT. Several actors were identified in the study and how they impacted the network and other actors. Apart from the ICT director represented by the university and EthERNet, other actors that were examined include institutional network/Campus network, NREN service, ICT Policy, and Strategy, Institutional ICT Policy, ICT strategy, retention mechanism for ICT staff, Skilled ICT staffs, inter-net bandwidth, support and maintenance, Budget for ICT and top-level management. The relationship between these actors and the ICT Directors and EthERNet is provided in Fig. 4 below.

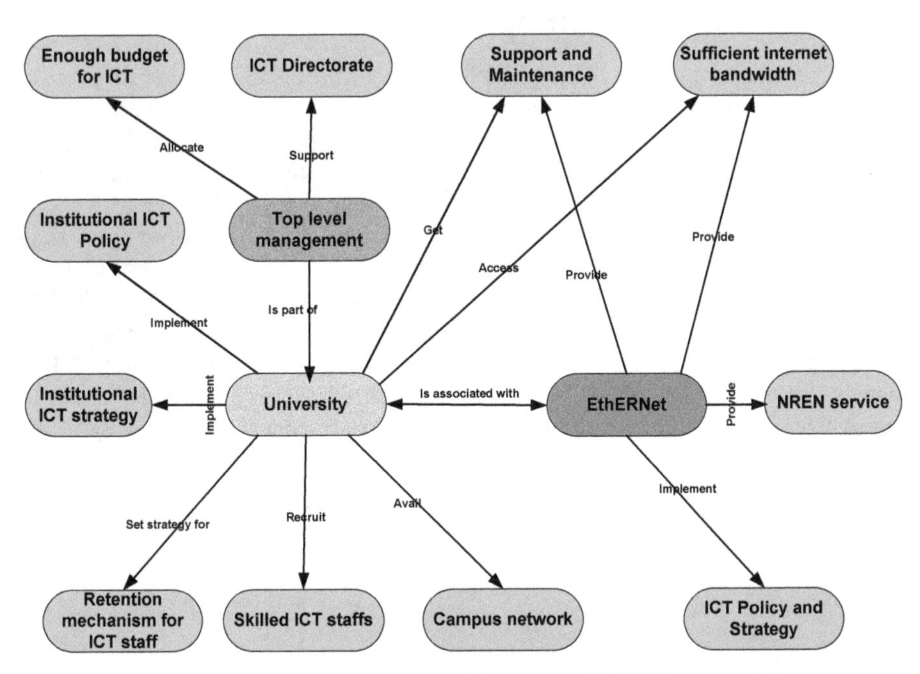

Fig. 4. Actors' relationships – ICT Directors and EthERNet perspective

The data revealed that there was a strong connection between the ICT Directors and all the identified actors above. Most of the respondents (95%) were interested in improving the institute's network infrastructure/campus network if they were granted access to a reliable, low-cost NREN. This suggests that the university is personally tasked with the responsibility of making the institutional ICT infrastructure available to the university communities (students, educators, and researchers). This includes services that are required by the end-user that can be provided by EthERNet, such as, but not limited to, video conferencing services, collaboration tools, online learning environment (LMS), MOOCs, and digital library. The ICT directors held the notion that the university needs to implement an ICT Policy and strategy as its absence would make it difficult for the university to allocate the required Budget for the implementation of the required ICT infrastructure and services.

The data also revealed that there was a strong relationship between the university, retention mechanism for ICT staff, and skilled ICT staff. This was despite the barriers which affected the ability of the EthERNet to offer high-quality ICT services for both research and education. The extent to which the challenges affected the quality of education, as indicated by the respondents, includes a lower retention mechanism for ICT staff (80%) and a lack of skilled ICT staff (76%). This suggests that ICT directorate need to set up a strategy on how to recruit and retain skilled ICT staff. Additionally, it was noted that a relationship was noted between EthERNet and support and maintenance as most of the ICT directors (64%) in the university claimed there was insufficient support from the network provider.

4.2 Challenges Affecting the EthERNet

The study also examined the challenges which affect the EthERNet from being able to provide ICT services that can support research and improve the quality of education. Some of the challenges which were examined include lack of an institutional ICT Policy, unavailability of institutional ICT strategy, and its alignment to the corporate strategy and difficulty of ICT strategy implementation, as presented in Fig. 5 below.

The results revealed that two main challenges affected EthERNet from providing ICT services. These included the unavailability of Institutional ICT Policy (80%), Institutional ICT Policy not being implemented (80%). Other challenges that were raised by the respondents include unavailability of Institutional ICT strategy (60%), Institutional ICT strategy not being aligned to the corporate strategy (52%), and difficulty of ICT strategy implementation (64%). This suggests that there is a stronger need for most of these universities to address most of these challenges to enhance the capacity of the EthERNet to provide ICT services for research and education to different actors.

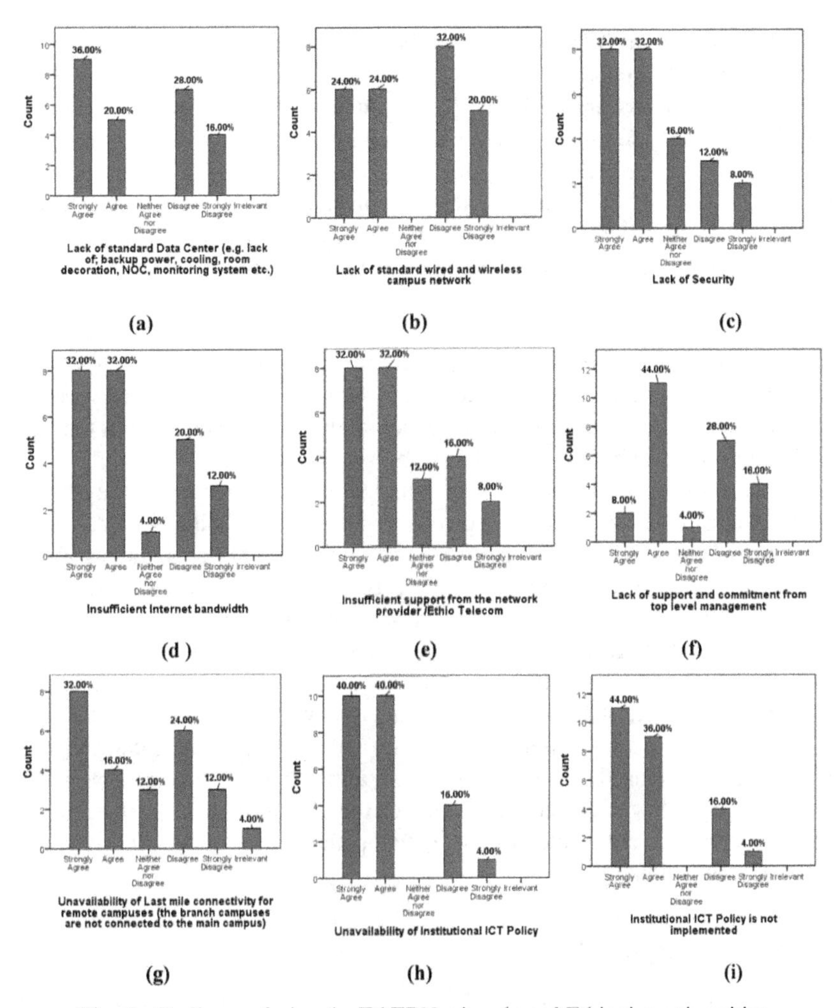

Fig. 5. Challenges facing the EthERNet in selected Ethiopian universities

5 Discussions

This study evaluates how the concept of ANT has been applied in meeting users' NREN service requirements within the Ethiopian institution of higher learning. It relied on the respondent's comments as the key players in NREN service requirements. Three major concepts relating to ANT that were examined in the study include OPPs recognition, the problematization, and the inscription process. It is noteworthy that the paper focuses only on NREN actors that are found in the university, namely ICT directors, researchers, learners, and educators.

Drawing from the findings, the enrolment of human actors such as learners, tutors, and the researchers have been identified. This suggests that there is a constant need for the enrolment of more human actors. These include management and ICT Directors, and the data also confirms the importance of non-human actors, which are NREN, EthERNet, and the University Campus.

From the ICT director and the EthERNet, it was noted that three main aspects affected the EthERNet from offering ICT services, namely the unavailability of Institutional ICT Policy, Institutional ICT Policy not being implemented, and lack of a retention mechanism for ICT staff. This suggests that the university, as one of the actors, to provide access to the NREN service provided by EthERNet, it must find ways that facilitate the end user's access to the resources and learning material through electronic devices inside the campus.

The results indicated that there was evidence which implied that there was an indirect relationship among the actors based on the researchers' comments on the role of other actors, which are not directly associated. This suggests that it is crucial to align the interest of the actors to those of the principal actor. This is expected to assist the network in making appropriate changes as well as stabilize the network.

To stabilize the network, there is a need for successful alignment of the actors' heterogeneous interests. This was findings concurred with the findings of Avgerou and McGrath [15], who emphasized the need for re-alignment of interest among all actors. This is despite that some researchers may freely opt out of the research activity [16]. There are nine attributes in the study that have been established, which can influence the re-alignment of interests of the principals, namely e-learning, awareness creation, understanding user requirements, Budget, power supply, off-campus accessibility, collaboration, and resource sharing, last-mile connectivity, and commitment of ICT staffs. This in line with the works of Bankole, Asse-fa, and Africa [18], who came up with an approach that can be used for achieving proper alignment of the heterogeneous interest. In their study, they suggested that some of the critical processes for re-alignment could include, but are not limited to, the actor's patterns, the role of the users, and the technological functionality.

5.1 Proposed ANT Model

This Chapter presents the proposed ANT model, which can illustrate the interactions of the various actors and how user requirements can be achieved. The proposed ANT model in Fig. 6 below could be used as a guide for the future for EthERNet and other NREN services, which can improve the quality of learning and research.

Research has suggested that the conventional adoption of technology has not exclusively been considered as the ultimate approach to evaluating technology use requirements [17]. ANT has been employed in this research to complement the views by providing dctailed systematic information.

Generally, the research has shown a wide spectrum demand for many educational and research networking services that have not yet been provided by the EthERNet. Given the wide range of possible service provisions and uses, EthERNet should focus on providing educational services (online resource access and MOOCs), online access to software used in education and research, roaming (inter-university login) and video/web conferencing services, and open access data repository services as the first set of services to achieve the required target, which can assist in improving the quality of education and increasing the research output at Ethiopian institutions of higher educations.

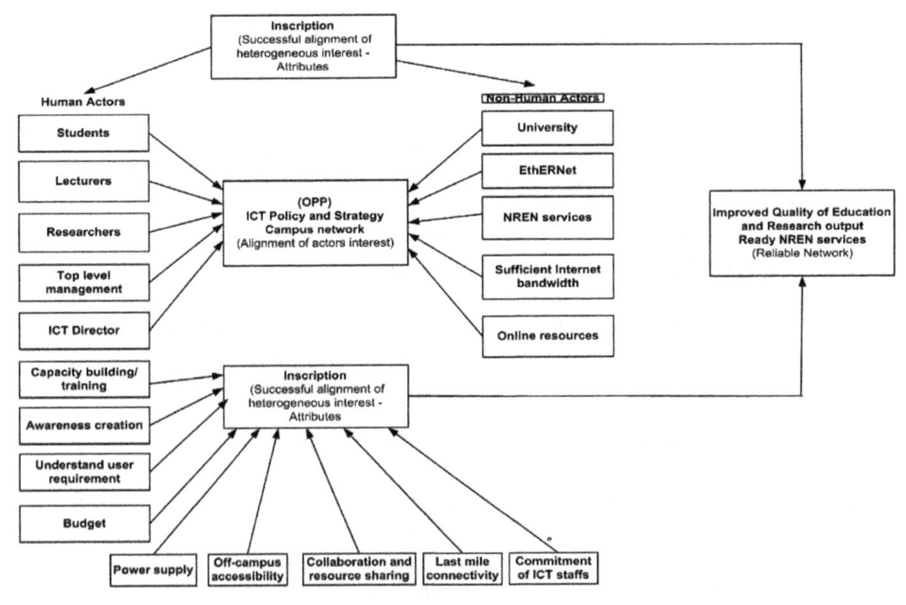

Fig. 6. The proposed ANT model

6 Conclusion and Recommendation

The study has explored the association between the various actors. They have influenced user requirements in the NREN services and EthERNet to improve educational quality and research output. The main actors, namely lecturers, researchers, ICT directors, and EthERNet, were identified from the respondent's comments as key players in the NREN service requirements.

The results indicated that there was a strong association between the main actors (lecturers, researchers, ICT directors, and EthERNet) and the other actors who used the NREN services. From the ANT, three concepts that were examined include OPPs recognition, the problematization, and the inscription process. Results also suggested that there an indirect relationship between the selected actors and others, even though they were not related.

From the results obtained, it was observed that actors could be able to influence other actors who were not related. Future studies could consider performing a comparison of the sensitive content held by different actors in NREN services and EthERNet. This can only be established if there is a holistic NREN service in which actor-network can be applied in various NREN settings. This is expected to eliminate the notion that there are different attributes and actors for different NREN.

References

1. Chipembele, M., Bwalya, K.J.: Assessing e-readiness of the Copperbelt University, Zambia: a case study. Int. J. Inf. Learn. Technol. **33**(5), 315–332 (2016)

2. Babbie Earl, R.: The Practice of Social Research. Nelson Education (2015)
3. Adam, L., Butcher, N., Tusubira, F.F., Claire, S., David, S.: Transformation-ready: the strategic application of information and communication technologies in Africa. Education Sector Study Final Report, 142 (2011)
4. Latour, B.: On actor-network theory: a few clarifications. Soziale Welt **47**, 369–381 (1996)
5. Cressman, D.: A brief overview of actor-network theory: punctualization, heterogeneous engineering & translation (2009)
6. Arnaboldi, M., Nicola, S.: Actor-network theory and stakeholder collaboration: the case of Cultural Districts. Tour. Manag. **32**(3), 641–654 (2011)
7. Rhodes, J.: Using actor-network theory to trace an ICT (telecenter) implementation trajectory in an African women's micro-enterprise development organization. Inf. Technol. Int. Dev. **5**(3), 1 (2009)
8. Arthur, T. (ed.): Social and Professional Applications of Actor-Network Theory for Technology Development. IGI Global (2012)
9. Myhra, A.H.: Cryptomarkets: an actor-network perspective on dark web-based black marketplaces. Master's thesis (unpublished), NTNU (2017)
10. Middendorf, M., Frank, R., Hartmut, S.: Information exchange in multi colony ant algorithms. In: International Parallel and Distributed Processing Symposium, pp. 645–652. Springer, Heidelberg (2000)
11. Tatnall, A., Burgess, S.: Using actor-network theory to research the implementation of a BB portal for regional SMEs in Melbourne, Australia. In: 15 the Bled Electronic Commerce Conference-'eReality: Constructing the eEconomy', Bled, Slovenia. University of Maribor (2002)
12. Calis, G., Orhan, Y.: An improved ant colony optimization algorithm for construction site layout problems. J. Build. Constr. Plan. Res. **3**(04), 221 (2015)
13. Dionys, D.: Introduction of ICT and multimedia into Cambodia's teacher training centers. Australas. J. Educ. Technol. **28**(6), 1068–1073 (2012)
14. Kunda, D.: Changing Faculty members' attitude towards the use of ICTs in teaching and research: the SIDS model (2014)
15. Avgerou, C., McGrath, K.: Power, rationality, and the art of living through socio-technical change. MIS Q. **31**, 295–315 (2007)
16. Gunawong, P., Gao, G.: Understanding eGovernment failure: an actor-network analysis of Thailand's smart ID card project. In: PACIS, p. 17 (2010)
17. Serdyukov, P.: Innovation in education: what works, what doesn't, and what to do about it? J. Res. Innov. Teach. Learn. **10**(1), 4–33 (2017)
18. Bankole, F., Assefa, Z., Africa, S.: Improving the quality of education and research output in Africa a case of Ethiopian education and research network (ethernet), vol. 10, no. 24, pp. 31–51 (2017)

Reading Network Peculiarity in Children with and Without Reading Disorders

Victoria Efimova[1], Artem Novozhilov[1], and Elena Nikolaeva[2](✉)

[1] Child Clinic "Prognosis", Paradnaja Street, 3(2), 191014 Saint-Petersburg, Russia
[2] Herzen State Pedagogical University, Nab. Reki Moika 48, 191186 Saint-Petersburg, Russia
klemtina@yandex.ru

Abstract. The reading of connected text has its own coordinated, distributed neural networks, often called the reading brain. Acquiring the ability to read calls on a child to develop not only certain cognitive functions but also the following background physiological capabilities: postural stability, maintenance of a certain muscle tone, automatic stabilization of the eyeballs and control over their movements, and motor planning skills as well. That is, the complex mechanisms of cognitive development are based on information received from structures that are at lower levels of the signaling process. In keeping with Bernstein's theory, these background abilities are related to levels A and B, whose. The purpose of this research was to study the state of the vestibular function and oculomotor activity in children with and without reading disorders. In all, 64 children and adolescents ranging in age from 6 to 16 were studied. The first group was composed of children and adolescents without any learning disabilities and without any reading disorders (WRD) – 32 individuals between the ages of 6 and 16. The second group consisted of children and adolescents of the same ages with reading disorders (RD) – 32 persons diagnosed with SLD Oculomotor Activity was assessed with Eyegaze Analysis System. The otolith function was evaluated using cervical myogenic induced potentials (cMIPs) using the Rehacor-T psychophysiological testing device (Medicom-MTD). Our experimental results reveal certain features in the oculomotor activity of children with learning disabilities, which may point to a non-optimal functional state or functional immaturity of the cerebellum and also the connections between the brainstem and the cerebellum. These features consisted of saccadic hypometria and a significant reduction in the amplitude and velocity of the saccades in comparison with the group of children that did not suffer from reading disorders.

Keywords: Reading brain · Reading disorder · Children and adolescents · Oculomotor activity · Otolith functions

1 Introduction

In the current classifications (ICD-10 and DSM-V), an impairment in the formation of scholastic skills is referred to as a specific learning disability (SLD). This emphasizes that it is a dysontogenetic disorder. The most frequent manifestation of SLD involves

K. Arai (Ed.): FICC 2021, AISC 1363, pp. 355–368, 2021.
https://doi.org/10.1007/978-3-030-73100-7_26

difficulties in learning how to read (dyslexia). In children, SLD rarely occurs in isolation; on the contrary, it is often accompanied by speech and attention disorders [1], which makes it difficult to analyze the psychophysical mechanisms of this disorder, since it is not clear whether it is a single diagnostic category or an umbrella term for a complex of dysfunctions.

The reading of connected text has its own coordinated, distributed neural networks, often called the reading brain. The left perisylvian region is connected with the temporal and frontal cortex through the arcuate bundle. These structures are engaged in reading and the generation of vocabulary [2]. But reading also has separate neural pathways: the dorsal phonological system and the ventral orthographic system [3]. The phonological pathway comprises (1) the left temporoparietal intersection, which decodes words by aligning graphemes to phonemes, and (2) Broca's area, which coordinates sounds with their phonological representations. The ventral pathway comprises the left occipital-temporal sulcus, or the fusiform gyrus.

This region of the brain is called the Visual Word Form Area, and it is thought to play a role in the fluent reading of whole words [4].

Up until very recently, models of the reading brain left out the right hemisphere [5, p. 89]. Ideas about including it in the reading process were limited to an understanding of its role in disturbances in this process, suggesting that the left hemisphere dominated reading. There are now two lines of thought associated with (1) the notion of a distributed brain network and (2) differences in neural resonance related to oscillation frequencies. During the process of reading, both hemispheres interact and contribute equally [6].

The brain networks can be described as topologically connected neural ensembles that are activated by particular perceptual/mental events and problems that call for attention [7, 8].

The networks differ in that they either process information or control other networks, basically predisposing neurons and neural pathways to a proactive or reactive function [9, 10].

The regions of the brain that are involved in reading are rated as multimodal, associative hubs [11]. Each region is located at the confluence of the posterior inferior parietal lobe, the posterior superior temporal sulcus, the lateral occipital cortex, the supramarginal gyrus and the angular gyrus. Each of them is connected with the perisylvian region through the arcuate bundle/the upper long bundle of each hemisphere. Each such temporoparietal junction contains intra-hemispherical corticocortical projections and extensive interhemispheric connections through the plexuses of the corpus callosum [12].

The left temporoparietal junction is a domain-specific processing hub, which functions in coordination with the left lower frontal cortex, with diversified language processing and deliberate attention [13]. In contrast, the right temporoparietal junction, in coordination with the right medial frontal gyrus (including the inferior frontal gyrus, the frontal operculum and the front of the islet) is a hub that may play a critical role in the processing of information, ascending and guided by stimuli to control behavior through descending nerves [14].

Bailey et al. [11] used fMRI on 1,000 test subjects who were in resting state, evaluating the state of the 7 cortical networks identified by Yeo et al. [15]. These were the visual, somatomotor, auditory, dorsal attention, ventral attention, default and frontoparietal control networks. The configuration of the networks corresponding to the best results for reading included the ventral and dorsal attention networks, and also the frontoparietal networks, which together contribute 56% of the total. Adding in visual activation (18%) increases the percentage of explained variance up to 74%.

Research shows that the right hemispheric networks should be incorporated into models of the reading brain; a deterioration in reading, after all, is evidently attributable to deterioration in control over attention and the hub functions.

In fNIR research, where the oxyhemoglobin concentration was assessed, right-hemispheric deterioration was more accurately localized in the supramarginal and angular gyrus, due to deterioration in the right-hemispheric networks of attention [16]. Research by Molinaro et al. [17] made it clear that deterioration in the right-hemispheric delta synchrony is imposed on the direct connection through the splenum of the corpus collosom, which worsens development of a higher order in the lower left frontal region. Finally, Di Liberto et al. [18] conducted an EEG study that confirmed atypical alpha- and theta-rhythms in children with dyslexia, and these impairments correlated with impairments in their phonological skills.

Acquiring the ability to read calls on a child to develop not only certain cognitive functions but also the following background physiological capabilities: postural stability, maintenance of a certain muscle tone, automatic stabilization of the eyeballs and control over their movements, and motor planning skills as well. That is, the complex mechanisms of cognitive development are based on information received from structures that are at lower levels of the signaling process.

In keeping with Bernstein's theory [19], these background abilities are related to levels A and B, whose anatomical substrates are the spinal cord, brainstem, basal ganglia and ancient regions of the cerebellum. In their turn, levels A and B function under the impulsation of the vestibular system, which generates the internal system of coordinates necessary for spatial orientation. When both the vestibular system and the background levels of the brain, which control activity, are not in an optimum state, this can cause disturbances in multisensory integration, which is critical for the acquisition of scholastic skills.

It is known that the background levels of activity control are already formed in the womb, and their development basically ends before the age of 7 [19]. As a result, if disturbances in levels A and B are identified in children with SLD, it can be assumed that the risk factors for learning disabilities will emerge in the early stages of ontogenesis.

The purpose of this research was to study, using instrumental methods, the state of the vestibular function and oculomotor activity in children with and without reading disorders.

2 Materials and methods

2.1 Subjects

In all, 64 children and adolescents ranging in age from 6 to 16 were studied. All of them attended a consultation with a neurologist and a speech therapist and underwent an instrumental examination.

The first group was composed of children and adolescents without any learning disabilities and without any reading disorders (WRD) – 32 individuals between the ages of 6 and 16 (the mean age was 10 ± 2), of which 70% were boys and 30%, girls. It was determined by a neurologist that in this group 45% of them had attention deficit disorder; 26%, phonemic disorders; 42%, asthenoneurotic syndrome; 10%, paralexia; and 16%, hyperactivity syndrome.

The second group consisted of children and adolescents with reading disorders (RD) – 32 persons diagnosed with SLD. It was ascertained by a neurologist that in this group 91% of them suffered from attention deficit disorder; 72%, from phonemic disorders; 38%, from asthenoneurotic syndrome; 19%, from paralexia; and 13%, from hyperactivity syndrome.

Thus, the WRD group included children and adolescents with neurological dysfunctions but without any learning disabilities, and the RD group included youngsters with neurological dysfunctions and learning disabilities, among which reading disorders were the most notable.

2.2 Evaluation of Oculomotor Activity

The children were asked to pay close attention to a round, yellow object with a diameter of 1.4 angular degrees on a computer screen. This object jumped back and forth, to the right and to the left, a total of ten times in each direction with equal amplitude from the central position. The target object would instantly change position on the screen, vanishing from the center and appearing on the periphery (without time delay). The distance between the central and peripheral positions of the target was 15.6 angular degrees. A second task involved using smooth pursuit eye movements while following a round, red object with a diameter of 0.5 angular degrees. This object made pendulum-like movements across the screen along a horizontal track (the amplitude being 33 angular degrees) at a constant velocity of 30 angular degrees a second, making six similar movements in each direction.

The gaze trajectory (Xt, Yt) was registered on the Eyegaze Analysis System, a device developed by LC Technologies, Inc. (USA) and equipped with NYAN 2.0XT software from Interactive Minds. The recording was done in monocular mode (right eye) with a sampling rate of Fd = 60 Hz. The young test subjects were seated in a comfortable position in front of the computer screen, and their head and neck were gently stabilized. The task was displayed on a 17-inch color LCD with a resolution of 1280 × 1024 pixels, and the size of a pixel was 0.264 mm. The distance between the eyes and the display screen was 65–70 sm.

Based on the results of the tasks, the following were assessed: the number of effective saccades in each direction, the amplitude of the saccades, the latency of the saccades (the reaction time), the peak velocity and the accuracy of the saccades (the distance between the fixation at the end of the saccade and the target). The function of smooth tracking was measured, determining the amplification coefficient according to the work of Tajik-Parvinchi [20].

Statistical processing of the of the data from the oculomotor activity was preceded by work with "raw" data, representing the gaze coordinates X and Y in screen pixels discretized at a rate of 60 samples a second, which was then converted from a text format into Excel format (Microsoft Office 2016). The beginning of a saccade/macro jerk was considered to be a change in the position of the gaze in 1/60th of a second, by one or more angular degrees.

2.3 Evaluation of the Vestibular Function

The functions of the semicircular canals and otolith organs of the vestibular system were evaluated separately by means of instrumental methods of examination.

The otolith function was evaluated using cervical myogenic induced potentials (cMIPs). The cMIPs were recorded with the help of a Neuro-MVP-4 electroneuromyograph (Neurosoft, Ivanovo). The latency of the cMIP p13 wave was evaluated. This is the interval of time from the delivery of an auditory signal to a slowdown in the action potential, and it was registered on the sternocleidomastoid muscle by two electrodes mounted on the tips of the muscle on the side from which clicks were presented (the sacculo-cervical reflex). Clicks of 130 dB ultrasound 0.5 ms long were presented through specially calibrated headphones, and 5–20 cMIPs in 10 series with superposition were averaged to assess the reproducibility of responses. The efferent pathways of the cMIP include the sacculus of the inner ear, the lower vestibular nerve, the vestibular nucleus, the vestibulospinal tract and the m. sternocleidomastoideus. When the ear is stimulated by a loud sound and a tonic contraction of the m. sternocleidomastoideus is registered, it is possible to evaluate the function of the lower vestibular nerve and the vestibule-cervical reflex. Recording the cMIPs ensures an accurate tonic diagnosis of the level of loss to the brainstem under normal parameters of brainstem auditory evoked potentials.

The functioning of the semicircular canals was assessed by registering the duration and nature of the post-rotational nystagmus and then using the Rehacor-T psychophysiological testing device (Medicom-MTD, Taganrog). The test subject was seated in a Barany rotational chair and asked to close their eyes or to put on a sleep mask, and their head was tilted forward from the vertical position of their torso by 30°. The chair was rotated clockwise at a speed of 10 rotations in 20 s and then counterclockwise, with a 5-min break between rotations. The horizontal component of the electro-oculogram (EOG) was recorded, using two EOG leads at the outer corners of the eyes and a neutral one in the center of the forehead. After an abrupt halt in the rotations of the chair, the test subject remained seated with their eyes closed or in a sleep mask, and, at this time, the post-rotational nystagmus was recorded, and the duration of its attenuation was measured for both directions.

Based on the findings for the cMIPs and the post-rotational nystagmus, the percentage of asymmetry for both indicators was calculated, in accordance with the work of Yakupov and Kuznetsova [21]: $(t_1 - t_2)|\cdot 100\% / (t_1 + t_2)$, where t1 and t2 are the latent period of contraction of the clavisternomastoid muscle in milliseconds after saccular stimulation left and right or the time of the postrotational nystagmus in seconds, after rotation to the left and to the right.

Statistical processing of the findings was carried out using the SPSS 21 set of programs. They are presented as the mean and the standard deviation. The null hypothesis was rejected at a level of $p < 0.05$, using the Student t-test. The values of the vestibular function are expressed as medians and deviations from the median, and the validity of the differences was determined by using the nonparametric Mann Whitney criterion.

The graphs are based on the mean, quartiles 1 and 3, and the minimum and maximum values. The correlation of the saccade rates within groups for each direction, and also between the oculomotor activity and vestibular function pairs, were evaluated using the Pearson criterion. The significance of the differences was evaluated on the basis of the Student t-test.

3 Results

3.1 Oculomotor Function Indicators

The indicators that were obtained for the amplitude of the saccades, the peak velocity and the latency time in the WRD group, and the indicators of the smooth tracking function correspond to the values seen in the literature [20, 22]. Some differences in the values are probably related to operational features of the eye tracker and inherent characteristics of the data processing.

On average, the children in both groups made the same number of effective saccades in both directions – 5–6 out of 10 (precisely from the center and in the proper direction). Altogether, in the WRD group, 165 saccades were analyzed to the left and 184 to the right, and in the RD group, 189 saccades to the left and 181 to the right.

One general pattern emerged: saccades to the right had significantly larger amplitudes than those to the left in both groups of participants ($p < 0.01$). The peak velocity for saccades to the left was also slower in both groups than for those to the right ($p < 0.001$). Likewise, in the RD group, there were more errors in saccades to the left than to the right ($p < 0.05$) (Table 1).

Table 1. The basic saccades' indicators (means and standard deviations).

Indicators	To the left		To the right	
	WRD	RD	WRD	RD
Amplitude of the saccades (angular degrees)	13.6 ± 0.9ΔΔ	12,9 ± 1,3***ΔΔ	14.1 ± 1.2	13.5 ± 1.4**
Errors in saccades/(angular degrees)	1.0 ± 0.9	1.6 ± 1.2***Δ	0.8 ± 1.1	1.2 ± 1.4*
The peak saccades velocity (angular degrees/s)	349.0 ± 48.0ΔΔΔ	333.0 ± 37.0**ΔΔΔ	382.0 ± 58.0	358.0 ± 53.0**
The latent period of the saccade ms	247.0 ± 56.0	239.0 ± 61.0	251.0 ± 51.0	242.0 ± 67.0
Amplification coefficient for the smooth tracking function	0.76 ± 0.13	0.70 ± 0.11 *	0.73 ± 0.11	0.72 ± 0.10

Note: differences between groups * - p < 0,05; ** - p < 0,01, *** - p < 0,001, t-тест; differences in group between left and right saccades direction: Δ - p < 0,05; ΔΔ - p < 0,01, ΔΔΔ - p < 0,001.

The youngsters in the RD group were found to have made saccades in both directions with significantly smaller amplitudes than those in the WRD group (Table 1). In the WRD group, the increase in the saccades to the left was 0.87 and to the right, 0.90; in the RD group, 0.83 and 0.87, respectively.

As for the accuracy of the saccades (the distance from the fixation of the gaze to the target after a saccade has been completed) in the WRD group, it was 1.0 ± 0.9 angular degrees in the left direction and 0.8 ± 1.1angular degrees in the right direction. In the RD group, when they were making saccades, the number of errors in both directions was significantly higher than in the WRD group: 1.6 ± 1.2 angular degrees to the left and 1.2 ± 1.4 angular degrees to the right (Table 1). The peak velocity of saccades in the RD group turned out to be significantly lower in both directions when compared with the WRD group. There were no differences in the latency of the saccades, both between the two groups and within the groups between movements to the left and to the right (Table 1).

The amplification coefficient for the smooth tracking function in the RD group was significantly lower when pursuit eye movements were performed to the left ($p < 0.05$) (0.76 to the left and 0.73 to the right in WRD; 0.70 to the left and 0.72 to the right in RD) (Table 1).

3.2 Vestibular System Indicators

Using the cMIPs method, a functional study of the otolith organ of the sacculus revealed no differences between the groups in the rate at which the nerve impulse traveled along the lower vestibular pathways, nor were there any differences in the asymmetry of this rate. The rates were moderately disrupted in both groups.

A functional study of the horizontal semicircular canals showed no differences in either group in the attenuation time/the length of the nystagmus after the rotational tests in the Barany chair. There was a marked tendency toward an increase in the duration of the post-rotational nystagmus in the RD group (11%, as against 7% in the control group, $p = 0.110$, the Manna-Whitney criteria) (Fig. 1). No predominant shift, to the left or to the right, was found (in both groups, the shift in either direction was approximately 1:1).

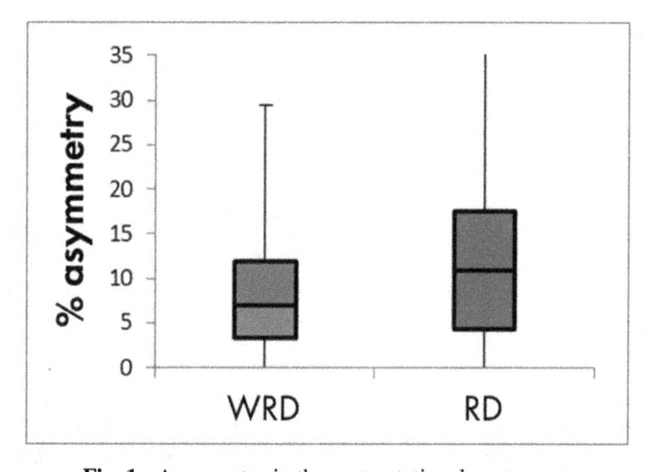

Fig. 1. Asymmetry in the post-rotational nystagmus

3.3 Correlation of the Indicators

A study of the correlation between corresponding left- and right-related indicators revealed special features in the WRD and RD groups. In terms of the amplitude of the saccades, the peak velocity, the smooth tracking function and the length of the post-rotational nystagmus, there were weak to moderate positive correlations. In the case of the post-rotational nystagmus, the correlation in the WRD group was very strong (0.838). Likewise, the correlation for the peak velocity of the saccades turned out to be more pronounced in that group (0.681).

A distinguishing feature of the WRD group was a weak to moderate negative correlation between the latency of the saccades and the FPS to the left and to the right (−0.456 and −0.500, respectively). Only in the RD group was there a moderate positive correlation between the latency of the saccades (0.517) and the cMIPs (0,623).

As for the correlation between oculomotor activity and vestibular function, a special feature of the WRD group was a moderate positive correlation between the amplitude of the saccades to the right and the post-rotational nystagmus in both directions (0.656 for the clockwise rotation and 0.515 for the counterclockwise rotation), and, in this group, there was also a weak negative correlation between the time of the saccade latency to the left and the cMIP to the right (−0.424) (Table 2).

Table 2. Pearson correlations of the basic saccades' indicators

Indicators	To the left		To the right	
	WRD	RD	WRD	RD
Amplitude of the saccades (angular degrees)	13.6 ± 0.9ΔΔ	12,9 ± 1,3***ΔΔ	14.1 ± 1.2	13.5 ± 1.4**
Errors in saccades/(angular degrees)	1.0 ± 0.9	1.6 ± 1.2***Δ	0.8 ± 1.1	1.2 ± 1.4*
The peak saccades velocity (angular degrees/s)	349.0 ± 48.0ΔΔΔ	333.0 ± 37.0**ΔΔΔ	382.0 ± 58.0	358.0 ± 53.0**
The latent period of the saccade ms	247.0 ± 56.0	239.0 ± 61.0	251.0 ± 51.0	242.0 ± 67.0
Amplification coefficient for the smooth tracking function	0.76 ± 0.13	0.70 ± 0.11*	0.73 ± 0.11	0.72 ± 0.10

Note: differences between groups * - p < 0,05; ** - p < 0,01, *** - p < 0,001, t-тест; differences in group between left and right saccades direction: Δ - p < 0,05; ΔΔ - p < 0,01, ΔΔΔ - p < 0,001.

3.4 Factor Analysis

Factor analyses were performed for each group separately and for the entire group of test subjects. Since the separate factor analyses have low Kaiser-Olkin-Meyer coefficients (0.503 and 0.522), the result of the factor analysis for the whole group is given here (Table 3). This factor analysis has a three-factor determination, KMO = 0.893, and the percentage of explained variance is 87.3%.

Table 3. A rotated component matrix

Indicators	Components		
	1	2	3
Maximal saccade amplitude L	.977	.106	.096
Saccade error R (% of saccade amplitude)	−.974	−.030	−.029
Maximal saccade velocity L	.973	.130	.099
Saccade error L	−.958	−.090	−.070
Saccade amplitude L	.951	.063	.041
Saccade amplitude R	.907	.183	.027
Post-rotational nystagmus L (s)	.867	.362	.161
Saccades: reaction L (ms)	−.866	.028	−.108
Saccades: reaction R (ms)	−.810	.092	−.418
Smooth tracking L	.640	.159	−.290
Post-rotational nystagmus R (s)	.104	.977	.055
Smooth tracking R	.088	.080	.933

Factor selection method: principal component method.

Rotation method: Kaiser normalized varimax. The rotation converged in four iterations.

As shown in Table 3, the first factor (67.3%) included almost all of the parameters, but they are separated into left and right components. The second factor was nystagmus to the right (10% dispersion), and the third factor was smooth tracking to the right (10%). Introducing both cMIPs (the left and the right) reduces the KMO, so they are not included in this version.

It is fair to assume that this determination indicates two factors that are significant for reading: post-rotational nystagmus, which occurs with rotation to the right, and smooth tracking to the right, which reflects the effectiveness of left-hemisphere reading mechanisms.

4 Discussion

Our research makes it clear that children with and without reading disorders differ significantly in their oculomotor activity, as can be seen from the particularities of the

RD group, where there were saccadic inaccuracies (with appreciably more errors in saccades to the left), considerably lower saccade velocity and lower amplitude in both directions.

There was no substantial difference between the groups in the vestibular indicators. In both groups, there were moderate abnormalities in vestibular sensory reactivity, both on the part of the otolithic segment and the semicircular canals. The prevalence of these dysfunctions among children with speech disorders, and attention-deficit/hyperactivity disorder is shown here by us and likewise in the work of other authors [23–26].

Accordingly, the difference between children with and without reading disorders consisted in the control of eye movement. In our research, nonverbal tests were used, and the findings are consistent with the literature [27–29] and with our previous works [30, 31].

It is shown in the literature that in more than 75% of children with learning disabilities, the following ophthalmological problems are brought to light: disturbance of binocular vision, oculomotor dysfunction, and impaired ability to follow moving objects [32]. The mechanisms behind these dysfunctions need further study [33].

Control over automatic eye movements is connected with the working of several structures in the nervous system. In our research, sensory hyporesponsiveness in the canal and otolith parts of the vestibular apparatus was detected in children from both groups. In the works of Wiener-Vacher [34], it is shown that in children the cerebellum can compensate for deficiency in the vestibular system. Since the difference between the WRD and RD groups in our study was only in the reading skills, it can be assumed that the functional state of the cerebellum in the children of the RD group did not make it possible to offset the deficit in processing the vestibular information at low levels of integration, which made reading difficult. This group also exhibited a more pronounced asymmetry in the post-rotational nystagmus, which is also associated with potential difficulties in stabilization of the eyeballs when the head is moving.

The difficulties that the children in the RD group have with reading may be related to inaccuracies in the saccades, as a result of which the gaze "overshoots" the target, and this makes reading tedious and ineffective. Inaccuracies in performing saccades are more often than not associated with the working of the cerebellum-brainstem circuit. The fastigial nuclei, acting as the conductor for the pre-motor burst neurons in the Paramedian pontine and the inhibitory neurons of the brainstem's rostral segment, on account of the precisely timed, reciprocal changeover of neuronal activity in their composition, determine the beginning and the end of saccades. At the same time, the activity thus produced is checked against an internal copy of a similar movement already built in during the development and interaction of the body with the surrounding environment. The forming of such commands in the brain can be accounted for by the coordinated functioning of the saccadic integrator, which includes the nucleus prepositus hypoglossi, the medial vestibular nucleus, the interstitial nucleus of the Cajal and the cerebellum. In addition, the activity of the pre-motor burst neurons during fixation is limited to the tonic activity of the omnipause neurons confined to the raphe nucleus of the pons Varolii. Any disruption in the interaction of the above-mentioned saccadic brainstem circuit with the omnipause neurons can trigger disruption of gaze fixation and the smooth tracking function [35, 36]. Excessive hypmetria of saccades is often associated with insufficient coordination in the functioning of the pre-motor burst neurons in the paramedian reticular function [36], and also with the cerebellum [37, 38].

A number of studies have shown that the cerebellum participates in learning [39, 40]. It is now known that it is involved not only in the automation of motor skills but also in the development of language skills. The hypothesis of the cerebellum's involvement in the mechanisms of dyslexia has been confirmed by studies using standardized clinical tests and fMRT [41, 42]. Changes have been detected in the morphology of the cerebellum that are characteristic of dyslexia [43].

It has been shown that the cerebellum participates in the following processes: the regulation of speed, strength, rhythm and precision in movement, memory retention, the results of motor learning and the control of habitual movement. It is also involved in the acquisition of phonological skills. Cerebellar cognitive affective syndrome has also been described [44].

Imamizu, Kuroda, and Miyauchi [45] hypothesize about the organization and preservation of activity models. They contend that this process takes place mainly in the cerebellum. It can be assumed that the reading skills of children whose cerebellum is in a non-optimal functional state are not automatic, since models of eye movements necessary for reading are initially distorted, due to vestibular dysfunction, or are not preserved, due to dysfunction of the cerebellum.

In our study, the children with learning disabilities were found to have a weak positive correlation between the latency of a saccade, on the one hand, and its amplitude and error, on the other. Saccadic latency reflects cortico-subcortical interplay, which includes visual processing, target selection and motor planning of a saccade [36]. This can also be seen as a minor disturbance in motor planning, in which an increase in the latency of a saccade – a slowdown in visual processing by the brain – is accompanied by a greater saccade error, an inadequate reflex response.

Our findings show how important low levels of sensory-motor integration are in the cognitive development of children. But they also speak to the role cooperation between the left and right hemispheres plays in reading. Both groups have neurological alterations, and they both show a decrease in amplitude with saccades to the left. At the same time, the children in the RD group committed more errors in saccades to the left than to the right. We can assume that general neurological disturbances are dependent on a reduction in the adaptive properties of the right hemisphere, attributable to developmental features in early ontogenesis, while problems directly associated with reading are brought about by a dysfunction of the left hemisphere reading network.

5 Conclusions

Our experimental results reveal certain features in the oculomotor activity of children with learning disabilities (reading disorders), which may point to a non-optimal functional state or functional immaturity of the cerebellum and also the connections between the brainstem and the cerebellum. These features consisted of saccadic hypometria and a significant reduction in the amplitude and velocity of the saccades in comparison with the group of children that did not suffer from reading disorders.

Throughout the world, the fMRT method is being used more and more often to evaluate the functions of the cerebellum; for many reasons, however, it cannot be used extensively with children. Our study shows that eye tracking, which is safe and non-intrusive, may be used to evaluate the psychophysiological mechanisms behind learning disabilities in children. Using simple, nonverbal tests, it is possible to study oculomotor activity in children who are unable to read. In this way, it is also possible to identify those children who are at risk of developing learning disabilities before they start school.

Correlation of eye tracking results with those from instrumental examinations of vestibular functions provide an opportunity to identify dysfunctions at lower levels of sensory-motor integration that can cause a child to be unsuccessful in school.

References

1. Schwenck, C., Dummert, F., Endlich, D., Schneider, W.: Cognitive functioning in children with learning problems. Eur. J. Psychol. Educ. **30**(3), 349–367 (2016)
2. Su, M., de Schotten, M.T., Zhao, J., Song, S., Zhou, W., Ramus, F.: Vocabulary growth rate from preschool to school-age years is reflected in the connectivity of the arcuate fasciculus in 14-year-old children. Dev. Sci. **21**(5), 1–26 (2018)
3. Vandermosten, M., Boets, B., Poelman, H., Sunaert, S., Wouters, J., Ghesquiere, A.: A tractology study of dyslexia: Neuroanatomical correlates of orthographic, phonological, and speech processes. Brain **135**, 935–958 (2012). https://doi.org/10.1093/brain/awr363
4. Dehaene, S., Cohen, L.: The role of the visual word form are a in reading. Trends Cogn. Sci. **15**, 255–261 (2011). https://doi.org/10.1016/j.tics.2011,04.003
5. Elliot, J., Gregorenko, D.: The Dyslexia Debate. Cambridge Press, New York (2014)
6. Kershner, J.R.: Neuroscience and education: Cerebral lateralization of networks and oscillations in dyslexia. Laterality **25**(1), 109–125 (2020). https://doi.org/10.1080/1357650X.2019.1606820
7. Fornito, A., Zalesky, A., Bullmore, E.: Fundamentals of Brain Network Analysis. Academic Press, San Diego (2016)
8. Medaglia, J., Lynall, M., Bassett, D.: Cognitive network neuroscience. J. Cogn. Neurosci. **27**(8), 1471–1491 (2015)
9. Spagna, A., Mackie, M., Fan, J.: Supramodal executive control of attention. Front. Psychol. **6**, 1–13 (2015)
10. Zanto, T.P., Rubens, M.T., Thangavel, A., Gazzaley, A.: Causal role of the prefrontal cortex in top-down modulation of visual processing and working memory. Neurosci **14**, 656–661 (2011)
11. Bailey, S., Aboud, K., Nguyen, T., Cutting, L.: Applying a network framework to the neurobiology of reading and dyslexia. J. Neurodev. Disord. **10**(37), 1–20 (2018). https://doi.org/10.1186/s11689-018-9251-2
12. Knyazeva, M.: Splenium of the corpus callosum: Patterns of interhemispheric interaction in children and adults. Neural Plast. **639430**, 1–12 (2013)
13. Dubois, J., Poupon, C., Thirion, B., Simonnet, H., Kuliloval, S., Leroy, F., Dehaene-Lambertz, G.: Explaining the early organization and maturation of linguistic pathways in the human brain. Cereb. Cortex **26**(5), 2283–2298 (2016). http://doi.org/10.1093/cercor/bhvo82
14. Wu, Q., Chang, C.-F., Xi, S., Huang, I.-W., Liu, Z., Juan, C.-H., Fan, J.: A critical role of temporoparietal junction in the integration of top-down and bottom-up attentional control. Hum. Brain Mapp. **36**, 4317–4333 (2015). https://doi.org/10.1002/hbm.229
15. Yeo, B., Krieman, F., Sepulcre, J., Sabuncu, M., Lashkar, D., Hollingshead, M., Buckner, R.: The organization of the brain cerebral cortex estimated by intrinsic functional connectivity. J. Neurophysiol. **106**, 1125–1165 (2011)

16. Cutini, S., Szucs, D., Mead, N., Huss, M., Goswami, U.: Atypical right hemisphere response to slow temporal modulations in children with developmental dyslexia. Neuroimage **143**, 40–49 (2016). https://doi.org/10.1016/j.neuroimage.2016.08.012
17. Molinaro, N., Lizarazu, M., Lallier, M., Bourguignon, M., Carreiras, M.: Out-of Synchrony speech entrainment in developmental dyslexia. Hum. Brain Mapp. **37**, 2767–2783 (2016). https://doi.org/10.1002/hbm.23206
18. Di Liberto, G., Peters, V., Kalashnikova, M., Goswami, U., Burnham, D., Laler, E.: Atypical cortical entrainment to speech in the right hemisphere underpins phonemic deficits in dyslexia. NeuroImage **175**, 70–79 (2018). https://doi.org/10.1016/j.neuroimage.2018.03.072
19. Bernstein, N.A.: Biomechanics and Physiology of Movements: Selected Psychological Works, 3rd edn. Ross. Akad. Education, Moscow, Voronezh (2008)
20. Tajik-Parvinchi, D.J., Lillakas, L., Irving, E., Steinbach, M.J.: Children's pursuit eye movements: a developmental study. Vis. Res. **43**(1), 77–84 (2003)
21. Yakupov, E.Z., Kuznetsova, E.A.: Diagnostic value of vestibular myogenic evoked potentials in vestibular-atactic syndrome of various etiologies. Bull. Int. Sci. Surg. Assoc. **5**(1), 17–21 (2010)
22. Hopf, S., Liesenfeld, M., Schmidtmann, I., Ashayer, S., Pitz, S.: Age dependent normative data of vertical and horizontal reflexive saccades. PLoS ONE **13**(9), (2018). https://doi.org/10.1371/journal.pone.0204008
23. Rine, R.M.: Vestibular rehabilitation for children. Semin. Hear **39**, 334–344 (2018). https://doi.org/10.1055/s-0038-1666822
24. Efimova, V.L., Reznik, E.N., Nikolaev, I.V.: Vestibular dysfunctions in children with ADHD symptoms. Bull. Psychophysiol. **3**, 38–43 (2019)
25. Efimova, V.L., Nikolaeva, E.I.: The role of the vestibular system in the formation of specific speech disorders in children. Hum. Physiol. **46**(3), 83–89 (2020)
26. Wiener-Vacher, S.R., Quarez, J., Priol, A.L.: Epidemiology of vestibular impairments in a pediatric population. Semin. Hear **39**, 229–242 (2018). https://doi.org/10.1055/s-0038-1666815
27. Pensiero, S., Accardo, A., Michieletto, P., Brambilla, P.: Saccadic alterations in severe developmental dyslexia. In: Case Reports in Neurological Medicine 2013, ID 406861 (2013). https://doi.org/10.1155/2013/406861
28. Romero, A.C., Stenico, M.B., de Oliveira, L.S., Franco, E.S., Capellini, S.A., Frizzo, A.C.F.: Vectoelectronystagmography in children with dyslexia and learning disorder. Rev. CEFAC **20**(4), 442–449 (2018)
29. Bucci, M.P., Nassibi, N., Gerard, C.-L., Bui-Quoc, E., Seassau M.: Immaturity of the oculomotor saccade and vergence interaction in dyslexic children, evidence from a reading and visual search study. PloS One **7**(3) (2012). https://doi.org/10.1371/journal.pone.0033458
30. Efimova, V.L., Novozhilov, A.V., Savelev, A.V.: Measurement and analysis of microsaccad trajectories in children with a specific disorder of school skills formation when tracking a moving stimulus. Neurocomput. Dev. Appl. **21**(4), 18–21 (2019)
31. Efimova, V.L., Novozhilov, A.V.: Neurocomputing in the study of features of oculomotor activity of children with attention deficit hyperactivity disorder and a specific disorder of school skills formation. J. Neurocomput. Dev. Appl. **5**, 35–40 (2018)
32. Everatt, J., Bradshaw, M.F., Hibbard, P.B.: Visual processing and dyslexia. Perception **28**, 243–254 (1999)
33. Ego, C., de Xivry, J.J.O., Nassogne, M.-C., Yüksel, D., Lefèvre, P.: The saccadic system does not compensate for the immaturity of the smooth pursuit system during visual tracking in children. J. Neurophysiol. **110**(2), 358–367 (2013)
34. Wiener-Vacher, S.R., Hamilton, D.A., Wiener, S.I.: Vestibular activity and cognitive development in children: perspectives. Front. Integr. Neurosci. **11**(7), 92 (2013). https://doi.org/10.3389/fnint.2013.00092

35. Termsarasab, P., Thammongkolchai, T., Rucker, J.C., Frucht, S.J.: The diagnostic value of saccades in movement disorder patients: a practical guide and review. J. Clin. Move. Disord. **2**(1), 14 (2015). https://doi.org/10.1186/s40734-015-0025
36. Leigh, R.J., Kennard, C.: Using saccades as a research tool in the clinical neurosciences. Brain **127**(3), 460–477 (2004)
37. Klyushnikov, S.A., Asiatskaja, G.A.: Oculomotor disorders in the practice of a neurologist. Nerv. Dis. **4**, 41–46 (2015)
38. Shurupova, M.A., Anisimov, V.N., Deveterikova, A.A., Kasatkin, V.N., Platonov, A.V.: Oculomotor and cognitive functions in children with cerebellar dysfunction. In: Shurupova, M.A., (ed.) Conference "Cognitive Science in Moscow: New Research, pp. 547–551, Buki Vedi, Moscow (2019)
39. Levinson, H.N.: The cerebellar-vestibular basis of learning disabilities in children, adolescents and adults; hypothesis and study. Percept. Motor Skills **67**, 983–1006 (1988)
40. Nicolson, R.I., Fawcett, A.J.: Dyslexia, dysgraphia, procedural learning and the cerebellum. Cortex **47**(1), 117–127 (2011). https://doi.org/10.1016/j.cortex.2009.08.016
41. Pernet, C.R., Poline, J.B., Demonet, J.F., Rousselet, G.A.: Brain classification reveals the right cerebellum as the best biomarker of dyslexia. BMC Neurosci. **10**(1), 67–78 (2009). https://doi.org/10.1186/1471-2202-10-67
42. Stoodley, C.J., Stein, J.F.: Cerebellar function in developmental dyslexia. Cerebellum **12**(2), 267–276 (2013). https://doi.org/10.1007/s12311-012-0407-1
43. Rae, C., Harasty, J.A., Dzendrowsky, T.E., Talcott, J.B., Simpson, J.M., Blamire, A.M., Dixon, R.M., Lee, M.A., Thompson, C.H., Styles, P., Richardson, A.J., Stein, J.F.: Cerebellar morphology in developmental dyslexia. Neuropsychologia **40**(8), 1285–1292 (2002). https://doi.org/10.1016/s0028-3932(01)00216-0
44. Schmahmann, J.D.: The role of the cerebellum in cognition and emotion: personal reflections since 1982 on the dysmetria of thought hypothesis, and its historical evolution from theory to therapy. Neuropsychol. Rev. **20**(3), 236–260 (2010)
45. Imamizu, H., Kuroda, T., Miyauchi, S., Yoshioka, T., Kawato, M.: Modular organization of internal models of tools in the human cerebellum. Proc. Nat. Acad. Sci. USA. **100**(9), 5461–5466 (2003)

A Development of Real-Time Failover Inter-domain Routing Framework Using Software-Defined Networking

Yoshiyuki Kido[1,2](✉), Juan Sebastian Aguirre Zarraonandia[2], Susumu Date[1,2], and Shinji Shimojo[1,2]

[1] Cybermedia Center, Osaka University, 5-1, Mihogaoka, Ibaraki-shi, Osaka 5670047, Japan
{kido,date,shimojo}@cmc.osaka-u.ac.jp

[2] Graduate School of Information Science and Technology, Osaka University, 1-5, Yamadaoka, Suita-shi, Osaka 5650871, Japan
j.sebastian@ais.cmc.osaka-u.ac.jp

Abstract. Internet Exchange Points (IXPs) play a crucial role in the interconnection between Internet Service Providers. Consequently, failures in the underlying IXP infrastructure impact the end-user experiences and can potentially translate into a financial loss for participants. Several research groups have attempted to override the Border Gateway Protocol (BGP) route selection process in IXP using a software-defined networking (SDN) approach to provide fast failover times, and to control how traffic flows between IXPs. In this research, data plane failures, such as physical disconnections between participant routers and the IXP switch, interrupt traffic flows until the configured BGP hold timer in the IXP route server expires. The objective of this research is to decrease the packets dropped during an IXP data plane failure event. We attempt to decrease the disconnection time between the IXP participants to a value lower than one second, which is the minimum configurable BGP Hold Time. In order to pursue this goal, we have developed a framework that reduces the failover process to the order of milliseconds. The results of the experiments registered an average failover time of 31 ms. As a consequence of improving the failover time, the average packet loss is reduced.

Keywords: Software defined networking · Stream processing · Internet Exchange Point

1 Introduction

Services that provide real-time communication such as VoIP, have stricter demands for continuous connectivity than during the origins of the Internet. Failing to do so could translate into a negative reputation and subsequently a large financial loss. At the same time, there has been an increase in the reliance of the scientific community on network infrastructures to share and store large

K. Arai (Ed.): FICC 2021, AISC 1363, pp. 369–387, 2021.
https://doi.org/10.1007/978-3-030-73100-7_27

amounts of data, which in most cases is geographically dispersed. For example, large-scale instruments like the Large Hadron Collider produce petabytes of data that are transmitted to regional locations and are finally distributed to researchers [7].

According to current trends, the data intensity of many disciplines is projected to increase by a factor of ten or more over the next few years [8]. Scientific de-militarized zones (DMZ) are not only transferring elephant flows through national research networks, but also storing large data sets in public cloud infrastructures. In order to support these operations cloud service providers peer directly with Research and Education networks in order to avoid congesting their own networks. In some cases, cloud service providers peer with commodity IXPs that serve the traffic of a large percentage of ISP customers, which are not designed to carry long-term data set file transfers [5]. As this trend continues, IXPs will play a crucial role in the interconnection between ISPs, CDNs, and research networks. Consequently, any failure in the underlying IXP infrastructure will impact end-user experiences and research endeavors, and can potentially translate into a financial loss for its participants.

As more heterogeneous participants around the world join the network, future IXP implementations will have to support several features in their network architectures, such as:

- Quickly identify connectivity disruptions within their infrastructure.
- Resolve connectivity failures faster than current routing protocols.

Guaranteeing continuous connectivity across the Internet is a large challenge for network operators, because exercising any resilience or traffic engineering technique beyond their administrative domain is not possible. Furthermore, there are several technologies and protocols implemented in Wide Area Networks (WAN) that are difficult to change (e.g., Border Gateway Protocol (BGP)) due to their strong standardization and widespread adoption during previous decades.

In order to enable fast failover and continuous forwarding between Internet domains, researchers from both academia and service providers have proposed traffic engineering approaches and inter-domain protocol enhancements that allow manipulation of the routing decision process. Still, these techniques have not achieved sufficient standardization to be widely adopted by all Internet participants, mainly because these techniques introduce routing incompatibilities and additional deployment and management complexities [26].

Based on the previous inter-domain scenario, we attempt to approach the identification and resolution of connectivity disruptions at IXPs. The objective is to decrease the number of packets lost during a routing protocol convergence to a new route. The disconnection time between IXP participants is reduced to a value lower than one second, which is the minimum configurable BGP hold timer. We explore the enhancement of an IXP architecture with software-defined networking (SDN) and stream processing technologies to provide fast failover on the order of milliseconds.

In Sect. 2, the technical issues to be solved and the derived research goal, as well as the research problem, are clarified. Section 3 formally introduces the concept, design, and implementation of the proposed IXP architecture as well as the technical issues derived in the previous section. In Sect. 4, evaluation experiments are conducted in order to verify the feasibility and correctness of the proposed method. Finally, Sect. 5 concludes this paper and suggests areas for future research.

2 Technical Issues and Related Research

This section presents the main challenges to reducing the failover time of data plane failures between Internet Exchange Point participants. An overview about how the Border Gateway Protocol works is presented, and the mechanisms that allow communication between network operators are provided.

2.1 Border Gateway Protocol (BGP)

Current Internet transport infrastructure relies on the BGP for interconnecting different administrative Internet domains, which are also technically referred to as autonomous systems (AS). As the de-facto standard for inter-domain routing for more than twenty years, the main function of the BGP is to exchange reachability information between ASes by means of BGP UPDATE or KEEPALIVE messages [9]. If a BGP speaker does not receive a KEEPALIVE or UPDATE message from a particular BGP neighbor for the duration of the Hold Time, then the speaker will close the BGP connection. By default, the Hold Time is set to 180 s on commercial routers. During the establishment of a BGP session, KEEPALIVE messages notify each BGP peer that the neighbor router is active. After a BGP session is established, the session is used as a mechanism to determine whether peers are reachable by being exchanged at a regular interval, which is usually one third of the configured Hold Time interval. KEEPALIVE messages must not be sent more frequently than once per second [24].

2.2 Internet Exchange Points (IXPs)

An IXP is a network infrastructure that facilitates the exchange of Internet traffic between ASes. Additional services offered at IXPs include the free use of route servers. Route servers provide participants with a scalable solution to peer with a large number of co-located ASes through a single BGP session, without having to manually configure individual adjacencies with each AS. Generally, participants in an IXP that want to exchange traffic must comply with some basic requirements:

1. Have a public Autonomous System Number (ASN).
2. Bring a router to the IXP. Connect one of its Ethernet ports to the IXP switch, and one of its WAN ports to the WAN media leading back to the network of the participant.
3. The router of the participant must be able to run the BGP.
4. Agree to the General Terms and Conditions (GTC) of the IXP.

Networks can peer publicly or privately in an IXP. Autonomous systems that publicly use an IXP network infrastructure for exchanging traffic make a one-time investment to establish a circuit from their premises to the IXP and pay a monthly charge for using an IXP port and an annual fee for membership to the operator of the IXP. Other than these costs, IXPs do not charge for exchanged traffic volume, unless there is a violation of the GTC (e.g., sending traffic to an AS without having received a route from that member). In addition, IXPs offer private peering interconnects that are separated from the public peering infrastructure and enable two ASes to directly exchange traffic and have a dedicated link that can handle stable and high-volume traffic (e.g., Internet2 Peer Exchange (I2PX) and Cloud Connect [14]).

2.3 Internet Exchange Point Data Plane Failure Detection Challenges

Data plane failures, such as physical disconnections between participants routers and the IXP switch, interrupt traffic flows until the configured BGP hold timer in the IXP route server expires. During this lapse of time, no failover mechanism is started, and, as a result, packets are dropped between the disconnected participants.

The goals of implementing a fast failover mechanism are to provide continuous service delivery after a failure has been detected and to optimize the performance of networks by controlling how traffic flows [28]. Among the motivations for providing fast failover and control to traffic that crosses multiple Internet domains are [10]:

Network Failures and Changes in Routing Policies: Neighbor domain failures and fluctuations degrade user performance and lead to inefficient use of network resources.

Violation of Peering Agreements: If the amount of traffic exchanged by two peer ASes is exceeded by one of the ASes, then an AS should have mechanisms to direct some traffic to a different AS without affecting its neighbor throughput.

Current limitations for providing fast failover at IXPs techniques represent a significant challenge for applications and collaborations groups between distributed domains that rely on underlying networks for services that include real-time data rendering and transfers, on demand communication tools, and processing data at high bandwidths for visual analysis. As pointed out in [15], each one of these applications has multiple network requirements, such as guaranteed bandwidth, traffic isolation, and time schedulers. Some of the limitations to providing fast failover at the IXP and performing traffic engineering in the inter-domain scenario derive from the BGP routing decision process, as presented in the following section.

The following are some of the limits that have been recognized based on the routing decision criteria of BGP that prevent fast reaction to data plane failure events and dynamic distribution of traffic between ASes at IXPs:

Single Next-Hop per Network Prefix: The BGP neighbor with the shortest AS path attribute is selected as the next-hop for forwarding all traffic to a given destination, discarding valid offers from other BGP neighbors.

Rigid Architecture: Widespread adoption of new routing features is difficult because contiguous BGP neighbors need to overcome implementation incompatibilities before agreeing on their deployment.

Slow Convergence: Border Gateway Protocol outages can take dozens of seconds to recover. Until the BGP Hold Time expires, a new route will not be selected for failover, and communication will not be resumed. This downtime impacts the performance of applications that rely on continuous connectivity.

Regarding the slow convergence of the BGP, the Hold Time parameter can be configured with low values. However, the problem with this is that if KEEPALIVE messages are buffered longer than the Hold Time, or if one of the BGP processes of the peering router cannot generate KEEPALIVE messages fast enough, then the BGP session is terminated. Moreover, BGP routers can protect themselves against aggressive KEEPALIVE timers from neighbors by refusing peering negotiation and never establishing a session [2].

2.4 Software-Defined Networking (SDN) in Inter-domain Scenarios

In network devices (e.g., routers and switches), the packet-forwarding task is performed by its data plane according to the rules and paths calculated by standard protocols running on its control plane. In a SDN architecture, the control plane of a network device is decoupled from its data, enabling programmability through an Application Programming Interface (API) such as OpenFlow or P4 [4,16].

OpenFlow is a communication protocol that enables a centralized SDN Controller to program the data path of a network device according to the decisions of applications and protocols running on the control plane or application layer. Among the features of the SDN architecture using OpenFlow, which is frequently used, is automatic isolation of traffic flows programming the match-action tables of the underlying network devices with OpenFlow rules.

The OpenFlow version 1.5 specification [21] supports the definition of multiple flow tables with rules that match flows to multiple parameters inside the packets header, such as the source and destination IP address, VLAN ID, and TCP or UDP port number. Inside the flow tables flow rules entries are sorted by priority, where the highest-priority rules are at the top of the flow table. Inbound packets are matched to the flow entries starting from the highest-priority rules. If there is a match, then the flow matching stops, and the set of actions for the matched flow entry are performed. Packets that do not match any flow entry are dropped or sent to the controller for further analysis.

Previous studies attempted to override the BGP route selection process in IXPs using SDN, provided fast failover times, and controlled how traffic flows between ASes. In this section, we present the more relevant proposals and some of their limitations.

Software-Defined IXP and iSDX. Gupta *et al.* [13] proposed a software-defined IXP (SDX) in which each participant AS interconnects to a shared data-plane and individually define forwarding policies relative to their current BGP routing table, enabling application-specific peering and fast failover.

In [12], how SDX can potentially fail at a large IXP as the compilation of BGP route updates scales between participant ASes is discussed. Policy compilation can take minutes as the number of participants exchanging traffic for hundreds of thousands of prefixes increases. In order to reduce the policy compilation time, Gupta *et al.* proposed a distributed model in which each participant AS runs locally an SDX controller that compiles routing policies. An IXP SDX controller collects the routing policies of all participants and updates the data plane accordingly.

Other Inter-domain SDN Approaches. A similar distributed approach to SDX was proposed by Wang *et al.* [27]. They introduced a framework for inter-domain path diversity based on SDN and the BGP. Each participating AS implements an enhanced SDN controller that exports and propagates to a peer AS SDN controller network functions that enable the setup of routing paths for particular applications. A richer criterion for the inter-domain route decision process is provided together with the BGP, as well as alternative routes to destinations through participant ASes.

Chen *et al.* [6] also extended SDN to an inter-domain network federation composed of several SDN autonomous domains. They proposed a network view exchange mechanism of routing criteria (e.g., Quality of Service attributes and application layer protocol numbers) across multiple SDN domains, enabling path diversity between ASes, and extending the BGP forwarding decision process. Based on the cited works in this section we recognize that the challenge of providing fast failover and application specific peering between ASes is only achievable only under the following assumptions:

- It is feasible for each participant AS to synchronously deploy the SDN technology locally and change their routing infrastructure.
- Autonomous systems with BGP routing implementation must be compatible with supplemental control-plane information and extensions, i.e., custom topology exchange algorithms.

2.5 Research Goal

The objective of this research is to decrease the packets dropped during an IXP data plane failure event. We attempt to decrease the disconnection time between the IXP participants to a value lower than one second, which is the minimum configurable BGP Hold Time. In order to pursue this goal, we propose a framework that reduces the failover process to the order of milliseconds. The proposed failover framework implements SDN technologies and stream processing concepts in an IXP. This failover mechanism detects in real-time data plane connectivity failure events that otherwise would have to wait until the BGP convergence process ends to finish. In addition, this mechanism provides the capability of reacting to BGP topology changes in order to avoid violations of inter-domain traffic forwarding between members of an IXP. Unlike previous studies, we look for a design that does not require the adoption of SDN technology in all participants of the IXP in order to not affect their throughputs. There are two main technical challenges to achieve the research goal:

Data Plane Failure Detection: Even though the BGP hold timer can be configured with a value of as low as one second, network device manufacturers recommend not configuring the hold timer below 30 s to avoid route flapping. In addition, in an IXP scenario, it is not feasible to demand that every participant configure a low BGP hold timer. Therefore, in order to achieve fast failover, it is not possible to rely on the BGP.

Correct Failover Path: This challenge derives from the previous challenge. If the BGP is not to be implemented in the failover process, it is difficult to guarantee that traffic redirection is to be done through an IXP participant that can reach the same routes that the compromised participant advertised. Lack of synchronization between the failover process and BGP route advertisements can lead to inter-domain policy violations.

Considering these technical challenges, the proposed failover framework should be designed to achieve a response time lower than the BGP Hold Time, and process BGP UPDATE messages to route traffic consistently between IXP participants.

3 Proposed Real-Time Failover Framework

The basic idea behind the proposal is to reduce the IXP failover time upon data plane failures by enhancing its architecture with SDN to execute the traffic redirection task as a control plane application. A data plane failure event is defined as the disconnection of the physical link between the border routers of the AS and the IXP data plane switch. The proposed framework adds software modules to the IXP in order to process these connectivity failure events in real time.

One of the most important advantages of adopting this framework is that it allows control plane applications to express their peering intents according to a set of rules or an algorithm, thus enabling control of the failover path. A hypothetical implementation of SDN in an IXP could be as follows. First, after a data plane failure event is detected, and an SDN controller is notified via a control channel. Then, an SDN application processes the event and proactively updates the data plane via the SDN controller northbound API. After the flow tables are updated, the failover process is completed and traffic continues crossing the IXP through an alternate AS. Even though this approach can failover to another path within a short time it overrides the next-hops calculated by the BGP. This can potentially violate BGP peering agreements and routing policies between IXP participants, so implementation becomes infeasible.

The main challenge in handling the failover process as an SDN application is that there is no direct channel or API with the IXP route server to learn valid alternative BGP neighbors to redirect traffic to. Network automation techniques allow applications to establish SSH sessions with a router, to collect and parse the output of commands that echo the Routing Information Base (RIB), and to store sessions in a database management system. However, collecting routing information in this fashion is prone to errors and lacks consistency in cases in which the result of the commands provide unexpected outputs or between routers operating systems. Moreover, BGP UPDATE messages that add or withdraw routes can arrive during the interval between SSH session establishment, making this approach ineffective in terms of correct packet forwarding to valid BGP neighbors. The failover process requires a mechanism that learns and considers valid BGP next hops from the IXP route server.

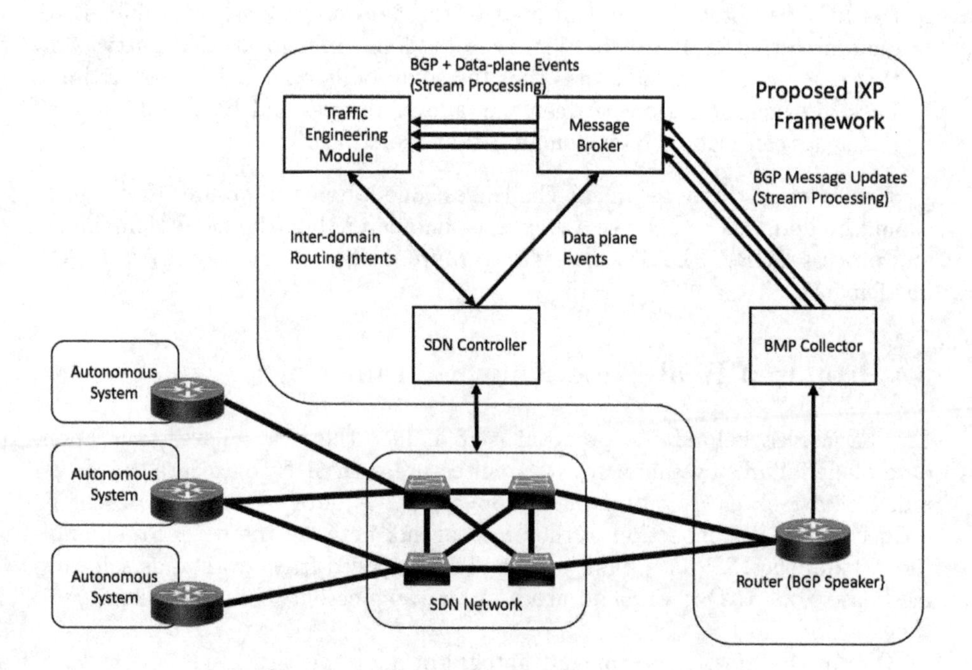

Fig. 1. Basic concept of the proposed framework.

We propose a real-time processing approach in which data plane failures and BGP updates are handled as a stream of events by a traffic engineering application. Figure 1 illustrates the concept of the proposed framework. Valid BGP alternatives for failover are learned from the IXP route server, bacause BGP UPDATE messages are handled as a stream of events and stored in a hash map data structure by the traffic engineering application. When a connectivity failure occurs in the IXP data plane, the event is detected by an SDN controller and published to a message broker system. Events are made available for the traffic engineering application to consume and make a failover decision.

3.1 Real-Time Failover Framework

Stream Processing and Messaging System. The failover processing time is proportional to the latency of the pipeline between event sources and processors. The proposed framework requires data plane failures and BGP route withdrawal events to be processed on the order of milliseconds. Because of this requirement a stream processing system is chosen due to its capability to process data continuously as it arrives at a system [17]. Two types of actors compose a stream processing system:

- Producers: write event records to the system datastore.
- Consumers: poll the system datastore periodically and read event records.

The consumer-polling task introduces overhead and delays into the system whenever records are written into the data store at a faster rate than they can be read. In order to overcome this drawback, a message broker server is implemented, and both producers and consumers connect to this server as clients. The message broker is an optimized database that stores producer events in memory and allows fast reads from connected consumers. Low latency and high bandwidth between the broker and its clients are recommended for implementations that require processing events at a fast rate [1].

Software-Defined Networking IXP. Implementing SDN and OpenFlow at an IXP enables external applications to manage the underlying switches forwarding tables. The BGP can run as an SDN application that synchronously configures the IXP data plane as BGP updates are received from participants. Apart from this, SDN features do not introduce any additional inter-domain benefits directly, but provide two additional capabilities that can be exploited for the failover process:

- A data plane failure detection mechanism based on port status notifications.
- Flow rules match inbound packets in priority order.

OpenFlow enables data plane switches to exchange messages with an SDN controller through an OpenFlow channel [21]. The channel is instantiated using Transport layer Security or a TCP connection and is only used to exchange control information (e.g., physical port status changes). As ports are added,

Table 1. OpenFlow Switch Port Status Messages.

Status	Description
OFPPR_ADD	A port was added
OFPPR_DELETE	A port was removed
OFPPR_MODIFY	An attribute of the port has changed

removed, or modified from data plane devices, notifications are sent to the SDN controller with an OFPT_PORT_STATUS message. Details describing the reason for this message are presented in Table 1. Whenever a loss of connectivity is detected, the OFPPS_LINK_DOWN flag of the status message is set to "true" in order to notify the SDN controller that there is no physical link present at the port.

Changes to the OFPT_PORT_STATUS message are useful in the proposed framework because such changes provide a relatively faster data plane failure detection mechanism than the BGP hold timer. OpenFlow switches send status messages asynchronously to the SDN controller without soliciting them.

In an IXP with OpenFlow switches, inbound packets header contents are matched to flow table entries. A flow table consists of entries defined by a match field and a priority. For the case in which there is a match for multiple flow entries, only the entry with the highest priority is considered [21]. For the proposed failover framework, once a data plane failure is detected it is necessary to install flow entries with a higher priority value than that installed by a BGP application. Following this, inbound traffic matching a BGP route that flows through a compromised BGP peer is redirected to another valid BGP next hop so that communication is not interrupted.

BGP UPDATE Message Ingestion. The proposed framework follows inter-domain traffic engineering guidelines by not redirecting flows to ASes that are not expecting to receive them (e.g., do not advertise routes to a compromised destination). Candidates for routing traffic between IXP peers should be valid according to their advertised BGP update messages. This principle is respected by the BGP Monitoring Protocol (BMP) [25]. The BMP allows a server to export an unfiltered BGP routing table from a router and to gather all possible routes that can be used to reach each destination.

The BMP is implemented by two entities, i.e., a BMP collector and a BMP client, that communicate over a TCP connection. Once the BMP session is opened, the BMP client will start dumping to the BMP collector Route Monitoring (RM) messages containing the Adjacent Routing Information Base Inbound (Adj-RIB-In). In addition, incremental updates to advertising or withdrawing routes are dumped. The peer session status is also dumped by a BMP client. Because of these features, we herein consider a BMP collector in order to guarantee that failover paths are valid according to the BGP. Moreover, whenever IXP participants add routes or stop advertising them, the proposed failover framework can automatically update the set of candidates considered for the traffic redirection task.

Traffic Engineering Module. As mentioned in Sect. 2, an IXP route server performs the best path selection process on behalf of all of its participants. In cases in which peering ASes implement route filters, these must be considered by the IXP as well. In order to facilitate both features, the proposed framework includes a Traffic Engineering module as an SDN application. Peering intents are pushed to the SDN controller through its exposed Northbound API. Route filters can be applied in the form of intents that explicitly specify the MAC address of the next hop for traffic being previously forwarded by an unreachable IXP participant. The Traffic Engineering module can also support custom traffic engineering algorithms that make decisions based on additional route attributes included in BGP update messages (e.g., LOCAL_PREF), or other criteria, maximally leveraging the capabilities of SDN technology.

3.2 Failover Mechanism Implementation

The message broker acts as a buffer between event producers and consumers and is implemented as a stream processing system that receives and makes available the following information with low latency:

- BGP update messages from the BMP collector.
- Communication failure between the IXP switch and participants.

In both operations, the consumer is the Traffic Engineering module. Upon receiving any routing update or host down events, the failover process is triggered. The message broker receives events related to connectivity failures, or BGP route updates. Depending on the case, the child node identifies the unreachable next-hop MAC address, or the IP address of an unreachable network. This information is used as input by the next child node, which selects an alternative next hop that can continue forwarding traffic. The final child node in the process installs flow rules through the Northbound API of the SDN controller using a flow priority value higher than that used by a BGP application.

Message Broker and Traffic Engineering Module. The Message Broker module implemented is Apache Kafka [17], which is chosen because of its capability as a stream processing platform for storing messages written by producers, and makes the messages available to consumer software components with low latency. Event records that arrive to the system are classified into topics and written into a database table residing on the memory of the Kafka broker. The Traffic Engineering module is an application implemented using the Kafka Java consumer API (Listing 1.1), which feeds from three topics, each one running as a separate thread:

- Data plane link failures.
- BGP Updates (new routes advertised or route withdrawals).
- BGP neighbor additions or deletions.

Listing 1.1. Traffic Engineering Module Main Thread

```
1 public class KafkaConsumerDemo {
2   public static void main(String[] args) {
3     boolean isAsync = args.length == 0 || !args[0].trim().
        equalsIgnoreCase("sync");
4     ConsumerHost thread1 = new ConsumerHost("pmacct.acct");
5     ConsumerMessage thread2 = new ConsumerMessage("pmacct.bmp_msglog")
        ;
6     HostDown thread3 = new HostDown("HOST");
7
8     thread1.start();
9     thread2.start();
10    thread3.start();
11  }
12 }
```

Each of these threads reads events and triggers the process for calculating a valid next hop for compromised traffic. The thread that reacts to data plane failures is shown in Listing 1.2. This thread starts executing by subscribing to the "HOST" Kafka topic, which is where the SDN controller commits records whenever it detects a link down in the data plane. The consumer poll method is an infinite loop that continuously polls the HOST topic to retrieve new commits at a frequency of 10 ms. Whenever a data plane failure event is committed to the HOST topic, a control loop is executed to perform the failover task in the following order:

- Identify the unreachable IXP participant MAC address using the ConsumerHost.get() method.
- Remove the compromised IXP participant from the Traffic Engineering module data structure using the alt.remove() method.
- Construct a POST message with a new peering intent and send it to the SDN controller through its REST API through the Postman.postman() method.

The desired method for the next-hop selection process is implemented on individual Java classes: one for pre-configured next hops, and another for a round robin algorithm that randomly selects a single next hop from among all valid options.

Listing 1.2. Data Plan Failure Reaction Thread

```
1 @override
2 public void doWork() {
3   consumer.subscribe(Collections.singletonList("HOST"));
4   ConsumerRecords<Integer, String> records = consumer.poll();
5   for (ConsumerRecord<Integer, String> record : records) {
6     if (!record.value().isEmpty()) {
7       alt = Consumerhost.get();
8       hostDown = (record.value().substring(record.value().length() -
            10)).substring(0, 8);
9       alt.remove(hostDown);
10    }
11  }
12  try {
13    Postman.postman();
14  } catch (IOException ioe) { ioe.printStackTrace(); }
15 }
```

For the failover process the Traffic Engineering module must identify the port of the unreachable IXP participant. To complete the task, the module must also select a valid participant. The reachability information of each IXP participant is stored in a hash map. The benefits of implementing this type of data structure are:

- Reachability information can be stored and retrieved in no particular order.
- Data can be stored based on a key-value pair: For this proposal the key-value pair consists of an IXP participant MAC address and the interface port number used by the IXP switch to reach the address.

Software-Defined Networking Controller. The Open Network Operating System (ONOS) is an SDN controller that implements OpenFlow as its primary southbound protocol [3]. Within ONOS features, there are two key functionalities required in an IXP environment that can be fulfilled:

- Physical layer failure notification to a message broker.
- Connection to external networks using BGP.

Section 3.2 introduced how the SDN controller detects data plane failure events. Next, we clarify how this type of event is handled for failover purposes. In order to share the failure events with other applications (e.g., the Traffic Engineering module), the SDN controller must expose an interface where the MAC address of an unreachable next hop can be propagated. It is desirable to transmit this information between software components with low latency. For this purpose, ONOS provides a Kafka Integration application [19] that serves as a data plane event producer. As soon as ONOS receives an OFPT_PORT_STATUS message with the OFPPS_LINK_DOWN flag set to "true", ONOS reacts by generating an event record containing the MAC and IP address of the now unreachable peer. These events are written into a Kafka topic, which is later consumed by the Traffic Engineering module.

In order to establish BGP sessions with other networks, ONOS provides the SDN-IP application [20]. When activated in an IXP, SDN-IP allows the ONOS controller to learn BGP routing information from the route server. The BGP routes are then translated by SDN-IP into OpenFlow rules that are installed on the IXP data plane.

Border Gateway Protocol Monitoring Protocol Collector. The implemented BMP collector is pmacct [23], an open-source telemetry tool that dumps BMP updates to a message broker like Kafka in JSON format. The IXP route server is configured as a BMP client that establishes a BMP session with pmacct. In the proposed framework it is of special interest when the IXP route server send pmacct the following messages:

- Peer Down notifications, which indicate that a BGP peering session was terminated.
- Route Monitoring messages, which provide a snapshot of the Adj-RIB-In of each monitored peer.

After pmacct receives both messages, the messages are published to the Kafka broker for the Traffic Engineering module to consume and make the necessary changes to the IXP data plane in order to maintain continuous communication between participants.

4 Evaluation

Three experiments have been conducted to evaluate the failover time and packet loss of the proposed real-time failover framework. The number of packets dropped and the processing time overhead of the framework are measured for two scenarios:

- PP: Failover to a pre-configured path in the Traffic Engineering module.
- RR: Failover to a path that is randomly selected by a round robin algorithm.

The third experiment is designed to estimate the number of packets forwarded incorrectly to an IXP participant after advertising a BGP route withdrawal.

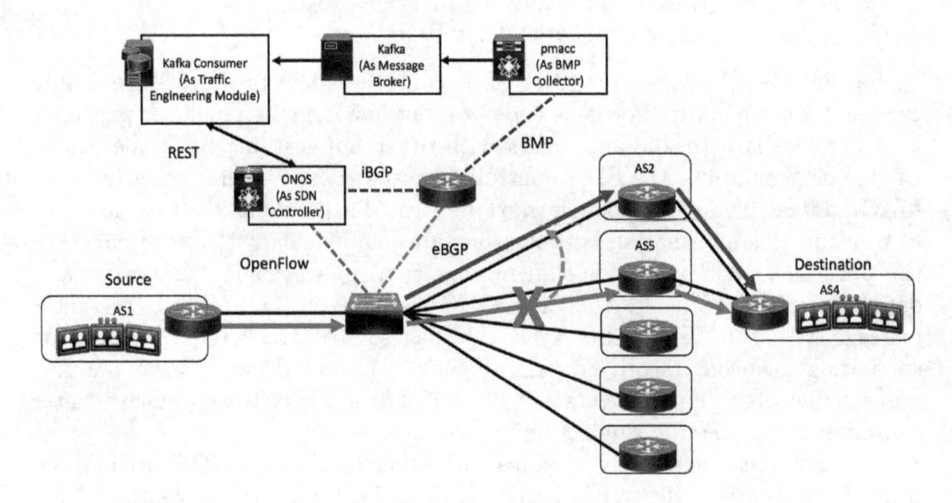

Fig. 2. Internet Exchange Points (IXP) testbed used for evaluation.

Figure 2 illustrates the proposed real-time failover framework testbed used for the experiments. Link failures are simulated between the IXP data-plane and the AS that is the next hop for traffic toward the destination host. The failover time is determined by measuring the processing time between the reception of a link failure event by the ONOS controller and the time at which the data plane intent is sent by the TE module.

Packets are captured in pcapng format using Wireshark [22]. These traces are used to perform live monitoring of the traffic flows between hosts before and after the failover process. In addition, Wireshark generates a timestamp of the time when packets are captured, which is later used to determine the number of packets dropped.

4.1 Experiment Environment

The IXP testbed has been emulated using GNS3 [11], which is a network emulator that supports the complete functionality of network devices operating system. The emulated routers run the Cisco Internetwork Operating System version 15.5 and are configured with 256 MB of RAM. The data-link layer is implemented using Ethernet, and network devices interfaces support a bandwidth of 10,000 Kbits/s, an MTU of 1,500 bytes, and a delay of 1,000 µs. The SDN IXP is responsible for translating BGP route entries to OpenFlow rules for each peer AS in the following form:

packets from peer A destined for network C → forward to peer domain B.

The SDN implementation includes the ONOS SDN controller and a single Open vSwitch. The SDN-IP and Kafka-integration applications are activated in the ONOS controller. Within the IXP, SDN-IP installs and updates the forwarding rules of the data-plane according to BGP route updates. The Kafka-integration application is also activated in the ONOS controller in order to generate connectivity failure events in real-time as a stream of records.

For evaluating the packets dropped during the failover process two virtual machines that exchange traffic are deployed in the testbed. As Fig. 2 shows, the source host is configured in the same network of AS1, which peers with the IXP route server. The destination host is configured in the network of AS4, which does not peer with the IXP directly. In this configuration, AS1 has multiple valid paths to reach AS4, but due to the BGP best-path selection algorithm, only one path is selected to forward the traffic between the two hosts. User Datagram Protocol traffic is generated using netsniff-ng [18] at four different rates, and the transmitted frames have a total length of 42 bytes.

For the first scenario, AS2 is pre-configured as the failover path and is selected as the next hop during the failover process. It is possible to implement an interface with the TE module that validates the pre-configured failover path against the hash map containing the list of valid BGP next hops. We have decided not to include such functionality because it will not impact the failover time or packets dropped during a data plane failure event. Moreover, if an OpenFlow rule is installed for forwarding traffic through a pre-configured next hop, and later a BGP update message withdraws the route, then the TE module detects the event and rollback to the route determined by the BGP best-path selection algorithm.

For the second scenario a round-robin function selects the best path from available options stored in a hash map. The data structure holds key value pairs corresponding to each IXP participant MAC address and port number at which the pairs connect to the IXP switch. For both cases, the TE module sends a new peering intent through the REST API exposed by the ONOS controller. A POST method is constructed with its content type defined as a JSON object that contains the MAC address, the port number of the next hop, and an alphanumeric tag that uniquely identifies the installed flow rule.

4.2 Experimental Results and Discussion

Failover simulations were performed five times for each transfer rate. The disconnection interval was registered by measuring the time difference between timestamps of data plane failures and peering intents sent by the TE module. The round-trip times between the SDN controller and the Open vSwitch in the data plane is not considered, because this time is on the order of microseconds. The round-trip time between the Open vSwitch and each IXP participant is also within the microseconds range, and so is discarded. Figure 3 shows traffic arriving at AS4 during one of the multiple experiments.

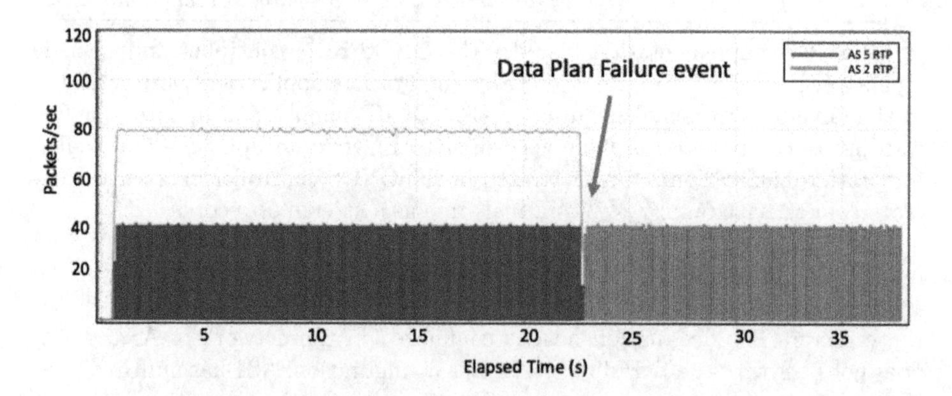

Fig. 3. Traffic forwarded through the IXP toward AS4.

Before the link failure, traffic toward AS4 is forwarded by AS5. When the link failure event is generated an interruption can be observed for a short interval of time until the proposed failover framework redirects the traffic to AS2 (PP case) or another IXP participant (RR case). Figure 4 illustrates the failover time for both pre-configured paths (PP) and round-robin (RR) selection. The results registered an average failover time of 31 ms for PP, and 617 ms for RR. Table 2 shows the average number of packets dropped at different transfer rates.

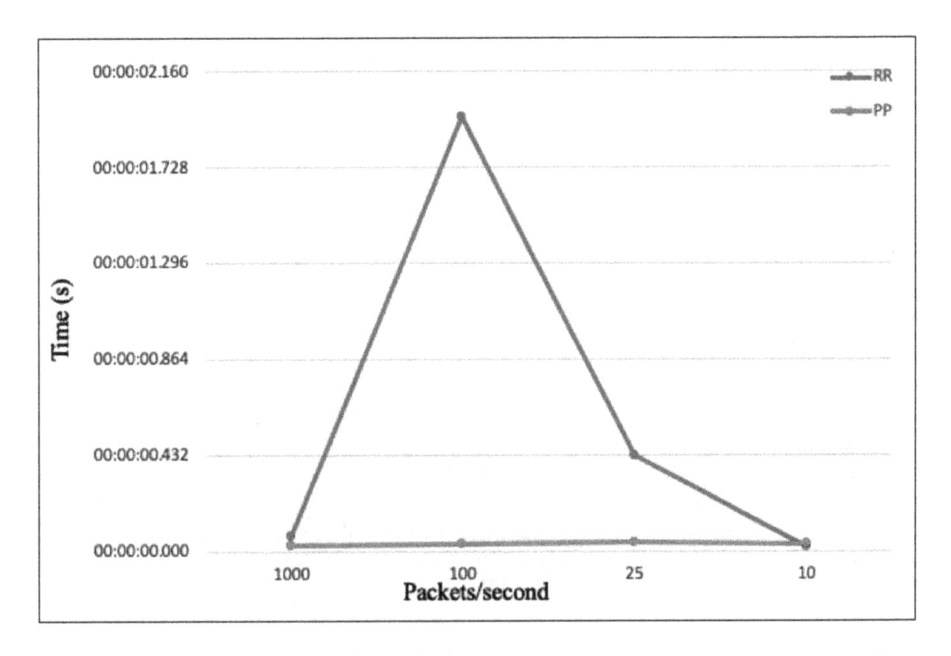

Fig. 4. Proposed framework failover time for pre-configured path and round-robin selection.

Table 2. Packets Dropped during Data Plane Failure Events.

Packets/s	Packets
10	0
25	1
100	2
1000	27

5 Conclusion

In the present study, we approach the challenge of reducing the failover time at an IXP by proposing a real-time failover framework that implements SDN and a stream processing system. Integrating SDN into the IXP architecture enables two features for improving the failover time provided by the BGP hold timer. The first feature is detection of connectivity failures between participants using OpenFlow, and the second feature is collection of the BGP routing tables of all the IXP participants through the BGP Monitoring Protocol (BMP). We evaluates the feasibility of the proposed framework by performing two experiments. These experiments investigate the failover time and packets dropped during data plane failure events. Traffic is exchanged between two hosts through an experimental IXP testbed, and a link failure event is simulated between the data plane switch and a participant AS. Although the proposed framework makes assumptions about the IXP underlying data plane infrastructure, we believe that its implementation in production has the potential to reduce the number of packets dropped between participants during data plane failures events, and the BGP Hold timer is configured with standard values (e.g., 90 s).

Acknowledgments. A part of this research was supported by the JSPS KAKENHI Grant Number JP18K11355.

References

1. Akidau, T., Chernyak, S., Lax, R.: Streaming Systems. O'reilly (2018)
2. Van Beijnum, I: BGP: Building Reliable Networks with the Border Gateway Protocol. O'reilly (2002
3. Berde, P., Gerola, M., Hart, J., Higuchi, Y., Kobayashi, M., Koide, T., Lantz, B., O'Connor, B., Radoslavov, P., Snow, W., Parulkar, G.: ONOS: towards an open, distributed SDN OS. In: Proceedings of the Third Workshop on Hot Topics in Software Defined Networking, HotSDN 2014, pp. 1–6, August 2014
4. Bosshart, P., Daly, D., Izzard, M., McKeown, N., Rexford, J., Talayco, D., Vahdat, A., Varghese, G., Walker, D.: P4: programming protocol-independent packet processors. ACM SIGCOMM Comput. Commun. Rev. **44**(3), 87–95 (2014)
5. Breen, J.: What happens when the science DMZ meets the commodity internet? In: The 2015 Internet2 Technology Exchange, October 2015
6. Chen, Z., Bi, J., Fu, Y., Wang, Y., Xu, A.: MLV: a multi-dimension routing information exchange mechanism for inter-domain SDN. In: 2015 IEEE 23rd International Conference on Network Protocols (ICNP), pp. 438–445, November 2015
7. Dart, E., Antypas, K., Bell, G., Bethel, E., Carlson, R., Dattoria, V., De, K., Foster, I., Helland, B., Hester, M., Klasky, S., Klimentov, A., Lehman, T., Livny, M., Metzger, J., Milner, R., Moreland, K., Nowell, L., Pelfrey, D., Pope, A., Prabhat, P., Ross, R., Rotman, L., Tierney, B., Wells, J., Wu, K., Wu, W., Yoo, B., Yu, D., Jason, Z.: Advanced Scientific Computing Research Network Requirements Review: Final Report 2015. Lawrence Berkeley National Laboratory Report, June 2016
8. Dart, E., Bell, G., Benton, D., Canon, S., Cowley, D., Dattoria, V., Fagnan, K., Foster, I., Giuntoli, P., Hester, M., Hettich, R., Hnilo, J., Howe, A., Jacob, R., Jacobsen, D., Love, L., Madupu, R., Metzger, J., Palinisamy, G., Rabinowicz, P., Rotman, L., Sears, R., Strand, G., Wehner, M., Williams, D., Zurawski, J.: Biological and Environmental Research Network Requirements Review 2015 - Final Report: Lawrence Berkeley National Laboratory Report, September 2015
9. Denning, P.J.: The ARPANET after Twenty Years. Computers under attack: intruders, worms, and viruses, pp. 11–19, February 1991
10. Feamster, N., Borkenhagen, J., Rexford, J.: Guidelines for interdomain traffic engineering. ACM SIGCOMM Comput. Commun. Rev. **33**(5), 19–30 (2003)
11. GNS3
12. Gupta, A., MacDavid, R., Birkner, R., Canini, M., Feamster, N., Rexford, J., Vanbever, L.: An industrial-scale software defined internet exchange point. In: 13th USENIX Symposium on Networked Systems Design and Implementation (NSDI 16), pp. 1–14. USENIX Association, March 2016

13. Gupta, A., Vanbever, L., Shahbaz, M., Sean Donovan, P., Brandon Schlinker, C., Feamster, N., Rexford, J., Scott Shenker, J., Russell Clark, J., Ethan Katz-Bassett, B.: SDX: a software defined internet exchange. ACM SIGCOMM Comput. Commun. Rev. **44**(4), 551–562 (2014)
14. Networking for Cloud, Internet2
15. Kiran, M., Pouyoul, E., Mercian, A., Tierney, B., Guok, C., Monga, I.: Enabling intent to configure scientific networks for high performance demands. Future Gener. Comput. Syst. **79**(1), 205–214 (2018)
16. McKeown, N., Anderson, T.E., Balakrishnan, H., Parulkar, G., Peterson, L.L., Rexford, J., Shenker, S.J., Turner, J.: "penFlow: enabling innovation in campus networks. ACM SIGCOMM Comput. Commun. Rev. **38**(2), 69–74 (2008)
17. Narkhede, N., Shapira, G., Palino, T.: Kafka: The Definitive Guide. O'reilly (2017)
18. netsniff-ng
19. Kafka Integration, ONOS Dashboard
20. SDN-IP Architecture, ONOS Dashboard
21. OpenFlow Switch Specification Version 1.5.1
22. PCAP Next Generation Dump File Format
23. pmacct project
24. Rekhter, Y., Li, T., Hares, S.: A Border Gateway Protocol 4 (BGP-4). RFC 4271, January 2016
25. Scudder, J., Fernando, R., Stuart, S.: BGP Monitoring Protocol (BMP). RFC 7854, January 2016
26. Silva, W.J.A., Sadok, D.F.H.: A survey on efforts to evolve the control plane of inter-domain routing. Information **9**(5), 125 (2018)
27. Wang, Y., Bi, J., Zhang, K.: A SDN-based framework for fine-grained inter-domain routing diversity. Mob. Netw. Appl. **22**, 906–917 (2017)
28. Xiao, X., Hannan, A., Bailey, B., Ni, L.M.: Traffic engineering with MPLS in the internet. IEEE Netw. **14**(2), 28–33 (2000)

Performance Improvement in NOMA User Rates and BER Using Multilevel Lattice Encoding and Multistage Decoding

Kwadwo Ntiamoah-Sarpong(✉), Parfait Ifede Tebe, and Guangjun Wen

University of Electronic Science and Technology of China, Chengdu 611731, Sichuan, China
kwadwons@std.uestc.edu.cn

Abstract. The encoding and decoding of construction D' lattices in a k-user downlink NOMA network are described. Conway and Sloane first described construction D' lattices obtained from LDPC codes. We extend the use of Construction D' lattices in a downlink NOMA. This paper considers the practical application of encoding and decoding these lattices and the improvement in spectral efficiency in NOMA networks. We evaluate the performance over the power-constrained AWGN channel, where near-optimum performance is demonstrated. The results from this study complement the existing body of knowledge; the relatively strong performance of lattices based on algebraic constructions provides excellent performance for many networks as well as the k-user NOMA network.

Keywords: Construction D' · Lattice codes · Non-orthogonal multiple access · 5G · Multi-level coding · Multi-stage decoding · Interference

1 Introduction

Non-orthogonal multiple access is considered as one of the key technologies to improve spectral efficiency and enable massive connectivity for 5G networks [1]. Through superposition coding at the transmitter, NOMA assigns the available transmission power to numerous users at different power levels within the same frequency band. However, the use of superposition coding results in significant inter-user interference at the receivers. To this end, NOMA employs successive interference cancellation (SIC) at the receiver to decode each user signal. Interference cancellation can be imperfect in practice and result in performance degradation due to channel estimation errors. Various solutions have been offered, including user clustering and multiple antenna NOMA, to minimize the interference [2]. Though NOMA's practical applications in wireless networks are comparatively new, related philosophies in information theory have been studied for quite some time.

Techniques proposed by academia and industry to improve NOMA performance include various multiple access methods [3], resource allocation, and user pairing strategies [4], different cooperative strategies to improve coverage and performance [5], to minimize interference [6] and the references within. However, in most of these works, it

K. Arai (Ed.): FICC 2021, AISC 1363, pp. 388–400, 2021.
https://doi.org/10.1007/978-3-030-73100-7_28

is not specified which precoding scheme is used. In [7, 8], the authors proposed a downlink non-orthogonal multiuser superposition transmission scheme. In this scheme, the base station transmits multi-level lattice codes to multiple users. It uses successive interference cancellation at the receiver to extract individual user messages. They, however, did not indicate how the lattice was to be constructed.

Precoding is considered an attractive technique to mitigate transmission over channels that suffer from interference and additive noise [9]. Algebraic lattice codes have been proven to achieve reliable communication with realistic encoding and decoding complexity [10]. However, accomplishing efficient decoding with excellent performance based on algebraic codes has not been comprehensively addressed in the literature and, particularly, NOMA networks and continues to be an essential area of study. In most recent reviews, the underlying lattices are constructed from A, D and D′ [11, 12]. These lattice constructions have proven beneficial and result in efficient encoding and decoding algorithms [13]. Construction D′ is a technique to produce lattices with high coding gains using sets of binary linear codes [14]. Construction D′ uses multiple binary codes, each with distinct scaling to form Euclidean-space codes [15, 16]. It has recently been used to form polar codes [13, 17]. The complexity of coding using construction D′ is significantly simplified to coding/decoding of a linear code over the AWGN channel over the finite field Z_p.

In [8], authors investigate and propose a new downlink NOMA scheme without SIC over slow fading channels where encoded users' signals correspond to algebraic lattice constellation over finite fields using single user decoding (without SIC). The scheme based on lattice partitions uses an arbitrary power allocation coefficient to determine the minimum product distance and derives its upper limit. In [13], authors propose encoding and decoding algorithms for low-complexity multi-level LDPC lattices using Construction D′. Results indicate that the performance under multi-stage decoding on the power-constrained channel is comparable to polar lattice codes.

This paper's motivation is to use construction D' lattices formed by LDPC codes to improve on user rates in a NOMA network. Different from [8] in which single user decoding is employed, in this contribution, we focus on practical aspects of precoding for NOMA k-users and multi-stage decoding (with SIC) at the receiver to achieve high spectral efficiency and reduce interference for the weak user(s). We thereby construct lattice codes using construction D′ and apply them to a k-user NOMA downlink network. A hallmark of our approach, similar to [18], is the insertion of the "dither" at the transmitter. The dither is shared with the receiver (common randomness).

To meet the power-constraint requirements, we apply hypercube shaping to achieve the Poityrev limit under multi-stage decoding. We show that for channel gains that are within 0.38 dB from the very strong interference regime, the weakest user can be decoded at signal-to-noise ratios within 1.51 dB from the Shannon limit. Construct D lattice codes can also be applied to compute-and-forward in cooperative relay NOMA networks. Multi-stage decoding considerably reduces decoding complexity because component decoders can be used at the user destinations.

Throughout the remainder of this paper, we express vectors in lower case boldface and matrices in uppercase boldface, respectively. This paper is further organized as follows. Section 2 presents the proposed system mode and encoding at the transmitter using construction D lattices. The end-to-end capacity using the proposed decoding with MMSE approximation is presented in Sect. 3. Simulation results demonstrating their excellent error-rate performance with practical decoding complexity is presented in Sect. 4. Section 5 concludes the paper and points out future work.

2 Multi-level Encoding and Transmission Scheme

Consider the NOMA downlink transmission scheme shown in Fig. 1, where a transmitter or base station *Bs* intends to send information to *k* users sharing the same frequency band.

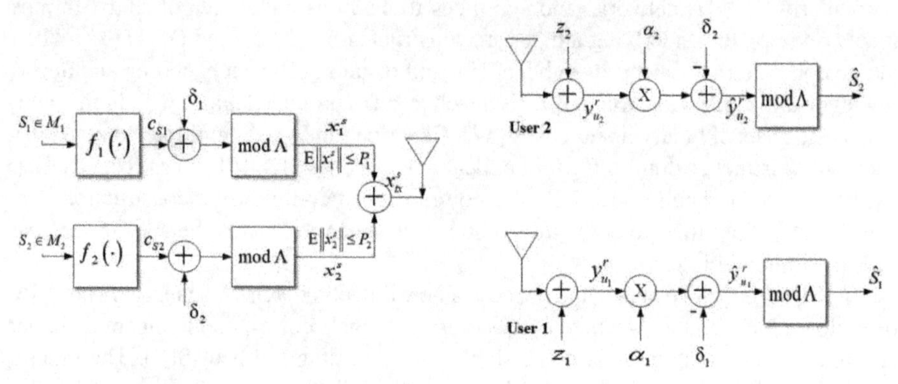

Fig. 1. Multi-level lattice encoding with multi-stage decoding scheme for downlink transmission with $k = 2$ users

For simplicity, we consider the case of a two-user NOMA, user U_1 is regarded as a robust user with excellent channel quality, and the user U_2 a weak user with poor channel quality. Consider that U_1 and U_2 are paired to perform NOMA among k users in a cluster. The channel coefficient between *Bs* and U_1 is denoted by h_{u_1} and between *Bs* and U_2 by h_{u_2}. We also assume that the *Bs* is under an input power constraint $\frac{1}{n}\mathrm{E}\left\|x_i^s\right\|^2 \leq P_{tx}$, $i = 1, 2$. We can thus transform the power-constrained channel to a modulo-lattice one, as we describe later.

Recall that Λ is a construct D′ dimensional lattice, let $\mathcal{V}_0$ be some fundamental region of Λ. Let the codewords $c_{S_1} \in \mathcal{V}_0$ and $c_{S2} \in \mathcal{V}_0$ be the information-bearing signal of the two users U_1 and U_2. Further, let δ_1 and δ_2, be independent dithers uniformly distributed over $\mathcal{V}_0$. The dithers are used to provide randomness to a lattice code to guarantee the desired power level and ensure that the transmitted signal is independent of messages S_2 and S_2. The codebook c_{si} is shared between the encoder and the decoder as well. The transmitter broadcasts to the users the signal with interference as

$$x_{tx}^s = \sqrt{\rho_1 P_{tx}} x_1^s + \sqrt{\rho_2 P_{tx}} x_2^s \tag{1}$$

Where ρ_1 and ρ_2 are the power allocation coefficients; $\rho_2 > \rho_1$ and $\rho_1^2 + \rho_2^2 = 1$.

At the user destinations, the interference is removed from the received data symbols. A modulo reduction is applied to the difference signal to generate the channel symbols x_i^s.

3 Construction D′ Preprocessing

For our coding purpose, we develop a generator matrix M out of the parity check matrix H_ρ for Construction D′ codes. We achieve this using modulo-2 arithmetic in the Gaussian elimination process. Generated once for a parity check matrix, M can be used for encoding all messages. The $1 \times k$ code vector c_ω is initially split into

$$C_\omega = [b : q] \tag{2}$$

where b is the $(n-k) \times 1$ parity vector and q is the message vector of size $(k \times 1)$. accordingly H_ρ is divided as

$$H_\rho^T = \begin{bmatrix} H_\phi \\ H_\theta \end{bmatrix} = 0 \tag{3}$$

where H_ϕ is a square matrix of size $(k-h) \times (k-h)$ and H_θ is a rectangular matrix of size $h \times (k-h)$. Assuming the constraint $C_\omega H_\rho^T = 0$, we write

$$bH_\phi + qH_\theta = 0 \tag{4}$$

where $b = qK$, and K is the coefficient matrix. For any non-zero q, K satisfies the relation

$$KH_\phi + H_\theta = 0 \tag{5}$$

$$K = H_\theta H_\phi^{-1} \tag{6}$$

The generator matrix M of Construction D codes is defined by

$$M = [K : I_j] = \left[H_\theta H_\phi^{-1} : I_j\right] \tag{7}$$

where I_j is a $j \times j$ unit matrix. The codeword is generated if $C = qM$

We outline the steps for efficient LDPC codes as follows:

Step 1; the H_p matrix is brought into the form

$$H_\rho^t = \begin{bmatrix} Q & R & Y \\ S & T & Z \end{bmatrix} \tag{8}$$

with a minimal gap g. where the matrix Q is () $(j-\acute{g})\times(k-j)$, R is $(j-\acute{g})\times\acute{g}$ matrix, Y is $(j-\acute{g})\times(j-\acute{g})$ matrix, S is $\acute{g}\times(k-j)$ matrix, T is $\acute{g}\times\acute{g}$ matrix and Z is $\acute{g}\times(j-\acute{g})$ matrix as shown in Fig. 2.

Step 2; pre-multiply H_ρ^t by $\begin{bmatrix} I_{j-\acute{g}} & 0 \\ -ZY^{-1} & I_0 \end{bmatrix}$

$$\begin{bmatrix} I_{j-\acute{g}} & 0 \\ -ZY^{-1} & I_0 \end{bmatrix}\begin{bmatrix} Q & R & Y \\ S & T & Z \end{bmatrix} = \begin{bmatrix} Q & R & Y \\ -ZY^{-1}Q+Y & -ZY^{-1}R+Y & 0 \end{bmatrix} \tag{9}$$

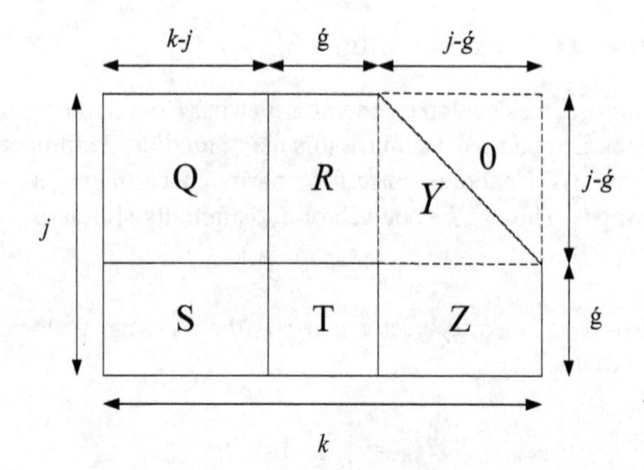

Fig. 2. H_{ρ}^{t} in proper lower triangular matrix form

Step 3; we obtain p_1 using

$$p_1^T = -\varphi^{-1}\left(-ZY^{-1}Q + S\right)\varsigma^T \tag{10}$$

where and ς is the message function.
Step 4; Let $\varphi = -ZY^{-1}R + T$ obtain p_2 with

$$p_2^T = -Y^{-1}\left(Q\varsigma^T + Rp_1^T\right) \tag{11}$$

Step 5; we construct the code vector c_v as

$$c_v = \left[\varsigma\; p_1\; p_2\right] \tag{12}$$

where p_1 and p_2 contains the first g parity and remaining parity bits, respectively.
From Fig. 1, the transmitter computes;

$$x_i^s = [c_{si} + \delta_i] \bmod \Lambda, \mathrm{i} = 1, 2 \tag{13}$$

It sends to the user destinations the signal

$$x_{tx}^s = x_1^s + x_2^s \tag{14}$$

where x_{tx}^s satisfies the power constraint,

$$\frac{1}{n}\mathrm{E}\left\|x_{tx}^s\right\|^2 \leq P_{tx} \tag{15}$$

3.1 Reception and Multi-level Decoding

The received signal at the user destinations can be generally expressed as

$$y_{u_i}^{(r)} = h_{u_i}x_{tx}^{(s)} + z_{u_i} \tag{16}$$

where z_{u_i}, is the additive white Gaussian channel noise experienced at the user U_i, which has zero mean and variance of σ_0^2 per dimension.

Thus, the signal received at user U_1 is expressed as,

$$y_{u1}^{(rx)} = h_{u_1} x_{tx}^{(s)} + z_{u1} \tag{17}$$

similarly, at U_2

$$y_{u2}^{(rx)} = h_{u_2} x_{tx}^{(s)} + z_{u2} \tag{18}$$

Upon reception of $y_{u_i}^{rx}$, $i = 1, 2$. The user U_i computes the signal $\hat{y}_{u_i}^{rx} = (\alpha_i y_{u_i}^{rx} - \delta_i) \bmod \Lambda$. Using p_1 and p_2 and manipulation of algebra, we obtain

$$\hat{y}_{u_1}^{rx} = \left(c_{s_1} + \alpha_1 z_{u_1} - (1 - \alpha_1) x_1^s\right) \bmod \Lambda \tag{19}$$

and

$$\hat{y}_{u_2}^{rx} = \left(c_{s_2} + \alpha_2 \left(z_{u_2} + x_1^s\right) - (1 - \alpha_2) x_2^s\right) \bmod \Lambda \tag{20}$$

Thus, the weak user U_2 expects the effective channel noise

$$z_{u_2}^{eff} = \left[\alpha_2 \left(z_{u_2} + x_1^s\right) - (1 - \alpha_2) x_2^s\right] \bmod \Lambda \tag{21}$$

and the robust user U_1

$$z_{u_1}^{eff} = \left[\alpha_1 z_{u_1} - (1 - \alpha_1) x_1^s\right] \bmod \Lambda \tag{22}$$

The MMSE coefficient for U_1 is expressed as

$$\alpha_1 = \frac{P_{rx} h_1^2}{P_{rx} h_1^2 + \sigma_{u_1}^2} \tag{23}$$

having a variance of

$$\frac{1}{n} E \left\| z_{u_1}^{eff} \right\|^2 = \frac{P_{rx} \sigma_{u_1}^2}{P_{rx} h_1^2 + \sigma_{u_1}^2} \tag{24}$$

and MMSE coefficient for U_2 is expressed as

$$\alpha_2 = \frac{P_{rx} h_{u_2}}{P_{rx} h_{u_2} + \sigma_2^2} \tag{25}$$

with a variance of

$$\frac{1}{n} \mathrm{E} \left\| z_{u_2}^{eff} \right\|^2 = \frac{P_{rx} \sigma_{u_2}^2}{P_{rx} h_2^2 + \sigma_{u_2}^2} \tag{26}$$

It can be seen that $\hat{y}_{u_1}^{rx}$ and $\hat{y}_{u_2}^{rx}$ has transformed into two modulo lattice AWGN channels with channel noise $z_{u_1}^{eff}$ and $z_{u_2}^{eff}$ respectively. It can be shown that when modulo reduction is with respect to a lattice Λ and a channel noise z is i.i.d Gaussian, then capacity in bits per g dimension can be written as

$$C_{\Lambda} = \frac{1}{n}\left(log_2(V) - h(z)\right) \tag{27}$$

here $h(\cdot)$ represents differential entropy. The maximally achievable rates are obtained by maximizing the expressions over α_1 and α_2 respectively. The corresponding achievable rate region

$$R_A = \bigcup_{0\le\gamma\le1} \left\{ \begin{array}{l} \left(\hat{R}_1, \hat{R}_2\right) : \hat{R}_1 \le \max\limits_{\alpha_1\in[0,1]} \frac{1}{n}\left(log_2\left(V(\Lambda)\right) - h\left(z_{u_1}^{eff}(\alpha_1, \gamma)\right)\right) \\ \hat{R}_2 \le \max\limits_{\alpha_1\in[0,1]} \frac{1}{n}\left(log_2\left(V(\Lambda)\right) - h\left(z_{u_2}^{eff}(\alpha_2, \gamma)\right)\right) \end{array} \right\} \tag{28}$$

With proper decoding decisions, we obtain the probabilities of decoding errors as:

$$P_r\left(\hat{c}_{s_1} \neq c_{s_1}\right) = P_r\left(z_{u_1} \bmod \Lambda \notin \mathcal{V}_0\right) \le P_r\left(z_{u_1} \notin \mathcal{V}_0\right) \tag{29}$$

$$P_r\left(\hat{c}_{s_2} \neq c_{s_2}\right) = P_r\left(z_{u_2} \bmod \Lambda \notin \mathcal{V}_0\right) \le P_r\left(z_{u_2} \notin \mathcal{V}_0\right) \tag{30}$$

Applying Theorem 3 in [19], the error probabilities vanishes as $n \to \infty$ if

$$\sigma^2(\Lambda) > \frac{1}{n} E\left\|z_{u_1}\right\|^2 \tag{31}$$

The rate of the end-to-end transmission for user U_2 is

$$R_{u2}^{s} < \frac{1}{2}\left(1 + \frac{P_{tx} h_2^2}{\sigma_{u2}^2}\right) \tag{32}$$

We observe that the effective noise consists of the noise and portions of the transmitted signal. Since x_{tx} is statistically independent of the information c_{s_i} to be decoded at the user destination, it does not result in a bias; portions of the desired signal are mistakenly allocated as noise [20]. It is significant to note that the effective noise ceases to be Gaussian and identically distributed. However, it is obtained by the convolution of an n-dimensional Gaussian pdf uniform over the Voronoi region. Therefore, channel detection based upon squared Euclidean distances ceases to be optimal. However, as n approaches infinity and its second moment approaches $1/2\pi e$, the low dimension projections tend to Gaussian pdf [21].

Proof

We exploit "good" nested codes according to that proposed in [22]. We assume that for $\varepsilon > 0$, there exist nested lattices $\Lambda_s \subseteq \Lambda_c$, such that the following conditions hold:

i. That the coarse lattice Λ_s is good for shaping (with the normalized second moment of $(2\pi e)^{-1} + \varepsilon$ and its Voronoi cell, $\mathcal{V}_0$ has second moment $\sigma^2(\Lambda) = P_{tx}$

ii. That the fine lattice Λ_c is "good" for channel coding (resilient to Gaussian noise of variance $PN/P+N$ while its fundamental cell has volume

$$V_0 \leq \left[\left(2\pi e \frac{PN}{P+N}\right)+\varepsilon\right]^{n/2} \tag{33}$$

According to Theorem 3 in [22], there exists for the AWGN channel a sequence of nested lattice codes that can achieve the following rate under multi-stage decoding

$$R_A = \frac{1}{2}\log\left(1+\frac{P_i}{N}\right)-\frac{1}{2}\log(2\pi e G(\Lambda_c))+\frac{1}{N}\mathcal{D}\left(z_{eff}\,\|z_{eff}^{+}\right) \tag{34}$$

here $\mathcal{D}(.\|.)$ is the *Kullback-Leibler* information divergence [23].

Proof: Lemma 4, [22], establishes that there exists a sequence of fine lattices Λ_a whose error probability can be made arbitrarily small under multi-stage decoding as $N \rightarrow \infty$ if

$$V(\Lambda)^{\frac{2}{N}} > 2\pi e\sigma_{eff}^2 \cdot 2^{-\frac{2}{N}\mathcal{D}\left(z_{eff}\,\|z_{eff}^{+}\right)} \tag{35}$$

There exists a series of proposed nested lattice codes with n-cube shaping that can achieve the rate above per real dimension is given by [22] as

$$\begin{aligned} R_{sc} &= \tfrac{1}{N}\log\left(\tfrac{V(\Lambda_s)}{V(\Lambda_c)}\right) \\ &= \tfrac{1}{N}\log(V(\Lambda_s)) - \tfrac{1}{N}\log(Vol(\Lambda_c)) \end{aligned} \tag{36}$$

$$= \frac{P}{G(\Lambda_c)} - \frac{1}{2}\log 2\pi e\sigma_{eff}^2 \cdot 2^{-\frac{2}{N}\mathcal{D}\left(z_{eff}\,\|z_{eff}^{+}\right)},\ (N \rightarrow \infty) \tag{37}$$

$$= \frac{1}{2}\log\left(1+\frac{P_i}{N}\right)-\frac{1}{2}\log(2\pi e G(\Lambda_c))+\frac{1}{N}\mathcal{D}\left(Z_{eff}\,\|z_{eff}^{+}\right). \tag{38}$$

We say that if Λ_s is good for MMSE quantization, then $G(\Lambda_s) \rightarrow (2\pi e)^{-1} = 0.0585498$ and

$$\frac{1}{N}D\left(z_{eff}\,\|z_{eff}^{+}\right) \rightarrow 0, \tag{39}$$

According to [22] if the coarse lattice $\Lambda_s = \gamma z^N$ is hypercube shaping, and $G(\Lambda_s) = 1/12$, we obtain the achievable rate as

$$R_A = \frac{1}{2}\log\left(1+\frac{P_i}{N}\right)-\frac{1}{2}\log\left(\frac{\pi e}{6}\right)+\frac{1}{N}D\left(z_{eff}\,\|z_{eff}^{+}\right) \tag{40}$$

substituting (39) into (40), we obtain the achievable rate as

$$R_A = \frac{1}{2}\log\left(1+\frac{P_i}{N}\right)-\frac{1}{2}\log\left(\frac{\pi e}{6}\right) \tag{41}$$

in the limit as $N \rightarrow \infty$. Here, $P_i \geq 0, P_i \neq 1$ are the constraints foisted upon the network. The actual transmit powers can be modified by the scalars ρ_1, ρ_2.

4 Results and Discussions

We present an analysis of the performance of the proposed encoding and decoding scheme for NOMA, using the system model in Fig. 1. The channel coefficients of the link between the transmitter and the users are modeled with i.i.d zero-mean Gaussian distribution whose variance is selected based on the strength of the corresponding link. More precisely, the channel coefficient for the link between the Bs and the users is modeled with a zero-mean Gaussian distribution with variance $\sigma_{u_i}^2$. Additionally we assume all terminals can estimate with reasonable accuracy the values of the channel coefficients at that time. We assume a two-user scenario in our analysis for simplicity in the discussion. It can be extended to a multiuser situation. All units of rates used are in bits/s/Hz. Monte Carlo simulations are performed to validate the analytical results. Figure 3 shows the achievable rate versus SNR for the weak user U_2. Note an excellent agreement between the analytical results and simulation is observed. It can be observed that for channel gains that are within 0.38 dB from the very strong interference regime, the weakest user can be decoded at signal-to-noise ratios within 1.51 dB from the Shannon limit. Also as the dimensionality of n approaches infinity, the Shannon capacity can be attained.

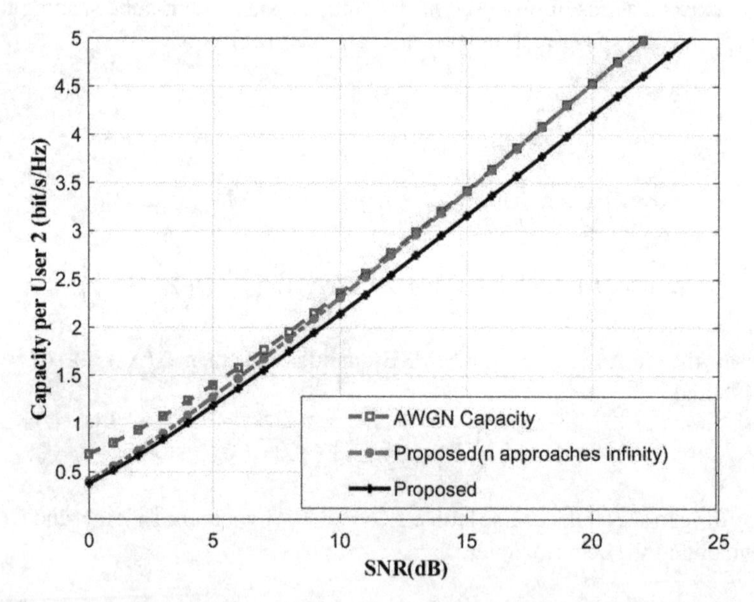

Fig. 3. Comparison between achievable capacity for proposed scheme and Shannon's capacity

Figure 4 depicts the achievable rates for the robust user U_1 and the weak user U_2 under different SNRs. We consider the case where the two users are the same same distances for the *Bs* with the same SNR 1 = SNR 2 = 20 dB. We also consider when the two users are at different distances from the *Bs*, with U_2 being the most distant (weakest user) and U_1 being the nearest (robust user) for SNR 1 = 20 dB and SNR 2 = 0.5 dB. The SNR for U_1 is maintained at 20 dB for high channel gain, while that of U_2 is varied from 20 dB to 0.5 dB indicating decreasing channel gain. It can be seen that higher rates is achieved for the NOMA weak user when there is a higher channel quality difference between the two (SNR 1 = 20 dB and SNR 2 = 2 dB), compared to when they are almost the same (SNR 1 = SNR 2 = 20 dB). This supports the existing knowledge that in NOMA for achievable high rates, user pairing is more advantageous when the two users have opposite channel qualities.

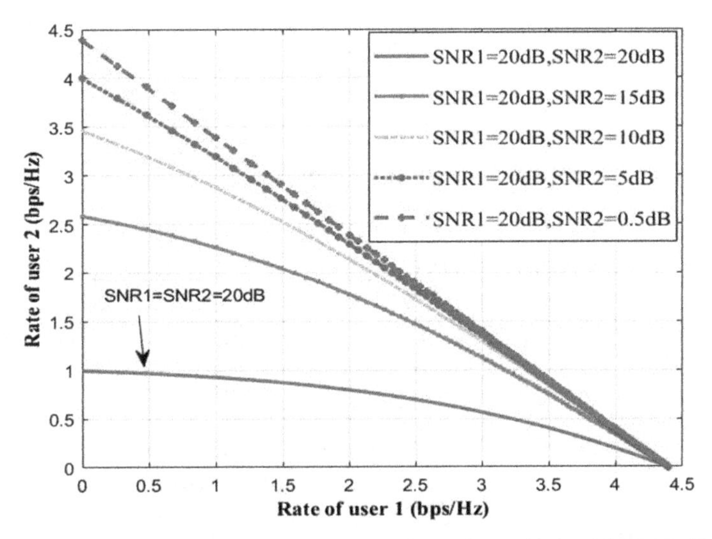

Fig. 4. Achievable rates for User 2 and User 1 for SNR1 = 20 dB, 15 dB,10 dB,5 dB

Figure 5 shows the achievable rates for the proposed NOMA scheme and conventional NOMA for SNR 1 = 20 dB and SNR 2 = 10 dB for users U_1 and U_2 respectively. It can be seen that the proposed scheme achieves a higher rate than conventional NOMA under the same channel conditions and the same power sharing coefficients. Despite the high rates achieved by the proposed scheme compared to conventional NOMA, as shown in Fig. 5, the proposed scheme imposes a higher decoding and encoding complexity than conventional NOMA.

We analyze the symbol error rate (SER) performance against the average SNR in Fig. 6 using QPSK modulation. For a power allocation factor of 0.82 for weak user U_2 and 0.18 for robust user U_1. With higher power allocation coefficient, the weak user is also able to achieve good error rates although it has poor channel quality compared to the robust user. This ensures user fairness and a higher throughput for NOMA networks. Though the SER performance analysis in Fig. 6, applies to quadrature phase shift keying (QPSK), high-order modulation methods such as quadrature amplitude modulation (QAM) or M-ary phase shift keying (M-PSK) can be similarly be applied.

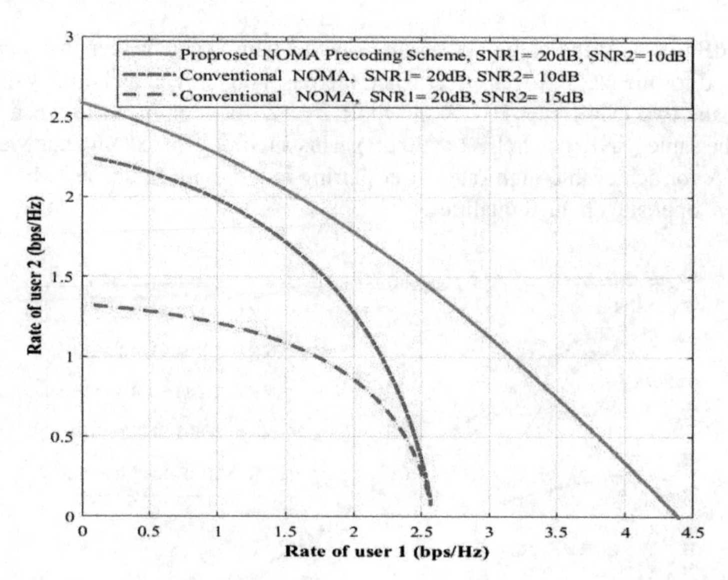

Fig. 5. Achievable rates for User 2 and User 1 for proposed NOMA precoding scheme and conventional NOMA

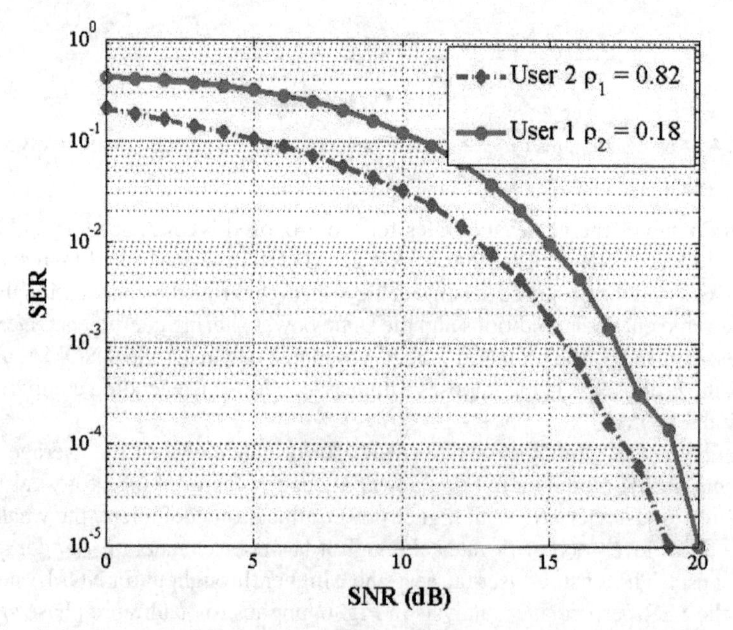

Fig. 6. Symbol error rate performance of proposed transmission scheme using QPSK for k = 2 users. Power allocation for weak and robust user are $\rho_2 = 0.82$ and $\rho_1 = 0.18$

5 Conclusion

This paper examined the performance improvement on the application of Construction D' lattices formed by LPDC codes on user rates in a k-user NOMA downlink system. We focus on practical a Construction D' lattice for precoding at the transmitter and multi-stage decoding at the receiver side to achieve high spectral efficiency for the weak user(s) and reduce inter-user interference. A hallmark of our approach is the insert of "dither" at the transmitter and the application of an MMSE estimator at the receiver to transform the power constraint AWGN channel into a modulo-lattice one. We show that for channel gains within 0.38 dB from the very strong interference regime, the weakest user can be decoded at signal-to-noise ratios within 1.51 dB from the Shannon limit. Additionally, simulation results show that the proposed scheme significantly exceeds the conventional NOMA scheme's performance, albeit with a higher encoding and decoding complexity. In future work, we will consider the use of our proposed scheme to cooperative relay NOMA network.

Acknowledgment. This work was supported in part by the National Natural Science Foundation of China Project contracts NO. 61701082, NO. 61701116, NO. 61601093, NO. 61971113 & NO. 61901095, in part by the National Key Research and Development Program under Project contracts N0. 2018YFB1802102 & 2018AA0103203, in part by Guangdong Provincial Research and Development Plan in Key areas under Project contracts N0. 2019B010141001 & NO. 2019B010142001, in part by Sichuan Provincial Science and Technology Planning under Project contracts N0. 2019YFG0418, NO. 2019YFG0120 & NO. 2020YFG0039, in part by the Ministry of Education - China Mobile Fund project under Project contract N0. MCM20180104, in part by Yibin Science and Technology Program under Project contracts N0. 2018 ZSF001 & NO. 2019GY001, and in part by the Fundamental Research Funds for the Central Universities under Project contract N0. ZYGX2019Z022.

References

1. Benjebbour, A., Li, A., Saito, K., Saito, Y., Kishiyama, Y., Nakamura, T.: NOMA: From concept to standardization. In: 2015 IEEE Conference on Standards for Communications and Networking, CSCN 2015, pp. 18–23 (2016)
2. Ding, Z., Lei, X., Karagiannidis, G.K., Schober, R., Yuan, J., Bhargava, V.K.: A survey on non-orthogonal multiple access for 5G networks: research challenges and future trends. IEEE J. Sel. Areas Commun. **35**(10), 2181–2195 (2017)
3. Vaezi, M., Ding, Z., Vincent Poor, H.: Multiple access techniques for 5G wireless networks and beyond (2018)
4. Islam, S.M.R., Zeng, M., Dobre, O.A., Kwak, K.S.: 'Resource allocation for downlink NOMA systems: Key techniques and open issues' IEEE Wireless Communications, 2018, 25, (2), pp. 40–47
5. Yang, Z., Ding, Z., Wu, Y., Fan, P.: 'Cooperative NOMA' IEEE Trans. on Vehicular Technology, 2017, 66, (11), pp. 10114–10123
6. Nazer, B., Gastpar, M.: 'Compute-and-forward: A novel strategy for cooperative networks'Conference Record - Asilomar Conference on Signals, Systems and Computers, 2008, pp. 69–73

7. Fang, D., Huang, Y.C., Ding, Z., Geraci, G., Shieh, S.L., Claussen, H.: 'Lattice partition multiple access: A new method of downlink non-orthogonal multiuser transmissions. In: 2016 IEEE Global Communications Conference, GLOBECOM 2016 – Proceedings (2016)
8. Qiu, M., Huang, Y.C., Yuan, J., Wang, C.L.: 'Lattice-partition-based downlink non-orthogonal multiple access without SIC for slow fading channels' IEEE Transactions on Communications, 2019, 67, (2), pp. 1166–1181
9. Fischer, R.F.H.: 'The modulo-lattice channel: The key feature in precoding schemes' AEU - International Journal of Electronics and Communications, 2005, 59, (4), pp. 244–253
10. Campello, A., Dadush, D., Ling, C.: 'AWGN-Goodness is enough: Capacity-achieving lattice codes based on dithered probabilistic shaping' IEEE Transactions on Information Theory, 2019, 65, (3), pp. 1961–1971
11. Kositwattanarerk, W., Oggier, F.: Connections between construction D and related constructions of lattices. Des. Codes Cryptograp. **73**(2), 441–455 (2014)
12. Strey, E., Costa, S.I.R.: Lattices from codes over Zq: generalization of constructions D, D′ and D¯. Des. Codes Cryptograp. **85**(1), 77–95 (2017)
13. Branco Da Silva, P.R., Silva, D.: Multilevel LDPC lattices with efficient encoding and decoding and a generalization of construction D. IEEE Trans. Inf. Theor. **65**(5), 3246–3260 (2019)
14. Barnes, E.S., Sloane, N.J.A.: New lattice packings of spheres. Can. J. Math. **35**(1), 117–130 (1983)
15. Kurkoski, B.M.: Encoding and indexing of lattice codes. IEEE Trans. Inf. Theor. **64**(9), 6320–6332 (2018)
16. Matsumine, T., Kurkoski, B.M., Ochiai, H.: Construction D lattice decoding and its application to BCH code lattices. In: 2018 IEEE Global Communications Conference, GLOBECOM 2018 - Proceedings, 2018
17. Liu, L., Yan, Y., Ling, C., Wu, X.: Construction of capacity-achieving lattice codes: Polar lattices. IEEE Trans. Commun. **67**(2), 915–928 (2019)
18. Erez, U., Shamai, S., Zamir, R.: Capacity and lattice strategies for canceling known interference. IEEE Trans. Inf. Theor. **51**(11), 3820–3833 (2005)
19. Nam, W., Chung, S.Y., Lee, Y.H.: Nested lattice codes for gaussian relay networks with interference. IEEE Trans. Inf. Theor. **57**(12), 7733–7745 (2011)
20. Forney, G.D.: On the role of MMSE estimation in approaching the information-theoretic limits of linear Gaussian channels: Shannon meets Wiener (2004)
21. Erez, U., Zamir, R.: Achieving 1/2 log $(1 + \text{SNR})$ on the AWGN channel with lattice encoding and decoding. IEEE Trans. Inf. Theor. **50**(10), 2293–2314 (2004)
22. Huang, Y., Narayanan, K.R.: Construction πA and πD lattices: Construction goodness, and decoding algorithms. IEEE Trans. Inf. Theor. **63**(9), 5718–5733 (2017)
23. Cover, T.M., Thomas, J.A.: Elements of information theory (2005)

iDATA - Orchestrated WiseCIO for Anything as a Service

Sheldon Liang(✉), Lance Mak, Evelyn Keele, and Peter McCarthy

Lane College, Jackson, TN 38301, USA
sliang@lanecollege.edu

Abstract. Integral digitalization aims to *liaise* with **Universal** interface for human-computer interaction, *assemble* **Brewing** aggregation via online analytical processing, and *engage* **Centered** user experience (UBC), which enables wiseCIO to orchestrate "Anything-as-a-Service" (XaaS). This paper presents three important concepts such as iDATA, iDEA and ACTiVE that together orchestrate XaaS on wiseCIO. ***iDATA*** stands for "integral digitalization via archival transformation and analytics" in support of content management, ***iDEA*** denotes "intelligence-driven efficient automation" for UBC processing with little coding required via machine learning automata, and ***ACTiVE*** represents "accessible, contextual and traceable information for vast engagement" with content delivery. Where iDATA is central to XaaS through computational thinking applied to multidimensional online analytical processing (mOLAP). Case studies are through discussed on the **massive** basis through iDATA over broad fields, such as *manageable* ARM (archival repository for manageable accessibility), *animated* BUS (biological understanding from STEM), *sensible* DASH (deliveries assembled for fast search & hits), *smart* DIGIA (digital intelligence governing instruction and administering), *informative* HARP (historical archives & religious preachings), *vivid* MATH (mathematical apps in teaching and hands-on exercise), and *engaging* SHARE (studies via hands-on assignment, review/revision and evaluation). As a result, iDATA-orchestrated wiseCIO is in favor of *archival* content management (ACM) and *massive* content delivery (MCD). Most recently, the comprehensive online teaching and learning (COTL) has been prepared and published as ACTiVE courseware with various multimedia and the student online profiles for paperless homework, labs and submissions. The ACTiVE courseware is integrated with a capacity equivalent to 10,000 + traditional web pages and broadly used for advanced remote learning (ARL) in both synchronous model and asynchronous model with great ease.

Keywords: wiseCIO: Web-based intelligent service engaging cloud intelligence outlet · iDATA: integral digitalization via archival transformation and analytics · iDEA: Intelligence-driven efficient automation · ACTiVE: accessible/available, contextual and traceable information for vast engagement · winCOM: Web-intensive composite · UBC: universal liaise, brewing assembly and centered engagement · ACM/MCD: archival content management/massive content delivery · COTL: comprehensive online teaching and learning

K. Arai (Ed.): FICC 2021, AISC 1363, pp. 401–424, 2021.
https://doi.org/10.1007/978-3-030-73100-7_29

1 Introduction: iDEA to Orchestrate XaaS

wiseCIO [1] takes a leadership towards the "Anything as a Service" or XaaS era that enhances organizational cloud service experience without needing to build their own data-centers and maintain Information Technology personnel [2].

The iDEA of orchestrating XaaS is to dedicate wiseCIO to various cloud services, such as PaaS (a platform for development) and SaaS (a software as a service) by introducing such conceptual models as *integral digitalization* via archival transformation and analytics [3–6], *efficient automation* via machine learning automaton [7–9], and *active servicing* against traditional web content management and delivery [10].

1.1 Cloud Service Needs Innovation

With wiseCIO initiating transitioning from "exhausted-ness" to excellence of web service [1], iDEA enables XaaS via integral digitalization, efficient automation and active servicing against traditional web service with following versus':

LIAR vs. **LIAiSE:** User interface (UI) is designed to liaise with the client and server (C/S) via enriched interactivity (request) and actionability (reply). However, an unfriendly interface sounds like a LIAR (layouts of interface for action and reaction) due to the "ad hoc creativity" applied to the UI design [11, 12], which may cause unclear actions and unpredictable reactions. The use of iDEA is to automate UI design with little coding required for universal interface via a LIAiSE (layout of interactivity and actionability via intelligent systems engineering).
AWK vs. **ACT:** Traditional web browsing by just simply downloading (from a remote website) and overlapping (current context on the client's screen) brings awkward experience to the user, which may cause the user to lose the context "like a chasing after wind (webpages)". The use of ACTiVE archives is to promote user engagement with ACT (accessible, contextual and traceable experience) via ubiquitous service, load-balancing and failover over iDATA.

1.2 iDEA for Intelligent Processing on UBC

The novel iDEA refers to two "persons" on the basis of iDATA: a "wise conductor" to orchestrate XaaS via archival content management, and a "CIO (chief-information-officer)" for massive content delivery of intelligence for business, education and entertainment (iBEE) through intelligent processing on UBC that collaborates three cloud services via *universal* interface, *brewing* aggregation, and *centering* user experience as a whole.

Integral Digitalization: Integral digitalization promotes transformational and analytical archiving for ACM/MCD [13–15] via computational thinking and manageable processing throughout FIAT approach as follows:

Feasible Decomposition represents a computational thinking process of breaking a big problem into smaller problems and archiving them hold their very own part of the whole.

Integral Pattern Recognition is a manageable analytical process of looking for a repeating sequence for "brewing" aggregation via reusable retrieval and assembly.

Agile Abstraction denotes an extracting and synthesizing process of removing unnecessary parts of a problem and creating a general solution for multiple problems via an agile and bidirectional approach: top-down analysis and bottom-up synthesis.

Tenable Algorithms reflect a sequential and concurrent process of step-by-step instructions to solve a primary problem, then other problems without needing additional solutions.

Efficient Automation: Efficient automation renovates a series of intelligent processing via three typical cloud services on *universal* interface, *brewing* aggregation, and *centering* user experience as a whole of UBC [16]. The brewing aggregation causes no contextual swapping, and is embodied via integral digitalization in depth as follows:

Universal Interface denotes an algorithmic model liaising with human-computer interaction for compatible interactivity and actionability to be automated with little coding required.

Brewing Aggregation models a synthesizable process for retrieval and assembly from remote servers via context-aware pervasiveness and OLAP [17, 18] that enables what to retrieve (brewing) and how to assemble (aggregation) through cryptography, availability, load-balancing, and failover (CALF).

Centered User Experience engages with hierarchical extensibility and contextuality in depth without context-changes to avoid driving the user like a chasing after webpages [19–21].

Active Servicing is central to ACTiVE archives that innovate web-intensive composite (winCOM) to engage with extremely-big size of customers (for the purpose of high availability) as follows:

Accessibility to the winCOM (or traditional websites) either uses hyperlinks, or commands in REST APIs [22] to "brew" (bring) out useful information to the users with ease.

Contextuality refers to assembling winCOM (depending on how to organize a cloud service in parts) or corresponding with traditional webpages. The winCOM's assemblability makes more sense with sensible contextuality.

Traceability over well-archived winCOM is queryable, assemblable and synthesizable (QAS) through absolute path (traditional hyperlink), shortcuts (under the context), and ubiquitous path (specified under control of load-balancing or failover), respectively.

1.3 Major Contributions and Organization

Integral digitalization is central to wiseCIO via iDATA that emerges from an innovative roadmap toward XaaS via archival content management and massive content delivery (ACM/MCD).

> An innovative roadmap from integral digitalization toward XaaS:
> iDATA: FIAT → iDEA: intelligent Processing on UBC → ACTiVE: XaaS

FIAT is the live soul of intelligent processing via digital archives [14] throughout universal liaise, brewing aggregation and centered experience with ACTiVE XaaS.

Major contributions are accountable as follows:

Integral digitalization via a FIAT approach including: *Feasible* decomposition, *integral* patterns, *agile* abstraction, and *tenable* algorithms to promote transformational and analytical archiving for archival content management and massive content delivery (CM/MCD).
Efficient automation throughout UBC processes as a whole to renovate universal liaise between the client and the server, brewing assembly for massive presentation, and centered user experience without being like a chasing after a website.
Active servicing in ACTs to innovate enhanced accessibility, seamless contextuality and queryable traceability over iDATA with vast engagement for better user experience.

The rest of the paper is organized as follows:

Section 2. LIAiSE in a FIAT approach to illustrate layouts of interactivity and actionability.
Section 3. iDEA for UBC universal liaise, browning aggregation, and centered experience.
Section 4. ACTiVE XaaS in use presented as typical applications via cloud services
Section 5. iDATA in transition to practical significance and application value, and discusses the scope of future work, and the future plan as well.

2 LIAiSE: Layouts of Interactivity and Actionability

Intelligent systems engineering [23] is an interdisciplinary approach on how to engineer, implement and manage complex systems over their life cycles, where iDATA (integral digitalization via archival transformations and analytics) represents the applied intelligence to the multi-perspective approach as illustrated in Fig. 1.

Figure 1 initiates LIAiSE with layouts of interactivity and actionability vai intelligent systems engineering to automate user interface design for user engagement in the exchange of information with computers through brewing aggregation and ACTiVE XaaS for user-centric experience.

2.1 Feasible Decomposition for Web-Intensive Composite

In terms of archival content management [24], we have introduced winCOM in support of integral digitalization in a feasible approach. The feasible decomposition represents breakdowns of a "giant" institution (say, Miami University, seemingly like a big bite we can't chew) into winCOM that are smaller and easy to manage.

Figure 2 illustrates one of the thumb-ups of computational thinking is its feasibility to break down a large university into separate colleges. If the college is still too large to handle, the breakdown process continues until easy enough to manage.

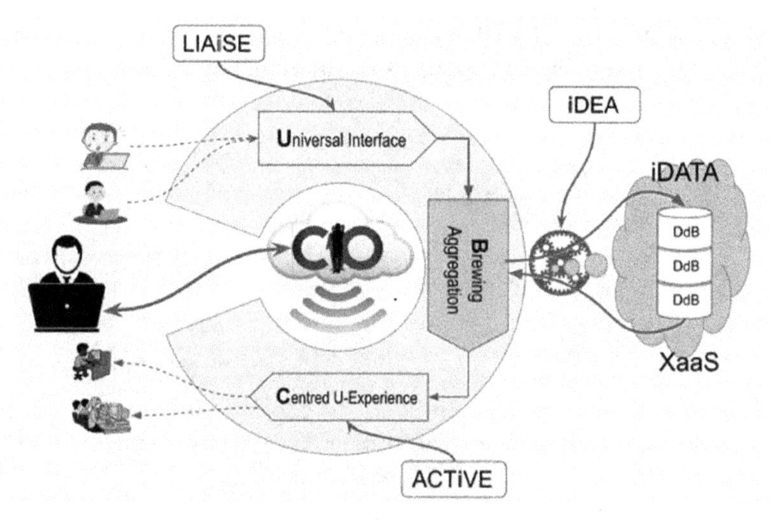

Fig. 1. Intelligent systems engineering via iDATA throughout LIAiSE, iDEA, and ACTiVE

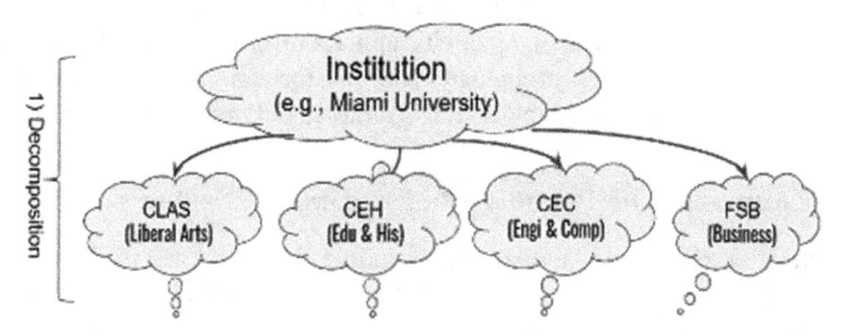

Fig. 2. Integral digitalization from feasible decomposition in principle of "divide-conquer"

The decompositional feasibility reflects an idea in engineering that a "college" should be highly cohesive (more interactions within) and lowly coupling (less ties to outside). So winCOM is used to represent logical organization and relational information groupings that are as a whole via QAS (*queryable*, *assemblable* & *synthesizable*).

Feasible decomposition is embodied as layouts of interactivity and actionability in support of the principle of "divide-conquer" for manageability and computability [25].

2.2 Integral Pattern Recognition for Iterative Processing Sequence

Integral pattern recognition comes from feasible decomposition via winCOM that does not treat the divided parts mysteriously, but meaningfully analyzed and recognized. The purpose of feasible decompositions is to identify and recognize parts as integral patterns [26] for a repeating sequence that is reusable for similar problems.

Let's take Miami University as an example – there are a couple of colleges underneath, an applicable "repeating sequence" means the similarity of logical organization and personnel in general: a dean, associate dean, an assistant to officials, offices and academic committee. The winCOM denotes a step forward from the "cloud" (unclear) to the "concrete" (integrally digitalized), as illustrated in Fig. 3.

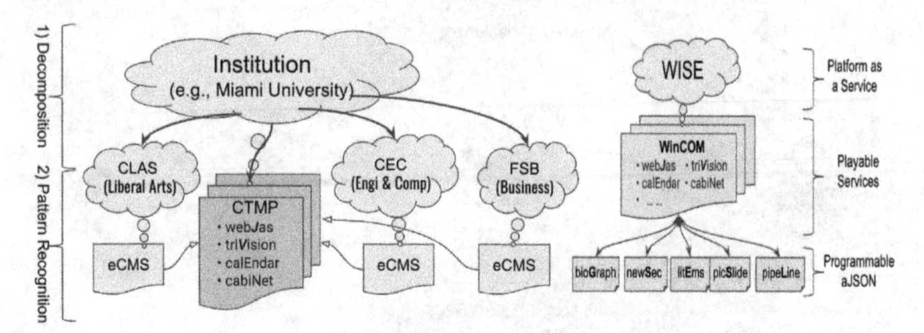

Fig. 3. Archival digitalization and transformation through integral pattern recognition

Figure 3 indicates the similar strategy applied to reusable patterns via winCOM that are recognized and reusable patterns in the FIAT approach from a concept (e.g., Miami University), a concrete solution - digital and transformational archives.

At this point, the layout of interactivity and actionability is enabled as a repeating sequence over all colleges through archival transformations and analytics by using playable services and programmable aJSON (sampled data for machine learning) [1].

2.3 Agile Abstraction for Essential Solutions Across Variations

Agile abstraction represents gradual understanding for digital reusability of archival components [14, 15]. wiseCIO may involve plenty of complex contents with winCOM to enhance human-computer interaction (*actionability*), engage user experience (*contextuality*) and assemble brewing aggregation (*ubiquity*).

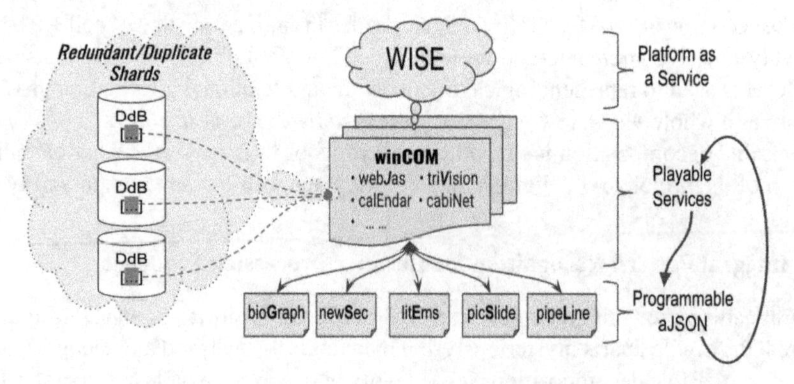

Fig. 4. Playable service via hierarchical winCOM and ubiquitous DdB

Agile abstraction also denotes scalability as the size of archival documents, which exponentially increased for digital networking service due to growing and increasing demand. Abstraction agility [27] can generalize a solution for multi-problems, and allow a specific solution by adding specific parts to the problem, as illustrated in Fig. 4.

wiseCIO represents the trinity of platform of web services, playable winCOM, and programmable aJSON (advanced JSON) in support of agile abstraction with following considerations to scalability [28]:

An Essential Solution Across Variations - in reality, there may be tons of problems, and suggested solutions. In a general sense, agile abstraction denotes logical organization and relational grouping in categorized “containers” for easy access, search and assembly such as “shelves,” “boxes,” “folders,” comparable to “situating archives (physical records), designated to hold their very own part of the whole. In a more significant sense, the agile abstraction allows those archival “containers” to be reused without introducing additional archival containers.

Recursive Containers via winCOM- a winCOM can be used as recursive containers via archival transformations. For instance, a winCOM can be seen as a top “shelf”, a middle level “box”, or a bottom “folder” recursively depending on the current context.

Distributed DocBases (DdB)- orchestrating winCOM with duplicates and redundancy across clustered servers for agility and scalability in order to grow digital networking service, and manageably increased demand via failover and load-balancing.

A playable service represents a winCOM and the associated archival storage on DdB. a winCOM reflects recursively support for hierarchical extensibility, and the archival DdB promotes analytical synthesis across clustered servers.

2.4 Tenable Algorithms via Step-by-Step Actionable Instructions

Tenable algorithms start in an agile developmental approach through step-by-step actionable instructions, then become more sophisticated via a test-driven analytical process. The algorithmic practice in favor of experimental “trial and error” processes via agile pattern recognition and recursive learning inference via archival transformation and analytics that is understood and usable by web-based intelligent service.

Digital archives are prepared for cloud content to be understood and used by a computer [29], and analytical transformation is to make web content useful (actionability), and usable (accessibility). At this point, not everything online is considered as “digitalized”. For instance, an uploaded PDF document would be considered a “deaf” content without actionable interaction. However iDATA can turn web content from “deaf” into “digitalized” by the use of machine learning automata [30].

Integral digitalization is the key to machine learning automaton through archival transformations and analytics, which denotes the tenable models to automate data analytics, identification of patterns, and decision making with minimal human intervention. In other words, computer systems are created to perform specific tasks without using explicit instructions, or relying on patterns and inference [31].

The tenable algorithm leads wiseCIO to its intelligent process automation based on feasible decomposition, integral pattern recognition and agile abstraction. This will be discussed in the next section of intelligent UBC processes.

3 iDEA: Intelligent UBC Processes

iDEA represents intelligence-driven efficient automation via digital integrity, archival transformations and analytics. The UBC process collaborates three essential cloud services as a whole, including *universal* interface liaising with human-computer interaction, *brewing* (retrieval) assembly from ubiquitously available service, and *centered* engagement for user experience [32] via context in depth and breadth.

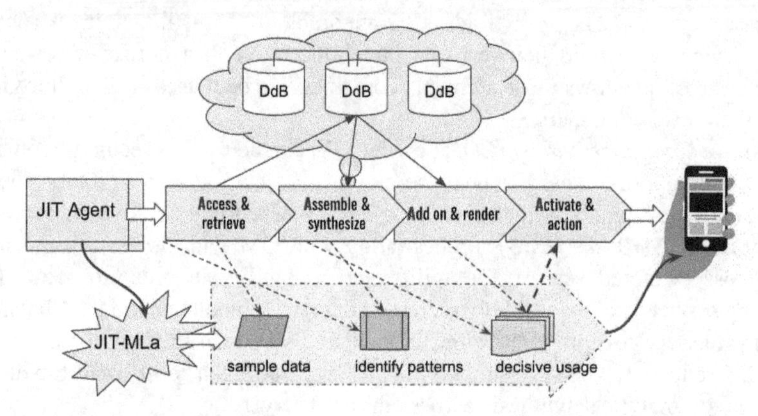

Fig. 5. JIT-MLa executes transformational analytics and logical inference over ubiquitous DdB

iDEA initiates a just-in-time machine learning automaton (JIT-MLa) and a series of AAAA tasks: access-assemble-addon-activate, which becomes a renderable presentation with HCI activated. iDEA highly relies on digital integrity through archival transformations, logical inference, and online analytical processing, as illustrated in Fig. 5.

Comprehensive UBC processes are automated to support following services:

- ACOM: archived content management
- COSA: context-oriented screening aggregation
- DASH: deliveries assembled for fast search and hits
- OLAS: online learning via analytical synthesis
- REAP: rapid extension (back-end) and active presentation (front-end)
- SPOT: special points on top on (or hotspot for short)

3.1 ACOM: Archived Content Management

Archived content management (ACOM) aims to discover, exhibit, express and present well-digitalized archives in the most convenient way based on feasible decomposition. ACOM is embodied as a twin browsing mechanism in support of both direct accessibility (like traditional webpage) and contextual extensibility as follows:

Direct Accessibility: Integral digitalization is always supposed to be sufficient and viewable. A browsing tree (similar to table of contents) is generated and maintained as the user surfs over. The user can always switch between the current context and table of contents to locate an individual winCOM that is shareable via a playable command interface (PLI), for instance:

```
?cMd=accKey@pathDdB&params        /// RESTful APIs
```

where: <u>*cMd*</u> queries via the access key and interprets the given DdB
<u>*accKey*</u> should be unique within the same DdB
<u>*params*</u> lists Key-Value pairs for possible modification

Contextual Extensibility: Well-digitalized archives enhances/strengthens archival transformation and online analytics. The JIT-MLa assembles intelligence via brewing aggregation over the big databases as an add-on subordinate to the context. By applying context-aware pervasiveness, the access key may be retrieved and assembled from multi-DdB, which demonstrates precisely on "brewing" aggregation in support of user-centric experience.

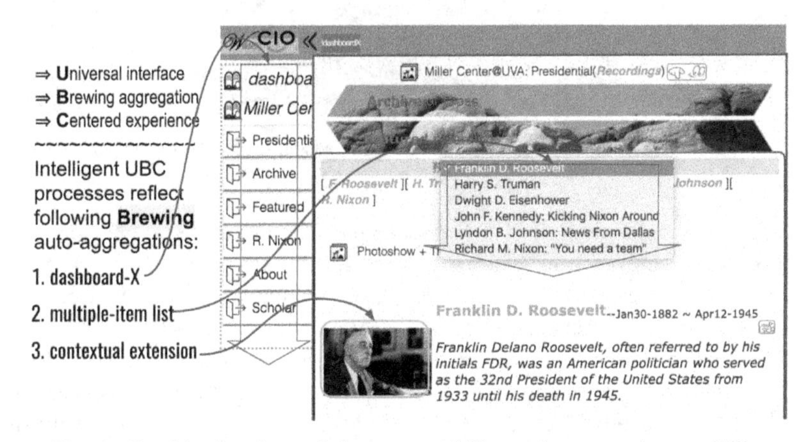

Fig. 6. Combinational use of direct accessibility and contextual extensibility

Figure 6 shows you an example of viewing a browsing tree and the brewing web content without swapping the current context. The direct accessibility favors rapid locating and easy entrance to individual winCOM (item: *Andrew Jackson*) while the contextual extensibility (*buttons are extensible*) is applied to embed content as a whole.

All winCOMs in iDATA are supportive for contextual extensibility and that offers user-centric experience via brewing aggregation. AwinCOM can work as an individual "dashboard", which is different from the traditional navigator. The "dashboard" is novel – an innovative navigator for user-centric experience with a highly cohesive context that avoids page swapping. More about contextual extensibility will be in Sect. 3.2 - DASH.

3.2 COSA: Context-Sensitive and Screening Aggregation

Context-oriented screening [33] was inspired by resume screening – a recruiter only spends a few minutes on reviewing each submitted resume by searching the predetermined keywords on the applicant's resume. If the matching rate is high, an applicant will be an initial consideration. The process may fail to find some great applicants, but it is very productive for preliminary screening.

Instead of trying to find any specific contents from a large search engine (Google, Bing, etc.), COSA encourages the users to stay with a subject-based context without swapping pages (websites), so the users can discover their interested contents by searching keywords and viewing the context via the underlying hyperlinks – those links won't cause context swapping.

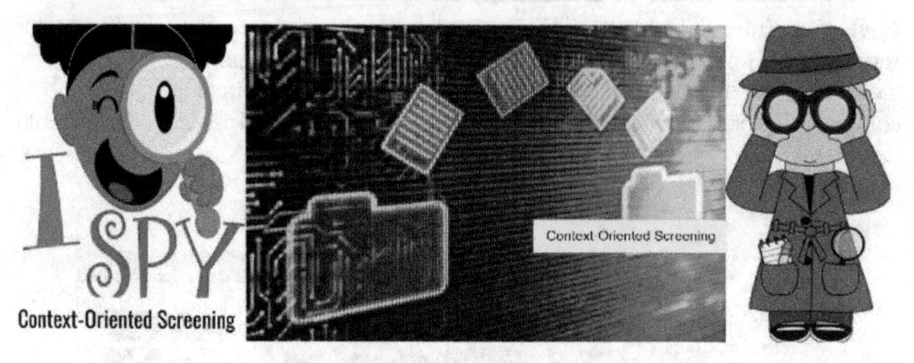

Fig. 7. COSA enables proactive brewing aggregation for "zoom-in/out"

COSA-based courseware [34]: For example, the user starts viewing a plain (hyperlink-free) context, so he won't be distracted by any unnecessary content). At his first glance, he can briefly view, and quickly figure out whether to explore the content in depth or to skip. Users can select a different context if they feel unfamiliar with any terminology by searching keywords associated with the desired context. After the JIT-MLa dynamically brews and aggregates content subordinate to the associated keywords, the user looks into the expected context via contextual extensibility as illustrated as Fig. 7.

COSA creates a proactive scenario that offers a better and user-centric experience: Users initially start in brief (without pre-layout of many hyperlinks), and intelligently discover and access interested subjects. As it says, "enough is enough", context-aware pervasiveness will offer "fairly good enough" reference triggered by keywords. Consequently, the user can be supported by dynamic brewing aggregation via JIT-MLa that effectively synthesizes useful and customizable context for individual users accordingly.

3.3 OLAS: Online Learning via Analytical Synthesis

Online learning via analytical synthesis (OLAS) is a twin service to COSA that aims to offer users proactive experience – users gets what they want. Cooperatively, OLAS supports COSA by archival transformations with preserved context-aware content reserved through the FIAT approach as follows:

Online Learning- basic use of the Internet to deliver online courseware at any time, in anywhere via XaaS. According to eLearning Industry [35], online education is totally worth the effort because a) you can learn whatever you want, b) self-paced learning, c) lower cost, d) comfort, etc. wiseCIO with its ubiquitous DdB helps create the online learning environment more friendly (context-oriented), more fostering (pondering and rethinking), and more fruitful (analytical synthesis), which would enhance the "readiness to learn without being taught"[1].

Ubiquitous winCOM- ubiquitous webcontent feasibly enables a holistic organization via scalable and digital archives from analytical synthesis over the DdB across a cluster of servers. Cloud services with wiseCIO are treated as winCOM of ubiquity that denote accessibility, actionability through archival transformations and analytics.

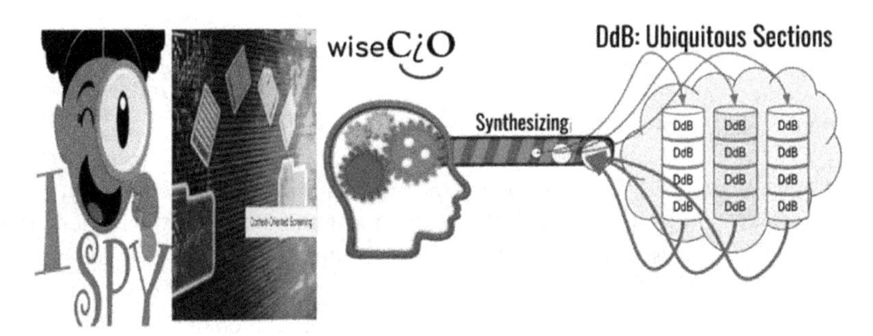

Fig. 8. OLAS propagates analytical synthesis over iDATA

Furthermore, the ubiquitous availability means CALF-cryptography, availability, load-balancing, and failover. A winCOM represents an essential unit of logical organization, and relational document groups by associating multiple shardings of content as a whole, as illustrated in Fig. 8.

3.4 REAP: Rapid Extension and Active Presentation

At the front-end with ACOM, wiseCIO *liaises* with universal interface to support both direct accessibility and contextual extensibility via winCOM. In addition, wiseCIO with REAP enables ubiquitous winCOM to *engage* user-centric experience in the back-end.

Viewing from the perspective of big databases, ACOM refers to archived content, and REAP refers to rapid extension and active presentation that assembles brewing aggregation through archival transformations and analytics, A winCOM is stored as shards in DdB for rapid assembly (transmissible content) and extension (contextual embedment), and active presentation on the client-side devices., as shown in Fig. 9.

According to the IBM, big data analytics [36] is the use of advanced analytic techniques against very large, diverse data sets that include structured, semi-structured and unstructured data, from different sources. The DdB is deployed on a cluster of servers that are integrally digitalized winCOM with CALF features.

Intelligent UBC processes are greatly orchestrated by iDATA to liaise with human-computer interaction (ACOM, DASH, and REAP), assemble brewing aggregation (OLAS), and engage centered user experience (COSA).

[1] Winston Churchill, "I am always ready to learn, although I do not always like being taught."

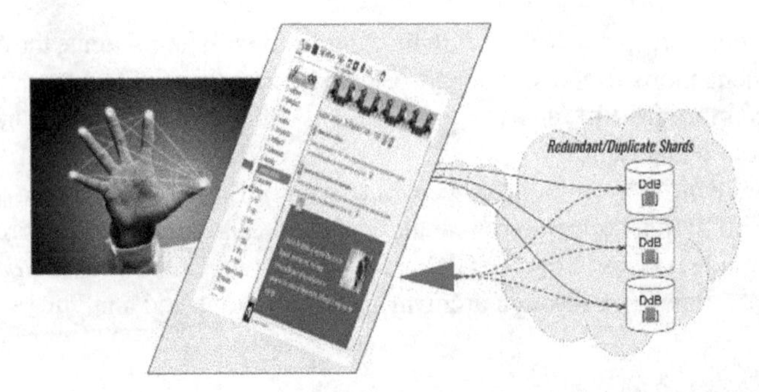

Fig. 9. Archival transformations start from actor, through interaction, to assembly

3.5 SPOT: Special Points on Top

Special points on top (SPOT) is a graphical user interface that provides at-a-glance views of key performance indicators (KPIs) relevant to a particular objective or business process [37] and extends on wiseCIO within the current context for user-centric experience in particular. wiseCIO has two types of SPOT: *favorite* spot and *thematic* spot as follows:

Favorite Spot- the primary winCOM, like a traditional homepage, will offer a favorite spot whose content (like some snacks) is either manually or automatically updated to help the user find things of common interest at a glance.

Thematic Spot- the primary winCOM, on the contrary, will also offer a thematic spot whose content (like a major meal) that is prepared, presented and published for users to explore things of deep interest at a fast access.

Fig. 10. SPOT: Special points on top allow one stop for all (zoom-in/out)

wiseCIO has constructed many dashboard features, such as customized user interface, database-embedded contents prioritized in several editable sections, and favorite functions and commands, etc. For example, after the users open the dashboard sections, they can add/remove/arrange any functions represented in icons, expand or shrink the section of the functions, and switch to another dashboard, etc., which helps users to locate and organize the information effectively. Within the dashboard section, users can have layouts with multimedia (docs, excels, ppt, audio, video, playlists, etc.).

Figure 10 shows you how wiseCIO provides an innovative user interface that helps enhance user-centric experience especially for young children and elders with little computer/programming experience. The SPOT embodies "one stop service online system" that prioritizes engagement with centered user experience and promotes efficiency and usability of the developed system in use [38].

The Miller Center[2] established the Presidential Recordings Program (PRP) to make these once-secret White House tapes accessible to all who have an interest or investment in the workings of American democracy. For instance, the PRP has five categories, such as "Archives of tapes", "Featured recordings", "Help and background", "About the program", and "Scholars". The thematic SPOT (*Recordings*) automates the best practice as a "one stop service online" that betters user experience of accessibility to citizens, journalists, policymakers, scholars, students, teachers with significant ease.

4 ACTiVE: XaaS in Use

wiseCIO uses iDATA to serve XaaS by archival transformations and analytics through intelligent UBC processes. As discussed in previous sections, Computational thinking is applied to integral digitalization via the FIAT approach (abstraction and decomposition, pattern recognition and algorithms), which provides feasibility and agility for a complex and difficult problem to be transformed into a solution.

iDATA supports bidirectional approach: top-down analytical (FIAT) processing and bottom-up transformational synthesis (brewing aggregation into a larger context) [38], as illustrated in Fig. 11.

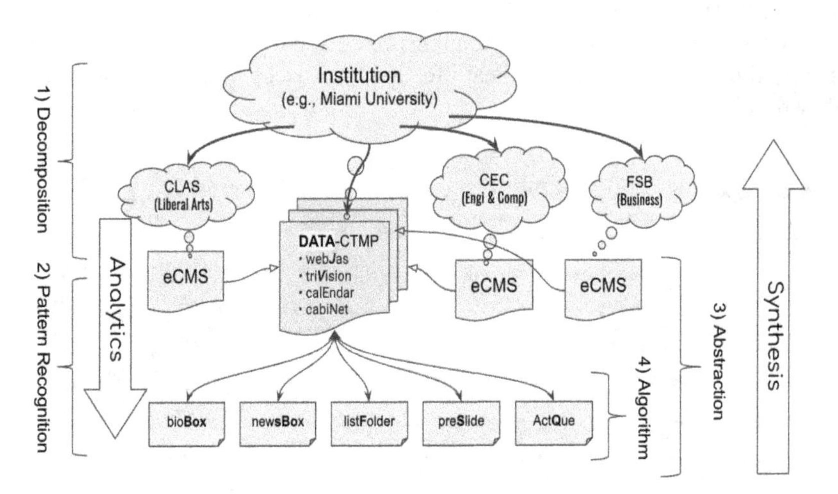

Fig. 11. The U-shape approach of digital analytics and archival transformation

[2] The Miller Center established PRP in 1998. Between 1940 and 1973, six consecutive American presidents secretly taped thousands of their meeting and telephone conversations.

Case studies in exploring possible and potential XaaS are discussed on a **massive** basis through ACTiVE mode, that is, accessible/available, contextual and traceable information for vast engagement with managed content for massive deliveries. This section covers efforts to orchestrate following ACTiVE XaaS over broad fields:

- **m**anageable ARM: Archival Repository for Manageable accessibility
- **a**nimated BUS: Biological Understanding from STEM Programs
- **s**entimental DASH: deliveries assembled for fast search & hits
- **s**mart DIGIA: Digital Intelligence Governing over Instruction and Administration
- **i**nformative HARP: Historical Archives and Religious Preachings/Presentations
- **v**ivid MATH: Mathematical Applications in Teaching and Hands-on exercises
- **e**ngaging SHARE: Studies via Hands-on Assignments, Review and Evaluation

4.1 ARM[3]: Archival Repository for Manageable Accessibility

Archival repository [39] and manageable accessibility (ARM) greatly impressed me because of wiseCIO's entrance-in-brief (one stop service), and exploration-in-depth (contextual extension). As an archivist, I have years of experience managing and categorizing materials in detail, so I am aware of how deep an archival box would be holding organized materials. An archival box will have a detail-oriented index, and may have more than a hundred folders. Each folder may have labels for the identification of its contents, which are authentic certifications and documents. Some of the papers can be very fragile.

Archiving materials in logical organization and relational document groupings by means of physical archiving and so does digital archiving - for instance, hierarchical layout via contextual expanding or extension for seventh United States President Andrew Jackson [40]. The capability of a digital archive for historical and biographical exploration allows for varied information groupings congruent to distinct aspects of Jackson's life: early life, military career, political life, and latter years through entrance-in-brief, as shown as Fig. 12.

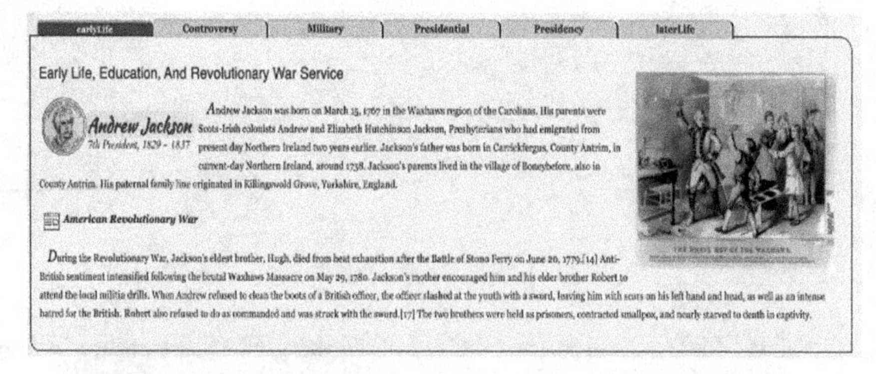

Fig. 12. The incredible entrance provides a spectrum through the whole life.

[3] Ms Evelyn Keele, Archivist, TN Room, Jackson-Madison, County Library, Jackson, TN .

Entrance-in-brief - wiseCIO lays out digital archives in a very brief layout by which a brief entry may take you through a person's whole life within the same context. For instance, six tabs almost cover (through) Andrew Jackseon's whole life from "early life" to "later life" with rich content associated in groupings, such as related video, websites, and so forth.

Explore-in-depth - wiseCIO also supports digital archives through logical organization with considerations to hierarchical depth (as deep as needed) and contextual breadth (as broad as you expect) without forcing the user like "a chasing after webpages". Why? It is all because that hierarchical extensibility allows the user to look into a "drawer" by opening it, and then surf the context by closing the "drawer", which is known as hierarchical extensibility.

In addition, information seekers view primary resource materials and special collections that have been digitalized and context-oriented to provide clear documents of Jackson's life relating to Tennessee State history and United States history. This learner-centered approach allows the researcher to explore those topics as needed. Accompanying digital images of primary resources, such as images of his personal correspondence, provide evidence and clarity of information while being highly interesting.

4.2 BUS[4]: Biological Understanding from STEM Perspectives

Computational thinking means not only how we are to think like computers, but also how to make good use of computers via MOLEC (modeling, observing, learning, experimenting and creativity) [41] in support of biological understanding from STEM perspectives.

iDATA is promising on how to digitally archive lecture and lab notes, and experimental labs via modeling, observing, (hands-on) learning with creativity promotes better and deeper understanding. Beta lactams are an important class of antibiotics that work by inhibiting cell wall synthesis in bacterial cells. Biochemistry tutorials help STEM students get better understanding visually by using 3D images, and animated GIF, as illustrated in Fig. 13.

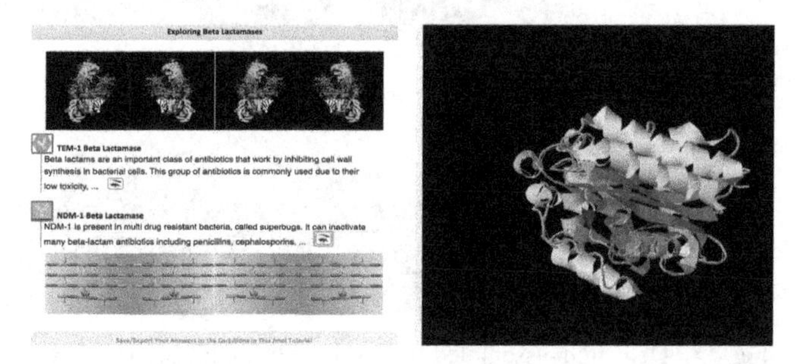

Fig. 13. Biological understanding via archived tutorials with Q/A on wiseCIO

[4] Dr Melanie Van Stry: Chair of STEM Division at Lane College, Jackson, TN. STEM stands for Science, Technology, Engineering and Mathematics.

4.3 DASH[5]: Deliveries Assembled for Fast Search and Hits

The DASH is sensible to the current context that a user is exploring intentionally or surfing flexibly managed content (winCOMs) and/or manifested content (traditional websites) of his own interest. *Acting* as a dynamic tracking service, the sensible DASH assembles the delivered web content in the meantime collectes its access-keys (similar to URLs) underneath the dashboardX for fast search and hits (used as a menu item to access), reflectively tracking the user's intention, which brings out user-centric experience and engagement through the traceable accessibility.

Managed content on DASH- The managed DASH is for winCOMs allowing the preparer of winCOMs to control subjective plants in advance and then dynamic popups as the user surfs to the point. Typically, a winCOM usually represents a one stop service for well-archived documents that supports hierarchical extendibility (*extending*) and contextual synthesis (*shrinking*) within the same context (witout often page-swapping), which reflects user-centric experience. However, the user may also like some traditional browsing experience with page swapping, the managed DASH will meet their needs of "menu-driven browsing".

Manifested content on DASH- The manifested DASH is used to support wiseCIO as a browser that allows the user to surf traditional websites. As a work in progress, the manifested DASH is enabled with chosen websites. The criteria for chosen websites are based on how the website is organized subject to the web expert or machine learning automata because the manifested DASH is a fully automated process. It would be risky if a website organized ad-hoc. For the sake of convenience, both manifested DASH and managed DASH can be enabled or disabled up to the user using wiseCIO as a browser.

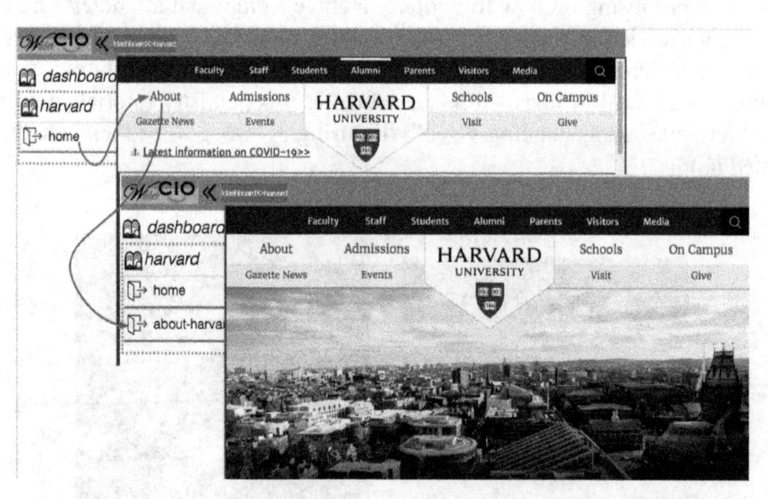

Fig. 14. Delivered–assembled for fast search and hits on wiseCIO

[5] Manifested DASH has chosen Harvard University https://www.harvard.edu, one of the best candidate websites, which inspired wiseCIO to do better on DASH for the convenience and user experience.

Figure 14 illustrates the use of Harvard University as a sentimental example - the DASH begins with empty items under dashboardX branch, then automated prompts for fast search and hits until web content is open to explore via wiseCIO.

In addition to excellent features on wiseCIO including intelligent UBC processing with more automation, the use of wiseCIO together with manifested DASH to explore traditional websites (if chosen) will provide new and better user experience than use of other browsers to explore traditional websites.

4.4 DIGIA[6]: Digital Intelligence Governing Instruction and Administration

With the demanding role of school leadership, school administrators often find themselves navigating multiple websites to accomplish tasks needed to support teachers and students. Websites inclusive of the district's student information management system, the state's online teacher evaluation system, and resources to help teachers improve instruction for students are among the many visited by administrators daily. Teachers supported by these administrators also find themselves responsible for managing many online resources to support students as well. However, current websites have all useful information archived in a way that the user has to surf around, which sometimes is quite segmentally distractive.

iDATA promises to digitalize archives for instruction and administration which have emerged from wiseCIO, to help prepare (well-archived), propagate (synthesis), and present (renderable & actionable) throughout manageable processing and quick approach. A web-based intelligence service and rich resources in the cloud environment will be useful and usable to support the use of digital intelligence to govern both instruction and administration, as illustrated in Fig. 15.

Remain focused on primaries - The scarcity of time in today's schools suggests that there is a need for a user-centred context for use, and wiseCIO emerges at digital age to help educators and administrators remain focused on their primary responsibility of instructing students without being distracted or exhausted in the "oceanic browsing". This interactive platform could also be used to engage students with content in an electronic format as well.

School leadership experience - Teachers struggle to organize online resources to incorporate technology into lessons on a daily basis. Though there is already a research

Fig. 15. Digital Intelligence Governing over Instruction and Administration

6 Ms Kimberly N. Quinn, Principal of Denmark Elementary School, Denmark, TN .

based established curriculum, often it lacks a daily technology component. However, there remains a need for student engagement. Despite the lack of planning time available to teachers, Herold notes 28% of principals surveyed thought the integration technologically for all is a transformative method to improve education, and an additional 23% saw it as a promising idea [42].

Integration of strong curriculum- wiseCIO may help to integrate well-established strong curriculum into an up-to-date digital platform via a transformatively manageable process that would provide positively profound effects on students' learning. Such a platform would also provide engaging learning opportunities, communication with a network of learners through the internet, innovative ways to assess student learning, and increase in digital fluency. Each of these are necessary skills as students enter the workforce. Additionally, providing students with technological and digital learning experience will also offer equitable learning opportunities for underserved populations [43].

It is the administrator's role to support teachers in the endeavor of integrating technology into their lessons, and it is the district's responsibility to allow autonomy for schools to do so with the understanding that technology based instruction will be paced to support the adopted curriculum. Flexibility of pacing lessons and units should be allowed in order for students to become digitally engaged in such learning.

4.5 HARP[7]: Historical Archives and Religious Preachings

As a scholar of religious history and a church historian, I have amassed many years of mining and researching at numerous archival sites and collections across the US. The majority of the physical sites have collections that are paper-based or analog-based, but are increasingly moving to include a wider diversity of digital-based resources.

The majority of the physical sites have collections that are paper-based or analog-based, but are increasingly moving to include a wider diversity of digital-based resources.

"No archives, no history."- Central to preserving and presenting history, archives are crucial for researching and writing a new denominational history. Transforming paper-based or analog-based collections (physical sites) into a wider diversity of digitalized resources holds promise to accessibility and availability in the cloud environment. However many of archives are operating under severe budget and personnel constraints that prevent them from expanding to new digital capabilities that have high value of incorporating emergent digital archiving training and technology.

"Digital archives, dicent history- The CME Church Archival collection in Memphis has identified a wide-range of sources ready for digitalization: denominational journals, minutes, reports, pictures, sermons, publications, musical recordings, etc. These sources not only considerably expand the scope of HARP to include more than the collection and preservation of preaching materials but also available, usable and useful online, as illustrated in Fig. 16.

[7] Dr. Raymond Sommerville, CME Church Historian/ Associate Professor of Religion, Lane College, Jackson, TN. He has conducted and experienced archival research at the CME Archival Room, the Tennessee Methodist Archives and Historical Library.

Fig. 16. Digital archiving becomes central to a new denominational history

iDATA promises to incorporate HARP in curricular and pedagogical developments at Lane College, particularly in the newly developed interdisciplinary Religion and Arts track. This track will draw on the disciplines of religion, music, and visual arts, proving both foundational courses and area-specific practicums (e.g., preaching in the religious track). An accessible, interdisciplinary archive of sources/resources for religion, music, and art would be useful for research, teaching, and learning. To enhance the curricular and pedagogical effectiveness of these collaborations, faculty and students alike will need some formal training and digital archiving understanding by use of wiseCIO on digital archives and analytical synthesis across multiple departments, such as the Religion and History Departments in HARP collaboration.

4.6 MATH[8]: Math Application in Teaching and Hands-On Exercises

In teaching, concrete materials and prompts that are real in students' world of significance help with student's learning. For instance, by making some smiley faces and bringing them to your classroom, you can start with addition, then connect it to multiplication – even if they did not know how to do multiplication, as a last resort, they could count it up, as shown in Fig. 17.

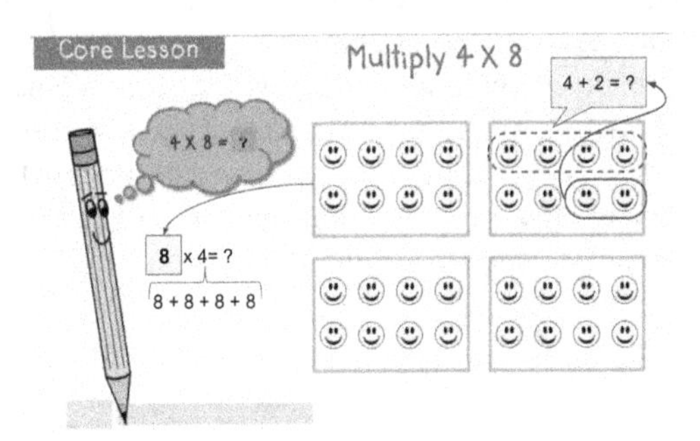

Fig. 17. Visual/concrete materials help to see multiplication as "repeated addition"

Concrete materials- Digital materials will greatly signify use of "concrete materials" in teaching via hands-on exercises to a visual, interactive and actionable (VIA) extent

[8] Dr Peter McCarthy, Associate Professor of Mathematics, Lane College, Jackson, TN.

that concrete materials can be customized at random. For instance, a girl may like flowers more than smiley faces. By name recognition via machine learning, on the girls screen, the same exercise can have flowers to replace smiley faces. By operating on the VIA materials on the screen acting as a digital "teaching assistant", the teacher may provide students with concrete experiences to help them model, describe, and explore mathematics.

Experimental illustration-Addition is essential to multiplication: Use of concrete materials makes learning mathematics experiential so that students can connect multiplication to addition as "repeated addition". They provide students the opportunities to interact with each other, and with the teacher as well.

Hands-on learning is important in the classroom learning experience, so web-based intelligent service helps orchestrate hands-on learning so as to engage students in kinesthetic learning. By operating (dragging and dropping on) visual user interfaces, students may experiment "trial and error", then learn from their mistakes. With wiseCIO the teacher may also create digitally effective learning environments that are valuable to students' prior knowledge. The teacher engages the students in meaningful math lessons and allows students sufficient time to think about, and discuss problems with their classmates.

wiseCIO will help in dynamical VIA means when students enter school seeing mathematics as an integral part of their world. Students' experiences are, therefore, much more global in essence. They can observe and make links to their prior knowledge when sharing their ideas.

wiseCIO allows flexible layouts of often-used icons that provides great convenience via a single block, say "Spring 2020", for instance, there is a set of buttons on the top so that we can conveniently access various online resources, as shown as follows:

wiseCIO provides excellent hands-on learning experience for students. On the top, a group of useful tools laid out as buttons, so students can find various online editor for different programming languages such as Python, Java, C ++, JavaScript and PHP, etc. which enhances students' hands-on programming experience by comparing and learning multiple languages with understanding of an algorithm. Through hands-on exercises in programming, we can write a program to run and test on the webpage. For instance, through hands-on learning, the professor may teach an algorithm in Python at beginning, then gradually transition to other programming languages such as Java, or C++, etc. As a result, "kill multi-birds with one stone", the students may master multiple languages at the same time,

Uwiragiye, Claudine, TA--Teaching Assistant to CSC 131
Claudine Uwiragiye is great and s/he is LOVE: love, objective, victory and enthusiastic; S/he always had a mind set on going out and making ways for study, family and friends (Video).

Fig. 18. The profile acts as an active section allowing paperless submission, online publishing.

The online service with wiseCIO is not only those posts by the professor, but also allows us to create our own web session within the profile framework so we students can customize our personal profiles. More significantly, we have flexible ways of submitting our work online, which helps us to keep up with our study progress because they are all online paperlessly.

Timely online sign-in is based on a calendar, and we are required to sign-in in the meantime individually by using computers in the lab. The date and time will be recorded when sign-in; also my photo in display indicates my success in sign-in shown in Fig. 18.

In addition, we can find almost all tools related to our studies within the current course modules, such as BlackBoard, Student/Faculty Portal. Worksheet and Schedule, and we're able to communicate with classmates easily because an email button is there.

5 Conclusion: iDATA in Transition to XaaS (Cloud Service)

This paper presents a central iDEA for wiseCIO over iDATA to orchestrate ACTiVE XaaS throughout intelligent UBC processes to *liaise* with **universal** interface for human-computer interaction, *assemble* **brewing** aggregation via online analytical processing and *engage* **centered** user exieruece.

What Has Been Achieved- wiseCIO as a cloud service platform has emerged with an innovative roadmap from integral digitalization (iDATA) that is intensive with archives and inline synthesis through intelligent processing (iDEA) toward ACTiVE cloud service as follows:

Digitalizing- **integral** digitalization reflects *practical significance* via archival transformation and analytics in computational thinking and manageable processing throughout a feasible FIAT approach to enable iDEA for efficient process automation.

Archiving- **intensive** archive aims to turn digital documents into a variety of labeled categories of such manageability as queryability, assemblability and synthesizability; where a labeled category could be a "shelf," "box," or "folder" for logical organization and relational groupings in support of archival management (ACM).

Transforming- **intelligent** transformations of (raw data) into useful and usable winCOMs that may generate high application values via massive delivery (MCD) of intelligence for business, education and entertainment (iBEE).

Analytics- **inline** processing of analytics is to examine information from distributed docBases using mathematical methods and machine learning techniques to find useful patterns and algorithmic fulfilment for information synthesis, which supports winCOM.

Application Scope and Limitation-In addition to ACTiVE XaaS (various cloud services) in Sect. 4, wiseCIO has been experimented and applied in broad application scopes, such as "Presidential Recordings Program" (*Miller Center, UVA*), "Parks Canada" (*Canadian National Parks*), Coach New York (*specializing in luxury accessories*), and Remote Online Teaching (*Comprehensive Online Teaching & Learning*), esp during the pandemic since March, 2020, which discloses a quite wide application scope. Although cryptography has "by nature" (considered initially) applied to storage and transmission of all digital documents, wiseCIO has not really been used in financial processing that requires high information security so never got experiment-on-attack against CIA (confidentiality, Integrity, and Availability).

Future Plan and Further Work-With fully digitalized documents as winCOMs, wiseCIO is particularly advantageous to archival management and massive delivery (ACM/MCD) through iDATA-based *integral* digitalization, *intelligence*-driven automation, and *information* for vast engagement.

In order to broaden application scope, more effort will be on wiseCIO as an intelligent browser to "brew" traditional websites for better user experience, where integral pattern recognition will play a key role in manifested DASH via machine learning automata.

Universal interfaces liaise with human-computer interaction with great ease. On one hand, intelligence-driven efficient automation is applied to universal interfaces with little coding required, but on the other hand, universal interfaces would be the "first buy of customer" (like or dislike), so some more deliberate work is needed to beautify the universal interface.

Last but not the least, further work will focus on connectivity and adaptability for enhanced cloud service (XaaS). Connectivity is about a "one stop service" that enables wiseCIO to pull existing websites on the platform so as to be beneficial to user-centric experience; adaptability is about capability of how easy to get existing apps/websites assembled as a part.

Acknowledgments. This work is partially supported by NSF DUE 1833960 – Any opinions, findings, and conclusions or recommendations expressed in this material are those of the authors and do not necessarily reflect the views of the National Science Foundation. Our sincere thanks also to Dr Melanie Van Stry (Chair of STEM Division at Lane College, Jackson, TN) for **BUS** (biological understanding from STEM programs), Ms Kimberly N. Quinn (Principal of Denmark Elementary School, Denmark, TN) for **DIGIA** (digital intelligence governing over instruction and administration), Dr. Raymond Sommerville (CME Church Historian/Associate Professor of Religion) for **HARP** (historical archives and religious preaching and presentations), Ms. Claudine Uwiragiye and Mr Michael Davis (students in CS, smart and very active as TAs in CSC program) for **SHARE** (studies via hands-on assignments, review and evaluation) and **COTL** (comprehensive online teaching and learning). Last but not least, I am deeply thankful to Angela for her encouragement, inspiration and loving of **wiseCIO** (!)

References

1. Liang, S., Lebby, K., McCarthy, P.: wiseCIO: web-based intelligent services engaging cloud intelligence outlet. In: Proceedings of the 2020 Computing Conference, vol. 1, pp. 169–195, London, UK, July, 2020
2. Verma, P., Kumar, K.: Foundation for XaaS: Service Architecture in 21st Century Enterprise One Edition
3. Blokdyk, G.: Digitalization Through Industrialization A Complete Guide - 2019 Edition
4. Wing, J.: Computational Thinking. https://www.cs.cmu.edu/~15110-s13/Wing06-ct.pdf
5. Denning, P.J., Tedre, M.: Computational Thinking (MIT Press Essential Knowledge series) 14 May 2019
6. NSF CSforAll:RPP. https://nsf.gov/pubs/2018/nsf18537/nsf18537.htm
7. Ameisen, E.: Building Machine Learning Powered Applications: Going from Idea to Product. O'Reilly Media, Inc., Sebastopol (2020)

8. Nilsson, N.J.: Introduction to Machine Learning - An Early Draft of a Proposed Textbook, Robotics Laboratory, Department of Computer Science, Stanford University. https://ai.stanford.edu/~nilsson/MLBOOK.pdf
9. SAS Insights: Machine Learning. https://www.sas.com/en_us/insights/analytics/machine-learning.html
10. Srivastav, M.K., Nath, A.: Web content management system, IJIRAE, Iss03 **3.** https://www.researchgate.net/publication/299438184_WEB_CONTENT_MANAGEMENT_SYSTEM
11. Galitz, W.O.: The Essential Guide to User Interface Design: An Introduction to GUI Design Principles and Techniques. 3 edn. Wiley, New York (2013)
12. (eBook PDF) Designing the User Interface: Strategies for Effective Human-Computer Interaction 6th Edition
13. Blokdyk, G.: Database Integrity A Complete Guide - 2020 Edition
14. Hodge, G.M.: Best Practices for Digital Archiving ~ An Information Life Cycle Approach, Information International Associates, Inc. http://www.dlib.org/dlib/january00/01hodge.html
15. Archives @ PAMA, Region of Peel: How do Archivists Organize Collections? https://peelarchivesblog.com/2015/08/26/how-do-archivists-organize-collections/
16. Leonard, A., Bradshaw, K.: SQL Server Data Automation through Frameworks: Building Metadata-driven Frameworks with T-SQL, SSIS, and Azure Data Factory 1st Ed
17. Beheshti, S., et al.: Process Analytics: Concepts and Techniques for Querying and Analyzing Process Data (2016)
18. Ranet OLAP Blog: OLAP Basics and Multidimensional Model. https://galaktika-soft.com/blog/overview-of-olap-technology.html
19. de Voil, N.: User Experience Foundations, BCS Learning & Development Limited, 1 edn., July 2020. ISBN: 9781780173511
20. Donoghue, K., Schrage, M.: Built For Use: Driving Profitability Through The User Experience 1st (eBook). ISBN-13: 978-0071383042, ISBN-10: 0071383042, Digital Format
21. Benyon, D.: Designing the User Experience - A Guide to HCI, UX and Interaction Design. Pearson Publishing, Inc. ISBN: 978-1-292-15551-7 (print)
22. IBM Cloud Learn Hub and Integration: REST APIs. https://www.ibm.com/cloud/learn/rest-apis
23. Sriram, R.D.: Intelligent Systems for Engineering - A Knowledge-Based Approach (1997)
24. Nakano, R.: Web Content Management: A Collaborative Approach (2002)
25. Mishra, D.D.: Divide and Conquer Paradigm. https://www.includehelp.com/algorithms/divide-and-conquer-paradigm.aspx
26. Gamma, E., Helm, R., Johnson, R., Vlissides, J.: Design Patterns: Elements of Reusable Object-Oriented Software, Computer Science Book. O'Reilly Media, Sebastopol (1994)
27. Aucsmith, D.: Information Hiding, Second International Workshop, Portland, Oregon, USA, April 1998
28. Liang, S., Puette, J., Luqi: Quantifiable Software Architecture of Dependable Systems of Systems. In: de Lemos, R. (ed.) Architecture, Dependable Systems II. Springer Verlag (LNCS) (2004)
29. Dobreva, M., Ivacs, G.: Digital Archives: Management, Use and Access (2001)
30. McCarthy, J.: Automata: Compiling State Machines. https://docs.racket-lang.org/automata/index.html
31. Kundan, A.P.: Intelligent Automation with VMware: apply machine learning techniques to VMware virtualization and networking Paperback, 30 Mar 2019
32. Benyon, D.: Designing the User Experience - A Guide to HCI, UX and Interaction Design. Pearson Publishing, Inc., London (2019). ISBN: 978-1-292-15551-7 (print)
33. Interaction design foundation - Context-Aware Computing. https://www.interaction-design.org/literature/book/the-encyclopedia-of-human-computer-interaction-2nd-ed/context-aware-computing-context-awareness-context-aware-user-interfaces-and-implicit-interaction

34. Liang, S.: COC: web-based intelligent services enabled comprehensive online courseware. In: CUR Biennial Conference, Crystal City, 30 Jun–03 Jul 2018
35. Norman, S.: Online Learning. https://elearningindustry.com/5-advantages-of-online-learning-education-without-leaving-home
36. IBM: what is big data analytics? https://www.ibm.com/analytics/hadoop/big-data-analytics
37. Few, S.: Information Dashboard Design - Effective Visual Communication of Data. 1st edn. O'Reilly, Sebastopol (2006)
38. Sriarunrasmee, J., Anutariya, C.: The Development of One Stop Service. https://www.researchgate.net/publication/341143924_The_Development_of_One_Stop_Service_Online_System_based_on_User_Experience_Design_and_AGILE_Method
39. Clobridge, A.: Building a Digital Repository Program with Limited Resources. https://www.thriftbooks.com/a/clobridge-abby/2951998/
40. Tennessee virtual Archive. https://teva.contentdm.oclc.org/digital/search/searchterm/Andrew%20Jackson
41. Biomolecules. https://www.biologydiscussion.com/biomolecules/biomolecules-top-4-classes-of-biomolecules/11169
42. Herold, B.: What principals really think about tech. Education Week. https://www.edweek.org/ew/articles/2018/04/18/what-principals-really-think-about-tech.html
43. Whitehead, B., Jenson, D., Boschee, F.: Planning for technology: a guide for school administrators, technology coordinators, and curriculum leaders (1st ed.)

Progressive Segment Routing for Vehicular Named Data Networks

Bassma Aldahlan(✉) and Zongming Fei

Department of Computer Science, University of Kentucky,
Lexington, KY 40506, USA
bal226@g.uky.edu, fei@cs.uky.edu

Abstract. Adopting the Named Data Networking (NDN) architecture for Vehicular Ad Hoc Networks (VANETs) can help improve the communication among vehicles because of the data-centric nature of NDN. Routing in vehicular NDNs is still a challenge because of high mobility and intermittent connections. This paper proposes a new progressive segment routing approach that takes into consideration how vehicles are distributed among different roads, with the goal of choosing well-traveled roads over less-traveled roads to reduce the failure rate of packet delivery. A novel criterion for determining progress of routing is designed to guarantee that the destination will be reached even if a temporary loop may be formed in the path. Simulation results demonstrate that the proposed progressive segment routing algorithm can achieve better performance than existing algorithms for vehicular NDNs.

Keywords: Vehicular networks · Named Data Networking · VANET · Source routing · Segment routing

1 Introduction

The Vehicular Ad Hoc Network (VANET) plays an essential role in future transportation systems and can provide services for improving the safety and the quality of experience using vehicles. There are many applications of using VANETs, from transportation related applications, such as road conditions, accident reports, parking availability, and toll information, to informational and entertainment applications, such as Internet access for web and email, video on demand, and video/audio conferencing.

We adopt Named Data Networking (NDN) architecture to VANET to take advantage of its focus on the name of the content that is better suited for the vehicular networks, where fixed IP addresses do not provide the same aggregation benefit as in the wired networks. In NDN, a consumer in need of some content will send out interest packets, specifying the name prefix of the content with parameters. Those producers of the content matching the interest packet, or some intermediate nodes that cache the content, will send back matched data packets to the consumer using the reverse path the interest packet traveled.

K. Arai (Ed.): FICC 2021, AISC 1363, pp. 425–437, 2021.
https://doi.org/10.1007/978-3-030-73100-7_30

Routing of interest packets in the vehicular NDN is still a challenge. Generally, we have to resort to broadcast of interest packets to neighbors to find where the content is located [5].

In this paper, we consider the category of VANET applications that are related to transportation. An observation is that the producers of the content for these applications are typically located in a certain geographical area. Using the location information of potential sources can direct the interest packets to certain direction. Therefore the flooding caused by broadcast of interest packets can be avoided. In particular, the application is equipped with a map of an area of interest (e.g. in urban areas). A simple approach is to use the Dijkstra's algorithm to calculate the shortest path from the consumer to the producer of the content as the routing path for forwarding interest packets.

We have to address several issues with regard to the routing of interest packets. First, we need to determine the metric of cost of edges used in Dijkstra's algorithm. The physical distance may be a good measure if we are figuring out a path for a vehicle to travel to the destination. However, we are interested in forwarding the interest packets to the destination. It is the nature of VANETS that communication may experience frequent disconnectivity due to fast mobility of vehicles. When selecting a path, we may want to have a path with least chance of being disconnected.

Second, the path condition may change over time. Letting the source node calculate the whole path to the producers of the content may lead to a subquality path. We need to make the decision on how far the source should dictate the forwarding of the interest packets.

Third, if multiple points along the path make decisions, we may end up with a loop in the forwarding process. We have to make sure the algorithm can avoid such situations.

To address these problems, we propose a progressive segment routing approach for handling interest packets in vehicular NDNs. The new routing algorithm takes into consideration how vehicles are distributed among different roads and chooses well-traveled roads over less-traveled roads. The goal is to reduce the failure rate of packet delivery. The whole routing path consists of multiple segments. The source will calculate the routing path to some intermediate target intersection as the first segment and the rest of the path will use the shortest path as the default. After reaching the intermediate target, the default shortest path can be re-calculated by relay nodes based on changed conditions. At this time, another segment can be formulated, followed by the another default shortest path to the destination. The process continues until the producer of the interest is reached.

One problem with the multi-segment routing is that it may generate loop because the intermediate targets calculate the path independently from the previous path. In order to avoid infinite loop in the path, we propose the idea of progressive segment routing. When a node generates a source routing path, not all nodes in this source routing path can re-calculate and update the path. Instead, only those nodes that have a shorter distance in the map graph than the current node can. By imposing this requirement, we can guarantee that the routing path makes progress in each step (segment) and finally the packet will reach the destination.

The rest of this is organized as follows. Section 2 describes related work. Section 3 presents background and an overview of the design idea. Section 4 describes progressive segment routing and gives the details of the algorithms for calculating density and calculating segments. Section 5 evaluates the proposed algorithms by comparing with existing routing algorithms in vehicular NDNs. Section 6 concludes the paper.

2 Related Work

Segment routing [2] is a special source routing paradigm that divides the whole source routing path into multiple segments. Each segment can be of different types, such as a single-hop tunnel, a multiple-hop tunnel (prefix segment) that uses shortest path links, and a segment applying the specified service. Our approach is based on the segment routing paradigm, but the segments are dynamically created/updated and the condition we imposed can guarantee that the routing process makes progress toward the destination.

Early work V-NDN explored the applicability of the Named Data Networking architecture to vehicular networks [5]. It took advantage of the content-centric approach and caching capabilities of intermediate nodes in NDN and made it work coherently with the vehicular networks. It studied how vehicles can use different channels to communicate among themselves and with centralized servers. It used implicit acknowledgment to reduce the broadcast traffic.

Last Encounter Content Routing [12] proposed an opportunistic routing method, which forwards interests to vehicles that have the content based on the information collected before. Example applications include retrieving music or sharing video. To decrease the content exploration overhead, it lets each node maintain the content locations using a data structure called "Last Encounter List", which contains the content list and content locations. These last encounter lists are exchanged when two vehicles encounters.

How to mitigate the flooding storm of interests and address link disconnectivity issues in Vehicular Content-Centric Networking was studied by Maryam, et al. [9]. They proposed an Intersection Based Forwarder Selection (IBFS) scheme, which tries to find the content within one hop, and if not found, the interest will be broadcast to the network. The method consists of two phases. The first phase is the intersection sharing phase, in which performance metrics are exchanged between neighbors within the transmission range. The second phase is the interest/data packet forwarding phase, in which a consumer selects a relay node whose trajectory information is the best. The goal of IBFS is to reduce the delay and minimize the number of forwarders in the forwarding path.

Geo-locations have been used to improve the forwarding of packets in vehicular NDNs [3]. This work focused on the issues caused by dynamic topology. It used a timer-based forwarding decision mechanism to promote reliability, multipath forwarding, and caching ability. It added a position field in both interest packets and data packets. The forwarding strategy is to select the furthest node from the sender to reduce the number of intermediate nodes.

GeoDTN-NDN [1] introduced hybrid routing and DTN support to vehicular Named Data Networking. It uses geo-location information to guide the forwarding of the interest packets. It also designed a data packet forwarding mechanism based on geographical information to deal with the mobility of the vehicles. In case the consumer vehicle moves, it will resend the interest packet toward the producer with the expectation that they will meet in the middle with data packets.

Navigo [6] exploited geo-location information to direct interest towards the destination to address the flooding problem of interest forwarding. It used the map data to find the best path for the interest packets to travel. Instead of using the shortest path generated by the Dijkstra's algorithm, it preferred more traveled roads than less traveled roads. The key is to prefer wider roads (with more lanes) over narrower roads (with fewer lanes), because they are likely to have more traffic. Therefore, it adjusted the costs of edges by making them inversely proportional to the number of lanes. In contrast, our approach measures the density of traffic on the roads and adjusted the costs of edges accordingly that can more accurately reflect the situation of road conditions.

3 Background and Design Overview

We focus on an urban environment where vehicles use vehicle to vehicle (V2V) communications. We do not consider using infrastructure-based communication such as Road Side Units (RSUs). We assume that each vehicle has a map to get the road information, including the attributes of roads and intersection locations. It is capable of WiFi Ad-hoc communication (IEEE 802.11p) and can communicate with neighboring vehicles. Also, each vehicle is equipped with the Global Positioning System (GPS) which can tell the vehicle's location.

The NDN architecture is a natural fit for vehicular networks because vehicles do not have pre-assigned IP addresses as computers and other equipment in the wired network and are more interested in getting content rather than setting up a communication with a specific target probably with a fixed IP address. NDN applications use the Application Data Unit [4], which allows the producer to publish its applications with a unique naming scheme, while the consumer can request the data content by using the application layer naming design. Using NDN in VANETs' applications can bring several benefits. The caching abilities of NDN allows the consumer of content to get the content from any vehicle that has content in its cache instead of requesting content from the publisher every time. In addition, NDN provides a naming scheme that supports requesting different types of applications based on these application names rather than requesting applications based on their addresses. When a consumer wants to get some content, it will send out interest packets to the network. When a node inside the network having the content specified in the interest packet, the content will be sent in data packets in the reverse path of the interest packet to the original consumer. Because there is no specific information about which nodes may have the content, interest packets are broadcast to some extent to different parts of the network. This may cause flooding of interest packets. This problem has been studied by some previous work [1,6]. They targeted at applications where the content is usually generated at or associated with a certain geographical location. The interest packets can be forwarded to those locations that have relevant information. We adopt a similar approach by adding geographical location information in the interest packet so that the location information can be used to direct where the packets will be forwarded to. From the routing perspective, the routing algorithm can use the information to figure out the path to the potential producers of the content.

Most routing problems can be reduced to finding the shortest path in a graph. If we can build a graph among all vehicles, called *vehicle graph*, we can use the same algorithm to find the routing path, except that this graph is constantly changing. Notice that this vehicle graph is different from the *map graph*, in which vertices are intersections and edges are the roads connecting these intersections.

The basic idea of our approach is to reduce the routing problem to the problem of finding the shortest path in the map graph. The starting step is to use source routing. When the consumer needs to send out an interest packet, it computes the shortest path using the map graph. The gap is how to forward a packet from one intersection to another. When a vehicle on one road approaching the intersection needs to relay packets to the next vehicle, it should choose one on the correct road, under the assumption that the GPS information is accurate enough and neighboring nodes have exchanged location information. Once the message is received by a vehicle on the new road, we assume that the restricted greedy forwarding mechanism [8] will be used. That is, nodes will always forward the packet along the road in the same direction the packet has traveled. This restricted greedy forwarding will continue until the node relaying the message is approaching the intersection.

As we mentioned before, the physical distance may be a good measure for finding a path for a vehicle to travel to the destination, and may not necessarily be the best for finding a path for packet forwarding. We need to take into consideration the frequent disconnectivity due to fast mobility of vehicles. The approach we will use is to avoid less traveled roads to reduce the chance of getting into the situation in which a vehicle cannot find the next relay node. To that end, we will estimate the density of vehicles on each road and adjust the distance by the density as the cost of edges used in the algorithm.

4 Progressive Segment Routing

Our proposed routing mechanism consists of two major steps, density estimation of roads and progressive routing.

4.1 Collection of Density Information

To achieve the goal of routing interest messages along the roads that have constant traffic so that a node can likely find the next relay nodes, we develop a mechanism to get an estimate of how frequently each road is traveled. We use the density as the measure of how busy a road is. It is defined as the number of vehicles traveled on a road per unit distance. We normalize it to the range from 0 to 1. We will give a definition next and explain how it will be used to adjust the distance information used for calculating shorted paths.

We do not expect to have density information of all the roads on a map for scalability reason. The pure number of roads may be too large to keep the status up to date. The states maintained at each node should also be minimal. Therefore, we will let each vehicle keep track of the numbers of vehicles encountered for the last n roads it traveled, where n is a design parameter and can be adjusted. For example, we can let a vehicle keep track of $n = 8$ roads most recently traveled. When a vehicle encounters another vehicle on a road, the exchange of beacon messages will allow each identify that another vehicle is on the same road. A vehicle maintains a list of IDs of other vehicles on the current road. Once it moves on to the next road, it only needs to remember the number of vehicles encountered. When a vehicle needs to record the number of vehicles on a new road and it has already had n road information, the oldest road information will be discarded. The stable state is that each vehicle has the information of the numbers of vehicles on last n roads it traveled. Note that the number recorded by a vehicle on a road is an undercount of the actual number of vehicles on the road because it definitely cannot meet every vehicle on the road. Nonetheless, it is the best estimate we can get that actually reflects how busy each road is.

For each road, the density is determined by the number of neighboring vehicles N_i, the length of the road L_i, and transmission range of wireless signal R. We can define the density of road i as follows.

$$Den_i = \min(\frac{N_i}{\beta * \frac{L_i}{R}}, 1) \tag{1}$$

where β is a parameter that defines the target of saturation level and $\beta \geq 1$.

Under the assumption of transmission range being R, for a road of length L, we need at least $\frac{L}{R}$ vehicles to cover the whole road length, if the vehicles are evenly distributed on the road with equal distance among themselves. However, that will not happen in reality. The redundancy parameter β states the target number of vehicles (i.e., β times $\frac{L}{R}$) to accomplish smooth communication on the road. In practice, β can be somehow in the range of 5 to 20. The ratio of the actual number (N_i) of vehicles on the road over this target number ($\beta * \frac{L_i}{R}$) is defined as the density of the road. To limit the density number to the range from 0 to 1, we put an upper limit of 1 to get the density value.

When a consumer node or a relay node needs to calculate the shortest path to the destination, it asks its neighbors about the numbers of the vehicles recorded on their last n roads traveled. The current node calculates the average number of vehicles for each road it has information. Using formula (1), it can calculate the density of roads it has information. If the node does not have information about the number of vehicles recorded, it just considers the density to be 0.

After getting the density information, the node can calculate the adjusted distance for each road as follows.

$$Dis_i = L_i * (1 - \alpha * Den_i) \tag{2}$$

where α $(0 \leq \alpha \leq 1)$ is a parameter that defines how aggressive we want to adjust distance values.

We can look more closely at how the values of α will adjust the distance values. If $\alpha = 0$, then $Dis_i = L_i$. The density information has no effect at all on the length value. In the unlikely case when $\alpha = 1$, we will have $Dis_i = 0$ if the density is $Den_i = 1$, that is, we will consider the length of the road is 0 if density is 1. The larger the values of α, the higher the effect of the density on the road length. More likely, we expect that α get a value close to 1, such as 0.8 or 0.9. This reflects our preference of higher-density roads over lower-density roads.

4.2 Progressive Routing

When a consumer (source) node needs to send an interest packet, it calculates the entire path using Dijkstra's algorithm on the underlying map with the length of road adjusted by the collected density information. Using source routing, it attaches the interest packet with the path, defined as a sequence of intersections that the interest packet should travel. The relay node selectively chooses a forwarder when approaching intersections to make sure that the packet is forwarded to the next node on the correct road. Assisted with the restricted greedy forwarding [8] on the road, the interest packet can be forwarded to the destination locations.

However, the source node only has a limited view of the whole network topology because its neighbors only store densities of up to n roads. It may find the better path close-by and choose sub-optimal path far away. Also the density information may change over time. Therefore, the source routing path may not necessarily be optimal for the interest packets to travel. To provide more flexibility, we allow the relay nodes on the source routing path to make new decisions to update the source routing path based on their density information collected at the location closer where the packet will travel with more up-to-date information. The path traveled so far from the consumer node to the relay node is called a *segment*. The rest of the path will be another segment, which can potentially be divided into more segments if other relay nodes downstream update the routing path.

For example, in Fig. 1, the source node A calculates the path to destination D is $[A, B, E, D]$. However, when the packet arrives at B, B finds that a better path $[B, C, E, D]$ based on its density information. It will replace the original path with $[B, C, E, D]$. Finally the packet will go through $[A, B, C, E, D]$ to get

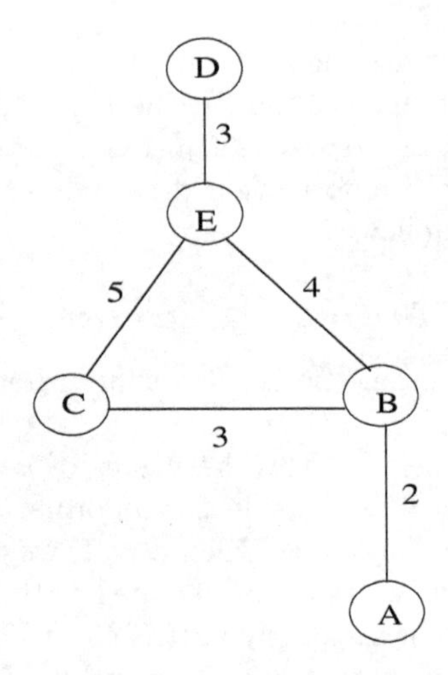

Fig. 1. Updating source routing path

to the destination D. We have two segments $[A, B]$ and $[B, C, E, D]$ in this routing path.

One possible problem with dynamic updating of the source routing path is that we may end up with a loop in the routing path. For example, in Fig. 1, after B updates the routing path $[A, B, E, D]$ with its newly found path $[B, C, E, D]$, it forwards the packet to C. However, C may find the path $[C, B, E, D]$ is the best path to destination D based on the information it collected. It will forward the packet back to B. This process can continue forever.

To deal with this problem, we propose a progressive approach for segment routing. While we still want relay nodes to be able to update the routing path, we will impose a condition on who can update, with the goal of making progress toward the destination at each step.

After calculating the source routing path from the current node to the destination with density-adjusted lengths, the progressive segment routing algorithm uses the unadjusted graph with the original lengths of roads to calculate the distance from each node in the path to the destination. In the source routing path, those nodes having a longer distance than the current node to the destination (both distances using the unadjusted lengths) will be marked. When approaching these marked intermediate nodes in the path, the relay node will not update the source routing path. Only when approaching unmarked intermediate nodes, the relay node may possibly update the source routing path.

We go back to the example in Fig. 1. At node A, all nodes in the source routing path including B, E and D have a shorter distance than A. So no nodes are marked. When approaching intersection B, the relay node will update the path because it finds a better path to D, i.e. $[B, C, E, D]$. However, because C has a longer distance to D than that from B to D in the unadjusted graph, C is marked. So the updated source routing path is $[C*, E, D]$. When the relay node approaches C, because C is marked, it will not update the routing path. Rather it will just forward the packet to a node on the road from C to E. When approaching E, a relay node will not be able find a better path to D. Finally the packet will be delivered to D. By imposing these restrictions, we can avoid the routing loop problem.

We present the progressive segment routing algorithm that a relay node will execute upon receiving a packet to forward it to the next node in Fig. 2. Step 2 determines whether to update the source routing path. Only unmarked intersections will trigger the update. Steps 9 to 15 decide which nodes in the updated path get marked based on the distance to the destination, compared to the distance from the current intersection to the destination.

Data: current node n_c, destination D, source routing path $P = [L_1, L_2, \cdots, L_k]$

1 **if** *the current node approaches an intersection* L_1 **then**
2   **if** *the intersection* L_1 *is marked in the source routing path* **then**
3   remove L_1 from the routing path
4   forward the packet to a node on the next road (L_1, L_2)
5   **else**
6   collect density information from neighboring nodes
7   calculate the adjusted length for each road edge based on densities
8   calculate the shortest path P' from L_1 to D using the adjusted lengths
9   **if** P' *is different from* P **then**
10   use the unadjusted map graph to calculate the distance d_c from L_1 to D
11   **for** *each node* n_i *in* P' *except* L_1 **do**
12   use the unadjusted map graph to calculate the distance d_i from n_i to D
13   **if** $d_i > d_c$ **then**
14   Mark n_i
15   replace P by P'
16   remove L_1 from P
17   forward the packet to a node on the road from L_1 to the first node in P
18 **else**
19   forward the packet to the next vehicle using restricted greedy forwarding

Fig. 2. Progressive segment routing algorithm

We will give another example in Fig. 3 to show that using the progressive segment routing can guarantee that the packet will be forwarded to the final destination, even if a node may be visited twice (a loop). When the original source A forwards the packet to D, the source routing path is $[A, B, C, F, D]$. Since no nodes in the path have a shorter distance than A, they are not marked.

Approaching intersection C, the updated routing path based on density can be $P' = [C, E, G, D]$. The distance of the current node C to D is $3+6=9$ using the map graph. We calculate the distances of nodes in P'. The distance from E to D is 7, which is smaller than C's distance to D. So E is not marked. Similarly G and D are not marked. So the updated routing path is $P = [E, G, D]$.

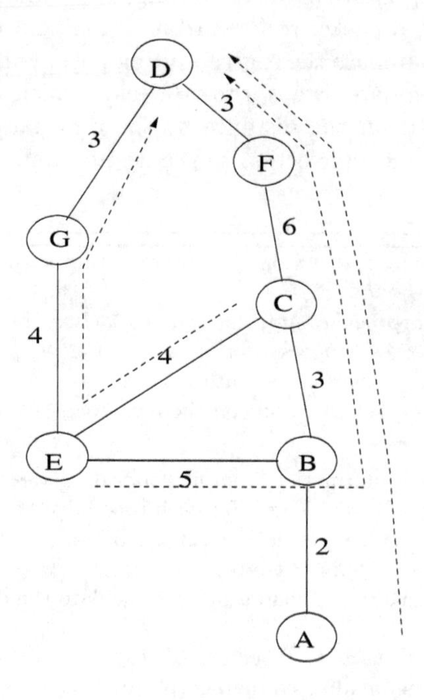

Fig. 3. An example of progressive segment routing

When the packet approaches E, the updated routing path based on density can be $P' = [E, B, C, F, D]$. The distance of the current node E to D is $3+4=7$ using the map graph. We calculate the distances of nodes in P'. The distance from B to D is 12, which is greater than E's distance to D. So B is marked. Similarly C is marked. However, F is not marked. So the updated routing path will be $P = [B*, C*, F, D]$. Based on the algorithm, when the packet approaches B, no update of the routing path will happen since B is marked. Similarly, when the packet approaches C, no update of the routing path will happen since C is marked. It is fine that the path may be updated when approaching F, but that will not cause any problem because we guarantee that progress is made when we do update each time. The routing path in this example consists of three segments, $[A, B, C]$, $[C, E]$ and $[E, B, C, F, D]$.

We will make one more observation. Just from the temporary loop ($[C, E, B, C]$) in which the packet travels, it seems that the routing path is not optimal. Actually, based on the states collected by the decision making relay nodes, the decisions are the best they can make at the time. The changes of the network states or the limitation of information they can get lead to inconsistent views of the network. However the algorithm will still make sure that the packet can be delivered to the final destination.

5 Simulation

5.1 Simulation Setup

The evaluation of the proposed algorithm uses an ns-3 based open-source NDN simulator ndnSIM [10] (Version 2.8) and the SUMO microscopic traffic simulation for urban mobility [7]. We use OpenStreetMap [11] to get a map of our University Campus that spans 1500×1000 m area. Vehicles move in both directions for all roads. The vehicle speed limit is 30–40 km/h. We assume that the transmission range of signals of all vehicles is 150 m. The number of vehicles varies from 50 to 900 with an increment of 50. We implemented the proposed algorithm and compared it with two other algorithms using ndnSIM.

The two previous schemes we compared with are V-NDN and GeoDTN-NDN. V-NDN [5] was the early work applying Named Data Networking architecture to vehicular networks. The pioneering work showed the promise of using a unifying architecture across wired infrastructure, ad hoc, or intermittent DTN. GeoDTN-NDN [1] introduced hybrid routing and DTN support to geographical routing for vehicular Named Data Networking. It also designed a data packet forwarding mechanism based on geographical information to deal with the mobility of the vehicles.

We use two metrics to compare these schemes. One is the packet delivery ratio, which is defined as the ratio of the number of packets successfully delivered to the destination over the total number of packets sent out. The other is the delay, which is defined as the average of delays of all packets that have successfully arrived at the destination.

5.2 Simulation Results

We first show the packet delivery ratios (PDR) of the three schemes, V-NDN, GeoDTN-NDN and progressive segment routing (PSR), when the number of vehicles increases from 50 to 900. In Fig. 4, we can see that the packet delivery ratio generally improves when the number vehicles increases. This is because when we have more vehicles, we have a higher chance to find the next vehicle to forward the interest packet. Therefore, we can reduce the drop rate. Our proposed PSR scheme has highest packet delivery ratios, compared with V-NDN and GeoDTN-NDN, mainly because the PSR scheme chooses roads with higher densities of vehicles, so that more likely a packet will be able to find a relay vehicle to keep forwarding process going. When the number of vehicles gets to 900, the performance of PSR is very close to GeoDTN-NDN, due to limited room of improvement.

Figure 5 shows the average delay of packets achieved by V-NDN, GeoDTN-NDN and PSR approaches. When the number of vehicles increases, the average delays for all three methods decreases. When we have more vehicles, we can use the shortest path in the map graph, rather than finding a longer path that the

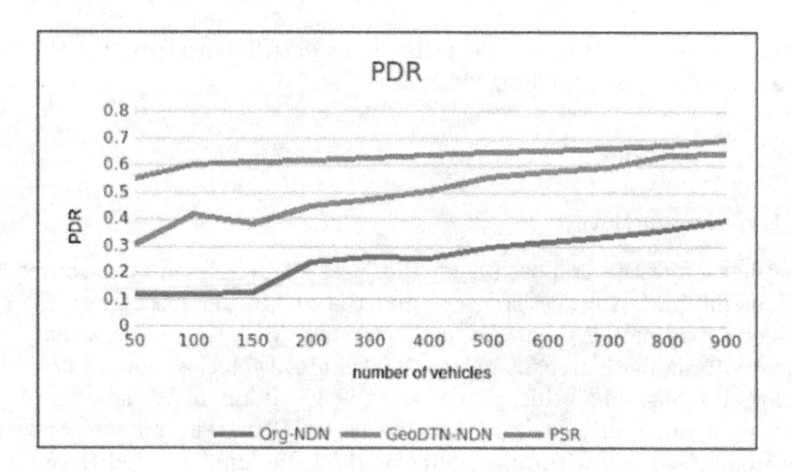

Fig. 4. PDR for the number of vehicles

packet can reach the destination. Again, the proposed PSR method performs better than both V-NDN and GeoDTN-NDN because the decisions made along the path can adapt to the changing conditions on the road and pick the best path for the packets to travel to the destination.

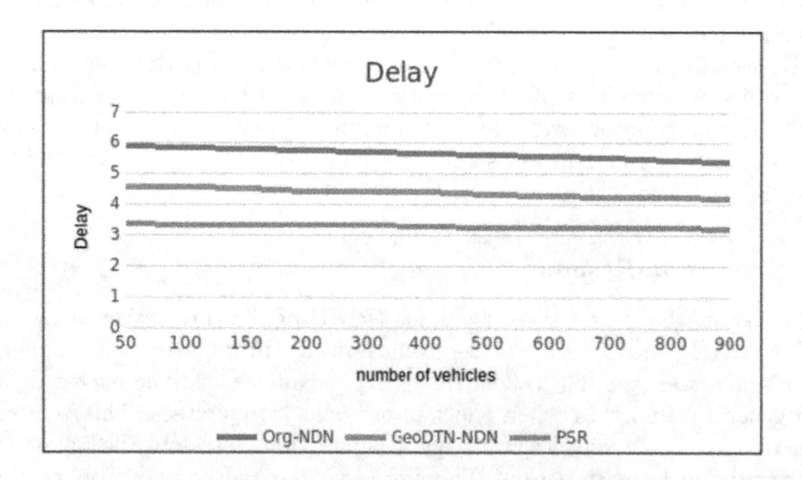

Fig. 5. Delay for the number of vehicles

6 Conclusion

We proposed a progressive segment routing approach for interest forwarding in vehicular Named Data Networking. Our approach is based on the source routing solution using the map graph, with the cost of the edges adjusted by the density information collected. By doing so, we can avoid empty or less traveled roads that have fewer vehicles to forward the traffic. We allowed intermediate nodes to update the routing path based on updated conditions in its proximity, thus creating multiple segments in the routing path. The progress condition imposed on who can update the source routing path guarantees that each routing segment makes progress toward getting to the final destination. Simulation results show that progressive segment routing can improve the packet delivery ratio and reduce the packet delivery time.

References

1. Aldahlan, B., Fei, Z.: A geographic routing strategy with DTN support for vehicular named data networking. In: 2019 IEEE International Conference on Computational Science and Engineering (CSE) and IEEE International Conference on Embedded and Ubiquitous Computing (EUC), pp. 361–366 (2019)
2. Bashandy, E.A., Filsfils, E.C., Previdi, S., Decraene, B., Litkowski, S., Shakir, R.: Segment Routing with the MPLS Data Plane. RFC 8660, RFC Editor (2019). http://www.rfc-editor.org/rfc/rfc8660.txt
3. Bian, C., Zhao, T., Li, X., Yan, W.: Boosting named data networking for efficient packet forwarding in urban VANET scenarios. In: The 21st IEEE International Workshop on Local and Metropolitan Area Networks, pp. 1–6 (2015). https://doi.org/10.1109/LANMAN.2015.7114718
4. Clark, D.D., Tennenhouse, D.L.: Architectural considerations for a new generation of protocols. Comput. Commun. Rev. **20**(4), 200–208 (1990)
5. Grassi, G., Pesavento, D., Pau, G., Vuyyuru, R., Wakikawa, R., Zhang, L.: VANET via named data networking. In: 2014 IEEE Conference on Computer Communications Workshops (INFOCOM WKSHPS), pp. 410–415 (2014). https://doi.org/10.1109/INFCOMW.2014.6849267
6. Grassi, G., Pesavento, D., Pau, G., Zhang, L., Fdida, S.: Navigo: interest forwarding by geolocations in vehicular named data networking. In: 2015 IEEE 16th International Symposium on A World of Wireless, Mobile and Multimedia Networks (WoWMoM), pp. 1–10 (2015). https://doi.org/10.1109/WoWMoM.2015.7158165
7. Krajzewicz, D., Erdmann, J., Behrisch, M., Bieker, L.: Recent development and applications of SUMO - simulation of Urban MObility. Int. J. Adv. Syst. Measur. **5**(3&4), 128–138 (2012). http://elib.dlr.de/80483/
8. Lochert, C., Mauve, M., Fussler, H., Hartenstein, H.: Geographic routing in city scenarios. SIGMOBILE Mob. Comput. Commun. Rev. **9**(1), 69–72 (2005). https://doi.org/10.1145/1055959.1055970. http://doi.acm.org/10.1145/1055959.1055970
9. Maryam, H., Wahid, A., Shah, M.A.: Mitigating broadcast storm in interest/data packet forwarding in vehicular content centric networking. In: 2017 International Conference on Communication Technologies (ComTech), pp. 162–167 (2017). https://doi.org/10.1109/COMTECH.2017.8065767
10. Mastorakis, S., Afanasyev, A., Moiseenko, I., Zhang, L.: ndnSIM 2: an updated NDN simulator for NS-3, NDN Technical Report NDN-0028, Revision 2 (2016). http://ndnsim.net/current/
11. OpenStreetMap contributors: Planet dump (2017). https://planet.osm.org. https://www.openstreetmap.org
12. Yu, Y.-T., Li, Y., Ma, X., Shang, W., Sanadidi, M.Y., Gerla, M.: Scalable opportunistic VANET content routing with encounter information. In: 2013 21st IEEE International Conference on Network Protocols (ICNP), pp. 1–6 (2013). https://doi.org/10.1109/ICNP.2013.6733679

Smart City Sensor Network Control and Optimization Using Intelligent Agents

Razan AlFar[1](✉), Yehia Kotb[2](✉), and Michael Bauer[1](✉)

[1] Computer Science Department, Western University, London, ON, Canada
{ralfar,bauer}@uwo.ca

[2] Computer Engineering Department, American University of the Middle East, Egaila, Kuwait
yehia.kotb@aum.edu.kw

Abstract. The sensor network is one of the most important layers in the smart city architecture. The major challenge for smart cities is processing the huge amount of information being transferred through the network. The data is usually generated from the sensor networks that build the physical layer of the smart city. In this paper, a new physical layer management framework is proposed to work as an edge layer that intelligently learns the behavior of the physical system and then optimizes it according to predefined objectives. The intelligence in the framework is divided into three modules. The first module learns the system states and the actions that can yield a state from another. The second module learns how to reach a certain objective expected from managing a sensor. The third module learns the overall system's behavior and performance. States are represented as the fuzzified values of the sensor readings.

Keywords: Smart City · Sensor Networks · Probabilistic Learning · Intelligent Systems

1 Introduction

Smart cities are one of the most active research fields today as it aims to provide solutions to problems in urban environments in order to help humans live better and increase efficiency in many aspects. Smart cities, like any emerging new topic, do not possess a clear definition. One simple definition was proposed by Celino and Kotoulas, which defined it to be a city that is able to effectively process networked information to improve outcomes on any aspect of city operations [1]. It should be noted that this definition is only from the perspective of internet computing and changing the perspective changes the definition. The city operations they mentioned included, but not limited to, providing information to authorities, e-commerce transactions and businesses, energy optimization and

K. Arai (Ed.): FICC 2021, AISC 1363, pp. 438–456, 2021.
https://doi.org/10.1007/978-3-030-73100-7_31

control, resource production, resource consumption, traffic management, public safety, emergency responses, medical applications, and many more.

Another attempt to provide a concrete definition for a smart city comes from Dameri [3]. Since the idea of a smart city is relatively a recent phenomenon, it suffers from ambiguity when it comes to a concrete definition. Not only that, but also the spread of the new trend world wide, and in different fields and industries, makes it very hard to come up with a definition that would define the purpose of a smart city in all sectors worldwide. The reason behind the spread of such a new phenomenon is due to the vital problems related to urban life in general including traffic, pollution, urban crowding, poverty, and so much more. Only few vital aspects were discussed by the authors to reach a definition for a smart city. The determining aspects are terminology, components, boundaries, and scope.

The word *smart* in a *smart city* is not the only word that describes the same concept. In literature, the phrase *intelligent city* is one of the terminology phrases that has been introduced and it refers to cities with several competencies. A city that is able to produce knowledge, use the produced knowledge to increase abilities and all the positive aspects of knowledge gain make the city an intelligent city. Digital city is another phrase that refers to the fact that the city is digitized, where it supports city wide data processing and information sharing. A third notion is the sustainable city. It is basically an environment-friendly city where reducing CO2 emissions with the help of technology is its main objective along with efficient energy production and improving buildings' energy consumption; it is also known as a *green city*. A different term is a *techno-city*, where technology is used to help infrastructure and services become more efficient and more effective. The focus in techno-city is on urban space quality, mobility, public transports, and logistics. The last (but not least) term used is the *well-being city*. A well-being city is a city that aims to guarantee the best quality of life for citizens and be attractive for businesses. This aspect not only depends on technology but also upon planning, culture, climate, history, and many other factors. It is obvious how the goal behind the application of smart cities dictates what its definition would be [3].

Components are the second vital aspect that was mentioned by Dameri in [3] for defining smart cities. They can be summarized as follows: land, technologies, citizens, and government. The land is the geographical area of the city used for the smart city project. Technologies provide feasible and sufficient infrastructure and services to serve people better. That includes government processes as well. Citizens are addressed by all smart initiatives and that is why they should feel the difference after the introduction of smart city projects. Lastly, the government makes decisions on the smart city projects that affect the citizens. Many other components were not mentioned in Dameri's work, like culture, laws, people's readiness, citizen privacy, and other views that can be considered as vital components, however, the topic is outside the scope of this paper.

Boundaries are the defined limits for the project whether it is geographical or procedural. In [3], only the geographical area and classifications like city, region, city networks, nation, and a global view were considered. The scope is also not well defined which makes finding a definition for the smart city even harder. To find a comprehensive definition for the term **smart city**, all elements discussed above need be taken into consideration.

Smart city research faces many challenges and those challenges vary from one application to another. For example, applications that deal with citizen's information suffers from privacy threats. That is why privacy in smart cities are grabbing attention since sectors, like the financial or healthcare sectors, will not fully participate in smart city initiatives before privacy and security are both guaranteed [1]. Other challenges are related to data itself. How diverse the data is? How noisy is it? What are the processes for data cleaning considering how huge the data volume is? The need to efficiently analyze collected data is one of the big challenges facing smart cities, especially when there is a necessity for real-time responses. Those challenges drive the research of big-data which is considered one of the components of smart city systems [1].

Pellicer *et al.* introduced two different definitions of *Smart City* in [14]. They defined it to be an urban system that uses information and communication technology (ICT) to make both its infrastructure and its public services more interactive, more accessible, and more efficient. This definition focuses on how infrastructure and services are interactive, accessible, and efficient, but it does not guarantee many other aspects concerning the reduction of energy consumption and being green. Also, it ignores privacy in many sectors like banking and healthcare. The other definition that they introduced defines a smart city to be "a city committed to its environment, both environmentally and in terms of the cultural and historical elements, and where the infrastructure is equipped with the most advanced technological solutions to facilitate citizen interaction with urban elements".

In this paper, a management system for the physical layer (sensor network) of smart cities is introduced. The idea is to simplify the management and operation tasks at the upper layers, e.g. the fog layer and cloud layer.

This paper is organized as follows: Sect. 2 presents some of the most important and latest work that has been done in the field of smart cities. Section 3 points out the contribution of this paper. Section 4 summarizes the contribution of the paper. Section 5 illustrates and describes the proposed mathematical model and algorithms. Section 6 proposed a set of equations to measure the success ratio of the proposed framework. Section 7 shows the simulated experiments and results. Finally, Sect. 8 concludes the work in this paper.

2 Literature Survey

Interest in smart cities has increased due to the growth of population in urban areas, as well as the many challenges being faced to meet the citizens' needs in areas such as education, healthcare, security, water, and far more. Smart cities aid in solving these issues by embracing the help of real-time sensors, cloud computing, and big data analytics. However, the adoption of smart city solutions comes with other challenges, including ensuring the existence of sufficient communication infrastructure, available funds, required skills, and legal aspects to ensure that the privacy of citizens is not violated. The smart city's infrastructure consists of sensors, data platforms, applications, and security. The sensor layer is considered as one of the most important layers in the smart city as it incorporates sensors and actuators. Sensors are distributed around urban areas and are used to collect data on events and conditions within the city. Actuators translate the received messages into actions [4]. Sensors are used in a variety of smart city applications, such as data collection on energy and water usage, environmental conditions, and traffic flow patterns [4]. As a result, there are many different types of sensors that can be utilized for different functionality. While the data collected and applications may make citizen's lives more convenient, more sensors also create additional energy consumption and create challenges in tracking and maintaining the physical assets.

Kang et al. [9] analyzed various types of sensors and described the sensor's importance from the smart home services perspective. To help understand and track the sensors and their behavior, they designed a sensor tree to map the sensors and their importance. From the tree, they were able to categorize the services of smart homes and discovered that motion sensors are the most valuable sensors to use and that the gas sensors are the least important ones when using them for smart home services.

Wireless sensors are being widely used because of their efficiency, availability, increased mobility, and low-cost nature. Wireless sensor networks are often preferred to be used in various smart city applications and wireless sensor networks (WSNs) have been extensively used in applications such as collecting data to maintain property security and enhance the quality of life [5]. The applications of WSNs can provide real-time data to city authorities to help operations [22]. The data collected from sensors can serve various objectives, such as health tracking, safety and control of industry, environmental monitoring, and prediction of disasters [6,12,16,20]. More complex sensors, such as accelerometers, cardio-tachometers, have been employed by Wang et al.[19] to detect elderly accidental falls. The authors in [2] presented a traffic light system using wireless communication where it delivers a support infrastructure for intelligent control in smart cities. The system relies on a ZigBee wireless structure. WSNs have been used for applications for automation, such as home automation [13] and industrial automation [15,23].

Understanding, processing, and deriving information from sensor data also represent significant challenges. Hidden Markov Models [18] and Conditional Random Fields [7] are known methods that can be implemented to understand latent random variables to supervise patterns of sensor-generated data. In WSNs, optimization plays an important role. WSN optimization can be loosely classified into a single and multi-objective optimization (MOO) problem. The difference between single and multi-objective optimization problems is the number of objectives to be targeted. The main goal is to maximize or minimize one objective at a time when dealing with a single optimization problem. Contrary to that, the optimization of multiple objectives takes place in a multi-objective optimization problem where several objectives get optimized simultaneously [8]. Accordingly, multi-objective optimization has caught the attention of researchers and engineers as it is considered a challenging and demanding task.

In MOO, problems are introduced when the objectives conflict with each other and it may not be possible to find the optimal solution to cater to all needs. To address this, the decision-makers may have to take a decision to reach the best possible solution. There are many approaches and algorithms used to solve this problem. The most common way is to measure the weights of the objectives and determine which objectives has the most significant weight. Other proposed methods and techniques exist to handle multi-objective optimization, including Min-Max, Pareto, Ranking, Goals, Preference, Gene, Sub-population, Lexicographic, Phenotype sharing function, and Fuzzy [17].

Olatinwo et al. [11] proposed a new approach to maximize the energy and throughput in a WSN. They used wireless information and power transfer method by harvesting energy from a dedicated radio frequency. The method was applied to water-quality monitoring where the improved energy efficiency addressed the problem of energy scarcity.

Our proposed model uses fuzzy states which is an effective technique, it utilizes approximate reasoning given the imprecise nature of information that is being manipulated. This is the main reason why research related to vagueness has been increasing [10]. Zadeh [21] presented the idea of a fuzzy set as a vague fact model. The fuzzy set theory is considered very convenient and credible, and it has been extensively used in control systems. It can be defined with membership function or commonsense and is a suitable optimization technique used for training online and offline machine learning algorithms.

In this paper, a framework that can combine various readings and optimize the performance of the physical layer is proposed. The way it functions is by moving the decision-making process from the cloud and fog layers to the edge and the sensor network layers.

3 Problem Statement

Smart cities are emerging and capturing the attention of researchers and citizens around the world. Extensive research has been done on approaches to support the development of smart city applications, specifically around handling sensors, their data, and associated data analysis. However, research done on addressing the management challenges of the smart city's environments is minimal. Research around smart cities is targeted to address various management issues and is focused on facilitating sensors.

This paper is a part of a project that focuses on solving the management issues that are contributed to sensors in smart cities. A smart city consists of several layers: a physical layer, an edge layer, and a cloud layer. All layers of this system should be well-connected to have faster communication, easier access, and a plethora of facilitating services. Generally, management layers send requests and tasks to their agents to be executed. Those agents happen to be mostly the fog nodes, which are the closest to the sensors and the hardware components.

In this paper, the focus is on the device or the physical layer. The device layer consists of sensors, actuators, etc. with different attributes and communication protocols. The idea is that smart cities depend on sensors to get readings to enable applications to manage houses, buildings, etc. When working in these areas, an abundance of problems may arise such as managing sensors with different attributes, functionalities, and dealing with heterogeneity. Accordingly, it is extremely important to make sure that each sensor is outputting the desired output and working efficiently to perform its task. One way to achieve this is by assuring each sensor is being utilized effectively and that sensors are working uniformly to leverage the efficiency of the whole system.

The main challenge is to deal with complexity when working with multiple applications. Many applications rely on many different sensing devices which can affect the emergence of smart cities. Therefore, a framework modeling the decision-making process that selects a set of actions from environment readings will be introduced to aid in solving this problem.

To run experiments, we utilize our simulator implemented using JAVA. Decisions can be done conforming to, for instance, event condition rules, objective functions, or prediction models based on techniques such as Bayesian networks, decision trees, or fuzzy logic. Probabilistic methods were used as an approach as they can produce probability functions, where they can be utilized to express the reason behind the failure of a sensor for instance. Such predictive mechanisms can keep learning and are being used to improve and optimize actions taken to manage sensors and help avoid errors in the future.

Furthermore, policies will be identified for different types of managed objects in a smart city, including sensors. For that, it is required to define the different kinds of policies that are reliable to optimize and monitor management systems using algorithms. We can also examine policies dealing with configuration management and adaptation as sensors may fail or stop working as expected, affecting the whole system or subsystem. Automating configurations and adaptation will be considered to introduce self-adaptation when making decisions, self-organization for routing protocols, self-optimization to define the optimal usage of the constrained resources, and many more that can be introduced as work progress.

In summary, in this paper, a framework for controlling the sensor's physical layer is proposed. The framework takes into consideration the set of integrated sensors and the data they produce. Additionally, it considers the events that get translated out of the sensor readings, the set of intelligent agents that control the proposed framework, the set of constraints to be met, and the goals to be reached throughout the process of framework execution.

4 Contribution

This paper proposes a novel approach to reduce the computational load on the cloud and fog layers. In addition, it looks to minimize the communication between sensors and the fog layer, and between the cloud layer and the fog layer.

This is achieved by optimizing the performance of the physical layer and moving some of the decision-making tasks from the cloud layer and the fog layer to the edge layer and the sensor network.

The physical layer is mathematically modeled and a proposed hierarchical approach is introduced for modeling the decision-making process by choosing a set of actions given some environment readings. The learning mechanism in the proposed framework is divided into three levels. The first level is to build what is called in this paper the *state event fabric*. They are the states that are achieved by certain sensor readings and the actions that will aid in moving the system from one state to another. The second level of the learning process is to learn the mapping between objectives for certain environmental property and the actions to be taken. The third level of the learning process is an overall learning mechanism that helps agents learn how to reach a global objective.

Moreover, the objective in level two is local and is simply an objective of one sensor. In level three, it is an objective for all the sensors connected, where they can affect each other as will be seen in this paper. The learning algorithm proposed in level one is a linear programming based learning algorithm that keeps trying alternatives until the system converges.

5 Proposed Model

The proposed model is a detailed mathematical model for the physical layer. The physical layer is described as follows:

5.1 Environment $\mathbb{E}$

In this paper, environment $\mathbb{E}$ is a set of values that describe the status of that environment. The length of this set depends on the values that need to be taken into consideration. Mathematically, the environment is modeled as follows:

$$\mathbb{E} = \langle v_1, v_2, \ldots v_n \rangle \tag{1}$$

where $n = \|\mathbb{E}\|$. An example related to those values is as follows: if the framework considers temperature, illumination, pressure, elevation, and pollution. Then $\mathbb{E} \equiv \langle v_1, v_2, v_3, v_4, v_5 \rangle$, where v_1 is the actual temperature measured in Celsius or Fahrenheit, v_2 is the illumination in Lux as an example, v_3 is the pressure measured in Pascal, v_4 is the elevation measured in meters, and lastly v_5 is the pollution measured in $\mu g/m^3$.

5.2 Sensors $\mathbb{S}$

For every $v_i \in \mathbb{E}$, there is a sensor that senses the environment for v and produces an equivalent reading $r \in \mathbb{R}$. Physical layer has a set of sensors $\mathbb{S}$ such that:

$$\mathbb{S} = \langle s_1, s_2, \ldots s_n \rangle \tag{2}$$

and $\|\mathbb{E}\| = \|\mathbb{S}\| = n$. Every sensor $s \in \mathbb{S}$ can be described as:

$$s_i \in \mathbb{S} = \langle v, \Psi_i, r, \tau \rangle \tag{3}$$

where $\Psi_i(v) = r$ is the transfer function that describes the sensor behavior. While v and r would seem the same, r is the numerical representation of v. This is crucial to be taken into consideration as the performance of the whole system could be sensitive to the transfer function error. In other words, if $\|v - r\| \geq T$, where T is the maximum acceptable error threshold.

$v \in \mathbb{E}$ is the actual level of the physical phenomenon under consideration. $r \in \mathbb{R}$ is the measurement of the level of that physical phenomenon, and $\tau = \frac{1}{f}$ is the periodicity of the sensor, which is the frequency of execution of Ψ.

5.3 Sensor Intelligent Agents $\mathbb{A}$

Every sensor is controlled by an intelligent agent that tries to bring the sensor output to meet a certain objective by applying an action to the environment. An agent has an objective Ξ and a learning algorithm Ω.

An agent accesses the reading history $\mathbb{H}$, actions Λ, sensor states Φ, learning algorithm Ω, objectives Ξ, and current reading r.

5.4 Reading History $\mathbb{H}$

History $\mathbb{H}$ is the previous readings that occurred in the past. Theoretically speaking, $\|\mathbb{H}\|$ could be infinite, however, in reality, and for the sake of performance and because of the limited memory capacity of embedded systems, history length has to be controlled. History can mathematically be represented as:

$$\mathbb{H} = \langle \hbar, \ell \rangle \tag{4}$$

Theoretically, $\hbar$ can be modeled as:

$$\hbar(t) = \int_0^t r \, dt \tag{5}$$

but since the history length is ℓ as explained previously. The equation is:

$$\hbar(t) = \left\{ \begin{array}{l} \int_0^t r \, dt, \quad \text{for } t \leq \ell \\ \\ \int_{t-\ell}^t r \, dt, \text{ for } t > \ell \end{array} \right\} \tag{6}$$

5.5 Actions Λ

Every agent has a predefined set of basic actions that a certain action can be chosen from. An action changes the environment from one state to another. The set of actions is countable and finite. Actions can be represented as follows:

$$\lambda = \{\lambda_1, \lambda_2, \ldots, \lambda_l\} \tag{7}$$

where $l = \|\Lambda\|$. As mentioned earlier, an action moves the environment from a state to another.

5.6 Sensor States Φ

For every sensor agent, there is a set of valid sensor states Φ. The agent tries to move the environment from one valid sensor state to another valid sensor state based on the objective the sensor agent is seeking. States can be mathematically modeled as:

$$\Phi = \{\phi_1, \phi_2, \ldots, \phi_{ssl}\} \tag{8}$$

where $ssl \geq 0$ is the number of sensor states the agent recognizes.

As mentioned earlier, every action leads the environment from one state to another. This makes the state space a connected fabric connected by actions. This is represented by:

$$\mathbb{F} = (\Phi \times \Phi) \tag{9}$$

where $\mathbb{F}$ is the state fabric that describes the state space. Every single transition between a pair of states Φ is done through an event in this way:

$$\forall \phi_i \in \Phi, \forall \phi_j \in \Phi, (\phi_i, \phi_j) \neq \emptyset \rightarrow \exists \lambda \in \Lambda \| \delta(\phi_i, \lambda) = \phi_j \tag{10}$$

where δ is the transition function that moves the system from one state to another when a certain action takes place. The equation means that if the sensor states ϕ_i and ϕ_j are connected, then there is an action λ that moves the sensor state from ϕ_i to ϕ_j. Point out that if $\delta(\phi_i, \lambda) = \phi_j$ that does not mean that $\delta(\phi_j, \lambda) = \phi_i$. That is:

$$(\delta(\phi_i, \lambda) = \phi_j) \not\equiv (\delta(\phi_j, \lambda) = \phi_i) \tag{11}$$

Mapping among states can be learned through interaction in an incident matrix shape as shown in Fig. 1.

$$\mathbb{M} = \begin{array}{c} \\ \phi_1 \\ \phi_2 \\ \phi_3 \\ \phi_4 \\ \phi_5 \end{array} \begin{array}{c} \begin{array}{ccccc} \phi_1 & \phi_2 & \phi_3 & \phi_4 & \phi_5 \end{array} \\ \begin{pmatrix} 0 & \lambda_1 & \infty & \infty & \lambda_4 \\ \infty & 0 & \infty & \lambda_3 & \lambda_1 \\ \lambda_2 & \infty & 0 & \infty & \lambda_3 \\ \infty & \infty & \lambda_2 & 0 & \lambda_3 \\ \infty & \infty & \lambda_1 & \infty & 0 \end{pmatrix} \end{array}$$

Fig. 1. Incident matrix $\mathbb{M}$ of the state space fabric

From the incident matrix shown in Fig. 1, the following can be concluded:

1. The incident matrix of a sensor state transition is always a square matrix.
2. The main diagonal of the matrix is always 0 since there is no event needed to move the system from a state to the same state.
3. If no event can move the system from ϕ_i to ϕ_j, then $\mathbb{M}(i, j) = \infty$.
4. If there is an event λ that can move the system from ϕ_i to ϕ_j, then $\mathbb{M}(i, j) = \lambda$.

Every sensor state is a fuzzy function and the system is in a certain state if and only if sensor reading has an acceptable membership in the fuzzy function that describes the state.

5.7 Incident Matrix Learning Algorithm

Incident matrix $\mathbb{M}$ is the description of the sensor state event fabric. It tells about the dynamic behavior of the system and how the state of the system changes when performing a certain action. At the beginning of the learning process, the incident matrix $\mathbb{M}$ is populated with ∞ except for the main diagonal. Initially, $\mathbb{M}$ looks as illustrated in Fig. 2.

$$\mathbb{M} = \begin{array}{c} \phi_1 \\ \phi_2 \\ \phi_3 \\ \phi_4 \\ \phi_5 \end{array} \overset{\begin{array}{ccccc} \phi_1 & \phi_2 & \phi_3 & \phi_4 & \phi_5 \end{array}}{\begin{pmatrix} 0 & \infty & \infty & \infty & \infty \\ \infty & 0 & \infty & \infty & \infty \\ \infty & \infty & 0 & \infty & \infty \\ \infty & \infty & \infty & 0 & \infty \\ \infty & \infty & \infty & \infty & 0 \end{pmatrix}}$$

Fig. 2. Incident matrix $\mathbb{M}$ of the state space fabric

The system then applies an event λ, and the new environment property is evaluated against the fuzzy sets describing the states. The state that will give the highest membership is the state that is chosen. Algorithm 1 is an algorithm that shows in detail how the incident matrix is learned. The assumption here is that the sensor state set are predefined and the actions the sensor agent would take is also predefined. Emphasize the fact that the actions are very basic. The agent can also take compound actions that are composed of several events defined in λ. Composed events do not appear in the incident matrix for two reasons:

1. They will lead to the same result as applying a series of simple actions in λ.
2. They are theoretically infinite since events can be repeated zero or more times.

Algorithm 1 works as follows: It takes the set of states and the set of events as input parameters. It creates the incident matrix $\mathbb{M}$ as a square matrix with a length equal to the length of the state set. That is because the incident matrix defines the relationships between the states with each other. Mathematically as mentioned before, $\mathbb{M}$ is nothing but $\Phi \times \Phi$. This cross-product defines the topology of the state connections.

After creating the incident matrix $\mathbb{M}$, it was initialized. The initialization process sets the main diagonal with 0s. As aforementioned, this is because the diagonal cells represent relationships between every state and itself. Since no event is needed to remain in the same state, the diagonal is set to 0s. Any other cell that is not a diagonal is set to ∞.

The process of filling up $\mathbb{M}$ starts by trying every possible state ϕ_i with every other state ϕ_j using every action λ_k and whenever a transition happens, λ_k is added to $\mathbb{M}(i, j)$.

Indicate that this algorithm works beforehand (offline). It learns the topology of the state event fabric before the system even starts. The online version of the algorithm works as follows:

Algorithm 1. Online Learning of $\mathbb{M}$

```
    INPUT   : φc, φg, Λ, M
    OUTPUT: M
    FUNCTION Online_Learn_M(φc, φg, Λ, M)
 4: Start_State_index ← indexOf(φc)
    Goal_State_index ← indexOf(φg)
    for i ← 1; i ≤ ‖ Λ‖; i ← i + 1; do
      if M( Start_State_index , Goal_State_index ) = ∞ then
 8:     r ← δ(φi, λk)
        if φc(r) ∈ φg then
          M( Start_State_index , Goal_State_index ) ← Λ(i)
        end if
12:   end if
    end for
    RETURN M
    END FUNCTION
```

Algorithm 1 is the online learning algorithm of $\mathbb{M}$. During run time, it takes the current state ϕ_c and the goal state ϕ_g as input parameters. It also takes the universal set of sensor agent actions Λ and the incident matrix $\mathbb{M}$. The algorithm tests every action, and the action that will lead to the goal state is set to the incident matrix. The complexity of this algorithm is $O(n)$ in its worst case.

It should be emphasized that Algorithm 1 is used to build the incident matrix $\mathbb{M}$, but it does not really find an optimal execution sequence to reach state b from state a when the two states are not directly linked through an action $\lambda \in \Lambda$.

If the two states a and b are not directly connected then the problem here is an $\mathbb{NP}$–Complete problem. The following is the approach to find the shortest sequence of actions that would lead from state a to state b when the two states are not directly connected through an action $\lambda \in \Lambda$. In other words, $\mathbb{M}(index(a), index(b)) = \infty$.

Assume that there is a sensor that has 5 states. After learning $\mathbb{M}$, the state action fabric is illustrated in Fig. 3.

$$\mathbb{M} = \begin{array}{c} \\ \phi_1 \\ \phi_2 \\ \phi_3 \\ \phi_4 \\ \phi_5 \end{array} \begin{array}{c} \begin{array}{ccccc} \phi_1 & \phi_2 & \phi_3 & \phi_4 & \phi_5 \end{array} \\ \begin{pmatrix} 0 & \lambda_1 & \infty & \lambda_7 & \infty \\ \lambda_2 & 0 & \lambda_3 & \infty & \infty \\ \lambda_1 & \infty & 0 & \lambda_6 & \infty \\ \infty & \lambda_4 & \infty & 0 & \lambda_5 \\ \infty & \infty & \infty & \lambda_9 & 0 \end{pmatrix} \end{array}$$

Fig. 3. Incident matrix $\mathbb{M}$ of the state space fabric

It is required to move the system from state ψ_2 to ψ_5. The following equation describes this situation:

$$\mathbb{P}(\phi_5|\phi_2) = \mathbb{P}(\phi_5|\phi_1) \times \mathbb{P}(\phi_1|\phi_2) + \mathbb{P}(\phi_5|\phi_3) \times \mathbb{P}(\phi_3|\phi_2) \tag{12}$$

That is simply because ϕ_2 is connected to ϕ_1 and ϕ_3. Knowing that ϕ_1 is connected to ϕ_2 and ϕ_4, and ϕ_3 is connected to ϕ_1 and ϕ_4, that leads to four new possibilities, which are: (Through ϕ_2 Through ϕ_1), (Through ϕ_4 Through ϕ_1), (Through ϕ_1 Through ϕ_3), or (Through ϕ_4 Through ϕ_3).

Since ϕ_1 and ϕ_2 are already visited, then they will not be added again to the equation. Hence, they would cause an infinite cycle. The equation then becomes:

$$\mathbb{P}(\phi_5|\phi_2) = \mathbb{P}(\phi_5|\phi_4)\times\mathbb{P}(\phi_4|\phi_1)\times\mathbb{P}(\phi_1|\phi_2)+\mathbb{P}(\phi_5|\phi_3)\times\mathbb{P}(\phi_4|\phi_3)\times\mathbb{P}(\phi_3|\phi_5) \tag{13}$$

From $\mathbb{M}$, it can be comprehended that ϕ_4 is directly connected to ϕ_5, this tells us that the probability for both passes is greater than zero, and it is possible to reach ϕ_5 from ϕ_2 through two different passes. Those passes can easily be found through reading, then by applying the probabilistic equation in reverse as follows:

1. $\phi_2 \xrightarrow{\lambda_2} \phi_1 \xrightarrow{\lambda_7} \phi_4 \xrightarrow{\lambda_5} \phi_5$
2. $\phi_2 \xrightarrow{\lambda_3} \phi_3 \xrightarrow{\lambda_6} \phi_4 \xrightarrow{\lambda_5} \phi_5$

Since the weight of executing the action itself is not taken into consideration, all actions have the same weight, and since the number of actions is the same in the two scenarios, then any of the two paths would be the solution. If actions have different weights, then the path that minimized the cost of execution would be the one to be selected.

After presenting the solution methodology. This is shown in Algorithm 2 as follows:

Algorithm 2. Optimal Path between Two States

```
    INPUT  : Φ, φ_c, φ_g, M
    OUTPUT: Path
    FUNCTION Find_Path(Φ, φ_c, φ_g, Λ, M)
    φ_w = φ_c
 5: create ω = ∅ as path space.
    create P(φ_g|φ_c) = ∅ as probability set.
    while φ_w ≠ φ_g do
      i ← 0
      for every φ ∈ φ_w• do
10:     P(φ_g|φ_c)[i] ← P(φ_g|φ_w) × →P(φ_w|φ_c)
        i ← i + 1
      end for
    end while
    END FUNCTION
```

5.8 Physical Layer Intelligent Agent $\mathbb{C}$

This agent can be seen as a global controller. While the model is a single point failure model, this problem can be solved by duplication or mirroring and it is out of the scope of this paper. This agent has a global objective like saving electrical energy as an example. The agent tries to manage sensor agents to apply actions so that the objective is met. This happens through reinforcement learning. This agent asks sensor agents to move to lower states. Note that the sensor agent senses and applies an action to change the environment. As an example and this will be seen in Sect. 7, the temperature sensor agent takes the reading from the sensor and then controls air conditioners and/or heaters as an instance. This agent is mathematically modeled as follows:

$$\mathbb{C} = \langle \mathbb{A}, \mathbb{O} \rangle \tag{14}$$

where $\mathbb{A}$ is the set of agents described in Sect. 5.3 and $\mathbb{O}$ is the set of overall objectives the framework needs to meet.

It is worth noting that $\mathbb{C}$ communicates to one or more $a \in \mathbb{A}$ commands to take actions $\lambda \in \lambda_l, \lambda_h$, where λ_l is the action to go to the lower state regardless what that state is and λ_h is the action to go to the higher state regardless what that state is. After every action λ, $\mathbb{C}$ evaluates the current overall state and compare it with the objective. If it is closer to the objective, then it is a success and $\mathbb{C}$ accepts the action applied. If not, then $\mathbb{C}$ rolls back the previous action and takes a new action. This is illustrated in Algorithm 3 as follows:

Algorithm 3. $\mathbb{C}$ behavior Algorithm

INPUT : $o \in \mathbb{O}, \mathbb{A}, \mathbb{E}_s$
OUTPUT: *Path*
FUNCTION Find_action($\Phi, \phi_c, \phi_g, \Lambda, \mathbb{M}$)
$\mathbb{E}_r = \mathbb{O} - \mathbb{E}_s$
$\lambda = \emptyset$
6: **if** $\mathbb{E}_r \geq \mathbb{T}_h$ **then**
  $\lambda = \lambda_l$
else
  if $\mathbb{E}_r \leq \mathbb{T}_l$ **then**
    $\lambda = \lambda_h$
  end if
12: **end if**
if $\lambda \neq \emptyset$ **then**
  choose $a \in \mathbb{A}$ that has a less negative effect of change.
  SEND λ to agent $a \in \mathbb{A}$
end if
END FUNCTION

---

6 Load Reduction of Network and Cloud Server

As mentioned earlier, the framework aims to allow the physical layer to be less dependent on the cloud layer. Specifically, to give the physical layer more autonomy and independence. As a result, this will reduce the load on the communication network and the cloud servers. This section measures the percentage of the load lifted from the network and servers.

Having n sensors each with its reading frequency f. Sending those messages to the cloud in separate messages means that the total length of messages that will be sent to the cloud servers are as follows:

$$\xi = \int_{x=1}^{n} f_x \times \ell_x \quad \partial x \tag{15}$$

where ξ is the total transmission length, ℓ_x is the average length of messages sent by sensor x, and f_x is the reading frequency of sensor x.

Messages that are being sent can be classified into two classes, **local** requests and, **global** requests. Local requests are messages that are being sent to ask for directions about sensor behavior, whereas global messages are messages that are being sent to ask for an overall objective. Assume that the percentage of local messages is β. This makes the percentage of global messages $1 - \beta$, where $0 \leq \beta \leq 1$. Equation 15 can be modified as follows:

$$\xi = \int_{x=1}^{n} \beta_x \times f_x \times \ell_x \quad \partial x + \int_{x=1}^{n} (1 - \beta_x) \times f_x \times g_x \quad \partial x \tag{16}$$

where ℓ_x is the average length of local messages sent by sensor x and g_x is the average length of global messages sent by sensor x.

If every local message initiates a process with processing cost p_l and every global message initiates a process with processing cost p_g, the processing cost on the server can be represented as follows:

$$p_c = \int_{x=1}^{n} \beta_x \times f_x \times p_\ell \quad \partial x + \int_{x=1}^{n} (1 - \beta_x) \times f_x \times p_g \quad \partial x \tag{17}$$

where p_c is the total processing cost imposed on the cloud server.

Now when delegating some decision-making responsibility to the physical layer through edge computing, this will result in reducing the amount of data that needs to be transmitted through the network and the processes that get initiated on the cloud server. The proposed framework processes all local decision-making. In addition to a portion of the global ones, which does not need any more information rather than what is already been collected by sensors. Assuming that the portion of global decision-making processes that need to be transmitted to the cloud is α, the network utilization reduction can be calculated by:

$$\nabla_c = \frac{\int_{x=1}^{n} \alpha_x (1 - \beta_x) \times f_x \times g_x \quad \partial x}{\int_{x=1}^{n} \beta_x \times f_x \times \ell_x \quad \partial x + \int_{x=1}^{n} (1 - \beta_x) \times f_x \times g_x \quad \partial x} \tag{18}$$

where ∇_c is the network utilization after delegating decision-making to the edge layer with respect to performing decision-making in the cloud layer.

In the same way, the reduction of server processing on the cloud was calculated as follows:

$$\nabla_p = \frac{\int_{x=1}^{n} \alpha_x (1-\beta_x) \times f_x \times p_g \quad \partial x}{\int_{x=1}^{n} \beta_x \times f_x \times p_\ell \quad \partial x + \int_{x=1}^{n} (1-\beta_x) \times f_x \times p_g \quad \partial x} \tag{19}$$

where ∇_p is the processing cost on the server after delegating decision-making to the edge layer with respect to performing decision-making in the cloud layer.

7 Simulation

Assume the subsequent scenario: A smart house with two sensors. One to sense the temperature of the surrounding environment and the other is to measure the illumination of the surrounding. The objective of the whole system is to minimize electricity consumption. In other words, there is a range of temperatures. When the actual temperature goes under this range, the heater starts up, and when the temperature goes above this range, the air conditioner goes on. Also, if the illumination is less than a certain predefined acceptable range, the light goes off and if it is above that acceptable range, the light goes on. If it is within the range, then it depends on how it is manually set. Temperature sensor has five fuzzy states: Hot state: $\phi_h \rightarrow 27C \leq t$, warm state: $\phi_w \rightarrow 22C \leq t < 27C$, normal state: $\phi_a \rightarrow 15C \leq t < 22C$, cool state: $\phi_o \rightarrow 10C \leq t < 15C$, and cold state: $\phi_c \rightarrow t < 10C$, where $\Phi_t = \{\phi_c, \phi_o, \phi_a, \phi_w, \phi_h\}$. Illumination sensor has five fuzzy states: Extremely bright state: $\phi_B \rightarrow 10000lux \leq \ell$, bright state: $\phi_b \rightarrow 1000lux \leq \ell < 10000lux$, normal state $\phi_n \rightarrow 100lux \leq \ell < 1000lux$, light saver state $\phi_s \rightarrow 10lux \leq \ell < 100lux$, and dark state $\phi_d \rightarrow 0lux \leq \ell < 10lux$, where $\Phi_t = \{\phi_B, \phi_b, \phi_n, \phi_s, \phi_d\}$. The temperature sensor agent has the incident matrix $\mathbb{M}_t$ shown in Fig. 4.

$$\mathbb{M}_t = \begin{array}{c} \\ \phi_c \\ \phi_o \\ \phi_a \\ \phi_w \\ \phi_h \end{array} \begin{array}{c} \begin{array}{ccccc} \phi_c & \phi_o & \phi_a & \phi_w & \phi_h \end{array} \\ \begin{pmatrix} 0 & \lambda_{co} & \infty & \infty & \infty \\ \lambda_{oc} & 0 & \lambda_{oa} & \infty & \infty \\ \infty & \lambda_{ao} & 0 & \lambda_{aw} & \infty \\ \infty & \infty & \lambda_{wa} & 0 & \lambda_{wh} \\ \infty & \infty & \infty & \lambda_{hw} & 0 \end{pmatrix} \end{array}.$$

Fig. 4. Incident matrix $\mathbb{M}_t$ of temperature

where λ_{co} is the action of raising the temperature state up from cold to cool, λ_{oc} is the action of lowering the temperature state down from cool to cold, λ_{oa} is the action of raising the temperature state up from cool to normal, λ_{ao} is the action of lowering the temperature state down from normal to cool, λ_{aw} is the action of raising the temperature state up from normal to warm, λ_{wa} is the action of lowering the temperature state down from warm to normal, λ_{wh} is the action of raising the temperature state up from warm to hot, λ_{hw} is the action of lowering the temperature state down from hot to warm. The illumination sensor agent has the incident matrix $\mathbb{M}_I$ shown in Fig. 5.

where λ_{bB} is the action of raising the illumination state up from bright to extremely bright, λ_{Bb} is the action of lowering the illumination state down from extremely bright to bright, λ_{nb} is the action of raising the illumination state up from normal to bright, λ_{bn} is the action of lowering the illumination state down from bright to normal, λ_{sn} is the action of raising the illumination state up from semi-dark to normal, λ_{ns} is the action of lowering the illumination state down from normal to semi-dark, λ_{ds} is the action of raising the illumination state up from dark to semi-dark, λ_{sd} is the action of lowering the illumination state down from semi-dark to dark.

$$\mathbb{M}_I = \begin{array}{c} \\ \phi_B \\ \phi_b \\ \phi_n \\ \phi_s \\ \phi_d \end{array} \begin{array}{c} \begin{array}{ccccc} \phi_B & \phi_b & \phi_n & \phi_s & \phi_d \end{array} \\ \begin{pmatrix} 0 & \lambda_{Bb} & \infty & \infty & \infty \\ \lambda_{bB} & 0 & \lambda_{bn} & \infty & \infty \\ \infty & \lambda_{nb} & 0 & \lambda_{ns} & \infty \\ \infty & \infty & \lambda_{sn} & 0 & \lambda_{sd} \\ \infty & \infty & \infty & \lambda_{ds} & 0 \end{pmatrix} \end{array}$$

Fig. 5. Incident matrix $\mathbb{M}_I$ of Illumination

Choosing the sensor agent that will have the least negative effect is not a trivial task, and it depends on the state $\phi \in \Phi$ that every sensor agent is in. For example, if the illumination is normal during the day, then it can be brought down to a lower state. However, if it is during the evening, then the state of the illumination or the temperature cannot be altered. The temperature cannot be kept at 40C in the summer without operating an air conditioner on to lower the state, however, if it is 24C, it may be possible to turn it off.

The algorithms that have been presented in this paper have been implemented using JAVA. The scenario explained here is simulated and the following are the results of the explained scenario of the simulation:

In Fig. 6, it is easy to notice that there are two sets of figures. The solid ones and the dashed ones. The solid ones represent the consumption change with respect to temperature change in the five different states of the illumination. The dashed ones represent the consumption change with respect to illumination change in the five different states of the temperature. Indicate that, while studying the case of temperature, if the illumination states changes, it will jump to another curve according to the new state of the illumination. Furthermore, the same effect will appear for the curves of the illumination when fixing the temperature. The purpose of this simulation is to show that the framework is functioning. Nevertheless, producing accurate results of consumption with respect to

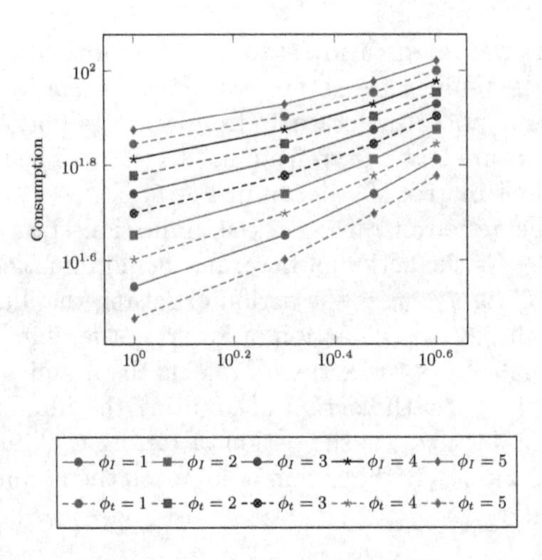

Fig. 6. Temperature and illumination VS consumption

temperatures or illuminations is not the scope of this paper since the equation of consumption was not taken into consideration.

Figure 7 shows the success of reducing costs through processing data locally. As shown in Fig. 7, the left-hand side presents the cost of sending messages to cloud servers, and how they change with the change of α.

In the transmission reduction curves, it can be seen that when $\beta = 0.9$, there is no reduction. This is because when $\beta = 0.9$, most of the messages are all processed locally anyways, and therefore, the amount of reduction is not even noticed. The red curve occurs when $\beta = 0.5$, that is when 50% of messages

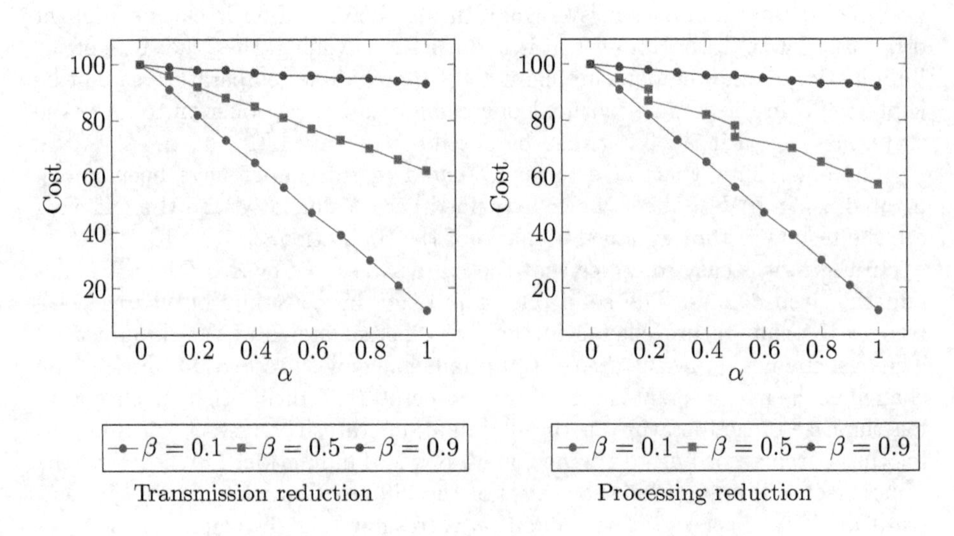

Fig. 7. Cost reduction with respect to change of α and β

can be processed locally and the other is on the cloud. In that case, reduction appears like the ones that used to be processed on the cloud, and are now being processed locally. It is noteworthy that the more α it is, the bigger the margin is for reducing cost. The last curve is when $\beta = 0.1$, which means that most of the messages are originally processed on the cloud. This is also a big margin for cost reduction since messages are being processed locally instead of being sent. When $\alpha = 1$, only 12.5% of messages are being sent. You can see from the curve that:

$$\frac{C(\alpha = 1, \beta = 0.1)}{C(\alpha = 0.1, \beta = 0.9)} = 12.5\%$$

where C is the cost of transmission. The same applies to the processing reduction on the servers.

8 Conclusion

In this paper, a framework for controlling and optimizing the performance of a smart city's physical layer is proposed. The framework provided complete mathematical modeling for the framework components and their relationships with each other. The main components that have been taken into consideration can be summarized as environment $\mathbb{E}$, sensors $\mathbb{S}$, intelligent agents that control sensors $\mathbb{A}$, reading history $\mathbb{H}$, sensor states Φ, and the physical layer intelligent agent $\mathbb{C}$, which works as a controller to lead every agent $a \in \mathbb{A}$ to reach a certain global goal. System states are presented in this paper as fuzzy sets. A set of algorithms have been proposed to describe the behavior of different framework components. An example was introduced to illustrate how the framework operates. The framework has been simulated and the results have been discussed. The results indicate that the load on the communication network and the cloud servers was reduced.

References

1. Celino, I., Kotoulas, S.: Smart cities [guest editors' introduction]. IEEE Internet Comput. **17**(6), 8–11 (2013)
2. Cunha, J., Batista, N., Cardeira, C., Melicio, R.: Wireless networks for traffic light control on urban and aerotropolis roads. J. Sens. Actuator Netw. **9**(2), 26 (2020)
3. Dameri, R.P.: Searching for smart city definition: a comprehensive proposal. Int. J. Comput. Technol. **11**(5), 2544–2551 (2013)
4. Dryjanski, M., Buczkowski, M., Ould-Cheikh-Mouhamedou, Y., Kliks, A.: Adoption of smart cities with a practical smart building implementation. IEEE Internet of Things Mag. **3**(1), 58–63 (2020)
5. Gungor, V.C., Lu, B., Hancke, G.P.: Opportunities and challenges of wireless sensor networks in smart grid. IEEE Trans. Ind. Electron. **57**(10), 3557–3564 (2010)
6. Hackmann, G., Guo, W., Yan, G., Sun, Z., Lu, C., Dyke, S.: Cyber-physical codesign of distributed structural health monitoring with wireless sensor networks. IEEE Trans. Parallel Distrib. Syst. **25**(1), 63–72 (2013)

7. Hsu, K.-C., Chiang, Y.-T., Lin, G.-Y., Lu, C.-H., Hsu, J.Y.-J., Fu, L.-C.: Strategies for inference mechanism of conditional random fields for multiple-resident activity recognition in a smart home. In: International Conference on Industrial, Engineering and Other Applications of Applied Intelligent Systems, pp. 417–426. Springer (2010)
8. Iqbal, M., Naeem, M., Anpalagan, A., Ahmed, A., Azam, M.: Wireless sensor network optimization: multi-objective paradigm. Sensors **15**(7), 17572–17620 (2015)
9. Kang, B., Kim, S., Choi, M., Cho, K., Jang, S., Park, S.: Analysis of types and importance of sensors in smart home services. In: 2016 IEEE 18th International Conference on High Performance Computing and Communications; IEEE 14th International Conference on Smart City; IEEE 2nd International Conference on Data Science and Systems (HPCC/SmartCity/DSS), pp. 1388–1389. IEEE (2016)
10. Mohamed, B., Abdelhadi, F., Adil, B., Haytam, H.: Smart city services monitoring framework using fuzzy logic based sentiment analysis and apache spark. In: 2019 1st International Conference on Smart Systems and Data Science (ICSSD), pp. 1–6. IEEE (2019)
11. Olatinwo, S.O., Joubert, T.-H.: Optimizing the energy and throughput of a water-quality monitoring system. Sensors **18**(4), 1198 (2018)
12. Oliveira, L.M.L., Rodrigues, J.J.P.C.: Wireless sensor networks: a survey on environmental monitoring. JCM **6**(2), 143–151 (2011)
13. Othman, M.F., Shazali, K.: Wireless sensor network applications: a study in environment monitoring system. Procedia Eng. **41**, 1204–1210 (2012)
14. Pellicer, S., Santa, G., Bleda, A.L., Maestre, R., Jara, A.J., Skarmeta, A.G.: A global perspective of smart cities: a survey. In: 2013 Seventh International Conference on Innovative Mobile and Internet Services in Ubiquitous Computing, pp. 439–444. IEEE (2013)
15. Silva, I., Guedes, L.A., Portugal, P., Vasques, F.: Reliability and availability evaluation of wireless sensor networks for industrial applications. Sensors **12**(1), 806–838 (2012)
16. Sung, W.-T.: Multi-sensors data fusion system for wireless sensors networks of factory monitoring via BPN technology. Expert Syst. Appl. **37**(3), 2124–2131 (2010)
17. Tan, K.C., Lee, T.H., Khor, E.F.: Evolutionary algorithms for multi-objective optimization: performance assessments and comparisons. Artif. Intell. Rev. **17**(4), 251–290 (2002)
18. van Kasteren, T.L.M., Englebienne, G., Kröse, B.J.A.: Human activity recognition from wireless sensor network data: benchmark and software. In: Activity Recognition in Pervasive Intelligent Environments, pp. 165–186. Springer (2011)
19. Wang, J., Zhang, Z., Li, B., Lee, S., Sherratt, R.S.: An enhanced fall detection system for elderly person monitoring using consumer home networks. IEEE Trans. Consum. Electron. **60**(1), 23–29 (2014)
20. Wu, C.-I., Kung, H.-Y., Chen, C.-H., Kuo, L.-C.: An intelligent slope disaster prediction and monitoring system based on WSN and ANP. Expert Syst. Appl. **41**(10), 4554–4562 (2014)
21. Zadeh, L.A.: Fuzzy sets. Inf. Control **8**(3), 338–353 (1965)
22. Zhang, T., Zhao, Q., Shin, K., Nakamoto, Y.: Bayesian-optimization-based peak searching algorithm for clustering in wireless sensor networks. J. Sens. Actuator Netw. **7**(1), 2 (2018)
23. Zhao, G., et al.: Wireless sensor networks for industrial process monitoring and control: a survey. Netw. Protoc. Algorithms **3**(1), 46–63 (2011)

Intelligent Collision Avoidance and Manoeuvring System with the Use of Augmented Reality and Artificial Intelligence

K. F. Bram-Larbi[1], V. Charissis[1(✉)], S. Khan[1], R. Lagoo[1], D. K. Harrison[2], and D. Drikakis[3]

[1] Virtual Reality Driving Simulator Laboratory, School of Computing, Engineering and Built Environment, Glasgow Caledonian University, Glasgow, UK
vassilis.charissis@gcu.ac.uk

[2] Department of Mechanical Engineering, School of Computing, Engineering and Built Environment, Glasgow Caledonian University, Glasgow, UK

[3] Defence and Security Research Institute, University of Nicosia, 2417 Nicosia, Cyprus

Abstract. The efficiency of collision-avoidance abrupt braking or manoeuvring is primarily based on a driver's response time. The latter is affected by the driver's spatial and situational awareness, which in turn is heavily depended on the driver's cognitive workload. Attention taxing, infotainment systems could dramatically reduce the driver's ability to respond effectively in an imminent collision situation . Current attempts to reduce this negative impact on driver's performance had limited success. To improve the driver's ability to perform a successful collision avoidance braking or manoeuvring, this paper presents the design considerations of a prototype system that employs Augmented Reality (AR) to overlay guidance information in the real-life environment. The proposed system will be further supported by an Artificial Intelligence (AI) system that will act as a co-driver, offering in real-time alternative options to the driver. Prior work for the development of a similar system for Emergency Services' (ES) vehicles sparked the idea to transfer and investigate the acceptance of this technology on a civilian vehicle domain. In conclusion, the paper presents the preliminary design for the development of the civilian version of the AR/AI system based on the feedback and suggestions of 20 drivers.

Keywords: Augmented reality · Intelligent transportation · HCI · AI · Collision avoidance · Head-up display · Smart Cities

1 Introduction

The typical vehicular interiors can be described as highly physically mutable environments with fluctuating noise, light levels and space availability. The external environment also presents variable levels of visual and auditory cues and subsequently, lead to increased cognitive loads reducing driver's ability to perform successfully the primary task of driving. Consequently, these conditions leave limited available attention

K. Arai (Ed.): FICC 2021, AISC 1363, pp. 457–469, 2021.
https://doi.org/10.1007/978-3-030-73100-7_32

resources for secondary tasks (i.e. infotainment, passenger discussions and navigation guidance amongst other) while driving.

Smart devices such as phones, tablets and wearables have created a new culture of constant provision and craving of information by the users. This contemporary trend has infiltrated and became a necessity also in the vehicular interiors. In turn, the vehicular environments responding to this user's requirement has been increasingly accommodating various infotainment devices that announce, project and otherwise call driver's attention to various pieces of information [1, 2]. An imminent consequence of these attention-seeking devices is the driver's distraction and increased probability of collisions [3, 4].

Despite legislative means of controlling the use of smart devices or other in-vehicle infotainment systems whilst driving having been introduced, it is expected that the profusion of such devices and their dangerous use will continue with a projected potential increase predicted in the following years [2].

As such, the proposed work examines the potential ways of fulfilling the prominent infotainment needs of modern drivers without jeopardising the safety of the driving process following previous studies examples [5–7]. In particular, this research is focusing on the safe guidance of the driver through traffic and the provision of safe manoeuvring options that will utilise better the road network and minimise abrupt lane changes that could result in a collision situation.

To achieve this, the proposed work employs emerging technologies such as Augmented Reality (AR) and Artificial Intelligence (AI). The latter will provide collisions avoidance guidance and manoeuvring options in the form of an AI co-driver suggesting in real-time the optimal speed and manoeuvring options that will enable the driver to avoid any potential collision incidents. This crucial information will be superimposed directly on the external environment and driver's field of view.

This AR representation of simplified options will utilise a full-windshield, prototype Head-Up Display (HUD) interface, avoiding any additional dashboard related screens. It was previously observed that the AR representation of data with the use of HUD conduits has resulted in faster driver responses and reduced accidents in various weather and traffic conditions [8–11].

Overall, the paper will present the initial design rationale and development process of the proposed AR/AI interface. In turn, the design aspects of the proposed interface will be contrasted against current systems through a comparative, qualitative study with 20 drivers. Their feedback will be presented and discussed in detail as it provides an understanding of the potential use of such emerging technologies in passenger vehicles. Finally, the paper will provide a tentative plan of future work which will explore further the interface development and the proposed technology acceptance from the drivers.

2 Current Navigation and Manoeuvring Issues

As aforementioned, the stream of incoming infotainment information and the interaction requirements with these devices have already reduced dramatically driver's attention. Additionally, variable traffic conditions, particularly in a motorway environment and unexpected fluctuation in the traffic flow could surprise the driver and result in a potential collision.

During these situations, the driver's decision-making process is slow, due to increased cognitive load. The driver's inattention to the road magnifies the difficulty to choose appropriate action. In milliseconds the driver has to identify potential options of braking and/or evading manoeuvres [12]. However, in such a demanding condition, the driver's spatial and situational awareness might be impeded and the choice of speed, lane, or braking might be obscured or inappropriate.

Typical navigation systems presented in the majority of passenger vehicles are offering limited information regarding the potential collision hazards lying ahead on the vehicle's path. Also, the generic traffic flow information provided, lack any visual linking to the external environment as typically are presented in a small-sized, dashboard screen.

Previous attempts of prototype systems to offer a real-time provision of crucial information to the driver for collision avoidance produced encouraging results [13–15]. These AR-HUD systems warned the driver for potential collision hazards related to the neighbouring vehicles and/or the road infrastructure as illustrated below (see Fig. 1) [11]. Similarly provision of information through full or small HUD displays presents various existing dashboard instrumentation and in some occasions, the optimal racing lines were presented for track-day/racing driving [16]. However, none of the aforementioned previous studies provides the optimal route or action (i.e. manoeuvring, braking) options for successful collision avoidance and a step by step guidance to avoid the accident.

Fig. 1. Previous collision avoidance systems which incorporated AR projection of data through a full-windshield HUD [11].

3 Proposed Solution

Following concurrent research in the development of intelligent collision avoidance and manoeuvring systems for Emergency Services' (ES) vehicles, it was observed that the bespoke system characteristics for collision avoidance could be adapted and transferred in the civilian vehicles [16, 17].

The feedback of the ES drivers was also in support of offering a simplified version for the passenger vehicles [18]. As such, instead of considering a more spartan infotainment environment or complete exclusion of smart devices (embedded or not on the vehicle), the current research is investigating the optimal way to control the incoming information whilst supporting in real-time the driver's decision making process. The proposed solution follows a two-fold approach with regards to the utilization of emerging technologies as described below.

3.1 Augmented Reality (AR – HUD)

The provision of crucial information to the driver in a timely and efficient manner is a challenging task. Current systems and applications require from the driver to divert his/her gaze from the road and interact with various devices and screens positioned typically in the dashboard area namely Head-Down Display (HDD) section of the vehicular cockpit. This contributes further to the time that the driver gazes away from the road and interacts with additional infotainment means [5, 9]. To counteract the above issue, this work follows previous studies which employed various types of HUD interfaces to present the required information directly to the driver's field of view [11, 14, 19–23].

The overall interface is an adaptation of the ES version, which maintains the primary functionality of real-time superimposed information for safe guidance. However, the civilian version complies fully with the speed and traffic regulations and does not offer unconventional manoeuvring options that apply only for emergency vehicles. The system will provide the best possible manoeuvre option given the vehicle's speed and proximity to the immobile obstacle (see Fig. 2). Yet if the collision is imminent the system will provide the best option to avoid or minimise the impact. This ultimate attempt will utilise every possible alternative to reduce the severity of the collision.

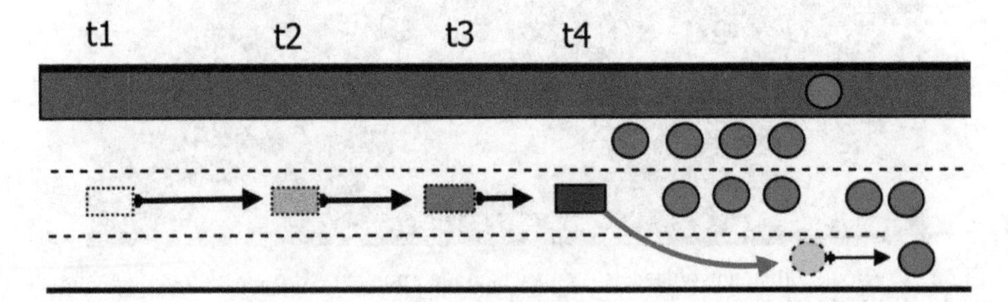

Fig. 2. Sudden braking scenario and manoeuvring options in an imminent collision case.

In a zoom-in, top view of the driving simulation, the red-vehicle driver is guided step by step towards the direction that will provide additional space for safer abrupt braking or escape route (see Fig. 3).

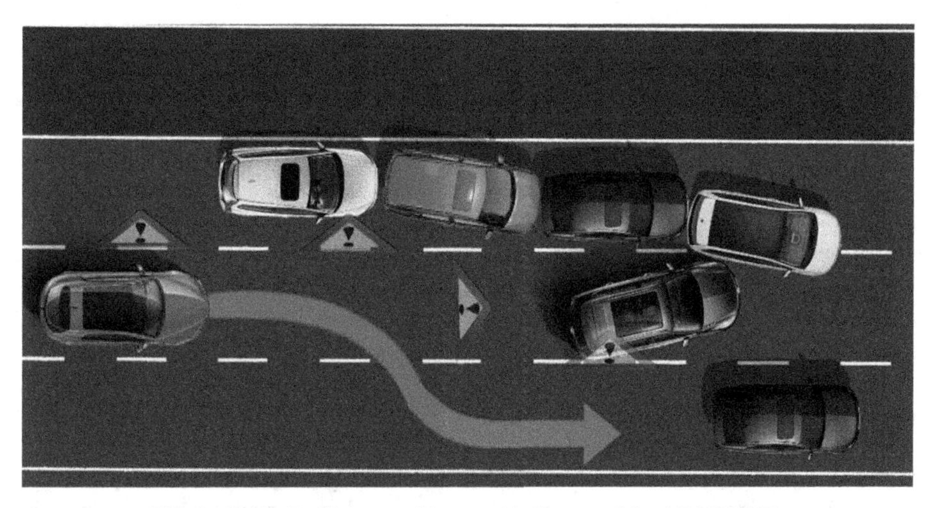

Fig. 3. Guided collision avoidance with the use of the AR/AI HUD.

Although on a top view, this option seems logical and obvious, in a real-life situation, the panic response of the driver typically results in a collision [5, 8]. The system's intervention is presented in a driving simulation environment (see Fig. 4).

Fig. 4. Suggested manoeuvre and speed by AR/AI HUD in a driving simulation environment.

3.2 Prototype Interface Design

Conveying the information to the user relies primarily on the type of information, interface design and conduit of presentation [11, 14, 19]. Simple and easy to follow multimodal interfaces reduce the cognitive load and the time of the decision-making process [11]. As this interface is intended to be activated before a potential collision, the information provided must be presented to the driver is a simplistic manner that enables to user to understand and respond instantly.

As such, the interface presents the guidance to the driver directly, superimposed in real-scale on the road ahead, as illustrated in the VR driving Simulator screenshot above (see Fig. 4). This approach imitates the naturalistic human hand directions provided typically by a co-driver. In this manner, the AI co-driver described below is materialized virtually with the use of the directional arrows instead of hands. The visual interface is complemented by a voice-over highlighting the directions and the actions. The interface follows a typical colour-coding approach of red, amber, green and blue, stating the risk level of the manoeuvring option with red being the riskier in contrast to blue which is the less urgent and easier to perform.

3.3 AI Co-driver

The effective operation of the AR HUD interface is based on the correct and timely information provision. This is a task that has been allocated to the second element of the proposed system namely Artificial Intelligence Co-Driver. The system under development aims to acquire information from the vehicles' sensors, the Vehicular Adhoc Networks, and the road infrastructure [11, 24, 25].

Additionally, the AI Co-Driver will be trained by previous accident data and through VR simulations performed by drivers in the VR Driving Simulator [26–29]. Initial feedback from focus groups has suggested that the AI Co-Driver should not be intrusive and offer only crucial information in challenging situations [17, 18].

The particular system design requirements will prevent the system's misuse (i.e. unsafe manoeuvring and speeding). As such the AR/AI HUD will strive to enable the driver to perform a necessary manoeuver safely and maintain the control of the vehicle in a challenging situation. The AI system is currently under development and a demo version has been utilised to simulate the AR/AI collaborative functionalities. As such is not in the remit of this paper to present further the AI prototype and the subsequent algorithms to be utilised. This paper is presenting primarily the overall system concept and focuses on the HCI development of the AR-HUD system. It also explores the drivers' attitudes towards the introduction of AR/AI systems in a form of a virtual co-driver.

4 Preliminary Evaluation and Feedback by 20 Drivers

A preliminary evaluation of a demo system was deemed essential for the further development and customisation of the proposed AR/AI for passenger vehicles.

4.1 Survey Rationale and Structure

This survey followed a similar structure to the ES drivers' survey to receive comparable information. As such the 20 drivers completed a pre-demo questionnaire aiming to acquire the demographic information of the group and their current experience and beliefs related to navigation and guidance systems.

In turn, the demonstration AR/AI HUD interface was presented to 20 drivers and contrasted to a default navigation system. A typical rear-collision scenario was employed to depict the conditions of the potential accident [11]. The provision of information with and without the AR/AI HUD was discussed. Finally, a post-demo survey was designed to obtain drivers' subjective feedback that could inform the future development and required adaptations of the proposed system for civilian, everyday use presented in Table 1. Also, this part of the survey aimed to identify the view of the typical driver on the technology. The second questionnaire served also as a public gauging for the use of emerging technologies to improve ES vehicles' performance for improving the speed and level of support to the public.

Table 1. Post questionnaire

Post-Questionnaire
Q11. Did you find the *interface design simple and clear*? ☐ 1 ☐ 2 ☐ 3 ☐ 4 ☐ 5 (1 Very Simple – 5 Very Difficult)
Q12. Do you think that *interface design and colour coding* would be useful to convey the manoeuvring information? ☐ 1 ☐ 2 ☐ 3 ☐ 4 ☐ 5 (1 Very Useful – 5 Not Useful at all)
Q13. Do you think that it would be useful to have ***AR*** *navigation/guidance system in the* ***ES vehicle***? ☐ 1 ☐ 2 ☐ 3 ☐ 4 ☐ 5 (1 Very Interested – 5 Not Interested at all)
Q14. Do you think that it would be useful to have ***AI*** *navigation/guidance system in the* ***ES vehicle***? ☐ 1 ☐ 2 ☐ 3 ☐ 4 ☐ 5 (1 Very Interested – 5 Not Interested at all)
Q15. Would you be interested to have ***AR*** *navigation/guidance system in the* **civilian *vehicles***? ☐ 1 ☐ 2 ☐ 3 ☐ 4 ☐ 5 (1 Very Interested – 5 Not Interested at all)
Q16. Would you be interested to have ***AI*** *navigation/guidance system in the* **civilian *vehicles***? ☐ 1 ☐ 2 ☐ 3 ☐ 4 ☐ 5 (1 Very Interested – 5 Not Interested at all)
Q17. Would you be interested to have *real-time guidance suggestions by an AI/AR system*? ☐ 1 ☐ 2 ☐ 3 ☐ 4 ☐ 5 (1 Very Interested – 5 Not Interested at all)
Q18. Do you think that the *AR/AI proposed system could replace other guidance systems*? ☐ Yes ☐ No
Q19. Do you think it would be a helpful system (AI/AR) to *integrate into future ES vehicles*? ☐ Yes ☐ No
Q20. Do you think it would be a helpful system (AI/AR) to *integrate into future civilian vehicles?* ☐ Yes ☐ No
Q21. Do you have *any other suggestions, comments or thoughts* regarding the proposed AR/AI system? If yes please use the space below to write your comments. ☐ Yes ☐ No

The questionnaires were mainly based on a Technology Acceptance Model (TAM) previously designed for the evaluation of emerging technologies in the vehicular environment and other consumer electronic market domains [21, 30, 31]. This was deemed essential to provide the validity and reliability for the particular survey.

As such the same survey could be repeated on a larger or different group of drivers and acquire quantifiable and comparable results whilst testing future versions of the proposed system. Notably, the TAM questionnaire was deployed selectively as the demo system could not reflect fully all the attributes of the final, future version of the system. As such, various constructs (i.e. Perceived Ease of Use (PEoU), Perceived Usefulness (PU) and others) were not included in this phase.

Also, for this stage of the project and due to the current pandemic limitations, 20 users were considered adequate for this preliminary evaluation. In the future development of the system, a larger cohort of users will be invited to test and experience the final versions.

4.2 Participants

The evaluation was performed by 20 users (5 female, 15 male) which held a valid driving licence and they were aged between 20 and 58 years of age. The participants were selected randomly and every effort was made to expand the sample pool. As mentioned previously, the current pandemic issue limited the availability of a larger sample of users. The 20 volunteers were of variable driving experience, occupations, computing experience and nationalities.

5 Data Analysis, Discussion and Recommendations

An indicative appraisal of the system was provided through the post questionnaire results presented below (see Fig. 5). Overall, the responses at this stage were in favour of a system. The minimalistic approach of the interface design was well perceived as 70% of the users considered it very simple and 20% simple enough to follow as presented in Question 11.

Similarly, the intended colour coding also received the users' approval with 90% in favour of the chosen scheme as presented from the results in Question 12. The system was deemed also logical and potentially very useful for ES vehicles for reducing potential collisions and the Estimated Time of Arrival (ETA) to the events as presented in the results of Question 13. However, the drivers raised some concerns regarding the potential intrusion of the AI Co-driver in an already challenging ES vehicle environment as seen from the responses in Question 14.

For the civilian use and in the particular example of the rear collision scenario, the drivers felt that such a system could have a major positive contribution in the safe manoeuvring and navigation through adverse traffic situations as illustrated by Question 15.

Question 16 results presented similar concerns to Question 14, as the drivers were unsure on the level of intrusion of the AI Co-Driver to the overall driving experience. The concerns on both questions stemmed from the experience of voice navigation systems

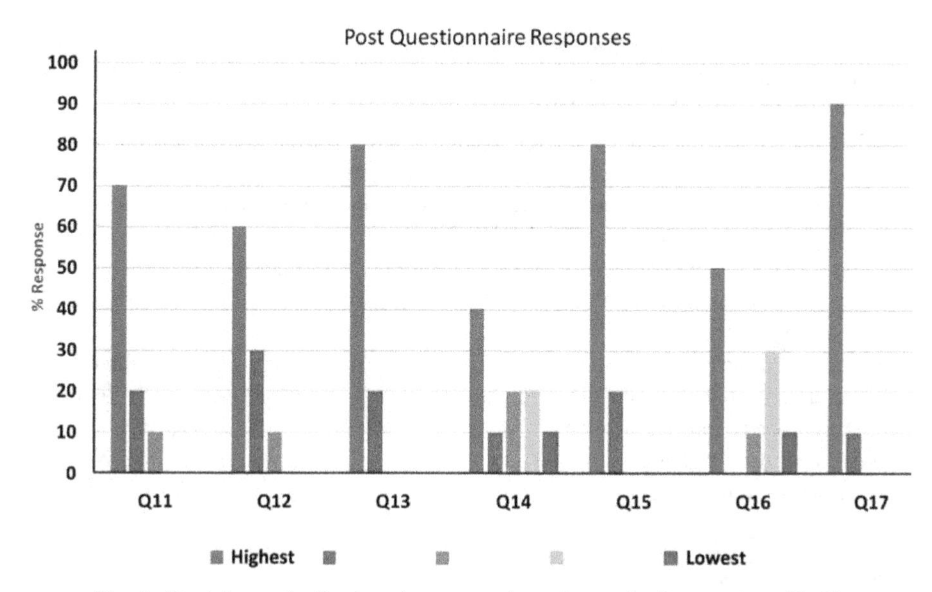

Fig. 5. Participants feedback and post-questionnaire results for questions 11–17.

and interfaces on contemporary vehicles. However, the combined approach of AI and AR HUD system scored 100% (see Fig. 5).

The remaining Questions (18,19,20) received also positive responses with 90%, 70% and 100% respectively highlighting the future expectations of the drivers.

Notably, the aforementioned results and drivers' feedback offers an overview appraisal of the users' tendency to accept this technology on their current and/or future vehicles. As the results were encouraging, further research will be commenced for the development of an updated version that will take into account the users' feedback. The future aim of this work is to produce a fully functional version which will be embedded on the VR Driving Simulator and utilised for a large cohort comparative evaluation between the proposed system and the traditional methods of guidance and navigation.

The derived results do not explore the users' performance improvements, they, however, offer an insight into the benefits and drawbacks of such systems according to user's feedback. Based on the aforementioned results and observations the following recommendations could support the Human-Computer Interaction and overall concept development of similar systems that embed emerging technologies on vehicular environments:

- Simple interface design: This is essential for accommodating the vast majority of users irrelevantly of their age, occupation, education, gender and nationality [5, 9]. Although this is considered common knowledge and part of the prevailing user-friendly design mantras, many interfaces are confined by the "family image" of each manufacturer resulting in unnecessary complex interfaces purely to comply with the branding identity.

- Prioritisation of the visual provision of essential information: The plurality of information provided and the number of devices that attract driver's attention should be planned to minimise the drivers' cognitive load [9, 21, 32]. Especially in navigation or collision avoidance systems, interrupt strategies should be put in place to control/mute non-crucial information. The current automotive trend is to provide an array of screens spanning across the dashboard providing a cinematic experience of Head-Down Displays (HDD) supporting the current infobesity issue that distracts the driver from the main task. In a similar manner various prototype HUD designs provide full windshield projections and numerous information without prioritisation or immediate relevance to the primary task.
- Human-Machine collaboration: Emerging technologies such as AI attract the interest of the drivers as they could offer an extra level of support on the driver's decision-making process and overall vehicle safety [33, 34]. Yet, in this stage, such systems should be introduced as complementary to the driver by supporting a human-machine collaboration until the drivers consider such systems trustworthy [9, 35].

6 Conclusions

This paper presented a prototype AR/AI system that aims to support the driver's decision-making process in challenging situations such as imminent collision scenarios. The system was designed to superimpose guidance information in a real-life environment and real-scale with the use of a full windshield HUD device. This work stemmed from a concurrent project related to the development of a similar AR/AI HUD system for the emergency response services. The proposed version capitalizes on the main design framework for the ES, yet it is customized to support the everyday driving of a typical civilian user.

Additionally, the paper described a preliminary evaluation of the prototype systems' functionalities by 20 drivers. Their feedback was in par to the previous studies aiming to identify the potential usability of such a system for ES vehicles. The derived results from the preliminary evaluation were analysed and presented offering an informative appraisal of the system's capabilities and users' expectations of the final output.

The drivers' feedback and suggestions highlighted also some concerns particularly for the AI component of the system that need to be addressed in the future versions. Furthermore, the future plans for this project entail the evaluation of a fully functional version, by a larger cohort of users, in our immersive VR driving simulation facility to achieve finer granularity of the results.

References

1. Morris, A., Reed, S., Welsh, R., Brown, L., Birrell, S.: Distraction effects of navigation and green-driving systems – results from field operational tests (FOTs) in the UK. Eur. Transp. Res. Rev. **7**(26) (2015)
2. Simons, S.M., Hicks, A., Caird, J.K.: Safety-critical event risk associated with cell phone tasks as measured in naturalistic driving studies: a systematic review and meta-analysis. Accid. Anal. Prev. **87**, 161–169 (2016)

3. Liao, Y., Li, G., Eben, S., Cheng, B., Green, P.: Understanding driver response patterns to mental workload increase in typical driving scenarios. IEEE Access **6**, 35890–35900 (2018)
4. Regan, M.A., Hallett, C., Gordon, C.P.: Driver distraction and driver inattention: definition, relationship and taxonomy. Accid. Anal. Prev. **43**(5), 1771–1781 (2011)
5. Charissis, V., Papanastasiou, S.: Human-machine collaboration through vehicle head-up display interface. Int. J. Cogn. Technol. Work **12**(1), 41–50 (2010). https://doi.org/10.1007/s10111-008-0117. P. C. Cacciabue and E. Hollangel (eds.)
6. Okumura, H., Hotta, A., Sasaki, T., Horiuchi, K., Okada, N.: Wide field of view optical combiner for augmented reality head-up displays. In: 2018 IEEE International Conference on Consumer Electronics (IEEE ICCE) (2018)
7. Charissis, V., Papanastasiou, S., Vlachos, G.: Interface development for early notification warning system: full windshield head-up display case study. In: Jacko, J.A. (ed.) Human-Computer Interaction. Interacting in Various Application Domains. HCI International 2009. Lecture Notes in Computer Science, vol. 5613. Springer, Heidelberg (2009)
8. Lagoo, R., Charissis, V., Harrison, D.K.: Mitigating driver's distraction: automotive head-up display and gesture recognition system. IEEE Consum. Electron. Mag. **8**(5), 79–85 (2019). https://doi.org/10.1109/mce.2019.2923896
9. Beck, D., Park, W.: Perceived importance of automotive HUD information items: a study with experienced HUD users. IEEE Access **6**, 21901–21909 (2018)
10. Hyungil, K., Xuefang, W., Gabbard, J.L., Polys, N.F.: Exploring head-up augmented reality interfaces for crash warning systems. In: 5th International Conference on Automotive User Interfaces and Interactive Vehicular Applications, ACM Automotive UI 2013, pp. 224–227. ACM (2013)
11. Charissis, V., Papanastasiou, S., Chan, W., Peytchev, E.: Evolution of a full-windshield HUD designed for current VANET communication standards. In: IEEE Intelligent Transportation Systems International Conference (IEEE ITS), The Hague, Netherlands, pp. 1637–1643 (2013). https://doi.org/10.1109/itsc.2013.6728464
12. Smith, D.L., Najm, W.G., Lam, A.H.: Analysis of braking and steering performance in car-following scenarios. In: Proceedings of the Society of Automotive Engineers World Congress (SAE 2003), Paper 2003-01-0283, Detroit, MI, USA (2003)
13. Weinberg, G., Harsham, B., Medenica, Z.: Evaluating the usability of a head-up display for selection from choice lists in cars. In: 3rd ACM International Conference on Automotive User Interfaces and Interactive Vehicular Applications, Automotive UI 2011, pp. 39–46. ACM (2011)
14. Okumura, H., Sasaki, T., Hotta, A., Shinohara, K.: Monocular hyperrealistic virtual and augmented reality display. In: 2014 IEEE Fourth International Conference on Consumer Electronics (ICCE-Berlin), Berlin, Germany, pp. 19–23 (2014)
15. Charissis, V., Papanastasiou, S., Mackenzie, L.: Evaluation of collision avoidance prototype head-up display interface for older drivers. In: 14th International Conference on Human-Computer Interaction, HCI International 2011, Orlando, 9–14 July 2011. Lecture Notes in Computer Science, vol. 6763, pp. 367–375. Springer, Heidelberg (2011)
16. Smith, N.: A National Perspective on Ambulance Crashes and Safety: Guidance from the National Highway Traffic Safety Administration on ambulance safety for patients and providers. NHTSA Report, EMS World, pp. 91–94 (2015)
17. Bram-Larbi, K.F., Charissis, V., Khan, S., Lagoo, R., Harrison, D.K., Drikakis, D.: Collision avoidance head-up display: design considerations for emergency services' vehicles. In: IEEE International Conference on Consumer Electronics (ICCE), Las Vegas, USA, pp. 1–7 (2020). https://doi.org/10.1109/icce46568.2020.9043068
18. Bram-Larbi, K.F., Charissis, V., Khan, S., Harrison, D.K., Drikakis, D.: Improving emergency vehicles' response times with the use of augmented reality and artificial intelligence. In: Lecture Notes in Computer Science (2020)

19. Zhang, Y., Yang, T., Zhang, X., Zhang, Y., Sun, Y.: Driver's visual attention analysis in smart car with FHUD. In: Stanton, N. (ed.) Advances in Human Aspects of Transportation. AHFE 2020. Advances in Intelligent Systems and Computing, vol. 1212. Springer, Cham (2020). https://doi.org/10.1007/978-3-030-50943-9_9
20. Beck, D., Jung, J., Park, J., Park, W.: A study on user experience of automotive HUD systems: contexts of information use and user-perceived design improvement points. Int. J. Hum. Comput. Interact. **35**(20), 1936–1946 (2019). https://doi.org/10.1080/10447318.2019.1587857
21. Lagoo, R., Charissis, V., Chan, W., Khan, S., Harrison, D.: Prototype gesture recognition interface for vehicular Head-Up Display system. In: IEEE International Conference on Consumer Electronics, Las Vegas, USA, pp. 1–6 (2018)
22. Wang, S., Charissis, V., Harrison, D.K.: Augmented reality prototype HUD for passenger infotainment in a vehicular environment. Adv. Sci. Technol. Eng. Syst. J. **2**(3), 634–641 (2017)
23. Charissis, V., Naef, M.: Evaluation of prototype automotive head-up display interface: testing driver's focusing ability through a VR simulation. In: IEEE Intelligent Vehicle Symposium (IV 2007), Istanbul, Turkey (2007)
24. Kianfar, R., et al.: Design and experimental validation of a cooperative driving system in the grand cooperative driving challenge. IEEE Trans. Intell. Transp. Syst. **13**(3), 994–1007 (2012). https://doi.org/10.1109/TITS.2012.2186513
25. Flanagan, S.K., He, J., Peng, X.: Improving emergency collision avoidance with vehicle to vehicle communications. In: 2018 IEEE 20th International Conference on High-Performance Computing and Communications; IEEE 16th International Conference on Smart City; IEEE 4th International Conference on Data Science and Systems (HPCC/SmartCity/DSS), Exeter, United Kingdom, pp. 1322–1329 (2018). https://doi.org/10.1109/hpcc/smartcity/dss.2018.00220
26. Vazquez, D., Meyer, L.A., Marın, J., Ponsa, D., Geronimo, D.: Virtual and real world adaptation for pedestrian detection. IEEE Trans. Pattern Anal. Mach. Intell. **36**, 797–809 (2014)
27. Frank, M., Drikakis, D., Charissis, V.: Machine learning methods for computational science and engineering. Comput. (J.) Comput. Eng. Sect. MDPI **8**(1), 15 (2020). https://doi.org/10.3390/computation8010015. ISSN 2079-3197, pp. 1–36
28. Charissis, V., Papanastasiou, S.: Artificial intelligence rationale for autonomous vehicle agents behaviour in driving simulation environment. In: Aramburo, J., Trevino, A.R. (eds.) Robotics, Automation and Control, pp. 314–332. I-Tech Education and Publishing KG, Vienna (2008). ISBN 953761916-8I
29. Li, J., Cheng, H., Guo, H., et al.: Survey on artificial intelligence for vehicles. Automot. Innov. **1**, 2–14 (2018). https://doi.org/10.1007/s42154-018-0009-9
30. Altarteer, S., Charissis, V.: Technology acceptance model for 3D virtual reality system in luxury brands online stores. IEEE Access J. **7**, 64053–64062 (2019). https://doi.org/10.1109/ACCESS.2019.2916353
31. Lee, H.-H., Fiore, A.M., Kim, J.: The role of the technology acceptance model in explaining effects of image interactivity technology on consumer responses. Int. J. Retail Distrib. Manage. **34**(8), 621–644 (2006)
32. Maroto, M., Caño, E., González, P., Villegas, D.: Head-up displays (HUD) in driving. Human-Computer Interaction, arXiv:1803.08383 (2018)
33. Dunbar, J., Gilbert, J.E.: The human element in autonomous vehicles. In: Harris, D. (ed.) Engineering Psychology and Cognitive Ergonomics: Cognition and Design, EPCE 2017. Lecture Notes in Computer Science, vol. 10276. Springer, Cham (2017). https://doi.org/10.1007/978-3-319-58475-1_26

34. Färber, B.: Communication and communication problems between autonomous vehicles and human drivers. In: Maurer, M., Gerdes, J., Lenz, B., Winner, H. (eds.) Autonomous Driving. Springer, Heidelberg (2016). https://doi.org/10.1007/978-3-662-48847-8_7
35. Souders, D., Charness, N.: Challenges of older drivers' adoption of advanced driver assistance systems and autonomous vehicles. In: Zhou, J., Salvendy, G. (eds.) Human Aspects of IT for the Aged Population. Healthy and Active Aging, ITAP 2016. Lecture Notes in Computer Science, vol. 9755. Springer, Cham (2016). https://doi.org/10.1007/978-3-319-39949-2_41

Analyzing Pedestrian Crossflow Through Complex Transfer Corridors

Emad Felemban, Faizan Ur Rehman(✉), Akhlaq Ahmad, and Shoaib Shahzad Obaidi

Umm Al-Qura University, Mecca, Saudi Arabia
{eafelemban,fsrehman,aajee}@uqu.edu.sa, s437036084@st.uqu.edu.sa

Abstract. Large gatherings which are commonly allied with spatial and temporal constraints possess several managerial challenges. As far as walking is considered as a sustainable mode of pedestrian movement with crowd safety as the core objective, hosting authorities put relentless efforts to analyze crowd dynamics and provide with pre- and during event mobility plans. In this paper, we present our ongoing research about analyzing pedestrian movement of a very large annual gathering, the Hajj, where topographical nature of the hosting city Makkah proffer complex network of straight, circular, Y, L, T shape transfer corridors. Moreover, Hajj crowd diversity in terms of race, age and language makes it more laborious for the hosting authorities to tackle pedestrians crossflow to avoid any possible stampede. As a part of our study, we have analyzed the mobility of a group of pilgrims approaching the Jamarat bridge through a less complicated route that contained a L-shape transfer corridor. Pilgrims' movement was recorded through GPS sensory data with a timestamp by deploying a mobile phone application, during the Hajj 2019 event. The initial outcomes led us to go on modelling and then simulating multiple scenarios of pedestrians' movement through the network of complex corridors. We aim that our findings will support in designing optimized mobility plans by considering routes' capacity and available time to accommodate all groups.

Keywords: Crowd management · Pedestrian movement · Pedestrian counterflow · Hajj · Modelling and simulation

1 Introduction

Increasing travelling facilities have mobilized a significant percentage of world population for sports events, leisure trips, education, adventure, medical treatments and pilgrimage. Mostly in sports or religion-related activities, people count reaches millions or even more. A few of the large gatherings[1] are Kumbh Mela (2013), gathering at the shrine of Husayn Ibn Ali, (2013), the funeral of C. N. Annadurai (1969), concert was given by Rod Stewart (1994), Hajj[2] the annual pilgrimage to Makkah, Dubai Exposition, protest in Circus Maximus, Rome against the government of Silvio Berlusconi (2002),

[1] https://en.wikipedia.org/wiki/List_of_largest_peaceful_gatherings.
[2] Fifth pillar of Islam.

K. Arai (Ed.): FICC 2021, AISC 1363, pp. 470–477, 2021.
https://doi.org/10.1007/978-3-030-73100-7_33

to name a few. For all such gatherings, pedestrian movement is mostly coupled with different spatio-temporal constraints.

Hajj is an annual pilgrimage to Makkah, where more than 2 million pilgrims from all over the world get together for a very short duration of about a week, to perform spatio-temporal rituals [1]. During the Hajj, pilgrims encamp for five days in Mina area (Fig. 1a), about 5 km^2 and visit other places Al-Haram Mosque, Arafat, Muzdalifah and Jamarat with temporal constraints [2]. Due to topographical nature of the Makkah city, pilgrims have to move through different types of the transfer corridors (uphill, downhill, zigzag, Y, L, T, and U-shape) (Fig. 1b–d) while they move in between the Hajj areas for performing Hajj rituals. Moreover, augmenting the diversified nature of the crowd in terms of race, age and language makes it backbreaking for organizers in providing pilgrims with efficient mobility plans and ensure pilgrims safety to avoid any crowd disasters [3] and massive causalities such as Mina stampede (2015) [4].

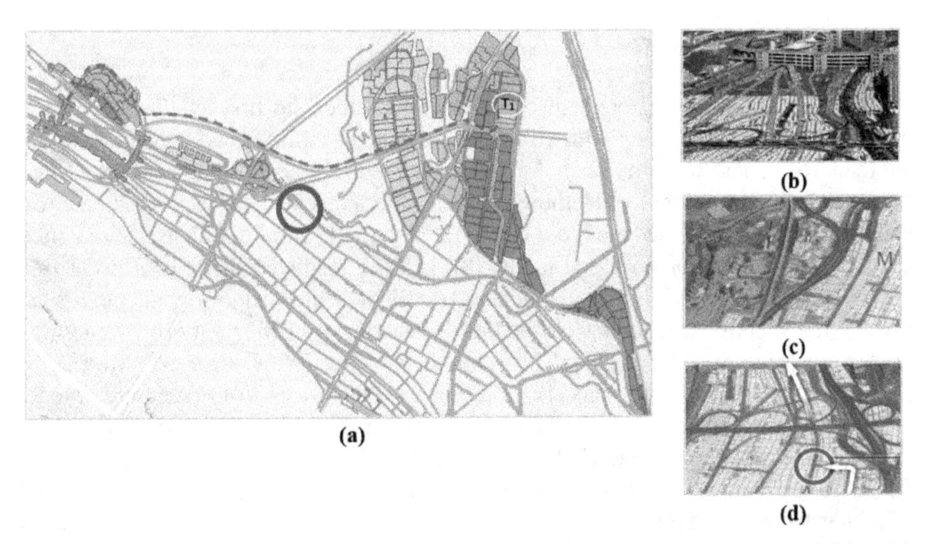

Fig. 1. (**a**) Mina area with multimodal transfer corridors, (**b**) Ramps towards Jamarat bridge, (**c**) A Y-shape corridor, and (**d**) Circular, T and U-shape corridors. The red circle is the location of Mina stampede (2015).

Considering walking, a sustainable mode of pedestrian movement and the crowd safety the core objective of the safe crowd management, organizing authorities are ones to completely analyze pedestrian behaviour, and study their mobility parameters like average speed/crowd density at the different hours of the day. Such information can lead the way to model and simulate complex situations when a massive crowd is moving through a complex network of available routes. In the same context, diversified nature of Hajj crowd, spatio-temporal constraints, available complex nature of pedestrians' walkways, and the possible crossflow of pilgrims through them, there is a need to analyze interconnectivity of Mina roads. Our ongoing research presents, a use case of pilgrims' movement of a group of pilgrims to Jamarat bridge for stoning the devil ritual, through a network of transfer corridors. Pilgrims position coordinates were captured with time stamp through a mobile phone application, and their average velocities were calculated while they moved through L, Y, U-shape, walkways with and without obstacles. We simulated this information to find the approximate time to pass through these corridors. We aim to extend our findings by considering the movement of multiple groups through these corridors and integrate the possible crossflow of pedestrians at the junctions. This will help us in designing optimized mobility plans by considering routes' capacity and available time to accommodate all groups.

The rest of the paper is organized as after the introduction, in Sect. 2 we added a brief literature review, followed by a simple scenario of pedestrian dynamics with the perspective of Hajj gathering in the next section. In Sect. 4, we present the modelling approach reflecting pedestrian movement through complex corridors. The next section presents a sample simulation activity and related results. Finally, we concluded our ongoing and future work.

2 Literature Review

Past Researchers from different disciplines allowed them to first analyze pedestrian movement and then come up with different models to describe crowd dynamics, which led them to propose mobility plans for safe crowd management. Lubas et al. presented a centrifugal force model [5], Helbing and P. Molnár presented social force model [6], and later Helbing presented the Fluid Dynamic Model [7]. Chraibi et al. used a modified spatially continuous forced-based model [8], Li et al. studied bidirectional pedestrian movement by image processing technique [9]. Dridi presented a microscopic model that uses PedFlow simulation to discuss the high-density crowd behaviour [10], which was earlier presented by Löhner [11]. Chen et al. extended LGM (Lattice Gas Model) for simulating pedestrian movements [12], Dietrich and G. Köster set of Force Based Model [13], Georgoudas et al. used a simulation tool for modelling pedestrian dynamics during the evacuation of large areas [14]. Kang and Han showcased a simple and Realistic Pedestrian Model for Crowd Simulation and Application [15], and Blue and Adler used "Cellular automata microsimulation for modelling bi-directional pedestrian walkways [16].

It has been observed that due to religious obligations pilgrims try their best to symmetrize their social interactions, which provides starting parameters like speed and crowd density helpful to connect our current study with social force models proposed for modelling crowd behaviour. However, crowd size and diversity along with temporal constraints suggests utilizing modern technologies which can track pilgrims in real-time and supportive in simulating the crowd mobility prior to the real event.

3 Pedestrian Dynamics

Hajj, a yearly activity, is a unique example of the heterogeneous crowd where more than 2 million attendees differing in race, age, language and culture get together with a common objective of performing religious rituals under spatial and temporal constraints. During their five days stay in Mina they move to other Hajj areas to perform Hajj rituals. Crowd nature impedes most of the crowd to utilize available transport and metro service, they prefer to walk through pedestrian walkways governed by pre-defined mobility plans. The following sub-sections will help readers to understand our ongoing research work to study Hajj crowd dynamics and the importance of applying modern methods to model, simulate and analyze, to explore and upgrade past events' mobility plans.

3.1 Pedestrian Movement to Jamarat

Figure 1 shows part of the Mina area and the highlighted camps that were allotted to south Asian pilgrims (209 tents and 693027 pilgrims) during the Hajj 2019 event. As a part of safe crowd management, these pilgrims were bound to only use the third floor of the Jamarat bridge for stoning the devil ritual. For this purpose, routes G (shown in green) and R (shown in red) was used by the pilgrims of tent T_1. Other than the large section of the route G (route R on return) that was commonly used by all the pilgrims of this section, there were several sub-routes (used by pilgrims of other tents) merging into this route, agitating movement of pilgrims of tent T_1. The similar can be imagined about the movement of pilgrims of other tents using different routes merging into route G.

3.2 Data Collection

A smartphone application was delivered to group leader of a group of 250 pilgrims (Fig. 2). The front-end of the application was developed using hybrid react JS and back-end using PHP Laravel with MS SQL database. The front-end communicated with the back end through HTTP RESTful APIs. The application can store GPS sensory data every two minutes and deliver to back-end. In case of network failure, the application can

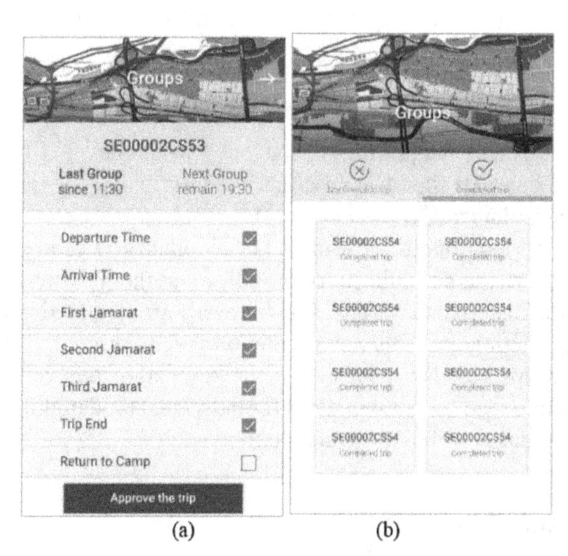

Fig. 2. Smart Phone Application used to Data collection (a) Check List before start of the journey, (b) Dashboard showing completed trips

store the information locally and forward to the relational database once get connectivity. Figure 2a shows the data collection checklist that group leader needs to click when he starts his trip till, he returns to his camp, and Fig. 2b shows a dashboard with the list of the completed trip by the same group leader.

4 Modelling Approach

We start with a simple use case that only pilgrims of a tent T_1 went to Jamarat for stoning the devil ritual through the rout G (shown in green) and came back through the route R (shown in red). The route G can be considered as a set of sub-routes g_1 (of length n units), g_2 (a curved route of radius r), g_3 a zigzag route etc. For some of the tents, these routes could be straight, but topographical nature of Mina area proffer route G for most of the tents were hybrid (the combination of straight, U, Y, T shape etc.). If the lengths of the routes $g_1, g_2, g_3, \ldots$ are taken $d_1, d_2, d_3, \ldots$ the average speed of pilgrims were $v_1, v_2, v_3, \ldots$, time taken by them can be measured as $t_1, t_2, t_3, \ldots$ for each sub route. Moreover, crowd densities $\sigma_1, \sigma_2, \sigma_3 \ldots$ can be measured at different hours of the day. By considering multiple similar sub-routes, average speed/density, etc., of each pilgrim (or groups) can be calculated that would lay the base for our mathematical modelling. We can introduce another variable α, the disturbance variable, by considering pilgrims from multiple tents were moving through and at junctions (Y, T, etc.) the speeds of pilgrims were affected. This could in general affect the total time taken by the pilgrims of each Tent in going and coming back to their tents. For the tent T_1, as;

$$T_1 = \{\mathrm{G}_1, \mathrm{R}_1\}, \quad \mathrm{G}_1 = \{g_1, g_2, g_3, \ldots\}, \quad R_1 = \{r_1, r_2, r_3, \ldots\}$$

$$\mathrm{g}_1 = \{d_1, t_1, v_1, \alpha_1, \sigma_1\}, \quad \mathrm{g}_2 = \{d_2, t_2, v_2, \alpha_2, \sigma_2\} \ldots$$

$$t_1 = \{d_1, v_1, \alpha_1\}, \quad t_2 = \{d_2, v_2, \alpha_2\} \ldots$$

The total time taken by the pilgrims of the tent 1 would be $t_{T1} = t_1 + t_2 + t_3 \ldots$.

So, in case we consider the ideal case that the pilgrims of the tent 1 are using the route g_1 only to go to Jamarat, $t_2 = t_3 = \ldots = 0$.

Considering the synchronous movement of few other groups of pilgrims with pilgrims of tent T1, there are possibilities of crossflows at the junctions of the corridors. This will trigger the disturbance variables α_i for every group affecting their times t_{T1}.

The following section explains our simulation results through different corridors, considering the groups speeds $v_1, v_2, v_3, \ldots$, the key parameters which upon analysis will suggest a suitable road (or combination) for individual (and multiple) groups so that all can be accommodated within the available time.

5 Crowd Simulation

Its known fact that once pilgrims from other tents also move towards the Jamarat Bridge, the crossflow at junctions will interrupt mobility of other pilgrims. Figure 3 shows six

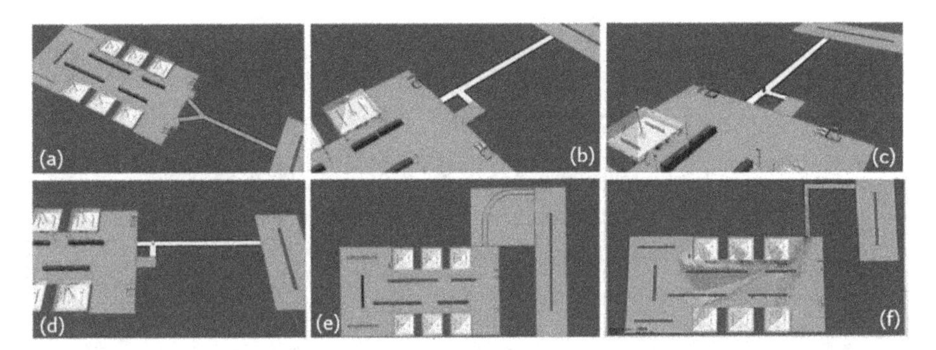

Fig. 3. Simulation Scenarios to analyze pedestrian movement (a) Y-shape Corridor, (b) Empty Intersection Corridor, (c) Intersection Corridor with a single 1 m Obstacle (length 1 m, width 1 m and height 1 m, (d) Intersection Corridor with a single 2 m Obstacle, (e) Smooth curved corridor, (f) L-Shape Corridor with simulation activity.

different types of scenarios that were chosen to simulate pedestrian movement clarifying the interruption and its effects in terms of travel times.

To start with the simple case study below is the detail of analysis of pedestrian movement through an L-shape corridor (Fig. 3f). The width of the corridor was measured as 5 m, and the length was 100 m. The simulation was carried out using a desktop computer, ensuring no other task was being processed by the machine. The software details are provided in Table 1 below.

Table 1. Software details

Software	Version	Manufacturer
MassMotion	10.5	Oasys
Revit	2018	AutoDesk
Operational dashboard	Online	ESRI

To find the speed and other necessary measures, we have captured pedestrians' spatio-temporal information through a mobile phone application. Our analysis showed that the average speed of pilgrims was 1 m/s in an ideal case. The capacity of the L-shape corridors calculated as if 1 agent crosses 1 m distance in one second so based on the width of the Corridor, 5 Agents cross the Corridor in 1 s.

5.1 Results and Discussion

After running the simulation for 10,000 Agents through a right angles detour corridor, the following results were found (Table 2).

The resulting flow rate was calculated as: $= 10000/30.16 = 331.6$ ped/min. The similar activity was carried out through other corridors (Fig. 3a–f), and the results are tabulated in Table 3.

Table 2. Simulation results

Activity	Time taken
First agent entered the corridor at	00:00:36
The last agent entered the corridor at	00:30:52
Time consumed by all agents to cross the corridor	30 min & 16 s
Time taken to finish the simulation	15 min & 45 s
Time taken to finish video rendering	25 min & 53 s

Table 3. Simulation through all six-shaped corridor-results

Corridor	Simulation time	Video rendering	Time consumed (10,000 agents)	Flow rate
Y-shape	12.33	23.26	28.05	356.5
Free intersection	13.34	24.15	28.53	350.5
Intersection with 1 Obstacle	17.37	25.55	29.5	339.0
Intersection with 2 Obstacles	21.02	31.06	35.42	282.3
L-shape corridor	15.45	25.53	30.16	331.5
Smooth corridor	15.03	24.58	29.54	338.5

The crossing time and the flow rates of Smooth and Y Shape corridor explain that there is an inverse correlation in such a way that by widening the width of the corridors or adding a second corridor, the crossing time will decrease and conversely the flow rate will increase. Y Shape is further explained as it has two 25 m corridor attached into a single 75 m long corridor, so in that at the first 25 m, it makes the flow rate of agents' a little higher compared to the smooth curved corridor.

6 Conclusion and Future Work

In this paper, we presented our ongoing research work about the pedestrian movement of a very large diversified gathering with a common objective of performing spatio-temporal religious rituals. Pedestrian movement through simple, L, Y, U-shape (or combination of) transfer corridors was discussed. We have simulated pilgrims, movement by considering their average velocities that were calculated by capturing position coordinates of a group of pilgrims while they travelled towards Jamarat bridge for stoning the devil ritual through a complex transfer corridor. Some other simulation results were presented that would be integrated by considering multiple groups movement to study the pedestrians' crossflow effects on their travel time through a complex network of available corridors. Our current work was limited to a group of pilgrims who travelled through a less complicated route that contained a L-shape transfer corridor. We aim that our findings will support in designing optimized mobility plans by considering routes' capacity and available time to accommodate all groups.

Acknowledgment. The authors extend their appreciation to the **Deputyship for Research & Innovation**, Ministry of Education in Saudi Arabia for funding this research work through project number 0909. Also, we would like to thank Ministry of Haj and Umrah, Saudi Arabia for providing the required data and resources.

References

1. Ahmad, A., et al.: A framework for crowd-sourced data collection and context-aware services in Hajj and Umrah. In: 2014 IEEE/ACS 11th International Conference on Computer Systems and Applications (AICCSA), pp. 405–412 (2014)
2. Ahmad, A., Rahman, M.A., Ridza Wahiddin, M., Ur Rehman, F., Khelil, A., Lbath, A.: Context-aware services based on spatio-temporal zoning and crowdsourcing. Behav. Inf. Technol. **37**(7), 736–760 (2018)
3. Haase, K., Kasper, M., Koch, M., Müller, S.: A pilgrim scheduling approach to increase safety during the Hajj. Oper. Res. **67**(2), 376–406 (2019)
4. Ganjeh, M., Einollahi, B.: Mass fatalities in Hajj in 2015. Trauma Mon. **21**(5) (2016)
5. Lubaś, R., Miller, J., Mycek, M., Porzycki, J., Wąs, J.: Three different approaches in pedestrian dynamics modeling – a case study. In: Advances in Intelligent Systems and Computing, October 2014, vol. 224, pp. 285–294 (2013)
6. Helbing, D., Molnár, P.: Social force model for pedestrian dynamics. Phys. Rev. E **51**(5), 4282–4286 (1995)
7. Helbing, D.: A Fluid Dynamic Model for the Movement of Pedestrians, pp. 1–23 (1998)
8. Chraibi, M., Seyfried, A., Schadschneider, A.: Generalized centrifugal-force model for pedestrian dynamics. Phys. Rev. E **82**(4), 046111 (2010)
9. Li, X., Ye, R., Fang, Z., Xu, Y., Cong, B., Han, X.: Uni- and bidirectional pedestrian flows through zigzag corridor in a tourism area: a field study. Adapt. Behav., 105971232090223 (2020)
10. Dridi, M.H.: Simulation of high density pedestrian flow: microscopic model. arXiv Prepr. arXiv:1501.06496, vol. XXI, no. 1, January 2015
11. Löhner, R.: On the modeling of pedestrian motion. Appl. Math. Model. **34**(2), 366–382 (2010)
12. Chen, T., Wang, W., Tu, Y., Hua, X.: Modelling unidirectional crowd motion in a corridor with statistical characteristics of pedestrian movements. Math. Probl. Eng. **2020**, 1–11 (2020)
13. Dietrich, F., Köster, G.: Gradient navigation model for pedestrian dynamics. Phys. Rev. E **89**(6), 062801 (2014)
14. Georgoudas, I.G., Sirakoulis, G.C., Andreadis, I.T.: A simulation tool for modelling pedestrian dynamics during evacuation of large areas. IFIP Int. Fed. Inf. Process. **204**, 618–626 (2006)
15. Kang, W., Han, Y.: A Simple and Realistic Pedestrian Model for Crowd Simulation and Application, pp. 1–6 (2017)
16. Blue, V.J., Adler, J.L.: Cellular automata microsimulation for modeling bi-directional pedestrian walkways. Transp. Res. Part B Methodol. **35**(3), 293–312 (2001)

A Secure and Privacy Preserving System Design for Teleoperated Driving

Stefan Neumeier[1(✉)], Christopher Corbett[2], and Christian Facchi[1]

[1] Technische Hochschule Ingolstadt, Ingolstadt, Germany
{stefan.neumeier,christian.facchi}@thi.de
[2] Universität Ulm, Ulm, Germany
christopher.corbett@uni-ulm.de

Abstract. Teleoperated Driving, where a human driver controls a vehicle remotely, has the ability to be a key technology for the introduction of autonomous vehicles in everyday's traffic scenarios. Already existing infrastructure like cellular networks have to be used to allow for an efficient use of such a system. Remote control is a sensitive subject and has high demands on security and, based on the fact that individuals are driven remotely, also on privacy. To take care of security and privacy, this paper introduces the minimal set of vehicle features, that are required for Teleoperated Driving. It also discusses a way of setting up a secure connection with valid and trusted remote operators, that can be selected taking into account various parameters. Involved parties are explained in detail. To allow for traceability, e.g. in case of an accident, by keeping a high level of privacy, a logging concept is introduced. Overall, this paper presents an initial approach to build a teleoperated system considering security and privacy as key factors, which can be used to build real-world systems.

Keywords: Teleoperated Driving · Remote control · Autonomous driving · Security · Privacy · Networking · System design · Concept

1 Introduction

Improving traffic safety is an ongoing effort since decades. Besides the introduction of essential safety-components like airbags and safety belts [27], the trend nowadays consists of driver assistant systems [4]. Improving such driver assistant systems enables vehicles to solve a growing number of situations independently in a safe way [3]. A definition of driver assistant systems is done by the SAE in [10], where levels between 0 and 5 were defined. Starting with level 3, the vehicle is responsible for taking care of the (isolated) driving task in very restricted situations. With level 4 vehicles are able to deal with a greater number of situations leading to fully autonomous vehicles reaching level 5. Unfortunately, even highly autonomous vehicles can face situations which they are not able to solve appropriately. Thus, autonomous test vehicles were already involved in

K. Arai (Ed.): FICC 2021, AISC 1363, pp. 478–496, 2021.
https://doi.org/10.1007/978-3-030-73100-7_34

accidents [16] and might face further issues. As shown in [20] such issues can be raised by complex road-side works. However, road-side works are not the only issues autonomous vehicles are confronted with. Kang et al. [24] lists further errors like system confusion or software/hardware failures. Human intervention is required to resolve such situations in a fast and safe manner. This requires a suitable human driver to be inside the vehicle and able to take over control. For autonomous vehicles this is not guaranteed anymore. It is more likely to happen that vehicles are empty [21] or humans inside are not able to take over control, e.g. in case of a medical emergency. Teleoperated Driving could be the solution in such situations. Teleoperated Driving is the remote control of a vehicle by a human in situations that require it. Such systems are already used and developed by different companies [12] and moreover also required by law in California if testing empty autonomous vehicles [13]. Thus, Teleoperated Driving is claimed to play an important role on the way to autonomous vehicles. Enabling Teleoperated Driving requires the utilization of wireless connections, to achieve high coverage, ideally with a technology that is already widely deployed [29]. Thus, cellular networks are the technology of choice. It is already widely deployed and with new technologies it is able to provide required demands regarding latency, bandwidth and packet loss [25].

Since controlling a vehicle remotely is a sensitive task, there are a number of specific requirements. In addition to safety-relevant requirements like emergency brake, driver support systems, etc. security and privacy must be taken into account. Security is one major factor, as several infrastructure components and external parties are involved in the communication and abusing Teleoperated Driving bears various dangers that finally might cause harm. Even if the connection is secure, there must be some kind of trust between the remote operator and the vehicle, which additionally considers privacy requirements of customers. Especially with already established European General Data Protection Regulation (GDPR/DSGVO) and the United Nations Economic Commission for Europe (UNECE) and ISO21434 [1] on the horizon to enforce security processes, mechanisms and monitoring for future vehicles.

Based on the fact that most recent work did focus on network and safety, but did neglect security and privacy, this work will focus on the latter two. It will present an initial security and privacy preserving concept that copes with the challenges of Teleoperated Driving.

Thus, this paper is structured as follows. Section 2 gives an overview of related work. Subsequent an introduction to Teleoperated Driving is given in Sect. 3. The proposed system design is presented in Sect. 4. Finally, Sect. 5 concludes the paper and draws an outline of future work.

2 Related Work

In case of teleoperated cars, a lot of research has been done, e.g. by [29,37,38]. Teleoperated cars are under development by various institutions [12,13,20]. The major part of research focuses on usage studies for teleoperated vehicles, the visualization of data at the teleoperator cockpit and the impacts of time delays and data rate when driving remotely.

A huge problem in remotely controlling a vehicle is the time delay between the teleoperator and the remote vehicle. Time delay leads to a lag in visualization at the teleoperator's side and to unsynchronized execution of commands at the remote vehicle. Wired connections would offer high data rates and low latency, but are unfeasible in typical road traffic. Therefore, wireless technologies have to be used. Utilizing Wi-Fi technology is a current approach, but has limited range and thus, without a costly infrastructure, is not able to provide the required long distance communication [7,9]. Another approach exploits current mobile communication technologies, e.g. 3G, LTE and 5G, but these connections suffer from high time delays, probably low bandwidth and potential packet loss. As an example, Chucholowski et al. [7] report time delays ranging from 65ms to 1299ms when transmitting pictures via 3G. Further research of Neumeier et al. [32] indicates lower latencies, but not at a continuous level. As such large delays are dangerous in a feedback control system, further approaches have to be applied to ensure safety. With the new standard, LTE-Advanced, the uplink rate is increased to up to 1.5 Gbps [39], which should be enough for transmitting the required data and commands. With 5G, even higher values are achievable [19]. Unfortunately, mobile connections suffer from potential high delays and packet loss [15]. Further, the data rate can drop sharply depending on the mobile cell workload. 5G could mitigate these problems, but only specific spots benefit from first 5G installations. Additionally, real-world measurements have to be applied to 5G to test its capabilities for Teleoperated Driving. Variable and fixed lags during the inconsistent wireless connection make it hard to safely control the vehicle [14]. Various research works have demonstrated several approaches to help mitigate this so-called time-lag problem. The main goal is to assist the driver so that he has the impression of physically sitting in the car. In [14] it has been shown that the use of a predictive display can mitigate the impacts of lags by representing the latency based state, e.g. foreshadowing the time delay based car position. In [8] various types of predictive displays have been compared in a study, showing that their usage can effectively assist the driver with his task. Using additional haptic feedback on the steering wheel to support the driver in challenging situations has been proposed by Hosseini et al. [23]. Most of those approaches aim for the situation awareness of the teleoperator. This situation awareness can be best achieved if the teleoperator is aware of its relevant environment [34], e.g. by having a suitable display of relevant data. In [22] the authors show that using virtual reality glasses and combining available sensor data to a representation of the environment can help to improve the situation awareness. This is achieved by combining video information from the cameras with 3D models built from LiDAR data. The authors in [17] show, that using a head-mounted display does not necessarily improve the driving performance. Exemplary setups of Teleoperated Driving systems are shown by Neumeier et al. [29], Gnatzig et al. [18] and Shen et al. [36]. Lichardopol [26] did a survey on teleoperation, whereas [6] analyzed more than 150 papers, identifying issues with teleoperation and proving some mitigations. All approaches shown in previous research try to address the drawbacks of non-deterministic network connections in various ways. However, all of those approaches put their focus on safety, but did not address the topics security and privacy elaborately. Thus, this paper wants to present a first overview of a system design for Teleoperated Driving, aiming for security and privacy.

3 Introduction of Terminology for Teleoperated Driving

Teleoperated systems, e.g. UAVs as in [41], usually consist of three crucial parts: teleoperated device, teleoperation work and the communication link in between [40]. An example setup of a teleoperated system as may be used for vehicles with road approval can be seen in Fig. 1.

Fig. 1. Overview of an teleoperated system (based on [37]).

One of the most crucial parts is the *Target Vehicle*. This is the vehicle that requires remote control, because it might not be able to solve a situation autonomously while there is no suitable driver available to take over control. Situations that can lead to such cases are for example complex road side works, unknown traffic situations or the failure of hardware or software.

When such a situation arises, a remote *Teleoperator* placed in a *Teleoperation Center* might utilize a Teleoperated Driving system and takes over control of the target vehicle. Teleoperators should be specially trained drivers [33], able to deal with the obstacles induced by the *Mobile Network* [32]. They additionally have to be trustworthy to ensure a safe and comfortable transportation of potential passengers. The teleoperation centers are institutions that provide the workspace for the teleoperator. They require a fast internet connection and further have to provide the required hardware, e.g. steering and pedals. In an ideal case, there are multiple teleoperation centers and the most suitable one is picked.

To allow the remote control of vehicles in a reasonable way, existing mobile networks have to be used. They provide the flexibility required due to the high mobility of vehicles, but, unfortunately, they induce obstacles such as latency, bandwidth and overall connectivity together with security issues. However, typical teleoperation sessions are required to overcome only a specific obstacle, i.e. the duration of a session is very short (<15 min).

3.1 Definition of the Requirements

To allow the remote operation, independent of security and privacy concerns, specific requirements need to be met. Features can be split into vehicle features, environmental conditions and remote workplace that must go hand in hand with a secure and safe vehicle status.

Required Vehicle Features: To enable the remote control of a vehicle, it requires some basic features such as the ones described by Neumeier et al. [29], i.e. it needs to allow an operator to remotely control it by providing required hardware and software. The most vital part besides the basic functionality is, that the vehicle is able to perform the transformation from the driving state to a safe state (usually stopped vehicle) autonomous and independent of the current state of the network. In addition to reaching the safe state on its own, the vehicle needs an automated emergency brake assist, that is able to deal with the sudden appearance of obstacles. This system can help to avoid crashes or reduce their impact even if the remote operator is not able to react appropriately. Further approaches, such as the ones shown in [28], might support the driver to safely control the vehicle under varying environmental conditions.

Required Environmental Conditions: For a safe Teleoperated Driving, the environmental conditions, especially bandwidth, latency and packet loss are crucial. These values need to be provided by cellular networks. Even though EDGE might be feasible under some circumstances, at least LTE should be available [32]. Based on existing research by Neumeier et al. [33], the overall Round Trip Time (RTT) should be at least below 300 ms to allow a safe remote control by trained operators. Following Kang et al. [24] and others, the latency should be kept at a fixed value. The required bandwidth depends on the weather conditions and the driving scenarios. Based on the findings in [31], using a single videostream under good environmental conditions (highway, day) it must always be above 280 KBit/s for the stream. This value is increased by additional requirements for logging and session handling. If more than one stream is required, e.g. in case of crossing an intersection or in a parking situation, the bandwidth requirement will further increase. Approaches, e.g. as shown by Schmid et al. [35], can be used to predict the network capabilities and adjust the route of the vehicle accordingly.

Required Operator Workspace Conditions: The ability to provide telepresence is an important enabler for Teleoperated Driving [5]. Thus, the teleoperator's workspace should follow specific requirements that reach beyond a good network connectivity. Following Hosseini and Lienkamp [22], Teleoperated Driving could be improved by virtual-reality systems. However, this is not necessarily the case. In [17] Georg et al. show that head-mounted displays allow for a better immersion, but—in contrast to the aforementioned results—do not necessarily improve the driving performance. A predictive display [8] or a combined approach with virtual-reality and augmented sensor data [22] could help to make driving safer.

Vehicle Status: Prerequisite for a remote connection is a valid and verified vehicle status. This includes both safety and security attributes.

Safety: To obtain a minimum level of safety, a bunch of different systems is required to allow the remote control of a vehicle. The most important part is, that the required sensors and actuators are working as expected. During the take-over phase of a remote operator, the required sensors and actuators have to be tested automatically. If required components fail, a takeover is not possible. Each remote system may require different sensors, e.g. LiDAR, ultrasonic, but remote driving can only be started if the required sensors are working. This can also be scenario dependent, e.g. if ultrasonic sensor are required only during parking, a remote control on a highway is possible even if those sensors do not work properly, as the sensors are not required then. Actuators are further required to work properly, e.g. steering, gas and braking are the most basic and required actuators that have to react on remote operator's commands. Further actions such as activating wipers are situation dependent and are required under specific constraints: If the rain sensor indicates rain or dirt and cameras for remote operation requires them they also needed to work properly. For safety reasons, a take-over will always need to happen from a non-driving situation, e.g. the vehicle needs to be in a "parking"-like mode.

Security: Vehicles are more and more in the focus of security researchers, hackers and criminals who impact or temper with the security or safety status of the vehicle. Therefore in-vehicle networks and respectively the network communication must comply with integrity, authenticity and confidentiality. Hence, the security testing of vehicle components [11] and the frequent monitoring of vulnerabilities is essential to detect and remedy potential attack vectors. Furthermore, technologies such as Intrusion Detection System (IDS) and Intrusion Detection and Prevention System (IDPS) are about to be integrated to monitor and detect any violation or misuse. If the vehicle was modified or tempered with—which has impact on the overall security state—Teleoperated Driving must not be enabled as the potential impact is unpredictable.

3.2 Parameters for Operator Selection

The basic idea behind the selection of suitable Teleoperator (TO) is motivated by the fact, that multiple TOs are required to provide a sufficient coverage for remote control. The following part describes an initial idea. The most important factor to select a remote teleoperation center is the network connectivity. This factor ensures that remote driving can happen in a way, that allows the remote operator to safely control the vehicle. Following Neumeier et al. [32], a major influence factor can be the distance between teleoperation station and the remote vehicle. Thus, the first step of selection is a selection based on the network connectivity, which often bears down to a distance-based selection.

If there is the case, that multiple operators are suitable, the selection might be based on:

Preference: The personal preference, e.g. the selected provider offering a payment model fitting the customer's needs or the preference of the manufacturer.

Available Operators: Which operator is available in the driving area.

Supported Service Providers: Which service is supported by X and is this compatible with Y?

Supported Languages: Which communication language is preferred, if interaction is required.

Trustability: Did something that happen in past, that may lowers the trust in a provider?

User Rating: Are the user ratings nicely and satisfying?

Based on those parameters, the selection of a suitable TO might happen.

3.3 Situations that Lead to Termination of Teleoperation

Teleoperated Driving, as a safety-critical process, has some situations at which the remote control of a vehicle must end. By ending a remote control, the vehicle will reach a safe state and control is handed over again either to the autonomous driving system, or if that is not applicable, to a person inside the remote vehicle. If there is no suitable person inside or the autonomous system can not takeover, the vehicle will remain in the safe state and further actions, such as towing away the vehicle, need to be undertaken. In some scenarios, e.g. emergency braking, it is possible to re-start remote driving, but a new initialization is required.

The most obvious scenario is the termination of a remote drive, either by the passenger or by the driver. If one of those parties decides to end the remote controlling, e.g. if the "tricky" driving scenario is solved, the remote drive will end. Another scenario consists of losing the required network connectivity. If the network connectivity of the remote vehicle will drop below a critical level regarding latency, jitter, bandwidth or packet loss, the vehicle will end the remote operation. A further scenario for ending teleoperation is that an integrated emergency assist had to conduct an emergency braking, e.g. avoid crashing into a pedestrian or cyclist. This immediately ends the remote operation.

The remote operator is required to keep the vehicle inside a pre-defined area. This allows him to drive around critical situations, but disables the ability to drive the vehicle somewhere. The pre-defined area lies inside the based on network connectivity [28] planned route. This also can support the prevention of car-theft utilizing Teleoperated Driving.

Finally, if there are any detected hardware or software failures that could disallow Teleoperated Driving, the system can not be used anymore and remote operation is not possible until the issue is solved.

4 Proposed System Design

Figure 2 shows the overall architecture of the proposed system design and gives a brief overview of the involved infrastructure and parties, which will be explained in detail in the following sections:

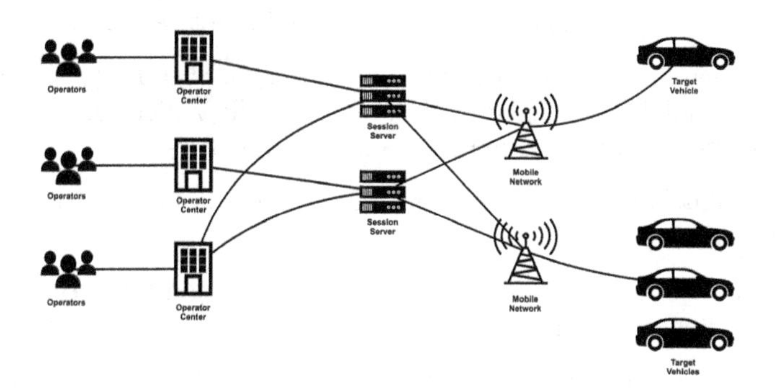

Fig. 2. Architecture overview of teleoperation infrastructure that interacts with vehicles.

The main architecture is meant to address potential attack vectors/problems within the communication between a vehicle and a teleoperator such as:

Car Theft: Operator drives unmanned vehicle to unknown destination for theft.

Take over: Control of an ongoing teleoperation by a third party.

Replay attack: Record and replay control information to misuse the vehicle or tamper with the vehicle.

Hopping criminal operator: Potential malicious teleoperator commits crime at several operator centers.

Control message tampering: Trust of messages between all stakeholders.

4.1 Connection Setup

A major process in Teleoperated Driving is connecting the operator and the vehicle. Therefore a 10 step connection setup, that allows for a secure connection process, is proposed and can be seen in Fig. 3.

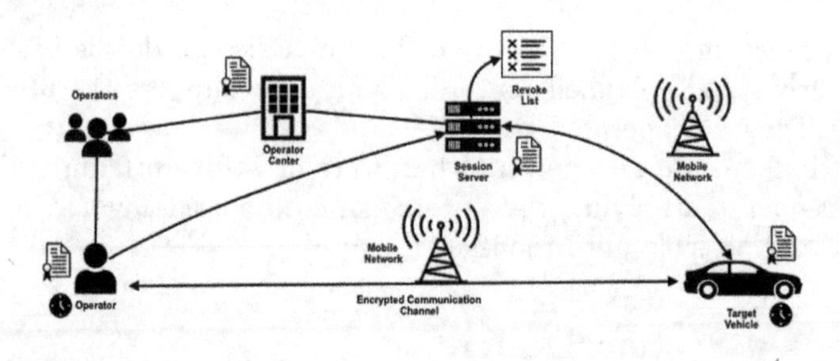

Fig. 3. Multi-step interconnection handling between operator and target vehicle.

Independent of the scenario, the trigger action for the remote control needs always to be sent by the customer, i.e. an operator can not select a vehicle randomly and choose to control it if the vehicle did not request this action.

Step1: A request for remote control was made by a customer (e.g. owner, car rental service, mobility service provider) for a specific vehicle. This request triggers the vehicle to start the connection setup process.

Step2: The vehicle registers a request for remote control and the destination at a designated session server.

Step3: At the session server several operation center and furthermore available operators register to signal availability.

Step4: Based on the parameters presented in Sect. 3.2 considering destination, latency, bandwidth, etc. and extended by a random selection process of compatible operators to prevent predictability the session server picks an operator to handle the request.

Step5: The vehicle receives a notification with the operator id that will handle the request.

Step6: The vehicle pulls the public key of the operator from the session server and generates a token that holds a random session key and a timestamp when the session key expires. To verify that the operator is valid and not banned the vehicle checks the revocation list before downloading.

Step7: The vehicle signs the token with its private key to enable verification if required later. The token with signature gets encrypted with the public key of the operator and is transmitted back to the session server along with the IP address of the vehicle.

Step8: The data is transmitted to the operator and decrypted. This way it is guaranteed that the information is only available to the specific operator and the involved vehicle.

Step9: The operator initiates an encrypted peer-to-peer connection. The foundation for the connection are the certificates from the chain of trust containing the roles and rights that are granted to the operator with the vehicle.

Step10: After the connection is established pre-tests are done to check the operability of the vehicles sensors and remote controls. If the tests are passed, the operator can send control commands (and e.g. receive the stream) to the vehicle that are signed with the session key to guarantee authenticity.

4.2 Chain of Trust

An important base for a system to provide a decent level of security is a chain of trust within the group of stakeholders. Such a trusted system could be inspired by a Car2X system as described in [2].

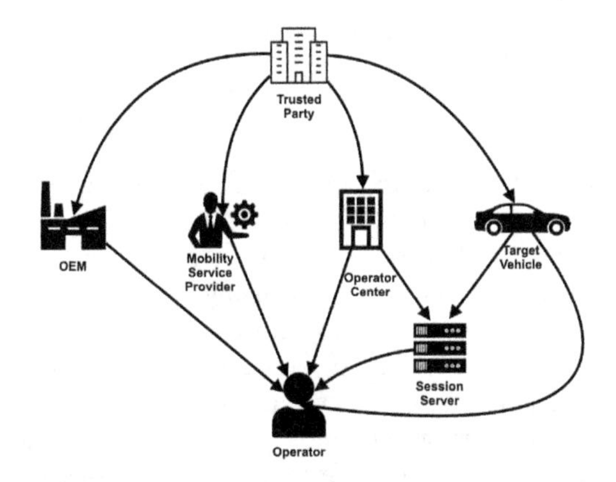

Fig. 4. Chain of trust within the group of stakeholders.

Key aspect of the proposed approach is a level of trust among all participating parties. As every stakeholder is either a service provider or service receiver none of the stakeholders by themselves is suitable as a provider of trust and an overall neutral entity is required to sustain an independent trust anchor. The chain of trust designed for the presented approach can be seen in Fig. 4. The identified stakeholders are described below in detail.

Trusted Party: Within this trusted party the Teleoperator Registry (TOR) is introduced. The TOR tracks information about a teleoperator such as overall rating, customer ratings and teleoperator center ratings. This is required to provide valid information to operation centers whether this teleoperator can be trusted. Only with a certain ranking a teleoperator is allowed to remote control vehicles.

Original Equipment Manufacturer (OEM): The original manufacturer of the vehicle, e.g. the party which is responsible for preparing the vehicle to allow Teleoperated Driving.

Mobility Service Provider: Party that provides a service with the vehicle involved, e.g. car rental, car sharing.

Operator Center: Service provider for remote vehicle control that provides operators.

Target Vehicle: Vehicle to take temporary control over, provided by other stakeholders.

Session Server: Trusted session server to relay requests and responses for Teleoperated Driving amongst stakeholders and customers (described in more detailed in the following).

Operator: Specific operator provided by an operation center to control the target vehicle.

With a superior trusted party as the anchor of trust, the trust relationship between the stakeholders is feasible with the teleoperator at the end of the chain.

4.3 Session Server

To work as expected, the approach presented in this paper requires a session server. The session server can be one or multiple trusted instances that vehicles and operators connect to. The selection of an appropriate session server can be based on factors such as customer licensing, geo position, bandwidth and latency (see Sect. 3.2). To prevent possible attack surfaces like Man-in-the-middle (MIDM) attacks or eavesdropping, any connection between the operator, the vehicle and the session server is encrypted. As a state of the art technology a certificate based TLS1.3 encrypted traffic with client authentication is suggested.

The trusted 3rd party holds a list of all existing sessions servers, the *Session Server Registry* and provides a downloadable list of servers for vehicles and operator centers based on criteria such as geo position, countries, etc.

To fulfill his task, the session server offers the basic functionality as described in the following.

Register New Teleoperator: Teleoperators that are ready for operation register at their designated session server.

Register New Vehicle: A vehicle can register for future requests and uploads its public key to the session server.

Connection Test to Vehicle: Send encrypted hello message (with vehicle public key) to Target Vehicle (TAV) and wait for reply. This way it can be verified that the correct cryptographic information is used.

Handle Vehicle Remote Control Request: Stakeholders that require the service of a remote controlling a vehicle (e.g. service providers, customers) send a request to the designated session server.

Notify Vehicle: Provide a target vehicle with necessary information.

Session Server Key Sync: Operation to sync operator public keys amongst other session servers. This can be either restricted to a customer or service provider subset or globally.

Notify Teleoperator: Channel to provide a teleoperator with necessary information.

Download Public Key: Download public key from the session server to the vehicle.

Upload Token: Upload encrypted token to session server.

When a teleoperator was randomly picked out of the list of available operators to process the vehicle request, a session key is generated by the teleoperator and submitted to the session server. The session server encrypts the session key with the public key of the vehicle and forwards the session key to the vehicle.

4.4 Exchanged Information Definition

Three different types of information need to be exchange to be able to safely communicate and remotely control the vehicle: vehicle information, control data and session control data.

Vehicle State Data: Information such as vehicle status, sensor and video data are transmitted from the vehicle to the operator. This information is required to allow the remote operator to sense the vehicle's environment and plan its actions

Table 1. Additional information sent from vehicle to teleoperator.

Name	Value	Type
Timestamp	div.	long
Gas Pedal	0–100	char
Brake Pedal	0–100	char
Steering Angle	0–180	char
Vehicle Speed	-20–235	char
Blinker State	Left, Right, Warn	char
Gear	Back, Forward	bool
Light State	Off, DRL, Normal, Distant	char
Wiper State	0–6 States	char
Lateral Acceleration	0–10	float
Longitudinal Acceleration	0–10	float
GPS Latitude	div.	double
GPS Longitude	div.	double
Temperature	0–100	char
Vehicle Error Codes	div.	char
X-Vec	div.	float
Y-Vec	div.	float
Z-Vec	div.	float

properly. Based on the specific implementation, this data can be manifold and may consist of one or more video-streams together with additional sensor data. However, there is a base set of additional information shown in Table 1 that is required to safely control a vehicle.

Vehicle Control Data: Information such as steering angle, actions (e.g. acceleration, breaking), activation of vehicle equipment (e.g. indicators) is transmitted from the operator to the vehicle. A minimum amount of control data that has to be send is listed in Table 2.

Table 2. Control Information Sent from the Teleoperator to the Vehicle.

Name	Value	Type
Timestamp	div.	long
Gas Pedal	0–100	char
Brake Pedal	0–100	char
Steering Angle	0–180	char
Gear	Back, Forward	bool
Blinker State	Left, Right, Warn	char
Light State	Off, DRL, Normal, Distant	char
Wiper State	0–6 States	char

Session Control Data: Information such as operator availability, vehicle control requests and session control protocol information are exchanged between the operator, target vehicle and session server.

As mentioned earlier, the establishment of a connection between the TO and the TAV is a multiple step process. When a dedicated operator was assigned to a vehicle request, the vehicle gets a notification with the id of the TO, the encrypted session key and awaits a connection attempt. As soon as a connection was established, control commands of the teleoperator are signed with the session key of the vehicle to provide authenticity of the messages. As speed is key in Teleoperated Driving, a User Datagram Protocol (UDP) based communication with Message Authentiation Code (MAC) is sufficient. Therefore a full encryption of the control messages is not required. To prevent replay attacks to the vehicle, session keys are utilized. Each session key needs a defined Time-To-Live (TTL) to prevent any further misuse. As remote control sessions are in average rather short (e.g. <15 min), a TTL of 30 min is proposed. The session key of the TAV always gets invalidated after the vehicle was completely switched off. Part of the authors proposal is a **MsgNr** at the start of each message (Fig. 5). This is primarily necessary due to possible UDP runtime and alternative packet path issues, so the receiver does not process older messages that are irrelevant by the time of reception.

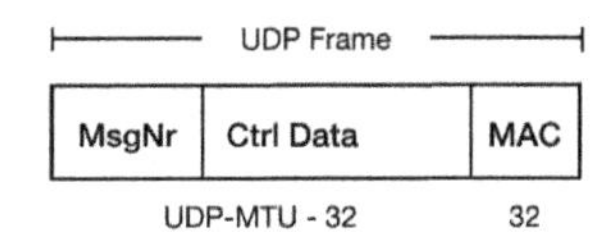

Fig. 5. UDP based control message frame with MsgNr and MAC.

For vehicle state data the same approach can be used. However, the final decision on how messages are structured depends on the specific implementation.

Session Keys: The authors propose the use of one session key shared amongst the TO and the TAV for every teleoperation request. For the session key a randomly generated symmetric key with a state of the art algorithm (e.g. SHA-256) is proposed. This session key later on represents the input for the Keyed-Hash Message Authentication Code (HMAC) algorithm to authenticate exchanged information with a MAC.

4.5 Logging/Event Recording/Legal Proceeding

Teleoperated Driving is a technology with potential legal, financial and health impact. In the case of such an event, the circumstances that led to the situation must be evaluated. This involves local authorities as well as mobility service providers and manufacturers. The authors therefore propose an event logging mechanism that supports any investigation by providing necessary information/data in compliance with General Data Protection Regulation (GDPR) and providing both—integrity and authenticity. The destination for any logging could be the already used session server instance or any other system. Figure 6 gives a brief overview of involved parties.

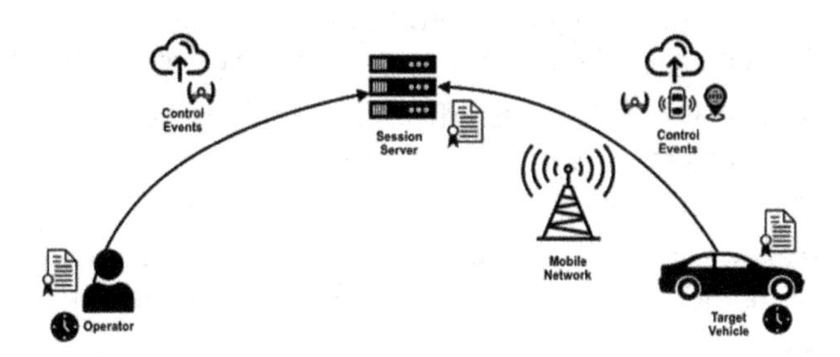

Fig. 6. Logging of events from vehicles and teleoperators during session.

Collecting data from each party on a central logging server involves frequent communication to the device. Hence, the reliability of the connection is bound to certain conditions each party is exposed to. The authors identified five key factors that have impact on the information exchange in the proposed scenario. In the following the impact for teleoperators and vehicles is examined.

Teleoperator

Bandwidth: Being part of a larger organizational unit and located at a static local facility, the bandwidth connection of a teleoperator is considered as high (above 1Gbit/s).

Latency: Given the assumption of a good connectivity the latency is considered low.

Synchronization Interval: Due to the high bandwidth and low latency connection to the logging server the interval can be considerably high (1 sync/1 s).

Time Accuracy: Due to the high bandwidth and low latency as well as the option to correlate several trusted time sources (NTP) the accuracy is considered high. GPS timestamps could be used instead, e.g. if target vehicle's network connectivity is poor.

Events/Content: The following data is considered relevant to be able to draw any conclusions on the origin of the event that led to a teleoperation malfunction.

- Teleoperator ID (A technical information not real name)
- Transmitted Control Command
- Teleoperation session identifier (technical)
- Timestamp
- Signed with the public key of the teleoperation center

Target Vehicle

Bandwidth: Relying on potentially limited wireless mobile broadband connections the bandwidth can heavily vary (100kbit/s to 1GBit/s, [32]).

Latency: Can vary from medium to high (e.g. from 65 ms to 1299 ms in LTE, following the results of [7], but can also vary heavily on both edges [32]).

Synchronization Interval: Due to the limited processing power and potential unstable connection to the logging server, the synchronization interval is considered as low (depending on the network parameters, ranging from 1–20 Hz).

Time Accuracy: In cases of high bandwidth and low latency combined with the option to correlate several trusted time sources (NTP) the accuracy is considered high. GPS timestamps could be used in cases with bad network connectivity.

Events/Content: The following data is considered relevant to be able to draw any conclusions on the origin of the event that led to a teleoperation malfunction:

- Teleoperation session identifier (technical).
- Vehicle Identifier (pot. VIN or pseudo technical information).
- Vehicle Status data (e.g. security status, sensor and actuator status, current speed).
- Current GPS position.

All this data needs to be stored to allow a potential investigation in cases such as accidents, etc. If data is stored encrypted on a privacy-conform server, where few people have access to, the proposed design is considered as privacy preserving.

4.6 Discussion of Potential Flaws and Limitations

The architecture described in this paper has some flaws that need to be discussed. Through the concept a designated teleoperator is bound to the operation request and as a result an instant handover to another teleoperator is not possible. A potential solution could be define a backup operator and assign a backup session with that teleoperator. There exist GDPR concerns when crossing country borders. Depending on the country laws, user or vehicle data are not allowed to leave the country and as such teleoperation across borders could become hard to manage or probably even impossible. With a single instance trusted 3rd party the state of the art, e.g. root ca in the internet is used. However, this might be a single point of failure. By randomizing the teleoperator in a session server a prevention against influencing who is controlling a dedicated vehicle exists. If a customer wants the same teleoperator multiple times, this can be hard to handle. Attack vectors such as jamming can not be overcome with this design, as lost packets can not be recovered and a high number of lost packets lead to end teleoperation.

Based on the design of the approach, the privacy is preserved strongly, e.g. teleoperators are selected randomly, etc. However, some flaws exist. If the teleoperation center is quite small, the randomisation might not work as expected. Additionally, a remote operator potentially may gain information on whom he is driving remotely. However, teleoperation usually happens only for a short amount of time (<15 min) and a specific areas, e.g. road-side works and as such, the loss of private information in considerably low. Saving of sensitive information (GPS, timestamp, etc.) and control data is required for traceability, e.g. in case of an crash, but can be handled in privacy-conform manner.

Based on the nature of this paper, its main limitation is the theoretical approach. However, as every implementation is slightly different, this paper can be seen as what it is meant to be: a general overview of an architecture that might serve as base for future implementations.

5 Conclusion and Future Work

Teleoperated Driving is a key technology for autonomous driving in future traffic scenarios. To function properly, it requires a security and privacy preserving concept to protect the customers as well as the service providers and comply with future legal requirements such as GDPR and UNECE. Based on the fact that security and safety were mentioned mostly incidental in previous works, this paper strikes into this notch by presenting a concept that allows for a safe and secure teleoperation. It addresses important factors that need to be considered when implementing a system for Teleoperated Driving and can serve as a base for future research and development.

In future work, an implementation of the proposed concept will be presented, allowing researchers to implement their own Teleoperated Driving systems and identify potential flaws. Especially a multi staged concept to implement an architecture that enables session transfers across international operation centers could be feasible. On a first step, the authors approach will be integrated in the Open-ROUTS3D driving simulator, an open-source driving simulator developed for the needs of Teleoperated Driving research [30]. This will allow researchers to directly use and test the concept in a virtual environment. First results of the concept will be present based on the findings of this implementation. In addition, the proposed concept will be implemented on a rc car based demonstrator. The specifications and the implementation of the rc car will be published. This allows for a safe investigation of the implementation and the gathering of first real-world data and measurements in a controlled environment.

References

1. Road Vehicles – Cybersecurity Engineering - ISO21434. Accessed 31 Jan 2020
2. Baldessari, R., Bödekker, B., Deegener, M., Festag, A., Franz, W., Christopher Kellum, C., Kosch, T., Kovacs, A., Lenardi, M., Menig, C., Peichl, T., Röckl, M., Seeberger, D., Straßberger, M., Stratil, H., Vögel, H., Weyl, B., Zhang, W.: Car-2-car communication consortium - manifesto. Technical report (2007)
3. Becker, J., Colas, M.-B.A., Nordbruch, S., Fausten, M.: Bosch's vision and roadmap toward fully autonomous driving, pp. 49–59. Springer, Cham (2014)
4. Bengler, K., Dietmayer, K., Farber, B., Maurer, M., Stiller, C., Winner, H.: Three decades of driver assistance systems: review and future perspectives. IEEE Intell. Transp. Syst. Mag. **6**(4), 6–22 (2014)
5. Chellali, R., Baizid, K.: What maps and what displays for remote situation awareness and ROV localization? In: Salvendy, G., Smith, M.J. (eds.) Human Interface and the Management of Information. Interacting with Information, pp. 364–372. Springer, Heidelberg (2011)
6. Chen, J.Y.C., Haas, E.C., Barnes, M.J.: Human performance issues and user interface design for teleoperated robots. IEEE Trans. Syst. Man Cybern. Part C (Appl. Rev.) **37**(6), 1231–1245 (2007)
7. Chucholowski, F., Tang, T., Lienkamp, M.: Teleoperated driving robust and secure data connections. ATZelektronik Worldwide **9**(1), 42–45 (2014)
8. Chucholowski, F.E.: Evaluation of display methods for teleoperation of road vehicles. J. Unmanned Syst. Technol. **3**(3), 80–85 (2016)
9. European Commission. A European strategy on Cooperative Intelligent Transport Systems, a milestone towards cooperative, connected and automated mobility, November 2016. http://ec.europa.eu/energy/sites/ener/files/documents/1_en_act_part1_v5.pdf. Accessed 30 Nov 2016
10. On-Road Automated Driving (ORAD) committee. Taxonomy and Definitions for Terms Related to On-Road Motor Vehicle Automated Driving Systems, January 2014
11. Corbett, C., Schmidt, K., Jakob, M.: Security testing for networked vehicles. In: 7th FKFS Autotest Conference, September 2018
12. Davies, A.: Nissan's Path to Self-Driving Cars? Humans in Call Centers, May 2017. https://www.wired.com/2017/01/nissans-self-driving-teleoperation/. Accessed 21 Oct 2018
13. Davies, A.: The War to Remotely Control Self-Driving Cars Heats Up, March 2019. https://www.wired.com/story/designated-driver-teleoperations-self-driving-cars/. Accessed 04 Apr 2019

14. Davis, J., Smyth, C., McDowell, K.: The effects of time lag on driving performance and a possible mitigation. IEEE Trans. Rob. **26**(3), 590–593 (2010)
15. Sichere Intelligente Mobilität Testfeld Deutschland. Deliverable D5.5 – Teil A TP5-Abschlussbericht – Teil A, June 2013. http://www.simtd.de/index.dhtml/object.media/deDE/8154/CS/-/backup_publications/Projektergebnisse/simTD-TP5-Abschlussbericht_Teil_A_Manteldokument_V10.pdf. Accessed 05 Dec 2016
16. Favarò, F.M., Nader, N., Eurich, S.O., Tripp, M., Varadaraju, N.: Examining accident reports involving autonomous vehicles in California. PLOS ONE **12**(9), 1–20 (2017)
17. Georg, J., Feiler, J., Diermeyer, F., Lienkamp, M.: Teleoperated driving, a key technology for automated driving? Comparison of actual test drives with a head mounted display and conventional monitors*. In: 2018 21st International Conference on Intelligent Transportation Systems (ITSC), pp. 3403–3408, November 2018
18. Gnatzig, S., Chucholowski, F., Tang, T., Lienkamp, M.: A system design for teleoperated road vehicles. In: Proceedings of the 10th International Conference on Informatics in Control, Automation and Robotics, pp. 231–238, July 2013
19. Gozalvez, J.: Samsung electronics sets 5G speed record at 7.5 GBS [mobile radio]. IEEE Veh. Technol. Mag. **10**(1), 12–16 (2015)
20. Harris, M.: CES 2018: Phantom Auto Demonstrates First Remote-Controlled Car on Public Roads, January 2018. https://spectrum.ieee.org/cars-that-think/transportation/self-driving/ces-2018-phantom-auto-demonstrates-first-remotecontrolled-car-on-public-roads. Accessed 28 Nov 2018
21. Hars, A.: Self-driving cars: the digital transformation of mobility, pp. 539–549. Springer, Heidelberg (2015)
22. Hosseini, A., Lienkamp, M.: Enhancing telepresence during the teleoperation of road vehicles using HMD-based mixed reality. In: 2016 IEEE Intelligent Vehicles Symposium (IV), pp. 1366–1373, June 2016
23. Hosseini, A., Richthammer, F., Lienkamp, M.: Predictive haptic feedback for safe lateral control of teleoperated road vehicles in urban areas. In: 2016 IEEE 83rd Vehicular Technology Conference (VTC Spring), pp. 1–7, May 2016
24. Kang, L., Zhao, W., Qi, B., Banerjee, S.: Augmenting self-driving with remote control: challenges and directions. In: Proceedings of the 19th International Workshop on Mobile Computing Systems & Applications, HotMobile 2018, pp. 19–24. Association for Computing Machinery, New York (2018)
25. Khan, A.H., Qadeer, M.A., Ansari, J.A., Waheed, S.: 4G as a next generation wireless network. In: 2009 International Conference on Future Computer and Communication, pp. 334–338, April 2009
26. Lichiardopol, S.: A survey on teleoperation. Technische Universitat Eindhoven, DCT report (2007)
27. Nayak, R., Padhye, R., Kanesalingam, S., Arnold, L., Behera, B.K.: Airbags. Textile Progress **45**, 209–301 (2013)
28. Neumeier, S., Facchi, C.: Towards a driver support system for teleoperated driving. In: 22nd Intelligent Transportation Systems Conference (ITSC), Auckland, New Zealand. IEEE, October 2019
29. Neumeier, S., Gay, N., Dannheim, C., Facchi, C.: On the way to autonomous vehicles - teleoperated driving. In: AmE 2018 - Automotive meets Electronics (2018)
30. Neumeier, S., Höpp, M., Facchi, C.: Yet another driving simulator openrouts3d: the driving simulator for teleoperated driving. In: 2019 IEEE International Conference on Connected Vehicles and Expo (ICCVE), Graz, Austria. IEEE, November 2019

31. Neumeier, S., Stapf, S., Facchi, C.: The visual quality of teleoperated driving scenarios - how good is good enough? In: 2020 International Symposium on Networks, Computers and Communications (ISNCC), Montreal, Canada, IEEE, October 2020 (Accepted Paper)
32. Neumeier, S., Walelgne, E., Bajpai, V., Ott, J., Facchi, C.: Measuring the feasibility of teleoperated driving in mobile networks. In: 2019 Network Traffic Measurement and Analysis Conference (TMA), France, June, Paris (2019)
33. Neumeier, S., Wintersberger, P., Frison, A.-K., Becher, A., Facchi, C., Riener, A.: Teleoperation: the holy grail to solve problems of automated driving? Sure, but latency matters. In: Proceedings of the 11th International ACM Conference on Automotive User Interfaces and Interactive Vehicular Applications, AutomotiveUI 2019, Utrecht, Netherlands. ACM (2019)
34. Nielsen, C.W., Goodrich, M.A., Ricks, R.W.: Ecological interfaces for improving mobile robot teleoperation. IEEE Trans. Rob. **23**(5), 927–941 (2007)
35. Schmid, J., Schneider, M., HöB, A., Schuller, B.: A deep learning approach for location independent throughput prediction. In: 2019 IEEE International Conference on Connected Vehicles and Expo (ICCVE), Graz, Austria. IEEE, November 2019
36. Shen, X., Chong, Z.J., Pendleton, S., Fu, G.M.J., Qin, B., Frazzoli, E., Ang, M.H.: Teleoperation of on-road vehicles via immersive telepresence using off-the-shelf components, pp. 1419–1433. Springer, Cham (2016)
37. Tang, T., Chucholowski, F., Lienkamp, M.: Teleoperated driving basics and system design. ATZ Worldwide **116**(2), 16–19 (2014)
38. Tang, T., Soto-Setzke, D., Kohl, C., Köhn, T., Lohrer, J., Betz, J.: EE-Architektur für mobile Dienste. ATZextra **19**(14), 40–45 (2014)
39. Wannstrom, J.: LTE-advanced. Third Generation Partnership Project (3GPP) (2012)
40. Winfield, A.F.: Future directions in tele-operated robotics. In: Telerobotic Applications (2000)
41. Zeng, Y., Zhang, R., Lim, T.J.: Wireless communications with unmanned aerial vehicles: opportunities and challenges. IEEE Commun. Mag. **54**(5), 36–42 (2016)

Towards Participatory Design of City Soundscapes

Aura Neuvonen[1](✉), Kari Salo[2], and Tommi Mikkonen[3]

[1] School of Media, Design and Conservation, Metropolia University of Applied Sciences, Helsinki, Finland
aura.neuvonen@metropolia.fi

[2] School of ICT, Metropolia University of Applied Sciences, Espoo, Finland
kari.salo@metropolia.fi

[3] Department of Computer Science, University of Helsinki, Helsinki, Finland
tommi.mikkonen@helsinki.fi

Abstract. Sonic environments of fast-growing urban areas are an integral part of the quality of our everyday living in cities. Due to the individual nature of the sonic experience, collecting and analyzing such experiences needs methods for gathering accurate and useful data about them. This paper describes how to incorporate the concept of soundscape into city planning processes. To achieve this, we propose creating participatory methods for gathering data from the citizens so that the data would be useful and relevant for the city planning professionals. Since the participatory planning process aims at in evolving the citizens, we suggest methods that utilize crowdsourcing, mobile technology and machine learning for presenting, workshopping, and designing soundscapes in the city context.

Keywords: Sound · Design · Soundscape research · Communicative planning · Smart cities · Urban planning · Tool support · Crowdsourcing

1 Introduction

Urbanization and fast-growing cities have catalyzed the importance of designing urban spaces that the citizens find pleasant, homey and that support the communal style of living. Unfortunately, tools and techniques that are suited for the task are rare, and even latest research focuses on noise abatement and preserving quiet areas [1–3], overlooking the design of our everyday sonic environment.

In his famous book "The Soundscape – Our Sonic Environment and the Tuning of the World", R. Murray Schafer asked if the soundscape is something over which we have no control, or are we its composers and responsible for giving it a form [4]. The design of an urban sonic environment soundscape should be a component of the urban planning process. In our visually orientated western culture, we tend to consider a city as a visual entity and the soundscape is a byproduct. Challenge in designing a high-quality sonic environment is the fact that different groups of people react to sounds differently [4]. Therefore, it is difficult to define a quality level for sonic environments.

K. Arai (Ed.): FICC 2021, AISC 1363, pp. 497–512, 2021.
https://doi.org/10.1007/978-3-030-73100-7_35

This paper proposes a smart, participatory method for presenting, workshopping, and designing soundscapes in city context. The method aims at serving the above objectives in city design. From the technical perspective, the method is based on previous development, where an audio platform containing sound mixing application for soundscape design was implemented to support crowdsourcing and data collection methods [5, 6]. In this phase, the research focuses on the development of the data collection, defining a common vocabulary for soundscape experience and the professional reuse of that data for planning purposes, where different stakeholders – citizens, municipal actors, and constructors – seek consensus on the designs.

The rest of this paper is structured as follows. In Sect. 2, we present the background and motivation for this work. In Sect. 3, we present the problem of expressing sonic memories and solutions how mobile technology could solve these issues. In Sect. 4, we discuss the possibilities that soundscapes offer to city planning. In Sect. 5, we present the lessons we have learned so far in the process. In Sect. 6, we draw some conclusions.

2 Background and Motivation

Soundscapes. A soundscape is any acoustic environment perceived by humans [4]. The term acoustic environment refers to sound as it is received from all sound sources modified by the environment [7]. Here, a soundscape is understood as an acoustic environment perceived or experienced and/or understood by a person or persons, in a specific context.

An urban soundscape – a soundscape that represents the sonic conditions of an urban area – is a complicated, multi-layered, multi-sensory experience. It is difficult to describe or define a city soundscape, since every city as well as different parts of them differ from each other, sometimes dramatically. Just by walking a kilometer or even less, the soundscape can change from heavy traffic noise to serenity of nature sounds, and vice versa. Furthermore, every component of the city soundscape is linked to another due to the nature of soundwaves. Therefore, a soundscape is constantly moving, breathing, changing both as an acoustic environment and as a sonic experience.

The soundscape is not only something that surrounds us but it also includes the listener's perception of the sounds. Us humans constantly produce, modify and change the soundscape and at the same time affect each other's experience of our sonic environment. This experience is not only dependent on the sounds and components of the soundscape but the subjective evaluation of acoustic phenomena [8].

Participatory Planning. Communicative or participatory planning is an approach to urban planning that aims to engage the citizens or other stakeholders into decision-making [9]. The theoretical conclusion of communicative planning is that in social, open and transparent processes the citizens or other stakeholders construct more reliable and influential knowledge [10]. In the so-called "communicative turn" in urban planning since 1990s, the role of the citizens has changed from the user of the residential areas to active participants of the planning process [2]. Even though the communicative planning process emphasizes citizen's trust to decision-making in create better environments, it is not trouble-free. Stakeholders have different approaches, interests and objectives, and they may not automatically serve the common good. Planner's professional role is to reconcile different viewpoints and information regarding planning and to offer participants an opportunity to reach a common understanding [11]. Due to a lack of systematic methods, gathering the data from the stakeholders is problematic.

The challenge in incorporating soundscape design in the participatory city planning process is that where buildings and plots have edges, soundwaves travel from one area to another as long as they have faded out. This means that there is no empty space or clean canvas when it comes to soundscapes. Introducing new buildings, streets, parks or changing the structure of the city in any way does affect the soundscape but it is hard for the public to imagine the changes that can be quite unpredictable sometimes. These changes are dependent on surface materials, structures and shapes of the buildings, the amount of traffic or people, and machinery such as air conditioning that are included in the city structures. Furthermore, animals, weather conditions, speed limits, special events and thousands of other little things change what the city sounds like.

Photography, maps, drawings and nowadays 3D models give us a living and accurate impression of a space or scenery. There is no corresponding method in sonography to describe the environment as well as any visual image can [4]. On linguistic side, we face the same problem: there is a lack of lexicalized terms and vocabulary for describing sounds [8].

Soundscapes in Participatory Planning. Describing a sound consists of the emotion and experience of the sound. A soundscape can be experienced negatively, positively or something in between and this is dependent on experiences, personal history, and preferences [12]. Therefore when the citizens are asked what they would like their sonic environment to sound like, the answers cannot be anything else than quite imprecise.

From an urban planning perspective, the soundscape is mainly studied as an acoustic space and the research focuses on noise abatement, noise pollution and protection of quiet areas. Noise levels are measured in decibels, which is important when the target is to reduce the overall noise level in urban areas. The importance of this approach is unquestionable as well as the fact that noise causes health problems, annoyance and lowers the inhabitants' positive relationship to their habitat [13, 14]. Yet measuring decibels does not tell much about the information or the individual's emotional experience of the sound in question [4].

Ever since composer and environmentalist R. Murray Schafer started his "World Soundscape Project" and stated that there should be a subject which we would call "acoustic design" [4], dozens of researchers from various fields of science have searched for a solution for designing and creating a better soundscape. Designing a sonic environment is a multi-disciplinary process that requires the knowledge and involvement of various professionals from planning, architecture, acoustics, noise abatement and so on. There are no official guidelines on how the design process should be carried out and how the involvement of different stakeholders should be done. The latest research shows that there is a need for more detailed and structured guidelines for soundscape planning [15]. The stakeholders should be involved during the whole planning process and that an appropriate engagement process with a relevant panel of representatives is crucial to the successful identification of the issues [1].

The current trend in soundscape research is to move from understanding towards designing the environments. There is a rapid expansion of research and the aim of this work is to provide policy-makers and practitioners with operative tools, standards and methods [3, 16, 17]. Wide range of research has been done in the field of noise abatement, noise monitoring, prediction models and auralization [2, 18, 19]. Kang et al. have proposed a model, which profiles recorded soundscapes, applies linear regression to soundscape profiles to predict suitable perceptual attributes related to each soundscape, and finally visualize perceptual attributes as layers in geographical maps (soundscape maps) [2]. While Kang would use a grid of small sensors to collect soundscapes, Zappatore would rely on crowdsourcing and mobile phones [19]. In Zappatore's approach, mobile devices would be used as recording decibel meters. Recordings, location info, info about user's perception about noise pollution, and other sensor data from a mobile device will be uploaded into backend service. Recordings are analyzed and visualized as noise maps. With all this research there is a common goal to understand how acoustic environments are perceived and thus enhance the sonic environment of urban areas.

Incorporating the concept of soundscape into the planning process is a fairly new idea. Soundscape expertise is not included in the planners' profession and training the planners is probably the first step towards better sonic environments [12]. Due to the individual nature of the sonic experience, the second step would be creating methods for gathering accurate and useful data. This paper aims to define how to incorporate the concept of soundscape into planning processes. This contains creating participatory methods for gathering data from the citizens so that the data would be useful and relevant for the city planning professionals.

3 Supporting the Sonic Memory with Audio Tools

In everyday listening, we focus on gathering relevant information about our environment. We sort, evaluate and describe the sounds according to our hedonic judgment and with spontaneous association [3]. Therefore, any public discussion about sonic environment rarely offers nothing more quite overall data. To assist this conversation we have created a mobile soundscape mixing application [4] in which the soundscape can be divided into pieces and re-arranged. In order to see how this changes the communication we ran a test to compare the different ways of expressing sonic experiences.

3.1 Mobile Mixing Tools

The idea of the mobile mixing tools was to create an easy and simple method for anyone to create and share their opinions about sonic environments. Our mobile soundscape mixing application is a part of an audio platform (Fig. 1). This platform consists of an audio digital asset management system (ADAM), a management application, and mobile applications. Soundscape management (soundscape mixing application), NFC tag management and audio story management applications (audio story sharing application) will run on smartphones. The admin console (management application) will run on the workstation's web browser. The data management and data storage modules (ADAM) run on application and database servers, which could be separate physical or virtual servers or one server combining both roles. The platform is modular so that a user can pick up only those mobile applications that they need. ADAM contains functionalities to manage the assets and an interface for the management application and mobile applications over the Internet. The management application is an administration console for managing the audio files and users.

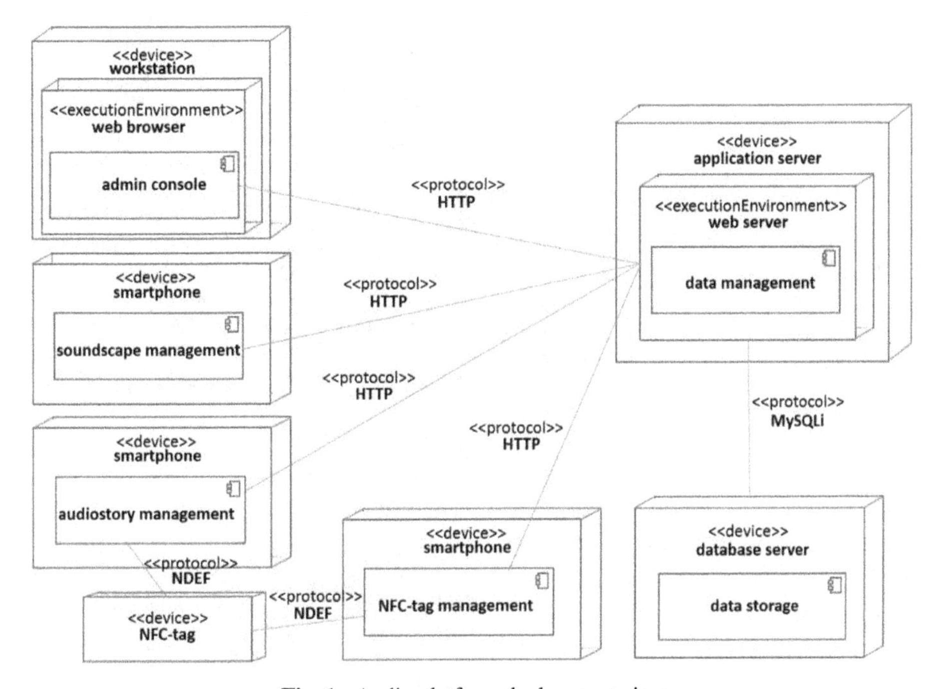

Fig. 1. Audio platform deployment view

We developed a soundscape mixing application called SoundSpace to increase user interaction by developing soundscapes from building blocks stored in ADAM [20]. The SoundSpace tool allows the user to test and play with soundscape elements [5]. The user can search audio files from ADAM, and listen to sounds before selecting them to create a soundscape segment by segment (Fig. 2). SoundSpace then plays the audio files together by looping them and thus giving an audible example of different kinds of soundscapes.

Such audio platform can be used in the participatory planning process in many ways. The mobile application could be offered for the citizens for free use and ask them to upload the kind of soundscapes they prefer. This could be done as a general enquiry or for a specific location. ADAM could be used to store and share uploaded soundscapes.

3.2 Sonic Memory and Useful Data

To observe how a mobile tool that creates hearable opinions affects the ways a person describes ones impressions about soundscapes we ran a test with group of university students. The aim of the test was to study the following topics:

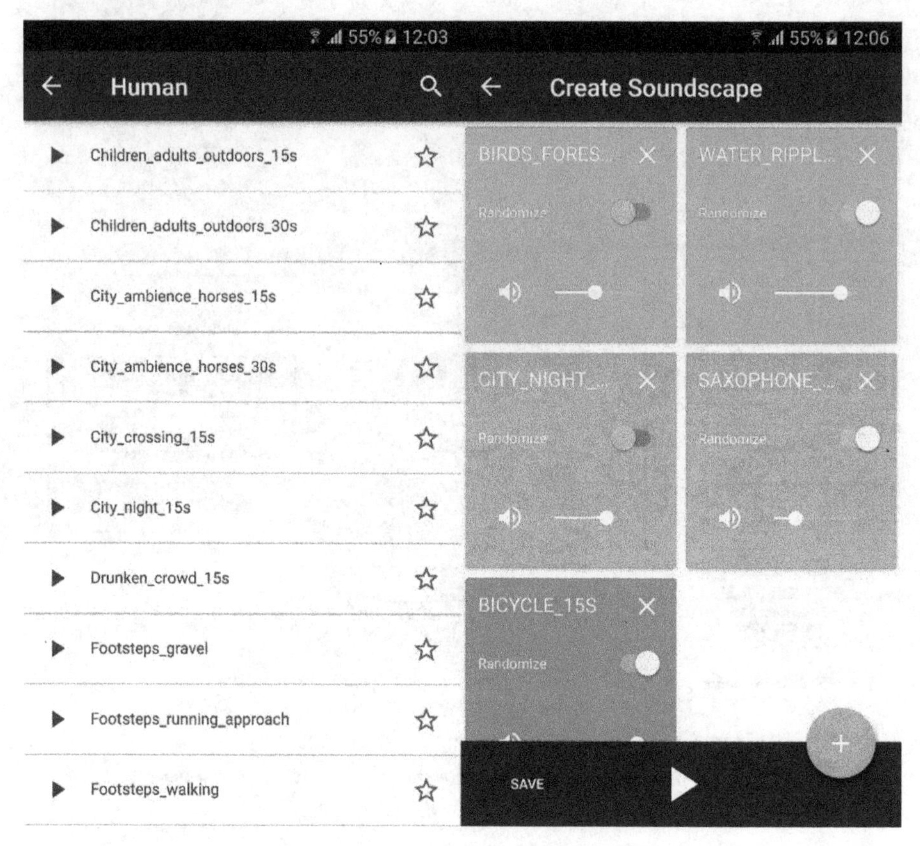

Fig. 2. The SoundSpace tool's user view

1. How well a person is able to recall a soundscape after a while.
2. What kind of verbalization/vocabulary the participants use to describe the sounds they heard.
3. Is there a difference between recollections that are memorized and written down, and memorized with the sonic mobile tools.

3.2.1 Method

Participants. The voluntary test group of 18 university students from Information and Communications Technology Department of Metropolia University of Applied Sciences.

Material. First soundscape resembled an urban nature environment (Fig. 3) and consisted four sound files from the sound collection:

- Finnish birds singing
- A summer forest ambience with mild wind and birds
- Lapping of light waves
- Distant discussion noise.

Fig. 3. Nature soundscape setup

The volume of the water sounds and discussion sounds were lower to make them more distant.

This soundscape could be called a 'hi-fi' soundscape as R. Murray Schafer defines [4]. In a hi-fi soundscape the background noise is low and even quieter and distant sounds can be heard. The listener is able to separate sounds from each other. On the contrary, in a lo-fi soundscape an individual sound disappears into a flood noise and only the most dominant and loudest sounds can be recognized.

The second 'lo-fi' soundscape sounded like a busy city with people and traffic (Fig. 4). This soundscape also consisted four sound files:

- Street noise with low frequency traffic sounds, tram rumbling and people walking and talking.
- City humming and a tram passing by
- A motorcycle passing by
- A person walking by from a close distance with high heels.

The volume of the street noise, city humming and the motorcycle was adjusted to lower to create a more balanced soundscape.

Schafer also presented a classification of the soundscape elements [4]:

- Keynote: ambient sounds (such as wind, traffic, humming, etc.) which are not actively listened because they are filtered out cognitively.

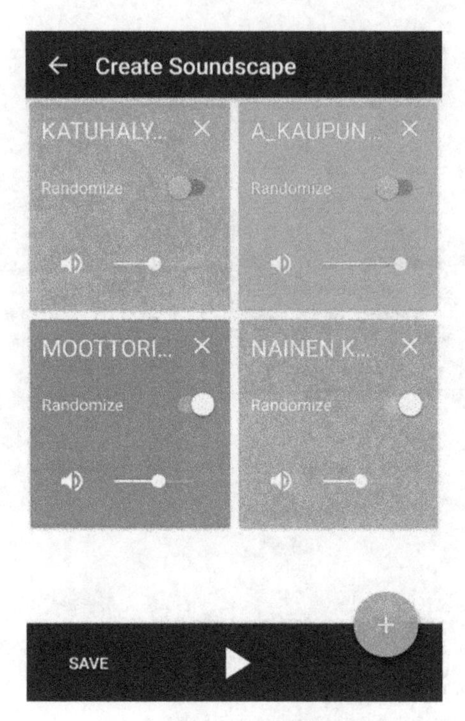

Fig. 4. City soundscape setup

- Soundmark: a sonic landmark; a sound which is characteristic of a place.
- Sound signal: a foreground sound that is listened actively. These sounds usually carry a signal with a message (car horn, dog barking, etc.).

The soundscape examples contained a keynote and several sound signals. To avoid too strong associations to real existing places there were no soundmarks. Tram sound is distinctive to some cities but still it is associated to several places around the globe.

Procedure. The participants first listened to two soundscape audio files in a classroom from loudspeakers. The participants were instructed to concentrate on listening and not to take notes or do anything else while listening. Both sound files were approximately 1.5 min long.

After 12 days, the test group was gathered in a classroom and divided randomly into to two groups. The first group stayed in the classroom and the second one was guided to another classroom. All the participants used their personal laptop to open an online form. Both groups were also given Android phones with the mobile mixing tool or instructions how to install the application to their own Android phone.

Group 1 was asked to create the two earlier heard soundscapes with the mobile mixing tool as well as they could one by one. They were then asked to take a screenshot of the soundscape created and evaluate how well they succeeded in building the same soundscape they had heard. Finally, the participants were asked how well they thought they remembered the soundscapes, if they found the application easy to use and if it helped them to remember the soundscapes.

Group 2 was asked to memorize and write down what sounds the two soundscapes contained one by one. Then the participants were asked how well they thought they remembered the soundscapes by memorizing. After this they were given the mobile devices and asked to test the mobile mixing tools. Then they were asked if they found the application easy to use and if they thought it would have helped them to remember the soundscapes.

3.2.2 Analysis

The two test groups both used approximately 40–60 min to complete the task and. The mobile mixing tool worked well and 84% of the participants evaluated it as easy to use. Only one participant found it difficult to remember the soundscapes well. Of the 18 participants, 14 remembered the listening order of the soundscapes somehow incorrectly. The nature environment soundscape was the first but most of the participants described the city soundscape as the first one. All the participants remembered that the other soundscape was somehow related to nature environment. Two did not describe a city environment at all, two described as some other kind of engine, and vehicle related sound source.

Group 1 was asked to build the soundscapes with the software. Five of the group thought they remembered the soundscapes well and four quite well.

The nature soundscape was easier to remember and most of the participants had picked the exact sounds that they heard 12 days earlier (Fig. 5). All the soundscapes the participants created sounded like a forest with birds. Seven of them had also the water sound in some form. None of the participants either heard or remembered the distant discussion sounds but this can also be due to the classroom listening conditions.

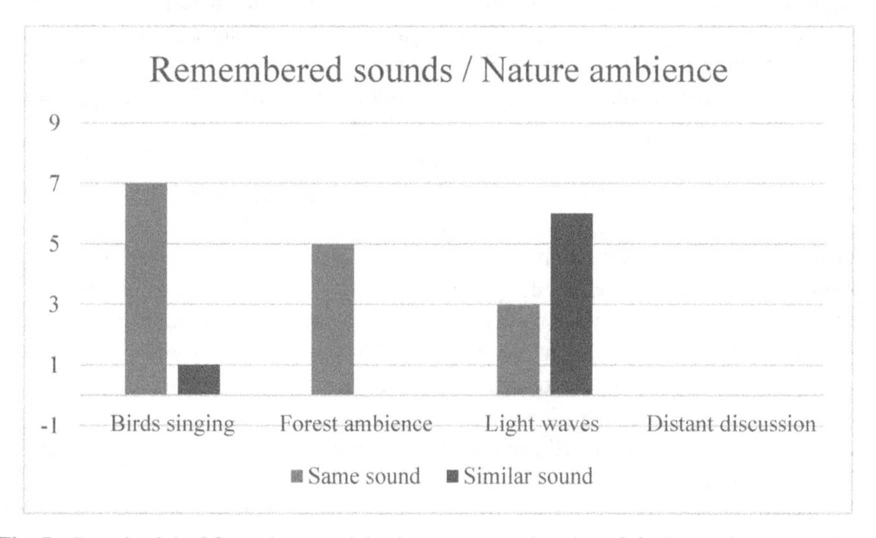

Fig. 5. Sounds picked from the sound database compared to the original soundscape number 1

The city soundscape is more complicated to perceive because there are several sounds constantly overlapping. There is also an overall background noise, the keynote sound, masking the more quiet sounds. The participants remembered the sound signals such as footsteps and tram passing. These were probably the most distinguishable sounds in this soundscape example. From the city soundscape eight of nine were able to pick at least one sound from the library that was exactly similar with the original soundscape (Fig. 6).

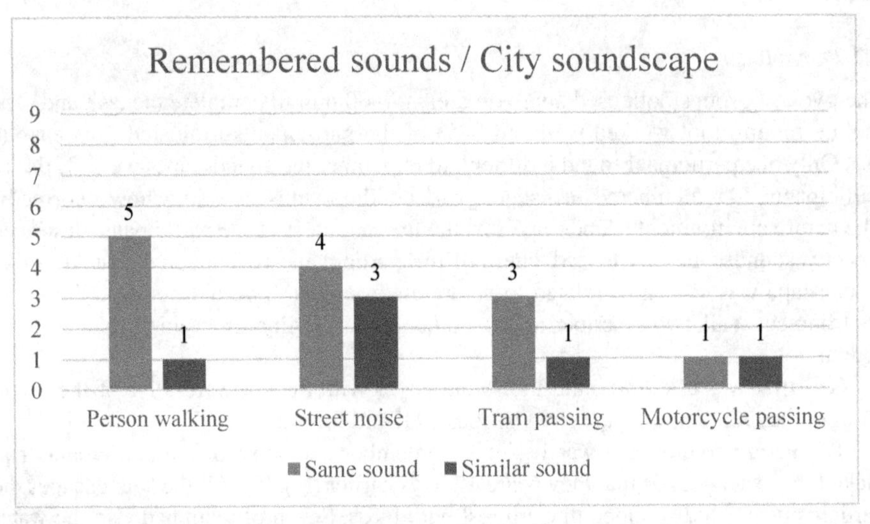

Fig. 6. Sounds picked from the sound database compared to the original soundscape number 2

Group 2 described the soundscapes as well as they remembered with no help of mobile tools. Only one participant thought it was difficult to remember the soundscapes well. Even though the participants were asked to list the sounds of each soundscape as precisely as possible the number of sounds mentioned varied from 1–5. Average amount of sounds described was 2,7.

The key elements of the nature soundscape had been memorized well. Compared to group 1 there is not much difference (Fig. 7). Eight participants mentioned birds, seven mentioned water sounds and five referred to nature or forest sound.

The difference occurs in expression and verbalization. Where group 1 was able to express their recollection with full soundscapes, group 2 used simple words such as “bird singing”, “wind” or “forest sounds”. Some participants verbalized the soundscape as sounds events such as “fishing on a rowing boat” or “nature scenery”. Some had associated the water to a spring, others to waves or water lapping.

The city soundscape seemed to be more difficult to remember and describe accurately (Fig. 8). Only two mentioned walking sounds, five referred to street noise, two had picked up a tram or a train and only one mentioned a motor vehicle.

The verbalization was more varied than with the nature soundscape. None of the participants mentioned street noise but used expressions like “city sounds” or “city noise”.

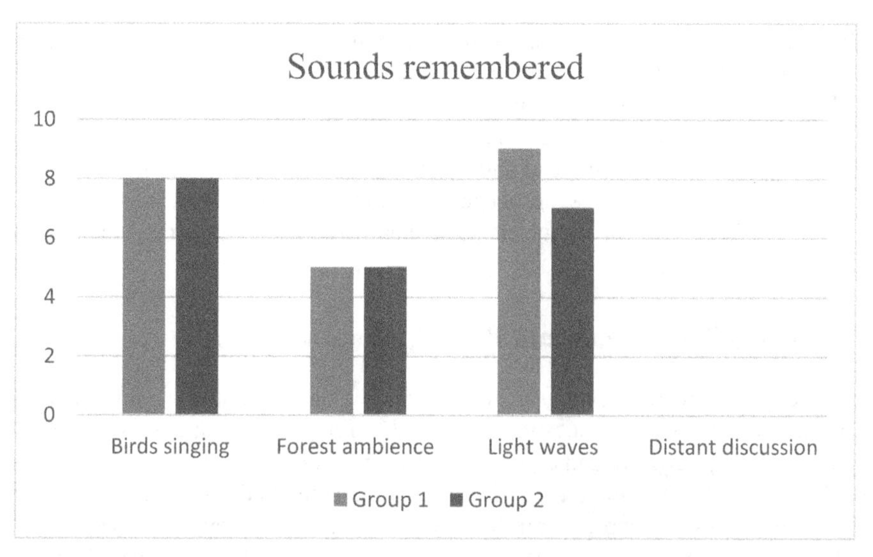

Fig. 7. Key sound elements of the nature soundscape remembered. Comparison between group 1 and group 2.

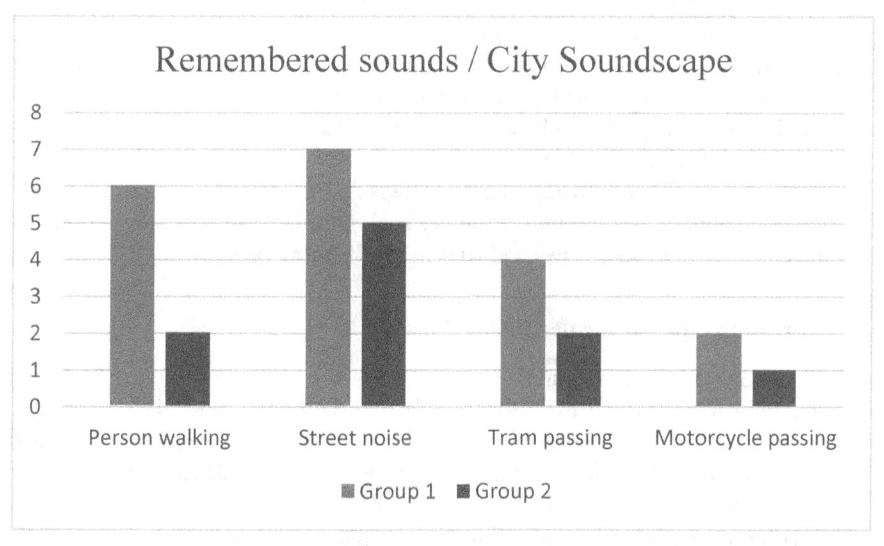

Fig. 8. Key sound elements of the city soundscape remembered. Comparison between group 1 and group 2.

Two participant mentioned human voices and one participant described the soundscape as “metro tunnel”.

After testing the mixing tools 55% of the group 2 thought that it would have helped them to remember the soundscapes. The participants knew at the beginning of the test that they will continue the task a week later. This probably affected the way they concentrated to listening. In a natural situation we listen to our environment more carelessly and therefore most of the sounds leave unnoticed.

With more simple hi-fi soundscapes there did not seem to be much difference between the two test groups' recollections. The second soundscape contained more sounds element and the group 2 remembered and described the sounds much less than what can be heard from the group 1's soundscapes. From the answers of the group 2 it can be observed that free verbal expression creates a remarkable possibility for misinterpretation and misunderstanding.

4 Mobile Mixing Tools and Participatory Methods

Since the participatory planning process aims to involve the citizens and utilize their knowledge about the area in question, the crowdsourcing requires methods and technology for gathering accurate and comparable data that is usable for the planning professionals. Our platform offers a user-friendly and smart method for the citizens or other stakeholders to discuss express and share ideas and opinions about soundscapes.

The soundscape design process and platform described here is based on three cornerstones:

1. Citizens of all ages can participate in the design process with mobile tools that are easy to use and offer a real hearable version of the soundscape in question. The citizens can express their opinion not only with words but also with sonic data.
2. The mobile tool gathers data from the design process. This data contains information about the sounds the citizen chose, deleted, and listened and so on. The data can be processed for various purposes.
3. The interaction with the citizens, the data that is collected and the sound components where the soundscapes are built from are such that they are equivalent to the real sonic environment and can be used later in the planning process.

4.1 Expressing Opinions with Mobile Tools

A sound can be described with numerous ways and sounds, sound sources and sound events are easily mixed up and used illogically. In the urban area context the soundscape is most likely dense and noisy which makes is difficult to remember, describe and separate all the various sound elements. With the mixing tool, it is possible to create a common method for expressing and discussing about sounds.

To incorporate soundscape design into the city planning it needs to be considered in every step of the planning process [1]. In most cases, the data that has been collected from the citizens during the planning process is based on written or oral opinions and feedback. Expressing a sonic experience is a complicated task and the result most likely leaves possibilities for misunderstanding. As R. M. Schafer writes: "To report one's impression of sound one must employ sound" [4]. This means that by representing sonic examples of the soundscapes and changing its components both citizens and the decision-makers can express their sonic ideas more specifically. By creating or modifying soundscapes element by element the user is creating a hearable opinion but also offering data by choosing and not choosing sounds.

4.2 Smart Data Collection and Sharing

In the new soundscape mixing prototypes, we need to collect and save two types of data in addition to audio data: logging data and metadata. Logging data will contain all user interaction with the application: what audio files user selected, in which order, which audio files were removed, how individual audio files were configured (for example volume level), how often user listened the soundscape created, etc. All these events will be time-stamped and saved into log storage, which is linked to the final soundscape file.

Soundscape-related metadata needs to be enhanced. So far, we have defined mainly metadata, which is compatible with unqualified Dublin Core [21]. We need also metadata that defines the structure of soundscape, i.e. what audio files are needed and how they are configured. Maybe we need metadata, which describes better the soundscape, i.e. adjective describing what kind of soundscape user has created and user's emotions related soundscape. One possibility is to define metadata describing if some audio files are a mandatory part of the soundscape. When the user has finalized the soundscape, then we need to upload soundscape file, related log file and metadata into ADAM. This means that also ADAM needs to be modified. Storing soundscape file, related log file and metadata into ADAM, enables detailed analysis of user interaction and behavior, and soundscape content.

4.3 Communicative Planning Process and Soundscape Design

Cities already able to smartly collect data about the locations of different soundscapes and what are the citizens' opinions about them. What is missing is the detailed data of the sounds that these locations contain. The application could also be used for demonstrating the effects of noise or other planned changes in the soundscape. The citizens could then modify or test different variations of these changes and express their opinions about it. This would, for example, give a possibility to define tolerable limits for certain sounds of suggested sounds that could be added or removed from the soundscape.

Cities arrange workshops and exhibitions as part of the interaction with the citizens. The audio platform could be available in an exhibition as audio-only or in later parts of the planning process combined with visual models of the area. Also, the soundscapes created by the citizens could be shared in exhibitions. One of the participatory methods used in city planning is guided walks in the location. Since the soundscape mixing application is mobile the soundscapes could be built on the location where the actual visual environment is fully available.

The further development will be about the data collection, recording and storing of the soundscapes and implementing them into augmented and virtual spaces. We have already developed new soundscape mixing prototypes, which address some of these possibilities. However, we need to further develop also ADAM to support all requirements.

5 Towards Smart Urban Soundscape Design

Urban areas are under a great interest at the moment. The planning methods and technologies are developing quickly but at the same time the process is becoming more and more complicated and multi-dimensional. Designing a soundscape is not a separate process from the rest of the planning process, and not least because the sonic environment is largely a result of material and living environment. Therefore it is important to think how the soundscape design process could be truly incorporated into to planning process and how it could benefit from other current inventions in the planning research.

Depending on the planned area in question, various professionals are involved in the planning process. Yet in most of the cases, the planning is presented visually but it can be a picture, map, model, 3D model etc. If the soundscape would be designed alongside with the visual and other environment there would need to be possibilities to transfer the soundscape design data from system to another. This is possible if visual environment planning platforms can be expanded to include audio elements.

In the method presented here ADAM has so far been designed to interact with mobile applications and management application. Thus, we have developed REST APIs for authentication, content search, content download, and content upload. Now it seems that we need to exchange information also with other IT systems, such as map based and 3D model based city planning systems. This means at least that we need to revisit existing APIs to enable searching and downloading soundscape files and related metadata. However, this approach is limited in the sense that it enables soundscape transfer to one direction only and requires that other IT systems probably need to be modified to access ADAM.

If we want to transfer soundscapes from a system to another we need a clear understanding or possibly a standard of the structure of urban soundscapes. As described earlier there is a great deal of theoretical information about soundscapes and various methods for analyzing and structuring them. These theoretical and technological inventions are an important part of the process that aim to create a smart working method for designing soundscapes. The method described here needs this knowledge in order to success.

Storing soundscape structure and other new metadata enables also new possibilities for analyzing data. Collecting soundscapes and the metadata with them would create in a long run knowledge about sonic experiences and opinions about soundscapes. This could be facilitated by using smart crowdsourcing methods for gathering soundscape data. Larger collected data would open a possibility to use Machine Learning approach for example. With machine learning, it would be possible to predict problematic sound components from urban soundscapes.

These are just few examples of co-operations and combinations that smart and structured audio data collection would open. Combining sound and picture would be a step towards a more comprehensive environmental planning process. Sonic environment affects the visual environment and vice versa. By adding audio to visual representations, the citizens would get a more realistic impression of the changes planned in their living environment.

6 Conclusions

Our experience of the world is a combination of our five senses but still our environment is mainly designed visually. It has been acknowledged that our sonic environment has a significant effect on our well-being and living conditions but due to the predominance of the visual planning, the methods of designing soundscapes are underdeveloped. There is a lack of comparable data of the citizens' sonic experiences, methods of collecting the data smartly and knowledge of how to implement the data into planning processes. A sonic experience contains a lot of tacit information that we have no words for. Writing down what an environment sounds like is a subjective interpretation and verbalization of something that we usually do not express that carefully. Therefore, we need a different kind of tools for interaction, expression and explanations.

In this paper, we have presented a concept for smart data collection of sonic experiences and methods for implementing them into city planning. The project aims to develop a method that could be transferred internationally to any city planning process. Furthermore, creating smart and standardized methods for data collection would open the possibility to use a machine learning approach for data analysis. This would create common knowledge about sonic experience and basis for the real design of soundscapes.

References

1. Xiao, J., Lavia, L., Kang, J.: Towards an agile participatory urban soundscape planning framework. J. Environ. Plann. Manag. **61**(4), 677–698 (2018)
2. Kang, J.B., Aletta, F., Margaritis, E.C., Yang, M.L.: A model for implementing soundscape maps in smart cities. Noise Mapp. **5**, 46–59 (2018)
3. Aletta, F., Kang, J.: Towards an urban vibrancy model: a soundscape approach. Int. J. Environ. Res. Public Health **15**(8), 1712 (2018)
4. Schafer, R.: Soundscape: The Tuning of the World. Destiny Books, Vermont (1994)
5. Neuvonen, A.: Experiencing the soundscape with mobile mixing tools and participatory methods. Int. J. Electron. Governance (IJEG) **11**(1), 44–61 (2019)
6. Salo, K.: Modular audio platform for youth engagement in a museum context. University of Helsinki, Helsinki (2019)
7. ISO: ISO 12913 - Acoustics – Soundscape (2014)
8. Guastavino, C.: Categorization of environmental sounds. Can. J. Exp. Psychol./Revue canadienne de psychologie expérimentale **61**(1), 54–63 (2007)
9. Healey, P.: Collaborative Planning: Shaping Places in Fragmented Societies. Macmillan, Basingstoke, Hampshire, Houndmills (1997)
10. Machler, L., Milz, D.: The evolution of communicative planning theory. In: AESOP Young Academics Booklet Series B - Conversations in Planning Theory, October 2015 (2015)
11. Puustinen, S.: Suomalainen kaavoittajaprofessio ja suunnittelun kommunikatiivinen käänne: vuorovaikutukseen liittyvät ongelmat ja mahdollisuudet suurten kaupunkien kaavoittajien näkökulmasta. Dissertation. Yhdyskuntasuunnittelun tutkimus- ja koulutuskeskuksen julkaisuja, no. 34 (2006)
12. Jennings, P.A., Cain, R.: A framework for improving urban soundscapes. Appl. Acoust. **74**(2), 293–299 (2013)
13. European Union (EU): Directive 2002/49/EC of the European Parliament and of the Council of 25 June 2002 relating to the assessment and management of environmental noise. Off. J. L **189**, 0012–0026 (2002)
14. Lercher, P., Schulte-Fortkamp, B.: The relevance of soundscape research to the assessment of noise annoyance at the community level. In: Proceedings of the 8th International Congress on Noise as a Public Health Problem, Berlin (2003)

15. Adams, M., Davies, B., Neil, B.: Soundscapes: an urban planning process map. In: INTER-NOISE and NOISE-CON Congress and Conference Proceedings (2009)
16. Kang, J.: From understanding to designing soundscapes. Front. Archit. Civ. Eng. China **4**(4), 403–417 (2010)
17. Schulte-Fortkamp, B., Jordan, P.: When soundscape meets architecture. Noise Mapp. **3**(1), 216–231 (2016)
18. Southern, A.: Sounding out smart cities: auralization and soundscape monitoring for environmental sound design. J. Acoust. Soc. Am. **141**, 3880 (2017)
19. Zappatore, M., Longo, A., Bochicchio, M.: Crowd-sensing our smart cities: a platform for noise monitoring and acoustic urban planning. J. Commun. Softw. Syst. **13**(2), 53–67 (2017)
20. Salo, K., Bauters, M., Mikkonen, T.: Mobile soundscape mixer. Ready for action. In: The International Conference on Mobile Web and Intelligent Information, Vienna (2016)
21. Salo, K., Mikkonen, T., Giova, D.: Backend infrastructure supporting audio augmented reality and storytelling. In: Human Interface and the Management of Information: Information and Knowledge in Context, Part II. LNCS, vol. 9735, Switzerland (2016)

Modeling Traffic Congestion in Developing Countries Using Google Maps Data

Md. Aktaruzzaman Pramanik[1], Md Mahbubur Rahman[2], A. S. M. Iftekhar Anam[3](✉), Amin Ahsan Ali[1], M. Ashraful Amin[1], and A. K. M. Mahbubur Rahman[1]

[1] Independent University, Bangladesh (IUB), Dhaka, Bangladesh
{aminali,aminmdashraful,akmmrahman}@iub.edu.bd
[2] Crowd Realty, Tokyo, Japan
[3] University of Wisconsin - Green Bay, Green Bay, USA
anami@uwgb.edu

Abstract. Traffic congestion research is on the rise, thanks to urbanization, economic growth, and industrialization. Developed countries invest a lot of research money in collecting traffic data using Radio Frequency Identification (RFID), loop detectors, speed sensors, high-end traffic light, and GPS. However, these processes are expensive, infeasible, and non-scalable for developing countries with numerous non-motorized vehicles, proliferated ride-sharing services, and frequent pedestrians. This paper proposes a novel approach to collect traffic data from Google Map's traffic layer with minimal cost. We have implemented widely used models such as Historical Averages (HA), Support Vector Regression (SVR), Support Vector Regression with Graph (SVR-Graph), Auto-Regressive Integrated Moving Average (ARIMA) to show the efficacy of the collected traffic data in forecasting future congestion. We show that even with these simple models, we could predict the traffic congestion ahead of time. We also demonstrate that the traffic patterns are significantly different between weekdays and weekends.

Keywords: Statistical modeling · Traffic congestion · Data collection · Intelligent transportation system

1 Introduction

Traffic congestion has been a growing concern for cities around the world. Rapid urbanization, limited space for expansion, and increasing demand for transportation are turning the congestion more vulnerable in the cities, especially in developing countries. Along with the factors mentioned above, the widespread availability of ride-sharing services and non-motorized vehicles are also responsible for the overcrowded streets in those cities. Traffic congestion affects the lives of the people living in urban areas; specifically, it increases mental stress and disrupts

K. Arai (Ed.): FICC 2021, AISC 1363, pp. 513–531, 2021.
https://doi.org/10.1007/978-3-030-73100-7_36

people's daily schedules, resulting in elevated blood pressure, increased negative mood states, and lowered tolerance for frustration. Moreover, traffic congestion has negative impacts on the economy. It increases business production costs due to longer travel times, missed deliveries, and increased fuel costs [1].

In recent years, transportation research focuses on traffic congestion control and minimization of congestion time, mostly to resolve the above problems. Developed countries invest in intelligent transportation research where researchers collect traffic data through loop detectors, Radio Frequency Identification (RFID), and sensor networks to model and predict the traffic pattern [28]. Such models can play a pivotal role in developing congestion prediction applications for commuters, travelers, and traffic management authorities.

The authors of [17] have collected traffic speed data using 39000 loop detector sensors from 12 roads in China (BJER4 dataset) with five-minute intervals. They have also used the PeMSD7 dataset [27] obtained from California Traffic Department to test their proposed traffic model. Another research group collected the vehicle speeds with five-minute intervals in Los Angeles County and California with 532 speed sensors [22]. Guo et al. have developed traffic models based on Dataset-I, Dataset-II, and PeMSD4 [27] that were collected from Washington DC (370 roads), Philadelphia (397 roads), and San Francisco Bay area (307 roads), respectively, using many speed sensors [19]. The authors of [20] accumulated traffic data from Chicago's 1250 arterial streets via the vehicles' GPS. Refer to Table 1 for a comparative summary of the studies discussed above.

Table 1. Traffic data collection scenarios in developed countries

Research	Dataset used	Location (Road Seg.)	Collection process	Collection period	Interval	Limitations
Yu et al. [17]	BJER4 PeMSD7 [27]	Beijing (12) California	Loop detectors Sensors	1 July-31 August, 2014, No weekends	5 min	39000 detectors, expensive
Chen et al. [22]	METR-LA PEMS-BAY	Los Angeles, California	Sensors	1 May-30 June, 2012 1 Jan-31 May, 2017	5 min	Total 532 Sensors expensive, not scalable
Guo et al. [19]	Dataset-I Dataset-II PeMSD4 [27]	Wash. D.C.(370) Philadelphia(397) SF. Bay Area(307)	Sensors	24000 measurements 24000 measurements Jan-Feb 2018, 15000 measurements	5 min	Enormous amount of sensors
Zhang et al. [20]	CTA Data [32]	Chicago's streets(1250)	GPS Tracing	6 months	10 min	Infeasible for Non-motorized vehicles in developing countries

It is easy to notice that the above works primarily focused on the developed countries where the transportation research division has a substantial budget to collect traffic data using loop detectors, sensors, and cameras. Deploying such an infrastructure could be prohibitively expensive for developing countries like Bangladesh, India, or Pakistan. Moreover, these techniques are not quite effective in the cities of developing countries with the narrow streets, the prevalence of non-motorized vehicles, and non-grid-like road structure. Some research studies attempt to collect traffic data in several developing countries, e.g., Bangladesh, Pakistan, Myanmar, and India. S.M. Labib et al. collected data with manual counts by deploying local surveyors for their case study area: Sheraton hotel junction in Dhaka, Bangladesh [23]. Another group collected traffic data by manually counting the vehicles at Science Laboratory Intersection in peak periods of the morning (9:00–10:00 AM) [24].

Some researchers collected traffic data using GPS for different purposes, such as GPS data from several taxis for 11769 road segments [25]. Salma et al. collected GPS data from a telecommunication company of Bangladesh with vehicles with GPS tracking devices [26]. However, they have collected traffic information with an interval of one hour. The interval is too high, and the number of vehicles is limited for modeling the traffic patterns.

Several studies focused on other developing countries of the South Asian region. Ali et al. collected traffic data of two peak periods in a day of nine different locations of Karachi, Pakistan [29]. They calculated the traffic delay manually based on the traffic volume at those locations. Another work collected GPS data for traffic congestion prediction [30]. For predicting the congestion on user's demand, the authors collected traffic speed, direction, timestamps, and other GPS data by tracing mobile phones on some predefined vehicles at Yangon, Myanmar. However, they could not perform large scale data collection as they focused on only android phones. In India, Sharma and his group used manual counting on vehicles from high-quality digital cameras mounted over a couple of two- and four-lane highways during two peak sessions of the working days [31]. They collected data from cities Roorkee, Haridwar, Delhi, and Muzaffarnagar. Then they used machine learning models to forecast traffic congestion. Please see Table 2 for the comprehensive summary. The table suggests that most of the data collection in developing countries did not target traffic congestion prediction, and the researchers relied mainly on manual counting or GPS data.

To summarize, the cited traditional data collection processes are tedious, require considerable human effort and instruments. Moreover, the approaches are not scalable for large collection scenarios for the whole city traffic. For developing countries, the traffic data needs to be collected in a configurable and scalable manner with minimal cost. In this paper, we propose to use traffic information from Google Map Service in a unique way so that the traffic data for any road segments/intersections can be collected seamlessly from any location in a city.

At present, the Google map provides the current situation of traffic congestion. We proposed an easy way of collecting traffic data. By extracting Google Map data, we can analyze the traffic condition for selected intersections. As

Table 2. Traffic Data Collection Scenarios in Some South Asian Developing Countries

Research	Location (Coverage)	Collection process	Collection period	Interval	Goal	Limitations
Labib et al. [23]	Sheraton Hotel Junction, Dhaka (1 intersection)	Manual counts	7:00 to 11:00 AM	15 min	Optimization of signal timing	Erroneous counting, non-scalable, time consuming
Roy et al. [24]	Science lab to Elephant road intersection, Dhaka (1 Intersection)	Manual counts	9:00 to 10:00 AM	15 min	VISSIM simulations	Erroneous and cumbersome manual counting
Rahman et al. [25]	11,769 road segments Dhaka	GPS Data	N/A	30 min	Learn traffic patterns	Not usable for traffic intensity prediction
Salma et al. [26]	Airport to Gulshan-2, Dhaka (9048 m)	GPS Data, Telecom company vehicle	working day (peak hour), weekend, Last day of week.	1 h	Predicting traffic intensity	No traffic intensity data, Interval is too high
Lwin et al. [30]	Yangon, Myanmar	GPS data, mobile phones on vehicle	4 months of GPS data	N/A	Probability of user's source and destination	Limited to Android only
Sharma et al. [31]	Roorkee, Haridwar, Delhi, and M.nagar India (3 Intersections)	Digital cameras, manual traffic volume count	9:00 am to 12:00 am 3:00 pm to 6:00 pm	5 min	Short-term traffic forecasting	Limited to small portion of highways
Ali et al. [29]	Karachi, Pakistan (9 locations)	Video camera, tracker, manual calculation	8:00 am to 1:00 pm 3:00 pm to 8:00 pm	N/A	Estimate traffic congestion cost	Manual calculation of traffic delay

Google Maps itself do not provide traffic data, we have collected the traffic condition of each road segment by observing their traffic color: green (no traffic), orange (moderate traffic), red (heavy traffic), and dark red (very heavy traffic). Using the Google Traffic Layer API, we have collected this information with a 30-second interval for six months: November 2019–April 2020. By analyzing the data from those intersections, we can develop a statistical traffic congestion model. Our data collection process is less expensive and more efficient compared to other approaches. Moreover, it is scalable to the entire city.

In Table 3, we have summarized the characteristics of the datasets collected by other research papers and our proposed one in terms of covered area, cost of deployment, scalability, interval flexibility, and remote data collection facility. It is easy to note that our proposed approach achieves most of the desired characteristics suitable for deployment in developing countries.

Table 3. Comparison of our dataset with developing countries

Researches	Larger Area Covering	Low Cost	Scalability	Flexibility in Adusting Interval	Remote Data Collection
Labib et al. [23]	✗	✓	✗	✓	✗
Roy et al. [24]	✗	✓	✗	✓	✗
Rahman et al. [25]	✓	✓	✗	✗	✓
Salma et al. [26]	✗	✓	✗	✗	✗
Lwin et al. [30]	✗	✗	✓	✗	✓
Sharma et al. [31]	✗	✗	✗	✓	✗
Ali et al. [29]	✓	✗	✗	✓	✗
Our Proposed Approach	✓	✓	✓	✓	✓

Finally, we have developed some well-known statistical (Historical average, ARIMA) and machine learning (Support Vector Regression (SVR), SVR with graph) predictive models to show the efficacy of the collected data. We used one month's data (November 2019) from Mirpur - Dhaka in our analysis. Therefore, our contribution includes:

- We propose a novel data collection method for traffic data collection that is scalable, low cost, and efficient for developing countries.
- We have demonstrated the efficacy of our traffic data in predicting traffic congestion in all intersections of the Mirpur area.
- We have shown the comparative analysis of how the history of traffic congestion length affects a predictive model's capability.
- Finally, we have shown the significant difference between traffic congestion patterns between weekdays and weekends in Mirpur, Dhaka.

The remainder of this paper is organized as follows: Sect. 2 describes the related literature about traffic data collection and forecasting. Section 3 introduces the proposed traffic data collection. Section 4 defines the statistical and machine learning models for traffic prediction. Section 5 presents the experimental setup and performance comparisons between different prediction models. The results also include performance analysis of traffic prediction in different time duration during weekdays and weekends. Finally, Sect. 6 concludes the paper and discusses future studies.

2 Related Works

Much research has been done in recent years on traffic congestion prediction and traffic forecasting based on time-series data [10,11]. In the case of developing countries, the variation of motorized and non-motorized vehicles is enormous [13]. For various reasons, the traffic-congestion in several areas has increased, the increasing traffic jam has become a threat to the major metropolitan cities, and its impact is observable [14]. In several works, traffic data was collected using GPS tracing or using the sensors located at several intersections [4,5,8,12].

Alghamdi and colleagues applied Auto-Regressive Integrated Moving Average(ARIMA)-based modeling in their short-term time series data with 13 attributes collected from traffic flow in four different lanes [2]. They have tuned different ARIMA parameters for better accuracy but did not show any comparison with other models. Iyer et al. proposed Deep Learning architecture(LSTM) for the prediction of congestion state of road segments using their dataset, which was collected by GPS tracing the specific vehicles [5]. Liu et al. applied Deep Learning models(CNN, RNN, SAE) on their mobility data collected by both some intelligent devices (e.g., loop detectors, traffic cameras) and a large number of public vehicles equipped with GPS devices [8]. Their data collection methods are quite expensive and complicated compared to the metropolitan cities in developing countries where vehicles have a considerable variety and density.

Zambrano-Martinez and colleagues applied logistic regression to characterize travel times and classified several street segments into three categories based on the measured number of vehicles from their Simulation of Urban MObility (SUMO [3]) simulations [4]. In [6], the authors proposed a neural network prediction model based on LSTM to predict traffic congestion at Shenzhen, China. The data set is collected by the traffic camera at all crossroads. This data collection process is expensive, and its almost impossible to implement in developing countries, i.e., Dhaka, Bangladesh. Research suggests that historical sensor-based data can provide less accuracy while predicting traffic state. For that reason, Ahmed and his group proposed the CTM-EKF-GA framework in which they did not use any historical traffic data; instead, they collected data from connected vehicles (CVs), which is considered real-time traffic data [9]. It is also an expensive way of collecting traffic data.

Yao et al. conducted their experiments on real datasets, such as taxi and bike trips records in New York City [16]. Collecting this type of data is impossible in cities like Dhaka, Karachi, or Delhi, as every route contains varieties of motorized and non-motorized vehicles, and they are equally responsible for traffic congestion. Some groups collected real-world datasets of two types using many sensors/loop detectors in Los Angeles County, California, Beijing City, and District 7 of California [15,17,21,22]. However, collecting large-scale data by setting up expensive loop detectors in highways is hardly feasible in developing countries. While in [7], they have shown an approach of using inexpensive Big Data from buildings in Hong Kong. By using building population data, they tried to predict the traffic density in the nearby area. They conducted a correlation between the ICC(Hong Kong International Commerce Centre) occupancy data and the traffic data for the roads of different distances from the ICC. Cui et al. have collected two real-world datasets; the first one contains traffic state data collected from 323 sensor stations in the Greater Seattle Area [18]. The second dataset contains road link-level traffic speeds aggregated from GPS probe data collected by commercial vehicle fleets and mobile apps provided by INRIX. While [20] have collected their dataset from Chicago Transit Authority(CTA) buses on Chicago's arterial streets in real-time by GPS traces, no organization in Bangladesh provides this kind of traffic data collected by GPS traces for any particular area.

There is a plethora of traffic research that has been done in developing countries. They have either done traffic simulation with vehicle counts or forecasting of traffic speed from GPS data. In [23], they have collected traffic data by manual counts employing local experts in several junctions and used traffic simulation at the micro level to create optimized signal timings for a congested intersection. Roy et al. collected traffic volume data manually in peak periods of the morning (9:00–10:00 AM) and collected geometric and control data for four intersections of Dhaka city and simulated the data using VISSIM optimization of existing traffic signals [24]. For identifying the traffic density patterns of Dhaka across different roads, [25] used GPS data for 11769 road segments. Furthermore, In [26], another group collected GPS data from a telecommunication company of Bangladesh who has vehicles with GPS tracking devices. The authors have accumulated traffic information with an interval of one hour, which is very high compared to the state of the art traffic data intervals because of traffic status changes in smaller intervals (Table 2).

For predicting traffic congestion probability on user's demand, [30] collected traffic speed, direction, timestamps, and other GPS data at Yangon, Myanmar, by tracing mobile phones on vehicles. In [31], the authors collected traffic data by manually counting vehicles on high-quality digital cameras, two or four-lane highway stretches of India. They have built a short-term traffic forecasting model by using the collected data. From the works mentioned in this section, we have identified some limitations in developing countries. In summary, they are

- Deploying sensors and loop detectors that are expensive.
- Collecting GPS data from a representative number of motorized vehicles, which is not feasible.
- Presence of a significant amount of non-motorized vehicles in the city streets.
- Manual counting is tedious, time-consuming, and non-scalable.
- Collecting traffic data for the whole city is practically impossible.

In the next section, we have described our proposed data collection process that overcomes the aforementioned limitations.

3 Data Collection

Figure 1 shows the steps in our proposed data collection and prediction framework. The following subsections describe each of the steps in more detail.

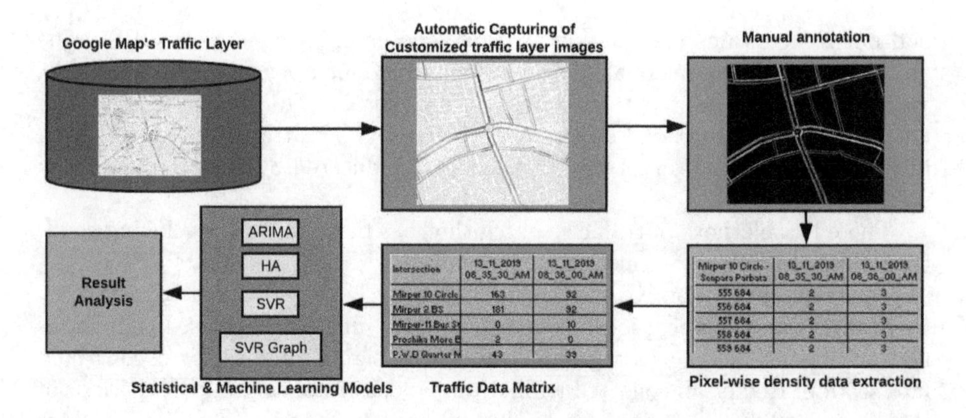

Fig. 1. Block Diagram of the proposed data collection and prediction

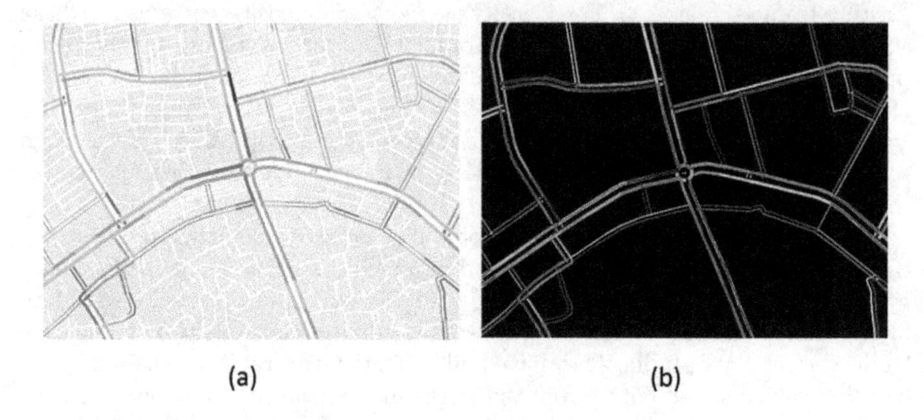

Fig. 2. Sample output of data capture (a) and preprocessing (b)

3.1 Automatic Google Data Capture

We have collected Google Map's traffic layer images of Mirpur, Dhaka (latitude: 23.8060493, longitude: 90.3712275) with a zoom level of 15 on Google Maps covering approximately 12.561 (km^2) area. Google does not provide traffic layer images of a specified time interval. So we developed a tool to capture traffic layer images automatically. We have used Google Maps' JavaScript API to customize the image, such as removing the labels of bus-stops, roads, transit, and administrative neighborhoods from the map. We need to provide three parameters for our data collection program: location(latitude-longitude), zoom level on Google Maps, and time interval for capturing each image. With these inputs, the tool automatically captures and saves images. Figure 2(a) shows a sample map image. For our selected area, we have captured images from 6:00 AM to 11:59 PM with an interval of 30 s, which stands 2160 images a day and 64800 images for a month. We have collected images of 6 months, from November 2019

	A	B	C	D
1	Mirpur 10 Circle- Senpara Parbata	13_11_2019 08_35_01_AM	13_11_2019 08_35_31_AM	13_11_201 08_36_01_A
2	555 684	2	3	2
3	556 684	2	3	2
4	557 684	2	3	2
5	558 684	2	3	2
6	559 684	2	3	2
7	560 684	2	3	2
8	561 684	2	3	2
9	562 684	2	3	2
10	563 685	0	0	0

(a)

	A	B	C	D
1	Intersection\Time stamp	13_11_2019 08_35_31_AM	13_11_2019 08_36_01_AM	13_11_2019 08_36_31_AI
2	Mirpur 10 Circle	163	92	164
3	Mirpur 2 BS	181	92	223
4	Mirpur-11 Bus Star	0	10	10
5	Proshika More Bus	2	0	2
6	P.W.D Quarter Mo	43	39	43
7	Mirpur Benaroshi F	84	36	84
8	Paris Road Mor	50	0	50
9	Mirpur College Uni	92	1	145
10	Avenue 5	87	0	87

(b)

Fig. 3. Sample data after of image processing (a) and Final data matrix for predictive model (b)

to April 2020. We have used an interval that is frequent enough to capture the traffic changes and does not cross the daily free quota of the Google Maps API.

3.2 Image Prepossessing

After capturing the images, we prepared them to extract traffic density data. We annotated each road segment in one image and assigned an ID and a color code (RGB) to each road segment. In this process, we have identified 155 road segments and 45 intersections in the Mirpur area. Figure 2(b) shows the color-annotated map image. Next, we extracted pixel locations for each road segment using their assigned RGB values. We adopted OpenStreetMap standard annotation to assign the IDs to each intersection. Thus, it can be extended to any arbitrary location.

3.3 Traffic Data Extraction

Using the pixel locations collected in the previous step, we extracted each road segment's pixel value. This time, we have pixel-wise traffic density information for each road segment(both incoming and outgoing) of an intersection. The numeric values 1, 2, 3, 4, and 0 mean no traffic delay, medium amount of traffic, near-heavy traffic, heavy traffic, and no traffic information. This information is assigned in this format because Google Map's traffic layer shows four colors(green, orange, red and dark red) to represent the amount of traffic density of a road segment[1].

3.4 Data Processing

In the final step of our data processing, we took every intersection and summed the intersection's incoming traffic density. We have only considered the traffic

[1] https://support.google.com/maps/answer/3092439?co=GENIE.Platform.

density values 3 and 4, which indicate near-heavy and heavy traffic density. As we know, incoming traffic is mainly responsible for creating traffic congestion in an intersection, so that, for a specific timestamp, we have used the summation of incoming traffic as a measure of traffic density of a particular intersection. Figure 3 shows the image data transformed into matrices for predictive modeling.

4 Statistical and Machine Learning Models for Predictions

4.1 Historical Average Forecast

Traffic flow in a particular area may follow some patterns because of some regular events occurring there. For that reason, we may expect that historical averages of traffic flow or density of a particular area at a particular time should provide us a significant traffic density forecast of that area at the same time and on the same weekday.

4.2 Support Vector Regression (SVR)

We have also used the Support Vector Regression (SVR) model for traffic density prediction. In Support Vector Regression, the training dataset includes multivariate sets of observations and observed response values.

While in linear SVR, if x_i is the ith traffic intensity of N observations with observed response traffic intensity values y_i and if we use non-negative multipliers α_i and α_i^* for each observation x_i, the prediction function becomes-

$$y = \sum_{i=1}^{N}(\alpha_i - \alpha_i^*) \cdot (x_i, x) + b \tag{1}$$

In the case of our traffic density dataset, we can not predict using a linear model, that is why we replace the dot product x_i, x_j with a nonlinear kernel function $K(x_i, x_j) = exp(-\frac{||x_i - x_j||^2}{2\sigma^2})$, which transforms the data into higher dimensional feature space and helps to perform the linear separation. So the equation becomes-

$$y = \sum_{i=1}^{N}(\alpha_i - \alpha_i^*) \cdot K(x_i, x) + b \tag{2}$$

4.3 Graph SVR

Sometimes, the traffic congestion in a particular intersection is highly influenced by the amount of traffic present in the neighboring intersections. We have used the Support Vector Regression (SVR) model in a different approach where we have considered the summation of the traffic density of neighboring nodes for each observation x_i. So this time, the SVR model predicts the future traffic density for a particular intersection considering the certain traffic status of neighboring intersections.

4.4 Autoregressive Integrated Moving Average (ARIMA)

ARIMA (also known as the Box-Jenkins model) is a simple but powerful model that we have used to train our time-series traffic density data for forecasting future traffic congestion. ARIMA model consists of autoregressive terms(AR), moving average terms(MA), and differencing operations(I).

Before applying ARIMA model, we must need to conform to the stationarity of the dataset. A degree of differencing reduces non-stationarity (seasonality) from the dataset. Such as, d = 1 means first-order differencing $Z_t = X_t - X_{t-1}$, again, when differencing order = 2, $z_t = (x_t - x_{t-1}) - (x_{t-1} - x_{t-2})$. In ARIMA(p,d,q) model the parameter d defines the differencing order. The parameter p defines the number of prior observations that have a significant correlation with the current observation. While q represents the moving window size in terms of error, which impacts current observation. The ARMA(p,q) equation-

$$Y_t = \Phi_1 Y_{t-1} + \Phi_2 Y_{t-2} + ... + \Phi_3 Y_{t-p} + \\ + \omega_1 \epsilon_{t-1} + \omega_2 \epsilon_{t-2} + ... + \omega_q \epsilon_{t-1} + \epsilon_t \quad (3)$$

where the weights $(\Phi_1, \Phi_2 ... \Phi_3)$ and $(\omega_1, \omega_2 ... \omega_3)$ respectively for the AR and MA are calculated depending on the correlations between the lagged observations and current observation. For our dataset, we have chosen ARIMA(1,0,0).

5 Experiments and Results

We performed some experiments to show the efficacy of the collected traffic data in predicting congestion intensity. Before going to the experimental setups, we would like to introduce various hyper-parameters: sampling rate, sequence length, and prediction length that impact the performance.

Sampling Rate denotes the time interval at which traffic intensities are used for each node.

Sequence Length represents the length of the input data across the previous timestamps from the current time. The sequence data is necessary to predict the traffic intensity in the future.

Prediction Length indicates the number of future timestamps when the models would predict traffic intensities.

5.1 Experimental Setup

Prediction of Traffic Intensity. In this experiment, we have used 20 days of traffic intensity data as training and ten days of data for testing. In these experiments, we have varied the hyper-parameters: sampling rate, sequence length, and prediction length to find what setting yields the best prediction results.

Traffic Analysis in Weekdays. Since the traffic patterns of Dhaka significantly vary on weekdays and weekends, we have performed an experiment that has taken these scenarios into account. In November 2019, there were 20 weekdays. We have trained the models with 14 days and tested using six days. In this experiment, we have used the hyper-parameters that yielded the best results in the earlier experiments.

Traffic Prediction: Weekdays vs. Weekends. In order to show the difference between traffic patterns on weekdays and weekends, we have performed a couple of experiments. In one experiment, we have taken 20 weekdays data as the training set and ten weekend days data as the test set. In the subsequent experiment, we have chosen seven days of weekend data as training and the last three weekday data as test data.

Traffic Analysis in Weekends. To analyze the model's performance during the weekends, we have designed this experiment where both training and test data are chosen from the weekends. In November 2019, we had a total of 10 days of weekends. Among them, we have chosen seven days of train and three days of tests.

5.2 Performance Metrics

Since the ground truth and predicted traffic intensities are all real numbers, we have used the following metrics to compare different statistical and machine learning models.

Root Mean Square Error (RMSE). This metric is calculated by taking the root mean square of the ground truth vector and the predicted vector by the following equation:

$$RMSE = \sqrt{\sum_{i=1}^{N} \frac{(\hat{y}_i - y_i)^2}{N}} \tag{4}$$

Here, y_i is the ground truth, and $\hat{y}_i$ is the predicted traffic intensity for i^{th} data sample. The lower value of RMSE represents better prediction performance.

Mean Absolute Error. This measure has been calculated by finding the mean absolute error between ground truth and predicted traffic intensity vector. The equation is given below:

$$MAE = \frac{1}{N} \sum_{i=1}^{N} |y_i - \hat{y}_i| \tag{5}$$

MAE's lower value is desirable because it indicates that the ground truth and prediction are close to each other.

Correlation Coefficient (*CORR*). We have used Pearson's Coefficient of Correlation for calculating the relationship between our ground truth and predicted traffic intensity values. The equation is

$$CORR = \frac{\sum(x-\bar{x})(y-\bar{y})}{\sqrt{\sum(x-\bar{x})^2 \sum(y-\bar{y})^2}} \tag{6}$$

Here x and y are respectively ground truth and predicted traffic intensity values, while $\bar{x}$ and $\bar{y}$ are the mean of the ground truth and predicted traffic intensity, respectively. Higher CORR means the predicted value follows the trends of the ground truth. Therefore, high CORR is desirable in the experiments.

5.3 Results and Discussions

In this section, we present the results of the experiments that have been designed in the last subsection.

Table 4. Different sample interval, sequence length, and prediction length used for the experiments

Index	Sample interval (Min)	Sequence length (Min)	Prediction length (Min)
1	0.5	15	5
2	1	30	15
3	5	45	30
4	–	60	45
5	–	–	60

Table 4 shows the list of parameters that we have exhaustively used in our experiments to find what combinations are the best for traffic estimation. We have used three different intervals between the samples (time interval), the sequence length of the input pattern (time length of the input traffic data), and the prediction length (the scheduled time from the current traffic where we need to predict). The interval between samples is 0.5 min (initial sampling rate: 1 sample in 30 sec), 1 min (1 sample in 1 min), and 5 min (1 sample in 5 min). The input sequence length differs from 15 min to 60 min, where the number of samples in the defined sequence length depends on the particular interval/sampling rate. For example, if the interval between samples is 1 min and the sequence length is 15 min, there are 15 samples inside the input data fed to the statistical models. The prediction length represents the scheduled time after the current traffic when we need to predict the traffic.

Therefore, we have run experiments for 60 parameter combinations ($3 \times 4 \times 5$) with HA, ARIMA, SVR, and SVR-Graph. We have calculated the average

Table 5. Eight Combinations of Sampling Interval, Sequence Length and Prediction Length with Top Performance

Sample interval	Sequence length	Predicted length	Models	RMSE	MAE	CORR
0.5 min	45 min	5 min	HA	**73.13**	**42.89**	**0.64**
			SVR	97.04	72.37	0.52
			SVR_GRAPH	95.82	72.18	0.52
			ARIMA	73.18	42.92	0.64
0.5 min	45 min	15 min	HA	75.44	44.28	0.62
			SVR	99.19	74.15	0.51
			SVR_GRAPH	96.94	73.01	0.51
			ARIMA	75.75	44.59	0.61
0.5 min	45 min	30 min	HA	78.03	45.89	0.59
			SVR	101.15	75.85	0.50
			SVR_GRAPH	98.14	73.91	0.50
			ARIMA	78.49	46.26	0.58
0.5 min	45 min	45 min	HA	80.37	47.31	0.57
			SVR	102.13	76.60	0.49
			SVR_GRAPH	98.86	74.53	0.50
			ARIMA	80.75	47.53	0.56
1.0 min	60 min	5 min	HA	73.61	43.02	0.64
			SVR	96.93	72.19	0.53
			SVR_GRAPH	95.03	71.23	0.53
			ARIMA	**73.26**	**42.70**	**0.64**
1.0 min	60 min	15 min	HA	75.71	44.27	0.62
			SVR	99.09	74.25	0.51
			SVR_GRAPH	96.27	72.20	0.52
			ARIMA	75.93	44.49	0.61
1.0 min	60 min	30 min	HA	78.28	45.79	0.59
			SVR	101.03	76.07	0.50
			SVR_GRAPH	97.47	73.20	0.51
			ARIMA	78.65	46.12	0.59
1.0 min	60 min	45 min	HA	80.28	47.01	0.57
			SVR	101.77	76.842	0.49
			SVR_GRAPH	98.19	73.79	0.50
			ARIMA	80.75	47.31	0.56

RMSE and MAE across all these techniques to sort out ten best performing combinations. Table 5 shows the detailed RMSE, MAE, and CORR for these eight best performing combinations. We have arranged them according to the sample interval 0.5 min and 1.0 min in the Table 5.

From Table 5, we observe that sampling interval 0.5 min and interval 1.0 min had yielded the best results among all combinations where prediction lengths are 5, 15, 30, and 45 min. Specifically, the combination

Table 6. Results for Weekdays and Weekend Experiments with 0.5 min Sampling Interval, 60 min Sequence Length, and 5 min Prediction Length

Train/Test data	Sample Interval	Sequence Length	Predicted Length	Model	RMSE	MAE	CORR
Train -14- weekdays Test -6- weekdays	0.5 min	60 min	5 min	HA	78.05	47.86	0.65
				SVR	91.83	67.37	0.58
				SVR_GRAPH	92.47	68.55	0.57
				ARIMA	78.09	47.85	0.65
Train -20- weekdays Test -10- weekend days	0.5 min	60 min	5 min	HA	72.82	42.15	0.63
				SVR	97.17	72.35	0.52
				SVR_GRAPH	95.41	71.74	0.53
				ARIMA	72.85	42.07	0.63
Train -7- weekend days Test -3- weekdays	0.5 min	60 min	5 min	HA	78.23	48.48	0.64
				SVR	85.12	58.57	0.58
				SVR_GRAPH	85.44	59.16	0.56
				ARIMA	78.35	48.56	0.64
Train -7- weekend days Test -3- weekend days	0.5 min	60 min	5 min	HA	**62.68**	**32.03**	**0.57**
				SVR	81.70	55.54	0.44
				SVR_GRAPH	78.81	53.77	0.44
				ARIMA	**62.68**	**32.02**	**0.57**

0.5 min/60 min/5 min achieved the best performance with the Historical Average (HA) model: 73.13 (RMSE), 42.89 (MAE), and 0.64 (CORR). On the other hand, 1.0 min/60 min/5 min combination with ARIMA yielded the second-best result: 73.26 (RMSE), 42.70 (MAE), and 0.64 (CORR). So, please note that these two combinations' performances are very close to each other.

While analyzing these two combinations' performances, we observed that the RMSE and MAE are large, whereas the CORR is very good. High CORR indicates that the trends between the ground truth and predicted traffic have matched with each other with a high positive correlation. To investigate the high RMSE and MAE, we have analyzed the performance for each intersection. For 0.5 min/60 min/5 min combination, we observed that two intersections (Kazi para overbridge and Senpara Parbata) have unusually high RMSE (201.92 and 126.88) and MAE (159. 32 and 88.58) for HA. If we disregard these two intersections, the average RMSE decreases significantly from 73.13 to 39.81, and the average MAE gets down from 42.70 to 35.34. The same situation is also true for 1-min/60-min/5-min combination, where Kazi para overbridge has RMSE= 202.37 and MAE = 159.52 for ARIMA. The numbers for Senpara Parbata are: RMSE= 127.07 and MAE = 88.74. After discarding these two intersections, the RMSE decreased from 73.26 to 39.81 and the MAE from 43.02 to 35.07. Therefore, the traffic pattern for these intersections might not be captured by HA and ARIMA, respectively.

Now, we run experiments to see the difference in traffic patterns between weekdays and weekends. To do that, we have run experiments using the following four different train and test data with the best combination from Table 5.

Table 7. Additional Results for Weekdays and Weekend Experiments with 1 min Sample Interval, 60 min Sequence Length, and 5 min Prediction Length

Train/Test data	Sampling Rate	Sequence Length	Predicted Length	Model	RMSE	MAE	CORR
Train -14- weekdays Test - 6 -weekdays	1.0 min	60 min	5 min	HA	78.04	47.58	0.65
				SVR	91.30	66.82	0.58
				SVR_GRAPH	90.76	66.88	0.58
				ARIMA	77.66	47.24	0.66
Train -20- weekdays Test -10 - weekend days	1.0 min	60 min	5 min	HA	73.03	42.36	0.64
				SVR	97.16	72.32	0.52
				SVR_GRAPH	94.00	70.22	0.53
				ARIMA	72.69	42.01	0.64
Train -7- weekend days Test -3- weekdays	1.0 min	60 min	5 min	HA	78.23	48.23	0.65
				SVR	84.81	58.09	0.58
				SVR_GRAPH	84.52	58.02	0.57
				ARIMA	77.97	47.98	0.65
Train -7-weekend days Test-3-weekend days	1.0 min	60 min	5 min	HA	**63.62**	**32.82**	**0.57**
				SVR	82.63	55.85	0.43
				SVR_GRAPH	78.55	52.99	0.44
				ARIMA	**63.44**	**32.69**	**0.57**

1. training with traffic data for 14 weekdays, testing with data for six weekdays.
2. training with data for 20 weekdays, testing with ten weekend days.
3. training with data for seven weekend days, testing with data for three weekdays.
4. training with data for seven weekend days, testing with three weekend days.

From the Table 6, it is easy to note that the 4th experiment has yielded the best result when both the train and test sets are from weekend data. HA and ARIMA models got very close performance scores in predicting the traffic. Therefore, we can conclude that traffic patterns on weekends are similar enough to be predicted by HA and ARIMA models.

In contrast, the 1st experimental result suggests that the weekdays' traffic patterns are very much dynamic and intermittent that the HA and ARIMA cannot capture them very well. The same result has been observed for 3rd experiments where the HA and ARIMA models have produced erroneous predictions for weekdays while they have been trained with data weekends. The result for the 2nd experiments lies in-between.

We can conclude from these experiments that the traffic pattern for weekends can be modeled with simple statistical and machine learning models. However, to learn the weekdays' traffic pattern with large variabilities, we need more sophisticated machine learning algorithms such as graph convolutional networks or recurrent neural networks. Since this paper focuses on the data collection, we plan to perform further analysis using graphical models in the future. With the proposed technique, we can collect data for the entire city in an automated manner. We also plan to incorporate a longer time frame to identify the traffic patterns for the weekday variations.

Similar results have been observed for combination 1 min/60 min/5 min combination as shown in Table 7.

6 Conclusion

This paper proposes a novel approach for traffic data collection in developing countries with minimal cost and human effort. We have proposed to use traffic information from Google Map Service in a novel way so that the traffic data for any road segments/intersections can be collected seamlessly from any location in a city. Therefore, the proposed approach is scalable from a small area to the entire metropolitan city. Further, we have implemented some statistical and machine learning models: historical averages (HA), Support Vector Regression (SVR), Support Vector Regression with Graph (SVR-Graph), and Auto-Regressive Integrated Moving Average (ARIMA) on the traffic data of November 2019 at Mirpur, Dhaka Bangladesh. The experimental results suggest that these simple models can effectively model the traffic congestion of Mirpur, Dhaka. Also, the results show the efficacy of the collected traffic data in forecasting future congestion.

In the future, we plan to incorporate more intersections to develop a comprehensive dataset of the city road network. As mentioned in the discussion, we will apply more sophisticated machine learning models such as Graph Convolution Network and Recurrent Neural Networks on the data to accurately estimate the traffic pattern.

Acknowledgment. This work is partially funded by the Independent University, Bangladesh, and the ICT Division of ICT Ministry, Bangladesh.

References

1. Falcocchio, J.C., Levinson, H.S.: The costs and other consequences of traffic congestion. In: Road Traffic Congestion: A Concise Guide, pp. 159–182. Springer (2015)
2. Alghamdi, T., Elgazzar, K., Bayoumi, M., Sharaf, T., Shah, S.: Forecasting traffic congestion using ARIMA modeling. In: 2019 15th International Wireless Communications & Mobile Computing Conference (IWCMC), pp. 1227–1232. IEEE (2019)
3. Behrisch, M., Bieker, L., Erdmann, J., Krajzewicz, D.: Sumo–simulation of urban mobility: an overview. In: Proceedings of SIMUL2011, The Third International Conference on Advances in System Simulation. ThinkMind (2011)
4. Zambrano-Martinez, J.L., Calafate, C.T., Soler, D., Cano, J.-C., Manzoni, P.: Modeling and characterization of traffic flows in urban environments. Sensors **18**(7), 2020 (2018)
5. Iyer, S.R., Boxer, K., Subramanian, L.: Urban traffic congestion mapping using bus mobility data. In: UMCit@ KDD, pp. 7–13 (2018)

6. Zhong, Y., Xie, X., Guo, J., Wang, Q., Ge, S.: A new method for short-term traffic congestion forecasting based on LSTM. In: Proceedings IOP Conference Series, Material Science Engineering (2018)
7. Zheng, Z., Wang, D., Pei, J., Yuan, Y., Fan, C., Xiao, F.: Urban traffic prediction through the second use of inexpensive big data from buildings. In: Proceedings of the 25th ACM International on Conference on Information and Knowledge Management, pp. 1363–1372 (2016)
8. Liu, Z., Li, Z., Wu, K., Li, M.: Urban traffic prediction from mobility data using deep learning. IEEE Netw. **32**(4), 40–46 (2018)
9. Ahmed, A., Naqvi, S.A.A., Watling, D., Ngoduy, D.: Real-time dynamic traffic control based on traffic-state estimation. Transp. Res. Rec. **2673**(5), 584–595 (2019)
10. Faloutsos, C., Gasthaus, J., Januschowski, T., Wang, Y.: Classical and contemporary approaches to big time series forecasting. In: Proceedings of the 2019 International Conference on Management of Data, pp. 2042–2047 (2019)
11. Kuznetsov, V., Mariet, Z.: Foundations of sequence-to-sequence modeling for time series. arXiv preprint arXiv:1805.03714 (2018)
12. Akbar, A., Khan, A., Carrez, F., Moessner, K.: Predictive analytics for complex IoT data streams. IEEE Internet Things J. **4**(5), 1571–1582 (2017)
13. Labib, S.M., Mohiuddin, H., Shakil, S.: Transport sustainability of Dhaka: a measure of ecological footprint and means for a sustainable transportation system. Labib, S.M., Mohiuddin, H., Shakil, S.H.: Transport sustainability of dhaka: a measure of ecological footprint and means for sustainable transportation system. J. Bangladesh Inst. Plan. **6**, 137–147 (2014)
14. Mahmud, K., Gope, K., Chowdhury, S.M.R.: Possible causes & solutions of traffic jam and their impact on the economy of Dhaka city. J. Mgmt. Sustain. **2**, 112 (2012)
15. Li, Y., Yu, R., Shahabi, C., Liu, Y.: Diffusion convolutional recurrent neural network: data-driven traffic forecasting. arXiv preprint arXiv:1707.01926 (2017)
16. Yao, H., Tang, X., Wei, H., Zheng, G., Yu, Y., Li, Z.: Modeling spatial-temporal dynamics for traffic prediction. arXiv preprint arXiv:1803.01254 (2018)
17. Yu, B., Yin, H., Zhu, Z.: Spatio-temporal graph convolutional networks: a deep learning framework for traffic forecasting. arXivpreprint arXiv:1709.04875 (2017)
18. Cui, Z., Henrickson, K., Ke, R., Wang, Y.: Traffic graph convolutional recurrent neural network: a deep learning framework for network-scale traffic learning and forecasting. IEEE Trans. Intell. Transp. Syst. **21**, 4883–4894 (2019)
19. Guo, K., Hu, Y., Qian, Z., Liu, H., Zhang, K., Sun, Y., Gao, J., Yin, B.: Optimized graph convolution recurrent neural network for traffic prediction. IEEE Trans. Intell. Transp. Syst. **22**, 1138–1149 (2020)
20. Zhang, Y., Wang, S., Chen, B., Cao, J., Huang, Z.: TrafficGAN: network-scale deep traffic prediction with generative adversarial nets. IEEE Trans. Intell. Transp. Syst. **22**, 219–230 (2019)
21. Guo, S., Lin, Y., Feng, N., Song, C., Wan, H.: Attention-based spatial-temporal graph convolutional networks for traffic flow forecasting. In: Proceedings of the AAAI Conference on Artificial Intelligence, vol. 33, pp. 922–929 (2019)
22. Chen, C., Li, K., Teo, S.G., Zou, X., Wang, K., Wang, J., Zeng, Z.: Gated residual recurrent graph neural networks for traffic prediction. In: Proceedings of the AAAI Conference on Artificial Intelligence, vol. 33, pp. 485–492 (2019)
23. Labib, S., Mohiuddin, H., Hasib, I.M.A., Sabuj, S.H., Hira, S.: Integrating data mining and microsimulation modeling to reduce traffic congestion: a case study of signalized intersections in Dhaka, Bangladesh. Urban Sci. **3**(2), 41 (2019)

24. Roy, K., Barua, S., Das, A.: A study on feasible traffic operation alternatives at a signalized intersection in Dhaka city. In: International Conference on Recent Innovation in Civil Engineering for Sustainable Development, Department of Civil Engineering, DUET, Gazipur, Bangladesh, pp. 656–660 (2015)
25. Rahman, M.M., Shuvo, M.M., Zaber, M.I., Ali, A.A.: Traffic pattern analysis from GPS data: A case study of Dhaka city. In: 2018 IEEE International Conference on Electronics, Computing and Communication Technologies (CONECCT), pp. 1–6. IEEE (2018)
26. Salma, F.N., Hosain, S.: Enhancing mixed road traffic forecasting method using vehicle tracking system in GPS. In: 2018 International Conference on Computing, Power and Communication Technologies (GUCON), pp. 14–19. IEEE (2018)
27. Chen, C., Petty, K., Skabardonis, A., Varaiya, P., Jia, Z.: Freeway performance measurement system: mining loop detector data. Transp. Res. Rec. **1748**(1), 96–102 (2001)
28. Chen, X., Chen, R.: A review on traffic prediction methods for an intelligent transportation system in smart cities. In: 2019 12th International Congress on Image and Signal Processing, BioMedical Engineering and Informatics (CISP-BMEI), pp. 1–5. IEEE (2019)
29. Ali, M.S., Adnan, M., Noman, S.M., Baqueri, S.F.A.: Estimation of traffic congestion cost- a case study of a major arterial in Karachi. Procedia Eng. **77**, 37–44 (2014)
30. Lwin, H.T., Naing, T.: Estimation of road traffic congestion using GPS data (2015)
31. Sharma, B., Kumar, S., Tiwari, P., Yadav, P., Nezhurina, M.I.: ANN based short-term traffic flow forecasting in undivided two-lane highway. J. Big Data **5**(1), 48 (2018)
32. https://data.cityofchicago.org/Transportation/Chicago-Traffic-Tracker-Historical-Congestion-Esti/77hq-huss/data

firstGlimpse: Learning How to Learn Through Observation via Memory Modeling with Reinforcement Learning

D. Michael Franklin(✉) and Ryan Kessler

Kennesaw State University, Marietta Campus, Marietta, GA, USA
mfranklin@kennesaw.edu, rkessle2@students.kennesaw.edu

Abstract. While there are many systems extant that take well-known models and refine how to learn with them, there are very few that deal with the initial stages of raw model formation. How do we, as human beings, learn? Can we teach machines to 'learn to learn'? In this paper, the authors introduce just such a system. firstGlimpse is a hierarchical memory-modeling technique that can learn from scratch via observation. firstGlimpse uses reinforcement learning to develop its models and reinforce the models that are most viable, most likely, etc. This hierarchical approach to memory modeling develops, through observation, basic symbols that comprise the elements of the learned material. These basic symbols are combined into words, these words combined into phrases, phrases into sentences, and so on, creating a modeling system for general learning that can be built from the ground up. As a new symbol is observed, it is recorded. Likewise, that symbol's connections with other symbols is also recorded in another layer of the memory. Next, these connected symbols are viewed as words, and the words are observed as being connected to other words. In this manner, phrases are added. This continues upwards to higher levels of the modeling hierarchy. This paper explains the process of firstGlimpse's reinforcement learning by using the game of Set as a simple example. We will show, starting from scratch, that it can learn to play by observing humans playing the game and recording its observations within its hierarchical memory. The experiments will show that this method is successful both in observing the links of the memory graph and in the results of successfully playing the game.

Keywords: Artificial intelligence · Reinforcement learning · Memory modeling · Learning from scratch

1 Introduction

One difficult problem in artificial intelligence is building an agent that is capable of domain-general learning and adapting in response to new stimuli. Sebastiani took such a general approach and noted that his work could be used for non-textual media [5]. However, even this agent is bound to classifying media, whether textual or otherwise. The agent being proposed would be capable of much more general learning, and not just in classification.

K. Arai (Ed.): FICC 2021, AISC 1363, pp. 532–540, 2021.
https://doi.org/10.1007/978-3-030-73100-7_37

One dimension of this difficulty is understanding the right first steps to a general knowledge learning system. Russell and Norvig describe four approaches to artificial intelligence: thinking humanly, acting humanly, thinking rationally and acting rationally [3]. We have decided to take the former two approaches. While eventually a unique approach may be found to create a general agent, we feel that looking closer to home, so to speak, can provide a wealth of knowledge upon which to build the agent.

In the introduction of his book [6], Ron Sun argues for the necessity of building comprehensive cognitive models. While our aim is not to advance computational cognitive modeling, it provided inspiration and grounding for our approach. Similar to cognitive modeling, our approach is to model the human mind and processes as closely as we can through observations on how we perform those tasks that we desire the agent to do perform. Thus, our approach is an intuitive one at this early stage. The agent's architecture is built around an abstract view of the functions performed by the mind. Specifically, memory, reasoning, emotions, input processing, and output processing (i.e., communication outside of the mind). Our agent contains a module that performs each of these functions to differing degrees.

2 Methodology

To help with testing the agent, we used the mathematical card game Set and wrote a program in Unity to model it. To keep the agent separated completely from the simulation, we wrote an interface between the agent and the simulation that would provide the agent with all relevant information and translate all output from the agent to actions within the simulation. These creates a virtual barrier between the actual data (contained within the simulation) and the agent's view such that the agent does not have access to any clues or hints other than the raw observations of the simulations. As it would take significant (and time-consuming) effort to have the AI learn against live human players we elected to build another AI in addition to the general agent being built. This brute force agent was built. This agent utilized exhaustive search (as Set is a finite game) to find any and all combinations of sets that were valid each round. As a result, this agent would always win or tie. This brute-force agent was then used to test the learning in the place of a highly skilled human player. Our assumption was that if the general agent was performing as intended against the brute-force agent it would achieve scores comparable to that of two brute force agents playing against each other (e.g., either a tie, or a win for the player who went first). Of course, initially, the agent will not perform at this level, but the goal would be for the general agent's performance to increase as it built up its memory, thus proving our learning technique.

While the complexity of our proposal far exceeds the simplicity of a game like Set, it is important at the early stages to prove the knowable for a smaller problem. We will use Set more as an example to ground our conversation rather than as an exhaustive proof of the completeness of the scenario. We will thus utilize this Set example to provide a narrative for describing our system so that the pieces and elements of the technique, the reinforcement learning, and the application of the firstGlimpse learning framework can be clearly seen.

The memory of the agent is built around a data structure called a symbol (Fig. 1). These symbols can represent anything from the color red to three cards that could be used as a valid set within the simulation. However, it is best to create a hierarchy to organize these symbols into larger and larger (i.e., more meaningful) collections.

```
{Symbol:
    public string,
    Int connectionNum,
    Array of size connectionNum[Connection]}

{Object:
    Int number = 3,
    Array of size number[Symbol]}

{Card:
    Int count [1..3],
    Array of size count[Object],
    Float emotion}
```

Fig. 1. Sample card data structures used by the simulation.

Before we march through the hierarchy, let us offer a quick word on the principle used to assure the separation of the simulation data from the learning data. To build these symbols and sets of symbols, we made use of C#'s reflection ability and a data structure in the simulation whose public properties are those of the physically observable properties of the cards.

By building the data structure of the simulation this way we can pass a list of objects to the agent and it can use reflection to observe the object. The class name represents what the object is, the names of the properties reflect the properties of each card, and the values of those properties reflect what makes each individual card unique. By using the class name and property names, the agent can build a model that represents the card class, and the values of those properties can be used to generate actual cards. However, it could also use this to observe other simulations such as Poker or Mahjongg without having to be changed, as described in a similar approach, in multiple ways, in their structure-Mapping Engine [1]. However, their focus is on the use of mapping knowledge as an ontology rather than a simple model. The author in [2] also provided a basis for this approach but with a more focused application on sentences.

Building on these ideas and innovating through the hierarchical idea from our own research, we need to apply these techniques to build our memory structure (used to hold the symbols and the aggregations of these symbols. The memory where it stores these symbols is separated into four tiers, as well as one separate tier to contain emotional memory (shown in Fig. 2). The first tier of memory holds primitive symbols such as: a shape (e.g., oval, squiggle, etc.), a fill (e.g., filled, empty, etc.), or a color (red, blue, etc.). Each symbol at this level has only its identifier and relevant connections. An example of a relevant connection at this level would be shape, fill, color and number being associated with card, or red being associated with color and so on. This level represents the alphabet that our agent can use.

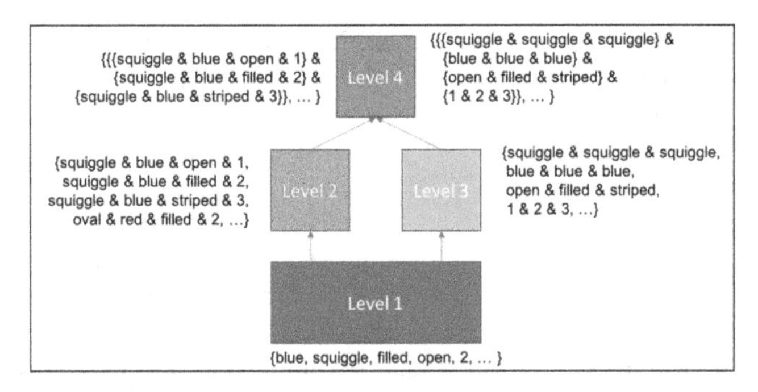

Fig. 2. firstGlimpse multi-layer memory model.

The second level of memory holds symbols that are an aggregate of these primitive symbols. This level would include the models built from data structures such a particular object on a card. These symbols also have an identifier, but instead of associated symbols, they contain a list of the primitive symbols that comprise their aggregate. An example of a level two symbol would be: {Oval, Filled, Red}. Models contain everything a level two symbol does. However, while the identifier of a level two symbol is the aggregate of the identifiers of its primitive symbols (i.e., Oval & Filled & Red), the identifier of a model would correspond to the name of the class, and the symbols within its aggregate correspond to the names of the properties of that class. In addition, models also contain a list of level two symbols that use the model as their basis. Level two symbols thus represent the words created from the alphabet represented by level one symbols.

The third level of memory holds symbols that are also aggregates of primitive symbols. For this simple example of Set, this would be a card (containing objects, each comprised of symbols). These are simple aggregations, so they may be valid cards or invalid cards (or generally, valid or invalid meta-objects). However, unlike level two symbols whose aggregate contains a member from each property of the model, the aggregate of a level three symbol contains primitive symbols all from the same property of the model. A few examples are: {Card:{Squiggle & Filled & Red}}, {Card:{Oval & Empty & Blue}, {Diamond & Filled & Green}}, and {Card:{Squiggle & Hatched & Blue}, {Squiggle & Empty & Red}, {Squiggle & Filled & Green}}. Level three symbols also have an emotional value (think of this as a score of the quality of the aggregation of symbols based on experience) attached to them that represents the sum of positive and negative feedback that the agent has received on this symbol. These can be considered the morphemes of the words.

The fourth level of memory contains symbols that are aggregates of level three memory. Level four symbols represent the sentences that can be formed from the words of level three memory and can also be considered patterns. In Set, such a pattern could be {Card 1, Card 2, Card 3} or {Card 9, Card 53, Card 19}. The emotion module currently houses the emotional values associated with level four symbols.

Now we will show how level three symbols relate to level four symbols. Level three symbols represent what we call semi patterns. While these are whole patterns, they represent the individual patterns that constitute the patterns represented in level four symbols. In another terminology, Level one symbols are letters, Level two symbols are words, Level three symbols are phrases, and Level four symbols are sentences. Each level can also be considered an aggregate of level before it. As an example, we will consider the example from level four memory above. For example, in the first level four symbol example above, if Card 1, 2, and 3 comprise a valid set, each semi pattern would receive positive feedback because it is valid according to the rules of Set. However, if the agent then used the second set, but here Card 9, 53, and 19 do not form a valid set, it would receive negative feedback regardless of the cards on the board as it is not a valid set. Semi patterns that are valid would return to zero, while the semi patterns that do not form valid sets would go to negative one and mixed sets would remain unchanged. Over a sufficient amount of simulation runs, the emotional value of valid semi patterns according to the simulation (whether Set or another) would tend towards infinity, and the emotional valid of invalid semi patterns according to the simulation would tend towards negative infinity. This approach was inspired by Quinlan's paper [4] who used the attributes of 'Saturday morning' along with feedback, indicating if a training set was positive or negative, to build a decision tree upon which his agent could work and learn.

The agent's reasoning module at the present only includes the ability to analyze level four symbols to look for level three symbols as well as housing the pattern-matching algorithm. The pattern-matching algorithm works by selecting an input to use as a reference, starting with the first input in the list. Next it will start to look through all known semi patterns and try to find one that contains the symbol of the input's first property in the first index of the semi pattern. Once a semi pattern is found, the algorithm will try to find more inputs that will fill out the semi pattern. If the pattern cannot be filled, the algorithm will move on to the next semi pattern, and if none are found, it will move on to the next input as the reference and start again. However, if the semi pattern is filled, then it will move to all subsequent properties and attempt to find semi patterns to fill those out. If all a semi pattern is found for each property then the symbols are added to a list of complete properties along with the sum of the emotional values of all the semi patterns that make up the full pattern. However, if at any level, a semi pattern is found, the pattern will return to the semi pattern for the first property and attempt to find a new group of inputs to fill it out and repeat the process if successful. If not, it moves on to the next semi pattern for the first property until all semi patterns have been looked at. It then moves on to the next input reference and repeats the process. When all inputs have been processed, the algorithm checks if any patterns have been found. If so, it will select the pattern that has the highest sum of emotional values and return that pattern. However, if none are found, it will look in level 3 memory for a semi pattern. If there is one, it will randomly select inputs until the number of inputs selected is equal to the number of symbols in the semi pattern. If no semi pattern is found, then it will choose a random amount of inputs to return.

The final module is for output processing. This module takes all information the agent wishes to communicate to the interface and processes it, so the interface receives information with the same encoding regardless of the symbols or simulation being played.

To see how these elements are all connected, we will look at an example of a turn as seen by the general agent. When the agent's turn comes up, it receives all the cards on the board as objects. These objects are passed into the input processor which first breaks them up into level one symbols and puts them into memory if they are not already there. Once this is complete, the processor then forms level two symbols from each of the cards and places these into level two memory. Now that the board observation phase is over, the inputs have been turned into level two symbols that the agent will send to the reasoning module. The reasoning module scans these symbols using the pattern matching algorithm to form patterns, or level four symbols. When a pattern has been selected, the reasoning module returns this, and it is sent to the output processor. The output processor turns it into a string that can be parsed by the interface and returns it to the interface. The interface will parse the string and turn the information into card objects and give these to the game to determine if it is a valid set. The game will then tell the interface whether the set is valid or not, and the interface will return this feedback, along with the set used, to the agent. Then agent will send the set to input processing and turn it into a level four symbol. This symbol is then sent to the reasoning module along with the feedback to analyze. The reasoning module will break the level four symbol up into its respective level three symbol parts. It will then try to place them into level three memory along with the feedback value if they are not already in memory. If they are in memory, it tells the symbol about the feedback and the emotional value is updated accordingly. Once all the level three symbols have been added or updated as necessary, the level four symbol is added to memory if needed, and its associated emotional value in emotional memory is added or updated. At this point, the agent has completed one round in the game. Each symbol viewed, each object observed, each card provided, and each set of cards matched is stored in memory alongside the feedback of how well these elements were received by the simulation. This forms the reinforcement loop the trains the framework on how to choose cards that will then maximize the score (thus trying to win the game).

3 Experimentation

Our experiments focused on allowing the general agent to play Set against both the brute force agent and against itself. When playing against the brute force agent, if the general agent was building an appropriate knowledge and emotion base and the pattern-matching algorithm is functioning as intended, then the gap between the individual scores should close until the agents either tied every simulation or the first player wins by one. However, when playing against itself, a limit had to be placed upon how many turns were allowed before the simulation would terminate and another one would start. This would keep the agent from getting into a permanent rut (due to no new information being gained).

The simulation was run with the brute force playing the general agent 500 times with a turn limit of 50. However, for the simulation with the general agent playing against itself, the simulation was given three different turn limits: 25, 30, and 50. For each limit, the simulation was run 10,000 times. Finally, there was a simulation run in which the agent observed the brute force agent playing against itself 500 times and then allowed to play against itself 500 times with a turn limit of 50. After each run, the scores of every simulation were output to a spreadsheet along with the number of turns for each simulation, the turn limit, and the number of symbols in each level of memory at the end of that simulation.

4 Analysis

In the simulation run against the brute force agent, the general agent learned to play optimally by run 6. It saw all level one symbols by run 3, all level two symbols by run 4, and reached the maximum number of level three symbols it would learn for the run by run 5. It saw linear growth in its level four symbols throughout the entire run.

In the simulation against itself with a turn limit of 25 (as shown in Fig. 3), the agent learned to play optimally by run 28. It saw all level one and two symbols by run 34, reached its maximum number of level three symbols observed by run 29, and showed rapid growth in the rate of level four symbols seen until run 39 at which point learning plateaued and only rose by 9 for the subsequent runs.

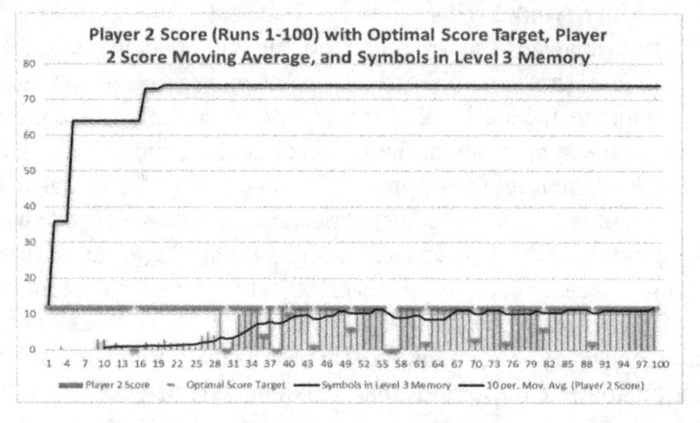

Fig. 3. Agent/Memory progress during simulation.

When the turn limit was raised to 30, the agent learned to play optimally by run 1962. However, the agent saw a steady growth in the number of points scored per game until run 310 in which it saw its first double digit score. From here until run 1962, the score fluctuated from mid-single digit scores to lower double-digit scores. By run 343 the agent had seen all level one and two symbols and reached the maximum number of level three symbols seen by run 335. Growth of level four symbols reach its plateau by run 1652 only growing by 10 for the remainder of the runs.

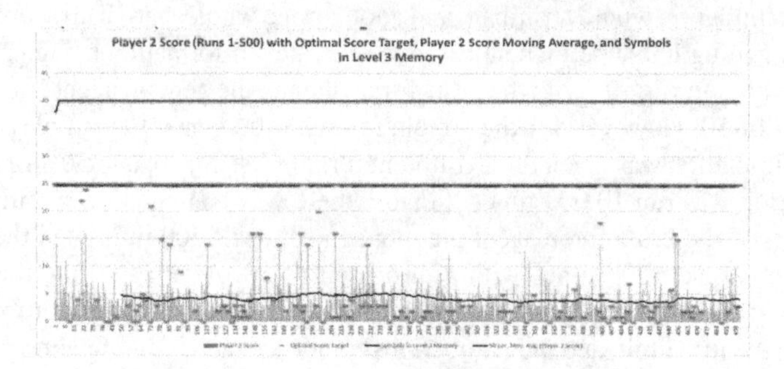

Fig. 4. Agent learning during simulation

When the turn limit was raised to 50, the agent learned to play optimally by run 26. It saw all level one and two symbols by run 20 and reached the maximum number of level three symbols it would see by run 20 as well. It reached the maximum number of level four symbols it would see by run 35.

Finally, during the run in which the agent observed two brute force agents play, it saw all level one and two symbols in the first game, and reached its maximum of 36 level three symbols by the second game. It saw consistent and rapid linear growth of level four memory throughout the entire game. When the agent then switched to playing against itself, the level three memory grew by 4, and the level four memory saw consistent, but much less rapid, linear growth.

Throughout every simulation set the growth of emotional memory mirrored the growth of level four memory almost exactly. In all simulations against itself, emotional memory was consistently only one below the number of symbols in level four memory. In the simulation against the brute force agent, the emotional memory ended up four below the final number of level four symbols.

The determining factor to playing optimally is the acquisition of a certain number of level three symbols. In the simulation versus the brute force agent, the number of symbols was 46. Against itself with turn caps of 25, 30, and 50, the number of level three symbols was 65, 75, and 74, respectively. Tied to this, however, are the emotional values associated with these symbols. By only observing the brute force agent playing, the general agent will never observe an invalid pattern and thus cannot form the proper emotional value required to dismiss those patterns when they come up. In the simulation run in which the general agent observed the brute force agent play against itself, as shown in Fig. 4, the general agent's level three memory reached a maximum of 36. However, when it then played against itself, it did not play optimally even after increasing its level 3 memory by 4 for the 500 runs it was allowed to play.

5 Conclusion

The current version of the agent meets the expectations set out for the simulation used to test it. The agent builds its memory as it observes actions made by itself and other agents and uses the feedback from these actions to determine how to perform optimally. The data shows that the ability of the agent to perform optimally is tied to the size of its memory and the quality of the feedback that it receives. It is our desire to make another simulation to which we can connect the agent and test its performance within the new arena without changing any code beyond the interface between the agent and simulation.

As for changes to the agent, we believe that how we make use of models will need to be refined and possibly expanded to include, at a minimum, level four symbols. The emotional memory needs to be refined as it is very basic in its current implementation and does not consider potential emotions about a given symbol and how that could impact the choices the agent might make. We also hope to give the agent the ability to reason beyond pattern-matching, including the ability to draw conclusions about similarities between symbols and models. An example would be the agent realizing that the semi pattern {Red & Green & Purple} in Set is the same as {Green & Red & Purple}.

References

1. Falkenhainer, B., Forbus, K.D., Gentner, D.: The structure-mapping engine: algorithm and examples. Artif. Intell. **41**, 1–63 (1989)
2. Lim, D.C.Y., Ng, H.T., Soon, W.M.: A machine learning approach to coreference resolution of noun phrases. Comput. Linguist. **24**, 521–544 (2001). Special Issue on Computational Anaphora
3. Norvig, P., Russell, S.: Artificial Intelligence: A Modern Approach, 3rd edn. Pearson Education Inc., Upper Saddle River (2010)
4. Quinlan, J.R.: Induction of decision trees. Mach. Learn. **1**, 81–106 (1986)
5. Sebastiani, F.: Machine learning in automated text categorization. ACM Comput. Surv. (CSUR) **34**, 1–47 (2002)
6. Sun, R. (ed.): The Cambridge Handbook of Computational Psychology. Cambridge University Press, Cambridge (2008)

Communicative Artificial Intelligence in Multi-agent Gaming

D. Michael Franklin(✉)

Kennesaw State University, Marietta, GA, USA
mfranklin@kennesaw.edu

Abstract. In many modern game implementations, the intelligent agents are abstracted from their context and from each other. This leads to a disconnected experience that causes information to be lost from each agent once they are defeated. We wish to show the Communicative Artificial Intelligence system for multiagent games. This system provides adaptive AI across generations of agents that the player will encounter. In this scenario the player encounters the first wave of enemy AI near the perimeter of the target zone. The player must choose which weapon or weapons to fight with. If they defeat the enemy agents, these agents pass on their acquired knowledge to the next wave of enemy combatants. This second wave now has developed immunity to the same form of attack that was previously used by the player. As a result, the player must now choose a different attack. This information is passed along to the third wave of enemy agents who now have immunity to all weapons that the player has used thus far (collected from each previous generation of enemy AI). In this manner we hope to create a more strategic approach to the game wherein the player must reason about which weapons to use knowing that they will lose effectiveness as they progress. The goal is for the player to have to use the least powerful weaponry that is sufficient for the task at each level or they will be hopelessly outmatched as the game progresses. This research aims to close a gap in current games that is created by agents not acting as a part of the greater whole. In real-world scenarios the teams would communicate with each other and pass along such vital information. We also hope to draw the conclusion that this system can also inform more complicated systems (such as threat detection, intrusion thwarting, and more capable firewalls). We imagine multi-tiered security systems that pass along the techniques and knowledge about the attack or attacker to each progressively more challenging system thus providing thus providing a much more solid defense. We model such a system in our game as a means of evaluating this type of communicative AI.

Keywords: Artificial intelligence · Communicative · Multi-agent systems · Team AI · Multi-level AI

1 Introduction

Games frequently implement enemy AI to keep the playing character from reaching the goal [3]. Game developers often attempt to layer multiple type of AI within the

K. Arai (Ed.): FICC 2021, AISC 1363, pp. 541–546, 2021.
https://doi.org/10.1007/978-3-030-73100-7_38

game context in an effort to create the appearance of more complicated AI. In fact, these multi-episodic journeys through the game reveal completely isolated AI schemas. These episodes are interesting for game players, but they do not really relate to each other. We wish to introduce Communicative AI to add a true progression of a learning AI. Communication is vital to effective artificial intelligence [4] but often neglected in implementation. Each AI in the domain is working independently and without communication or centralized coordination. Communicative AI (CAI) adds a new layer to existing AI schemas by sending information learned by one AI to the next level AI. As an example, suppose the player encounters a first round of AI and uses a particular weapon against them. The player defeats the first round of AI, but not before they pass this information along to the next layer of AI agents who are now prepared for the player with a defense against this particular weapon. They now hold this information in their collective memory along with the information they gather from their encounter with the player. Again, the player defeats this level of AI, but has had to choose a different weapon to use since the previous one is now of little effect. The player dispatches this enemy as well, but not before that information is passed to the next level AI. This third level of AI agents now has the collective memory of the previous two and has prepared itself for this confrontation appropriately. This third level enemy is now going to be more difficult. Our goal in introducing this multi-level generational AI is to deepen the play experience in gaming without having to create significantly more intricate AI. We believe that CAI will add a richer depth to the game without requiring the equivalent extensive coding that would otherwise be required. This will require that the player develop a deeper strategy to defeat each enemy and cannot choose the simpler path of perfecting the use of a single favorite weapon and using it in every encounter. This will make for a richer and deeper game-playing experience that will require critical thinking, planning, and strategic reasoning to handle this incrementally increasing challenge. Further, the player will have to conquer a broader range of the game's weapon choices to complete the game leading to better exploration of the entire game space.

An additional application of this technology will be to apply this to the security/defense realm that is dominated by hackers that develop a favorite attack scheme. With an AI in place to defeat these attackers it can adapt itself at each level and pass along what it learns in one layer of the defense to the next layer. This should increase the ability of the system to repel attacks. The system would learn the attacker's methodology in the first layer and pass that information along to the next layer of security. Even if the attacker gets through this first layer they will have to contend with a more capable and better-informed AI in the next layer. There are many more applications of this Communicative AI in realms such as these where the constant relaying of information learned can help to make the next layer of security even stronger. This layered approach provides a deeper security by varying many different layers of security rather than one monolithic layer.

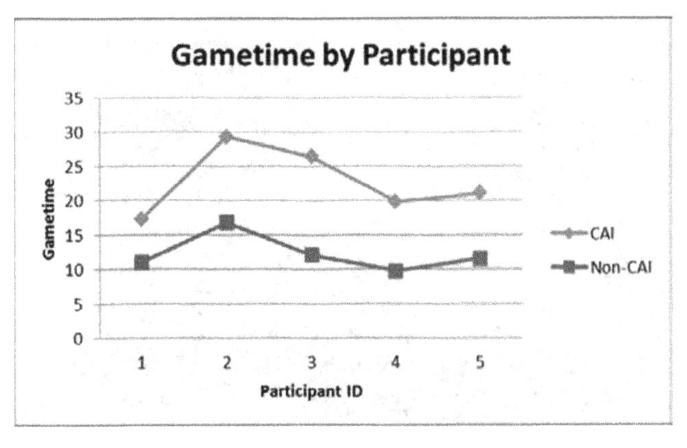

Fig. 1. Increased challenge from communicative AI

2 Methodology

To test this Communicative AI we devised a simple game that has the player face three different waves of enemy agents.

There are a variety of weapons to choose from while playing the game. The player must choose wisely to avoid trouble in the later rounds. The first round has the player face an enemy force of five warriors. They are lightly armed and lightly armored. The player must choose from the dagger (a thrown weapon), a sword, or the flaming sword (see Figs. 2, 3 and 4). The player may use any one weapon, any combination of weapons, or all of the weapons in this first round. As the player dispatches the enemies, they make note of the weapons used and pass that information along to the second-round enemies. Presuming the player makes it to the second round, they must face three warriors, but each is armored against any weapons that the player used in the first round. They are configured to be the same as the first-round enemies except for this armor. The player must once again choose which weapon or weapons to use in this battle. Once this battle

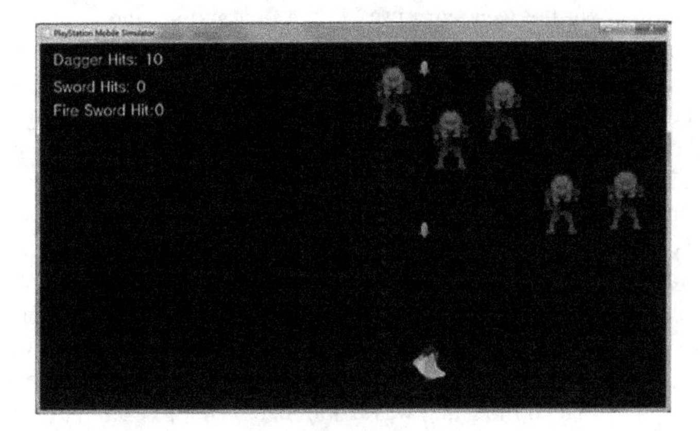

Fig. 2. Scenario 1

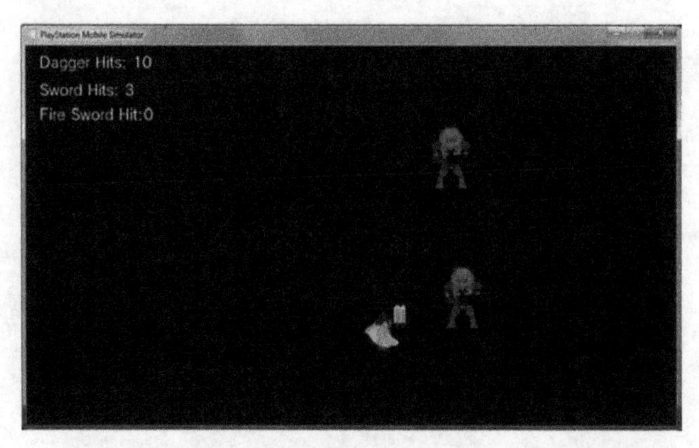

Fig. 3. Scenario 2

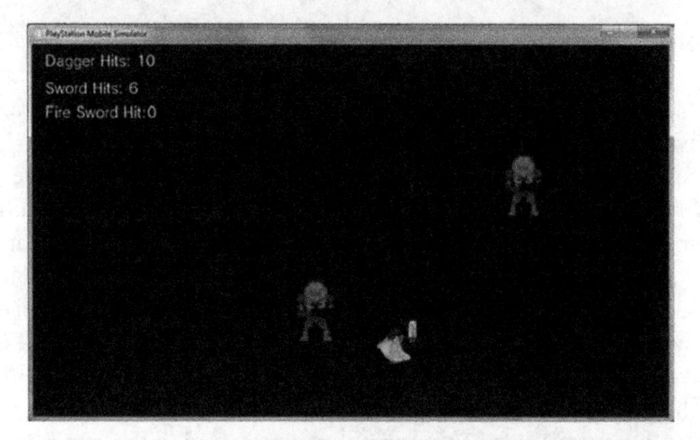

Fig. 4. Final scenario/result

is over, presuming another victory for the player, the second-round enemies pass along their information about the weaponry used in the first and second rounds to the third-round enemy. In the third round the player must face only one enemy, but this enemy is also armored against any and all weapons seen by any enemy in the game thus far. This final enemy is configured just like the other except for this armoring. The player must then choose which weapon or weapons it will use for the final battle and the round commences. If they have chosen wisely, they will have the right weapon or weapons left to dispatch this final enemy and achieve victory.

The trials for this research involved playing this exact game in both scenarios, with standard AI and with the communicative AI, and comparing the delta of the scores. Scoring is measured in total enemies dispatched minus the amount of damage sustained. Additionally, the total game time is measured. Finally, the participants were asked to evaluate their perceived difficulty in playing the game. To remove experience bias, the order in which they played the game was randomized. Before each participant played, they were briefed on the controls, the concept of the game, and the general idea of communicative AI. They were allowed to play each round three times and their cumulative scores were averaged.

3 Analysis

The early testing showed that the CAI made the game more difficult and more enjoyable. The players commented that they enjoyed the challenge. The trials were run with each participant playing both versions of the game and using runtime as the indicator of the level of challenge presented by the gameplay. The results show that the average length of the game using no Communicative AI was 12.2 s (std. dev. 2.46). This shorter gameplay reveals the lack of complexity and challenge as opposed to the 22.8 s (std. dev. of 4.4) average game length under the CAI. The game was considerably more engaging with the CAI in force. The participants had varying levels of experience in gaming, so their survival times varied, but the delta between the non-CAI and the CAI was clear (Fig. 1). The participants also noted their increased satisfaction with the CAI version, even though it was harder, because it made the game more interesting.

The game performed as desired. It made the player make a strategic choice at the first level about which weapon they would choose. If they did not choose wisely their progress was blocked. The weapon choices were checked at the next level and that opponent was impervious to the first weapon chosen. This continued on as planned to the next level. The players were challenged to think through their overall plan as they progressed. There were several scenarios wherein the player was barred from progress because of squandered resources along the way. Many games ended in failure because of this. It was interesting to note the number of players who commented on this intriguing development or remarked with an inquisitive brow-furrowing about the need to plan better. While such qualitative analysis may seem out of place in serious research, it is, in fact, quite relevant to the idea of this research.

The game was simple, but it was intended only to showcase the added quality of gaming provided by a communicative AI system that could transmit its learning from one generation to the next. Further, and worthy of more study, is the initial finding that the satisfaction of the game designer is increased by added complexity and challenge to the game. This makes the time and effort of game design more worth it for the developer by providing more challenge and depth without a corresponding increase in programming time.

4 Conclusions

The study showed that the game was made more engaging and interesting by the addition of the Communicative AI. Further, it showed that CAI was both viable and credible. The implementation of this technique made the game more enjoyable and the challenge greater. This served to increase the satisfaction of the players as well as the game's developer and acted as a force multiplier for the effort put in to create a video game.

The success of proving the efficacy of CAI in this simple context leads to the hope that, in future research, this same principle can be applied to security research in the information technology discipline. This research indicated that a layered approach to intrusion detection and prevention could prove to be more secure than one monolithic firewall. This principle will be challenging in that there must be several smaller levels of security to act as probes into the attacker's methods. Once these levels have done their part, gathering data on the attacker's tricks and techniques, they can forward that on to the more secure layers that can be better prepared for the attacker. It has been shown in [1] that attackers tend to have a single favorite technique for attacking, or a finite set of such techniques. It could be possible to create a database of these attacks and defeats and pass that along not just from one layer to the next but from one system to the next. This could provide a valuable resource for researchers and security experts alike and become analogous to the virus databases that serve as the underpinning for antivirus software [2]. We will continue this research with colleagues from IT to verify and expand these ideas and concepts.

5 Future Work

This work will be extended by researching broader application for in game AI, increased attention to the developer's perceptions of the deeper and more involved gameplay, and the applications to the security industry.

References

1. Russ, R., Ed, F., Greg, M., Bryan, C.: Network Security Evaluation Using the NSA IEM. O'Reilly Media (2005)
2. Kirwan/Power: Cybercrime. Cambridge University Press, Cambridge (2013)
3. Edward, M., Brian, E.: Encyclopedia of Quantitative Risk Analysis and Assessment, vol. 1. Wiley Publishers, Chichester (2008)
4. Nobuyoshi, T.: Intelligent Communication Systems: Toward Constructing Human Friendly Communication Environments. Academic Press, Cambridge (2002)

The Adaptive Affective Loop: How AI Agents Can Generate Empathetic Systemic Experiences

Sara Colombo[1](✉), Lucia Rampino[2], and Filippo Zambrelli[2]

[1] Eindhoven University of Technology, Eindhoven, The Netherlands
S.colombo@tue.nl
[2] Politecnico di Milano, Via Candiani 72, 20158 Milan, Italy
lucia.rampino@polimi.it, filippo.zambrelli@mail.polimi.it

Abstract. Affect and emotions play a crucial role in any human experience. However, these components are often overlooked in the design of user experiences with virtual AI agents. In this paper, we investigate the possibility of AI agents - in particular virtual assistants, to adapt to users' emotional states in different interaction scenarios. Currently, AI agents such as Google Home, Alexa, and Siri, support very limited forms of affective interactions, despite showing great potentials for their implementation. This gap reflects the lack of specific theoretical models for affective and empathetic interactions with AI assistants, as well as current technological limitations. In this work, we address the first issues, i.e. the lack of theoretical models, from an experience design perspective. We present the Adaptive Affective Loop, a revised version of the Affective Loop model from social robotics, where we introduce two new concepts. First, the use of distributed interfaces for AI agents, where all IoT elements controlled by the agent can be leveraged to generate affective interactions. Second, the integration of learning and adaptive features, which allow the AI agent to assess the effectiveness of its affective response, and to adapt it over time, in order to generate custom empathetic responses for each user. We apply the model to three interaction scenarios: direct, pre-defined, and indirect interactions with the AI agent. We discuss benefits and limits of our model, and we address the challenges designers will face while envisioning experiences for such new affective scenarios.

Keywords: AI agents · Virtual assistants · Adaptive affective loop · Empathetic interaction · Distributed interface · Experience design

1 Introduction

Affect is a fundamental aspect of human intelligence. It is widely recognized that emotions play a crucial role in our decision-making processes [1]. Furthermore, affective states like anger, anxiety, and joy set the tone to our days and greatly influence our social interactions. Although emotions are essential elements of our relationships and experiences, they are rarely addressed in our interactions with the AI agents and virtual assistants that surround us. Such systems are mainly conversational, and the design effort so far has been focused on the creation of cognitive interactions aimed at retrieving

K. Arai (Ed.): FICC 2021, AISC 1363, pp. 547–559, 2021.
https://doi.org/10.1007/978-3-030-73100-7_39

information or executing commands, rather than at generating fulfilling user experiences [2, 3].

The importance of considering emotional aspects in the interaction between humans and intelligent systems is becoming more and more evident to all disciplines involved in the design of AI agents: "While A/IS [Autonomous/Intelligent Systems] have tremendous potential to effect positive change, there is also potential that artifacts used in society could cause harm either by amplifying, altering, or even dampening human emotional experience." [4, p. 90].

In the past decades, these considerations have brought to the development of research fields such as social robotics and affective computing, which focus on affective interactions between humans and robots, or machines in general [5, 6]. A number of prototypes and commercial examples of affective and social robots have emerged as outcomes of research in these fields. Some examples are described in Sect. 2.1.

Nevertheless, although both theoretical models and tangible experiments of human-machine affective interactions have been developed in the field of affective computing, the way we currently interact with virtual AI agents, or assistants, is still far from any truly affective or empathetic experience. This gap is due to both the lack of theoretical models specifically focused on this kind of virtual agents, and the technological limits of such systems, which are still struggling with making the interaction efficient and completely satisfying from a user experience perspective [3]. One attempt in this direction is Amazon's release of new voice styles for Alexa - excited and disappointed – that can be used by developers of Alexa Skills. Despite being very useful tools, such speaking styles represent just a first step towards the design of comprehensive and consistent affective experiences.

The use of human-like features (e.g. voice tone, facial expressions, body language) is often seen as the only way to create empathy between humans and artificial agents. In this view, AI agents need to acquire the status of sentient beings able to both understand user's emotions and express their own feelings in ways that resemble human-human interactions. In addition to raising ethical questions regarding the personification of AI agents, this also poses a limitation to transferring social robotics affective models to the field of virtual assistants. Indeed, virtual AI agents lack the physical presence - often anthropomorphized - of robots, therefore lacking the ability to use their bodily or facial expressions to convey emotions.

In this paper, we argue that AI virtual agents can generate empathetic experiences even when they lack the physical presence of robots, and beyond the use of human-like voices. We present a model that shows how an empathetic interaction between humans and AI agents can unfold in different scenarios thanks to a systemic approach, where IoT elements connected to the AI agents are used as an extension of their own interfaces. Finally, we present the possibility for AI agents to become truly empathetic by learning to adapt their affective reactions to users' individual emotional profiles over time, therefore creating personalized experiences for each user.

2 Related Work

2.1 Social Robots

How to create affective interactions between machines and humans is an issue that research in affective computing and social robotics has extensively investigated. Commercial outcomes of these studies include robots such as Jibo [7]. Its facial and voice recognition technology allows it to recall up to 16 different users, so to create personalized experiences in each interaction. Another example is Moxie [8], a child development-focused robot with machine learning technology that allows it to perceive and respond to natural conversation and other behaviors as well as to create individualized experiences. Perhaps the most advanced example of commercial affective robots is Pepper, a social humanoid robot able to recognize not only faces but also basic human emotions [9]. Pepper is optimized for human interaction and can adapt its behavior to the mood of the human beings around it, thus creating personalized experiences based on the affective states of the users.

In all these three examples, the robots use anthropomorphic features such as eyes, mouths, or simplified bodies, to mimic or express human emotions.

2.2 Personalized Affective Experiences

From a theoretical standpoint, Egger et al. [10] demonstrated how human-machine interactions follow both social and natural principles based on human-human interactions. According to these authors, in order for an intelligent system to perform a correct affective computing process, it should follow three steps: recognizing the user's emotional state, generating an emotional behavior profile, and finally eliciting emotions in the user.

In the first step, the user's emotional state needs to be detected and classified. Emotion recognition can be performed through a multimodal analysis that involves detecting visual, auditory, and physiological signals [10] and subsequently classifying such signals to identify the affective state of the user [11]. In order to classify emotions correctly, machine learning algorithms are trained on signals collected through different sensors and technologies [10, 12, 13]. In the second step, the emotion is evaluated, and an emotional profile for the intelligent system is generated, in response to the emotion identified in the user [10, 14]. In the third step, the focus is on expressing such emotional profile, in order to elicit one or more emotions in the user. Since each user has a highly personal and non-generalizable emotional reaction, it is essential to understand how the intelligent system can generate an emotional reaction suitable for each individual user. In trying to answer this question, recent studies have focused on creating embodied conversational agents (ECA) and affective robots [15].

In literature, there is a lack of theoretical studies connecting intelligent agents' empathic responses to the user's affective states they will more likely trigger. For this reason, so far the field of social robotics has proceeded through empirical analysis in which the robot regulates its behavior based on individual user's preferences. In a recent study, for instance, it emerged that the presence of adaptive empathic behaviors in a robot led children to perceive the robot as a social support to an extent similar to how they perceive their peers [16]. In this specific study, the robot was able to use a 'no-regret'

Reinforcement Learning (RL) algorithm to learn the supportive behaviors that were most effective for each child (i.e. the behaviors that were most likely to increase the positive affective state in the child). The reward function of the RL algorithm was based on the child's non-verbal responses, captured a few seconds after the robot employed a certain support strategy.

In general, as discussed by Magyar et al. [17], robots can create a personalized affective experience by generating adaptive interactions. In adaptive interactions, the system can assess the effectiveness of its emotional response at the end of each affective cycle, by detecting the new emotional state of the user.

2.3 The Affective Loop

In 2008, Höök introduced the concept of Affective Loop [18] as a basis to design embodied affective experiences. The Affective Loop consists of an interactive process in which the artificial system detects the users' emotions expressed, for instance, through body language and movements and responds by generating affective expressions through its own physical and sensory features. These embodied affective expressions engage the user, making them feel more and more involved with the system. Therefore, an affective loop experience is triggered by systems that can both influence and be influenced by users. Such systems leverage the physical and cultural features of our embodied experiences, creating events that have the power to deeply affect us. Besides augmenting engagement during human-robot interactions, the Affective Loop can be adopted to create a personalized affective experience [15]. According to Paiva et al. [14], to establish such an Affective Loop with users, robots require an affect detection system that understands whether the user is experiencing positive or negative sentiments, and an action selection mechanism that chooses the preferable emotional response to display. As explained by Carberry et al. [19], the robot should consider the context in which the emotion was recognized, thus trying to infer the reasons for the user's emotional state. Then, the way in which the robot shows its affective response should be easily perceivable by the user.

In brief, it emerges that the main feature of the Affective Loop as defined by Höök is its being based on physical and embodied interactions, both on the user's and system's sides.

3 Affective Experiences with AI Agents

Differently from social robots, virtual AI agents and assistants do not have an artificial body or facial features that facilitate the generation of empathetic interactions through movements and expressions. However, adding an affective component to the interaction with AI agents that goes beyond conversational elements may greatly improve the user experience [20].

We started investigating the question of how to elicit such an affective experience and the feeling of empathy in the interaction with body-less AI agents, from an experience design perspective. To this end, we defined the Adaptive Affective Loop for AI agents, a revised and extended version of the Affective Loop model proposed by Höök and visualized by Paiva, based on the concept of AI agent's distributed interfaces. In our model, we also integrate empathetic adaptation and personalization through continuous learning, as a possible future step in the generation of truly empathetic virtual AI agents.

3.1 Empathetic Interactions

In the context of this paper, we focus on the generation of empathetic interactions between humans and virtual AI agents. We define empathetic interactions as interactions having two possible aims: i) mimicking the emotion expressed by the user, to make them feel understood, or to reinforce positive feelings (e.g. if the user is happy or excited, the system can reinforce such emotions by mimicking them) [17]; ii) eliciting new desired affective states, for instance relaxing the user if they are stressed, or energizing them if they are bored. This can be related to what [17] define as a "social supportive behavior", i.e. a behavior performed by a robot in order to motivate the user to achieve specific goals. In our case, the goal is shifting the emotional state of the user from a negative - or unwanted, to a positive – or desired, one.

4 A Model of Adaptive Affective Loop for AI Agents

Our model of Adaptive Affective Loop is based on the concept of *distributed interface.* We introduce this concept in the next section, followed by the description of three scenarios of human-agent interactions, which we will use as a basis to explain our model.

4.1 AI Agents' Distributed Interfaces

AI agents such as Google Home, Alexa, and Siri have little to almost no physical presence. However, they can be connected to home appliances (e.g. ambient lights) or devices (TV, smartphones, music streaming systems), which generate an interconnected system of physical elements through which the AI agent can ideally manifest its presence to the person. In our view, this physical IoT system becomes the AI agent's distributed interface, through which empathy can be conveyed. All elements connected to the AI agent can proactively be controlled by the agent itself, in order to generate an affective, empathetic interaction.

In this light, the AI agent can activate empathetic responses to the user by acting on:

- its own interface and physical/sensory appearance (voice tone or volume, lights, sounds emitted by the interface, and any combination of these).
- connected ambient elements and home appliances (e.g. ambient lights, music speakers, etc.). These elements are characterized by modifying the sensory quality of the environment.
- connected digital devices and content platforms (e.g. TV, smartphones, game consoles). These platforms are characterized by providing access to multiple contents.

The sensory appearance of the AI agent's physical interface and the connected ambient elements (e.g. audio systems and lights) can be used to either mimic the user's emotions or to modify their emotional states.

Content platforms do not have specific sensory qualities that might affect the emotional state of the user. However, the interaction with these systems can be made more empathetic by i) suggesting specific contents that adapt to the user's emotional state, ii) adjusting the responsiveness of the digital device or the platform interface (e.g. the interface colors and sounds, the amount of contents displayed, haptic feedbacks, reaction speed, etc.).

The concept of distributed interface can be adopted also in the first step of the Affective Loop, i.e. emotion recognition. Indeed, sensors required for emotion recognition (e.g. microphones, cameras, heart rate sensors, etc.) are already present in several domestic environments and IoT devices and could be then employed to collect data about the user's emotional state through a systemic approach.

4.2 Three Interaction Scenarios with AI Agents

In our model, an empathetic experience can take place in three different human-agent interaction scenarios: direct, pre-defined, and indirect interactions.

Direct Interaction. The user initiates an interaction with the system to perform a task or just to start a conversation. The task can involve either a core function of the AI agent (e.g. retrieving information, setting a timer), or a function of any of the elements connected to it (e.g. turning on the light in the kitchen, or the TV).

Pre-defined Interaction. The AI agent performs a task according to a pre-defined routine (e.g. turn the music on at 7 pm) or a previous user's input (e.g. appointment reminder). Pre-defined tasks can involve either the AI agent's core functions or any element controlled by the agent.

Indirect Interaction. The user interacts with a connected digital device or platform to access contents (e.g. TV shows, news section on the smartphone). The AI agent does not directly take part in the interaction, but it has the ability to filter or select the displayed contents, and to transform the interface of connected devices.

For each scenario, we show how the Affective Loop can take place through the AI agent's distributed interface, where both the AI agent's own interface and the connected elements are used by the agent to convey an affective, empathetic response.

4.3 Three Models of Affective Loops for AI Agents

Affective Loop in Direct Interactions. When the user directly interacts with the AI agent, the agent can collect information on the users' emotional state by embedded sensors (e.g. microphone, camera), as well as by processing data coming from external devices, such as wearables, or smartphones (Fig. 1). For instance, the agent can directly detect the user's facial expressions, voice, body gestures, and movements during the interaction. If a touch sensitive interface is embedded, the system may be able to analyze touch gestures and applied pressure while interacting. Wearable devices can detect other physiological data such as electrodermal activity and heart rate. Moreover, they can collect useful data on the quality of sleep or physical activity for that day.

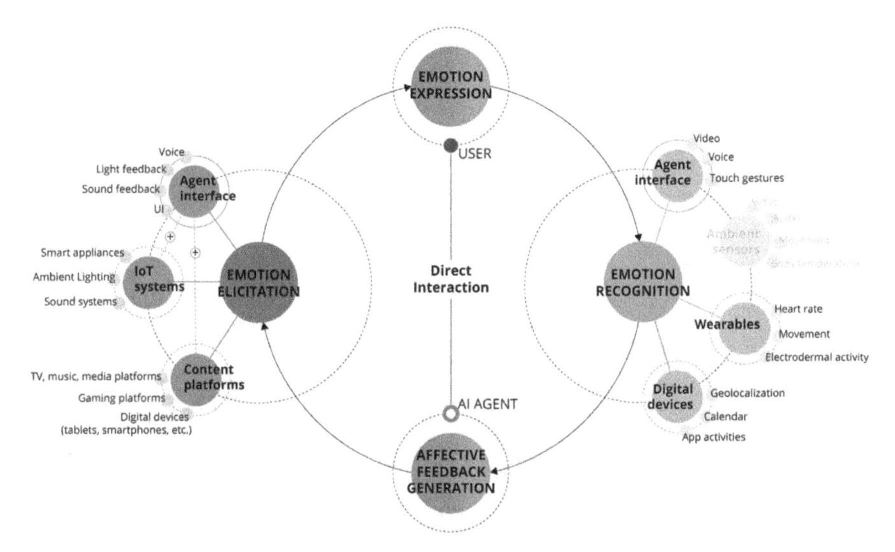

Fig. 1. Affective loop in direct interactions.

In this case, the AI agent's feedback is adjusted to the user's emotions through, for instance, the voice tone used to notify that it received the message or performed the required task. If the user initiates a conversation with the AI agent to add an entry to its calendar, and the AI agent detects stress in their voice, it will be able to modulate its response accordingly: the agent's voice tone will be softer, and the lights used as feedback will have a calmer pace. In case the user also asks to play music, the agent will select relaxing playlists and it will start it at a lower volume than the pre-set one, still interacting through a calm and soft voice.

Affective Loop in Pre-defined Interactions. In this second case, there is no direct interaction with the AI agent, therefore the user's emotions are collected through other channels than embedded microphones or cameras. In this case, the AI agent may detect and assess the user's emotional state in advance, through data coming from ambient sensors, wearables, and other digital devices (Fig. 2). For instance, it can infer the user's stress level by merging ambient cameras data, heart rate analysis, and information coming from the user's digital agenda.

Routines (e.g. alarms, lights turning on/off) or previously defined inputs (e.g. reminders and notifications) can either adapt to the user's emotional state, or try to elicit a different affective response, based on pre-defined goals. For example, if the AI agent detects that the user is bored, but they have an upcoming work appointment, the pre-set reminder can be changed into an 'activating' one. In a similar way, depending on the ecosystem elements that the routine engages, the AI agent might select an energizing light hue, or a more dynamic ambient music.

Affective Loop in Indirect Interactions. In this third case, there is neither direct interaction with the AI agent, nor previously programmed actions performed by the agent. However, by interacting with devices or content platforms connected to the AI agent,

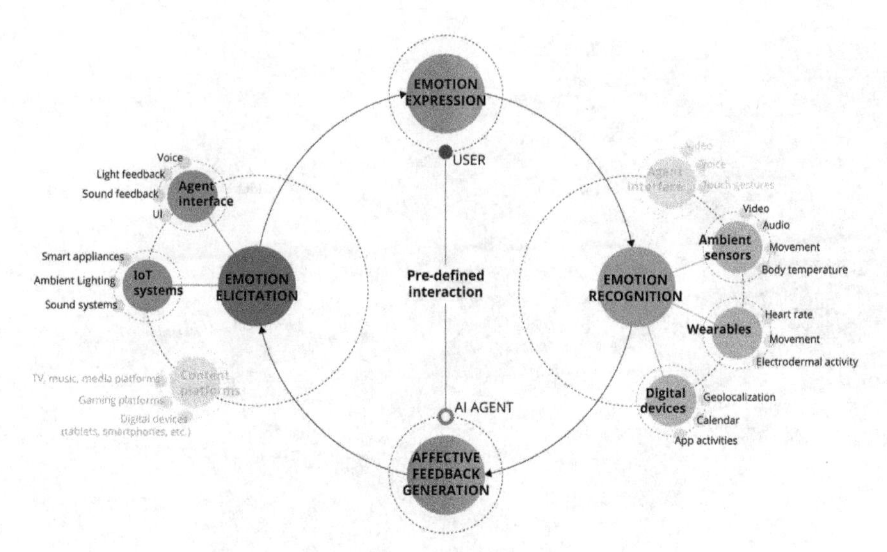

Fig. 2. Affective loop in pre-defined interactions.

the user can still trigger an affective loop. Indeed, the agent may select the contents displayed by digital devices or platforms (e.g. movies streaming, news platforms, etc.), or affect the way such devices react to user's inputs, according to the user's emotional state. In this scenario, the affective state is identified through data coming from ambient sensors, wearables, and other connected digital devices (Fig. 3).

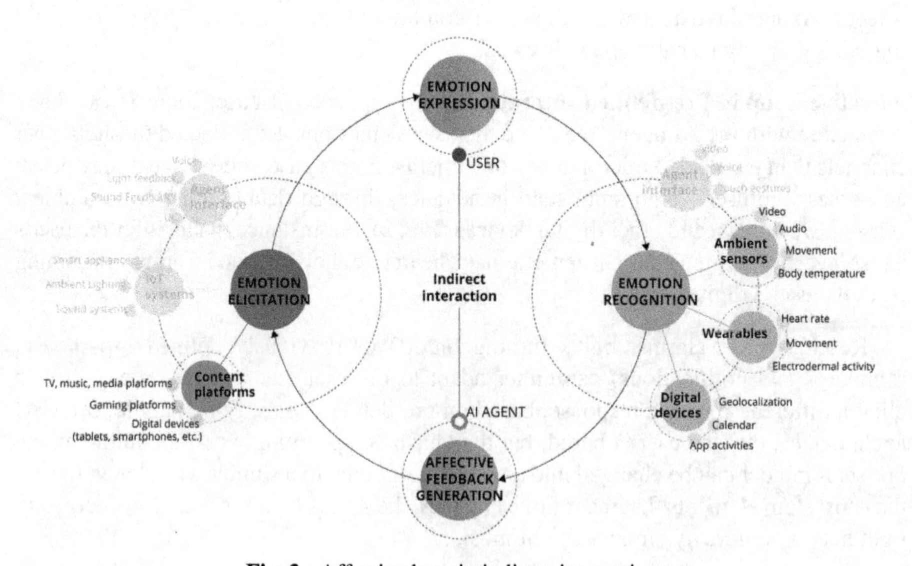

Fig. 3. Affective loop in indirect interactions.

For example, if the user decides to watch TV and accesses their streaming platform, the AI agent might suggest the most suitable contents based on their current emotional state. A similar case is the interaction with a news or music platform. In all these cases, the AI agent can display and suggest contents based on the emotional state of the user. Ideally, the system might also modify the interface elements of the connected devices or platforms, so that they respond through colors, sounds and animations to the user's affective state.

In our vision, the goal of the affective interaction with AI agents shifts from creating empathy between the person and the agent itself (as it happens with social robots) to creating empathetic experiences for the user *through* AI agents. Such experiences affect all the elements of the agent-controlled IoT system, as empathy is achieved through the interaction with different types of appliances, platforms, and interfaces.

In addition to the systemic approach we integrated in the original Affective Loop, our model can also be extended in another direction. It can indeed become iterative, or recursive, meaning that the system can assess the emotional reaction elicited in the user at the end of one loop, and adapt its subsequent interaction in turn. This idea is similar to the one presented above for the creation of personalized affective experiences with social robots. In that case, the affective loop was used to create engaging interactions where the robot keeps adapting to the user response at the end of each loop, starting a new one until the desired affective state or behavior is reached in the user. However, we believe this iterative approach can be stretched even further, to achieve a truly personalized empathetic interaction, through an Adaptive Affective Loop.

We envision agents that can learn over time what affective stimuli are effective in creating empathy or in triggering specific emotional responses in each user. Indeed, every person can react differently to the same stimuli conveyed through the AI agent's distributed interface. Being able to assess if the stimuli elicited the intended reaction or not is essential to create an AI agent that can learn over time how to best respond to the emotional state of a specific user. This way, the affective response is not predefined, but it can be generated by the AI agent by mixing a number of possible stimuli, and by learning what works best in a specific situation.

From a technological standpoint, machine learning allows the generation of such intelligent and adaptive systems. By performing repeated cycles of affective loops with an individual, the agent may be able to predict the best way to achieve its affective goal. To give an example, we can assume that the AI agent used a softer notification sound when the user was stressed. Did it really lower their stress level? If yes, the agent will consider that stimulus as effective for the user. If not, next time the system will not use that stimulus anymore, but it will try a different sound. The system can also leverage the knowledge built through many interactions with other users. Such knowledge can be leveraged by the AI agent to predict what stimuli might better work in the future, because they were effective in the interaction with other people. By merging the personal data with the ones generated by interactions with a huge number of other users, the system can more quickly go through the trial and error process and acquire data useful to create personalized empathetic interactions.

5 Discussion

5.1 Benefits

Our Adaptive Affective Loop extends the idea of the AI agent's interface, and the ways the agent is able to generate affective interactions with the user. Such repeated affective interactions ultimately lead to the creation of custom emotional experiences. In our model, empathy is not necessarily generated by human-like features, but it can be achieved beyond anthropomorphism or personification.

Our model suggests an alternative approach to the current one based on human-like interactions, where agents are personified. Indeed, what creates empathy is not the mimicking of human language, but the sensitiveness of an artificial system, which adapts to the user's emotions to generate the experience the user needs in a specific moment. This helps to avoid the problem of seeing the AI agent as a human companion, and the ethical issues connected to it. Indeed, in the interaction with existing vocal assistant, ethical concerns arise for instance when technology imitates human-human conversation – the human voice in particular – to such an extent that users are driven to wonder whether they are interacting with a real person or a machine [5].

5.2 Limits

Along with the foreseen benefits, the model we present in this paper shows a number of limits concerning both user experience and technical aspects.

Data Privacy. Collecting data on users, especially related to their emotional state, through multiple sensors can generate the feeling of "being monitored". Moreover, the concern connected to the interaction with existing vocal assistant, for their being 'always-on' and 'always-listening' would remain a problem. This issue should be addressed in order to create trust between the person and the agent, without which real empathy cannot be achieved. Moreover, integrating machine learning systems that can learn and generate a personalized experience over time would require a great amount of data to be collected and analyzed. This could possibly generate a feeling of distrust, as much data would need to be collected, stored, and analyzed remotely. Anonymity would be essential, and the use of such data by companies should be clearly defined and made transparent.

User Experience. In our model, the agent is envisioned to learn the best way to react to the user's emotional state over time. During the learning process, however, the user might not experience a proper and personalized affective response. Moreover, the user would need to set a number of goals for the interactions – e.g. when they would prefer the agent to mimic and adapt to their emotion, or to change their current feelings by eliciting a different emotional reaction (e.g. making them feel excited if they are bored). Therefore, even though the final goal of our model is to significantly enhance the user experience, during the initial phase of the interaction, mismatches and misunderstandings may likely occur.

Technical Limits. Our theoretical model is based on the potentialities of technologies that are currently in development and are not reliable enough to guarantee proper emotional recognition. This represents the main limitation of the model. Moreover, making connected IoT elements flexible enough to interface with and be controlled by a single AI agent, adapting their contents and UI to the user's state, might require a significant effort.

5.3 The Designer's Role

In the model we presented, the designer's role will still be a central one, but designers will need to focus on elements that differ from the ones they are used to design in AI agents. The mindset and approaches adopted to design empathetic experiences will have to change, for two main reasons. The first one is connected to the fact that designers will have to shift from designing the agent's personality, conversational contents, and physical interface, to designing systems of interfaces, where the agent can manifest itself through a number of connected IoT elements. The design of the affective, empathetic response will have to assume a systemic nature.

The second reason concerns the autonomy of the intelligent agent. Indeed, the AI agent will have the ability to modify its affective responses according to what it learns about the user's individual emotional profile over time, i.e. what set of stimuli work in specific circumstances to generate empathy or to elicit the intended emotion. Instead of designing pre-defined feedback in response to users' interactions with the agent, designers will have to determine a set of elements that can be dynamically adapted and autonomously mixed by the agent. This will be necessary to adapt the experience and the empathetic response to each individual user. The AI agent will have the ability, for instance, to adjust a sound or a light feedback to the specific sensitivity of the user, by experimenting with subtle changes on a range of variables, until the intended emotional response is achieved.

Despite the agent's autonomy, consistency in the language of these elements, and in the way they can change over time (speed, rhythm, color range, etc.) will need to be carefully designed. Designers will have to give the agent the freedom to adapt and merge different elements of its language – and the languages of connected elements - in order to respond to a detected emotion, but they will also have to set some boundaries or rules, to guarantee quality in the user experience, e.g. avoiding a mix of stimuli that might generate a negative or unpleasant experience from a sensory or aesthetic viewpoint.

Designers will also still have to design ideal scenarios that the AI agent can use as starting points for its learning process. They will have to ideate sets of affective responses to each detected emotion that will be initially used by the AI agent, before they are adjusted to each specific user over time.

6 Conclusions

We presented the Adaptive Affective Loop, a theoretical model for the generation of empathetic interactions between AI agents and humans. Despite its limits, mainly related to technology and privacy, our model intends to propose a new design perspective. We started from the existing Affective Loop for social robots and we adapted it to body-less AI agents, by extending their interfaces to a distributed and systemic level. We also introduced the idea of personalized empathetic interactions, based on the possibility to create a set of custom affective responses that can effectively adapt to the user's emotions, or elicit new intended affective states.

Our goal is to provide a model focused on human-agent interactions in real-life scenarios, in order to steer technological development towards the creation of meaningful experiences with AI agents.

Although implementing such a model in reality would require emotion recognition techniques to reach a higher level of reliability, we believe having a reference model for the design of personal empathetic experiences with AI agents will be useful to guide future work in this field.

References

1. Damasio, A.R.: Descartes' Error. Random House (2006)
2. Aylett, M.P., Clark, L., Cowan, B.R.: Siri, echo and performance: you have to suffer darling. In Conference on Human Factors in Computing Systems – Proceedings (2019)
3. Velkovska, J., Zouinar, M.: The illusion of natural conversation: interacting with smart assistants in home settings. In: Proceedings of the 2018 CHI Conference on Human Factors in Computing Systems (2018)
4. IEEE: Advancing Technology for Humanity. Ethically Aligned Design. A Vision for Prioritizing Human Well-being with Autonomous and Intelligent Systems, 5th edn. (2019)
5. Picard, R.W.: Affective Computing. MIT press, Cambridge (2000)
6. Tao, J., Tan, T.: Affective computing: a review. In: International Conference on Affective computing and intelligent interaction, pp. 981–995. Springer, Berlin (2005)
7. Breazeal, C., Faridi, F.: U.S. Patent Application No. 29/491,780 (2016)
8. Hurst, N., Clabaugh, C., Baynes, R., Cohn, J., Mitroff, D., Scherer, S.: Social and Emotional Skills Training with Embodied Moxie. arXiv:2004.12962 (2020)
9. Pandey, A.K., Gelin, R.: A mass-produced sociable humanoid robot: Pepper: The first machine of its kind. IEEE Robot. Autom. Mag. **25**(3), 40–48 (2018)
10. Egger, M., Ley, M., Hanke, S.: Emotion Recognition from Physiological Signal Analysis: A Review, vol. 343, pp. 35–55. AIT Austrian Institute of Technology GmbH, Vienna, Austria (2019). https://doi.org/10.1016/j.entcs.2019.04.009
11. Den Uyl, M.J., Van Kuilenburg, H.: The FaceReader: online facial expression recognition. In: Noldus, L.P.J.J., Grieco, F., Loijens, L.W.S., Zimmerman, P.H. (eds.) Proceedings of Measuring Behavior 2005, pp. 589–590. Wageningen, 30 August – 2 September 2005
12. Lang, P.J., Bradley, M.M., Cuthbert, B.N.: International affective picture system: affective ratings of pictures and instruction manual. Technical Report A-8 University of Florida, Gainesville (2008)
13. Silva, C., Ferreira, A.C., Soares, I., Esteves, F.: Emotions under the skin autonomic reactivity to emotional pictures in insecure attachment. J. Psychophysiol. **29**, 161–170 (2015)
14. Paiva, A., Leite, I., Ribeiro, T.: Emotion modeling for social robots. The Oxford handbook of affective computing, 296–308. Oxford University Press, USA (2014)
15. Calvo, R., D'Mello, S., Gratch, J., Kappas, A., Arkin, R., Moshkina, L.: The Oxford Handbook of Affective Computing. Oxford University Press, USA (2015)

16. Leite, I., Castellano, G., Pereira, A., Martinho, C., Paiva, A.: Modelling empathic behaviour in a robotic game companion for children: an ethnographic study in a real world settings, In: Proceedings of the HRI, ACM, pp. 367–374 (2012)
17. Vircikova, M., Magyar, G. and Sincak, P.: The affective loop: A tool for autonomous and adaptive emotional human-robot interaction. In: Robot Intelligence Technology and Applications, vol. 3, pp. 247–254. Springer, Cham (2015)
18. Höök, K.: Affective loop experiences–what are they?. In: International Conference on Persuasive Technology, pp. 1–12. Springer, Berlin, Heidelberg (2008)
19. Carberry, S., de Rosis, F.: Introduction to special Issue on 'Affective modeling and adaptation'. User Model. User-Adap. Inter. **18**(1–2), 1–9 (2008)
20. Braun, M., Alt, F.: Affective assistants: a matter of states and traits, In: Conference on Human Factors in Computing Systems - Proceedings (2019)

From What Is Promised to What Is Experienced with Intelligent Bots

Sandrine Prom Tep(✉), Manon Arcand(✉), Lova Rajaobelina(✉), and Line Ricard(✉)

ESG-UQAM, 315 Rue Sainte-Catherine E, Montréal, QC H2X 3X2, Canada
{promtep.sandrine,arcand.manon,rajaobelina.lova,
ricard.line}@uqam.ca

Abstract. Based on survey research, this study, aims to identify the gap between what is promised (usefulness) and what consumers experience (usability and enjoyment) from virtual assistants such as voice- or chat-based bots. Our empirical results compare these three factors applied to AI bots and their adoption for customer service. They were collected with a sample of 148 participants in the financial sector (utilitarian experience), and with an exploratory sample of 112 exposed to a tourism voice bot offering (hedonic experience). Interesting comparisons (similarities and differences) are highlighted to provide managerial user experience (UX) guidance for the design and evaluation of virtual assistants from a user/customer-centric perspective.

Keywords: Chatbot · Voice bot · Virtual assistant · User experience · Tourism · Insurance · FinTech

1 Introduction

Experts predict that by 2025 over 50% of companies will invest in chatbots in preference to mobile apps. Fueled by promises of instant and personalized customer experience, the chatbot market is growing at an annual growth rate of 24% [1]. Service bots represent a new type of interaction through which companies can create value for customers by providing them with additional resources driven by artificial intelligence [2]. However, there are mixed reports regarding the experience when using chatbots: for example, according to [1], over 60% of consumers did not complete their intended task when using chatbots due poor usability resulting in frustration and deception. When gathering consumer feedback from different sectors such as insurance (utilitarian) and tourism (hedonic), it is highly relevant to compare how users experienced each sector offered through chatbot and voice bot service, respectively [3]. In both contexts, the virtual assistants (VA) are service-oriented, but one is more commerce-oriented (insurance) while the other, is more geared toward entertainment [4]. The need for more research on the successfulness of (ro)bots has been demonstrated [5], particularly for private stakeholders since their own financial and human resources can be involved in these new service channels. Also, in tourism most private stakeholders are small to medium-sized enterprises (SMEs), who need to know precisely where to focus, as they have

K. Arai (Ed.): FICC 2021, AISC 1363, pp. 560–565, 2021.
https://doi.org/10.1007/978-3-030-73100-7_40

limited resources available for time-consuming projects. The idea is to ensure a more natural interface and improve efficiency for simple tasks [1], assisting human agents as a tool (see Fig. 1) to enhance the customer experience [6].

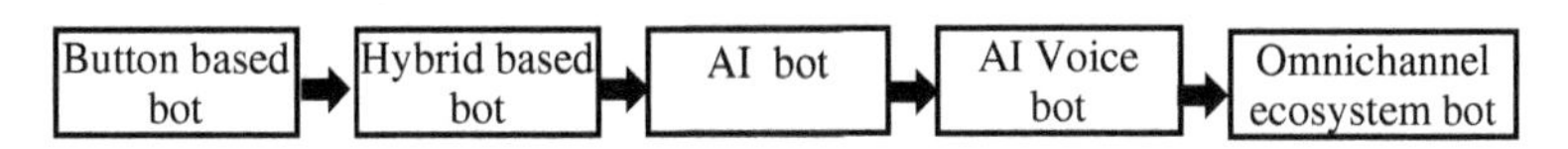

Fig. 1. The complexity continuum of chatbots and smart speakers [7].

2 Research Question and Literature Review

They are many UX challenges associated with virtual assistants (VA). UX is a specific form of human experience that comes from the interaction with a new product, a new service or a new technology [8]. The key dimensions of UX are usefulness (PU), usability (PEOU) and enjoyment (PE) [9]. But most chatbots are not useful or easy to use [1]. The pragmatic (utilitarian) vs hedonic aspects depend on the use context and the user's age category [10].

To address the UX challenges, best practices are crucial in developing a new intelligent product and to help to optimize the user experience based on user-centered design approaches [11] and agile (vs waterfall) development [12]. In that context, it is important to consider consumer behavior in relation to technology, since differences exist between millenials and baby-boomers [9], along with the utilitarian and hedonic use contexts [10].

Hence, our research question is: "How can the adoption of a virtual assistant in tourism be fostered by leveraging the UX dimensions during its development?"

3 Methodology

To answer the research question, a survey was conducted online with 112 domestic tourists from the Montreal metropolitan area in the province of Quebec, Canada. To be eligible, respondents had to be between 18 and 45 years old and own a smartphone. The study took place in the winter of 2020 (Pre COVID-19). Respondents were recruited through social networks (i.e. LinkedIn, Facebook) and in the academic community according to a convenience sampling procedure. Perceptions (PEOU, PU and PE) and usage intentions regarding the VA were measured after viewing an online ad promoting Google's virtual assistant (i.e. A restaurant reservation according to, see Fig. 2). Other questions about their preferences/opinions regarding virtual assistant and tourism-related options in the Montreal area were asked as well as familiarity with VA.

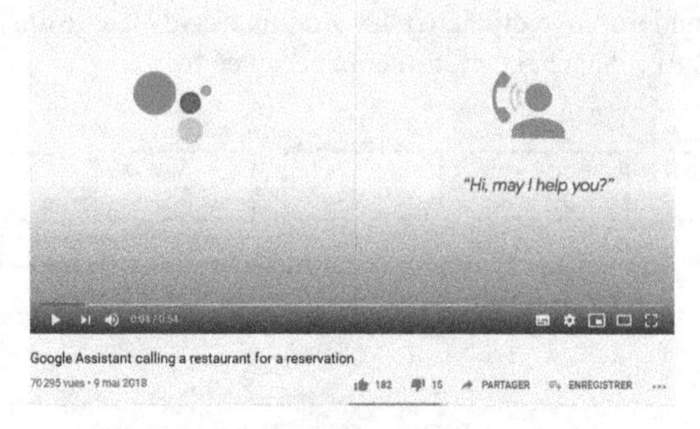

Fig. 2. Google VA demo on Youtube.

Turning to the respondents' profile, a little more than half of the sample was female (58%), and consumers of all ages were represented with 43.8% of respondents aged 18-25, 36.6% between 26 and 35 and 19.6% between 36 and 45. Eighty-nine percent were from the metropolitan Montreal area and 9.8% were international students. Note that 58% of respondents declared no previous experience with a virtual assistant, while 30.4% had previous exposure to one type of VA and only 11.6% were familiar with two or more types of assistants. PEOU, PU and PE scales (3 items each) were adapted from [9] and TAM [13] using 5-point Likert scales. Usage intention was measured by 2 items, whether info is provided by locals or experts using a bipolar scale (Very unlikely and Very likely) in accordance with TAM.

In order to contrast the perceptions and intentions of consumers regarding VA with another activity sector (utilitarian), results were compared with findings of a similar survey conducted regarding a Fintech chatbot using similar sampling criteria (45 and below, mobile device owners) with 148 respondents. However, in this study, respondents assessed PEOU, PU and PE and usage intention after using a realistic prototype of car insurance service chatbot (see Fig. 3). Measures were similar as the ones used in the tourism study though (with 5-point scales). Usage intention was measured when offered by the insurance company whether on Alexa or Google Home.

Fig. 3. Car insurance chatbot.

4 Results

Table 1 shows results for PEOU, PU, PE and usage intention regarding the domestic tourists sample as well as the correlations between constructs. With means close to 4/5, usability (3.99) of the virtual assistant is quite good and usage intentions (3.91) are high. Perceived enjoyment associated with the VA is however average (3.08/5). Interestingly, PU and PE are significantly and positively ($p < 0.01$) associated with usage intentions (adoption), while PEOU is not ($p > 0.10$). Further, we note that usefulness and enjoyment are strongly correlated ($r = 0.73$) in the tourism context.

Table 1. Descriptive results and correlations tourism (*p < 0.01).

Construct	Mean/5	Std-dev	Correlations*		
			Ease of use	Usefulness	Enjoyment
Ease of use	3.99	1.05	–	0.47*	0.39*
Usefulness	3.69	0.92		–	**0.73***
Enjoyment	3.08	1.17			–
Usage intention	**3.91**	**1,02**	**0.15 (ns)**	**0.56***	**0.57***

Further, results show that the preferred type of VA while visiting a location (tourism context) is a voice and text combination (hybrid) for 60.9% of the respondents. When asked what is missing when looking for info on events and attractions in Montreal, 62% mentioned information centralized on a single digital platform, while 53% indicated that it is from "difficult to very difficult" to find information about those events favored by the locals. It is therefore not surprising to find that usage intention of a virtual assistant is a bit higher (76.8%) if information is based on the locals' opinions than if it is based on the experts' opinions (73.2%). Finally, additional analyses (t-tests) showed that familiarity with VA is driving all variables. Therefore, the more exposure to VA for one consumer, the higher his/her PEOU, PU, PE and usage intentions ($p < 0.05$).

By comparison, results regarding the chatbot in the insurance sector (see Table 2) showed that while perceived usability is similar, it is perceived as being a little bit more useful ($\bar{x} = 3.81_{\text{insurance}}$ vs 3.69_{tourism}) but less fun ($\bar{x} = 2.34_{\text{insurance}}$ vs 3.08_{tourism}). Further, usage intention - if offered on the insurance website - is below average (2.81/5), well below intention in the tourism context (3.91/5).

Opinions are also more divided on this question in the insurance ($\sigma = 1.52$). According to correlation tests, results show that PEOU, PU and PE are all significantly associated with usage intention. In this context, PU is strongly associated with usability ($r = 0.87$), and less correlated with PE than in the tourism context. Interestingly, if the chatbot was to be offered on Alexa or Google, usage intentions drop sharply to 1.17/5 and only PE is then associated with adoption intention ($r = 0.30$). Further, familiarity with VA was associated with usage intention ($p < 0.05$) but not with PEOU, PU and PE in the insurance study (all $p > 0.10$), which are next explained through methodological considerations.

Table 2. Comparative results: car insurance (*p < 0.01).

Construct	Mean/5	Std-dev	Correlations*		
			Ease of use	Usefulness	Enjoyment
Ease of use	4.02	0.96	–	**0.87***	0.49*
Usefulness	3.81	0.90		–	**0.53***
Enjoyment	**2.34**	1.22			–
Usage intention (insurance website)	**2.81**	**1,52**	**0.55***	**0.60***	**0.56***
Usage intention (Alexa/Google)	**1.17**	**0.84**	**0.12 (ns)**	**0.15 (ns)**	**0.30***

5 Discussion and Managerial Implications

The results of this comparison allowed us to assess the two distinct bot performances in each industry. These basic standard variables for technology adoption are in accordance with human-centered design, and the literature proved TAM to be a good indicator of an interactive system/service scoring well with users, based on their user experience [8, 14]. Further, this comparative analysis showed that PU is associated with PEOU in the insurance sector but with PE in tourism. This might be because the task involved in the insurance sector is goal-oriented or utilitarian while in the tourism sector it is hedonic. PE is also the least important factor among the three for users in both sectors, whereas the importance of PEOU and PU differ between sectors. This difference might be explained by the experimental procedure (testing a real chatbot vs. watching a promotional demo for V.A.). Familiarity with virtual assistants has a positive effect on usage intention in both sectors, consistent with the results from study [15]. They showed that individuals' differences in their level of familiarity with robotic systems play a crucial role in the adoption of robot-advisors.

We believe these results are useful, especially in the Covid-19 crisis context, in which VA can play a role helping with local tourism. Our results show that VA is a tool with great potential for tourism-related information and to help discover local favorites (high interest area). The findings of the study provide managers with insights for evaluating whether they should be moving on with designing a full (ro)bot service or not for their existing offering, based on prototype user evaluation. Managers can also use this approach to audit an existing bot design offering, in order to improve its user experience, and better understand which factor to focus on to improve their ongoing user research process depending on the industry and the type of service context they offer (hedonic vs utilitarian). Applied to the tourism industry, some key practical implications could be offered. In the light of our results, it is relevant to focus on usefulness and enjoyment to foster VA adoption by users < 45 years old. Also, since our results indicate that familiarity is significantly associated with PEOU, PU, PE and usage intentions, it would be appropriate to target the "early adopters" segment who might have experience with virtual assistants. Finally, various bot project stakeholders can use these findings to plan the implementation of a future VA service offering, insisting on a large UX budget based on the study results - knowing that "tech novelty" is never a guarantee of UX success, and that adoption and repeated use can only derive from a user-centric approach.

6 Limits and Future Research

Research works are bound to have limitations. A small exploratory-size sample was used for the tourism voice bot evaluation, in comparison to the one used for the insurance chatbot. Both samples are restricted to Quebec (Canada), where VA familiarity is poor in comparison with the Japanese context for instance [16]. The sample for domestic tourism targets only respondents 45 years old or less. These limits prevent generalization of the results to the Canadian population. A post-COVID future research, taking into consideration the social distance constraints and targeting international tourists would be interesting to conduct and to compare the implementation challenges of VA in the insurance sector again.

References

1. Ask, J., Facemire, M., Hogan, A.: The State of Chatbots. Forrester Research. [Report] (2016)
2. Riikkinen, M., Saarijärvi, H., Sarlin, P., Lähteenmäki, I.: Using artificial intelligence to create value in insurance. Int. J. Bank Mark. **36**(6), 1145–1168 (2018)
3. De Lacoste, A.: The Social Client Tout comprendre sur les chatbots. Livre blanc (2017)
4. Hassenzahl, M.: Funology: from usability to enjoyment. In: Blyth, M.A., et al. (eds.) Springer, Netherlands (2004)
5. Bordeau, J.: Comment le commerce vocal révolutionne le langage? Chroniques d'experts marketing. Harvard Business Review France (2019)
6. Automation Edge. https://automationedge.com/ai-chatbot-transforming-customer-care-industry/. Accessed 20 May 2020
7. Medium. https://medium.com/voiceui/five-different-types-of-chatbot-17bb255b23b4. Accessed 20 May 2020
8. Lallemand, C., Gronier, G., Dugué, M.: Méthodes de design UX: 30 méthodes fondamentales pour concevoir des expériences optimales, 2nd edn. Eyrolles, France (2018)
9. Bruner, G.C., Kumar, A.: Explaining consumer acceptance of handheld Internet devices. J. Bus. Res. **58**(5), 553–558 (2005)
10. Følstad, A., Brandtzaeg, P.B.: Users' experiences with chatbots: findings from a questionnaire study. Qual. User Exp. **5**(3) (2020)
11. Norman, D.A., Draper, S.W.: User Centered System Design; New Perspectives on Human-Computer Interaction. L. Erlbaum Associates Inc., USA (1986)
12. Study.com. https://study.com/academy/lesson/agile-vs-waterfall-project-management.html. Accessed 20 May 2020
13. Davis, F.: Perceived usefulness, perceived ease of use, and user acceptance of information technology. MIS Q. **13**(3), 319–340 (1989)
14. Bevan, N., Carter, J., Harker S.: ISO 9241-11 Revised: What Have We Learnt About Usability Since 1998? In: Kurosu, M. (eds.) Human-Computer Interaction: Design and Evaluation 2015, Lecture Notes in Computer Science, vol. 9169, pp. 143–151. Springer, Cham (2015)
15. Belanche, D., Casaló, L.V., Flavián, C.: Artificial Intelligence in FinTech: understanding robo-advisors adoption among customers. Ind. Manage. Data Syst. **119**(7), 1411–1430 (2019)
16. New York Times. https://www.nytimes.com/2019/12/31/world/asia/japan-robots-automation.html. Accessed 15 May 2020

Cognitive Advantage

Richard J. Carter[1,2(✉)]

[1] University of Bristol, Bristol, UK
rich.carter@bristol.ac.uk
[2] Mayhill Management Consultancy Ltd., Longhope, UK

Abstract. We are moving into a new paradigm of hyper-accelerated decision making, where artificial intelligence and machine learning facilitate automated action at a pace and scale that human minds alone cannot keep up. We need to respond to this challenge by combining artificial and human intellect to build a cognitive advantage, which is the demonstrable superiority gained through comprehending and acting to shape a competitive environment at such a pace that an adversary's ability to comprehend and act is compromised. In the longer term, cognitive advantage provides the means to cultivate the broader sustainability of the environment on which the organization depends. This white paper provides context for this trend, summarizes relevant scientific understanding and codifies the distinguishing features of the type of organization that will succeed and lead in this new competitive landscape – the cognitive enterprise.

Keywords: Artificial intelligence · Decision-making · Strategy

1 Introduction

1.1 The Shifting Sands of Competition

There are many examples of AI out-competing human intellect: beating world champion chessmasters [1] and Go players [2], detecting breast cancer [3], through to autonomous cyber warfare [4, 5]. The quality exhibited by AI that allows it to out-compete humans are the scale and speed of its data processing capacity. Testing vast numbers of alternative hypotheses – primarily in synthetic digital environments – can yield successful strategies that would simply be too expensive, time-consuming or dangerous for humans experimenting in the real world to discover.

In the same way that it would be unthinkable now to run an organization without computers – whereas 20 years ago that would still have been possible – in another decade it will be unthinkable for an organization to function without an army of AI that enhances human capabilities.

The penetration of AI across industry and government is already significant and contributed to a global digital economy that was valued at $11 trillion in 2018, fueled by the generation of 33 zettabytes of data (the equivalent of 660 billion Blu-ray discs) [6]. By 2025, with the expansion of the Internet of Things, enabled by adoption of 5G and cloud technologies, the digital economy is forecast to reach a value of $23 trillion

K. Arai (Ed.): FICC 2021, AISC 1363, pp. 566–576, 2021.
https://doi.org/10.1007/978-3-030-73100-7_41

– with AI as a major driver of growth [7] - and to be generating nearly 175 zettabytes of data annually [8].

Within this context 'data is the new oil' [9] holds some truth as the driver of the digital economy. Insights from data can provide an understanding of what is happening with customers, infrastructure and competitors; but it takes work to extract such value, and simple understanding is not enough. Businesses and governments also need to apply data-driven insight to inform action at sufficient pace to generate precise, high impact effects in their operating environment. AI has a significant role in augmenting and automating such insight and decision-making.

AI is already taking us from the existing paradigm – of seeking an information advantage as the basis of competing successfully – to one of hyper-accelerated decision-making. In a world where everyone is investing in AI to make better and quicker decisions, how is it possible to compete?

I advocate that the main driver for success in this new paradigm is the ability to shape the environment that is generating the data flowing within the digital economy. The focus therefore needs to be on building and maintaining operational capabilities that enable deep, insightful action.

A cognitive advantage is the new basis for power in an AI-rich world and I define it as:

the demonstrable superiority gained through comprehending and acting to shape a competitive environment at a pace and with an ingenuity that an adversary's ability to comprehend and act is compromised

Consequently, the competitor that has an insufficient understanding of their environment will make poor decisions that are little better than random chance. As time goes by, and without drastic intervention, this will lead to further regression of their ability to act with impact.

There is a duality to holding a cognitive advantage: the power to compete and win can also be used to cultivate a sustainable operating environment for the organization.

This latter point can be described as acting with sagacity. Sometimes this can lead to action that appears counterintuitive, for example, where an organization is both competing against but also cooperating with another organization (see 'Samsung vs Apple' [10]).

Both of these aspects of cognitive advantage will now be described.

1.2 Out-Pacing and Out-Innovating Adversaries

Cognitive advantage is about understanding the operating environment to enable rapid exploration of options, and to then take precise and deliberate action to successfully shape the environment in the organization's favor; and to do so at pace. Cognitive advantage builds on the concept of information advantage by explicitly including elements of decision-making such as comprehension, prediction, goal setting, evaluation of alternatives and action. It is an approach that requires the effective fusion of artificial and human intellects.

Example 1. Amazon Web Services (AWS) is the market leader in the cloud service sector with a 33% share [11]. The platform was launched in 2002 and AWS have continued to build rapidly on their first mover advantage to become the dominant solution in the cloud services market; they benefit from the 'network effects' [12] of facilitating exchanges between a growing community of customers and suppliers. Key to their success is their extraordinary rate of innovation. In 2019 AWS was releasing ~2,000 new capabilities onto their platform every day. By comparison, AWS' nearest competitors – Microsoft Azure, IBM and Google – do not publicize their rate of innovation in this way and are focused on growth through acquisition of other cloud companies. Time will tell which strategy will win out. For now, AWS continue to leverage their ability to comprehend (use of machine learning to analyze and report trends on platform usage), to act (a skilled workforce, they own the platform) and to do so at pace (they have a culture relentlessly focused on time-to-market).

Example 2. During the UK's 2016 EU referendum, the Leave campaign employed a team of data scientists to analyze the behaviors of sections of the voting population on social networks [13]. Using Facebook's analytical services, they were able to identify opportunities to influence the voting behavior of individuals. They targeted these individuals with messaging that favored a Leave vote, testing a variety of content. Crucially, they were able to monitor and assess the impact of this content in real-time and adjust their messaging accordingly. Their extensive use of data analytics to inform precise, high impact interventions proved to be their 'secret weapon' to counter the Remain campaign who relied on more traditional approaches to communicating with the public, such that their ability to comprehend the impact of their efforts, and therefore adapt, was limited.

In 2017, the US Air Force's General Goldfein succinctly captured the transformation underway in defense and national security when he stated that "... we're transitioning from wars of attrition to wars of cognition..." [14]. Taken to its logical conclusion, cognitive advantage leads to an adversary experiencing cognitive fatigue, which renders them incapable of keeping up with, or effectively responding to, their environment. Of course, the exit of an adversary from an environment may have its own consequences; the cognitive enterprise would have already prepared for such an event.

1.3 Acting Sagaciously

Sagacity is variously defined as: having or showing understanding and the ability to make good judgments [15]; keen discernment and farsightedness [16]; a shrewd and judicious understanding of human motives and actions [17]; and acute insight and wisdom [18]. Acting sagaciously is about thinking and acting in the best interests of the long-term viability of the operating environment in which the organization exists. Sagacity is not about altruism (although that may play a part); it is the holistic view of the role and the impact that an organization has in its environment and the interdependency it has with its environment. A form of reflexivity occurs where the actions that an organization takes to change the environment, also change the organization too.

Every person, team, company or government has the possibility of bringing wisdom – sagacity – to their action. Collectively, an organization can make astute assessments of what is going on and form wise and humane views on what it should do next, on what would have the most beneficial impact in society. An organization that exhibits sagacity – through the collective behavior of its artificial and human intellects – recognizes that what benefits society also benefits itself.

In the field of biology, it has been observed that healthy ecosystems are neither overly efficient nor excessively diverse [19]. They are sufficiently efficient and sufficiently diverse to allow them to persist – to sustain – over long periods of time. There is a balance between efficiency and diversity – a 'window of viability'. Too efficient, and the ability to adapt is limited; too diverse and resources are wasted. Species that endure exist within this window of viability. The long-term success of any company or government relies on the ability to maintain itself in a window of viability. A short-term, 'winner takes all' strategy may ultimately cause the entire ecosystem to fail.

Social, political and commercial ecosystems are constantly being disrupted because there is constant innovation, there is constant striving to outcompete and out-maneuver competitors and adversaries. It is critical to understand the environment and to predict how it is likely to change. That gives the ability to plan and prepare, and beyond that, the ability to shape the environment to the advantage of the organization and the people it serves.

Successful organizations not only respond and adapt to disruption in their environment – they create disruption to give themselves an advantage; and the pace and scale of that disruption will increase exponentially as automated decision making becomes mainstream.

Example 3. The nascent field of distributed applications, enabled by Distributed Ledger Technology [20], may have a profound impact on the global economy. The ability to transact peer-to-peer, without the need for a trusted third party to facilitate (such as a bank), could disrupt many business models across all industrial sectors; and as some organizations embrace the transparency proffered by such technology, for the benefit of their customers, those who do not will begin to stand out for the wrong reasons. One example of an organization intending to increase adoption of decentralized applications is Holo with its development of the Holochain protocol [21]. This protocol, still in development, is designed to enable distributed applications to run with thousands of times less overhead than the current leading protocol for distributed applications, Ethereum. Benefits to users include owning their own data and controlling their own identity, and promotion of democratic, co-operative platforms.

Example 4. China's sometimes surprising actions in the wider world indicate a more profound disruption. As part of their Belt and Road initiative [22], China is investing billions into infrastructure projects around the world. They are playing a long game, and it may be a smart move - infrastructure shapes our operating environment. It dictates the movement of people and of goods. As a country they appear to be pursuing a cognitive advantage trajectory and are positioning themselves to shape the world in unforeseeable ways. However, such global expansion is generating consequences such as de-forestation and a rise in CO_2 emissions; and humanitarian issues remain a significant challenge. The level of sagacity they bring to this endeavor remains to be seen.

2 Orchestrating a Collective Intelligence

2.1 The Cognitive Enterprise

The structural element of building and maintaining a cognitive advantage is the cognitive enterprise. This is a form of organization that is centered on informed dialogue and decision-making by the right people, equipped with unique data insights and augmented by artificial intelligence.

A key motivation is to reserve human time and effort to that which we are best suited: purpose, ingenuity, curiosity, wisdom and compassion. People are good at knowing what questions to ask; AI is good at answering questions but not in evaluating whether an action is 'a good idea'. Humans remain exclusive in that regard. Therefore, a human-centric ethos is pre-eminent in designing the cognitive enterprise. To make the best use of human time, artificial and human intellects must be combined to form a collective intelligence. This intelligence will drive the most efficacious collective action to maintain the organization within a window of viability.

The critical center of the cognitive enterprise is the synthetic core – an encoded representation of everything that the organization has learned and continues to learn about itself and its environment. This body of knowledge and meaning is queried, explored and updated by the collective intelligence of the organization. An effective synthetic core will require the integration of a diverse range of capabilities in artificial intelligence, behavioral science, cloud computing, complexity science, data science, decision science, knowledge management and simulation environments.

Individuals perceive and comprehend things differently from each other; in groups, comprehension changes and knowledge are created due to a dialectic process [23], with people informing and being informed by each other's opinions. For much of the second half of the 20th century, the majority of organizations have cultivated a 'hero culture', with everyone looking at the leader as an all-knowing mind. In the new, fast-moving world of automated systems, this will no longer work. As Eddie Obeng has said, 'Our world changes faster than we can learn about it' [24]. Google, for example, takes a more de-centralized approach to leadership. The founders recognized that for Google to really flourish and grow, they needed to cultivate a culture where everyone was encouraged to openly innovate and to lead in their own unique way [25].

As we pass information and knowledge between individuals and teams, up to the more ephemeral organizational level, there is dialectic influence here as well. The integration of these information feeds gives rise to a mutual understanding of the current state of the operating environment. This is where collective intelligence arises. That collective intelligence is greater than the sum of the parts – there is emergence, in the complexity science sense. From that collective intelligence emerges understanding, realization and insight that no individual in that organization could have reached alone. Organizational understanding, or corporate understanding, comes from the collective understanding of all elements.

The architecture of the cognitive enterprise is centered on cognitive functions that optimize the informational and operational limits of the organization. The cognitive enterprise encodes these insights into the architecture of the organization, meaning the design of processes; the types of people and skills employed; the type of work that humans do vs. the work that machines do; the software; the hardware. It involves consciously designing an organization to maximize the collective intelligence of human and artificial workforces as a critical, protected asset. The interdependency and interflow of ideas and knowledge and information between all human and artificial entities is what gives the edge to the organization as a whole – and the edge here is about having a cognitive advantage.

2.2 What Does a Cognitive Enterprise Look Like?

Customer/Citizen. From the perspective of a customer, or citizen, a cognitive enterprise is in tune with what they are thinking at that time, taps into their belief system and provides the products and services they need, when they need them. This comes from having the right amount of information about them, and with that power and that deep insight into the behavior and experience of a single individual human, comes responsibility and accountability. Trust will become increasingly important and a vital asset for any organization.

Competitor/Adversary. From an adversary's viewpoint, the actions of a cognitive enterprise may be an enigma. It won't be until after an organization has shifted the environment beyond recognition that competitors may begin to understand what has happened. Competitors will act but with reduced confidence in achieving their intended effect. This may lead to competitors over-analyzing to the point of procrastination and subsequently a lack of action, or the opposite, an explosion of a full spectrum of actions that will be of questionable benefit and almost certainly unsustainable. Conversely, the cognitive enterprise may be seen as excessively open and transparent, almost to the point of foolhardiness. For example, the owner of Tesla Inc. communicates openly and publicly about his intent to disrupt the automotive industry. By shifting consumer expectations, for example that electric cars can be luxurious and high performance, Tesla are successfully re-shaping the operating environment of the automotive industry.

Partners/Allies. From the viewpoint of a partner organization, a cognitive enterprise is loyal, trusted, has a clear vision and is good at communicating it to its partners. Also, because a cognitive enterprise organization excels at understanding its environment, predicting and anticipating and making active moves to shape that environment, those organizations tend to know more about their partners, and how they are performing in that environment than the partners even know themselves. They can hold up a mirror and point out to trusted partners where they have blind spots, such as a competitor they are not paying enough attention to. The cognitive enterprise looks after its partners because it is dependent on them; it understands its dependencies in the ecosystem and invests in managing them, which has enormous value to the partner.

Workforce. People who work within a cognitive enterprise love it. In the same way that the organization excels at giving customers what they need before they even know they need it; the cognitive enterprise applies that same power of insight to giving the best to its people. The way to get the best from people is to make them feel valued, trusted and understood.

3 The Challenges of Building and Maintaining a Cognitive Advantage

The challenge to building a cognitive advantage can be summarized as getting the right data, to the right intellect, at the right time, guided by the right values, to have the right impact.

3.1 Data, Data Everywhere But Not a Byte to Eat

Every organization has an informational limit [26] which is the amount of data that it can gather and process to inform action. There is a finite limit to the physical resources (people, money) and the technical resources (data analysis capabilities, knowledge) available and so the challenge is in optimizing the data aperture of the organization. This is an iterative process, requiring capabilities in data analysis, in understanding how, why and where data is generated, and interpreting the value of the data in terms of impact. Some investment is required to start the process, with initial trial and error followed by continuous learning to inform and refine what data is needed.

The needs of customers/citizens can change on a daily, even hourly, basis. Real-time, continuous feedback of large amounts of data from which to identify patterns and emerging trends, to comprehend what is happening and what may happen next, is the hallmark of a digitally capable organization. One only needs to look at the empty supermarket shelves that appeared in March 2020, as the Covid-19 lockdown began, to see the impact that a lack of foresight has, and an over-emphasis on prediction and just-in-time operations.

The classic example of mishandling informational limits is Kodak [27]. They were defining their informational limits to their existing knowledge of their market and investing their resources on unchallenged assumptions about the future direction of their industry. At the organizational level they didn't understand the emergence of digital technologies – even those they invented themselves, including the digital camera – so they were completely blindsided. Kodak's data aperture was too small.

There are three main approaches to gathering data: directly, with some sort of platform or device in the environment that is capturing information; buying data, whether that's ad tech, or a commercial data analysis company that scrapes the internet on a daily basis and does its own analysis; or the vast repository of open-source data available on the internet. Whatever it is, a data strategy will be required that clearly outlines how the economic and information constraints of the organization will be optimized – for example, the investment in proprietary data vs. acquired data - to achieve the purpose of the organization.

3.2 Using Our 50 Bits Per Second Wisely

It is estimated that when we are reading a book or solving a puzzle our conscious mind processes a maximum of just 50 bits a second [28]. This is a very slow, very expensive (albeit powerful) cognitive process, so we should seek to optimize the use of this precious resource.

The functions of cognition – sensory input, encoding, processing, meaning – can be viewed as a hierarchy: data at the bottom, then information, knowledge and finally wisdom at the top; the scarcest and most valuable process. At the moment, many humans are using their 50 bps down at the data level. In an AI world, that no longer needs to be the case. Ideally the genius of our über-expensive but talented, compassionate and ethical humans should be focused on acquiring and acting on wisdom - where humans can add the most value - and let AI do as much as possible of the rest.

Very large amounts of information and knowledge will need to be encoded in a form that is accessible to both artificial and human intellects alike. Data-linked models (semantic graphs) deployed on a cloud infrastructure provide the flexibility and the scalability required. These models are an encoding of the collective understanding that a collective intelligence has about the organization and its operating environment. Cognitive functions such as reasoning, learning, and knowledge retrieval provide access into this knowledge store. As a whole, this knowledge store – and the means to query, explore and update it – are called the synthetic core.

To fully utilize the sophistication and complexity of the synthetic core it needs to be accessible. As such, the human-machine interface will take on an even greater priority if we are to extract the maximum value from the collective intelligence of our artificial and human intellects. This is where we can learn from the videogames industry; they are, after all, experts in creating large-scale virtual worlds that equip humans – often with AI working alongside them – with the ability to act and to problem solve to win in a rapidly changing simulated world, competing against other artificial and human players.

Embracing technology developments such as augmented and virtual reality, haptics, data visualization, simulations and natural language processing will enable humans to fully utilize the synthetic core.

3.3 Mastering Deep Insight

A key difference between the general intelligence that humans possess, and the narrower intelligence of AI, is the ability to take learning from one context and apply it to another. For example, if a deep learning algorithm is taught to recognize dogs, it can't simply transfer that knowledge to immediately begin recognizing cats; but humans can transfer knowledge about dogs to cats somewhat successfully.

This is because humans can conceptualize. We can infer relationships between concepts either through experience or through inference. From our ontological model we can anticipate how our environment might change. This requires us to seek to understand cause-and-effect in our world – no matter how rudimentary – and we do this by testing hypotheses. We imagine and creatively link concepts that we have yet to experience. We then seek data to assess the credibility of such connections.

This ability to think laterally, to make new connections through a creative union of causal concepts, can unlock new, previously hidden knowledge about our environment. From this deep insight we can synthesize precise, deliberate action to inject influence into cause-and-effect pathways. To achieve such deeper insights, we will need to fuse data science with causal analysis and complexity science.

Causal analysis [29] goes upstream of the data itself and seeks to understand the system that has generated the data; to seek out the answers to why and how the data was created in the first place. However, many real-world processes are non-linear - predicting an outcome from the occurrence of an event is generally poor due to small variations in the conditions that caused the event leading to large changes in the effect. This makes inferring causality hard.

To help resolve this, we need the multi-disciplinary approach of complexity science [30] which brings with it the tools of information theory, statistics, dynamical system modelling, agent-based modelling, and large-scale computer simulations. This will allow us to explore, experiment and problem solve across a wide range of scenarios in a synthetic environment so that, in an uncertain world, we can begin to anticipate likely outcomes. To do this at a scale and a speed required to pursue a cognitive advantage trajectory, AI will be required to automate such reasoning and to automatically update the synthetic core – the knowledge store - of the cognitive enterprise.

There is a more profound aspect to causality and AI too, as Judea Pearl stated, '... AI can't be truly intelligent until it has a rich understanding of cause and effect ...' [31]. Investing in causal analysis of an organization's body of knowledge may also enhance that organization's proprietary AI capabilities too. Which, in turn, will improve the cognitive capabilities of the organization as a whole.

3.4 Developing Sagacious Leaders

In the latter half of the 20th century, the ability to win at all costs was the main compass by which we lauded and rewarded leaders. Maximizing shareholder value or winning votes has been the largely unquestioned aim and a board that is responsible to its shareholders or to party members need the best person for that job. Action happened, but was it wise action? Ethical work practices, caring about people from a more human perspective, have become secondary, or fallen away altogether. Using an algorithm to judge a human workforce – such as Amazon's infamous 'the rate' algorithm [32], which has led to mental health issues in the workforce – is a stark reminder that AI can introduce unintended and negative consequences.

As AI becomes widely used to augment and significantly increase the cognitive capabilities of decision-makers, it becomes supremely important that our leaders of tomorrow have a strong foundation in ethics and morality. The Greeks and the Romans placed emphasis on the qualities of a good leader. In Meditations, the Roman Emperor Marcus Aurelius said, "Don't think about being a good person, just be that good person" [33] whilst citing attributes such as wisdom, morality, moderation and courage. We need to set greater expectations for our leaders along these lines.

A sagacious leader is someone who embodies such values whilst also having the innate ability to process complex information [34] – ambiguous, volatile, incomplete – to inform action that has profound impact. Oftentimes such decisions may be tactical, short-term and immediate. But the sagacious leader is also comfortable with thinking strategically, long-term, and sitting with ambiguity. Sagacious leaders are comfortable with developing strategy shaped by ethos [35].

With the right mix of minds, better decisions can be made; a more well-rounded leadership team gives a deeper, more insightful, more morally balanced view of the major strategic decisions that their organization needs to make. Cultivating different viewpoints is critical to arrive at the most sagacious actions [36].

4 Conclusion

To build and maintain a cognitive advantage requires:

- Sensing, thinking and moving faster than your adversaries and maximizing the use of AI to do this

- Being sufficiently self-aware of the fit of the organization to the operating environment; continually adapting to optimize the collective intelligence of the organization
- Relentlessly shaping the operating environment to maintain a cognitive advantage

Whenever there is a paradigm shift in technology – the integrated circuit, the internet, mobile communication technology – there are those organizations that fail to adapt. The new paradigm of hyper-accelerated decision-making enabled by AI is no different. However, this is not about responding and adapting to changes in the operating environment that this brings. Ultimately, this is about shaping the operating environment, enabled through having a cognitive advantage. This introduces a shift in how we think about the ideal organization.

Whilst agile organizations are currently seen as the exemplar for being competitive, in an AI rich 'cognitive' world, being responsive to change is not enough. Where agile organizations respond, sagacious organizations shape.

To become a sagacious organization requires foresight and a willingness to act, even when there is no immediate threat to the organization. It is about asking:

"If we were starting from a blank canvas and we knew about the AI capability that is coming, how would we organize ourselves to best take advantage of this? How would we organize ourselves to make the most of what's now possible, so that we can launch ourselves onto a cognitive advantage trajectory? And how can we do all that in such a way that we shape the environment we operate in so that it thrives and, subsequently, so do we?"

References

1. Porter, J.: Kasparov v Computer: When Deep Blue Beat The Grandmaster (2020). www.thesportsman.com
2. Reynolds, M.: DeepMind's AI beats world's best Go player in latest face-off (2017). www.newscientist.com
3. Thomson, A.: Google shows AI can spot breast cancer better than doctors (2020). www.bloomberg.com
4. Panda, J.: The Weaponization of Artificial Intelligence (2019). www.forbes.com
5. Norton, S.: Era of AI-Powered Cyberattacks Has Started (2017). http://blogs.wsj.com/cio/2017/11/15/. Accessed 15 Oct 2020
6. Digital Economy Compass 2019, Statista
7. Powering Intelligent Connectivity with Global Collaboration: Mapping your transformation into a digital economy with GCI 2019. Global Connectivity Index, Huawei (2019)
8. Reinsel, D., Gantz, J., Rydning, J.: The Digitization of the World: From Edge to Core, IDC White Paper, US44413318 (2018)
9. Attributed to Clive Humby
10. 'Samsung vs Apple: Competition or collaboration? (2018). http://descrier.co.uk. Accessed 15 Oct 2020
11. Richter, F.: Amazon Leads $100 Billion Cloud Market (2020). Statista
12. Parker, G.G., Val Alstyne, M.W., Choudary, S.P.: Platform Revolution, W.W. Norton & Company (2016)
13. Moore, M.: Democracy Hacked: Political Turmoil and Information Warfare in the Digital Age (2018)

14. Friedl, M.: Goldfein delivers Air Force update (2017). www.af.mil. Accessed 15 Oct 2020
15. Cambridge Dictionary, University of Cambridge
16. The American Heritage Dictionary of the English Language, 5th edn
17. The Century Dictionary. http://www.global-language.com/CENTURY. Accessed 15 Oct 2020
18. WordNet3.0 www.wordnet.princeton.edu. Accessed 15 Oct 2020
19. Ulanowicz, R.E., Goerner, S.J., Lietaer, B., Gomez, R.: Quantifying sustainability: resilience, efficiency and the return of information theory. Ecol. Compl. **6**(1), 27–36 (2009)
20. Distributed Ledger Technology (2018). www.investopedia.com. Accessed 15 Oct 2020
21. www.holochain.org. Accessed 15 Oct 2020
22. Belt and Road Initiative (2018). www.worldbank.org. Accessed 15 Oct 2020
23. Stacy, R.: Complex Responsive Processes in Organizations: Learning and Knowledge Creation. Routledge, London (2001)
24. Obeng, E.: The World After Midnight, TEDGlobal (2012)
25. Smithson, N.: Google's Organizational Structure and Organizational Culture (An Analysis) (2019). www.panmore.com
26. Vakarelov, O.: 'The Cognitive Agent: Overcoming informational limits (2011). www.u.arizona.edu. Accessed 15 Oct 2020
27. Gottschall, J.: The Innovator's Dilemma: Lessons from Kodak (2017). www.innovationmanagement.se. Accessed 15 Oct 2020
28. Markowsky, G.: Information Theory – Physiology (2017). Encyclopaedia Britannica. Accessed 15 Oct 2020
29. Pearl, J.: Causality: Models, Reasoning and Inference. Cambridge University Press, Cambridge (2009)
30. Mitchell, M.: Complexity A Guided Tour. Oxford University Press, New York (2013)
31. Bergstein, B.: What AI still can't do. MIT Technology Review (2020)
32. Lecher, C.: How Amazon automatically tracks and fires warehouse workers for 'productivity' (2019). www.theverge.com. Accessed 15 Oct 2020
33. Aurelius, M.: Meditations, Penguin Classics
34. Jaques, E., Cason, K.: Human Capability: A Study of Individual Potential and its Application. Cason Hall & Co. (1994)
35. Cummings, S.: ReCreating Strategy. SAGE Publications (2002)
36. Graham, P. (ed.): Mary Parker Follett – Prophet of Management, Harvard University Press (1995)

Dynamic Formation Reshaping Based on Point Set Registration in a Swarm of Drones

Jawad N. Yasin[1(✉)], Sherif A. S. Mohamed[1], Mohammad-Hashem Haghbayan[1], Jukka Heikkonen[1], Hannu Tenhunen[2], Muhammad Mehboob Yasin[3], and Juha Plosila[1]

[1] Autonomous Systems Laboratory, Department of Future Technologies, University of Turku, Vesilinnantie 5, 20500 Turku, Finland
{janaya,samoha,mohhag,jukhei,juplos}@utu.fi
[2] Department of Industrial and Medical Electronics, KTH Royal Institute of Technology, Brinellvägen 8, 114 28 Stockholm, Sweden
hannu@kth.se
[3] Department of Computer Networks, College of Computer Sciences and Information Technology, King Faisal University, Hofuf, Saudi Arabia
mmyasin@kfu.edu.sa

Abstract. This work focuses on the formation reshaping in an optimized manner in autonomous swarm of drones. Here, the two main problems are: 1) how to break and reshape the initial formation in an optimal manner, and 2) how to do such reformation while minimizing the overall deviation of the drones and the overall time, i.e. without slowing down. To address the first problem, we introduce a set of routines for the drones/agents to follow while reshaping to a secondary formation shape. And the second problem is resolved by utilizing the temperature function reduction technique, originally used in the point set registration process. The goal is to be able to dynamically reform the shape of multi-agent based swarm in near-optimal manner while going through narrow openings between, for instance obstacles, and then bringing the agents back to their original shape after passing through the narrow passage using point set registration technique.

Keywords: Autonomous swarm · Multi-agent systems · Point set registration · Agent-based modeling · Swarm intelligence · Collision avoidance

1 Introduction

Unmanned aerial vehicles (UAVs) are gaining more attention of the researchers due to their numerous advantages, such as robustness, agility, cost-effectiveness, over ground vehicles. These benefits have created even more demand for the autonomous UAVs and especially their swarms in various applications and fields,

K. Arai (Ed.): FICC 2021, AISC 1363, pp. 577–588, 2021.
https://doi.org/10.1007/978-3-030-73100-7_42

for instance threat detection, intelligent transportation systems, search and rescue, military purposes, surveying, and mapping [1–3].

Drones and swarms of drones, despite getting more sophisticated, still face several mission or design limitations and challenges, such as optimal navigation and obstacle avoidance, dynamic reformation, payload, flight time due to limited battery life, design of state-of-the-art stability controllers, and optimal resource allocation [4–6]. UAVs can encounter stationary or dynamic obstacles while navigating and therefore require a reliable collision avoidance system on-board for safe navigation [7,8]. Different types of collision avoidance methods can be generalized into four categories [9,10]: potential field based methods [11,12]; geometric methods [2,10,13]; sense and avoid methods [6,14,15]; optimization based methods [16,17]. Furthermore, the methodologies for formation maintenance can be categorized into three approaches [18,19]: leader follower based [20,21]; virtual structure based [22,23]; behaviour based [24,25].

One of the biggest limitation for UAVs is the mission life due to their limited battery capacity [5,21]. More deviation from the path would result in more power consumption leading to shortened mission life. Similarly, if the swarm slows down or hovers at a certain location for reshaping before continuing, it increases the mission time meaning more power consumed and shortened mission life on a single charge. Maintaining formation and only reshaping the formation while continuing towards the destination can lower the unexpected deviations and power consumption or losing a drone, as while navigating through the obstacles (e.g. going through the window of a building) breaking the formation to bypass the obstacles in different directions may cause inter-drone communication failure or excessive deviation, resulting in more power consumption.

In this paper, for maintaining the formation, we utilize the leader-follower based approach due to its simplicity, non-complex implementation, reliability, and scalability [6]. Keeping the above mentioned limitations and time criticality, we propose our algorithm where a swarm formation comes across multiple obstacles and can go through the gap between them while dynamically changing the formation and reforming to original shape without slowing down. In the proposed algorithm, upon detection of the obstacles and the available gap between them, the swarm reshapes into a queue formation. In order to do so, the agents in the swarm select their temporary leaders and navigate to reach and maintain the minimum defined distance between two agents. Once the agents come out of the obstacles, to reform the original formation shape, they immediately start navigating to the best possible position determined through the point-set registration technique.

The rest of the paper is organized as follows. Section 2 provides the development of the proposed approach. Section 3 is based on simulation results. And finally, conclusions, discussions, and future work is provided in Sect. 4.

2 Proposed Approach

In this section we describe the proposed Dynamic Formation Reshaping Based on Point Set Registration (DFRPSR) algorithm in a Swarm of Drones. The overall strategy is to be able to navigate through the available gap between multiple obstacles while reshaping the formation in a near-optimal manner, without slowing down, and after passing through the obstacles bringing the agents back into formation dynamically while navigating towards the destination in order to minimize the overall time it takes to reach the destination. A novel top-level algorithm is developed to accomplish this, i.e. composed of partial feedback algorithms: one for realigning the agents into a queue formation for passing through the narrow openings between the obstacles and one for bringing them back into the original formation shape. Upon detection of the available gap between the obstacles in the planned route, the algorithm starts evaluating the possible reshaping solutions based on the calculated gap between the obstacles and the angle at which the available gap for safe passage lies. If the detected obstacles already lie in such a manner that the leader does not require any deviation from its original path, then the leader continues to navigate forward while its followers reshape into a queue formation by merging both *Left Leg* and *Right Leg* of the formation while communicating with each other to allow for an efficient integration of both legs of the swarm as shown in Fig. 1(a). After successfully passing the obstacles, the *Turn-Back* function brings the agents back into the initial formation shape in an effective manner by utilizing the point set registration technique, as shown in Fig. 1(b).

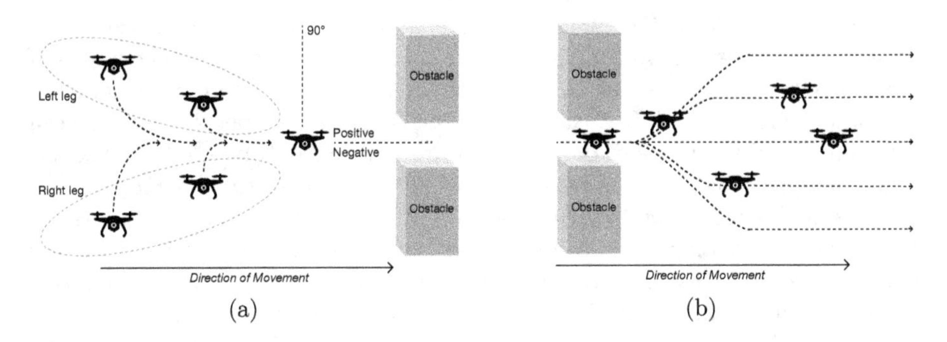

Fig. 1. Illustration of the *Left and Right Legs* and angles and transition phases (a) formation reshaping process while going through the obstacles, (b) Transitioning back to initial formation

2.1 Navigation

The general pseudo-code of the top-level routine is given in Algorithm 1. As our initial setup, we presume that the agents are already in the defined formation shape, and the leader is dedicated and the ID assignment to the agents is

Algorithm 1. Global Routine

```
   procedure NAVIGATION
2:     ObsFlag ← False;
       PSR ← False;
4:     TShape ← Initialized based on the current state;
       while True do
6:         ObsFlag, ObsDist, ObsAng, ObsNum ← Obstacle_Detection();
           if ObsFlag then
8:             Reshape&Avoidance(ObsDist, ObsAng, ObsNum);
               ObsFlag ← False;
10:            PSR ← True;
           else
12:            Update TShape;
           end if
14:        Update CShape;
           if PSR then
16:            Turn-Back(TShape);
           end if
18:    end while
   end procedure
```

completed before the mission is started. This top-level algorithm is executed by every agent locally. A Boolean variable (*ObsFlag*) is initialized, whose role is to indicate the presence of an obstacle *True* or absence *False* (Line 2). Similarly, a Boolean variable (*PSR*) is initialized (Line 3), whose role is to allow (if *True*) the algorithm to call the *Turn-Back* function after successful reshaping and collision avoidance or bypass the *Turn-Back* function (if *False*). Then based on the current position of the every agent, the target shape of the swarm *TShape* is then set up, i.e., initialized, (Line 4). This shape is calculated for every agent at every time interval for the next interval where they have to navigate to. After the initialization, *Obstacle_Detection* procedure is called (Line 6). And based on the presence of an obstacle in the vicinity, the information of the characteristics of the obstacle, such as, distance, angle at which the obstacle is detected, number of obstacles are updated. If *ObsFlag* == *True*, indicating an obstacle(s) has been detected, the Reshape&Avoidance procedure is invoked (Lines 7–8). Once the reshaping and avoidance has been successfully performed, the *ObsFlag* is reset to *False* and *PSR* is set to *True* (Lines 9–10), indicating that the formation shape has been changed due to avoidance maneuver and now initial formation shape needs to be restored. On the other hand, if after the *Obstacle_Detection* procedure, the condition *ObsFlag* == *False* (Line 7), indicating that no obstacle is detected, then only the *TShape* is updated (Line 12). Then the current shape (*CShape*) of the swarm is updated (Line 14). Finally, if *Reshape&Avoidance* procedure was invoked earlier, i.e., *PSR* == *True*, then *Turn-Back* procedure is called for bringing the agents back into the original formation shape (Line 15–16).

2.2 Obstacle Detection

The pseudo-code for the *Obstacle_Detection* procedure is given in Algorithm 2. The node, while being in this procedure, continuously keeps on scanning for the obstacles in the vicinity, and as soon as an obstacle is detected the obstacle detection flag *ObsFlag* is set to *True* (Lines 2–3). After the detection of the obstacle, the parameters of the detected obstacle(s) are calculated and updated (Line 4), such as, the distance of the obstacle (*ObsDist*), the angle at which the obstacle is detected (*ObsAng*), and the number of detected obstacles (*ObsNum*).

Algorithm 2. Obstacle Detection

```
1: procedure OBSTACLE_DETECTION()
2:     if obs in Detection_Range then
3:         ObsFlag ← True;
4:         ObsNum, ObsDist, ObsAng ← Calculate number of obstacles, the distance and angles
           at which the edges lie;
5:     end if
6: end procedure
```

2.3 Formation Reshaping and Avoidance

After the detection of obstacles, the *Reshape&Avoidance* procedure is called, pseudo-code given in Algorithm 3. This procedure is responsible for reshaping the formation, according to the situation, and successful collision avoidance. As soon as this procedure takes control, the node calculates the *Gap* between the detected obstacles, in case of multiple detected obstacles (Lines 2–3). This algorithm works on three cases (Lines 5–18). If only one obstacle was detected, i.e. $ObsNum \ngtr 1$, in this case the algorithm simply calls for collision avoidance procedure, developed and presented in [6], to perform the avoidance maneuver (Lines 7–8). If more than one obstacles were detected, i.e., $ObsNum > 1$, but the calculated *Gap* between the obstacles is less than the defined safe distance $Safe\_dist$ indicating the gap is not wide enough for the agent to go through. In this case, the algorithm treats them as a single obstacle to perform the collision avoidance maneuver (Lines 9–10). And the third case, the condition $ObsNum >$ 1 holds True, i.e. more than one obstacles detected and the gap between the obstacles more than the defined safe distance for the agents to pass through, i.e. $Gap >= Safe\_dist$ holds True (Line 11). The algorithm then checks if the *ObsAng* is negative, as shown in Fig. 1(a). In this case, the leader has to move towards its right to align itself to be able to pass through the gap between the obstacles without colliding. And consequently, the *Right leg*, i.e. the agents with "odd" IDs (ID= 3, 5, 7,.., n), will have to move less to be able to come into the queue formation as compared to the agents with "even" IDs (ID = 2, 4, 6,.., n+1), i.e., the *Left leg*, will have to move more towards their right to align properly. And therefore, the *Left leg* is merged into the *Right leg*, (Lines 12–13) as shown in Fig. 1(b). In this case, every agent follow the following protocol:

1. if SELF.ID is even number
2. make SELF.ID - 1 as temporary leader
3. if SELF.ID is odd number
4. make SELF.ID - 1 as temporary leader

On the other hand, if the *ObsAng* is positive, then *Right leg* is merged into the *Left leg* (Lines 14–16). And the agents follow the following protocol:

1. if SELF.ID is odd number
2. make SELF.ID - 3 as temporary leader
3. if SELF.ID is even number
4. make SELF.ID + 1 as temporary leader

However, if the *ObsAng* is neither negative or positive (Lines 17–18), meaning that the leader is already well aligned in the gap, in this case *Left leg* is merged into the *Right leg* as well. It is important to note here that in this case either leg can be merged into the other without any efficiency or energy difference, but instead of randomizing the routine, the routine has been hard-coded for easier function.

Algorithm 3. Formation Reshaping and Collision Avoidance

```
    procedure RESHAPE&AVOIDANCE(ObsNum, ObstDist, ObsAng)
2:      if ObsNum > 1 then
            Gap ← calculated gap between the obstacles;
4:      end if
        STATE ← (Gap >= Safe_dist, ObsNum > 1);
6:      switch STATE do
            case (-, False)
8:              collision avoidance();                    ▷ single obstacle
            case (False, True)
10:             collision avoidance();                    ▷ Treated as single obstacle
            case (True, True)                             ▷ Multi-obstacle
12:             if ObsAng is negative then
                    Left leg merges into the right leg;
14:             else
                    if ObsAng is positive then
16:                     Right leg merges into the left leg;
                    else
18:                     Left leg merges into the right leg;   ▷ Priority assignment
                    end if
20:             end if
        end switch
22: end procedure
```

2.4 Turn-Back Function

Once the original formation shape is distorted or reshaped, into a queue formation in our case, a procedure for bringing the swarm back into their original formation shape in a near-optimal manner is presented in this part. While bringing the swarm back to the original shape we utilize the point set registration technique [26,27] that is based on the thin-plate spline [28] and is a well known methodology in data interpolation and smoothing. This technique is used as due to its non-complicated construction, complicated shapes are approximated easily using splines [28]. Here we provide and discuss for 2-dimensional formulation, where we have two sets of correlating points X (x_i, i = 1, 2, . . . , n) and V (v_i, i = 1, 2, . . . , n), for current shape $CShape$ and the original/target shape $TShape$

respectively. Here the coordinates of the location of a point are represented by $x_i = (1, x_{ix}, x_{iy})$ and $v_i = (1, v_{ix}, v_{iy})$. The mapping function $f(v_i)$ can be found by minimizing the following:

$$E(f) = \sum_{i=1}^{n} ||x_i - f(v_i)||^2 + \lambda \iint [(\frac{\partial^2 f}{\partial x^2})^2 + 2(\frac{\partial^2 f}{\partial x \partial y})^2 + (\frac{\partial^2 f}{\partial y^2})] dxdy \tag{1}$$

where E is the energy function, i.e. the amount of disturbance in the formation, the integral part of the equation represents the mapping of the corresponding point sets to the correlating ones while considering the intended formation, and the scaling factor is provided by the λ. Here we set λ to zero, as our intention is to map one point set over the other without considering the $CShape$, and therefore the closest points are mapped accordingly. Once the mapping process is performed, every node navigates towards its calculated position by utilizing the shortest path approach.

Algorithm 4. Turn-Back

```
   procedure TURN-BACK(TShape)
2:     while PSR do
           NLoc ← Calculate the new coordinates for each agent;
4:         Temperature minimization function(NLoc);
           Update CShape;
6:         if TShape == CShape then
               PSR ← False;
8:         end if
       end while
10: end procedure
```

An overview of our *Turn-Back* function is given in Algorithm 4. First the next location (*NLoc*) of every agent is calculated, i.e. the coordinates where every respective node should be in order to maintain the initial formation shape (Line 3). These values are then fed to the temperature minimization function for bringing the nodes to their new locations as optimally as possible (Line 4). Then the current shape (*CShape*) of the swarm is updated (Line 5). Finally it is checked if the swarm has reached its intended formation shape, if the current shape matches with the target shape, the PSR flag is set to False and the control is returned to the main function (Lines 6–7).

3 Simulation Results

The initial conditions defined for our work are as follows:

1. all UAVs/agents travel with constant ground speeds
2. the communication between UAVs is ideal, i.e., without information loss
3. all UAVs are at the same altitude

4. on-board localization techniques are utilized by the UAVs for obtaining their respective positions

The simulation results are shown in Fig. 2. Figure 2(a) shows the overall trend of the trajectories of the swarm from mission start till the destination, including the transition phases while going through the narrow opening between the obstacles and returning back to the initial formation shape. In Fig. 1(a), at around $X - Axis = 85\,\text{m}$ the obstacle is detected by the leader, i.e., UAV 1 (green trajectory). Upon calculating the gap between them, UAV 1 realigns itself to safely pass through the obstacles and at the same time the followers are notified of the detected obstacle. As it can be seen from the figure, the followers nodes start rearranging and realigning themselves, even before the obstacle is visible to them, by merging both *legs* of the swarm and following the optimal path to now follow their temporary respective leaders in a queue formation. At $X - Axis$ around 116m, the swarm coming out of the gap between the obstacles and starting the reformation process to the initial formation shape at the same time.

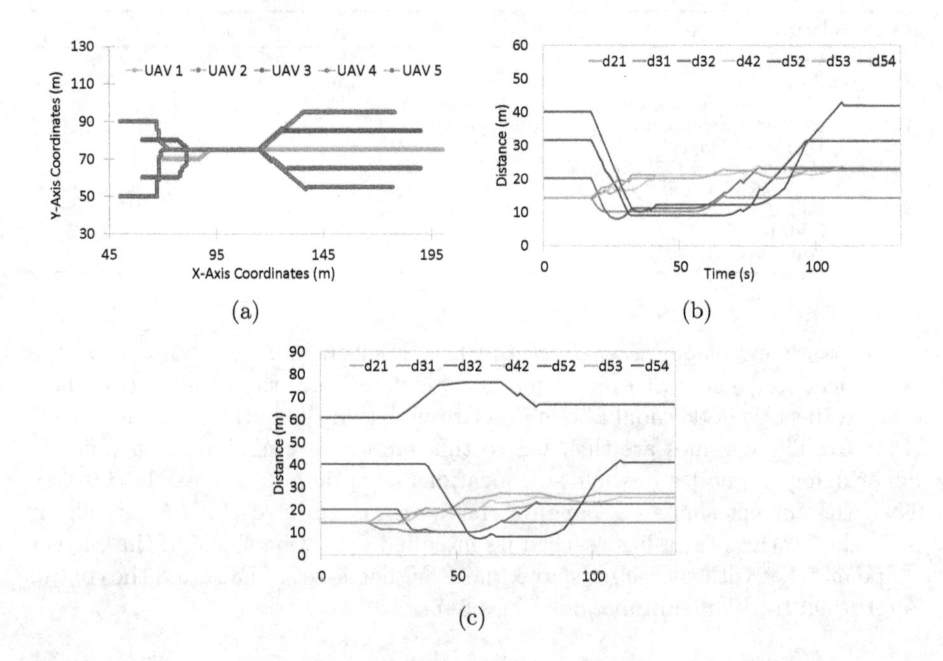

Fig. 2. Simulation results (a) Overall trend of the trajectories of all UAVs. (b) Distances maintained by UAVs from their immediate leaders and temporary leaders utilizing DFRPSR algorithm. (c) Distances maintained by UAVs from their immediate leaders and temporary leaders utilizing sense&avoid only

The distances maintained by the UAVs, with their immediate leaders and the temporary leaders (while in transition shape), through out the mission is shown

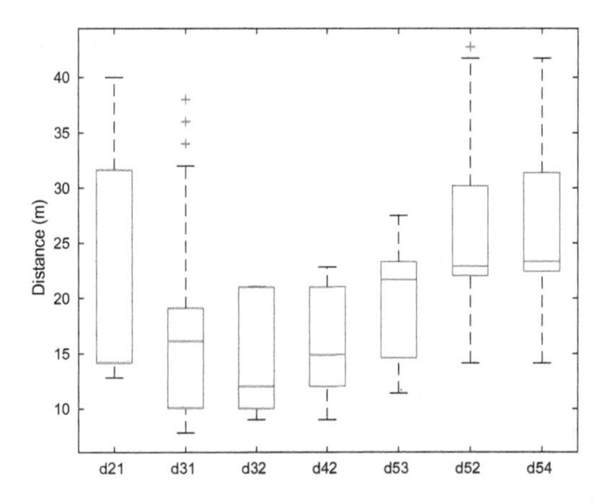

Fig. 3. Distance relationship characteristics

in Fig. 2(b), where distance maintained between UAV 2 and UAV 1 is represented by *d21*, distance maintained between UAV 3 and UAV 1 *d31*, distance maintained between UAV 3 and UAV 2 *d32*, and so on. The overall distance maintenance between the nodes and their immediate leaders without utilizing the proposed DFRPSR algorithm is shown in Fig. 2(c). In Fig. 2(b), we show the trend of distance maintenance between the nodes and their immediate leaders only, for easier visualization and cleaner graphs. As it can be seen that as the obstacles come in the vicinity, at around $t = 17$ s, the drones started reshaping, to be able to pass through the narrow passage between the obstacles, by temporarily changing their respective leaders to form a queue. From $t = 31$ s to $t = 56$ s, the swarm in reformed queue shape is navigating through the gap between the obstacles. And as soon as the nodes emerge from the end of the obstacles, they start navigating to regain the initial V-shaped formation as can be seen from their trajectories and the changed inter-node distances. However, Fig. 2(c) shows that utilizing standard sense&avoid algorithm with formation, every agent started the reshaping process based on their respective local detection. Utilizing the proposed DFRPSR approach, the swarm came back to the initial formation at $t = 97$ s, whereas while local sense&avoid methodology was applied the swarm came back to the initial formation at $t = 111$ s. It shows the efficiency of the DFRPSR approach in reducing the disturbance in the formation shape more efficiently.

Figure 3 shows the overall trend of the minimum, maximum, and the percentile of the distances maintained by the respective nodes. Minimum distance, while navigating between UAV 1 and 2 is 12.8 m, whereas the maximum distance is 40 m that is due to the reshaping process and changed leaders while in transition phase, i.e. queue formation. Similarly, UAV 3 to UAV 1 the minimum distance maintained is 7.81 m due to the compression between the agents while dynamic reformation process and the maximum is 38 m, with three outliers. The minimum distance between UAV 4 and 2 is 9 m, maximum distance was 22.82 m, and the median 14.86 m, that is due to the fact that UAV 2 remained the leader for UAV 4 even while in transition phase.

4 Conclusions

In this paper, we developed an algorithm for smooth and dynamic transformation from initial formation shape into a queue formation, by smoothly merging the agents together, to enable the swarm to efficiently pass through the gap or opening between obstacles. Upon detection of the obstacles, the leader of the swarm calculates the gap between the obstacles, aligns itself to pass through the gap, and sends that information to the followers. At the same time, the followers, start realigning and reforming the swarm by re-selecting their temporary leaders respectively. Upon successful avoidance, the nodes rearrange themselves to regain the initial formation shape as optimally as possible. The simulation results show that the proposed method works reliably and efficiently in static environments. All the agents maintained the respective defined distances between them, reshaped to maintained the minimum defined distances while being in queue formation, and expanded back to attain the original formation effectively.

The proposed approach is limited to 2-dimensional movement, with fixed altitude, in its current form and its efficiency needs to be evaluated by extending the work by introducing the third dimension. Moreover, the effectiveness of the proposed approach still needs to be tested by introducing the environmental effects such as air drag on individual drones as well as on the overall shape of the swarm. In the future work, we plan to further investigate the effectiveness of this approach for multi-layered formations. Another interesting analysis will be the optimization of resource management in the swarm. This will be very interesting to analyze especially when considering other environmental effects, for instance air drag on the layers, and dynamic swapping of the outer layered drones while reforming to increase the mission life.

Acknowledgment. This work has been supported in part by the Academy of Finland-funded research project 314048 and Nokia Foundation (Award No. 20200147).

References

1. Ladd, G., Bland, G.: Non-military applications for small UAS platforms. In: AIAA Infotech@ Aerospace Conference and AIAA Unmanned... Unlimited Conference, p. 2046 (2009)
2. Shakhatreh, H., Sawalmeh, A.H., Al-Fuqaha, A., Dou, Z., Almaita, E., Khalil, I., Othman, N.S., Khreishah, A., Guizani, M.: Unmanned aerial vehicles (UAVs): a survey on civil applications and key research challenges. IEEE Access **7**, 48572–48634 (2019)
3. He, L., Bai, P., Liang, X., Zhang, J., Wang, W.: Feedback formation control of UAV swarm with multiple implicit leaders. Aerosp. Sci. Technol. **72**, 327–334 (2018). http://www.sciencedirect.com/science/article/pii/S1270963816309816

4. Campion, M., Ranganathan, P., Faruque, S.: A review and future directions of UAV swarm communication architectures. In: 2018 IEEE International Conference on Electro/Information Technology (EIT), pp. 0903–0908 (2018)
5. Tseng, C.M., Chau, C.K., Elbassioni, K.M., Khonji, M.: Flight tour planning with recharging optimization for battery-operated autonomous drones. CoRR, abs/1703.10049 (2017)
6. Yasin, J.N., Haghbayan, M.H., Heikkonen, J., Tenhunen, H., Plosila, J.: Formation maintenance and collision avoidance in a swarm of drones. In: Proceedings of the 2019 3rd International Symposium on Computer Science and Intelligent Control. ISCSIC 2019, Association for Computing Machinery, New York (2019). https://doi.org/10.1145/3386164.3386176
7. Seo, J., Kim, Y., Kim, S., Tsourdos, A.: Collision avoidance strategies for unmanned aerial vehicles in formation flight. IEEE Trans. Aerosp. Electron. Syst. **53**(6), 2718–2734 (2017)
8. Lin, Y., Saripalli, S.: Collision avoidance for UAVs using reachable sets. In: 2015 International Conference on Unmanned Aircraft Systems, ICUAS 2015, pp. 226–235. Institute of Electrical and Electronics Engineers Inc. (2015)
9. Yasin, J.N., Mohamed, S.A.S., Haghbayan, M., Heikkonen, J., Tenhunen, H., Plosila, J.: Unmanned aerial vehicles (UAVs): collision avoidance systems and approaches. IEEE Access **8**, 105139–105155 (2020)
10. Payal, A., Singh, C.R.: A summarization of collision avoidance techniques for autonomous navigation of UAV. In: Jain, K., Khoshelham, K., Zhu, X., Tiwari, A. (eds.) Proceedings of UASG 2019, pp. 393–401. Springer, Cham (2020)
11. Choi, D., Lee, K., Kim, D.: Enhanced potential field-based collision avoidance for unmanned aerial vehicles in a dynamic environment (2020). https://arc.aiaa.org/doi/abs/10.2514/6.2020-0487
12. Senanayake, M., Senthooran, I., Barca, J.C., Chung, H., Kamruzzaman, J., Murshed, M.: Search and tracking algorithms for swarms of robots: a survey. Robot. Auton. Syst. **75**, 422–434 (2016). http://www.sciencedirect.com/science/article/pii/S0921889015001876
13. Alexopoulos, A., Kandil, A., Orzechowski, P., Badreddin, E.: A comparative study of collision avoidance techniques for unmanned aerial vehicles. In: 2013 IEEE International Conference on Systems, Man, and Cybernetics, pp. 1969–1974 (2013)
14. Prats, X., Delgado, L., Ramirez, J., Royo, P., Pastor, E.: Requirements, issues, and challenges for sense and avoid in unmanned aircraft systems. J. Aircr. **49**(3), 677–687 (2012)
15. Albaker, B.M., Rahim, N.A.: A survey of collision avoidance approaches for unmanned aerial vehicles. In: 2009 International Conference for Technical Postgraduates (TECHPOS), pp. 1–7 (2009)
16. Zhang, X., Liniger, A., Borrelli, F.: Optimization-based collision avoidance. arXiv preprint arXiv:1711.03449 (2017)
17. Smith, N.E., Cobb, R., Pierce, S.J., Raska, V.: Optimal collision avoidance trajectories via direct orthogonal collocation for unmanned/remotely piloted aircraft sense and avoid operations. https://arc.aiaa.org/doi/abs/10.2514/6.2014-0966
18. Ren, W.: Consensus based formation control strategies for multi-vehicle systems. In: 2006 American Control Conference, p. 6 (2006)
19. Oh, K.K., Park, M.C., Ahn, H.S.: A survey of multi-agent formation control. Automatica **53**, 424–440 (2015)
20. Shen, D., Sun, Z., Sun, W.: Leader-follower formation control without leader's velocity information. Sci. China Inf. Sci. **57**(9), 1–12 (2014)

21. Yasin, J.N., Mohamed, S.A.S., Haghbayan, M.H., Heikkonen, J., Tenhunen, H., Plosila, J.: Navigation of autonomous swarm of drones using translational coordinates. In: Demazeau, Y., Holvoet, T., Corchado, J.M., Costantini, S. (eds.) Advances in Practical Applications of Agents, Multi-Agent Systems, and Trustworthiness. The PAAMS Collection, pp. 353–362. Springer, Cham (2020)
22. Li, N.H., Liu, H.H.: Formation UAV flight control using virtual structure and motion synchronization. In: 2008 American Control Conference, pp. 1782–1787. IEEE (2008)
23. Dong, L., Chen, Y., Qu, X.: Formation control strategy for nonholonomic intelligent vehicles based on virtual structure and consensus approach. Procedia Eng. **137**, 415–424 (2016). Green Intelligent Transportation System and Safety
24. Lawton, J.R., Beard, R.W., Young, B.J.: A decentralized approach to formation maneuvers. IEEE Trans. Robot. Autom. **19**(6), 933–941 (2003)
25. Balch, T., Arkin, R.C.: Behavior-based formation control for multirobot teams. IEEE Trans. Robot. Autom. **14**(6), 926–939 (1998)
26. Guo, P., Hu, W., Ren, H., Zhang, Y.: PCAOT: a Manhattan point cloud registration method towards large rotation and small overlap. In: 2018 IEEE/RSJ International Conference on Intelligent Robots and Systems (IROS), pp. 7912–7917 (2018)
27. Myronenko, A., Song, X.B.: Point-set registration: coherent point drift. CoRR abs/0905.2635 (2009). http://arxiv.org/abs/0905.2635
28. Chui, H., Rangarajan, A.: A new algorithm for non-rigid point matching. In: Proceedings IEEE Conference on Computer Vision and Pattern Recognition. CVPR 2000 (Cat. No. PR00662), vol. 2, pp. 44–51 (2000)

Customer Driven Bargain on Multi-agent System

P. L. A. U. Rathnasekara(✉) and A. S. Karunananda

Department of Computational Mathematics, University of Moratuwa, Moratuwa, Sri Lanka
asokakaru@uom.lk

Abstract. The modern E-commerce incorporates online merchants and digital payment modes, offering benefits to the customers. However, the research conducted in the area of customer driven bargaining is limited and it is hard to find a single e-commerce platform to facilitate customers for ordering some product by negotiating the prices with the identified best merchants in the echo system by using past data except traditional auction supporting merchants. The study was conducted with the assumption that some customers are more important than others, so that that they should not be treated equally with common rewards. The research executes Multi Agent Based merchant-customer bargaining solution for online transactions. This solution contains Agents, namely, Customer Agents, Merchant Agents, Manager Agents and Search Agent. Upon the request by a customer agent with the product name, customer location, history of purchase pattern of the customer, multiple merchant agents will be activated. Consequently, agents initiate communication about the product and the negotiation ultimately sets out terms on an agreeable price to both parties. The solution has been implemented on JADE and the results yielded can imply avenues for further research.

Keywords: Communication · Bargaining · Negotiation · Coordination · Multi agent systems

1 Introduction

E-Shopping has become a fad during the last few decades owing to its great flexibility of buying from home or office. Software agent technologies have taken the new-generation e-commerce to a different level, saving more time of customers who would have otherwise taken the hassle of making visits to physical stores by automating most time-consuming stages [1, 2]. As per the definition of Electronic Commerce Association (ECA), electronic commerce means any form of business or administrative transaction or information exchange that occurs within the broad spectrum of Information and Communication Technology [3]. Agent based e-commerce has come into being and is now the focus of the next generation e-commerce [2]. Bargaining is meant to negotiate a price between two parties. People use bargaining in any kind of market ranging from a rural level boutique to the international market. In the present context, there is a widely shared enthusiasm over online shopping, so that numerous digital payment platforms are

K. Arai (Ed.): FICC 2021, AISC 1363, pp. 589–604, 2021.
https://doi.org/10.1007/978-3-030-73100-7_43

available to online shoppers. Even though there are many digital payment methods and online merchants, it is an arduous task to look for a merchant who facilitates bargaining except the traditional auction supporting merchants.

In a competitive market, sellers make their products and services available, allowing the customers to select from a variety of options with the best prices. Simultaneously, the merchants should make sure to make sales without any loss, still retaining the customer base. Among the strategies to maintain a win-win situation are offering discounts and promotions. Hence, it is useful to create a payment platform that allows both the seller and the buyer to bargain online. In that type of a scenario, Multi Agent based solutions (MAS) can provide with smart solutions.

Agents have been identified as the next breakthrough in software development, resulting in powerful multi-agent platforms and innovative e-business applications. However, it should be noted that MAS is not yet advanced enough to cater to the sophisticated e-business applications [4]. As a team lead of a similar project, the researcher has identified the needs, challenges and opportunities in the market. So far, little progress has been made in developing a system incorporating bargaining facilities.

It is a possibility to request details of any product or service in a certain area. Taxi ride service apps which are widely in use across the country is one such instance. Nevertheless, that kind of simple solution cannot yield benefits to the customers. As a payment platform system can store more data of customers and merchants for a significantly longer period, it is viable to trace the patterns of customer and merchant behavior. The study discusses that use of AI technologies can facilitate customers to find the best merchants in an efficient way.

The rest of this paper is organized as follows: Sect. 2 provides a critical review of the developments and issues in the area of digital payment platform by defining the research problem. Section 3 is on technology adapted to building the multi agent solution for digital payment platforms. Section 4 then presents our approach to multi-agent based solution for ordering product in digital payment platform. Section 5 is about the design of the proposed solution and it describes the modules and the connections. Section 6 presents the implementation details of each module and connections in the design. Section 7 gives the evaluation details such as experimental design, text cases, evaluation strategy and data correction. Finally, Sect. 8 concludes with a note on the possible further work.

2 Review on Digital Payment Platforms

2.1 MAS in Industrial Landscape

Among AI technologies, Multi-Agent Systems (MAS) are more popular in the industry. MAS have been widely implemented in business domains of manufacturing, electrical engineering, electronic, commerce, graphics (e.g., computer games and movies), transportation, logistics, software development, robotics, telecommunications and energy. Most of the giants in IT industry are using those technologies to develop commercial products to identify more opportunities in the competitive market.

Multi-agent technology has been used in many areas however industry applications have taken the earliest advancement of agent technology when compared to others. It has been proved that the features of MAS technology can be used to handle complexity in communication [5]. MAS originated from distributed artificial intelligence and that allows a substitute way of designing design distributed control systems based on autonomous and exhibiting modularity, cooperative agents, flexibility, robustness and adaptability. The most important thing is that MAS technology has proven how complex systems can be modeled to generate smart solutions, which may not be done easily by using classical computing technologies nevertheless they have shown a little bit success in the recent history [5].

MAS are using in many areas like production planning, inventory management, vehicle routing, electronics & electrical productions and software development industry [6]. It has been showcased in a considerable amount of literature that the capability of multi-agent system (MAS) technology to effectively use the power of communication a problem-solving strategy in complex systems.

Many electrical engineering products are benefited from the advantages of this technology in many ways. Mini/Micro grids are one of the best applications of MAS in the electrical engineering world. Mini/micro grids are a potential solution being studied for systems relying on distributed generation. MAS have been able to maximize the power generation of the micro-grid and reduce the cost of the operation. In this manner, MAS in micro-grid helps to best use of monetary and environmental resources [7].

Agent technology is used in many real word applications in traffic management systems in dynamic environments to handle the complexity in an easiest manner. The domain of traffic and transportation systems is another classic domain for an agent based approach owing to its geographically distributed and dynamic changing nature. Moreover, with MAS there is an opportunity to improve the ability of the traffic management systems to deal with the uncertainty in a dynamic environment and this one of the crucial features of Multi Agent Systems [8].

OASIS is an agent based air traffic control system used in Sydney airport, Australia which was introduced in early nineties [9]. Which has been developed to assist alleviate air traffic congestion by maximizing runway utilization, achieved through arranging landing aircraft into an optimal order, assigning them a landing time, and then monitoring the progress of each individual aircraft in real time.

One of the interesting areas in which MAS used is 3D games. Designing complex and reasonable 3D environments for modern 3D games is one of the time consuming challenges faced by present video game industry. Each 3D model in a 3D game environment can be associated with an agent with simple rules and system can allow users to introduce new 3D models and associate them with agent types. By using MAS it can build a 3D game environments with self-organized 3D models positioned and oriented in most suitable places [10].

MAS solutions can be applied in the wireless sensor network environments. It can be implemented by combining the use of intelligent sensors and middle agent architecture. Intelligent sensor nodes may exploit as autonomous agents which may monitor the incidents in the environment. For retrieving the data from the target nodes, mobile software agents can be used in an energy saving manner. It's been proven that using

MAS solution is highly suitable to access and distribute sensory data efficiently and timely manner [11].

With the development of software industry, software architectures and tools are also heavily influence the future trends in deployment of agent base solutions [12]. Agent based software developments has the potential to significantly improve the designing and implementing systems in disruptive manner. Agent based software systems are significantly popular for the purpose of solving complex real world problems. Especially, as MAS has the capability of being robust, scalable and working as autonomous agents who can achieve their objectives in an uncertain environment by communicating, coordinating and negotiation with each other [13].

2.2 Development and Challenges in Digital Payment Platform

It is a known factor that the business environment is highly connected in the present context and there is a great competition with each other. There is a rising demand for greater responsiveness to market challenges and opportunities. Multi Agent Systems are the ideal solution for creating responsive and adaptable business solutions to achieve these demands in this dynamic environment. Considerable amount of work has already been done for couple of decades since agents have been identified as the next breakthrough in software development, resulting in powerful multi-agent platforms and innovative e-business applications [14].

One of the most advantages of online shopping is saving our time especially during rush hours and a holiday season. Otherwise we need to wait in long lines or search from store to store for a particular item. The rapid growth of the Internet users in world opened a new business opportunity to the whole world. It has shown an exponential growth of shopping activities over the internet in the last few years. It is hard to find a proper mechanism for facilitating online transactions and automate shopping process on behalf of customers. So a human buyer is still responsible for gathering commodity information from multiple suppliers on Internet, making decisions about each commodity, then making the best possible selection, and ultimately performing the e-payment. So it takes lot of time to buy things over the Internet. Hence, to reduce the time and to enhance the automation of the e-shopping system a multi agent environment is used [1].

Even though there are many different shopping sites on the public internet, most of these sites haven't a proper user-friendly interface design, which is essential for the success of online software [1]. The interface of e-shopping systems must be able to attract the eyes of a customer, easy to understand and easy to use [1]. Otherwise, people will likely become less interested in e-shopping applications.

2.3 Future Trends

Owing to the numerous reasons agent technologies have not achieved yet that level of maturity which industry expected to cater the business needs of modern world. There are still remain some research challenges and much work is to be done adapt relevant existing agent technologies to the requirements of the new generation e-business systems. The frameworks such as FIPA agent system platforms and communication languages have arisen that implementation of common standards and these frameworks would help immensely to reduce the valuable time of developers [14].

There are many future trends in this area. An intelligent shopping system can be developed using a multi-agent system to provide shopping service for the commodities that a consumer does not buy frequently. The system integrates built-in expert knowledge and the customer's current needs, and recommends optimal products based on multi-attribute decision making method [1].

Software agent technologies provide a new scenario that is used to develop the new-generation e-commerce system, in which the most time-consuming stages of the customers' shopping process will be automated [14]. At now on the internet most of the e-shopping systems are using the normal webpage implementation. That will take the more heavy work for the web servers. Intelligent e-shopping systems can be developed using data mining. But it can be too slow. The agent technology can be used to reduce the enormous time taken by the e-shopping system based on web application and data mining [5].

2.4 Problem in Brief

The exponential development of the Internet has changed the way enterprises do business. Electronic commerce is becoming an attractive way of conducting business transactions. However, the progress of e-commerce seems to be hindered by the lack of a widely accepted payment standard which is suitable for e-commerce. Lack of intelligence is another factor stymieing electronic commerce is also emerging to the surface. The vast size of information on the Internet also means that it is inconvenience for potential customers to locate products that they are interested in. Therefore, e-commerce demands advanced technologies as support. Agent technology seems to be an excellent candidate with its properties of intelligence, autonomy, and mobility [2].

Currently there are many digital payment platforms in the market and innovative features bring customer attraction and the more advantages in the competitive market to the platform owners. Even though there are many challenges and opportunities in this filed, this research focus on the below business opportunity currently available in the market.

A web based or mobile based system option for putting an order and bring something through a platform in a secure manner by negotiating with the best merchants in the echo system. Merchants are also benefited from this kind of solution to serve in a better way and increase their revenue in an easiest manner.

3 Technology and Methodologies

3.1 Multi Agent System Technology

In the present context the business environment is highly dynamic. However, the majority of the people in this competitive world expect their daily requirements to be met at shorter agreed service level. The Internet and mobile computing means that the majority of business processes are "always-on" and increasingly interconnected. Agent based solutions are designed to handle these kind of complex and dynamic environments. Multi-agent systems have proven to offer measurable business benefit for many organizations in the recent past [5].

The study of multi-agent systems focuses on systems in which many intelligent agents interact with each other through coordination, communication and negotiation. Agents can be defined as software programs which running on a system or network and implemented to achieve a set of specific. They are designed to be autonomously process in the platform to achieve their goals by interacting with other agents, either collaboratively or in competition. This showcases that groups of agents can work together, each carrying out their own visions and processes. This provides a powerful technique for designing software applications for complex or distributed business processes [5].

Several researchers have attempted to provide a meaningful classification of the attributes that agents might have. A list of common agent attributes is shown below [13]:

Adaptively: The ability to learn and improve with experience.
Autonomy: The ability to act without any interference from the outside.
Activity: The ability to show its own initiative.
Collaborative behavior: The ability to work with other agents to achieve a common goal.
Mobility: The ability to migrate in a self-directed way from one host platform to another.
Reactivity: The ability to selectively sense and act.
Temporal continuity: Persistence of identity and state over long periods of time.
Multi-agent systems are designed to handle changing and dynamic business processes. They continue to operate even when there may be incomplete information, but a decision must still be made by the most effective method. Properly integrated into a business process they can offer flexible and intelligently adaptable systems support [5].

3.2 JADE

JADE (Java Agent DEvelopment Framework) is a Java base software Framework which simplifies the implementation of multi-agent systems through a middle-ware that complies with the FIPA specifications (IEEE Computer Society standards organization that promotes agent-based technology and the interoperability of its standards with other technologies) and through a set of graphical tools that support the debugging and deployment phases. A JADE-based system can be distributed across machines (which not even need to share the same OS) and the configuration can be controlled via a remote GUI. The configuration can be even changed at run-time by moving agents from one machine to another, as and when required.

3.3 Genetic Algorithm Technology

Genetic algorithms (GAs) are a combination of heuristic search and optimization technique inspired by natural evolution. This technique has been successfully applied in wide range of real-world complex problems. By starting with a randomly generated population of chromosomes, a GA carries out a process of fitness based selection and recombination to produce a successor population, the next generation. Parent chromosomes are selected at the recombination and their genetic material is recombined to produce child chromosomes. These then pass into the successor population. This is an iterative process and a sequence of successive generations evolves and the average fitness of the chromosomes tends to increase until some stopping criterion is reached. Therefore, GA provides an optimum solution to a given problem [15].

John Holland was the person who proposed this solution to the world at first as a means to find good solutions to problems that were otherwise computationally intractable. Holland's Schema Theorem, and the related building block hypothesis, provided a theoretical and conceptual basis for the design of efficient GAs. It also proved straightforward to implement GAs due to their highly modular nature. This results in, the exponential growth of the field and the technique was successfully applied enormous practical problems in science, engineering and industry [7].

GAs are specially suitable for the problems where traditional optimization techniques break down, either due to the irregular structure of the search space (for example, absence of gradient information) or because the search becomes computationally intractable [7].

4 Novel Approach to Customer Driven Bargain

For addressing such kind of problem which explained in the above, Multi-Agent based solution is proposed according to the study done in the previous days. Our new approach to customer bargaining is named as MASCB acronyms for Multi Agent System for Customer Bargaining and Fig. 1 showcases the process in high level.

Based on the requested product, customer location, his or her expenditure pattern and other considerable factors, any product order request will sent to the search agent for selecting suitable merchants in that area. The search agent is using a Genetic algorithm process and the distance between the merchant and the customer, previous customer reviews for the particular merchant is considered in the fitness function. That process will propose set of merchants who satisfy the fitness function and customer agent send a request to all these merchants for offering a price for the particular product. Any desired merchant is allowed to offering a price the merchant who sends the minimum price will introduce for the customer for accepting it.

The customer and merchant will be allowed to communicate with each other by negotiating the price through multi agent system and incase of rejecting the request another merchant will be proposed by the system and above process will repeat until customer find the suitable merchant who satisfy with his/her requirements. If the customer accept that order it will inform to the merchant and the merchant will pass that order to the delivery module. This will open another business opportunity as well. System can be modified to host an API for any third party delivery partner to process the delivery process. The security level of this kind of system must be satisfied into an industry standard level with zero bugs.

The inputs of the proposed system are the product needs to order, location of the customer and expect price. The main two outcomes are to introduce the most suitable merchants in the platform and order the products according to the negotiated price. And the expected users of the MASCB are mainly the customers in a payment platform and its registered merchants. MASCB will contain the below main features:

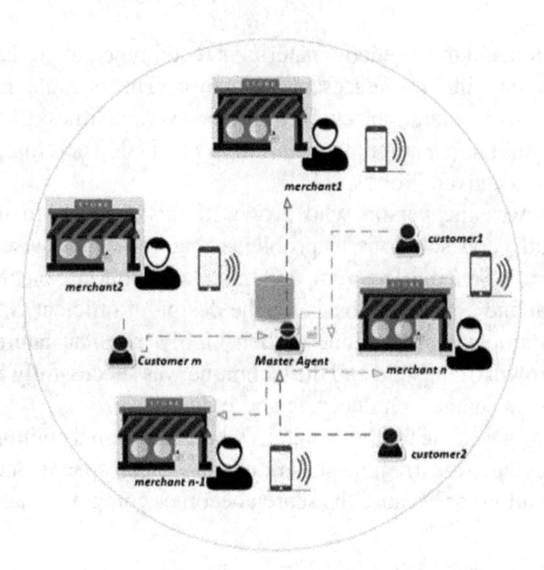

Fig. 1. Process diagram

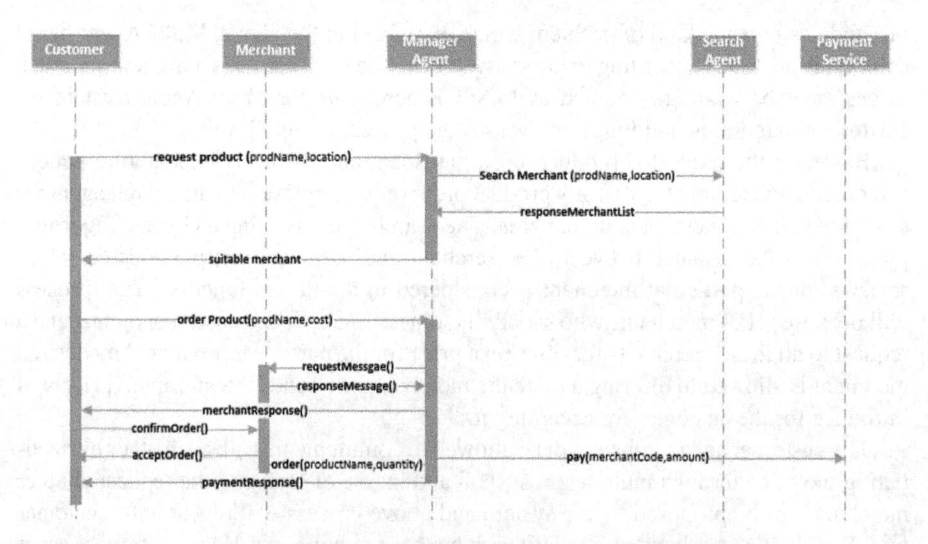

Fig. 2. Merchant customer interaction sequence diagram

- Analyze the past data and select the most suitable merchant in the area.
- Facilitate the customer and merchant to negotiate about the price.
- Accuracy.
- Time saving/efficiency.

5 Design

This system consists of four types of agents as described below. These agents access a common ontology of knowledge of payment platform rule and practices.

The sequence diagram in Fig. 2 provides the processes flow and the objects involve and the messages exchange between them to perform. Customer, merchant, Manager and search Agent interact with each other agent according to the mentioned sequence in this diagram. The process starts with the "request product" request and ends with the "payment response" and the flow of the process happen from top to bottom & left to right order.

Describe the following agent types by indicating their roles/task and their connection with the other agents.

Manager Agent
This agent can communicate and even interrupt the other agents at his wish. This does happen with other agents unless in very exceptional situations. This agent is doing the main coordination part of the whole process. The main role of this agent is initialize agents, support execution, suspend and terminate the other agents.

Customer Agent
This agent represents the customer who initiates the request for ordering product. Agent needs to communicate with manager agent and has to communicate with the merchant agent though the manager agent for negotiating prices.

Merchant Agent
Merchant agent is another key agent in this system. This agent will communicate with the manager agent to make success this process.

Search Agent
The responsibility of this agent is to find the more suitable merchants related to the product request sent by the customer. In this case search agent has to work with Manager Agent and customer agent to check whether the results satisfy the system defines constraints.

Figure 3 showcases the architecture of this system. UI denotes User Interfaces of the application and once any request comes from the customers though the UI it directs into the service layer of the application which handles all the business logics of the system. Service layer connects with the Database (DB) through the Model layer using Java Persistence API (JPA). DB connects with Multi Agent System module and it connects with the Genetic Algorithm (GA) module which uses to identify the set of most suitable merchants in the platform.

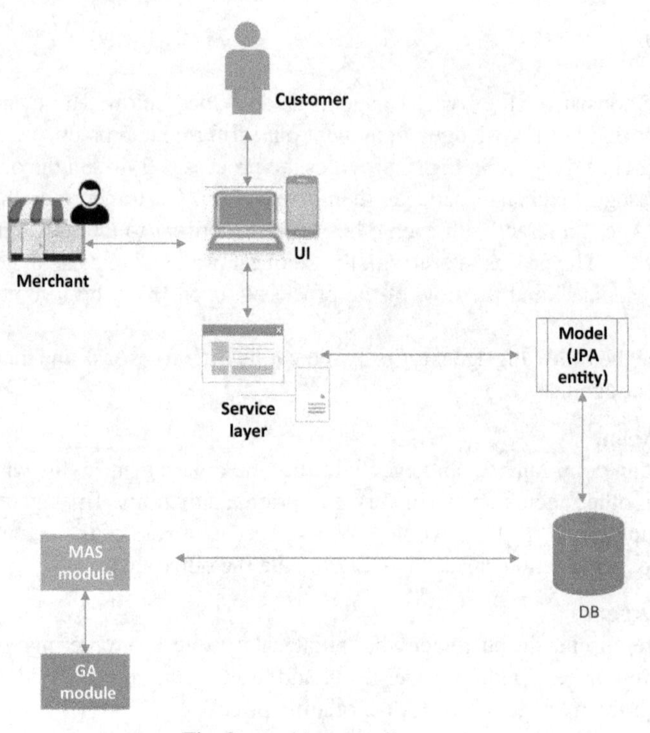

Fig. 3. Architecture diagram

6 Implementation

6.1 Agent Development Pseudo Code

The following pseudo codes are used for implementation of MAS module and which is the core module of this system.

Manager Agent

AGENT INITIALIZATIONS
START SETUP
Create customer agents

```
        Create merchant agents
      END SETUP
```

Customer Agent

```
      AGENT  INITIALIZATIONS
      START SETUP
            ADD BEHAVIOR RequestMerchant
      END SETUP

  // This is the behaviour used by customer agents to request product from
merchant agents
      START BEHAVIOR RequestMerchant
            Send the CFP to Search Agent
            Receive the list of suitable merchants
            Send the CFP to suitable merchants
            Receive all proposals/refusals from merchant agents
            Send the purchase order to the merchant that provided the best
   offer
  Receive the purchase order reply
      END BEHAVIOR
```

Merchant Agent

```
      AGENT  INITIALIZATIONS
      START SETUP

   // Add the behaviour serving queries from customer agents
                    addBehaviour(new OfferRequestsServer());

   // Add the behaviour serving orders from customer agents
                    addBehaviour(new    PurchaseOrdersServer());

      END SETUP
  // This is the behaviour used by customer agents to request product from
merchant agents

      START BEHAVIOR OfferRequestsServer
            Receive CFP messages from customers
            If requested item available
            ACLMessage.PROPOSE
               Else
            ACLMessage.REFUSE
      END BEHAVIOR
```

Search Agent

```
      AGENT  INITIALIZATIONS
      START SETUP
   // Add the behaviour serving queries from customer agents
                    addBehaviour(new proposeMerchants());
```

```
END SETUP
START BEHAVIOR proposeMerchants
      Receive CFP messages from customers
      If suitable merchants available
            ACLMessage.PROPOSE
      Send the merchant list
   Else
      ACLMessage.REFUSE

END BEHAVIOR
```

6.2 UI Implementation

User Interfaces is developed using JSP pages which support for Spring MVC architecture. Figure 4 displays the main screens of this system and customers have to use this UI face to initiate their orders.

Fig. 4. Customer order main UI

7 Evaluation and Discussions

7.1 Evaluation

It is very important to follow a proper mechanism to evaluate the system to keep the quality and deliver an error free solution to the end users. So experiments were designed to evaluate the accuracy of the system while developing the system. As MASCB is a new system module development it was hard to compare it with an existing this kind of a module. Unit testing was done while developing the system and integration & UI testing was conducted at the end of the development process.

Number of testing cycles was conducted for evaluating this solution. The following test scenarios are explaining the success of the solution. Different customers will get different prices for the same product based on the conditions like quantity, number of times earlier visited, loyalty points, etc. The following scenario is for ordering a same product from merchants. Figure 5 displays the customers' order details and Fig. 6 showcases the relevant merchants in the platform.

id	first_name	user_credentials_id	ordrProduct	ordrProductType	quantity	ordrStatus
2	Kamal	1,008	sandwitch	3	1	I
3	Rangamal	1,084	sandwitch	4	1	I
4	Kamal	1,008	sandwitch	3	1	I
5	Achira	1,104	Anchor	4	6	I
6	anushka	938	Anchor	4	1	I
7	nelara	527	Anchor	4	6	I
8	Kamal	1,008	sandwitch	3	1	I
9	anushka	938	Anchor	4	1	I
10	Kamal	1,008	sandwitch	3	1	I
11	Rangamal	1,084	Anchor	4	1	A
12	Achira	1,104	Anchor	4	6	A
13	anushka	938	Anchor	4	1	A

Fig. 5. Customer order details1

id	name	latitude	longitude	merchant_category_id
92	Pthum Food Store	6.9392	79.4136	4
166	Dilan Cooolspot	6.1292	79.5436	4
168	Disny food city	6.3254	79.1436	4
171	Mithuru Cafe	6.7758	79.1036	4
172	Araliya Food Center	6.2534	79.2336	4
184	DJ Cafe	6.9301	79.3136	4
185	Dilan Coffee Shop	6.5292	79.6336	4
187	Anu Cafe	6.2542	79.8556	4
191	Supun Trade Center	6.1235	79.4236	4
228	gamage Cafe	6.9352	79.3236	4
233	Ravi Trade Center	6.9336	79.1136	4
245	SMC	6.9374	79.9936	4

Fig. 6. Relevant merchants

Even though there are busying the same product a merchant in the platform offering the same for different prices as showing in Fig. 7.

```
Anchor sold to agent Cus_Rangamal_1547302232053@192.168.195.1:1099/JADE
==============================
Cus_Rangamal_1547302232053@192.168.195.1:1099/JADE Anchor successfully purchased for 741.0 from agent Mer_Supun Trade Center
==============================
bestMerchant = ( agent-identifier :name "Mer_Supun Trade Center_1547302232299@192.168.195.1:1099/JADE"  :addresses (sequence
bestMerchant = ( agent-identifier :name "Mer_Supun Trade Center_1547302232236@192.168.195.1:1099/JADE"  :addresses (sequence
Customer-agent Cus_Rangamal_1547302232053@192.168.195.1:1099/JADE terminating.
Anchor sold to agent Cus_Achira_1547302232189@192.168.195.1:1099/JADE
==============================
Cus_Achira_1547302232189@192.168.195.1:1099/JADE Anchor successfully purchased for 680.0 from agent Mer_Supun Trade Center_
==============================
Anchor sold to agent Cus_anushka_1547302232236@192.168.195.1:1099/JADE
==============================
Cus_anushka_1547302232236@192.168.195.1:1099/JADE Anchor successfully purchased for 717.0 from agent Mer_Supun Trade Center_
==============================
```

Fig. 7. Results

The Following scenario is for ordering a Sandwich from the merchants in the platform. Figure 8 displays the customer's order detail, Fig. 9 exhibit the relevant merchants in the platform and Fig. 10 showcases the inventory details of the merchants. The search agent proposes set suitable merchants based on the selection conditions like distance, customer ratings and etc. These selected merchant agents propose different prices and the system select the merchant with the best offer as showing in Fig. 11.

id	first_name	user_credentials_id	ordrProduct	ordrProductType	quantity	ordrStatus	buyMerchantId
14	Kamal	1,008	sandwitch	3	1	A	(NULL)

Fig. 8. Customer order details2

id	name	latitude	longitude	merchant_category_id
74	Pathum Food Store	6.9392	79.8435	3
90	The Fruit Shop	6.7321	79.2245	3
132	Dinara Food Center	6.1452	79.6436	3
158	Arpico Foods	6.9392	79.2541	3
160	Keels	6.3214	79.3101	3
177	Durdans Hospital Shop	6.4251	79.4652	3
188	Cargills	6.2245	79.3452	3
192	Perera Cafe	6.3653	79.1245	3
205	Divya Foods	6.3625	79.6435	3
226	Eleven01 Cafe	6.6524	79.3245	3
229	Ruwan Foods	6.1145	79.2543	3
230	Ravi Cafe	6.9377	79.3214	3

Fig. 9. Relevant merchants

id	prod_cat	prod_code	merchant_id	prod_name	price	quantity	status
1	3	ric	132	rice	250.00	25	A
2	3	bak	132	bun	200.00	20	A
3	3	rty	132	roty	100.00	10	A
4	3	bak	90	bread	70.00	40	A
5	3	bak	90	sandwitch	400.00	15	A
6	3	bak	132	bread	65.00	29	A
7	3	bak	74	pitza	750.00	19	A
8	3	ric	74	rice	175.00	29	A
9	4	MIL	191	Anchor	780.00	1	A
10	3	bak	132	sandwitch	275.00	22	A
11	3	bak	74	sandwitch	325.00	19	A

Fig. 10. Merchant inventory details

```
sandwitch sold to agent Cus_Kamal_1547305705576@192.168.195.1:1099/JADE
=============================
Cus_Kamal_1547305705576@192.168.195.1:1099/JADE sandwitch successfully purchased for 261.0 from agent Mer_Dinara Food Center.
=============================
```

Fig. 11. Results

7.2 Discussions

As explained in earlier scenarios, this proposed solution helps the customers to order product online from the merchants in the platform in an easier manner. While developing the system, incremental evaluation is done but formal evaluation due and we are expecting to do it as follows:

- Retention of customers after some time
- Merchant revenue incensement
- Customer satisfaction
- Merchant satisfaction.

The research conducted in the area of customer driven bargaining is limited and it was hard to find an e-commerce platform to facilitate customers for ordering some product by negotiating the prices with the identified best merchants in the echo system by using past data except traditional auction supporting merchants. Therefore, it is hard to compare the results with a previously developed model.

8 Conclusion and Future Work

8.1 Conclusion

The above-mentioned customer driven bargaining system can bring more benefits to customers and sellers by making their lives easier. The present system stands out from the rest of the bargaining systems which are merchant driven. MAS based solution can ideally handle the complexity of negotiation, coordination and communication of the dynamic environment. The platform may provide ideal solutions and recommendations based on the past transaction history and related data stored in the system. This can inevitably lead to greater levels of customer satisfaction and help the seller achieve a competitive edge over other competitors.

8.2 Limitations of the System

The developed MASCB system is a web based system and this should be provided through a mobile application and it may bring more convenience to the both customers and the merchants to use the application in a simpler manner. Customer location should automatically get through the browser or from the mobile app and at present, users need to input their location to the system and this may be a hassle for the customers. These are the identified limitations of this project and this may overcome with future works to deliver a better service to the end users.

8.3 Future Work

In future, MASCB can be integrated with any third part delivery platform through the APIs and this has opened a new business opportunity as well. So this module can be expanded while increasing the scope and it will be more helpful for both customers and merchants in an echo system. And also to overcome the above mentioned limitations providing the solution through a mobile application or integrate with an existing mobile application would improve the user experience to a greater extent.

References

1. Zhang Yuheng, S.K.: A multi agent based e-shopping system. J. Glob. Res. Comput. Sci. **2**(4), 1–11 (2011)
2. Guan, S.-U., Feng, H.: A multi-agent architecture for electronic payment. Int. J. Inform. Technol. Decis. Mak. (2011)
3. Wautelet, Y., Heng, S., Kiv, S., Kolp, M.: User-story driven development of multi-agent systems: a process fragment for agile methods. Comput. Lang. Syst. Struct. **50**(Suppl. C), 159–176 (2017)
4. Leitao, P.: Multi-agent Systems in Industry: Current Trends & Future Challenges. https://pdfs.semanticscholar.org/234c/6a74f071697baf3f8211ba5244e111971850.pdf. Accessed 27 Oct 2017
5. Karunananda, A., Perera, L.: Using a multi-agent system for supply chain management. Int. J. Des. Nat. Ecodyn. **11**, 107–115 (2016)
6. Moyaux, T., Chaib-draa, B., D'Amours, S.: Supply chain management and multiagent systems: an overview. In: Chaib-draa, B., Müller, J.P. (eds.) Multiagent based supply chain management. SCI, vol. 28, pp. 1–27. Springer, Heidelberg (2006). https://doi.org/10.1007/978-3-540-33876-5_1
7. Kulasekera, A., Gopura, R., Hemapala, K.T.M.U., Perera, N.: A review on multi-agent systems in microgrid applications. In: 2011 IEEE PES International Conference on Innovative Smart Grid Technologies-India, ISGT India 2011 (2011)
8. Integrating mobile agent technology with MAS. http://citeseerx.ist.psu.edu/viewdoc/download?doi=10.1.1.467.2880&rep=rep1&type=pdf. Accessed 27 Oct 2017
9. Lucas, A.: The OASIS air traffic management system. https://www.researchgate.net/publication/2702604_The_OASIS_Air_Traffic_Management_System. Accessed 28 Oct 2017
10. Multi agent based approach to assist the design process of 3D game environments. https://www.researchgate.net/publication/261449497_Multi_agent_based_approach_to_assist_the_design_process_of_3D_game_environments. Accessed 27 Oct 2017
11. Tapia, D.I., Alonso, R.S., García, Ó., Corchado, J.M.: Wireless_Sensor_Networks_Real-Time_Locat
12. Pěchouček, M., Mařík, V.: Industrial deployment of multi-agent technologies: review and selected case studies. http://citeseerx.ist.psu.edu/viewdoc/download?doi=10.1.1.145.9080&rep=rep1&type=pdf. Accessed 28 Oct 2017
13. Bradshaw, J.M.: An Introduction to Software Agents, pp. 3–46. AAAI Press, Menlo Park (1997)
14. Solodukha, T.V., Sosnovskiy, O.A.: Multi-Agent Systems for E-Commerce
15. McCall, J.: GA for modelling & optimisation. J. Comput. Appl. Math. **184**, 205–222 (2005)

Digital Ecosystems Control Based on Predictive Real-Time Situational Models

Alexander Suleykin(✉) and Natalya Bakhtadze

V.A. Trapeznikov Institute of Control Sciences, RAS 117997 Moscow, Russia

Abstract. In this paper, the architecture of Digital Ecosystems Control based on Predictive Real-Time Situational models is presented. It has been reviewed Data Fusion issues in Digital Ecosystems and proposed a methodological approach for solving such issues based on modern big data technologies, investigated the Digital Ecosystem control in real-time as a situational management , modeled the architecture of the real-time DES control systems with in-memory technologies, message-oriented systems, virtualization, containerization and ability for horizontal scaling. Finally, the paper proposes the principles of predictive models developing for situational control. The methods are based on the intelligent analysis of DES statistical data and knowledge bases development. Retraining the model is carried out in real-time regime.

Keywords: Digital ecosystems · Data fusion · Big data architecture · Real-time predictive Situational-based identification · Associative search methods

1 Introduction

Nowadays digitalization is taking more and more active role in the development of global economy, and industry and intersectoral digital platforms are transforming into digital ecosystems, which, in particular, allow companies develop innovations, new services and increase their income [1]. The last period of management of developing industrial production was characterized by the development of digital platforms, in particular - the expansion of their functionality [2]. In recent years the trend for development of digital ecosystems has become worldwide in a way that new ecosystems are more and more connected with other external for company ecosystems, sharing data between them and increasing income from different digital services. The pandemic and post pandemic COVID-19 global digital market has been already revealing the high increase in the creation of new digital services for months, generating even more new digital products that lead to high growth of data exchange and need for horizontally scalable, sustainable, reliable, fast-computing, fault-tolerant and self-organizing architecture of such new ecosystems.

Moreover, the data increase generated by new digital services has shown high complexity between ecosystems communication. Thus, it is used different approaches and methodologies, different data volumes, formats and protocols for data transmissions.

K. Arai (Ed.): FICC 2021, AISC 1363, pp. 605–623, 2021.
https://doi.org/10.1007/978-3-030-73100-7_44

Applying control modeling in real time regime generates even higher complexity, which requires application of modern big data technologies to overcome such issues.

In this paper, we will review Data Fusion issues in Digital Ecosystems and propose a methodological approach for solving such issues, investigate the Digital Ecosystem control in real time as situational management, model the architecture of the real time DES control systems with in-memory technologies, message-oriented systems, virtualization, containerization and ability for horizontal scaling. Moreover, the paper proposes the principles of predictive models developing for situational control. The methods are based on the intelligent analysis of DES statistical data and knowledge bases development. Retraining the model is carried out in near-real time regime.

This paper is organized as follows: first, we review the literature of current state of the Digital Ecosystems development. Next, we investigate Data Fusion issues for Digital Ecosystems and review relevant publications. After that, main Big Data methods for the creation of real-time architecture are considered such as in-memory computing, message-oriented systems, virtualization and containerization. Next, it is formulated an approach for Situational modeling as an Identification method. In addition, Associative search identification method for time-varying objects is described as well as the Conditions of the associative model stability in the aspect of the analysis of the spectrum of multi-scale wavelet expansion. Finally, main Digital Ecosystem architecture is modeled and described the role of each Digital Agent and its properties.

2 Related Work

By digital ecosystem we consider a distributed socio-technical system with the properties of adaptability, self-organization and sustainability, functioning in a competitive environment and cooperation between various subjects of this system (automated systems and economic entities) for the exchange of knowledge in the context of the evolutionary development of the system. The digital ecosystem is functioning based on computer network infrastructure using intelligent control technologies, for example, multi-agent technologies [3]. The formation of national digital ecosystems based on global digital space today is necessary a condition for faster growth of the economy of states [1]. Breakthrough economic development can be achieved through high-quality changes in the structure and management system economic assets. One of the most significant assets become voluminous diversified information data. Priority becomes Big Data Management (Big Data) and real-time predictive models. In-memory computing technologies speed up processing large amounts of data and therefore, as such phenomena as large data are becoming increasingly popular among enterprises.

Models of objects and their dynamics, as well as models of dynamic processes, customizable in real time, use patterns recoverable through historical and current data. These patterns are the essence of the term inductive knowledge [4]. Under inductive knowledge understanding of the laws, derived from statistical data that characterize the operation of the object, its “dynamic traces”.

In [5], by a digital ecosystem the authors mean the union of IT-networks and social networks, implementing the exchange of knowledge. In [6] the authors have identified the concept of ecosystems with e-learning ecosystem and digital ecosystem. Digital

Ecosystem provides access to knowledge, global added value cost chains, specific services, the introduction of new business models. In this case, the economy is no longer considered as a fully managed system for which the operation is made up of: its functioning in the framework of the specific situation determines the interaction of the active elements.

In [7] under the DES refers to the temporary pooling of resources and the actions of competitors and partners on specific projects. Within the framework of a specific project, they cooperate and exchange information, and outside the project, competition and autonomous functioning take place. In [8] is reviewed the emphasis on the interpretation of information as a resource that can be used, produced and transformed in the same way as material resources. Ecological idea applies to maintaining and growth of the cost of the user information. In [9] it is concluded that with the development of information and communication technologies at the same time people began to live in a digital and ecological environment, i.e. in a dual environment. In the digital ecosystem, the economy spontaneously transforms into a networked one, that is, into a "continuously flowing space of flows", gaining the ability to continuously update. Nonlinear forms of communication with blurred spatial and temporal boundaries appear [10].

Combining various aspects of the cited definitions and descriptions, we will understand the digital ecosystem as a DT of a distributed social-and-technical system that has the properties of adaptability, self-organization and sustainability, operating under conditions of competition and cooperation between various subsystem (automated systems and economic entities) that exchange knowledge to improve control efficiency and evolutionary development of the system.

All authors note that the digital economy really motivates the development of digital ecosystems, their interconnections, new types of interactions and new intelligent models of DES control. Research continues into the digital ecosystem and expand their applications in such areas as manufacturing and supply chain management, energy, transport, education and health care [6]. In [11], the authors present an opportunity to develop a digital ecosystem for transport and warehouse logistics.

In [5–11], and other works a lot of attention is paid to the principle of multi-agent ecosystem control, the establishment of their relations and to ensure the stable operating. The principles and semantics used in digital ecosystems are formulated in [12].

The functioning of a DES is characterized not only by a large number of stakeholders and the complexity of their interaction, but also by the fragmentation and asynchrony of large amounts of information coming from different sources. The need for on-line analysis of scattered data, efficient utilization of used computing resources, the ability to ensure sustainability and sustainable development require the use of the most modern approaches to the development of the architecture of the DES. Author in [13] considered the problem of system creating for digital ecosystem control based on predictive models in real time.

3 Data Fusion Issues for Digital Ecosystems Development

One use of big data to control the digital ecosystem is the scenario of using these data to create predictive models. For example, based on historical smart grid data, a model can be built that can predict the energy consumption profiles of the next day, which will help better plan the distribution of energy to consumers and therefore provide cheaper energy for the end user. To create the best possible forecasts, it is necessary to take into account not only the current values of power consumption, but also historical data, various flow aggregates and possible derivatives, as well as be able to expand the current data streams with relevant contextual information - weather data, weather forecasts, news services and on-the-fly text analysis, data from social networks based on keywords or other filters, data on human behavior, data from CCTV cameras, and any open sources as appropriate. Thus, to build the most accurate predictive model, it is necessary to have a large amount of scattered data - in different formats, arriving at different intervals, of different volumes and types. To retrain the predictive management model, all data must first be preprocessed to meet the necessary data markup requirements, aggregated, combined and combined with historical data of the same structure. The problem arises of defining such a methodology and architecture, in which a set of architecture components will be clearly defined, the main subsystems and their interaction are defined, the architecture and information flows between subsystems are modeled, in which the system will be stable, scalable, reliable, with high-performance and capable of retraining with minimal delays. Models based on both new input vectors and historical data, followed by seamless switching of the control model working at the current time.

The two main reasons that such a data fusion task is not the trivial one are the heterogeneity of IoT data and its temporal inconsistency. According to [14], heterogeneity is a property of big data. Many definitions of heterogeneous data can be found in the literature, and there is no consensus among different researchers about the definition. Properties illustrating the problem include: multimodality of data (even taking into account a combination of continuous and categorical features and structured and unstructured data) and technical aspects (i.e. data format), the degree of independence of some data from others, concept bias, dynamics of change and security. The temporal aspect of data inconsistency manifests itself in the following properties:

- sampling frequency;
- temporary delay;
- data availability.

The sampling rate differs from sensor to sensor. Some sensors provide a constant sample rate. Different sensors within a setup can implement constant but different sample rates. Many sensors implement approximately constant sample rates. Some sensors can use arbitrary sampling rates (that is, they can only report an event). Time delay correlates with data transmission delays, legacy systems, privacy and access issues. Measurements can be delayed by several milliseconds or more, depending on various conditions.

Latency is closely related to data availability. A data stream is a sequence of values with a corresponding time stamp (in the IoT, a time stamp indicates the time when a measurement was made). A sequential time series is defined in a stream to create a sequence where each subsequent measurement in the series was taken later than the previous one.

Merging data from different sensors is necessary to enrich the data and improve the efficiency of the predictive models. Data fusion is used for various cases in industry -

for positioning and navigation [15, 16], for actions recognition [17, 18], in monitoring systems and diagnostics of various malfunctions [19–24], in transport [25], in healthcare [26] and in other industries. For example, in healthcare, data fusion is used in IoT settings such as remote patient monitoring systems. Here, the patient is monitored using various body and environmental sensors, whose signals are processed and used to inform doctors about the patient's condition. Data fusion is used at various levels: the raw data level - combining raw and raw data from sensors, the functional level - combining functions provided by various methods, and the decision level - combining several expert decisions. Fusion is also used to compute context awareness, which is used to assign dynamic roles to doctors. The use of fuzzy logic in the context of understanding is discussed in [27]. Integration of the proposed methodology and architecture in this paper, for example, into the medical field of remote patient monitoring can provide additional improvements - the inclusion of historical data in modeling, as well as their use in combination with the current streams of real data from sensors.

Most of the sensor data fusion techniques mentioned assume that all data will be consistent, available immediately, and arriving in the correct order. This is true in many proprietary systems, but in IoT scenarios, data availability exacerbates most of the problems. Access control plays an important role in data management and is a prominent topic in recent scientific publications [28–30]. The methodology and architecture proposed in this paper is based on mechanisms described in the literature and can take advantage of recent findings in work, especially related to streaming platforms. The processing of delayed or inconsistent measurements is discussed in [31]. Many of the systems also include significant domain knowledge (model-based approaches) in the data fusion models [32], in which the models lose their generalizability potential. Frameworks and architectures that can be applied in different use cases should be domain independent and solely data driven [33]. The idea was further developed using machines for fusion of heterogeneous features [15], however, the task of creating a methodology and architecture to solve data fusion issues in real time remains outside the scope of the work.

In large datasets and streams of data, data preprocessing and data reduction becomes critical for knowledge extraction [34, 35]. In these papers, the authors discuss the important role of the effectiveness of machine learning systems for real time. Basic preprocessing functions include drift detection and adaptation, missing data search, noise processing, data reduction, and efficient and accurate sampling of the algorithm stream.

Automatic data analysis is useless if data preprocessing requires manual intervention [35]. The authors present adaptive preprocessing that improves the final prediction accuracy based on real sensory data. In our paper, it is proposed to use the same strategy, however, the paper focuses on building a scalable system architecture and methodology for solving data fusion issues and retraining predictive models based on a constantly updated knowledge base in real time.

According to the reviewed publications, Data Fusion issues are highly important and are still ongoing research subject. In order to create an efficient high-speed scalable real-time control models with constantly increasing data flows and historical database, special DE architecture is needed based on big data technologies. In the next chapters we will regard main methods for the creation of such architecture: in-memory computing, message-oriented systems, virtualization and containerization. Finally, we will present the architectural model based on such technologies.

4 In-Memory Computing

In-memory database management systems (in-memory database systems, IMDS) are a growing segment of the global database market. The creation of an in-memory DBMS was a response to the emergence of new challenges facing applications, new system requirements and the operating environment.

An in-memory DBMS is a database management system that stores information directly in RAM [36]. This contrasts sharply with the approach of traditional DBMSs, which are designed to store data on stable media. Since data processing in RAM is faster than accessing and reading information from the file system, an in-memory DBMS provides an order of magnitude higher performance of software applications. Because the design of in-memory DBMSs is much simpler than traditional design, they also have much lower memory and CPU requirements.

Also, the in-memory DBMS does not bear any load from data input/output operations. Initially, the architecture of such databases is more rational, and the memory load processes and processor cycles are optimized.

In-memory DBMSs are typically used for applications that require ultra-fast data access, storage, and manipulation, as well as systems that do not have a disk but still need to manage a significant amount of data. In this paper, this is the main class of the systems, which is proposed to be used for the possibility of retraining models in a real time mode. The data will be stored in memory, which will allow quick access to them, and performance issues are proposed to be solved using scale-out technologies. In-memory DBMSs can scale perfectly well beyond terabytes. Based on the analysis of the report [37] on the use of DBMS in memory, we can conclude that they show almost linear horizontal scalability, and many companies already use such systems for fast data processing and calculations in memory, where minimum latency is required. In addition, in-memory calculations are also used for deep learning [38], but this thesis proposes to constantly replenish the knowledge base (in-memory storage) and retrain the model only when it is necessary based on the preliminary analysis of the incoming data.

Thus, in-memory computing technologies are the main method of storing data for fast retraining of the model. Only on the basis of fast access to data - reading and writing - it is possible not only to retrain the model in near real time, but also to do it on the constantly increasingly updated knowledge base.

5 Message-Oriented Systems

Messaging systems are a distributed system based on asynchronous messaging between system components, while message-oriented middleware is the product on which the messaging system is built. Unlike traditional systems, in messaging systems, applications do not interact directly, but through MOM (Message Oriented Middleware) - specialized middleware, which is asynchronous systems of interaction between the server and the client. If one component of the system wants to send a message to another component, it sends the message to the MOM, and then the MOM forwards it to the recipient [39].

The application that sent the message does not have to wait for a response and can continue its current activity. Neither the application sending the message nor the recipient of the message should be active at the same time. If the destination of the message is not active, MOM ensures that the message is delivered as soon as the destination becomes active. System components are not directly connected to each other (disconnected), and therefore, it is possible to transfer components from one server to another at runtime without compromising system performance.

Earlier studies have compared different messaging systems [40], and analysis shows that most systems are infinitely scalable and fault-tolerant. At the same time, these systems also have significant performance, which can be increased based on needs. These characteristics of messaging systems will make the system high-performance, fault-tolerant and horizontally scalable. There are two "basic" messaging models [39]:

- point-to-point;
- publication-subscription (pub-sub).

The point-to-point model is used when one or more components (called senders) need to send a message to one receiver component. The model is based on the concept of a message queue. Senders send messages to the queue, and the recipient reads messages from the queue (Fig. 1):

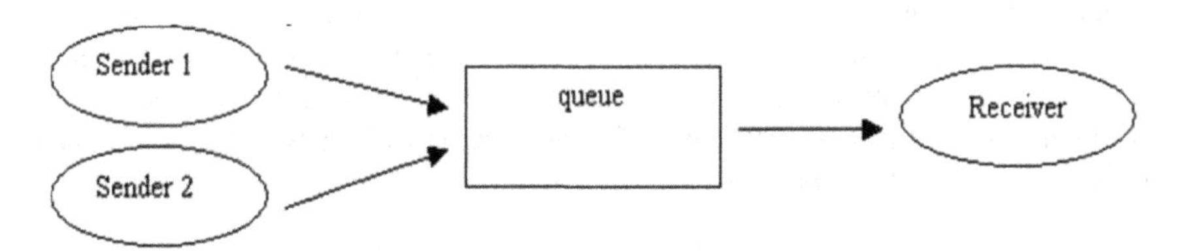

Fig. 1. Point-two-point model for messaging systems

It is often assumed that there is one and only one receiver in a point-to-point model. However, this is not quite true. Several recipients can be connected to one queue. But the system will deliver the message to only one of them, and who exactly depends on the implementation. Some MOMs deliver the message to the first registered recipient, while there are implementations that loop through.

The publish-subscribe model is used when one or more components (publishers) need to send a message to one or more subscribers. This model is based on the concept of a message subject.

Publishers send messages to a specific topic, and all specified topics receive these messages (Fig. 2):

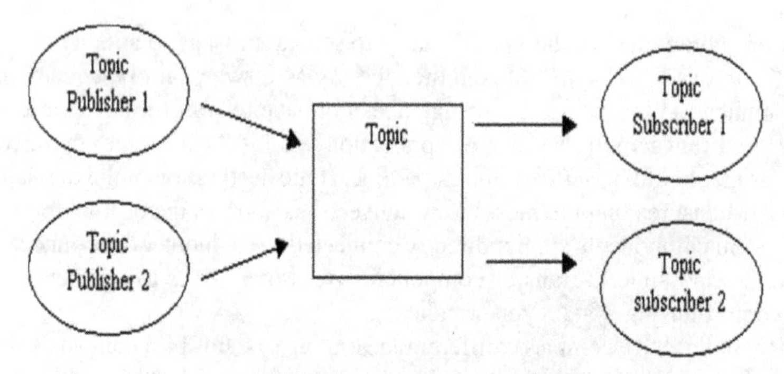

Fig. 2. Publish-subscribe model for messaging systems

Which model to use depends on the solution to a specific business problem. In this paper, a universal model is proposed that is suitable for solving most business problems - the publish-subscribe model. This model is used to exchange data between different subsystems in real time - firstly, it is the main entry point of all data into the system for further processing; secondly, this model is used for further transfer to other subsystems only the necessary cleaned, aggregated, normalized and preprocessed data; thirdly, with the help of this model, interaction occurs within each subsystem (for example, when more than 1 model is sequentially applied for cleaning, aggregation, filtration, etc.); fourthly, sending a signal about the need to retrain the predictive model also comes from this model.

Thus, messaging systems are very important for high-performance processing of data in real time. This class of systems will allow the exchange of data between various system components and within the components themselves in real time, and the high performance and horizontal scalability of such systems will allow the system to scale based on the necessary needs.

6 Virtualization and Containerization

Virtualization is a technology for creating virtual instances of computer resources for multiple uses of the same physical resource. There are several virtualization technologies that are capable of virtualizing server, storage, networks, and operating systems. Virtualization technology is known for being easy to configure and optimizing the cost of setting up an environment with higher processing efficiency. The virtualization environment separated the actual components of resources, such as memory (RAM), disk space, and network, as a separate group of resources. Virtualization helps organizations scale compute resources. Virtual machines (VMs) are a popular use of hardware virtualization that is managed through a hypervisor [41]. It is one physical machine with multiple virtual machines. Large companies with tens or even hundreds of servers are using server virtualization to improve their efficiency and consolidate the number of working computers. This can help businesses by allowing fewer devices to be serviced, better utilized by those devices, and more reliable backup and recovery procedures. After configuring the virtual server, it is possible to set up new virtual servers using the same configuration, which only takes a few minutes.

The following advantages of virtualization can be distinguished [41, 42]:

- saving space and operating costs;
- simple management of data centers;
- increasing the productivity of IT departments;
- increasing the reliability and fault tolerance of the system in case of failures;
- reducing the time of resource allocation when users request it;
- no need to install hardware components;
- automatic updating of hardware and software by providers;
- reduced energy consumption because less energy is consumed on physical servers;
- efficient utilization of all computing resources.

Virtual machines come with a host of challenges in terms of managing dedicated resources. They simplify the task of working in ever-changing data centers, but they are difficult to scale. The ability to run four virtual machines on the same dedicated hardware is a boon for power-hungry workers and space teams.

For the past five years, teams have addressed these issues by implementing containerization [43]. In many ways, containerization is best seen as the natural evolution of virtualization. Virtualization treats each virtual machine as its own logical (but not physically) separate server. Container handling treats each application as its own, logically separate server. Multiple applications will use the same underlying operating system. These containers are unaware that other containers are running on their dedicated hardware. If they want to communicate with another server, they need to communicate through a network interface as if they were on different physical devices [44].

Thus, the advantage of containerization is that there is no need to allocate hardware resources for redundant operations. Rather than allocating operating system cores and memory for each virtual machine, a server is created with one underlying virtual machine, and cores and memory are allocated for only one application that runs in each container. Application containers work by virtualizing the operating system, in particular the Linux kernel. Each application container thinks it is the only application running on the computer. Each container is defined as one running application and a set of supporting libraries. One of the significant differences between containers and virtual machines is that containers are immutable. That is, they cannot be changed by any action that is taken. This is done on purpose - this means that every time a container is created it will be the same.

No matter where it is created, on what hardware or what operating system, the container will perform the same. This consistency eliminates a whole class of bugs. While many developers are used to taking snapshots of virtual machines, it's easy to imagine a container this way. Each time the container is started, it is restored to its original snapshot, regardless of what actions were last performed.

7 Situational Control in Digital Ecosystems

Situational control is understood as a methodology for complex semi-structured technical, organizational and socio-economic objects control using artificial intelligence methods. The approach was proposed in 1958 by H. Simon and A. Newell [45]. Later, this approach became the basis of the control methodology, within the framework of which various specific methods were proposed.

"Identification" of the control object is carried out on the basis of comparing the state of the object, represented as a features set, with different reference situations (reference features set), in most cases - using the "nearest neighbor" method. Control algorithms in the corresponding technical (and not only) problems use, as a rule, not the traditional approach of integra-differential and difference equations, but logical-linguistic, fuzzy, neuro-fuzzy and other approaches based on the artificial intelligence methods.

The literature also contains numerous examples of this approach application.

The direction of research is, of course, constructive and the further, the more promising, and for a large class of control objects - the most (and sometimes the only) acceptable.

Recently, in the international scientific literature there are quite a lot of works devoted to this area, which is called the synthesis of systems with situational awareness (Situational Awareness) [46].

8 Predictive-Based Identification Model Development

8.1 Data Normalization for Coding Situational Data Flows

The method for constructing a situation model by means of binary coding of key features presented in Sects. 7.1 and 7.2' is not always convenient. In particular, if some of the features characterizing the situation are categorical. For example, a color characteristic (categories: "red", "yellow", "blue", ...). In some cases, features can be described by fuzzy variables. Otherwise, it is possible to encode categorical features in other ways, starting with primitive numbering within the permissible set of values, and ending with the use of all kinds of projection methods on the real axis.

Traditional normalization of data values can be an efficient way of encoding. The most famous methods are: Standard Scaling (Z-score normalization) and MinMax Scaling.

$$\tilde{x}_{ik} = \frac{x_{ik} - x_{\min i}}{x_{\max i} - x_{\min i}}$$

$$y_{jk} = y_{\min j} + \tilde{y}_{jk}\left(y_{\max j} - y_{\min j}\right)$$

Linear algorithms are for normalization. Among nonlinear algorithms, the most popular algorithms are those using the sigmoid logistic function or the hyperbolic tangent. The transition from traditional units of measurement to normalized and back during normalization and denormalization within [0, 1] is carried out according to the formulas:

$$\tilde{x}_{ik} = \frac{1}{e^{-a(x_{ik} - x_{ci})} + 1};$$

$$y_{jk} = y_{cj} - \frac{1}{a}\ln\left(\frac{1}{\tilde{y}_{jk}} - 1\right),$$

where $x_{c\,i}, y_{c\,j}$ are the centers of the normalized intervals of the input and output variables variation:

$$x_{ci} = \left(x_{\min i} + x_{\max i}\right)/2, \;\; y_{cj} = \left(y_{\min j} + y_{\max j}\right)/2;$$

8.2 Situational-Based Predictive Model Development as Identification Approach

The digital twins (of both control objects and situations to be controlled) can be interpreted as digital identification models. Accordingly, their values at specific points in time can be considered as dynamic traces used in the associative search algorithm.

The construction of digital identification (including predictive) models of a certain situation, in particular, for assessing the state of the DES, is carried out as follows:

- A set of indicators and/or signs is formed that characterize the situation for the control object.
- The values of numerical indicators and/or attributes are normalized, as a result we obtain a projection of these values on the segment [0,1].
- Values of binary signs are considered.
- Coding categorical attributes carried by the projection on the interval [0,1], or by projecting the real axis, followed by normalization.
- The identification model of the situation (in particular, the predictive one) is developed for the control object, the input variables of which are encoded signs of the situation. The associative search algorithm is used as identification algorithms.

When it is solved a specific problem, the data is normalized using standard functions of integrated statistical analysis environments. It can be, for example, IBM SPSS, SAS or special python libraries for ML and data analysis: Scikit-learn, pandas, numpy, etc. It is possible to write original code in R, Python or any other language.

8.3 Associative Search Identification Method for Time-Varying Objects

What if the object is time varying? In this case, both for the case of non-stationary input variables and for the unknown internal dynamics of the object, it is advisable to carry out research in the space of wavelet transformations.

Within the last two decades, applying the wavelet-transform to the analysis of non-stationary processes has been widely used. The wavelet-transform of signals is a generalization of the spectral analysis, for instance, with regard to the Fourier transform. First papers on the wavelet analysis of time (spatial) series with a pronounces heterogeneity have appeared in the middle of 1980s [46]. The method was positioned as an alternative to the Fourier transform, localizing the frequencies but not providing the time extension of a process under study. In sequel, the theory of wavelets has appeared and is developed, as well as its numerous applications.

Today the wavelet analysis is used for processing and synthesis of non-stationary signals, solving problems of compressing and coding information, image processing, in the theory and practice of the pattern recognition, in particular, in the medicine, and in many other branches. The practice of applying wavelets was found effective for studying geophysical fields, time meteorological series, forecasting earth quakes. The approach is effective for research of functions and signals being non-stationary in the time are heterogeneous in the space, when results of the analysis are to contain not only the frequency signal characteristics (distribution of the signal power over frequency components), but also information on local coordinates, at which certain groups of the frequency components are manifested, or at which fast changes of the frequency components of the signal are the case.

The wavelet analysis is based on applying a special linear transform of processes to study real data interpreted by these processes, characterizing processes and physical properties of real plants, in particular, technological processes. Such a linear transform is implemented by use of special soliton-like functions (wavelets) forming an orthonormal basis in L^2.

In comparison to the Fourier transform apparatus, when a function is used generating the orthonormal basis of the space by use of the scale transform, the wavelet transform is formed by use of a basis function localized in a bounded domain and belonging to the space, i.e. to the all numerical axis. In comparison to the "window" Fourier transform, to obtain the transformation on one frequency all time information is not already required. The Fourier transform does not provide information on local properties of a signal under fast enough changes in the time of its spectral make-up. Thus, the wavelet transform may provide one with frequency-time information on a function, which, in many practical situations is more actual than information obtained by the standard Fourier analysis [47]. Wavelets also provide a powerful tool of approximation, which may be used for synthesis of functions that are poorly approximated by other methods, with a minimal number of the basic functions.

The wavelet transform serves, basically, to analyze data in the time and frequency domains. The wavelet theory may be used for the system identification. An investigation of the interaction of the identification theory and wavelet analysis was, to some extent, presented in [48–50]. Besides the direct wavelet analysis, for system identification there may be used bi-orthogonal wavelets, wavelet-frames, wavelet- networks. Preising [50] studied the identification of systems with a specific input/output structure, in which the parameters are identified via spline-wavelets and their derivatives. In [51] an extended by use of an orthonormal transformation least squares method is presented in order to revealing useful information from data.

8.4 Conditions of the Associative Model Stability in the Aspect of the Analysis of the Spectrum of Multi-Scale Wavelet Expansion

Let a predicting associative model of a non-linear time-varying plant meet Eq. (3). For the selected detail level L for the current input vector $x(t)$, we obtain the multi-scale expansion [47]:

$$x(t) = \sum_{k=1}^{N} c_{L,k}^{x}(t)\varphi_{L,k}(t) + \sum_{l=1}^{L}\sum_{k=1}^{N} d_{l,k}^{x}(t)\psi_{l,k}(t),$$
$$y(t) = \sum_{k=1}^{N} c_{L,k}^{y}(t)\varphi_{L,k}(t) + \sum_{l=1}^{L}\sum_{k=1}^{N} d_{l,k}^{y}(t)\psi_{l,k}(t), \quad (6)$$

where: L is the depth of the multi-scale expansion ($1 \le L \le L_{max}$, where $L_{max} = \lfloor \log_2 N^* \rfloor$ and N^* is the power of the set of states of the system in the System Dynamics Knowledge Base); $\varphi_{L,k}(t)$ – are scaling functions; $\psi_{l,k}(t)$ are the wavelet functions that are obtained from the mother wavelets by the tension/combustion and shift

$$\psi_{l,k}(t) = 2^{l/2}\psi_{\text{mother}}\left(2^{l}t - k\right)$$

(as the mother wavelets, in the present case we consider the Haar wavelets); l is the level of data detailing; $c_{L,k}$ are the scaling coefficients, $d_{l,k}$ are the detailing coefficients. The coefficients are calculated by use of the Mallat algorithm.

In [47, 52] it was shown that a sufficient condition of the stability of plant (3) (and, hence, also (11) is as follows: for $\forall k = \overline{1,N}$ meeting the inequalities is to be provided:

- if $m > R$, $R = \max_{s=\overline{1,S}} r_s$, then the condition for the detailing coefficients:

$$\left|a_m d_{lk}^{y}(t-m)\right| < \left|d_{lk}^{y}(t)\right|,$$

  for the approximating coefficients:

$$\left|a_m c_{Lk}^{y}(t-m)\right| < \left|c_{Lk}^{y}(t)\right|,$$

- if $m < R$, $R = \max_{s=\overline{1,S}} r_s$, then the condition for the detailing coefficients:

$$\left|\sum_{s=1}^{S} b_{sR} d_{lk}^{s}(t-R)\right| < \left|d_{lk}^{y}(t)\right|,$$

  for the approximating coefficients:

$$\left|\sum_{s=1}^{S} b_{sR} c_{Lk}^{s}(t-R)\right| < \left|c_{Lk}^{y}(t)\right|,$$

– if $m = R \neq 1, R = \max_{s=\overline{1,S}} r_s$, then the condition of the stability for the detailing coefficients:

$$\left| a_m d_{lk}^y(t-m) + \sum_{s=1}^{S} b_{sm} d_{lk}^s(t-m) \right| < \left| d_{lk}^y(t) \right|,$$

for the approximating coefficients:

$$\left| a_m c_{Lk}^y(t-m) + \sum_{s=1}^{S} b_{sm} c_{Lk}^s(t-m) \right| < \left| c_{Lk}^y(t) \right|,$$

– if $m = R = 1, R = \max_{s=\overline{1,S}} r_s$, then the condition of the stability for the detailing coefficients: $\left| a_1 d_{lk}^y(t-1) + \sum_{s=1}^{S} b_{s1} d_{lk}^s(t-1) \right| < \left| d_{lk}^y(t) \right|$,
for the approximating coefficients:

$$\left| a_1 c_{Lk}^y(t-1) + \sum_{s=1}^{S} b_{s1} c_{Lk}^s(t-1) \right| < \left| c_{Lk}^y(t) \right|.$$

9 Digital Ecosystem Architecture

To solve described problems, it is necessary to build an architecture that can collect, aggregate, normalize and combine data streams in real time as well as conduct constant retraining of the predictive model. Thus, the methodology proposed in this paper includes the following Digital Agents:

- Data sources Agent. It is represented by real-time measurements, which are various data from sensors of production processes, IoT sensors (sensors), data on the metrics of the functioning of any real system – situational data. This type of data is mainly characterized by a large continuous flow, sometimes with a rather low degree of structure (for example, logs). Data from sensors contains metrics of production processes, IoT metrics, which contain information about the functioning of a particular production process, metrics of equipment, installations, etc. Data is transmitted in a real-time manner.
- Preprocessing and Aggregation Agent. An Agent that is responsible for the formation of basic sets of incoming data for subsequent retraining. In this subsystem, the main logic of all aggregations, merges of data streams, transformations, cleansing, normalization, data caching and any other operations take place, which are aimed at converting all raw data streams into a storage structure, based on which predictive models can be retrained in real time. How quickly the predictive model can retrain on an ever-increasing amount of data depends on data preprocessing and efficient storage organization. It is also worth noting that constant interaction with the next Agent - the messaging Agent - allows to build an infinite number of different solutions in stream mode, which, when executed sequentially in real time, from the raw stream will

make a new stream for efficient storage and fast access. This Agent is responsible for building auxiliary models, which does all the necessary preprocessing, and many operations can be combined into one auxiliary model (for example, merging two streams and removing duplicates and normalization), and some operations require a separate auxiliary model (for example, serialization). The required number of auxiliary models is determined based on the business problem, and from the interaction occurs in streaming mode through the messaging system. The main technology requirements for auxiliary models are fault tolerance, horizontal scalability, in-memory computing, reliability, and high performance.

- Digital Twins Agent. An Agent for real object modeling in particular time period mostly based on situational data. This Agent models the current state of the real object forming a digital copy.
- In-memory storage Agent. An Agent for storing "hot" data - data that needs to be quickly accessed. This subsystem stores the basis for predictive models - a knowledge base that is constantly updated with the help of auxiliary models, on the basis of which a new virtual model is built in a specific period of time. This subsystem is characterized by extremely fast data access (read speed), as well as fault tolerance, horizontal scalability, in-memory computing, reliability and high performance. Without an efficient data storage system for fast access, it is impossible to quickly recalculate the predictive model due to computational constraints.
- Messaging Agent. An Agent for exchanging messages in streaming mode (Real-time data streams) between all components of the architecture as well as the input subsystems for all incoming data flows in real-time mode. With the help of this Agent, it is possible to build any logic of interaction of auxiliary models for preparing data for storage with the lowest latency, since messages appear in streaming mode. This layer is the basis of the architecture, since it is this layer that allows universally interconnect all components - any auxiliary models, raw data from sensors, the Agent of "hot" data storage, as well as the simulation subsystem itself. Very common cases when the auxiliary model, after processing and combining several streams from the data from the sensors, will put the results of the data fusion back into the messaging subsystem, where the new vector will be immediately processed by the hot storage subsystem, and then, also immediately, the data will be read by a predictive model, then retrained and built a new one. Thus, the messaging system binds together all the components of the architecture, being a kind of a clipboard and the main level of communication between Agents. It should be noted that this Agent must be horizontally scalable, fault-tolerant, reliable and high-performance.

The Digital Ecosystem Architecture Control based on Predictive Real-Time Situational models is shown in Fig. 3:

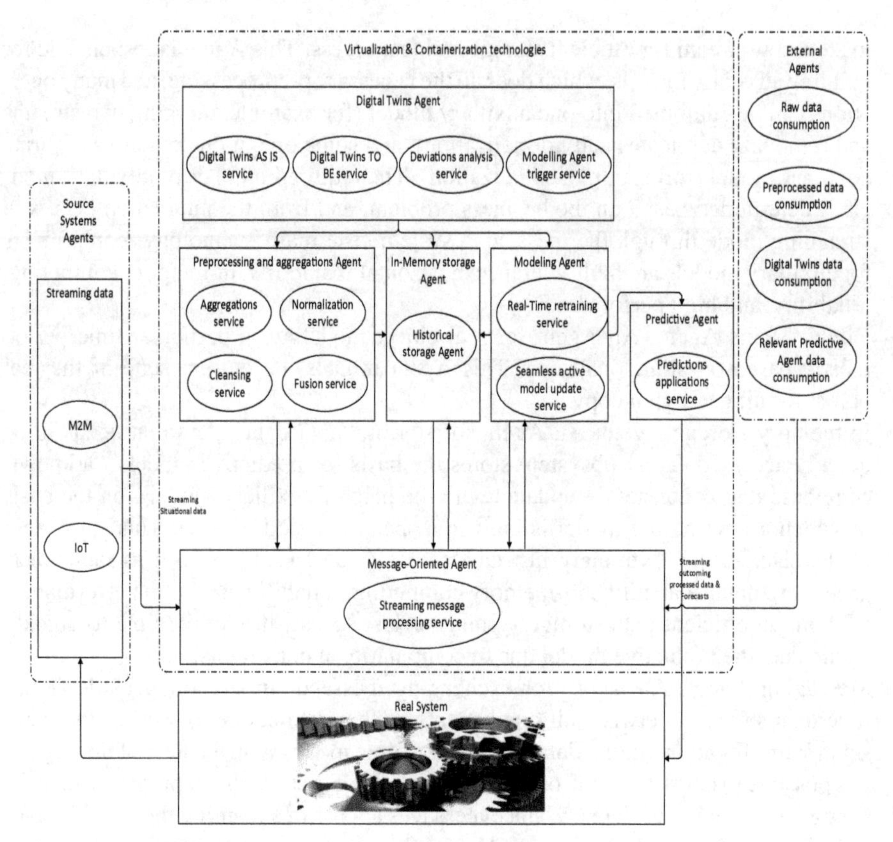

Fig. 3. The digital ecosystem architecture control based on predictive real-time situational models

10 Conclusion

In this paper, we have reviewed data fusion issues and literature for development of digital ecosystems. It is proposed in-memory computing methods, messaging systems, virtualization and containerization technologies for development of digital ecosystems. The use of such technologies will make all digital agents stable, reliable, horizontally-scalable, fault-tolerant and self-organizing. Proposed architecture is capable with streaming data volumes from IoT and M2M sensors, different data formats, asynchrony with incoming data flows and high-performance data processing in real-time. Also, it is formulated an approach to the management of digital ecosystems control based on predictive models of the situation in real-time mode. The situation-based modeling represents the state of the system and the state of environmental factors, which at a given time can have a significant impact on the system.

Also, methods for constructing predictive models of nonlinear and non-stationary systems using associative search algorithms are proposed on the basis of constantly increasing knowledge base. In addition, the stability criterias are presented by studying the spectrum of the multiple-scale wavelet decomposition.

Acknowledgments. The reported study was funded by RFBR according to the research project № 20-31-70001.

References

1. © 2018 International Bank for Reconstruction and Development/ World Bank 1818 H Street NW, Washington DC https://openknowledge.worldbank.org/bitstream/handle/10986/30584/AUS0000158RU.pdf?sequence=4&isAllowed=yMark
2. Babiolakis, M.: Forget Products. Build Ecosystems. How products are transforming to open interconnectable interfaces. https://medium.com/@manolisbabiolakis/forget-productsbuild-ecosystems-792dea2cc4f2
3. Senyo, P.K., Liu, K., Effah, J.: Understanding behaviour patterns of multi-agents in digital business ecosystems: an organisational semiotics inspired framework. In: Kantola, J.I., Nazir, S., Barath, T. (eds.) AHFE 2018. AISC, vol. 783, pp. 206–217. Springer, Cham (2019). https://doi.org/10.1007/978-3-319-94709-9_21
4. Vapnik, V.N.: Statistical Learning Theory. John Wiley & Sons Inc., New York (1998)
5. Nachira, F., Dini, P., Nicolai, A.A.: Network of Digital Business Ecosystems for Europe: Roots, Processes and Perspectives. Digital Business Ecosystems. European Commission, Bruxelles (2007)
6. Chang E., West, M.: Digital ecosystems: a next generation of the collaborative environment. In: iiWAS, pp. 3–24 (2006)
7. Baker, K.S., Bowker, G.C.: Information ecology: open system environment for data, memories, and knowing. J. Intell. Inf. Syst. **29**(1), 127–144 (2007)
8. Kastel's, M., kul'tura. M.: Informatsionnaia epokha. Ekonomika, obshchestva, GU VShE. 129 s. (2000). (in Russian)
9. Fuller, M.: Media Ecologies: Materialist Energies in Art and Technoculture (Leonardo Books). The MIT Press, Cambridge (2007)
10. Papaioannou, T., Wield, D., Chataway, J.: Knowledge ecologies and ecosystems. An empirically grounded reflection on recent developments in innovation systems theory. Environ. Plan C: Govern. Policy **27**(2), 319–339 (2009)
11. Camarinha-Matos, L.M., Afsarmanesh, H., Galeano, N., Molina, A.: Collaborative networked organizations - concepts and practice in manufacturing enterprises. Comput. Ind. Eng. **57**(1), 46–60 (2009)
12. Chang, E., West, M.: Digital Ecosystems and comparison to existing collaboration environment. WSEAS Trans. Environ. Dev. **2**(11), 1396–1404 (2006)
13. Seuring, S.: A review of modeling approaches for sustainable supply chain management. Decis. Support Syst. **54**(4), 1–8 (2012)
14. Kolomvatsos, K., Anagnostopoulos, C., Hadjiefthymiades, S.: Data fusion and type-2 fuzzy inference in contextual data stream monitoring. IEEE Trans. Syst. Man Cybern. Syst. **47**, 1839–1853 (2017). https://doi.org/10.1109/TSMC.2016.2560533
15. Zhang, L., Xiao, N., Yang, W., Li, J.: Advanced heterogeneous feature fusion machine learning models and algorithms for improving indoor localization. Sensors **19**, 125 (2019). https://doi.org/10.3390/s19010125
16. Bouguelia, M.R., Karlsson, A., Pashami, S., Nowaczyk, S., Holst, A.: Mode tracking using multiple data streams. Inf. Fus. **43**, 33–46 (2018). https://doi.org/10.1016/j.inffus.2017.11.011
17. Kong, J.L., Wang, Z.N., Jin, X.B., Wang, X.Y., Su, T.L., Wang, J.L.: Semi-supervised segmentation framework based on spot-divergence supervoxelization of multi-sensor fusion data for autonomous forest machine applications. Sensors **18**, 61 (2018). https://doi.org/10.3390/s18093061

18. Wu, J., Feng, Y., Sun, P.: Sensor fusion for recognition of activities of daily living. Sensors **18**, 4029 (2018)
19. Ma, M., Song, Q., Gu, Y., Li, Y., Zhou, Z.: An adaptive zero velocity detection algorithm based on multi-sensor fusion for a pedestrian navigation system. Sensors **18**, 3261 (2018). https://doi.org/10.3390/s18103261
20. Zhou, Y., Xue, W.: A multisensor fusion method for tool condition monitoring in milling. Sensors **18**, 3866 (2018). https://doi.org/10.3390/s18113866
21. Shi, P., Li, G., Yuan, Y., Kuang, L.: Data fusion using improved support degree function in aquaculture wireless sensor networks. https://doi.org/10.3390/s18113851
22. Zhou, F., Hu, P., Yang, S., Wen, C.: A multimodal feature fusion-based deep learning method for online fault diagnosis of rotating machinery. Sensors **18**, 3521 (2018). https://doi.org/10.3390/s18103521
23. Lu, K., Yang, L., Seoane, F., Abtahi, F., Forsman, M., Lindecrantz, K.: Fusion of heart rate, respiration and motion measurements from a wearable sensor system to enhance energy expenditure estimation. Sensors **18**, 3092 (2018). https://doi.org/10.3390/s18093092
24. Hu, J., Huang, T., Zhou, J., Zeng, J.: Electronic systems diagnosis fault in gasoline engines based on multi-information fusion. https://doi.org/10.3390/s18092917
25. Wu, B., Huang, T., Jin, Y., Pan, J., Song, K.: Fusion of high-dynamic and low-drift sensors using Kalman filters. Sensors **19**, 186 (2019). https://doi.org/10.3390/s19010186
26. Akbar, A., Kousiouris, G., Pervaiz, H., Sancho, J., Ta-Shma, P., Carrez, F., Moessner, K.: Real-time probabilistic data fusion for large-scale IoT applications. IEEE Access **6**, 10015–10027 (2018)
27. Kayes, A., Rahayu, W., Dillon, T., Chang, E., Han, J.: Context-aware access control with imprecise context characterization for cloud-based data resources. Future Gener. Comput. Syst. **93**, 237–255 (2019). https://doi.org/10.1016/j.future.2018.10.036
28. Colombo, P., Ferrari, E.: Fine-grained access control within NoSQL document-oriented datastores. Data Sci. Eng. **1**, 127–138 (2016). https://doi.org/10.1007/s41019-016-0015-z
29. Kayes, A.S.M., Rahayu, W., Dillon, T.: Critical situation management utilizing IoT-based data resources through dynamic contextual role modeling and activation. Computing (2018). https://doi.org/10.1007/s00607-018-0654-1
30. Colombo, P., Ferrari, E.: Access control enforcement within MQTT-based Internet of Things ecosystems. In: Proceedings of the 23rd ACM on Symposium on Access Control Models and Technologies, Indianapolis, IN, USA, 13–15 June 2018, pp. 223–234. ACM, New York, NY, USA (2018). https://doi.org/10.1145/3205977.3205986
31. Zhang, K., Li, X.R., Zhu, Y.: Optimal update with out-of-sequence measurements. IEEE Trans. Signal Process. **53**, 1992–2004 (2005)
32. Khaleghi, B., Khamis, A., Karray, F.: Multisensor data fusion: a data-centric review of the state of the art and overview of emerging trends. In: Fourati, H. (ed.) Multisensor Data Fusion: From Algorithms and Architectural Design to Applications, pp. 15–33. CRC Press, Boca Raton (2015)
33. Tu, D.Q., Kayes, A.S.M., Rahayu, W., Nguyen, K.: ISDI: a new window-based framework for integrating IoT streaming data from multiple sources. In: Barolli, L., Takizawa, M., Xhafa, F., Enokido, T. (eds.) AINA 2019. AISC, vol. 926, pp. 498–511. Springer, Cham (2020)
34. García, S., Ramírez-Gallego, S., Luengo, J., Benítez, J.M., Herrera, F.: Big data preprocessing: methods and prospects. Big Data Anal. **1**, 9 (2016). https://doi.org/10.1186/s41044-016-0014-0
35. Zliobaite, I., Gabrys, B.: Adaptive preprocessing for streaming data. IEEE Trans. Knowl. Data Eng. **26**, 309–321 (2014). https://doi.org/10.1109/TKDE.2012.147
36. Shirinzadeh, S., Drechsler, R.: In-Memory Computing: The Integration of Storage and Processing. In: Große, C.S., Drechsler, R. (eds.) Information Storage, pp. 79–110. Springer, Cham (2020). https://doi.org/10.1007/978-3-030-19262-4_3

37. Luo, L., Liu, Y., Qian, D.-P.: Survey on in-memory computing technology. J. Softw. **27**, 2147–2167 (2016). https://doi.org/10.13328/j.cnki.jos.005103
38. Lung, H.-L.: AI: from deep learning to in-memory computing. **53** (2019). https://doi.org/10.1117/12.2517237
39. Yongguo, J., Qiang, L., Changshuai, Q., Jian, S., Qianqian, L.: Message-oriented middleware: a review. 88-97 (2019). https://doi.org/10.1109/bigcom.2019.00023
40. Sachs, K., Kounev, S., Appel, S., Buchmann, A.: Benchmarking of message-oriented middleware (2009). https://doi.org/10.1145/1619258.1619313
41. Pujolle, G.: Virtualization (2020). https://doi.org/10.1002/9781119694748.ch1
42. Piper, B.: Network Virtualization (2020). https://doi.org/10.1002/9781119658795.ch10
43. Pahl, C., Jamshidi, P., Zimmermann, O.: Microservices and Containers (2020)
44. Lim, J., Nieh, J.: Optimizing nested virtualization performance using direct virtual hardware. 557–574 (2020). https://doi.org/10.1145/3373376.3378467
45. Simon, H.A., Newell, A.: Human problem solving: the state of the theory in 1970. Am. Psychol. **26**(2), 145–159 (1971). https://doi.org/10.1037/h0030806
46. Jajodia, S., Albanese, M.: An integrated framework for cyber situation awareness. In: Liu, P., Jajodia, S., Wang, C. (eds.) Theory and Models for Cyber Situation Awareness. LNCS, vol. 10030, pp. 29–46. Springer, Cham (2017). https://doi.org/10.1007/978-3-319-61152-5_2
47. Bakhtadze, N., Sakrutina, E.: The intelligent identification technique with associative search. Int. J. Math. Models Methods Appl. Sci. **9**, 418–431 (2015)
48. Ghanem, R., Romeo, F.: A wavelet-based approach for the identification of linear time-varying dynamical systems. J. Sound Vibr. **234**, 555–576 (2000)
49. Tsatsanis, M., Giannakis, G.: Time-varying system identification and model validation using wavelet. IEEE Trans. Signal Process. **41**(12), 3512–3523 (2002)
50. Preisig, H.: Parameter estimation using multi-wavelets. Comput. Aid. Chem. Eng. **28**, 367–372 (2010). Elsevier
51. Carrier, J., Stephanopoulos, G.: Wavelet-based modulation in control-relevant process identification. AIChE J. **44**(2), 341–360 (1998)
52. Yadykin, I., Bakhtadze, N., Lototsky, V., Maximov, E., Sakrutina, E.: Stability analysis methods of discrete power supply systems in industry. IFAC Paper OnLine. **49**(12), 355–359 (2016)

A Multi-layered Approach to Fake News Identification, Measurement and Mitigation

Danielle D. Godsey, Yen-Hung (Frank) Hu(✉), and Mary Ann Hoppa

Norfolk State University, Norfolk, VA 23504, USA
yhu@nsu.edu

Abstract. Technology and the Internet of Things (IoT) has changed the way the world communicates and shares information. Emergent technologies, with their elaborate infrastructures for uploading, commenting, liking, and sharing, have created an almost ideal environment for news manipulation and abuse. Fake news stories in the media have real impacts on society and politics. The ease and speed of obtaining content through media networks have made information consumers more susceptible to falling prey to fake news. Critics across the political spectrum claim that fake news and cyber-attacks are playing an increasingly significant role in determining the course of world events. Unfortunately, solutions to this issue remain unclear: existing algorithms are not good enough to filter out blatant lies with 100% accuracy. This paper aims to discuss the mechanisms used to spread fake news and the algorithms used for identifying and measuring it. Multiple cyber security-enabled fake news case studies are analyzed, along with the attack mechanisms that led to those incidents. A new algorithm is proposed for identifying fake news, along with identification strategies to thwart fake news attacks.

Keywords: Fake news · Cyber security · Social media · Internet of Things (IoT)

1 Introduction

Fake news is information that may not be verifiable and is possibly untrue. Currently running rampant across all media outlets, fake news is a content-based social engineering attack that "hacks the human" with serious security implications. Fake news can exploit human behavior that can lead to the psychological manipulation of people into acting or performing how threat actors want them to.

Fake news is not a new phenomenon: satire is arguably the oldest and most-studied form of fake news. While the original purpose of satire was entertainment, poor media literacy has blurred the line between satire and deliberate misinformation. Some websites claim to be satire as a defense against potential legal action. Still, their content may mislead readers into believing it is true [1]. Compared to what people see as traditional communication, there is enhanced attention, enhanced interest, and enhanced processing that occurs related to the mental processing of satire. Polarized times have contributed to a lack of critical thinking that also has led people to lose their ability to tell the truth from fiction.

K. Arai (Ed.): FICC 2021, AISC 1363, pp. 624–642, 2021.
https://doi.org/10.1007/978-3-030-73100-7_45

No one is immune to the threat of fake news. Fake news stories – specifically those tied to real events – creates significant cyber security risk. Today's technologies, with their elaborate infrastructures for uploading, commenting, liking, and sharing, have created an almost ideal environment for manipulation and abuse of news. Trusted and fake news stories are merging as one, creating the perfect setting for fueling previously held ideas and reinforcing personal bias that leads to psychological based social engineering.

Social media accounts are a common means to spread fake news stories. Smith et al. explored the anatomy of the information space on Facebook by characterizing on a global scale, the news consumption patterns of users over six years [2]. Their research looked at 376 Facebook users' interactions with over 900 news outlets and found that people tend to seek information that aligns with their views, thus making them vulnerable to accepting and acting on fake news and misinformation [3].

According to BBC News, cybercrime on social media is most often deployed in broad-sweep scams that try to get users to visit pages that push malware to their devices. Users who click on fake news articles and then share them on social media may also expose their online connections to potential hacks. From a cyber-security standpoint, the security implications of fake news are of significant concern to information technology (IT) and security professionals. Greg Mancusi-Ungaro from BrandProtect emphasized that false and misleading information online can affect your company and should be on the radar of IT security [4].

There are quantitative and qualitative damages caused by the spread of fake news. Problematic in numerous ways, fake news can cause irreversible damage to public and private sector entities. The weaponization of fake news has become a national security concern, as several countries and leaders have used it as ammunition to oppress the media and control public opinion. Often politically or financially motivated, fake news has the potential to cause significant economic damage. Due to a trickle-down effect, the real-world consequences of the spread of fake new can and will last for generations.

Unfortunately, solutions to the fake news issue remain unclear as algorithms are not good enough to filter out blatant lies with 100% accuracy, and the line on what is constitutionally protected free speech and decide what is and is not acceptable is questionable [5]. In September 2017, after much scrutiny, Facebook CEO Mark Zuckerberg admitted that Facebook possibly influenced the 2016 U.S. presidential elections [6]. In January 2020, ahead of the 2020 US presidential election, Facebook announced that it would evaluate how it handles misinformation (fake news) posts on its platform. According to Zuckerberg, Facebook's current approach to dealing with fake news problems is employing 35,000 people to review online content and suspend fake accounts daily. Besides the aforementioned information, full mitigation specifics have not been released, and many top-ranking US politicians are not convinced that the company's approach will stop the spread of fake news [7].

The goal of this research was to conduct a qualitative survey and analyze previous studies to recommend a multi-step plan to thwart the spread of fake news. The loopholes of social media companies like Facebook and Twitter, who are still dealing with fake news issues, will be researched to create universal mitigation strategies. Analyzing the cyber security vulnerabilities of the enablers of fake news, will lead to a multifaceted recommendation of permanent solutions. Individuals can then use these countermeasures to safeguard themselves from the ill effects of fake news. Educating the public at large on legitimate sourcing and teaching them techniques to cross-reference information likewise will help reduce the effectiveness of fake news [8].

This project surveyed fake news incidents reported in the media using a multilayered approach by:

1. Creating a report of noteworthy fake news incidents reported in the media.
2. Characterizing the incidents using the following information: who was the attacker, the target, goal, the popularity/population, and the effects.
3. Surveying tactics and techniques to thwart the spread of fake news.
4. Finding solutions to help determine the kinds of fake news that occur.
5. Providing provide tools to evaluate news for its reliability and truth.

First, the background and mechanisms for spreading fake news are described. Next related work along with an in-depth analysis of several types of fake news incidents – including but not limited to satire, news fabrication, photo/video manipulation, click-bait headlines, and propaganda – are summarized. Then tactics, techniques and tools will be surveyed to thwart the spread of fake news, to find solutions to help determine the kinds of fakes news that exits, to evaluate news for its reliability and truth. These general findings will be analyzing according to the technology, people and processes involved in several noteworthy fake news incidents reported in the media. These events will be interpreted into a table using information like who was the attacker, the target/victim, goal, popularity or population, and estimated damages. This paper will include information to restructure existing literature and future research efforts dealing with the two-dimensional phenomenon of the fake news genre and fake news label at large. Finally, this paper will recommend tools and countermeasures for evaluating news for its reliability and truth.

The remainder of this paper is organized as follows: Sect. 2 summarizes fake news background. Section 3 introduces related work. Section 4 describes existing algorithms for identifying and measuring fake news. Section 5 introduces the research methodology. Section 6 analyzes representative case studies. Section 7 explains the proposed Fake News Identifier (FNI) algorithm proposed for analyzing news for characteristics that suggest it may be fake. Section 8 concludes the paper with some reflections on findings and suggestions for future work to build upon them.

2 Fake News Background

2.1 Fake News Overview

The topic of fake news is as old as the news industry itself, and the internet is the latest means for abusing communication to spread lies and misinformation [9]. Technology takes misinformation, hoaxes, propaganda, and satire to a new and dangerous level. Social media users in particular are being targeted by phony stories and advertisements designed to undermine faith in American institutions, with everything circulating through an extensive network of outlets that spread partisan attacks and propaganda with minimal regard for conventional standards of evidence or editorial review. Trending news stories, both fake and real, buy into what is called the attraction of the economy, which posits that whenever people pay attention to a topic, more information on that topic will be produced.

Satire and propaganda have been around for centuries. Satire is arguably the most prevalent kind of fake news and, arguably, the best studied [10]. Compared to what people see as traditional communication, there is enhanced attention, enhanced interest, and enhanced processing that happens concerning the mental processing of satire. During the 2014 Ebola crises, websites that self-identified as political satire began to resemble legitimate news sources. Many stories from these sites were shared hundreds of thousands of times on Facebook. During the 2016 United States (U.S.) Presidential election, social media spam had evolved into political clickbait using fabricated money-making posts that lured millions of Facebook, Twitter, and YouTube users into sharing provocative lies [5].

2.2 Common Mechanisms to Spread Fake News

News Parody/Satire. Parody is used to generate fake news and as an antidote to it [11]. Parody mimics mainstream news sources with fictitious news stories by relying on humor as a means of drawing in an audience. Both satire and news parodies share similar deception roles, but parodies use non-factual information to inject humor [12]. Satirical journalism or news satire is presented in such a way that it mimics genuine news stories, but instead has fictional content based on follies, vices, abuses, or shortcomings. Satire uses techniques such as reversal, parody, incongruity, and exaggerations to arouse the intended audience's disapproval of a topic by holding it up to ridicule. The last step is to use ambiguity to advance the satire topic. Satire is very popular on the web, due to its ease in creation. There are many advantages and disadvantages to satirical news and journalism. In journalism, satire most commonly pokes fun at the news or uses parody portrayed as conventional news. A potential advantage of satirical news is that is has made politics more accessible, leading to more informed viewers who have the potential to form more educated opinions and discuses those views with others [13]. One of the most significant disadvantages of satire is that it uses deadpan humor to create what would typically be called fake news, making it fundamentally biased. The impact of satirical news is that it does affect political efficacy. Satire can influence opinions and preconceived biases that lead to personal and political change [14].

News Fabrication. News fabrication refers to articles that have no factual basis but are published in a style of news articles to create legitimacy [12]. News fabrication is intended to deceive and do harm, as the creator has the intention to misinform readers to draw on pre-existing biases. Technology has added a new dimension in news fabrication exploits as the development of news bots are creating an easily accessible environment for cybercriminals to spread fake news. To successfully execute the news fabrication mechanism, the creator must draw on pre-existing memes or partialities under the veneer of authenticity. The first step in news fabrication is to create a false (fake) news story that appears to be rea but is intended to deceive. Next, that story is shared on social media platforms to generate traffic and social shares among users. From the information consumer's perspective, there are no advantages to news fabrication; whereas the cons of news fabrication have lasting effects. In the past four years, news fabrication has been injected into society as a weapon of political warfare. By manipulating audiences and shaping public opinion, news fabrication has been a catalyst for spreading misinformation and bolstering tensions and conflicts. Twitter news bots also have contributed to the spread of news fabrication. The more widely circulated these stories are by multiple sources, the more the unsuspecting public tends to view them as genuine.

Photo/Video Manipulation. Manipulation of real images and videos can create a false narrative. This so-called deepfake technology is being used to create fake videos that look and sound authentic [15]. By creating seemingly realistic videos, the manipulated deepfake video is used to broadcast fabricated images and sounds that influence and persuade viewers. Several states in the U.S. are introducing bills to criminalize deepfake when it is used in pornography and to interfere with elections [15]. However, regulations are still lacking to prohibit the use of this technology for entertainment and business purposes. The first step in creating a deepfake is to scan a target video to isolate phonemes spoken by the subject. After matching the phonemes with corresponding visemes, a three-dimentional model of the lower half of the subject's face is created using the target video. The last step is to use deepfake software that combines all the data to construct new footage that matches the desired text and deceptive video inputs. The quality of deepfakes is rapidly improving, increasing the use and impact of this technology. A positive implication of deepfakes is that they can be positive for humanity. For example, the ability of deepfakes to generate realistic simulations using artificial intelligence (AI) can be beneficial in fields like medical education and computer sciences. The downside of deepfakes is the potential for this technique to be linked with misinformation and malicious hoaxes. In other words, these videos – just like other mechanisms of spreading fake news – are another form of manipulation to influence society and politics.

Clickbait Headlines. Clickbait refers to using sensationalist headlines to share deliberately fabricated news and unsubstantiated rumors. Often shared through social media feeds, clickbait at times resembles advertising on side banners, which has led to an increase in the ways cyber attackers can amplify their target reach. By sharing exaggerated and attention-grabbing stories thousands of times over social media channels, attackers can rapidly spread content that can lead to a major cyber security breach. The first step in this mechanism is to create a melodramatic attention-grabbing title for an online article. Next, people are manipulated into clicking the link, reading the content, and sharing the content. Clickbait headlines have many advantages and disadvantages, depending on their intended use. Clickbait headlines specifically benefit content creators, since they generate more page views and a greater potential for social shares. Disadvantages of clickbait headlines include the negative effects they may have on branding. A headline that does not match the corresponding article, thereby misleading the audience, it diminishes trust. This bait-and-switch tactic may initially draw more visitors to a web page, but ultimately will erode their trust. Although clickbait headlines can influence subsequent attitudes and beliefs, effects also can be highly contextualized based on the issue [16].

Propaganda. Propaganda is fake news stories created by a political entity to influence the public's perception. The primary attack vector of propaganda is to try to influence opinion. Specifically, cyber propaganda can be broadly defined as the use of modern electric means to manipulate an event or influence public perception toward a point of view. Depending on the end goal, to influence perception, propagandists use varied techniques, including stealing private information and releasing it to the public, hacking machines directly, and creating spreading fake news [17]. The first step in this mechanism is to conduct a hyper-target audience analysis. Next select inflammatory content based on its persuasive effect. After, inject the selected content into echo chambers identified through audience analysis. The last steps are to mobilize followers to action and win media attention to be a trend or stage a scandal. Oddly enough, the advantage for propaganda is also one of its biggest disadvantages. Since its creation centuries ago, governments and political leaders have used propaganda as a tool to sway public opinion. Another advantage is that when propaganda is used to device the public, the creators will lose their credibility. The digital age and emergence of social media networks have had a significant impact on propaganda. What used to be a tool for changing opinions, is now a pathway to instantaneous participation in political conflicts right from the safety of your own home [18].

3 Related Work

Fake news is everywhere, especially on social media platforms. The prevalence and ease of using fake news on social media sites like Facebook, Twitter, and Reddit have spread to the ways news is accessed. According to a 2016 Pew Research Center survey, conducted in association with the John S. and James L. Knight Foundation, about 6-in-10 Americans get their news from social media platforms. With over 62% of adults now accessing their news via social media platforms as of 2020, users are more susceptible to being exposed to and believing misinformation [19].

Fake news has quickly become a contender for the most famous phrase of the last few years. There are genuine consequences of fake news stories in the media. Cybercriminals are using real and fake news to grab people's attention through social engineering campaigns that employ psychological tricks to cause excitement or fear and spur action. The damages caused by fake news are not only monetary but also reputational. Fake news contributes to billions in global financial loss yearly. Creating and sharing unflattering and artificial fake news across social media platforms has the potential to diminish consumer trust and damage a company's brand [20]. While social media platforms have become more vigilant about the spread of misinformation since the 2016 presidential campaign, there is no full approach to mitigate the issue.

Maasberg Ayaburi, Lui and Au, analyzed data to survey the trend of fake news in cyberspace. The purpose of their research was to understand how different the phenomenon of fake cyber news is and to determine the process and antecedents of its propagation [21]. They emphasized the importance of combating fake news in the cyber security context based on its use as a content-based social engineering attack, or weaponization of information to compromise corporate information assets. By using the Engagement Theory to investigate the dual cognitive and affective process of information, their findings suggested that the possibility of propagation of cyber news is strikingly dependent on how it attracts people and less on how intellectually enhancing it is.

Researchers Lion Gu, Vladimir Kropotov, and Fyodor Yarochkin wrote a research paper titled *"The Fake News Machine: How Propagandists Abuse the Internet and Manipulate the Public,"* which explores the techniques and methods used by actors to spread fake news and manipulate public opinion [9]. Their paper also discusses some of the signs of fake news with the hope that readers will be able to determine for themselves how to spot fake news. Although in-depth, their contributing study regarding the impact of fake news requires further research on how to thwart these incidents. This suggests that recommended countermeasures need to consider the full process; technology, people, and processes, of the fake news cycle to warrant success.

Shivam B. Parikh and Pradeep K. Atrey recognized that media-rich news ads been a topic of concern in the research community due to its challenges in detecting fake news [22]. To study the problem, they presented a paper aimed at characterizing news stories in the modern diaspora combined with the different content types of news stories and their impact on readers. Several categories, including visual-based, user-based, post-based, network-based, knowledge-based, style-based, and stance-based types of fake news, were textually analyzed. After presenting an overview of comparing existing fake news detections methods from different perspectives, Parikh and Atrey highlighted critical research challenges for future research.

The spread of fake news has become a crucial problem for our information-driven society. Gravanis et al. proposed a model for fake news detection using content-based features and Machine Learning (ML) algorithms [23]. These researchers were motivated by the need for novel and reliable solutions to detect fake news through datasets in ML algorithms. They proposed an enhanced set of linguistic features with robust capabilities for discriminating fake news from real articles. The results of this comprehensive study proved that the classification of articles according to their truthfulness is possible by selecting proper features and suitable ML algorithms.

Gupta et al. wrote a research paper called *"Faking Sandy: Characterizing and Identifying Fake Images on Twitter During Hurricane Sandy"* to study the roles of social media during crisis events [24]. The paper aimed to highlight the role of Twitter during Hurricane Sandy by performing a characterization analysis to understand temporal, social reputation, and influence patterns in how fake images are spread. Through analysis, these researchers found that retweets accounted for most of the Hurricane Sandy fake image tweets. Very few tweets were the original tweets by users. Their results through analysis provided proof of concept that automated techniques can be used to identify malicious or fake content spread on Twitter during real-world events.

4 Existing Algorithms for Identifying and Measuring Fake News

4.1 Media Literacy and the CRAAP Test

Part of media literacy is the process of being able to evaluate and separate fake news from real news. Social media platforms have changed the use of old media literacy tools like source checking and relying on known outlets. Originally intended to check the objective reliability of sources across academic disciplines, the Currency, Relevance,

Authority, Accuracy, and Purpose (CRAAP) test can be used to verify if news is fake or real [25].

The CRAAP test uses an evaluation checklist as a shortcut to determine source credibility. By asking questions about currency, relevance, authority, accuracy, and purpose, researchers will develop more informed opinions regarding the truthfulness of information sources they are consuming. The anarchistic nature of social media demands that researchers using the CRAAP test account for contextual considerations, and additionally use critical thinking skills. Although there is a fair amount of controversy over the use of checklists to evaluate information and sources, the CRAAP test provides an effective approach [26]. As a mechanistic evaluation tool, the CRAAP test is extensible to content beyond social media.

4.2 Online Fact-Checking Applications and Websites

Stony Brook University's Center for News Literacy defines news literacy as "the ability to use critical thinking still to judge the reliability and credibility of news reports, whether they come via print, television, or the Internet [27]." Credibility is no longer a binary situation. Traditional news sources need to be questioned as well, as rigorous fact-checking is necessary to adhere to professional ethics in journalism. Several online fact-checking applications and websites gather the credibility scores of news to thwart fake news. The following are some noteworthy sites:

- FactCheck.org
- Washington Post's Fact Checker
- PolitiFact.com
- Snopes.com
- Hoax-Slayer.com

Online fact-checking applications and websites algorithms are effective because of their low level of bias and real-time fact-checking abilities. A common criticism is that fact-checking has failed to eradicate deceptive and misleading claims by politicians and is therefore ineffective. But without fact-checking websites, it would be much more difficult to find accurate and reliable information about claims made by political leaders [28, 29].

An unfortunate limitation of fact-checking websites is that they are under-utilized. Cognitive biases will lead social media users to perceive any information they consume there as facts, even if it is fake. Lack of consistency among different sites also can make it seem as though the fact-checking tools are not reliable. This issue would cause more confusion among users about the truthfulness of the content they are researching. Ultimately, good guiding principles include questioning everything that is shared on social media networks and using fact-checking techniques before sharing.

5 Research Methodology

The following are cases selected from media for research:

1. Spear phishing campaign targeting Lithuanian Defense Minister Raimundas Karoblis.
2. AP Twitter tweet about President Obama injured in a White House explosion.
3. Deepfake video of President Trump offering advice to Belgium citizens about climate change.
4. Hackers Post Fake News Story About Infected U.S. Soldier in Lithuania.
5. Lithuanian Fake Press Release States About U.S Nuclear Weapons.

These cases were chosen for analysis because of their cyber security-based attack methods and their connection to the three pillars of cyber security: people, process, technology.

In this research, a new comprehensive method will be proposed to identify fake news (details are in Sect. 7). The case studies and approaches will be used to analyze and interpret the subjective effects of fake news in the media. Discussion and recommendations resulting from the case analysis also will be covered.

6 Case Studies

Misinformation of fake news is a huge security concern. Hackers construct fake accounts and create hyper-tailored lures to target social media connections, thus opening users to potential hacks and malware attacks. The growth of Internet of Things (IoT) devices has helped cyber adversaries to capitalize on real and fake news. An abundance of information, and ways to access it, has led to information spread without the verification of its accuracy.

This project will survey five newsworthy cases of fake news incidents. Each case explains who was targeted, attacker, attack method, goal, and the effects of the incident. The following five cases are examples of real cyber security-based fake news attacks.

6.1 Spear Phishing Campaign Uses Misinformation Tactics Aimed at the Lithuanian Defense Minister

On April 10, 2019, attackers launched a spear-phishing campaign targeting the Lithuanian Ministry of Defence from a spoofed email address. The email that also included a malicious link, spoofed to look like it was an employee of the Ministry of Defence, used misinformation to target the Lithuanian Defence Minister Raimundas Karoblis. With the intent of discrediting Minister Karoblis, the emails were spreading fake news regarding alleged corruption within the Ministry of Defence. It was claimed that Minister Karoblis took a bribe of USD 586,000 in the process of weapons procurement and that one of the banks in Lithuania has documents proving it [30, 31].

The spoofed email was sent to several dozen email addresses at the Government, Parliament, and the President's office. Attackers also managed to inject fake news stories into legitimate news portals in Lithuania and across the Baltics. As the fake news contained in the email spread across social media sites, the National Cyber Security Centre (NCSC) warned the public not to forward the spoofed emails of open the link in the email. Essentially, the incident was a social engineering cyber warfare-based attack. At the same time, a shady U.S. news outlet also promoted the story [32]. The NCSC attributed the attack to a foreign government. An advisory, published by the Ministry of Defense, called the attack a complex cyber information attack that affected the Kasvyksta Kaune news portal and also spread across social media [32]. A similar email hacking-based attack against Lithuania leadership occurred in 2018. During that attack, fake quotes were attributed to Defense Minister Karoblis about Crimea belonging to Russia [31]. As with other cyber-attacks, the NSCS urged citizens to think critically and not give in to manipulation [32].

6.2 AP Twitter Account Makes False Claim of Explosions at the White House

The official Twitter account of the Associated Press (AP) sent out a fake tweet on Tuesday, April 23, 2013, that President Barack Obama was injured in an explosion at the White House. The incident came after hackers made repeated attempts to steal the passwords of AP journalists using phishing emails [33]. In this incident, tweets sent out by the official Associated Press Twitter account and Matt Moore, who was at the time an AP foreign correspondent. As a result, social media users had retweeted the information over a thousand times. The progression of tweets shows that internal sources notified AP followers after they realized the official account had been hacked.

Two million of the AP's Twitter followers had access to the misinformation, and several news outlets retweeted the fake news. Although the fake news tweet was dispelled almost immediately by White House Press Secretary Jay Carney, the damage was already done. Within minutes of the tweet, the market temporarily dropped 145 points. $136.5 billion of the S&P 500 index was wiped out in a three-minute plunge. The AP Twitter accounts were shut down after the incident, only to return after they could be assured of their security. A group calling itself the Syrian Electronic Army, which is supportive of Syrian leader Bashar Al-Assad, has claimed responsibility for the hacked tweet. The Syrian Electric Army announced their ties via their Twitter account along with the hashtag #ByeByeObama. After the incident, Twitter suspended that account associated with the Syrian Electronic Army.

The Dow Jones index plunged in the moments after hackers sent the message about the explosion at the White House from the Twitter feed of the AP. Then the market recovered within a few minutes. The immediate plunging and recovery of the Dow Jones demonstrates how tightly intertwined Wall Street has become with social networks [34]. Although brief, the incident left traders speculating about their vulnerability and questioning Twitter's security procedures. Although losses were rapidly recovered once it became clear that it was a false alarm [35], this incident should be a cause for security concerns.

6.3 Donald Trump Offers Advice to the People of Belgium About Climate Change

AI is now at the forefront of central automated cyberattacks. First emerging around 2017, deepfakes are videos doctored to look and sound just like real videos. In other words, these videos use AI to make it appear that someone has said or done something they have not [36]. The AI technology used to create deepfakes makes propaganda easy and cheap. The digital disinformation shared in deepfakes has the potential to manipulate and spread propaganda, making it a frontline tool for misinformation warfare. Social media companies are struggling to find ways to deal with this growing form of disinformation.

In May of 2018, President Donald Trump became the center of a deepfake campaign directed at Belgian citizens. A deepfake video was created a posted by Belgian political party, Socialistische Partij Anders (sp.a), offering advice by President Trump to the Belgian people on the issue of climate change. Sp.a is a social-democratic Flemish political party in Belgium. In the deepfake video, President Trump stated, "as you know, I had the balls to withdraw from the Paris climate agreement, and so should you." Sp.a intended to use the deepfake video to grab people's attention, then redirect them to an online petition calling on the Belgian government to take more urgent climate actions [37]. The video, which was viewed on Twitter and Facebook by over 25,000 of sp.a's followers, also called for the signing of a petition that urged investing in renewable energies. Instead of revealing the hoax, sp.a collected over 900 of the required 25,000 signatures for their climate change petition. After public scrutiny, the creators of the video went into damage control after they realized Belgian citizens took what they called a joke, legitimately. The video creators of the video tried to copy President Trump's inflections and style of speaking to make it appear real. Sp.a tried to downplay their video, but it shows how deepfakes can be a threat to democracy and national security.

6.4 Hackers Post Fake News Story About Infected U.S. Soldier in Lithuania

In January 2020, the U.S. military was the target of a fake news campaign in the Baltics. The Lithuanian news website Kauno.diena.lt, was hacked and posted a fake news story from the Baltic News Service (BNS) alleging that a U.S. military officer deployed to Lithuania was infected with COVID-19 coronavirus. The fake news story, which resembled a BNS report, was well written but featured some style details that differed from the site's original content. Hacking into the site's content control system allowed the fake news story to be visible on the website Kauno.diena.it for about ten minutes before it was taken down. The fake news incident was debunked by a Lithuanian army spokesman who said at that time there were no cases COVID-19 among U.S troops deployed to the country.

According to the same spokesman, the fake news information sought to discredit North American Treaty Organization (NATO) and cause a little panic, and mistrust among the society [38]. Lithuania is the subject of thousands of cyber-attacks annually. The Lithuania Ministry of Foreign Affairs said, "NATO and the presence of allied troops in Lithuania continue to be the target of disinformation and cyberattacks by NATO adversaries and their troll army that seek to dilute and diminish Lithuanian society's support for NATO and show mistrust of Lithuania's allies [39]."

6.5 Lithuanian Fake Press Release About US Nuclear Weapons

In October 2019, the Lithuanian Ministry of Foreign Affairs was the target of a cyber-attack. A fake news press release was circulated by way of email and later, social media outlets, alleging the US was planning to build a military base that would host US nuclear weapons transferred from the Incirlik Air Base in Turkey. The release appeared to be from a spoofed media@urm.lt Lithuanian government email address and said 500 US soldiers recently deployed were on a mission to lay the groundwork for a base in the country [40].

The email stated that the Ministry of Foreign Affairs of the Republic of Lithuania confirmed the fact that the President of Lithuania Gitanas Nauseda had proposed to build a US military base in Lithuania and to transfer tactical nuclear weapons stored at Incirlik Air Base. Included in the email was legit contact information from the Department of Communication and Cultural Diplomacy of the Ministry of Foreign Affairs. In what appeared to be a complex cyber-attack, the fake press release was issued after a US official made the rare acknowledgment that the Air Force keeps nuclear weapons in Turkey, and shortly before NATO forces arrived in Lithuania [41]. Several media outlets, diplomatic missions, and NATO member countries were targeted. Lithuania issued a public clarification shortly after the incident explaining that the press release was a cyber-attack that was under investigation by Lithuanian authorities. The attack was likely an attempt to discredit NATO. Quick clarification from the Ministry of Foreign Affairs minimized any damages that could have occurred because of the attacks.

7 Algorithm Development

This section shows how to use media and information literacy tactics to create a new multi-layered method to identify fake news. The proposed FNI model is grounded in key identification markers attributed to the common mechanisms for sharing fake news in the media.

7.1 FNI Algorithm

The FNI model is based loosely on the academic CRAAP test. The model includes following characteristics to analyze news for its truthfulness:

1. Consider the source and check the URL
2. Confirm author credentials
3. Check against other sources
4. Check headlines and text for grammatical and formatting errors
5. Recognize media manipulation.

Figure 1 shows the step-by-step procedures of the FNI model. The model includes the best courses of action to determine if media is real or fake. The FNI model, along with critical thinking, can help information consumers avoid believing fake news. Using the key identifiers of fake news – including source information, author information, grammatical errors, and media manipulation – users of the proposed algorithm will be able to use a multilayered technique to combat misinformation. This model will be broadly applicable because the majority of news and information can be tested using one or all of its pertinent categories to measure truthfulness.

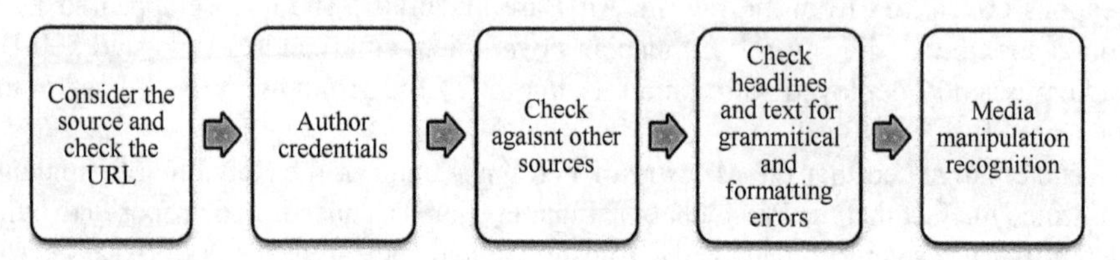

Fig. 1. Flowchart of the FNI model

Step 1: Consider the Source and Check the URL. Fake news sites can masquerade as legitimate news and information sources. Online news consumers must be alert to this and should review the *Contact* and *About* sections of the websites they visit. Omitting information these sections is a red flag that should prompt questions about the website's validity. Some websites include explicit disclaimers that the content they contain is satire. Table 1 is a list of some popular fake news websites and some specifics about them. The URL also can be a red flag regarding a site's truthfulness. Legitimate news websites use tactics like adding '.co' or names similar to their news agencies in the URL.

Table 1. List of Fake News Websites

Now8news.com	Designed to look like a real website. Several fake news stories from the site have spread on social media.
Burrandstreetjournal.com	Satirical website designed to look like a real news website that shares headlines based on current news stories
Realnewsrightnow.com	Designed to look like a real website. The site consists of stories about U.S. news, world news, politics, science & technology, entertainment, and sports. The about section of the website does not include that stories are based on satire, but all stories are satire
Infowars.com	The website is run by conspiracy theorist Alex Jones, who alleged the Sandy Hook school shooting was a hoax

Step 2: Author Credentials. The author's educational and professional credentials should be vetted by checking the following criteria:

- Educational - college, degree programs, collegiate affiliations
- Professional - LinkedIn profile and connections, job history, organizations, memberships

- Past publications - journals, conferences, articles
- Reputation - social media profiles

Step 3: Check Against Other Sources. Other sources and websites should be consulted to confirm whether information is factual. Trying to find the original source of posted content is an essential step useful for weeding out fake news from real news. Well-established, authoritative websites rarely contain information of unknown origins. Table 2 lists fact-checking websites useful for debunking misinformation, especially that shows up in social media.

Table 2. List of Fact-Checking Sources

Factcheck.org	A project of the Annenberg Public Policy Center of the University of Pennsylvania. The website is a nonpartisan advocate that monitors the factual accuracy of U.S. political players
Politifact.com	Pulitzer Prize winning fact-checking website that rates the accuracy of claims by elected officials. PolitiFact uses a Truth-O-Meter rating to reflect the relative accuracy of statements
Snopes.com	One of the internet's oldest fact-checking websites. Uses investigative reporting and evidence-based contextualized analysis to rate facts so users can ascertain the credibility of a claim
Botsentinel.com	A website that utilizes machine learning and artificial intelligence to identify and track troll-bot accounts on Twitter

Step 4: Check for Spelling and Formatting Issues. Spelling and formatting issues are red flags in fake news. Clickbait headlines are also a signal of fake news content meant to persuade its audience. Credible news sources will almost never contain spelling or grammatical errors, nor will they use caps lock for headlines. Newsrooms employ skilled copy editors to check and double-check for these types of errors. For example, a news story titled "FINAL ELECTION 2016 NUMBERS: TRUMP WON BOTH POPULAR (62.9 M - 62.2 M) AND ELECTORAL COLLEGE" appeared on a now defunct site, the 70News blog [42]. Without even reading the headline, the all caps lock should be a sign to informed consumers that the information might be fake.

Step 5: Media Manipulation Recognition. Digital content can be easily edited to mislead. The following is a list of tests that can be used to spot fake media created through this kind of manipulation:

- Drop the image into Google's reverse image search (images.google.com) to find out if it has been recirculated on other websites.
- Upload the image to a photo inspector tool to strip the metadata information. The information that can obtained from this includes global positioning system (GPS) coordinates, the make of camera used, the date the image was taken, and photoshop information.

- Check for inconsistent lights and shadows that reveal the image is edited.
- Check videos to see if the audio and video are in sync.

For example, on May 9, 2015, North Korea's state shared a photo of a successful firing of a cutting-edge anti-ship missile. News sources, including CNN, shared the story with the public [43]. Experts later suggested that the photo was significantly altered based on the reflection of the missile exhaust flame on the water. In the picture of North Korean Supreme Leader Kim Jong-un observing the missile launch, the "red shadow" – meant to be a reflection of the missile – turns out to be a "red flag" that the photo might be photoshopped.

7.2 Analysis and Interpretation

The fake news case studies summarized in Sect. 6 showcase the serious impact fake news can have on national security and society. Bad actors were at the root of all the attacks. In each case, technology was used to share misinformation, hack media, and to manipulate people to spread fake news as part of a larger campaign. Monetary losses were measurable only for the AP Twitter White House explosions tweet. The purpose of the other attacks was mainly to discredit national security or to influence society. Research showed that the Baltic States are often victims of fake news campaigns from Russia. Multi-layered countermeasures are the only way to thwart attacks by agents of the Russian governments or other countries.

In the post-truth era, where news and information are difficult to verify, the universal set of directives in the FNI model can help filter out fake news. Table 3 shows how the case studies in Sect. 6 can be analyzed using the FNI model. In each case, a discrepancy is found that should serve as a red flag that further investigation is needed. This suggests how the model can be used as a useful tool to detect fake news content when consuming and before sharing news and information.

Table 3. FNI model case study table

FNI	Step 1	Step 2	Step 3	Step 4	Step 5
Case 1	OpEdNews posted an article about the incident. They are known for spreading fake news	The alleged author of the OpEdNew article is Vytautas Benokraitis. He is CEO at DELFI Lithuania and never wrote material like the Defence Minister allegations on OpEdNews			
Case 2			Other sources including the AP's official Facebook page and White House Press Secretary Jay Carney debunked the original tweet	A tweet was sent from the same AP account before the explosion tweet that contained capitalization errors, sending a red flag the account might have been hacked	
Case 3			English speakers would have recognized the video was fake because it was stated at the end. Native Dutch only speakers had to read the captions, that omitted that the video was fake		The color of the video is abnormal, and it looks to be heavily edited. It is obvious that the audio is not that of President Trump, and instead a personal who is attempting to use vocal modulation techniques to sound like him
Case 4				The published news report featured some style details that differed from Kauno Diena's original content	
Case 5				The Lithuanian Foreign Ministry tweeted that they did not issue the press release about the intention to establish and U.S base in Lithuania	

8 Conclusion and Future Directions

Free enabler technology has made it easy for both bad actors and social bots to spread fake news. Fake news affects us all, even when inadvertently exposed to it. Social media networks like Facebook and Twitter can conveniently hide behind the First Amendment to protect their bottom lines and to avoid responsibility for the consequences of fake news gone viral. This paper presented a comprehensive view of the mechanisms used to spread fake news, the effects of fake news, and countermeasures useful to mitigate it. Cyber security-based fake news cases were analyzed to further define the technology, processes and people necessary to spread fake news to its target audiences. Models were proposed to thwart fake news.

Real-world examples demonstrate that spotting fake news is a complicated task. The case studies discussed here highlight how politics and national security are frequent targets of cyber security-based fake news attacks. Several social media companies are engaged in creating algorithms to detect fake news, but show uncertain confidence in potential reliability. Individuals must go beyond relying on pre-generated social media algorithms to protect themselves from fake news threats. Common characteristics of fake news content – especially shared on social media networks – mean consumers must be skeptical, dig deeper, and ask questions.

This research determined that a universal tool is needed to address identification based on the characteristics of fake news. The proposed FNI model includes means to assess the truthfulness of news and information and to spot fake news, such as considering the source, checking the URL, checking the author's credentials, checking against other sources, checking headlines and text for grammatical and formation errors, and recognizing media manipulation. The FNI model should be applied before sharing news and information online or offline.

Because fake news research –especially its connection to cyber security – is limited, more work is needed. An extensive study should be conducted about the long-term effects of fake news campaigns in the media to enrich the findings presented here. Because fake news and misinformation can affect behavior, a study is needed on the impacts of fake news in the workplace and the possible damages it causes to organizational learning cultures.

Acknowledgments. "This work was supported [in part] by the Commonwealth Cyber Initiative, an investment in the advancement of cyber R&D, innovation and workforce development. For more information about CCI, visit cyberinitiative.org."

References

1. Satirical News Sources. https://guides.ucf.edu/fakenews/satire
2. Schmidt, L., Zollo, F., Vicario, M.D., Bessi, A., Scala, A., Caldarelli, G., Stanley, H.E., Quattrociocchi, W.: Anatomy of News Consumption on Facebook. Proc. Nat. Acad. Sci. US Am. **114**(12), 3035–3039 (2017)
3. Anderson, J., Rainie, L.: The future of truth and misinformation online. 19 Oct 2017. https://www.pewresearch.org/internet/2017/10/19/the-future-of-truth-and-misinformation-online/
4. Fruhlinger, J.: What fake news means for IT - and how IT security can help fight it, 23 Dec 2016. https://www.csoonline.com/article/3153358/what-fake-news-means-for-it-and-how-it-security-can-help-fight-it.html
5. Waldrop, M.M.: News feature: the genuine problem of fake news. Proc. Nat. Acad. Sci. US Am. **114**(48), 12631–12634 (2017)

6. Levin, S.: Mark Zuckerberg: i regret ridiculing fears over facebook's effect on election, 27 Sept 2017. https://www.theguardian.com/technology/2017/sep/27/mark-zuckerberg-facebook-2016-election-fake-news
7. Ghaffary, S.: A senator is demanding to know how facebook will stop misinformation from spreading online, 24 Feb 2020. https://www.vox.com/recode/2020/2/24/21147428/facebook-2020-elections-misinformation
8. Matthews, K.: What does fake news have to do with cybersecurity? A lot, 19 June 2019. https://securityboulevard.com/2019/06/what-does-fake-news-have-to-do-with-cybersecurity-a-lot/
9. Gu, L., Kropotov, V., Yarochkin, F.: The Fake News Machine How Propagandists Abuse The Internet and Manipulate The Public, TrendLabs Trend Micro (2017)
10. Akpan, N.: The very real consequences of fake news stories and why your brain can't ignore them, 5 Dec 2016. https://www.pbs.org/newshour/science/real-consequences-fake-news-stories-brain-cant-ignore
11. Sinclair, C.: Parody: fake news, regeneration and education. Postdigit. Sci. Educ. **2**, 61–77 (2020)
12. Tandoc, J.E.C., Lim, Z.W., Ling, R.: Defining "fake news" - a typology of scholarly definitions. Digit. J. **6**(2), 137–153 (2018)
13. Thai, A.: Political satire: beyond the humor, 6 Feb 2014. https://www.thecrimson.com/article/2014/2/6/harvard-political-satire
14. Knobloch-Westerwick, S., Lavis, S.M.: Selecting serious or satirical, supporting or stirring news? selective exposure to partisan versus mockery news online videos. J. Commun. **67**(1), 54–81 (2017)
15. Adee, S.: What are deepfakes and how are they created? 29 Apr 2020. https://spectrum.ieee.org/tech-talk/computing/software/what-are-deepfakes-how-are-they-created
16. Scacco, J., Muddiman, A.: Investigating the influence of "clickbait" news headlines, 8 2016. https://mediaengagement.org/research/clickbait-headlines/
17. Trend Micro: Cyber Propaganda 101, 10 Mar 2017. https://www.trendmicro.com/vinfo/us/security/news/cybercrime-and-digital-threats/cyber-propaganda-101
18. Asmolov, G.: The effects of participatory propaganda: from socialization to internalization of conflicts. J. Des. Sci. **6**, 1–25 (2019)
19. Gottfried, J., Shearer, E.: News use across social media platforms 2016, 26 May 2016. https://www.journalism.org/2016/05/26/news-use-across-social-media-platforms-2016/
20. Atkinson, C.: Fake news can cause 'irreversible damage' to companies — and sink their stock price, 11 Apr 2019. https://www.nbcnews.com/business/business-news/fake-news-can-cause-irreversible-damage-companies-sink-their-stock-n995436
21. Maasberg, M., Ayabur, E., Liu, C.Z., Au, Y.A.: Exploring the propagation of fake cyber news: an experimental approach. In: Proceedings of the 51st Hawaii International Conference on System Sciences (HICSS 2018), Big Island, Hawaii (2018)
22. Parikh, S.B., Atrey, P.K.: Media-rich fake news detection: a survey. In: 2018 IEEE Conference on Multimedia Information Processing and Retrieval (MIPR), Miami, Florida (2018)
23. Gravanis, G., Vakali, A., Diamantaras, K., Karadais, P.: Behind the cues: a benchmarking study for fake news detection. Expert Syst. Appl. **218**, 201–213 (2019)
24. Gupta, H. L., Kumaraguru, P., Joshi, A.: Faking sandy: characterizing and identifying fake images on twitter during hurricane sandy. In: WWW'13 Companion: Proceedings of the 22nd International Conference on World Wide Web, Rio de Janeiro, Brazil (2013)
25. Blakeslee, S.: The CRAAP test. LOEX Q. **31**(3), 6–7 (2004)
26. Folk, M., Apostel, S.: Online Credibility and Digital Ethos: Evaluating Computer-Mediated Communication. IGI Global (2012)
27. Center for News Literacy, Stony Brook University, School of Journalism, What Is News Literacy? https://www.centerfornewsliteracy.org/what-is-news-literacy/

28. Walter, N., Cohen, J., Holbert, R.L., Morag, Y.: Fact-checking: a meta-analysis of what works and for whom. Polit. Commun. **37**(3), 350–375 (2020)
29. Altmire, J.: The Importance of Fact-Checking in a Post-Truth World, 18 Sept 2018. https://thehill.com/opinion/technology/405429-the-importance-of-fact-checking-in-a-post-truth-world
30. New Cyber-Information Attack, Targets Minister of National Defence, 11 Apr 2019. https://kam.lt/en/news_1098/current_issues/new_cyber-attack_targets_minister_of_national_defence.html?__cf_chl_jschl_tk__ = e5da6e7a83203ea10bfd47b5c2808dba0ae44415-1593587264-0-AaRFo9qgFwMgngZ_XyBlNMsrv_ArF86ym_sQHkocpYzV2cDMULfcpNwR-_U-o1qfYYUdHgHwFjMtEGp
31. Saldžiūnas, V.: How Lithuanian Defense Minister Became A Target: Cyber and Fake News Attack Was Just the Beginning 8, 13 Apr 2019. https://en.delfi.lt/politics/how-lithuanian-defense-minister-became-a-target-cyber-and-fake
32. Paganini, P.: Major Coordinated Disinformation Campaign Hit The Lithuanian Defense, 14 Apr 2019. https://securityaffairs.co/wordpress/83813/cyber-warfare-2/attack-hit-lithuanian-defense.html
33. AP Twitter Account Hacked in Fake 'White House Blasts' Post, 24 Apr 2013. https://www.bbc.com/news/world-us-canada-21508660
34. Moore, H., Roberts, D.: AP Twitter Hack Causes Panic on Wall Street and Sends Dow Plunging, 23 Apr 2013. https://www.theguardian.com/business/2013/apr/23/ap-tweet-hack-wall-street-freefall
35. Foster, P.: 'Bogus' AP Tweet About Explosion at The White House Wipes Billions off US Markets, 23 Apr 2013. https://www.telegraph.co.uk/finance/markets/10013768/Bogus-AP-tweet-about-explosion-at-the-White-House-wipes-billions-off-US-markets.html
36. Guynn, J.: Fake Trump Video? How to Spot Deepfakes on Facebook and YouTube Ahead of the Presidential Election, 8 Jan 2020. https://www.usatoday.com/story/tech/2020/01/08/deepfakes-facebook-youtube-donald-trump-election/2836428001/
37. Schwartz, O.: You Thought Fake News Was Bad? Deep Fakes Are Where Truth Goes to Die, 12 Nov 2018. https://www.theguardian.com/technology/2018/nov/12/deep-fakes-fake-news-truth
38. Vandiver, J.: Hacking Leads to Fake Story Claiming US Soldier in Lithuania Has Coronavirus, 3 Feb 2020. https://www.stripes.com/news/europe/hacking-leads-to-fake-story-claiming-us-soldier-in-lithuania-has-coronavirus-1.617404
39. Keller, J.: No, A US Soldier Deployed to Lithuania Hasn't Come Down With The Wuhan Coronavirus, 3 Feb 2020. https://taskandpurpose.com/news/army-coronavirus-lithuania-fake-news
40. Vandiver, J.: Lithuania Says Statement About Accepting US Nuclear Weapons Is Fake, 18 Oct 2019. https://www.stripes.com/news/lithuania-says-statement-about-accepting-us-nuclear-weapons-is-fake-1.603616
41. Gehrke, J.: Urgent Information': Lithuania Denounces Fake Press Release About US Nuclear Weapons, 22 Oct 2019. https://www.washingtonexaminer.com/policy/defense-national-security/urgent-information-lithuania-denounces-fake-press-release-about-us-nuclear-weapons
42. Gore, L.: Did Donald Trump Win The Popular Vote? Google's Top Election Results Site Is A Fake, 14 Nov 2016. https://www.al.com/news/2016/11/did_donald_trump_win_the_popul.html
43. Ripley, W., Castillo, M.: Report: North Korea Tests Ballistic Missile, 9 May 2015. https://www.cnn.com/2015/05/08/asia/north-korea-missile-test/index.html

Introducing the TSTR Metric to Improve Network Traffic GANs

Pasquale Zingo(✉) and Andrew Novocin

University of Delaware, Newark, DE 19711, USA
patzingo@udel.edu

Abstract. Labeled network traffic is a vital resource for improving anomaly-based Intrusion Detection Systems (IDS). However, a majority of the real-world traffic data sets are antiquated compared to modern traffic patterns, or are too small. To overcome this, efforts have been made to create synthetic network traffic with Generative Adversarial Networks (GANs). These efforts have been hindered by the lack of effective metrics for evaluating the appropriateness of the model structure, hyperparameters, or for comparing traffic GAN architectures. This work proposes the application of the Train Synthetic, Test Real (TSTR) metric for evaluating traffic GANs in terms of their ability to generalize patterns that are relevant in the Anomaly-based IDS, as well as preliminary results applying TSTR to an existing traffic GAN model.

Keywords: GAN · IDS · TSTR · Network security

1 Motivation

The need for of automated network security has increased as voting, banking, energy infrastructure, agriculture, and national defense have become increasingly networked, and so are threatened by network attacks [1]. Recent events have driven a substantial portion of information work to a work-from-home paradigm which has served to further decentralize corporate networks. While the world finds itself more networked than ever, the rate and cost of Cybersecurity incidents are on the rise. Existing Intrusion Detection Systems (IDS), which almost exclusively use human-defined signatures to evaluate traffic buckle under the modern threat landscape. To solve these problems, anomaly-based IDS have been proposed to detect malicious traffic in ways that are more flexible and can update quickly as new attacks become identified. However, the limitations of existing traffic data sets have hindered the performance of anomaly-based IDS, since they depend on machine learning methods and so are limited in their effectiveness to the data available to train them. To bridge this gap, efforts have been made to create synthetic network data to train anomaly-based IDS for use in the real world, though so far synthetic data sets have failed to generate realistic enough traffic for training anomaly-based IDS systems. Our contribution is the new application of the TSTR method to network traffic generating Generative Adversarial Networks (GANs), as well as a new perspective on TSTR to improve traffic generation systems.

K. Arai (Ed.): FICC 2021, AISC 1363, pp. 643–650, 2021.
https://doi.org/10.1007/978-3-030-73100-7_46

The next section details the state of signature-based and anomaly-based IDS, as well as prior work on network traffic generation and the TSTR metric. Section 3 details the implementation and methodology of our experiments, with results and discussion in Section 4. Section 5 details future work and use cases for the TSTR metric.

2 Background

2.1 Intrusion Detection Systems (IDS)

IDS create alerts when traffic meets certain conditions which indicate the presence of abnormal traffic, generally focused on malicious use and attacks. IDS are closely related to intrusion prevention systems, which outright block traffic which meets certain conditions as opposed to simply informing the administrator. Blocking suspicious traffic can stop attacks, but if the system has a high false positive rate it will interrupt benign user behavior. Conversely, over correcting leads to weighting towards false negatives, where the traffic is malicious and goes undetected, which poses security risks. Figure 1 describes the confusion matrix of possible outcomes for an IDS traffic classification.

		Predicted Class	
		Normal	Attack
Actual Class	Normal	True Negative (TN)	False Positive (FP)
	Attack	False Negative (FN)	True Positive (TP)

Fig. 1. Confusion matrix for IDS classification.

2.2 Signature-Based IDS

Signature-based IDS uses human-designed patterns to classify packets as malicious or benign. In recent years their utility has fallen precipitously [2]. Their increasing tendency to fail is in part due to the rise of predominantly encrypted traffic and in part due to the increasing rate of attacks. IDS designers put weight on catching potential threats, which naturally results in an increased false positive rate. However, IDS require an administrator to take action based on alerts created by traffic events, so a high false positive rate can render the system useless by creating too much noise for the administrator to react to the real threats. Signature based IDS require an expert to analyze traffic, detect threat patterns, and synthesize their intuition as **signatures**. New attacks which have never been observed will not be detected, the administrator is not alerted, and so the system fails silently. Additionally, attacks can be modified in insignificant ways (such that the attack behavior goes unchanged) while its signature differs enough that it is misclassified as normal traffic. This is possible due to the brittle nature of the signature-based paradigm, which does not operate on the level that traffic and attacks do, but rather by abstractly capturing a human-expert level of intuitive detection. This same tension is paralleled in the fascinating history of algorithmic image recognition.

2.3 Anomaly-Based IDS

Anomaly-based IDS are proposed as a solution to many of the problems outlined for signature-based IDS. Anomaly-based IDS operate based on statistics and patterns learned from labeled traffic examples, and so they can update continuously with less labor from the administrator. Since the anomaly-based paradigm uses features closer to the actual traffic than in the signature-based paradigm, it is more flexible in detecting modified attacks. An attractive feature of anomaly-based IDS is the richness of the possible models, which can be significantly more complicated than signature-based ones, allowing for models to better differentiate between normal traffic and attack traffic thus reducing false-positive and false-negative error.

2.4 Network Data Shortcomings

Despite their potential, anomaly-based IDS have yet to see widespread adoption in practice. The primary challenge in deploying anomaly-based IDS is the difficulty training them. Without sufficient data to tune the model, it will underperform compared to signature-based IDS, and traffic data can be difficult to obtain. The majority of open traffic data sets are antiquated [3]. As a result they are both too small to work with modern data techniques, and too old to effectively represent patterns of modern traffic that differentiate between normal traffic and attacks. Making private traffic public poses security risks to the organization of origination, so new real-world data sets are uncommon.

2.5 Network Traffic GANs

To account for this lack of real-world data sets, several attempts to synthesize network traffic using GANs have been made. These methods have gone some way to recreate synthetic traffic that is realistic, but to this date the evaluation of their output has not measured their applicability to training anomaly-based IDS. Some examples of architectures are listed below:

B-WGAN-GP & E-WGAN-GP. B-WGAN-GP and E-WGAN-GP [4] are GAN architectures designed to generate NetFlow traffic. B-WGAN-GP was trained on (and recreated) a binary representation on the NetFlow data from week 1 of the CIDDS-001 data set [5], and E-WGAN-GP works on an embedding of the same data using IP2Vec [6] for the categorical NetFlow features. Both are based on the Wasserstein GAN with Gradient Policy [7] and use the two time-scale update rule (TTUR) [8]. Other than the data representation and hidden dimension of the generator and discriminator, the models are designed identically and trained for 5 epochs.

Evaluation. The authors cite the difficulty of evaluating generated traffic, and used two strategies for measuring the similarity between weeks 2-4 of the original data set, a benchmark, and the output of their GANs. The benchmark used the empirical distribution of the data and sampled each feature independently.

- Euclidean Distances: the Euclidean distance between single features of the empirical distribution of the of weeks 2–4 and the synthetic data sets (benchmark and *-WGAN-GP's).
- Domain Knowledge Checks: behaviors expected of network traffic flows were checked for consistency.

1. No TCP flags for UDP packets
2. At least one of Source IP and Dest IP must be internal
3. Normal user behavior using ports 80 and 443 must use TCP
4. Normal user behavior using ports 53 must use UDP
5. Only destination IP can be broadcasts
6. Netbios flows the source must be internal and the dest IP must be external
7. TCP, UDP, ICMP must not exceed maximum packet size

PAC-GAN. PAC-GAN [9] is a model which generates packet-level traffic using a Convolutional Neural Network (CNN) GAN. The first 784 bits of each packet was encoded as a 28×28 matrix, leveraging the natural data structure used in CNN models which are normally used in the image domain. To account for assumptions of continuity of values and the insignificance of small variations in the image domain that hindered performance for generating packets, the values encoded were duplicated to 'create clusters of similar converted packet byte values.

Evaluation. PAC-GAN was evaluated by attempting to send the decoded packets generated by the model over a network, and measuring the Success Rate and Byte Error of the packets.

2.6 Train Synthetic, Test Real

TSTR [10] is a method to evaluate the performance of GANs generating time series data. The algorithm (see Fig. 2) works by training classifier models on data produced by the GAN and evaluating based on the performance of the classifier on the real data. This has several nice features:

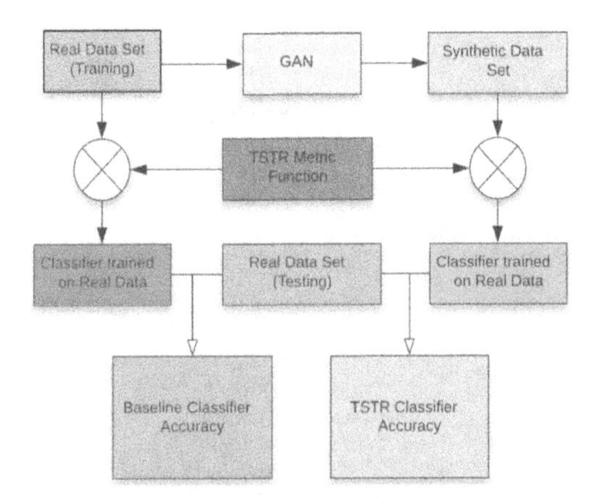

Fig. 2. How the TSTR classifier accuracy is calculated.

- The TSTR value after training a model can inform a hyperparameter search (not addressed in either traffic GAN paper referenced above).
- Comparing the synthetically trained classifier to a classifier trained on the original data set gives some indication as to the success the GAN generator had in generalizing the data set. If the GAN-trained classifier significantly under performs the real-trained classifier, then we can conclude the traffic GAN has not generalized well. If the classifiers perform similarly, or if the GAN-trained classifier performs better, we conclude that the GAN is generalizing well.

2.7 The Coburg Intrusion Detection Data Set

The Coburg Intrusion Detection Data Set (CIDSS-001) is a synthetic NetFlow traffic data set [5]. It contains traffic from a network of virtual machines with scripted traffic generating behaviors, as well as flows from hosts used by the creators to execute attacks. The data set contains flows paired with labels indicating the presence and type of attack. See Table 1 for the NetFlow data description and for the encoding described in [4].

Table 1. NetFlow variables and binary encodings

Attribute	Example	Binary encoding	
		Attr.	Value
Date first seen	2020-07-04 17:28:21.123	isMonday isTuesday isWednesday isThursday isFriday isSaturday isSunday daytime	0 0 0 0 0 1 0 $\frac{62901}{86400}$
Duration	1.503	norm dur	$\frac{1.503-dur_{min}}{durmax-durmin}$
Transport protocol	TCP	isTCP isUDP isICMP	1 0 0
Source IP address	192.168.210.5	Ip_1 to ip_8 Ip_9 to ip_16 ip_17 to ip_24 Ip_25to ip_32	1,1,0,0,0,0,0,0 1,0,1,0,1,0,0,0 1,1,0,1,0,0,1,0 0,0,0,0,0,1,0,1
Source port	445	pt 1 to pt 8 pt 9 to pt 16	0,0,0,0,0,0,0,1 1,0,1,1,1,1,0,1
Destination IP address	192.168.220.16	Ip_1 to ip_8 Ip_9 to ip_16 ip_17 to ip_24 Ip_25 to ip_32	1,1,0,0,0,0,0,0 1,0,1,0,1,0,0,0 1,1,0,1,1,1,0,0 0,0,0,1,0,0,0,0
Destination port	445	pt 1 to pt 8 pt 9 to pt 16	0,0,0,0,0,0,0,1 1,1,0,1,1,1,0,0
Bytes	144	byt 1 to byt 8 byt 9 to byt 16 byt 17 to byt 24 byt 25 to byt 32	0,0,0,0,0,0,0,0 0,0,0,0,0,0,0,0 0,0,0,0,0,0,0,0 1,0,0,1,0,0,0,0
Packet	1	pkt 1 to pkt 8 pkt 9 to pkt 16 pkt 17 to pkt 24 pkt 25 to pkt 32	0,0,0,0,0,0,0,0 0,0,0,0,0,0,0,0 0,0,0,0,0,0,0,0 0,0,0,0,0,0,0,1
TCP flags	A...S	isURG isACK isPSH isRES isSYN isFIN	0 1 0 0 1 0

3 Method

Our experiment recreated the B-WGAN-GP [4], trained it as is described in their original paper, and created data sets to evaluate under TSTR. A training set of 150,000 samples and a testing set of 50,000 samples were created from week 1 of CIDDS. An additional test set of 50,000 samples was created from week 2 of CIDDS. All hyperparameters followed the description in the original paper, and the B-WGAN-GP model was trained for 100 epochs. The model was implemented in the TorchGAN framework.

The classifiers used for TSTR were K Nearest Neighbors, Gaussian Naïve Bayes, Random Forest, and Adaboost. The KNN TSTR score for the last synthetic sample was not calculated due to computational restrictions. All classifier models used sci-kit-learn implementations, using all the binary features as were transformed for training

B-WGAN-GP to predict the binary class value of the traffic flow (1 if the flow was part of an attack, 0 in all other cases).

For each classifier type, three classifiers were trained and scored: one trained on the original training set, another trained on 150,000 samples from the trained B-WGAN-GP model (reflecting the original size of the data set), and a third trained on 1,500,000 samples from the B-WGAN-GP model (oversampling the model). Each classifier was tested on the week 1 test set (Table 2 and the week 2 testing set (Table 3). The method for calculating the scores is reflected in Fig. 2.

Table 2. Real and TSTR scores - trained on week 1, tested on week 1

Classifier	Trained on real	B-WGAN-GP (150k)	B-WGAN-GP (1.5M)
KNN	0.99728	0.99058	
RFC	0.99984	0.99732	0.24826
ABC	0.99884	0.99374	0.31588
GNB	0.96536	0.82744	0.48216

Table 3. Real and TSTR scores - trained on week 1, tested on week 2

Classifier	Trained on real	B-WGAN-GP (150k)	B-WGAN-GP (1.5M)
KNN	0.99156	0.97258	0.2426
RFC	0.91066	0.90976	
ABC	0.91034	0.90814	0.29524
GNB	0.89458	0.8564	0.478

4 Results and Discussion

In all cases the classifiers trained on synthetic samples performed worse than classifiers trained on the original data set, though the classifiers were equally impacted. When tested on the week 1 testing set, KNN trained on the smaller B-WGAN-GP sample achieved over 99% of the accuracy of the classifier trained on the real training set. RFC and ABC performed similarly. GNB did not perform as well, but unlike other classifiers, when trained on the small synthetic sample it's performance *increased* from testing on week 1 to week 2, even when the GNB classifier trained on real data decreased in performance.

5 Future Work

5.1 Improved Hyperparameter Search

Given TSTR as a quantifiable metric for the performance of a traffic GAN model, the possibility arises for hyperparameter search to optimize performance on that model.

5.2 Direct Model Comparison

As in the case of hyperparameter search, TSTR allows for direct model comparison. This will enable demonstrable improvement from one architecture to another, which the authors believe will enable more research in this area.

5.3 Determining Optional Set of TSTR Classifiers

The TSTR classifiers here were illustrative, though in practice the optimal TSTR classifier may depend on the intended use (actual anomaly-based IDS algorithm) or a useful ensemble may approximate general performance.

References

1. Critical infrastructure sectors. https://www.cisa.gov/critical-infrastructure-sectors
2. Naylor, D., et al.: The cost of the "s" in https. In: Proceedings of the 10th ACM International on Conference on Emerging Networking Experiments and Technologies. CoNEXT'14. New York, NY, USA: Association for Computing Machinery, pp. 133–140 (2014). https://doi.org/10.1145/2674005.2674991
3. Ring, M., Wunderlich, S., Scheuring, D., Landes, D., Hotho, A.: A survey of network-based intrusion detection data sets. Comput. Secur. **86**, 06 (2019)
4. Ring, M., Schlor, D., Landes, D., Hotho, A.: Flow-based network traffic generation using generative adversarial networks. *CoRR*, abs/1810.07795 (2018). http://arxiv.org/abs/1810.07795
5. Ring, M., Wunderlich,S., Grüdl, D., Landes, D., Hotho, A.: Creation of flow based data sets for intrusion detection. J. Inf. Warfare **16**, 40–53 (2017)
6. Ring, M., Dallmann, A., Landes, D., Hotho, A.: Ip2vec: learning similarities between ip addresses. In: 2017 IEEE International Conference on Data Mining Workshops (ICDMW), pp. 657–666 (2017)
7. Gulrajani, I., Ahmed, F., Arjovsky, M., Dumoulin, V., Courville, A.C.: Improved training of wasserstein gans. In: Guyon, I., et al. (eds.) Advances in Neural Information Processing Systems 30. Curran Associates, Inc. pp. 5767–5777 (2017). http://papers.nips.cc/paper/7159-improved-training-of-wasserstein-gans.pdf
8. Heusel, M., Ramsauer, H., Unterthiner, T., Nessler, B., Hochreiter, S.: Gans trained by a two time-scale update rule converge to a local nash equilibrium. In: Advances in Neural Information Processing Systems, pp. 6626–6637 (2017)
9. Cheng, A.: PAC-GAN: packet generation of network traffic using generative adversarial networks. In: 2019 IEEE 10th Annual Information Technology, Electronics and Mobile Communication Conference (IEMCON), pp. 0728–0734 (2019)
10. Esteban, C., Hyland, S.L., Rätsch, G.: Real-valued (medical) time series generation with recurrent conditional GANs (2017)

HyPA: A Hybrid Password-Based Authentication Mechanism

Saroj Gopali[1], Pranaya Sharma[1], Praveen Kumar Khethavath[2(✉)], and Doyel Pal[2]

[1] Department of Computer Science, Texas Tech University (TTU), Lubbock 79409, USA
[2] Mathematics Engineering and Computer Science Department, LaGuardia Community College, CUNY, New York, NY 11101, USA
{pkhethavath,dpal}@lagcc.cuny.edu

Abstract. User authentication is the process of verifying identity of a user. A user's identity can be verified by using different types of authentication mechanisms such as text-based password, graphical password, and biometrics. Password-based authentication is the primary line of defense against intruders and cyber-attacks. Text-based password is widely used for authentication purpose, but it is vulnerable to different kind of security attacks, such as brute force attack, dictionary attack, shoulder surfing attack etc. To overcome such risks, we propose a hybrid password authentication (HyPA) mechanism in this paper. In our proposed method we combine graphical password with text-based password. Using HyPA user needs to provide text-based password (alphanumerical character) along with image-based password. The analysis proves the efficiency, effectiveness, and security of HyPA. We also show that our proposed mechanism is secure against brute force attack, dictionary attack, and shoulder surfing attack.

Keywords: Password · Authentication · Text-based · Graphical based · Biometric · Security · Attacks

1 Introduction

Password authentication is one of the effective and convenient authentication mechanisms over insecure networks. Password is the protection key which protects the data and information of any individual or any organization from being stolen. Authorized user can only get access to his/her resources by using password. Most of the online services such as college learning management systems, social networking sites, twitter, database management systems, bank account, etc. are based on the password authentication. Besides the numerous advantages of password authentication mechanisms, it is vulnerable to various attacks such as replay attack, dictionary attack, guessing attack, shoulder surfing etc. Password authentication mechanisms can be categorized into three different categories: text-based, graphical, and biometric.

K. Arai (Ed.): FICC 2021, AISC 1363, pp. 651–665, 2021.
https://doi.org/10.1007/978-3-030-73100-7_47

1.1 Text Based Password

Text-based authentication method is a traditional authentication method in which credentials are directly inserted into the login field in the form of alphanumerical. Most importantly text-based passwords are inexpensive and consumes less time than any other authentication method [1]. Text-based password does not need professional people to set it up. Despite its significant benefits, text-based authentication has drawbacks [2, 3, 14]. Since the credentials are directly inserted, text-based authentications are vulnerable to spyware attack and shoulder surfing [3]. Users usually tend to have short and easy password so that they can remember it easily at any time. Easy and same text-based password is convenient for user to remember, but it becomes easily vulnerable to dictionary attack, brute force attack and surfing attacks [4]. To overcome the risk and vulnerabilities to dictionary attack and brute force attack in text-based password authentication, Graphical Password was developed [4].

1.2 Graphical Password

Graphical password schemes offer a good alternative to text-based passwords in terms of memorability and security. Graphical passwords are user friendly and provide good resistance to attacks like dictionary attack and brute force attack [4, 6]. Though graphical password seems more effective than text-based password, it has some key challenges like time consuming and easy shoulder surfing [5, 6]. Graphical passwords are more likely better at memorizing compared to text-based passwords. Graphical password can be categorized into three different types:

Recognition: Most of the graphical passwords' authentication are based on the Recognition. User is provided with a set of images from which user must select the correct image which he/she has selected during registration phase. Example: image-based authentication, Déjà vu, Pass face.
Recall: User is required to produce something that was created during registration process. In this method, User does not get any hint for the login process.
Cued-Recall: User is provided with some hint so that user can recall the password that he/she selected during registration phase.

1.3 Biometric Authentication

Biometric uses unique physical characteristics and behavior of user for authentication. Static characteristics such as, fingerprint, iris, retina, face, and dynamic characteristics such as, voice print, and signature are the types of biometrics. It is secure and less likely to be copied and counterfeited. In most biometric authentication mechanisms, users are needed to be available physically to authorize their identity. Users do not need to remember any pattern or characters for authorization. Though biometric authentication is secure and convenient for users however, it has a lot of disadvantages also. This mechanism is not possible in all the currently used operating system. It can be expensive based on password choice made by users (for devices or software to read voice, face, iris). Moreover, it is also difficult to replace.

In the field of authentication. Applications more likely focus on the single authentication method (e.g., text-based password or graphical password or biometric authentication [2–4]. In our study, we found that there are various ways of attack for the currently used single authentication method [2–6, 8, 9, 16]. Though text-based password (single authentication method) are easy to remember, attacker can impersonate a user by intercepting the user's device by using keylogging malware, brute force attack. Similarly, Graphical password most likely has a shoulder surfing problem. Images can be easily remembered and seen from few meters of distance. In Biometrics authentication mechanisms, user does not need to remember pattern or character, but it is difficult to replace [1]. Moreover, biometric authentication with Iris Recognition requires expensive hardware and it needs to keep the user's head still during authentication process [15]. So, to make the strong platform of password authentication, there must be creative changes on the currently using authentication mechanism.

In this paper we proposed HyPA: a hybrid password authentication technique that combines text-based and graphical password mechanism. The aim of the proposed scheme is not only to make user authentications simple but also to add strong defense layer to make it secure. The registration and login process of proposed scheme is easy and simple for users to follow. During sign-up and log in process, user needs to provide text-based password for each selected image. HyPA resist the cyberattacks like brute force, shoulder surfing and dictionary attacks.

The rest of the paper is organized as follows. In Sect. 2 we discuss about the existing related work. We describe how the HyPA mechanism works in Sect. 3. We explain the experimental analysis, and security analysis in Sect. 4 and 5, respectively. The conclusion of the proposed scheme is in Sect. 6.

2 Related Work

Though single form of authentication has lot of advantages, lot of existing single authentication schemes [2–6] are not liable to shoulder surfing, brute-force, and dictionary attacks. Combined form of authentication is developed to overcome the risk and vulnerabilities of single form of authentication. Combined form of authentication adds new level of security to protect user's credentials. Wantong Zheng and Chunfu Jia [7] proposed online password authentication mechanism, CombinedPwd, which makes the password mechanism stronger by inserting separators into the password. This method adds a middleware between user input and website database. In this scheme, for the registration process, user must input password into password field with including some separators. In the back end of this mechanism, code will hold several counters to count the separator in every position. Though this scheme resists brute force and dictionary attack, it is vulnerable to SQL injection or XSS attacks leading to data theft.

Yi-Lun Chen, Wei-Chi Ku, Yu-Chang Yeh and Dun-Min Liao [2] proposed an improved text-based shoulder surfing resistant graphical password scheme which is improved form of scheme proposed by Zhao and Li. This scheme uses color triangle to make more efficiency than SP3PAS scheme. In this scheme, For the registration process user has to select one color as a pass color from 8 colors designed by the system and also user has to enter text password into the password field of registration screen similarly as another textual password. In the login phase, the system displays a circle composed of 8 equally sized triangle. Though this scheme resists attacks like shoulder surfing, keystroke logger and mouse logger attacks, this is also time-consuming scheme since user must look for the password triangle among 8 of the triangles Sand rotates it into the actual pass color triangle.

Shah et al. [4] proposed a text-based user authentication which improves the security of textual password scheme by modifying the password input method and adding a password transformation layer. The purpose of this scheme is most likely to resist shoulder surfing, spyware, and key logger attacks. Though the registration process is same as ordinary textual password scheme, user inserts a new set of decimal numbers in the password field during login process. Entered decimal number represents a password which will go for the password matching process in the database. Though, this scheme has some security advantages, the login time is higher than the traditional textual password scheme.

Fawaz A. Alsulaiman and Abdulmotaleb El Saddik [5] proposed a scheme called "Three-Dimensional Password for More Secure Authentication" which is a multifactor authentication. These authentication scheme includes text-based, graphical, and biometric authentication. At first, user needs to provide a username and moves to next step where user is presented with 3-D virtual environment in the user registration screen. User needs to navigate and interacts with various object which are recorded in the database. User must follow the same sequence and type of actions and interactions toward the objects for the user's original password to authenticate login screen. The time required for the legitimate user to login is vary from 20 s to 2 min varies on the number of interactions.

Zhao and Li [8] proposed a scheme called Scalable Shoulder Surfing Resistant Textual- Graphical Password Authentication Scheme(S3PAS). This scheme integrates both graphical and textual password schemes and provides nearly perfect resistant to shoulder surfing, spyware attacks and hidden camera. This scheme consists of two different passwords: original password and session password. Every time the user tried to login; user must select the session password rather than original password. The specialty of the S3PAS is letting the user to input session password so that original password kept hidden. In the login process, alphanumeric characters are shown in the image format which consist the password character that was entered during registration process. However, this scheme provides resistance to the shoulder surfing and keystroke logger, this is not efficient because of the amount of time it takes.

Lopez et al. [9] proposed a scheme called "Even or Odd: A Simple Graphical Authentication System" which presents a user with a row of typically three faces and needs to decide whether the number of friends is even or odd. Though this schemes challenges shoulder surfing and spyware attack, the number of images that being used in this scheme is small which results in vulnerable to brute force attack. Moreover, this scheme found to be failed after multiple times of key logger and screen scrapper attack.

Akpulat et al. [10] present a hybrid graphical password scheme called Type&Click (T&C). This scheme uses the combination of text password and graphical passwords. Type&Click itself defines how the system work in this scheme. User need to provide a text password at first and then user need to make a single click on the image to complete the registration and login process. Since the passwords are directly inserted into the password field just like other textual password authentication, this scheme is also vulnerable to brute force and dictionary attack. At first, attacker can find the password using brute or dictionary attack, and then they can click on the different part of image every time they tried to login until they get the actual click point of image.

Similarly, P.C van Oorschot and Tao Wan [11] proposed a scheme called "TwoStep: An Authentication Method Combining text and Graphical Passwords" also a combination of textual password and graphical password authentication. A user is needed to provide the username and text password at step 1 and then user is asked to select images sequentially in the registration process. During login phase, user needs to provide the same text-password and needs to select images in the same sequences that was used during registration process. In this scheme, even if the text-based password got stolen, it does not compromise an account since actual user can be alerted if the sequence of graphical password does not match.

Sung-shiou et al. [12] suggested a graphic pattern protection mechanism for enhance authentication level in the keypad lock screen app field. This scheme is mostly for the smart android devices. Most of the android device with pattern unlock need same sequences of pattern that was used in registration process. The graphics of the pattern are fixed which makes the user to provide same pattern to unlock. User can draw different graphic pattern since the position of digital graphics (characters provided by user) changes every time user tries to login. This scheme provides strong layer of defense to shoulder surfing attacks because of the uncertain position of character in graphical pattern.

Jitendra et al. [13] suggest a scheme as an improved form of text-based shoulder surfing resistant graphical password by using colors. This scheme has challenged both shoulder surfing and key logger attacks by combining the text-based password and graphical password. User must register an email address and his/her phone number and text-based password in the password field during sign-up process. Most importantly, user must select only one out of 8 colors shown in the sign-up screen. During login process, user will be shown a circle composed of 8 equally sized sectors with 9 characters on each of them. User must rotate the sector containing the ith pass character of his password K and press confirm till he/she reach all the character of password. After completing this process, user will be provided with OTP (One Time Authentication) which he/she must enter to finish the login process. This scheme also prevents attacks like dictionary and brute force attack by adding OTP to complete login process.

Though proposed scheme in [2–5, 8–13] has several benefits, they are vulnerable to attacks like dictionary, brute force, shoulder surfing. Spyware attack etc. Our study concludes that existing combined form of authentication either need whole text password or graphical password at first step and then move to next step for remaining part of combined authentication. This results an attacker to attack the text-based password first and tries multiple rounds of guessing attack on graphical password authentication. A scheme proposed by the Lopez et al. [9] is found to be failed after multiple times of key logger and screen scrapper attack. Some of existing authentication need [2, 5, 8] user to have a look for the logical triangle which only consist of password element which user never want to do that because of the time it takes and some of them need advanced device which can support the 3-D virtual environment. Moreover, some of the existing combined authentication are easily vulnerable to dictionary attack and brute force attack because they have used the single text-based password directly inserted to the password field [1]. To overcome such vulnerabilities and drawbacks, our proposed scheme approached in a new way that has never used before to the best of our knowledge. In our proposed scheme user needs to enter the text-based password for every image he selects and every time images appears in random order which makes our scheme secure against different types of attacks such as dictionary attacks, brute force attacks, shoulder surfing attacks, key logger attacks and spyware attacks.

3 HyPA: Hybrid Password-Based Authentication Scheme

In this section, we describe our proposed Hybrid Password-Based Authentication (HyPA) mechanism scheme which reduces the risk of password protection and security weakness of currently used single password mechanism. It is designed in a way which is friendly for user to use in different online services. We use image-based authentication as a base of hybrid password mechanism since image is easy to remember rather than alphanumerical character. Additionally, we aimed to satisfy user requirement by providing the freedom of selection. User can select images as many as they want but less than or equal to nine. User also have freedom to enter any kind of alphanumerical character without any limitation on length. HyPA is easy and convenient for users to remember, user friendly and most importantly it provides security of user's credential since it uses unique combination of random graphical images and text-based password for authentication. Users to need to set up their hybrid password during sign-up phase and use the same password every time they log in. The overall flow of the HyPA scheme is shown in Fig. 1. The next two subsections describe the sign-up phase and log in phase of HyPA.

3.1 Sign-Up Phase

In the sign-up process, user needs to enter his email address. HyPA uses user's email address to save the user details in the database.

To set up the password using HyPA users are provided with 9 random images displayed on the sign-up screen as shown in Fig. 2. Users can choose any number of images from those 9 images. At the same time, user also needs to enter a text-based password after selecting each image. There is no restriction on the length of a text password so user even have a choice to put a single character after selecting image which would be easier to remember. User needs to remember the sequence of images and corresponding text-based password that he/she select during sign-up process, because the index of image sequence and content of images both are stored in database which is used to authenticate during the log in process. Since images appear in random order every time and each selected image are stored with different text password entered by user, HyPA is not vulnerable to brute force, dictionary attack, shoulder surfing and spyware attack. To sign-up, user must complete the following steps:

Steps to sign-up using HyPA

Step 1: The user must enter an email address.
Step 2: The system displays a 3*3 grid-based table of 9 images. User needs to select first image and system will prompt user for text-based password where user must enter text as shown in Fig. 3. The combination of image and text-based password will be used as password during login process.
Step 3: Every time user selects an image and enter corresponding password, the screen with image grid (Fig. 2) appears again on the screen. User can keep repeating step 2 at most nine times since HyPA allows combination of at most nine image and text pairs. To complete the sign-up process user must select the "Sign-Up" button. There is no restriction on selecting number of images so that user can select as many images as he wants.

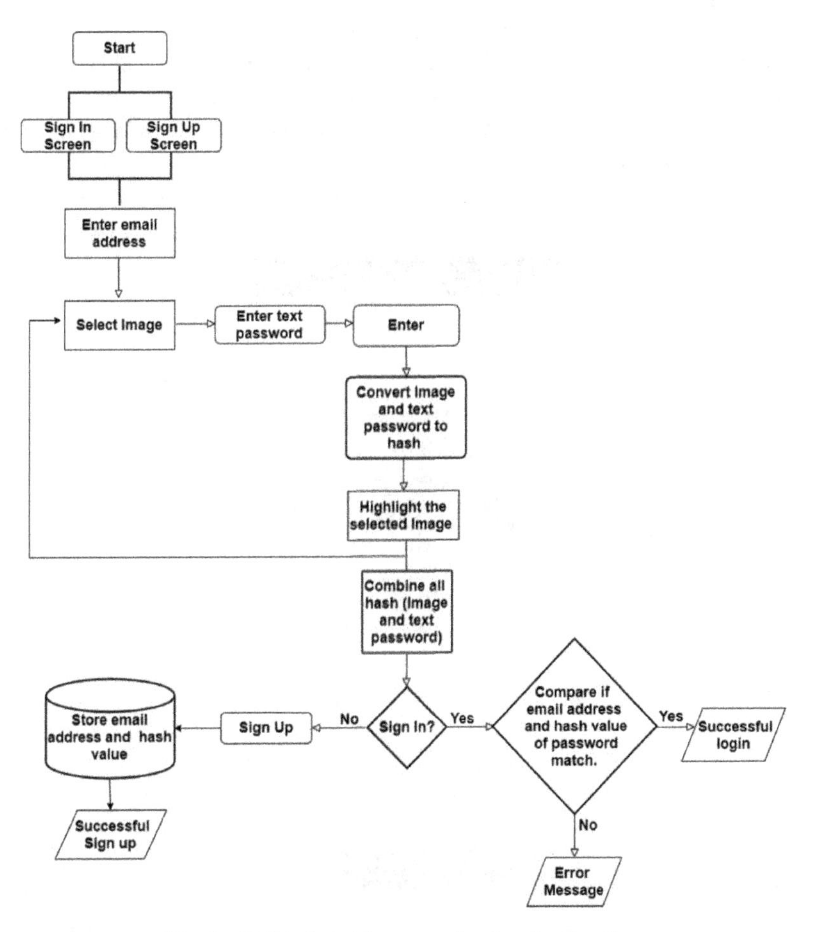

Fig. 1. HyPA: hybrid password authentication flow

3.2 Sign-In Phase

In the sign-in process, user must enter an email address that was provided during sign-up process. User will be asked to select image/s and enter text-based password/s which was used during sign-up process maintaining the same sequence. Positions of images change every time user tries to log in. Each Image is stored in the database with its respective text. User must remember the order of image and its corresponding text. To sign-in, user must complete the following steps:

Steps to sign-in using HyPA

Step 1: The user needs to provide the email address as username. Click Next to enter passwords (Fig. 4).
Step 2: Images will appear on the screen like sign-up page (Fig. 2). User needs to select image/ images that have been used to register/ sign-up.
Step 3: For each image user needs to enter the corresponding text-based password.
Step 4: Steps 2 and 3 will continue depending on the number of image-text pair user used to sign-up. Correct username, and correct image, text-based password pair/s authenticates the user (Fig. 5).

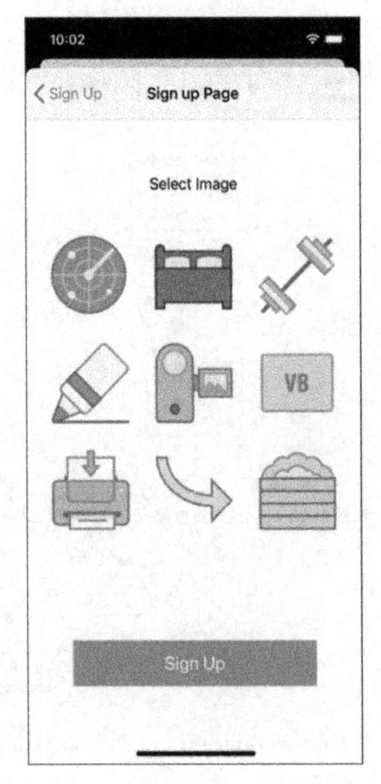

Fig. 2. HyPA sign-up screen with a grid of random images.

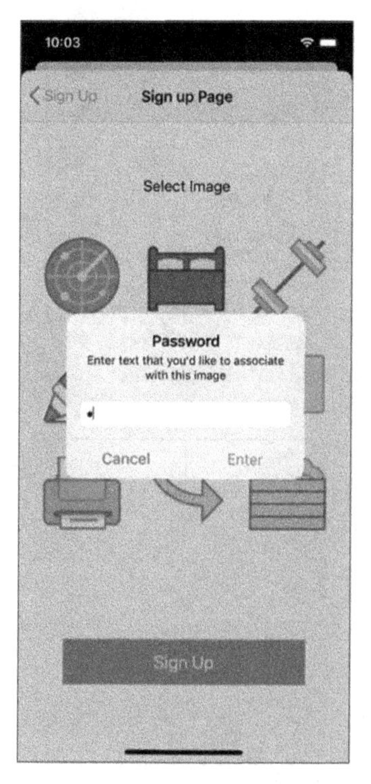

Fig. 3. HyPA sign-up screen with prompt for text password

3.3 Handling Combination of Images and Text Password in the Back End

Users' credentials such as email address, combined image, and text-based password pair/s are stored into the database. Instead of saving users' password in raw format, HyPA generates the hash value of user's image-text password pair/s and saves it in the database. Figure 1 depicts the flowchart of the procedure to save users' password in the database. Also, our application uses a pool of images which is random for every user. Moreover, during sign-in phase the images that are displayed are random but will includes all the images users have selected as their password.

Fig. 4. Sign-In: user needs to provide email address

4 Experimental Analysis

4.1 Experiment Set Up

For the implementation of our proposed HyPA mechanism, we implemented an iOS mobile application. We have used "XCode 10.2.1" software which is a developer tools for creating apps for mac, iPhone, apple-watch, and apple Tv. We used Mac book Pro 15 inch with 16 GB memory 2400 MHz DDR4, intel core i7 processor with 2.6 GHZ speed. Swift programming language [17] has been used for the implementation process to create the application. We used online database platform, Firebase Realtime Database [18] to store and sync data between user in real time on the application.

4.2 Performance Analysis

In this paper, we have developed and implemented an iOS mobile application for the proposed HyPA mechanism. Using the iPhone emulator of the iPhone X model, we ran the prototype and obtained the data of sign-up and sign-in phase.

For experiments we used various number of image-text password pairs for sign-up and sign-in. All text-based passwords are of length 8 characters and are required to have at least a special character and alphanumerical characters. We measured the time it takes for sign-up and sign-in after the user inputs their credentials. So, the times shown below

Fig. 5. Successful login

are processing times not taking time users take. Different users might take different times for entering the inputs and is the reason we used processing times for the algorithms to run.

The time taken for sign-up with 1 image-text password pair was at 3.1 microseconds. The graph in Fig. 6 shows sign-up times for various image-text password pairs. We took average of 3 readings for each setting. The graph shows there is a spike for 4 image-text password pairs. But if we look at overall graph, the spike is negligible. The sign-up times for various password pair are always between 2.5–3.5 microseconds except for 4 password pairs. Since, the difference is very negligible, we can say that the time for sign-up for various number of image-text password pairs is almost same.

For sign-in, we measured the time for one through nine image-text password pairs with 10 and 100 users in the Firebase database. We used Firebase authentication for sign-in. The graph in Fig. 7 shows the sign-in time lies between 2.7 to 3.5 (10 microseconds) from images count 1 to 9 for 10 users and 100 users. Each image has a text-based password associated with it. For example, one image count has one text-based password of 8 characters, where nine images have nine different eight characters text-based passwords. The time taken for sign-in for various image-text password pairs with 10 and 100 users in the database is almost the same. The difference is always in between 0.8 microseconds for various number of image-text password pairs. From the result we can say that the

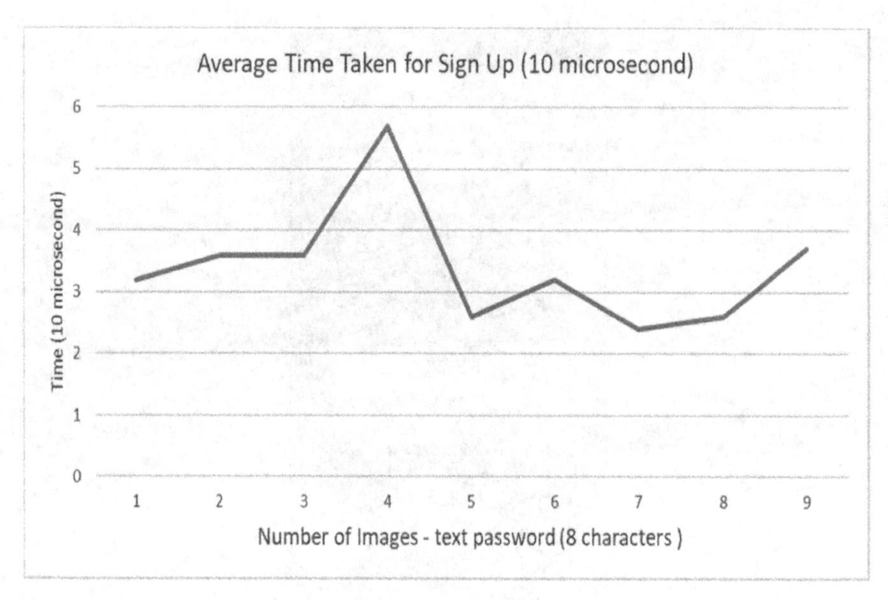

Fig. 6. Sign-up time

system is scalable and can accommodate large number of users with various number of custom image-text password pairs without effecting the sign-in time.

To validate that the time difference for sign-in with various number of users in database remains negligible, we ran experiments with 3, 4 and 5 image-text password pairs. We calculated average time for sign-in with 20, 40, 60, 80 and 100 users in database using 3, 4 and 5 image-text password pairs. We chose between 3 and 5 passwords because those are the numbers where we saw some spikes. From Fig. 8. we see that the time difference at any point is within 1 microsecond. Even though there are highs and lows at certain points the average time is still very low, and the differences are very negligible.

5 Security Analysis

In our proposed HyPA scheme, user needs to provide text-based password for each randomly selected image which improves the security of user authentication. In this section, we discuss how our proposed scheme is secure against different security attacks.

5.1 Brute Force Attack

In brute force attack, attacker tries different combinations of password until the correct password is discovered. HyPA can effectively resist brute force attack since user has to select the correct image and must provide the corresponding correct text- based password every time. If user choose multiple image – text password pairs, then attacker not only has to provide correct image-text password pairs but also in proper sequence which makes it hard to crack the correct password using brute force attack. Also, most of the systems have a mechanism to lock out the attacker or manually verify if it is actual user after few failed attempts.

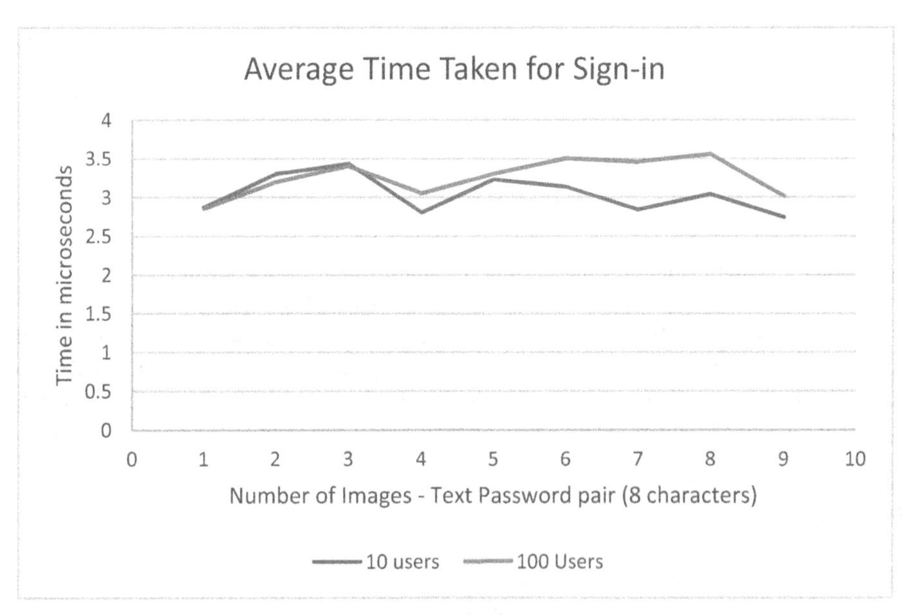

Fig. 7. Sign-in time for 10 and 100 users with various image-text password pairs

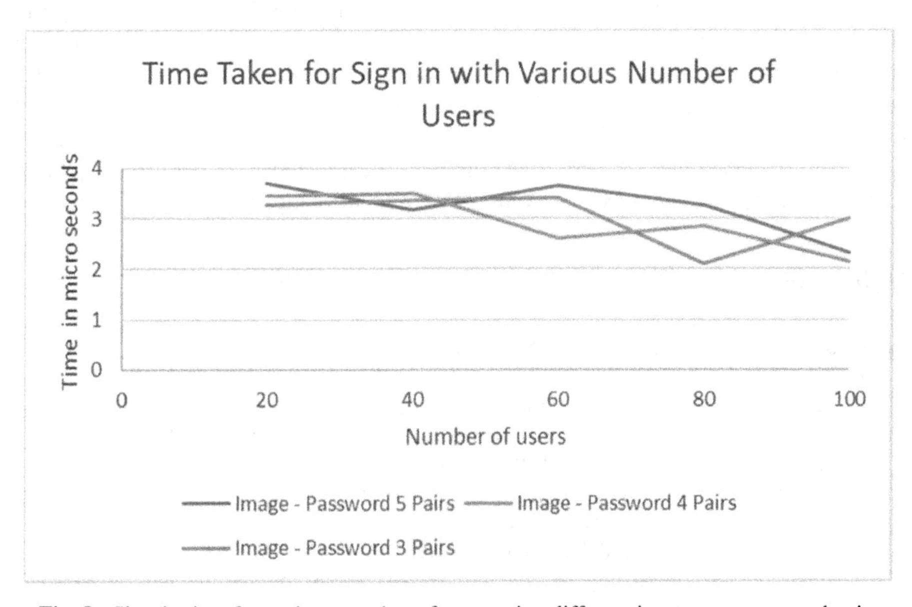

Fig. 8. Sign-in time for various number of users using different image-text password pairs

5.2 Dictionary Attack

In the dictionary attack, a list of passwords is generated, and attacker tries them systematically to crack the correct password. HyPA resists dictionary attacks in the similar way it resists the brute force attack. All text-based passwords are stored along with its corresponding selected images in the database. Even if the attacker succeeds to get the text-based passwords, they also need to get the image for which the password is matched. It is difficult to get the right sequence of selected images and their respective passwords unless the attacker visually sees the users entering the password.

5.3 Shoulder Surfing Attack

In shoulder surfing attack, attacker tries to steal users' password by observing their keystroke, shoulder movement etc. In our proposed HyPA scheme, shoulder surfing attack is not possible because every time user gets different and random images in random position on the login screen. Images and sequences of images get changed every time. In addition, user needs to enter text-based password for each selected image which makes it more difficult for the attacker to steal correct password.

6 Conclusion

In this paper, we have proposed HyPA: a hybrid password authentication mechanism which is a combined form of the graphical and text-based password mechanism. The authorized user can easily access to system by selecting correct image-text password pair. The process of sign-up and login using HyPA mechanism is simple and easy. Hybrid password mechanism can add appropriate layer of defense in addition to the existing password mechanisms that are currently being used. From the performance analysis we can say that our proposed scheme is scalable and can accommodate large number of users with various number of custom image-text password pairs without effecting the sign-up and sign-in time. Our proposed scheme is secure against different cyber-attacks such as, brute force attack, dictionary attack, shoulder surfing attack.

References

1. O'Gorman, L.: Comparing passwords, tokens, and biometrics for user authentication. Proc. IEEE **91**(12), 2021–2040 (2003)
2. Chen, Y.-L., Ku, W.-C., Yeh, Y.-C., Liao, D.-M.: A simple text- based shoulder surfing resistant graphical password scheme. In: 2013 IEEE International Symposium on Next- Generation Electronics (ISNE), pp. 161–164. IEEE (2013)
3. Yan, Q., Han, J., Li, Y., Deng, H., et al.: On limitations of designing usable leakage-resilient password systems: attacks, principles and usability (2012)
4. Shah Zaman, N., Tariq Jamil, K., Syed Rahul, H., Mohd Zalishman, J.: A text based authentication scheme for improving security of textual passwords. Int. J. Adv. Comput. Sci. Appl. **8**(7), 513–521 (2017)
5. Alsulaiman, F.A., El Saddik, A.: Three-dimensional password for more secure authentication. IEEE Trans. Instrum. Meas. **57**(9), 1929–1938 (2008)
6. Arti, B., Bhavika, M., Harshika, V., Hetal, B., Poonam, B.: Comparison of graphical password authentication techniques. Int. J. Comput. Appl. **116**(1) (2015)
7. Zheng, W., Jia, C.: CombinedPWD: a new password authentication mechanism using separators between keystrokes. In: 2017 13th International Conference on Computational Intelligence and Security (CIS), Hong Kong, pp. 557–560 (2017)

8. Zhao, H., Li, X.: S3pas: a scalable shoulder-surfing resistant textual graphical password authentication scheme. In: 21st International Conference on Advanced Information Networking and Applications Workshops, AINAW 2007, vol. 2, pp. 467–472. IEEE (2007)
9. Lopez, N., Rodriguez, M., Fellegi, C., Long, D., Schwarz, T.: Even or odd: a simple graphical authentication system. IEEE Lat. Am. Trans. **13**(3), 804–809 (2015)
10. Akpulat, M., Bicakci, K., Cil, U.: Revisiting graphical passwords for augmenting, not replacing, text passwords. In: Proceedings of the 29th Annual Computer Security Applications Conference, pp. 119–128. ACM (2013)
11. Van Oorschot, P.C., Wan, T.: TwoStep: an authentication method combining text and graphical passwords. In: Babin, G., Kropf, P., Weiss, M. (eds.) E-Technologies: Innovation in an Open World. MCETECH 2009. Lecture Notes in Business Information Processing, vol. 26. Springer, Heidelberg (2009)
12. Shen, S., Kang, T., Lin, S., Chien, W.: Random graphic user password authentication scheme in mobile devices. In: 2017 International Conference on Applied System Innovation (ICASI), Sapporo, pp. 1251–1254 (2017)
13. Neerukonda, J., Vinay, N.S., Ram, P.S., Sidhardha, P.N., Deepthi, D.: Text-based shoulder surfing and key logger resistant graphical password. J. Eng. Sci. (JES) **11**(3), 214–223 (2020)
14. Roshni, R., Bhavna, G., Asmita, R.: Textual and graphical password authentication scheme resistant to shoulder surfing. Int. J. Comput. Appl. **114**(19), 26–30 (2015)
15. Maria, H., Jafar, A.: A proposed password-free authentication scheme based on a hybrid vein- keystroke approach. In: Conference: The International Conference on New Trends in Computing Sciences (ICTCS 2017) (2017)
16. Mudassar, R., Muhammad, I., Muhammad, S., Waqas, H.: A survey of passwords attacks and comparative analysis on method for secure authentication. World Appl. Sci. J. **19**(4), 439–444 (2012)
17. Swift Developer page. https://developer.apple.com/swift/. Accessed 15 July 7 2020
18. Firebase documentation. https://firebase.google.com/docs/database. Accessed 15 July 7 2020

Human Factors in Biocybersecurity Wargames

Lucas Potter[1(✉)] and Xavier-Lewis Palmer[1,2]

[1] Department of Engineering and Technology, Biomedical Engineering Institute, Old Dominion University, Norfolk, VA 23529, USA
{Lpott005,Xpalm001}@odu.edu

[2] School of Cybersecurity, Old Dominion University, Norfolk, VA 23529, USA

Abstract. Within the field of biocybersecurity, it is important to understand what vulnerabilities may be uncovered in the processing of biologics as well as how they can be safeguarded as they intersect with cyber and cyber-physical systems, as noted by the Peccoud Lab, to ensure not only product and brand integrity, but protect those served. Recent findings have revealed that biological systems can be used to compromise computer systems and vice versa. While regular and sophisticated attacks are still years away, time is of the essence to better understand ways to deepen critique and grasp intersectional vulnerabilities within bioprocessing as processes involved become increasingly digitally accessible. Wargames have been shown to be successful within improving group dynamics in response to anticipated cyber threats, and they can be used towards addressing possible threats within biocybersecurity. Within this paper, we discuss the growing prominence of biocybersecurity, the importance of biocybersecurity to bioprocessing, with respect to domestic and international contexts, and reasons for emphasizing the biological component in the face of explosive growth in biotechnology and thus separating the terms biocybersecurity and cyberbiosecurity. Additionally, a discussion and manual are provided for a simulation towards organizational learning to sense and shore up vulnerabilities that may emerge within an organization's bioprocessing pipeline. This article includes an introduction to Biocybersecurity (BCS), highlights games in the methodology used, retrains focus on the biological component, and includes exact instructions for doing a war game on one's own.

Keywords: Human Factors · Cybersecurity · Systems engineering · Biosecurity · Biocybersecurity · Cyberbiosecurity · Wargames · Bioprocessing

1 Introduction: Importance of Biocybersecurity to Bioprocessing

Work governing the processing of living biological systems and their components falls under bioprocessing. This stretches from the initial research to the development and manufacturing, and eventually the commercialization of products [1]. Bioprocessing constitutes trillions towards the world economy and continues to grow as novel uses of bioproducts increase in demand, covering food, fuel, cosmetics, drugs, construction, packaging, and more [1]. As the means of biocomputing, bioprocessing, and storage merge and become increasingly accessible and feasible, paired with growing appetites for green technology, the growth of the bioeconomy can be expected to further expand

K. Arai (Ed.): FICC 2021, AISC 1363, pp. 666–673, 2021.
https://doi.org/10.1007/978-3-030-73100-7_48

[1–11]. The works of numerous researchers demonstrate that the entire bioprocessing pipeline is vulnerable to attacks at multiple steps along the way where bioprocessing equipment converges with the internet, which calls for the need for additional scrutiny in the design and monitoring of organization bioprocessing pipelines [1, 6, 12–14]. To do so otherwise invites potential disruptions in the world's functioning as the bioeconomy grows [1, 4].

Much of the progress in bioprocessing will hinge on increased automation aided by advanced algorithmic processes, increasing engagement with Information Technology, or IT, for short. IT spending, as with that of bioprocessing, has also reached trillions in worldwide spending, with the expectation to continue [15, 16]. Paired with open-source methodologies being both profitable and adapted worldwide, economies worldwide are witnessing unprecedented growth in communication and digital technological development [15–22]. Further paired with biological computing and storage, technologies within the bioprocessing pipeline may witness shifts in development that will require new lines of expertise in operation and defense. Nonetheless, the basic means of protecting data and processes remain essential [15–22]. As entities engage in advanced bioprocessing, to run lean operations, it is reasonable to expect that they will employ networks of connected bioprocessing infrastructure, which will need a great deal of IT expertise for both management and security [1]. For this, it is important that war games be employed by such IT teams to simulate risks posed in their infrastructure.

2 Research Gaps

The research gaps in this endeavor are numerous. It is important to note that this exercise was not and is not at all meant to be a comprehensive exhibition of all possible vectors or aspects of a BCS threat. It was meant to be a primary investigation of the factors that would come into play while attempting to defend an organization from a combined BCS threat. A complete review of all BCS aspects would be unfeasible for two primary reasons. First is that some of the information required for this would (and frankly should) be classified. Secondarily, a war game that included all of these risks would essentially be a handbook to committing a major BCS threat. This is, hopefully for obvious reasons, not the author's goal.

3 Importance of Emphasizing the Bio Component in Biocybersecurity

The primary cause for the development and demarcation of a new field of study is that biological processes are no longer simply the result of data processing. Biological data is being used as interlocks (retina, fingerprint scanners), to inform decisions (health monitors), and even as the data processing techniques themselves (biocomputing). Thus, in a biocybersecurity context, biological phenomena can act as interlocks, and even as facilitatory steps in a cybersecurity system. Biology and its manifold aspects introduce numerous targets that can be exploited such as in simple behaviors exhibited by organisms down to the qualities of organic compounds that comprise said organisms. That is, the transport of biomolecules within or between organisms can be monitored precisely as well as the targeting of specific genes within an organism [4, 5]. These can be examined and targeted on both macro and microscales, requiring creativity, but yielding potentially profitable results. For example, the actor can examine the data from someone's Fitbit to determine their night and daytime activities, for deciphering their target's lifestyle and how to take advantage of that [5, 22–24].

The same can be done for an entire community, in which a company might want to gain additional data on diet or work patterns, to better plan food truck locations. In terms of monitoring and targeting qualities, an insurance company can examine one's genome to glean and set new rates for individuals based on projected diseases that an individual may develop [25]. Although flawed, it is expected that the tools of analysis and their predictive power will strengthen [26]. In time, as bioprocessing infrastructure leans on biocomputing, our societies will need to examine such biological systems in an analogous way. In terms of the bioprocessing pipeline, DNA can be used to encode weapons to attack machinery that interfaces with it, but also can be a means of smuggling [13, 27]. As Peccoud and others rightly point out, a naivety in design and protocol can prove disastrous for an organization's enterprise [4, 12, 13, 27]. Thus, teams are encouraged to engage in a rigorous examination of the pipeline.

4 War Game Simulation

4.1 Description of and Manual for Wargame Activity for Bioprocessing Teams

The following guide is a fast and simple guide to doing a preliminary biocybersecurity analysis for a bioprocessing center, or any facility that uses biosecurity measures or uses large amounts of biological data. Figure 1 contains the common routes of attack and defense for these types of organizations:

Preliminary Step 1 - Selection: Elect or appoint an individual to run the activity. It is recommended that this person has knowledge of Information and cybersecurity, but any person with active interpersonal skills should do. Then gather any individuals interested or with a vested interest in security. Additionally, be sure to screen and update participants on knowledge of biology and biosecurity concepts.

Preliminary Step 2 - Review: Review the current, active security procedures and evaluate participation in them. For instance, investigate how many unsecured devices are in the facility, or how many people leave passwords out or computers unlocked. Now the activity can begin.

Step 1 - Training: Divide the participants into two groups. The data defenders and the data hackers. After reviewing the current security protocols as a group, a hacker and defender will pair up. The hacker should attempt to identify a given vulnerability in the current system. The data defender should then attempt to find a way to fix or "patch" that vulnerability. For instance, a facility that has users that frequently use exposed USB SSD devices can be made vulnerable by a malicious actor distributing flash drives with malware on them. Potential defenses include having these devices in the facility, educating users as to the hazards they pose or disabling inactive USB ports. This section

Common Routes of Attack

- Social Engineering
- Hardware modification
- Bio-Cyber Physical Interfaces
- Supply-Line Corruption
- Phishing

Common Routes of Defence

- Mandated Cybersecurity Training
- Segmented Privileges
- Use of Red/Blue Teams
- Ban of personal electronics
- Regular Security Patches

Fig. 1. A review of attack and defense routes to consider

is meant to be an introduction to the key methods used and should only last 5–10 min. Then have the participants switch partners, switch roles, or both. Table 1 describes the common "tiers" of attackers and defenders for this activity.

Table 1. Potential teams: data defenders vs. data hackers

Data defender	Data hacker
University/community bio lab	Company or rival lab looking to gain lab/research/material access or impede research
Government lab/regulatory agency	State-level hacker from rival country looking to gain access to privileged research or product information
Company lab	Rival company looking to steal/overcome IP and gain a market edge or sabotage company assets
Hospital, Prison,	Ransomware hackers looking to attack patients/equipment to derive ransoms or compromise medical equipment to harm an individual
Insurance	Insurance Fraudsters looking to increase costs, skim resources, and or disrupt healthcare access

Step 2 - Group Ideation: After a few rounds of training, several key ideas or vulnerabilities may become evident. Instead of patching them immediately, at this time, two groups should be formed- again Data Defenders and Data Hackers. This time, they should both be isolated and allowed to envision defenses against the vulnerabilities or exploits pertaining to vulnerabilities of the facility (respectively). Self-selection into the groups is encouraged, as long as both are roughly equal. Allow sufficient time to ideate strategies, and supply materials like adhesive notes, whiteboards, and other common office supplies. The time spent on this stage should be slightly longer than the training stage.

Step 3 - War Game: This is the terminal stage of the activity. This section should be recorded in some way in order to analyze the weaknesses and potential plans that could ameliorate security vulnerabilities. The activity will take a call-and-response style. Align the groups opposite of each other. The hacker group will announce a plan to exploit a security vulnerability, and the data defender group will have to state a plan to implement mitigation strategies for that plan. This will continue until an impasse is reached or the realistic time limit is reached. The flow of this stage of the activity is illustrated in Fig. 2.

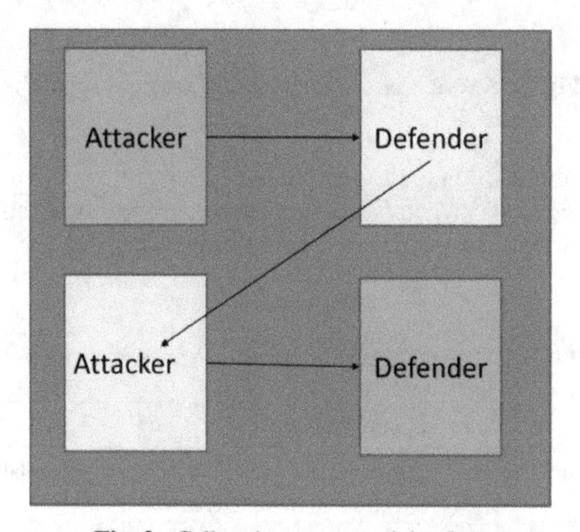

Fig. 2. Call and response activity flow

5 Discussion

Common exploit sources found in trial tuns of this activity have been the following:

1. The inefficiency and potential exploitations of security theater
2. The security implications of underpaid or unpaid workers
3. Miscommunications of conventional security threats
4. Lack of knowledge of novel types of threats
5. Security implications of sub-standard resources

All of these combined exploit sources illustrate that considerable gaps in knowledge exist among potential staff concerning firms that deal with some degree of bioprocessing operations. Regardless of subindustry, it is important that members who take part in such wargames switch sides and remain updated on new trends within biocybersecurity as well as separate developments within cybersecurity and biosecurity, as both fields may have developments that do not yet materially overlap. Further, these war games should be run frequently so as to keep staff in practice and cognizant of dangers that may exist. It is important to note that the style and order of wargaming can be varied to suit the organization's needs. That being said, it is suggested and encouraged that teams vary their playstyle and introduce different scenarios. For example, the question of a wargame involving three or more opposing groups is worthy of exploration as can be imagined from the interplay of state-level actors, corporate actors, internal actors, and ethical hackers within. As it is increasingly apparent that IT and Bioprocessing possess a shared destiny in the years to come, operations both mental and physical should reflect that for optimal security.

6 Importance and Conclusion

The primary importance of doing this as a group and as a facility includes the fast-paced nature of development in biology and bioprocessing. Different labs tend to have different requirements, tools, cyber-physical interactions, and workarounds. Lower funded labs may be vulnerable to more conventional means of penetration. Labs with greater amounts of funding could be specifically targeted. Coupled with the fact that almost no research facility will have a complete end-to-end supply of anything except the most basic of supplies (for instance ethanol or double-distilled water), interactions that could be used and vulnerabilities and exploited for penetration will always exist. Only by the use of representation from all of the invested groups will more complete security coverage be achieved.

7 Future Directions

The primary future direction of this activity was already exhibited at the MIT Global Community Bio Summit 4.0 over October 10th and 11th, 2020. The war game was modified for accessibility and took the form of a card game similar to the classic card game "War", but with multivariate "suits" (namely biological, cyber, and human) which took the different kinds of threats previously identified and coalesced them for use in a less educated audience. Notes on such work may be found in the Proceedings of the Global Community Bio Summit, to be released later within 2020.

Acknowledgments. We obtained no external funding for this work.

References

1. Murch, R.S., So, W.K., Buchholz, W.G., Raman, S., Peccoud, J.: Cyberbiosecurity: an emerging new discipline to help safeguard the bioeconomy. Front. Bioeng. Biotechnol. **6** (2018). https://doi.org/10.3389/fbioe.2018.00039
2. Yong, E.: Synthetic double-helix faithfully stores Shakespeares sonnets. Nature (2013). https://doi.org/10.1038/nature.2013.12279

3. Liverpool, L.: 3D-printed bunny contains DNA instructions to make a copy of itself, 9 December 2019. https://www.newscientist.com/article/2226644-3d-printed-bunny-contains-dna-instructions-to-make-a-copy-of-itself/
4. Dieuliis, D.: Biodata risks and synthetic biology: a critical juncture. J. Bioterrorism Biodefense **09**(01) (2018). https://doi.org/10.4172/2157-2526.1000159
5. Arriba-Pérez, F.D., Caeiro-Rodríguez, M., Santos-Gago, J.: Collection and processing of data from wrist wearable devices in heterogeneous and multiple-user scenarios. Sensors **16**(9), 1538 (2016). https://doi.org/10.3390/s16091538
6. Pauwels, E., Dunlap, G.: The Intelligent and Connected Bio-Labs of the Future: The Promise and Peril in the Fourth Industrial Revolution 2017. The Wilson Center , Wilson Briefs. Washington, DC (2017)
7. Podolsky, I.A., Seppälä, S., Lankiewicz, T.S., Brown, J.L., Swift, C.L., O'Malley, M.A.: Annual Review of Chemical and Biomolecular Engineering **10**(1), 105–128 (2019)
8. Ubando, A.T., Felix, C.B., Chen, W.-H.: Biorefineries in circular bioeconomy: a comprehensive review. Biores. Technol. **299**, 122585 (2020). https://doi.org/10.1016/j.biortech.2019.122585
9. Agapakis, C.M.: Designing synthetic biology. ACS Synth. Biol. **3**(3), 121–128 (2013). https://doi.org/10.1021/sb4001068
10. Wilbanks, R.: Real vegan cheese and the artistic critique of biotechnology. Engaging Sci. Technol. Soc. **3**, 180 (2017). https://doi.org/10.17351/ests2017.53
11. Amos, M.: DNA computing. In: Adamatzky, A. (eds.) Unconventional Computing. Encyclopedia of Complexity and Systems Science Series. Springer, New York (2018)
12. Ney, P., Koscher, K., Organick, L., Ceze, L., Kohno, T.: Computer Security, Privacy, and DNA Sequencing: Compromising Computers with Synthesized DNA, Privacy Leaks, and More (2017). https://www.usenix.org/conference/usenixsecurity17/technical-sessions/presentation/ney
13. Peccoud, J., Gallegos, J.E., Murch, R., Buchholz, W.G., Raman, S.: Cyberbiosecurity: from naive trust to risk awareness. Trends Biotechnol. **36**, 4–7 (2017). https://doi.org/10.1016/j.tibtech.2017.10.012
14. O'Neill, P.H.: Data leak exposes unchangeable biometric data of over 1 million people, 14 August 2019. https://www.technologyreview.com/f/614163/data-leak-exposes-unchangeable-biometric-data-of-over-1-million-people
15. Shea, V.J., Dow, K.E., Chong, A.Y.-L., Ngai, E.W.T.: An examination of the long-term business value of investments in information technology. Inf. Syst. Front. **21**(1), 213–227 (2017). https://doi.org/10.1007/s10796-017-9735-5
16. Dehgani, R., Jafari Navimipour, N.: The impact of information technology and communication systems on the agility of supply chain management systems. Kybernetes **48**(10), 2217–2236 (2019). https://doi.org/10.1108/K-10-2018-0532
17. Rumbley, L.E.: Intelligent internationalization. In: Intelligent Internationalization. Brill|Sense, Leiden (2019). https://doi.org/10.1163/9789004418912_001
18. Elder-Vass, D.: Lifeworld and systems in the digital economy. Eur. J. Soc. Theory **21**(2), 227–244 (2017). https://doi.org/10.1177/1368431017709703
19. Agrawal, A., Gans, J., Goldfarb, A.: Economic policy for artificial intelligence. Innov. Policy Econ. **19**, 139–159 (2019). https://doi.org/10.1086/699935
20. Ghafele, R., Gibert, B.: Open growth: the economic impact of open source software in the USA. In: Khosrow-Pour, M.D.B.A. (Ed.) Optimizing Contemporary Application and Processes in Open Source Software, pp. 164–197. IGI Global, Hershey (2018). https://doi.org/10.4018/978-1-5225-5314-4.ch007

21. Winter, L., Pellicer-Guridi, R., Broche, L., Winkler, S.A., Reimann, H.M., Han, H., Benchoufi, M.: Open source medical devices for innovation, education and global health: case study of open source magnetic resonance imaging. Manage. Prof. Co-Creation 147–163 (2018). https://doi.org/10.1007/978-3-319-97788-1_12
22. Greenwald, M.: Cybersecurity in Sports (2017)
23. Tanev, G., Tzolov, P., Apiafi, R.: A value blueprint approach to cybersecurity in networked medical devices. Technol. Innov. Manage. Rev. **5**(6) (2015)
24. Langone, M., Setola, R., Lopez, J.: Cybersecurity of wearable devices: an experimental analysis and a vulnerability assessment method. In: 2017 IEEE 41st Annual Computer Software and Applications Conference (COMPSAC), vol. 2, pp. 304–309. IEEE, July 2017
25. Barry, P.: Seeking genetic fate: Personal genomics companies offer forecasts of disease risk, but the science behind the packaging is still evolving. Sci. News **176**(1), 16–21 (2009). https://doi.org/10.1002/scin.5591760123
26. Fumagalli, E.: Direct-to-Consumer DNA Testing and Product Personalization: One Size Does Not Fit All Genes (2019). https://doi.org/10.4135/9781526463951
27. Andy, E.: How DNA could store all the world's data. Nature **537**(7618) (2016). Gale Academic OneFile, Accessed 24 Mar 2020

The Effects of Data Breaches on Public Companies: A Mirage or Reality?

Bright Frimpong[1,2](✉) and Lei Chen[1,2]

[1] Department of Enterprise Systems and Analytics, Georgia Southern University, Statesboro, GA, USA
{bright_frimpong,LChen}@georgiasouthern.edu
[2] Department of Information Technology, Georgia Southern University, Statesboro, GA, USA

Abstract. Public companies are often criticized for their lackluster approach to cybersecurity investments. Critics assert that there are no financial incentives for public companies to invest in cybersecurity. Previous studies have mostly focused on the effects of data breaches on the market value of public companies. Nonetheless, this study focused on the impact of data breaches on revenue, net income and marketing expenses of public companies. Results of the study indicated that data breaches have significant impact on revenue but not on net income and marketing expenses. These results affirm the assertion that public companies do not have sufficient financial incentives to invest against data breaches. Therefore, we assert that public companies not only need to look at their bottom line but also consider non-financial incentives when making decisions regarding cybersecurity investments.

Keywords: Data breach effects · Revenue · Cybersecurity investment · Marketing expense · Public companies

1 Introduction

Traditional information systems have phased out as most companies now depend on some form of digital information systems for their business operations. Powered by the internet, digital information systems provide companies with benefits such as improving daily operational efficiency, creating new business models and products, making enhanced decisions and providing sustainable competitive advantages. Digital data is rapidly expanding as the amount of data stored online has increased at an unparalleled rate. The United States continues to experience a growing number of online businesses. In 2013, there were approximately 102,728 e-commerce retailers in the country [1]. It is estimated that global retail e-commerce sales will reach $4.5 trillion by 2021 which is not surprising considering companies' increasing dependency on digital information systems [8]. However, one pertinent drawback associated with using digital information systems is the threat of data breach. Companies must relentlessly safeguard their data from potential threats and breaches. Data breaches have become a headache for both large and small companies as the number of breached records continues to soar in the

K. Arai (Ed.): FICC 2021, AISC 1363, pp. 674–683, 2021.
https://doi.org/10.1007/978-3-030-73100-7_49

United States above the global average. For instance, public companies like Equifax, Target, eBay, and Uber have all experienced massive data breaches in recent years.

Data breaches are bad publicity for affected companies as they affect brand reputation, goodwill, consumer trust and most importantly, financial performance [15]. Also, businesses are most likely to experience a high turnover of customers due to present notification laws. Notification laws are rules and regulations which require businesses to inform and educate customers about a data breach event. These laws have been designed to give customers considerable awareness and expectations regarding how data breaches should be handled [15]. After eBay suffered a massive data breach in 2014, the company was required to notify customers about the breach and convince them to reset their passwords. eBay observed a decline in user activity after the data breach due to a loss of consumer trust and goodwill. Also, KPMG reported that data breach attacks led to a 31% decrease in brand image and 30% loss of clients [5].

Public companies have two options when it comes to dealing with data breaches. First, they can take proactive measures by making substantial investments in their cybersecurity. Also, they can forgo such cybersecurity investments and deal with breach issues as and when they occur. It has been asserted that public companies have very little financial incentive to invest in cybersecurity because the real cost of data breaches is very insignificant and barely affect their bottom line. Apparently, the costs of data breach make up 1% to 2% of public companies' revenue or net profit when scaled [16]. However, it is also argued that public companies need to consider the hidden and unquantifiable costs of data breaches such as declining goodwill and reputation. In any case, businesses have to determine whether they are going to take a proactive or reactive approach to cybersecurity investment. This paper attempts to explain the rationale behind both approaches and how it affects various businesses' approach to cybersecurity investment.

One of the fundamental objectives of every company is to maximize profits for its shareholders. Profit is a function of revenue and cost. This implies that a change in revenue or cost elements should result in change in profit. A lot of studies have confirmed the effect of data breaches on cost [2–4]. These effects have been found to be direct like legal fees, court fines, and database repair costs [2–4]. Recently, several public companies have coughed up huge amounts of money due to data breaches. Hilton Hotels was slapped with $700,000 in fines in 2017, Home Depot paid $15.3 million in legal fees and $46.75 million in settling class action lawsuit from customers and banks while Equifax recorded $113.3 million in data breach-related expenses [6, 11, 16]. While there is a high level of certainty that a company will incur some form of breach-related costs, the same cannot be said for its impact on revenue. A review of a few of the biggest data breach scandals over the years provides mixed results on the effect of a data breach on revenue [12, 16]. For instance, Target experienced a 46 percent decline in revenue the next quarter following their 2013 breach whilst eBay missed its revenue target by $200 million [11, 16]. Whilst both Target and eBay experienced a dip in revenue after their reported data breach, Equifax's experience was somewhat opposite. In a massive data breach that affected approximately 148 million people, Equifax reported a 7% and 20% increase in both revenue and net income respectively [6]. This shows how data breaches can have contrasting yet significant effects on the fortunes of different companies. The purpose of this study is to examine the relationship between data breach events and revenue, net profit, and marketing expenses among public companies in the U.S. The researchers attempt to investigate whether data breach events have a significant impact on any of these financial accounting measures and subsequently, put forward recommendations regarding how public companies should approach cybersecurity investments.

This study proceeds as follows: The next section discusses the related literature and develops our theoretical framework with specific hypotheses. The proceeding section covers our research design and methodology including an evaluation or prior measurement techniques. Finally, results and findings are presented, and the implications of this research are discussed together with the limitations and recommendations for future work.

2 Literature Review

The leading cause of data breaches is attributed to hacking and malware attacks. A study found that 47% of all data breaches were caused by malicious attacks from hackers [5]. Data breaches may also take the form of denial of service attacks, computer viruses and unauthorized access to confidential information such as credit card data. To gain access to confidential information, hackers normally employ the use of phishing. Phishing is the act of sending sham emails or making impostor phone calls to deceive customers. Additionally, companies that rely on time-sensitive data like hospitals and banks are also likely to be attacked by ransomware, which is a form of a computer virus. A ransom note is typically attached to this ransomware with the sole intent to extort money from breached companies before the database system can be unlocked for normal business operations. While data breaches are mostly intentional, they can also be accidental by means of employee error, improper disposal of data systems and unintended data disclosures.

Public companies are likely to suffer a loss in market value when a data breach is disclosed to the public [3]. Whenever a publicly traded company discloses a data breach, statistical evidence suggests there is a significant chance that stock prices will fall [3]. This is probably because investors fear that the breach will have a negative impact on the future financial performance of the company. Interestingly, such decline in stock prices has been found to last for a short period of time between 1–5 days after which the stock prices start rising again [3]. Campbell et al. found that data breaches had insignificant long-term economic effects on public companies [2]. The study further explained that public companies were more likely to make additional security investments to prevent future breaches, therefore, limiting the impact of the breach to only the short-term. Investor confidence has been found to be linked to how firms handle data breach issues [2]. Gatzlaff & McCullough found that refusal of a public company to disclose details about a breach led to a strong negative market reaction [4]. As such, public companies ought to provide details about breach events if they want to boost investor confidence and prevent drastic loss of market value. Few studies have also sought to quantify the amount of impact data breaches have on market value in the short term [2–4]. Cavusoglu et al. concluded that breached companies experienced a market value loss of 2.1% over 2 days, Garg et al. found that the overall loss in market value was 5.3% over a 3-day window while Campbell et al. gathered that the loss of firm value was approximately 5.5% [2–4] (Table 1).

Table 1. Summary of previous research findings

Researchers	Findings
Cavusoglu et al. (2004)	1. Breaches result in overall loss of 2.1% of value over 2 days following event 2. Breach costs are higher for Internet firms 3. Costs not related to breach type 4. Breach costs increase over time 5. Negative correlation between size and stock market response
Hovav and D'Arcy (2003)	1. Breach costs higher for Internet firms 2. No overall significant market impact for denial of service attacks
Garg et al. (2003)	1. Security attacks result in overall loss of 5.3% of value over 3-day event window 2. Internet security vendors experience positive returns of 10.3% over the same window when security attacks are reported 3. Property–casualty insurers experience a loss of 2.0% over the same window when security attacks are reported
Campbell et al. (2003)	1. Breaches result in no statistically significant loss for entire sample 2. Breaches involving unauthorized access to customer personal data or firm proprietary data result in an average loss of firm value of 5.5%
McCullough and Gatzlaff (2010)	1. Negative association between market reaction and firms that are less forthcoming about the details of the breach 2. Firms with higher market-to-book ratios experience greater negative abnormal returns associated with a data breach 3. Firm size and subsidiary status mitigate the negative effect of a data breach on the firm's stock price and that the negative market reaction to a data breach is more significant in the most recent time periods of the sample

*An updated review table extracted from McCullogh and Gathzalog (2010)

While the impact of a data breach on stock prices have been quantified in several studies, the true cost of data breaches has been more difficult to quantify. Most studies and industry reports have resorted to measures of estimation or approximation in the attempt to measure the cost of data breaches [7]. In the attempt to measure the cost of data breaches, such studies tend to evaluate and estimate factors such as breach detection expenses, notification and public disclosure expenses, legal expenses and court fines, damaged systems, investigative teams and disruption of business. For instance, TJ Maxx paid up to $256 million in costs when the company experienced a data breach attack in 2007 [14]. This figure included notification costs, credit monitoring and court settlements

[14]. A report found that U.S companies incurred the highest data breach costs although the global average cost of a data breach had decreased by 10% [9, 15]. The report further revealed that US companies were likely to incur more costs on post-breach responses than companies from other countries. Additionally, results from several studies indicate that there is a positive significant relationship between the number of records breached and the cost of the data breach (2,7,9). It has also been found that companies are likely to incur lesser costs if a data breach is discovered earlier than when discovered later [9]. Specifically, another study found that companies that discovered a breach in less than 100 days were more likely to save more than $1 million than rival firms that took the average of 197 days [9].

On the consumer side of things, findings from a Branden Williams study revealed that consumers were quick to return to companies that had suffered a data breach [10]. The study claimed that most of these customers were either not aware of the data breach or completely unfazed about it. The study also concluded that while breached companies were not resistant to sales or revenue reduction, any decline in their sales or revenue was only temporary.

Reviewed literature seems to downplay the true impact of data breaches on public companies, therefore the need to examine the relationship between data breach events and revenue, net income and marketing expenses of public companies. Fundamentally, the researchers attempt to study these relationships through extensive secondary data collection and rigorous statistical tests. As already stated, the primary objective of the study is to determine if there is enough evidence to suggest that data breach events have direct significant impact on revenue and net profit among publicly traded companies in the U.S. Also, public companies are most likely to increase their marketing and advertising expenses after a data breach event. Several industry reports have indicated that public companies spend more money on sales promotion, advertising and public relations in the bid to restore consumer trust and salvage brand reputation after a data breach event [12]. If so, how significant is such an increase? To achieve our research objective, our study is guided by the following hypotheses:

H1 - There is a significant relationship between data breach event and revenue.
H2- There is a significant relationship between data breach event and net income.
H3- There is a significant relationship between data breach event and marketing expenses.

3 Research Methodology

The study evaluated the financial impact of data breaches on public companies using various accounting measures. The use of accounting measures is the most popular method of assessing company performance [8]. We collected a sampled list of 50 public companies that suffered a data breach attack between 2014 and 2018 from Privacy Rights ClearingHouse. Privacy Rights ClearingHouse is a nonprofit organization that keeps records of reported data breach events in its database. We focused on only public companies in the U.S because these companies are legally required to provide their financial statements to the public which makes their financial data readily available. The financial data of the 50 sampled public companies were extracted from S&P Net Advantage, a database that stores financial reports of public companies. Quarterly financial performance data were collected for each public company on revenue, net income, and marketing expenses. Therefore, performance measures were collected for "four quarters prior to the breach event" and the "four quarters immediately after the breach event". All the companies in the sample suffered a data breach attack at some point between 2014 and 2018.

We used normality plots to examine the distributions of the variables. The data were all close to normal except for few outliers in the variables. However, due to the small sample size and potential effects of these outliers, nonparametric tests were conducted. A nonparametric test is most appropriate when the sample size is small [18]. Therefore, the Wilcoxon signed rank test was used to test the various hypothesis designed for the study. This nonparametric test is much less sensitive to outliers and very satisfactory when comparing before-and-after observations on the same subjects [18]. The Wilcoxon signed rank test was used to test for the average differences between the before-and-after in the variables being observed. The researchers also employed the sign test as a diagnostic test to compare with the results of the Wilcoxon's. The primary objective of both tests was to determine whether the statistical average difference between the before-and-after differed from zero.

To test H1, average revenue for the 4 quarters prior to the breach was compared with the average revenue for the 4 quarters after the breach. The same procedure was carried out for the remaining two hypotheses also. This way, the study was able to determine if there was an average difference between revenue, net income, and marketing expenses before-and-after the data breach event at a 95% significance level.

4 Results and Discussion

Results from the analysis show that there is statistical difference between revenue levels before-and-after data breach at a 95% significance level. The P-values computed for both tests indicate that the average revenue levels before and after a data breach do differ from zero since the p-value is below 0.05. This means that the null hypothesis should be rejected as the relationship between data breach event and revenue is significant among public companies in the U.S. Findings for net income and marketing expenses were insignificant as the P-values for H2 and H3 were above 0.05. Therefore, H2 and H3 were not confirmed as there is no evidence to suggest that data breaches significantly impact revenue, net income, and marketing expenses.

	Revenue	Marketing expenses	Net income
Wilcoxon Signed Ranks Test			
Z	-2.058^{b}	$- 956^{b}$	-1.373^{b}
Asymp. Sig. (2-tailed)	.040	.339	.170
Sign Test			
Z	−2.155	−1.323	−1.486
Asymp. Sig. (2-tailed)	.031	.186	.137

Results from the data analysis deviate from the general perception that data breaches have a significant impact on financial performance. In fact, these findings indicate that the effects of data breach on public companies are illusions and might have been grossly overrated by media outlets and industry analysts. Although the study found a significant relationship between data breach events and revenue, previous studies have successful argued and demonstrated that potential effects of data breach events are very momentarily. That is to say, although public companies might notice a change in their revenue trend after a data breach, such a change is very brief [10]. Research shows that consumers are quick to return to companies that have suffered a data breach. Most consumers simply do not care about data breaches and will remain loyal to the company while other consumers might also be totally unaware of the data breach. Customers who choose to remain loyal after a data breach might do so because of high switching costs, contract lockups and high discounts from loyalty programs. As a result of this phenomenon, public companies rarely experience a reduction in their sales volume after a breach event.

At a closer look, there are a lot of potential factors that have the potential to mitigate the impact of data breaches on firm performance. Public companies in the U.S are typically huge enterprises with many subsidiaries. Therefore, a data breach event at one subsidiary might not have the cascading effect on general firm performance as expected. Simply because, the financial performance of the other subsidiaries will strongly nullify any weak performance from the breached subsidiary.

Also, public companies are more likely to have insurance policies covering data breaches. For instance, Equifax received $125 million from their insurance package to cover part of the $439 million cost of the massive 2017 data breach. Target also received $15 million under its insurance coverage for a data breach cost of $43 million. These insurance payouts reduced Equifax's cost by 28% and that of Target by more than 30%. Also, breached companies have the option to use tax deductions to offset a huge chunk of the remaining cost. These two factors mitigate the negative effects of data breaches on companies as they provide indirect revenue injections into breached companies.

Another factor to consider is that public companies normally anticipate huge cash flows from legal fines, court settlements, and cybersecurity investment after a data breach. We argue that these anticipated cash flows make companies skeptical to increase their marketing and advertising expenses after a data breach. Companies are most likely to reserve funds to tackle the legal fees, subsequent fines and additional cybersecurity investments that will accrue after the breach. At the end of the day, the impact of data breaches on revenue, net income and marketing expenses are more likely to be rendered insignificant by these factors.

5 Evaluation

In the table below (Table 2), we present a list of the various methodologies that have been used in previous literature. We devote this section to compare these methodologies and explain the need to experiment with additional methodologies to explore the relationship between data breaches and firm performance.

Table 2. Comparison framework

Researchers	Methodology	Variables
Hovav and D'Arcy (2003)	Event study	Stock price
Garg et al. (2003)	Event study	Stock price
Campbell et al. (2003)	Event study	Stock price
Cavusoglu et al. (2004) McCullough and Gatzlaff (2010)	Event study	Stock price
Ko and Dorantes (2006)	Matched-sample analysis	Operating income, Return on Assets and Sales
Makridis and Dean (2018)	Econometric models	Revenue per worker, capital per worker
*Current study	Wilcoxon/ Sign t-test	Revenue, Net income, Marketing expenses

The event study methodology is adopted to investigate the effect of an event on a specific dependent variable. For instance, it can be used to examine the response of stock price around the announcement or occurrence of an event. Studies that have used event studies tend to depend on the use of stock prices as indicators for firm performance. [12] argue that stock prices do not fully reflect the risk inherent in cyber security breaches because stock prices have been found to bounce back to normal trends after two days. Also, detailed examination of prior studies that have reported a negative association between breach incidents and stock prices reveal a wide dispersion in their estimated conditional correlations between breaches and stock prices [12]. Additionally, event studies depend on the assumption of an efficient market, an assumption which is not always valid. In spite of these limitations, event studies have a powerful and easy design which makes it easy to interpret and detect abnormal performance. Meanwhile, the use of matched-sample analysis allows the researcher to control for the effects of other unwanted variables [18]. However, they can also be time-consuming and impossible to match subjects perfectly. Econometric models also assist in reducing risk and predicting outcomes with some probability to make decision process easier. However, they share similar weaknesses with event studies as they depend on a lot of assumptions which are not always valid [12].

In this study, we used the much simpler Wilcoxon Signed Rank test because it does not depend on the data distribution or its parameters. The Wilcoxon test does not require any assumptions about the shape of the data distribution [18]. Also, the test is fairly robust, easy to calculate and interpret. The test is suitable for small data with sensitive outliers making it compatible with data used in this study. We also employed the sign test as a diagnostic test to compare with the results of the Wilcoxon's. One major disadvantage with the Wilcoxon test is that one cannot control for the effects of the environment. In this study, we were not able to control for potential factors that could have affected the revenue, net income and marketing expenses of the companies in the sampled data. However, our study provided significant results which are line with previous studies [2,

10]. [12] challenges the validity and reliability of samples and measurement techniques used in previous data breach related studies. This challenge is not surprising considering that all of the measurement techniques are fraught with several limitations. Therefore, we recommend future studies to explore additional effective techniques or perhaps, a combination of prior adopted techniques in assessing the relationship between data breach incidents and firm performance.

6 Conclusion

Several analysts and researchers have asserted that public companies lack the financial incentive to invest in cybersecurity [13, 16]. Results gathered from this study support this notion as the impact of breaches on net income and marketing expenses were found to be insignificant. To get public companies to invest more in cybersecurity, the focus must be on the non-financial effects of data breaches. Public companies need to be made aware of how data breaches erode brand reputation, goodwill and customer trust and how these metrics indirectly affect company performance in the long run. Although, these are non-quantifiable metrics and hardly show up on financial statements, these metrics are leading indicators which gives a prediction of future performance. Public companies need to stop looking at their bottom-line but rather focus on the hidden cost of data breaches like soaring insurance premiums, third-party damages, negative brand image and declining goodwill and trust among customers. Further, regulatory bodies like the government and the SEC need to impose heavier fines on negligent public companies to ensure a more proactive attitude towards cybersecurity investment.

Public companies need to adopt a more proactive approach towards cybersecurity investment even though data breaches have insignificant financial impacts on their bottom-line. While companies might not have the financial incentive to invest in cybersecurity, the non-financial repercussions associated with data breaches can have lasting effects on company brand and reputation and indirectly affect financial performance in the long run.

7 Limitations and Future Research

The sample used for the study was limited to public companies therefore affecting the generalizability of the results to private companies. We recommend that future studies consider larger sample sizes comprising both public and private companies. Also, accounting metrics may not the best standards to evaluate data breach impacts even though they are widely used. We recommend that future studies adopt a case study approach to offer additional insights on the effects of data breaches on individual companies. Future studies need to consider diving deeper to analyze data breaches by type and the affected companies by industry. We believe that doing so will provide more insights on how different types of data breaches affect companies from different industries. Like previous data breach studies, this study encountered difficulties in its sampling and measurement techniques. Whiles some companies simply do not report data breach incidents, there is no proven measurement tool to measure the effects of data breach.

Makridis and Dean published a study challenging the validity and reliability of samples and measurement techniques used in previous data breach related studies [12]. The authors call for the development of comprehensive data on cybersecurity incidents at the firm level and proper measurement technique devoid of sampling selection problems as decision decision-makers require proper evidence to make well-informed decisions regarding information security investments.

References

1. Ablon, L., et al.: Consumer Attitudes Toward Data Breach Notifications and Loss of Personal Information. Rand Corporation, Santa Monica (2016)
2. Belicove, M.: How many U.S. based online retail stores are on the internet? Forbes (2013)
3. Campbell, K., et al.: The economic cost of publicly announced information security breaches: empirical evidence from the stock market. J. Comput. Secur. **11**(3), 431–448 (2003)
4. Cavusoglu, H., et al.: The effect of internet security breach announcements on market value: Capital market reactions for breached firms and internet security developers. Int. J. Electron. Commer. **9**(1), 70–104 (2004)
5. Dean, B.: Sorry consumers, companies have little incentive to invest in better cybersecurity (2015)
6. Fowler, K.: Data breach preparation and response (2016)
7. Gatzlaff, K.M., McCullough, K.A.: The effect of data breaches on shareholder wealth. Risk Manag. Insur. Rev. **13**(1), 61–83 (2010)
8. Ko, M., Dorantes, C.: The impact of information security breaches on financial performance of the breached firms: an empirical investigation. J. Inf. Technol. Manag. **17**(2), 13–22 (2006)
9. Kerber, R., Globe, B.: Cost of data breach at TJX soars to $256 m. Boston Globe (2007)
10. K.P.M.G.: Small business reputation and the cyber risk (2015)
11. Moore, T.: On the harms arising from the Equifax data breach of 2017. Int. J. Crit. Infrastruct. Protect. **19**(C), 47–48 (2017)
12. Makridis, C., Dean, B.: Measuring the economic effects of data breaches on firm outcomes: challenges and opportunities. J. Econ. Soc. Meas. **43**(1–2), 59–83 (2018)
13. Manworren, N., et al.: Why you should care about the target data breach. Bus. Horiz. **59**(3), 257–266 (2016)
14. Martin, K.D., et al.: Data privacy: effects on customer and firm performance. J. Mark. **81**(1), 36–58 (2017)
15. Orendorff, A.: Global ecommerce: statistics and international growth trends [Infographic. Enterprise Ecommerce Blog-Enterprise Business Marketing, News, Tips & More (2017)
16. Ponemon Institute: The impact of data breaches on reputation and share value (2017)
17. PricewaterhouseCoopers: The Global State of Information Security (2017)
18. Hair, J.F., et al.: Multivariate Data Analysis. Prentice Hall, Upper Saddle River (1998)
19. Garg, A., et al.: Quantifying the financial impact of IT security breaches. Inf. Manag. Comput. Secur. **11**, 74–83 (2003)
20. Hovav, A., D'Arcy, J.: The impact of denial-of-service attack announcements on the market value of firms. Risk Manag. Insur. Rev. **6**(2), 97–121 (2003)

Trust Models and Risk in the Internet of Things

Jeffrey Hemmes[1(✉)], Judson Dressler[2], and Steven Fulton[2]

[1] Regis University, Denver, CO, USA
jhemmes@regis.edu
[2] United States Air Force Academy, Colorado Springs, CO, USA

Abstract. The Internet of Things (IoT) is envisaged to be a large-scale, massively heterogeneous ecosystem of devices with varying purposes and capabilities. While architectures and frameworks have focused on functionality and performance, security is a critical aspect that must be integrated into system design. This work proposes a method of risk assessment of devices using both trust models and static capability profiles to determine the level of risk each device poses. By combining the concepts of trust and secure device fingerprinting, security mechanisms can be more efficiently allocated across networked IoT devices. Simultaneously, devices can be allowed a greater degree of functionality while ensuring system availability and security. This paper describes the integration of risk assessment into a prototype IoT network. The purpose of this prototype is to explore whether finer-grained security policies based on risk can adequately protect the network while also allowing for efficiency and system functionality to a greater extent than traditional security protocols permit. Furthermore, we demonstrate how identification, trust, and risk can be synthesized to provide a finer degree of control over system security.

Keywords: Security · Access control · Trust · Trust models · Device fingerprinting · Risk · Profiling · Device characterization · Internet of Things · IoT security

1 Introduction

In recent years, IoT concepts and implementations have rapidly matured towards a vision of a broad eco system of devices, collaborating and exchanging information to accomplish useful and beneficial tasks. However, the massively heterogeneous nature of IoT poses unique and significant security challenges. Devices can join different network enclaves or from multiple access points, while having varying capabilities in terms of computing power. Although these characteristics offer tremendous opportunities in terms of functionality, designing an effective and efficient security architecture for such systems is quite difficult. With the growing reliance on smart devices, the need for robust security protocols is ever more apparent. However, security protocols for IoT must be sufficiently flexible to accommodate a wide range of devices with varying capabilities.

This work was supported by the Air Force Office of Scientific Research (AFOSR) Summer Faculty Fellowship Program (SFFP).

K. Arai (Ed.): FICC 2021, AISC 1363, pp. 684–695, 2021.
https://doi.org/10.1007/978-3-030-73100-7_50

The introduction of new frameworks and architectures for functional purposes has revealed unique design challenges that must be addressed going forward. Limited capabilities, legacy, and/or low-powered devices cannot always be upgraded with state-of-the-art security mechanisms due to a number of reasons including costs and non-availability of newer devices with the same functionality. Any framework, therefore, must account for such limitations in its security architecture.

Conversely, with a substantial array of computing devices on the network, it is assumed any node potentially could be compromised to send malicious information or denying service, causing disruption in the overall system with real-world consequences. In this work, we propose accounting for variations in device capability through a combination of trust modeling and device fingerprinting in a security architecture. Such an approach could be incorporated into the design of IoT middleware or a gateway security policy, and is based on both the verified capabilities of connected devices as well as their demonstrated behavior.

The goal of this work is to balance flexibility and efficiency with security considerations such that available security mechanisms can be directed towards and focused on those devices more capable of causing harm on the network. That is not to say that less powerful devices are ignored; rather, a tiered approach can be employed based on a combination of capability and trust. More powerful and less trusted devices may be subjected to restrictive security controls such as sandboxing, bandwidth throttling, etc., until its behavior over time establishes a high level of trust. The device may then be permitted to access additional services.

This paper is organized as follows: Sect. 2 discusses the most relevant prior works related to trust-based security and trust modeling. Section 3 outlines the basic trust model used in this work. Section 4 describes methods for device characterization necessary for assessing risk. Section 5 presents the system architecture of our prototype. Section 6 describes the experimental setup. Section 7 through Sect. 9 present evaluation results, conclusions, and the future direction of this work.

2 Related Work

The need for trust-based security in mobile IoT networks is well established and acknowledged [18]. Consequently, trust models have long been a robust area of study in wireless network security [3, 6–8, 10, 12, 18–20, 22]. With respect to IoT, the notion of trust appears in many recent works. However, it takes on several distinct forms. This is in large part due to a general lack of consensus across the research community of what trust really means and what it ought to look like in systems exemplified by IoT [10].

Most systems employing trust models involve some element of reputation that is developed based on a demonstrated pattern of behavior over time. Most commonly, trust is handled in a decentralized fashion, with nodes developing a measure of trustworthiness of their immediate neighbors through direct interactions. For more distant nodes, trust can be developed through the receipt of trust evaluations from others, which can subsequently be modified based on additional interactions. With most work related to trust in IoT systems, trust is generally a standalone concept often used as the sole criterion for making access control or routing decisions. For instance, in [8], trust is used for adaptive network protection with the goal of allocating security services efficiently based on trust level of devices, much like this work. However, the use of trust is implemented in a decentralized manner, with each node making local decisions about whether to authenticate messages received from neighboring nodes. More significantly, trust is the sole criterion for making decisions about whether to authenticate traffic. Many other works take a similar approach, e.g., [2, 3, 11, 16, 22].

As for how trust models are defined for IoT networks, there are a number of distinct approaches. Several of the most relevant works are cited in this paper, but the work proposed here would be generally compatible with a wide variety of different trust models.

3 Trust Models

One approach to security that has been well established in the literature is the use of trust models [14]. In essence, a trust model provides an indication of the extent to which an entity operates within the bounds of specified parameters. These models have broader application in more traditional wireless ad-hoc networks. However, given that incorporating wireless sensor networks into the IoT is not uncommon [10], trust models have seen wider application in IoT systems in recent years. Typically, the relationship between trust models and IoT has been with single-purpose applications, e.g., [7, 12]. Applying trust models in a more general and heterogeneous network beyond what has been previously explored is the primary focus of this work. The purpose is to be able to prioritize and efficiently allocate security mechanisms towards those devices which pose the greatest threat based on both their capabilities as well as their trustworthiness. This differs from traditional reputation-based systems, which do not address capabilities in any meaningful way [14].

Beginning with a general trust model such as that described and presented by Liu et al. [13], we extend it by incorporating capability characteristics into assessments. This information informs decisions regarding what actions each device is permitted to take on the network. For example, a low-powered sensor is less capable than a laptop computer with less potential to cause significant damage, although permitting any unknown device to connect to a network carries with it some degree of risk. A device with the computational power of a FitBit has less potential to cause a denial of service attack across a network than a laptop, although it certainly can cause some disruption under certain conditions. Given that any device is inherently untrusted when it joins the network, the question is, which device poses the greater potential threat?

To explore this idea, existing trust modeling can be expanded towards a more comprehensive risk assessment mechanism that takes into account device capabilities. The basic trust model used is described by Liu et al. [13] as:

$$TL_i(j) = \frac{\sum_{k=0}^{n} TL_k(j)}{n} \tag{1}$$

Where $TL_i(j)$ is the estimated trust level of another node on the network, node j, as determined by node i, and n is the total number of nodes that submitted trust reports about node j to node i. We can similarly estimate the reliability of the reports using the method described in [13] as:

$$TL_i(j) = \frac{\sum_{k=0}^{n} TL_k(j) \div TL_{req}(j)}{n} \quad (2)$$

where $TL_i(k)$ represents the level of trust node i maintains for node k, $TL_k(j)$ represents the level of trust for node j as estimated through observation by node k, and TL_{req} represents the minimum threshold of trust to permit message transmission. Unlike Eq. 2 as presented in [13], for a more heterogeneous network with devices having different roles, TL_{req} is generalized to specify different trust thresholds for different services requested.

This trust model is augmented with information about device capability. In order to provide for this, the system must have a mechanism for securely identifying device characteristics and assessing capabilities. Static device profiling is one possible means of accomplishing this and has been studied for various applications such as multimedia [8]. Once a static profile for a newly joined device is established, its behavior on the network is then monitored and trust values computed both by other devices and at the access point based on its expected and observed behavior. This is strictly a design choice. Monitoring could be centralized for each IoT enclave, or decentralized as is typically the case with intrusion detection systems in ad-hoc networks [2]. As devices join the network and share or access data or services, a trust value for each device is computed and refined based on the extent to which its behavior deviates from its established behavioral profile using Eqs. 1 and 2. For decentralized systems, a reliability measure for trust reports can be computed as described above and originally presented in [13].

Trust values are combined with device capability indicators based on static profiles to produce a taxonomy of values similar to a risk assessment matrix. Risk is categorized into three discrete categories: high (HR), medium (MR), and low risk (LR), with the specific delineation between risk levels determined by the system owner as is traditionally done in any risk assessment. Trust levels range from 0 to 5, with 0 representing a node that is known to be compromised, then values from 1 to 5 represent unknown to fully trusted, respectively. Similarly, capability values range from 1 (least capable) to 5 (most capable).

Table 1: Risk assessment matrix

Trust level	Device capability				
	1	2	3	4	5
0	HR	HR	HR	HR	HR
1	M/HR	M/HR	HR	HR	HR
2	L/MR	L/MR	MR	M/HR	M/HR
3	L/MR	L/MR	MR	MR	MR
4	LR	LR	L/MR	L/MR	L/MR
5	LR	LR	LR	LR	L/MR

Combining these in the risk assessment matrix produces an estimate of risk, which is shown in Table 1. It is important to point out that machines can never accept risk. Risk can only be accepted by the owner of the resources. Therefore, risk values in Table 1 may be different based on policy, but an automated system such as that proposed in this paper can execute in accordance with that policy.

With a determination of the level of risk posed by each device, services may be accessed or denied accordingly. As trust levels increase or decrease, the corresponding level of risk will fluctuate accordingly. Because capabilities are immutable characteristics of each device, they can be captured in a static profile such that trust level is the only variable that impacts access control decisions.

4 Device Characterization

In our prototype system, security has a significant foundation on identifying device characteristics. This requires mechanisms for fingerprinting each device connected to the network and constructing a device profile. The notion of trust inherently maintains a close relationship to identification, and fingerprinting is commonly used for identification purposes and identity management. Therefore, there is a natural interrelationship between trust and device profiling. Traditionally, fingerprinting has been accomplished using active approaches, which require each device to self-report information regarding its own hardware and software capabilities. No security mechanism can rely on well-behaved clients providing accurate information, so any fingerprinting technique used for security purposes must be resistant to forgery and spoofing.

There are a number of ways in which secure device fingerprinting can be accomplished [1, 5, 11, 15], primarily through passive techniques that examine network traffic from each device and characterize them based on vendor-specific implementation details. Passive approaches do not require devices to voluntarily provide identifying information beyond what they ordinarily would connecting to a network, to have knowledge of the fingerprinting technique being used, or even have awareness that fingerprinting is occurring at all.

The literature shows numerous existing and complementary methods for fingerprinting devices using passive techniques, e.g., the use of network parameters and features of 802.11 implementations are described in [15]. Features such as random back-off timers, rate switching, and duration field values in 802.11 frames for both data transmission and network management purposes provide sufficient information to identify device type in a manner difficult to spoof. For other types of wireless connectivity such as Bluetooth Low Energy, similar fingerprinting techniques can be applied [5]. To further characterize devices, it is possible to securely fingerprint operating system type and versions using spoof-resistant passive techniques based on differences in TCP/IP and SSL/TLS protocol implementations described in [1].

Such techniques may be used individually to classify devices by type using a single attribute. Alternatively, a combined approach as described in [11] could be used to achieve greater fidelity in secure device profiling. Because of the dependence placed on secure device characterization in this work and the requirement to ensure device profiles cannot be maliciously forged or otherwise manipulated, we assume the use of a combined approach would be preferred in any implementation [23].

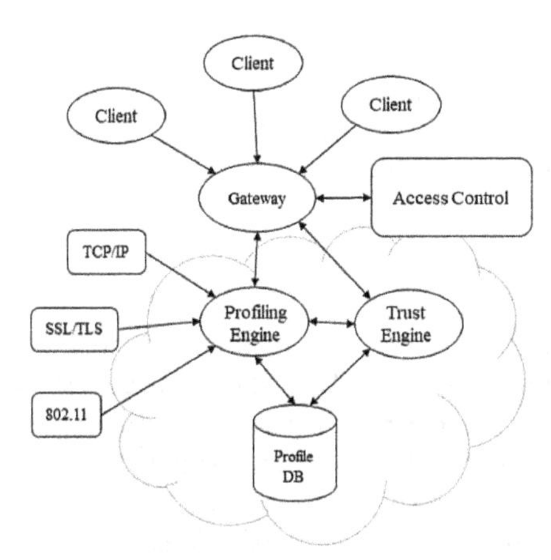

Fig. 1. System architecture

5 System Design

We have a prototype design developed that will form the basis for future research and evaluation. Originally implemented as a simulated network using the NS-3 network simulator [21], using the Whitefield simulator to represent wireless sensor network components [17]. Whitefield is layered on top of both NS-3 and Contiki OS [4]. In such a setup, NS-3 provides the network component, particularly wireless RF transmissions, while Contiki provides an implementation of the limited network stacks commonly found in IoT devices. We found this configuration to be somewhat lacking and challenging to work with, which resulted in the subsequent development of a prototype system. The basic structure of the prototype was originally presented in [9] to describe the fingerprinting mechanisms but has since been expanded to include the trust model and risk assessment components, which are new to this work. The overall architecture is shown in Fig. 1.

A single gateway might provide network connectivity to a range of devices with varying capabilities. These capabilities are captured in device profiles that reflect the characteristics of each type of networked device. Device profiles are then used in conjunction with stochastic behavioral profiles, which form the bases of trust evaluation. Each new device that joins the network is immediately characterized based on its type and capabilities, which could be unknown. Its network behavior is then monitored and a profile constructed based on the traffic it initiates. This profile reflects the overall trust level.

In the system packet transmissions are evaluated based on frequency of occurrence and expected destination, and compared against profiles of devices in a similar class. For instance, one would expect temperature sensors to behave similarly. Behavior indicative of a significant outlier based on comparable devices or its own past behavior will reduce the trust level. Additionally, any malicious behavior detected by an intrusion detection system will immediately flag the responsible node and reduce its trust level to zero, causing its quarantine or eviction from the network. A node known to be malicious or compromised cannot be readmitted to the network using automated means in this work.

Nodes not necessarily known to be compromised but with varying levels of associated risk, security policy may dictate that network communications be limited or that access to particular services be restricted. With restoration of trust, greater network communication traffic may be permitted or access to services granted.

6 Experimental Setup

The initial configuration of the system is designed with a notional mechanism for detecting anomalous behavior in devices based on patterns of network traffic. It should be noted that any number of anomaly detection mechanisms could be used. For simplicity, we constructed a statistical model of devices that transmit UDP packets at regular time intervals, which is representative of many types of sensor motes. A Raspberry Pi is used as a representative IoT sensor device for this work. To construct models of representative network traffic patterns that accounts for variance in packet arrival times due to delays, we began by recording inter-arrival times of UDP packets transmitted over wireless Ethernet. Detection of statistically significant variations in packet arrival times relies on a X^2 test with one degree of freedom and a critical value of 3.84.

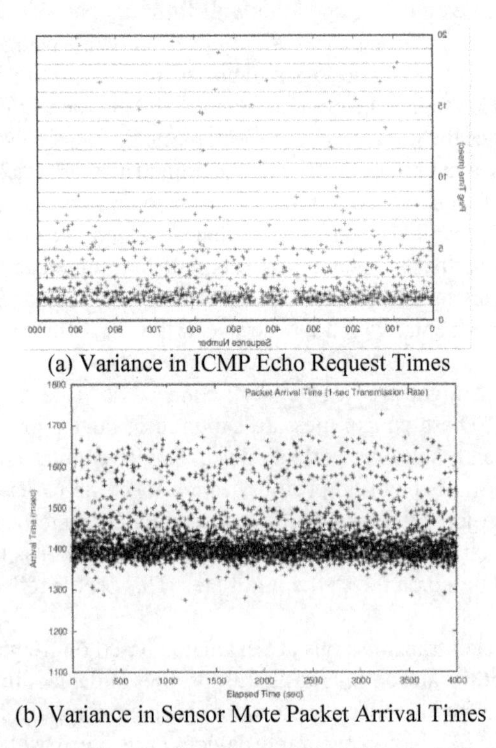

(a) Variance in ICMP Echo Request Times

(b) Variance in Sensor Mote Packet Arrival Times

Fig. 2. Empirical measurement of variance in packet arrival times

As a preliminary step, we first measured ping round-trip times between the sensor device and a virtualized network router. The variation in the arrival times for ICMP echo

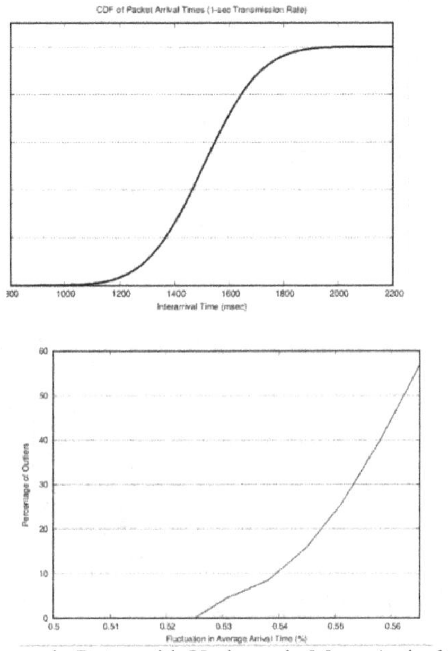

(b) Anomaly Rates with Variance in Mean Arrival Times

Fig. 3. Experimental data for system evaluation

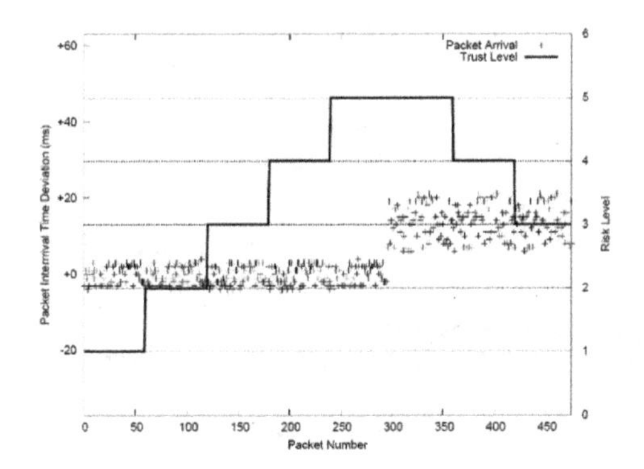

Fig. 4. Trust levels with high- or low-capability devices

request packets is shown in Fig. 2a. For the sensor mote device, the arrival times between packets sent at $1-s$ intervals were recorded over a period of 30 min. The variation in packet arrival times is shown in Fig. 2b and is consistent with the results in Fig. 2a, and the CDF of the packet arrival times is shown in Fig. 3a. This empirical data was used to construct a statistical model of the variations in packet arrival times that is used to determine expected versus anomalous behavior. The results in Fig. 2a and 2b represent expected behavior.

The purpose of the experiment is to evaluate the synthesis of statistical anomaly detection, trust values, and risk assessment. In the experiments, the sensor mote device transmitted UDP packets at regular intervals maintaining interarrival times consistent with the statistical model. The X^2 test determines whether arrival times were consistent with the model, or if there were statistically significant deviations. At 30-s intervals, the trust level of the device was assessed, and behavior consistent with the model resulted in an increased level of trust, with the trust levels ranging from 0 to 5. Each device began the experiment with a low initial trust value based on the assumption that it is a new and unknown device joining the network for the first time. In this experiment, trust is evaluated using Eq. 1 with $n = 1$.

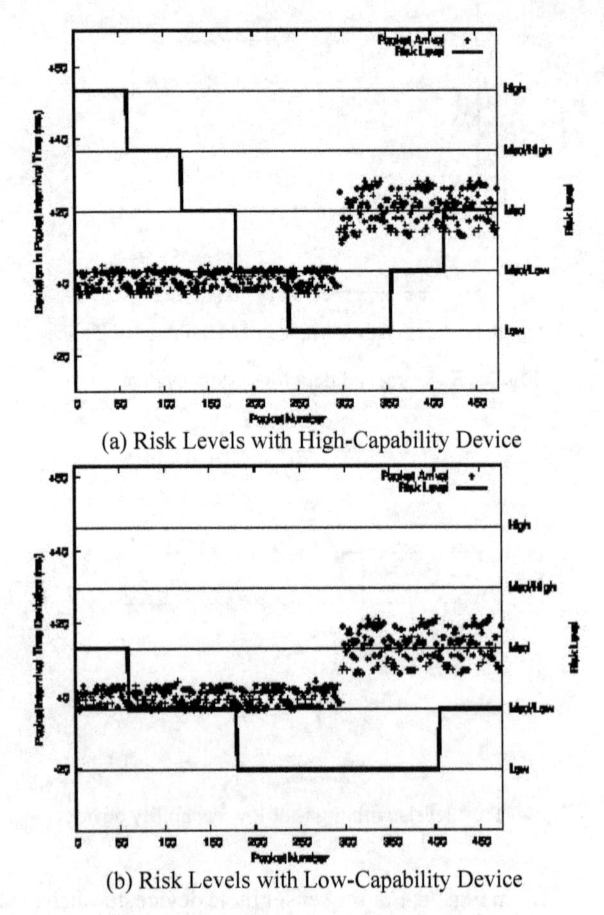

(a) Risk Levels with High-Capability Device

(b) Risk Levels with Low-Capability Device

Fig. 5. Risk levels for varying capability devices

After 5 min, the behavior of the network traffic is modified in a statistically significant way by inserting network delays that cause packet interarrival times to deviate significantly from the model. This was accomplished by injecting random delays to the packet transmission times such the X^2 method would detect the change. Because this new behavior deviates from the model and is considered anomalous, the trust level for the device decreases with each additional assessment as long as the anomalous behavior continues.

This experiment is repeated with devices considered to be less capable or low powered, and more capable or high powered. The high-powered device used is a Raspberry Pi. For the low-powered device, we emulated a less sophisticated type of sensor mote with an alpha processor that is representative of many types of older IoT sensor motes. An older, lower-powered device is assumed to pose less of a threat to the system and would therefore constitute a lower risk based on its capabilities.

7 Discussion

In experimenting with the system, we observed that anomaly detection rates increase significantly with changes in the mean packet arrival time. Figure 3b shows the increase in anomalous behavior as the mean packet arrival time increases. The sensitivity to mean arrival time shown in the figure suggests that an accurate statistical model of expected device behavior is of utmost importance with respect to correct system behavior. As mentioned previously, the a X^2 method using packet arrival times is only one of several possible alternative approaches to anomaly detection. It is certainly possible for a machine learning approach to be used instead, for example, but the use of such techniques is beyond the scope of this work.

With both high-powered and low-powered devices, as they transmit UDP packets in a manner consistent with the expected pattern captured in the models, their trust level increases. Trust levels in this system are a set of discrete values that increase or decrease in a step-wise manners, as shown in Fig. 4. Once network traffic patterns deviate from the models in a statistically significant manner, trust levels decrease with each periodic trust assessment. If anomalous behavior continues, eventually the trust level diminishes to a minimal level that indicates a misbehaving device that should not be trusted with critical functions. Regardless of the capability of the device, trust levels are entirely a function of behavior. The pattern of trust levels subsequent to either expected or anomalous behavior is the same for all categories of devices. This result illustrates a weakness of trust models as a security mechanism. Trust alone does not permit tailoring of security policies since the only conclusion that can be drawn from trust values is whether a device is behaving correctly. Trust gives no indication of threat. However, trust values are a good indicator of the potential threat posed by an unknown device. For this reason, the trust values assigned to devices in this system based on behavior are used to assess risk.

To illustrate the difference between trust and risk, consider both a high-powered device and a low-powered device behaving in exactly the same manner. Each device transmits packets consistent with the model for a period of time, then the behavior becomes anomalous. Figure 5a shows the assessed risk level for such a device over time as its behavior changes from expected to anomalous. In this case, the risk levels correspond inversely to the trust levels in Fig. 4. For a low-powered device exhibiting the same behavior pattern and being assessed the same trust levels as the high-powered device, the risk levels are much different. A low-powered device does not pose the same threat as a higher-powered device, and even with anomalous behavior does not constitute the same risk.

Over time as trust values increase, the level of risk associated with a device's presence on the network may decrease in accordance with network security policy. If the trust value diminishes to zero, the device is known malicious or compromised, and should be evicted from the network or quarantined. For other trust values, the action taken might depend on the capability of the device involved. For instance, a low-powered sensor that reports data to a server might be considered lower risk, even if it has not yet established a high level of trust. Similarly, a more powerful device may be considered to be of some risk, even if its behavior indicates a higher level of trust, because a device may behave correctly for some period of time before a malicious program such as a logic bomb is executed. By associating higher risk with a more powerful device, the device can be sequestered from certain functions and services or be subjected to additional security monitoring. This risk assessment is the metric by which security controls could then be applied. Different risk levels based on the same pattern of trust differentiates devices based on potential threat and can focus attention and resources on those devices more likely to cause harm to the system. Traditional behavior-based trust modeling does not provide any capability to identify varying degrees of potential threats.

This approach improves existing trust models and the way they are employed in sensor networks and reflects that not all devices are the same and do not necessarily pose the same threat. Thus, a spoof-resistant device identification method is critically important to proper functioning of an automated risk assessment.

8 Conclusions

The purpose of this work is to understand whether a finer degree of control over network access based on risk provides a benefit beyond a uniform and highly restrictive security policy. While resources must be protected, maximizing utility and functionality of IoT networks could be very beneficial. This work aims to understand how to maximize system functionality while simultaneously balancing security requirements. Using behavior and capability-based risk to tailor security policies and mechanisms provides additional information that can be used by security administrators for more efficient resource allocation.

9 Future Work

The planned direction of future work involves better understanding the relationship and interaction between trust models and device profiling in terms of efficiencies gained through an adaptive approach to deploying security mechanisms. Experiences in practice using the system could provide additional lessons learned which would be fed back into the system design and implementation. This approach to risk assessment is currently implemented in a test IoT network and is being subject to ongoing study.

References

1. Anderson, B., McGrew, D.: OS fingerprinting: new techniques and a study of information gain and obfuscation. In: IEEE Conference on Communication and Network Security (CNS), Las Vegas, NV, pp. 1–9 (2017)

2. Bakar, K., Daud, N., Hasan, M.: Adaptive authentication: a case study for unified authentication platform. In: World Congress on Computer and Information Technology (WCCIT 2013), pp. 1–6 (2013).
3. Bao, F., Chen, I.: Dynamic trust management for internet of things applications. In: 2012 International Workshop on Self-Aware Internet of Things (Self-IoT 2012), San Jose, pp. 1–6 (2012)
4. Contiki: The Open Source OS for the Internet of Things. https://www.contiki-os.org
5. Gu, T., Mohapatra, P.: BF-IoT: securing the IoT networks via fingerprinting-based device authentication. In: 15th IEEE International Conference on Mobile Ad-Hoc and Sensor Systems, Chengdu (2018)
6. Gutscher, A., Heesen, J., Siemoneit, O.: Possibilities and limitations of modeling trust and reputation. In: CEUR Workshop (2008)
7. Han, G., Jiang, J., Shu, L., Guizani, M.: An attack-resistant trust model based on multidimensional trust metrics in underwater acoustic sensor network. IEEE Trans. Mobile Comput. **14**(12), 2447–2459 (2015)
8. Hellaoui, H., Bouabdallah, A., Koudil, M.: TAS-IoT: trust-based adaptive security in the IoT. In: IEEE 41st Conference on Local Computer Networks (LCN), Dubai, 2016, pp. 599–602 (2016)
9. Hemmes, J., Dressler, J.: Work-in-progress: IoT device signature validation. In: IEEE 10th Conference on Information Technology, Electronics, and Mobile Computing (IEMCON), Vancouver (2019)
10. Hudson, F.: Enabling trust and security: TIPPSS for IoT. IT Prof. **20**(2), 15–18 (2018)
11. Kamvar, S., Schlosser, M., Garcia-Molina, H.: The Eigen-trust algorithm for reputation management in p2p networks. In: 12th International Conference on World Wide Web (WWW 2003), New York, USA (2003)
12. Kerrache, C., Calafate, C., Cano, J., Lagraa, N., Manzoni, P.: Trust management for vehicular networks: an adversary-oriented overview. IEEE Access **4**, 9293–9307 (2016)
13. Liu, Z., Joy, A., Thompson, R.: A dynamic trust model for mobile Ad Hoc networks. In: 10th IEEE Workshop on Future Trends of Distributed Computer Systems (2010)
14. Neumann, C., Heen, O., Onno, S.: An empirical study of passive 802.11 device fingerprinting. In: 32nd International Conference on Distributed Computer Systems Workshops, Macau, pp. 593–602 (2012)
15. Nitti, M., Girau, R., Atzori, L., Iera, A., Morabito, G.: A Subjective model for trustworthiness evaluation in the social Internet of Things. In: IEEE 23rd International Symposium on Personal Indoor and Mobile Radio Communications (PIMRC), Sydney, pp. 18–23 (2012)
16. NS-3 Consortium, NS-3 Network Simulator. https://nsnam.org
17. Sharma, V., You, I., Andersson, K., Palmieri, F., Rehmani, M.H.: security, privacy and trust for smart mobile Internet of Things (M-IoT): a survey (2019). https://arxiv.org/abs/1903.05362.
18. Sicari, S., Rizzardi, A., Grieco, L., Coen-Porsini, A.: Security, privacy and trust in Internet of Things: the road ahead. Comput. Netw. **76**, 146–164 (2015)
19. Trivedi, A., Kapoor, R., Arora, R., Sanyal, S., Sanyal, S.: RISM - reputation based intrusion detection system for mobile Ad-Hoc network. In: 3rd International Conference on Computers and Devices for Communication (2006)
20. The Whitefield Simulation Framework for Wireless Sensor Networks. https://github.com/whitefield-framework/whitefield.
21. Xiong, L., Liu, L.: Peertrust: supporting reputation-based trust for peer-to-peer electronic communities. IEEE Trans. Knowl. Data Eng. **16**, 843–857 (2004)
22. Xu, Q., Zheng, R., Saad, W., Han, Z.: Device fingerprinting in wireless networks: challenges and opportunities. IEEE Commun. Surv. Tutor. **18**(1), 94–104 (2015)
23. Yan, Z., Zhang, P., Vasilakos, A.: A survey on trust management for Internet of Things. J. Netw. Comput. Appl. **42**, 120–134 (2014)

Mobile Per-app Security Settings

Carlton Northern[1(✉)], Michael Peck[1], Johann Thairu[1], and Vincent Sritapan[2]

[1] The Homeland Security Systems Engineering and Development Institute (HSSEDI), which is operated and managed by the MITRE Corporation, McLean, VA, for the U.S. Department of Homeland Security, McLean, VA, USA
cnorthern@mitre.org

[2] U.S. Department of Homeland Security, Science and Technology Directorate, Washington, DC, USA

Abstract. Mobile apps should be deployed with care because they can pose substantial risk to enterprise organizations due to their potential to contain exploitable vulnerabilities, malicious code, or privacy-violating behaviors. Even apps from the Apple App Store or Google Play are not free of these risks. Mobile app vetting solutions can automate analysis of third-party mobile apps to help enterprises determine whether an app is safe to deploy on their mobile devices, but this is primarily a human-driven process which is time consuming. A new, automated approach called *continuous app vetting* is emerging that attempts to automate this entire process through use of app behavior rulesets and enforcement via enterprise mobility management (EMM) solutions. This study sought to develop a set of configurations and rulesets for continuous app vetting to be used by enterprises to identify potentially malicious, exploitable, or privacy-violating behavior of apps; define rulesets governing acceptable/unacceptable mobile app behavior; and describe how to apply mitigations individually to apps via EMM.

Keywords: Mobile · App vetting · Enterprise mobility management

1 Introduction

Mobile devices and their applications (hereafter "apps") have completely transformed the way enterprises work and conduct business. The ability to communicate, collaborate, and access data via a networked hand-carried device empowers workers to conduct business from any location. However, apps should be deployed with care because they can pose substantial risk to enterprises due to their potential to contain exploitable vulnerabilities, malicious code, or privacy-violating behaviors. Even apps from the Apple App Store or Google Play are not free of these risks. Numerous high-profile examples have shown that offerings in both app stores may include malware, even though the apps themselves went through Google's and/or Apple's vetting processes. Furthermore, researchers have discovered mobile apps sometimes request unnecessary permissions or collect and share more personal information than is necessary for the app to function.[1]

[1] https://www.symantec.com/blogs/threat-intelligence/mobile-privacy-apps.

K. Arai (Ed.): FICC 2021, AISC 1363, pp. 696–715, 2021.
https://doi.org/10.1007/978-3-030-73100-7_51

During installation of a new app, mobile device users are often ill-equipped to judge whether they should grant permissions to gather data, and too often carelessly click through the permission request prompts to use the app. Therefore, enterprise organizations should not rely solely on the vetting capabilities of these commercial app stores in a time of constant cybersecurity threat.

Mobile app vetting solutions can automate analysis of third-party mobile apps to help enterprise organizations determine whether an app is safe to deploy on its mobile devices. Traditionally, app vetting takes place before deployment or at update time, but not all steps in the vetting process can be automated. It can take time to review the findings and decide to accept or reject deployment of the app, resulting in considerable delay from the time the app is requested until it is approved for use. This is an especially burdensome process given the frequency of updates to mobile apps.

However, a new approach called continuous app vetting is emerging that attempts to automate this entire process through use of app behavior rulesets and enforcement via enterprise mobility management (EMM) solutions. This approach is described in a previous work [1]. While that report explored app vetting capabilities and their integration with EMM for the purposes of providing continuous app vetting, it did not recommend a configuration of such tools to identify mobile app risks and proposed mitigations. The present report recommends configuration of these tools and explores the ability to apply appropriate mitigations on a per-app basis and how best to apply them with EMM.

1.1 Purpose

The study described in this report seeks to develop a set of configurations and rulesets for continuous app vetting for enterprises that identifies potentially harmful, exploitable, or privacy-violating behavior of apps. It also attempts to define rulesets governing acceptable/unacceptable mobile app behavior and describe how to apply mitigations individually to apps via EMM. This will allow multiple levels of heightened or relaxed security in apps, for instance, in the case of devices used during international travel to locations of concern.

1.2 Background

In [2], the authors provided an in-depth evaluation of mobile app vetting tools. The study focused on the ability of the tools to find vulnerabilities and privacy-violating and malicious behaviors in mobile apps. It did not describe how such tools would be employed in practice, merely their ability to detect these problems in a given mobile app. The study cited a need for future work, which prompted a follow-on study [1] to examine app intelligence services that gather app/developer reputation and known threat information and to incorporate results from those services into the vetting process [1]. It also considered the ability of the app vetting solution to integrate with EMM solutions but did not identify or examine what capabilities should be present in such a configuration.

MITRE's Adversarial Tactics, Techniques, and Common Knowledge® (ATT&CK) framework details specific techniques that adversaries can use to obtain unauthorized access to enterprise resources or achieve other objectives [3]. ATT&CK for Mobile[2] describes techniques that adversaries can employ to gain access to mobile devices and then make use of that access to achieve their objectives. The entry for each technique typically includes a detailed technical description, mitigations, detection analytics, examples of use by adversaries, and references. The techniques are organized into tactic categories, which represent higher-level adversarial objectives.

As described in ATT&CK, mobile apps present a significant attack vector against mobile devices, as they can enable an adversary to execute arbitrary code. Access to mobile devices could give adversaries access to enterprise data stored either on the device itself or on backend enterprise systems. It could also give the adversary access to other valuable information, for example data obtained from the variety of sensors present on mobile devices (e.g., microphone, camera, location data).

App vetting provides some form of mitigation against 34 out of the 76 techniques currently included in ATT&CK for Mobile [3]. The techniques mitigated by app vetting are spread throughout ATT&CK for Mobile's tactic categories. Many of them fall into the "Collection," "Credential Access," and "Discovery" categories. Adversaries can use Collection techniques to gather data stored on the mobile device, transmitted through the mobile device, or obtained through the mobile device's sensors. They can use Credential Access techniques to gather passwords, tokens, cryptographic keys, or other values to gain unauthorized access to resources. Discovery techniques allow adversaries to gain knowledge about the characteristics of the mobile device and potentially of other networked systems.

By default, mobile devices are configured to only allow apps to be installed from the device platform's official app store (i.e., Google Play Store or Apple App Store). This default behavior provides significant protection, as these app stores perform some level of vetting (of both the developer's identity and of the apps themselves). Also, requiring the apps to be published openly increases the potential of detecting an adversary's actions. However, examples of malicious apps that have been distributed through the official app stores are common and many occurrences have been documented [4–6]. This implies that relying on this default behavior is not by itself sufficient to ensure device security, although the presence of sideloaded apps (installed from sources other than the official app store or an authorized enterprise app store) should be treated as an indicator of higher potential risk.

2 Per-app Mobile Security Use Cases

In [1], the authors presented a new approach to mobile app vetting known as continuous app vetting, in which the vetting process is streamlined by integrating app vetting tools with EMM. This continuous approach aims to strike a balance between maintaining security and allowing enterprise employees the freedom to use apps they need to conduct business and accomplish the organization's mission.

Continuous app vetting uses the analysis capabilities of app vetting tools to periodically inspect apps installed on enterprise devices for security issues and relay the

[2] ATT&CK for Mobile provides a capability to query a specific mitigation such as app vetting and retrieve all of the techniques that list that mitigation.

results to the EMM solution for potential action. In some cases, the app vetting tool categorizes or scores the findings (categories and scoring are configurable by the organization). Among the actions that the EMM solution can take are notifying an administrator to perform further inspection, notifying the user that an app violates enterprise policy, restricting use of the app, restricting access to enterprise resources, restricting access to specific device profiles (e.g., Bluetooth), or prohibiting use of the employee's device completely until the issue is mitigated. Figure 1 depicts this process.

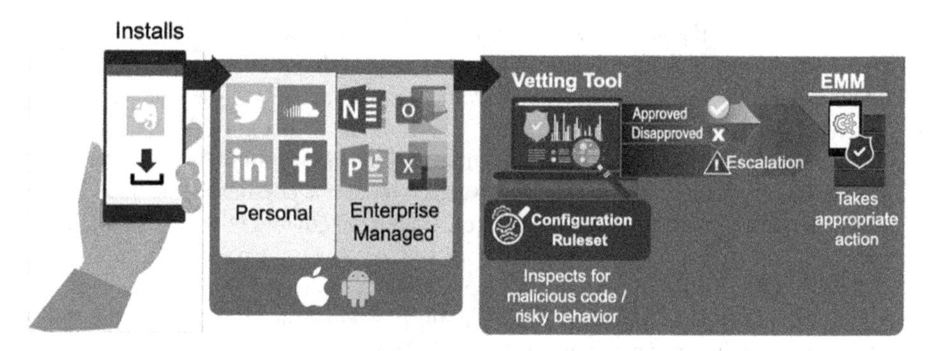

Fig. 1. Continuous app vetting process.

The ability to vet apps in an automated fashion depends on a configuration ruleset that looks for specific risky behaviors and provides appropriate mitigations. Rulesets are organization-specific based on the mission and risk tolerance of the organization. A ruleset can also be dynamically applied based on heightened security levels (e.g., in the case of international travel). This process differs from traditional app vetting in that rather than simply denying apps that do not meet organizationally defined rules, organizations can take actions on an app-by-app basis, removing potentially risky aspects (e.g., permissions) of an app while allowing use of the non-risky app features. Consider, for instance, a game app that has been installed on the personal segment of the device. If the app uses an ad network library to serve ads, and that ad network happens to communicate to a country of concern, one mitigation could be to remove the app's ability to communicate via the internet. This would prevent the app from communicating with the country of concern while still potentially allowing the user to play the game until a human security analyst could make a use determination.

Rulesets can be complex or simple depending upon the complexity of the organization. An organization with differing missions and user profiles may choose to deploy different rulesets based upon the type of user (e.g., there may be different security concerns for field workers than for users who work in an office setting). A highly sensitive organization may simply choose to put any app that has a security finding of significant risk on a deny list, whereas a low-sensitivity organization may choose not to act on most findings as long as they do not indicate flagrant violations of policy (e.g., known malware). Knowing what security findings to look for, how to respond appropriately, and under what conditions to enforce them is not a trivial undertaking, and organizations should have a rigorous process for putting these rulesets in place. Section 4 of this report provides a starting point for organizations to use based on low, medium, and high security levels of the organization.

3 Methodology

The authors employed the following methodology to make recommendations for a mobile per-app security implementation:

1. Identify specific risky or dangerous app behaviors. The authors achieved this by analyzing risk factors that mobile app scanners such as Kryptowire[3] use to determine how dangerous an app is. These risk factors include security issues/vulnerabilities, such as an app not properly verifying Transport Layer Security (TLS) server certificates (making the app's network communications susceptible to interception and modification); potential privacy risks, such as an app accessing the device camera; and potentially malicious behaviors, such as an app showing windows over another app's user interface (UI), which could be used to carry out a phishing attack.
2. Identify all possible mitigations to these behaviors. To do this, the authors first inspected the administrative options available in Android and iOS, such as the ability to revoke runtime permissions on Android. The authors then analyzed the administrative implementation offerings from two leading EMM providers, to determine the implementations actually available to be used in practice. The authors did this because there are some differences between what Android and iOS offer and what has been implemented by commercial EMM providers.
3. Create a ruleset that includes the action to be performed by the EMM to mitigate a problematic behavior based on whether an organization is operating at a low, medium, or high security level. The authors determined this classification by analyzing the risk posed by the behavior and the impact that the resulting mitigation would have. For example, if the risky app behavior included connecting to a country deemed "suspicious," a possible mitigation would be to block the offending app's internet access, or to block all of Android's "dangerous" permissions,[4] effectively neutering the app. While this makes sense for organizations operating at a medium or high security level, it may be deemed too restrictive for those operating at a low security level.
4. Develop a risk model in which multiple app vetting findings would result in a compound score that could be used to determine whether or not to apply a specific mitigation. For example, an app may request access to the device camera, which would not be considered a risky behavior in and of itself, but if the app also requested a location permission, and/or access to the contacts list, an organization might want to block these permissions until a human security analyst could make a risk determination for a Go/ No-Go decision.
5. Investigate the ability to apply the per-app settings in a real-world example. The authors did this by developing a proof-of-concept prototype that demonstrates risky/dangerous behaviors that are mitigated at the app level on an individual basis

[3] "Kryptowire" - https://www.kryptowire.com/.

[4] https://developer.android.com/guide/topics/permissions/overview#dangerous_permissions.

and in a completely automated fashion. The prototype was further configured with the ruleset and risk model developed in Steps 3 and 4. Apps were identified that demonstrate use cases that utilize the ruleset and risk model and demonstrate the prototype's capabilities with those apps and the EMMs.

4 Per-app Settings

The authors followed the methodology summarized in Sect. 3 to develop a listing of risky app security behaviors and mitigations, and to develop a mapping between the two based on the severity of the threat and security levels of the organization. This section documents these behaviors, mitigations and mappings.

4.1 App Security Behaviors

The authors generated a list of app security behaviors by inspecting findings of various mobile app vetting tools as well as by analyzing security requirements in the National Information Assurance Partnership Protection Profile for Application Software. The list that follows is an initial list; it does not represent all potential mobile app risky behaviors. Additionally, some of these findings might not always present a security issue; some may represent appropriate app behavior depending on the context. However, this list serves as a starting point for examining the feasibility of automated enterprise mitigations.

- Contains cryptographic vulnerabilities
  - Hardcoded IV (Initialization Vector)
  - Hardcoded encryption key
  - No data in transit encryption
  - No data at rest encryption
- Uses TLS improperly
  - Accepts all TLS certificates without proper verification
  - Bypasses TLS
- Checks for root
- Leaves app as debuggable
- Dynamically loads Java classes
- Contains hard-coded credentials
- Has extra (unused) permissions
- Loads third-party libraries – includes accessing shared libraries on iOS
- Stores files insecurely
- Is vulnerable to Structured Query Language (SQL) injection
- Exposes user password
- Has device administrative privileges
- Can remain persistent in memory

- Can kill background processes
- Can show windows over UI
- Contains malware
- Attempts to gain root access
- Accesses sensitive information repositories
    - Contacts
    - Calendar
    - SMS/MMS (Simple Message Service/Multimedia Message Service)
- Accesses hardware resources
    - Camera
    - External storage
    - Bluetooth
    - Location
    - Microphone/audio
    - NFC (Near-field communication)
- Uses cloud storage
- Uses social network
- Improperly exposes personally identifiable information (PII)
- Connects to countries of concern (e.g., China, North Korea)
- Integrates advertising network
- Can call outside numbers
- Can get user accounts
- Collects device identifier information
    - Device ID collection
    - SIM (Subscriber Identity Module) serial number
    - Phone number

4.2 Mitigations

The authors created an initial listing of mitigations by inspecting the administration options available in Android and iOS. The mitigations are grouped by revoking runtime permissions, enacting a device-wide policy, or taking an action such as notifying a user of a violation. The authors then analyzed the offerings from two different leading EMM providers to determine what configuration options were currently available in commercial offerings.

Figure 2 depicts the various runtime permission mitigations that can be managed through the EMM providers analyzed. The types of permissions are classified into sensitive information repositories or hardware (HW) resources to help inform users about the impact on usability from implementing the various mitigations.

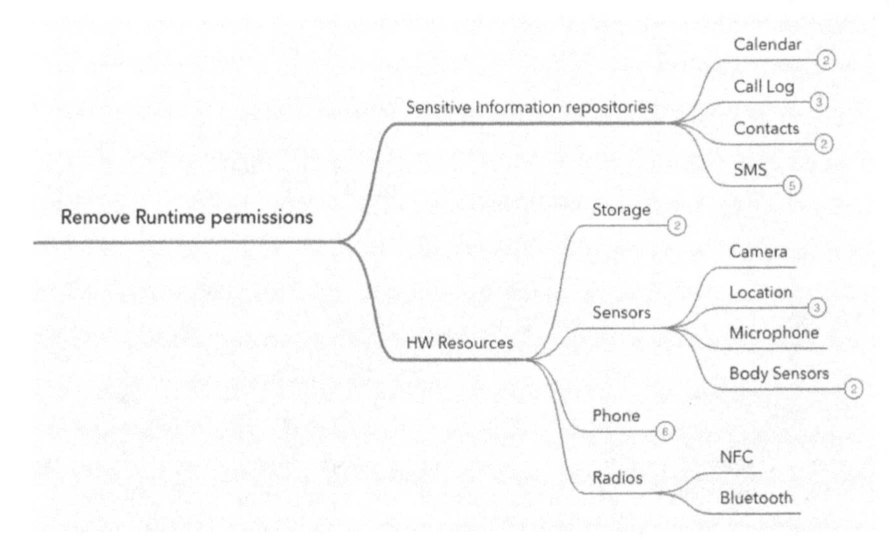

Fig. 2. Runtime permission mitigations.

Figure 3 presents device policy mitigations that the EMM can apply. These mitigations block access to enterprise resources, enforce virtual private network (VPN) usage, or place the apps on a deny list. These mitigation techniques are typically employed as part of an escalation plan once an application has been determined to be malicious.

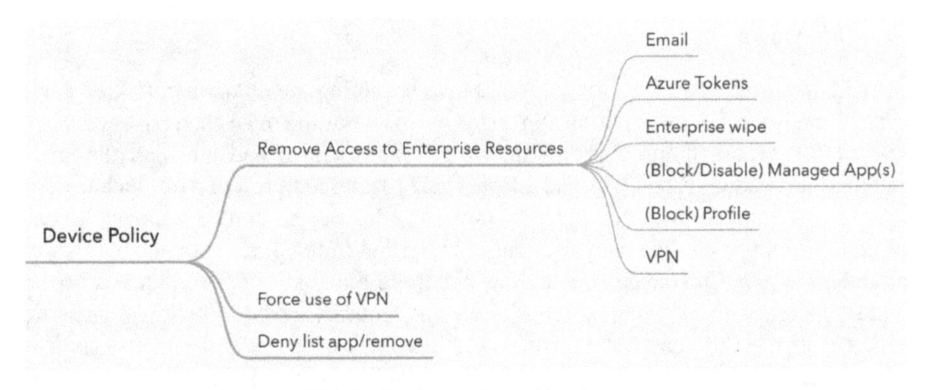

Fig. 3. Device policy mitigations.

Figure 4 depicts mitigations intended to either notify various user types (general users, administrators, and supervisors), or to notify administrators of specific issues that require their review and possible removal of access to the work profile. An organization can apply these actions in various ways depending on its security level. For example, instead of immediately blocking a permission, an organization operating at a low security level might flag a behavior as suspicious and have an administrator review it.

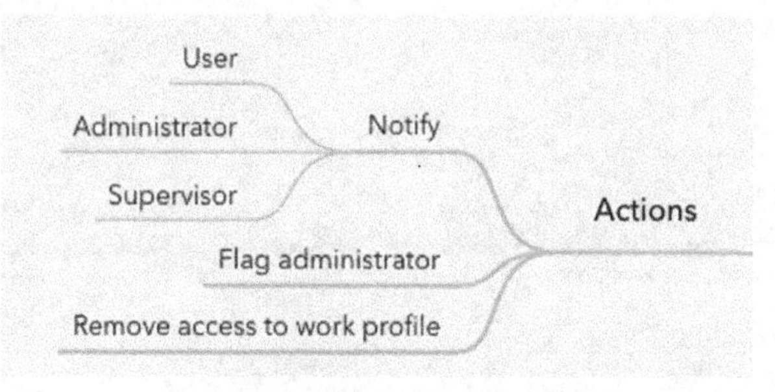

Fig. 4. Mitigation actions.

4.3 Rulesets

The authors developed an initial mapping of app behaviors to mitigations at low, medium, and high security levels to accommodate the needs of organizations with differing sensitivity/security concerns. Given the cumulative nature of these behaviors (i.e., an app may exhibit multiple behaviors and the overall risk level will compound with each additional behavior), an overall risk score is needed to determine if a specific behavior should be allowed or disallowed and subsequently analyzed by a security analyst. If the overall risk score is low, and the mitigation is listed as "None," then the organization need not do anything, but if the risk score is above an acceptable level (defined in Sect. 4.4) and the mitigation is listed as "Block Permission," then the permission should be revoked at runtime until a security analyst can make a Go/No-Go decision regarding use.

4.4 Risk Model

The authors developed a risk model to aid in determining the compound risk an app's behavior presents. For example, an app on the personal portion of an enterprise-managed device whose security findings are minimal (e.g., it was found to load third-party libraries, contains hard-coded credentials, and allows backup) does not present a significant security risk given that apps on the personal portion of the device do not intermingle with the enterprise apps on the enterprise-managed portion of the device and the number of infractions is low. Conversely, an app that has those findings, and also requests access to multiple sensitive information repositories and requests access to multiple hardware resources such as the camera or microphone, presents a greater risk to the enterprise. This occurs even though the app resides on the personally segmented portion of the device, because of the chance it could be used as a surveillance target or for other potential nefarious purposes. Thus, any one of those findings individually may not be a reason for concern, but when aggregated would present a potential security concern.

The risk model is fairly simplistic, relying on a point system to provide an indicator of trust. To calculate an app's risk score, the authors simply assigned one point for low-risk behaviors, three for medium-risk behaviors, and five for high-risk behaviors. The authors then took the cumulative score of all the app behaviors found and subtracted it from 100. For example, an app whose cumulative behaviors amounted to 25 would have a risk score of 75 (i.e., $100 - 25 = 75$). To adjust for the differing levels of sensitivity of

Table 1. App behavior mapping to organization security level.

App behavior	Risk	Organization security level		
		Low	Medium	High
Allows backup	Low	None	None	None
Leaves app left as debuggable	Low	None	None	None
Attempts to gain root access	Unacceptable	Deny-list, flag admin	Deny-list, flag admin	Deny-list, flag admin
Can call outside numbers	Medium	None	Block Permission	Block Permission
Can get user accounts	Low	None	None	None
Can remain persistent in memory	Low	None	None	None
Can show windows over UI	High	None	Flag Security Analyst	Flag Security Analyst
Checks for root	Low	None	None	None
Collects device information	Low	None	None	None
Connects to countries of concern	Medium	None	Block Internet	Block Internet
Contains hard-coded credentials	Low	None	None	None
Creates resources accessible 3rd party	Low	None	None	None
Cryptographic vulnerabilities	Low	None	None	Notify User
Dynamically loads Java classes	Medium	None	None	None
Executes native code	Low	None	None	None
Hardware resources - Unacceptable Risk Score	High	Block Permission	Block Permission	Block Permission
Hardware resources - Acceptable Risk Score	High	None	None	None
Has device admin privileges	Unacceptable	Deny-list, flag admin	Deny-list, flag admin	Deny-list, flag admin
Has extra permissions	Low	None	None	None
Improper PII exposure	Low	None	None	None
Improper SSL Usage	Low	None	Notify User	Notify User
Integrates ad network	Low	None	None	None
Can kill background processes	Low	None	Block Permission	Block Permission
Loads a third-party library	Low	None	None	None
Malware signature detected	Unacceptable	Deny-list, flag admin	Deny-list, flag admin	Deny-list, flag admin
Sensitive information repositories - Unacceptable Risk Score	High	Block Permission	Block Permission	Block Permission
Sensitive information repositories - Acceptable Risk Score	High	None	None	None
Stores files insecurely	Low	None	None	None
User password exposed	Medium	None	None	None
Uses cloud storage	Medium	None	None	None
Uses social network	Low	None	None	None
Vulnerable to SQL injection	Low	None	None	None

an organization, the authors propose use of a similar low, medium, and high value system in which a low-sensitivity organization can tolerate a score of 70 or higher; a medium-sensitivity organization can tolerate a score of 80; and a high-sensitivity organization can tolerate a score of 90.

To provide a real-world example, the authors used Kryptowire to analyze the Twitter Android application (v7.93.4-release.05) and seven behaviors were identified:

Low-risk behaviors (total points 5):

- Has extra permissions
- No data at rest encryption
- No data in transit encryption
- Allows backup
- Access to user accounts

High-risk behaviors (total points 25):

- Writes to external storage
- Reads external storage
- Accesses location
- Accesses contacts
- Accesses camera

Cumulative risk behavior value: 30
Risk score: 70

For a medium-sensitivity organization, the Twitter app, with a risk score of 70, may present too much risk, whereas a low-sensitivity organization may find the risk acceptable. For a high-sensitivity organization, appropriate mitigations may be to block the app's ability to write to external storage and to block access to the camera until a security analyst can make a risk determination regarding use of Twitter in the organization. The analyst can either decide that the organization can accept the risk and allow the app to be used as-is, or that the organization should block the app entirely (i.e., put it on a deny list). A third option could be to allow use of the app while blocking those two high-risk permissions, which may provide use of most of the app's intended functionality while preventing the app's riskier behaviors.

5 Per-app Settings Proof of Concept Prototype

The authors developed a proof-of-concept prototype to show how a per-app security settings deployment could operate. The prototype utilizes Android, given its open nature and access to device administration application programming interfaces (APIs) that are not available openly with iOS. After examining five leading commercially available EMM products, the authors determined that app-specific permission blocking was not available on the personal profile of the corporately-owned, personally-enabled device used for testing. The ability to block permissions of individual apps is available on the enterprise-managed portion of the device, but this is not useful because a traditional app vetting approach is more appropriate on the enterprise-managed profile due to the increased access to enterprise data and the risk of it being compromised.

Figure 5 provides an architectural depiction of the prototype. An EMM Agent on the device provides an up-to-date list of apps installed on the device to an EMM server. The EMM is included in this architecture solely to provide an up-to-date app list. It has been integrated with an instance of Kryptowire to provide the app list. Kryptowire then scans the apps to determine their behaviors and security findings. Reports on the apps are sent to a Python server that the authors created to calculate a risk score for each app and to block permissions for apps as dictated by the ruleset. App permission blocks are pushed to the device using Google Cloud Messaging (GCM). A custom-developed Android Device Policy Controller (DPC)[5] app receives those messages and performs the per-app, runtime-permission blocking. In a real-world deployment, the EMM would perform the runtime-permission blocking, but this functionality is not presently available in any of the EMM's analyzed by the author team.

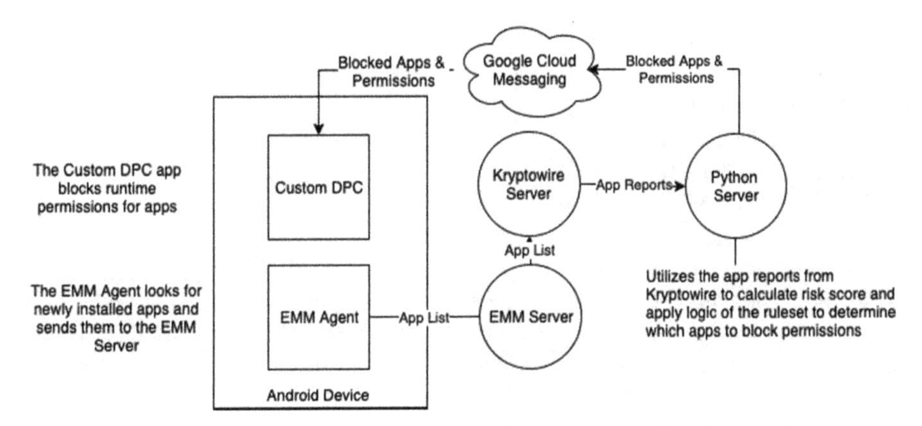

Fig. 5. Proof-of-concept per-app settings architecture.

Two use cases are demonstrated with the prototype. The first, called *Country Blocker*, looks for any installed apps that communicate with countries of concern. If such communications are found, it will block all of the permissions that Android deems "dangerous,"[6] namely:

- ACCEPT_HANDOVER
- ACCESS_BACKGROUND_LOCATION
- ACCESS_COARSE_LOCATION
- ACCESS_FINE_LOCATION
- ACCESS_MEDIA_LOCATION
- ACTIVITY_RECOGNITION
- ADD_VOICEMAIL
- ANSWER_PHONE_CALLS
- BODY_SENSORS
- CALL_PHONE
- CAMERA
- GET_ACCOUNTS
- PROCESS_OUTGOING_CALLS

[5] https://developer.android.com/work/dpc/build-dpc.

[6] https://developer.android.com/reference/android/Manifest.permission.

- READ_CALENDAR
- READ_CALL_LOG
- READ_CONTACTS
- READ_EXTERNAL_STORAGE
- READ_PHONE_NUMBERS
- READ_PHONE_STATE
- READ_SMS
- RECEIVE_MMS
- RECEIVE_SMS
- RECEIVE_WAP_PUSH
- RECORD_AUDIO
- USE_SIP
- WRITE_CALENDAR
- WRITE_CALL_LOG
- WRITE_CONTACTS
- WRITE_EXTERNAL_STORAGE.

As a first attempt, the authors tried blocking the INTERNET permission, but for unknown reasons this did not work. The authors succeeded in blocking all other permissions. More work should be performed to determine if blocking the INTERNET permission is achievable as this would be an important permission to block. As a second attempt, the authors decided to block all dangerous permissions instead. Blocking these dangerous permissions means that the application is effectively neutered and would be unable to effect serious harm.

The app used to demonstrate this use case was Pitu[7] (6.2.2.2610), as shown in Fig. 6. Pitu was developed by Tencent, a Chinese-owned company, which communicates with servers in China. Tencent also develops the WeChat application that is extremely popular in China.

The figures below use screenshots to depict the dangerous permissions blocked at runtime. The three screenshots in Fig. 7 show the app requesting permissions to take video, which requires:

- CAMERA
- RECORD_AUDIO
- READ/WRITE_EXTERNAL_STORAGE.

After running the Python Server application with the Country Blocker use case the following messages are displayed on the terminal:

```
2020-07-30 13:50:31-0700 [-] Suspicious application found: Pitu
2020-07-30 13:50:31-0700 [-] Application contacts the following suspicious countries:  ['China']
2020-07-30 13:50:35-0700 [-] Pitu requests the following dangerous permissions['record audio', 'course location',
'phone state', 'camera', 'write storage', 'fine location', 'read storage']
2020-07-30 13:50:35-0700 [-] Adding permission(s) block to dictionary
```

[7] https://play.google.com/store/apps/details?id=com.tencent.ttpic&hl=en_US.

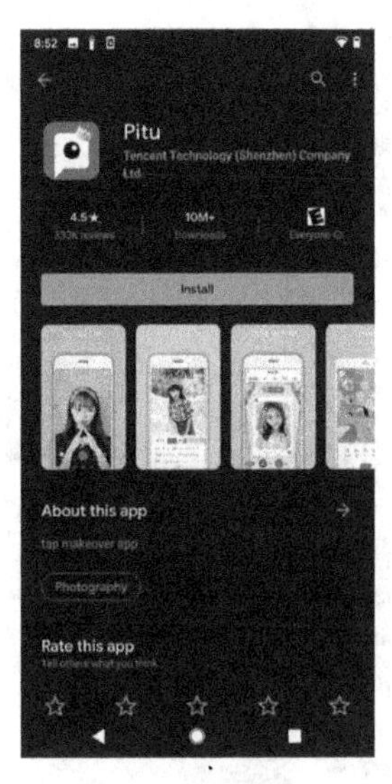

Fig. 6. Pitu App.

The terminal then displays the resulting permission blocks:

```
2020-07-30 13:50:35-0700 [-] Sending blocks to all devices.
2020-07-30 13:50:35-0700 [-] Send message to [com.tencent.ttpic] to block permision [record audio]
2020-07-30 13:50:36-0700 [-] Send message to [com.tencent.ttpic] to block permision [course location]
2020-07-30 13:50:36-0700 [-] Send message to [com.tencent.ttpic] to block permision [fine location]
2020-07-30 13:50:37-0700 [-] Send message to [com.tencent.ttpic] to block permision [camera]
2020-07-30 13:50:38-0700 [-] Send message to [com.tencent.ttpic] to block permision [write storage]
2020-07-30 13:50:39-0700 [-] Send message to [com.tencent.ttpic] to block permision [phone state]
2020-07-30 13:50:39-0700 [-] Send message to [com.tencent.ttpic] to block permision [read storage]
```

The app then shows the messages depicted in Fig. 8 requesting the permissions that were taken away. Clicking on either "Confirm" or "Grant Permissions" takes the user to the App Permissions screen. This occurs as part of the system app settings that show the permissions have been "Disabled by admin" and therefore users are not allowed to grant permissions. It should be noted that the user can still use the app without these permissions enabled. Almost all of the functionality is inoperable, but for some apps they could still prove useful. Take for instance an app that allows the user to view and create video content. While the video feature may be disabled, the user can still view others' videos.

The second use case, called Risky Permissions, utilizes the risk model to determine if access to sensitive information repositories or HW resources should be allowed. The sample app chosen for this use case was Flashlight – Bright LED Light[8] (v1.0.4), as shown in Fig. 9. In this case, the app has a risk score of 75. If the organization had a medium level of sensitivity, it would deem the app as 'risky' and would disallow access to any of the permissions requested from the Sensitive Information Repositories or Hardware Resources.

In this particular case, the app had the following findings:

[8] https://play.google.com/store/apps/details?id=com.crusader.flashlight&hl=en_US.

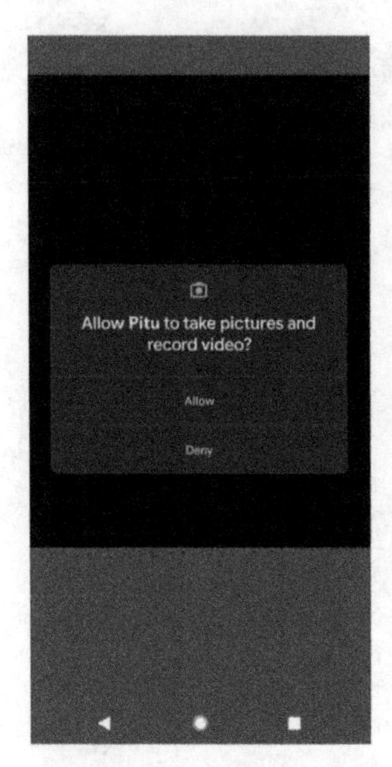

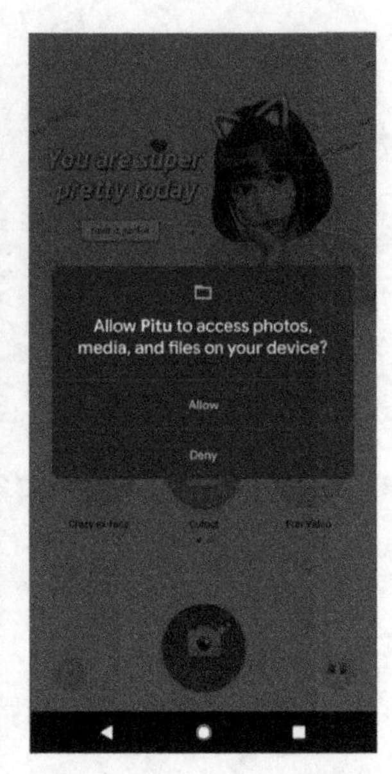

Fig. 7. Pitu request for video and audio.

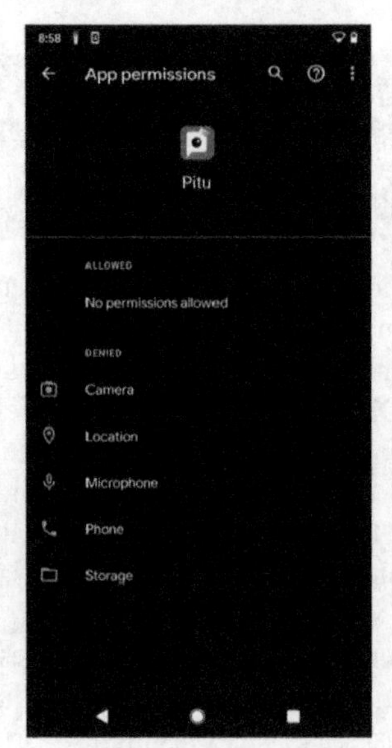

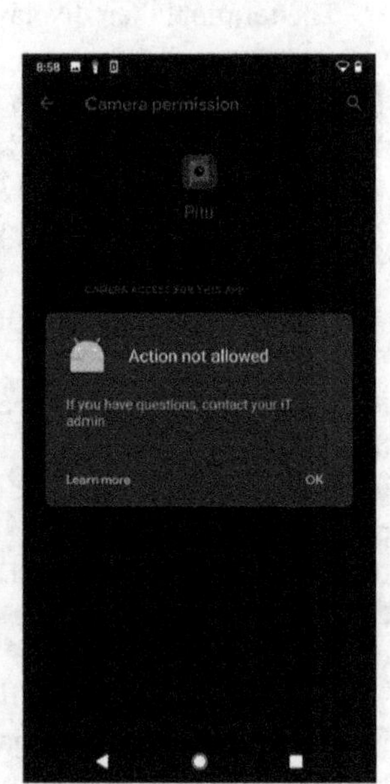

Fig. 8. Pitu demonstrating blocked permissions.

Low-risk behaviors (total 5):

- Has extra permissions
- No data at rest encryption

- No data in transit encryption
- Allows backup
- Integrates with a social network

High-risk behaviors (total 20):

- Writes to external storage
- Reads external storage
- Accesses location
- Accesses camera

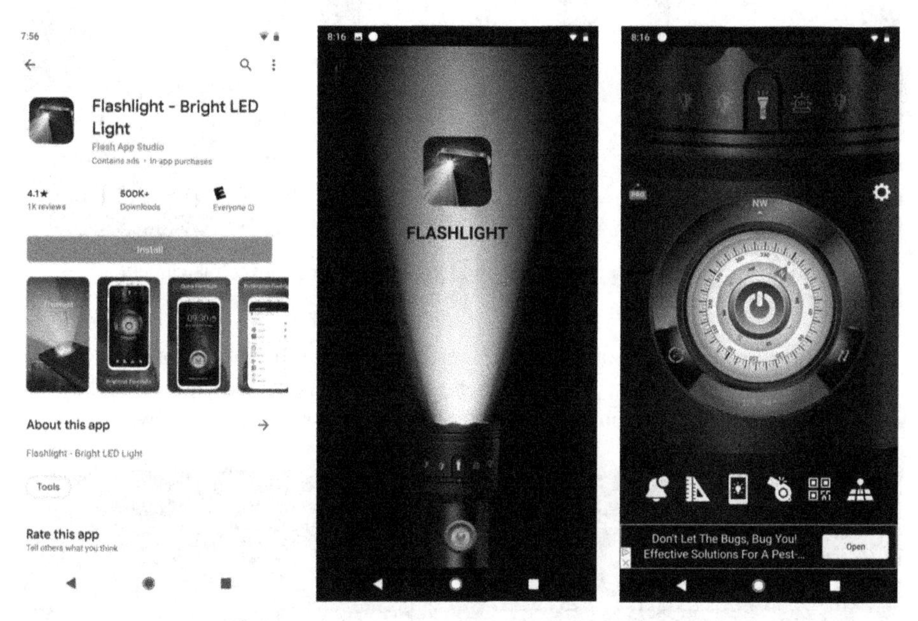

Fig. 9. Flashlight – Bright LED light screenshots.

Cumulative risk behavior value: 25.
Risk score: 75.

It is suspicious that the app needs access to the high-risk behaviors in order to provide flashlight functionality utilizing either the hardware flash on the back of the device or the light from the screen. As part of Google's review process, the app must have functionality in place that utilizes the requested permissions. In this case the app has a Quick Response (QR) code reader function that utilizes the camera and external storage. Figure 10 shows the QR code function in screenshots.

The app also has a compass and mapping function that uses the location permission. It is unclear if these permissions are used elsewhere in the code, but it is conceivable that this app, or other apps like it, could do so, especially if an app utilizes dynamic loading of Java classes that can be downloaded and executed at runtime, potentially weeks after installation. Figure 11 shows the compass and mapping functions as well as the ability to operate from the device's lockscreen while also showing ads.

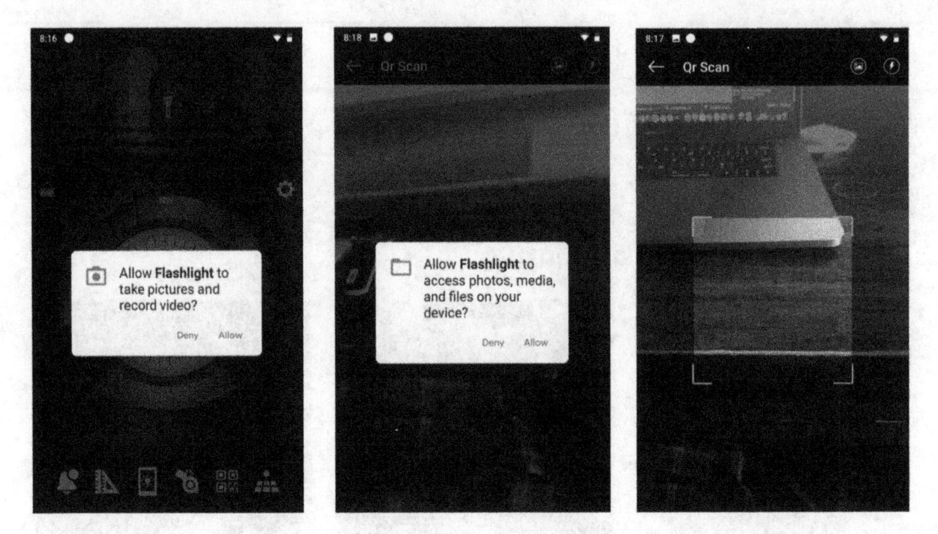

Fig. 10. Flashlight app requesting and granting permissions to the camera.

In this example, the authors blocked Sensitive Information Repositories and Hardware Resources permissions, which happen to be the same as the high-risk behaviors listed above. This is done from the python app; the process is shown below.

```
2020-04-01 17:02:42-0400 [-] Flashlight - Bright LED Light( version: 1.0.4): high risk (score=76)
2020-04-01 17:02:43-0400 [-] Flashlight - Bright LED Light requests the following dangerous permissions['course location',
phone state', 'camera', 'write storage', 'fine location', 'read storage']
2020-04-01 17:02:43-0400 [-] Adding permission(s) block to dictionary
2020-04-01 17:02:43-0400 [-] Sending blocks to all devices.
2020-04-01 17:02:43-0400 [-] Send message to [com.crusader.flashlight] to block permision [course location]
2020-04-01 17:02:43-0400 [-] Send message to [com.crusader.flashlight] to block permision [fine location]
2020-04-01 17:02:44-0400 [-] Send message to [com.crusader.flashlight] to block permision [camera]
2020-04-01 17:02:44-0400 [-] Send message to [com.crusader.flashlight] to block permision [write storage]
2020-04-01 17:02:45-0400 [-] Send message to [com.crusader.flashlight] to block permision [phone state]
2020-04-01 17:02:45-0400 [-] Send message to [com.crusader.flashlight] to block permision [read storage]
```

As a result, the functionality of the QR code reader and the mapping functionality were disabled, and those activities of the app will no longer execute. In addition, the camera's flashlight on the back of the device will no longer work because it needs the camera permission; however, the flashlight capability of the screen will still work. Similar to the Pitu app, the authors blocked the permissions as shown in Fig. 12.

6 Recommendations

Enterprises can only utilize per-app security settings if the capability is present in commercially available EMM solutions. Given that this capability is not presently available, it would behoove enterprise organizations to request that vendors add this functionality; specifically, the ability to revoke runtime permissions on an individual app basis to the personally managed portion of the device. Revoking permissions only accounts for

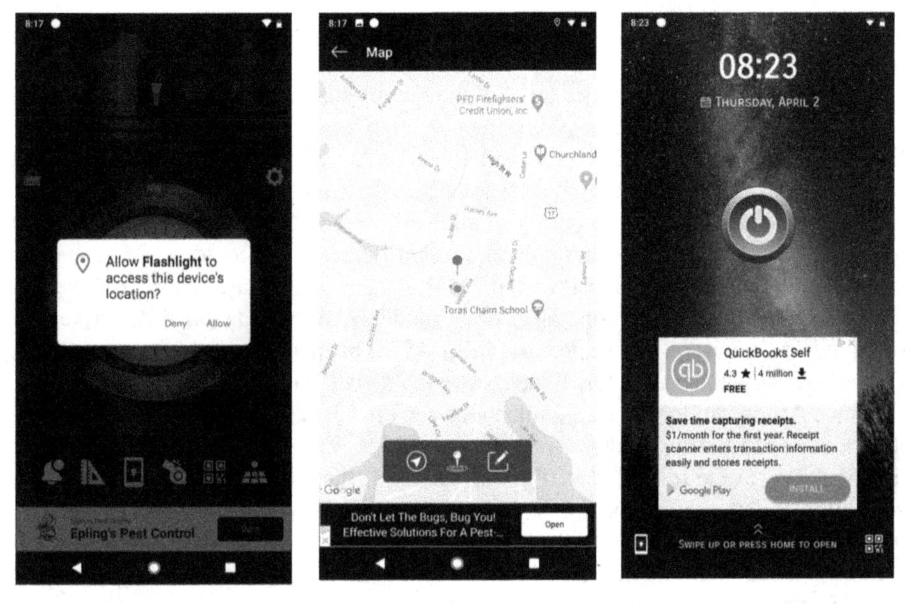

Fig. 11. Flashlight app requesting permissions for location and lockscreen operation.

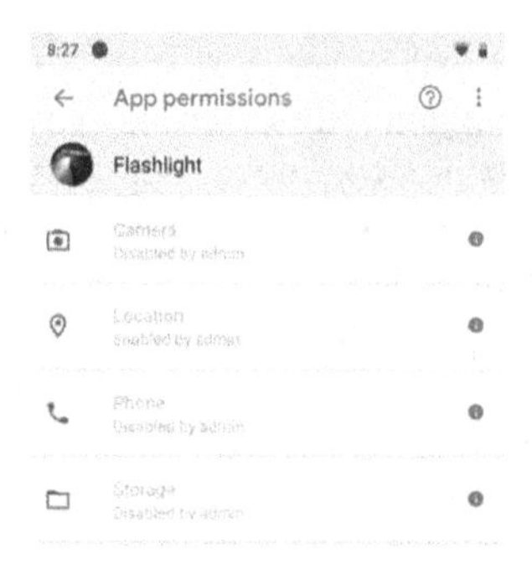

Fig. 12. Unable to grant permissions to flashlight app.

four of the 30 rules (as shown in Table 1). However, when accounting for the number of permissions to be revoked if an app violates one of these four rules – four sensitive information repository permissions, and 11 HW resource permissions – the importance increases.

For vendors, the needed functionality may require greater integration with mobile app vetting providers than current offerings permit. Current solutions do not enable the EMM to programmatically interpret the findings of the vetting report, which is needed for the EMM to act appropriately when a particular app behavior is found. Another option could be to have the app vetting solution utilize the EMM API to perform per-app actions. In other words, the ruleset configuration functionality could reside in the app vetting solution and it would instruct the EMM to carry out actions as needed, based on the findings. In the latter option, the EMM would have to provide the ability to revoke runtime permissions on a per-app basis in its API. This may be done in either case, as EMMs typically make most functionality available via their APIs.

Once this capability exists, organizations can take the configuration ruleset and use it as a basis for starting their own enterprise ruleset. Each organization should make modifications to the ruleset (described in Sect. 4.3) that consider its mission and the sensitivity of data used on its mobile devices. In some cases, organizational policy may dictate certain rules that can be added as needed. The overall goal is to improve the security posture of the organization's mobile deployment while not restricting employee use of personal or non-sanctioned apps for work purposes.

7 Summary

This document describes a configuration ruleset composed of app behaviors and mitigations that enterprises can use in a continuous app vetting process that makes use of per-app security settings for mobile devices. The work builds on previous studies that investigated the effectiveness of mobile app vetting products as well as their ability to integrate with EMM solutions to provide a continuous app vetting capability.

The authors reviewed and carefully inspected Android and iOS documentation as well as the security findings of commercial mobile app vetting vendors and used those findings to generate configuration rulesets that can be applied automatically or with minimal human intervention. The authors developed a proof-of-concept prototype to demonstrate the feasibility of this process utilizing commercial off-the-shelf mobile security products that could be employed by enterprise organizations.

Acknowledgement. The MITRE authors conducted this work under Homeland Security Systems Engineering Institute (HSSEDI) Task Order 70RSAT19FR0000019. The MITRE Corporation operates HSSEDI under Department of Homeland Security (DHS) contract number HSHQDC-14-D-00006. Approved for Public Release; Distribution Unlimited. Public Release Case Number 20-2309.

References

1. Homeland Security Systems Engineering and Development Institute: Evaluating Mobile App Vetting Integration with Enterprise Mobility Management in the Enterprise. https://www.dhs.gov/publication/st-evaluating-mobile-app-vetting-integration-enterprise-mobility-management-enterprise

2. Peck, M., Northern, C.: Analyzing the effectiveness of app vetting tools in the enterprise. MITRE Technical Report 160242. https://www.mitre.org/publications/technical-papers/analyzing-the-effectiveness-of-app-vetting-tools-in-the-enterprise
3. MITRE ATT&CK website. https://attack.mitre.org/mitigations/mobile/. Accessed 11 Mar 2018
4. Banking Trojan Attacks European Users of Android Devices. https://news.drweb.com/show/?i=12940&lng=en. Accessed 16 Nov 2018
5. Goodin, D.: 22 apps with 2 million+ Google Play downloads had a malicious backdoor. https://arstechnica.com/information-technology/2018/12/google-play-ejects-22-backdoored-apps-with-2-million-downloads/. Accessed 6 Dec 2018
6. Stefanko, L.: Scam iOS apps promise fitness, steal money instead. https://www.welivesecurity.com/2018/12/03/scam-ios-apps-promise-fitness-steal-money-instead. Accessed 3 Dec 2018

Blockchain Technology for Improving Transparency and Citizen's Trust

Naresh Kshetri(✉)

Department of Computer Science, University of Missouri - St. Louis, St. Louis, MO 63119, USA
nkbgy@umsystem.edu

Abstract. Public service organizations are always driven by goals to uplift the living standard of citizens. Transparency of information and procedures plays vital role for gaining the citizens' trust. This article is a study about Blockchain Technology's roles in managing the transparency, trust, citizen's satisfaction and minimizing corruption so as to improve the efficiency of public service delivery. Transparency and trust have close relation to corruption and citizen's satisfaction with public services. Various case studies and analysis clarifies the necessity of additional layer in the e-services delivered by public organization. Based on some case studies of ongoing Blockchain-based public services, the potential benefits as well as the existing challenges when introducing blockchain technology in the public sector is presented. Findings show that blockchain technology has the capacity to enable transparency and build citizen's trust in public service delivery while maintaining a sufficient level of privacy. Therefore, keeping the transparency and trust as an important driver, blockchain technology could be the promising technology.

Keywords: Blockchain technology · Citizen's trust · Transparency · Public service

1 Background

Blockchain technology is the technology behind the world's popular cryptocurrency i.e. Bitcoin; since the invent of Bitcoin, number of applications of blockchain technology has been invented. Satoshi Nakamoto's white paper [1], published on SourceForge.net on 2008 has brought this burning technology in the limelight. These days the popular network architecture in almost all the areas is Client-Server architecture; where a central server remains the powerful computer that bears the capacity to control the overall transactions or activities. Intermediation involves the third party handling the transfer of information between the transacting bodies. Main advantage brought by the existing technology was the possibility of allowing the exchange of information between the parties without the physical presence. Major drawbacks of the existing system are the risk of third party being compromised i.e. then presence of single point of failure.

In contrast to Client-Server architecture, blockchain technology is based on P2P architecture. Here the power or authority doesn't rest on single computer. Rather the

K. Arai (Ed.): FICC 2021, AISC 1363, pp. 716–735, 2021.
https://doi.org/10.1007/978-3-030-73100-7_52

authority rests on all the participants of networks. Distributed peer-to-peer nature of blockchain-based technologies has no single point of failure. It uses the renowned technical advancements like hashing, digital signatures, consensus mechanisms and difficulty adjustment. Every computers of the network, controls every computer in the network. Absence of trusted third-party has removed the single point of failure. Whereas the consensus algorithm employed by Blockchain-based technologies has proved the security, scalability and privacy of the platform. Blockchain-based technologies based on PoW Consensus Algorithm (e.g. Bitcoins) operate on 100% verification and 0% trust.

1.1 Distributed Ledger Technology (DLT)

It refers to the protocols and supporting infrastructure that allow computers in different locations to propose and validate transaction and update records in a synchronized way across a network. The technology stores transactions in a decentralized way. Transactions that are assigned for value-exchange are executed directly between connected computers/peers and verified consensually using several algorithms over the network.

Every actor in the network has a copy of record of the transaction, and the change in the ownership of digital asset required validation from the users. So, DLT is supposed to address the 'Double Spending' problem [2, 3]. It is a good alternative to centralized database system in terms of transparency and the trust. Blockchain, Directed Acyclic Graph (DAG), Holochain DLT, Radix, etc. are few examples of DLTs. Major components of Distributed Ledger Technologies are:

Peer to Peer Network (Fig. 1)
Unlike Client-Server network, P2P network is created by the mutual connection among all the participating computing devices. Here the data communication occurs without going through the central server. This type of network doesn't possess single point of failure as in Client-Server.

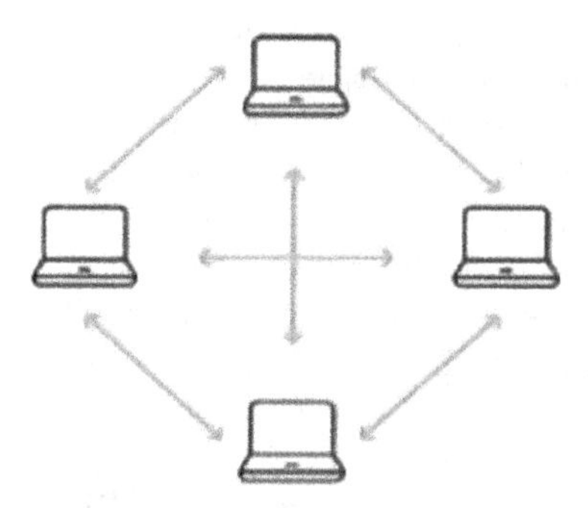

Fig. 1. Peer-to-peer network

Nodes
Nodes are independent computers that act as a communication point. Higher the number of nodes, greater is the security of the network. Nodes preserve the all copy of transactions

separately. Based on the varieties of DLT, one may employ the concept of full node (or master node) as well. The connection of Nodes in DLT takes the form of P2P network.

Consensus Mechanism

Consensus mechanism is the protocol that determines the procedure in the network for reaching the agreement regarding the changes in the ledger. The consensus algorithm defines whether the changes occurred are valid or not. Only the valid changes or transactions are subjected to be the part of distributed ledger. For instance, if a node request for a change in transaction or simply request for any service. Then the verifiability of the request is not under the control of single computer rather it is determined through the consensus among participating nodes. Proof-of-Work (PoW) and Proof-of-Stake (PoS) are the popular consensus algorithm popular on blockchain network.

2 Introduction

Blockchain is used interchangeably with DLT. But the fact is that Blockchain is a type of Distributed Technology. Increased recognition of it leads to more widespread usage, even in cases where the underlying technology being used is quite different than Blockchain. Similar to Blockchain there are different types of DLT that varies according to the number of transactions per second, consensus mechanism, data structure and transaction validation process.

BCT comprises blocks of data (transactions) which are ordered in a linear sequence. The sequence is maintained as according to the verification of new transaction. Each block contains previous block header hash, its payload, timestamp, nonce, difficulty and Merkle root. Hash of the current block is based on payload (transaction) it comprises. Due to the fact that the upcoming block contains the signature of previous one it creates interlinked chain that connects the newest block with the previous one. These blocks are immutably recorded across a peer-to-peer network, using cryptographic trust and assurance mechanism. Tampering the block would cost a lot since, all the blocks are immutably interlinked with each other. Blockchain finds its origin in a paper published by an anonymous author(s) known as Satoshi Nakamoto [1]. Nakatomo has introduced Bitcoin as a purely Peer-To-Peer electronic transaction network. This network allows for direct financial transaction rather than through a third-party intermediation. The consensus mechanism used in the network helps to establish verifiability of the data in the system which was traditionally established by an intermediary of a centralized system. Even in the presence of trustless environment in the network, secure transactions can be made. This was not possible before [4]. Blockchain technology is the most commonly known distributed ledger technology.

There is clear distinction in these two technologies. Blockchain is a type of DLT that stores transaction details in block that are sequentially linked, whereas in other DLTs this does not necessarily have to be the case [5].

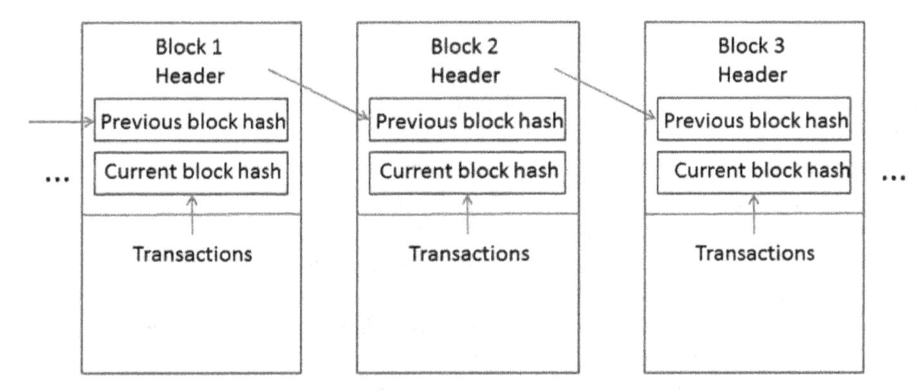

Fig. 2. Simple blockchain

Figure 2 is the simple interlinked sequence of blocks. Block 2 contains the hash of Block 1, Block 3 contains the hash of Block 2 and so on. This way, all the upcoming blocks are linked with the previous blocks. The interlinked nature of blocks and consensus algorithm makes it immutable and tamper-resist. Meanwhile two different models of Blockchain Technology have been popular nowadays.

Permissionless (Public) Blockchain
Bitcoin is the good example of public Blockchain. Accessibility is open to any participants willing to transact within the network. It is not owned by any superior body/entity. Participants of the network can create and validate transactions and hold an identical copy of the complete ledger.

Permissioned (Private) Blockchain
In contrary to public blockchain, permissioned blockchain are owned by certain organization or group of stakeholders. This nature of distributed technology results in simplicity and time-effectiveness of operations which ultimately leads to increased efficiency compared to public blockchain. Further, private blockchain can't be considered as fully decentralized one.

Section 1 presents the background, DLT and objectives of the research. Section 2 presents the literature review of the proposed work. Section 3 introduces the methodology using traditional and existing system. Section 4 presents results and findings of the study. Section 5 concludes the paper.

2.1 Rationale and Objectives

Government of Nepal is prioritizing Information Technology as the sector of national interest. In the era of technological innovation, the advancement of technology is far beyond the pace of contemporary Nepalese pace. Every day newer inventions are brought to replace the age-old technology. Government is investing a lot in the implementation of webpages and applications which are not based on recent technologies like AI, ML or IoT. Unlike to our activities, the developed countries are focusing on automation of existing technologies. So, we need to incorporate the advancement with no delay in order to grab the benefits of IT. e-Government implementations in developed as well as developing countries are facing large difficulties. The case study regarding the failure of e-Government projects in Egypt suggests the role of User Satisfaction for the evaluation of project's success/failure. The quality of information, system and service determines the level of user satisfaction which is directly linked with the citizen's trust. So, citizen's trust can be regarded as the major metrics behind the success of technologies. If the project failed to achieve the citizen's trust or failed to deliver the services with the promising level of transparency, then the investment will go in vain.

Blockchain technology, which is based on Distributed Ledger, would be appropriate way for gaining the trust and transparency in public service delivery. The hype of BTC, implementation of Blockchain based system on aforementioned projects and ongoing projects of Blockchain are the indicators of this technology.

In the research project, it is attempted to cover the major public service sector. In-depth study regarding the feasibility and comparison with existing technology has been prioritized throughout the study.

The proposed research has the following objectives:

- To find the metrics of public service delivery for efficiency.
- To derive the feasibility and suitability of employing Blockchain-based technology at the public services and propose a model.

3 Literature Review

From the advent of the concept of *peer-to-peer form of electronic cash* in 2008, and driven by the success of Bitcoin, high expectations are on the transformative role of Blockchain for the industry and public sector. The analysis of a group of pioneering developments of public services shows that Blockchain technology can reduce bureaucracy, increase the efficiency of administrative processes and increase the level of trust in public recordkeeping [5]. The technologies based on Distributed Computing or Distributed Ledgers are about to change the role of web from centralized document sharing platform to generic de-centralized platform. Distributed Ledger Technologies using specific consensus algorithm can be assumed to bring the openness by disintermediation and without compromising the security. Blockchain and related technology can provide more secured and verifiable transacting environment [6].

Cong & He define a blockchain as a 'distributed database that autonomously maintains a continuously growing list of public records in unit of 'blocks', secured from tampering and revision [7], while Atzori describes it as an 'irreversible and tamper-proof public records repository for documents, contracts, properties, and assets [that] can be used to embed information and instructions, with a wide range of applications [8]. The combination of a distributed, append-only ledger and consensus mechanism employed by Blockchain technology is supposed to confirm disintermediation. Thus, eliminating the middlemen and brokers, results in removal of middle-man or broker related transactional costs [5]. Decentralization of power and authority to the peers of networks boosts the trustworthiness of any services delivered.

Innovation and transformation of governmental processes is the core question that needs to be addressed regardless of any sort of technologies being used. A critical assessment regarding the benefits of blockchain technology and their implications for governmental organization and processes is a must. Svein Olnes' paper has plead for a shift from a technology-driven to need-driven approach in which blockchain applications are customized to ensure a fit with requirements of administrative processes and in which administrative processes are changed to benefit from the technology. Deeper dives into the blockchain technology haven't been made yet in various potential areas. Potential applications of Blockchain have still remained unexplored beside cryptocurrency. The new technology has potential benefits in e-governments to comply with societal needs and public values [9]. BTC has shown the promising potential in bringing the revolution in financial sectors by the removal of trusted third party in payment solutions. Innovations of BTC in financial sectors have depicted both challenges and opportunities for enhancing digital public services.

Implementation of Blockchain on few public services like Voting, Land Reform Management, Supply Chain Management, Health Care, etc. has proved an existence of great potential for reforming and even transforming public service delivery. It is also argued that open blockchain technology is best understood as a possible information infrastructure, given its universal, evolving, open and transparent nature [10]. Blockchain Technology can be understood even as an Information Infrastructure. It could also be understood as Information Superhighway (I-Way) serving as a backbone network for transferring the information across the geographical ends. Numerous research and innovations have been done for exploring the potential of Blockchain technology in various sectors like Digital Agriculture, Food Supply Chain, Democracy, Supply Chain Management, Health Care and so forth.

Public service organization of any country is driven by the main purpose of delivering the services to the people so as to promote a better living standard. Despite the holy purpose of public service organizations, there are several threats that prevent the achievement of better living standard goals of citizens. The threats are corruption and lack of accountability. Corruption free public institution that are accountable to government and public are the fundamental building blocks for increasing the citizen's trust and participation so as to ultimately uplift the living standard of citizens. To improve the efficiency of public services, one must build the citizen's trust on service delivery. One of the ways of improving the citizen's trust is through transparency and accountability. F. Rana [11] has suggested accountability in public organization as an important metric to enhance efficiency level of public service delivery. The conclusion was taken through the statistical analysis of data of Pakistan. An efficient and effective public administration is an essential precursor for socio-economic development. However, it is recognized that many regimes in developing countries are partly open to public scrutiny and influence. Due to the fear of citizens are informed about the flow of ideas and information in the government, few non-democratic countries have been restricting the openness in the flow of information. Despite the fact regarding the various practices of e-Governance, citizens are always eager to adopt the more convenient, efficient and transparent form of services [12].

Inefficiency of public service is contributed by lack of transparency and lack of accountability. The worse consequences of these inefficient practices are corruption and deterring level of citizen's trust towards public service. This ultimately leads towards the failure of governance. Client-Server architecture which is the existing backbone of almost all e-government services has a single point of failure and it contains a third-party with whom all the transaction relies on. Any breaches on server-side could unprecedented loss of e-government services. This is why, digitization of governance through existing client-server infrastructure is not sufficient to get rid of corruption and non-transparency of public services. The new technological advancement, i.e. Distributed Ledger Technology, could offer potential benefit at public service level to get rid of the long-lasting problem of public services in most of the developing countries like Nepal as well as in developed countries. The study of blockchain initiative in D5 (Digital 5) countries has come with the good findings that recommends for policymakers on emerging governance topics that require investigation in order to realize the full potentials of blockchain innovation in public administration and the government domain. Large number of research topics and concepts are being derived from the interaction of BCT and Governance. The study of D5 countries has revealed blockchain as a secure information management and provenance infrastructure, authentication and validation infrastructure, financial settlement infrastructure and transaction governance infrastructure [13, 14]. When, all the stakeholders are able to maintain the copy of the entire ledger from the very beginning and the validity of transaction is done through the consensus algorithm, then we can re-create the real e-government with maximum efficiency, 100% transparency, no single point of failure and no corruption. The blockchain technology can reshape the way governments interact with citizens and with each other. Distribution of all the information, inclusion of all the stakeholders, transparency of all the activities, protection of users' privacy, anonymity of the users, verifiability and immutability of the transaction can be achieved using Blockchain-based technologies [15]. The paper studies the various metrics of public service delivery which are considered as the major parameter for improving the citizen's trust. Further the Blockchain's potential in achieving those parameters is studied.

4 Methodology

Identity Management System has been taken into consideration during the explanation of existing traditional and blockchain-based system. Review of literatures regarding the DLT, BCT, Public Services and Metrics of Public Services to meet the objectives of the research project. Data required for the research project are collected through verifiable and trusted sources from online databases. Following are the few topics that are explored for the achievement of the aforementioned objectives of the study.

4.1 Traditional Existing System

Existing System comprises of a dedicated server through which the clients request for services. Advancement on this technology has brought distributed Client-Server architecture in limelight as well (Fig. 3). But despite the increase in number of server either for backup or recovery purpose, these types of system have a single point of failure.

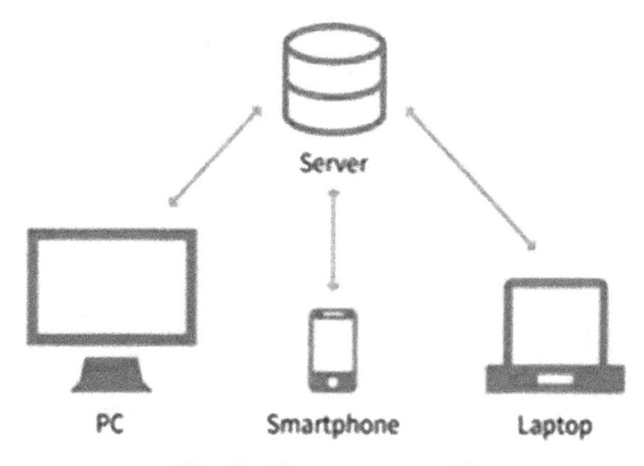

Fig. 3. Client-server model

The authority rests on the powerful machine. Any breaches of information from the server remain untraced instantly by the client. Attackers only need to exploit the server machine. Their' unauthorized access to sever machine would create unprecedented loss to the system. The major threats on existing non-blockchain based client-server e-governance system are unauthorized access, malicious damage, data intercepts, etc. Non-transparency in overall flow of information and dictating mode of transaction verification can created doubt on the efficiency of these types of intermediated system.

One cannot have 100% trusts regarding the transparency on the system being operated by some powerful machine. Furthermore, news on data breaches, loss of privacy, potential vulnerabilities, etc. has contributed for preventing citizens to trust the public services to the fullest.

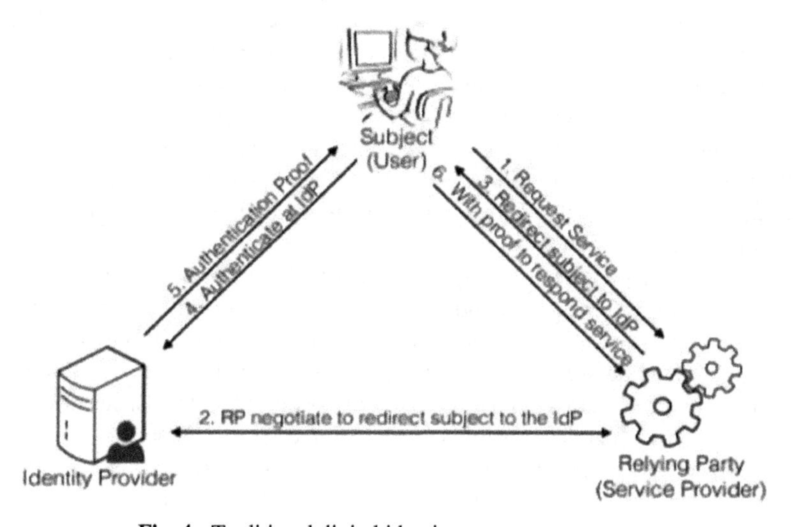

Fig. 4. Traditional digital identity management system

Figure 4 illustrates the model of traditional Digital Identity Management System taken as a reference from the research paper of Xiaoyang Zhu [16].

4.2 Blockchain-Based Existing System

The ability of blockchain based technologies to eliminate the need of intermediaries has sought the attention of tech enthusiasts. BCT is proving as a promising technology to tackle the threats of cyberspace through 100% transparency and disintermediation. Simple figure of blockchain is stated in Fig. 2. It consists of interlinked blocked that are linked together by the cryptographic hash which is difficult to tamper.

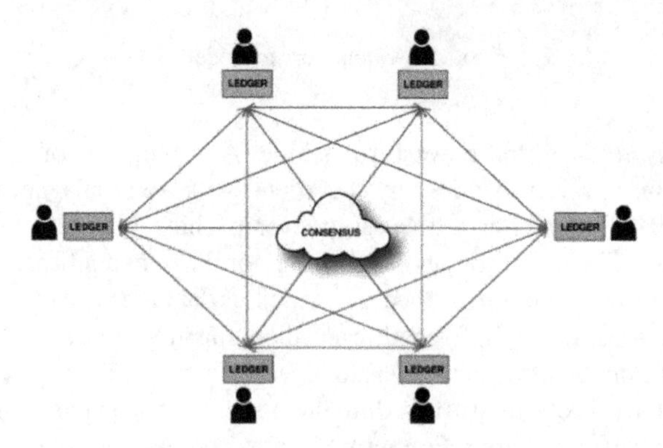

Fig. 5. Interconnected nodes in BC

Each node in Blockchain consists of identical copy of ledger. When all node operators agree to the change and consensus is reached, the entire network will update their own ledgers. This ensures the immutability of records for network participants and end-users. Since, all the node shares the identical ledger, any tampering activities or misbehaves can be easily caught. Fraud activities within the network are not liable to be the part of original chain and thus are thrown away from the blockchain. Considering the Xiaoyang Zhu's Identity management system as illustrated in Fig. 5, the research paper has also presented the alternative of traditional system. The Blockchain based Identity management System shown in the following figure (Fig. 6):

Here the user, do not need to rely on third-party to claim his/her identity. He/She could easily verify his/her identity and claim the services or benefits offered by any governmental organizations or public service organizations. In contrast to it, the traditional system would force the user to submit all the private information and should rely on the permission of intermediaries to have valid claims. Moreover, the chances of fraudulent claims and activities are relatively higher in traditional system. Absence of central point of failure, distributed nature and presence of consensus algorithm along with secured cryptographic hashes has proved the transparency, verifiability, and tamper-resist and secured nature of Blockchain-based system.

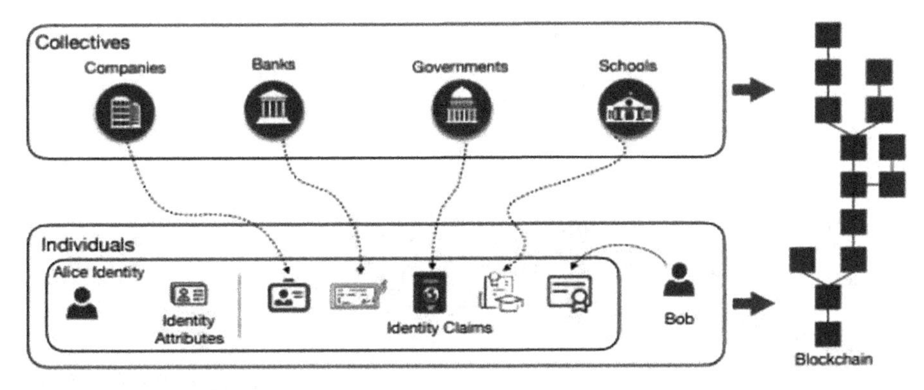

Fig. 6. Blockchain based identity management system

4.3 Smart Contracts

Built upon the BC, a smart contract is usually:

- a pre-written logic in the form of computer code.
- stored and replicated on the blockchain.
- executed and run by the network of computers running the blockchain.
- can result in updates to accounts on the ledger.

Smart contracts can be understood as autonomous control platforms such as software applications that can be used to validate transaction without the necessity of human interference [17]. They are self-executing after meeting the certain criteria. By offering unparalleled transparency as well as elimination of third parties, smart contract could ultimately have profound impact on governance. These days, a risk in contract occurs due to asymmetric information. Blockchain based smart contract reduces asymmetric information because they are open and is available to all parties. In public service, the information is supposed to be available to entire society.

4.4 Feasibility Study of BCT

By nature, BC cannot be applied to all systems as a modular off-the-shelf solution. The following feasibility flowchart (Fig. 7) [18] clarifies the usefulness of BC. If Blockchain is not useful for a project, then it cannot be feasible either.

The characteristics of BC that have added value in it are; disintermediation, auditability, transparency, cost saving, immutability, remote participation and verifiability.

Economic Feasibility

Automating the traditional system will undoubtedly lower the administrative costs in long run. Replacing the existing system will obviously result in higher initial investment costs but in a long run, if properly aligned, almost all the technical innovations offers good benefits for both public as well as private sectors. Cost of existing non-blockchain based digital technologies includes the development of software, integration of hardware

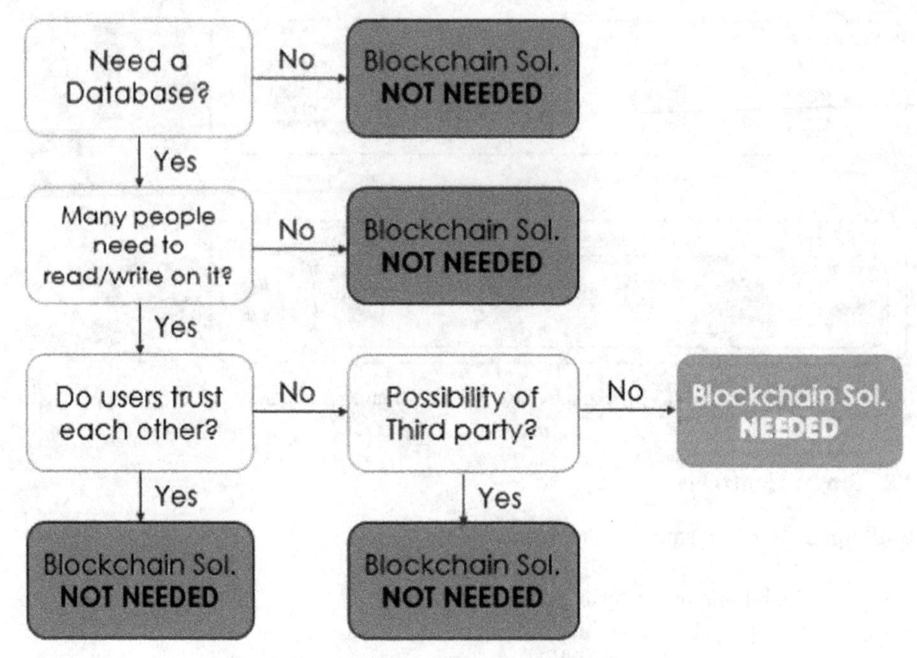

Fig. 7. Blockchain feasibility chart

and operation & maintenance cost. Replacing these technologies with blockchain-based solution will even lower the cost as most blockchain are open-source projects and come with customizable APIs.

Technical Feasibility

The network of BC consists of powerful nodes that keep the records of all the transaction from the very beginning. They are also called miners, who are responsible to solve the complex mathematical puzzle. The availability and fault tolerance of the system is very high. Integrity of records, availability, fault tolerance and privacy of BCT is unbeatable in comparison with existing technologies. Blockchain provides anonymity along with transparency. Identity verification (authentication) is done through public-key cryptography. Elliptic-Curve Cryptography (ECC) is widely used in this process. Thus, from the technical perspective, a country needs to setup basic network infrastructure. Infrastructure for the distribution of public and private keys. The energy consumption is comparatively higher than the other technologies. So, the basic energy requirements for the operation of nodes of the network must be ensured. Addressing the basic infrastructure and energy requirements, the application of Blockchain is feasible from technical point of view as well.

Socio-Political Feasibility

BC stores all the transaction immutably. The consensus algorithm ensures the verifiability and accuracy of the transaction. It is built upon 100% verification and 0% trust. This makes the technology, suitable from the socio-political perspectives as it addresses the major socio-political problem brought in the society by virtue of misuse of technology. This time, the technologies are being used to manipulate the information by misusing the power of servers. Due the fact, the software application being deployed in public services are losing trust of the citizens. To address the privacy, transparency and verifiability is the socio-political need of contemporary world. This proves the socio-political feasibility of BCT/DLT.

4.5 Public Service

The quality of public service is increasingly important for the general public as well as policymakers. With the expectation of raising the living standards, the demand will tend to shift towards higher qualitative services rather than quantitative ones. Government should focus towards efficient and effective service delivery. The promise of good governance is often threatened by the corruption and lack of accountability which worsens the situation. Ultimately the trust of citizens towards the public services diminishes. This will result in reduced efficiency of any forms of governance i.e. traditional or modern e-governance [11]. The research has selected the few metrics behind the success of public service delivery in terms of efficiency.

Transparency
Being transparent is being open to all. A transparent network is open to all its participants. Blockchain offers this transparency. In public service delivery, the transparency reduces the attempt of corruption. The traditional client-server-based e-governance is ruled by the powerful server that might be corrupted (less likely). Despite the several preparations, the non-blockchain based systems are vulnerable to attack and possess a central point of failure. Transparency is one of the most sought metrics behind the success of any public service.

Citizen's Trust
The success of public service or simply the weight of efficiency if highly contributed by the citizen's trust towards it. Trusting the service ensures the suitability and sustainability of it. Blockchain technology being distributed open ledger, collaborates with all the nodes of the network in deciding its' own future. Being ruled by every participants of network, the BCT can be termed as fully democratic system. Having no central point of failure, 100% verifiability, and immutability of transactions, anonymity and transparency are the features that can promise anyone to have faith towards the service.

Other Metrics
Accountability is also one of the major metrics of measuring the sustainability of public service delivery. It is defined as being answerable for our actions or taking responsibility for our action.

Privacy is also the most sought parameter of any kind of services. Citizen wants their information not to be leaked by any purpose without their consent. For the services to be more efficient, it must endorse the citizen's privacy at any cost. Breaches or any attacks on the privacy of user's information may diminish the efficiency of the services. Similarly, Responsiveness, Participation, Impact, Resource Optimization, etc. are also the metrics on public service that can be taken into consideration while having comprehensive study on efficiency of public services [19].

4.6 Blockchain in Public Service

Ability to trust government data and services in any situation is one of the fundamental capabilities of any country. The ability to verify the integrity of public data independently of its home database, in real time, enables data interoperability between systems and across boundaries [20]. The security mechanism of BC enables implementation in a wide range of processes. Most of the developed countries are piloting the blockchain-based projects while few of them had it already implemented successfully. Application area of Blockchain in public services includes digital identity, the storing of judicial decisions, financing of school buildings and tracing money, marital status, e-voting, business licenses, passports, criminal records and even tax records and many more.

A November 2017 report by Reform (the independent non-party thinks tank) states that only "13% of people trust government to use their data appropriately, while 46 per cent do not". The report asserts that current models of "identity" cause duplication and friction and suggests that blockchain could simplify the management of trusted information, making it easier for government agencies to access and use critical public-sector data while maintaining the security of this information. Some records exist only in paper form, and if changes need to be made in official registries, citizens often must appear in person to do so. Individual agencies tend to build their own silos of data and information-management protocols, which preclude other parts of the government from using them. The Reform report suggests that blockchain could enable a shift ownership of personal data from the government to the citizen and proposes a new identity management model powered by blockchain [21].

Estonia - Secured by Blockchain

Pioneer of E-Government services, Estonia has been the first Nation-State in the world to deploy blockchain technology in production system in 2012 [20]. The services backed up by BC are Healthcare Registry, Property Registry, Business Registry, Succession Registry, Digital Court Registry and State Gazette. Since, they had a matured digital government infrastructure; the citizens were used-to with the internet-based services. With substantial investments in cyber-security infrastructure, Estonia has developed extensive expertise in this area. Higher digital literacy and readiness in almost all the sector of digitization has helped Estonia for the successful implementation of Blockchain in public services. On top of the X-Road system [22] the Estonian government has built transparent services that lets the citizens not only easily access their own data, but also see who else has accessed their data and when. This increased the transparency that led to improve in efficiency and citizen's trust in public service delivery.

The government has employed KSI blockchain. Unlike traditional digital signature approaches, e.g. Public Key Infrastructure (PKI), that depend on asymmetric key cryptography, KSI uses only hash-function cryptography, allowing verification to rely only on the security of hash-functions and the availability of a public ledger commonly referred to as a blockchain. Benefits imposed by KSI blockchain are Massive Scale, Portability, Quantum Immunity, Independent Verification and Data Privacy. Data never leaves the system; only hash is sent to blockchain service. As no data is stored on the KSI Blockchain, it can scale to provide immutability for petabytes of data, every second. The lesson learned from Estonia is that speed is essential for citizen experience (Fig. 8).

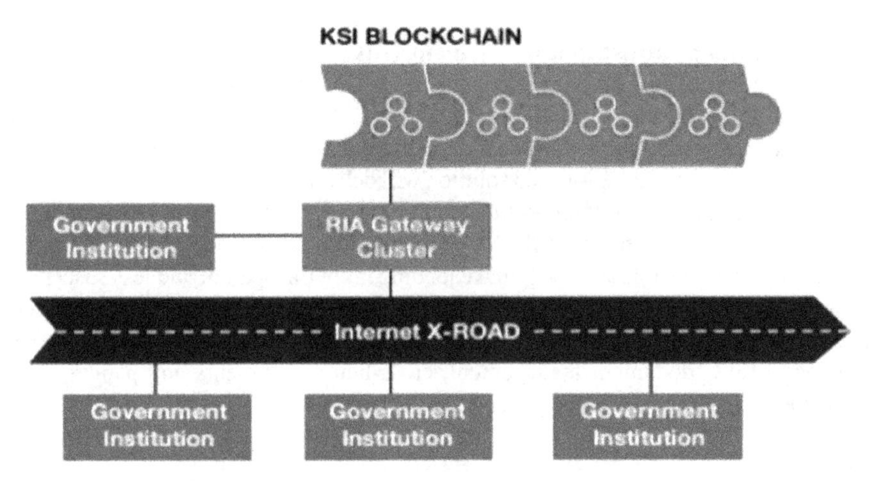

Fig. 8. Estonian blockchain implementation framework

Malta - Blockchain Island

Malta has taken a significant step toward becoming a "Blockchain Island" by passing bills regarding cryptocurrencies, blockchain and distributed ledger technology (DLT) marking it as one of the first jurisdictions in the world to pass specific legislation around the tech. U.K. middleware blockchain startup Omnitude has been working to improve the Maltese Public Transport Service using DLT. Malta has developed a regulatory framework and technological infrastructure favorable for blockchain-based projects. These attempts were to ensure transparency and certainty to the companies or organizations. Ultimately, the act ensures the delivery of public services using DLT/BCT.

Blockchain in United Kingdom

Major undertakings of UK Government are the partnership with IBM, Nestle and Unilever to improve the traceability of contaminated food. Actions like this have the capability of making huge improvements to peoples' lives. Also, National Archives and the University of Surrey on 'Archangel' are collaboratively pursuing a project aiming to preserve digital archives through blockchain technology. UK government had successfully tested the use of a Blockchain-based system to distribute welfare payments through the Department of Work and Pensions, explore the use of Blockchain as a service for each governmental department, available as of August 2016, and saw the Financial Conduct Authority (FCA) permit Blockchain startup Tramonex to issue its digital currency to UK citizens. Investment of over £10M in Blockchain projects focused on energy distribution, clean water provision, electoral systems and charitable giving clarifies the necessity felt by UK government. UK comprises huge number of Companies, Investors, Influencers,

R&D Centers, Hubs, Conferences and Accelerators in the field of BCT/DLT. The necessity of BC has been proved through the success of earlier projects. Major benefits are integrity, consistency, immutability and security of data that has gained the citizen's trust and improved the transparency of public services as well [23].

India - Blockchain Infrastructure Company (BIC)

A consortium (i.e. BIC) of 11 big lenders set to launch the India's first blockchain-linked funding for small and medium enterprises (SMEs). It is driven by the motive of removing any communication hurdle among different bank. The consortium is hoping a new industry-wide blockchain-based solution will deliver a number of benefits including cutting timeframes in supply-chain financing, reduced costs, deepen credit catchment area and increase the number of SMEs integrated into the formal credit system. It is expected that this new initiative will make lending more transparent and less susceptible to fraud [24]. Also, BCT is engendering significant attention within India across a wide range of industrial sectors for a range of use cases including trade finance, supply chain financing, e-KYC document management, cross-border payments and patient record management.

Others

Beside aforementioned countries and their brief implementations, there are many completed and ongoing projects on BC/DLT. Implementation of BCT on e-Voting, Healthcare management, SCM, Identity Management, etc. are driving this technology towards the maturity. Larger companies like Walmart, IBM, Maersk, Luxoft, etc. are investing huge amount of resources for the deployment of Blockchain-powered services in the field of Identity Management, Health Care, SCM, IoT and Energy Industry. Moreover, the application of BC on public service is exploring the newer way of consensus that provide efficient and fast means of services promoting the citizen's satisfaction and trust.

4.7 Benefits Brought

Strategic

BCT democratizes the access to data. Interlinked sequence of chains remains visible and every node has an identical copy of ledger. Unnoticeable changes are difficult to commit. Further, the storage of information in distributed ledger prevents from fraudulent or corrupting activities. Changing the ownership of any assets (information) can't be manipulated beyond the consensus mechanism. So, from the strategic point of view, the BCT invite **transparency** and **reduces corruption and frauds**.

Organizational

Immutability and verifiability of the BCT increases the trust among every participants of the network. The technology allows us to track the changes or updates from the very beginning of the network. Thus, from organizational point of view, **trust, auditability, predictability** are the prime benefits.

Economical
BCT is tamper-resist through cryptographic consensus mechanism. Immutability and consensus algorithm have provided higher level of resilience and security, thus reducing the costs of measures to prevent attacks. Involvement of humans is discouraged, rather the transactions are verified with the help of consensus algorithms among nodes. Thus, the cost of conducting and validating a transaction is reduced. From economic point of view, **reduced costs** can be regarded as the benefits of BCT.

Informational
Automatic operations discourage the human-made error in the output. Distributed nature of network allows easy and quick access of information. Users have their own control over their information i.e. shouldn't rely on third-party to access the information. Thus, BC ensures **integrity, accessibility and reliability** of the information assets.

Technological
Activities within BC network are resilient to malicious behavior. It has no single point of failure. Hacking is less likely as well. So, Resilience, Security, Immutability are the technological benefits promised by BCT.

4.8 Cons of BCT

Redundancy
Complete ledger is shared with all the nodes of the network. This creates redundancy in BC network driven only by the purpose of removal of intermediation. This redundancy offers increased costs.

Low Scalability
The ledger grows faster than the number of network members. This will invite computational and storage burden on the member of distributed network barring to effective scaling.

Privacy and Decentralization
Whenever the public services are powered by Blockchain technology esp. Open blockchain, then the information are open to all the participants. However, the organization could have really critical information not to be made open to all the general public. This might impose a challenge in the public service. As a remedial action, we may use permissioned blockchain that prevents the data from being open to all the general public in the expense of reduced decentralization.

51% Attack
Despite the fact that the economic incentives are heavily aligned against 51% attack, there is the theoretical feasibility of it. It is the method of using large amount of hash rate to generate fraudulent transactions.

5 Results and Findings

Through the review of several case studies and applications of blockchain in emerging nations on various dimensions of public service, the study has brought some findings. The findings of the study have highlighted the importance of transparency, immutability and distributed nature of BCT for improving the efficiency of public service. Anonymity is achieved through cryptography. Not all blockchain uses anonymity however because, it poses a problem when it comes to trust. Private blockchain would be useful when we want an extra layer of transparency and higher level of security. Being decentralized, the BCT requires a consensus algorithm to reach the agreement/solution of any problem. PoW and PoS are the popular consensus algorithms having their own benefits and limitations. PoS solves the problem of energy requirement associated with miners and offers more speed and scalability. Going through the statistics, it is found that 67% of Americans are more worried about electronic machines getting hacked or manipulated than about tampering with paper ballots. Also, 75% Americans trust their votes will be accurately counted with paper ballots, while 68% trust votes to be accurately counted using e-voting [25]. This statistic shows the necessity of fully decentralized, immutable and transparent Blockchain-powered technologies in public services. With blockchain-powered services, the user has their own control over their data. This ensures the privacy as well.

Furthermore, according to WTO, reducing barriers within the international supply chain could increase worldwide GDP by almost 5% and total trade volume by 15%. These are the statistics that reveals the economic as well as social importance of Blockchain technologies. Successful implementation of LifeID for Identity Management System, MedChain for medical record keeping, Blockchain-powered e-Voting System, Blockchain-powered Land Records and major undertaking of Luxoft and other companies like IBM has proved the importance of employing Blockchain Technology in public services. The reason behind the success of the technology is due to the feature of Transparency and Immutability nature of BC which guarantees the trustful environment. More particularly, a person is ensured that his/her private information are in his/her own control, a voter is ensured that his/her vote hasn't been tampered, etc. In public services, the citizens require the transparency of all the activities. They search for the traceability of all the activities. The public service organizations are accountable to their services if and only if their activities are transparent and immutable from the people's view. The guarantee of transparency, immutability and privacy can't be guaranteed by the existing client-server-based technologies. It was found through the study that the BCT has the potential to provide the transparency, immutability and privacy sought by the citizens.

5.1 Proposed Model

The user is a citizen, bearer of a mobile application. The application comprises of his/her credentials recorded in a Blockchain Network in the form of cryptographically secured Identifiers. Citizen could use the country's backbone network to access the public services. Moreover, the important records of the public services are immutably recorded in the Blockchain network. This way, we could ensure trust and transparency in the absence of third party which will ultimately improve the efficiency of public services. The proposed model is shown in Fig. 9.

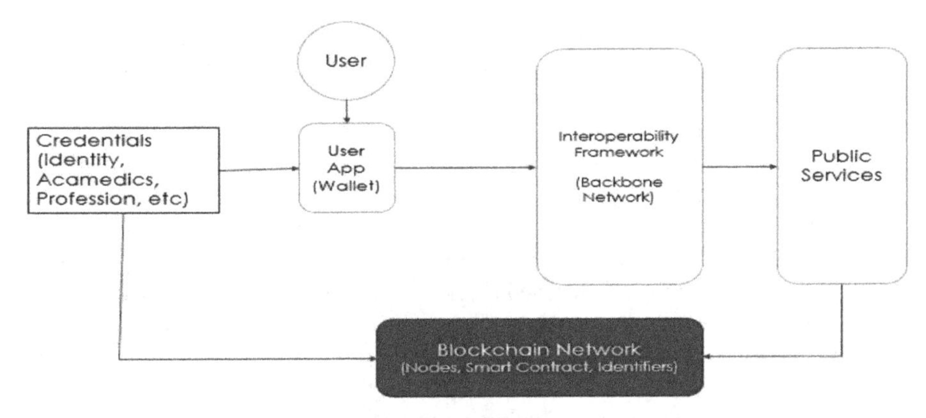

Fig. 9. Proposed model for blockchain-based public services

5.2 Opinions on Digital Nepal Framework (DCN)

The findings of the research project have come up with some suggestions or opinions regarding Digital Nepal Framework as well. DNF has defined 8 sectors and 80 digital initiatives covering almost all the public services offered at different level of government [26]. In all those projects, the application of blockchain could offer advantage on following sectors as identified on DNF: *Agriculture, Energy, Tourism, Financial Services, Digital Foundation, Urban Structure, Education, Health.* Most importantly, the management of National Identity, Citizen's credentials related to several sectors and transparency of information through immutable ledger could drive the digital initiatives as designed in the framework for gaining the trust of Citizen more quickly. We could employ the similar architecture as of Estonia's (i.e. KSI Blockchain) to add an additional layer of trust as well.

5.3 Challenges and Limitations

Adoption
Technology Adoption is the major challenges after the deployment of e-services. Making the general public aware regarding the usability and suitability of the application always hinders the success of Technology. In context of BCT, still being in infancy stage, especially in developing countries like Nepal, where the government hasn't yet envisioned the use of DLT in public services, the Adoption of this technology is the major challenges.

Laws and Regulations
Due to the higher energy consumption brought by the mining of BTCs, Nepal and most of the countries don't have any legal regulations regarding this technology. In the absence of laws and regulations, the deployment of blockchain-powered technology becomes illegal act. This prevents the innovation in this emerging technology. Regarding

the energy consumption issues, there could be many alternative consensus algorithms suitable to employ in the expense of certain features of PoW-based algorithms.

Permissioned vs. Permissionless BCT

These two flavors of Blockchain has their own pros and cons. Public Blockchain are fully decentralized while it doesn't provide the privacy of sensitive and critical information regarding the organization and its' strategies. Further, the public blockchain operates on energy inefficient PoW consensus but they offer 100% transparency.

On the other hand, the permissioned blockchain or private blockchain have the pre-assigned nodes. Not all the participants can have the authority of node. It contains the extra layer of transparency. This type of BC can't be considered as fully decentralized. Rather, it can classify the information and assign the access level as according to users. In most of the application of public service, these type of Blockchain could be beneficial from the organizational perspective. Moreover, private blockchain are energy efficient and more scalable than public. The clear distinction between private and public blockchain has led the companies to re-think and re-design the appropriate architecture in order to deploy public services through Blockchain-based tools.

6 Conclusion and Future Work

In this paper, a new model is proposed for blockchain based public services. Thus, we have the problems of scalability and adoption model of newer technology. Further, the public blockchain which relies on PoW consensus is not energy efficient and private blockchain is not fully decentralized. However, millions and trillions of investments has been done in this sector by larger group of companies. Many developed and developing countries of the world are pioneering in several dimensions of BCT. Moreover, the success of LifeID, MedChain, Estonia's KSI Blockchain Model, Blockchain-powered SCM of IBM, etc. has proved its innovations. Still, the technology is in infancy. More research and in-depth exploitation of the technology is must.

Keeping the transparency and trust and the important driver for the success of public service delivery, blockchain technology could be the promising technology to improve the efficiency of public services by virtue of its decentralized, immutable and transparent nature.

References

1. Satoshi, N.: Bitcoin: A Peer-to-Peer Electronic Cash System (2009)
2. EVRY: Blockchain - Powering the internet of value (2016). https://www.evry.com/en/about-evry/media/white-papers/blockchain-powering-the-internet-of-value/
3. Swan, M.: Blockchain: Blueprint for a New Economy. O'Reilly Media, Inc. (2015)
4. Basnet, S.R., Shakya, S.: BSS: Blockchain security over software defined network (2017)
5. Allessie, D., Sobolewski, M., Vaccari, L., Pignatelli, F.: Blockchain for digital government (2019)
6. Kadam, S.: Review of Distributed Ledgers: The technological Advances behind cryptocurrency (2018)

7. Cong, L.W., He, Z.: Blockchain disruption and smart contracts. Rev. Financ. Stud. **32**, 1754–1797 (2019)
8. Atzori, M.: Blockchain technology and decentralized governance: is the state still necessary? J. Gov. Regul. **6**, 02–03 (2017)
9. Ølnes, S., Ubacht, J., Janssen, M.: Blockchain in government: Benefits and implications of distributed ledger technology for information sharing. Gov. Inf. Q. **34**, 355–356, 363 (2017)
10. Ølnes, S., Jansen, A.: Blockchain technology as infrastructure in public sector: an analytical framework (2018)
11. Rana, F., Ali, A., Riaz, W., Irfan, A.: Impact of accountability on public service delivery efficiency. J. Publ. Value Adm. Insights **2**, 7–9 (2019)
12. Sharma, G., Xi, B., Wang, Q.: E-Govemment: public participation and ethical issues. J. E-Gov. **35**, 195–204 (2012)
13. Ojo, A., Adebayo, S.: Blockchain as a next generation government information infrastructure: a review of initiatives in D5 countries. In: Public Administration and Information Technology, pp. 283–298. Springer International Publishing (2017)
14. Davidson, S., De Filippi, P., Potts, J.: Disrupting governance: the new institutional economics of distributed ledger technology. SSRN Electron. J., 15–16, 18 (2016). https://doi.org/10.2139/ssrn.2811995
15. Zhu, X., Badr, Y.: A Survey on Blockchain-Based Identity Management Systems for the Internet of Things (2018)
16. Infographic: The Power of Smart Contracts on the Blockchain, Visual Capitalist. https://www.visualcapitalist.com/smart-contracts-blockchain/. Accessed 22 Apr 2020
17. Çabuk, U., Adiguzel, E., Karaarslan, E.: A Survey on feasibility and suitability of blockchain techniques for the e-voting systems. Int. J. Adv. Res. Comput. Commun. Eng. (IJARCCE) **7**, 124–134 (2018)
18. Smith, P., Mayston, D.: Measuring efficiency in the public sector. Omega **15**, 181–189 (1987)
19. PwC: Estonia-the-digital-republic-powered-by-blockchain. pwc.com (2020). https://www.pwc.com/gx/en/services/legal/tech/assets/estonia-the-digital-republic-secured-by-blockchain.pdf. Accessed 20 Apr 2020
20. Borrows, M., Harwich, E., Heselwood, L.: The future of public service identity: Blockchain. Reform Research Trust, London (2017)
21. Paide, K., Pappel, I., Vainsalu, H., Draheim, D.: On the Systematic Exploitation of the Estonian Data Exchange Layer X-Road for Strengthening Public-Private Partnerships (2018)
22. Laurie, C.: How is the UK Government Using Blockchain? https://www.computerworld.com/article/3427866/how-is-the-uk-government-using-blockchain.html. Accessed 20 Apr 2020
23. ICICI, Kotak, Axis among 11 to launch blockchain-linked funding for. https://economictimes.indiatimes.com/markets/stocks/news/icici-kotak-axis-among-11-to-launch-blockchain-linked-funding-for-smes/articleshow/67718025.cms. Accessed 22 Apr 2020
24. Vavra, S.: Exclusive Poll: Majority Expects Foreign Meddling in Midterms, Axios, 04 June 2018. https://www.axios.com/exclusive-poll-majority-expects-foreign-meddling-in-midterms-515c9556-be3e-4d51-b575-ff3ae47af205.html. Accessed 23 Apr 2020
25. Government of Nepal, Ministry of Communication and Information Technology: Digital Nepal Framework 2019. Government of Nepal, Ministry of Communication and Information Technology (2019)
26. Shiang, J., Lo, J., Wang, H.-J.: Transparency in E-Governance. In: Proceedings of the 4th International Conference on Theory and Practice of Electronic Governance, pp. 268–273. Association for Computing Machinery, Beijing (2010)

SCADA Testbed Implementation, Attacks, and Security Solutions

John Stranahan, Tapan Soni, Jacob Carpenter, and Vahid Heydari(✉)

Rowan University, Glassboro, NJ 08028, USA
{stranahaj3,sonit9,carpentej8}@students.rowan.edu, heydari@rowan.edu
http://cybersecurity.rowan.edu/index.html

Abstract. Supervisory Control and Data Acquisition (SCADA) systems have been in use for decades. They provide remote management and monitoring capabilities for Industrial Control Systems (ICS) such as power plants, trains, water treatment plants, and dams. In recent years, SCADA systems have been targeted by malicious attackers. The Modbus TCP/IP protocol, which is the standard communication protocol used by many SCADA systems for network communication, is unencrypted and therefore it is insecure by design. In this research, cost-effective design and implementation of a custom SCADA testbed is proposed to assess common vulnerabilities and exploits in real-world Industrial Control Systems. A solution is then proposed to prevent these types of vulnerabilities from being exploited on real-world systems by implementing a secure tunnel.

Keywords: SCADA · Modbus · ICS · Testbed · Denial-of-Service · Man-in-the-middle · IPsec · VPN

1 Introduction

Supervisory Control and Data Acquisition (SCADA) systems are widely used for monitoring, controlling, and gathering data in Industrial Control Systems (ICS) such as power plants, train systems, and water treatment plants. With the implementation of these SCADA systems, operators can remotely manage and monitor the physical systems without having to be present at the site locations. In a SCADA system, there are three parts: the Human Machine Interface (HMI), the physical system that includes the Programmable Logic Controller (PLC) assembly and the physical devices being controlled, and the network which connects the physical systems to the HMI. The HMI displays data from the physical system in a human-readable format. Additionally, through the HMI, an operator can monitor and control the machines, sensors, and other industrial equipment in the physical system. The PLC in the physical system controls the machines via a pre-installed program, written with Ladder Logic, which is stored inside the memory of the PLC. In a typical implementation, both the HMI and the PLC are connected to a wide area network so that they can communicate

K. Arai (Ed.): FICC 2021, AISC 1363, pp. 736–755, 2021.
https://doi.org/10.1007/978-3-030-73100-7_53

with one another. Because of the network connection, the HMI does not have to be in the same building as the physical system. The HMI can be operated at remote management facilities, located in a different building, city, state, or even country. Since SCADA systems are designed for remote control and management of critical infrastructure systems; they are a prime target for cyber-attacks. A famous example of a cyber-attack on a SCADA system is Stuxnet. Stuxnet is a computer worm, developed to target PLCs that automate many industrial processes. Stuxnet propagates into the target system via infected USB drives and looks for Siemens Step 7 software on computers which are controlling the PLCs [17]. Once the target machine is found, the malware injects a rootkit into the PLC, the Siemens Step 7 software modifies the program code, and then it sends malicious commands to the PLC all while showing normal operation on the HMI. Another cyber-attack on industrial control systems was the Ukraine Power Plant attack [4,5]. In December of 2015, the Ukrainian power grid was compromised by a cyber-attack in which approximately 260,000 residents were left without power for a period of one to six hours. The attack started with a spear-phishing campaign that targeted the IT employees working in the power plants. Spear phishing is a form of electronic communication that targets specific people in an organization with the intent of stealing data or gaining access [15]. The emails were infected with a Trojan called BlackEnergy [1,5] which provided the attackers with VPN credentials to access the network of the power plants. After that, the attackers disabled the backup power and overwrote the serial to Ethernet firmware which is used to control the grid breakers. Then the attackers used the KillDisk software to delete the master boot record, the logs, and the software that was used to communicate with the grid breakers. In one final act, they performed a Denial-of-Service (DoS) attack on the call centers of the power plants after the attack. The sheer volume of residents affected by this attack is an indication of the severity of these types of attacks. Another type of threat against SCADA systems is malware that targets the safety instrumented system. Safety instrumented systems are the last line of defense against life-threatening disasters [19]. Triton [18] is a piece of malware discovered in 2017 which was specifically designed to disable and tamper with these systems. For the first time in the world of cybersecurity, there has been code which was explicitly designed to put lives at risk. In the Ukrainian power plant attack and the Stuxnet attack, the goal was to disrupt and cause harm to the systems, but in the case of Triton, the intent was purely malicious. It is crucial to model and examine the vulnerabilities in a controlled environment so that mitigation techniques can be developed, tested, and then implemented on vulnerable real-world systems. To put that plan into action, a custom SCADA testbed was designed and implemented [31] to assess the different types of attacks that can be conducted in the real-word [32][1]. The testbed was completed over six months and the total materials budget was less than $6,000.00. It is a good example for

[1] This paper is the extended version of the two conference papers with more details, the source codes, the list of materials for building the testbed, and a new security solution to prevent attacks.

other schools who are interested in having a cost-effective custom ICS testbed for research and educational purposes to sparks students' enthusiasm for learning and build a strong foundation that will lead to further interest in the subject matter. The source codes and the list of materials for building the testbed can be found on GitHub [3].

To model a real-world system, it is necessary to build a system that uses the Modbus TCP/IP protocol to communicate from the HMI to the physical system over a network that simulates a direct Internet connection. The Modbus TCP/IP protocol was designed in 1979 by Modicon (now Schneider Electric). Modbus TCP/IP is a variant of the Modbus protocol in which the messages are sent over a TCP/IP stack through the standard port 502 [11]. Modbus TCP/IP is a free and open protocol, meaning industrial manufacturers and companies can use the protocol in their own equipment without having to pay any royalty fees. Modbus TCP/IP can be used on equipment from many different vendors which has made it the standard communication protocol for the industry [23]. The Modbus TCP/IP protocol uses the master-slave communication structure. The device which is requesting information from other devices is called the Modbus Master, and the device(s) providing data back to the Master are called the Modbus Slave(s). The standard Modbus network can have up to 247 Modbus Slaves and 1 Modbus Master. Each of the Modbus Slaves has a unique address ranging from 1 to 247. The Modbus Master, in addition to receiving all the data from the Modbus Slaves, can also write to the Modbus Slaves [23]. However, only efficiency and reliability requirements were considered when designing the protocol; security was not a consideration. Modbus TCP/IP lacks these critical security features [33]:

1. No certification mechanism: In Modbus TCP/IP, there is no packet authentication. An attacker can set up a Modbus TCP/IP communication session and initiate a man-in-the-middle attack. A potential attacker only needs to intercept the packets, modify the data and forward the modified packets to the original destination address to send malicious commands to the PLC.
2. No privilege management: Modbus TCP/IP does not utilize privilege separation. In other words, any user can perform arbitrary functions without the permission of any other user. This is a high-security risk since any user can potentially send any command to a critical system that utilizes Modbus TCP/IP.
3. No encryption mechanism: Encryption is critical when transmitting critical information as it prevents third parties from reading the transmitted information. Modbus TCP/IP does not have any encryption, and as a result, all messages, commands, and addresses are passed in the packets as plain text which can easily be read by any network packet sniffer.

SCADA security research is very important because SCADA systems are the foundation of our modern lives. They control critical infrastructure systems such as power and water systems, both of which are crucial to every aspect of our lives. Without power, clean running water, and other critical services, our daily lives would be severely impacted. This is one of the many reasons why SCADA research is important because of the gravitas of these systems, how ingrained they are in our society, and because of the impact they will have if they are taken offline by a cyber attack.

The remainder of the paper is organized as follows. Section two reviews related work, section three details the design of the testbed, section four outlines the attack tools and scenarios, section five presents exploit mitigation techniques, and section six concludes the research and outlines future work.

2 Related Work

There are many different types of SCADA testbed implementations. The testbeds reviewed are primarily small scale physical models, virtually simulated testbeds or hybrid models which use a combination of simulated and physical parts.

2.1 Small Scale Physical Models

The National SCADA Test Bed (NSTB) [13] is a testbed which was designed in 2003. The NSTB was developed by multiple national laboratories in the Department of Energy to address different cybersecurity challenges and to conduct state-of-the-art research in critical vulnerabilities in SCADA systems.

The University of New Orleans built a small scale SCADA testbed [25] which consists of a power distribution model, a gas pipeline, and a wastewater treatment plant. Each site is connected to a PLC which is routed through a switch to the HMI. The testbed uses several different physical components in addition to a PLC including transformers, aerators, and an air compressor. The protocols used in this testbed are the Modbus protocol, EtherNet/IP protocol, and the PROFINET protocol.

The Critical Infrastructure Cybersecurity Laboratory at the University of Leon is a similar implementation of the SCADA testbed [29]. The lab took into account four different scenarios for flexibility and scalability: industrial control systems, building management systems, energy management systems, and smart city sensor networks. The industrial control systems use PLCs connected to input and output devices and they use the Modbus TCP, Ethernet/IP, and PROFINET protocols. The building management system consists of several integrated technologies including Zigbee, EnOcean, etc. The energy management subsystem is designed to emulate a real-world electricity grid with functionality such as distribution and use of electricity. The smart city subsystem uses physically distributed sensors to measure different variables in the lab.

Korkmaz et al. [30] describe a physical testbed designed to emulate a power station. The physical system is designed using two blower motors which are used to generate electricity. The HMI, designed using the Proficy HMI/SCADA - iFIX software, is deployed onto a computer with an LCD monitor. The testbed has a network which uses Rockwell Automation Company's EtherNet/IP Modules and RSLinx Enterprise.

2.2 Simulated Models

Morris et al. [26] proposed a simulated SCADA system using virtual simulation. By segmenting the parts of a SCADA testbed, it allows for the virtualization of the SCADA testbed implementation. This results in a reduction of size and cost for the testbed.

Chabukswar et al. [28] present a SCADA testbed simulation which uses the Command and Control WindTunnel application. The simulated testbed also utilizes Simulink and network simulators (OMNet++, etc.) in the implementation.

2.3 Hybrid Models

Del Canto et al. [27] developed an addition to the University of Leon's testbed. This was comprised of a hybrid system of virtual machines which controlled the physical PLCs and other equipment. A web interface was also developed for students to monitor the testbed and to perform vulnerability scanning with OpenVAS.

3 Design of the SCADA Testbed

Fig. 1. Physical model of the SCADA testbed.

The design of the SCADA testbed, seen in Fig. 1, went through many revisions. Initially, a traffic signal model implementation with moving cars was considered. The idea of a more complex industrial system such as a dam was entertained but it was decided that moving water was overly complicated to implement using the necessary plumbing. Therefore, a model of a nuclear power plant surrounded by a train system was finally decided. With the combination of these two industrial control systems, the effects of malicious cyber-attacks on two separate systems can be accomplished simultaneously without having to build multiple models.

3.1 Physical Model Design

The physical model is built on top of a 30-inch by a 36-inch wooden platform. The platform is mounted on top of a movable cabinet. All of the power and data cables run through the wooden platform into the lower cabinet for discrete functionality. As seen in Fig. 1, The physical model can be divided into two parts:

1. Nuclear Power Plant: The Nuclear Power Plant is the centerpiece of the SCADA testbed. It is made up of the central cooling tower assembly with a 3D printed cooling tower, an industrial office building, an external cooling fan assembly, and auxiliary industrial power plant buildings. To replicate the cooling tower functionality, a 110 V DC heater was installed inside the tower which sits on top of a 24 V DC fan. The purpose of the fan is to transfer the heat from the heater to the temperature sensor that is mounted above the heater. Inside the cooling tower, there is a red 110 V DC LED warning light which blinks whenever the heater is turned on. The power plant is enclosed inside a fence to separate it from the surrounding train system.
2. Train System: The Train system is the second industrial control system that was implemented in the SCADA testbed. It surrounds the nuclear power plant in an oval track. The track runs through a tunnel and past a 3D printed train station. The train system also includes an automatic railroad crossing which is activated when the train drives over infra-red sensors that are embedded in the track. The train is an N-Scale Kato USA Santa Fe USA Super Chief. The automatic crossing gate assembly is the North American Style Automatic Crossing Gate from Kato USA. The crossing gate includes infra-red sensors that detect the train and respond accordingly depending on which infra-red sensor is triggered. The crossing gate has lights which flash from side to side and alarms that go off whenever the train goes over the activating infra-red sensors located behind the nuclear cooling tower. The train station also has an infra-red sensor which signals the train to stop at the station.

3.2 PLC Control Panel

The PLC Control Panel consists of numerous modules that are responsible for capturing input signals and controlling output signals. Each of the components is mounted to the control panel using a 35 mm DIN rail mount. The components used are the following:

1. Circuit Breaker: The circuit breaker used on the PLC Control Panel is the Eaton FAZ-D15-1-NA-SP Miniature Circuit Breaker. This circuit breaker acts as a power switch for the PLC. Using a circuit breaker instead of a fuse is more beneficial for industrial control systems because they are more sensitive to faults, they make it easy to find fault locations and they make it easier to restore power.
2. 24 V DC Power Supply: The 24 V DC Power Supply used on the PLC Control Panel is the RHINO PSB24-060-P switching power supply. This powers the ZIPLink 24 V DC Sensor Input Module, the ZIPLink 12–24 volt DC Output Module and the ZIPLink Analog Input Module.

3. 12 V DC Power Supply: The 12 V DC Power Supply used on the PLC Control Panel is the RHINO PSB12-015-P switching power supply. This powers an Azatrax MRD1 model train detector that is connected to an infra-red sensor.
4. Direct Logic 205 PLC: The PLC used to control the testbed is the Direct Logic 205 Koyo D2-04BDC1-1. Connected to the PLC are input and output modules that are used to control the testbed.
5. Duplex-receptacle Outlet: This outlet drives power to the Kato Train power pack that controls the voltage, direction, and speed of the train.
6. ZIPLink 24 V DC Sensor Input Module: This 24 V DC Sensor Input Module takes in input from an Azatrax MRD1 model train detector that senses the presence of a train at a single location (the train station).
7. ZIPLink 12–24 V DC Output Module: The ZipLink 12–24 V DC Output Module has most of the connections to the physical model. The Kato Train power pack is wired into the module to allow the control of the direction of the train (forward, brake, and reverse). The tower fan, cooling fan, heater, and lights are connected to this module.
8. ZIPLink Analog Input Module: The ZipLink Analog Input Module has an analog to digital converter connected to it that is connected to a Resistance Temperature Detector (RTD) sensor.

3.3 The HMI Design

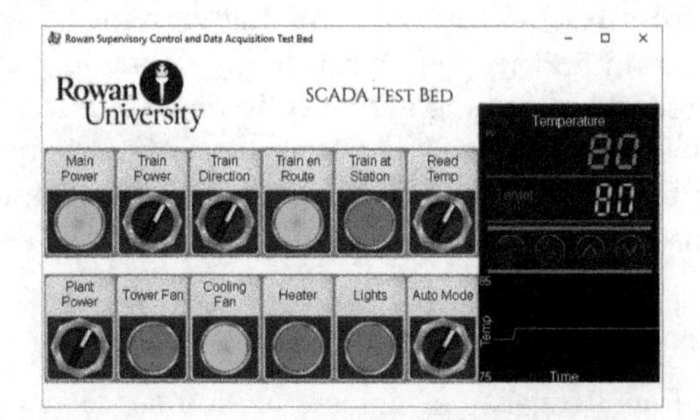

Fig. 2. HMI panel used in the SCADA testbed.

The HMI is the interface that the operator uses to control and monitor the SCADA testbed. In the testbed, the AdvancedHMI [8] software was used to build the HMI. AdvancedHMI offers robust HMI solutions to the user with rich graphics and an array of switches, monitoring screens, and controls, all in an easy to program Visual Studio environment. Using the Visual Studio Interactive Development Environment, the user can drag and drop different components onto the screen and link each button to its corresponding output in the ladder logic program code. Therefore, if a toggle light button is pressed then light is toggled either on or off, and users can see that change graphically on the HMI. The AdvancedHMI software is loaded onto a Raspberry Pi 3 which is connected to a 7-inch touch screen to emulate a real-life control panel. The HMI is then connected to the network to remotely monitor and control the physical system. Figure 2 shows the HMI panel used to control the SCADA testbed. The button operations from the top left to the bottom right are as follows:

1. Main Power (Toggle Indicator Button): This button controls the power to the entire PLC assembly. When the color is GREEN (on), it powers the train assembly and the nuclear power plant assembly. When the color is RED (off), all power is cut from the inputs and outputs.
2. Train Power (Two-Way Selector Switch): This selector switch controls the main power for the train assembly only. When the switch is in the RIGHT position (on), it allows power and control to the train assembly. When the switch is in the LEFT position (off), no power can flow through the train assembly. This toggle switch does not affect the nuclear power plant at all.
3. Train Direction (Three-Way Selector Switch): This three-way toggle switch controls the direction of the train. The train has three statuses; Forward, backward, and stop. When the switch is in the RIGHT position, the train is going forward. When the switch is in the MIDDLE position, the train stops. When the switch is in the LEFT position, the train goes in the reverse direction.
4. Train en Route (Status Indicator Light): This status indicator light shows the operator if the train is traveling to the train station. When the switch is in the LEFT position (off), the train is at the train station. When the switch is in the RIGHT position (on), the train is going towards the train station.
5. Train at Station (Status Indicator Light): This status indicator light shows the operator if the train has stopped at the station or not. When the color is GREEN, it means that the train has stopped at the train station and will resume normal functionality after 5 s. When the color is RED, it says that the train has left the train station.
6. Read Temp (Two-Way Selector Switch): This selector switch enables the temperature sensor to read, scale, and filter the data read from the temperature sensor if the plant power is on. This helps avoid interference and temperature spikes.
7. Plant Power (Two-Way Selector Switch): This selector switch controls the power for the plant only. When the switch is in the RIGHT position (on), it allows power and control to the Power Plant assembly. When the switch is in the LEFT position (off), no power can flow through the Power Plant assembly. This toggle switch does not affect the train system.
8. Tower Fan (Toggle Indicator Button): This button controls the operation of the fan which is mounted inside the cooling tower assembly. When the color is GREEN (on), the fan is turned on. When the color is RED (off), the fan is turned off.
9. Cooling Fan (Toggle Indicator Button): This button controls the operation of the cooling fan that is mounted outside the cooling tower. It provides cold air to reduce the temperature of the cooling tower once the temperature

crosses a predefined threshold. When the color is GREEN (on), the fan is turned on. When the color is RED (off), the fan is turned off.

10. Heater (Toggle Indicator Button): This button controls the operation of the heater. The heater is mounted on top of the tower fan. When the color is GREEN (on), the heater is powered and heating the surrounding air. When the color is RED (off), the heater is turned off and the cooling fan is turned on to cool the tower.
11. Lights (Toggle Indicator Button): This button controls the operation of the building lights. The lights are mounted in each of the buildings and the surrounding terrain. When the color is GREEN (on), the lights are on. When the color is RED (off), the lights are off and not powered.
12. Auto Mode (Two-Way Selector Switch): This toggle switch controls the mode for the nuclear power plant. When the switch is in the RIGHT position (auto mode), the heater, tower fan, cooling fan, and the lights turn on and off automatically based on the built-in ladder logic. When the switch is in the LEFT position (manual mode), the heater, tower fan, cooling fan, and the lights must be manually turned on or off. The user can also change the temperature threshold from the HMI.
13. Temperature (7 Segment Display): This 7-segment display shows the current temperature reported by the temperature sensor.
14. Target Temperature Selection (Temperature Controller): The target temperature can be incremented and decremented by the user to change the temperature that will cause the cooling fans to activate. This is available when the Auto Mode is on or off.
15. Temperature Graph (Basic Trend Chart): The graph shows the temperature data read from the sensors over time. This allows the user to see the temperature relative to past temperature readings within a small time frame.

3.4 The Network Architecture

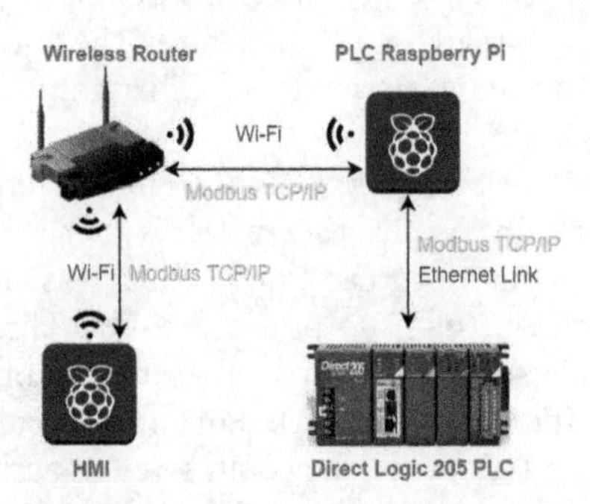

Fig. 3. Network architecture of the SCADA testbed.

The network architecture of the SCADA testbed is shown in Fig. 3. The HMI and the PLC Raspberry Pi communicate wirelessly through the router. The PLC

Raspberry Pi communicates to the Direct Logic 205 PLC via an Ethernet link. The PLC is assigned a static IP address so that the HMI can map it's controls to the PLC.

3.5 Normal Functionality of the SCADA Testbed

It is important to define the normal operations of the SCADA testbed in order to show how compromising the network changes the functionality of the system.

Train System Functionality: The normal functionality of the train starts with the main power switched to the ON position. When the main power and train power are both switched on, the train starts to move on the train track. When the train triggers the infra-red sensor embedded in the track behind the nuclear cooling tower, the crossing gate lights begin to flash, and the crossing gate arms come down. The crossing gate alarms also go off indicating to the user that the train is about to cross. Once the train successfully passes through the crossing gate and the last crossing gate infra-red sensor (embedded in the track after the crossing gate arms), the crossing gate arms rise, the lights stop flashing, and the alarm stops. After the train passes the crossing gate, it triggers the train station infra-red sensor. Three seconds after the train triggers the train station infra-red sensor the train stops so that the passenger car stops directly in front of a train station. After five seconds, the train starts to move again and heads for the first crossing gate sensor located behind the nuclear power plant. As discussed in the HMI Design section, the train direction and power can be controlled from the HMI remotely.

Nuclear Power Plant Functionality: The normal functionality of the nuclear power plant starts in the same way as the train system, with the main power switched to the ON position, and the nuclear power plant powered on. If the temperature of the heater inside the cooling tower is below 80° Fahrenheit (or whatever the user sets it to), then the heater is turned on, and it keeps heating until the temperature goes above the target value (by default it is set to 80° Fahrenheit). There is a temperature sensor directly on top of the heater which measures the temperature and sends that data to the PLC. A red LED light is placed inside the cooling tower and it flashes whenever the heater is turned on. When the temperature goes above the target value, a fan which is mounted underneath the heater, as well as the external cooling fan, are both turned on. These fans cool the heater until it falls below the set limit and then resumes heating to the target value again. There are LED lights inside the industrial buildings that can be controlled by the user in manual mode. In automatic mode, the LED lights turn on when the heater is on and turn off when the fans are cooling the tower. As discussed in the HMI Design section, the tower fan, the heater, the external cooling fan, and the lights can all be controlled remotely from the HMI.

4 Attack Tools and Scenarios

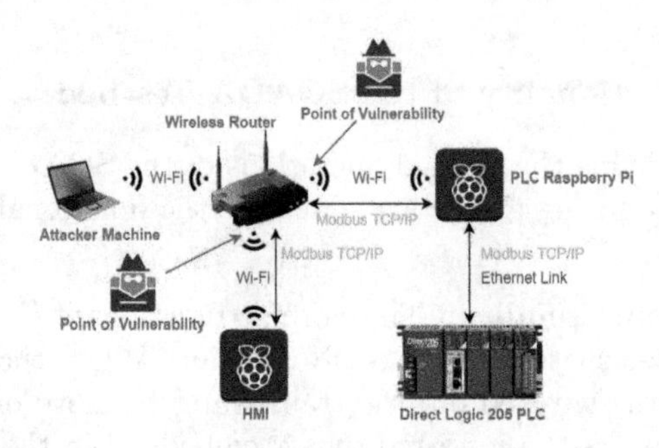

Fig. 4. Points of vulnerability in the SCADA testbed's network.

In this section, the various tools and attacks used against the SCADA testbed are described. First, it is necessary to understand where the points of vulnerability in the system are. In Fig. 4, the points of vulnerability are shown in the network. The points of vulnerability in the SCADA network are between the wireless router to the PLC Raspberry Pi and the wireless router to the HMI. These points of vulnerability can be exploited by an attacker because the Modbus TCP/IP protocol lacks certification, privilege management, and encryption. An attacker can set his/her own Modbus TCP/IP session and send malicious commands to the PLC. The attacker can also use a network sniffing tool to monitor the plain text data within Modbus TCP/IP packets.

4.1 Attack Tools

When penetration testing the SCADA testbed, open source tools like Wireshark, Ettercap, Metasploit, and Nmap are useful for active reconnaissance and exploitation.

Wireshark: Wireshark [24], developed by The Wireshark Team, is a free and open-source program that is used to sniff and analyze packets. Because the Modbus TCP/IP protocol has no security or encryption, this tool allows attackers to completely dissect such packets for further analysis and use the gathered information for active attacks.

Ettercap: Ettercap [6] is a utility included in Kali Linux that is used to execute man-in-the-middle attacks on networks [6]. It includes features such as sniffing for live connections, content filtering on the fly, active and passive dissection of protocols and many other features for network and host analysis [6]. By using

Ettercap, the attacker can take down an entire network or manipulate information being sent from a host to host. In the SCADA testbed implementation, Ettercap was used to perform an Address Resolution Protocol (ARP) poisoning attack.

Metasploit: Metasploit [7] is a beneficial penetration testing tool that is used for hacking into systems for testing and security purposes. Metasploit is a working collection of exploits that is available to anyone. By using this tool, there are multiple data manipulation and scanning attacks that can be performed on the PLC that impact the safe operation of our SCADA testbed. The malicious attacks possible on our system with Metasploit are fabrications, interceptions, interruptions, and modifications.

Nmap: Nmap [14] is a network mapper that is used to scan for devices on a network. It is possible to run many types of scans which range from stealthy to highly detectable. Nmap also has powerful and flexible scripting capabilities where users can write their own scripts which can be used for network discovery, vulnerability detection and exploitation, and more.

4.2 Scanning/Discovery

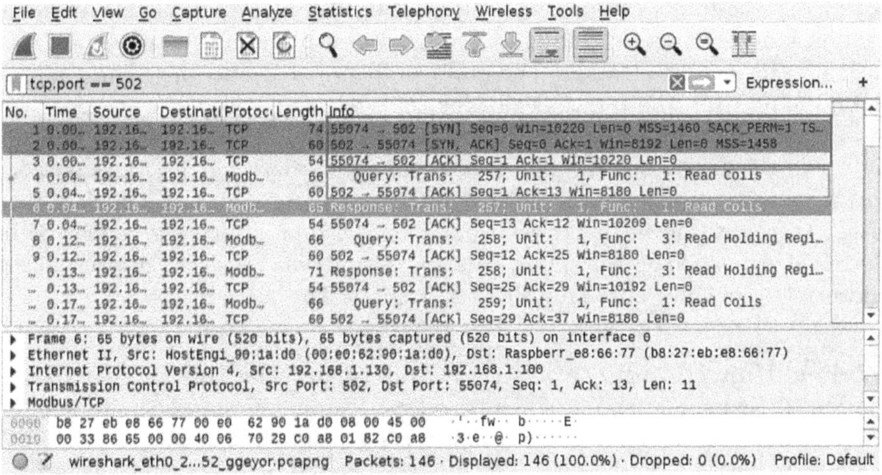

Fig. 5. Captured Modbus TCP/IP 3-way handshake.

Wireshark Sniffing: When using Wireshark, the user can see traffic on the network related to the PLC and strictly view Modbus TCP/IP packets when the port number is filtered to port 502. When Wireshark is running and a command is sent to the PLC for the first time, the TCP 3-way handshake followed by the first Modbus TCP/IP packet is captured. After this occurs, the first Modbus TCP/IP packet is sent. If a READ_COIL query is sent to the PLC, Wireshark will allow the user to see the contents of the packet such as the unit identifier, function code, and bit count. With the Direct LOGIC 205 Koyo PLC, the unit identifier is always "1" and the bit count is the number of bits that it is asked to read. The function code can vary depending on the function requested, and the READ_COIL function code is 01. As shown in Fig. 5, the 3-way TCP handshake is shown in the red box. Inside the orange box, is the query to the PLC which is the Read function. After the "Query", the "Response" is sent which contains the actual data that is read. The different coil values can be read in plain text because Modbus TCP/IP does not have an encryption mechanism to encrypt data which is being passed [33].

Modbus Version Scanner: The Modbus Version Scanner [12] is a Metasploit module that detects if the target address uses the Modbus TCP/IP service. The module is found in the Metasploit framework console in the "auxiliary/scanner/scada/modbusdetect" module. The scanner identifies if the device employs the Modbus TCP/IP protocol by sending a Modbus TCP/IP request to it and verifying if the endpoint returns with the same transaction-id, and protocol-id. The transaction identifier allows the synchronization between messages of a server and client. When the Modbus Version Scanner runs, it sends a query to the server and checks the transaction identifier of the query is sent as well as the transaction identifier of the response it receives - and if they do not match, the target does not use the Modbus TCP/IP protocol. The protocol identifier must match as well. It is composed of a block of 2 bytes, 0x00 0x00, that are sent by the attacker. The protocol identifier in the query must match the response from the target device for the scanner to tell the attacker if the device uses the Modbus TCP/IP protocol. When a PLC is connected to the Internet, the device has the potential to be exposed to all the same security threats that a computer will. This tool allows attackers to locate devices that utilize this protocol. In the attack simulation, once an attacker is connected to the SCADA testbed network, they can run this module on all the devices on the network to see which ones are using the Modbus TCP/IP protocol.

Modbus Unit ID and Station ID Enumerator: The Modbus Unit ID and Station ID Enumerator [21] is a Metasploit module that checks the unit ID of the PLC endpoint. It is found in the "auxiliary/scanner/scada/modbus_findunitid" module. In most cases, the correct device is addressed by referencing the devices IP address without the need for the unit ID. However, in cases where multiple PLCs share the same IP address, the unit ID must be set to distinguish the devices. When the unit ID is 0, all Modbus TCP/IP slave devices can accept that message. The attacker can use the "UNIT_ID_FROM and UNIT_ID_TO" variable to select the range of the search (within 1 to 254). The BENICE variable allows the hacker to set the number of seconds to wait between probes. Once the attacker knows what Unit Ids are valid, they can attack specific devices rather than every device using the IP address.

Nmap - Modbus-Discover.Nse: Nmap, as mentioned above, has a script available by the name of Modbus-discover.nse [10]. It is a vulnerability and port scanner that enumerates SCADA Modbus TCP/IP slave IDs and collects device information. The Modbus-discover.nse script tries to find devices that use the Modbus TCP/IP service and displays their device information. Users are able to determine the model of the enumerated devices. After the execution of the Nmap script, the MAC address is shown and using the MAC address, the user can discover more attributes about the PLC.

4.3 Data Manipulation

Modbus Client Utility: When running Metasploit, the attacker can use the Modbus Client Utility module [21]. It is located at "auxiliary/scanner/scada/modbusclient". This module allows the attacker to read and write to the coils and registers of the PLC using the Modbus TCP/IP Protocol. This is achievable once the attacker is connected to the same local network. The module options that can be set are: RHOST (the target address), RPORT (the target port), UNIT_NUMBER (Modbus Unit Number), DATA_ADDRESS (Modbus Data Address), and DATA (Data to Write). The module actions that can be set are: Read Coil, Read Coils, Read Register, Read Registers, Write Coil, Write Coils, Write Register, and Write Registers. This allows the attacker to have full control over the Power Plant and Train System.

(i) Power Plant: In the power plant system of the SCADA testbed, an attacker can read the values of every coil and register and see whether there is the power running to the mode selection, heater, cooling fans, and lights. If the coil for the heater reads a zero that means it is off, if the coil for the mode selection is 0 then the manual mode is activated. The attacker can configure what components they want to have on or off. The HMI will show these changes as they occur, and the HMI operator can fix the changes; therefore this attack is not extremely detrimental unless combined with different attacks.
(ii) Train System: In the train system of the SCADA testbed, an attacker can read the coil values of the Train Forward, Train Brake, Train Reverse, Train Station Stop, and Train Crossing Signal. The attacker can change the values of the coils to direct the train to stop, go forward, go backward, disable the infra-red sensor that stops the train at the station, and the turn off the power to the crossing signal.

4.4 Man-in-the-Middle Attacks

ARP Poisoning with Ettercap: Address Resolution Protocol (ARP) Poisoning is a type of network attack in which a malicious attacker sends fake ARP packets over a LAN [16]. By broadcasting fake ARP packets, the attacker can link their MAC address with the IP address of a legitimate host on the network which results in data being received by the attacker instead of the target host [16]. ARP Poisoning allows attackers to intercept, modify or stop data packets which are in transit [16].

To perform this attack on the SCADA testbed, the Ettercap utility in Kali Linux was used. By scanning the different hosts on the network interface, the attacker is able to target the HMI and the router. The attacker is then able to drop packets that are caught in transit between the HMI and the PLC. With this type of attack, the sensor data from the PLC can be controlled, modified, and forwarded to the HMI. The physical systems could be malfunctioning, but the HMI will not receive any warning messages or abnormal register values because the attacker is blocking or modifying them. Figure 6 shows how ARP Poisoning was implemented on the SCADA testbed's network.

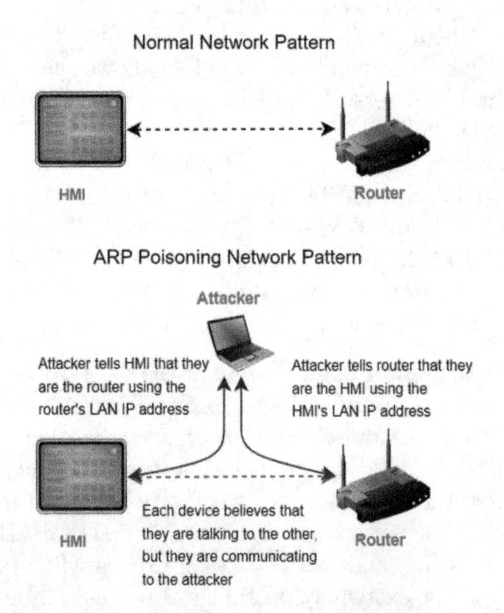

Fig. 6. ARP poisoning implementation on the SCADA testbed's network.

4.5 Denial-of-Service Attacks

TCP SYN Flood Attack: A TCP SYN Flood Attack [21] is a type of Denial-of-Service (DoS) attack that exploits the method of setting up a TCP/IP connection over an Internet Protocol based network. This attack consumes resources on the target PLC, makes the HMI unresponsive to reading and writing, and causes the PLC to become unresponsive and ultimately reboot. Using Metasploit's auxiliary module called "auxiliary/dos/tcp/synflood", an attacker can send TCP connection requests faster than the PLC can process them. The attacker uses the module to send synchronize messages to port 502 on the PLC to establish communication. Next, the PLC sends a synchronize-acknowledge message back to the attacker, or a different IP address entirely. Lastly, the PLC waits for an acknowledgment message back from the device looking to connect, yet it never receives it. Before the connection request times out, the PLC is sent another synchronize message, and so on, to create more half-open connections. Eventually the overflow tables will fill, and this causes the connection to the HMI or other legitimate devices to be denied, and subsequently crash the system [22].

(i) The Nuclear Power Plant in the SCADA testbed can be attacked using this method and cause the system to change to an unwanted state. Before running the TCP SYN flood attack, the attacker can modify the coils that control the logic of the mode selection, heater, cooling fans, and lights. First, the attacker can turn on the manual mode to turn off the automatic mode that keeps the temperature at a safe level. Once in manual mode, the values of the coils in the system can be altered to what the attacker would like. For example, the attacker can now turn on the heater, turn off the cooling fans and turn off the lights. The temperature will increase beyond the limit that would normally shut off the heater and turn on the cooling fans. Once the heater is turned on, the attacker can run the TCP SYN flood attack. The PLC and HMI are now unresponsive, and the operator is unable to fix the issues by pressing the buttons on the HMI. The HMI can now no longer write to the coils of the PLC or show the state of each component of the Power Plant. The temperature will rise above the threshold and the cooling fan will not start.

(ii) The Train System in the SCADA testbed can be attacked and create dangerous conditions. The coils in the PLC that can be modified are the Train Forward, Train Brake, Train Reverse, Train Station Stop, and Train Crossing Signals. For example, the attacker can reverse the train, turn off the train station infra-red sensor, and turn off the train crossing signals. Once these values are changed the attacker can run the TCP SYN flood attack. Now the HMI operator cannot control the train from the HMI.

5 Exploit Mitigation

Internet Protocol Security (IPsec) is a suite of protocols which ensures private communication over a network [9]. IPsec provides a means to protect data authenticity, integrity and confidentiality [9]. Data authenticity is accomplished by means of authentication and integrity. Authentication verifies that data comes from a trusted source and integrity is the assurance that the data is accurate and consistent. Confidentiality provides protection from network sniffers and other parties attempting to monitor the traffic between two hosts on a network. IPsec, as the name implies, is implemented in the network layer; therefore it is transparent to the application layer which allows it to be application agnostic [9].

In the SCADA testbed, the OpenVPN protocol is used to handle IP security. OpenVPN is a portable, open-source, cross-platform package that is used to create a Virtual Private Network (VPN) between two systems [20]. OpenVPN is used to provide secure communications between the HMI and the PLC. As shown in Fig. 7, an IPsec tunnel was implemented between the HMI and the PLC Raspberry Pi. To create the tunnel, the PLC Raspberry Pi was configured as the VPN server and the HMI was configured as the VPN client to create a persistent connection to the server. Once the VPN is working, the Modbus TCP/IP packets from the HMI are encrypted and passed through the IPsec tunnel to the OpenVPN server where they are decrypted and forwarded to the PLC via the attached Ethernet link. In this IPsec implementation, if an attacker attempts to sniff packets for information, they would only get the encrypted packets sent between the HMI and the OpenVPN server. The Modbus TCP/IP packets are only sent as plaintext via a direct, hardwired connection between the OpenVPN server and the PLC. For additional traffic management, iptables are used to configuring the network traffic rules. From the HMI all incoming and outgoing traffic is routed through the IPsec tunnel. On the PLC Raspberry Pi, all incoming traffic from the VPN tunnel is routed to the Ethernet interface that is connected to the PLC. All outgoing traffic from the PLC is routed via the Ethernet link to the PLC Raspberry Pi where it is routed back through the VPN tunnel to the HMI. By using these rules, the network traffic is isolated from third parties on the network by enforcing the data path of the network.

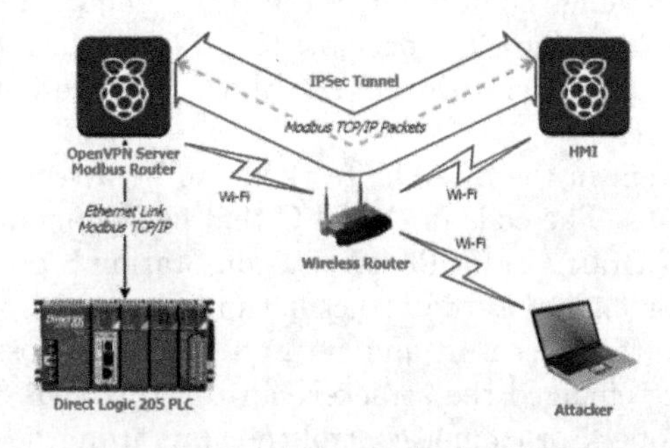

Fig. 7. IPsec implementation in the SCADA testbed.

5.1 Scanning/Discovery

Now that the IPsec VPN Tunnel is implemented, using Nmap to run the modbus-discover.nse script against the PLC shows that the host is down which means that the attacker can no longer get device information from the PLC such as the MAC address, open port numbers or what brand of PLC the device is. Some attacks are specific to certain brands of PLCs due to different system architecture. For example, an attack that was made for a Seimens PLC may not necessarily work on an Allen-Bradley PLC. Prevention of this attack increases the confidentiality of device information.

5.2 Data Manipulation

Using the auxiliary "scanner/scada/modbusclient" module in Metasploit to overwrite the values in the PLC does not work when the IPsec tunnel is implemented as the PLC located at the target IP address is no longer accessible to the attacker because the only way to access the PLC is through the tunnel, which is encrypted and only the HMI has access to it. Previously an attacker could change the direction of the train, disable the train station sensors, turn off the cooling fans in the tower, and more: but now data manipulation is not possible because the PLC is now inaccessible by unauthorized parties. Additionally, the data is encrypted, so decryption is not possible without the correct key and altering the data is not possible because IPsec protects against data manipulation by providing data integrity.

5.3 Denial-of-Service

Although the IPsec tunnel provides confidentiality, authenticity, and integrity [9], it does not protect a node from a DoS attack. When running a DoS attack with the tunnel implementation, the HMI and the OpenVPN server, although unreachable, can still be overwhelmed with data packets and become unresponsive for a certain amount of time.

5.4 Man-in-the-Middle

With the IPsec tunnel implementation, the attacker can no longer read and inject malicious data between the router and the HMI or the OpenVPN server. Even though the attacker can still drop packets using filters in Ettercap and perform a DoS attack, decryption of the data and modification of the data is not possible because the data is encrypted and IPsec provides authentication and integrity to the network [9].

6 Conclusion and Future Work

In this paper, a custom SCADA testbed is proposed for research and education. SCADA systems are used throughout the world and they play a critical role in monitoring and controlling critical infrastructures such as power plants, train systems, and water dams. Because of their essential role and unencrypted communication protocol Modbus TCP/IP, SCADA systems are a prime target for malicious hackers. In this research, cost-effective design and implementation of a custom SCADA testbed is proposed. Several types of attacks have been performed on the testbed including denial-of-service, packet sniffing, ARP poisoning, man-in-the-middle, and data manipulation. For preventing these types of attacks, an IPsec VPN tunnel has been integrated into the testbed's network architecture. By implementing the IPsec VPN tunnel, the impact of these types of attacks is significantly reduced.

The most important part of the future work is to develop a moving target system which prevents system attacks, including Denial-of-Service attacks by creating a moving target. The moving target is always harder to hit than one that is standing still. Taking a system with a static IP address and continuously moving that system by way of changing its IP address would make it harder to attack. Changing the IP address of a target system at regular intervals, say every five minutes, will add several more layers of security to protect the transfer of data. Additionally, cracking a 2048-bit SSL certificate would require the factorization of a 617 digit number. It is estimated that it would take a standard desktop computer 4,294,967,296 $\times$ 1.5 million years, or about 6.4 quadrillion years to perform the required calculations [2]. If a new certificate is created every time an address change occurs, not much time is allowed for any effort to break the encryption mechanisms. The change in IP address would also inhibit an attacker from targeting the system for very long before that system "moves" again.

References

1. BlackEnergy APT attacks | what is BlackEnergy? | kaspersky lab US. Accessed 23 Feb 2019
2. Check our numbers. Accessed 29 Mar 2019
3. Code for the HMI and PLC. Accessed 29 Mar 2019
4. E-ISAC_sans_ukraine_duc_5.pdf. Accessed 23 Feb 2019
5. EBOOK_cyberattacks-AGAINST-UKRAINIAN-ICS.pdf. Accessed 23 Feb 2019
6. Ettercap home page. Accessed 23 Feb 2019
7. Getting started - metasploit. Accessed 23 Feb 2019
8. HMI software by AdvancedHMI, application creation framework. Accessed 23 Feb 2019
9. Introduction to cisco IPsec technology - cisco. Accessed 23 Feb 2019
10. modbus-discover NSE script. Accessed 23 Feb 2019
11. Modbus TCP/IP overview. Accessed 23 Feb 2019
12. Modbus version scanner | rapid7. Accessed 23 Feb 2019
13. National SCADA test bed
14. Nmap: the network mapper - free security scanner. Accessed 23 Feb 2019
15. Spear phishing definition and prevention kaspersky lab US. Accessed 23 Feb 2019
16. Spoofing attack: IP, DNS & ARP. Accessed 23 Feb 2019
17. Stuxnet. Accessed 23 Feb 2019
18. Trisis malware. Analysis of safety system targeted malware. Accessed 10 Mar 2019
19. Triton is the world's most murderous malware, and it's spreading. Accessed 10 Mar 2019
20. VPN software solutions & services for business. Accessed 23 Feb 2019
21. Vulnerability & exploit database | rapid7. Accessed 23 Feb 2019
22. What is a TCP SYN flood | DDoS, attack glossary | incapsula. Accessed 23 Feb 2019
23. What is modbus and how does it work? Accessed 23 Feb 2019
24. Wireshark · go deep. Accessed 23 Feb 2019

25. Ahmed, I., Roussev, V., Johnson, W., Senthivel, S., Sudhakaran, S.: A scada system testbed for cybersecurity and forensic research and pedagogy. In: Proceedings of the 2Nd Annual Industrial Control System Security Workshop, ICSS 2016, pp. 1–9. ACM, New York (2016)
26. Alves, T., Das, R., Werth, A., Morris, T.: Virtualization of scada testbeds for cybersecurity research: a modular approach. In: 2015 Joint International Mechanical, Electronic and Information Technology Conference (JIMET-15). Atlantis Press (2015)
27. Del Canto, C.J., Prada, M.A., Fuertes, J.J., Alonso, S., Domínguez, M.: Remote laboratory for cybersecurity of industrial control systems. IFAC-PapersOnLine **48**(29), 13–18 (2015)
28. Chabukswar, R., Sinopoli, B. Karsai, G., Giani, A. Neema, H., Davis, A.: Simulation of network attacks on SCADA systems. In: First Workshop on Secure Control Systems, Cyber Physical Systems Week 2010, April 2010
29. Domínguez, M., Prada, M.A., Reguera, P., Fuertes, J.J., Alonso, S., Morán, A.: Cybersecurity training in control systems using real equipment. IFAC-PapersOnLine **50**(1), 12179–12184 (2017)
30. Korkmaz, E., Dolgikh, A., Davis, M., Skormin, V.: Industrial control systems security testbed (2016)
31. Stranahan, J., Soni, T., Heydari, V.: Supervisory control and data acquisition testbed for research and education. In: 2019 IEEE 9th Annual Computing and Communication Workshop and Conference (CCWC), pp. 0085–0089, January 2019
32. Stranahan, J. Soni, T., Heydari, V.: Supervisory control and data acquisition testbed vulnerabilities and attacks. In: SoutheastCon 2019 (to appear), April 2019
33. Qu, W., Wei, W., Zhu, S., Zhao, Y.: The study of security issues for the industrial control systems communication protocols. In: 2015 Joint International Mechanical, Electronic and Information Technology Conference (JIMET-15). Atlantis Press (2015)

D3CyT: Deceptive Camouflaging for Cyber Threat Detection and Deterrence

Kuntal Das(✉), Ellen Gethner, Ersin Dincelli, and J. Haadi Jafarian

University of Colorado Denver, Denver, CO, USA
{kuntal.das,ellen.gethner,ersin.dincelli,haadi.jafarian}@ucdenver.edu

Abstract. Even the most secure cyber systems could be compromised, and their data could be stolen. Once the data is stolen, even if it is encrypted or hashed, the attackers can conduct offline brute-forcing on it to recover the plaintext without being disrupted or detected. In this paper, we propose D3CyT, a simple, deceptive approach to camouflage sensitive data against such data thefts. In this approach, we transform a sensitive data value (which could be encrypted or hashed) to a deceptive value, called honeyvalue. The honeyvalue is stored instead of the original value, and the key to retrieve the original value from the honeyvalue is stored on a dedicated and secure server. If the data is stolen, the adversary would only attain the honeyvalues. The honeyvalues would either dissuade the attackers from using them by making stolen data look unimportant, or enable detection of data theft in case the attacker uses them. Through three different case studies focused on camouflaging passwords, QR codes, and logged IP addresses, we show the broad usability of our approach in different domains. We also show that even if our deception fails, the system is still technically more secure and computationally as secure as the original system.

Keywords: Data theft · Camouflage · Deception · Proactive defense · Password security

1 Introduction

Computer systems and networks can never be assumed to immune to compromises. In the cybersecurity arms race, adversaries are constantly inventing new ways to intrude into systems, ranging from exploiting zero-day vulnerabilities to social engineering attacks. Recent data breaches like the Equifax breach with 147.9 million consumers data stolen or Canva breach resulting in the disclosure of 137 million user accounts are just a few examples of the prevalence and the grave consequences of such data thefts [10].

This is why cryptographic systems, as one of the main building blocks of security, are used in cyber systems to ensure secrecy of stored data. These approaches transform plaintexts into arbitrary values, using a cryptographic hash, or a keyed

K. Arai (Ed.): FICC 2021, AISC 1363, pp. 756–771, 2021.
https://doi.org/10.1007/978-3-030-73100-7_54

cipher. This ensures that if the data is stolen, its secrecy is not (immediately) violated.

However, these ciphertexts, once obtained by an attacker, are very distinguishable from a plaintext, and reveal information about the underlying cryptographic system (for example, if the length of a hashed value is 160 bits, the used cipher is probably SHA-1). This encrypted nature of data immediately incentives attackers to try breaking the cipher or reversing the hashed value. While the cryptanalysis attacks on modern ciphers and cryptographic hash functions are not viable, with the growing available computational power, brute-forcing is becoming a major risk. This threat is more serious when the brute-force could be performed offline or away from the site; *e.g.*, when a database of hashed passwords is leaked [1]. In this case, attackers can use various techniques like high-performance cluster computing (HPCC) or on a smnaller scale: brute-force using multiple cores and threads running in parallel or by simply using online lookup tables [1] like rainbow tables, to invert the hashed passwords [16]. Since the attack happens away from the source or offline, the site owners have no mechanism to protect their data against them.

Motivated by *cyber deception* paradigms and systems [3], in this paper, we propose, D3CyT, a novel approach for bridging this gap by wrapping a deceptive layer around sensitive, and potentially encrypted/hashed, values. Deception refers to a controlled act to feed information that deliberately misleads the enemy decision-makers. It is identified as a core cyber defense tactic, and has been widely studied in the context of physical (military) conflict [3,24] and cybersecurity extensively [18,21]. A well-known cyber deception technique called *camouflaging*, is based on disguising sensitive information or objects by changing their appearance [3]. In the literature of cybersecurity, several works used camouflaging techniques for disguising critical assets or parameters of the system [7,19,20,22].

In this paper, we propose a novel camouflaging approach to enable the detection of stolen sensitive data. Our approach uses a special property of XOR that could be used for camouflage:

$$(p \oplus q = k) \leftrightarrow (q \oplus k = p) \tag{1}$$

Building on this property of XOR, D3CyT works as follows: to camouflage a real value p (which could be encrypted, hashed, or even hashed with a salt), we first select a camouflaging value q. We call this value *honeyvalue*. Then, we do a bit-wise XOR operation between p and q to generate a camouflaging key ($camo\_key = p \oplus q$). This key is stored on a dedicated server, called *camouflaging key server* or key server for short. All computer systems can communicate with the key server through encrypted and authenticated channels, and the key server has extensive instrumentation to detect anomalies and intrusions of various sorts.

The goal of this camouflage is to disguise the real value, the underlying cipher system and even the fact that the data is encrypted, thus either (1) dissuading attackers by making the data look unimportant, or (2) detecting the attack if the attacker believes that the honeyvalue is a real (truthful) value.

For legitimate users, to de-camouflage a honeyvalue q to its original value p, the system first retrieves the camouflaging key for q from the key server (*camo_key*). Then, the system conducts a bit-wise XOR between the camouflaging key and q to retrieve the original value, p ($p = camo\_key \oplus q$).

For malicious actors, three adversarial scenarios may occur when facing this honeyvalue:

- The attacker believes our camouflaging and wrongly concludes that the honeyvalue is real, and continues the attack based on this. In this case, we have the potential to detect the attacker. For example, suppose the sensitive data item is an IP address that has been camouflaged to a honeypot-monitored IP address (see case studies in Sect. 4). Then, once the attacker buys our deception and probes this honey IP, they could be trapped in the honeypot and potentially identified and characterized.
- The attacker suspects that since the data is not encrypted, then it could be decoy data that has been retrieved from a honeypot. For example, assume in a Web application, a hashed user password is camouflaged as a honeypassword that is in plaintext (see Sect. 4). Once the user data is compromised, the attacker may suspect that the data is fake since the user passwords are not hashed. This potentially dissuades the attacker from acting upon this information and continuing the attack.
- The attacker discovers our deception scheme and realizes that the data is camouflaged. In this case, the problem is not any simpler than when the conventional approaches are used: the attacker has to compromise the highly-secured and dedicated key server to discover both the camouflaging key; then since the real value is encrypted/hashed, brute-force the underlying cipher as before to retrieve the unencrypted real value. Therefore, the compromise of the camouflaging key server at worst only reduces security to the level it was at before the deployment of our camouflaging approach.

The rest of the paper is organized as follows. Section 2 discusses the related work. In Sect. 3 we introduce our architecture and algorithms for camouflaging. Section 4 proposes three motivating use cases for our proposed approach, called D3CyT. Section 5 evaluates the overheads of the approach and discusses the main challenges in its adoption for new domains. Section 6 concludes the paper.

2 Related Work

Deception is a core tactic for cyber defense [12,21]. Defense techniques based on the notion of cyber deception have long served as a robust defensive layer in the defense-in-depth strategy of enterprises and networks [11,14]. The goal of these deception-based techniques is to detect attackers and characterize them or deter their attacks by distracting them from real targets. A quintessential example of deception systems are honeypots, which are computing resources that are placed in sensitive locations in the computing system or networks to be probed, attacked, and compromised by the attackers [18].

Deception tactics are broadly classified into A-Type and M-Type [3]. A-type or ambiguity-increasing tactics are designed to create confusion and to distract an adversary by making noise, while M-type or misleading variety deception is more ambitious in that it is designed to mislead an enemy into believing a specific deception plan [3]. Camouflaging is a well-known M-type deception tactic [3] that is used for distracting adversary's attention away from the area of interest or attack [3]. Camouflaging could be passive or active. In passive camouflaging, we disguise real capabilities to prevent their detection by an adversary, while active camouflaging entails artificial creation of an image or impression that you have a capability that does not actually exist.

Camouflaging techniques have been used in the cybersecurity domain by malwares to hide their signature [20], as well as for the detection of attacks on cyber infrastructures, including wireless sensor networks [22], smart grid [17], and energy harvesting motivated networks [19]. A highly relevant work to ours is a passive camouflaging technique proposed by Fu *et al.*[7] that deceptively hides the patterns (packet sizes, timings) of an encrypted network flow by embedding (camouflaging) its packets into the packets of another flow. These patterns, if not camouflaged, could be exploited by ML-based traffic analysis techniques to identify certain properties of the flow (even though it is encrypted) like which website is being visited [5].

In the rich literature of defensive cyber deception, few works have focused on the idea of integrating deception with cryptography. A work close to our approach is Honeywords [14], which is proposed by Juels and Rivest and has gained notable popularity in both academia and industry. Honeywords are a group of fake passwords (*e.g.*, 19) that are associated with a user account along with the real password. An adversary who steals a file of hashed passwords and brute-forces the hashed passwords cannot tell if they have found the password or a honeyword. The attempted use of a honeyword for login sets off an alarm. This approach uses active camouflaging (creation of things that do not exist) while our approach uses passive camouflaging (obscuring things that are real). Several other works have investigated the effectiveness of Honeywords [8,23] or proposed extensions and improvements [4,6]. In Sect. 4, we compare Honeyword with our passive camouflaging technique and discuss our advantages.

Password-based authentication protocols often use collision-resistant one-way hash functions to make inversion infeasible. However, collisionful hash functions [9] with selective collision could be used as an alternative where the password hash has collisions with some fake passwords. During brute-forcing of the hashed password, if the attacker discovers one of the fake passwords and uses that for login, the intrusion is detected.

Honey Encryption [13] is another close like of work. The primary idea behind Honey Encryption is to generate a fake but plausible-looking plaintext upon decryption by certain (controlled) wrong keys. An attacker trying to iterate through several keys will end up getting an equal number of genuine-looking plaintexts, and it becomes impossible to distinguish the real one. To accomplish the above functionality, the authors used an intermediate block called Distribution Transforming Encoder (DTE) [13]. The DTE maps the messages to multiple seeds that correspond to different outputs. But since each message is mapped to multiple seeds, it is non-deterministic with some local randomness. The closest works to D3CyT are Honeywords [14] and Honey Encryption [13]. The main similarity between the three approaches is the use of deception for making brute-forcing detectable. However, while Honey Encryption and Honeyword achieve this by giving attackers multiple choices (*e.g.*, multiple passwords), we achieve this goal by pretending that the data is not encrypted while hiding the underlying cipher system. In other words, while Honeyword and Honey Encryption aim to confuse attackers (*e.g.*, about which password or key is real or not), we aim to deceive them through camouflaging (*e.g.*, hiding the real password and showing a fake one instead). Moreover, while Honeyword and Honey Encryption are only effective for specific threat models (*e.g.*, offline password cracking), our approach is more generic and could be adapted to various domains, as shown by the use cases in Sect. 4.

3 Methodology

Our goal is to protect a sensitive data item on a machine (*e.g.*, a Web server), which we simply refer to as the *main server*. We assume that the adversary does not compromise this server on a persistent basis, directly observing and capturing newly created honeyvalues. However, the adversary can compromise the main server at least once, and steal the sensitive data (*e.g.*, user passwords, log files, emails).

To realize our camouflaging, in addition to this main server, an auxiliary, dedicated, and secure server, called *Camouflage Key Server*, assists with the use of honeyvalues. Camouflaging keys are stored and retrieved from the key server, using authenticated and encrypted communication channels. The key server only includes the processes relevant to the key encryption/decryption, storage, and retrieval with no other services running on it; its OS and services are hardened with the latest security patches, and it is protected with various firewalling and intrusion detection systems at network and host level.

Next, we discuss the algorithms for camouflaging and de-camouflaging based on this architecture.

3.1 Camouflaging Algorithm

Assume our goal is to camouflage a sensitive value p to a honeyvalue q. This sensitive value could itself be a hashed or encrypted value, and q is a honeyvalue in plaintext that is believable in place of p.

Given p and q, the main server converts p and q to their binary encoding *p_bv* and *q_bv* respectively. Then, it does a bit-wise XOR between them to obtain the camouflaging key *camo_key*. Then, this *camo_key* is sent to the key server through an authenticated and encrypted channel.

The key server stores the key-value pair $(q, camo\_key)$ in a dictionary data structure:

$$dict[q] = camo\_key$$

Algorithm 1 denotes the steps of the algorithm.

To ensure consistency and robustness of the approach, the selected honeyvalue must satisfy two constraints:

- It must be a unique value.
- Its length must be at least as large as the real value $(len(q) \geq len(p))$ to ensure that no part of p is revealed in *camo_key*.

In addition to these constraints, the honeyvalue must be chosen such that it enhances the likelihood that adversaries believe it and act upon it. In Sect. 5, we discuss two heuristic metrics for selecting desirable honeyvalues.

Algorithm 1. Camouflage(p, q)

Input: p: real value, q: honeyvalue

$p\_bv := string\_to\_binary(p)$
$c\_bv := string\_to\_binary(c)$
$camo\_key := p\_bv \oplus c\_bv$
store key-value pair $(q, camo\_key)$ on key server

3.2 De-camouflaging Algorithm

To de-camouflage q, the main server provides q to the key server through the authenticated and encrypted channel to retrieve the camouflaging key, *camo_key*, through the same channel.

The main server first converts q to its binary encoding and then performs a bit-wise XOR between q and *camo_key* to retrieve *q_bv*. Finally, the main server converts this binary encoding back to its string representation, p.

Algorithm 2 denotes the steps of the algorithm.

Algorithm 2. De-Camouflage(q, *camo_key*)

Input: q: honeyvalue, *camo_key*: camouflaging key
Output: p: real value

retrieve *camo_key* for q from key server
$q\_bv := string\_to\_binary(q)$
$p\_bv := q\_bv \oplus camo\_key$
$p := binary\_to\_string(p\_bv)$

4 Use Cases

In this section, we discuss three potential use cases of our camouflaging method in order to highlight its wide applicability and valuable advantages in different defense scenarios.

4.1 Password Camouflaging Against Offline Password Cracking

In traditional password protection techniques, user passwords are not stored in the database or documents in plaintext; rather, the password is hashed using a secure hash function like SHA-2, and the hashed password is stored instead. Since the hash functions are irreversible, even if hashed passwords are stolen, attackers still need to brute-force the hashes to discover the plaintext password.

Although cryptanalysis attacks on modern hash functions are not viable, the increasing computational power and collective password recovery efforts [1] have simplified the process to a great extent [15].

Using our camouflaging technique, we propose a novel mechanisms for password protection. In D3CyT, given a user's password, p, we first hash the password, $h(p)$ as in a conventional system. Then, we choose a plaintext-looking string that is believable as the password for the user as the honey-password, q. This honey-password could be algorithmic generated based on user data [14] or chosen from the previous passwords of the same or other users. Next, we generate the camouflaging key ($camo\_key = h(p) \oplus q$). The key-value pair (q, $camo\_key$) is sent to the key server through an authenticated channel.

In the password field of the user, we store the honey-password, q. Figure 1 shows this modified sign-up procedure.

Figure 2 shows the modified sign-in process. Once a user tries to sign in, they provide their user ID and password. The Web server first retrieves the honey-password, q, for this user from the main database. Then, it provides this honey-password to the key server through the authenticated and encrypted channel. The key server retrieves the camouflaging key for q and sends it back. The main server retrieves the real hashed password by XORing camouflaging key with the honey-password. Then, it hashes the input password given by the user and compares it with the retrieved hashed value. If both are equal, then the user access is authorized.

Now, assume the user data is leaked, for example, through a SQL injection vulnerability on the Web application. If the attacker uses the honey-password to log in, the Web server detects that the user passwords have been leaked and will take necessary countermeasures, like informing admins and users to change their passwords and finding the intrusion point.

In the first look, our camouflaging password protection may look similar to Honeyword [14]. In fact, similar to Honeyword [14], our approach uses deception to make offline password cracking detectable. However, the two approaches realize this goal through different mechanisms. Honeyword extends user credentials with a number of honey passwords, while in D3CyT, we essentially replace real passwords with honey passwords. The strength of Juels and Rivest's Honeyword

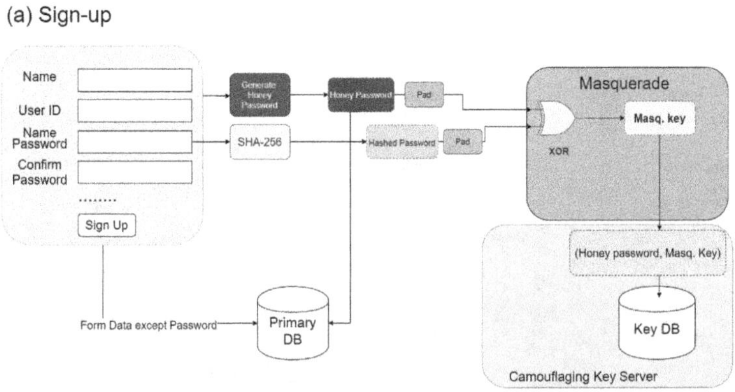

Fig. 1. A flowchart of modified sign-up using the camouflaging technique

[14] approach depends on the indistinguishability of honey and real passwords, and the main deployment challenge is devising methods for automatic generation of indistinguishable honey passwords. Evaluation of the honey password generation algorithms proposed by Juels and Rivest shows that for 19 honey passwords and one real password, a basic attacker can correctly guess the real password with probability 29.29%–32.62% which is far from the expected 5% (1/20) [23].

In contrast, being a passive camouflaging technique, the strength of D3CyT comes from deceptively obscuring the fact that the data is even hashed or encrypted. While the generation of a plausible honey password is a challenge for our approach as well, the strength of our technique does not completely rely on it. If the user database is disclosed, then it only includes a bunch of dummy texts as passwords. Without the camouflaging keys, the attacker has no mechanism to discover the real password hashes.

If the attacker believes that these honey-passwords are real passwords and tries them, then the compromise is detected. Similar to Honeywords, generating indistinguishable honey passwords can enhance the likelihood of success for our approach. However, the real password is not stored anywhere in the user data file, and the attacker's main challenge is not distinguishing the real password from the honey ones.

Thus, if the attacker discovers that the plaintext-looking passwords are only deceptive, then they still have no clue about what the real password could be. To obtain the real password, they first need to compromise the key server (which is a hardened and very secure server) to retrieve the camouflaging key. Once the attacker has both the honeyvalue and camouflaging key, they can retrieve the real value, but since this value is encrypted/hashed, they still need to brute-force the hashed password, as before.

In a nutshell, our approach increases the security level of password-based online systems by either (1) enabling the detection of password compromises, (2) dissuading attackers by convincing them that the compromised user data is dummy, or (3) making the brute-forcing process more challenging.

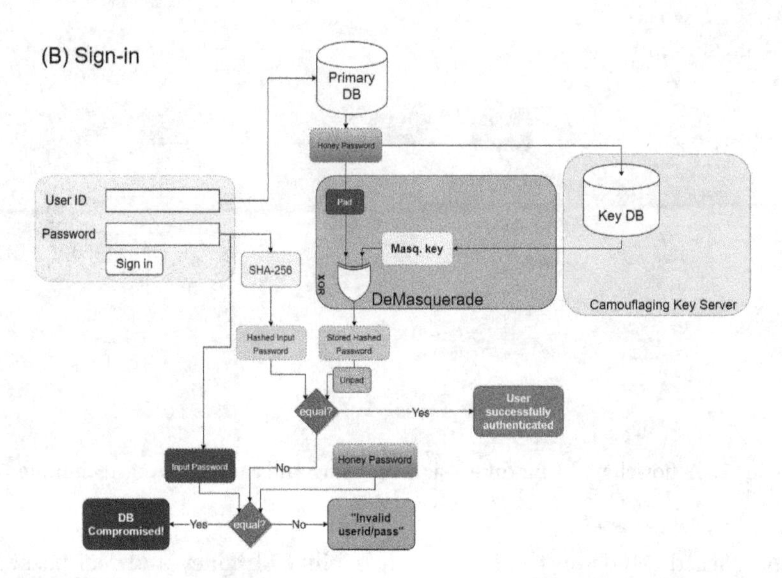

Fig. 2. A flowchart of modified sign-in algorithms using the camouflaging technique

4.2 Camouflaged QR Codes

QR codes are becoming one of the most extensively used encodings of modern times. Having the ability to store more information within a very compact space, it is getting preference over barcodes. QR codes represent textual information in the form of a monochromatic image that can be decoded using a QR-code scanner. With smartphones and cloud-based services becoming popular, everyone can encode and decode these image-based encodings. To take advantage of its widespread usage, we propose our idea of using camouflaged QR-codes as a means of deception.

To begin with, let us assume a scenario where a company (say KamSecure) uses a cloud-based service that maintains several user accounts. The company, depending on its profit, offers gift cards to its employees at the end of each week. Each employee/user receives a unique voucher code via email that can be used for online shopping. Over the years, this policy started attracting more and more attackers to compromise their accounts and steal the voucher codes. To detect compromised accounts, the company finally adopted the idea of camouflaging QR codes as shown in Fig. 3.

To this aim, given a real voucher code of an employee (*i.e.*, real value), the system first generates a unique decoy voucher code for each employee (*i.e.*, honeyvalue). Next, as per Fig. 3, in step 1, the unique decoy voucher code encoded in the form of a QRcode is sent to each employee separately through their email addresses. Then, the system generates the camouflaging key for the employees by applying the camouflaging Algorithm 1 on the real and the decoy voucher codes, and stores the key-value pairs (decoy voucher code, camouflaging key) on the key server. Then, in step 2, the server sends out a text message containing a URL to all employee's smartphones via SMS. The URL has the following format. "https://www.kamsecure.com/redeem/ <*random string*>". For example, we use: "https://www.kamsecure.com/redeem/53c8a4222785637c8fc8fc1c8ee24dec" as the URL. This is to ensure that the URL can not be discovered through brute-forcing. Although the example shows a 128 bit random string, we recommend using one of length 512 bits. Each week a new URL (random sequence) is generated for new gift cards, and the same URL is sent to all employees.

In step 3, once the user clicks on the link in the SMS, the web application launches an online scanner. The user uses the smartphone camera to scan the QRcode sent via email. This QRcode includes the decoy voucher code. Next, they can redeem their code on the designated online website for shopping, as shown in step 4. The web application uses the decoy voucher code to retrieve the camouflaging key for the user from the key server, and then generates the real voucher number using the de-camouflaging algorithm (Algorithm 2), and displays it to the user.

From the perspective of the users, it is the same as just scanning a QRcode, with the only difference that they cannot use any other scanning application. We assume that the company trains its users separately, and none of them will be using a different scanner.

Now, suppose there is a user account that has been compromised by an attacker through phishing or any other social engineering techniques. The attacker also has access to the same email as the employee but does not have access to employee's smartphone. Seemingly genuine, the attacker decodes the voucher by scanning the QRcode, believing it is meant for the user, as depicted by step 5 in Fig. 3. Since this code is unique to every account, once the code is redeemed, the compromised account will be revealed. Note that the same QRcode that generates a real voucher code for the genuine user will act as a trap for the attacker.

4.3 Data Camouflaging Against Stealthy Intrusion Attacks

Attackers use a myriad of techniques to hack into machines for information theft or espionage. For example, in recent years, we have witnessed an unprecedented increase in a novel class of attacks called Advanced Persistent Threat (APT), which relies on sophisticated social engineering attacks, such as spear-phishing messages (emails with malicious attachments) to take over sensitive machines and steal information [2].

Once a system is compromised, attackers will look through existing files and documents on the compromised machine to extract sensitive information about users or other devices in the network. To complicate this process and make such compromises potentially detectable, one may use our technique to deceptively camouflage sensitive data in documents and files.

To make our example concrete, without loss of generality, we focus on a specific threat model in which the attacker, after compromising a web application running on a server (like Apache), retrieves the server access log file. For example,

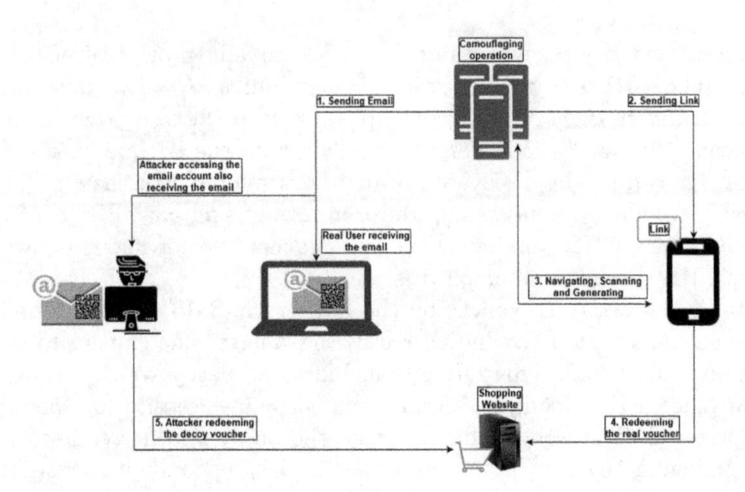

Fig. 3. A schematic of the qr-code camouflaging technique to discover compromised email accounts.

if the Web application has a directory traversal vulnerability, an attacker can use variations of `../apache/logs/access.log` to traverse the filesystem and retrieve the `access.log` file. Once the access log is retrieved, the attacker parses it to extract useful information, including IP addresses of clients.

The following shows a typical log message of Apache HTTP Server Version 2.4 generated for an internal client:

```
[Fri Sep 09 10:42:29.902022 2019][core:error][pid 35708:tid 4328636416]
[client 192.168.100.112] File does not exist:
/usr/local/apache2/htdocs/favicon.ico
```

Using D3CyT, we provide a robust deception scheme against such potential exfiltration of sensitive data. Now, instead of putting the actual IP addresses of internal hosts in the access log, we camouflage each of these real IP addresses into a honeypot IP address. Assume that "192.168.110.50" is the IP address of a honeypot (*e.g.*, an SSH honeypot like `Kippo` within our network), which is used to trap insider threats (*e.g.*, SSH credential brute-forcing attacks).

Using the camouflaging technique, instead of storing the real clients' IP address in the log file, the Apache server replace it with the IP address of the honeypot. To achieve this, the Apache Web server uses Algorithm 1 to generate the encrypted camouflaging key:

$$camo\_key = Camouflage(realIP = '192.168.100.112', honeyIP = '192.168.110.50') \tag{2}$$

Next, the following key-value pair is added to the camouflaging key dictionary on the key server. Finally, in the access log, the Apache Web server would store the IP address of the honeypot ('192.168.110.50') instead of the real client's IP address ('192.168.100.112'). If this access log is leaked, an unsuspecting attacker will extract the honeypot IP address from the access log. Probing the extracted IP addresses would only direct the attacker to our honeypot, thus enabling the characterization of the attack (*e.g.*, attacker's IP address). While the same property can be achieved by Honeywords, using our approach, we do not need to keep multiple instances of the same log forged with different decoy IP addresses. Since the internal systems of the server have access to the *camo_key*, they perceive the logs with the real IP addresses. However, the same log file shows the decoy IP addresses to the attackers. This raises low or no suspicion as compared to keeping multiple instances together, like Honeywords. Moreover, even if the attacker learns about our deception scheme, they still need to discover the camouflaging key to retrieve the original IP address.

In order to retrieve the original IP address from the honey IP in access log, the Apache server first needs to retrieve the camouflaging key from the camouflaging server and then use Algorithm 2 to retrieve the original real IP address:

$$camo\_key = dict[honeyIP]$$
$$realIP = De\text{-}Camouflage(honeyIP, camo\_key)$$

While our discussion has been focused on a specific case study, the same notion could be applied to camouflage other sensitive information like domain names, file names, *etc.* as well.

5 Evaluation

In this section, we evaluate the overheads of this approach in terms of execution time and space.

5.1 Execution Overhead

In conventional encryption, a given plaintext is directly encrypted and later decrypted into a ciphertext. In contrast, in our camouflaging technique, after encryption, we only perform an extra XOR operation to calculate the camouflaging key. In de-camouflaging, we first retrieve the camouflaging key and then perform an XOR operation with the honeyvalue to retrieve the original plaintext. The only extra operation required for both camouflaging and de-camouflaging as compared to conventional encryption and decryption is a Boolean XOR operation. As a binary operator, bit-wise XOR is one of the fastest operators because it is hardware supported and parallelizable (each bit could be calculated in parallel). This overhead, compared to the execution time of the block cipher, is negligible. This is confirmed through our evaluation results achieved via implementation.

Figure 4 compares the execution time of a conventional encryption approach (without camouflaging) with that of our camouflaging approach applied after encryption. For both techniques, AES-128 in CBC mode is used as the symmetric block cipher. We implemented our algorithm and evaluation testbed in Python

3.6 and used standard pycrypto library for AES. The times in the figure are calculated on an Intel Core i5, 2.6 GHz, with 4 GB RAM. Each data point is averaged over 1,000 execution to reduce the error margin. On an average it took $8 * 10^{-4}\%$ more time per camouflage operation. This confirms our expectation that the overhead of XOR is negligible compared to the overhead of encryption and decryption functions.

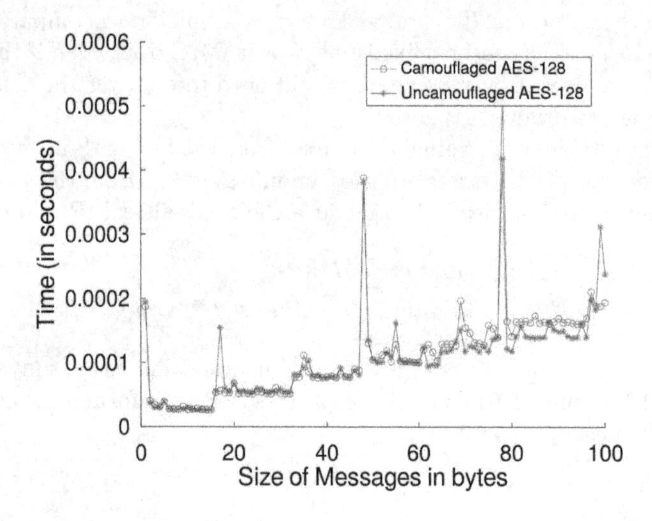

Fig. 4. Computation time for conventional encryption vs. encryption followed by camouflaging for different input sizes

5.2 Resources and Space Overhead

Our camouflaging technique requires an additional secure server with additional storage space for the encrypted camouflaging key. In practice, this server could be deployed on a secure virtual machine or docker container for efficient deployment. In addition, a secure communication channel such as SSH Server needs to be setup on the key server for secure communication.

The camouflaging keys also need to be stored on the key server. Assuming that the ciphertext has the same size as the plaintext, the camouflaging key will be of the same or smaller size as of the plaintext. In the worst case, the camouflaging operation increases space overhead by 100%. However, the use case of this camouflaging is not to encrypt the bulk of data, but to protect sensitive attributes like passwords or IP addresses.

5.3 Metrics for Generation of Decoys/Honeyvalues

The main challenge in the adoption of our approach in a new domain is the choice of decoys or honeyvalues. This problem has been investigated in the context of honey passwords. Juels and Rivest propose four techniques for generating honey passwords for users [14]. They define a metric called ϵ-flatness. A honey-password generation method is ϵ-flat if the maximum value calculated over all adversaries of the adversary's winning probability is ϵ. For example, if 19 honey passwords are generated for each user (thus having a total of 20 real and honey passwords), and the maximum correct guessing probability for adversaries is 5% (1/20), then the deception has *perfect flatness*. Several other works have shown generating perfectly-flat honey passwords is a very challenging task [6,23]. In this section, we define two heuristic metrics for the generation of honeyvalues, based on the literature of cyber deception. In future works, we provide a more in-depth study of these metrics in various defense scenarios and with human users and through red-teaming experiments to identify fine-grained principles for realizing them:

- (1) Plausibility: a honeyvalue must be believable as a real value, and thus it must conform to the expectations of attackers regarding potential values for that field. For example, the honeyvalue generated for an IP address field must be another IP address, or a honey-password must be believable as the password for that user.
- (2) Playability: the honeyvalue must be chosen such that it maximizes the likelihood that attackers believe and act based on this deceptive information, thereby enabling their detection. For example, when camouflaging an IP address, we can choose the honeyvalue to be the IP address of a honeypot device. So, if a potential attacker attains this honeypot IP address since the IP address belongs to a live machine, it increases the likelihood of engagement, and once the attacker probes the honeypot machine, we can characterize the attack (*e.g.*, identify the source IP address, potential motives, *etc.*.).

5.4 Limitations of D3CyT

Like all other approaches, D3CyT also has its own limitations. We already talked about the space overhead in Sect. 5.2. It implies that creating a decoy and a camouflaging key pair of substantially large amounts of data makes it quite inefficient and consequently confines its usage to small data chunks only. Furthermore, storing the camouflaging key in most cases will require an additional level of security in place, as the element of deception loses its meaning as soon as it gets revealed. As XOR is a very fast operator, the decoy and the camouflaging key for small texts must be periodically updated to keep the system protected against offline brute-force attacks.

6 Conclusion and Future Work

In this paper, we present a novel method for camouflaging sensitive data, which adds a layer of deception on top of conventional encryption mechanisms to potentially detect or deter stealthy data thefts as a result of potential compromises. We present the necessary algorithms and architecture for realizing the working principle of D3CyT. We also discuss three different use cases of our approach to highlight its broad advantages and applicability. Through implementation, we also show that the execution time overhead of D3CyT is negligible, and the space overhead is reasonable. As a part of our future work, we will extend the use of camouflaging to other kinds of cybercrimes like illegal software re-distribution. A lot of the single-user license activation codes are being leaked/circulated online and used by multiple parties. To prevent this, we can create camouflaged codes, using a real key and a decoy key. When a genuine user buys software, a real key is generated against their online account. Next, a random $(decoy\_key, camo\_key)$ pair will be generated. The decoy key will be embedded into the software and will be unique to the user. The camouflaging key will be given to the user as a part of the activation key. Upon activation, the real key will be generated using the algorithm of D3CyT and matched online. Upon successful completion, the software gets activated. Immediately the embedded key in the software is changed to *camo_key*. Now if the user redistributes their copy of the software and the *camo_key*, the result of the de-camouflaging will result in 0. The software will get locked and a warning message will be displayed.

References

1. Hashes.org - shared community password recovery. https://hashes.org/leaks.php. Accessed 15 June 2019
2. Butavicius, M., Parsons, K., Pattinson, M., McCormac, A.: Breaching the human firewall: social engineering in phishing and spear-phishing emails. arXiv preprint arXiv: 1606.00887 (2016)
3. Caddell, J.W.: Deception 101-primer on deception. Technical report, Army War Coll Strategic Studies Inst Carlisle Barracks PA (2004)
4. Chakraborty, N., Mondal, S.: On designing a modified-UI based honeyword generation approach forovercoming the existing limitations. Comput. Secur. **66**, 155–168 (2017)
5. Dyer, K.P., Coull, S.E., Ristenpart, T., Shrimpton, T.: Peek-a-boo, i still see you: why efficient traffic analysis countermeasures fail. In: 2012 IEEE symposium on security and privacy, pp. 332–346. IEEE (2012)
6. Erguler, I.: Achieving flatness: selecting the honeywords from existing user passwords. IEEE Trans. Dependable Secure Comput. **13**(2), 284–295 (2015)
7. Fu, X., Guan, Y., Graham, B., Bettati, R., Zhao, W.: Using parasite flows to camouflage flow traffic. In: Proceedings of 3rd Annual IEEE Information Assurance Workshop 2002. Citeseer (2002)
8. Genc, Z.A., Kardaş, S., Kiraz, M.S.: Examination of a new defense mechanism: honeywords. In: IFIP International Conference on Information Security Theory and Practice, pp. 130–139. Springer (2017)
9. Gong, L.: Collisionful keyed hash functions with selectable collisions. Inf. Process. Lett. **55**(3), 167–170 (1995)
10. Gressin, S.: The equifax data breach: What to do. Federal Trade Commission **8** (2017)
11. Heckman, K.E., Stech, F.J., Schmoker, B.S., Thomas, R.K.: Denial and deception in cyber defense. Computer **48**(4), 36–44 (2015)

12. Jafarian, J.H.: Cyber agility for attack deterrence and deception. Ph.D. thesis, The University of North Carolina at Charlotte (2017)
13. Juels, A., Ristenpart, T.: Honey encryption: security beyond the brute-force bound. In: Annual International Conference on the Theory and Applications of Cryptographic Techniques, pp. 293–310. Springer (2014)
14. Juels, A., Rivest, R.L.: Honeywords: making password-cracking detectable. In: Proceedings of the 2013 ACM SIGSAC Conference on Computer & communications Security, pp. 145–160 (2013)
15. Kelley, P.G., Komanduri, S., Mazurek, M.L., Shay, R., Vidas, T., Bauer, L., Christin, N., Cranor, L.F., Lopez, J.: Guess again (and again and again): measuring password strength by simulating password-cracking algorithms. In: 2012 IEEE Symposium on Security and Privacy, pp. 523–537. IEEE (2012)
16. Kim, K.: Distributed password cracking on GPU nodes. In: 2012 7th International Conference on Computing and Convergence Technology (ICCCT), pp. 647–650. IEEE (2012)
17. Lu, Z., Wang, W., Wang, C.: Camouflage traffic: minimizing message delay for smart grid applications under jamming. IEEE Trans. Dependable Secure Comput. **12**(1), 31–44 (2014)
18. Provos, N., et al.: A virtual honeypot framework. In: USENIX Security Symposium, vol. 173 (2004)
19. Pu, C., Lim, S.: Spy vs. spy: camouflage-based active detection in energy harvesting motivated networks. In: MILCOM 2015-2015 IEEE Military Communications Conference, pp. 903–908. IEEE (2015)
20. Rad, B.B., Masrom, M., Ibrahim, S.: Camouflage in malware: from encryption to metamorphism. Int. J. Comput. Sci. Netw. Secur. **12**(8), 74–83 (2012)
21. Rowe, N.: A taxonomy of deception in cyberspace. In: International Conference on Information Warfare and Security, pp. 173–181 (2006)
22. Ubaid, S., Shafeeq, M.F., Hussain, M., Akbar, A.H., Abuarqoub, A., Zia, M.S., Abbas, B.: SCOUT: a sink camouflage and concealed data delivery paradigm for circumvention of sink-targeted cyber threats in wireless sensor networks. J. Supercomput. **74**(10), 5022–5040 (2018)
23. Wang, D. Cheng, H., Wang, P., Yan, J., Huang, X.: A security analysis of honeywords. In: NDSS (2018)
24. Whaley, B.: Toward a general theory of deception. J. Strategic Stud. **5**(1), 178–192 (1982)

Homomorphic Password Manager Using Multiple-Hash with PUF

Sareh Assiri(✉) and Bertrand Cambou(✉)

Northern Arizona University, Flagstaff, AZ 86011, USA
{sa2363,Bertrand.Cambou}@nau.edu

Abstract. In the proposed homomorphic methods, the server authenticates clients without ever knowing their passwords. During enrollment, the users subject their passwords to multiple hashing cycles, typically 1000 times, and communicate the resulting message digests to the server. Rather than storing these message digests, the server uses them to find addresses in the physical unclonable functions, which generate data streams that are stored for future authentication. The authentication cycles use the following steps: i) The users hash their passwords multiple times, at levels lower than the one used during enrollment; ii) The server generates data streams from the physical elements at the address extracted from the message digest and compares it to the data streams stored during enrollment, and iii) The server reiterates the previous step by incrementally hashing the resulting message digest to find a match, or it rejects the password. During subsequent authentication cycles, the users again hash their passwords multiple times, but at levels lower than the ones used during the previous cycles. Thereby it becomes pointless for third parties to intercept previously hashed passwords; they are never used twice. Hacking a database containing the data streams extracted from the physical unclonable functions during enrollment is also pointless without also having access to the devices. In this entire homomorphic protocol, the users are the only ones who know their passwords. This paper presents a prototype demonstrating the functionality of an example of a homomorphic password manager protocol with SHA-3–512 hashing algorithm exploiting the physical randomness of static random-access memories.

Keywords: Homomorphy · Password management · Physical unclonable function · Hash function · Authentication

1 Introduction

A password-based authentication offers access to computer systems, applications, or online accounts. The password is a mix string of characters, numbers, and symbols used to verify the identity of a user during the authentication phase. The passwords are used with a username (USER ID). One way the USER ID and PW can be intercepted is by keeping watching the network flow. Monitoring the network flow assists the hacker to get the user's information. Usually, the eavesdroppers infer out from watching the same network flow several times, the passwords, or keys of encryption or so on [1–4]. By

K. Arai (Ed.): FICC 2021, AISC 1363, pp. 772–792, 2021.
https://doi.org/10.1007/978-3-030-73100-7_55

sending different network flow each time, it might be mitigated the risk of inferring some user's information. Also, one way of steal the passwords which the hacker usually used is to hack the database (DB). If the password that has been stored in the DB is not encrypted in a strong way, it might be exposed. The easiest manner to store passwords in the DB is to create a table in the DB, which contains the USERID and PW; this table keeps all of the USER IDs and PWs in a human-readable format [5–7]. Therefore, if hackers are able to hack the DB, they can easily read all the content stored in DB. So, all data that has been stored in plain text format in a DB is vulnerable to be readable by hackers. Consequently, one of the worst possible methods, in security terms, used by some websites and applications is storing a USER ID and PW in a human-readable format [5, 7, 8]. The weakness of DBs containing USER IDs and passwords is of major concern for cybersecurity developers. It prompts investigation of a solution that will help make the content of DBs encrypted to forbid hackers from understanding DB content [9–11]. As a solution, it has been suggested that replacing the output of the hash function (message direct (MD)) of the password in DB [7, 12, 13]. The MD format replaces the content of DB instead of the real plaintext USER ID and PW. As known, hashing is one-way encryption after hashing the input; it is inaccessible. Whereas, by inventing the rainbow table, the hash function becomes no more secure to be used. The Rainbow table is a huge number of passwords with their outputs of the hash. The hackers can use the rainbow table to try many different passwords until one matches the hash output [14–17].

In this paper, we are presenting two additional lines of defense; the first one is in the way we handle the password at the client level, the second one is to eliminate the need to store the passwords or the hashing of the password in the database of the server. On the client side, the users subject their passwords to a great number of multiple hashing and send over the last MD to the server. The server converts the MD into data streams generated with physical unclonable functions (PUFs) [9, 18–21], and store these streams in DB. The PUFs hardware components work like human fingerprints; in this work, we use commercially available components, static random access memories (SRAMs) as PUFs [22, 23]. The organization of this work is the following.

Section 2 presents a brief overview of the background important for this work; the hash functions, which are one-way cryptographic function widely used in the proposed protocol; how the hash functions are used to protect mainstream password managers; the physical unclonable functions (PUFs) that we also use in the proposed protocol; the addressable PUF generators (APGs) that are used to generate data streams and cryptographic primitives; and how APGs can be used to design secure password managers.

Section 3 presents the main objectives of the work and security threats. We worry about insider attacks stealing databases of passwords (or message digests generated from the passwords), and man-in-the-middle attacks intercepting the passwords. This paper shows how the servers authenticate the users without knowing the passwords.

Section 4 presents the detailed architecture of the proposed password manager both at the client and server level. This paper shows how the DB of the hash of passwords is replaced by the output of physical unclonable functions (PUFs). Phrased differently, the addressable PUF generators (APGs) are implemented to "encrypt" the DB of the password manager scheme.

Section 5 presents the implementation of the architecture and the design of a prototype.

2 Background Information

2.1 Hash Function

The hash function is defined as one-way encryption; by hashing the input, you will not be able to obtain it again. The hash function, such as sha1, sha2, and sha3, has been used in several cryptography applications. For the digital signature, there is a scheme called Winternitz one-time signature (W-OTS) that has used the hash function. The W-OTS idea is that it hashes several times the private key. In this study, we can take advantage of the W-OTS idea, hashing the password several times, to be implemented in the homomorphic password scheme to obtain each time different password [12, 15, 24].

2.2 Hashing the Passwords

As mentioned in Sect. 2.1, the hash function is one-way cryptographic methods. We can use the hash function because it is similar to encryption in the sense that it turns the password into a long string of letters and numbers to keep it hidden. For example, if we used SHA-1 to hash "the password." After hashing it the output will be like this(e38ad214943daad1d64c102faec29de4afe9da3d). A hash function can be used to protect, or replace, the tables containing USERIDs with their corresponding passwords by lookup tables, as shown in the block diagram of Fig. 1 [11, 15, 19]. In [11, 13, 25, 26] it is shown that:

- The hashing of the password results in the first message digest h(PW).
- The USER ID, USER ID, and PW are XORed.
- The hashing of USER ID ⊕ PW results in the second message digest h(USER ID ⊕ PW).
- The message digest h(USER ID ⊕ PW) is used as an address in DB.
- The message digest h(PW) is stored in the DB at that address.

2.3 Physically Unclonable Functions (PUFs)

The PUFs hardware components work like the fingerprints of a human. The output of PUF is unclonable and random for each different input; therefore, PUFs have used to strengthen the level of security in the cybers. Later on, for the authentication purposes and several fields of security have started using the PUFs. According to [9, 22, 23, 27, 28], Several types of PUFs are excellent elements to generate strong PUFs: ring oscillators, SRAM, DRAM, ReRAM, and. Also, the output of PUF has two types,

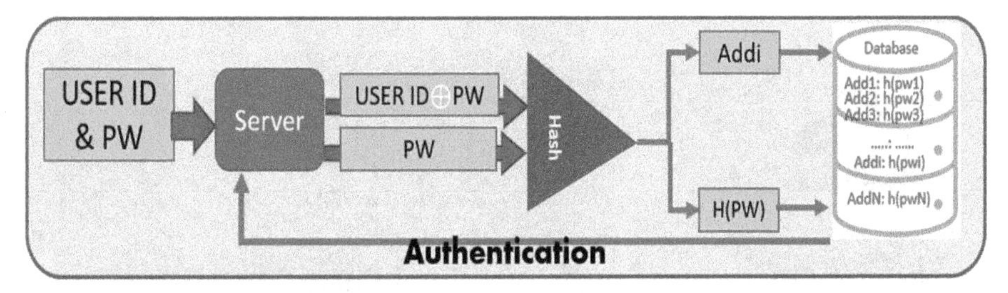

Fig. 1. The diagram is describing the data flow for authentication. On the left, PW and USER IDs are converted into H(PW) and addresses = (h (USER ID ⊕ pw)). On the right, the database store the message digests of Add and PW [11, 19].

which are binary or ternary. The outputs of PUFs are called challenges or responses. The challenges are generated upfront from the PUF, whereas the responses are generated during access control rounds or authentication rounds. Each time the PUF is read, it is slightly mismatched, so it complicates authentication [29]. However, as noted in [9, 10], during authentication, when we read the PUF to generate the responses, we must calculate the rate of matching challenge-response-pairs (CRP). If the matching CRP is high enough with a low error rate, then we can accept; otherwise, we must reject the user [9]. Northern Arizona University (NAU) has a disclosure PUF based password generation scheme [10], and the Addressable PUF generators (APG) disclosure number [9].

2.4 Addressable PUF Generators (APG)

There are two similar operation modes of the APG, which are the PUF challenge generation and the PUF response generation. The PUF challenge is to extract a reference pattern when you read the PUF first time. Whereas, the PUF response is when you read the PUF to authenticate what you already have. In general, the APG has two main parts, which are the hash function, and the PUF. APG works as following:

1) The input which might be a random number, or something entered by the users (e.g., a password), then
2) The input is fed to the hash function, and the hash function gives the MD,
3) MD is divided into several couples of bits; every two groups of bits hold the value of coordinate x and y.

The block diagram in Fig. 2 shows that the CRP generation varies with the relative value of the parameters within the multiple cells that are selected at a particular address. We give a value of 'x' and 'y,' and then we find a particular address in the PUF. Therefore, as a particular cell in the PUF could be a "0" when part of one group of cells, and a "1" when part of a different group, or when read with different instructions.

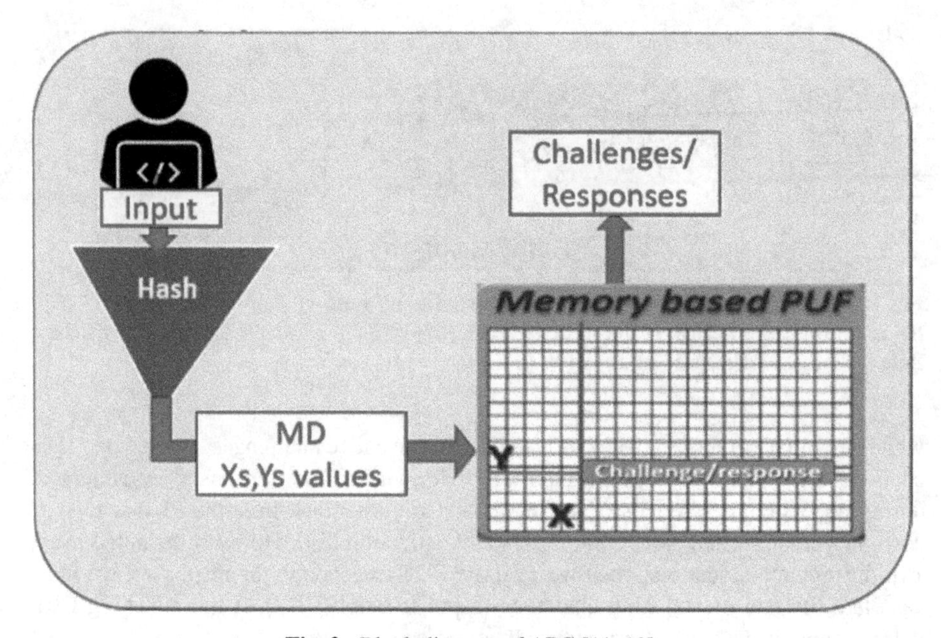

Fig. 2. Block diagram of APG [11, 19].

2.5 Password Manager with Binary or Ternary APG

By using the hash function and APG, we will be able to generate the passwords and also to authenticate users. In [30] it has shown a diagram of a temporary password that is generated by using the APG. Also, in [19, 30], it shows how the hash function and the APG are used in the password manager scheme. In the beginning, the password and USER ID are XORed. Both the password and result of XOR are going to feed to the hash function separately to obtained two MDs. The MD that coming from (UID ⊕ password) will feed to the PUF to extract the address challenge, which will point to a

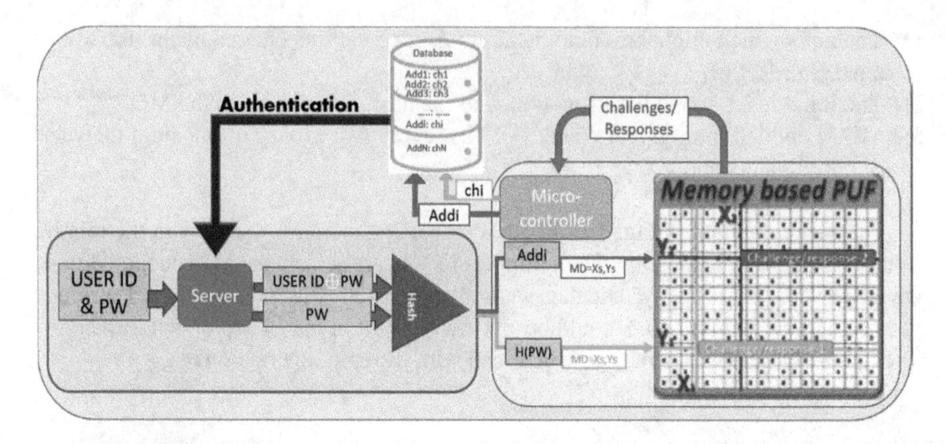

Fig. 3. Block diagram showing a password manager with APG at the server [19].

specific location in the DB. Whereas the MD that coming from the password will find the addresses (position of x, y) in the APG. After that the challenge will be extracted from the APG and will be stored in the database corresponding to its address. See Fig. 3.

3 The Objectives: Mitigating Security Threats

For many applications, current password managers with salted hashing are good enough; however, the level of risk is now higher; organized opponents now have access to large organizations working multiple years to successfully penetrate the network of their victims by finding ways to hack the passwords. The list of possible attacks includes sophisticated password guessing attacks, insiders accessing databases of users ID and sometimes passwords, man-in-the-middle (MIM) attacks generating illegitimate passwords, and distributed denied of service attacks forcing users to try new passwords to login [31–33]. The work presented in this paper cannot alone mitigate such an exhaustive list of threats and should be combined with other remedies. The three main objectives of this work are first to add a level of protection against insider and password guessing attacks, second to mitigate man-in-the-middle attacks, third to offer a practical and low-cost solution.

3.1 Protecting the Password Manager Against Attacks on the DB

This study aims to address methods and solutions that help to avoid guessing the password and understanding the content on the database when lost to the opponent. One of the weaknesses that we wish to address is the one where the hacker has access to the database of hashed passwords and the salting of the method. In such a case, a sophisticated guessing method can uncover many passwords, in particular the weakest ones. To address the hash function weakness gap, we propose the use of ternary SRAM-based PUFs, and to convert the message digests of the passwords into data streams stored in the DB. To have access to the content of the DB, the hackers need to understand how the ternary PUF operates, which vary PUT-to-PUF and get access to the PUF itself, which is buried in the overall hardware of the server. Generic methods on how to protect the PUFs are part of this work; see examples of analysis in the references [21–24].

3.2 Protecting the Password Manager Against MIM Attacks

In this paper, we analyze the benefit of the hash function when it uses in both client and server to generate a one-time password each time wanted to log in to the system. The benefit of that is to confuse the man in the middle, which mean that the man in the middle won't be able to learn from the network flow because each authentication time, new encryption of the hash of the password will be sent, see Fig. 4 [18, 34–36].

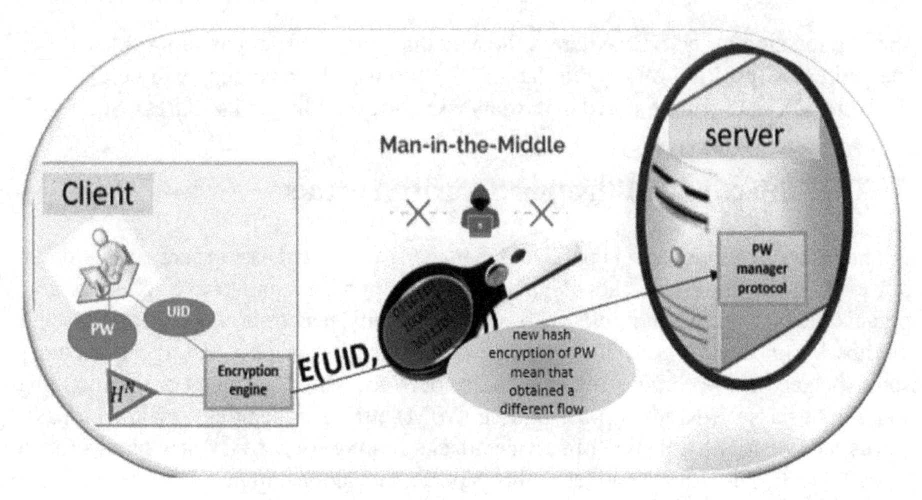

Fig. 4. Man in the middle try to learn from the flow.

3.3 Cost-Effective and Practical Implementation

In this work, mature and commercially available technologies were selected. We use the standard hash algorithm SHA-3 as hash functions. If further cost reductions are needed, SHA-3 can be replaced by SHA-2, which is secure enough for most implementation. The SRAMs are excellent components to prove the architecture due to their wide availability. If further protections are needed, they can be replaced by components such as MRAMs or ReRAMs. That could be more tamper resistants.

4 Methodology and Designing the Prototype

This paper shows how the hash function and APG are used in password management. The hash function will be used in both client and server-side, wherein the APG will be used in the server-side [18, 28]. There are two modes in the password management scheme, which are a new user and exist. Each mode has several steps that must be done on both the client and server-side. In this section, both modes protocol will be explained in more detail.

4.1 New User

a) **At Client Side**

For new users:

- The users will enter their user ID and password the first time on the client-side, and the password will be hashed a specified number of times (N).
- The user ID is named the address.

- The output of the last hash password is referred to as a hash password (H^N(PW) = hash password).
- The Address, hash password, and the number of times the password is hashed (N) are encrypted [E (Address, hash password, N)] and sent to the server, as shown in Fig. 5 and Table 1-part A [18].

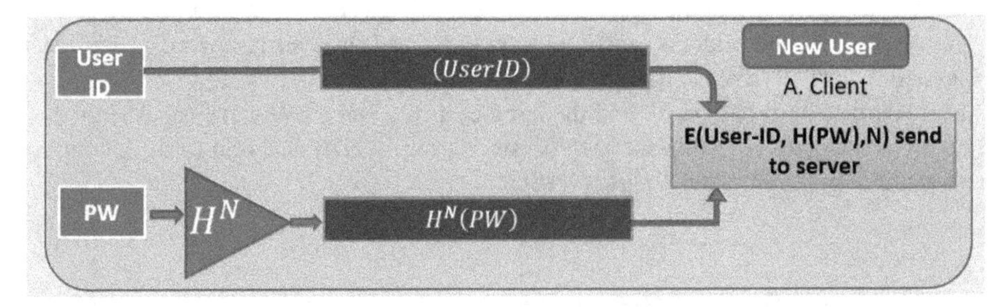

Fig. 5. The new user at client side

b) **At the Server Side**

On the server-side:

- The encrypted information [E (Address, hash password, N)] is decrypted.
- The last output of the hashed password will be XORed with the user ID.
- The output of this is named the address($UserID\, XOR\, H^N(PW)$= Address). And independently hashed an additional time to be H^1(Add), and H^{N+1}(PW).
- The output of the hash the address will be inputted into the PUF; the challenge received from this input in the PUF is stored in the address column in the database.
- The output of the hash password with the additional hash will be inputted into the PUF; the challenge received from this input in the PUF will be stored in the password column in the database, as shown in Fig. 6 and Table 1-part B [18].

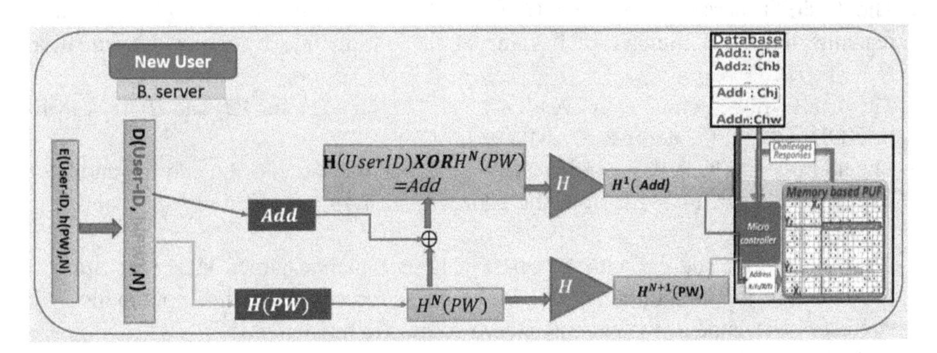

Fig. 6. The new user at server side

4.2 Returning User

a) At Client Side

For existing users:

- The users will enter their user ID, and password in the client-side and the password will be hashed less than the specified number of times ($H^{M<N}(PW)$).
- The user ID is the address, and the output of the last hash password is the hash password ($H^{M<N}$(PW) = hash password).
- The Address, hash password, and the number of times the password is hashed (M < N) are encrypted [E (Address, hash password, (M < N))] and sent to the server, as shown in Fig. 7 and Table 1 part A' [18].

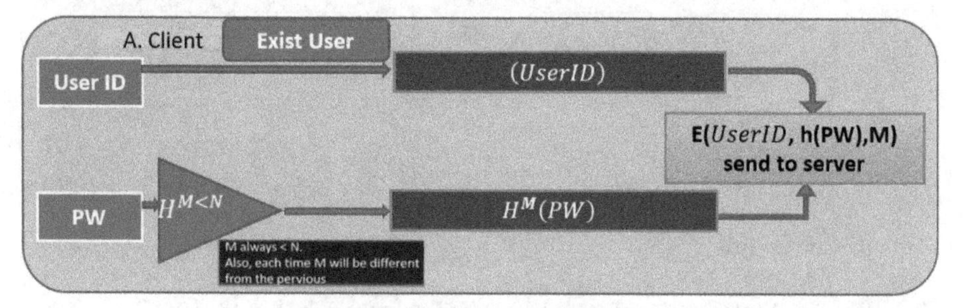

Fig. 7. Exist user at client side

b) At the Server Side

For existing users on the server-side:

- The encrypted information [E (Address, hash password, M < N)] is decrypted.
- The hash of password m time (H^M(PW)) is continue hashing H^M(PW) until reaching the same number of hashing at the initial time by using this equation H^{N-M}((H^M(PW)).
- The result will be getting H^N(PW). Then, the Username (User-ID) and H^N(PW) will be XORed to get the address (UserID $\oplus$ H^N(PW)).
- The address and hash password H^N(PW) must be the same as the address and hash password in Sect. (4.1. b). Both the address and hash password are independently hashed one time.
- The output of the address with the hash one times which is named MD1 ($H^1(Add) = MD1$) will be inputted into the PUF; the response received from this input in the PUF is must be the same as the existing stored address in the database.

- The output of the hash password with the additional hash which is named MD2($H^{N+1}(PW) = MD2$) will be inputted into the PUF; the response received from this input in the PUF must be the same as the existing stored password the database, as shown in Fig. 8 and Table 1 part B' [18].

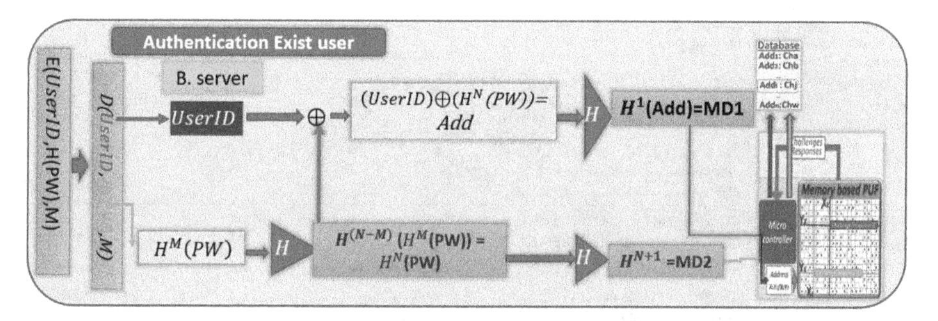

Fig. 8. Exist user at server side

5 Implementation

For implementing the homomorphic password manager with multiple hashes, we need two important tools, which are the hash function and PUF. Also, for the protocol implementation, we need to have two sides, which are the client and server-side. In this work, the hash function that has been used is SAH-3, and for the APG PUF, we have used SRAM PUF. The programming language that has been used for coding this protocol is C++. The next section is going to describe why SHA-3 has been selected and how the SHA-3 has been used in both side client and server.

5.1 SHA-3

According to [30, 19], the hash function accepts arbitrary input, whereas it gives a fixed output length. Besides, they are three main attributes for the hash function point to the level of security in the hash function. According to [37], the attributes of a hash function are Pre-image resistance, Second pre-image resistance, Collision resistance. Pre-image resistance means that it is extremely hard to reverse a hash function computationally, whereas the second pre-image resistance means that having MD and input, but it is hard to find a different input with the same hash. Finally, the collision resistance means If there are two different inputs of any length, it should be hard to find the same MD. Also, According to [37], The Secure Hashing Algorithm (SHA), which is a family of cryptographic hash functions, is recommended to be used. SAH family has several types, which are SHA-1, SHA-2, and SHA-3, so in this work, we decided to used SHA-3. Therefore, the Cryptopp has been used to provide the hash function software and some encryption software. The Cryptopp package is an open-source C++ class library of cryptographic algorithms, and schemes have been written by Wei Dai [38]. So, we built a

Table 1. Shown all steps of the protocol

Steps	Mode	Description of the instruction	Data stream/information	where
A	New User	The password hash N times	$H^N(PW) = H(PW)$	Client
A	New User	The USER-ID named as address	$UID = ADD$	Client
A	New User	Address, H(PW), and number of hashing are encrypted	$E\{Add, H(PW), N\}$	Client
B	New User	Address, H(PW), and number of hashing are decrypted	$D\{Add, H(PW), N\}$	Server
B. 1	New User	(The last time of hashing of PW) XOR with (USER-ID) hashed one time to obtain the message digest which is called MD1	$H\left(UID \oplus H^N(PW)\right) = H^1(ADD) = MD1$	Server (Hash)
B. 2	New User	The password which has been hashed N times is hashed one more time to obtain the message digest which is called MD2	$H^1(PW) = H^{N+1}(PW) = MD2$	Server (Hash)
B. 1. a	New User	Extracting from MD1(X, Y) Then, send as input to PUF to tell which the location in database DB	$MD1(X_i, Y_i)$ *input to the PUF to get the challenge for the Address*(tell the location in DB)	Server (APG)
B. 1. a	New User	The obtained challenge is stored in DB instead of the real User ID	The challenge = Add_i in database column	Server (DB)
B. 2. a	New User	Extracting from MD2(X, Y) Then, send as input to PUF	$MD2 = (X_i, Y_i)$ *as input to the PUF to get the challenge for the PW*	Server (APG)
B. 2. b	New User	The obtained challenge of password(cha(pw)) is stored to its corresponding address in the database	$Add_i : cha(pw)_i intheDB$	Server (DB)
A'	Exist User	The password hash M times where M must be less than N	$H^{M<N}(PW) = H^M(PW)$	Client
A'	Exist User	The USER-ID named as Address	$UID = ADD$	Client
A'	Exist User	Address, $H^M(PW)$, and number of hashing(M) are encrypted	$E\left\{Add, H^M(PW), M\right\}$	Client
B'	Exist User	Address, $H^M(PW)$, and number of hashing(M) are decrypted	$D\left\{Add, H^M(PW), M\right\}$	Server
B'. 1	Exist User	The initial password was hashed M times. Subsequent PW are hashed (N-M) times to obtain the message digest $H^N(PW)$	$H^{(N-M)}\left(H^M(PW)\right) = H^N(PW)$	Server (Hash)
B'. 2	Exist User	(The last time of hashing of $H^N(PW)$) XOR with (USER-ID) to get address	$\left(UID \oplus H^N(PW)\right) = add$	Server (Hash)
B'. 3	Exist User	Hash (USER-ID is XORed with $H^N(PW)$ one time to obtain the MD1'	$(H^1(UID \oplus H^N(PW)) = MD1'$	Server
B'. 3.a	Exist User	Extracting from MD1(X, Y) Then, send as input to PUF to get the response	$MD1' = (X_i, Y_i)$*as input to the PUF to get the response for the* Add'_i	Server (APG)
B'. 3.a.1	Exist User	The obtained response is compared with the challenge that has been stored in DB as address	If (The challenge = Add_i) = new (response = Add'_i) - > Go to next step, otherwise reject	Server (DB)
B'. 4	Exist User	Hashing $H^N(PW)$ one more to get $H^{N+1}(PW)$ which is called the MD2'	$H^1\left(H^N(PW)\right) = MD2'$	Server
B'. 4.a	Exist User	Extracting from $MD2'$ (X, Y) Then, send as input to PUF to get the response	$MD2' = (X_i, Y_i)$ *as input to the PUF to get the responce for the* $response(pw)_i$	Server (APG)
B'. 4.a.1	Exist User	The obtained response must be same as the Exist password(cha(pw)) in the DB, otherwise reject the user	Is $(response(pw)_i = cha(pw)_i$ *in the database) then accept* otherwise *reject*	Server (DB)

function that can call the hash function from the CryptoPP. The hash function that we used can accept the user password as input and provided 512 bits as output, see Fig. 9.

```
byte* Homomorphic_PW_SRAM::hash_512MessageDigest(byte* result, int size){
   byte* digest512 = new byte[64];
   SHA3_512().Calculate Digest(digest512, result, size);
   return digest512;
}
```

Fig. 9. Shows the method that has been used to call the hash function from Crypto++ package.

The second important tool is PUF. The next section is going to describe how the SRAM read and how the challenges and responses have been extracted from the SRAM PUF.

5.2 SRAM-Based PUF

In this work, we got the benefit of the SRAM PUF that has been developed by a cybersecurity lab at Northern Arizona University (NAU) [19, 30]. To read the SRAM PUF, we used prior work that has been used [39, 40]; three main routine methods, which are "generatePubPriKeys", "getPriKey", and "exec_bin" have been built to send a "command read" to read the SRAM, see Fig. 10.

```
vector<byte> Homomorphic_PW_SRAM::exec_bin(const char* cmd) {
   vector<byte> output(PUF_SIZE);

   FILE* pipe = popen(cmd, "r");
   if (!pipe) throw runtime_error("popen() failed!");
   try {
      size_t status = fread(output.data(), sizeof(byte), output.size(), pipe);
      if(status < output.size()) {
         throw;
      }
   } catch (...) {
      pclose(pipe);
      throw;
   }
   pclose(pipe);

   return output;
}
```

Fig. 10. Shows the method that has been used to run the command to read the SRAM ship.

By having the hash function and APG work, we can start building the protocol stages. The next section will describe the implementation of protocol steps and what is the strategy that we have followed to simulate the protocol.

5.3 The Implementation of Protocol Steps

To implement the protocol, we have decided to simulate the protocol steps and parties by using the C++ programming language. We have built a code for the new user in both side client and server, as well as we have built a code for the registered user on both sides as well. In the beginning, the user will declare if he is a new user or not. So, in order to solve user diagnostic, we have used the SWITCH statement. By choosing a new user, the code will run the part of the new user at the client-side; and after that, the server-side will run to finish the rest of the procedures. So, for the new user, there are two work parts must be accomplished, one at the client-side and the rest at the server as follows:

A New User at Client Side

- The code will hash the password n times.
- The hash number n with user-ID is going to be sent to the server.
    - o To do that we built a function that received the password as input and keeping the output of each time hash as input again to this function n time. So, the hash number n, which is the last hash, will be the returned value from that function.
- Both the output of hash n, and user-id XORing with the output of hash n is going to be encrypted and sent to the server, see Fig. 11.

```
/////////////// Client side  for NEW USER ///////////////////
Enter an userID: Hello
Enter a password: pass123
===============================================
 *****beginning Hash N times****
Enter the number of Hash: 10
=====================================================================================
The new message digest hashed(10 times):
97A6B6CF634F786FC5DE9CD7E54DB9E93A5C58767EE55E50395E498E46C57C7A3E9F82FF28593F73E416395D
62C93DF26A9E41B0242AA43EFC2203DC76F80606AC346897ABA8AF3AE922EE1844638E58A97A442574E59B41
27FEFA8BA1E495D269A700765ECECD0E9520688805612DECD7ECA9AD0A88B186AD28BCA1A1480E2A6EEAEC0C
BB85920E3C1F319DC98996DCB73B9F4C7A9357DE0BAE1057A5885F0EAB58810B07D9A2AD86FBF122E4566496
7326EF564F9BFCD5E4E5435583F1D3B302B40B725DF92185E67758A04D85CBF9DA068002FE58AD447C52E228
534C52C18A9E66AD04C732DD613DEACF279991CC4261F8EC17CF27CB42A2C7A5D9B35B9F47182ED83666F6BF
9B467510F1BBE4E3CE1EFADB1AFCA404266960ECE716C7D786B600794CBADD5D4965BE1C3F656B79D50A087F
98F88D73BAD07EE1C1546626F43174D9BE0DF3576B8C29AFF8C3667E990119E1462F2F4943ABDB1D4432A8CE
2E8A50FFEA8F379CA342EC795E93E3388AA511EA3B70622FB15414186932291BADA1A342483F35FD872B7E0B
909DCF83789E0BC68F370239AE560D8203B432BF743D0F3C403DBD22CFDD7636571F83AE3BEA03E9931836FE

The last hash (H(PW)^N times) 10:
909DCF83789E0BC68F370239AE560D8203B432BF743D0F3C403DBD22CFDD7636571F83AE3BEA03E9931836FE
The userID⊕lasthashPW( Add ) :
F1EEBCEA0AF70BC68F370239AE560D8203B432BF743D0F3C403DBD22CFDD7636571F83AE3BEA03E9931836FE
```

Fig. 11. Illustrated both values of last has of password H(PW)10, and the value of XORing User-ID with H(PW)10

Figure 11 shows the example we have run through the simulations. We decided to go 10-time repeating the hash. The output of the hash at 10 times is XORed with user-ID to hide the user-ID as well. Both the output of hash n and the result of XORing are going to be encrypted and forwarded to the server.

A New User at Server Side

When the server received the output of hash n and XORing the user-ID with the output of hash n:

- It is going to decrypt the XORing by XOR the result of XORing with the output of hash n again to obtain back the user-ID in the server-side.
- By retrieving the user-ID, both values of hash 10 times and user-ID, they are going to be fed to the hash function again. And then, the two new hash values 512 bits will be obtained. One for user-ID and one for the last hash of the password.
- Every 16 bits from 512 bits value of the hash are going to point to one cell in the SRAM PUF.
- The value of the cell and its following 15 cell values is going to be read.
- These 16 bits are stored in the first chunk (2 bytes) of the challenge array.
- Repeating that 32 times will give us new 32 different chunks, (32* 16 bits) means that we received a new 512 bit come from the SRAM PUB, see Fig. 12.

```
///////////////// server side  for NEW USER ///////////////////
The address before sending to the PUF :
5DD28D47A79051FF6FBDA9D43D8A33C61F69A2DC01F3FA720F31535B523BB3923FFE6E2ACA9DA0EADC549F6E8A7DAC4307B2C7E2FA0B3938E0BF4B4A7224420
The password before feed to the PUF :
007330C112CD03E6510C37BB3986E99AA686D75FA06972655EE5E1E25AEA94F1E0E23B2D4D923607DE58FD25DD759C7FE9BA1A6AD0208251361A8C88816631C

The SRAM_PUF has been read
 Done**
The addresses that have been extracted from the SRAM (in stream of byte form):
E4E73D7BA0601200C9FBDFFBA7354128F3CE40850440BBBF6FF61B09270045B9BFFFFC6FF37324127F8CDFFFF3BF977F41A88AC08557AB3F100D9F5F6381823C
The password that have been extracted from the SRAM (in stream of byte form):
002828E03186216583060312BEFFFBFA3C8A20B035E00480FC6712357DF820001102DE3F772F1100BFBBF9D9FCEDFDE6FA273D7B0C9722040401FF1D3401056C
```

Fig. 12. Every 2 bytes in both the address stream and password stream pointed to a specific cell in SRAM. Reading the cell and the 16th cells after will provide new 2 bytes again. These 2 bytes come from the SRAM's cells values.

According to [18], after extracting the challenges from the SRAM PUF, they are going to be stored in the database for later authentication purposes, see Fig. 13.

Figure 13 illustrates how the first byte of the challenge of address is going to tell us about the location in the DB. E4 was the first-byte value, which means that in the location of E4 in the DB table, the challenge of the password must store corresponding to E4. After storing the hash of the password to its location in the database, we are on the last phase of new user procedures. After that, we can move to the registered user or

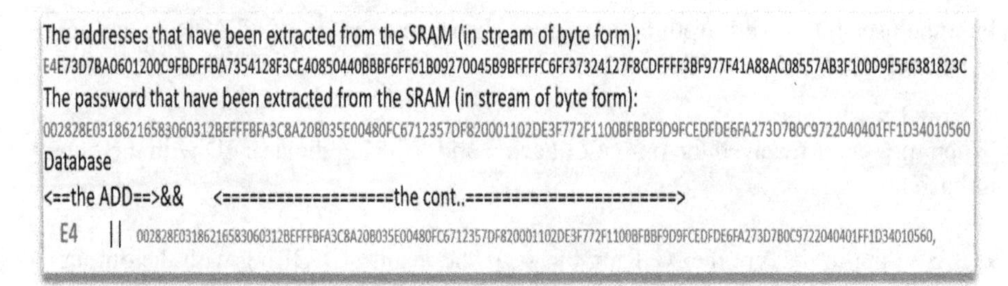

Fig. 13. The First byte from the address challenge point to the location number in the database, and the challenge of password hash is stored corresponding to its location.

the user already in the system. By choosing a registered user, the code will run the part of the registered user at the client-side and, after that, will move to the server side to finish the authentication of procedures. So, there are two phases for the registered user must be accomplished, one at the client-side and the rest at the server as follows:

A Registered User at Client

Same as what occurs in the client-side for the new user will happen for registered users with slight differences:

- The code will hash the password m times instead of n time in the new user phase.
- After that, the hash number m with user-ID is going to be sent to the server.
- The same function that has been used for the new user will be used for the registered user as well. So, the hash number m will be the returned value from that function.
- Both the output of hash m, and user-ID XORing with the output of hash m is going to be encrypted and sent to the server, see Fig. 14.

Figure 14 shows the example we have run through the simulations. For the registered user, we must go for the number of hashes less than the time of registration of the new user. Here in the example, we decided to go 5-time repeating the hash. The output of the hash at 5 times will be XORed with user-ID to hide the user-ID as well. Both the output of hash 5 and the result of XORING are going to be encrypted and forwarded to the server.

To control the value of m, we can use several ways:

- One of them is that the value of M at first login will equal to N – 1, and for the second login, it will be N – 2 and so on. When the M = (N – (N – 1)), so the server automatically will upgrade the number of N.
- The second way for controlling the M is to choose the number for M randomly among N and 2. The value for M must not be repeated.

In this work, we used a table to store the number of M that already has been chosen to avoid using them in upcoming logins. If the server receives the same number of hashes that have already been used, then the server can consider it an abnormal event.

```
//////////////// Client side  for Existing USER ///////////////////
Enter an userID: Hello
Enter a password: pass123
==============================================
 *****beginning Hash M times****
Enter the number of Hash must : 5
=========================================================================================
The new message digest hashed(5 times):
CBF1E1E9AE0B78E9F85F22454E2B89253368E170396CE2CBFAEA8F81F389B370F00CEAB5DC92F1F42A8D8FE1
4B40E0AA676F34EA3F7766AE54FAF206F45C42B6931E91C924447E40BD47DC274CEF41ECC6C075541FD9DCDB
5EB7B60819F870512ECCA936DDD3A72F82EF1A81F7B73C0F3966CFD4A7B8821EE47F4F28BD6989D1878B9464
CCA61C4A61CCC363E4A074AED5C22E9592DB7D9AD91912069BA3EB4FBECBF06CA170221EF0D38CFB89138A8A
3B4BCFA51D46C1F4782060BC69A5CC677DF0ECB38B226900909D3BEEF31F6971F4CA3DD1F839EBBD7B3A5C04
=========================================================================================
The hash number (M) 5:
3B4BCFA51D46C1F4782060BC69A5CC677DF0ECB38B226900909D3BEEF31F6971F4CA3DD1F839EBBD7B3A5C04

The userIDXorlasthashPW(Add) :
5A38BCCC6F2FC1F4782060BC69A5CC677DF0ECB38B226900909D3BEEF31F6971F4CA3DD1F839EBBD7B3A5C04
```

Fig. 14. Illustrated both values of last has of password H(PW)5, and the value of XORing User-ID with H(PW)5

Registered User at Server
In server:

- We are going to XOR the result XORing the output of hash at 5 times with user-ID to retrieve the user-ID, see Fig. 15.
- Also, we are going to hash n-m times to reach n times hash result. By having user-ID and result of n time hash, see Fig. 16.

```
//////////////// server side  for Existing USER ///////////////////

The (user-ID XOR last hash(PW)^5 XOR with hash Password 5 times : 6173736972690000 ( retrieve USER-ID )

The hash number (M) 5:
3B4BCFA51D46C1F4782060BC69A5CC677DF0ECB38B226900909D3BEEF31F6971F4CA3DD1F839EBBD7B3A5C04AB11A54B310752AFC72C8AB48EF1D4142F4B8126
=========================================================================
```

Fig. 15. Shows how can the user-ID is retrieved.

- We will be able to do the procedures as a new user server-side phase.
- By reading the SRAM PUF and extracting the responses, we can make a comparison among the result that already existed in the database to finish the authentication for the user.

```
//////////////// server side for Existing USER ////////////////////
The new message digest hashed(5 times):
E5DE3620D3E9106F3536B98D2F1693C0D9641A04B17414CC00E93CE55D2D0002A8867C3A385BE3CE6ADE1B063D272AA74A5BEF2395589273CF12C7F1EB277BCF
19F8EDC6ED616488ECDC1FFC5B0777DCE2E7D16B8AAACD950A8B22F2254F12E0107D8413B7CF0BD8ABC46C15CA55598E20350186E222E5C654AEC9D7F59961A5
105DC5FBCFF5E0EE30529929C5F50AC6D25168C435F9938080D8119F1DF742FE2C5C79380D723F22C6396C6BB405C4C7BF7C23E0F915311F9EBDC41356ABEB58
85E7A96EBFE0AA7B3B59D803B8ABA496D8CA60891623374BFCAB64F7D7644DD7CFFBAD00DDFC5A03F210F5C73A21F2562EE9A4CD9E4F7FAC32C09FBC615BE4F1
E38BCD62F6F8EB38EA0A409A0065D74409948CDD424D4A80522AC2A12C1DCEB6578BCB674C382F6DBE15ECCD3B1395A0008C68C60635CFE642EE46963C1872D8
The last hash number from (M)= (5) to N= (10):
E38BCD62F6F8EB38EA0A409A0065D74409948CDD424D4A80522AC2A12C1DCEB6578BCB674C382F6DBE15ECCD3B1395A0008C68C60635CFE642EE46963C1872D8
The New user-ID XOR last hash PW after reaching the N level :
82F8BE0B8491EB38EA0A409A0065D74409948CDD424D4A80522AC2A12C1DCEB6578BCB674C382F6DBE15ECCD3B1395A0008C68C60635CFE642EE46963C1872D8
```

Fig. 16. Shows how can move the hash number (M) = (5) to N = (10).

According to [19], we have the same approach to solve the problem of the SRAM Cells that are unsustainable. If the mismatches among the challenge and response less than five percentages, we are going to accept as a normal user, otherwise we will reject, see Fig. 17.

```
The address before sending to the PUF for Exist User:
5DD28D47A79051FF6FBDA9D43D8A33C61F69A2DC01F3FA720F31535B523BB3923FFE6E2ACA9DA0EADC549F6E8A7DAC4307B2C7E2FA0B3938E0BF4B4A7224420D
The password before feed to the PUF for Exist User:
007330C112CD03E6510C37BB3986E99AA686D75FA06972655EE5E1E25AEA94F1E0E23B2D4D923607DE58FD25DD759C7FE9BA1A6AD0208251361A8C88816631D7
The SRAM_PUF has been read
The addresses that have been extracted from the SRAM (in stream of byte form):
E4E73D7BA0601200C9FBDFFBA7354128F3DE40850440BBBF6FFE1B09250045B9BFFFFC6FF3732C127F9CDFFFF3BFD77F51888AC08D5FAB3F128D9FDF6381823C
The password that have been extracted from the SRAM (in stream of byte form):
006828E0318601658307011 2BEFFFBFA3C8A20B035E00480FC6712357DF820001102DC3F772F11029FBBF9D9FCCFFFEEFA273D6F2C9722048401FF1534010560

**** Find if there is matching in the system among the challenge and response :

The exist challenge  :002828E03186216583060312BEFFFBFA3C8A20B035E00480FC6712357DF820001102DE3F772F1100BEBBF9D9FCEDEDE6FA273D7B0C9722040401FF1D34010560
The Mismatch         :00 00000000 0000 000000000000000000000000000000000000000000 00000000 0000000 - 000000 000000 0000 00000000
The new response     :006828E0318601658307011 2BEFFFBFA3C8A20B035E00480FC6712357DF820001102DC3F772F11029FBBF9D9FCCFFFEEFA273D6F2C9722048401FF1534010560
The number of mismatch is: 14
*************:

The smallest number of mismatches is : 14
**************************

* Hello exists in the system    *
**************************
```

Fig. 17. Illustrates the authentication process happens, if the number of mismatches less than 5%, then the user will be accepted; otherwise, reject the user.

At the end of illustrating the implementation of protocol stages, we can recognize that each time the existing user wants to do authentication, a different hash of password will be sent. This hash of password value will be a value among the hash of number one and hash of n − 1. We can infer that the man in the middle each time will see a different stream of bits. By sending each different time stream via network makes learning from network flow impossible. Also, the server will not store or know what the original password is, because it just knows the hash of it.

6 Conclusion and Future Work

This paper describes the implementation of homomorphic password managers using multiple hashing cycles, and SRAM-based PUFs. The work is divided into separate parts, at the client and the server sides. Both client and server have their own processes to implement the protocol whether a new user is enrolled, or an existing user tries to be authenticated. Examples of methodologies to register new users and to authenticate returning users are demonstrated with full hardware and software implementations. The pivotal elements of the protocol are the multiple hash function algorithm combined with SRAM-based PUF to generate a database that does not disclose user IDs and passwords.

To enroll as a new user, the password is subject to a great number of multiple hashing, and only the last MD is sent to the server as reference. Rather than storing MD, the server uses MD to find addresses in PUF, which generate data streams that are stored in the DB for future authentication. To enroll as an existing user, the password is hashed multiple times (at levels lower than the one used during registration time), and the final MD is sent to the server for authentication. The server hashes the MD to get the level of hash at registration time; after that, the server extract data streams from the PUF and compares it to the data streams stored during registration to find a match, or it rejects the password. During subsequent authentication cycles, the users again hash their passwords multiple times, but at levels lower than the ones used during the previous cycles. Thereby it becomes useless for third parties to intercept previously hashed passwords, because hashing functions are not reversible, and the previous message digests are never used twice. Hacking a database containing the data streams extracted from the PUF during registration is worthless without also hacking the PUF, thereby offering an additional level of protection.

The password managers that we are proposing are relatively light and inexpensive to implement. Hardware implementation of hash algorithms are commercially available at a low cost. We estimate that the latencies of protocols incorporating one thousand SHA-3 cycles can be less than 10 ms with commercially available microcontrollers powering client devices such as cell phones. The 1 Mbit SRAM devices that were used in this work to design the prototypes are also commercially available at less than $5 each.

Two important issues remain outstanding and will be addressed by future work: i) the resilience of the architecture if the SRAM fails due to quality problems or malicious interventions, and ii) limited bandwidth in term of the number of possible user authentications per second due to the delays in converting the message digests into data streams generated from the PUFs. Two methods are under investigation to mitigate these two issues and build redundancy; the first one is to use larger SRAM devices that can incorporate multiple addressable PUFs, the second one is to include routers and two-dimensional arrays of addressable SRAM devices. To address the resilience issue, two (or more than two) PUFs can be used concurrently to generate two data streams from a message digest that are stored as a reference in the DB. If the first PUF fails during authentication, which is highly infrequent, the probability of having a concurrent failure of the second PUF is negligible. Such multi-PUF architectures also address the bandwidth issue by implementing concurrent authentication cycles for multiple clients.

We are also studying other methods of homomorphy in addition, or in replacement, of the multi-hash methods used in this work. We are considering using variations of the post-quantum cryptographic (PQC) protocols under evaluation by the National Institute of Standards and Technology (NIST) [41–43].

Acknowledgments. The author is thanking the contribution of several graduate students at the cyber-security lab at Northern Arizona University, in particular, Christopher Philabaum, Vince Rodriguez, Ian Burke, and Dina Ghanaimiandoab. Also, the author is thanking the contribution of Jazan University.

References

1. Jeong, Y.-S., Park, J. S., Park, J.H.: An efficient authentication system of smart device using multi factors in mobile cloud service architecture. Int. J. Commun. Syst. **28**(4), 659–674 (2015)
2. Saxena, N., Choi, B.J.: State of the art authentication, access control, and secure integration in smart grid, vol. 8, MDPI AG, pp. 11883–11915 (2015)
3. Zhang, M., Zhang, J., Zhang, Y.: Remote three-factor authentication scheme based on Fuzzy extractors. Secur. Commun. Netw. **8**(4), 682–693 (2015)
4. US20050125699A1 - Sarts password manager - Google Patents. https://patents.google.com/patent/US20050125699A1
5. Coates, M.: darkreading.com, Safely Storing User Passwords: Hashing vs. Encrypting, 4 June 2014. https://www.darkreading.com/safely-storing-user-passwords-hashing-vs-encrypting. Accessed 20 Dec 2018
6. Gordon, W.: Life hacker, How Your Passwords Are Stored on the Internet (and When Your Password Strength Doesn't Matter), 20 June 2012. https://lifehacker.com/how-your-passwords-are-stored-on-the-internet-and-when-5919918. Accessed 28 Aug 2018
7. Higgins, K.J.: Dark reading, 8 5 2008. https://www.darkreading.com/risk/hackers-choice-top-six-database-attacks/d/d-id/1129481. Accessed 25 Oct 2018
8. Hari Balakrishnan, B.M., Raluca Ada Popa, C.M.: Methods and apparatus for securing a database. USA Patent US13/357,988, 25 1 (2012)
9. Cambou, B.: Physically Unclonable Function Based Password Generation Scheme. United States of America Patent D2016–011, Sept 2016
10. Cambou, B.: Password management with addressable PUF generators. USA Patent **D2018–040**, 04 (2018)
11. Cambou, B.: Addressabke PUF generators for database-free password management system. In: Advances in Intelligent Systems and Computing, Flagstaff (2018)
12. Tsai, J.L.: Efficient multi-server authentication scheme based on one-way hash function without verification table. Comput. Secur. **27**(3–4), 115–121 (2008)
13. Zen, J.: Iterated password hash systems and methods for preserving password entropy (2007)
14. (Rainbow Table). https://www.windowsecurity.com/uplarticle/Cryptography/LSO-RainbowCrack.pdf
15. Arias, D.: auth0.com Hashing Passwords: One-Way Road to Security, Hashing Passwords: One-Way Road to Security, 25 April 2018. https://auth0.com/blog/hashing-passwords-one-way-road-to-security/. Accessed 4 Feb 2019
16. US8291491B2 - Password system, method of generating a password, and method of checking a password - Google Patents
17. US Patent for Systems and methods for providing a covert password manager Patent (Patent # 9,571,487 issued February 14, 2017) - Justia Patents Search

18. Assiri, S., Cambou, B.: Homomorphic Password Manager Using Multiple-Hash with PUF. USA Patent **07**(05), D2019–D2045 (2019)
19. Assiri, S., Cambou, B., Duane Booher, D., Mohammadinodoushan, M.: Software implementation of a SRAM PUF-based password manager. In: Advances in Intelligent Systems and Computing 2020 Computing Conference, London (2020)
20. Gao, Y., Ranasinghe, D., Al-Sarawi, S., Kavehei, O., Abbott, D.: Emerging physical unclonable functions with nanotechnology (2016). ieeexplore.ieee.org.
21. Herder, C., Yu, M.D., Koushanfar, F., Devadas, S.: Physical unclonable functions and applications: a tutorial, vol. 102, Institute of Electrical and Electronics Engineers Inc., pp. 1126–1141 (2014)
22. Maes, R., Tuyls, P., Verbauwhede, I.: A soft decision helper data algorithm for SRAM PUFs. In: IEEE International Symposium on Information Theory - Proceedings (2009)
23. Holcomb, D.E., Burleson, W.P., Fu, K.: Power-Up SRAM state as an identifying fingerprint and source of true random numbers. IEEE Trans. Comput. **58**(9), 1198–1210 (2009)
24. Robust, Q.: asecurity site.com, Winternitz one-time signature scheme (W-OTS). https://asecuritysite.com/encryption/wint. Accessed 17 Jan 2019
25. Forler, C., List, E., Lucks, S., Wenzel, J.: Overview of the candidates for the password hashing competition and their resistance against garbage-collector attacks. Lecture Notes in Computer Science (including subseries Lecture Notes in Artificial Intelligence and Lecture Notes in Bioinformatics) (2015)
26. Zhang, Z., Yang, K., Hu, X., Wang, Y.: Practical anonymous password authentication and TLS with anonymous client authentication. In: Proceedings of the ACM Conference on Computer and Communications Security, New York (2016)
27. Paral, Z., Edward, G., Thomas, S., Ras, C., Devadas, R.N., Handelval, V.: Authentication with physical unclonable functions, patent, 19 9 2007
28. Dong-gyu, K.: Puf-based hardware device for providing one-time password, and method for 2-factor authenticating using thereof. Korean Patent KR20140126787A, 22 4 2013
29. Becker, G.T., Wild, A., Guneysu, T.: Security analysis of index-based syndrome coding for PUF-based key generation. In: Proceedings of the 2015 IEEE International Symposium on Hardware-Oriented Security and Trust, HOST 2015 (2015)
30. Cambou, B.: Password manager combining hashing functions and ternary PUFs. In: Intelligent Computing-Proceedings of the Computing Conference., London (2019)
31. 427 million Hacked Myspace Passwords Get Dumped Online | Digital Trends. https://www.digitaltrends.com/social-media/myspace-hack-password-dump/
32. Cybercrime Damages $6 Trillion by 2021. https://cybersecurityventures.com/hackerpocalypse-cybercrime-report-2016/
33. Target: Data stolen from up to 70 million customers. https://www.usatoday.com/story/money/business/2014/01/10/target-customers-data-breach/4404467/
34. Wang, D., Zhang, Z., Wang, P., Yan, J., Huang, X.: Targeted online password guessing: an underestimated threat. In: Proceedings of the ACM Conference on Computer and Communications Security, New York, NY, USA (2016)
35. Bonneau, J., Van Oorschot, P.C., Herley, C., Stajano, F.: Passwords and the evolution of imperfect authentication (2015)
36. Tsai, C.-S., Lee, C.-C., Hwang, M.-S.: Password Authentication Schemes: Current Status and Key Issues (2006)
37. N.-. H. function, NIST - information technology labortory Computer security resource center, Hash function, 04 01 2017
38. Dai, W.: Crypto++. https://en.wikipedia.org/wiki/Crypto++
39. Booher, D.D., Cambou, B., Carlson, A.H., Philabaum, C.: Dynamic key generation for polymorphic encryption. In: IEEE 9th Annual Computing systems and Conference (CCWC), Las Vegas (2019)

40. Assiri, S., Cambou, B., Booher, D.D., Ghanai Miandoab, D.: Key exchange using ternary system to enhance security. In: IEEE 9th Annual Computing and Communication Workshop and Conference (CCWC), Las Vegas, NV, USA (2019)
41. US20040193925A1 - Portable password manager - Google Patents. https://patents.google.com/patent/US20040193925A1/en
42. US20070226783A1 - User-administered single sign-on with automatic password management for web server authentication - Google Patents. https://patents.google.com/patent/US20070226783A1/en
43. Blocki, J., Harsha, B., Zhou, S.: On the economics of offline password cracking. In: Proceedings - IEEE Symposium on Security and Privacy (2018)

An Assessment of Obfuscated Ransomware Detection and Prevention Methods

Ahmad Ghafarian(✉), Deniz Keskin, and Graham Helton

Department of CSIS, Mike Cottrell College of Business, University of North Georgai, Dahlonega, GA 30597, USA
ahmad.ghafarian@ung.edu

Abstract. Ransomware is a special type of malware, which infects a system and limits user's access to the system and its resources until a ransom is paid. It does that by creating a denial of service of a system to its own user by encrypting files and/or locking the machine. The malware takes advantage of people's fear of revealing their private information, losing their critical data, or facing serious hardware damage. Some of the most recent well-known ransomware include WannaCry, Petya, Bad Rabbit, and Baltimore City. Ransomware creators use a special technique called obfuscation to evade detection by antivirus software. The degree of antivirus detection depends on the complexity of the obfuscation process. The purpose of this research is to assess the efficiency of current malware analysis methods and technologies in the detection of ransomware as well as prevention methods to keep files safe using cloud computing. The experiments presented here were performed using antivirus engines and dynamic malware analysis against live obfuscated ransomware samples.

Keywords: Ransomware · Obfuscation · Detection · Prevention · Antivirus · Cloud storage · Tools

1 Introduction

Ransomware is a special type of malware, which infects a system and limits user's access to the system and its resources until a ransom is paid. It does that by creating a denial of service of a system to its user by encrypting files and/or locking the machine. In the past few years, this malware has become popular among cybercriminals and it is regarded as a billion-dollar industry. Cybercriminals launch ransomware attack to extort money. Some of the most recent well-known and destructive ransomware include WannaCry [1], Petya [2] and Bad Rabbit [3]. WannaCry attacked known Windows network vulnerabilities using various exploits, which allowed an intruder to execute arbitrary code on a targeted system by transmitting customized data packets. WannaCry made global headlines after infecting more than 230,000 systems in over 150 countries and causing an estimated $5 billion in damages. Like WannaCry, Petya used Windows vulnerabilities to propagate itself. It impacted large organizations in multiple countries with billions of dollars damage. Bad Rabbit appeared shortly after WannaCry and Petya

K. Arai (Ed.): FICC 2021, AISC 1363, pp. 793–811, 2021.
https://doi.org/10.1007/978-3-030-73100-7_56

ransomware families, which targeted Ukraine's Ministry of Infrastructure and Kiev's public transport system.

There are two main types of modern ransomware i.e. locker and crypto [4]. The infection for both kinds of malicious software happens in similar way. These attacks may include drive-by-download, phishing, spam, social engineering, etc. In the case of locker ransomware, the user is denied access to an infected machine. However, in the case of crypto ransomware attack, the victim is prevented from accessing her/his vital files or data by using some form of encryption. Generally, the victim must pay a ransom in the form of cryptocurrency to obtain the decryption key. In general, ransomware detection is evaded by using obfuscation [5]. In this process the ransomware authors obfuscate the program to protect the inner workings of the malware from security researchers, malware analysts, and reverse engineers. These obfuscation techniques make it difficult to detect/analyze the binary, extracting the strings from such binary results in very fewer strings, and most of the strings are obscured. There are several obfuscation techniques. One of the most common methods of obfuscation is to use existing tools. The current published research on ransomware analysis mostly focus on identifying the behaviors of ransomware once they infect a system. As a result, the infected files and folders are not useable after the system being infected. We believe it is necessary to implement a method to upload the system files securely and automatically using cloud e.

With the above in mind, in this research we assess the efficiency of current malware analysis methods and technologies in the detection of ransomware as well as prevention methods to keep files safe using cloud computing. The experiments presented here were performed using antivirus engines and dynamic malware analysis against live obfuscated ransomware samples. We will use VirusTotal [6] to compare the normal ransomware detection rate against the obfuscated ransomware.

The rest of this paper is structured as follows: Sect. 2 provides literature review and background. Section 3 presents scope and research methodology. Section 4 covers ransomware obfuscation experiment. In Sect. 5, ransomware detection and prevention approaches is presented. The paper concludes in Sect. 6.

2 Background and Literature Review

Although ransomware analysis is a relatively new, but many researchers and professionals have contributed to the field of ransomware detection and prevention. Surati and Prajapati [7] reviewed ransomware detection and prevention research papers for the period of 2010 to 2016. The authors concluded that significant number of ransomware families exhibit similar characteristics. Kharraz et al. [8] report on their review of ransomware for the period of 2006 to 2014. The authors present an early-warning detection system that alerts a user during suspicious file activity. A brief history of ransomware that includes suggestions to prevent an infection as well as recovering from an infection has been performed by Richardson and North [9]. The authors cover the evolution of ransomware from 1989 to 2016 and how the hackers sophisticated the ransomware effectiveness. In a similar work, a detailed ransomware attack lifecycle and its characteristics is presented in [10]. The authors reviewed the existing techniques for ransomware detection, as well as the pros and cons of each technique. A study of 36 academic publications between 2006 and 2011 about methodologies and techniques of malware detection and prevention is covered by Kok et al. [11]. Windows API calls that differ significantly in their usage between normal non- malicious operations and ransomware activities is presented by Hampton et al. [12]. The authors conclude that the study of low-level API calls will help in the analysis of malware detection and prevention.

In another study, Kardile [13] presents a dynamic approach to ransomware analysis. The authors use Cuckoo Sandbox for isolating ransomware and monitoring system behavior using Process Monitor to analyze ransomware in its early stages. Their model also keeps a record of file access rates and other file-related details. The use of machine learning in ransomware detection is studied by Jethva [14]. The author presents a dynamic ransomware detection model with grouped registry key operations and combined file entropy features. The author analyzes the features by exploring different linear machine learning techniques. Fedler et al. [15] conducted various tests on several antivirus apps for Android, in which antivirus software is tested for recognizing known malware samples. The author's test setup cope with typical malware distribution channels, infection routines, and privilege escalation techniques. The authors conclude that through obfuscation techniques the malware can evade detection by most antivirus software. In an experimental research, Sechel [16] assessed the performance of antivirus in detection of obfuscated ransomware. Their experiments show that by using various obfuscation techniques on known ransomware samples the detection and classification can be hindered by antivirus engines.

An evaluation of commercial antimalware effectiveness has been reported by Morales et al. [17]. Their method concentrates on detection and treatment components against malicious objects by partitioning true positives to incorporate detection and treatment. This new measurement is used to evaluate the effectiveness of four commercial antimalware programs in three tests. Kharraz et al. [18] evaluate the behaviors of ransomware vs. General malware. They suggest that defending ransomware should be more straight forward than general malware. They argue that the goal of ransomware is often a reversible DoS attack on data availability. In practice, this means first, performing cryptographic operations on user data, and second modifying many data files. Defenders can use these features to enhance both the detection of and protection against ransomware in ways that are not applicable to malware in general.

3 Scope and Research Methodology

This research aims to assess the efficiency and the effectiveness of ransomware analysis and prevention methods. In addition, we propose a cloud computing approach for preventing users' files and data from possible ransomware attack. In our experiment, we use some of the popular antivirus engines and dynamic malware analysis against live obfuscated ransomware samples. Our research experiment complements the existing methods in various ways including environment, tools, dataset, and the structure. The details of the experiment environment is described below:

- Lab Environment
- Tools and Technology

- Sample Ransomwares
- Ransomware Behavior

3.1 Lab Environment

Since we use live ransomware, it is important to protect the system and perform the testing in a sandbox. This is because analysis of the ransomware without sandboxing may infect the network system. In practice, there are several ways of sandboxing including use of Cuckoo Sandbox, using simulation tools, VMware virtual machine network, etc. In this work, we will use VMware virtual machine network.

3.2 Tools and Technology

This subsection highlights the software tools and technology we used in this research. Most of the tools listed here are downloadable for free.

1. **VMware Workstation Player:** VMware Workstation player [19] is a hosted hypervisor for Windows and Linux systems. We used VMware Workstation 16, to create multiple virtual machines to dynamically test ransomware samples.
2. **IDA:** IDA [20] is a disassembler software for translating machine code into assembly language. The disassembly process allows software engineers to analyze suspected programs such as spyware or ransomware. For our research, we used IDA to analyze the process ransomware uses to encrypt and decrypt files.

a. **Ghidra:** Like IDA, Ghidra [21] is a software reverse engineering framework. It helps analyze malicious code and malware like ransomware. Ghidra can give cybersecurity professionals a better understanding of potential vulnerabilities in their network. For our research, we also used Ghidra to analyze the process Ransomware uses to encrypt and decrypt files.

b. **VirusTotal:** VirusTotal [6] is a website created by the Spanish security company Hispasec Sistemas. VirusTotal aggregates 70 antivirus engines to check for viruses that the user's own antivirus may have missed, or to verify against any false positives. Files up to 550 MB can be uploaded to the website or sent via email (max. 32 MB). Users can also scan suspect URLs and search through the VirusTotal dataset. VirusTotal for dynamic analysis of malware uses the Cuckoo Sandbox. For our purpose, we used VirusTotal to test our obfuscated malware detection rates against the VirusTotals detection rate.

c. **Themida:** Themida [22] is an advanced Windows software protector system. When a protected application is going to be run by the operating system, the software protector will first take control of the CPU and check for possible cracking tools that may be running on the system. If everything is safe the software protector will proceed to decrypt the protected application and giving it the control of the CPU to be executed as normal. For our purposes, we used Themeda to obfuscate our ransomware files.

d. **theZoo Malware Repository:** theZoo [23] is a repository of live malwares for download. TheZoo is a project created to make the possibility of malware analysis open and available to the researchers. For our purposes, we used theZoo repository to download and test selected ransomware.

e. **Cloud Storage OneDrive:** Microsoft OneDrive is a file hosting and synchronization service that allows users to store files. For our purpose, we used OneDrive to upload/download ransomware files for sharing among the members of this research team. In addition, we use OneDrive to protect user's files from Ransomware attacks.
f. **Antivirus Software:** Windows Defender, ESET, and Malwarebytes

3.3 Sample Ransomwares

We selected/downloaded 16 ransomware specimens from theZoo malware Github repository. Each ransomware sample was executed in a sandboxed virtual machine to validate the encryption behavior of malwares. Table 1 shows the selected ransomwares.

Table 1. List of Selected Ransomwares

No	Ransomware	Remarks
1	Cerber	Ransomware-as-a-service (raas), which means the attacker licenses Cerber ransomware over the Internet and splits the ransom with the developer
2	Cryptowall	Scans the infected computer looking for files to encrypt
3	Jigsaw	Spreads via email attachment and is most recognizable for its picture of Billy the Puppet from the popular horror movie Saw. It attempts at scare users for paying the ransom. However, the decryption key had been left in the source code and thus allowed the ransomware to be easily reverse engineered
4	Locky	Uses social engineering to get users to install ransomware. It is spread via spoofed emails. Once the victim opens the document, it would execute macros that encrypts files
5	Mamba	It forces the computer to reboot and upon reboot, a full disk encryption programmed called DiskCryptor is implemented which immediately begins to encrypt the hard drive
6	Matsu	This malware acts more of a backdoor software. Once installed on a system it connects back to a command and control server, allowing for remote code execution. This code can be used for a variety of malicious purposes such as encrypting files as well as stealing user's data
7	PetrWrap	It is a variant of the well-known Petya ransomware and uses many of its functions. PetrWrap 'wraps' the Petya ransomware in such a way that it utilizes its encryption process. This is done to allow PetrWrap to use its servers for handling encryption keys and payments
5	Petya	It encrypts a hard drives file system and propagates via the EtenralBlue exploit. This ransomware is most easily identified by the flashing skill that is seen upon restarting the infected computer
9	Radament	It is a malware that masquerades as directx.exe. While this phony directx is running in the background it will begin to quietly encrypt all your files using AES-256 encryption. Once the files are encrypted it will open a web page that instructs the infected user on how to send an anonymous payment of bitcoin
10	Rex	Rex was first found in May of 2016 and targeted Linux web servers. It did this by scanning for known yet unpatched vulnerabilities. Once it was able to take control of a server it would then lock the website until a ransom was paid
11	Satana	Once a computer is infected with the Satana ransomware it begins to encrypt the Master Boot Record (MBR) which contains information about the computer file system and partitions. Once encrypted the Satana ransomware demands a payment of bitcoins in exchange for the user's files

(*continued*)

Table 1. (*continued*)

No	Ransomware	Remarks
12	TeslaCrypt	It is distributed through phishing emails with malicious attachments. Once the user is infected a web page is pulled up stating how to pay for the decryption key
13	Unnamed_0	A newer version of TeslaCry with additional features
14	Vipasana	Another variation of WannaCry ransomware
15	Vipasana	It is one of the most well-known ransomwares because of its impact on the British National Health Service. What makes the WannaCry debacle so infuriating for many security researchers is the fact that the exploit was not disclosed to Microsoft
16	WannaCry_Plus	It behaves like Wannary

3.4 Ransomware Behavior

In this subsection we demonstrate the behavior of the ransomware after attacking a system. Ransomware is a malware that uses strong encryption to encrypt victims' files and various operating system functions until a ransom is paid to the attacker. The most common way is the server and client asymmetric encryption with symmetric encryption [24]. This method allows attackers to infect and encrypt victims without the need for Internet connection. Internet connection is only needed in the decryption part of the process. The process we followed is described below:

1. Start the Virtual Machine
2. Open Process Hacker/Task Manager
3. Take a screenshot of the process hacker
4. Start the Ransomware (Let it run on the background for 5–10 min)
5. Record the CPU usage and RAM usage
6. Take note of the new processes started
7. Take another screenshot of the Process Hacker
8. Cancel Reservation or Revert to the Last Snapshot

Figure1 shows malware Cerber running as a background encrypting files. The network tab shows that Cerber Ransomware is sending and receiving data (private/public keys) over the network.

After encryption of the user's files, the malware generates a notepad window with instructions on how to pay the ransom for decrypting the encrypted files. As shown in Fig. 2, users are greeted with Read me Notepad message and a Windows background change instructing user on how to pay and decrypt the files. Fig. 3. Shows the user's Windows background telling users that their data is encrypted.

Based on our observation, once the ransomware is in the system, it is uncommon for ransomware to establish persistence, since it increases the chance of being detected by antivirus software. Ransomware detection presents a unique challenge as often the only indicator of the machine being infected is the discovery of files being encrypted and a ransom note appearing. In general, it is recommended to focus more on the early prevention of ransomware, as opposed to detecting it after a network being compromised. The

Task Manager

File Options View

Processes | Performance | App history | Startup | Users | Details | Services

Name	S...	2% CPU	17% Memory	0% Disk
Apps (3)				
Process Hacker		0%	12.8 MB	0 MB/s
Task Manager		0%	23.5 MB	0 MB/s
Windows Explorer		0%	39.3 MB	0 MB/s
Background processes (36)				
Application Frame Host		0%	5.4 MB	0 MB/s
cerber.exe (32 bit)		1.2%	2.8 MB	0.1 MB/s
COM Surrogate		0%	2.3 MB	0 MB/s
COM Surrogate		0%	3.3 MB	0 MB/s
Cortana (2)		0%	7.0 MB	0 MB/s
CTF Loader		0%	3.3 MB	0 MB/s

Fewer details

Fig. 1. Shows ransomware Cerber running in the background

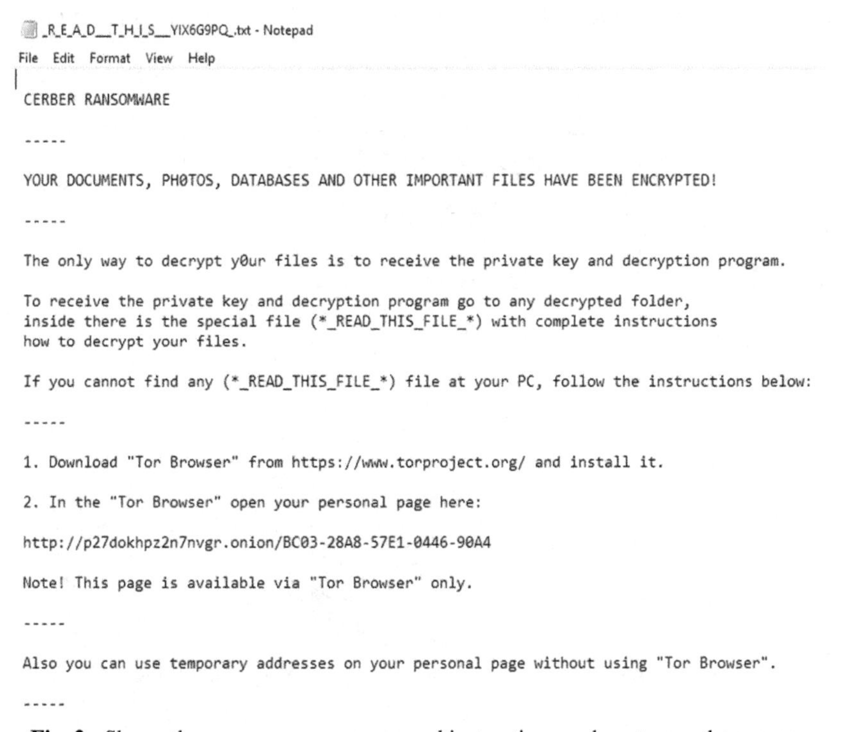

_R_E_A_D__T_H_I_S__YIX6G9PQ_.txt - Notepad

File Edit Format View Help

CERBER RANSOMWARE

YOUR DOCUMENTS, PH0TOS, DATABASES AND OTHER IMPORTANT FILES HAVE BEEN ENCRYPTED!

The only way to decrypt y0ur files is to receive the private key and decryption program.

To receive the private key and decryption program go to any decrypted folder,
inside there is the special file (*_READ_THIS_FILE_*) with complete instructions
how to decrypt your files.

If you cannot find any (*_READ_THIS_FILE_*) file at your PC, follow the instructions below:

-----

1. Download "Tor Browser" from https://www.torproject.org/ and install it.

2. In the "Tor Browser" open your personal page here:

http://p27dokhpz2n7nvgr.onion/BC03-28A8-57E1-0446-90A4

Note! This page is available via "Tor Browser" only.

Also you can use temporary addresses on your personal page without using "Tor Browser".

Fig. 2. Shows the ransomware message and instructions on how to pay the ransom.

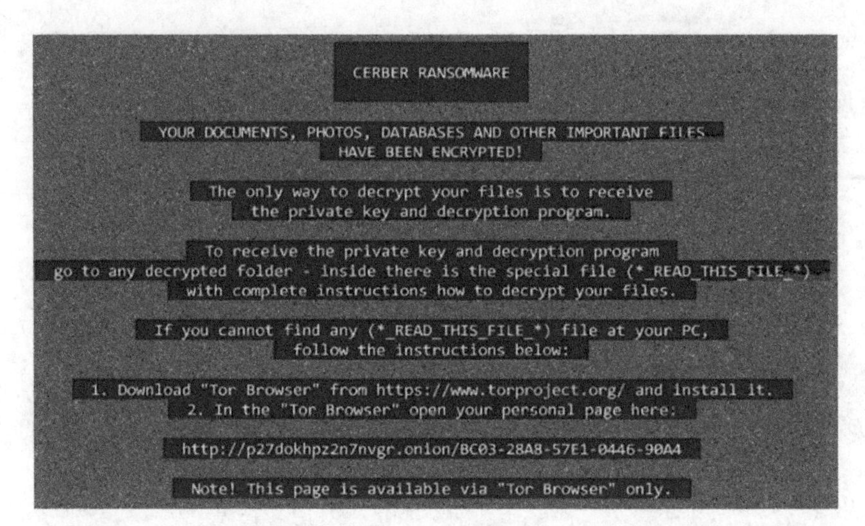

Fig. 3. Shows the user's Windows background telling users that their data is encrypted.

successful infection of a network with ransomware is highly dependent on the transmission vector used. Masquerading ransomware as inconspicuous files such as Microsoft Word documents is a common approach to deceive users into opening malicious files. Exploiting vulnerabilities in unpatched software is another approach. Another common technique used by attackers is to obfuscate ransomware or to hide it in another program, such as a calculator. Figure 4 shows ransomware TeslaCrypt disguised as an ordinary

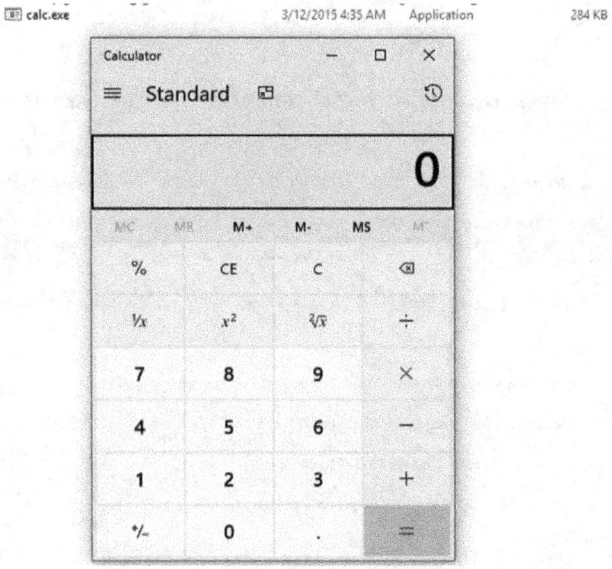

Fig. 4. Shows TeslaCrypt disguised as a calculator application.

calculator application [25]. When the user starts the Calc.exe program, it works as a calculator application while running the encryption application as a hidden background process.

For ransomware to evade antivirus, attackers commonly use Packers and Crypters to hide the ransomware [26]. A Packer is a program that uses compression to obfuscate the executable content. It takes the contents of the software to decompress it while changing the structure of the ransomware. This process allows antivirus software to ignore it until its execution, which decompresses the program back to its original self. Crypter is a software that uses encryption to obfuscate the executable's content. Tools like Themida allows executables to be encrypted in a new executable file. Upon execution, encrypted programs decrypt itself to the original binary code of the executable [27]. Figure 5 shows common ransomware obfuscation technique of packing and unpacking software.

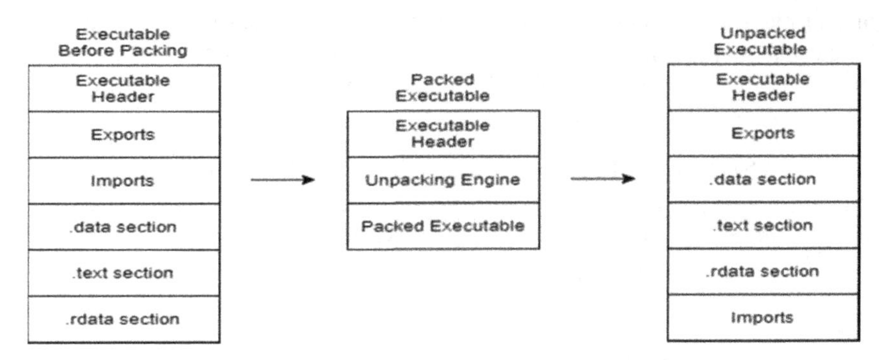

Fig. 5. Shows common ransomware obfuscation technique.

See Fig. 8 below for the original dissembled version of ransomware Wannacry in Ghidra. This version is susceptible to reverse engineering and antivirus detection. See Fig. 9 below for the packed version of the Wannacry ransomware which reveals the code for the decompression process.

4 Ransomware Obfuscation Experiment

Now that we have demonstrated how ransomware behaves, we are ready to apply obfuscation to 16 ransomware datasets. We do that to evaluate the effectiveness of some of the existing antivirus engine detection technologies against obfuscated ransomware. During the experiment, we recorded the detection rate of both obfuscated and original ransomware for each ransomware.

In software development, obfuscation is used to deliberately change the source code or assembly code of the software to make reverse engineering harder [28,29]. Ransomware obfuscation is very popular among malicious hackers. We used VMware Pro as discussed above for our experiment and built a network Virtual Machines. In its simplest from we have a host VM and a server VM. We then performed the following obfuscation experiment.

1) Start host Virtual Machine
2) Start Themida
3) Adjust Themida Protection Options
4) Obfuscate Ransomware using Themida
5) Run the Obfuscated files through VirusTotal
6) Tabulate the detection results on server virtual machine.

Ransomware obtained by the theZoo repository went through the above obfuscation process to increase the difficulty of detection and reverse engineering. Obfuscated ransomware was then analyzed using Virustotal. Testing of the malware was done on a Windows 10 platform with modified security settings (Windows Defender and Ransomware Protection turned off). For the obfuscation process, Themida x34 3.0.4.0 was used to modify ransomware samples. Themida packer has options like string encryption, anti-file patching, and anti-virtualization techniques, it makes reverse engineering and detecting malware much harder. Table 2 shows the options that we selected for this experiment. See Fig. 6 and Fig. 7 for Themida options.

Table 2. Shows Themida Options Selected in the Obfuscation Process.

No	Themida Options	Selection
1	Anti-Debugger Detection	Enable
2	Advanced-API-Wrapping	Enable
3	Compress and Encrypt	Application
	Compress and Encrypt	Resources
	Compress and Encrypt	Security Engine
4	Encrypt Strings in VM macros	ANSI String
	Encrypt Strings in VM macros	UNICODE string
5	Extra Protection Options	Delete File/Registry
	Extra Protection Options	Monitors
	Extra Protection Options	Entry Point
	Extra Protection Options	Virtualization
	Extra Protection Options	Anti-File Patching
	Extra Protection Options	Anti-Sandbox
	Extra Protection Option	Allow execution of VMware/Virtual PC

After we obfuscated the sample ransomwares, they were analyzed using VirusTotal and sandboxed virtual machine to detect behavior and analyze the damage caused by the malware. Sandboxing ransomware allows ransomware to be dynamically analyzed in a secure space.

To analyze the obfuscation process and behavior of the ransomware, Ghidra was used to disassemble the popular ransomware WannaCry (see Fig. 8 and 9).

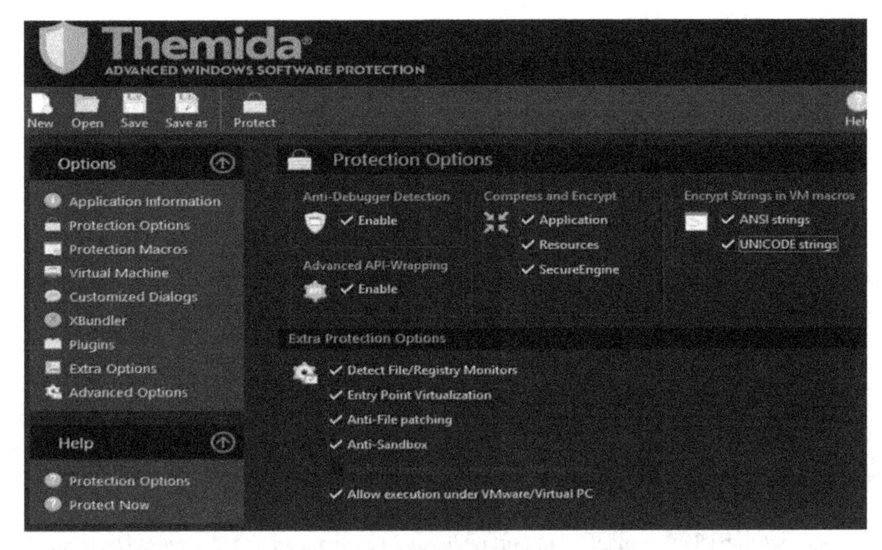

Fig. 6. Shows Themida packer protection options

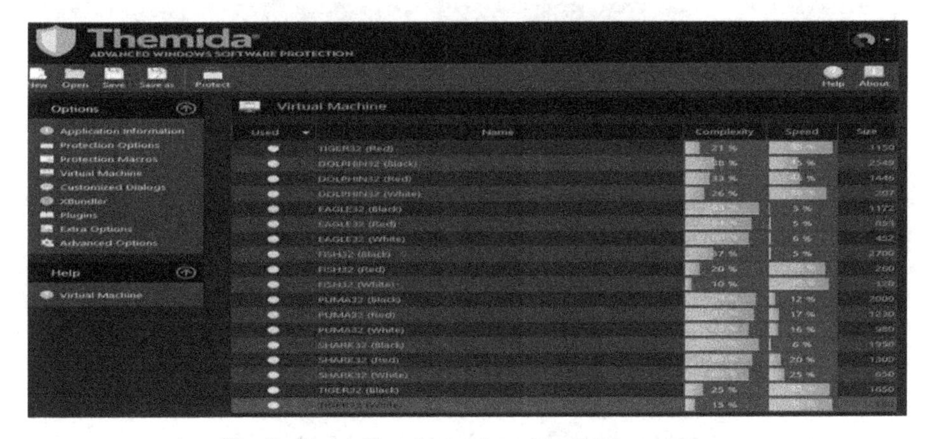

Fig. 7. Shows Themida packer virtualization options

Themida packer was successfully managed to obfuscate the ransomware which based on Fig. 8 and 9 made reverse engineering and the detection of the ransomware impossible. This is further verified by the detection rates of VirusTotal analysis. Running 16 obfuscated ransomware files through VirusTotal's detection engines resulted in a reduction of 36% detections rates compared to 87% (see Table 3).

In summary, using various obfuscation techniques like packing, encrypting, and anti-debugging on ransomware samples, hinder the detection and analysis of the ransomware by antivirus engines. After running through obfuscation process the average rate of detection among 16 ransomwares was 50.88%. This is a decrease of 35.71% from 86.71% original ransomware. Despite the malware being older than 36 months, using a simple software packer managed to bypass heuristic antivirus engines with ease.

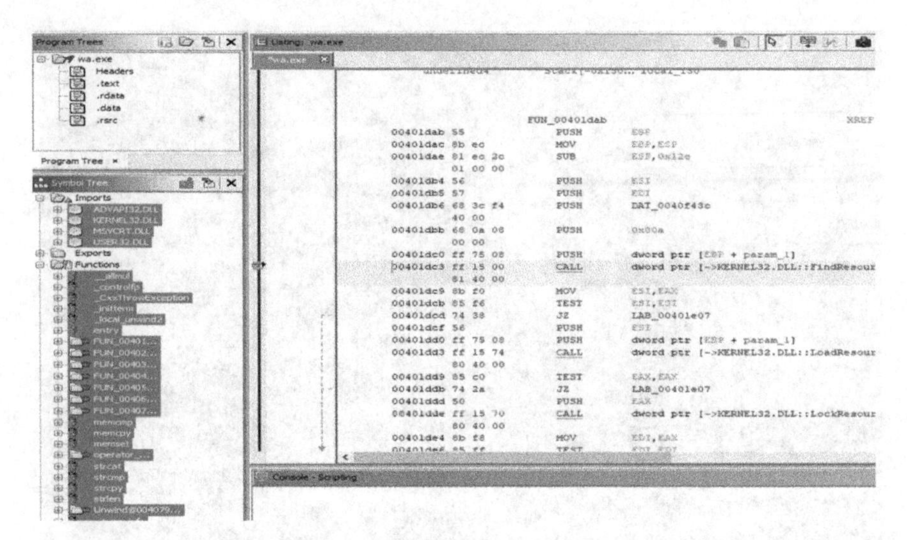

Fig. 8. Shows unpacked WannaCry import tables and functions loaded in Ghidra Disassembler

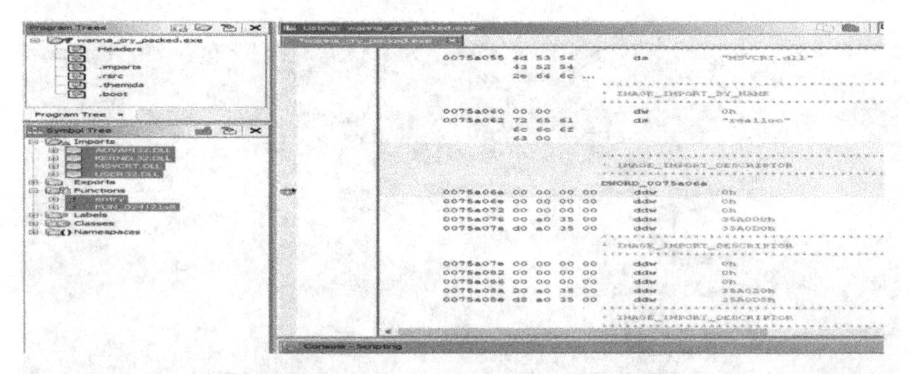

Fig. 9. Shows packed WannaCry import tables and functions loaded in Ghidra disassembler.

5 Ransomware Detection and Prevention Methods

Preventing ransomware infections can be a daunting task because new strains of ransomware are constantly being developed. Despite this challenge, there are ways to harden a network to reduce the risk of a ransomware attack. This section will go over the methods of Ransomware prevention and mitigation methods. First, we will test various antivirus software to determine their effectiveness. Second, we will examine cloud backup of the user's file for mitigating the attack.

5.1 Effectiveness of Antivirus Software

Early prevention of ransomware attack is crucial. That is because once a computer has been infected with ransomware, it can be too late unless the computer is running an antivirus software that can stop the ransomware before any encryption of files takes

Table 3. Percentage of VirusTotal protection rate

No	Ransomware	VirusTotal detection	VirusTotal detection Rate after obfuscation
1	Cerber	90.27%	40.57%
2	Cryptowall	91.66%	30.98%
3	Jigsaw	93.05%	50.11%
4	Locky	95.83%	43.66%
5	Mamba	89.04%	36.23%
6	Matsnu	87.67%	47.88%
7	Petrwrap	86.30%	25.35%
8	Petya	92.95%	43.05%
9	Radamant	83.33%	50.00%
10	Rex	62.71% t	null
11	Satana	90.14%	44.44%
12	TeslaCrypt	89.04%	45.83%
13	Unnamed_0	80.08%	32.85%
14	Vipassana	85.91%	43.39%
15	WannaCry	84.72%	45.71%
16	WannaCry_Plus	84.72%	39.43%
	Total Detection Rater	**86.71%**	**50.88%**

place. In this section will examine the effectiveness of Windows Defender and other common antivirus software against ransomware attack. The steps for this experiment described below:

1) Create a brand-new Windows 10 Virtual Machine
2) Install or turn on antivirus software
3) Infect the Virtual Machine
4) Observe the behavior of the antivirus software

To observe the actions taken by various antivirus engines, a fresh Windows 10 virtual machine was created to create an isolated sandbox environment for testing. Once creating the virtual machine, ransomware was loaded on to the system inside of zip folders. Three different antivirus solutions were employed one at a time to gauge their effectiveness. These are Windows Defender, Malwarebytes and ESET. Once the tests were completed for one of the antivirus software, the machine was reimaged, and another antivirus software was tested.

5.1.1 Windows Defender

Our experiment showed that Windows Defender is the most lightweight of the three antivirus options observed. By default, it did not allow ransomware such as WannaCry and Petya to run when clicked on, instead, displaying a large red warning information the user that the application you are trying to run is likely malicious and may cause damage to your device (see Fig. 10). However, it does allow the user to click through and run the file which results in the machine becoming infected.

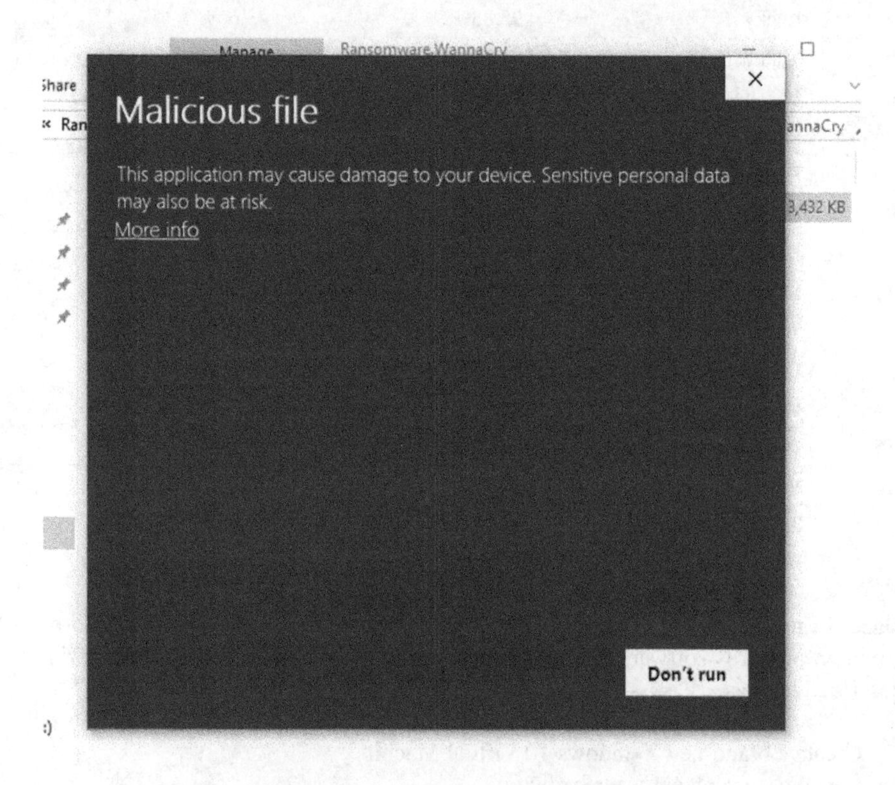

Fig. 10. Windows Defender warning pop-up

5.1.2 Malwarebytes

Malwarebytes is a popular commercial antivirus engine that offers a lightweight look and feel as well as an easy-to-use control panel. When this antivirus installed the ransomware tested was immediately stopped and a large notification appeared in the lower right of the screen informing the user that an action had been taken to quarantine the ransomware (see Fig. 11). Once quarantined, the user can go to the Malwarebytes dashboard and reinstate the removed ransomware; however, Malwarebytes ensures that the user knows the file is most likely malicious.

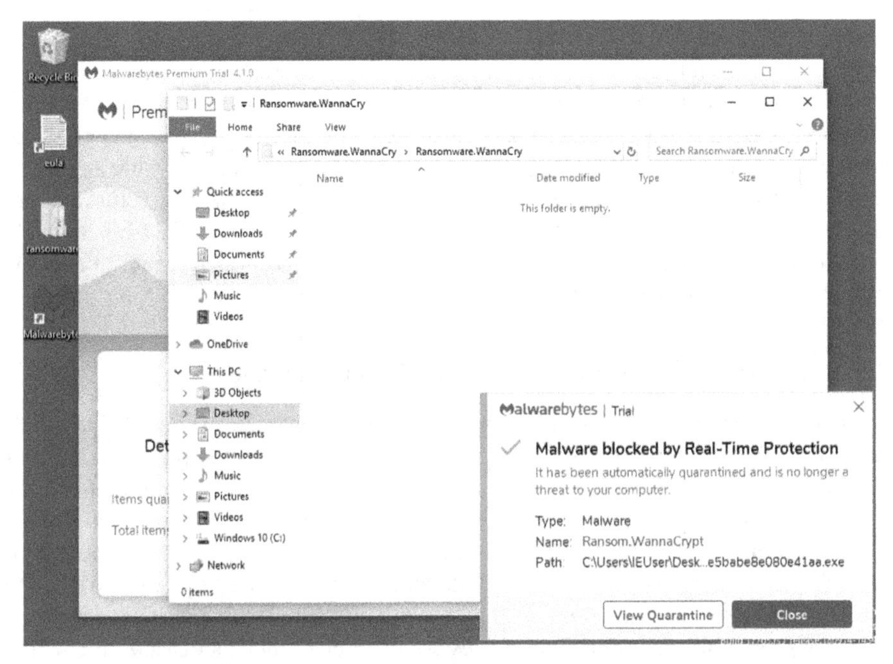

Fig. 11. Malwarebytes pop-up notifying users about blocked ransomware.

5.1.3 ESET

ESET is another powerful antivirus engine that, like Malwarebytes. Once installed, ESET antivirus prompts the user with a plethora of information promising real-time protection. With the antivirus installed, the first ransomware sample was taken out of the ZIP file,

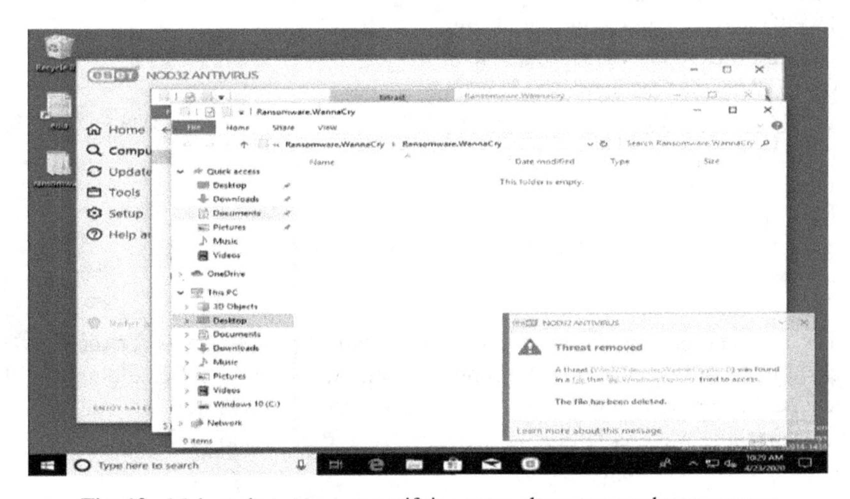

Fig. 12. Malwarebytes pop-up notifying users about removed ransomware

and immediately the user is notified that a malicious file has been removed from the system and placed into quarantine (see Fig. 12).

In summary, all the tested antivirus solutions were able to successfully stop known strains of ransomware such as WannaCry and Petya with varying degrees of aggression. Windows defender was the least aggressive, simply warning the user of a file's possible malicious properties. Malwarebytes was slightly more aggressive with an actual quarantine of the malicious file. ESET Antivirus was the most aggressive. ESET immediately removed the tested ransomware files from the computer and placed it into quarantine without the ransomware even needing to be run.

5.2 Mitigating Ransomware Attack using Cloud Computing

Using cloud backup devices has seen a rise in popularity over the last decade, primarily due to the price of cloud storage decreasing. Many of these cloud platforms such as amazon web services (AWS), Microsoft Azure, and Google's Cloud platform offer large amounts of storage space for a relatively low cost. The demand for more cloud storage space has brought down the price of cloud storage space to a point where it is well suited to handle large backups of files. Mitigating damage caused by ransomware can be accomplished by enabling routine cloud backup files and documents. Cloud backups allow for version history control so if files are encrypted, they can easily be restored to their previous versions. However, this technique is not effective against ransomware that locks the computers. In this part of the experiment, Microsoft's OneDrive [30] will be installed onto the virtual environment. Once installed, the machine will be infected with various strains of ransomware. Using Microsoft's web interface for OneDrive will determine if the files are recoverable to their original content. The step-by-step process is shown below:

1) Start virtual machine
2) Install OneDrive on the virtual machine
3) Add test files to the virtual machine
4) Open Windows OneDrive and Sign In
5) Create a folder in OneDrive and populate it with different file types (word file, ppt,.txt, etc.)
6) Infect the computer
7) Restore the VM via snapshot
8) Try access and download infected files from OneDrive cloud storage
9) Observe the results.

The first step is creating a sandboxed environment with test files for ransomware attacks. This was done in a virtual machine running windows 10. After adding test files to the virtual machine, a snapshot was created, and various ransomware was run to ensure that this machine was susceptible to being infected. Fig. 13 shows the virtual machine being infected with WannaCry ransomware.

After this test, the virtual machine was reverted to its previous state and OneDrive was installed which automatically began backing up files on the computer to OneDrive. Next, the WannaCry ransomware was loaded onto the machine which encrypted all the

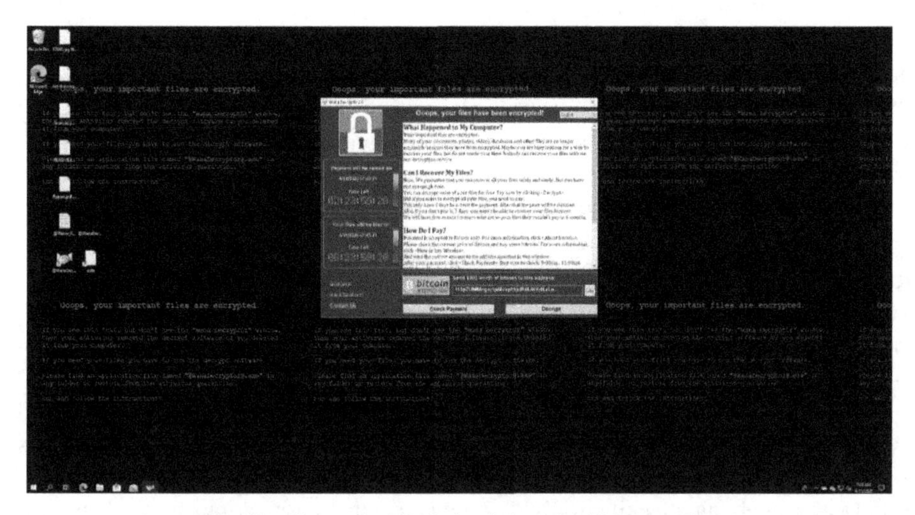

Fig. 13. Shows the VM being infected with WannaCry.

photos and PDF files on the machine. From there, looking at our Microsoft OneDrive web interface, users can see many files that have been added, including the files that were encrypted. Once verifying the presence of our file on Microsoft OneDrive, users can go ahead and restore our machine to a known safe state and download our once infected files back to our machine by using the restore to previous versions option. Restoring infected files to their previous versions is not difficult using Microsoft OneDrive since they can be download from the version history tab (see Fig. 14). This proves to be a successful mitigation technique, as all files were able to be restored to their original state.

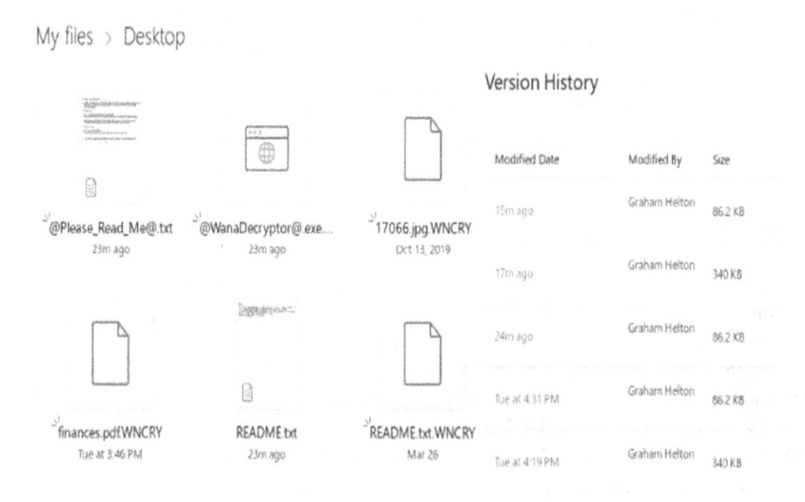

Fig. 14. OneDrive's web interface and version of files

In summary, Microsoft OneDrive is a cloud backup service integrated into the Windows 10 machines. Once activated and on the network, OneDrive uploads user-selected files to the cloud over the internet. This allows for simple yet effective backups and version history controls. Allowing users to revert previous states of a file which is ideal for mitigating the damage that ransomware can cause to user's machine, as the user can simply reimage the machine and download the unencrypted versions of the files.

6 Conclusions

Due to the destructive nature of ransomware, early detection and prevention methods are crucial to prevent or mitigate risks possessed by ransomware. In this research, we used VirusTotal antivirus engines and dynamic malware analysis against 16 live obfuscated as well as original ransomware samples. The results showed that VirusTotal antivirus engines can detect up to 86.71% of 16 sample ransomware selected in our experiment. Due to ever-evolving attacking tools ransomware authors use various methods to pass antivirus detection engines. One of the most popular ways to get past detection is to obfuscate ransomware using Packers and Crypters. Application of this method of obfuscation to sample ransomware showed that it reduces antivirus detection from 86.71% down to 50.88% which is a reduction of 35.71%. This indicates that ransomware authors apply sophisticated techniques to evade detection, which implies we should use stronger antivirus engine. To this end, we tested three proprietary antivirus software to determine their effectiveness in detecting and preventing ransomware attack. These are Windows defender, Malwarebytes and ESET antivirus. Out of these three ESET showed to be the strongest in preventing ransomware attack. Experiment of using OneDrive cloud storage demonstrates that implementing automatic cloud backup method is a very effective way of mitigating ransomware attack.

This research can be extended in several ways. First, analyze ransomware using memory forensics. Second, perform the experiment on a larger sample of ransomware. Third, apply different methods and tools for obfuscation of ransomware. Fourth, use mobile devices as the target of the attack. Finally, fully integrating the cloud storage to the network system for automatic backup.

References

1. Kaspersky. What is WannaCry Ransomware. https://www.kaspersky.com/resource-center/threats/ransomware-wannacry (2020)
2. Norton. What you need to know about the Petya ransomware outbreak (2017). https://us.norton.com/internetsecurity-emerging-threats-what-to-know-petya-ransomware.html
3. Bad Rabbit: A new ransomware epidemic is on the rise (2020). https://www.kaspersky.com/blog/bad-rabbit-ransomware/19887/
4. Cabaj, K., Gregorczyk, M., Mazurczyk, W.: Software-defined networking-based crypto ransomware detection using HTTP traffic characteristics. Comput. Electr. Eng. **66**(C), 1–14 (2018)
5. Okane, P., Sezer, S., McLaughlin, K.: Obfuscation: the hidden malware. IEEE Secure. Priv. **9**(5), 41–47 (2011)
6. VirusTotal. https://www.virustotal.com/gui/. Accessed 14 Feb 2020

7. Surati, S.B., Prajapati, G.I.: A review on ransomware detection & prevention. Int. J. Res. Sci. Innov. (IJRSI) **IV**(IX), 86–91 (2017)
8. Kharraz, A., Robertson, W., Balzarotti, D., Bilge4, L., Kirda, E.: Cutting the gordian knot: a look under the hood of ransomware attacks. In: DIMVA, pp. 1–20 (2015)
9. Richardson, R., North, M.: Ransomware: evolution, mitigation and prevention. Int. Manag. Rev. **13**(1), 0–22 (2017)
10. Kok, S., Abdullah, A., Jhanjhi, N., Supramaniam, M.: Ransomware, threat and detection techniques: a review. IJCSNS Int. J. Comput. Sci. Netw. Secur. **19**(2), 136–146 (2019)
11. Rossow, C., Dietrich, C., Grier, C., Kreibich, C., Paxson, V., Pohlmann, N., Bos, H., Steen, M.V.: Prudent practices for designing malware experiments: status quo and outlook. In: IEEE Symposium on Security and Privacy, pp. 66–79 (2012)
12. Hampton, N., Baig, Z., Zeadally, S.: Ransomware behavioral analysis on windows platforms. J. Inf. Secur. Appl. **40**, 44–51 (2018)
13. Kardile, A.B.: Crypto ransomware analysis and detection using process monitor. MS Thesis Presented to the University of Texas at Arlington (2017)
14. Jethva, B.A.: New Ransomware Detection Scheme based on Tracking File Signature and File Entropy. MS Thesis presented to the University of Victoria (2014)
15. Fedler, R., Schuttle, J., Kulike, M.: On the Effectiveness of Malware Protection on Android: an Evaluation of Android Antivirus Apps. Fraunhofer AISEC, 2–35 (2013)
16. Sechel, S.: A comparative assessment of obfuscated ransomware detection methods. Informatica Economică **23**(2), 45–62 (2019)
17. Morales, J.A., Sandhu, R., Xu, S.: Evaluating detection and treatment effectiveness of commercial anti-malware programs. In: Fifth International Conference on Malicious and Unwanted Software, pp. 31–38 (2010)
18. Kharraz, A., Robertson, W., Kirda, E.: Protecting against ransomware: a new line of research or restating classic ideas? IEEE Secur. Privacy **16**(3), 103–107 (2018)
19. VMware. https://www.vmware.com/. Accessed 15 Jan 2020
20. IDA Freeware. https://www.hex-rays.com/products/ida/support/download_freeware/. Accessed 14 Feb 2020
21. Ghidra. https://ghidra-sre.org/. Accessed 24 Feb 2020
22. Themida. https://themida.en.softonic.com. Accessed 23 Feb 2020
23. theZoo https://github.com/ytisf/theZoo. Accessed 11 Mar 2020
24. Agrawal, M., Mishra, P.: A comparative survey on symmetric key encryption techniques. Int. J. Comput. Sci. Eng. (IJCSE) **4**(1), 877–883 (2012)
25. Lemmou, L., Souidi, E.M.: Infection, self-reproduction and over infection in ransomware: the case of teslacrypt. In: International Conference on Cyber Security and Protection of Digital Services, pp. 212–220 (2018)
26. Reaves, J.: A Case Study into solving Crypters/Packers in Malware Obfuscation using an SMT approach. academia.edu (2018)
27. Yan, W., Zhang, Z., Ansari, N.: Revealing packed malware. IEEE Secur. Privacy Mag. **6**(5), 65–69 (2008)
28. Lau, B., Svajcer, V.: Measuring virtual machine detection in malware using DSD tracer. J. Comput. Virol. **6**(3), 181–195 (2008)
29. Song, S., Kim, B., Lee, L.: The effective ransomware prevention technique using process monitoring on Android platform. Hindawi Publ. Corporation Mob. Inf. Syst. **16**, 1–9 (2016)
30. Microsoft OneDrive. https://www.microsoft.com/en-us/microsoft-365/onedrive/online-cloud-storage. Accessed 10 Apr 2020

Effective Detection of Cyber Attack in a Cyber-Physical Power Grid System

Uneneibotejit Otokwala(✉), Andrei Petrovski(✉), and Harsha Kalutarage(✉)

School of Computing, Robert Gordon University, Aberdeen, UK
{u.otokwala,a.petrovski,h.kalutarage}@rgu.ac.uk

Abstract. Advancement in technology and the adoption of smart devices in the operation of power grid systems have made it imperative to ensure adequate protection for the cyber-physical power grid system against cyber-attacks. This is because, contemporary cyber-attack landscapes have made devices' first line of defense (i.e. authentication and authorization) hardly enough to withstand the attacks. To detect these attacks, this paper proposes a detection methodology based on Machine Learning techniques. The dataset used in this experiment was obtained from the synchrophasor measurements of data logs from snort, simulated control panels and relays of a smart power grid transmission system. After the preprocessing of the dataset, it was then scaled and analyzed before the fitting of - Random Forest, Support Vector Machine, Linear Discriminant Analysis and K-Nearest Neighbor algorithms. The fitting of the different classifiers was done in order to find the algorithm with the best output. Upon the completion of the experiment, the results of classifiers were tabulated and the result of the Random Forest model was the most effective with an accuracy of 92% and a significantly low rate of misclassification. The Random Forest model also shows a high percentage of the true positive rate that is critical to the security issue.

Keywords: Cyber-attack detection · True positive rate · Smart grid system

1 Introduction

The Purdue model for Industrial Control System (ICS) has bridged the gap between Information Technology (IT) and Operation Technology (OT) through the deployment of Wireless Sensor Network (WSN) and robots. As a result, the cyber-physical power grid system which is also known as the smart grid system has witnessed a tremendous advancement as Intelligent Electronic Device (IED) and other internet enabled devices have been incorporated into its structure for effective monitoring and value addition in its operations [1]. In fact, Cedric et al. [2] had proposed that "*next generation of electric power grid system and other critical infrastructures will rely mainly on advanced technologies such as: industrial automation control systems, error diagnostics, preventive maintenance, automatic safety switching, advance metering infrastructure, and synchrophasor systems*". These advancements however, have exposed the system to a new vista of cyber-attack landscape which are clearly intended to undermine the smart grid system, cause system misuse and obviate it from the critical role it plays in the society.

K. Arai (Ed.): FICC 2021, AISC 1363, pp. 812–829, 2021.
https://doi.org/10.1007/978-3-030-73100-7_57

Cyber-attacks on the smart grid system occur when an unauthorized user leverages on the flaws and vulnerabilities of the devices to gain access to the internet enabled device. Some of the vulnerabilities include: weak passwords, unpatched firmware, weak encryption, insecure web links, etc. [3]. According to Alasdair Gilchrist [3], hackers have in recent times resorted to looking for older firmware to perform their attacks especially for versions with known vulnerabilities. For example, the power grid infrastructure system which were isolated and only run on proprietary softwares are now running on Commercial-of-the-Shelf (COTS) components and according to reports [4, 5], several cyber-attacks have been targeted against it because the COTS are not resilient enough and because the built-in safeguards against cyber-attacks are not properly hardened, maintained or updated [6]. It is also noteworthy that before now, most cyber-attacks were restricted to the IT infrastructure of critical organizations; however, with the convergence of OT and the IT infrastructure, there has been a significant shift in cyber-attacks to OT infrastructures [7] and these breaches often results in: reset of the phasor parameters, system shutdown, and disruption of the power grid system [6]. Usually, the Operating System (OS) provides the abstraction and support mechanism for the protection of hardware and application from misuse [8]; however, the cyber-attacks and threats especially, from non-state actors have assumed some level of sophistication in recent times. This therefore, makes the effective detection and prevention of cyber-attacks on the smart power grid system very important [9, 10].

1.1 Structure of a Smart Grid System

A typical structure of a cyber-physical power grid system is shown in Fig. 1 with the components.

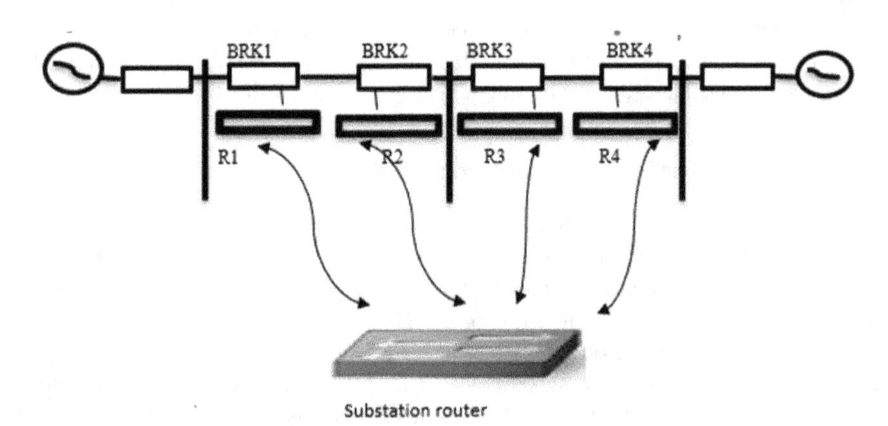

Fig. 1. Structure of a cyber-physical power grid system [11]

A typical structure of a power grid system has power generators on both ends to supply electricity to the grid. The devices labeled R1, R2, R3 and R4 are Intelligent Electronic Devices (IEDs) which are connected to each circuit breaker, BRK1, BRK2 through BRK4. The role of the IED is to monitor events on the grid and to switch on or off the circuit breakers. According to the authors [11], there are two events that can cause the circuit breakers to trip and the events are: (a) an alert within the line (L1 and L2) that could initiate the IEDs to cause the breakers to trip (b) the operators manually issuing a command to the IED to break the circuit. In both instances, the intelligent devices, use a distance protection algorithm which enables the circuit breakers to trip irrespective of the cause of the command, i.e. whether it is a valid or invalid cause. Below is a list of events scenarios from the two mentioned above that can result in line tripping:

(a) Short-circuit fault – This is when there is a short circuit between two lines (two or more lines touching each other). This often results in very high voltage that could lead to massive damages.
(b) Line maintenance – This is when the line is intentionally disabled to allow for line maintenance.
(c) Remote tripping command – This is a possible attack in which an attacker breaches the device's defense and sends a command to a relay thereby causing a breaker to open.
(d) Relay setting change – This is another form of attack in which the attacker upon penetrating the device's defense, reconfigures the relay's setting and disables the relay function such that the relay will not trip even for a valid fault or a valid command.
(e) Data Injection – This is another form of attack in which the attacker upon entry, initiates a seeming valid fault by changing the phasor values of current, voltage, and other parameters just to ensure that the line trips.

It is apparent that from the scenarios highlighted above, successful attacks against the power system has the propensity to obliterate and render the power grid system incapable of providing efficient power. With these inadequacies and the insufficient scalability of the smart power grid system to mitigate the cyber challenges [12], there is a need to identify the cyber-attacks and secure the power grid system infrastructure.

1.2 Objective, Contribution and Structure of Paper

The objective of this paper is to find an effective cyber-attack detection model by fitting different machine learning classifiers on a simulated smart power grid system dataset. The results will then be compared and the most effective of the models will be tested for effective performance using different metrics. The effectiveness of the performance of our model will therefore be our contribution for effective intrusion detection of cyber-attacks in the smart power grid system. The rest of this paper is organized as follows. Section 2 related literatures. Section 3 discusses data analysis. Section 4 model fitting and performance evaluation. Section 5 is conclusion.

2 Related Literatures

While a lot of research papers have been written on the subject of intrusion detection in the smart grid system, a number of them appears static in their approach to intrusion detection especially in looking out for particular anomalous deviant behaviors. Considering that contemporary attacks on smart grid system have become dynamic, it therefore, requires that approaches should be dynamic and holistic such that detection could be effective even in multiclass situations. For example, cyber-attacks such as Relay Setting Change are common in smart grid system and they are often subtle and obfuscated in order to anonymize the attack. This kind of attack may likely not display an anomalous deviant behaviour to enable some of the proposed IDS detect. Here are some of the literatures:

2.1 Wide Area Monitoring System (WAMS)

This system was adapted through a concerted effort by several organizations to widely monitor power grids system in real-time within a "neighboring grids cluster". Basically, WAMS monitors the cyber-physical system parameters such as phasors of voltage, current, and the status of the IEDs, relays, circuit breakers, etc. [2]. The real-time data so generated from the multiple remote points are then synchronized by the WAMS and then transmitted for measurements by the Phasor Measurement Unit (PMU). The PMU is a device used to estimate the magnitude and phase angle of the phasor parameters (voltage, current, etc.) in the electricity grid. The monitoring and synchronization is done in order to ensure accuracy whilst looking out for deviations and malicious values that could lead to down time resulting from attacks [13, 14].

2.2 Specification-Based Intrusion Detection System

Unlike the signature-based and anomaly-based Intrusion Detection Systems, the Specification-based IDS is a behavior-rule specification-based technique for intrusion detection that was introduced by Ko in 1996. It has its application mostly in medical cyber-physical systems, electrical cyber-physical grid system, software engineering and in network protocol of some critical infrastructures [2, 15]. In this IDS, the rules work by representing the system behavior of the state machine at every instance of time. According to Pan et al. [2], the state machine behaviors are represented by a sequence of states according to the policies specified. The devices are then monitored and tracked for intrusion, changes and anomalous behaviors that could drive the system state from safe to unsafe state. Any noticeable sequence of behaviors that are outside the predefined specifications are flagged as intrusion. In a nutshell, the authors averred that the Specification-based IDS can be likened to a complement of the anomaly-based IDS.

2.3 Semi-supervised Anomaly IDS

This is another form of behavior-based IDS which was proposed by Park et al. [16]. Though this model was targeted at the Medical Cyber-Physical devices (MCPD) for assisted living environments, it could as well be adopted to detect anomalous behavior in the power grid system. Basically, this Semi-Supervised Anomaly IDS audits a series of events called, episodes. These episodes are sensor ID, start time and duration of events. In using the Hidden Markov Model (HMM) technique, a comparative analysis is then done to determine the current state of events and what happens thereafter. Based on the noticed behaviour, classification is then done by classifying the behaviour as low-level state or high-level state in order to be able to infer whether it is an abnormal or normal behavior.

2.4 Data-Stream-Based Mining IDS

Faisal et al. [17] proposed the use of Data-Stream-Based Mining IDS for the monitoring of intrusion in smart grid Advanced Metering Infrastructure (AMI). The structure of this IDS is similar to the anomaly-based IDS, but it selects a stream of data as against the conventional static mining techniques often observed in the anomaly technique. This proposal is, however, very limited in application to the smart meter. Therefore, this model is not suitable for intrusion detection of cyber-attacks in the cyber-physical smart power grid system.

3 Data Analysis

3.1 Description of Dataset

The dataset used in this paper is the power system dataset [2, 18]. It is made of 129 variables (128 predictors and 1 response variable of 3-classes). The dataset contains the measurement of electric transmission on a smart power grid system. These measurements were done using 4 synchrophasors which measures 29 features of the events in each Phasor Measurement Unit (PMU) totaling 116 features. The PMU uses a common time source to synchronize the various measurements and the features so measured were classified as attacks and benign data. The benign data is consisted of Normal traffic and NoEvents. These measurements were obtained using: snort, a simulated control panel, and relays. The parameters measured are: the voltage phase angle (PA1:VH – PA3:VH), the voltage phase magnitude (PM1:V – PM3:V), the current phase angle (PA4:IH – PA6:IH) and the current phase magnitude (PM4:I – PM6:I). Others are: the zero voltage phase angle (PA7:VH – PA9:VH), zero voltage phase magnitude (PM7:V – PM9:V), the zero current phase angle (PA10:VH – PA12:VH) as well as the zero current phase magnitude (PM10:V – PM12:V). In addition, there were also other parameters that were measured, and they are: frequency for relay (F), frequency delta (DF), appearance impedance for relays (PA:Z), appearance impedance angle for relays (PA:ZH) and status flag for relays (S). Other descriptions in the dataset are fault location, line maintenance and load condition. The entire setup was aimed at measuring both the normal traffic transmission in the grid as well as the attacks (cyber intrusion) that could impact the power grid system.

3.2 Dataset Class Distribution

To enable us visualise the distribution of the instant classes in the response variable of our dataset, using RStudio Integrated Development Environment (IDE), we plotted a barplot of the values. See the plot in Fig. 2 and the R code in Appendix A.

A code snippet of the proportional representation of the classes in the response variable are as shown in Appendix B. Though the class representation and barplot shows

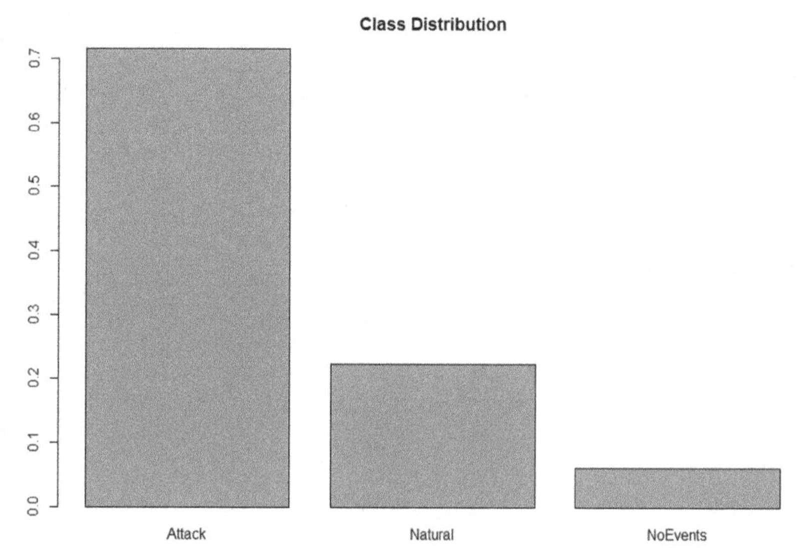

Fig. 2. Barplot showing the distribution of the instances of the response variable

the Attack class as the majority class over the benign class, the dataset does not fit into the description of an imbalanced dataset in cybersecurity considering the ratio between the classes. If we consider the dataset as binary (attacks and benign) then the ratio is 1:2. For attack to Natural it is 1:3 and for attack to NoEvent it is 1:11. In typical intrusion dataset, a ratio of 1:10 and above for a majority to minority class is expected before a dataset could be classified as an imbalanced dataset. More importantly, since our target class is the attack class, and it is a majority class, we elected to proceed with the dataset but with a view to ensuring that a higher recall rate is achieved and that the Area Under the Curve (AUC) for the ROC curves is high.

3.3 Data Pre-processing

Data cleaning and pre-processing is a way of preparing the dataset for eventual use and to also ensure that all the data points contribute to the model without bias. It involves outlier removal, feature selection and data normalisation. However, in our experiment, we only performed outlier removal and data normalisation using the scaling function.

Outlier Removal. While summarising and visualizing the dataset, we observed that the dataset was fraught with outliers that needs to be removed. However, further introspection into the dataset shows that the anomaly was caused by fault of 10%–19% on the relay of Line 1 which results in "Inf" values. The same outliers were found in Line 2, and relays number 3 and 4 of the power line. In addition, our observation also gave credence to the fact that these outliers may have been as a result of either the disabling of a single relay for line maintenance, remote tripping command of a single relay or a fault. In all the cases, the percentage of the disabling function lies between 10% and 29%. In view of these and the need to visualise the data points that clearly deviate from the others, we decided to use boxplot package of RStudio to visualise the outliers in the dataset [19].

From the plot (Fig. 3), data points that were discovered to significantly deviate from the rest of the points were identified and removed. As could be seen in the figure, the outliers are "Inf" and they were found in the following variables: "R1.PA.Z", "R2.PA.Z", "R3.PA.Z", and "R4.PA.Z".

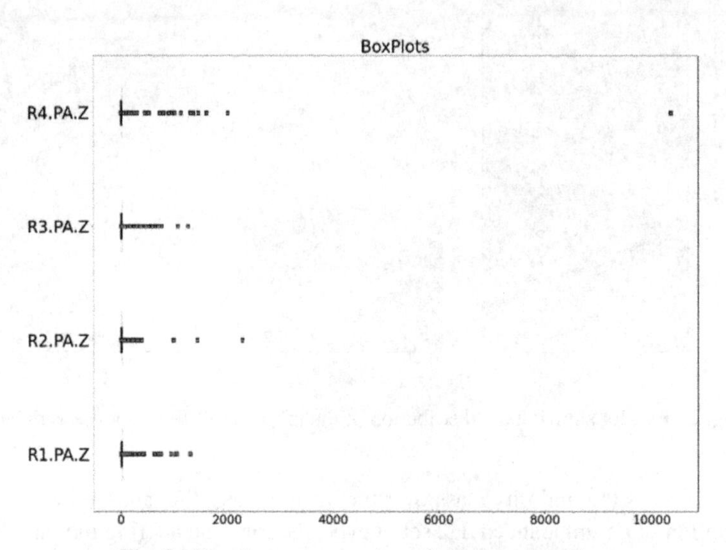

Fig. 3. Boxplot representation of values and outliers

Data Normalization. Data normalization during multivariate analysis is to enable each variable to contribute equally to the analysis. Therefore, the normalization method we used was scaling, and we scaled from the first to the 128th variable leaving out the response variable which is a factor variable. Upon completion of the scaling, we then appended the response variable before we commenced the application of the classifier for modeling. See Appendix C for the code snippet on data scaling.

4 Model Fitting and Performance Evaluation

At this stage of our experiment, using a Windows 10 computing machine with intel core i5 processors and RStudio IDE, we applied some machine learning algorithms on the dataset. The essence was to fit several models and then compare the results of the models in order to determine which of them has the best accuracy, sensitivity and specificity. Also, our reason for using both linear and non-linear classifiers to fit the models was because, we observed that a few of the classifiers are highly likely to be biased toward the majority class in their output. However, before we applied the classifiers, we ensured that the dataset was clean of all factors that might affect our output. At this point, the total number of observations and variables after data pre-processing was, 52,885 – observations, and 129 - variables. We then partitioned the dataset into training and testing data and assigned 37,000 of the observations which constitute about 70% to training of the classifier. The remainder of the dataset which constitutes 30% of the observations was then used for validation. After the splitting, we went on to fit the model using the different classifiers.

4.1 Linear Discriminant Analysis (LDA)

The LDA [20] was the first classifier we used. It is a linear classifier that is robust and good at performing dimension reduction in the course of its application on datasets. It mostly works by dividing the data space into N number of disjoint regions such that probability densities are calculated with the assumption that the data is Gaussian with each attribute having same variance close to the mean. This classifier produced an accuracy of 71% with a high percentage of misclassification rate. Table 1 contains the values of the sensitivity and specificity of this classifier. Also find the R-code snippet for the model in Appendix D.

4.2 Support Vector Machine (SVM)

Support Vector Machine (SVM) [21] is a non-linear classifier that is used for both regression and classification problems. SVM produces significant accuracy with less computation power. To maximize the output and margin, SVM uses decision boundaries to classify data points that are closer to the hyperplane. These data points then influence the number of data points closed to the hyperplane, position and the orientation of the hyperplane. Our accuracy while using this classifier to fit our model was 72%. This model also showed a high percentage of misclassification rate hence our desire to tune the kernel parameters in order to ensure improved performance. See Appendix E for the R-code and Table 1 for the value of sensitivity and specificity.

SVM Tuning

Since the accuracy of our SVM model was not very high especially considering the high rate of misclassification, we decided to tune our SVM kernel parameters in order to improve the accuracy as well as reduce the Cost Matrix [21]. Usually, the SVM kernels takes data points as inputs and outputs similarity score that affects the class boundaries. The measure of the closeness on both sides of the hyperplane is the similarity and the nearer the data points are to the hyperplane, the higher the similarity score. We knew that to achieve a better SVM classifier output, it would require a better measure of closeness which can only be achieved through the right values of the kernel parameters. At this point, we then proceeded to try several values for gamma and cost with a view to having an optimal value that will yield a better accuracy and recall rate. We also applied the different kinds of kernel: Radial kernel, Polynomial kernel, Sigmoid kernel and Linear kernel. In the end, we were able to obtain a gamma value of 0.1 and cost parameter value of 20 in the radial kernel. With these values, we were able to tune the kernel parameters and obtained a better accuracy and a little reduction in the misclassification rate. With this tuning, we were able to improve the accuracy from 72% to 77%. However, we observed that the misclassification rate was still high hence the need for us to further apply some other non-linear classifiers. The sensitivity and specificity values have been provided in Table 1 and the R-code snippets are in Appendix F.

4.3 K – Nearest Neighbour (KNN)

The K-Nearest Neighbor (KNN) [22] is another non-linear classifier that we also used to model our work. KNN uses Euclidean distance to measure the distance between one data point and its neighbor. Based on the size of our dataset, we calculated the value of K as 192 and 193 (nearest neighbour), we then fit in the model and computed the confusion Matrix. The accuracy of the KNN model when it was fitted was 71% with a very high misclassification rate as the sensitivity and specificity were very low. See Table 1 for the values and Appendix G for code.

4.4 Random Forest

Random Forest (RF) [23] uses decision trees that are randomly created from selected data samples to make its predictions on each tree and then selects the best solution by means of voting. Usually, the more trees the classifier can create, the more robust the forest is. Its method of data splitting is an ensemble approach based on divide and conquer method. Individual trees are usually generated by the classifier using an attribute selection indicator. The application of Random Forest classifier to fit the model improved the accuracy of the model to 92% at 95% CI. Also, the model detection rate of the true positives (sensitivity) and specificity also improved. The improved accuracy makes the model quite relevant for the detection of instances of attacks in a multiclass dataset as the one we are using. Furthermore, the balanced accuracy across the three instances were also very high which is an indication of suitability of the classifier for our experiment. It is also worthwhile to add that with a Kappa value of 82%, the model could be said to have performed very well in the identification and detection of the attack classes. See Table 1 for more on the detected values and Appendix H for a snippet of the code.

4.5 Experimental Result Comparison

We computed the Confusion Matrix of each of the classifiers and tabulated the values of the classes in Table 1. For the purpose of this experiment, we restricted the values to the computed Accuracy, Sensitivity and Specificity.

Table 1. Outputs of the confusion matrix of each of the models

	Accuracy	Sensitivity			Specificity		
		Attack	Natural	NoEvent	Attack	Natural	NoEvent
LDA	71%	99%	1%	6%	3%	99%	99%
SVM	71%	99%	0	4%	1%	99%	99%
Tuned SVM	77%	94%	28%	70%	39%	65%	98%
KNN	71%	100%	0	0	0	100%	100%
RF	92%	98%	68%	91%	73%	98%	99%

4.6 Confusion Matrix of Best Model

From the comparison of the values in Table 1, the output of the Random Forest model gave the best result of all the classifiers. In addition, the RF model also gave the lowest misclassification rate of all the models hence the confusion matrix in Table 2. The numbers along the diagonal represent the correct decisions made, and the numbers on the left and right of the diagonal represent the errors otherwise known as misclassification of the various classes. The confusion matrix code is in Appendix I.

Table 2. The confusion matrix of the random forest classifier

Predicted Values	ACTUAL VALUES			
		Attacks	Natural	NoEvents
	Attacks	11202	984	56
	Natural	142	2592	6
	NoEvents	9	4	890

Recall and Precision. The recall otherwise known as Hit Rate or sensitivity is one of the metrics of measurement of the performance of a model. It is the number or proportion of the correctly predicted positive values divided by the total number of positive values (TP/(TP + FP)). False positives are values that our model incorrectly classified as positives but are actually negative values.

Attack- from the confusion matrix, from first column/row is

$$\text{Recall} = \frac{11202}{11202 + 142 + 9} = 98\%$$

$$\text{Precision} = \frac{11202}{11202 + 984 + 56} = 92\%$$

Natural - from the confusion matrix, from second column/row is

$$\text{Recall} = \frac{2592}{2592 + 984 + 4} = 72\%$$

$$\text{Precision} = \frac{2592}{2592 + 142 + 6} = 95\%$$

NoEvent - from the confusion matrix, from third column/row is

$$\text{Recall} = \frac{890}{890 + 56 + 6} = 94\%$$

$$\text{Precision} = \frac{890}{890 + 9 + 4} = 99\%$$

To further explain the value of our recall and precision - given all the predicted labeled class called Attack, the number of instances that were correctly predicted has a **precision = 0.92** (92%). Also, a **recall = 0.98** shows that for all instances that should have label Attack, our model correctly captured 98%.

F – Measure. F- Measure also known as F-score or F1 is another metric for the measurement of the accuracy of a classifier especially a dataset whose distribution of classes in the dataset is slightly skewed towards the majority class. Our dataset fits into this category hence our desire to also compute the F-score of our model. It is described as the harmonic mean of the precision and recall as it is the most common metric that is used on an uneven or imbalanced classification problem.

$$F = 2 \times \frac{Precision * Recall}{Precision + Recall}$$

$$F = 2 \times \frac{92 * 98}{92 + 98} = 2 \times 47.45 = 94.9 \approx 95\%$$

An F-score value of 1 indicates that the variance among the class mean is exactly what is expected given the within-classes variance and not by chance. Therefore, with our model's F-score tending to 1 (F1≈1), we can infer that the RF model was able to classify and detect the attacks. Also, considering our Confidence Interval of 95% with a significance of 0.05, the value of our computed P-value (see Appendix I) is less than the significance level (0.05) therefore, we can also infer that the value is statistically significant and supports the adoption of the RF model as suitable for detection of attacks.

4.7 Cutoff Value, Receiver Operating Characteristics (ROC) and AUC

Cut-off value - The ROC curve is used to determine the optimum cut-off value especially as it shows the trade-off between the true positives and the false positive at different cut-off marks. Basically, it evaluates the hit rate and false alarm rate at varying thresholds (Fig. 4) [24].

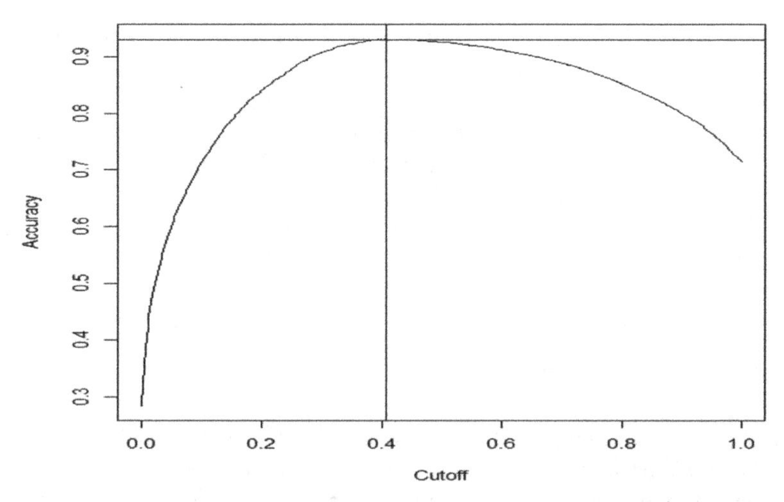

Fig. 4. A plot showing the overall Accuracy values against several Cutoff values of the RF model.

From Fig. 4, it can be observed that the accuracy of our model tends to increase with an increase in the cutoff values. However, at a maximum threshold value before the default cutoff (0.5), the model was able to achieve the maximum accuracy. The code snippet is in Appendix J.

ROC Curve - The ROC curve is a veritable tool for visualizing and evaluating classifiers performance accuracy and it is independent of the class distribution. ROC curve's ability to tend to the top-left corner of the graph indicates a better performance. Our RF model ROC from the graph (Fig. 5) tends to the top left corner of the graph which is a pointer to the ability of the model to predict the true positive rates more correctly.

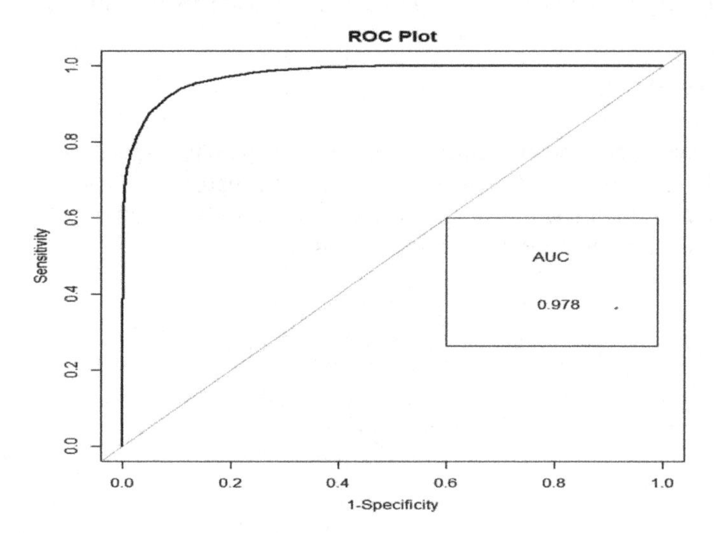

Fig. 5. Showing ROC graph of sensitivity against 1-specificity and the area under the curve.

AUC - The area of ROC graph is 1 and its scale ranges from points 0.0 – 1.0. To measure the predictive accuracy of a model, the AUC of the curve needs to be computed as it is the probability that a given randomly chosen value is a positive instance of higher rank. An AUC of 0.5 indicates that the ROC curve lies on the baseline (the diagonal) where FPR = TPR which indicates that the predictive value of the AUC in the ROC curve is less accurate or at best the detection could only happen by chance. However, with our model AUC = 0.978, it is indicative that our RF model has a higher chance of detecting high positives.

5 Conclusion

In dealing with the growing integration and complexity of cyber-physical smart grid system, there is a necessity to explore an effective approach to detection, monitoring, optimizing, and more importantly, securing the smart power grid system. This paper has proposed an effective anomaly detection method against cyber-attacks in a smart grid system. Because the dataset we used has multiclass response variable, our focus was more on how to correctly classify and detect the true positive rate (Attacks) with a commensurate value of accuracy. The methodology we adopted to achieve this objective involved the application of several machine learning classifiers that will be able to provide a high accuracy as well as a high detection rate of the true positive rate. The classifiers we applied after necessary data cleaning and preparation were: Linear Discriminant Analysis, Support Vector Machine, K-Nearest Neighbor and Random Forest. Of all the classifiers, the Random forest model gave us the highest accuracy, a better detection rate of the true positives and also the specificity. We then went further to evaluate the performance of our model using metrics like precision, recall rate, F-score, ROC and Area Under the Curve. It is interesting to point out that all the metrics supported our model with very high probability for the detection of anomaly in a smart grid system.

6 Future Work

The smart power grid system is experiencing a number of domain specific forms of cyber-attacks. These attacks include: data injection, remote tripping command injection, relay reset and others. Future works should look at identifying and classifying these forms of cyber-attack against the smart grid system infrastructure.

Appendix

A.

```
> boxplot(new_Data)
> #ploting  the response variable
> barplot(prop.table(table(new_Data$marker)),
+         col = rainbow(2),
+         ylim = c(0,0.7),
+         main = "Class Distribution")
```

B.

```
> cbind(freq=table(new_Data$marker),
+       percentage=prop.table(table(new_Data$marker))*100)
          freq percentage
Attack    37851  71.572279
Natural   11809  22.329583
NoEvents   3225   6.098137
```

C.

```
> #scaling of dataframe
> Norm_Data <- as.data.frame(scale(new_Data[1:128]))
> Norm_Data$marker <-new_Data$marker
```

D.

```
> set.seed(222)
> part <- sample(1:52885, 37000, replace = F)
> LDAtraining <- Norm_Data[part,]
> LDAtesting <- Norm_Data[-part,]
> fit_LDA <- lda(marker~.,data = LDAtraining)
Warning message:
In lda.default(x, grouping, ...) : variables are collinear
> fit_LDA_predict <- predict(fit_LDA, LDAtesting)
> confusionMatrix(table(fit_LDA_predict$class, LDAtesting$marker))
```

E.

```
> #SUPPORT VECTOR MACHINE (SVM)
> fit_svm <- svm(marker~., data = training)
Warning message:
In svm.default(x, y, scale = scale, ..., na.action = na.action) :
  Variable(s) 'control_panel_log3' and 'control_panel_log4' constant. Cannot scale data.
> pred_svm <- predict(fit_svm,testing)
> confusionMatrix(table(pred_svm,testing$marker))
```

F.

```
> tuneSVM <- tune.svm(marker~., data = training, gamma = seq(0.1,1,by=0.2),
+                     cost =c(1,20, by=2))

There were 31 warnings (use warnings() to see them)
> newsvm <- svm(marker~., data = training, gamma=0.1, cost=20)
```

G.

```
> caret::confusionMatrix(table(fit_KNN192, testMarker))
Confusion Matrix and Statistics

           testMarker
fit_KNN192 Attack Natural NoEvents
  Attack    11359    3565      961
  Natural       0       0        0
  NoEvents      0       0        0

Overall Statistics

               Accuracy : 0.7151
                 95% CI : (0.708, 0.7221)
    No Information Rate : 0.7151
    P-Value [Acc > NIR] : 0.504

                  Kappa : 0

 Mcnemar's Test P-Value : NA

Statistics by Class:

                     Class: Attack Class: Natural Class: NoEvents
Sensitivity                 1.0000         0.0000          0.0000
Specificity                 0.0000         1.0000          1.0000
Pos Pred Value              0.7151            NaN             NaN
Neg Pred Value                 NaN         0.7756          0.9395
Prevalence                  0.7151         0.2244          0.0605
Detection Rate              0.7151         0.0000          0.0000
Detection Prevalence        1.0000         0.0000          0.0000
Balanced Accuracy           0.5000         0.5000          0.5000
> |
```

H.

```
# using RandomForest
fit_randforest <-randomForest(marker~., data = training)
pred_randforest <-predict(fit_randforest,testing)
confusionMatrix(pred_randforest, testing$marker)
```

I.

```
> # using RandomForest
> fit_randforest <-randomForest(marker~., data = training)
> pred_randforest <-predict(fit_randforest,testing)
> confusionMatrix(pred_randforest, testing$marker)
Confusion Matrix and Statistics

          Reference
Prediction Attack Natural NoEvents
  Attack    11202     984       56
  Natural     142    2592        6
  NoEvents      9       4      890

Overall Statistics

               Accuracy : 0.9244
                 95% CI : (0.9202, 0.9285)
    No Information Rate : 0.7147
    P-Value [Acc > NIR] : < 2.2e-16

                  Kappa : 0.8142

 Mcnemar's Test P-Value : < 2.2e-16

Statistics by Class:

                     Class: Attack Class: Natural Class: NoEvents
Sensitivity                 0.9867         0.7240         0.93487
Specificity                 0.7705         0.9880         0.99913
Pos Pred Value              0.9150         0.9460         0.98560
Neg Pred Value              0.9586         0.9248         0.99586
Prevalence                  0.7147         0.2254         0.05993
Detection Rate              0.7052         0.1632         0.05603
Detection Prevalence        0.7707         0.1725         0.05685
Balanced Accuracy           0.8786         0.8560         0.96700
```

J.

```
A performance instance
  'Cutoff' vs. 'Accuracy' (alpha: 'none')
  with 502 data points
> max <- which.max(slot(eval, "y.values")[[1]])
> max
[1] 298
> acc <- slot(eval, "y.values")[[1]][max]
> acc
[1] 0.9305005
> cut <- slot(eval, "x.values")[[1]][max]
> cut
57449
0.408
> print(c(Accuracy=acc, Cutoff=cut))
   Accuracy Cutoff.57449
  0.9305005    0.4080000
```

References

1. Escudero, C., Sicard, F., Zamai, E.: Process-aware model based IDSs for industrial control systems cybersecurity: approaches, limits and further research. In: Emerging Technologies and Factory Automation ETFA, vol. 2018, pp. 605–612 (2018). https://doi.org/10.1109/ETFA.2018.8502585
2. Pan, S., Morris, T., Adhikari, U.: A specification-based intrusion detection framework for cyber-physical environment in electric power system. Int. J. Netw. Secur. **17**(2), 174–188 (2015)

3. Gilchrist, A.: IoT security issues. Walter de Gruyter GmbH & Co KG (2017)
4. Dondossola, G., Szanto, J., Masera, M., Fovino, I.N.: Effects of intentional threats to power substation control systems. Int. J. Crit. Infrastructures **4**(1–2), 129–143 (2008). https://doi.org/10.1504/IJCIS.2008.016096
5. Morris, T., et al.: Cybersecurity risk testing of substation phasor measurement units and phasor data concentrators. ACM Int. Conf. Proceeding Ser. (2011). https://doi.org/10.1145/2179298.2179324
6. Haber, M.J., Haber, M.J.: Privileged Attack Vectors (2020)
7. Maglaras, L.A., et al.: Cyber security of critical infrastructures.pdf. Elsevier, vol. ICT Expres, pp. 42–45 (2018). https://doi.org/10.1016/j.icte.2018.02.001
8. Mollus, K., Westhoff, D., Markmann, T.: Curtailing privilege escalation attacks over asynchronous channels on Android. In: 14th Int. Conf. Innov. Community Serv. "Technologies Everyone", I4CS 2014 - Conf. Proc., pp. 87–94 (2014). https://doi.org/10.1109/I4CS.2014.6860558
9. Wilhelm, T.: Chapter 10 - Privilege Escalation _ Elsevier Enhanced Reader.pdf. In: Professional Penetration Testing, Elsevier, pp. 271–306 (2013)
10. Conteh, D.N.Y., Royer, M.D.: The rise in cybercrime and the dynamics of exploiting the human vulnerability factor. Int. J. Comput. **20**(1), 12 (2016). https://www.ijcjournal.org/index.php/InternationalJournalOfComputer/article/view/518/374
11. Events, N., et al.: Power System Attack Datasets - Mississippi State University and Oak Ridge National Laboratory - 4 / 15 / 2014, no. 8, pp. 1–3 (2014)
12. Mo, Y., et al.: Cyber-physical security of a smart grid infrastructure. Proc. IEEE **100**(1), 195–209 (2012). https://doi.org/10.1109/JPROC.2011.2161428
13. Bakken, D.E., Bose, A., Hauser, C.H., Whitehead, D.E., Zweigle, G.C.: Smart generation and transmission with coherent, real-time data. Proc. IEEE **99**(6), 928–951 (2011). https://doi.org/10.1109/JPROC.2011.2116110
14. Liu, W., Lin, Z., Wen, F., Ledwich, G., Member, S.: A Wide area monitoring system based load restoration method. IEEE Xplore **28**(2), 2025–2034 (2013). https://doi.org/10.1109/TPWRS.2013.2249595
15. Mitchell, R., Chen, I.R.: Behavior rule specification-based intrusion detection for safety critical medical cyber physical systems. IEEE Trans. Dependable Secur. Comput. **12**(1), 16–30 (2015). https://doi.org/10.1109/TDSC.2014.2312327
16. Park, K., Lin, Y., Metsis, V., Le, Z., Makedon, F.: Abnormal human behavioral pattern detection in assisted living environments. ACM Int. Conf. Proceeding Ser. (2010). https://doi.org/10.1145/1839294.1839305
17. Faisal, M.A., Aung, Z., Williams, J.R., Sanchez, A.: Data-stream-based intrusion detection system for advanced metering infrastructure in smart grid: A feasibility study. IEEE Syst. J. **9**(1), 31–44 (2015). https://doi.org/10.1109/JSYST.2013.2294120
18. Pan, S., Morris, T., Adhikari, U.: Developing a hybrid intrusion detection system using data mining for power systems. IEEE Trans. Smart Grid **6**(6), 3104–3113 (2015). https://doi.org/10.1109/TSG.2015.2409775
19. Aggarwal, C.C.: Outlier analysis, Second Edn., vol. 9781461463. Springer, Heidelberg (2017)
20. Gaber, T., Tharwat, A., Ibrahim, A., Hassanien, A.: Linear Discriminant Analysis : A Detailed Tutorial. Univ. Salford, Manchester, pp. 0–22 (2017). https://doi.org/10.3233/AIC-170729
21. Schlkopf, B., Smola, A.J., Bach, F.: Learning with Kernels: Support Vector Machines, Regularization, Optimization, and Beyond. The MIT Press, Cambridge (2018)
22. Thanh Noi, P., Kappas, M.: Comparison of random forest, k-nearest neighbor, and support vector machine classifiers for land cover classification using Sentinel-2 imagery. Sensors **18**(1), 18 (2018)

23. Van Essen, B., Macaraeg, C., Gokhale, M., Prenger, R.: Accelerating a random forest classifier: multi-core, GP-GPU, or FPGA In: 2012 IEEE 20th International Symposium on Field-Programmable Custom Computing Machines, pp. 232–239 (2012)
24. Fawcett, T.: An Introduction to ROC Graphs, pp. 861–874 (2005). https://doi.org/10.1016/j.patrec.2005.10.010

On the Design and Engineering of a Zero Trust Security Artefact

Yuri Bobbert[1,2](✉) and Jeroen Scheerder[2]

[1] University of Antwerp, Antwerp, Belgium
yuri.bobbert@on2it.net
[2] ON2IT, Zaltbommel, Netherlands
jeroen.scheerder@on2it.net

Abstract. Adequately informing the board of directors about operational security effectiveness is cumbersome. The concept of Zero Trust (ZT) approaches information and cybersecurity from the perspective of the asset, or sets of assets, to be protected, and from the value that it represents. Zero Trust has been around for quite some time. This paper continues on the authors previous research work on the examination of Zero Trust approaches, what is lacking in terms of operationalisation and which elements need to be addressed in future implementations and why and how this requires empirical validation. In the first part of the paper, we summarise the limitations in the state of the art approaches and how these are addressed in the Zero Trust Framework developed by ON2IT 'Zero Trust Innovators'. Then we describe the design and engineering of a Zero Trust artefact (dashboard) that addresses the problems at hand, according to Design Science Research (DSR). The last part of this paper outlines the setup of an empirical validation trough practitioner-oriented research, in order to gain a better implementation of Zero Trust strategies. And how this validation was conducted in 2020 with 73 security practitioners. The final result is a proposed framework and associated technology which, via Zero Trust principles, addresses multiple layers of the organization to grasp and align cybersecurity risks and understand the readiness and fitness of the organization and its measures to counter cybersecurity risks.

Keywords: Zero Trust security · Architecture · Cybersecurity · Digital Security · Managed Security Services (MSS) · Security Operation Centre (SOC) · Security strategy · Design Science Research (DSR) · Group Support System (GSS) · Platform technology · Security Orchestration · Automation and Response (SOAR)

1 Introduction

These days it's impossible to imagine business without technology. Most industries are becoming "smarter" and more tech-driven — ranging from small individual tech initiatives to complete business models with intertwined supply chains and "Platform" based business models [1]. New ways of working such as Agile and DevOps are introduced and thereby new risks arise [2, 3]. Not only technology risks, but also risks that are

K. Arai (Ed.): FICC 2021, AISC 1363, pp. 830–848, 2021.
https://doi.org/10.1007/978-3-030-73100-7_58

caused by teams working together at high pace and autonomy [4, 5]. Where decisions on risk acceptances or security measures most of the time take place in the team itself rather than looking at the bigger picture of accumulated risks [6]. According to CRO-Forum[1] this is an increasing "silent risk" [3]. This autonomous way of working in agile teams – in most case in a distributed manner- is needed to enable speed, quality and craftmanship and there is a quicker time to market [7]. For policymakers and business leaders technology is no longer a domain that is shrouded in mystery; rather it's an essential business discipline that is here to stay, and it's taught at business schools all over the world. It's also a professional discipline that has won the attention of analysts and supervisory boards. However, at the same time, nefarious nation -state activity and organized crime have arrived on the scene in a big way. Through hacks and denial-of-service attacks, all sorts of malicious actors are infiltrating our 'digital' society. They can easily take advantage of systems that are sloppily designed, built and configured and they frequently use advanced "socially engineering" techniques to trick their way into organizations. Platform oriented businesses are typically built on api-based-ecosystems of data, assets, applications and services (DAAS). These hybrid technology landscapes, most of the time built-in software defined networks in clouds [8], lack real-time visibility and control when it comes to their operations [9, 10]. This makes it hard for boards to take ownership and accountability of cyber risks [11]. In practice, we have seen the application of security and privacy frameworks falter because they tend to become a goal on their own rather than a supporting frame of reference to start dialogues with key stakeholders [12]. Kluge et al. [13] for example also noted that the use of frameworks as a goal on its own does not support the intrinsic willingness and commitment to improve. This is especially the case for mid-market organizations that lack dedicated security staff, capabilities and/or sufficient budgets [14]. Puhakainen and Siponen [15] noted that information security approaches are lacking not only theoretically grounded methods, but also empirical evidence of their effectiveness. Many other researchers [16–18] have also pointed out the necessity of empirical research into practical interventions and preconditions in order to support organizations improve the effectiveness of their security. These theoretical voids, as well as the practical observation of failing compliant-oriented approaches, widen the knowledge gap [19]. This "knowing-doing gap" [20] is also perceived in the current Zero Trust approaches, which predominantly aim at the technology or by the technology industry. In our previous paper published in June, titled "Zero Trust Validation: From Practical Approaches to Theory" [21] we have described the several streams of Zero Trust, such as security vendors aiming to deliver point solutions for Zero Trust security and why ON2IT developed a Framework that addresses the problems we describe in the next section.

2 Problem

Although the term "Zero Trust" can be perceived that individuals, i.e. human beings cannot be trusted, Zero Trust actually implies humans can be trusted, but always need

[1] Chief Risk Officer Forum; The CRO Forum's Emerging Risk Initiative continually scans the horizon to identify and communicate emerging risks.

to be verified *before* access and authorization is granted. Jagasia quotes; *"perimeter-based security primarily follows "trust and verify," which is fundamentally different from ZTA's paradigm shift of "verify, and then trust."* Kindervag formulates it more strongly: we have to get rid of the concept of trust: *"The point of Zero Trust is not to make networks, clouds, or endpoints more trusted; it's to eliminate the concept of trust from digital systems altogether.* Kindervag proceeds with; *"We've injected this concept of trust into digital systems, but it should have never been there, because trust represents a vulnerability for digital systems"* [22].

Since its Zero Trust inception in 2010, research and consulting firm Forrester puts forward the thought leadership of Kindervag [22] in their approaches, focusing mainly on managerial level but lacking operational detailing that DevOps teams and engineers can get proper guidance from. Most of the security measures are derived from the control objectives in control frameworks and are not directly aligned with security measures prescribed by tech vendors. Consequently, linking the strategic objectives to operational security measures is complex and is rarely implemented [18]. The problem with an approach that lacks alignment with strategic goals lies in the limitations of mainly IT-focused security and security experts. Bobbert refers to operating in silo's without any reflection outside the silo [23]. The security experts operate in silos with limited view on the world and the business drivers and business context [19]. This is important, as information security is subject to many different interpretations, meanings and viewpoints [24], especially since major breaches can have serious impact on the continuity of the firm as well as their individual board members [25]. Bobbert states in his research into improving Business information security that it needs to be a collaborative effort between Technology, Business (Asset Owners) and risk management to establish and maintain a proper and -near- real-time Cyberrisk and security administration. From strategic level towards the operations and vice versa. To effectively link the strategic level of the organization to the operational level in the organization, we need to have a proper level of awareness and understanding on *how* to do this. We explore this challenge based on

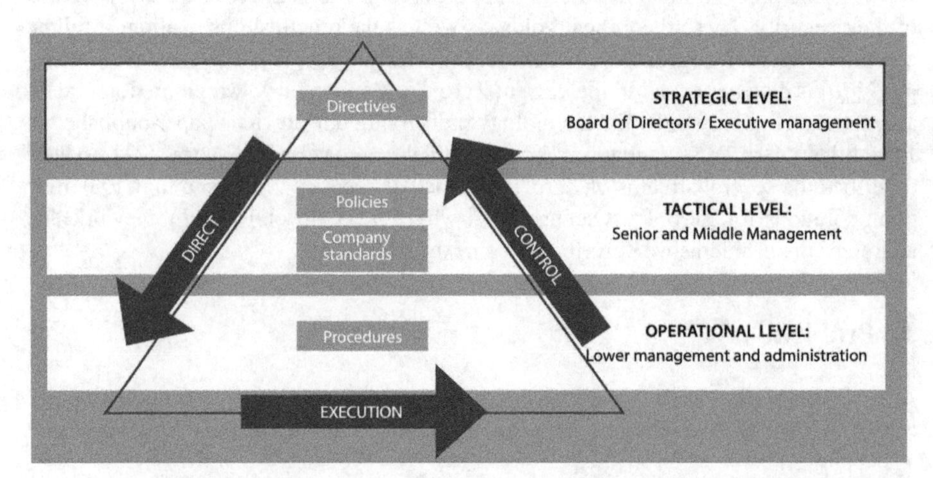

Fig. 1. The IS Governance Direct Control Cycle taken from Von Solms and Von Solms [26] and applied in ISACA's COBIT5 Framework for Enterprise Governance of IT.

earlier research in this domain, distinguishing per organizational level the processes and data.

2.1 Business Information Security Processes and Data

The Information Security Governance (ISG) layers to bridge the so called knowing-doing [20] gap and to gain the integral view the Von Solms brothers developed the Direct Control Cycle [26] (Fig. 1). The authors distinguish three levels of the organisation: Governance, Management and Operational level.

We elaborate on each level, including some examples. The directive-setting objectives stem from the strategic level. Risk appetite and accompanying policies are communicated to senior management in the form of requirements. Senior management is then mandated to put these policies into standards (e.g. technical, human and process requirements). These standards are applied in terms of all kinds of risks (e.g. through maintenance of risk logs) and security (e.g. security action plans, advisories) processes and controls (e.g. general IT controls). Processes and controls depend on underlying processes such as operational services processes like: change management, configuration management, incident management and problem management. All with clear requirements. Due to changes in legislation, technological trends (Cloud, IoT, OT, Big Data) and changing business environments, the subsequent security requirements also change. In many organizations, these requirement-setting documents reside on personal laptops, fileshares (e.g. sharepoint), desktops in spreadsheets [27]. This Excel spreadsheet based way of working generates an administrative burden to maintain and becomes a risk on its own since there is no single, authoritative place of truth [2].

2.2 Problem Statement

This problem becomes bigger with the growth of all sorts of smart devices and data sitting all over the place. Regulated companies perform better in this respect, since managing information risk and security is part of their license to operate and losing that poses a business continuity risk. Continuous measurement and reporting on the performance of risk and security processes is needed in order for senior business leaders to take ownership of assets and risks, due to rotation of personnel, the introduction of new tech-services without IT involvement, formal procurement processes (vendor vetting, etc.), mergers and acquisitions, rough and orphan assets become the new standard rather than an exception. Accurate administration of critical assets, the value they represent, CIA ratings etc. is not in place nor centrally administered. This wood of security tooling causes decision latency[2], during a hack, due to inefficient security operations that has limited interaction. The tools are owned, consumed, managed and measured by multiple stakeholders like auditors, managers, security staff, IT, business users. This brings us to the main problem statement, which is:

[2] The Standish Group: Decision latency theory states: "The value of the interval is greater than the quality of the decision." Therefore, to improve performance, organizations need to consider ways to speed-up their decisions.

"Current emphasis of Zero Trust lies on architecture principles that are understood only by insiders. The current approaches and documents lack alignment with risk management, existing frameworks and associated processes. Board and business involvement are not addressed and ownership of data, risks, security controls and processes is limited. And the main focus is on the change and not on the run and its value contributors".

2.3 Research Questions

Pondering the issues mentioned above there is a need to establish a more collaborative way of working among stakeholders when addressing the dynamics of the environment and the organization, gain a more qualitative and integral view based on facts related to tactical and operational data, to secure an increase in awareness at board level, to cultivate a certain level of reflection and self-learning and improvement to use recognised best-practice frameworks produced and maintained by existing communities. Therefore, the aim of this research is to answer the following main research question "*How can we establish a method which utilizes best practices and collaboration for improving Zero Trust security implementations?*".

In order to answer this main research question, we follow Wieringa [28] to distinguish Knowledge Questions (KQ) and Design Questions (DQ). Knowledge questions provide us with insights and learnings that together with Design Questions contribute in the construction of the design artefact (later referred to as Portal) since the artefact will be integrated in the exiting Managed Security Service Portal (MSSP) of ON2IT. This means that during the Design and development stages separate –requirement- design questions are formulated with the objective to design artefact requirements. The Design Science Research Framework of Johannesson and Perjons [29] is adopted and visualized in Fig. 2. This approach follows earlier design and engineering efforts at the University of Antwerp and Radboud University [9, 2].

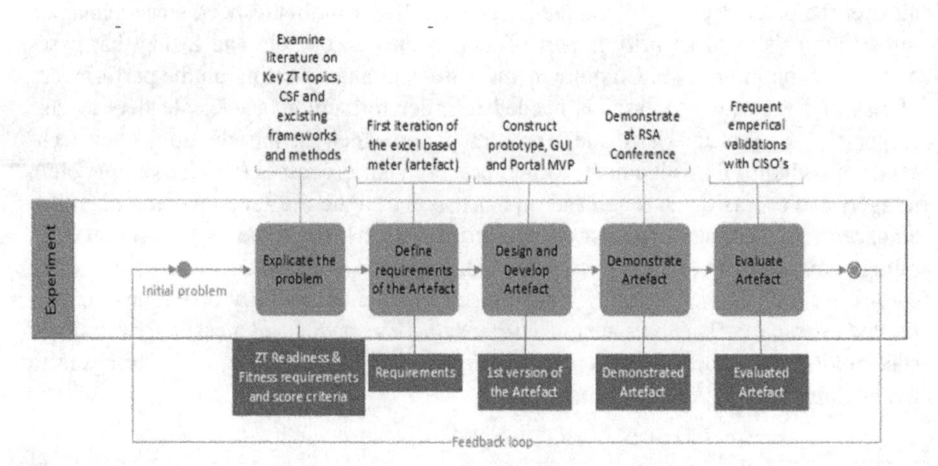

Fig. 2. Conceptual model for the Zero Trust Framework and artefact based on Design Science Research (Taken from Perjon and Johannesson) as proposed in the authors earlier research work [21]

In our previous publication we have formulated the following research questions:

1. What are Critical Success Factors for drafting and implementing a ZTA?
2. What is an easy to consume capability maturity -readiness- model and its associated portal technology that enables the adoption of ZTA and guides boards and management teams and facilitates collaboration and ownership?
3. How does the future empirical validation of the framework and the associated portal look like and how does it provide feedback to relevant stakeholders?

3 Research and Development Methodology

Design Science Research (DSR) has attracted increasing interest in the Information System research domain. March and Mith initiated important DSR work with their early paper on a two-dimensional framework for research on information technology [30]. The objective of DSR research is to establish artefacts that solve real-life problems. The collective set of requirements within the DSR artefact should contribute in this goal. Frequent validation involving stakeholders, such as users, engineers and customers to confirm that the artefact requirements actually help solve the problem at hand is necessary [31]. Wieringa [31] refers to using the regulative cycle to determine the right set of artefact requirements and to validate if it contributes to solving the problem.

Hevner et al. [32] produced a broad framework which is used worldwide to perform and publish DSR work. This framework is visualized in see Fig. 3 contrasts two research paradigms in information system research: *behavior sciences* and *design sciences.* Both domains are relevant for Information Security because the first is concerned with soft aspects such as the knowledge, attitudes and capabilities required to study and solve problems. The second is concerned with establishing and validating artefacts. To put it more precisely, Johannesson and Perjons distinguish between the design, development, presentation and evaluation of an artefact [29]. Wieringa distinguished many methods for

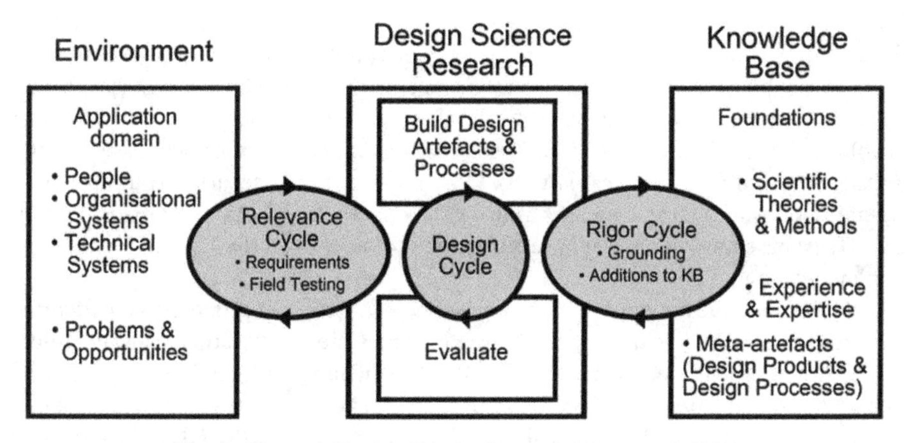

Fig. 3. Hevner's Design Science Research Framework [32].

examining numerous types of problems, e.g. design problems and knowledge problems [33]. In this Zero Trust project we used Hevner's work as a frame of reference for the entire DSR project and potential later validation by practitioners and we use Wieringa's approach to address the challenges and technical requirements we encounter during the current and future journey of portal development.

Design Science Research in a Business Context

Like any other longitudinal research new insights emerged from the problems we encountered in real life environments during the performance of our research project. The complete project, specifically the design of the artefact, is done in a practical business setting. We applied the research strategy displayed in Fig. 4, departing conducting literature research on main ZT topics and shortcomings. This was largely published in our research paper: "Zero Trust Validation: From Practical Approaches to Theory". That research paper mainly focusses on phase 1 and 2. This research paper manly focusses on 3, 4 and 5.

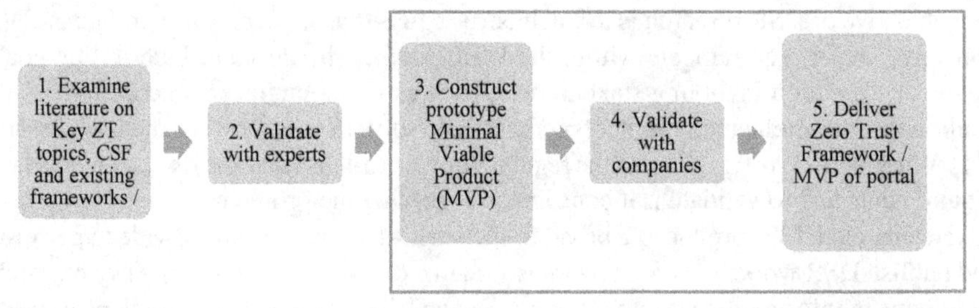

Fig. 4. Scope of the research project and strategy to design and build the Zero Trust Framework and portal technology (artefact) based on DSR.

4 Results of the Research

Based upon the above-mentioned insights from the literature and experiences we have detailed the Critical Success Factors in our previous publication, being; *Engage and collaborate with relevant stakeholders* on the value of Zero Trust for the business. *Alignment with existing control framework* and their scaling, metrics and taxonomy so it enables collaboration. *Complete and accurate administration* of critical assets (Data, Assets, Applications, Services: DAAS), CIA rating and their security requirements in a central repository (one source of truth). Establish a *Clear technology roadmap* with Zero Trust based measures that have a clear definition of done (DoD) and timelines for implementation.

As a result of the previous publication, Bobbert & Scheerder proposed a longitudinal research methodology to do empirical research with Chief Information Security Officer (CISO's) and Data Protection Officer (DPO's) on validating this Framework collectively due to group collaboration in small groups [34]. This research paper continues on that research proposition of longitudinal research. Table 1 shows the ON2IT Zero Trust Framework and the Three Organizational Levels.

Table 1. The ON2IT Zero Trust Framework and the Three Organizational Levels According to the Direct Monitor and Control Cycle

Strategic/Governance	*Know your environment and capabilities*	Determine to what extend context analysis, leadership capabilities, roles, and accountabilities are in place to execute a Zero Trust strategy
Managerial	*Know your risks*	Determine to what extend structures, processes, and relational mechanisms (reporting, roles, tone at the top, and accountabilities) are in place to execute the strategic zero trust objectives. E.g. Capabilities for logical segmenting[a]
Operations	*Know your technology*	Determine to what extent current or future technology is able to utilize zero trust measures. (Fitness)

[a]Segment being "A logical part of the environment which consist of collaborating data, assets, applications and services that represent a certain value, business dependency and exposed to certain risks".

5 Alignment of the ON2IT Zero Trust Framework

To answer research question two;

> *"What is an easy to consume capability maturity -readiness- model and its associated portal technology that enables the adoption of ZTA and guides boards and management teams and facilitates collaboration and ownership?"*

Rendering to the aforementioned shortcomings, obtained from the literature, the improved framework has the objective to function as guidance for senior managers and boards, before they start a Zero Trust strategy. In this section we recall the improvements put forward in our previous paper, in the next section we demonstrate how the Framework and Portal artefact addresses just that:

- A common language is used by making use of existing control Framework. This makes it an easy to consume model.
- The Framework enforces strict sign off for asset owners and board members on preconditions that are required before you can implement Zero Trust.
- Segmenting the environment based on data flows; Each Data, Application, Asset or Service (DAAS) element in a segment requires ownership and annotation of the level of Confidentiality, Integrity and Availability (CIA) in a central repository to ensure sufficient asset qualification so security measures can be assigned to these assets.
- By assessing the readiness and technological fitness to utilize Zero Trust there is transparency in the level of a successful ZT implementation, the "progress monitor" in the framework monitors the progress.

6 From Zero Trust Strategy to Operations

In the Zero Trust architecture, measures are implemented to limit the attack surface, and to provide visibility and, hence, swift and to-the-point incident response.
How we allocate measures to segments is described below. First, by identifying traffic flows relevant to a (closely coupled) application. In physical networks, the notion of segmentation takes a step further; the term 'microsegments' has been coined. By intention, such segments contain a (functional) application or a set of closely associated applications. By this additional segmentation, a 'microperimeter' is formed that can be leverage to exert control over, and visibility into, traffic to/from the contained (functional) application. A policy governs the traffic flows, inspects those flows and thereby the Zero Trust architecture not just prescribes defence in depth by isolation but also by inspection, response and reporting. We will elaborate this in more specific detail;

- Policy regulating traffic to and from a Zero Trust segment

  o s *specific* and *narrow*, satisfying the 'least privilege access' principle: it allows what's functionally necessary, and *nothing more*;
  o is, whenever possible, related to (functional) *user groups*
  o enforces that traffic flows contain *only* the network applications that are defined for that specific segment;
  o enforces content inspection to enable threat detection and mitigation on;
  o visibility is ensured;

- Logs are, whenever possible, related to *individual users*;
- Presence and conformance of policy is operationally safeguarded;
- Policy is *orchestrated*, if applicable, across multiple components in complex network paths;
- Operational state and run-time characteristics (availability, capacity instrumentation) are structurally monitored.
  The very same concepts applied above to physical networks are used, unchanged, in virtual-, container-, cloud- or other software defined networks. In all cases, a way is found to create a logical point of 'visibility and control' that enables insertion and safeguarding of the appropriate measures.
  Extending the Zero Trust architecture to endpoints is a step that is conceivable as well, considering the endpoint itself as a complex collective of potentially unwanted (malicious) processes to be safeguarded. At endpoint level, an agent can be introduced to detect and mitigate malicious processes. When doing this, fine-grained endpoint behaviour extends the visibility beyond the network layer, and 'large data' analysis of (user) behaviour becomes viable, further deepening both visibility and defence in depth. Extracting the telemetry data -near- real time from these technological measures is needed to feed this data back to tactical and strategical levels and promptly respond and telecommand back[3]. This relates to the increasing question; "*how to inform the CEO in minutes after a breach?*

[3] Telemetry is the collection of measurements or other data at remote or inaccessible points and their automatic transmission to receiving equipment for monitoring. The word is derived from

7 Deliverables

- Due to practical experiences we see that an important factor for Zero Trust success is to start with assessing the organizations readiness and technological fitness to adopt and execute Zero Trust. Therefore the "Framework" should follow a certain sequence of application, as displayed in Fig. 5, initially consist of:
- A Readiness assessment to determine how ready and fit you are as a company on the strategic level and managerial level.
- An assessment to determine your technological fitness compared to objectives. This fitness level represents to what extent an organization is technically capable of utilizing the required Zero Trust measures and understand their limitations. Starting with five segments to aid the learning curve and understand the earlier mentioned Knowing Doing Gap.
- A process of labelling these five segments with meta-data such as CIA ratings, Relevance scores, regulatory-compliance tags. This is desired to determine the technical policies and guidelines that should be applied to the segments. Table 2 displays the Relevance score and associated DAAS label. Determining the relevance score can be done by making use of existing CIA rating methods. Where CIA normally rates the asset, will the Relevance score rate the entire segment with potentially multiple assets.

Table 2. Overview of the Relevance Score and the DAAS Labels in the Artefact (Portal)

Relevance score	Detailing	DAAS Label
0–25 (CI11)	No personal data, no sensitive data, limited number and amount of financial data. No possible legal, contractual or regulatory impact. Minor local reputational damage possible. Medium financial impact	
25–50 (CI22)	Limited amount of personal data (<4 different data types) and a limited amount of data subjects, no sensitive data, limited number and amount of financial data. No possible legal, contractual or regulatory impact. Minor local reputational damage possible. Medium financial impact	Subject to audit, Core processing 3rd party access to data Personal data (PII)
50–75 (CI33)	Personal data or financial data available, Legal, contractual, or regulatory impact possible. Reputational damage can be locally impacted. High financial impact. Industrial Control Systems with High availability. (with business case and all applies with potential waiver process)	Personal data (PII) Core processing 3rd party access to data Confidential data
75–100(CI44)	Special personal data or sensitive financial data available. Serious Legal, contractual, or regulatory impact (serious fines, suspension or loss of license) possible. Risk of sustained (inter)national reputational damage. Industrial Control Systems with High availability requirements. Very high financial impact. (At any cost)	Special personal data, Industrial Control Systems Personal data (PII) Core processing 3rd party access to data Confidential data

Greek the roots tele, "remote", and metron, "measure". Systems that need external instructions and data to operate require the counterpart of telemetry, telecommand. Source.

- A Progress Monitor to report to boards and regulators on a periodical basis and thereby involve them in the required decision making and avoid decision latency. An additional portal functionality built into the ON2IT Managed Security Services Platform Portal (also referred to as an *artefact*) that captures; a. the readiness assessment results on the three levels of strategy, management and operations, b. the Zero Trust segments with meta-data and c. the fitness score of the segments extracted from the operational technology by ingesting logs of technology such as segmentation gateways a.k.a. firewalls, end points and other security measures.

The operationalization of the ON2IT Zero Trust Framework is done in the portal. We detail per Framework component "how" this is operationalized and evidenced in this paper by making use of portal screenshots.

Figure 6 displays the Zero Trust overall score, the progress of ZT implementation and the Readiness maturity level on strategic and tactical level.

Figure 7 shows the screenshot of the Readiness assessment results on Strategy (including example of "Ownership and sign off" criteria and score).

Figure 8 shows the screenshot of the Fitness assessment results for segment ATMS, including status per measure and in the upper right corner the ZT Heatmap. This heatmap displays the relevance score compared to the security gap (amount of security measures implemented in the specific segments). Segments with a high relevance score but large security gap are calculated in "red zone" of this heatmap. This enables boards and senior managers to have direct insights into weak spots and where to take action.

Figure 9 shows the screenshot of the Fitness assessment results for segment ATMs, including status per measure (e.g. Encryption) and in the lower left corner, the tags for the application of various Standards and frameworks.

Figure 10 shows the screenshot of the operational status of segment "Insurance" with ownership, relevance score and cyber events; exfiltrations, investigations, intrusions, threats and advisories in October 2019.

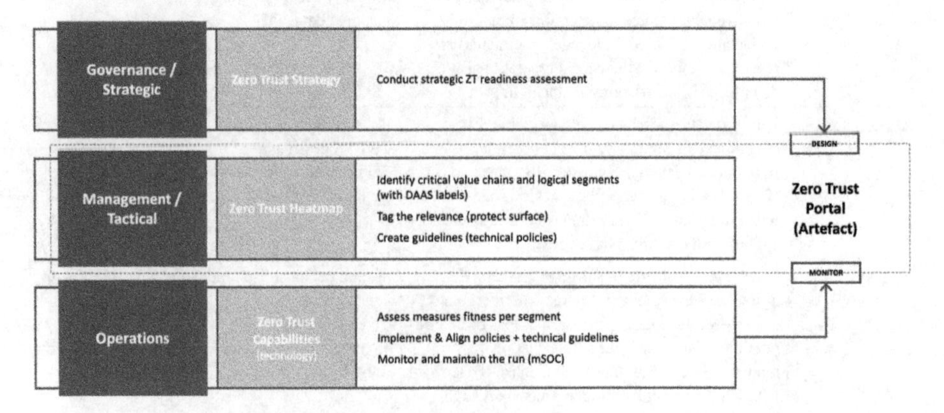

Fig. 5. The ON2IT Zero Trust Framework Approach; from Strategy to Operations

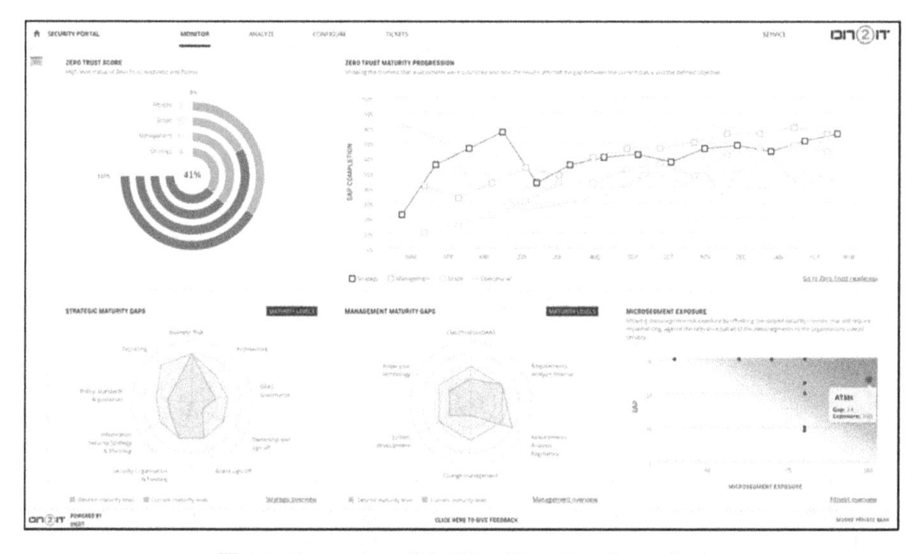

Fig. 6. Screenshot of the Zero Trust Portal (artefact)

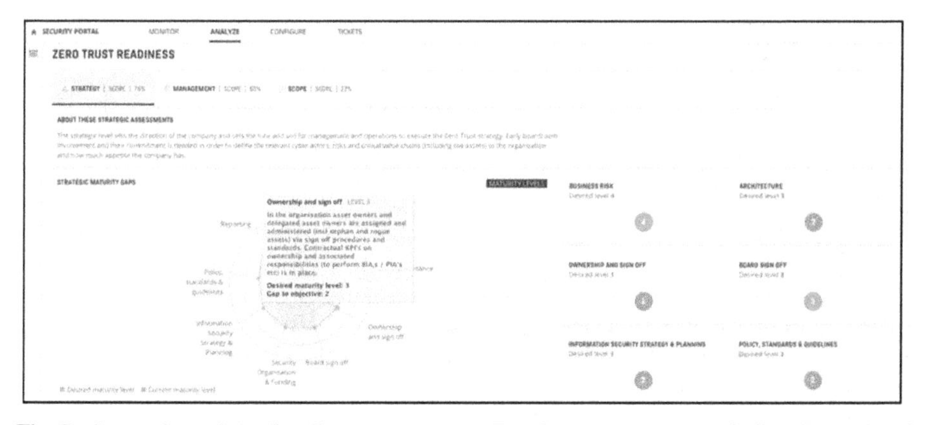

Fig. 7. Screenshot of the Readiness Assessment Results on Strategy (Including Example of "Ownership and Sign Off" Criteria and Score)

8 Preliminarily Results of GSS Validation Sessions

To execute this empirical validation with practitioners, sessions are held and facilitated via Group Support System [9]. The opportunity for larger scale longitudinal research lies specifically in gaining knowledge at the organizational level and using that data, collected with GSS system technologies, to establish a collective knowledge base. This larger set of data can then form a frame of reference for a certain industry, country or community and thus contribute to other sectors, countries or communities. The application of GSS for such large-scale longitudinal research has been identified by De Vreede.

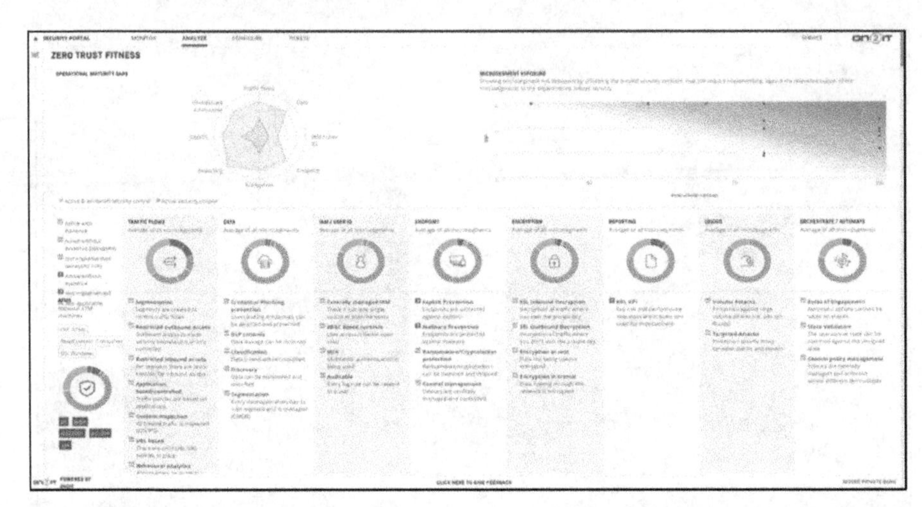

Fig. 8. Screenshot of the Fitness Assessment Results for Segment ATMS (**ATM is** an abbreviation for automated teller machine. The ATM segment is part of the Moore private bank organization used in this demonstration of the artefact), Including Status Per Measure and in the Upper Right Corner the ZT Heatmap

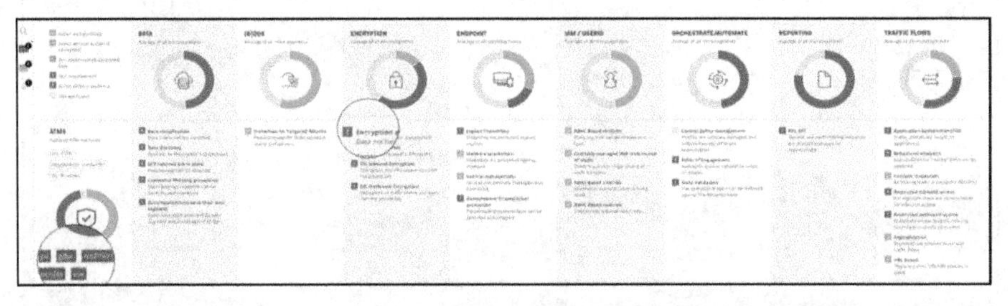

Fig. 9. Screenshot of the Fitness Assessment Results for Segment ATMs, Including Status Per measure (e.g. Encryption). In the Lower Left Corner, the Tags for the Application of Various Standards and Frameworks.

A Group Support System (GSS) method was applied over a period of January 2020 to December 2020 to gain a deeper understanding of the topic, validate the questionnaire questions on clearness and completeness and gather additional viewpoints on Zero Trust. The multiple GSS sessions enabled peer-reflection, which generates new knowledge on the Zero Trust topics. Each group of practitioners developed new insights that were taken into consideration by the next group. This "double-loop learning" [35] provides additional scrutiny to the latter and thereby contributes to the overall quality of the ON2IT Zero Trust Framework. The sessions were executed by a professional GSS moderator, which, according to Hengst, is key [36]. All steps, scores and arguments are recorded in the GSS software to assure objectivity, transparency, controllability and repeatability. A predefined agenda, clear introduction and the readiness questionnaire were shared prior to the meetings so the participants can individually prepare the GSS session.

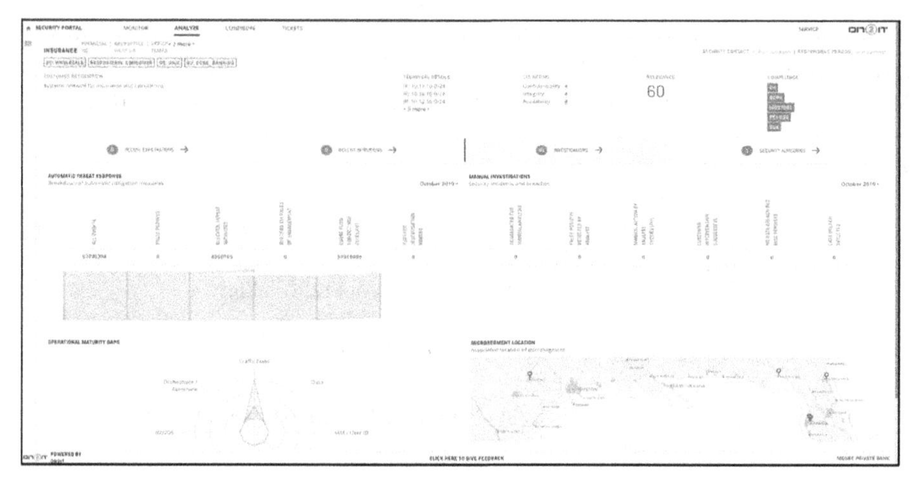

Fig. 10. Screenshot of the Operational Status of Segment "Insurance" with Ownership, Relevance Score and Cyber Events; Exfiltrations, Investigations, Intrusions, Threats and Advisories in October 2019

Process of the GSS Meetups with CISO's and Security Professionals
As proposed in earlier research [21] GSS was used to validate the framework and the associated topics relevant for determining the Readiness maturity level. These questions can function as a questionnaire to diagnose organisations and identify gaps. During 2020 73 participants in 10 sessions validated the framework questions on:

- Completeness, do you have something to complement to the current question set?
- Clearness, is the question easy to understand and without ambiguity?
- Validity, does the set of questions represent topics related to Zero Trust and present in contemporary environments?
- Priority, which one do you think has higher validity over the other and can be prioritised as a core pre-requisite?

Before each session participants were instructed via a clear instruction that included; a video, letter with guidance (including the questionnaire) and once confirmed a phone call to explain:

- Agenda of the session
- Objective of the session
- Expectations of the participants
- Explanation of the process and timelines and required preparation
- The end results

One week prior to each session participants were called to confirm their participation and if preparations were made. Each session was moderated by a professor of Antwerp Management School and professional GSS facilitator. Each of the GSS reports detail:

- The name of the participants;
- Introduction of the session objectives;
- The score of the relevance of the topics and questions, ranking from 1 to 10. 1 being not relevant and 10 being very relevant;
- Comments to the scores per strategic, tactical and operational level;
- Answers to additional questions;
- End evaluation of the session, to verify if objectives are met.

The results of the 10 GSS sessions held among 73 participants with GSS are detailed in the Table 3.

Table 3. Overview of the total amount of GSS participants at strategic level

Validation Session dates in 2020	#Participants
Feb	5
March	7
April	6
May	9
July	4
August	4
September	5
October	6
October	9
Nov	8
Total GSS participants	**73**

Improvements were made to the Framework based upon the empirical validations via GSS. The ON2IT Zero Trust Framework has the objective to act as a guide for boards and managers prior to starting a Zero Trust strategy and during the implementation. Below, we list the major findings for improving the Framework.

- A common language is needed by making use of existing control framework as of level > 3, for example, ISF, NIST Cybersecurity Framework, NIST privacy Framework, PCI DSS or ISO27000 controls. CMMI based maturity levels on a 1 to 5 scale are applied, based on audit terminology (such as Test of Design (ToD), Test of Implementation (ToI) and Test of Effectiveness (ToE)) that NOREA is using.
- Most pf the GSS panel participants acknowledge that Zero Trust can help to inform the CEO quicker, more granular. But Zero Trust can also be viewed as something negative

due to the naming "Zero Trust" that can have a negative perception or aftertaste by boards.

- Following category 1 (Know your environment and capabilities) you identify whether Business, Privacy and IT alignment takes place, threats and trends are identified that influence the enterprise risk management (ERM) and assign appropriate ownership at board and managerial level (according to the COBIT EDM model). On a managerial/tactical level, NIST can be used and on an operational level ISO can be used.

  o COSO / COBIT – Strategic (Enterprise-Level Approach to Risk Management)
  o ISO – Operational (Initiative / Program-Level Approach to Risk Management)
  o NIST – Tactical (Asset / Project-Level Approach to Risk Management)

- A future research project was defined based upon the feedback of the GSS to map all Zero Trust Measures to the SCF framework, that captures all frameworks for Digital Security.
- Each DAAS element requires ownership and CIA annotation in a repository (e.g. CMDB) to ensure adequate asset qualification and even quantification so security measures can be assigned to these assets. A Relevance Score on scale 0–100 combines the standard Business Impact Assessments (BIA) and Privacy Impact Assessment methods (PIA). The presence of personal data, its importance for the Business-critical processes and the type of personal data are all factors that affects the Relevance score, and therefore the type of Security measures to be applies. 0 being a segment with low exposure and 100 with high exposure.
- The Framework encourages strict sign off for board members on preconditions that are required before you can implement Zero Trust. Organizations that use the COBIT5 or COBIT2019 processes and design principles can plot these to the EDM layers of Governance, Management and Operations. This brings the required common language on technical and organizational security measures.
- By assessing the readiness of the organization in terms of processes and structures as well as the technological fitness to utilize Zero Trust transparency is given into the current and desired states. Some participants raised the concerns that in large environments, this segmenting and putting measures in place might take years. Monitoring the progress is vital not to lose attention and urgency.

9 Future Research

To answer research question three: "*How does the future empirical validation of the framework and the associated portal look like and how does it provide feedback to relevant stakeholders?*" can be answered in the following ways:

Assessing an organizations' posture with respect to Zero Trust viability requires evaluating these three levels, *and this ON2IT framework.* We propose four research areas:

1. Validation of the Zero Trust Readiness framework (pre- and post-implementation progress monitor).

2. Assessing the presence and relevance of strategic capability attributes (strategic level).
3. Assessing the presence and relevance of executive capability attributes (managerial level).
4. Assessing the presence and relevance of adequate technical capabilities (operational level).

These assessments determine the relevance, coverage, depth and actionability of the controls/objectives (at their respective level) needed to successfully implement and maintain Zero Trust security strategies.

10 Conclusions

For answering our Research Question; *How can we establish a method which utilizes best practices and collaboration for improving Zero Trust security implementations?"* the ON2IT Zero Trust Framework explicitly recognizes all major shortcomings in the current approaches. Such as the lack of board and business involvement and explicit sign-off to ensure commitment. The presented framework of ON2IT assigns and organises the ownership and responsibility for segment and asset risks and their measures, aka controls. These assets and controls are clearly defined in the 'classic' Zero Trust concepts of segments and transaction flows. By forcing the Zero Trust concept of *segmentation* 'up' into the boardroom strategic risk level, the alignment between risk and the required measures becomes more tangible and manageable than in existing frameworks. Mainly due to the fact that names are ascribed to assets.

A key design objective of the ON2IT Zero Trust Framework is to formalize the involvement of organization asset owners from a business perspective, yielding in more insightful interpretations of concepts such as *recovery time objectives* and *risk appetite.*

The framework obviously addresses the readiness necessities at the three separate organizational levels of cybersecurity and provides insight and control across these levels with a common language and metrics for relevant measurements. Because the effectiveness of operational measures is -near realtime- assessed in relation to the Zero Trust segments defined at the upper levels, the alignment of risk and technology can be designed and measured with greater precision. The 'relevance score', derived from traditional CIA ratings, of every individual segment, a concept integrally embedded in the methodology and Zero Trust orchestration and automation portal, drives the required controls and the required dynamic feedback on their effectiveness. This is a near real-time process. This simply cannot be a static or manual process else you cannot inform "upper" levels with proper and actual information.

Further Design Science Research based research and development for both the framework as well as the portal technology will continue and is needed in order to improve organization's security maturity, the security and risk administration, decrease risks and lower the operational cost of information security to focus on what really matters. Empirical demonstrations and evaluations of the artefact with industry professionals (CISO's, Security managers, architects) are -again- planned for 2021–2022 to continue the longitudinal research.

References

1. Betz, C.: The Impact of Digital Transformation, Agile, and DevOps on Future IT Curricula (2016)
2. Bobbert, Y., Ozkanli, N.: LockChain technology as one source of truth for Cyber, Information Security and Privacy. In: Computing Conference, London (2020)
3. CROForum. Understanding and managing the IT risk landscape: A practitioner's guide (2018). https://www.thecroforum.org/2018/12/20/understanding-and-managing-the-it-risk-land-scape-a-practitioners-guide/
4. Kumar, T.: What is the impact of distributed agile software development on team performance? Antwerp Management School, Antwerp (2020)
5. Lencioni, P.: The Five Dysfunctions of a Team; a Leadership Fable. Wiley Imprint Jossey Bass, SA USA (2002)
6. Ozkanli, N.: Implementation of Continuous Compliance; Automation of Information Security Measures in the software development process to ensure Continuous Compliance, Utrecht: Open University Press Netherlands (2020)
7. Forsgren, N.: Accelerate: The Science of Lean Software and Devops: Building and Scaling High Performing Technology Organisations. IT Revolution Press, United States (2018)
8. McCarthy, M.A.: A compliance aware software defined infrastructure. In: Proceedings of IEEE International Conference on Services Computing, pp. 560–567 (2014)
9. Bobbert, Y.: Defining a research method for engineering a Business Information Security artefact. In: Proceedings of the Enterprise Engineering Working Conference (EEWC) Forum, Antwerp (2017)
10. Hilton, M.N.N.: Trade-offs in continuous integration: assurance, security, and flexibility. In: Proceedings of the 2017 11th Joint Meeting on Foundations of Software Engineering (2017)
11. ITGI, Information Risks; Who's Business are they?, United States: IT Governance Institute (2005)
12. Kuijper, N.: Effective Privacy Governance and (Change) Management Practices (Limited to GDPR Article 32) A View on GDPR Ambiguity, Non-Compliancy Risks and Effectiveness of ISO 27701 as Privacy Management System. Antwerp Management School, Antwerp (2020)
13. Kluge, D., Sambasivam, S.: Formal information security standards in German medium enterprises. In: Conisar, Phoenix (2008)
14. Siponen, M., Willison, R.: Information security management standards: problems and solutions. Inf. Manag. 46 (2009)
15. Puhakainen, P., Siponen, M.: Improving employees compliance through information systems security training; an action research study. MIS Q. **34**(4), 757–778 (2010)
16. Workman, M., Bommer, W., Straub, D.: Security lapses and the omission of information security measures: a threat control model and empirical test. Comput. Hum. Behav. **24**(6), 2799–2816 (2008)
17. Lebek, B., Uffen, J., Neumann, M., Hohler, B., Breitner, M.: Information security awareness and behavior: a theory-based literature review. Manag. Res. Rev. **12**(37), 1049–1092 (2014)
18. Yaokumah, W., Brown, S.: An empirical examination of the relationship between information security/business strategic alignment and information security governance. J. Bus. Syst. Governance Ethics **2**(9), 50–65 (2014)
19. Flores, W., Antonsen, E., Ekstedt, M.: Information security knowledge sharing in organizations: investigating the effect of behavioral information security governance and national culture. Comput. Secur. **2014–43**, 90–110 (2014)
20. Pfeffer, J., Sutton, R.: The Knowing-Doing Gap: How Smart Companies Turn Knowledge into Action. no. Harvard Business School Press (2001)

21. Bobbert, Y., Scheerder, J.: Zero trust validation: from practical approaches to theory. Sci. J. Res. Rev. **2**(5) (2020). https://doi.org/10.33552/SJRR.2020.02.000546
22. Kindervag, J.: Build Security Into Your Network's DNA: The Zero Trust Network Archit Security (2010)
23. Bobbert, Y.: Improving the Maturity of Business Information Security; On the Design and Engineering of a Business Information Security Artefact. Radboud University, Nijmegen (2018)
24. Van Niekerk, J., Von Solms, R.: Information Security Culture; a Management Perspective, pp. 476–486. Elsevier (2010)
25. Papelard, T.: Critical Succes Factors for effective Business Information Security. Antwerp Management School, Antwerpen (2017)
26. Von Solms, R., Von Solms, B.: Information security governance; a model based on the direct–control cycle. Comput. Secur. **2006**(Elsevier) Comput. Secur. **25**, 408–412 (2006)
27. Volchkov, A.: How to measure security from a governance perspective. ISACA J. **5** (2013)
28. Wieringa, R.: Design Science Methodology: For Information System and Software Engineering. Springer, Berlin (2014)
29. Johannesson, P., Perjons, E.: An Introduction to Design Science. Springer, Stockholm University (2014)
30. March, S., Smith, G.: Design and natural science research on information technology. Decis. Support Syst. **15**, 251–266 (1995)
31. Bobbert, Y.M.J.: Enterprise engineering in business information security; a case study & expert validation in security, risk and compliance artefact engineering. In: Aveiro, D. et al. (eds.) EEWC 2018. LNBIP 334, pp. 1–25. Springer, Heidelberg (2019). https://doi.org/10.1007/978-3-030-06097-8_6
32. Hevner, S., Park, J.M., Ram, S.: Design science research in information systems. Manag. Inf. Syst. Q. **28**(1), 75–105 (2004)
33. Wieringa, R.: Design science as nested problem solving. In: Proceedings of the 4th International Conference on Design Science Research in Information Systems and Technology, New York (2009)
34. Straus, D.: How to Make Collaboration Work; Powerfull Ways to Build Consensus, Solve Problems and Make Decisions. Berrett-Koehler Publishers Inc, San Franciso (2002)
35. Argyris, C.: Double-loop learning, teaching, and research. Acad. Manag. **1**(2), 206–218 (2002)
36. den Hengst, M., Adkins, M., Keeken, S., Lim, A.: Which Facilitation Functions are Most Challenging: A Global Survey of Facilitators. Delft University of Technology, Delft (2005)

A Compact Quantum Reversible Circuit for Simplified-DES and Applying Grover Attack

Ananya Kes[1], Mishal Almazrooie[2](✉), Azman Samsudin[1], and Turki F. Al-Somani[3]

[1] School of Computer Sciences, Universiti Sains Malaysia (USM), 11800 Gelugor, Pulau Pinang, Malaysia
[2] Khulais, Saudi Arabia
[3] Department of Computer Engineering, Umm Al-Qura University, Makkah, Saudi Arabia

Abstract. Security of symmetric cryptosystem in post quantum era can be evaluated by implementing the algorithm on quantum platform. The number of qubits and ancilla bits used to design the quantum circuit is questionable as it impacts the cost of the circuit. This study proposes a compact and efficient quantum circuit for simplified DES which utilizes the lowest number of qubits and ancilla bits. The key generation steps are fused so as to maintain the essence of the algorithm and at the same time make the circuit compact. Key and plaintext is mixed without the help of ancilla bits. The results show that nineteen qubits and seven ancilla bits along with polynomial number of quantum gates are sufficient to implement simplified DES on quantum circuit. Quantum attacks are conducted on the proposed circuit using Grover's algorithm for key search as it acts as the black box for quantum search.

Keywords: Block cipher · Grover search · Quantum simulation · Quantum cryptanalysis

1 Introduction

With the advent of the World Wide Web, the Internet is widely used across the globe. With the increase of Internet usage, the importance of information or data security has also increased. The science of cryptography helps us to achieve data or information security by encrypting the data using strong encryption methods by applying complex mathematics and logic. The security and strength of the existing cryptographic algorithms against any possible attacks must be ensured. Asymmetric and symmetric cryptography are believed to be secure against any attack using classical computers. Unfortunately, this view is no longer valid in the presence of quantum mechanics, where the calculations are performed based

M. Almazrooie—Independent Researcher.

K. Arai (Ed.): FICC 2021, AISC 1363, pp. 849–864, 2021.
https://doi.org/10.1007/978-3-030-73100-7_59

on the behavior of particles at subatomic levels. Thus, quantum computing poses threats to asymmetric and symmetric cryptography. Various studies have been published on quantum number factorization [1,2], along with studies on alternative solutions such as code-based cryptography and lattice-based cryptography [3]. Some solutions to the key distribution problem have come from quantum mechanics and opened the field of quantum cryptography [4].

The only known and clear quantum threat to symmetric algorithms is that the exhaustive key search can be performed more efficiently on the quantum platform with quadratic speedup using Grover's algorithm [5]. The study of quantum cryptanalysis on a block cipher was performed with the assumption that the block cipher was already implemented on a quantum platform [6]. For example, the security of a block cipher could be evaluated by using Brassard, Hoyer, and Tapp's quantum algorithm [7], but the block cipher should be implemented as a quantum reversible circuit. Block ciphers such as Data Encryption Standard (DES) [8] and Advanced Encryption Standard (AES) are widely studied cryptographic algorithms. Though these are not inordinately complicated, it would be best understood if it can be worked through an example by hand [9]. Thus, a simplified version of the algorithm DES such as Simplified DES (S-DES) was devised [10] to conduct studies and perform a cryptanalytic attack so as to obtain an estimate of the cost of the attack on the original version of the algorithm. S-DES, being a miniature version of the original DES, retains the flavor of DES and at the same time is useful for the study. Recent studies on quantum circuits for block ciphers [11] have designed an effective quantum circuit for block cipher S-DES. However, the existent circuit is not cost-efficient as it has a large workspace of ancilla bits and qubits. The extra workspace created using ancilla bits allow for simpler circuit construction. This simplicity comes with a cost involved. The addition of ancilla bit increases the computation cost and also makes the circuit workspace bigger in size, which leads to a complex circuit design with a high cost. However, in the field of computer science, focus is to not only achieve simplicity in a particular design, but the simplicity achieved should also not come with a higher complexity. An effective circuit is a circuit which is adequate to accomplish the purpose of designing a working quantum circuit for block cipher S-DES. An efficient circuit is such that performs or functions in the best possible manner with the least cost. This study aims to design a compact and efficient quantum circuit for S-DES by reducing the number of qubits and ancilla bits to build the circuit and finally apply Grover's algorithm to perform a quantum cryptanalysis of the algorithm.

Classical S-DES will be discussed in this paper followed by an introduction to the world of quantum circuits and its components. A detailed explanation of the various components of the proposed quantum compact circuit for S-DES is provided followed by the circuit design. The quantum resources of the circuit are calculated, and Grover's algorithm is applied on the circuit for exhaustive key search, and the results are compared with classical S-DES and previous studies.

2 Classical S-DES

S-DES is a simple version of the well-known cipher DES developed by Schaefer [10]. With small parameters, S-DES has similar properties and structure to DES. In the key generation of S-DES, two 8-bit sub keys are generated from the main 10-bit secret key after passing through the stages of permutation, shift, and mixing followed by final permutation. The S-DES encryption algorithm consists of only two rounds of encryption. Each round is composed of permutation of 8-bit plaintext, splitting the plaintext into two halves, plaintext expansion, XOR-ing (plaintext with the key) and substitution (S-Box) and switching. This research focuses on implementing S-DES on a quantum circuit with minimal usage of qubits and ancilla bits. A detailed explanation of S-DES is provided in [11].

3 Introduction to Quantum Circuit Components

A quantum circuit is a model for quantum computation in which a computation is a sequence of quantum gates that are reversible transformations on a quantum mechanical analogue of n-bit register also known as n-qubit register. Quantum information and unitary transformation are discussed in the quantum computing introductory book in [12].

3.1 Quantum Bit (Qubit)

A qubit or quantum bit is a unit of quantum information. It is the quantum analogue of a classical bit. Quantum mechanics allow the qubit to be in superimposition of both states at the same time. A qubit has a few similarities to a classical bit but is overall different. Two possible outcomes for the measurement of a qubit –usually 0 and 1– such as a classical bit. The difference is that whereas the state of a classical bit is either 0 or 1, the state of a qubit can also be a superposition of both [11]. A pure qubit state is a linear superposition of the basis states. This means that the qubit can be represented as a linear combination of $|0\rangle$ and $|1\rangle$: $|\psi\rangle = \alpha|0\rangle + \beta|1\rangle$, where α and β are probability amplitudes and are generally complex numbers. In classical computation, any memory bit can be turned on or off (i.e., the state can be changed from 0 to 1 and vice versa without any loss of information about the initial value of the bit being changed). This is not the case in quantum computation, as operations in these models are reversible, while turning a bit on (or off) would lose the information about the initial value of that bit. For this reason, in a quantum computation, there is no way to deterministically put qubits in a specific prescribed state unless one is given access to bits whose original state is known in advance. Ancilla bits are generally used to copy or store qubit information and simplify complicated quantum gates to simple gates.

3.2 Quantum Gate

Quantum gates are the basic building blocks of a quantum circuit. It is a basic quantum circuit operating on a small number of qubits. Various quantum gates are used in this study to build the proposed circuit. An **X-Pauli** gate is applied on a single qubit. An **X-Pauli** is the quantum equivalent of a NOT gate and is sometimes known as bit flip. **Hadamard** gate acts on a single qubit. It maps the basis state $|0\rangle$ to $\frac{|0\rangle+|1\rangle}{\sqrt{2}}$ and $|1\rangle$ to $\frac{|0\rangle-|1\rangle}{\sqrt{2}}$ which means that a measurement will have equal probabilities to become 1 or 0. A controlled gates act on 2 or more qubits, where 1 or more qubits act as a control for some operation. Controlled-NOT or **CNOT** gate acts on 2 qubits and performs the **CNOT** operation on the second qubit only when the first qubit is $|1\rangle$, else leaves it unchanged. The **Toffoli** gate, also known as Controlled-CNOT gate, is a 3-qubit gate, which is universal for classical computation. It is defined for 3 qubits. If the first two bits are in the state $|1\rangle$, it applies **X-Pauli** on the third bit, else it does nothing. This study involves usage of the described quantum gates. For more insight on quantum gates, readers may refer to advanced study on quantum gates [13].

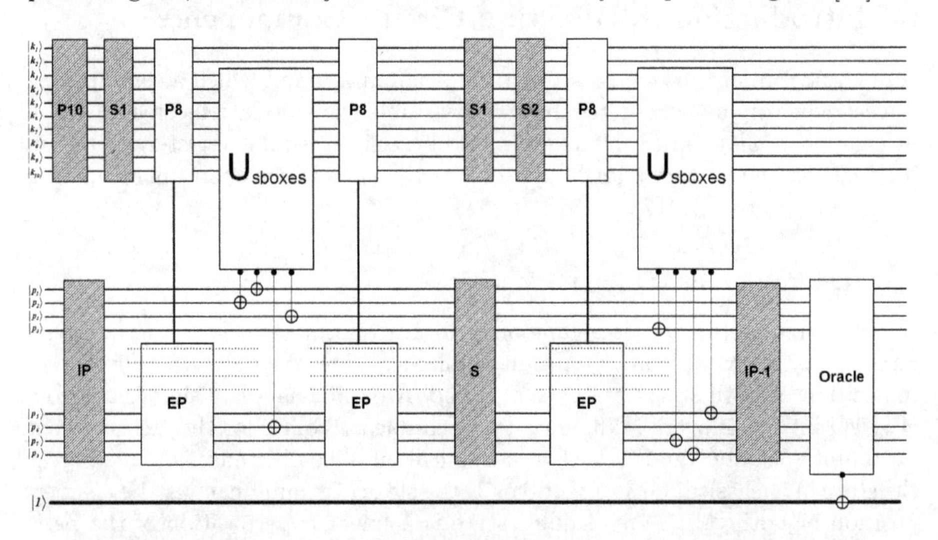

Fig. 1. The proposed block diagram of the compact quantum S-DES. This proposal makes use of 19 qubits preserved for the key ($|k_1\rangle$ to $|k_{10}\rangle$), plaintext ($|p_1\rangle$ to $|p_8\rangle$), and the Oracle qubit. A total of 7 ancilla qubits are utilized for the workspace. The circuit of S-boxes can be found in Fig. 2. **S** refers to the switch function that occurs at the end of the first encryption round. **IP** and **EP** refer to initial permutation and expansion permutation respectively. Note that, all the highlighted boxes in this figure refer to the virtual permutation; that is, these steps are not implemented.

4 Proposed Compact Circuit Design for S-DES

To design the proposed compact quantum circuit for S-DES, various components of S-DES must be configured and designed. Figure 1 shows the proposed compact quantum circuit for S-DES. The figure gives an insight of each and every component of S-DES, including the virtual boxes. Figure 3 provides the circuit architecture and focuses on the circuitry such as the quantum gates, qubits, and ancilla bits. Both figures will be discussed in details in subsequent subsections.

4.1 Substitution Boxes (S-Boxes)

A substitution box or S-Box plays a pivotal role in block ciphers as it satisfies the Shannon's property of confusion. It obscures the relationship between the secret key and ciphertext. The S-Boxes are represented by $\mathbf{U_{sboxes}}$ in Fig. 1. In this study, two different quantum designs of S-Boxes from previous studies are discussed. The first quantum S-Box design was presented by Almazrooie et al. [11], and the second quantum design was demonstrated by Denisenko et al. [14]. Almazrooie et al. [11] presented a systematic quantum circuit design for Handcrafted S-Boxes (predefined S-Boxes). We reproduce the quantum design of Almazrooie et al. as an algorithm as shown in Algorithm 1. This algorithm can produce a quantum circuit for any given S-Box. The inputs of the algorithm are S-Box inputs and output, and accordingly, the quantum circuit of the S-Box is designed.

The second quantum design by Denisenko et al. [14] is an efficient quantum design that makes use of less quantum gates compared with the first design. We also reproduce the quantum circuit of the two S-Boxes from the enclosed source code in the paper of Denisenko et al. as shown in Fig. 2. Both of the quantum designs of the S-Boxes are implemented in this work. More discussions are conducted about the two S-Boxes quantum designs in Subsect. 4.5 and in the future work in Sect. 6.

Algorithm 1: QUANTUM S-DES S-BOX$_0$

Constant: $\text{SBox}_0^{in} = \{2, 7, 8, 0, 5, 11, 12, 3, 6, 10, 15, 1, 4, 9, 13, 14\}$
Constant: $\text{SBox}_0^{out} = \{0, 0, 0, 1, 1, 1, 1, 2, 2, 2, 2, 3, 3, 3, 3, 3\}$
Assumption: Two ancilla qubits a_1, a_2

```
2  Function QSBox0(in0, in1, in2, in3, out0, out1)
3    for i < 2^4 do
4      temp ← SBox_i^in
       for j < 4 do
         // To check the state of the j^th bit in temp
5        if ¬(temp ∧ (1 <<j)) then
6          X-Pauli(in_j)// If the j^th bit in temp is 0, apply
             X-Pauli to the j^th bit in in.
7      TOFFOLI(in0, in1, a1)
1      TOFFOLI(in2, a1, a2)
       temp2 ← SBox_i^out
       for k < 2 do
8        if (temp2 ∧ (1 <<k)) then
9          TOFFOLI(in3, a2, out_k)
10     TOFFOLI(in2, a1, a2)
       TOFFOLI(in0, in1, a1)
       temp ← SBox_i^in
       for j < 4 do
         // To check the state of the j^th bit in temp
11       if ¬(temp ∧ (1 <<j)) then
12         X-Pauli(in_j)// If the j^th bit in temp is 0, apply
             X-Pauli.
```

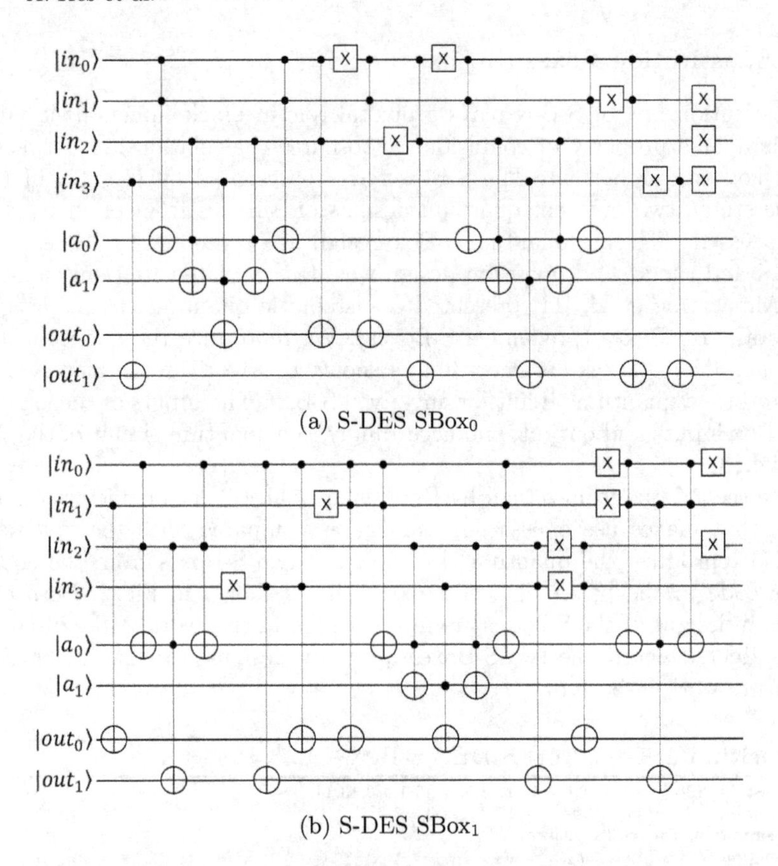

(a) S-DES SBox$_0$

(b) S-DES SBox$_1$

Fig. 2. The quantum circuit of S-Box0 and S-Box1 of S-DES. These two circuits are reproduced from the source code included in the work of Denisenko et al. [14]. Four qubits ($|in_0\rangle$ to $|in_3\rangle$) are for the S-Box inputs and two qubits ($|out_0\rangle$ and $|out_1\rangle$) are for the outputs. Two ancilla qubits ($|a_0\rangle$ and $|a_1\rangle$) are used for the workspace.

4.2 Initial Permutation and Expansion

Permutation and Expansion can be achieved in a quantum circuit using the conventional fan-out circuit [15]. However, using a fan-out circuit involves usage of more ancilla bits [11]. To make the circuit compact, the initial permutation (IP) of the plaintext and the expansion permutation (EP) are evaluated based on the IP and EP matrices. The virtual boxes as shown in Fig. 1 are applied directly on the plaintext. A total of Eight qubits, which are denoted by $|p_1\rangle$ to $|p_8\rangle$ in Fig. 1, are used to represent the plaintext. Ancilla bits are initialized to $|0\rangle$. **X-Pauli** gates are applied on the circuit where the plaintext bit is 1, and plaintext bit 0 remains untouched. Thus, the number of **X-Pauli** gates used to represent the plaintext depends on the number of plaintext bit with value 1. For example, if the plaintext is 1001 1001, then **X-Pauli** gates are applied on the qubit numbers 0, 3, 4, and 7 while the remaining qubits are kept untouched. Once the plaintext is configured on the circuit, the IP is performed on the plaintext. S-DES has a predefined function, which is used to permute the plaintext. The plaintext is split into two halves. The right half of the permuted plaintext is subjected to EP as per S-DES design, and the resultant expanded right half of the plain text is XOR-ed with the key bits. All these steps, which include IP and EP are done virtually by identifying which output bits from a stage forms the input bits for the next stage. With the IP and EP in hand, the virtual implementation becomes easy as they clearly state which bit of plaintext is permuted to which bit and how to achieve it. With this technique, only 8 qubits are used for the plaintext instead of 8 qubits and 8 ancilla qubits as used in the previous circuit [11].

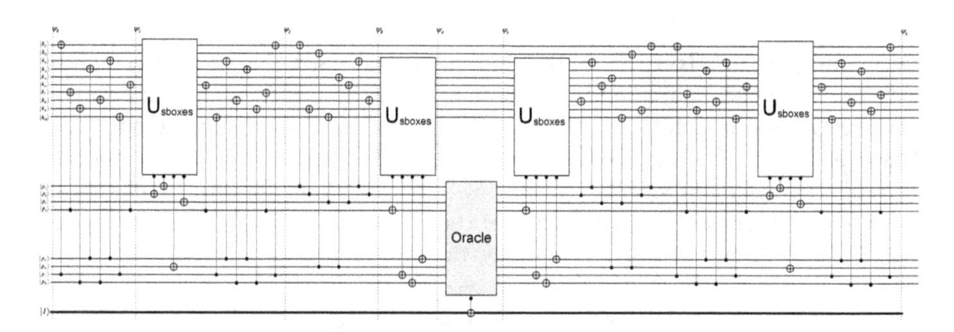

Fig. 3. The complete reversible circuit of the compact quantum S-DES. At the initialization stage $|\psi_0\rangle$, the 10 key qubits are initialized to the state $|0\rangle^{\otimes 10}$ and then the 10 sets of **Hadamard** gates are applied to acquire an equally uniform superposition. The plaintext qubits are initialized to $|0\rangle^{\otimes 8}$ and a set of **X-Pauli** gates are used according to the 1's in the plaintext. The Grover oracle qubit is set to $|1\rangle$ and a **Hadamard** to put it in a superposition. At $|\psi_1\rangle$, after tracing the virtual permutations P10, S1, IP, and EP shown in Fig. 1, the qubits $|p_1\rangle$ to $|p_8\rangle$ are XOR-ed accordingly with the key qubits using a set of **CNOT** gates. Note that $\oplus$ is a commutative function; hence, in this work, the plaintext is XOR-ed with a key, while in the previous works, [11,14] is the opposite action. The previous processes are reversed to clear the key qubits to generate the second sub-key at $|\psi_3\rangle$. At $|\psi_4\rangle$, the ciphertext is generated, which inputs to the net of **Toffoli** gates ($\mathbf{C^8(X)}$), which is called oracle in this figure. From $|\psi_5\rangle$ to $|\psi_6\rangle$, all the processes are reversed so that the Grover Conditional Phase Flip **CPF** is applied.

4.3 First Sub-Key Generation and Key Mixing

S-DES involves two 8-bit keys that are generated from a 10-bit key. This is achieved by applying a 10-bit permutation and splitting it into two halves, left shift and an 8-bit permutation at the last [11]. In this study, 10 qubits, represented by ($|k_1\rangle$ to $|k_{10}\rangle$) in Fig. 1, are used for the key. All the steps involved in sub-key generation are done virtually, similar to Initial IP and EP stage. In this stage, the permutation function P10, which is represented by a virtual box P10 in Fig. 1, is defined for the initial 10-bit key, and the permutation function P8 which is represented by a virtual box P8 in Fig. 1, is defined for the 8-bit permutation of the final 8-bit key generated, which is XOR-ed with the plaintext. A detailed explanation on the permutation functions and shift operation is provided in the previous work [11].

Once the key, the plaintext, and the S-Boxes are ready to be used, all these components are put together to build the proposed compact quantum circuit for S-DES, which utilizes the lowest number of qubits and ancilla bits so as to reduce the overall cost of building a quantum circuit for S-DES, leading to a simplified and compact quantum circuit with better complexity. The proposed circuit in this study is used for exhaustive key search using Grover's algorithm given a pair of plaintext and ciphertext.

First Encryption Round. A total of 8 qubits are used to represent the plaintext. The plaintext is applied on the circuit with the help of **X-pauli** gates, as described in Subsect. 4.2. The designed circuit at this stage is for the particular chosen plaintext. The circuit can be easily modified at this stage to accommodate different plaintexts, only by altering the position of **X-pauli** gates based on the position of the bit with value 1 of the chosen plaintext. This step involves 8 qubits and **X-pauli** gates depending on the number of plaintext bits with value 1.

The IP and EP of the plaintext is done virtually, as explained in Subsect. 4.2, and the final output is directly applied on the plaintext; thus, no extra quantum gates or qubits are involved in this stage. The sub-key generation and key mixing is also performed virtually; thus, no quantum gates or qubits other than 10 qubits for the initial 10-bit key are required in this stage. With the plaintext and key in place, the expanded left half of the plaintext is XOR-ed with the key with the help of 8 **CNOT** gates. Here, the plaintext bits act as the control bits and modify the key qubits accordingly. Figure 3 shows how the expanded plaintext bits are XOR-ed with the key qubits. For example, key qubit $|k_1\rangle$ is XOR-ed with plaintext bit $|p_7\rangle$ and so on.

The modified key qubits are passed to the S-Boxes which have a placeholder for 4 input bits. The S-Box takes 4 qubits as input and stores the output in 2 qubits as shown in Fig. 3. The 4 input qubits registry location is fixed, and the plaintext is fed into the registry locations meant for the input qubits for the S-Box. The 2-qubit output from the S-Box is also stored in predefined registry locations meant for the output. The output of the S-Box is 2-bit, which is stored in 2 qubits (the plaintext quibits as shown in Fig. 3). This resulted in reducing the usage of ancilla bits by 6 per S-Box which were used in the previous circuit [11] for manipulating the input plaintext and storing the output within the S-Box. This is where the S-Boxes from the previous work [11] have been modified by making the S-Boxes accept runtime values for the input into the specific placeholder for the input bits and store the output into the specific placeholders for the output bits. A detailed breakup of how the S-Box and the quantum gates are designed to build the S-Boxes, is available in [11].

Switch Function. The Fiestel structure, which governs the design of S-DES, involves operations to be done on one half of the plaintext, and the other half remains untouched in a round that leads to the introduction of a switching function at the end of each round, where the two halves of the plaintext are switched or permuted. The permutation involved in switching can be done virtually, represented by virtual box **S** in Fig. 1, with the help of the S-DES design and directly implemented on the plaintext. The circuitry for the reversal is shown in Fig. 3. This approach is followed in designing this circuit; thus, this step does not involve any quantum gates or qubits unlike the previous implementation [11], which involved 12 **CNOT** gates for the switch function. This marks the end of the first round of encryption.

Second Encryption Round. The second round of encryption is similar to the first round. The second round starts with EP unlike the first round, which starts with IP and ends with inverse permutation unlike the switching function of first round. Before the start of the second round, because of the reversible nature of the circuit, all the qubits and ancilla bits are reversed and set to their initial state except the ancilla bits holding the intermediate ciphertext (the plaintext qubits). The last step, which involves inverse permutation, is achieved virtually with the help of Inverse Permutation virtual box IP1 in Fig. 1. Once all the functions are performed, at the end of the second round, all the steps are reversed.

4.4 Grover Attack on QS-DES

The key used in S-DES is of 8 bits, which is derived from an initial 10 bits key. Since the intention of this study is to retrieve the secret key, **Hadamard** gates are applied on the 10 qubits used for the key, one gate for each qubit. 10 **Hadamard** gates are used to put the 10 key qubits in superposition such that all the possible ($2^{10} = 1024$) combinations of the key are acquired with equal probability ($\frac{1}{2^{10}} = \frac{1}{1024}$ each) at the same time. For instance, if one performs any operation on the key qubits, the operation will be performed on the 1024 possible keys simultaneously. In contrast to classical computing, one needs to invoke such operation 1024 times compared with one invoke on quantum computing.

Grover's algorithm is used for quantum exhaustive key search and is applied on a black box, which is assumed to have the implementation of block cipher. The proposed compact quantum circuit for S-DES is built so that it can be used as the black box or oracle for Grover's quantum search. The circuit involves multiple levels of reversibility, wherein the qubits and ancilla bits are returned to their original states. At this phase, an oracle qubit, represented by oracle in Fig. 1, which is an ancilla bit, is required. It is set to 1 using an **X-Pauli** gate. A detailed explanation on the working of the oracle bit and how it is used for quantum search is provided in [11].

4.5 Quantum Resources

The quantum resources in terms of the total number of qubits and quantum gates are calculated for the proposed quantum S-DES (QS-DES) circuit. As mentioned in Subsect. 4.1, two different approaches of S-Boxes design are adopted. Hence, the quantum resources of the compact QS-DES circuit are evaluated for the two designs.

First, the resources needed in the QS-DES circuit are calculated without considering the design cost of the S-Boxes. The total number of the quantum gates (excluding the S-Boxes) used in designing this compact circuit is: 10 **Hadamard** gates for the key qubits, 1 **Hadamard** gate for the oracle qubit, and 4 **X-Pauli** gates (depends on the plaintext). A total of 48 **CNOT** gates are needed for mixing, including the reversal process. In addition, 13 **Toffoli** gates for the Grover oracle.

Almazrooie et al. [11] proposed a circuit design for S-Boxes as shown in Algorithm 1. The proposed approach can be used to design a quantum circuit for any Handcrafted S-Box (predefined S-Box). This approach makes use of 4 qubits for the input, 2 qubits for the output, and 2 ancilla qubits for the workspace. However, the number of the quantum gates needed in this approach is directly proportional to the input size of the S-Box. Every S-Box in QS-DES has 16 states that correspond to the size of the input to the S-Box which is 2^4. Each state of them needs **X-Pauli** gates to implement the zeros. Thus, approximately 32 **X-Pauli** gates are needed for the 16 states. In addition, every state needs 5 **Toffoli** gates. Therefore, for the 16 states of one S-Box, 16×5, 80 **Toffoli** gates are used.

Recently, Denisenko et al. [14] proposed a more efficient quantum circuit dedicated for QS-DES S-Boxes as shown in Fig. 2. Similar to Almazrooie et al., this approach makes use of 4 qubits for the input, 2 qubits for the output, and 2 ancilla qubits for the workspace. In contrast, a smaller number of quantum gates is needed in this approach compared with Almazrooie et al.'s. It can be seen in Fig. 2 that 1 **CNOT** gate, 15 **Toffoli** gates, and 8 **X-Pauli** gates are needed to design SBox_0.

To conduct a fair comparison between our proposed QS-DES circuit and the previous work by Denisenko et al., some graphical notations of the quantum gates should be clarified. Figure 4 shows the graphical notations of **SWAP** gate and the multiple-controls Toffoli gates ($\mathbf{C^n(X)}$) where n is the number of the controls. It should be noted that, Denisenko et al. refer to the multiple-controls Toffoli gate ($\mathbf{C^n(X)}$) as "*generalized CNOT(n) gates*" in their paper.

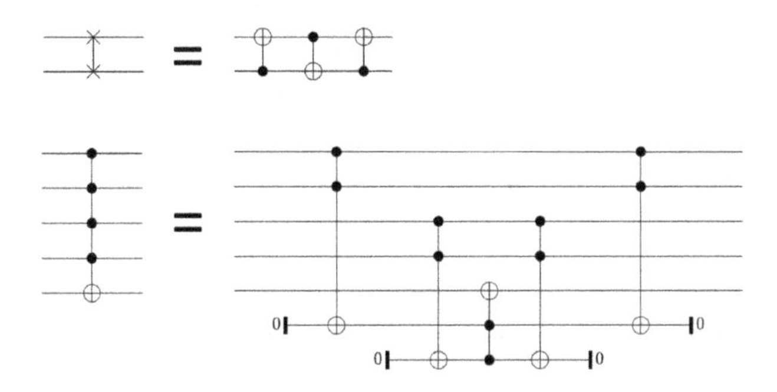

Fig. 4. The graphical notations of some quantum gates and their implementation. The notation of **SWAP** gate is on the top left, and its implementation in terms of **CNOT** gates is on the top right. The notation of $\mathbf{C^n(X)}$ gate is on the bottom left, and its implementation in terms of **CNOT** and **Toffoli** gates is on the bottom right. One may notice that the implementation of $\mathbf{C^n(X)}$ requires extra ancilla qubits.

As can be seen from Fig. 4, the graphical notation of the **SWAP** gate (at the top left of the figure) is represented as a single quantum gate; however, the implementation of **SWAP** is composed of 3 **CNOT** gates as can be seen at the top right of the figure. Similarly, the graphical notation of the multiple-controls Toffoli gate ($\mathbf{C^n(X)}$) (at the bottom left of the figure) is also represented as a single quantum gate, while at the implementation level (at the bottom right if the figure), $\mathbf{C^n(X)}$ is actually composed of 5 **Toffoli** gates and 2 anchilla qubits, where $n = 4$. The number of anchilla qubits needed to implement $\mathbf{C^n(X)}$ is linear to the number of the controls of $\mathbf{C^n(X)}$, which can be calculated as $(n - 2)$, where n is the number of controls qubits [16, Chap. 4, p. 183]. Thus, using multiple-controls Toffoli gate ($\mathbf{C^n(X)}$) in the design of QS-DES circuit whether in this work or the previous work by Deniseko et al. involves using ancilla qubits for the workspace regardless the graphical notations.

Moreover, since Denisenko et al. have used *Quipper Quantum Simulator* [17] to implement their proposed quantum circuit of S-DES, we had to verify the implementation of $\mathbf{C^n(X)}$ in the Quipper simulator. Deniseko et al. have used the Quipper command: (`qnot_at qubit`$_0$ `'controlled' [qubit`$_1$, `qubit`$_2$, `qubit`$_3$, `qubit`$_4$`]`) as for example in [14, Page 38, Line 24] to implement $\mathbf{C^4(X)}$ in which `qubit`$_1$, `qubit`$_2$, `qubit`$_3$, `qubit`$_4$ are the controls, and `qubit`$_0$ is the target. Hence, we have installed the Quipper simulator and implemented the same command shown above, and then we used the Quipper Decomposition Library to decompose the command. The decomposition of the gate shows that 5 **Toffoli** gates and 2 ancilla qubits are needed to implement $\mathbf{C^4(X)}$. The implementation and results of this simple experiment are shown in Appendix A.

In this work, the quantum resources (the numbers of qubits and quantum gates) will be evaluated in terms of **CNOT** and **Toffoli** gates to facilitate the comparison between this work and the previous ones. It should be noted that even **CNOT** and **Toffoli** gates can be decomposed further to elementary quantum gates in terms of **Chifford** and **T-gates** [18]. Table 1 shows the comparison between our proposed quantum circuit and the previous work.

Table 1. Comparison of the quantum resources in the proposed QS-DES versus the previous study [14].

	Proposed QS-DES	Previous QS-DES [14]
Qubits	19	19
Ancilla	7	7
Hadamard	34	34
X-Pauli	62	88
CNOT	52	96
Toffoli	90	148
Quantum simulator	Libquantum	Quipper

5 Experiments and Results

In this section, the coding method and the simulation of the quantum computer are discussed in the first subsection. Then the verification of the proposed quantum circuit of S-DES, as well as the application of the Grover attack to recover the secret keys, are discussed in the second subsection.

5.1 Coding and Simulation

Simulation is a mechanism to reproduce or imitate an operation, real-world environment, process, or system over time. To simulate, a model has to be developed which represents the key characteristics or behavior of the system or process that is simulated. Libquantum is a C library simulator of a quantum computer that provides all the vital operations and simulation of a quantum computer [19]. It is used to simulate the quantum S-DES and apply Grover's algorithm for quantum search. Libquantum offers a high-performance and low-memory consumption. Since *libquantum* is a C library simulator, C language is used for coding the circuit. The source code of the proposed compact QS-DES can be found in [20]. All the experiments are conducted on a machine equipped with Intel Core i5-4470 3.5 GHz processor.

5.2 Verification and Grover Attack

The functionality of the proposed quantum circuit is verified against classical S-DES. Since the aim of the study is to build a quantum circuit, the key qubits will be in superimposition. Thus, the selected plaintext 1001 1010 is encrypted with all the possible keys at the same time with only one query of the proposed quantum S-DES. Table 2 shows all 1024 (2^{10}) possibilities for the 10-qubit key. Each state has an equal probability of 9.765623×10^{-04}. Data references related to plaintext-ciphertext pairs are drawn from the previous work [11] to justify the working functionality with the proposed compact and efficient quantum circuit for S-DES.

Table 2. Proposed compact quantum S-DES results when keys are in superposition

	Plaintext	Key and ciphertext	Probability
0	1001 1010	$\|0000000000\ 11111001\rangle$	9.765623×10^{-04}
1	1001 1010	$\|0000000001\ 01010001\rangle$	9.765623×10^{-04}
2	1001 1010	$\|0000000010\ 01101001\rangle$	9.765623×10^{-04}
⋮	⋮	⋮	⋮
1022	1001 1010	$\|1111111110\ 11100110\rangle$	9.765623×10^{-04}
1023	1001 1010	$\|1111111111\ 00001011\rangle$	9.765623×10^{-04}

Grover's algorithm is used to perform a cryptanalytic attack on the designed quantum S-DES circuit by performing an exhaustive quantum key search for a known plaintext/ciphertext pair. To test the working and functionality of the designed compact quantum circuit for S-DES, a plaintext/ciphertext pair is implemented on the circuit, and then Grover's algorithm is applied on the circuit to retrieve the secret key for that particular plaintext/ciphertext pair. A detailed explanation of Grover's algorithm is provided in the previous study [11]. The equation used to calculate the number of Grover iterations to find the key for a particular plaintext/ciphertext pair is: $R = \frac{\pi}{4}\sqrt{N}$, which is derived from [11, Eq. 5]. Thus, around 25 Grover iterations are required to find the key for a particular plaintext/ciphertext pair. Table 3 shows the results of quantum exhaustive key search for the key used to encrypt the known plaintext 0001 0000 to the ciphertext 0011 0011. For the mentioned plaintext/ciphertext pair, there is only one cryptographic key that can perform the encryption. Table 3 shows that the state $|01100010011\rangle$ has the highest probability of 0.9994553, whereas the remaining states have low probabilities of 5.266347×10^{-07}. Thus, the secret key 11 0001 0011 is found in 25 iterations.

There might be cases where there can be multiple keys for a single plaintext/ciphertext pair, specifically in those cases when the key size is more than the plaintext size. In those cases, the number of Grover iterations required to find the keys is $R = \frac{\pi}{4}\sqrt{\frac{N}{M}}$ which is derived from the previous work [11, Eq. 6]. For example, for the plaintext 1010 0101 and ciphertext 0011 0110 as shown in Table 4, two states, $|00010010111\rangle$ and $|00011011111\rangle$, have the highest probability of 0.4997221 each, whereas all the remaining states have a probability of 5.400958×10^{-07}. In this case, only 17 Grover iterations are required as per the equation to find the set of secret keys 00 1001 0111 and 00 1101 1111. Finding out the key in these collision cases in a classical computer is difficult.

Table 3. Quantum exhaustive key search on proposed compact quantum S-DES circuit in case of single solution

	Ciphertext	Oracle qubit and key and plaintext	Probability
0	0011 0011	$\|1\ 0000000000\ 00010000\rangle$	5.266659×10^{-07}
1	0011 0011	$\|1\ 0000000001\ 00010000\rangle$	5.266659×10^{-07}
⋮	⋮	⋮	⋮
787	0011 0011	$\|1\ 1100010011\ 00010000\rangle$	0.9994553
⋮	⋮	⋮	⋮
1022	0011 0011	$\|1\ 1111111110\ 00010000\rangle$	5.266659×10^{-07}
1023	0011 0011	$\|1\ 1111111111\ 00010000\rangle$	5.266659×10^{-07}

Table 4. Quantum exhaustive key search on proposed compact quantum S-DES circuit in case of multiple solutions

	Ciphertext	Oracle qubit and Key and Plaintext	Probability
0	0011 0110	$\|1\ 0000000000\ 10100101\rangle$	4.118168×10^{-06}
1	0011 0110	$\|1\ 0000000001\ 10100101\rangle$	4.118168×10^{-06}
⋮	⋮	⋮	⋮
151	0011 0110	$\|1\ 0010010111\ 10100101\rangle$	0.4978935
⋮	⋮	⋮	⋮
223	0011 0110	$\|1\ 0011011111\ 10100101\rangle$	0.4978935
⋮	⋮	⋮	⋮
1022	0011 0110	$\|1\ 1111111110\ 10100101\rangle$	4.118168×10^{-06}
1023	0011 0110	$\|1\ 1111111111\ 10100101\rangle$	4.118168×10^{-06}

It should be noted that Denisenko et al. [14] reported that the execution times for 18 Grover iterations is $\sim$6, 593 s and for 25 Grover iterations is $\sim$7, 665 s. Whereas in this work, it takes $\sim$0.034 s for 18 Grover iterarions, and $\sim$0.05 s for 25 Grover iterations.

6 Conclusion and Future Work

Quantum cryptanalytic attacks on a symmetric cryptosystem have not been studied extensively because of the precondition of implementing the symmetric algorithm on a quantum platform. In this study, an efficient, simple, and compact quantum circuit for block cipher S-DES is built, with the help of a minimum or polynomial number of quantum gates, qubits, and ancilla bits. The goal of the proposed circuit design in reducing the number of ancilla bits and quantum gates used is achieved as it has successfully reduced the number of ancilla bits used during the various phase of S-DES design. The proposed circuit uses a minimal number of ancilla bits (only where it is a design requirement). A reduced number of quantum gates used for circuit design affect the complexity of the circuit (computed in terms of the number of quantum gates used). A polynomial number of quantum gates are used in designing the circuit. This overall affects the circuit design as it will be a much smaller and compact and efficient circuit.

All these factors combined open up a wide opportunity for researchers to study the effect of cryptanalysis on a symmetric cryptosystem in the post-quantum era, an area of study that has not been sufficiently explored the lack of an efficient circuit quantum circuit design for block ciphers. This will help in designing a robust cryptosystem in a quantum environment, resilient to quantum cryptanalytic attacks.

In this study, two different approaches for the quantum S-Boxes design were discussed. First, Almazrooie et al.'s [11] approach can be used to design a quantum circuit for any predefined S-Box. Second, a more efficient quantum circuit dedicated for S-DES S-Boxes was presented by Denisenko et al. [14], however the team have not shown any method for designing the S-Boxes. Since the size of S-DES is small, they might try manually till they come up with the proposed circuits of the S-Boxes. In the future, it would be great if such a method as efficient as Denisenko et al.'s can be developed and generalized to be used in designing any predefined S-Boxes such as the ones in 3DES, Twofish, AES, or any block cipher.

A Multiple-control Toffoli Gate ($C^n(X)$) Decompostion

In this simple experiment, the multiple-control Toffoli gate ($\mathbf{C^n(X)}$) is decomposed to **CNOT** and **Toffoli** gates using Quipper quantum simulator. The Quipper library `QuipperLib.GateDecompositions` contains functions for decomposing multiple controls. Listing 1 shows the Quipper script of decomposing $\mathbf{C^6(X)}$ and the output circuits are shown in Fig. 5.

Listing 1: Quipper script of decomposing $\mathbf{C^6(X)}$.

```
import Quipper
import QuipperLib.Decompose

toffoli::(Qubit,Qubit,Qubit,Qubit,Qubit,Qubit,Qubit)->Circ (Qubit,Qubit,
          Qubit,Qubit,Qubit,Qubit,Qubit)
toffoli (q1,q2,q3,q4,q5,q6,q7) = do
         qnot_at q7 `controlled` (q1,q2,q3,q4,q5,q6)
         return (q1,q2,q3,q4,q5,q6,q7)

main :: IO ()
main = do
         print_simple Preview toffoli
         print_simple Preview (decompose_generic Toffoli toffoli)
         --print_simple Preview (decompose_generic Binary toffoli)
```

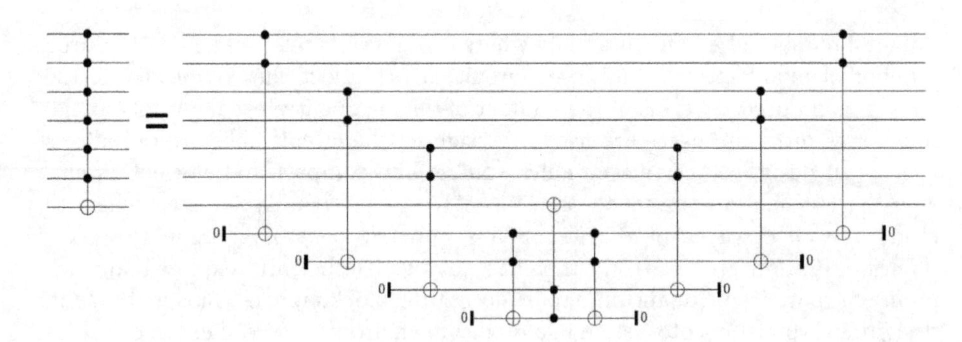

Fig. 5. The decomposition output of Quipper script in Listing 1

References

1. Markov, I.L., Saeedi, M.: Phys. Rev. A **87**(1), 012–310 (2013)
2. Martin-Lopez, E., Laing, A., Lawson, T., Alvarez, R., Zhou, X.-Q., O'Brien, J.L.: Nat. Photonics **6**, 773–776 (2012). https://doi.org/10.1038/nphoton.2012.259
3. Bernstein, D.J., Buchmann, J., Dahmen, E.: Post Quantum Cryptography. 1st. edn Springer (2008). ISBN 3540887016
4. Nicolas, G., Gregoire, R., Wolfgang, T., Hugo, Z.: Rev. Mod. Phys. **74**(1), 145–195 (2002). https://doi.org/10.1103/RevModPhys.74.145
5. Christof, Z.: Phys. Rev. A **60**(4), 2746 (1999)
6. Akihiro, Y.: Algebraic Systems, Formal Languages and Computations, vol. 1166, pp. 235–243. RIMS Kokyuroku (2000)
7. Boyer, M., Brassard, G., Hoyer, P., Tapp, A.: Fortschritte der Phys. **46**, 493–505 (1998)
8. Stallings, W.: Cryptography and Network Security: Principles and Practice, 5th edn. Prentice Hall Press, Upper Saddle River (2010)
9. Musa, M., Schaefer, E.F., Wedig, S.: Cryptologia **27**(2), 148–177 (2003)
10. Schaefer, E.F.: Cryptologia **20**(1), 77–84 (1996)
11. Almazoorie, M., Samsudin, A., Abdullah, R., Mutter, K.: SpringerPlus **5**, 1494 (2016). https://doi.org/10.1186/s40064-016-3159-4
12. Mermin, N.: Quantum Computer Science: An Introduction. Cambridge University Press, Cambridge (2007). https://doi.org/10.1017/CBO9780511813870
13. Williams, C.P.: Texts in Computer Science. Springer, London (2011). https://doi.org/10.1007/978-1-84628-887-6_2
14. Denisenko, D.V., Nikitenkova, M.V.: J. Exp. Theor. Phys. **128**, 25 (2019). https://doi.org/10.1134/S1063776118120142
15. Hoyer, P., Spalek, R.: Theory Comput. **1**(5), 81–103 (2005). ISSN 1557-2862
16. Nielsen, M.A., Chuang, I.L.: Quantum Computation and Quantum Information: 10th Anniversary Edition, 10th edn. Cambridge University Press, New York (2011)
17. The Quipper Language. https://www.mathstat.dal.ca/~selinger/quipper/. Accessed 24 Oct 2019
18. Amy, M., Maslov, D., Mosca, M., Roetteler, M.: Trans. Comput. Aided Design Integr. Circ. Syst. **32**(6), 818–830 (2013)
19. Simulation of quantum mechanics. http://www.libquantum.de/. Accessed 24 Oct 2019
20. C Implementation of Quantum Simplified-DES. https://github.com/mishal131/QSDES. Accessed 06 Dec 2019

A Review of Time-Series Anomaly Detection Techniques: A Step to Future Perspectives

Kamran Shaukat[1(✉)], Talha Mahboob Alam[2], Suhuai Luo[1], Shakir Shabbir[2], Ibrahim A. Hameed[3], Jiaming Li[4], Syed Konain Abbas[2], and Umair Javed[2]

[1] The University of Newcastle, Newcastle 2308, Australia
kamran.shaukat@uon.edu.au
[2] University of Engineering and Technology, Lahore 54890, Pakistan
[3] Norwegian University of Science and Technology, 7491 Trondheim, Norway
[4] Data61, Commonwealth Scientific and Industrial Research Organisation, Canberra, Australia

Abstract. Anomaly detection is a significant problem that has been studied in a broader spectrum of research areas due to its diverse applications in different domains. Despite the usage of modern technologies and the advances in system monitoring and anomaly detection techniques, false-positive rates are still high. There exist many anomaly detection algorithms, among them few are domain-specific, and others are more generic techniques. Despite a significant amount of advance in this research area, there does not exist a single winning anomaly detector known to work well across different datasets. In this paper, we review the literature related to types of anomalies, data types of anomalies, data types of time-series, components of time-series data, classification of anomalies context, and classification methods of time-series anomalies detection. We presented a taxonomy to characterize the various aspects related to time-series anomaly detection. One of the key challenges in current anomaly detection techniques is to perform anomaly detection with regards to the type of activities or the context that a system is exposed. We hope that this investigation gives a more remarkable ability to understand the evolving methods of time-series anomaly detection and how computational methods can be applied in this domain in the future.

Keywords: Anomaly detection · Anomalies context · Anomaly types · Anomaly data types · Machine learning · Review · Time-series

1 Introduction

Anomaly detection is a technique used in the field of statistics to determine outliers from data [1]. An anomaly in time-series data is defined as a point or sequence of points that deviate from the normal behavior of the data. Time-series data that has values associated with timestamps. A sudden increase in the temperature of a room in a factory can signal that there is a fire. It would be beneficial to have a system that can detect and flag anomalies, giving administrators the ability to minimize losses [2]. If we were to plot the amount of data being generated within organizations, it would be an exponential curve. To make sense of all this data, companies need to hire analysts to match the

K. Arai (Ed.): FICC 2021, AISC 1363, pp. 865–877, 2021.
https://doi.org/10.1007/978-3-030-73100-7_60

tremendous rate of growth of the data. However, organizations can at best only afford to hire linearly due to financial constraints. There is a growing gap between the amount of data being generated and the number of personnel available to go through this data. This gap causes an intelligence gap, and it represents an enormous problem that companies in the digital age are facing. Namely, it is too expensive to hire humans to review data at the rate it is being generated [3]. This is where computers can contribute. Machine intelligence programs can be built to monitor information from critical systems and flag anomalies as they occur. Human experts can then view these warnings, and decide to deal with them on a case-by-case basis. This would decrease the amount of work that human experts have to do, thereby decreasing the burden on companies to hire more and more experts. Building machine learning systems that can process information and identify anomalies is more cost-effective than creating human-based teams. It can also detect changes in the signal that are too subtle for humans to identify [1].

There are a variety of methods and measures that can be used to differentiate the normal points from anomalous ones that depend on what is being predicted. Using machine learning for detecting anomalies in time-series data has been approached from both an unsupervised and supervised manner. The easiest anomaly to detect is extreme values or outliers that exceed the standard operating range of the process. Limits can be placed on each sensor channel to automatically detect these point anomalies when the specified threshold is violated [4, 5]. In this paper, we review the literature related to types of anomalies, data types of anomalies, data types of time-series, components of time-series, classification of anomalies context, and classification methods of time-series anomalies detection. We presented a taxonomy to characterize the various aspects related to anomaly detection, as shown in Fig. 1.

2 Types of Anomalies

2.1 Point Anomalies

These are the simplest type of anomalies in which each data point can be analyzed by the anomaly detector without considering any other data points in the input dataset. Point anomaly is an observation x or y that deviates remarkably from X according to some predefined criteria, where $x \in X$ and $y \notin X$ [6]. For instance, the absence of a student in the Math lesson on Monday morning during the term time is a point anomaly, because, different from other students, the student is absent. Limits can be placed on to automatically detect these point anomalies when the specified threshold is violated.

2.2 Contextual Anomalies

For detecting a contextual anomaly, an anomaly detector has to consider the context in which the data point occurred. In practice, this type of anomaly can occur, especially in sequence or spatial. A contextual anomaly, however, is anomalous only in the context of the data around it. A more significant challenge is contextual anomalies that occur within the normal operating range but which are not conforming to the expected temporal pattern [7, 8].

2.3 Collective Anomalies

Collective anomalies are notable because only sets of data points can be labelled as collective anomalies (not individual data points). The data points in the anomalous set themselves can be normal, but together, they represent an anomaly.

In some cases, algorithms designed for point anomalies can be used for contextual anomalies if we include context as new features, such as month number [8]. To summarize, the most straightforward form of an anomaly is point anomaly. It is also the simplest one to detect since there is no need to find what sequence or context. When detecting sequential anomalies, it is common to transform subsequences of the data to points containing information about the subsequence. Then the techniques for finding point anomalies can be used to find sequential anomalies. However, how to construct subsequences is a problem on its own and can be challenging to solve.

3 Types of Data

A key and fundamental aspect of any anomaly detection technique is the nature of the target dataset. Essentially, a dataset is a collection of data instances or observations. According to specific application scenarios, a data instance can be a number, record, video, song, graph, image, event, and profile. All these disparate forms of data should be transformed into general data types for anomaly detection. Temporally, general data types comprise scalar, vector, matrix as well as a tensor, and their elements are known as data attributes, features, or fields, which can be numerical or categorical. In literature, the scalar is called univariate, while the vector, matrix, and tensor are all multivariate data types.

4 Components of Time-Series Data

Time-series data often includes trend, seasonal and cyclic variations [9]. The dynamic nature of time series can be related to the different components comprising time series, namely: the trend, seasonality, cyclic, and irregular components. The trend component is usually referred to as a long-term change in the mean level of the data. If the trend of the time series is of no interest, it can be removed. When time-series experiences regular and predictable changes in fixed periods, it is said to contain a seasonal component. The monthly variation can be estimated, or removed if it is of no interest. Besides seasonal effects, sometimes time series show other predictable oscillations, but which do not have a fixed period. This type of variation is a cyclic pattern. The typical example of a cyclic pattern is a business cycle, in which the economy experiences periods of growth and periods of recession.

After removing the three above components from the time series, the remaining part is known as the irregular component or residuals. Mathematically, these components can be explained by a trend component, T_t, a cyclic component, C_t, a seasonal component, S_t, and a reminder, R_t, denoting the stochastic variations not captured by the other components. These components can be combined by an additive decomposition or by a multiplicative decomposition. The additive decomposition assumes that the components

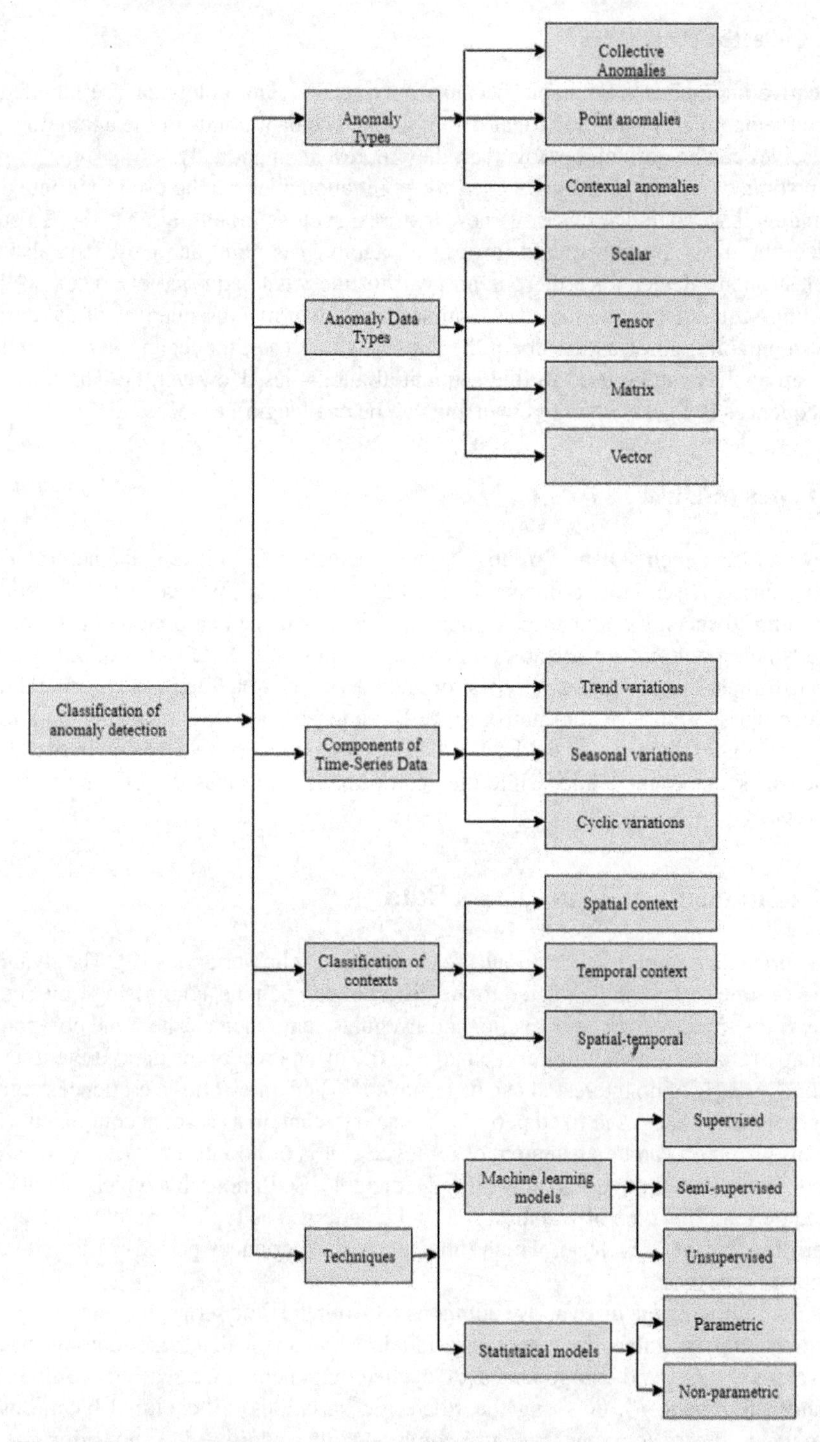

Fig. 1. The taxonomy based on anomaly detection to characterize the various aspects

are independent of one another, such that $Y_t = T_t + S_t + C_t + R_t$, whereas the multivariate decomposition assumes that the components are dependent, $Y_t = T_t * S_t * C_t * R_t$ [10, 11].

5 Classification of Contexts

The contexts under which the data instances are observed are another crucial source of information that is helpful in detecting abnormal events. Usually, the contextual information is distinctive and not measured or recorded explicitly in different applications [12]. The context that is helpful for anomaly detection is always obscure. Therefore, we only considered ubiquitous contexts, i.e., spatial context, temporal context, and spatial-temporal context. Specifically, the spatial context concentrates on the location where a data instance is observed, while the temporal context reveals the sequential information among observations. The integration of the two contexts motivates the spatial-temporal context, which has attracted much attention in recent years. There are two strategies to analyze contextual information in anomaly detection.

On the other hand, feature engineering is adopted to consolidate the contextual information and the original information. A well-developed example is the time embedding technique for time-series analysis in which temporal information is critical. On the other hand, contexts are analyzed separately for contextual anomalies [13]. For instance, the year-on-year growth of the financial income in a company instantiates the contextual information of previous incomes and can be investigated for contextual anomalies.

6 Machine Learning Techniques for Anomaly Detection

Under the umbrella of anomaly detection, the solutions are typically categorized into three aspects according to the type of input, including supervised, unsupervised, semi-supervised anomaly detection. Therefore, the essence of a supervised anomaly detection problem is a classification problem that endeavors in distinguishing abnormality from normality. On the other hand, unsupervised anomaly detection gains no access to the exact labels of the given dataset. It achieves anomaly detection by identifying the shared patterns among the data instances and observing the outliers. Besides, Semi-supervised anomaly detection accepts normal or abnormal datasets and determines the concept of normality or abnormality for anomaly detection [14]. Machine learning techniques are widely used in the domain of social media [15], healthcare [16, 17], surveillance [18], robotics [19], and business [20].

6.1 Supervised Anomaly Detection

Supervised methods require having a label for each data point in the training dataset. They use this information to learn the differences between normal and anomalous instances using discriminative or generative algorithms [9]. Theoretically, supervised anomaly detection is superior in its overall accuracy due to the clear understanding of normality and abnormality. Neural network models are considered more robust and highly accurate in applications such as forecasting medical time-series data [21] and stock market prices [22]. The Multi-Layer Perceptron (MLP), a simple feed-forward neural network with multiple hidden layers, can learn complex time dependencies and correlations within the data. Recurrent neural networks (RNNs) are also a topic of interest in time-series forecasting, as the inherent structure of the RNN enables it to retain a memory state from each time step to the next. The Long Short Term Memory (LSTM) model, a variant of the RNN structure, introduced an innovative solution to the problem. LSTM models have been widely researched and have been found to perform exceptionally well in forecasting time-series data with long term temporal dependencies [23, 24]. Except for the popular SVM-based classification models, other widely adopted models like neural networks, Bayesian networks, and decision trees are also applied to anomaly detection. Previous studies [25–28] showed that LSTM provides better prediction accuracy than many other approaches and neural networks.

Nevertheless, some practical problems immensely hinder its utilization. Firstly, the data labels are usually not available or too costly to obtain under many scenarios, e.g., the label of a configuration of a network is not entirely clear unless the network is practically run and examined. Secondly, the data labels may not be balanced. Typically, in practical anomaly detection problems, the normal samples greatly outnumber the abnormal ones, which results in a prominent bias in the classification model that may degrade the performance of anomaly detection. Lastly, the involvement of both normal and abnormal data may introduce more noise into the model; hence the performance may be disgraced.

6.2 Semi-supervised Anomaly Detection

Semi-supervised techniques endeavor the model to a single concept and achieve anomaly detection according to the fitness of the data in the concept. Therefore, semi-supervised anomaly detection constitutes a one-class classification problem. In comparison to the classification problem, one-class classification requires only normal or abnormal samples, which is more feasible in reality. Also, due to the sole type of samples, one-class classification negates the problem of an imbalanced dataset.

The goal of clustering is to group similar data into multiple clusters depending on the machine learning tasks. In this general type, the normal class is characterized by several prototype points in the input data set. The class label of a test point is quantified by the distance to the center of the nearest cluster. In anomaly detection, the normal data with similar behavior will be in the same cluster, while outliers will show distinct differences in the criteria. Depending on the application scenario, the data will be partitioned into multiple groups. The classic k-means clustering is the most representative and intuitive in this category of algorithms [29, 30]. The k-means algorithm works by choosing k random initial cluster centers, computing the distances between these cluster centers and each point in the training set, and then identifying those points that are closest to each cluster center. Many variations based on the k-means are also developed, among which the most popular is the combination with fuzzy theory, such as the fuzzy c-means and probabilistic c-means [31]. Other algorithms like Self-Organizing Maps(SOM), Expectation-Maximization, Find Out, CLAD, and CBLOF [32] also belongs to this category.

6.3 Unsupervised Anomaly Detection

Unsupervised techniques are typically employed in a situation where no prior knowledge of the dataset is known. In other words, no label information is presented. An anomaly detection method has to analyze the dataset to infer the real concept of abnormality or make an assumption of the concept. A concrete example of this type is the set of clustering-based anomaly detection methods which presume the data that rest inside small clusters are prone to be anomalous. Unsupervised anomaly detection enjoys similar merits to semi-supervised anomaly detection, while it is always criticized because of the validity of the assumptions made in related tasks. Pang et al. [33] evaluated Hierarchical clustering as a method of unsupervised feature selection. In this study, the final cluster represents variables that are important to each other. For clustering approaches, a chosen number of variables are examined, often a maximum of two or three. However, the clustering approaches require little modeling effort and are known to be relatively easy to build and implement [31]. The anomaly detector based on residuals has the advantage that the algorithm easily can be trained to capture seasonal and cyclic variations, where the relationship between the target variable and input features for different seasons is captured, assuming that the data is available [34].

7 Statistical-Based Models

Statistical models [35] were the earliest adopted for anomaly detection. They are mostly based on the comparison of some statistical properties to test whether outliers exist. During the training phase, the distribution parameter is optimized under specific evaluation criteria. The resultant distribution will define the boundary with a probabilistic threshold. Furthermore, to test the generalization of unseen data, the learned model is assumed to have potential outliers to lie in the low-probability density region [36]. As can be seen, for training, a suitable probabilistic model should be defined. Thus the fitting could perform.

7.1 Parametric Models

Parametric models [35] assume that data is sampled from a known distribution. The training phase then is to estimate the parameters of the model with training data through statistics. These models are suitable for extensive data set since the model complexity is irrelevant to the data size, and the test evaluation is performed rapidly. However, the accurate assumption of distribution is the key issue for this type of model, leading to the requirement of a priori knowledge of data. Gaussian distribution is frequently assumed [37] for both univariate and multivariate continuous data. Training involves estimations of mean and covariance using Maximum Likelihood Estimation. For categorical data, the commonly used model is multinomial distributions. Markovian model is suitable for sequential data [38]. In real applications, many datasets do not follow one specific distribution model and are randomly distributed; thus, a single model is not enough to capture true underlying statistical characteristics. In these cases, a mixture of multiple models is often assumed [39], then the parameters and model weight coefficient for each model are optimized jointly.

7.2 Non-parametric Models

This type of model [35] is opposite to the parametric ones as they make no assumptions on the data distributions. The complexity of these models grows to be adaptive to fit the data in size and complexity. One common thing about these methods is that they are distance-based. The training phase mainly reorganizes the data set through the transformation to better demonstrate structure information. Despite being more flexible and autonomous than a parametric model, non-parametric models also require meaningful a priori knowledge to help the learning of distance-based similarity. This set of models could be applied to both categorical and numeric features. One branch of representative models is histogram analysis [40], in which the normal input feature represented by the frequency of occurrences will be used to maintain the normal behavior pattern.

Histogram analysis is based on the frequency by counting. When the number of feature dimensions is low, it is very efficient for univariate data but suffers as dimensions increase. The kernel density estimators (KDE) is used to estimate the probability density function using a large number of kernels distributed over the space. The algorithm estimates the density at each location in the data space in a localized neighbourhood of the kernel. Parzen window methods [41] hires a set of Gaussian kernels to place them on each data and sums the local contributions from each kernel to estimate the actual density, with a single shared variance hyper-parameter.

8 Challenges in Anomaly Detection

In short, anomalous behaviour is a deviation from normal behaviour. Therefore, to detect deviations from normal behaviour, a criterion or measurement is needed to define normal behaviour. However, this seemingly simple task of defining a range or criterion for normal behaviour and distinguishing it from abnormal behaviour is very challenging in practice. We outline these challenges as the following:

8.1 Defining Normal Region

Defining proper criteria that indicate the range of normal data instances is often a difficult task. This is especially challenging when we need to draw a line between normal data instances and anomalous data instances. In fact, in many cases, likely, data in a normal region that is close to a borderline is an indicator of an anomaly, and vice versa.

8.2 Detecting Malicious Actions

Malicious actions may adapt themselves to mimic normal behaviour in a system, thereby making it difficult to distinguish normal behaviour from non-normal behaviour.

8.3 Evolution of Normal Behaviour

If the characteristics of the current notion of normal behaviour change, then the normal behaviour needs to be updated as normal behavior changes.

8.4 Different Domains Require Different Anomaly Notions

The exact notion of an anomaly varies from one application domain to another. For example, while fluctuations in the value of a stock market may be accounted for as normal, in the medical field, a minimal deviation from the normal value of a blood test can be considered anomalous. This adapts an effective anomaly detection technique of one domain not suitable in another domain.

8.5 Lack of Labelled Data

Many anomaly detection techniques rely on labelled data, while for much training data, such labelling is not available.

8.6 Distinguishing Noise from Anomalies

There is a chance that noise is similar to anomalous data instances, thereby making a differentiation between these two a challenging task.

8.7 Availability, Reliability, Low Latency

In critical applications like healthcare, the reliable detection of anomalies is fundamental. The detection systems have to make sure that each data point is considered and none of them is lost or omitted. Sometimes the delivery guarantee is even more strict, and it is also necessary to ensure that no data point is delivered more than once. This so-called exactly-once delivery guarantee, often required, e.g. in the finance industry, is in general very hard to achieve in distributed systems. In e-commerce, on the other side, the requirements for delivery are lower. However, the minimization of latency is extra important because it has been experimentally shown that the latency correlates with a decrease in revenues [42].

In reality, the requirements for low-latency or availability and reliability are inherently contradictory due to the CAP theorem [43]. Therefore, tradeoffs like sacrificing reliability in favour of low-latency or the other way around must be made in the architecture and implementation of stream processing systems. Also, the source systems (especially IoT devices) have often minimal capabilities. These devices cannot effort to wait long periods to complete sending a data point, nor can they buffer the data points in the case that a stream processing system is unavailable. Therefore, any outage of the stream processing systems results in a loss of data.

8.8 High Throughput, Parallelization, Distribution, and Scalability

To handle multiple large data streams, the underlying computer systems must be able to parallelize or distribute the workload to more than one physical machine. This distribution requirement represents a challenge for the design and implementation of anomaly detection algorithms. The algorithm must be able to efficiently exchange the gained knowledge about data points between compute instances. The load of the detection system may also vary over time. The data stream can grow, which requires more compute instances. On the other side, there might be some periods of low activity of the source systems, and in this situation, it is not feasible to run the same number of machines as at peak hours (the number of compute instances must be scaled down). Because of the reasons above, the detection systems must be able to dynamically scale-out and down.

The challenges outlined above make the detection of anomalies a challenging task. As a result, most anomaly detection techniques are designed for a particular domain, and choosing a proper anomaly detection method requires consideration of a set of factors such as the field of work, the characteristics of the available data, the type of anomalies to be detected, and data dependency and data volume.

9 Conclusion

Anomaly mining is an exciting research area that has extensive use in a wide variety of application domains. An overview of outlier detection techniques in time-series data has been proposed. Moreover, a taxonomy that categorizes these methods depending on the input data type, the anomaly type, and the nature of the detection method has been included. This section first discusses some general remarks about the analyzed techniques, and it then introduces the conclusions regarding the axes of the taxonomy. There are several promising directions for further research in anomaly detection. In the following, we describe some specific directions for future research works.

Online Algorithms: Nowadays, most real-world application domains generate data that is dynamic. Mining anomalies in such dynamic datasets with online techniques required to execute the detection algorithm on the whole extended dataset as new data arrives. As the size of the data increases rapidly with time, running online methods periodically is not feasible. Most of the existing online anomaly detectors use some supervision, and there are very few unsupervised online detectors. Hence, designing online versions of our ensemble and multimodal learning approaches will be an interesting future research direction.

Scalability: There is continuous advancement in data collection and storage as real-world data keep growing. It needs petabytes of storage to process such big data. Therefore, the major bottleneck of any data mining algorithm design is the scalability. Previously, the main focus of an algorithm design was the accuracy of the algorithm. However, nowadays, researchers focus on both accuracy and running time as the algorithms need to process data that does not fit into the main memory. As such, exciting future research would be to design practical approximation algorithms that require linear time and space. Moreover, designing anomaly ensemble algorithms that exploit parallelism can also be interesting future work.

Anomaly Ensemble for Complex Heterogeneous Graphs: The majority of the existing works in graph anomaly mining are designed to work for simple undirected homogeneous graphs. However, real-world graphs are much complex having various types of nodes, edges with weights, node and edge attributes, etc. Most of the social media graphs, such as Facebook, Twitter, and Pinterest have a heterogeneous structure, and representing them as plain undirected graph results in losing much information. There is a lack of work in finding anomalies from such complex graphs. Therefore, designing effective and robust ensemble approaches for complex heterogeneous graphs can be a fascinating future research direction.

References

1. Wu, H.-S.: A survey of research on anomaly detection for time series. In: 2016 13th International Computer Conference on Wavelet Active Media Technology and Information Processing (ICCWAMTIP), pp. 426–431 (2016)
2. Blázquez-García, A., Conde, A., Mori, U., Lozano, J.A.: A review on outlier/anomaly detection in time series data, arXiv preprint arXiv:2002.04236 (2020)
3. Yeoh, W., Koronios, A.: Critical success factors for business intelligence systems. J. Comput. Inf. Syst. **50**, 23–32 (2010)
4. Al Mamun, S.A., Valimaki, J.: Anomaly detection and classification in Cellular Networks using automatic labeling technique for applying supervised learning. Procedia Comput. Sci. **140**, 186–195 (2018)
5. Landauer, M., Wurzenberger, M., Skopik, F., Settanni, G., Filzmoser, P.: Time series analysis: unsupervised anomaly detection beyond outlier detection. In: International Conference on Information Security Practice and Experience, pp. 19–36 (2018)
6. Teng, X., Lin, Y.-R., Wen, X.: Anomaly detection in dynamic networks using multi-view time-series hypersphere learning. In: Proceedings of the 2017 ACM on Conference on Information and Knowledge Management, pp. 827–836 (2017)
7. Fahim, M., Sillitti, A.: Anomaly detection, analysis and prediction techniques in iot environment: a systematic literature review. IEEE Access **7**, 81664–81681 (2019)
8. Foorthuis, R.: On the nature and types of anomalies: a review, arXiv preprint arXiv:2007.15634 (2020)
9. Esling, P., Agon, C.: Time-series data mining. ACM Comput. Surv. (CSUR) **45**, 1–34 (2012)
10. Talagala, T.S., Hyndman, R.J., Athanasopoulos, G.: Meta-learning how to forecast time series. Monash Econometrics and Business Statistics Working Papers, vol. 6, p. 18, (2018)
11. Idrees, S.M., Alam, M.A., Agarwal, P.: A prediction approach for stock market volatility based on time series data. IEEE Access **7**, 17287–17298 (2019)
12. Zhou, S., Shen, W., Zeng, D., Fang, M., Wei, Y., Zhang, Z.: Spatial–temporal convolutional neural networks for anomaly detection and localization in crowded scenes. Signal Process. Image Commun. **47**, 358–368 (2016)
13. Liu, C., Liu, J., Wang, J., Xu, S., Han, H., Chen, Y.: An attention-based spatiotemporal gated recurrent unit network for point-of-interest recommendation. ISPRS Int. J. Geo-Inf. **8**, 355 (2019)
14. Boukerche, A., Zheng, L., Alfandi, O.: Outlier detection: methods, models, and classification. ACM Comput. Surv. (CSUR) **53**, 1–37 (2020)
15. Alam, T.M., Awan, M.J.: Domain analysis of information extraction techniques. Int. J. Multidiscip. Sci. Eng. **9**, 1–9 (2018)
16. Ghani, M.U., Alam, T.M., Jaskani, F.H.: Comparison of classification models for early prediction of breast cancer. In: 2019 International Conference on Innovative Computing (ICIC), pp. 1–6 (2019)
17. Ali, Y., Farooq, A., Alam, T.M., Farooq, M.S., Awan, M.J., Baig, T.I.: Detection of schistosomiasis factors using association rule mining. IEEE Access **7**, 186108–186114 (2019)
18. Baig, T.I., Alam, T.M., Anjum, T., Naseer, S., Wahab, A., Imtiaz, M., et al.: Classification of human face: asian and non-Asian people. In: 2019 International Conference on Innovative Computing (ICIC), pp. 1–6 (2019)

19. Kamran, S., Farhat, I., Talha Mahboob, A., Gagandeep Kaur, A., Liton, D., Abdul Ghaffar, K., et al.: The impact of artificial intelligence and robotics on the future employment opportunities. Trends Comput. Sci. Inf. Technol. **5**, 5 (2020)
20. Alam, T.M., Shaukat, K., Mushtaq, M., Ali, Y., Khushi, M., Luo, S., et al.: Corporate bankruptcy prediction: an approach towards better corporate world. Comput. J. **63** (2020)
21. Cerqueira, V., Torgo, L., Soares, C.: Layered learning for early anomaly detection: predicting critical health episodes. In: International Conference on Discovery Science, pp. 445–459 (2019)
22. Golmohammadi, K., Zaiane, O.R.: Sentiment analysis on Twitter to improve time series contextual anomaly detection for detecting stock market manipulation. In: International Conference on Big Data Analytics and Knowledge Discovery, pp. 327–342 (2017)
23. Le, X.-H., Ho, H.V., Lee, G., Jung, S.: Application of long short-term memory (LSTM) neural network for flood forecasting. Water **11**, 1387 (2019)
24. Gao, S., Huang, Y., Zhang, S., Han, J., Wang, G., Zhang, M., et al.: Short-term runoff prediction with GRU and LSTM networks without requiring time step optimization during sample generation. J. Hydrol. **589**, 125188 (2020)
25. Singh, A.: Anomaly detection for temporal data using long short-term memory (LSTM) (2017)
26. Hundman, K., Constantinou, V., Laporte, C., Colwell, I., Soderstrom, T.: Detecting spacecraft anomalies using LSTMs and nonparametric dynamic thresholding. In: Proceedings of the 24th ACM SIGKDD International Conference on Knowledge Discovery & Data Mining, pp. 387–395 (2018)
27. Lee, T.J., Gottschlich, J., Tatbul, N., Metcalf, E., Zdonik, S.: Greenhouse: a zero-positive machine learning system for time-series anomaly detection, arXiv preprint arXiv:1801.03168 (2018)
28. Lee, M.-C., Lin, J.-C., Gran, E.G.: RePAD: real-time proactive anomaly detection for time series. In: International Conference on Advanced Information Networking and Applications, pp. 1291–1302 (2020)
29. Basora, L., Olive, X., Dubot, T.: Recent advances in anomaly detection methods applied to aviation. Aerospace **6**, 117 (2019)
30. Shaukat, K., Luo, S., Varadharajan, V., Hameed, I.A., Chen, S., Liu, D., et al.: Performance comparison and current challenges of using machine learning techniques in cybersecurity. Energies **13**, 2509 (2020)
31. Kacprzyk, J., Owsiński, J.W., Viattchenin, D.A., Shyrai, S.: A new heuristic algorithm of possibilistic clustering based on intuitionistic fuzzy relations. In: Novel Developments in Uncertainty Representation and Processing. Springer (2016)
32. Heryadi, Y.: The effect of several kernel functions to financial transaction anomaly detection performance using one-class SVM. In: 2019 International Congress on Applied Information Technology (AIT), pp. 1–7 (2019)
33. Pang, G., Cao, L., Chen, L., Liu, H.: Unsupervised feature selection for outlier detection by modelling hierarchical value-feature couplings. In: 2016 IEEE 16th International Conference on Data Mining (ICDM), pp. 410–419 (2016)
34. Munir, M., Siddiqui, S.A., Chattha, M.A., Dengel, A., Ahmed, S.: FuseAD: unsupervised anomaly detection in streaming sensors data by fusing statistical and deep learning models. Sensors **19**, 2451 (2019)
35. Kourtis, M.-A., Xilouris, G., Gardikis, G., Koutras, I.: Statistical-based anomaly detection for NFV services. In: 2016 IEEE Conference on Network Function Virtualization and Software Defined Networks (NFV-SDN), pp. 161–166 (2016)
36. Garcia-Font, V., Garrigues, C., Rifà-Pous, H.: A comparative study of anomaly detection techniques for smart city wireless sensor networks. Sensors **16**, 868 (2016)

37. Luo, H., Zhong, S.: Gas turbine engine gas path anomaly detection using deep learning with Gaussian distribution. In: 2017 Prognostics and System Health Management Conference (PHM-Harbin), pp. 1–6 (2017)
38. Lüdtke, O., Robitzsch, A., West, S.G.: Regression models involving nonlinear effects with missing data: a sequential modeling approach using Bayesian estimation. Psychol. Methods **25**, 157–181 (2019)
39. Ahmed, M., Mahmood, A.N., Hu, J.: A survey of network anomaly detection techniques. J. Netw. Comput. Appl. **60**, 19–31 (2016)
40. Bansod, S.D., Nandedkar, A.V.: Crowd anomaly detection and localization using histogram of magnitude and momentum. Vis. Comput. **36**, 609–620 (2020)
41. das Chagas, J.V.S., Ivo, R.F., Guimarães, M.T., de A. Rodrigues, D., de S. Rebouças, E., Rebouças Filho, P.P.: Fast fully automatic skin lesions segmentation probabilistic with Parzen window. Comput. Med. Imaging Graph. **85**, 101774 (2020)
42. Gharaibeh, A., Salahuddin, M.A., Hussini, S.J., Khreishah, A., Khalil, I., Guizani, M., et al.: Smart cities: a survey on data management, security, and enabling technologies. IEEE Commun. Surv. Tutor. **19**, 2456–2501 (2017)
43. Abadi, D.: Consistency tradeoffs in modern distributed database system design: CAP is only part of the story. Computer **45**, 37–42 (2012)

Tapis: An API Platform for Reproducible, Distributed Computational Research

Joe Stubbs[1(✉)], Richard Cardone[1], Mike Packard[1], Anagha Jamthe[1], Smruti Padhy[1], Steve Terry[1], Julia Looney[1], Joseph Meiring[1], Steve Black[1], Maytal Dahan[1], Sean Cleveland[2], and Gwen Jacobs[2]

[1] Texas Advanced Computing Center, Austin, TX, USA
{jstubbs,rcardone,mpackard,ajamthe,spadhy,sterry1,
jlooney,jmeiring,scblack,maytal}@tacc.utexas.edu

[2] University of Hawaii - System, Honolulu, HI, USA
{seanbc,gwenj}@hawaii.edu

Abstract. Modern computational research increasingly spans multiple, geographically distributed data centers and leverages instruments, experimental facilities and a network of national and regional cyberinfrastructure (CI). Tapis is an open-source API platform developed at the Texas Advanced Computing Center at the University of Texas at Austin to increase reproducibility and minimize time-to-solution for distributed computational experiments. Core features of Tapis include data management and code execution, a fine-grained permissions system enabling objects to be saved privately, shared with individuals or "published" to a community, and provenance endpoints exposing the detailed history Tapis collects on analyses, enabling workflows to be repeated and results reproduced. In this paper, we describe the evolution of the Tapis platform, from its origins in 2008, and discuss the growth and success of the project as well as challenges and limitations that have led to a new design effort, funded by the National Science Foundation in September of 2019. We present a detailed overview of the new system, including reference architecture and new features such as support for streaming/sensor data, and we discuss some of the early science use cases driving its design. We conclude with the roadmap for future work.

Keywords: HPC · Cloud · Virtualization · Middleware

1 Introduction

Modern computational experiments are becoming increasingly distributed, particularly those designed by interdisciplinary teams. These efforts span geographically distributed data centers and leverage instruments, experimental facilities, and national and regional cyberinfrastructure to address fundamental problems in science and engineering. For example, investigators might first apply machine learning techniques to raw data from remote sensors, or they might perform genetics analysis against sequencing data produced in a robotics wet

K. Arai (Ed.): FICC 2021, AISC 1363, pp. 878–900, 2021.
https://doi.org/10.1007/978-3-030-73100-7_61

lab. Next, they might pass those results to a simulation that runs on a traditional HPC supercomputer. The simulation results in turn help to calibrate the remote instruments and processes for the next iteration, and the cycle repeats. Programming the execution of such experiments, much less in a scalable, reproducible way, presents a formidable challenge to investigators and prevents many such experiments from ever being run. Furthermore, the operation of the advanced cyberinfrastructure to accommodate such experiments requires specialized knowledge and experience, creating another obstacle for many regional academic providers.

Tapis is an open source, hosted Application Program Interface (API) platform for distributed computation that enables researchers to manage data and execute codes on a wide range of remote systems, from high-speed storage and high-performance computing systems to commodity servers. Tapis has experienced tremendous growth and success since its initial incarnation as the "Foundation API" for the iPlant Collaborative project in 2008 [23]. In the ensuing twelve years, more than 14 independently funded projects have leveraged the Tapis platform across a wide variety of domains of science and engineering, and many smaller projects interact with Tapis without a formal engagement. Two major version releases of the framework as well as numerous minor releases were made over this period in response to community usage and feedback.

In 2019, the Nation Science Foundation funded a five-year project to extend and enhance Tapis, officially the third major version of the platform ("v3"). This new work will deliver production-grade capabilities to enable researchers to 1) securely execute workflows that span geographically distributed providers, 2) store and retrieve streaming/sensor data for real-time and batch job processing, with support for temporal and spatial indexes and queries, 3) leverage containerized codes to enable portability, and reduce the overall time-to-solution by utilizing data locality and other "smart scheduling" techniques, 4) improve repeatability and reproducibility of computations with history and provenance tracking built into the API, and 5) manage access to data and results through a fine-grained permissions model, so that digital assets can be securely shared with colleagues or the community at large.

Tapis is itself distributed: a hosted RESTful API platform, deployed at various institutions including the Texas Advanced Computing Center (TACC) at the University of Texas at Austin, the University of Hawaii (UH), Manoa, and others. End users and applications interact with Tapis by making authenticated HTTP requests to Tapis's public endpoints. In response to requests, Tapis's network of microservices interact with a vast array of physical resources on behalf of users: high performance and high throughput computing clusters, file servers and other storage systems, databases, bare metal and virtual servers, etc. The goal of Tapis is to provide a unified, simple to use API enabling teams to accomplish computational and data intensive computing in a secure, scalable, and reproducible way so that domain experts can focus on their research instead of the technology needed to accomplish it. Working alongside researchers from various fields to drive real-world use cases, Tapis aims to be the underlying cyberinfrastructure for a diverse set of research projects: from large scale science gateways built to serve entire communities, to smaller projects and individual labs wanting to automate one or more components of their process.

The Tapis v3 design is motivated by lessons learned from developing and operating the current production platform and draws on a number of prior NSF investments, including the Agave API [23], the Abaco (Actor Based Containers) functions-as-a-service project [39], and the CHORDS (Cloud-Hosted Real-time Data Services for the Geosciences) platform [26]. The primary Tapis APIs include the Files API for managing data on remote storage, the Systems, Apps and Jobs APIs for registering and executing software and research codes, the Actors API for registering small executables that run with very low latency in response to messages sent over HTTP, the Meta API, a high performance document store for scaling research data collections to billions of documents serialized using formats such as JSON or XML, and the Streams API for storing and retrieving sensor data. Underlying all of these services is a set of authorization APIs which comprise the Tapis Security Kernel. The Tapis Security Kernel is a unique, decentralized solution to authorization and security management with APIs that enable trust federation across physical and institutional boundaries.

A recurring technique leveraged throughout the platform is the use of container technology to enhance portability and reproducibility, not just for the end user's research computations, but for all Tapis execution. Each Tapis microservice is packaged into a Docker image, and, for the new Tapis v3, the official deployment tooling targets the Kubernetes platform. This approach simplifies operations across datacenters while simultaneously increasing the uniformity of deployed Tapis components, whether they run at academic institutions, nationally funded cloud providers or the commercial cloud.

The rest of the paper is organized as follows: in Sect. 2, we discuss in detail the background, motivation and related work; Sect. 3 describes the existing production platform including successes and lessons learned. The paper transitions to the new v3 system in Sect. 4 where we describe the primary Tapis v3 capabilities including ones provided by the new design; in Sect. 5 we highlight some projects that have agreed to be early adopters of the new platform, and describe their use cases and requirements; in Sect. 6 we provide a high-level overview of the Tapis architecture; finally, in Sect. 7 we conclude with a road map for the five funded project years.

2 Background and Related Work

2.1 API Platforms

Microservice architectures, JSON and OAuth2 have greatly reduced barriers to distributed application development and have enabled new usage patterns across industry and academia. Microservices typically are HTTP-based APIs built on open standards such as REST. All leading cloud providers including Amazon AWS, Google Cloud Platform and Microsoft Azure provide such services. In collaboration with its partner institutions, the Texas Advanced Computing Center builds and operates a number of APIs as part of ongoing projects. The Chameleon and JetStream systems, cloud computing resources for the national research community, offer APIs for managing infrastructure-as-a-service, similar to the commercial cloud providers. The Tapis API offers data management and code execution. The more recently funded Abaco API supports functional programming models and event driven architectures on cloud infrastructure.

The ubiquity and popularity of web-friendly APIs is hard to refute. While obtaining official numbers is difficult, according to some reports AWS now supports over one million active users [18]. Though academic clouds are significantly smaller, they still service thousands of users and millions of requests a year. Nearly 50,000 OAuth clients have been registered to the Tapis framework alone. See Sect. 3 for a more in-depth discussion of Tapis usage. The relatively recent Abaco platform already supports several projects that have registered over 44,000 functions after less than two years in production.

2.2 Containers and Distributed Computations

The number of scientific workflows making use of distributed computational experiments is increasing. Advances in technologies related to small devices, instruments, robotics and other forms of experimental facilities have led to an explosion in large, real-time data sources. Examples include genomic sequencing facilities, survey telescopes, climate sensors, shake tables, wind tunnels, and laser light sources. Computations often leverage a mix of AI/Machine Learning techniques in combination with model simulations to derive new insights. As a result, these experiments require a mix of computational resources, data access methods and management techniques.

Researchers analyze sensor data using batch and real-time processing. With batch processing, investigators write programs to analyze a static set of data defined at the start of the analysis. These programs may utilize traditional HPC machines scheduled via a batch scheduler such as SLURM, or high-throughput and/or cloud resources, and they may be written in a variety of languages and frameworks. Such frameworks include traditional model simulations that leverage MPI, AI/ML code that depend on CUDA or higher-level libraries such as TensorFlow, or other data-intensive frameworks like Hadoop and Spark. With real-time processing, codes analyze new streaming data points as they arrive. In the simplest cases, these analyses check basic conditions to determine if interesting or unusual criteria are met; in the general case, real-time processing can involve many of the same kinds of analysis as batch processing.

Finding a mechanism to enable computational portability can be a great simplification if not an out right requirement for successfully orchestrating such workflows. Steps in the workflow that analyze streaming data often have a real-time, low-latency requirement that can only be achieved by running code near the data. Traditional simulation steps often require running codes across large clusters that may only be available in world-class HPC centers. For other steps with less resource requirements, minimizing time-to-solution may involve leveraging high throughput systems and machines with shorter queue times than highly-demanded supercomputers. If the workflow application (or individual step components) can be easily moved to different computing resources, different steps can leverage the most appropriate resource available at the time of execution.

Over the last five years or so, software teams have started adopting Linux container technologies, chiefly the Docker platform, to improve application portability. Container best practices encourage practitioners to bundle all application assets, including software libraries and other dependencies, into the container image to produce an independent package to minimize dependencies on the execution environment. The goal is to enable containers to be executed from the image on any machine where the container runtime is available.

While container technology significantly improves the portability and reproducibility of general applications, adoption within the scientific computing community has lagged behind industry. The primary reason is that containers cannot easily encapsulate all dependencies of scientific code running in customized, heterogeneous environments. Scientific computations rely on specialized hardware and custom software installed across a computing infrastructure, including MPI, networking, mathematical and GPU libraries, and specialized workflow managers, that cannot be captured within a container. Additionally, the container runtime itself becomes a dependency of containerized applications, which may rely on specific versions, configurations, plugins or optional features of that runtime.

Our strategy with Tapis therefore is to augment container technology with metadata captured by the platform that describes requirements of applications as well as capabilities of different execution environments. These abstractions, described in more detail in Sect. 4.1, are based on our experience running containerized workloads in a variety of contexts. The Singularity [27] runtime is available on all major HPC systems at TACC, where over 4,400 unique containerized applications are available for use. In its first two years of operation, nearly 25,000 Docker images representing actor functions were registered with the Abaco platform. TACC manages many other containerized applications across its various cloud offerings such as its custom JuptyerHub clusters.

A growing body of evidence suggests that time-to-solution can be reduced using these techniques. For instance, a 2018 paper entitled "Virtualizing the Stampede2 SuperComputer with Applications to HPC in the Cloud" [34] describes methods for building a virtual HPC resource in the JetStream cloud that shares many properties of the Stampede2 supercomputer. The performance of a number of popular scientific applications was measured on both the virtual and real Stampede2 clusters. The paper presents a profile for scientific applications where performance in the virtual cluster is similar to that on the real Stampede2 cluster, suggesting that such applications are amenable to "cloud bursting", i.e., scheduling on alternative resources to bypass potentially long queues on HPC systems.

2.3 Distributed Security

Distributed applications are often constrained by rigid authentication, authorization and secrets management requirements. Some applications, for instance, build in their authentication mechanisms, making integration into enterprises with existing security infrastructures and policies difficult. The practice of inventing ad-hoc access control is even more pervasive, where the introduction of administrative users, permissions, groups and roles occurs incrementally as user requirements evolve. Distributed applications almost always have to manage passwords, keys and other secrets; these secrets often get embedded in configuration files, databases, build scripts, deployment scripts, container definition files, etc. All of these practices make application integration and management more difficult and less secure.

The security subsystem is among the most basic and critical of all modules in a distributed application, one on which almost all other subsystems depend, and one that does not depend on other subsystems. Many distributed applications face the same security design issues as Tapis, yet there is not an off-the-shelf package that delivers a robust, lightweight, flexible, high performance solution.

The goal is for the Tapis Security Kernel to provide a complete security API for distributed applications with the following characteristics:

- Easy integration with any authentication/identity manager.
- Fast, fine-grained authorization checking that scales to millions of objects.
- A secure, highly available, multi-tenant secrets store.
- Management of all secrets through the store.
- Support for on-premises or remote installation.

Our approach incorporates hardened open-source packages into a portable system unified by a simple, high-level API. A key simplification is that we use signed tokens created outside of our system as proof of identity. Any authentication system that can create and properly sign a token with a trusted key can be integrated into Tapis—no further specification required. The community version of HashiCorp's Vault [9] product provides a mature, robust secrets store with numerous management features. Apache Shiro [4] provides the basis for a simple, powerful authorization mechanism that we extend (a) with hierarchical roles for improved user management and (b) with a permissions model that allows file path names to be succinctly represented for scalability.

2.4 Software Comparison

We briefly survey the landscape of software systems available for computational research.

Gateway Frameworks. Gateway frameworks provide components for building web-based computational science portals with intuitive user interfaces to advanced cyberinfrastructure. Apache Airavata [3], the Agave project [23], Galaxy [7], Globus [8], HubZero [10] and WS-PGRADE [25] are among the most widely used, domain agnostic projects, and while all have enjoyed undeniable success, none currently attempt to address the aforementioned challenges of distributed experiments. For example, none of these frameworks provide APIs for sensor data. With the exception of Galaxy, which provides support for tools packaged as Docker images, none provide first-class support for containerized application workloads. With Dockerized Galaxy tools, the abstractions do not include notions of capabilities such as specialized hardware and system libraries, required to achieve portability of high-performance codes. Additionally, among these major gateway frameworks, only Globus Auth attempts to provide a decentralized security system; however, Globus Auth does not provide a secrets store for securing arbitrary data such as keys, certificates, database passwords, etc. Moreover, it is neither open source nor free to use, and the charge model isn't publicly available. This is a very different model from that implemented for Tapis.

Gateway Tools. Gateway tools comprise another class of comparable software including workflow managers such as Pegasus [22] and Taverna [42], real-time data mediation services like Brown Dog [32] and SciServer [16], and infrastructures for streaming data such as CHORDs and Data Turbine [5]. We assessed each project for features that would be of potential benefit for distributed workflows and discovered that most would require a heavy development investment to utilize. A detailed discussion of the assessment is beyond the scope of this manuscript.

Commercial Platforms. Tapis draws comparison to a number of commercial cloud offerings. Amazon Web Services (AWS) provides a suite of offerings for storing and processing streaming data including IoT, Kinesis, SQS, Lambda, etc. as well as IAM and secrets management offerings. Google Cloud Platform and Microsoft Azure provide similar if less mature offerings; However, in each case, usage is restricted to the specific commercial platform where costs can be prohibitive for researchers. Moreover, these services lack any significant integration into the national cyberinfrastructure provider fabric, and the closed, proprietary nature of these platforms makes the prospect of future integrations unlikely.

Application Security Software. Distributed applications implement security using different underlying technologies, each offering its own mix of features, compliance, performance, availability, cost, deployment options, etc. Technologies such as OAuth2, Kerberos, PERMIS and Amazon Key Management Service [2,12,14,15] focus on one aspect of security, where others such as Apache Fortress and Microsoft Active Directory [6,13] incorporate several aspects. As a group these implementations assume sole control of the security fabric, which makes integration and management in existing environments difficult.

In contrast, Tapis decouples application code from authentication method, using robust open source authorization and secrets management tools, and packaging those tools as an easily deployable system accessible through a unified API available to any application that wishes to incorporate it. This enables distributed implementation and management while offering fine grained controls for administrators and flexibility for developers and researchers.

3 Tapis in Production: Achievements and Lessons Learned

In this section, we describe the history of growth and success of the Tapis project, and identify the key challenges facing future growth.

3.1 Growth and Success

The evolution of the current production Tapis framework dates back to the initial release in 2008 of the Foundation API, a single server deployment supporting only the iPlant project. Today, the production Tapis system is a multi-tenant platform-as-a-service deployed across more than 50 servers at TACC supporting thousands of users. Additionally, separate "official" Tapis installations run at the University of Hawai'i and the Centers for Disease Control.

At TACC alone, 14 different independently funded projects representing a total of more than 30 million dollars in annual investments leverage Tapis to manage data, run jobs on HPC and HTC systems, and track provenance and metadata about computational experiments. Official collaborations include projects across a wide range of scientific/engineering domains and NSF directorates, including CyVerse [31] (formerly, the iPlant Collaborative) [24,30], DesignSafe [35], 3DEM Hub [29], the Science Gateways Community Institute [41], and VDJServer [17]. Non-NSF projects include the iReceptor portal for immune genetics research; the CNAP Center of Biomedical Research Excellence at Kansas State University; and Planet Texas 2050, funded by Texas and the University of Texas, Austin. An analysis of usage data suggests that many more projects leverage these platforms in an unofficial capacity, with over 53,000 OAuth client applications having been registered. The remote deployments at University of Hawai'i and CDC also support substantial projects such as the "Ike Wai Gateway" for water sustainability [20].

At the end of 2019, annual usage reached 30 million total API requests, nearly one Petabyte of data transferred, and 70,000 jobs launched. Aggregate usage numbers for the entire lifetime of the platform are provided in Table 1.

Beyond the obviously compelling usage numbers, Tapis consistently commands large turnouts at workshops and tutorials at conferences such as Practice And Experience in Advanced Research Computing (PEARC), Gateways and SuperComputing (SC). Over the years, a significant number of students have been trained as part of Research Experience for Undergraduates (REUs) fellowships sponsored by the Tapis project. Additionally, a number of successful "hack-a-thon" events established how productive researchers could be on the platform. For example, as part of the kickoff meeting of the Synergistic Discovery and Design project in 2018, a three day hack-a-thon of roughly 100 new Tapis users managed to analyze 10 Terabytes of raw data across 3,000 jobs.

3.2 Challenges

Despite its undeniable successes, a number of challenges threatened to prevent Tapis from reaching additional orders of magnitude of usage.

Table 1. Aggregate Tapis usage metrics

Metric	Value	Description
Official Installations	3	Total official, independent installations of the platform
Official Projects	14	Independently funded projects leveraging the platform
Systems	8,700	Number of systems registered
Apps	5,150	Number of unique application codes registered
Clients	53,126	Number of unique OAuth client applications registered
Jobs	285K	Total number of batch jobs run
Metadata items	442K	Total number of metadata item stored across all projects
Postits	400K	Total number of API objects shared and retrieved with postits
Data moved (files/size)	1.15B/3.5 PB	Total number of files transferred and total number of bytes transferred

Centralized Deployment Model. Many aspects of the current Tapis architecture make it difficult or even impossible to run in a decentralized configuration, where a subset of components run at different institutions or sites. As a result, the three official Tapis installations (at TACC, CDC and UH) are completely independent. This translates to two sub-optimal conditions. First, each site must run a complete Tapis platform, including all APIs, databases and third-party services. Running Tapis is a serious undertaking—not only the initial installation but the ongoing maintenance—as a minimum of three servers is required, but in practice, several more are needed. This requirement precludes all but the most dedicated institutions from actually running Tapis. Additionally, the independent nature of the instances means transfer performance suffers significantly when agents running at a given site are transferring data between systems at another site. The lack of any possible coordination between Tapis components means the Tapis agent must first transfer data from the source system at the remote site to its home institution, and then from its home institution to the destination system. This so called "two-legged" transfer can be very costly when sites are thousands of miles apart.

Expanding Footprint and Maintenance Burden. The current Tapis installation at TACC spans approximately 50 servers, constituting a significant maintenance burden. One reason for the expanding footprint has to do with the tenancy model in the current architecture. Tapis deploys an isolated "authorization server" for every tenant supported, and these authorization servers alone account for 15 servers in the deployment. And while possibly slightly more secure than an authorization server capable of servicing multiple tenants, in practice the projects derive little to no benefit from having an independent server, and many of them are configured very similarly. Deployment footprints of other components of Tapis, such as the Jobs and File transfers components, have also grown substantially to keep up with usage and provide dedicated capacity to projects. Many of these shortcomings can be attributed to fundamental architecture choices made at a time when the possibility supporting 20 concurrent projects seemed unimaginable. Today, however, that possibility is very real.

System-Dependent App Model. Finally, the "app" model for executable codes in the current Tapis system is limited by the fact that a given app must be tied directly to a specific execution system. If a user wants to run a code on Stampede, Lonestar and Frontera, she must register three different copies of the app. On traditional HPC systems, which were the primary target of the current Tapis app model, where hardware and operating system environments vary greatly from machine to machine, this may not seem like a significant issue. But on cloud and high-throughput systems, this becomes an overwhelming obstacle to getting work done, particularly in light of the explosion in usage of containers for application portability.

4 Tapis v3: Primary Capabilities

In this section we detail the primary capabilities of the new Tapis v3 framework. While this new version of the framework is under active development and targeting an initial public release during the summer of 2020, it draws upon architecture, ideas, lessons learned, and, in some cases, actual code, developed for Tapis as well as a number of other projects.

4.1 Data Management and Code Execution

Fundamental to all computational research is the ability to manage data and execute codes or *apps* to analyze such data. The Tapis Files API enables users to manage data on remote storage including SFTP and iRODS storage servers and S3-compatible object stores. Synchronous endpoints exist for listing, uploading, downloading and renaming data, and an asynchronous endpoint provides a managed transfer capability between two storage resources. Entire storage resources as well as individual files and directories can be registered as private to a single API user or shared with one or more users.

The Tapis Apps API provides a catalog of executable software which, like the resources within the Files API, can be private to a specific user or shared. The Tapis Jobs API is then used to execute an instance of an app on a remote execution resource. These execution resources can themselves be scheduled using a traditional HPC scheduler such as SLURM, in which case Tapis can inject the necessary scheduler directives based on metadata provided in the job request and in the app definition; otherwise, the Jobs service can start the app by directly forking a process on the underlying operating system.

The app model centers around containers to improve portability and reproducibility, and to enable flexible scheduling of computational workloads across geographically distributed providers. We achieve this flexibility by introducing execution system *capabilities* and application *requirements*. Based on our experience running both native and containerized applications, we have identified an initial list of capabilities required for determining whether a specific execution system can support a given application. The initial list of capabilities can be grouped into the following types:

- *Container runtime capabilities.* These capabilities include type and version of the runtime as well as optional features of the runtime (e.g., Singularity bind mounts) that may be enabled or disabled through configuration by the system administrator.
- *MPI capabilities.* The capabilities include the MPI version as well as the associated networking technology (e.g., Mellanox Infiniband, Intel Omni-Path, etc.)
- *GPU capabilities.* These capabilities include the GPU API type and version available (e.g., CUDA, OpenGL, etc.).

Additionally, the Jobs service will use these enhanced app definitions to execute jobs in a distributed manner to take advantage of data locality and, optionally, to schedule jobs on underutilized systems. An instance of the Tapis Jobs service running at a given institution can be configured to utilize a local security kernel for system credentials, enabling a datacenter to keep all sensitive credentials on premise; see 4.2 for more details.

4.2 Identity, Authorization, Security and Tenancy

Tapis provides a modular authentication subsystem with the goal of achieving sufficient flexibility to enable institutions to integrate their existing identity providers and related systems. Fundamental to the platform is the notion of *tenancy*; a *tenant* in the Tapis framework represents a logical separation of Tapis entities (i.e. apps, jobs, actors, etc.) as well as a high-level authentication and security configuration. A key simplification in our approach is to leverage signed JSON Web Tokens (JWT, [1]) as the single, sole mechanism for proving identity to any Tapis API. As each tenant is configured with its own public key for token signatures, the entire authentication system can be customized on a per-tenant basis.

The Tapis authentication subsystem is comprised of the following components:

- *Tokens API.* A stateless microservice and reference implementation for generating a properly formatted, signed JWT.
- *Authentication Server.* An OAuth2 and OIDC compliant web server capable of integrating with LDAP servers for authenticating end users and generating signed JWTs using a token API.

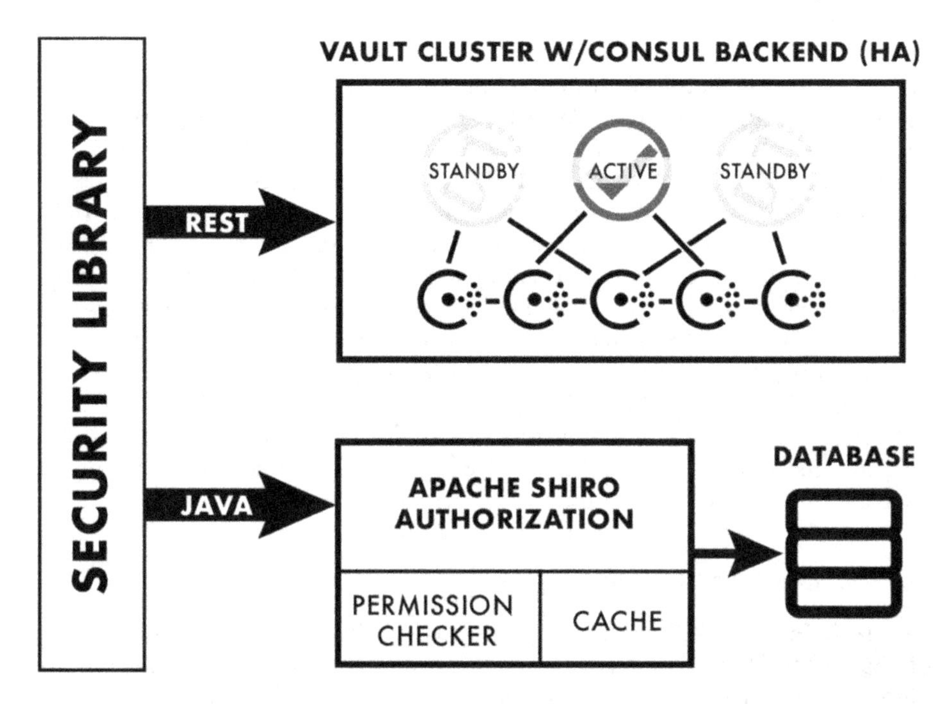

Fig. 1. Tapis Security Kernel. The Security Kernel presents an REST API over HTTP to other Tapis microservices while leveraging the open source Vault and Shiro projects for its implementation.

- *Tenants API.* Administrative API for managing the registry of tenants globally in a distributed Tapis installation.
- *API Router.* A load balancer and "edge router" capable of routing requests to back-end services, potentially across sites.

Tapis administrators manage the registry of tenants via the Tenants API, and it provides public (even anonymous) endpoints for discovering fundamental properties of a tenant, such as the public key used to sign JWTs for the tenant and the location of the tenant's security kernel. The rest of the components can be viewed as optional, but are used, for example, to provide an OAuth2-compliant authentication system against the primary TACC LDAP server.

The Tapis Security Kernel is a distributed subsystem comprised of open-source software components tied together by a unifying API. Figure 1 shows the main components of the subsystem. The security kernel starts with its inclusion of a secrets store, Vault [9], and an authorization service, Shiro [4] and, significantly, its exclusion of an authentication component. The security kernel provides an API that the platform's other microservices use to securely interact with users' storage and execution resources and with each other. Vault manages all secrets in the system, including all passwords and keys, using a fault-tolerant, scalable, highly available cluster of VMs.

The Tapis Security Kernel builds upon Shiro's security framework to create a scalable, fine-grained authorization facility by extending its permissions model with representations of file path names. This extension, along with proper caching, provides a scalable solution to the problem of fine-grained authorization checking across a virtually unlimited, distributed namespace.

The security kernel's API presents a unified interface to secrets and permissions management. Secrets are only accessible to a microservice if the user on behalf of whom the microservice is acting is authorized via Shiro. The API operates on users in multiple tenants, all of which have previously authenticated to the microservice. Multiple instances of the security kernel can exist in the system. Each tenant is assigned a kernel, which may be shared with other tenants or be exclusive to itself. In addition, a kernel instance can reside in a tenant's data center and be locally administered while the rest of the system runs and is administered in a central location. By configuring a local kernel, organizations have the option to keep and administer their secrets in-house.

4.3 Support for Streaming Data

The Tapis Streams API provides a production-quality service that builds on top of the CHORDS project for real-time data services and extends the primary data models including *site*, *instrument* and *variable*, with additional metadata including adding spatial indexing and permissions; Fig. 2 presents the high-level workflows and architecture. The Streams API also integrates Tapis event-driven functions (see Sect. 4.4) and data management and code execution capabilities (see Sect. 4.1) to provide an analysis capability on streaming data sources. Support for streaming data includes the following capabilities:

- Storing and retrieving streaming data for batch job processing, with support for temporal and spatial indexes and queries.
- Automated, event-driven data stream processing workflows with integration into Tapis functions as well as other streaming frameworks.
- Automated data management and scheduled archiving based on programmable policies.

Batch job processing is supported by allowing a Tapis stream, specified as a query, to be the input to a job. In such a case, the Jobs service will schedule the stream data to be transferred to the execution system, manifested as a JSON, CSV or similarly formatted file, prior to launching the app.

Real-time, event-driven workflows can also be supported by the Streams API. Tapis provides this capability through integration with its functions-as-a-service, jobs service and other systems or services that support web-hooks. For these use cases, analysis of incoming data in real time is critical to the success of the experiment. The Streams API will provide the following capabilities to support various levels of analysis based on their computational needs:

- *Alerts as Events*—Third-party processing engines will receive real-time notifications when measurement data hits predefined criteria. Subscribers will receive an HTTP POST request containing JSON data detailing the event that triggered the alert.
- *Processing Events with Tapis Functions*—Developers will be able to register an Abaco function with an alert to perform scalable processing with no infrastructure to maintain.
- *Scheduled Relays to Third-Party Systems*—Analogous to the transfers support for batch applications, the streaming data API will be capable of supporting scheduled relays for fixed time intervals to remote services, such as, message brokers and streaming frameworks capable of handling large quantities of data.

In order to support science use cases involving vast amounts of generated data (e.g., many GB/sec), the streaming data API will also provide endpoints for robust data management and archiving policies. For each stream, the data management APIs will support an archiving threshold and a policy. Thresholds will include time-to-live and max sizes (in bytes), and policies will include compressed transfer to a remote storage system and physical delete.

4.4 Functions-as-a-Service and Events-Driven Workloads

Tapis will adopt and evolve the NSF funded Abaco (Actor Based Containers) project to provide functions-as-a-service to support event-driven workloads. Abaco is based on the Actor Model of concurrent computation and Docker; users define computational primitives called *actors* with a Docker image, and Abaco assigns each actor a unique URL over which it can receive messages. Users send the actor a message by making an HTTP POST request to the URL. In response to an actor receiving a message, Abaco launches a container from the associated image, injecting the message into the container. Typically, the container execution is asynchronous from the message request, though Abaco does provide an endpoint for sending a message to an actor and blocking until the execution completes, providing synchronous execution semantics. Abaco maintains a queue of messages for each actor, and is capable of launching containers in parallel for a given actor when the actor is registered as *stateless*.

A primary use case is to allow users to develop and register actors to process events in Tapis. Examples of events include new files arriving on a storage system, a job completing on an execution system, either normally or abnormally, or alerts from the streaming data service indicating measurements have hit predefined thresholds, as described in Sect. 4.3.

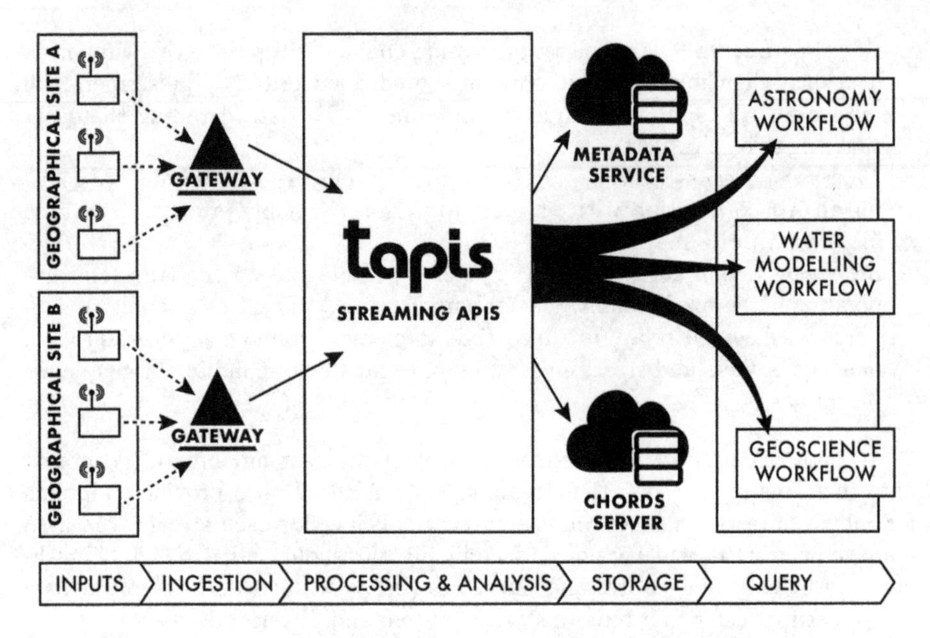

Fig. 2. Tapis Streaming APIs. Raw data are streamed from sensors to the Tapis Streams API, potentially first through a gateway node. The time series data are stored in CHORDS while associated metadata are stored in the Tapis Metadata service. Workflows from various domains can process data in real-time or batch.

4.5 Container Registries and Advanced Job Scheduling

One of the chief accomplishments of the Docker platform has been to build a catalog of millions of containerized applications freely available for download from its Docker Hub. Other community efforts such as BioContainers and Singularity Hub have followed in Docker's footsteps to provide registries of containerized software for research computing. The Tapis Apps and Jobs services will capitalize on these efforts so that applications stored in these third-party registries will be available for use through the platform.

Leveraging the computational portability features of containers and the system capabilities and application requirements described in Sect. 4.1, the Jobs service provide an optional "smart scheduling" capability to users interested in minimizing time to solution for a given workload. The smart scheduling option, when enabled for a specific job, will examine the execution systems available to the user whose capabilities meet the requirements of the application. Additionally, the user can provide a whitelist or blacklist of systems to consider or ignore, respectively. For systems meeting the requirements, the Jobs service will initially consider two factors as part of its scheduling decision: the queue time for the execution system and the time to transfer any input data. A mix of real-time data (current queue lengths) and historical data (average transfer rates to the execution system from the storage systems) will be used to estimate the job start time on each system, and the job will be scheduled on the system with the earliest start time. We will consider evolving the scheduling algorithm to use additional information over time.

4.6 Highly Scalable Document Store and Metadata API

A growing number of research projects curate and analyze large collections of highly structured data, with performance requirements regarding time to access, search and move the data. The Tapis Meta API provides a scalable storage solution for structured documents serialized with formats such as JSON or XML. The service aims to support a wide variety of data collections and usage patterns by providing users with a high degree of customization of the backend data store structure. Within a tenant, *collections* can be registered that define their own schema and indexes. Support for geospatial and hashed indexes is included.

In addition to basic create, read, update and delete functionality for documents, the Meta API integrates with the Tapis Security Kernel (see Sect. 4.2) to provide a permissions system over the documents themselves. Like other Tapis objects, Meta documents can be private to a single API user or shared with one or more users. Backing the Meta API is a managed MongoDB cluster running on dedicated hardware with high-speed attached storage. The service scales to support large collections—a research group at UT Southwester Medical School is planning to scale an adaptive immune repertoire sequencing data repository to 10B records as part of the iReceptor Plus project [11], a consortium of institutions funded by EU and the Canadian Institutes of Health (see Sect. 5.2 for more details).

4.7 DevOps Tooling for Automated Platform Management

Tapis is providing automated deployment and system management to simplify the administration of components at a given site. Based on experience developing the TACC Deployer project [40], [19], the Tapis DevOps tools will:

- Deploy and configure the security kernel.
- Deploy and configure individual Tapis services.
- Provide detailed, actionable logging information.
- Monitor the status of system components.
- Provide an audit trail for security-sensitive operations.

Each Tapis service is packaged into a Docker image, and the official deployment tooling targets the open source Kubernetes container orchestration system. Kubernetes provides a number of important features, including: service deployment scheduling onto one or more compute nodes, networking and discovery between services in the same Kubernetes namespace, service configuration management via Kubernetes Configmaps, persistent storage via Kubernetes Persistent Volume Claims (PVCs), and basic service health monitoring and restart policies. Moreover, Kubernetes serves as a common denominator across different datacenters and deployment environments, and is emerging as the dominant container orchestration technology. As an open source project, Kubernetes is easy to install in an academic datacenter, but at the same time, all major commercial cloud platforms, including AWS, Azure and Google Cloud Platform, provide a "Kubernetes-as-a-service" solution. The Tapis project is planning to utilize a small cluster in the Azure cloud for testing our provisioning tools in a sandbox environment.

5 Early Science Adopters

5.1 Real Time Climate Data for the Hawaiian Islands

Climate observations in Hawai'i are critically important to understand past, current, and future climate of the nation and the globe. Despite Hawai'i's importance, national climate data compilations and analyses include only sparse and inadequate coverage of the Hawaiian Islands. As the only US state within the tropics, Hawai'i also provides a model system for studying the functional responses of tropical terrestrial ecosystems, including tropical rain forests, dryland forests, and agriculture, to changes in atmospheric CO2 and climate. Up to date, high quality, reliable climate information in Hawai'i is of great value to the nation.

Tom Giambelluca's research group at the University of Hawaii is working on automated workflows to provide ready access to a spatially comprehensive, high quality, reliable climate data set and data analysis products covering the Hawaiian Islands. The climate researchers will use Tapis to: (1) gather and centralize climate data from all known sources and new monitoring stations using Tapis data streams; (2) update the data archive to maintain near-real-time status with Tapis triggered events for Tapis functions and containerized process workflows; (3) automate basic data screening, homogeneity testing, and gap filling using Tapis functions; (4) automatically update statistical analysis of available data by site and produce daily near-real-time, high-resolution, gridded digital map products and reports of climate variables (e.g., precipitation, temperature, humidity, wind speed, solar radiation) using Tapis containerized workflows scheduled on HPC or cloud resources; (5) share data and products with the wider community through Tapis data management and sharing services.

5.2 iReceptor Plus: Human Immunological Data Storage, Integration and Controlled Sharing

Increasingly, scientists depend on next generation sequencing (NGS) data to understand the immune response in autoimmune and infectious disease as well for developing vaccines. The iReceptor Plus platform provides a science gateway linking to federated, geo-distributed repositories of NGS data conforming to adaptive immune receptor repertoire (AIRR) community standards. Co-funded by the European Union and the Canadian Institutes of Health Research, iReceptor Plus constitutes an international consortium of institutions providing AIRR datasets as well as an API for access.

The Cowell Lab at the University of Texas Southwestern Medical Center is leveraging the Tapis Meta service for storage and management of their AIRR dataset. An initial prototype effort in which 350 million records were created in the Tapis Meta service has yielded promising results. The Cowell group plan to grow the collection to ten billion records during the four year iReceptor Plus project. They are also considering using Tapis for high-throughput analysis jobs at TACC and Compute Canada.

5.3 Planet Texas 2050

The Planet Texas 2050 group wants to leverage spatially dense and ever-increasing Geoscience temporal datasets generated by Arduino-based microcontrollers as ground-truth inputs for integrated models of water-land-atmosphere-urban systems [33]. Deployments collect hydrological and atmospheric measurements to feed models describing various processes related to flooding and aquifer recharge.

Work is already underway to leverage the Tapis Streams service for real-time data from the Adruino-based microcotrollers using the microcontroller API associated with Particle Argon and Xenon mesh series devices. Future work involves hardening this preliminary effort into a production-grade service and scaling up the size of the sensor deployment. Additionally, a workflow will be built to automate the process of converting raw stream data to dataframes for integrated modelling analysis.

5.4 Ocean Wave Fingerprinting and Automatic Wave Forecasting

Oceanic and atmospheric processes leave their imprint, much like a fingerprint, on the ocean surface. These fingerprints can be used to understand air-sea interaction, ocean wave propagation, wave-ice interaction, and small-scale atmospheric weather phenomena. Justin Stopa, UH Ocean Engineering uses high resolution, synthetic aperture radar (SAR) from the Sentinal-1 satellites to study sea surface roughness at very fine spatial resolution (>10 m). SAR provides high-resolution data regardless of environmental conditions and provides information at a global scale capturing approximately 200–400 Gb per platform/day. Manual classification is impractical for the S1 database (120,000 images per month) and machine learning techniques and efficient data workflows are required to make best use of this data. The Stopa team is also developing a real-time wave forecasting system for mitigation of damaging events such as tropical cyclones or large swells, trans-oceanic commerce, and recreation activities. This system requires information from satellites, in-situ sensors, and models.

In-situ sensors include buoy observations that measure wind speeds, wave spectra (energy as a function of frequency and direction), or air-sea temperature difference. Model information includes wave forecast sea state parameters which are usually gridded in time and space. Tapis will be used to support both the classification of satellite images via containerized ML workflows and the automatic wave forecasting workflow. The forecast model is launched when new data is posted by the major weather forecast offices that run global atmospheric and oceanic simulations. Wind speeds, ice concentrations, air-sea temperature differences, and sometimes ocean currents are required forcing fields that drive a spectral wave model such as WAVEWATCH3 (WW3). Regional (e.g. state of Hawaii) and local (e.g. Oahu, island-scale) information, is needed to resolve the wave transformations around the intricate bathymetry of Hawaii (e.g. [36], [28]). WW3 (> V5.0) is extremely efficient at linking and nesting multiple wave grids internally (e.g. global and regional models). The Simulating Waves Nearshore (SWAN) spectral wave model is more efficient in coastal environments (island scales 1–100 km) when the spatial scales become small due to its implicit numerical scheme to solve the governing equation. The spectral wave information and forcing fields must be passed from WW3 to SWAN. Since the wave model data could miss details of the wave field, the observations provide complementary information. Tapis functions can be triggered to pull auxiliary datasets from models, other satellite observations or sensors as additional inputs to the appropriate algorithm based on the earlier classification. The results of these calculations can become secondary Tapis data streams that provide geophysical parameters [38] that will be used to build automated algorithms using Tapis containers and HPC jobs to move the data from the storage systems to the compute system to generate new products such as swell fields by back-propagating swell components (direction and wavenumber) and the forward associated swell components with a storm source (e.g. [21,37]).

6 System Architecture

Tapis capabilities are organized into a set of microservices, where each service provides an independent API over HTTP, designed in a RESTful style. API contracts are codified using OpenAPI v3 definitions, and these definitions are used to generate live docs and initial client libraries in Python and Java. Most services also communicate with one or more databases and/or message queues using direct connections, and with other Tapis services over HTTP. All service requests—both external requests and requests from other Tapis services—are authenticated using a JSON Web Token (JWT). As part of initialization, a Tapis service retrieves a JWT representing itself from the Tapis Tokens API.

Tapis supports deployment of its components to multiple sites or physical locations. The components at these sites "work together" to comprise a single distributed Tapis installation. Each distributed installation will have a single *primary site* and zero or more *associate sites*. The primary site has the following roles and responsibilities:

- The primary site runs one or more instances of every (primary) Tapis service.
- The primary site runs a copy of each database needed by the primary Tapis services. The Tapis services running in the primary site communicate with these databases and not any other databases running at external associate sites.
- The primary site runs the Tenants API (see description of multi-tenancy below); no other site in the installation runs the Tenants API.

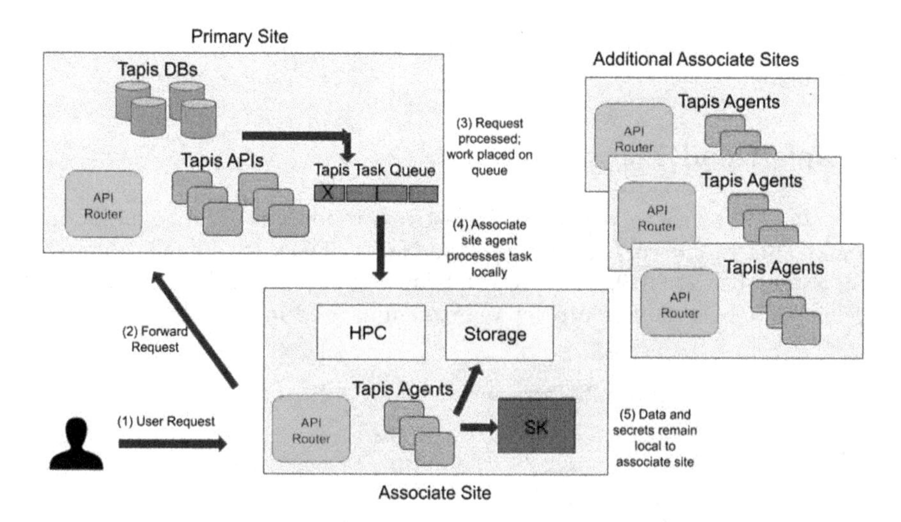

Fig. 3. Multi-Site Tapis Deployment. Each site represents a physical datacenter. (1) A user's request is first resolved to a site based on the *base URL*. Based on configuration, the site's API Router routes the request to the Primary Site (2) which processes the request and puts work on a shared Task Queue (3). An agent at the Associate Site takes the work unit (4) and consults its own Security Kernel while processing the task (5).

In addition to one or more Tapis services, each site runs a Tapis API Router (or "edge router") component. The Tapis API Router is responsible for forwarding requests to specific Tapis microservices.

Tapis also supports multi-tenancy: a *tenant* is a logical separation of Tapis objects for a specific project or group. Tapis objects (e.g. systems, apps, jobs, streams, etc.) in one tenant cannot be accessed from another tenant. Tenants have the following properties:

1. A universe of user accounts, ultimately linked to an LDAP query or some other identity store.
2. A base URL for all (public) API requests (e.g., https://api.sd2e.org).
3. A base URL for the location of any services running at the primary site (e.g., https://sd2e.api.tapis.io).

Tenants and associate sites are related through a hub and spoke model (see Fig. 3) based on which site is hosting the tenant's base URL. There are two cases: If the primary site hosts the tenant's base URL, then all Tapis services for that tenant are hosted at the primary site. Alternatively, an associate site can host the tenant's base URL. In this case, the associate site's API Router is responsible for routing service requests either to a service running locally at the associate site or to a service running at the primary site. To route to the primary site, the associate site configures its API Router to route requests to the tenant's primary site base URL, defined in 3) above. Thus, all services for a given tenant are hosted either exclusively at the primary site, or between the primary site and one associate site. No cross-service requests or forwarding between associate sites occurs.

7 Tapis Road Map

Tapis is funded by a five year Cyberinfrastructure for Sustained Scientific Innovation (CSSI) Framework grant from the National Science Foundation, which began September 1, 2019. The project team has defined specific milestones to be delivered across the five project years, summarized in the following Table 2:

Table 2. Tapis annual project milestones

Year	Real-time capabilities	Batch capabilities	Security and authorization
Year 1	Production release of streaming API	System, Application and Job abstractions	Security Kernel initial release
Year 2	Data management and archiving features	Support for third-party registries	Automated devops tools for security kernel
Year 3	Support for Alerts as Events	Gateway in a Box release	Local caching solution for security kernel
Year 4	Support for Abaco functions	Smart Jobs scheduling, initial release	Automated devops tools for hybrid and HA deployment
Year 5	Support for scheduled relay streams	Expanding support and features for existing services	Devops for monitoring and log aggregation

Additionally, an Early Adopters workshop is planned for the end of the first project year, targeting the July, 2020 time frame. During the workshop, Tapis core team members will present talks on the system capabilities developed, and early adopters will be invited to present their use cases, data sets, and adoption of the platform to date.

Acknowledgment. This material is based upon work supported by the National Science Foundation Office of Advanced CyberInfrastructure, grant numbers 1931439 and 1931575.

References

1. Json Web Token (JWT) (2018). https://tools.ietf.org/html/rfc7519. Accessed 17 Mar 2018
2. Amazon Key Management Service (2019). https://aws.amazon.com/kms/. Accessed 30 Oct 2019

3. Apache Airavata (2019). https://airavata.apache.org/index.html. Accessed 30 Oct 2019
4. Apache Shiro (2019). http://shiro.apache.org/. Accessed 30 Oct 2019
5. Dataturbine (2019). http://dataturbine.org/. Accessed 30 Oct 2019
6. Fortress (2019). https://directory.apache.org/fortress/. Accessed 30 Oct 2019
7. Galaxy community hub (2019). https://galaxyproject.org/. Accessed 30 Oct 2019
8. Globus (2019). https://www.globus.org/. Accessed 30 Oct 2019
9. HashiCorp Vault (2019). https://www.vaultproject.io/. Accessed 30 Oct 2019
10. HubZero (2019). https://hubzero.org/. Accessed 30 Oct 2019
11. iReceptor Plus (2019). https://www.ireceptor-plus.com/. Accessed 30 Oct 2019
12. Kerberos (2019). https://web.mit.edu/Kerberos/. Accessed 30 Oct 2019
13. Microsoft Active Directory (2019). https://azure.microsoft.com/en-us/services/active-directory. Accessed 30 Oct 2019
14. OAuth2 (2019). https://oauth.net/2/. Accessed 30 Oct 2019
15. PERMIS (2019). http://www.openpermis.info/. Accessed 30 Oct 2019
16. Sciserver (2019). http://www.sciserver.org/. Accessed 30 Oct 2019
17. VDJServer (2019). http://vdjserver.org. Accessed 30 Oct 2019
18. Who's Using AWS (2020). https://www.contino.io/insights/whos-using-aws. Accessed 08 May 2020
19. Cleveland, S.B., et al.: Building science gateway infrastructure in the middle of the pacific and beyond: experiences using the Agave Deployer and Agave platform to build science gateways. In: Proceedings of the Practice and Experience on Advanced Research Computing. PEARC 2018 (2018a)
20. Cleveland, S.B., et al.: The 'Ike Wai Gateway - A science gateway for the water future of Hawai'i. In: Proceedings of Science Gateways 2018, Austin TX, USA, September 2018. Science Gateways Community Institute (2018b)
21. Collard, F., Ardhuin, F., Chapron, B.: Monitoring and analysis of ocean swell fields from space: new methods for routine observations. JGR-Oceans **114**(C7) (2009). https://doi.org/10.1029/2008jc005215
22. Deelman, E., et al.: Pegasus: a workflow management system for science automation. Future Gener. Comput. Syst. **46**, 17–35 (2015). http://pegasus.isi.edu/publications/2014/2014-fgcs-deelman.pdf
23. Dooley, R., et al.: Software-as-a-Service: the iPlant foundation API. In: 5th IEEE Workshop on Many-Task Computing on Grids and Supercomputers (MTAGS). IEEE (2012)
24. Goff, S.A., Vaughn, M., McKay, S., Lyons, E., Stapleton, A.E., Gessler, D., Matasci, N., Wang, L., Hanlon, M., Lenards, A., et al.: The iPlant collaborative: cyberinfrastructure for plant biology. Front. Plant Sci. **2**, 34 (2011)
25. Gottdank, T.: Introduction to the WS-PGRADE/gUSE science gateway framework, pp. 19–32. Springer, Cham (2014). https://doi.org/10.1007/978-3-319-11268-8_2
26. Kerkez, B., et al.: Cloud hosted real-time data services for the Geosciences (CHORDS). Geosci. Data J., 2–4 (2016)
27. Kurtzer, G.M., Sochat, V., Bauer, M.W.: Singularity: scientific containers for mobility of compute. PloS One **12**(5), e0177,459 (2017)
28. Li, N., Cheung, K., Stopa, J., Hsiao, F., Chen, Y.L., Vega, L., Cross, P.: Thirty-four years of Hawaii wave hindcast from downscaling of climate forecast system reanalysis. Ocean Model. **100**, 78–95 (2016). https://doi.org/10.1016/j.ocemod.2016.02.001

29. Litvina, E., Adams, A., Barth, A., Bruchez, M., Carson, J., Chung, J., Dupre, K., Frank, L., Gates, K., Harris, K., Joo, H.: BRAIN initiative: cutting-edge tools and resources for the community. J. Neurosci. **39**(42), 8275–84 (2019). https://doi.org/10.1523/JNEUROSCI.1169-19.2019
30. Merchant, N., Lyons, E., Goff, S., Vaughn, M., Ware, D., Micklos, D., Antin, P.: The iPlant collaborative: cyberinfrastructure for enabling data to discovery for the life sciences. PLoS Biol. **14**(1), e1002,342 (2016a)
31. Merchant, N., et al.: The iPlant collaborative: cyberinfrastructure for enabling data to discovery for the life sciences. PLOS Biol. (2016b). https://doi.org/10.1371/journal.pbio.1002342
32. Padhy, S., Jansen, G., Alameda, J., Black, E., Diesendruck, L., Dietze, M., Kumar, P., Kooper, R., Lee, J., Liu, R., et al.: Brown dog: leveraging everything towards autocuration. In: 2015 IEEE International Conference on Big Data (Big Data), pp. 493–500. IEEE (2015)
33. Powell, J., Stubbs, J., Cleveland, S., Pierce, S., Daniels, M.: Streamed data via cloud-hosted real-time data services for the geosciences as an ingestion interface into the Planet Texas science gateway and integrated modeling platform. In: Proceedings of Science Gateways, San Diego, CA, USA, September 2019, p. 2019. Science Gateways Community Institute (2019)
34. Proctor, W.C., Packard, M., Jamthe, A., Cardone, R., Stubbs, J.: Virtualizing the Stampede2 supercomputer with applications to HPC in the In: Proceedings of the Practice and Experience on Advanced Research Computing (2018). https://doi.org/10.1145/3219104.3219131
35. Rathje, E.M., Dawson, C., Padgett, J.E., Pinelli, J.P., Stanzione, D., Adair, A., Arduino, P., Brandenberg, S.J., Cockerill, T., Dey, C., et al.: DesignSafe: new cyberinfrastructure for natural hazards engineering. Nat. Hazards Rev. **18**(3), 06017,001 (2017)
36. Stopa, J., Cheung, K.F., Chen, Y.L.: Assessment of wave energy resources in Hawaii. Renew. Energy **36**(2), 554–567 (2011). https://doi.org/10.1016/j.renene.07.014
37. Stopa, J., Ardhuin, F., Husson, R., Jiang, H., Chapron, B., Collard, F.: Swell dissipation from 10 years of Envisat ASAR in wave mode. GRL (2016). https://doi.org/10.1002/2015GL067566
38. Stopa, J.E., Mouche, A.: Significant wave heights from Sentinel-1 SAR: validation and applications. J. Geophys. Res. Oceans **122**, 1827–1848 (2017). https://doi.org/10.1002/2016JC012364
39. Stubbs, J., et al.: Rapid development of scalable, distributed computation with Abaco. In: 10th International Workshop on Science Gateways. Science Gateways Community Institute (2018a)
40. Stubbs, J., et al.: TACC's Cloud Deployer: automating the management of distributed software systems. In: The 2nd Industry/University Joint International Workshop on Data Center Automation, Analytics, and Control (DAAC). Supercomputing (2018b). https://drive.google.com/file/d/1oORwQdQEWTHLpARVJPzQqR_0OY5SOrfg/view
41. Wilkins-Diehr, N., Zentner, M., Pierce, M., Dahan, M., Lawrence, K., Hayden, L., Mullinix, N.: The science gateways community institute at two years. In: Proceedings of the Practice and Experience on Advanced Research Computing, p. 53. ACM (2018)
42. Wolstencroft, K., et al.: The Taverna workflow suite: designing and executing workflows of web services on the desktop, web or in the cloud. Nucleic Acids Res. **41**, W557–W561 (2013)

Towards the Evaluation of the Performance Efficiency of Fog Computing Applications

Wilson Valdez[1(✉)], Priscila Cedillo[1(✉)], Kevin Chávez-Z[1(✉)], Sebastián Espinoza-A[1(✉)], Ana Barzallo[1(✉)], and Lenin Erazo[2(✉)]

[1] Computer Science Department, University of Cuenca, Cuenca, Ecuador
{wilson.valdezs,priscila.cedillo,kevin.chavezz, sebastian.espinoza,ana.barzallo}@ucuenca.edu.ec

[2] Laboratory for Research and Development in Computer Science – LIDI, University of Azuay, Cuenca, Ecuador
lerazo@uazuay.edu.ec

Abstract. Nowadays, Cloud Computing represents an effective solution for organizations that need to adopt a model of elastic computing services and pay-as-you-go. However, in new domains, such as Cyber-Physical Systems and the Internet of Things (IoT), which manage large volumes of data, the time of latency in communication between the Cloud server and users, as well as data processing costs, can be a limitation. Therefore, it is proposed the Fog Computing architecture, which is used as a mediator among IoT devices that generate and demand data (e.g., cellphones, video cameras, sensors) and Cloud infrastructure. The purpose of Fog Computing is to relieve the Cloud server load and to delegate it to intermediate devices. In this context, it is essential to measure the performance to evaluate the solution and exploit all the advantages provided by Fog nodes. This paper presents a performance efficiency quality model for Fog Computing applications. The model breaks down in sub-characteristics and attributes with defined metrics and thresholds that should be used to evaluate these kinds of applications to understand how efficient is the use of computational resources (i.e., time, processing, memory). Also, it is offered an evaluation method that is supported by the quality model to assess the software development process for Fog Computing applications. The evaluation method is aligned with the ISO/IEC 25040 standard and indicates the developer steps towards a good quality Fog Computing solution.

Keywords: Fog Computing · Evaluation · Performance efficiency · Measurement · Metric · Quality model

1 Introduction

Internet of Things (IoT) is conceptualized as a new paradigm in which every device is digitally connected, regardless of their function, and can communicate with other devices and people over communication protocols [1]. According to the previous definition, IoT refers to all electronic devices identified uniquely, which use the Internet to send data [2]. Therefore, the primary purpose of IoT is to empower the connected things or devices

K. Arai (Ed.): FICC 2021, AISC 1363, pp. 901–916, 2021.
https://doi.org/10.1007/978-3-030-73100-7_62

(e.g., sensors, mobile devices, security cameras) with new capabilities and to provide infrastructure, communications, interfaces, protocols, standards, and data management to process and analyze the obtained information [3].

In this sense, IoT and Cloud Computing are closely related, since all data produced by IoT devices require to have virtual resources to process and store data, and it is possible by integrating it into the Cloud [4]. Therefore, the elasticity property provided by the Cloud provider may be useful for increasing or decreasing computing capacity according to the demand of the IoT architecture. However, it is essential to consider the geographic distribution of the servers, as it can lead to excessive bandwidth consumption and high latency times between the server and IoT devices. Consequently, Cloud Computing may be incompatible with Cyber-Physical Systems and IoT architectures due to their large volumes of data, and thus, the challenge created by the time of latency in communication between the Cloud and users, and data-processing costs [5]. To aboard this issue has emerged the Fog Computing paradigm.

Fog Computing is defined by Vaquero et al. [6], as: "a scenario where a huge number of heterogeneous (wireless and sometimes autonomous) ubiquitous and decentralized devices communicate and potentially cooperate among them and with the network to perform storage and processing tasks without the intervention of third parties." Thus, Fog Computing is a middle layer between the IoT devices and Cloud Computing service that provides data processing through devices called Fog nodes. These hardware devices are found or included in the IoT end-nodes. Indeed, Fog nodes are devices with limited resources such as routers, set-top boxes, switches, embedded servers, and industrial controllers, or high-performance machines as the called Cloudlets. These Cloudlets are data centers at a small scale that provide data processing with short latency times [7].

The main reason for this paper comes at the absence of models and methods aligned to quality standards, which provide a systematic way to assess the performance efficiency of Fog Computing applications. Furthermore, Fog Computing is a trending architecture with several studies about its performance; thus, this paper is a contribution for researchers and development organizations in this Computing paradigm. Hence, this paper presents a performance efficiency quality model for Fog Computing applications.

The presented model is divided into sub-characteristics and attributes with defined metrics and thresholds that should be used to evaluate these kinds of applications to understand how efficient is the use of computational resources (i.e., time, processing, and memory). This paper presents two contributions: on the one hand, it proposes a performance and efficiency quality model, and on the other hand, it provides a method for the evaluation of the sub-characteristics of quality included in the model. These contributions are aligned with standards ISO/IEC 25010 [8] and ISO/IEC 25040 [8], respectively. Moreover, the presented solution is focused on supporting the improvement of the performance efficiency quality during the software development process of Fog Computing applications.

This paper is organized as follows. Section 2 describes the related work. Sect. 3 provides the quality model of the efficiency of Fog Computing applications. Section 4 provides the evaluation method for Fog Computing applications using the quality model of the previous Section. Finally, Sect. 5 presents conclusions and further work.

2 Related Work

Nowadays, there are not software quality models, neither methods that allow the evaluation of the performance efficiency for Fog Computing applications, which are aligned with a quality standard such as ISO/IEC 25000, and that enables the assessment of the quality of software products in this kind of architectures. Nevertheless, there are some previous studies about Quality of Experience (QoE) and the time behavior measurements that have been considered to carry out this aim.

Some studies about QoE to improve performance in Fog Computing applications are described. First, the research presented by Mahmud et al. [9] proposes a QoE-aware application placement policy by giving priority to different requests of placement. Those requests, according to user expectations in a Fog Computing environment to suitable Fog instances so that user QoE is maximized in respect of utility access, resource consumption, and service delivery. The study uses two separate fuzzy logic models to simplify the mapping of applications to compatible instances by calculating the application's Rating of Expectations from the users and the capacity class score of the instances. Finally, a linear optimization problem ensures the best convergence between user expectations and scope within the Fog environment that maximizes the QoE.

Moreover, the study presented by Xiao et al. [10] investigates two performance metrics for Fog Computing networks: (i) user QoE, and (ii) the power efficiency in Fog nodes. These are useful to affront the workload offloading problem for Fog Computing networks. A Fog node cooperation strategy referred to as the offload forwarding is used to distribute the workload correctly. In this strategy, each Fog node can forward part or all of its offloaded workload to other local Fog nodes to further improve the performance and QoE of users. However, both presented studies are not aligned to any standard to bring performance efficiency and quality to the application.

On the other hand, the research addressed by Zeng et al. [11] about the optimization of task scheduling in Fog Computing architectures offers algorithms to optimize their performance. Through the execution of these algorithms in a simulator, researchers present statistics about the efficiency improvement in the response time. Therefore, that study becomes a method to evaluate the performance efficiency of a Fog Computing application by using a set of metrics. However, it is not based on any standard for software product quality.

Sarkar et al. [12] proposes another study about Fog Computing efficiency. In that paper, the researchers assess the suitability of Fog Computing in the IoT context. One of the characteristics evaluated in that study is the service latency (transmission and processing time), which is analyzed in a simulation environment through a comparison between a Cloud server and a Fog node. As a result of the study, they concluded that the latency time in Fog Computing is better than in Cloud Computing when there is a high percentage of IoT applications that demand real-time services. This result offers another reason for the importance of assessing a Fog Computing application using a quality model and an evaluation method.

On the other hand, In [13] is presented a discussion about a load balancing based smart gateway with Fog and Cloud environment where analysis of load balancing environment with regards to processing delay and network delay is performed. The authors have compared different scenarios with and without a load balancing algorithm in the projected architecture to evaluate the performance of the smart gateway in a Fog Computing application. As a result, in a few cases analyzed, an intelligent gateway can be used as a smart load balancer. Still, in the application, the smart gateway doesn't need to trim and preprocess data every time. However, the analysis presented is focused on a specific component in the network, and it is not aligned with a quality standard to evaluate the performance.

Moreover, the study presented in [14] shows a queuing model to predict the minimum required number of computing resources (both Fog and Cloud nodes) to meet the Service Level Agreement (SLA) for response time. The analytical model is validated through a discrete event simulation. The results of the analysis of the model show that it can correctly and effectively predict the number of computing resources needed for health data services to achieve the required response time under different workload conditions. However, beyond successful results, these performance models have not been aligned to a quality standard, which is a gap in the presented study.

In summary, the related studies presented, although they have shown solutions to evaluate the Quality of Service (QoS), Quality of Experience (QoE), or efficiency, are not aligned within any regulation or standard. Thus, they do not provide technical support that guarantees that an analyzed product and service have the desired quality.

3 Quality Model of the Performance Efficiency of Fog Computing

Aligning models to standards always brings essential benefits: they allow minimizing costs, avoiding errors, have consistent measurement criteria and indicators, being more competitive, and increasing productivity. They are also relevant to the customer: it increases their satisfaction, guarantees product quality, reliability, and safety [15–17].

Therefore, this paper proposes to make use of ISO/IEC 25000 quality standard family [18] to carry out quality analysis of (i) the performance efficiency of Fog Computing applications by following the ISO/IEC 25010 and (ii) an evaluation method of Fog Computing applications aligned by following the ISO/IEC 25040. This standard is a validated model that provides step-by-step guidelines for creating the performance and evaluation method of the Fog Computing applications, conceived as the software product to be evaluated.

Then, here is defined as a hierarchical quality model composed of a set of sub-characteristics and attributes relevant to the Fog Computing applications domain. Also, several metrics have been included from third part studies for the corresponding quality measurement. This model is based on the characteristics and sub-characteristics proposed by the quality standard ISO/IEC 25010 [8], which allows assessing the properties of a software product. The standard is made up of eight characteristics, which have been selected only the *performance efficiency* characteristic since Fog Computing focuses on the efficient use of resources.

3.1 Time Behavior Sub-characteristic

The first sub-characteristic of the quality model for Fog Computing applications is the *time behavior*; its purpose is to measure the degree with which the communication time between the IoT device and the Fog node fulfills the requirements. This sub-characteristic is composed of four attributes:

1) *Request submission from the IoT device:* What is the request sending time from the IoT device to the Fog node?
2) *The response of the request from the Fog node:* What is the time it takes for the Fog node to process the request and send the answer to the IoT device?
3) *Communication latency between the IoT device and the Fog node:* What is the sum of the request sending time plus the response time between the IoT device and the Fog node?
4) *The simultaneity of requests processed by the Fog node:* How many requests are accepted at the same time?

Table 1 presents the metrics designed for the attributes of the *Time Behavior* sub-characteristic. The evaluation thresholds are based on task completion time proposed by

Table 1. Metrics of the attributes of time behavior.

Attribute	Metric/Thresholds
Request submission from the IoT device	**RQST** = Request sending time (data and specification of the operations to be performed) from the IoT device to the Fog node **Unit:** milliseconds (ms) **Excellent:** less than 50 (ms) **Good:** between 50–120 (ms) **Bad:** greater than 120 (ms)
Response of the request from the Fog node	**RPT** = Request processing time **RPST** = Response sending time to the IoT device from the Fog node **TRT** = Total response time **TRT = RPT + RPST** **Unit:** milliseconds (ms) **Excellent:** less than 1050 (ms) **Good:** between 1050–2550 (ms) **Bad:** greater than 2050 (ms)
Communication latency between the IoT device and the Fog Computing node	**RQST** = Request sending time from the IoT device to the Fog node **TRT** = Total response time **TCT** = Total communication time ***TCT = RQST*** + **TRT** **Unit:** milliseconds (ms) **Excellent:** less than 1600 (ms) **Good:** between 1600–2720 (ms) **Bad:** greater than 2720 (ms)
Simultaneity of requests processed by the Fog node	Probing of the number of average requests that the Fog node process in a certain time-slot **Unit:** number of average requests per minute **Excellent:** between 5–10 average requests per minute **Good:** between 2–5 average requests per minute **Bad:** less than two average requests per minute

the research work of Zeng et al. [11] about the experiments of task scheduling in Fog Computing.

3.2 Resource Utilization Sub-characteristics

The second sub-characteristic of the quality model for Fog Computing applications is the Resource Utilization, whose purpose is to measure the degree to which the amounts of resources (CPU) used by the Fog node when performing its functions meet with the preset requirements. This sub-characteristic is composed of three attributes:

1) *Performance of the CPU processing in the Fog nodes:* What is the efficiency degree of the Fog nodes CPU to process requests from IoT devices?
2) *Performance of the search algorithms in the Fog node:* What is the performance offered by each of the search algorithms in the Fog node?
3) *Performance of the server when receiving requests from the IoT device:* What is the level of performance provided by the server of a Fog node at the moment of receiving the requests?

Table 2 presents the metrics designed for the attributes of the *resource utilization* sub-characteristic. These metrics are based on the formulas proposed in [19].

Table 2. Metrics of the attributes of resource utilization.

Attribute	Metric/Thresholds
Performance of the CPU processing in the Fog nodes	**CPUCN** = CPU cores number **CPUPC** = CPU performance coefficient **CPUP** = CPU performance **CPUP = CPUCN * CPUPC * 100** **Unit:** percent (%) **Excellent:** less than 50% **Good:** between 50–120% **Bad:** greater than 120%
Performance of the search algorithms in the Fog node	**DCT** = Data collection time **ST** = Synchronization time between the IoT devices and the Fog node **TCT** = Total communication time **TCT = DCT + ST** **Unit: milliseconds (ms)** **Excellent:** less than 1050 (ms) **Good:** between 1050–2550 (ms) **Bad:** greater than 2550 (ms)
Performance of the server when receiving requests from the IoT device and the Fog Computing node	**NC** = Number of confirmations per second **NNR** = Number of new requests **NA** = Number of actions per minute **PSRR** = Performance of the server when receiving requests from the IoT device **PSRR = NC + NNR + NA** **Unit:** requests number **Excellent:** 50–80 requests **Good:** between 30–50 requests **Bad:** less than 30 requests

3.3 Capacity Sub-characteristics

The third sub-characteristic of the quality model for Fog Computing applications is the *capacity*, whose purpose is to measure the concurrence capacity in the Fog node. This sub-characteristic is composed of four attributes:

1) *Performance of the Fog node data transfer:* What is the data transfer time from the IoT device to the Fog node?
2) *The Number of requests to the Fog node:* How many requests are made to the Fog node when it is running or waiting?
3) *Fog node success rate:* What is the success degree of the user when performing tasks with the operations available by the Fog node?

Table 3 presents the metrics designed for the attributes of the *capacity* sub-characteristic. These metrics are based on the formulas proposed in [20–22].

4 Evaluation Method for Fog Computing

This section provides information about the methodology to be followed for using the quality model above mentioned and assess a Fog Computing application. The method is aligned with the evaluation reference guide of the standard ISO/IEC 25040. This standard provides three types of evaluation: process for developers, acquirers, and independent assessment. The procedure chosen is for developers because the selected characteristic *performance efficiency* has more relevance within the software development process.

Table 3. Metrics of the attributes of capacity.

Attribute	Metric/Thresholds
Performance of the Fog node data transfer	**DS** = Data size **BW** = Bandwidth **TT** = Data transfer time $TT = DS/BW$ **Unit:** milliseconds (ms) **Excellent:** less than 50 (ms) **Good:** between 50–120 (ms) **Bad:** greater than 120 (ms)
Number of requests to the Fog node	**NR** = Count of the number of incoming requests to the Fog node **Unit:** requests number **Excellent:** 3000 or greater to 5000 requests **Good:** between 1000 and 3000 requests. Bad: less than 1000 requests
Fog node success rate	**SR** = Success rate **NCT** = Number of completed tasks **NHCT** = Number of halfway complete tasks **NUT** = Number of uncompleted tasks **TNT** = Total number of tasks $SR = \frac{(NCT + NHCT * 0{,}5 + NUT * 0)}{TNT} * 100\%$ **Unit:** percent (%) **Excellent:** between 80%–100% tasks **Good:** between 30%–80% tasks **Bad:** less than 30% of tasks

4.1 Evaluation Requirements Specification

This phase allows getting to know the way the quality requisites, software product components, and the strictness with which the developers must perform the evaluation.

1) *Software product requirements:* The software requirements, considering the internal quality of the Product, are as follows:

   - The processing algorithms must be practical with the lower time consumption in the execution.
   - The response times must be lower at the moment of communicating with other systems.
   - The source code must be well documented to understand the complexity of the algorithm.
   - The communications with other systems for the data exchange must not consume too many computational resources.
   - The system must allow concurrent access and parallel transaction execution.

2) *Software product components:* The software components that are part of the evaluation of the quality are listed below:

   - The source code (algorithms).
   - Communication interfaces.
   - The software product (running on the Fog node).
   - The technical documentation.

3) *Definition of the Evaluation Strictness.* The chosen sub-characteristics define the evaluation strictness due they represent the levels to consider, which are:

   - Strictness according to the algorithm execution time
   - Strictness according to algorithm complexity in the source code.
   - Strictness according to the capacity tests over the hardware.

4.2 Specification of the Evaluation

In this phase, the characteristics, sub-characteristics, attributes, metrics, and thresholds to be used by the developers are specified.

1) *Selection of sub-characteristics and attributes:* The sub-characteristic attributes to be used are specified in Sect. 3.
2) *Selection of the quality metrics:* The quality metrics to be used are defined in Tables 1, 2, and 3.
3) *Definition of the decision criteria for the quality metrics:* The decision criteria for quality metrics to be used are specified by the thresholds shown in Tables 1, 2, and 3.
4) *Specification of the decision criteria for the Evaluation:* The decision criteria for the evaluation are shown in Table 4.

Table 4. Attributes and sub-characteristics qualification.

Criteria	Description	Procedure
Attributes qualification	The attributes should be qualified according to the range achieved by these	The qualifications of the attributes are: *1:* If the range is Excellent *0,5:* If the range is Good *0:* If the range is Bad
Sub-characteristic qualification	The qualification is classified in ranges according to the sub- characteristic score The score of the sub-characteristic is equal to the sum of the qualifications of all its attributes	The ranges considered for the qualifications are: *Excellent:* If the summed the score is greater than 2,5 *Good:* If the summed score is between 1,5 - 2,5 *Bad:* If the summed score is less than 1,5

4.3 Evaluation Design

This phase specifies the activities to be performed to evaluate the software product as a developer. Also, there are indicated the artifacts to be evaluated, constraints, and the order to be assessed. Thus, to better present this section, it is followed a Fog Computing application in the healthcare domain (see Fig. 1).

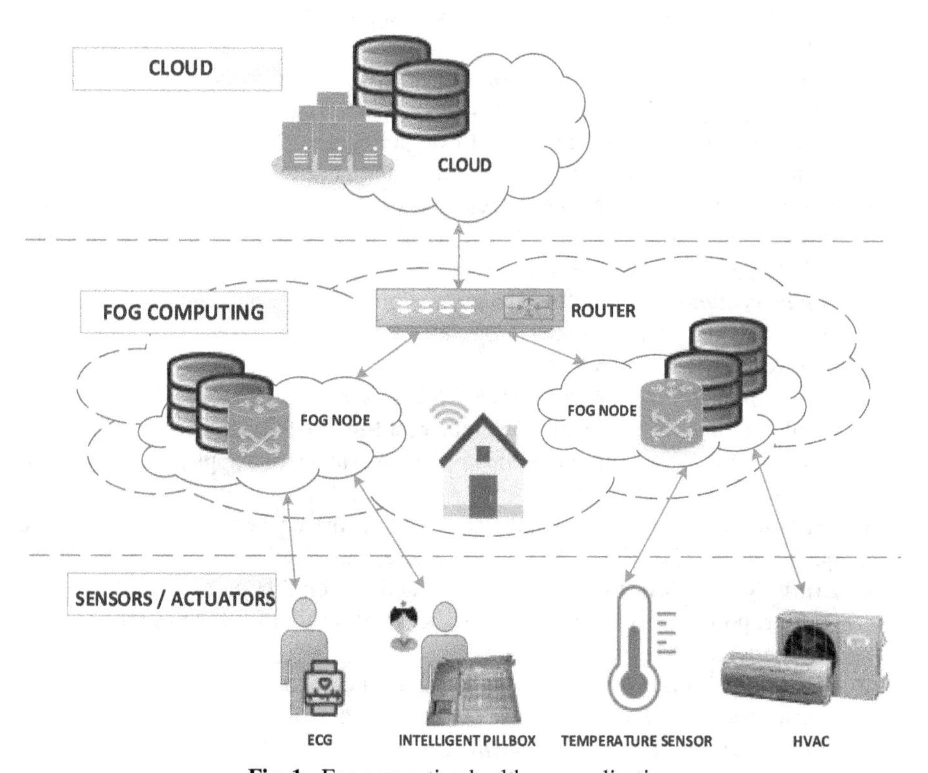

Fig. 1. Fog computing healthcare application.

Figure 1 presents a Fog Computing solution for healthcare where a patient is monitored. The application is divided into three sections:

- *Sensors actuators or Edge layer.* The first layer represents the edge of the application, which presents electrocardiogram (ECG) monitoring and a medication tracking supported by an Intelligent Pillbox [23, 24]. Moreover, a temperature sensor and a fan to keep the room temperature.
- *Fog Computing layer.* It contains two Fog nodes that collect, process, store data. Then, significant data is sent to the Cloud.
- *Cloud Computing layer.* It represents the Cloud services (i.e., SaaS, IaaS, PaaS) of the healthcare application.

Then, in the lines below is described the evaluation of the performance of this application.

1) *Planning of the evaluation activities:* The plan to achieve a correct evaluation has the following activities:

   - ***Evaluation purpose.*** The objective is to evaluate the performance of the Fog Computing application presented in Fig. 1, determine the time, CPU resources used, and capacity of this application as determined in the quality model shown in Sect. 3.
   - ***Organization involved in the evaluation (developer and evaluator enterprise).*** In this case, it could be considered as a developer, a consulting firm with expertise in IoT, Fog Computing, and Cloud Computing solutions, and as an evaluator, another consulting firm confirmed by the authors of this paper.
   - ***The expected result by the end of the assessment.*** As an application in the healthcare domain, the desired results are mainly related to better time performance and efficient use of CPU resources.
   - ***Stages evaluation schedule.*** The evaluation will be scheduled in three stages aligned to the quality model proposed in Sect. 3 to analyze the time, CPU resources used, and capacity.
   - ***The responsibilities of the involved parts in the assessment***. The developer should provide the application with the best functionality and performance possible. On the other hand, the evaluator has to evaluate the application aligned to the proposed quality model.
   - ***Evaluation scope.*** The evaluation should cover the whole parts presented in the quality model (i.e., time, CPU resources use, capacity).
   - ***Evaluation tools to use.*** Tables 1, 2, 3, and 4 should be used to evaluate each item proposed. Besides, in this case, the iFogSim tool [25] is used to simulate the operation of the application.
   - ***Metrics decision criteria.*** The evaluation of the application has to consider the metrics shown in Tables 1, 2, 3, and 4 of the quality models.
   - ***Decision criteria for the assessment quality product.*** Tables 1, 2, 3, and 4 present the levels to determine the quality of the product. In this case, those criteria determine the final decision.

- ***Adopted standards.*** The standard which supported the quality model presented in Sect. 3, the ISO/IEC 25010, and the ISO/IEC 25040.

2) *Constraints definition:* The constraints that could restrict the evaluation are:

- ***The evaluation budgets.*** It depends on the tools needed and the professionals evaluating the application regarding to the evaluation case.
- ***The tools regarded for the evaluation.*** The tools required should be selected according to the size and type of application. In this case, to evaluate the example solution, it is used the iFogSim free software tool.
- ***The evaluation methods and standards.*** The evaluation methods are presented in Sect. 3 aligned to ISO/IEC 25010 standard to determine the quality of the software product.
- ***The reach and time defined in the planning.*** It is determined to cover the three components which compose the quality model of Sect. 3.
- ***The evaluation activities as the milestones, also including both the human and material resources.*** Each activity proposed in the quality model of Sect. 3 contains the tables, materials, and the domain expert evaluating the software product.

3) *Implementation of the evaluation:* This phase specifies the steps to execute the evaluation. In order to better explain the steps, the Fog Computing healthcare application has been simulated in the iFogSim tool based on the network topology (see Fig. 2). The steps of this phase are as follows:

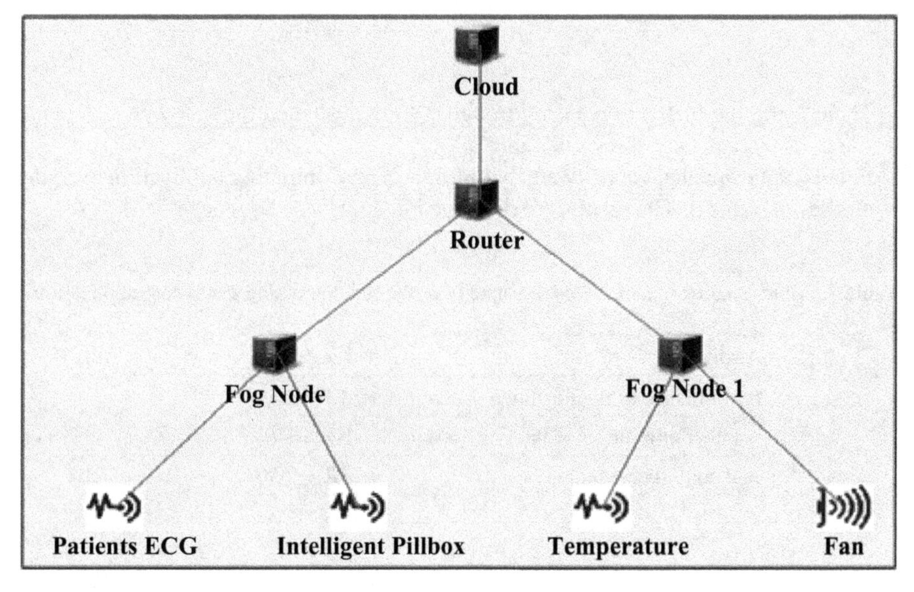

Fig. 2. Fog computing healthcare application network topology in iFogSim.

- ***Measurement execution:*** The measures should be carried out according to the quality metrics proposed in Tables 1, 2, and 3.

First, are evaluated the metrics of the attributes of time behavior by following the presented in Table 1. The results obtained are presented in Table 5.

Table 5. Metrics of the attributes of time behavior of the fog computing healthcare application.

Attribute	Metric
Request submission from the IoT device	**RQST** = *55* ms
Response of the request from the Fog node	**RPT** = *1200* ms
Communication latency between the IoT device and the Fog Node	**TCT** = *1500* ms
Simultaneity of requests processed by the Fog node	***Average:****5 requests/min*

Table 6. Metrics of the resource utilization attributes of the fog computing healthcare application.

Attribute	Metric
Performance of the CPU processing in the Fog node	**CPUP** = *59%*
Performance of the search algorithms in the Fog node	**TCT** = *1100* ms
Performance of the server when receiving requests from the IoT device and the Fog node	**PSRR** = *45 requests*

Then, using the metrics presented in Table 4 are evaluated the resource utilization attributes. The results are shown in Table 6.

Finally, are evaluated capacity attributes of the Fog Computing application by using the metrics of Table 3. The results are displayed in Table 7.

Table 7. Metrics about the attributes of capacity of the fog computing healthcare application.

Attribute	Metric
Performance of the Fog node data transfer	**TT** = *45* ms
Number of requests to the Fog node	**NR** = *4000 request*
Fog node success rate	**SR** = *68%*

- ***Use of the decision criteria for measurements:*** The decision criteria should be carried out according to the levels purposed in Tables 1, 2, and 3.

Based on the measurement execution results, Table 8 summarizes the decision criteria.

Table 8. Decision criteria from measurement execution of fog computing healthcare application.

Time behavior attribute	Decision criteria
Request submission from the IoT device	*Good*
Response of the request from the Fog node	*Good*
Communication latency between the IoT device and the Fog node	*Excellent*
Simultaneity of requests processed by the Fog node	*Good*
Resource utilization attributes	**Decision criteria**
Performance of the CPU processing in the Fog node	*Good*
Performance of the search algorithms in the Fog node	*Good*
Performance of the server when receiving requests from the IoT device and the Fog node	*Good*
Capacity attributes	**Decision criteria**
Performance of the Fog node data transfer	*Excellent*
Number of requests to the Fog node	*Excellent*
Fog node success rate	*Good*

- ***Use of the evaluation decision criteria:*** The decision criteria for the evaluation should be carried out according to Table 4. About the Fog Computing healthcare application, this evaluation is presented in Table 9.

The output of this phase is a report with the performance efficiency problems of the Fog Computing application.

4.4 Finishing the Evaluation

This phase concludes the evaluation through a report that provides the results on the quality achieved by the software product (Fog Computing application). Therefore, the report should be written so everybody involved can understand it.

1) *Review of the evaluation results:* The people involved in the evaluation should write a detailed report. Then, they must socialize the results, give the report to the respective authorities for review, and approve it. In an adverse response scenario, a new version of the report should be made until the authorities are satisfied with the report. Overall, as presented in Table 9, the analyzed Fog Computing healthcare application offers a good quality after being evaluated the time, resources utilization, and capacity of the application through the quality model of the performance efficiency of Fog Computing.

Table 9. Evaluation decision criteria of the fog computing healthcare application.

Time behavior attribute	Qualification	Total
Request submission from the IoT device	*0.5*	*2.5 ≈ Good*
Response of the request from the Fog node	*0.5*	
Communication latency between the IoT device and the Fog node	*1*	
Simultaneity of requests processed by the Fog node	*0.5*	
Resource utilization attributes	**Qualification**	**Total**
Performance of the CPU processing in the Fog nodes	*0.5*	*1.5 ≈Good*
Performance of the search algorithms in the Fog node	*0.5*	
Performance of the server when receiving requests from the IoT device and the Fog node	*0.5*	
Capacity attributes	**Qualification**	**Total**
Performance of the Fog node data transfer	*1*	*2.5 ≈ Good*
Number of requests to the Fog node	*1*	
Fog node success rate	*0.5*	

2) *Disposition of the evaluation results:* Once the organization involved has finished the evaluation, there should be a suitable treatment of the data and information obtained by the report. Thereby, it is assured the evaluation process confidentiality and accessibility.

5 Conclusions and Further Work

In this paper, it has been presented both a quality model and an evaluation method for Fog Computing applications, which can be used along the software development process. This contribution is aligned with the ISO/IEC 25010 and ISO/IEC 25040 standards, so it helps to the Fog Computing development community with issues about quantifying and detecting performance efficiency during the development. The quality model of the performance aligned to ISO/IEC 25010 is based on three sub-characteristics Time behavior, resource utilization, and capacity, which summarize the features of the Fog Computing network. Then, the evaluation method aligned to ISO/IEC 25040 describes the process to evaluate the performance of Fog Computing Applications by using the quality model. Finally, the viability of the model has been assessed through a case study by using a simulated Fog Computing network topology from a healthcare application. This case study presents the advantages of the quality model and its benefits to the developers in order to develop robustness applications adequate to the domain solution and needs.

As further work, it is possible to present the queueing model, which is commonly applied for performance evaluation and QoS studies. Moreover, to develop a Fog Computing application and then use the quality model and the evaluation method to assess it. Besides, it is planned to create and apply an evaluation process for Fog Computing application acquirers. Therefore, two kinds of assessment for the performance efficiency of Fog Computing could be presented, one focused on the developers and another for the acquirers.

Acknowledgment. The authors would like to thank to *"Corporación Ecuatoriana para el Desarrollo de la Investigación y Academia (CEDIA)"* for the partial financial support given to the present research, development, and innovation work through its CEPRA program, especially for the *"CEPRA XIV-2020-07 Adultos Mayores"* fund. Moreover, to the research projects *"Fog Computing applied to monitor devices used in assisted living environments; study case: a platform for the elderly"*, winner of the call for research projects DIUC XVII, and *"Run-time model-based self-awareness middleware for Internet of Things (IoT) eco-systems"*. Hence, the authors thank to *"Dirección de Investigación de la Universidad de Cuenca (DIUC)"*, and "Laboratorio *de Investigación y Desarrollo en Informática (LIDI) - Universidad del Azuay"* for its academic and financial support.

References

1. Silva, F., Analide, C.: Sensorization to promote the well-being of people and the betterment of health organizations. In: Applying Business Intelligence to Clinical and Healthcare Organizations, pp. 116–135 (2016)
2. Thien, A.T., Colomo-Palacios, R.: A systematic literature review of fog computing, vol. 24, no. 1, pp. 28–30 (2016)
3. Li, S., Da Xu, L., Zhao, S.: The internet of things: a survey. Inf. Syst. Front. **17**(2), 243–259 (2015). https://doi.org/10.1007/s10796-014-9492-7
4. Aazam, M., Khan, I., Alsaffar, A.A., Huh, E.: Cloud of things: integrating Internet of Things and cloud computing and the issues involved. In: Proceedings of 2014 11th International Bhurban Conference on Applied Sciences Technology (IBCAST), Islamabad, Pakistan, 14th–18th January 2014, pp. 414–419 (2014). https://doi.org/10.1109/IBCAST.2014.6778179
5. Hosseinpour, F., Westerlund, T., Meng, Y.: A review on fog computing systems. Int. J. Adv. Comput. Technol. Hannu Tenhunen **8**(5), 48–61 (2016)
6. Vaquero, L.M., Rodero-Merino, L.: Finding your way in the fog: towards a comprehensive definition of fog computing. SIGCOMM Comput. Commun. Rev. **44**(5), 27–32 (2014).https://doi.org/10.1145/2677046.2677052
7. Willis, D., Dasgupta, A., Banerjee, S.: ParaDrop: a multi-tenant platform to dynamically install third party services on wireless gateways (2014). https://doi.org/10.1145/2645892.2645901
8. International Organization for Standardization: Software engineering - Software product Quality Requirements and Evaluation (SQuaRE) – System and software quality models. ISO/IEC 25010:2011, vol. 2, no. Resolution 937 (2011)
9. Mahmud, R., Srirama, S.N., Ramamohanarao, K., Buyya, R.: Quality of Experience (QoE)-aware placement of applications in fog computing environments. J. Parallel Distrib. Comput. **132**, 190–203 (2019). https://doi.org/10.1016/j.jpdc.2018.03.004
10. Xiao, Y., Krunz, M.: QoE and power efficiency tradeoff for fog computing networks with fog node cooperation. In: IEEE INFOCOM 2017 - IEEE Conference on Computer Communications, pp. 1–9, May 2017. https://doi.org/10.1109/INFOCOM.2017.8057196
11. Zeng, D., Gu, L., Guo, S., Cheng, Z., Yu, S.: Joint optimization of task scheduling and image placement in fog computing supported software-defined embedded system. IEEE Trans. Comput. **65**(12) (2016). https://doi.org/10.1109/TC.2016.2536019

12. Sarkar, S., Chatterjee, S., Misra, S.: Assessment of the suitability of fog computing in the context of internet of things. IEEE Trans. Cloud Comput. **6**(1) (2018). https://doi.org/10.1109/TCC.2015.2485206
13. Sarma, B., Kumar, G., Kumar, R., Tuithung, T.: Fog computing: an enhanced performance analysis emulation framework for IoT with load balancing smart gateway architecture. In: Proceedings of the 4th International Conference on Communication and Electronics Systems. ICCES 2019, pp. 1–5, July 2019. https://doi.org/10.1109/ICCES45898.2019.9002172
14. El Kafhali, S., Salah, K., Ben Alla, S.: Performance Evaluation of IoT-Fag-Cloud Deployment for Healthcare services, July 2018. https://doi.org/10.1109/CloudTech.2018.8713355
15. Laporte, C.Y., April, A.: Software engineering standards and models. In: Software Quality Assurance, First. Wiley Ed., New Jersey. IEEE Computer Society Press (2017)
16. Cedillo, P.: Monitorización de calidad de servicios cloud mediante modelos en tiempo de ejecución, pp. 1–361 (2016)
17. Calidad del software. Ventajas de los Modelos/Estándares de calidad del Software (2017). https://dankocs2012.blogspot.com/2012/12/ventajas-de-los-modelos-estandares-de.html
18. ISO/IEC 25000. Software engineering - Software product Quality Requirements and Evaluation (SQuaRE) – System and software Quality Requirements and Evaluation (SQuaRE), ISO/IEC 25000:2011, vol. 2 (2011)
19. Hyper-V. Hyper-V Performance. https://ciudadanozero.azurewebsites.net/hyper-v-performance-parteiconsumodecpu/
20. Microsoft: Performance and capacity requirements calculation for Windows SharePoint Services collaboration environments (2012). https://docs.microsoft.com/es-es/previousversions/%0Aoffice/sharepoint-2007-products-andtechnologies/%0Acc261795(v=office.12)
21. GeekRed: Data Transmission Calculation, Data Transmission Calculation. https://geekred.blogspot.com/2008/05/calculo-de-transmicion-dedatos.%0Ahtml
22. Flores, K.: Success Metrics, Success Metrics. https://es.scribd.com/document/327668739/Metricas
23. Parra, J.M., Valdez, W., Guevara, A., Cedillo, P., Ortíz-Segarra, J.: Intelligent pillbox: automatic and programmable Assistive Technology device. In: 2017 13th IASTED International Conference on Biomedical Engineering (BioMed), pp. 74–81 (2017). https://doi.org/10.2316/P.2017.852-051
24. Wilson, V.S., Orellana, I.C., Parra, J., Guevara, A., Ortiz, J.: Intelligent pillbox: evaluating the user perceptions of elderly people. In: International Conference on Information Systems Development, September 2017. https://aisel.aisnet.org/isd2014/proceedings2017/HCI/3. Accessed 29 Aug 2020
25. Gupta, H., Vahid Dastjerdi, A., Ghosh, S.K., Buyya, R.: iFogSim: a toolkit for modeling and simulation of resource management techniques in the Internet of Things, Edge and Fog computing environments. Softw. Pract. Exp. **47**(9), 1275–1296 (2017). https://doi.org/10.1002/spe.2509

Auto Attendance Smartphones Application Based on the Global Positioning System (GPS)

Mahmoud Abdul-Aziz Elsayed Yousef(✉) and Vishal Dattana

Middle East College, Muscat, Sultanate of Oman
Vishal@mec.edu.om

Abstract. For several years, the mechanism of recording attendance has evolved from traditional manual systems, such as recording in daybooks, to electronic systems, where modern systems have included the integration of fingerprint devices and data management systems, including ERP systems. Earlier, the manual attendance method used, which is not only consuming the time, but it also gives erroneous results. The attendance controlling system provides many benefits to organizations. This diminishes the need of a pen-and-paper-based manual attendance system. Following this thought, this paper has proposed a smartphone auto attendance application based on the global positioning system (GPS). We are going to use GPS, wireless fidelity (Wi-Fi) and network data to determine the location of the mobile device with the desired accuracy. The solution is implemented and tested on an Android and iOS device, which results in "no need of additional biometric scanner devices and other readers". The application can be configured with the organization locations that were identified in the system, which can be determined by GPS. Each user's location can be logged by GPS using a smartphone. This location is declared as a key of time and attendance tracking in this paper.

Keywords: Mobile attendance system · GPS · Wi-Fi · Data connection · Android applications · iOS applications · Location-based service

1 Introduction

Nowadays, monitoring attendance is one of the basic parameters of controlling work in organizations, whether private or public. Attendance systems and working hour calculations are very important for any organization. It effects many results, whether the quality of work or salaries of the employees, because of the additional wages that must be calculated correctly in order to prevent losses that may be borne by the organizations.

Usually, there are two types of attendance systems existing in most of organizations: paperwork systems and automated systems. A paperwork system contains the use of paper sheets, including Excel sheets and notebooks, in recording attendance where employee or student fill out and managers take care of accuracy. This method could be lead to incorrect result because of many reasons, such as incorrect taking time or user ID or even due to damaged or lost sheets. The extraction of user's data and the manual computation of working time is consuming hug time. In addition, hiring a worker to check and enter the manual computation data is costing as well [1].

K. Arai (Ed.): FICC 2021, AISC 1363, pp. 917–934, 2021.
https://doi.org/10.1007/978-3-030-73100-7_63

The second method, automated attendance systems, involves the use of magnetic stripe cards, electronic tags, and biometrics devices instead of paper sheets and electronic sheets [2]. In these aforementioned methods, users touch or swipe in order to provide their identification to enter the start-time and end-time to calculate their working hours. The data captured by any of these devices is recorded in a device database and then automatically transferred to the database server. Using an automated system for time and attendance monitoring reduces the errors of manual systems and saves time. However, these automated system devices have others issues.

Moreover, this paper discusses related-works in the issues of using these biometrics devices while considering the wide popularity of smartphones. The paper introduces the auto attendance application on smartphones for recording attendance purposes as a mobile application that communicates with the server [3].

Internet-connectivity (Wi-Fi/4G) is-needed for connecting to a database-server and getting a current location. The auto attendance application does not require any exterior device other than the user smartphone, which will reduce computational time and cost of engaging an extra device.

2 Literature Review

In demand to improve the attendance system framework, the scholars cut-away at the change from the alternate perspective [4] with the evaluation that attendance administration is missed by the organizations'-administration frameworks, focusing just-on record administration, training method and so on; thus, they established an attendance administration framework utilizing Java and Oracle [5]. They outlined a novel finger impression gadget that utilizes a part of a unique finger attendance framework.

The users check their attendance by putting their finger on the sensor of the biometric device. However, this framework has a lack-of-feasibility, because a fingerprint biometric-scanner cannot always sense the user fingerprint, and it can waste a lot of time to verify the employee's fingerprint. In addition, NFC-based applications arrange different human ordinary experience by touching a thing settled or coordinated with an NFC tag. For example, Smart Touch is one of the early NFC projects that spotlights NFC innovation, which introduced by Technical Research Centre TRC in Finland, the Applications in different zones were produced [6]. However, the system is highly expensive. In addition, the system faces some other kinds of issues and insufficiency of honesty and probability of forgery, and therefore, the inability to confirm the identity of the user [7].

In addition, some devices came with face detection systems and face recognition systems to achieve the target goals, but because of the COVID-19 pandemic, employees are forced to wear facemasks; this will limit the uses of these features [8].

Soewito et al. proposed the Soewito attendance-application using fingerprints and GPS technology via smartphone [9]. The application is able to collect data, but it cannot export reports as.pdf-or.xlsx-files. According to a Soewito paper, the application is an Android application; users can hack the location using a fake GPS location application. Moreover, the application cannot use the auto attendance feature and mobile service-provider internet data.

M. Noor et al. [10] developed an application for student attendance automation. Each student has an ID assigned with barcode, which smartphone application can read it. One student can carry the barcode of the others students, which will mislead the system.

3 Smartphone vs. Barcode Systems and RFID

There are many differences between biometric and auto attendance smartphones. From accuracy to cost savings, smartphone attendance systems are definitely a better choice for small and big companies.

No matter the number of users in your organization, the cost of installation is high because of the biometric equipment and software.

Rather than capturing data at two points, a smartphone attendance system records data during the working hours in a day. In fact, when applying the attendance smartphone application, you will dispense with the current devices installed in the organization building, and if you decide to expand your business by operating new branches for your organization, you will not need other connections or a hole in a wall in the organization to install these devices.

Data is power today. Moreover, you can only imagine how a detailed and accurate user data can help companies take empowered HR decisions. Consider two systems with identical mechanisms but with different structure functions.

Next Table 1 is a comparison between an auto attendance application and RFID/barcode readers.

Table 1. Smartphone application vs. RFID/barcode readers

Comparison	Smartphone application vs. RFID/barcode readers	
	Auto attendance – smartphone	*RFID and BARCODE*
Requirements	Smartphone, database server	RFID, barcode, biometric readers, and database server
Cost	Cost-effective and faster since the application does not require any additional devices	Expensive, since it requires RFID/barcode readers
Authentication	Each user must carry his/her smartphone device, which is easy to do always	Each user must bring RFID card or barcode label, which is hard to do always. Moreover, a biometric fingerprint reader does not always work and can cause delay of data transferring
Authorization	User can carry only one unique smartphone UUID	One user can carry another's card
Different work locations	Easley can define many work location for each user and trace the data	Easy to define many work locations for each user and trace the data
Verification	Verification can be done by face ID/fingerprint/username and password	Only the cardholder and the reader can do verification
Availability	The probability of losing the phone is very low	The probability of losing the card is very high
Time	Smartphone requires minimum time and there is no need for an extra operator to operate	Reading from RFID or barcoding is consuming time and sometimes it needs an employee to operate
Transparency	Attendance data can be accessed in the smartphone application itself	Need s third-party application to access the attendance data

*·The table contents are based on my personal test and experience

4 Methodology

The users-authentications and authorization is one of the main key in the proposed mobile application. Each user is authenticated based on unique user identification code and device UUID. This unique identification code is the number given by the Information

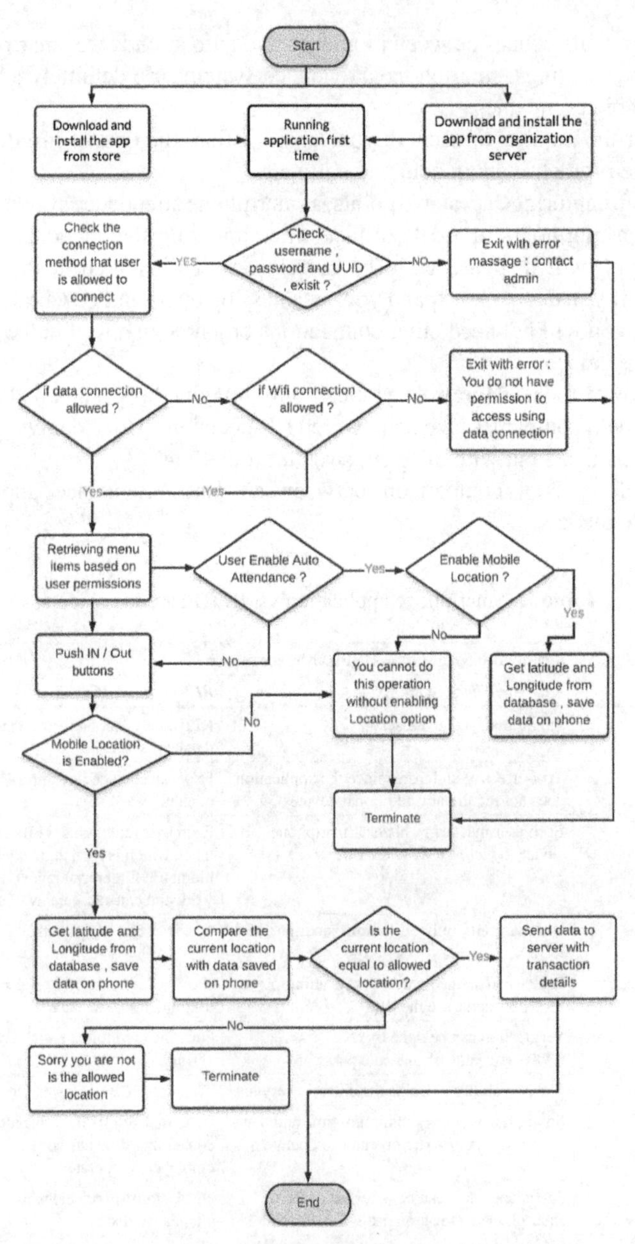

Fig. 1. The workflow of the app with manual attendance and enabling the auto attendance method.

Technology department of the organization. The identification number, along with device UUI and other information, is saved in the user device and database. Initially, the users have to download the application APK file or iOS file into their Android/iOS smart phone. The location service should be always on during usage of the mobile application. If location service is turned off, the application will request the user to turn the location on, Fig. 1 shows the workflow.

Location service option in the smartphone has an effective role to trace the user location. Once the user enters the range area that allows to record attendance from it, there are two ways to record the attendance: the manual method is required from the employee to choose the type of the attendance. The automatic record attendance automatically records his attendance without the need to press any button.

The user smartphone is directly connected to the organization's Wi-Fi internet. He can use mobile data as an alternative option, then the app will send the record including the date, time, location, user-ID to the organization sever; this is parameters consider as the login date and time of that user. While leaving the company area, a record is sent to the server with the user-ID and local-time, which is consider as the exit time. The app also give a nice feature to allow users to check and monitor their daily record on their smartphone without the need to contact the human resource department.

5 System Overview

The proposed system provides a solution to solve attendance issues with GPS as a location service implemented in the smartphone application, to service the attendance control system or the enterprise system, which is based on the concept of web applications. An auto attendance application improves the automated recording of users. It also avoids any loss of data that proves the attendance and departure of the user, and records the attendance automatically without the need for the user to follow up to prove his/her attendance condition. At the same time, the user can view his daily report from his/her smartphone without contacting the human resource department, below Fig. 2 showing the auto attendance method road map.

Using smartphones to achieve these valuable goals, the application can connect and handle data whether the phone connected to the wireless network or data network. Initially, it is important to save the organization coordinates by inserting these parameters (latitude, longitude) and radius of target area into the attendance control system (ACS). The ACS has great features such as entering different latitude and longitude for different branch location, and name the location as the branch location address, which make it easy later on to assign it to the user. The users' needs to download and install the auto attendance APK or iOS file on their smartphone devices.

For the first-time use, the user must register his/her device with the IT department in the organization. The information technology department will give to the user a username and password. The user will sign manually into the app using the username and password only the first time; if the user device supports face recognition or fingerprint technology, then the application will ask the user to scan his face/fingerprint to confirm the identity. If the user wants the application to register the attendance automatically, he must activate that by enabling this option on his smartphone, so he won't need to open

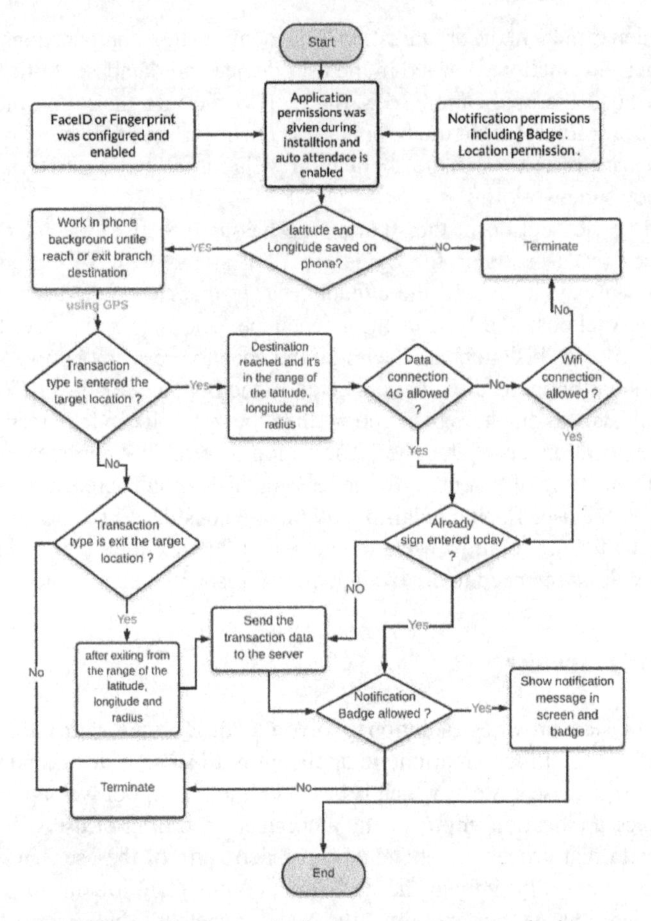

Fig. 2. Road map of using auto attendance method.

the application every time he reaches the target location. The app will sign him/her IN and OUT automatically.

If the user doesn't want to enable auto attendancc, he/she will have to open the app every time they reach the target location and login with FACE ID or fingerprint ID in the smartphone then choose the attendance option such as IN, OUT, etc.

The application checks the user location and compares it with the location, which the organization assigned to the current user. Once the user enters the radius, the application will register the transaction as "IN Duty" automatically in the database, and vice versa.

One of the interesting application features is the application can compare the current user location with the target location, even if the user dos not have internet, by using only GPS, but it will not send the data to the server without internet connectivity.

6 System Design

The auto attendance application is a client>server approach and follows a standard hardware and software architecture. The integration of the software components is the main challenge, as the application is calling also APIs, which are needed to verify user's data.

The system is designed in a great way, which easily can be configured with other devices if they are available in the workplace, such as biometric devices and readers.

The system consists of two main components: smartphone application and desktop web application.

6.1 Software Architecture

The software architecture consists of the mobile application, desktop web application (ACS), database and the server.

Mobile Application

The application was developed and built with a Flutter framework, the code was written in an Android studio using the Dart language and configured to run in XCODE for iOS version. Flutter count on the mobile operating system for numerous capabilities and configurations. Flutter one of the best way to build a beautiful UIs for mobile. If the developer is an expert with Android, the developer does not have to relearn everything to use Flutter. The smartphone application sends the data to the database server via an integrated development environment (IDE).

Desktop Web Application (ACS)

Attendance control system desktop application developed with SQL/PL and Java language using APEX 19 environment. The application provides user-friendly interface to the admin to configure the mobile application and to print out reports in many different file types, such as Excel and PDF and it even can automatically send emails with daily reports to the users' managers.

The application also offers many types of reports, such as daily late and early out, reports between two dates and it calculates percentage automatically.

Database

The database contains a number of tables, procedures and functions to store records. The Oracle database has been used in this project, which is fast, efficient and can store a large number of records. You may use any other database brand.

Server

The server is deployed on a VMware machine using IDE and apache With 16 GB ram and a 100 GB hard disk, the system can work perfectly with 120 users.

6.2 Hardware Architecture

The hardware requirements to run the auto attendance smartphone application is just an Android device or an iOS device, which will run the application; the user will mark their attendance by taking their IN attendance OUT time automatically without any hassle. The server side requirement depends on the way the information technology department wants to implement it. The standard requirement is a two VMware machine (in one physical machine). The first machine is for the database and the second is for the application server. If the information technology department wants to implement the application with their existing database, then one VMware machine could be used as the application server.

6.3 Technical Implementation

Database Entity Relationship Diagram (ERD)
ERD is a structural diagram used for database design. An ERD contains different symbols

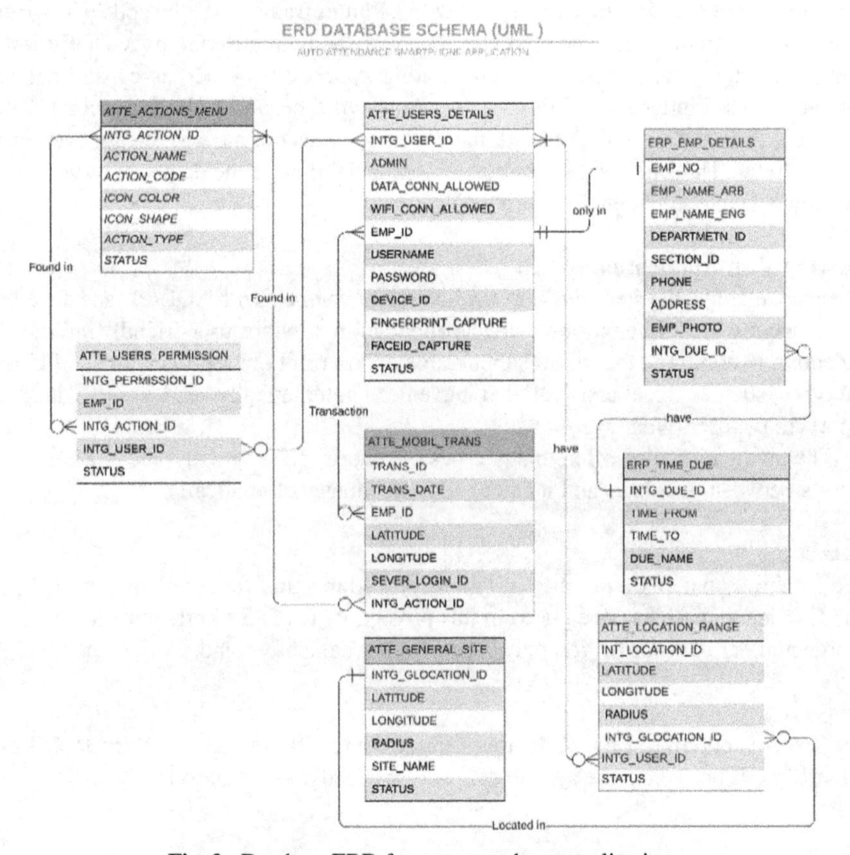

Fig. 3. Database ERD for auto attendance application.

and connectors that visualize two important. Figure 3 shows the ERD diagram, which used to design the auto attendance application.

Auto Attendance Application Background Activities

In main dart, we are going to call Geofencing API to get the current location in both method auto attendance, which works in the background and manual attendance.

Latitude (LAT) and longitude (LON) coordinates assume the earth is sphere in shape (i.e., not flat) (1). Therefore, since the earth is a sphere, you cannot use the distance formula that works for two points in a plane as shows in Fig. 4. Instead, using the formula that approaches and assumes the earth is a sphere. First, let us have a look at how to calculate the distance between two points on a flat plane using the distance formula [11] .

$$Distance = \sqrt{(x2 - x1)^2 + (y2 - y1)^2} \quad (1)$$

For example:

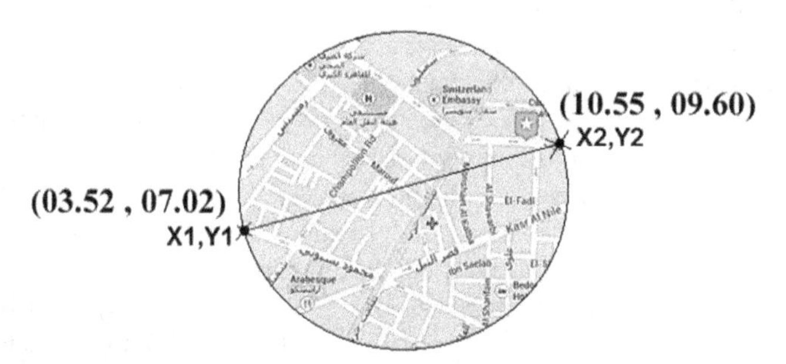

Fig. 4. Distance between two points on a flat plane.

$$Distance = \sqrt{(x2 - x1)^2 + (y2 - y1)^2}$$
$$= \sqrt{(10.55 - 3.52)^2 + (09.60 - 07.02)^2}$$
$$= \sqrt{49.42 + 6.6} = 7.49 \text{ Meter}$$

Haversine formula – get the distance between two points' latitude and longitude that was given by Haversine formula:

$$\mathrm{a} = \sin^2(\Delta\varphi/2) + \cos\varphi 1. \cos\varphi 2. \sin^2(\Delta\lambda/2)$$

$$\mathrm{c} = 2 \cdot \mathrm{a}\tan 2\left(\sqrt{a},\ \sqrt{}(1 - \mathrm{a})\right)$$

$$\mathrm{d} = \mathrm{R} \cdot \mathrm{c}$$

Where φ is latitude, λ is longitude, R is earth's radius (mean radius = 6,371 km). Figure 5 shows how you use the formula as script.

```
1  from math import sin, cos, sqrt, atan2, radians
2
3  # approximate radius of earth in km
4  R = 6371.0
5
6  lat1 = radians(3.52)
7  lon1 = radians(7.02)
8  lat2 = radians(10.55)
9  lon2 = radians(9.60)
10
11 dlon = lon2 - lon1
12 dlat = lat2 - lat1
13
14 a = sin(dlat / 2)**2 + cos(lat1) * cos(lat2) * sin(dlon / 2)**2
15 c = 2 * atan2(sqrt(a), sqrt(1 - a))
16
17 distance = R * c
18
19 print("Result:", distance)
20
```

Fig. 5. Distance between two lat/lng coordinates in km using the Haversine formula.

For best practice, the minimum radius of the geofence must be set between 60 and 150 m. Therefore, when Wi-Fi signal is available, accuracy is regularly between 30 and 60 m. In addition, when the user location is inside the target location, the accuracy range can be as small as 5 m. If you know that the indoor location is available inside the geofence, assume that Wi-Fi location accuracy is about 60 m.

When Wi-Fi signal is not available, you will use data connection (4G) and the location accuracy is not bad. The radius accuracy range can be as large as several hundred meters. In this case, geofences should use a larger radius between 60 and 100 m.

We are going to use GeofencingAPI and use these values (enter, exit). These values will identify the type of the location action, which helps to identify the type of the record in the database. The application allows the administrator to add many types of actions from the desktop interface, but each action in the database must be specified whether it is "enter" or "exit." For example, in the database ERD, there is a table named "AT TE_ACTIONS_MENU." The table has an attribute called "Action_Type." In that attribute, we should identify the type of the record that will be used in the transaction. Below, Table 2 is clarifying the attribute data style.

Table 2. Action type in database master table.

Code	Action name	Action type
IN	Attendance	Enter
OUT	Departure	Exit
F1	Request for delayed arrival	Enter
F2	Request for early exit from work	Exit

Now, the events which we used here as mentioned it above are to describe the log and transaction type which is mean if the user entering the target location, the application will send the record to transaction table in database as "IN" record, moreover the application will notify the user instantly that his attendance has been recorded, Fig. 6 shows the event overview as script.

```
GeofenceEvent intToGeofenceEvent(int e) {
  switch (e) {
    case _kEnterEvent:
      return GeofenceEvent.enter;
    case _kExitEvent:
      return GeofenceEvent.exit;
    case _kDwellEvent:
      return GeofenceEvent.dwell;
    default:
      throw UnimplementedError();
  }
}
```

Fig. 6. Event type overview

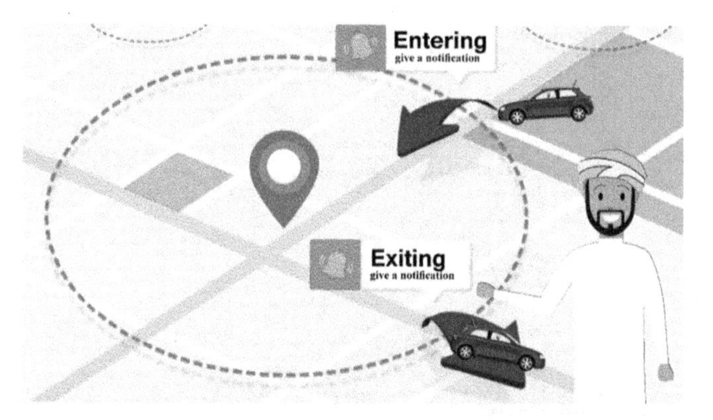

Fig. 7. Entering and Exiting location range

To apply this method which shows in Fig. 7 and use the advantage of the Flutter framework, we going to import the plugin and start to call GeofenceAPI in main dart; we need to make sure that the function is running before the splash screen. As explained before in 0, latitude and longitude must be saved on the device to avoid constantly requesting data from the server. Figure 8 shows the code that is used in the dart.

After that, we will start to compare the event type and send the transaction to the database via IDE in application server as showing in Fig. 9. Flutter notification also executes the notification message and badge as shown in Fig. 10.

```
1  try {
2    if (setting.geofanceEnabled) {
3      AppUtils.setDeviceId( await AppUtils.getUUID());
4      PermissionStatus permissionStatus = await PermissionHandler()
5          .checkPermissionStatus(PermissionGroup.location);
6      if (permissionStatus != PermissionStatus.granted) {
7        Map<PermissionGroup, PermissionStatus> permissions =
8        await PermissionHandler().requestPermissions([
9          PermissionGroup.location,
10         PermissionGroup.locationAlways,
11         PermissionGroup.locationWhenInUse
12       ]);
13       permissionStatus = permissions[PermissionGroup.location];
14       if (permissionStatus != PermissionStatus.granted) {
15         throw Exception(AppLocalization.of(context)
16             .text("required_location_permission"));
17       }
18     } else {
19       await geofencing.GeofencingManager.removeGeofenceById('mtv');
20       if (setting.permission == null)
21         throw Exception(AppLocalization.of(context)
22             .text("validate_permission_for_auto_action"));
23       User user = await AppUtils.getUserSession();
24       Location location = user.location;
25       if (user.location == null)
26         throw Exception(AppLocalization.of(context)
27             .text("location_not_configured_for_user"));
28       await geofencing.GeofencingManager.registerGeofence(
29           geofencing.GeofenceRegion(
30               'mtv',
31               location.latitude,
32               location.longitude,
33               location.radius,
34               GeofenceTrigger.intriggers,
35               androidSettings: GeofenceTrigger.inandroidSettings),
36           geofenceCallback);
37       print("Geofence mtv registred");
38       AppUtils.setGeofenceEnabled(true);
39
40     }
41   } else {
42     await geofencing.GeofencingManager.removeGeofenceById('mtv');
43     AppUtils.setGeofenceEnabled(false);
44   }
45   final completer = snackBarCompleter(
46       context, AppLocalization.of(context).text("refresh_complete"));
47   store.dispatch(UpdateSettings(completer, true, setting));
48   return completer.future;
49 } catch (ex) {
50   Scaffold.of(context).showSnackBar(SnackBar(
51       content: SnackBarRow(
52         message: ex.message,
53       )));
54   return Future<Null>(null);
55 }
56 }
```

Fig. 8. Adding and removing geofence coordinates

6.4 Graphical User Interface (GUI)

The Smartphone Application

The smartphone application interface is very simple and modern; the user can do everything just by clicking some buttons, as Fig. 11 shows the loading screen and login screen. At first, the user should get the user credentials along with registering his device with the information technology department. They will need to have the UUID number, which is shown below the login button.

Once the user registers the UUID and gets user credentials, the user needs to verify the face ID or fingerprint ID if the smartphone has these features. Figure 12 shows the fingerprint authentication request.

After authentication is complete, the user will be able to see the home page screen, which includes all the buttons the user is allowed to see that have been assigned to him by the system administrator via the web application. Figure 13 shows two small icons in the top right screen: the gear icon and the shield icon.

```
//import 'package:flutter_background_geolocation/flutter_background_geolocation.dart';
void geofenceCallback(List<String> ids, Location l, GeofenceEvent e) async {
  print('Fences: $ids Location $l Event: $e');

  FlutterLocalNotificationsPlugin notification = initialisedNotification();
  print("notificatio nobject created");
  Permission permission = null;
  print("enter event executed");
  TouchApi api = TouchApi();
  print("tuch api object created");
  User userSession = await AppUtils.getUserSession();
  print("got the user session");

  if ((e == GeofenceEvent.enter))
    permission = await AppUtils.getAutoPermission();
  if ((e == GeofenceEvent.exit))
    permission = await AppUtils.getAutoOutPermission();
  print("got the permission");
  Locale locale = await AppUtils.getLocale();
  print("got the locale");
  var languageCode = locale.languageCode;
  String deviceId = await AppUtils.getDeviceId();
  print("got the device id");
  try {
    if (userSession == null || permission == null)
      throw Exception("No user configuration found");
    await api.setAuthorizationWithDeviceId(
        userSession.username, userSession.password, deviceId);

    print("API Autherization setup");
    await api.markAttendanceBackground(Transaction((b) => b
      ..action.replace(permission.action)
      ..latitude = l.latitude
      ..longitude = l.longitude));
    print("got the api response");
    if (languageCode == AppConstant.LOCALE_ENG_CODE)
      showNotification(
          "Action :" +
              permission.action.nameEn +
              " attendance has been successfully recorded.",
          notification);
    else
      showNotification(
          "تم تسجيل :" + permission.action.nameAr + " بنجاح.", notification);
  } catch (ex) {
    print("exceptiontion executed");

    print(ex.message);
    if (languageCode == AppConstant.LOCALE_AREBIC_CODE &&
        ex.message == "you already marked attendance for this day")
      showNotification("لقد تم تسجيلكم سابقاً اليوم", notification);
    else
      showNotification(ex.message, notification);
  }
}
```

Fig. 9. Checking the event type and database if the event was already inserted on the same day.

The gear icon is a settings icon that allows the user to choose the attendance method whether manual or automatic; the user can choose the event separately (Fig. 14), while the shield icon is shown only to the administrator. The shield icon allows the administrator to control all user permissions from his smartphone instead of the web application; this is a nice feature of the app.

```
1   void showNotification(
2       String message, FlutterLocalNotificationsPlugin notification) {
3     var androidPlatformChannelSpecifics = new AndroidNotificationDetails(
4         'touch_app_id', 'Touch App', 'Touch App Description',
5         importance: Importance.Max, priority: Priority.High);
6     var iOSPlatformChannelSpecifics = new IOSNotificationDetails();
7     NotificationDetails notificationDetails = NotificationDetails(
8         androidPlatformChannelSpecifics, iOSPlatformChannelSpecifics);
9     notification.show(
10        Random(100).nextInt(100), "Auto Attendance", message, notificationDetails);
11  }
12
13  FlutterLocalNotificationsPlugin initialisedNotification() {
14    var initializationSettingsAndroid =
15        AndroidInitializationSettings('@mipmap/launcher_icon');
16    var initializationSettingsIOS = IOSInitializationSettings();
17    var initializationSettings = InitializationSettings(
18        initializationSettingsAndroid, initializationSettingsIOS);
19    var flutterLocalNotificationsPlugin = FlutterLocalNotificationsPlugin();
20    flutterLocalNotificationsPlugin.initialize(initializationSettings);
21    return flutterLocalNotificationsPlugin;
22  }
23
```

Fig. 10. Using Flutter plugin to show notification.

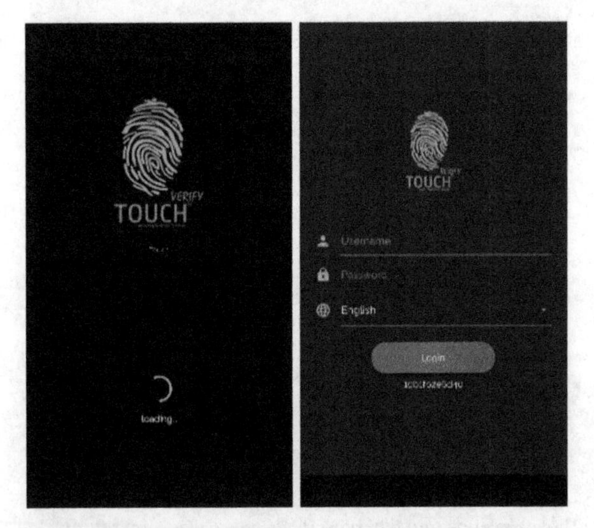

Fig. 11. Loading and login screen including UUID number

The home page has two tabs available. The first tab shows the user profile, such as name, department, job title, etc. while the report tab allows the user to view his/her transaction history, daily or monthly.

Desktop Web Application (ACS)

The web application is used to setup the user details and to operate the mobile application. In addition, the operator must login to the web application to register the user's devices, and to create the user's credentials (Fig. 15). Printing out reports and viewing data can done through the web application and reports menu (Fig. 16). HR can easily print out the reports or send reports by email. The application is able to generate.pdf/xlsx and xml.

Fig. 12. Fingerprint authentication request

Fig. 13. Home page

Assigning the users permissions and locations can be done simply by choosing from a list of values. Google Maps has been implemented in the web application to easily choose the location and range from the map and to define the location as a site location, then simply assign the site name to each user or group of users (Fig. 17).

Fig. 14. Settings screen

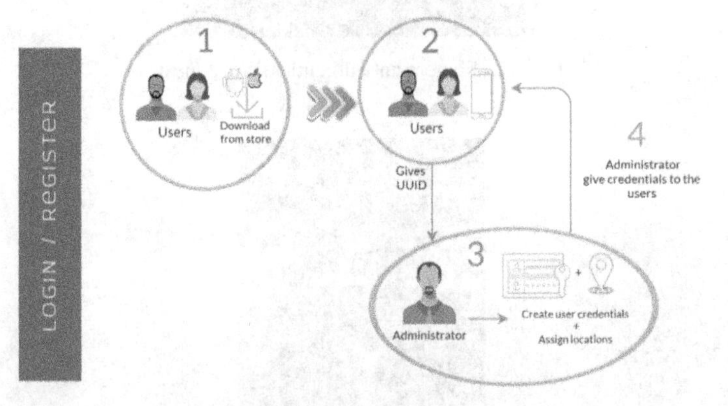

Fig. 15. Admin creating user credentials

7 Application Features

7.1 Taking Attendance and Percentage Calculations

If the time remaining on the location/site is related to the requested time, the attendance will be referred to as "present." The method can be changed easily from the web application settings, as it is possible once the user enters the site to be considered "present." In addition, the system calculates the different in time between "entering" and "exiting" the location. There is no opportunity to duplicate the system; the only "OUT" events can be recorded more than one time. For example, the user went out to pray or eat; the system will record that, and then the user returns to the office and then goes home, so there will be two records of "OUT," but only one record will be calculated, which is the

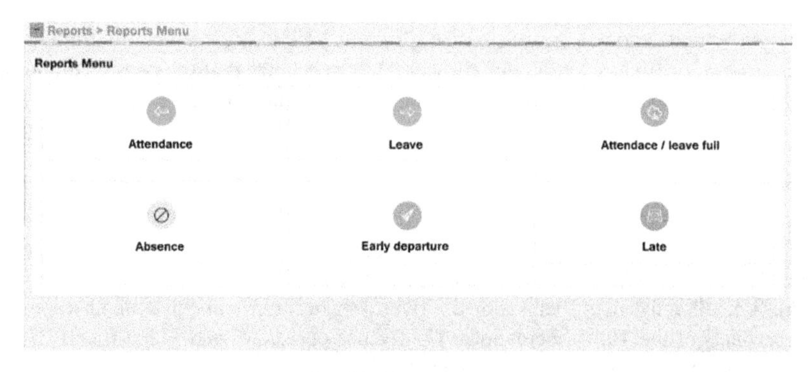

Fig. 16. Reports menu

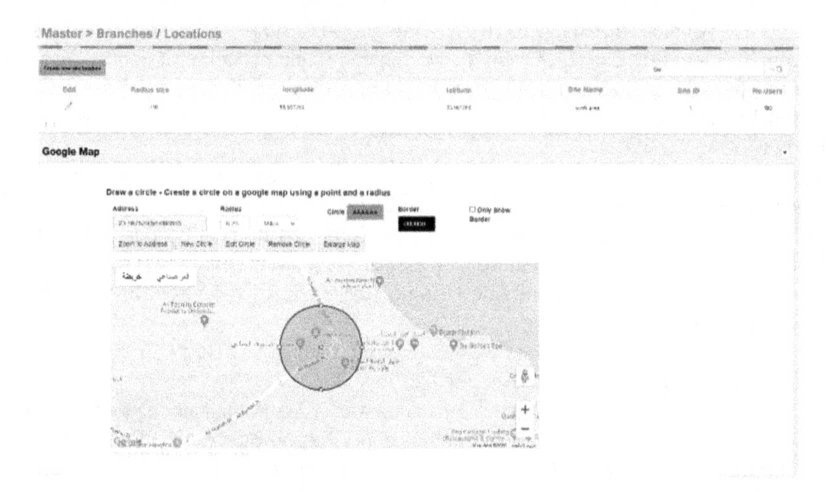

Fig. 17. Add new site / view existing site

last transaction time. The percentage is calculated automatically by the web application for each user in each department.

7.2 E-Mailing System

In case of low attendance percentage, the web application will send an e-mail to the user including the number of delayed hours as well as a warning letters. Moreover, the HR department can set the e-mail settings to send daily reports to the user manager with full daily reports.

8 Conclusion

This paper introduced a smartphone auto attendance application, which uses GPS as the core component of attendance tracking. It is a computerized information accumulation technology, which leads to more precise data entry. The area is set for tracing using GPS and user coordinates inside the allowed area to be sure that the user is attend in the organization. The application is available for Android and iOS platforms.

References

1. Islam, Md., Hasan, Md., Billah, Md., Uddin, Md.: Development of smartphone-based student attendance system (2017). https://doi.org/10.1109/R10-HTC.2017.8288945
2. Jain, A.K., Prabhakar, S., Pankanti, S.: Twin test: on discriminability of fingerprints. In: Proceedings of the Third International Conference on Audio- and Video-Based Biometric Person Authentication, pp. 211–216, June 2001
3. Park, J., An, K., Kim, D., Choi, J.: Multiple human tracking using multiple kinects for an attendance check system of a smart class. In: 2013 10th International Conference on Ubiquitous Robots and Ambient Intelligence (URAI), Jeju, p. 130 (2013)
4. Zhang, Z., Gong, P., Cao, L., Chen, Y.: Design and implementation of educational administration attendance management system based on B/S and C/S. In: 2007 First IEEE International Symposium on Information Technologies and Applications in Education, Kunming, pp. 606–609 (2007)
5. Mohamed, B.K.P., Raghu, C.V.: Fingerprint attendance system for classroom needs. In: 2012 Annual IEEE India Conference (INDICON), Kochi, pp. 433–438 (2012)
6. Strommer, M., et al.: Smart NFC interface platform and its applications. In: Tuikka, T., Isomursu, M. (eds.) Touch the Future with a Smart Touch (2009)
7. Bermejo, C., Hui, P.: Steal Your Life Using 5 Cents: Hacking Android Smartphones with NFC Tags (2017)
8. BBC News: Why Some Countries Wear Face Masks and Others Don't (2020). https://www.bbc.com/news/world-52015486. Accessed 28 June 2020
9. Soewito, B., Gaol, F.L., Simanjuntak, E., Gunawan, F.E.: Attendance system on Android smartphone. In: 2015 International Conference on Control, Electronics, Renewable Energy and Communications (ICCEREC), Bandung, pp. 208–211 (2015)
10. Noor, S.A.M., Zaini, N., Latip, M.F.A., Hamzah, N.: Android-based attendance management system. In: 2015 IEEE Conference on Systems, Process and Control (ICSPC), Bandar Sunway, pp. 118–122 (2015)
11. Larson, R.: Precalculus with Limits. Function and their Graphs, 2nd edn., p. 4. Brooks Cole, Boston (2017)

Transforming Audience into Spectator/Actor: Assimilating VR into Live/Theatre Performance

Saikrishna Srinivasan(✉)

University of Waikato, Hamilton, WA 3216, New Zealand

Abstract. Virtual Reality (VR) is a medium that is used to teleport audience virtually from one place to another in the physical world or into a virtual world. VR does prove to bring a different and more enhancing perspective to the audience by immersing them into the performance. This paper is a continuation of a prototype project on the early development and use of a three-dimensional theatre prototyping in order to explore the technical requirements for the application to a VR theatre experience, which is a part of more extensive research to use VR as an assistive technology in order to enhance audience access and propinquity to a live performance. This paper explores and investigates the relative position of the audience on the stage (virtually transported to the stage using VR) with respect to their traditional off-stage seating position. The evaluation between the two situations is based on perceptual experience along the lines of the field of view of the audience, obstruction and framing. This experiment is a part of a larger research and its evaluation will be critical to the result of integrating VR inside theatre performance. The study of the larger research was to introduce VR into the theatre and induce a state of deeper engagement from bodily presence. However, this experiment not only introduces VR into the theatre space but also helps to evaluate how the virtual presence of an audience can induce a sensation of being inconspicuous yet close at hand and quite removed from the audience. The project aims to exploit the idea of having the stage cameras either mounted on a pole (stage) or on the actor (participant) and comparing them with the traditional audience perspective using the technique of video analysis. This paper will account for the findings and comparison between the audience position outside the stage and on the stage using VR technology, providing the necessary foundation of assimilating VR inside of live performance.

Keywords: Virtual Reality · 360-degree camera · Assistive technology · Emancipation · Google cardboard · Perspective · Theatre · Invisible

1 Introduction

1.1 Project (Scope)

Virtual Reality is primarily defined as a medium used to virtually teleport audience into a virtual world with the help of VR headsets and software [12]. The primary aim of the experiment is to virtually teleport the audience from their traditional off-stage

K. Arai (Ed.): FICC 2021, AISC 1363, pp. 935–944, 2021.
https://doi.org/10.1007/978-3-030-73100-7_64

spectator space in a proscenium theatre and gradually introduce them into the performance arena. The experiment aims to compare the two positions of the audience and evaluate based on perceptual experience along the lines of the field of view of the audience, obstruction and framing. This experiment constitutes the initial phase of a larger research to have VR stimulate new approaches to theatre practice and performance by offering and perhaps empowering theatre practitioners with a means to experiment with closer audience proximity and interactive set projection. This may help in developing or devising a novel approach or branch in the field of theatre performance. Affordable VR technologies, such as Google cardboard [6] crafted from low cost cardboard have been used in this experiment. The overall aim of the research is to exploit the cheaper cost of the VR technologies and help the audience to get a novel experience of the live performance. In order to achieve the required result, the different available camera angles are being experimented and evaluated in this experiment to give the best experience to the audience.

1.2 Experiment

Studies in cinema show that camera angles play a major role in determining the perception of any character or object in the audience's mind [5]. Supporting the same fact Joan Meyers-Levy, and Peracchio Laura experimented to study the effect of camera angle on the audience and how it affects the audience's perception [8]. The experiment involved participants viewing a product (object - an Everex 386/20 personal computer) from three different camera angles. Their result suggested that the camera angle at which a product is shown in advertising can affect product evaluations. When observed at eye level, the product was regarded less favorably than when it was viewed from a low camera angle (looking up at the product) but more favorably than when it was observed from a high camera angle looking down at the product.

It provides a valid conclusion on Joan Meyers-Levy, and Peracchio Laura statement that camera angle plays a major role in affecting the audience's perception [8]. Similarly in this experiment the audience is made to experience the same scene in a different camera angle (person's perspective) to determine the impact of the scene on the audience.

This experiment is a progressive development to the previous prototypic experiment, which was focused on the production of a 3-D mock-up of a theatre setting in order to identify possible areas for camera (and in doing so, audience) placement on stage [13]. This experiment is aimed to evaluate the camera position similar to Joan Meyers-Levy, and Peracchio Laura's [8] approach using the method of video analysis to evaluate the camera positions based on perceptual experience along the lines of the field of view of the audience, obstruction and framing. The prototype experiment helped to find out the suitable camera positions for the audience on the stage, however it is required to test out the camera positions in real time to find the viability of the previous findings.

This experiment uses four actors inside a proscenium rehearsal space, which will be further conducted in a black box theatre atmosphere with a higher number of participants. The experiment consists of showcasing a theatre performance scene of a monologue and a duologue between two characters, which will be captured and recorded using a 360 degree camera (Insta X 360). Figure 1 and 2 showcases the performance in different angle with the same actors performing the same duologue scene. The theoretically derived camera positions from the previous experiment are implemented here to capture the performance and evaluate them. Since this is the first experiment of its kind, only three camera positions were tested. Figure 1 shows the actors performing from the audience's perspective, while Fig. 2 shows the same performance on the stage from the actor's perspective. However, it does give rise to a lot of novel camera positions and ideas which can pivot theatre performance to a whole new interactive medium using Virtual Reality. The entire scene in this experiment is enacted by the four actors for four minutes. The reason for the selection of such a short scene is due to the fact that theatre audiences need to get accustomed to this new technology, also it should not be a hindrance for the actors and theatre producers. The experimentation on a short timed scene of four minutes might be a step behind the work of Charlton and Moar experiment [2], in which the 360° camera was placed on the center of the stage for the entire length of the play (nearly 20 min). However, the justification for taking a step behind and re-examining the effect of using 360° cameras in a theatre space is because theatre audiences need to gradually be introduced to the use of this new technology. Compelling the theatre audience to use VR headset and gadgets for an entire long performance is very questionable.

In the previously concluded experiment "Virtual Teleportation of a Theatre Audience Onto the Stage: VR as an Assistive Technology as an assistive medium" [13] I have recreated a duologue sequence in an theatre environment, in 3D using 3D modelling softwares and calculated possible desirable camera positions for the audience. The experiment was concluded with six major impactful camera positions, which would be very apt and delightful for the audience in a duologue sequence. The concluded six positions were two positions on top of each of the actors, four positions at an angle 45° and −45° from each actor.

Fig. 1. Audience perspective from the camera on the actor

1.3 Method (Procedure)

Every camera position from the previous experiment is tested in different scenes and character setup to maximize the randomness of the result and to conclude an accurate result. The video analysis of the same has been performed using a suitable method of Grounded Video Analysis (Strauss, A. and J. Corbin). These recorded scenarios will be later showcased to the participants in the upcoming experiments. Both the theoretical result from this experiment and the practical conclusion from the participants' showcase will be compared to get a better and self-standing result.

The experiment is conducted in a proscenium space for rehearsal of a theatre play in order to get the better out of the lighting setup. As lights play a major role when experimenting with cameras and its placement. A group of 4 theatre artists have been employed as volunteers for the experiment. For this experiment they were asked to enact a short scene of Shakespeare's writing 'Constellations' [11] to test the various camera angles. To organize the experiment and data collection in a linear framework, a monologue and a dialogue scenes were enacted by the employed volunteers. These enacted scenes were captured and recorded using the 360-degree camera which were placed in the theoretically calculated positions (in front, behind, in between and physically on the actor). This experiment is in line with the experiment conducted by Charlton and Moar [2]. Unlike Charlton and Moar's experiment of using a single camera at the centre of the proscenium space, here the various available and enticing camera placement are being experimented and devised upon.

There are various recent studies and research being undertaken in the field of performance art and VR like Professor Gareth White [16] conducted an experiment by devising a cleverly structured theatre interior which helped the audience to feel their bodily presence in the environment. Pearlman [10] developed a brain opera which had cleverly structured confined interior space with display monitors on all the surrounding walls with the audience in the center. This experiment will be a successor to those experiments in technology and artistic innovations, focusing more on the capabilities of

Fig. 2. Audience perspective from the camera on the actor in the stage

the technology. However it focuses majorly on the artistic innovations available using technological advancements. The technologies being used in this experiment are a highly stabilizable 360-degree camera, Insta 360 X along with cheapest possible VR headgear, the google cardboard [6] which is clearly an enhancement from the previous experiments and research and providing the audience and artist their personal space.

2 Technical Evaluation

2.1 Evaluation Criteria

The evaluation is performed by comparing the two camera positions (one on the actor and the other from the traditional audience position). By this way, the perceptual experience like distance, angle, lights, obstructed vs clear view, objects in a frame, performance expression will be evaluated through video analysis.

This experiment was planned and conducted for a blackbox theatre for an arrangement of proscenium space with the audience seated from a single direction. The experiment is performed to compare and evaluate the audience vision from the tradition off stage seated perspective to the virtually teleported audience vision using 360° camera. The evaluation will study if the virtually teleported audience will experience the same level of audio and visual experience to that of the naked eye viewing from their seats. This experiment concentrates on the visual detailing, lucidity from the 360 camera, obstructions and deviations due to the other technicalities such as varied lighting, audio and actor interactions and movements.

2.2 Process of Evaluation for the Video Analysis

Whole to Part Analysis (Grounded Video Analysis)
The first part of the analysis will be following the method of Whole to Part analysis. This method consists of breaking down the video as a whole into individual smaller pieces and analyzing more [4]. This will help in finding and analysing all the data from the video recorded and categorize the necessary data accordingly.

The data (video) is split into smaller parts and put back into new categories and it will be open coded. From the smaller parts the concepts are identified separately, which are further grouped into categories. From the separate categories the properties and dimensions are identified. This is a very useful method of video analysis, as it helps to analysis the small details and parameters in the video [14].

In this particular research the concepts will consist of events or happenings or instances (scenes). Categories will consist of the group of concepts (type of scene). The properties will be the unique characteristic of the category (visible elements and actions in the scene). The dimension will be the frequency of the properties happening (how often the elements and actions are visible or repeated).

The dimensions measured in this experiment are basically camera properties (dimensions) that are necessary to provide a high dynamic range video experience to the audience similar to naked eye view. Image/visual quality is very important for the reproduction of the same performance from a different angle [9]. Hence the sub dimensions of the visual quality: lighting, sound and dynamic range effects are examined in this experiment. Light is an important factor to be considered when experimenting with cameras. Since a 360 degree camera captures all the surrounding space, lighting will be very much vital even in the corner of the proscenium space. The physical measure of light that is the most appropriate for imaging systems is luminance [9]. The luminance of the light is kept constant throughout the experiment by using a bright light source from the ceiling to have the consistent luminance throughout the performance space. Sound is the next important factor after light, since it can be used either as a cue to the audience or a distraction. The digital video revolution has given rise to cameras and camcorders with greater image resolution and cleaner sound reproduction than that of older analog formats [Holman, Tomlinson, and Arthur Baum]. The final measurement is the signal-to-noise ratio (SNR) is most often used to express the dynamic range of a digital camera. In this context, ratio is between the brightest and the darkest parts of a scene given in luminance [9]. In order to minimize the dynamic range of the camera the intensity of the light is kept constant throughout the experiment.

In addition to these dimensions, the visual distractions by the ambience and the immersive are noted by examining the tone reproducing ability of the camera. If the camera could not reproduce the same tone due to the external lighting or luminance factor, distortion. The lighting setup in a theatre performance is very vital to convey the mood or tone of the entire play [15]. When recording or watching the play through a camera lens, the camera should be able to convey the same mood and tone without distractions like lens flare (unless it is necessary for the performance itself). Similarly, sound plays a major role in any video to convey the mood and tone of the video [7]. Hence, these parameters for the various camera angles are noted and tabulated in Table 1 and Table 2 for analysis. Table 1 shows the parameter captured from the traditional audience view, while Table 2 tabulates the parameter captured from the position of the actor.

3 Findings

The data findings from the above concept of duologue, compares the camera position on the stage/actor to the traditional audience seating position. A consistent lighting set up is used in this experiment for the overall comparison between the two positions. It can be noted that in this experiment, only the technical variables of the 360-camera, the theatre space, lighting setup are examined. The primary aim of the experiment is to compare the two viewpoints of the audience: on the stage/actor and traditional seating arrangement. The primary goal of the total research is to bring the audience into the stage virtually using a camera on the stage.

Camera angle plays a major role in conveying a story to any audience [1]. Hence, it has been considered as the primary factor for experimentation and various camera positions are recorded for the same performance of a duologue, under the same lighting condition. The camera placed on an actor shows more likeness to the audio-visual technicalities and thus encapsulates the best capabilities of the technology. It provides a better visibility to the character B's expressions, gives a feel of better ambience with comparatively fewer visual distractions. The visibility of the character's facial expression and gestures are

Table 1. Video analysis of various dimensions from traditional audience view

Concept	Category	Property	Dimension	Range
A conversation between two characters	Character's self - explanation about his/her past	Character A's facial expression	Visibility	Poor - Medium (depending on the audience seat)
			Lighting	Constant white light from ceiling
			Intensity of the visual	Medium intense white light
			Sound	Dialogue exchange
			Dynamic range	Consistent due to the usage of same luminance of light
			Lucidity of the visual	Continuous flow of performance
		Character B's facial expression	Visibility	Poor- Medium (depending on the audience seat)
			Lighting	Constant white light from ceiling
			Intensity of the lights	Medium intense white light
			Sound	Dialogue exchange
			Dynamic range	Consistent due to the usage of same luminance of light
			Lucidity of the visuals	Continuous flow of performance
		Ambience (stage and set)	Visibility and Auditory Cues	Not necessary for this set up
			Immersiveness	Cannot achieve visual immersion
		Distractions	Visual and Auditory distractions	Depends on the seating placement
			Maintaining the tone of the play	Yes

very clear and captured with detail using the camera on the actor/stage. The technical capabilities of the camera on the stage provides the audience with the same visual impact from that of the view from a seated position. It in fact enhances and captures a lot of minor details like the actor's gestures and facial expression. The tone or the mood of the entire performance is kept intact by controlling the luminance, dynamic range of the camera so that it does affect the performance. The other major takeaway from this experiment is the use of sound. Sound constitutes of both the dialogues delivered by the actors and also the background music or the ambience noise inside the performance. Sound can be used as an audio cue for the audience to look at a specific direction, if they are overwhelmed or lost in the 360-degree space. At the same time, it can be used as a distraction if there needs to be any in the performance. Similarly, by experimenting with

Table 2. Video analysis of various dimensions from the position of actor

Concept	Category	Property	Dimension	Range
A conversation between two characters	Character's self - explanation about his/her past	Character A's facial expression	Visibility	Poor
			Lighting	Constant white light from ceiling
			Intensity of the visual	Medium intense white light
			Sound	Dialogue exchange
			Dynamic range	Consistent due to the usage of same luminance of light
			Lucidity of the visuals	Continuous flow of performance
		Character B's facial expression	Visibility	Medium
			Lighting	Constant white light from ceiling
			Intensity of the lights	Medium intense white light
			Sound	Dialogue exchange
			Dynamic range	Consistent due to the usage of same luminance of light
			Lucidity of the visual	Continuous flow of performance
		Ambience (stage and set)	Visibility and Auditory Cues	The intensity /luminance of the lights and the background sounds can be used as cues
			Immersiveness	Audience can see the environment in 360 degree providing with immersion
		Visual Distractions	Visual and Auditory distractions	It can be added or removed as per necessity
			Maintaining the tone of the play	Yes the tone is maintained by adjusting the camera settings and lights

the luminance or the intensity or the color of the light audience can be provided with visual cues. If these technicalities are used as per need the performance can contain a lot of visual and audio Easter eggs in it.

4 Results and Future Work

The findings suggest that each camera position has its own pros and cons depending on the nature of the scene. It is similar to the idea of cinematography in movies and games where you need the specified camera angle in order to provide the best visual and audio experience to the audience [3]. The primary aim of this experiment was to find if an audience is allowed to view a theatre performance through a 360-degree camera on the stage/actor, will they have the same audio and visual experience compared to their traditional off stage seating position. The finding suggests that using the 360-degree camera on the stage/actor and the VR technology audience can virtually teleport themselves onto the stage and the visual experience can be maintained on par with their traditional naked eye experience from off stage. In fact this on stage camera enhances their visual experience by being amidst the performance and helping them to capture minute details of the performance.

However, this idea will not be suitable for all theatre plays and performances. As a starter, this idea could be suitable for performances and theatre styles, that can be devised and re-constructed based on what the VR technology could offer to the audience's virtual participation and the actor's performance. This idea of having various camera positions and placement available, can provide the audience with a choice of changing their field of view to different cameras or to the traditional naked eye view providing them to be the story creator or the editors of their own. However this idea does come with a shortcoming of the relay and lag between the live performance and performance viewed through the Google Cardboard [6], which are proposed to be addressed in the upcoming experiments and research. The other future work from this experiment would be to study the effects of placing a camera on the stage/actor. How does the camera on the stage affect the performance of the actor and virtual realities overall effect in the field of theatre performance.

Even though there is a room for a lot of experiments, breakthroughs and applications using virtual reality, like the virtual teleconferences, virtual gaming and virtual training experience, it is high time that the focus of VR should also fall on other mediums such as arts, theatre and movies. This experiment stands as implementing one of the first and novel ideas of using VR in a theoretical experience and addresses its effects on the audience. This could be one of the first steps in illustrating theatre performance to a new branch, where environment and set properties can be virtually superimposed on the stage and the audience could visualize it using the VR headset. This could result in reducing the time, money and physical efforts involved in creation and designing of set properties and helping the actors and producers to focus more on the content. This opens the scope of the research to a wider prospect and it will be addressed in future projects.

References

1. Baranowski, A.M., Hecht, H.: Effect of camera angle on perception of trust and attractiveness. Empirical Stud. Arts **36**(1), 90–100 (2018)
2. Charlton, J.M., Moar, M.: VR and the dramatic theatre: are they fellow creatures? Int. J. Perform. Arts Digit. Media **14**, 187–198 (2018)
3. Creighton, T.: Camera rolling and action! Metaphor **4**, 38–40 (2014)

4. FitzGerald, L.: Video analysis techniques (2016). https://www.slideshare.net/ejfitzgerald/video-analysis-techniques. Accessed 26 Jan 2020
5. Giannetti, L.D.: Cinematic metaphors. J. Aesthetic Educ. **6**(4), 49–61 (1972)
6. Google. Google Cardboard (2016). https://arvr.google.com/cardboard/
7. Holman, T., Baum, A.: Sound for Digital Video. Taylor & Francis Group (2013). ProQuest Ebook Central. https://ebookcentral.proquest.com/lib/waikato/detail.action?docID=1221469
8. Meyers-Levy, J., Peracchio, L.A.: Getting an angle in advertising: the effect of camera angle on product evaluations. J. Mark. Res. **29**(4), 454–461 (1992)
9. Myszkowski, K., Krawczyk, G., Mantiuk, R.: High Dynamic Range Video. Synthesis Lectures on Computer Graphics and Animation, 1st edn. Morgan & Claypool, San Rafael (2008)
10. Pearlman, E.: 3-D opera. PAJ: J. Perform. Art **40**(1), 43–46 (2018)
11. Shakespeare, W.: Constellation: A Shakespeare Anthology. Farrar Straus & Giroux, New York (1968)
12. Sherman, W.R., Craig, A.B.: Understanding Virtual Reality Interface, Application, and Design. Morgan Kaufmann Series in Computer Graphics and Geometric Modeling, 2nd edn. (2003)
13. Srinivasan, S., Schott, G.: Virtual teleportation of a theatre audience onto the stage: VR as an assistive technology. In: Arai, K., Kapoor, S., Bhatia, R. (eds.) Advances in Information and Communication. FICC 2020. Advances in Intelligent Systems and Computing, vol. 1129. Springer, Cham (2020)
14. Strauss, A., Corbin, J.: Basics of Qualitative Research: Techniques and Procedures for Developing Grounded Theory. Sage Publications, Thousand Oaks (1998)
15. Strong, J., Association of British Theatre Technicians: Theatre Buildings a Design Guide. Routledge, New York, Oxon [England] (2010)
16. White, G.: On immersive theatre. Theatre Res. Int. **37**(3), 221–235 (2012)

Automated Ontology Instantiation of OpenAPI REST Service Descriptions

Aikaterini Karavisileiou, Nikolaos Mainas, Fotios Bouraimis, and Euripides G. M. Petrakis(✉)

School of Electrical and Computer Engineering, Technical University of Crete (TUC), Chania, Crete, Greece
{akaravasileiou,fbouraimis}@isc.tuc.gr, {nmainas,petrakis}@intelligence.tuc.gr

Abstract. Web services are published in service registries on the Web by various software vendors and cloud providers so that can be easily discovered and used by the users. The need for standardizing technologies for service publishing and discovery is of crucial importance for their adoption and market success. Web services need to be described in a way that eliminates ambiguities so that they can be uniquely identified by users or machines. OpenAPI Specification is a powerful framework for the description of REST APIs. OpenAPI standard aims to provide service descriptions which are understandable by both humans and machines. Despite its rigorous service language format, OpenAPI service descriptions can be vague. In our previous work we identified the causes of ambiguity in service descriptions and showed that ambiguities can be eliminated by associating ambiguous OpenAPI properties to concepts in a semantic model. Each service is then uniquely described by a JSON object representing its identity, properties, purpose and functionality. Further to the above objective, the Semantic Web vision defines the requirements for unifying the world of Web services and suggests representing the services as semantic objects accessible on the Web. All services should be published on the Web by means of Semantic Web tools (e.g. ontologies), become discovered by Web search engines (e.g. SPARQL) and be reused in applications. Leveraging latest results for hypermedia-based construction of Web APIs (i.e. Hydra) and the latest update of the OpenAPI specification (OpenAPI v3.0), the present work proposes a reference ontology for REST services along with a formal procedure for converting OpenAPI service descriptions to instances of this ontology.

Keywords: Web service · Ontology · OpenAPI · REST · SHACL · Hydra

K. Arai (Ed.): FICC 2021, AISC 1363, pp. 945–962, 2021.
https://doi.org/10.1007/978-3-030-73100-7_65

1 Introduction

OpenAPI[1] provides a basic format based in JSON (or YAML) for describing RESTful services. The specification offers a syntactic service description that is enriched with text descriptions so that users may easily discover and understand the service and interact with it. OpenAPI Specification[2] is a powerful framework for the description of REST APIs which is now being supported by a large users community and well known software vendors and cloud providers like Google, Microsoft, IBM, Oracle and many others. OpenAPI 3.0[3] is the latest update of the specification. Despite its user friendliness, OpenAPI 3.0 service descriptions can be vague: the same property may appear with different names within the same OpenAPI document or, its meaning may not be defined. In our recent work [3,5] we analyzed the causes of ambiguity and we concluded that, in order to eliminate these ambiguities, OpenAPI properties must be semantically annotated and associated with entities of a semantic model (e.g. in www.schema.org shared vocabulary). Then, it is plausible to represent OpenAPI descriptions using ontologies and we proposed a reference ontology for OpenAPI service descriptions. This is the first work in the literature that handles ambiguities in OpenAPI version 3.0 descriptions.

The present work forwards this approach in certain ways: (a) presents the Beta (stable) version of the OpenAPI ontology and (b) proposes a formal methodology and an algorithm for converting OpenAPI version 3.0 service descriptions to instances of the ontology. A related Web application[4] which supports uploading of OpenAPI descriptions of REST services, instantiation to the ontology and downloading of the resulting ontology, is available on the Web for evaluation by all users. The new (Beta version)of the ontology takes full advantage of Hydra [4] and SHACL[5]. Hydra is a promising technology towards understanding and constructing Web services that meet the HATEOAS requirement of REST architectural style. OpenAPI 3.0 ontology incorporates features of Hydra for modeling service operations along with models not foreseen in Hydra (e.g. REST security features, headers, constraints). Classes together with constraints on class properties are described using SHACL allowing for service descriptions to be validated against the ontology. The OpenAPI ontology combines the advantages of all and supports the efficient representation and dynamic discovery on the Web of hypermedia driven APIs.

Related work on services description languages (with emphasis to REST services) is presented in Sect. 2. The OpenAPI 3.0 approach is discussed in Sect. 3. The instantiation of service descriptions to the OpenAPI 3.0 ontology is discussed in Sect. 4, followed by issues for future work in Sect. 6.

---

[1] https://www.openapis.org.
[2] https://swagger.io/specification/.
[3] https://blog.restcase.com/6-most-significant-changes-in-oas-3-0/.
[4] http://www.intelligence.tuc.gr/semantic-open-api/.
[5] https://www.w3.org/TR/shacl/.

2 Related Work

OpenAPI is a fairly new technology and despite its high impact and adoption by the industry, it has not been explored in depth in the related (Web services) literature. As a result, the proposed model is also novel and the related work is very limited.

Syntactic description languages describe the requirements for establishing a connection with a service and the message formats to successfully communicate with it. WSDL[6] for SOAP, WADL[7], RAML[8] for REST, are popular mainly due to their simplicity and compatibility with common machine readable formats like XML and JSON. They are not specifically designed for hypermedia-driven APIs (such as REST) that call for the dynamic discovery of resources at run-time (referred to as HATEOAS). Their reputation has been overshadowed by OpenAPI Specification[9], formerly known as Swagger.

Semantic approaches build-upon the idea of describing services by means of semantic models (i.e. ontologies) and are more capable of supporting automated service discovery and composition. WSMO[10] defines a conceptual model and WSML language for the semantic description of Web services. OWL-S[11] is an upper ontology for Web services but, similar to all other methods, do not support the dynamic discovery of resources at runtime.

Hydra [4] simplifies the construction of hypermedia-driven APIs. The Hydra vocabulary defines concepts that a server can use to advertise valid state transitions to a client as a result of a sequence of service invocations. Server responses are provided as JSON-LD which a client can use at run-time to discover the available actions and resources, in order to formulate new HTTP requests and achieve a specific goal. Hydra is a promising technology towards understanding and constructing Web services that meet the HATEOAS requirement of REST architectural style.

OpenAPI 2.0. Musyaffa et al. [6] introduce annotations in *Schema* and *Parameter* objects for OpenAPI 2.0. They do not handle all causes of ambiguity nor do they handle OpenAPI 3.0 descriptions. The annotations appear within text properties and cannot be interpreted by a machine without pre-processing. Schwichtenberg, Gerth and Engels [7] map OpenAPI 2.0 service descriptions to OWL-S ontology but do not deal with any causes of ambiguity in service descriptions, nor do they handle security properties. Their approach attempts to find a mapping of OpenAPI 2.0 *Schema* objects to OWL-S using heuristics and name similarity matching techniques. The mapping is error prone and need to be adjusted manually. Most important, their choice of OWL-S model for representing REST services is controversial: OWL-S is good for SOAP services but

[6] https://www.w3.org/TR/wsdl.html.

[7] https://www.w3.org/Submission/wadl/.

[8] https://raml.org.

[9] https://www.openapis.org.

[10] https://www.w3.org/Submission/WSMO/.

[11] https://www.w3.org/Submission/OWL-S/.

not good for hypermedia-driven Web services like REST. Finally, they do not handle OpenAPI 3.0 descriptions.

In a recent contribution, Hamza et al. [1] propose a example-driven approach and a representation of REST APIs for discovering OpenAPI compliant REST Web APIs. This facilitates the API discovery favoring software reuse. The approach does not deal with ambiguity in OpenAPI and is complementary to our work, in the sense that, our proposed ontology representation is a far more powerful tool for supporting services discovery enhanced with reasoning for service synthesis and integration of existing APIs.

3 Annotated OpenAPI Descriptions

Figure 1 illustrates the structure of an OpenAPI 3.0 service description[12]. OpenAPI 3.0 service descriptions comprise many objects. Each object has a list of properties which can be objects as well.

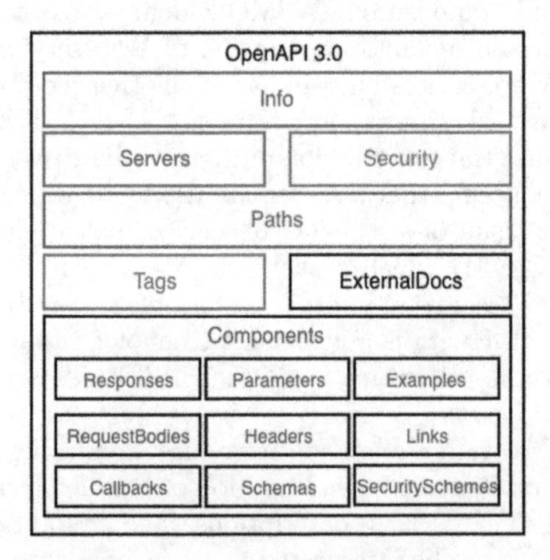

Fig. 1. OpenAPI document structure.

The *Info* object provides non-functional information such as the name of the service, service provider and terms of use. The *Servers* object provides information on where the API servers are located (i.e. multiple servers can be defined). The *Security* object contains the security schemes that the service uses for authentication (i.e. API keys, OAuth2.0 common flows[13] and OpenID Connect[14]). The *Paths* object contains the relative paths of the service endpoints. Each *Path* item describes the available operations based on HTTP methods (e.g. get, put, post).

[12] https://blog.readme.io/an-example-filled-guide-to-swagger-3-2/.
[13] https://oauth.net/2/.
[14] https://auth0.com/docs/protocols/openid-connect-protocol.

The core of an OpenAPI document is the *Operation* object that provides the needed information for expressing HTTP requests to the service. It provides information regarding the HTTP responses of the service. The *Components* object holds a set of reusable objects. These objects can be responses, parameters, schemas, request bodies and more. The *Responses* object specifies the expected responses of an *Operation* and maps each response to an HTTP status code and to HTTP Headers that an operation's response may return. The *Parameters* object specifies the parameters that operations can use. These, can be either path parameters (i.e. specified in operation's path), query parameters (which are appended to the URL when sending a request), header or cookie parameters.

The *Schemas* object specifies the data type that can be used to describe the request and response messages. It can be a primitive (string, integer), an array or a model. For the definition of *Schema* objects, the specification is based on JSON Schema[15]. A *Schema* object definition can be be enhanced with XML data types. Finally, new data types can be defined as a combination or a specification of existing ones (i.e. using the *allOf* property or, the *oneOf* property, respectively).

Listing 1.1 demonstrates how polymorphism is expressed using the discriminator property: the model of a *Person* uses the *gender* property as a discriminator in order to determine whether this is a model of a male or a female person. In the same example, the *allOf* property is used to specify that a *Male* or a *Female* model has the properties of a *Person* model and additional properties.

Listing 1.1. A *Person* Model Polymorphism Example.

```
schemas:
  Person:   # An Person model that will be extended
    type: object
      properties:
        firstname:
          type: string
        lastname:
          type: string
        gender:
          type: string
        required:
          - firstname
          - lastname
          - gender
        discriminator:
          propertyName: gender
  Male: # A Male model extending the Person Model
    description: A representation of a male person
    allOf:
      - $ref: '#/components/schemas/Person'
      - type: object
```

[15] http://json-schema.org/latest/json-schema-core.html.

```
    properties:
      height:
        type: integer
      description: height in cms
        weight:
          type: integer
      description: weight in kgs
      required:
        - height
        - weight
  Female: # A Female model extending the Person Model
    description: A representation of a female Person
    allOf:
      - $ref: '#/components/schemas/Person'
      - type: object
    properties:
      eyesColor:
        type: string
    required:
    ...
```

In an OpenAPI document, there can be properties that share the same meaning (although they are defined using different names) or, their meaning is ambiguous. Probably, a human can easily resolve ambiguities either by the element names or by the description that may be provided with the properties. However, in order for a machine to act similarly to a human it is necessary to provide additional information, that clarifies ambiguous properties in an OpenAPI service description. In our recent work [3,5] we introduced extra properties (referred to as *extension properties*) to annotate OpenAPI properties which are ambiguous. Table 1 summarizes the extension properties, their scope and their meaning.

Table 1. OAS extension properties

Property	Applies to	Meaning
x-refersTo	Schema object	The concept in a semantic model that describes an OAS element
x-kindOf	Schema object	A specialization between an OAS element and a concept in a semantic model
x-mapsTo	Schema object	An OAS element which is semantically similar with another OAS element
x-collectionOn	Schema object	A model describes a collection over a specific property
x-onResource	Tag object	The specific *Tag* object refers to a resource described by a *Schema* object
x-operationType	Operation object	Clarifies the type of operation

The *x-refersTo* extension property specifies the association between an OpenAPI element in *Schemas* object and a concept in a semantic model. The property accepts a URI, that represents the concept in the semantic model. Listing 1.2 demonstrates how the property *x-refersTo* is used to semantically annotate a *Person* model and its properties. However, it is not always possible for a model to have a relation with an equivalent semantic concept (e.g. model may have a narrower meaning over the referenced semantic concept). For example, if the model *Person* model describes a specific group of people (e.g. teenagers), it would be inappropriate to associate the model with a generic concept such as the *Person* type of the www.schema.org vocabulary and, a more specific type would be used instead. In this case, the *x-kindOf* extension property is used to denote that the model is actually a subclass of the referred semantic concept. In addition, the *x-mapsTo* extension property can be used to define elements that share the same semantics (e.g. in cases where the same property is referred with different names across the OpenAPI document).

Listing 1.2. Semantic Annotations of Schema Objects.

```
schemas:
Person:
  type: object
  x-refersTo: http://schema.org/Person
  properties:
    firstname:
      type: string
      x-refersTo: http://schema.org/givenName
    lastname:
      type: string
      x-refersTo: http://schema.org/familyName
    gender:
      type: string
      x-refersTo: http://schema.org/gender
    required:
      - firstname
      - lastname
      - gender
      ...
```

The *x-collectionOn* extension property is used to indicate that a model in *Schemas* object is actually a collection. Typically, a collection (or a list) of resources is described using the *array* type. However, the definition of a collection can be encapsulated within an *object* type with additional properties so that, it is not easy to distinguish the type of the objects that form the collection. Then, the *x-collectionOn* property is used to denote the data types of the objects of the collection. Listing 1.3 defines a model as a collection of *Person* objects (*totalItems* property denotes population).

Listing 1.3. Model Definition Representing a Collection.

```
...
schemas:
  PersonCollection:    # A groupd of Persons
  x-collectionOn: persons
  type: object
  properties:
    persons:
    type: array
    items:
    $ref: '#/components/schemas/Person'$
  totalItems:
    type: integer
```

The *x-onResource* extension property is used in *Tag* objects to specify the resource that a tag refers. In OpenAPI, tags are used to group operations either by resources or by any other qualifier. If the tag is used to group operations by resources, a human may recognize that the referred resource is described by an object in *Schemas* but, a machine cannot. The *x-onResource* property is used to associate the tag with an object that describes a specific resource. Finally, the *x-operationType* extension property is used to semantically specify the type of an *Operation* object. A request (i.e. an operation) is characterized by the HTTP method it uses. However, the semantics of the HTTP methods are too generic and in the context of a service they may have a more specific meaning (e.g. a GET request on a path is in fact a search operation). The value of the property is a URL pointing to the concept that semantically describes the operation type. The *Action*[16] type of the *www.schema.org* vocabulary provides a detailed hierarchy of *Action* sub-types that can be used by the property.

4 OpenAPI Ontology

The OpenAPI ontology of Fig. 2 captures all information of an OpenAPI description of [3]. At the heart of the ontology is *Hydra* core vocabulary. In this work, *Hydra* is enhanced with additional semantic models representing information about security, headers and constraints.

The *Document* class provides general information about the service (in *Info* class) and, specifies service paths, the entities and the security schemes that it supports. The *Path* class represents (relative) service paths (in *pathName* property). The *Operation* class provides information for issuing HTTP requests. Request bodies are represented by the *Body* class, while, responses are declared in *Response* class specifying the status code and the data returned. The entire range of HTTP responses is represented. The *MediaType* class describes the format (the most common being JSON, XML) of a request or response body data. Class *Operation* refers to a security scheme in *SecurityRequirement* class.

Figure 3 shows the security schemes supported by OpenAPI. Class *Security* has security schemes as sub-classes. Class *OAuth2* has different flows (grants) as

[16] http://schema.org/Action.

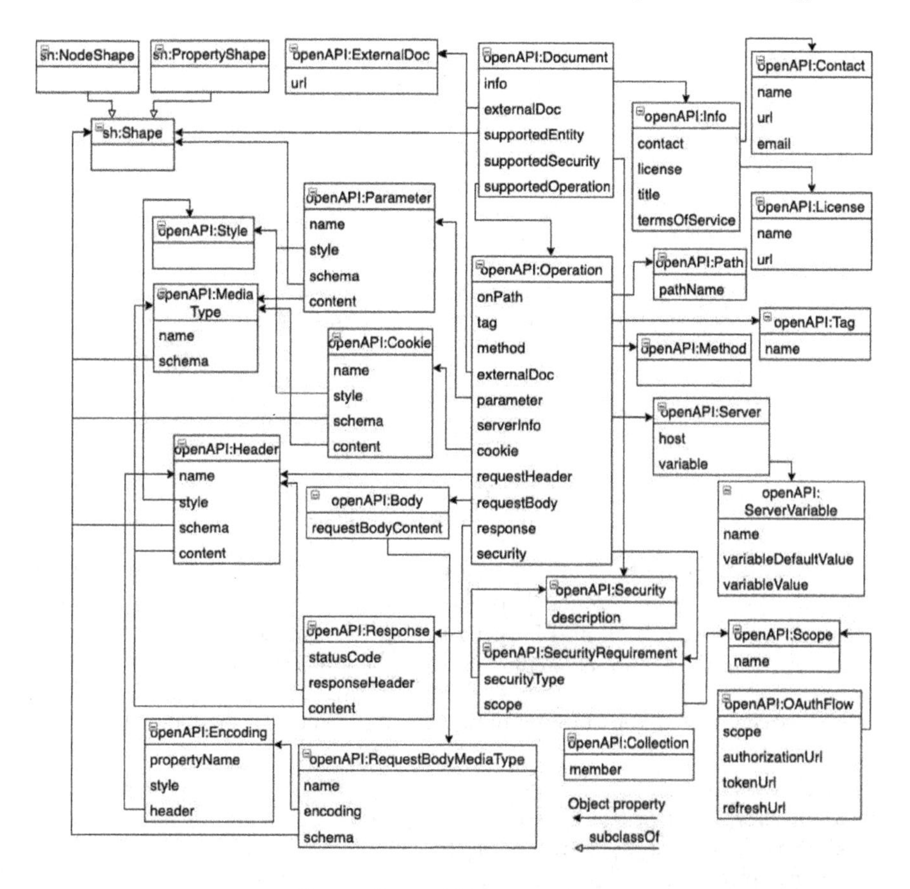

Fig. 2. OpenAPI 3.0 ontology.

sub-classes. If the security scheme is of type *OAuth2* or *OpenID Connect*, then scope names are defined as properties.

Schema objects are expressed as classes, object and data properties using SHACL. SHACL is an RDF vocabulary that can be used to define classes together with constraints on their properties. It provides built-in types of constraints (e.g. cardinality: *minCount, maxCount*) and expr the *Shape* class. The *NodeShape* class defines the properties of a class and specifies whether a class may contain additional properties (*additionalProperties*) of a specific type. Additionally, it represents operations related to a class (*supportedOperation*), which come from *x-onResource* extension property. Class *PropertyShape* represents the properties of a class, their datatype and restrictions (e.g. a maximum value for a numeric property) and indicates whether the supported property is required or read-only. Table 2 shows how Schema object properties are associated with properties of the SHACL vocabulary.

Listing 1.4 shows how the *Person* model of Listing 1.1 is represented in the OpenAPI ontology. The model contains references to the www.schema.org vocabulary using the *x-refersTo* extension property. The SHACL class *Person-Shape* is now defined according to the definition of *Person* object with the addi-

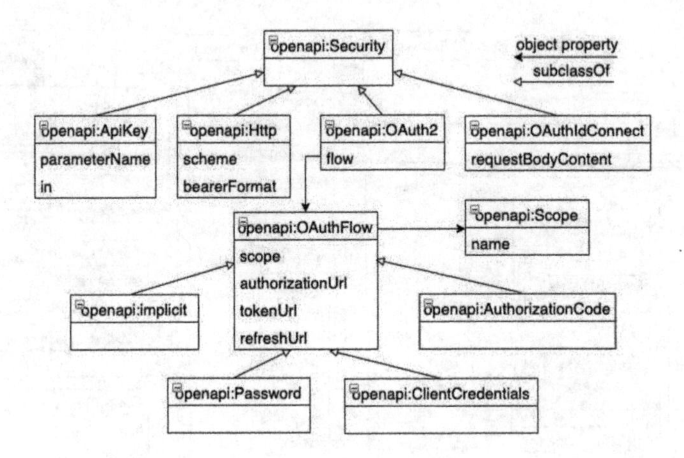

Fig. 3. OpenAPI 3.0 security class.

tion of new data properties and constraints (e.g. each person has exactly one first name, last name and gender). If the *x-refersTo* extension property is missing from a model definition and its properties, then the corresponding class and its properties must be defined explicitly in the OpenAPI ontology.

Listing 1.4. Representation of an OpenAPI Model in the Ontology.

```
...
ex:PersonShape
  a sh:NodeShape;
  sh:targetClass schema:Person;
  sh:property [
    sh:path schema:givenName;
    sh:name "firstname";
    sh:datatype xsd:string ;
    sh:minCount 1;
    sh:maxCount 1;
  ];
  sh:property [
    sh:path schema:familyName;
    sh:name "lastname";
    sh:datatype xsd:string;
    sh:minCount 1;
    sh:maxCount 1;
  ];
  sh:property [
    sh:path schema:gender;
    sh:name "gender";
    sh:datatype xsd:string;
    sh:minCount 1;
    sh:maxCount 1;
  ] .
...
```

Table 2. Mapping OpenAPI Schema Object Properties to SHACL.

Schema object property	SHACL property
maximum	sh:exclusiveMaximum if openAPI exclusiveMaximum is true
exclusive Maximum	sh:inclusiveMaximum if openAPI exclusiveMaximum is false
minimum	sh:exclusiveMinimum if openAPI exclusiveMinimum is true
exclusive Minimum	sh:inclusiveMinimum if openAPI exclusiveMinimum is false
maxLength	sh:maxLength
minLength	sh:minLenght
pattern	sh:pattern
maxItems	sh:maxCount
minItems	sh:minCount
enum	sh:in
allOf	sh:and
oneOf	sh:xone
anyOf	sh:or
not	sh:not
default	sh:defaultValue

A model defined using the combination of *allOf* property and *discriminator* property, is represented in the OpenAPI ontology as a subclass of the model that is extended. A subclass defined using *x-kindOf* become a sub-class of the referenced semantic concept. Listing 1.5 demonstrates how the models of Listing 1.1 are represented in the OpenAPI ontology. The *Male* and *Female* classes are defined as sub-classes of the *Person* class due to the existence of the *allOf* property in model definitions. In addition, they inherit the properties of the *Person* model.

Listing 1.5. Representation of OpenAPI Models Defined using the *allOf* Property in the Ontology.

```
...
ex:Person
  a rdfs:Class .
ex:Male
  rdfs:subClassOf ex:Person.
ex:Female
  rdfs:subClassOf ex:Person.
ex:PersonShape
  a sh:NodeShape;
  sh:targetClass ex:Person;
```

```
    sh:property ex:FirstNameShape, ex:LastNameShape, ex:
        GenderShape;
  openapi:discriminator [
    openapi:discriminatorProperty "gender";
  ] .
  ex:MaleShape
    a sh:NodeShape;
    sh:targetClass ex:Male;
    sh:and (
      ex:PersonShape
      ex:HeightShape
      ex:WeightShape
    ) .
  ex:FemaleShape
    a sh:NodeShape;
    sh:targetClass ex:Female;
    sh:and (
      ex:PersonShape
      ex:EyesColorShape
    ) .
  ...
```

Collections are represented in the OpenAPI ontology through the *Collection* class by specifying the members (*member*) of a collection. Listing 1.6 demonstrates the *PersonCollection* of Listing 1.3 representation in the OpenAPI ontology. A *PersonCollection* class is defined in the OpenAPI ontology as a subclass of the *Collection* class. The *PersonCollection* model in Listing 1.3 is defined as an object, and the *x-collectionOn* extension property is used in order to specify that the model is in fact a collection whose members are *Persons*. Without the *x-collectionOn* extension property, the *PersonCollection* model would be defined as a simple class without any reference of being a collection.

Listing 1.6. Representation of Collections in the OpenAPI Ontology.

```
  ...
  ex:PersonCollection
    rdfs:subClassOf openAPI:Collection.
  ex:PersonCollectionShape
    a sh:NodeShape;
    sh:targetClass ex:PersonCollection;
    sh:property [
      sh:path openapi:member;
      sh:name "persons";
      sh:class ex:Person;
    ];
    sh:property [
      sh:path ex:totalItems;
```

```
      sh:name "totalItems";
      sh:datatype xsd:integer;
    ].
  ...
```

Finally, OpenAPI parameters are represented as separate classes for every parameter type in the OpenAPI ontology. The *Header* class contains all the definitions of header parameters that are used in HTTP requests and responses. The *Cookie* class defines the cookies that are sent through HTTP requests and responses. In addition, the *Parameter* class defines all parameters that are attached to operation's URL. The class is further organized in *PathParameter* and *Query* classes that refer to the corresponding path and query parameters of the specification.

A request or response body (defined using *content* property) is used to send and receive data via the REST API respectively (a response contains also a response code (e.g. 200, 400, etc.). Media type is a format of a request or response body data in different formats, the most common being JSON, XML, text and images. They are typically defined in *Paths* object; however re-usable bodies can also be defined in *Components* object. Each media type includes a *Schema* property, defining the data type of the message body. Request and response bodies are represented as properties of class *Operation*. Request and response bodies in particular, are defined as classes and so is defined their media type. Class *Encoding* defines keywords denoting serialization rules for media types with primitive properties (e.g. *contentType* for nested arrays or JSON).

5 Instantiating OpenAPI Descriptions

Instantiating OpenAPI descriptions and services is a rather complicated process. The reasons can be any or all the following: the same property may appear many times in the same document (with the same or different meaning), with different scopes (local property declarations overshadow global ones) or, it can be nested inside other properties. In addition, a *Schema* object can be defined as an extension or the composition of existing models (e.g. using the *allOf* property) which add extra complexity to the instantiation process. The proposed algorithm handles all these issues.

The algorithm presented hereafter is described in detail in the author's thesis [2]. Algorithm 1 scans the OpenAPI document of a service and instantiates *OpenAPI* objects to classes of the ontology. In particular, after uploading the ontology in the memory, the algorithm will scan the *OpenAPI* file to extract *info*, *servers*, *securitySchemes*, *securityRequirements*, *tags* and *paths* objects. These objects will become individuals of their corresponding classes.

Algorithm 1. Instantiating OpenAPI Object to Ontology.

```
1: procedure ParseDocumentObject(OpenAPI doc)              ▷ Main body of algorithm:
2:     ont =InitializeOntologyModel();
3:     InfoObject(ont, doc);
4:     Read servers property of OpenAPI object;
5:     globalServerList= ServerObject(ont, servers);
6:     SecuritySchemeObject(ont, doc);
7:     Read security property of OpenAPI object;
8:     globalSecurityReqList= SecurityRequirementObject(ont, security);
9:     tagShapeMap = TagObject(ont, doc);
10:    PathObject(ont, doc, globalServerList, globalSecurityReqList, tagShapeMap);
11: function InfoObject(Ontology ont, OpenAPI doc):          ▷ Function definitions:
12:    Info object becomes individual of class Info;
13: function ServerObject(Ontology ont, List servers):
14:    Initialize list containing server individuals (serverList);
15:    for Server object ∈ servers do
16:        Server object becomes individual of class Server;
17:        add individual in serverList;
18:    return serverList;
19: function SecuritySchemeObject(Ontology ont, OpenAPI doc):
20:    Each Security Scheme object becomes individual of class Security;
21: function SecurityRequirementObject(Ontology ont, List security):
22:    Initialize list containing Security Requirement individuals (securityReqList);
23:    for Security Requirement object ∈ security do
24:        Security Requirement object becomes individual of class SecurityRequirement;
25:        add individual in securityReqList;
26:    return securityReqList;
```

The *OpenAPI* object is mapped to class *Document*. There may exist more than one appearances of *servers* or *securityRequirements* in an *OpenAPI* file. Property *servers* declares server information which applies across the description (global servers). This will be overwritten by server information defined in *Path* or *Operations* objects. Similarly, *Security* property declares security requirements. Security requirements declared by an operation will also override global declaration of security requirements. Property *Tags* contains the *Tag* objects for operations which are grouped. Through the *x-onResource* property *Tag* objects can associate operations with *Schema* objects.

Algorithm 3 illustrates the instantiation of *Tag* objects in the ontology and how *x-onResource* relations are handled. In Algorithm 4, *Tag* names and their associated *Shapes* are kept in a Map structure (*tagShapeMap*) that will be used when instantiating *Operation* objects. *tagShapeMap* defines a mapping *(key, value)* between a string and an individual. An entry to the *tagShapeMap* will contain the tag name and the corresponding *Shape* individual. This will take place

Algorithm 2. Instantiating *Path* Objects in OpenAPI Ontology.

```
1: function PathObject(Ontology ont, OpenAPI doc, List globalServerList, List
   globalSecurityReqList, Map tagShapeMap)
2:     for Path Item object ∈ doc do
3:         Create an individual of class Path;
4:         Select the server list that will be used in operations (pathServerList);
5:         if servers property in Path Item object is empty then
6:             pathServerList = globalServerList;
7:         else
8:             Read Server objects of Path Item object (path level);
9:             pathServerList = ServerObject(ont, servers);
10:        Read Parameter objects of Path Item object (path level);
11:        pathParameterList= ParameterObject(ont, parameters);
12:        for Operation object (opObj) ∈ Path Item object do
13:            operationInd = OperationObject(ont, opObj, pathServerList, glob-
   alSecurityReqList, pathParameterList, tagShapeMap);
14:            add property onPath to operationInd associating Path and Operation
   individuals;
15: function ParameterObject(Ontology ont, List parameters)          ▷ Function
   definition:
16:    Initialize list containing parameter individuals (parameterList);
17:    for Parameter Object ∈ parameters do
18:        Create individual of corresponding class (PathParameter, Query, Cookie,
   Header);
19:        add individual in parameterList;
20:    return parameterList;
```

if an *x-onResource* annotation has been defined on a tag object (i.e. describing a *Schema* object). The *Schema* object will be converted to an individual of the *Shape* class and will be added inside the corresponding mapping (e.g. *(tag.getName(), SchemaInd)*).

Algorithm 2 shows how *Path* objects are converted to individuals of class *Path*. This is where *Operation* individuals are created with their respective properties (i.e. *tag*, *security*, *parameter*, *server*). This is done after creating individuals for Server and Parameter objects that may be declared in a Path. *Server* objects declared in *Path* (as stated in OpenAPI Specification) will override global declaration of *Server* Objects.

Algorithm 3. Create Instances of *Tag* Class.

```
1: function TagObject(Ontology ont, OpenAPI doc)
2:     initialize a Map structure (tagShapeMap) to hold the entries of Tag and Shape
   individuals;
3:     for Tag Object ∈ doc do
4:         create an individual of class Tag;
5:         if Tag object contains the x-onResource property then
6:             find the Schema object it refers to;
7:             convert Schema Object into an individual of class Shape;
8:             add individuals (Tag and Shape) into tagShapeMap;
9:     return tagShapeMap;
```

Algorithm 4 shows how the individuals of class *Operation* are created. When the *x-operationType* annotation is used, the *Operation* individual is also considered as an individual of the provided Semantic entity. The structural elements of the operation (e.g. *responseObject*, *requestBody*, *security* and *tag* become properties of the *Operation* individual (e.g. *operationInd property:tag tagInd* is an example triple). Property *method* defines the type of an operation (*get*, *put*, etc.) which are already defined in the OpenAPI ontology. *Parameter* objects can be any of type *Path*, *Query Header* or *Cookie* and are instantiated to the corresponding classes (i.e. *PathParameter*, *Query*, *Header* and *Cookie*). Then, parameter individuals are associated to the *Operation* individual using the respective properties. Finally, property *onPath* associates an *Operation* individual with a *Path* individual.

The OpenAPI Specification allows the combination and extension of model definitions using the *allOf* property. However, in order to support polymorphism, the *discriminator* property is used to determine which *Schema* definition validates the structure of the model. A model defined using the *allOf* property becomes a subclass of the model it extends. In the example of Listing 1.1, classes *Male* and *Female* become sub-classes of *Person*. OpenAPI service description may have *Schema* objects sharing common properties. Instead of describing these properties for each *Schema* repeatedly, these *Schema* objects are described as a composition of common properties and schema-specific properties. If these *Schema* objects are not be related to a semantic entity, we suggest annotating *Schema* objects with the property *x-refersTo: none*, and the algorithm should not attempt to relate these models with any semantic entity.

Algorithm 4. Converting *Operation* Object to Instances of Class *Operation.*

```
1: function OperationObject(Ontology ont, Operation Object opObj, List servers,
   List securityReqList, List parameters, Map tagShapeMap)
2:     Create individual (operationInd) of class Operation;
3:     Check for the x-operationType property;
4:     if opObj contains the x-operationType property then
5:         Add operationInd as member of the class the property refers;
6:     for Tag ∈ opObj do
7:         Add property Tag to operationInd;
8:         if Tag exists in tagShapeMap then
9:             Get the corresponding Shape Individual;
10:            Add property supportedOperation to Shape individual;
11:    if servers property in opObj is not empty then
12:        for Server Object ∈ servers do
13:            Server object becomes individual of class Server;
14:            Add property server to operationInd;
15:    else
16:        for server individual in pathServerList do
17:            Add property server to operationInd;
18:    if security property in opObj is not empty then
19:        for Security Requirement object ∈ security do
20:            Security Requirement object becomes individual of class SecurityRe-
   quirement;
21:            Add property security to operationInd;
22:    else
23:        for security requirement individual in globalSecurityReqList do
24:            Add property security to operationInd;
25:    for Parameter object ∈ parameters property of opObj do
26:        Parameter object becomes individual of the corresponding class (PathPa-
   rameter, Query, Cookie, Header);
27:        Add property parameter to operationInd for PathPrameter and Query indi-
   viduals;
28:        Add property cookie to operationInd for Cookie individuals;
29:        Add property requestHeader to operationInd for Header individuals;
30:    for parameter individual in pathParameterList do
31:        if parameter individual doesn't exist in parameters of operationInd then
32:            Add property parameter to operationInd for PathPrameter and Query
   individuals;
33:            Add property cookie to operationInd for Cookie individuals;
34:            Add property requestHeader to operationInd for Header individuals;
35:    if RequestBody property is not empty in opObj then
36:        RequestBody object becomes individual of class Body;
37:        Add property requestBody to operationInd;
38:    for Response object ∈ responses property of opObj do
39:        Response object becomes individual of class Response;
40:        Add property response to operationInd;
41:    return operationInd;
```

6 Conclusions

Instantiating the RESTful APIs of GURU[17] catalogue is an issue for future work. The instantiated ontology will be available on the Web. Future improvements to the ontology and algorithm will incorporate comments received by its users worldwide. Query formulation on the catalogue can be facilitated by Graphical User Interfaces allowing users to select service properties from pull-down menu. Alternatively, a dedicated query language in the spirit of SOWL-QL [8] will be designed so that the user need not be familiar with the peculiarities and syntax of the underlying representation.

References

1. Hamza, E., Izquierdo, C., Luis, J., Jordi, C.: Example-driven web API specification discovery. In: Modelling Foundations and Applications (ECMFA 2017), Marburg, Germany, pp. 267–284, July 2017
2. Karavasileiou, A.: An ontology for OpenAPI version 3 services in the cloud. Technical report TR-TUC-ISL-07-2019, Diploma Thesis, School of Electrical and Computer Engineering, Technical University of Crete (TUC), Chania, Crete, November 2019
3. Karavasileiou, A., Mainas, N., Petrakis, E.G.M.: Ontology for OpenAPI REST services descriptions. In: International Conference on Tools with Artificial Intelligence (ICTAI 2020), Baltimore, MD, USA, pp. 35–40, November 2020
4. Lanthaler, M., Gütl, C.: A vocabulary for hypermedia-driven web APIs. In: Workshop on Linked Data on the Web (LDOW 2013), Rio de Janeiro, Brazil (2013)
5. Mainas, N., Petrakis, E.G.M.: SOAS 3.0: semantically enriched OpenAPI 3.0 descriptions and ontology for rest services. In: IEEE International Conference on Semantic Computing (ICSC 2020), San Diego, California, pp. 207–210, February 2020
6. Musyaffa, F.A., Halilaj, L., Siebes, R., Orlandi, F., Auer, S.: Minimally invasive semantification of light weight service descriptions. In: IEEE International Conference on Web Services (ICWS 2016), San Francisco, CA, USA, pp. 672–677, June 2016
7. Schwichtenberg, S., Gerth, C., Engels, G.: From open API to semantic specifications and code adapters. In: IEEE International Conference on Web Services (ICWS 2017), San Francisco, CA, USA, pp. 484–491, June 2017
8. Stravoskoufos, K., Petrakis, E.G.M., Mainas, N., Batsakis, S., Samoladas, V.: SOWL QL: Querying spatio-temporal ontologies in OWL. J. Data Semant. **5**(4), 249–269 (2016)

[17] https://apis.guru/browse-apis/.

High Level Software Separation: Experience Report for e-Health and Legal Metrology

Patrick Scholz[1], Daniel Peters[1](✉), Jörn Berger[2], and Florian Thiel[1]

[1] Physikalisch-Technische Bundesanstalt (PTB), Abbestr. 2-12, 10587 Berlin, Germany
{patrick.scholz,daniel.peters,florian.thiel}@ptb.de
[2] Xiralite GmbH, Robert-Koch-Platz 4, 10115 Berlin, Germany
berger@xiralite.com

Abstract. We give a practical example of transferring High Level Software Separation and Risk Analysis Techniques initially developed for measuring instruments in Legal Metrology, into the medical device domain. The concepts are used to separate medically sensitive, and hence regulated software, from the dynamic, unregulated software parts. In that way, the highly valuable patient data are securely separated from potential security threats, ensuring greater IT protection of the medical device and hence the patient data. Concretely, virtualization with the Xen hypervisor has been implemented, to show how IT security for Windows software can be enhanced by separating software parts and restricting access to hardware. The knowledge gained is presented as an experience report and compares the concept with the already known software separation mechanisms for Legal Metrology. Additionally, an experimental test was done, to analyze how a second graphics card, which is forwarded to the medically relevant software parts through PCI passthrough, can significantly improve performance.

Keywords: Hypervisor Xen · Legally and medically relevant software · Software separation · Virtualization

1 Introduction

As part of e-health in the German healthcare system, digital technologies are addressing challenges of treating increasingly elderly and chronically ill people, financing costly medical innovations, and providing medical care to rural, underdeveloped areas. As a key technology of the 21st century, information and communication technology is becoming increasingly important in health care as well, enabling more efficient care and broader access to medical expertise for people who previously did not have access. In medicine, the health and protection of the patient are the most important, see §1 in the German Medical Devices Act [1]. The German Federal Office for Information Security (BSI) names the healthcare system as a critical infrastructure in the national IT security law [2], thereby emphasizing the need for protection in this area. In the European General Data Protection Regulation [3] and in the German Federal Data Protection Act [4], health data is listed as a special category of personal data, which emphasizes a higher

K. Arai (Ed.): FICC 2021, AISC 1363, pp. 963–975, 2021.
https://doi.org/10.1007/978-3-030-73100-7_66

level of protection of this data. Therefore, the protection of sensitive patient data against unauthorized disclosure as well as tampering is the number one priority in IT security for the health care industry.

In order to improve IT security for medical devices, we turned our attention towards software separation through virtualization. Concretely, we used a hypervisor that enables the operation of several isolated virtual machines (VMs) as an additional software abstraction layer that also controls and monitors the software within those virtual machines. There are two major types of hypervisors. The bare-metal type I runs directly on the hardware, while a type II hypervisor is a program running on a host operating system. In our software architecture we decided to use Xen. The open source type I hypervisor Xen has been developed by the Linux Foundation, since 2013, and has many benefits [5]. The architecture separates the hypervisor from the Linux kernel, so attacks on the Linux kernel normally should have no effect on Xen itself. The Xen Security Modules define separate VMs with limited rights. Xen virtual machines can run at almost native performance through their paravirtualization capabilities.

1.1 Overview

First, we present the hardware basis for our experiments, the Xiralite X5 camera, and analyze possible security risks of the given device. Second, we describe our security architecture with which we improve the security of the given device. In Sect. 3, we focus on the devised Xen framework in detail, especially the start-up procedure. Test results can be found in Sect. 4, e.g. impact of certain visualization components on the system performance. We summarize our results in Sect. 5 and finish with a conclusion in Sect. 6.

2 The Medical Device

2.1 The General Outline of the Device

Xiralite GmbH is a Berlin-based medical technology company for optical imaging in rheumatology [6]. The company is the manufacturer of the Xiralite series of fluorescence cameras, like the version X4 [7] and its successor X5, shown in Fig. 1.

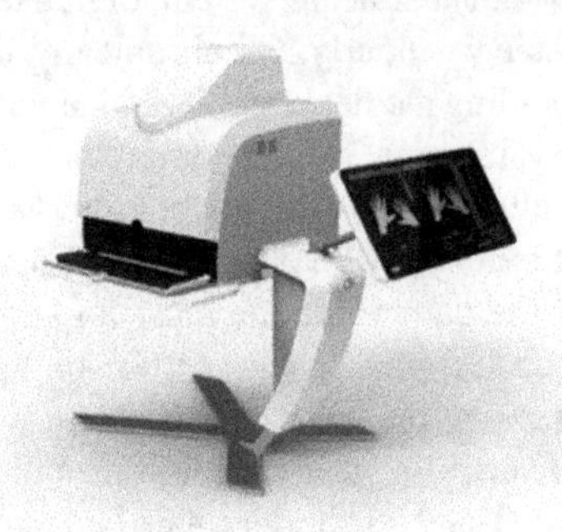

Fig. 1. Xiralite X5 fluorescence camera systems of the Xiralite GmbH.

Our test was implemented on the Xiralite X5. In general, these devices display microcirculations on up to 30 joints of the hands to detect inflammatory foci such as arthrosis or rheumatoid arthritis. The device uses imaging techniques to identify disturbances of the blood circulation in the hands. The kind of disturbance gives a clear indication of a possible disease. For an examination of a patient, the fluorescent dye "indocyanine green" is injected into the blood stream, which is approved for microcirculation diagnostics in Europe.

During a clinical routine diagnostic, the dye is stimulated by light-emitting diodes in the dark red-light spectrum during the approximately 360 s long examination, and the light signal intensity of the hands is recorded with a high-sensitivity CCD camera. Throughout the examination a picture is taken every second. The diagnostic software XiraView 4.0 controls the radiation-free examination and assists in the subsequent evaluation. Since the first serial approval in Europe in 2009, more than 40 medical devices have been installed in clinics, hospitals and research facilities in Germany alone. Thus, 13.000 patients have already been examined with the Xiralite procedure.

2.2 Hardware Specifics

For conceptual design, development, establishment and subsequent practical evaluation of the IT security software architecture presented here, a modified version of the Xiralite X5 was used. This test platform consisted mainly of the computer control hardware of the X5, without the actual medical measuring hardware, thus without a CCD camera or LEDs. The core of the hardware of the X5 test system was the FUJITSU motherboard D3433-S2 [8], which used the Intel Q170 chipset. It leveraged the VT-d Intel Directed I/O virtualization technology, which expanded the Intel virtualization solution VT-x with support for I/O device virtualization, as well as improving the security and reliability of systems and the performance of I/O devices in virtualized environments. The X5 test system included 8GB of RAM, 2 ethernet ports and a 4-core (8-threaded) CPU, Intel Core i7-7700T CPU @ 2.90 GHz. There were also 5 USB 2.0 ports and 5 USB 3.0 ports available. In addition to the on-board processor graphics card Intel HD Graphics 630, an NVIDIA GF106GL Quadro 2000 1 GB PCI-Express graphics card was installed. This was passed through to the virtual machine and used as a dedicated graphics adapter for the VM in order to yield the necessary performance. A more in-detail discussion can be found below. Additionally, two 1 TB hard drives have been installed in the medical measure device, which were combined to a RAID 1 system to redundantly store the highly sensitive patient and examination data.

2.3 Identifying the Risk

In order to determine what quality our software architecture should at least achieve and what challenges it would face, a risk analysis was initially performed. The described device can be classified as a measuring instrument for the medical domain. In 2015, the Physikalisch-Technische Bundesanstalt (PTB) developed a risk analysis method based on ISO/IEC 27005 [9] for Legal Metrology [10, 11], which since then has been used to assess the software quality and IT security of measuring instruments within the process of conformity assessment according to 2014/32/EU [12]. The PTB is Germany´s National

Metrology Institute and is tasked with legal regulations on the national and European level, e.g. in Legal Metrology. In this framework the PTB guaranties consumer protection and trust in measuring instruments and the measurement results. Additionally, to the risk analysis method two other documents are of interest concerning software in Legal Metrology.

In Europe, the Measuring Instruments Directive (MID) 2014/32 / EU [12] discusses requirements for measuring instruments in general and more specific for software. The WELMEC cooperation formulates guidelines such as the WELMEC 7.2:2019 [13] Software Guide. In this guide, all software that is used for the function of a measuring device is considered legally relevant. This software is under legal supervision and is subject to a certification process for possible approval of the measuring device. In addition, any software that is not separate from the legally relevant software and thus can influence the legally relevant software is considered to be legally relevant. Using software separation, software that is not necessary for the operation of a measuring device can be separated and thus released from legal monitoring. This legally non-relevant software can therefore be changed without re-certification. In the past, two distinctions were helpful, i.e. low and high-level separation. Here, the former is a software separation at the programming language level and is described by combining program parts, i.e. by means of subroutines, procedures, functions or classes. We go one step further and concentrate on high level separation. Hereby, the WELMEC 7.2 recommends that legally relevant software parts should be compiled into one object, or program that runs separated by means of the underlining operating system from legally irrelevant programs that are unrelated with the measurement. Specifically, in our architecture the distinct virtual machine is used for this purpose.

The risk assessment procedure developed and used by PTB can be applied for a wide range of software and is not limited to the unique constrains in Legal Metrology. At the beginning of the assessment the assets that need protection must be identified. In our case those assets are health care assets. Furthermore, we formulated possible threats and attack vectors for the Xiralite X5. For those threats we developed countermeasures in our IT security architecture. Since we are dealing with a medical device, certain assets are already defined by laws and official regulations such as the German industry standards for medical devices [14, 15] and medical software [16], recommendations of the US Food and Drug Administration [17, 18] as well as the German Medical Devices Act [1] and the Federal Data Protection Act [4] were used. Protective goods would be the protection of the patient and the patient and examination data (see [1] in Sect. 1 and [4] in Sect. 1). Therefore, the integrity, authenticity, confidentiality, and availability of the data, as well as availability of the medical device have been identified as targets of a threat, as shown in Table 1.

In addition to a denial-of-service (DOS) or distributed-denial-of-service (DDOS) attacks, attack vectors have also been analyzed for zero-day exploits against the Linux Ubuntu operating system or the Xen hypervisor. We have also considered that an attacker could access the BIOS or UEFI of the medical device or access one of the operating systems of a virtual machine by "guessing" the password, as well as the unwanted installation of malware by a user of the medical device by opening an infected web site or by opening a compromised USB device or a malicious e-mail attachment. Also, a banality such as the failure of a hard drive and the concomitant loss of patient and examination data has been considered.

Table 1. Threats against the protected goods.

ID	Threat target	Description
B1	Data integrity	Attacker changes data on the medical device
B2	Data authenticity	Attacker generates fake data that looks real
B3	Data confidentiality	Attacker copies or steals data
B4	Data availability	Attacker manipulates medical device so that data access is denied
B5	Device availability	An attacker succeeds in denying device assess

3 Our Framework

3.1 High Level Software Separation

Our main contribution is the modification of the Xiralite X5 with regards to the separation of the individual software components using virtual machines. Emphasis is placed on the availability of the device in the clinic and its data protection, because the Xiralite device collects and stores highly sensitive patient data which requires the highest level of protection in health care. Theft or manipulation of patient data could lead to misdiagnosis, followed by mistreatment and high claims for damages. Therefore, a robust IT security construct is necessary. At the same time, we want a system that can be easily implemented, and whose software components have already proven themselves in the field, to ensure both performance and reliability, and ultimately, security. Hence, we decided to use the open source type 1 hypervisor Xen in version 4.15.0 together with a minimal Linux Ubuntu operating system core in version 18.04 for virtualization, collectively referred to as the first domain of virtualization, i.e., domain 0, shown in Fig. 2. Dom0 represents an administrative layer that is used to manage real and virtual hardware resources and assigning them to the VMs as well as monitoring the VMs. We chose Ubuntu because it is one of the broadly used and well-supported Linux distributions and it is Secure Boot compatible. We deliberately wanted to have few functionalities in Ubuntu, because more lines of code can mean more potential errors due to a larger attack surface [19].

The medically relevant virtual machine, also known as DomU1, accommodates patient and exam data as well as the Xiralite X5 XiraView control software and the CCD camera driver. At the same time network access to the clinical system and the data archiving programs is necessary and a certain USB stick type, which is intended explicitly only for medically relevant data exchange. In the medically non-relevant virtual machine, which we refer to as DomU2, full Internet access is available, administered and monitored by Dom0 and a strict iptables configuration, so that the medical staff can rely on research from the network for an extensive diagnosis after an examination.

This function was specifically placed in the DomU2 in order to prevent potential security attacks against the medical device and its data. Additionally, all USB mass storage devices not reserved for the medically relevant VM, were passed into the DomU2. The medical staff should have access to publications stored on private data storage and can call them up in the non-relevant VM. Windows 10 is installed in both virtual machines, the medically relevant and the medically non-relevant. In the medically relevant VM, Windows was chosen as the hypervisor guest because the control program XiraView of the Xiralite X5 is only available for Windows.

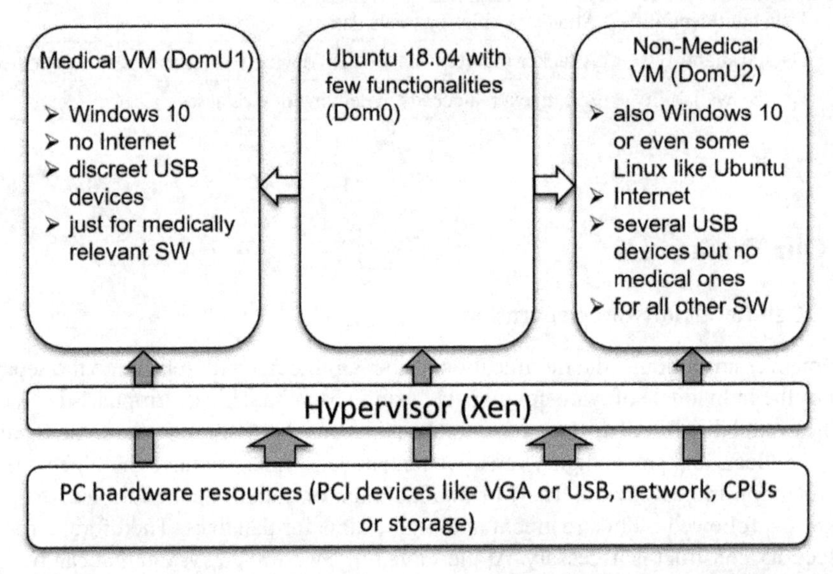

Fig. 2. High level software separation through hypervisor Xen in a minimal Linux Ubuntu.

Dom0 provides and controls the network access for both VMs. A selective USB device passthrough was established by self-written scripts, which will be discussed later. The medically relevant VM received access to a physical graphics card through the hypervisor to enable higher display performance through PCI VGA passthrough. Another script automatically started the system directly in the medically relevant VM, from which the medically irrelevant VM can be called up as a service by a remote session with the VNC protocol. RAID 1 contributes to the availability of the Xiralite X5 device. If one of the Xiralite X5's hard drives fails, the system can continue to function, therefore providing redundancy for patient exam data. For Dom0, password-protected servicing access through ssh in the local network is set up for remote maintenance. It is also possible for the medical staff to set up remote diagnostics.

3.2 Initial Software System

The basis of our system was a minimal Ubuntu 18.04 installation. To use the UEFI mode for Secure Boot and the GUID partition tables (GPT), we installed the system in

EFI-Mode. Secure Boot and RAID 1 were set-up in the BIOS. The Secure Boot chain established by Secure Boot started from the UEFI with a Platform Key (PK) and a Key Exchange Key (KEK) and executed the signed, minimal Bootloader Shim by storing so-called Machine Owner Keys (MOKs). Shim loads another boot loader Grub, which is signed with a MOK. Then grub loads the Xen hypervisor which in turn verifies and executes Dom0 Linux kernel.

The hard drive was split into three partitions, 200 MB for an EFI partition, 50 GB for Dom0 in the Ubuntu Xen as an ext4 file system, and more than 900 GB as a physical volume for a logical volume manager (LVM) for our virtual machines. With the LVM, we created a Physical Volume, fasting it into a Volume Group, and created three Logical Volume (LV) partitions. One each for a medically relevant and a medically irrelevant VM of 45 GB size for installing Windows 10, and a further LV as a separate approximately 800 GB partition in the medically relevant VM for the patient and examination data. For network configuration we used Network Address Translation (NAT). We had to pay attention to the NAT solution on postrouting masquerade of the iptables and the link to the actual network, where in Xen eth0 is expected.

3.3 Selective USB Device Passthrough

In order to provide the DomUs with USB devices such as mouse, keyboard and USB sticks, we used USB passthrough via vendor and product ID with the qemu monitor command in Xen, as they remained the same even when the devices were replugged, and were also unique for each device class per manufacturer. Thus, a particular type of USB flash drive was declared medically relevant, exclusively for DomU1. All other USB sticks were exclusively made available to the medically not relevant DomU2. To achieve this, Xen provided each VM with a hot plug USB3.0 controller with 10 ports at startup. For use in DomU's Windows, the Renesas USB 3.0 host controller driver was installed. Through scripts in Dom0, the USB passthrough for both DomUs has been automated. While it was fixed, always the same device for the medically relevant DomU1, DomU2 had to react dynamically to new devices. Other USB passthrough methods including the Xen hotplug commands like usbdev-attach or usbdev settings in the config file of a DomU used either only bus and device ID, what resulted in a USB passthrough log-down because of a change of the replug ID, which could only be solved by restarting the VM, or offered only USB2.0.

3.4 System Startup

Each time the medical device is switched on, a script is executed directly, which is shown in Fig. 3 as a program flow chart. First of all, it is checked whether the hypervisor Xen and thus the starting point of our software separation is active. If Xen is not loaded, an immediate hardware shutdown takes place. If the Xen kernel module is loaded successfully at system startup, it is then checked whether the medically irrelevant virtual machine (DomU2) exists. If not, it is created and booted by an automatic command, and scripts for network, USB passthrough initialization and selective USB passthrough for DomU2 are executed, see Fig. 3, block A. Next, it is checked whether the medically relevant virtual machine (DomU1) is active. Together with the DomU1 scripts for network and exclusive USB passthrough for the medically relevant VM DomU1 are started, if it did not exist yet, see Fig. 3, block B.

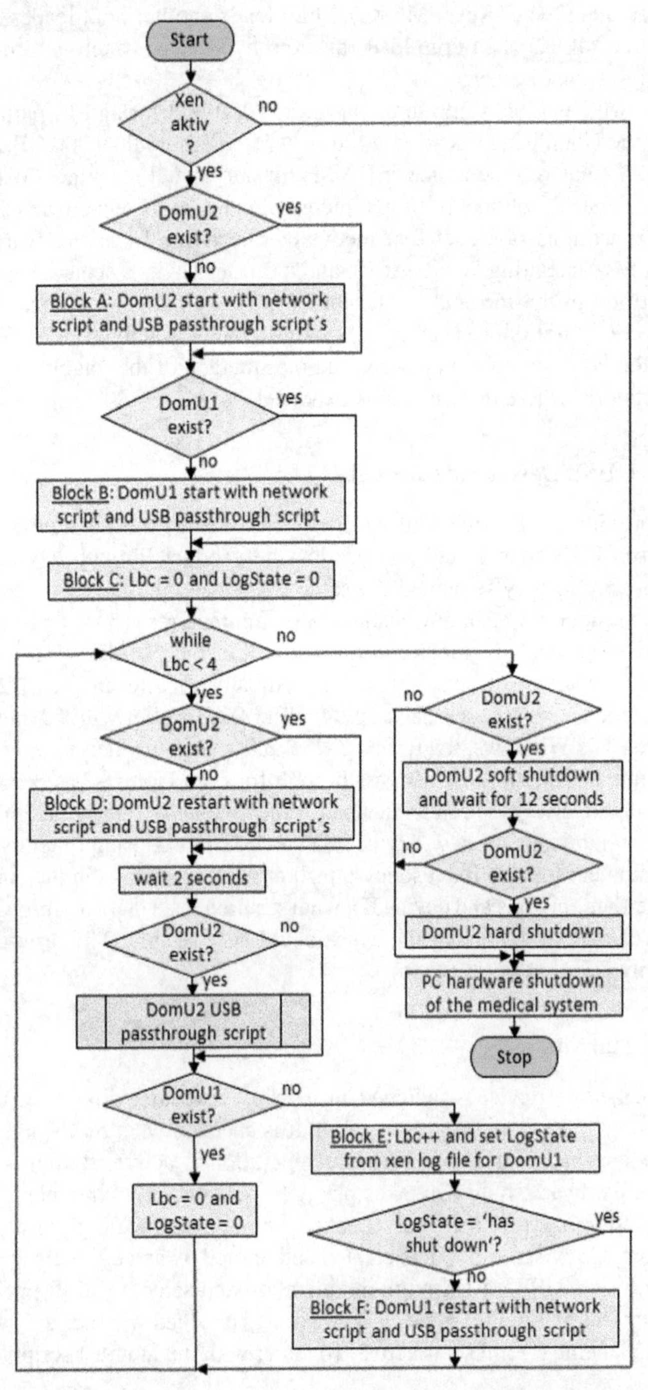

Fig. 3. Program flow chart of the script for startup, restart, and shutdown of DomU1 and DomU2.

After both VMs started up, two parameters are necessary to control the subsequent loop, see Fig. 3, block C. The leaky bucket counter Lbc counts every 2 s as soon as DomU1 is no longer available until it reaches 4 to perform a subsequent DomU2 followed by system hardware shutdown. The parameter LogState, holds the information passed, whether DomU1 should be restarted by the user, or shut down. For this purpose, a Xen log file is evaluated. As long as the leaky bucket counter is less than 4, a while loop is executed that confirms whether a restart or shut down of DomU1 is performed. Once a subsequent check detects that DomU2 is no longer available, the medically irrelevant VM, together with the associated network and selective USB passthrough scripts, is automatically rebooted (see Fig. 3, block D) to have DomU2 always available as a service for the user of the medical device. By a subsequent wait command of 2 s, the loop is slowed down. After successfully checking whether DomU2 is active, the while loop ran the selective USB passthrough script for DomU2 at least every 2 s. This script captures all USB devices connected to the medical device, checks whether they are exclusively intended for the medically relevant VM (DomU1) or are already been passed on to DomU2 based on the vendor and device ID, and thus passes newly recognized USB Devices that are not intended for DomU1 automatically through to DomU2. Subsequently, it is checked whether DomU1 was still active. If so, the two parameters Lbc and LogState are set to 0, and a new loop cycle begins. If DomU1 is no longer active, the leaky bucket counter is incremented and the LogState set by the contents of a xen log file for DomU1, see Fig. 3, block E. In this log file two codes, $0\ 0 \times 0$ for shutdown and $1\ 0 \times 1$ for restart, are used by Xen to differentiate between the corresponding actions. When DomU1 is deliberately shut down by the user, LogState contains the content 'has shut down', which causes the loop to be executed 4 more times by the leaky bucket counter until the loop is exited. If DomU1 is deliberately restarted by the user, the LogState contains the content 0, which causes DomU1 to restart and the scripts for network and exclusive USB passthrough for DomU1 are executed, see Fig. 3, block F. If it is a shutdown of DomU1 by the user, the loop is left after 4 cycles. If DomU2 is active at that time, a soft shutdown of DomU2 takes place after an activity check. Following a wait command of 12 s, DomU2 is checked for activity once more. If the non-legally relevant VM is still active, an automatic command results in a hard shutdown of DomU2. Finally, the medical device itself shut down automatically, without the user ever getting into the administrative software.

3.5 Remote Access

For remote maintenance, a password-protected access was established via ssh in the Dom0. With this both DomUs could be shut down and restarted in an emergency. To train medical staff and to realize remote diagnostics, DomU1 is externally accessible via VNC.

4 PCI Passthrough Tests

At the beginning, we displayed the DomUs with a remote viewer via VNC and the SPICE protocol. By using an X server and with the respective settings in the config file of the DomUs, we were able to use a standard VGA (stvga) with the VNC protocol and a more powerful qxl emulated graphics card with the SPICE protocol. SPICE even provided features such as copying host-guest information, setting screen resolutions, and custom USB passthrough for up to 4 devices with the spice agent (vdagent). To get an approximate of the graphics and system performance with the VNC protocol, we observed the system load with the tool htop in version 2.1.0 running in Dom0. With an idle desktop the system load was around 1% but raised to around 30% when moving a window in the VNC session. The SPICE protocol imposed even higher at 70% for the same action. Therefore, we decided to directly pass an Nvidia Quadro 2000 via PCI VGA passthrough to the medically relevant DomU1. This required support for IOMMU IO virtualization (VT-D) from the motherboard chipset, BIOS and CPU. For passthrough, the physical graphic card in Dom0 had to be assigned to the xen-pciback driver. The average CPU load compared with this solution remained below 5%. For DomU2 we still used the remote solution with VNC protocol.

Table 2. Threats benchmark of graphics cards for DomU1.

Parameter	Emulated qxl	Emulated stvga	NVIDIA Quadro 2000
Total score	1605.0	1748.0	3229.0
Video playback workload in fps	2.77	4.87	23.98
Video transcoding workload in fps	2647.82	2532.98	3044.86
Image manipulation in Mpx/s	16.31	18.40	18.41
Web browsing in pages/s	10.44	10.86	10.72
Data decrypting in MB/s	94.38	95.31	94.06

To verify our estimates, we checked all three graphics card variants (emulated stvga, emulated qxl and real NVIDA Quadro 2000) for the medically relevant VM (DomU1) within the VM and thus within the operating system Windows 7 with the Benchmark Tool PCMark 7 Basic Edition v1.4.0 [20], as shown in Table 2. In addition to values such as the video playback workload, which measured the performance (frame rate without glitches) when playing a video file in frames per second (fps), or the video transcoding workload, which measured the video's resolution, frame rate, and bit rate, this tool gave us a total score, which could be considered as a performance index of the graphics card. Here our observation was confirmed. The qxl over remote connection and SPICE protocol had the worst performance. The physical NVIDIA Quadro 2000 was by far the best choice concerning performance.

5 Discussion and Further Work

In Legal Metrology, a distinction can be made between low and high-level software separation [13]. There, the low-level software separation can be implemented at the level of the programming language by combining program parts, including subroutines, procedures, functions or classes. The high-level software separation is achieved by combining all legally relevant software parts into an object or program that runs alongside the other, legally irrelevant programs in the same operating system. This has been discussed in more detail in [20]. At the operating system level, without the use of virtualization, software separation utilizing Linux can also be realized by using SELinux or AppArmor and in Windows by using Mandatory Integrity Control (MIC) [21]. Each data folder, every program and every user can be assigned a role and a security level. Legally relevant software is set to a high level, not legally relevant to a low level. An operating system-wide policy ensures that the low level has no rights and therefore no impact on the high level.

To secure the highly sensitive patient and exam data of a medical measurement device from loss, tampering and theft and also to protect a patient during an examination, we realized high-level software separation through virtualization, by separating medically relevant software and medically irrelevant software into various, separate and strictly hypervisor-controlled virtual machines with a type I hypervisor running directly on the hardware. With a hypervisor, we establish an additional layer of software security that closely monitors all system hardware resources, including network access to the hospital network, to the Internet or also to VMs among themselves, remote VM sessions, and most importantly, PCI devices such as graphics cards and USB devices. With the separating into two VMs, we were able to outsource threats and attack surfaces, revealed by the risk analysis, from the medical device that emanates from inadvertently opened email attachments or infected USB sticks, into the medically irrelevant VM. Thus, the user still has the convenience to use both without endangering medically relevant programs and patient data. If a VM falls into unauthorized hands due to malicious code, it can easily be shut down and replaced, and does not endanger the rest of the system. Another benefit of this software separation implementation has been in the area of Legal Metrology, where typically every software change leads to a recertification of the software conformity of a meter. Due to the separation, legally irrelevant software can be changed arbitrarily without losing the certificate for the legally relevant part [13].

The high-level software separation discussed here can be modified by using additional VMs and outsourcing functions, as described in the software reference architecture from [20] for Legal Metrology. The software separation is implemented and administered by a microkernel, which sits directly on the hardware and performs the function of a hypervisor, e.g., see [22]. Through a Communication Supervisor in another VM, [21] management functions for Key & Signature, Download, Storage, Connection and Communication, as well as a Secure GUI are provided.

Our aim was to achieve sufficiently high security, which could be adapted by profiled open source applications by anyone, to ensure reliable performance in everyday practice. By integrating Secure Boot and a RAID 1 (hard disk mirroring) we achieved additional security and data redundancy. Through remote maintenance and the possibility of treatment and diagnosis from a remote system, we followed the current digitalization future trends. In further work the collected examination data of the Xiralite X5 devices from the practices and clinics will be summarized, anonymized and pseudonymized in a database initially in Germany, later also in Europe and finally in a global network. The aim of this "big data" collection is to extract smart data by means of machine learning tools in order to obtain characteristics and patterns for patients' clinical pictures. This should lead to an improvement of prevention, diagnostics and therapy. Additionally, the hypervisor can easily manage the Internet access from and to the device and make sure that only authorized access is granted.

6 Conclusion

We presented an experience report in health care and Legal Metrology of the implementation of high-level software separation by separating medically relevant software from medically non-relevant software using virtualization, i.e. an additional administrative software layer, the hypervisor. Software groups were separated with different virtual machines. These machines are monitored, controlled, and protected, like their network connections and access to PCI devices such as graphics card or USB devices. Dangers due to infected USB sticks and manipulated e-mail attachments are outsourced to the non-relevant VM. Thus, for patient and examination data of a medical device, which have been identified by a separate risk analysis as particularly valuable assets in addition to the protection of the patient, a higher level of protection is realized. By eliminating unnecessary functionalities, the probability of code errors and therefore the attack surface has been reduced. By implementing Secure Boot and a RAID 1 further security measures and redundancies were established. At last, by using well-supported open source software, we created an architecture that can be adapted by everyone.

References

1. Deutscher Bundestag der Bundesrepublik Deutschland: Gesetz über Medizinprodukte (Medizinproduktegesetz - MPG) - Medizinproduktegesetz in der Fassung der Bekanntmachung vom 7. August 2002 (BGBI. I S. 3146), das zuletzt durch Artikel 11 des Gesetzes vom 9. August 2019 (BGBI. I S. 1202) geändert worden ist (2002)
2. Bundesamt für Sicherheit in der Informationstechnik (BSI): Das IT-Sicherheitsgesetz - Kritische Infrastrukturen schützen (2016)
3. Official Journal of the European Union: Regulation (EU) 2016/679 of the European parliament and of the council of 27 April 2016 on the protection of natural persons with regard to the processing of personal data and on the free movement of such data, and repealing Directive 95/46/EC (General Data Protection Regulation) (2016)
4. Deutscher Bundestag der Bundesrepublik Deutschland: Bundesdatenschutzgesetz (BDSG) - Bundesdatenschutzgesetz vom 30. Juni 2017 (BGBI. I S. 2097) (2017)
5. Barham, P., Dragovic, B., Fraser, K., Hand, S., Harris, T., Ho, A., Neugebauer, R., Pratt, I., Warfield, A.: Xen and the art of virtualization. University of Cambridge Computer Laboratory (2003)
6. Bauer, D.F.W.: Vorstellung eines neuen Verfahrens zur Diagnostik der Rheumatoiden Arthritis: ein Nahinfrarot-Fluoreszenzbildgebungssystem, Inauguraldissertation, Gießen (2015)
7. Xiralite GmbH: Gebrauchsanweisung zum Xiralite Fluoreszenzbildgeber X4, Revision 22 (2017)
8. Fujitsu Technology Solutions GmbH: Data Sheet - FUJITSU Mainboard D3433-S Mini-ITX - Industrial Series (2017)

9. International Organization for Standardization: ISO/IEC 27005:2011(e) Information technology - Security techniques - Information security risk management, Geneva, CH, Standard (2011)
10. Esche, M., Thiel, F.: Software risk assessment for measuring instruments in legal metrology. In: Proceedings of the Federated Conference on Computer Science and Information Systems, Lodz, Poland, vol. 4, pp. 1113–1123 (2015). https://doi.org/10.15439/978-83-60810-66-8
11. Esche, M., Thiel, F.: Incorporating a measure for attacker motivation into software risk assessment for measuring instruments in legal metrology. In: Proceedings of the 18th GMA/ITG-Fachtagung Sensoren und Messsysteme 2016, Nuremberg, Germany, pp. 735–742 (2016). https://doi.org/10.5162/sensoren2016/P7.4
12. Official Journal of the European Union: Directive 2014/32/EU of the European Parliament and of the Council; European Commission: Brussels, Belgium, (2014)
13. WELMEC (European Cooperation in Legal Metrology): Softwareleitfaden (Europäische Messgeräterichtlinie 2914/32/EU), WELMEC 7.2 (2015)
14. DIN Deutsche Institut für Normung e.V.: DIN EN ISO 14971:2013-04 Medizinprodukte - Anwendung des Risikomanagements auf Medizinprodukte (ISO 14971:2007, korrigierte Fassung 2007-10-01); Deutsche Fassung EN ISO 14971:2012 (2013)
15. DIN Deutsche Institut für Normung e.V.: DIN EN ISO 13485:2016-08 Medizinprodukte - Qualitätsmanagementsysteme - Anforderungen für regulatorische Zwecke (ISO 13485:2016); Deutsche Fassung EN ISO 13485:2016, Berichtigung zu DIN EN ISO 13485:2016-08; Deutsche Fassung EN ISO 13485:2016/AC:2016 (2017)
16. DIN Deutsche Institut für Normung e.V.: DIN EN 62304:2016-10 Medizingeräte-Software - Software-Lebenszyklus-Prozesse (IEC 62304:2006 + A1:2015); Deutsche Fassung EN 62304:2006 + Cor.:2008 + A1:2015 (2016)
17. U.S. Food and Drug Administration: Postmarket Management of Cybersecurity in Medical Devices - Guidance for Industry and Food and Drug Administration Staff (2016)
18. U.S. Food and Drug Administration: FDA Sefety Communication: Cybersecurity for Medical Devices and Hospital Networks (2013)
19. Chelf, B.: Measuring software quality – A study of Open Source Software. Chief Technology Officer, San Francisco, CA, USA (2011)
20. PCMARK 7 - Technical Guide - Updated 24 August 2018, UL (2018). https://benchmarks.ul.com/. Accessed 27 Aug 2020
21. Peters, D., Scholz, P., Thiel, F.: Software separation in measuring instruments through security concepts and separation kernels, Acta IMEKO, vol. 7, no. 1, article 4, identifier: IMEKO-ACTA-07 (2018)-01–04 (2018)
22. Nordholz, J.: Design and provability of a statically configurable hypervisor. Dissertation, TU Berlin, Germany (2017)

Making Financial Sense from EaaS for MSE During Economic Uncertainty

P. S. JosephNg[1(✉)] and H. C. Eaw[2]

[1] Institute of Computer Science and Digital Innovation, UCSI University Kuala Lumpur, Kuala Lumpur, Malaysia
josephng@ucsiuniversity.edu.my
[2] Faculty of Business and Management, UCSI University Kuala Lumpur, Kuala Lumpur, Malaysia
eawhc@ucsiuniversity.edu.my

Abstract. Economic uncertainty and financial senses impacts have affected IT infrastructure implementation as Medium Size Enterprise goes bleeding for extravagant technology pursuing. Antecedents during stable market were not for optimization used during business uncertainty condition is been re-questioned for its relevancies. This mixed-mode with double data collection sequential strategy methodology lessens shallow perceptions or outlooks that simplify each phenomenal activities that merit the necessity for competitive innovation. The integrated framework applied to sampling users and tested using ordinal regression relationship that yields strong low relationship that provokes the null hypothesis. The embracing of BOINC shareware grid technology with consolidated computing configurations are applied to develop the system's productivity by encouraging positive monetary prospect via synchronized interworks hence better team accomplishment and branding respect. The study proposed an elastic internal IT infrastructure alongside the market condition with the development of EaaS framework beyond the legacy TAM implementation.

Keywords: BOINC · Cloud computing · Economic uncertainty · Exostructure · Framework · Grid computing · IaaS · model · Optimization · Reusability · Resource pooling · Small and medium-sized enterprise · Virtualization

1 Introduction

The Medium Size Enterprise (MSE) has been a prevailing segment, comprising twenty-three percentage consolidate Malaysia's Gross Domestic Product (GDP) development with thirty three percentage of occupation business. Unfortunately, the MSE is stuck among the non economic of scale of Small and Micro Enterprise (SME) as compare with bigger unpredictable Multi-National Corporation (MNC). The significant challenges emerges from the promise from the MSE, despite the fact that they simply provided 4% of the total Malaysia actual economy substances, MSE delivered estimated two billions for Malaysia's GDP in 2019 [1]. Such companies will continue competing with the fight with bigger enterprise, and they could be hesitant or ill-suited to place

K. Arai (Ed.): FICC 2021, AISC 1363, pp. 976–989, 2021.
https://doi.org/10.1007/978-3-030-73100-7_67

assets for each improvement tasks and practices, besides just new advancement [2, 3]. This is a result from of smaller production capabilities, monetary resources, knowledge and data, and human resource. Notwithstanding, Asian MSE that have embraced innovation before has now made the most of their change benefits when contrasted with the undertaking that was not receiving innovation before. In any case, [4] have featured that “venture abilities, limit and production growth will lessen the probability of disappointment”. Data communication has empowered innovative administrations towards advanced industry working within supply chain for upgrade execution, mobility and improving their abundance. These are additionally in accordance for their aggressive technology development of the company and industry and enter-prise. The company’s implementation in IT framework deployment with incredible yet easy configuration equipped with versatile resources integrated into requirements from the new IT framework activity. While MSE to extend domestically and all around the world, so is the normal for IT development and its challenges, particularly in the accompanying areas [5–7]:

- Escalating capital consumption and working costs.
- Inconsistent organic market financial rewards from infrastructure expectation.
- Unable to maximize infrastructure development with business cycle.
- Volatile, vulnerability, multifaceted nature and uncertainty business climate.

The MSE is influenced by individualized IT activity that zoomed in on the regular basic operational methodologies. As business digitize their activity, this has made IT foundation all the more expensive [12–15]. The MSE is additionally quiet in related past keep secret at the misuse of foundation administrations and registering progression in view of the apparent low business esteems rather than greater multinationals advancement. Past examinations have indicated that various MSE has exceeded their spending on assets for IT framework at the same time increasing their investment and influencing IT reasonability [8–11]. With the movement in innovation reappropriating, Medium Size Enterprise should have the ability to get a handle on the IT framework speculation choice fundamentally much better. Thusly, it has been foreseen that while MSE starts to extend, they ought to gain by data frameworks that grant the company to build up extra data but avoiding the dependencies on the innovation imperatives. Khan, Z.H. [13] suggestion that company could decrease about five millions of IT costs if these assets are improved. Past literatures reveals that IT expenditure uses major portion of company’s maintenance costs. While it is a recognized issue where IT framework speculation eventually causes negative pay back on investment, these inheritance mentalities requires a thump on the paradigm change to push ahead. However, for these starter study have demonstrated where critical IT ventures are ineffective. Interestingly, most of ventures are as yet wanting to in-wrinkle their IT speculation more than their last yearly spending plan.

As the business market turns out to be more powerful yet convoluted and rivalry among industry has become amazingly threatening. The monetary vulnerability is an unsteady economic situation where business deals income at that specific second is lacking or faulty for business congruity. Throughout the financial vulnerability cycle likewise, it makes deeper IT framework venture decision more troublesome because of restricted income to help higher working costs. However, company need to have a more extensive psyche to comprehend its estimation infrastructure and commitment towards moving and orchestrating into the standard mindset of computerized economy. Sadly, IT foundation additionally haul together a colossal initial cost and costly support. Moreover, the spotlight has been more on boosting innovation advancement, instead of expanding the innovation business measure redesigning. The authors in [16] focused on that "*there is imperative information on the commitment where innovation highlights shapes in embellishment the strategic sense-production from imaginative IT framework, and how the strategic reaction aided in it sense-production*". Even though the anticipated shareware network arrangement is now been recognize for its steady innovation, the selection of such innovation has not been completely used for business benefits.

Moreover, with limited literature regarding the utilization of shareware matrix answer at company stage, particularly the Medium Size Enterprises. Depicted innovation among the executives is a fundamental mechanism to improve intensity and to direct the company in the purchasing, usage and supporting of innovation to support the initiative that should be viable with the company's essential objective [17]. Although innovation development is a critical discovery to advance work measure, innovation isn't the lone alternative to improve the work cycle. All the more significantly is the way any innovation was really being utilized to develop the work cycle, unessential if the innovation sent as an innovation advancement or a basic matured innovation [18, 19]. With these issues, the difficult assertion has created the exploration speculation:

H_R1: EaaS optimizes available resources when buying new hardware.
H_R2: EaaS avoids unnecessary additional maintenance when buying new hardware.
H_R3: EaaS shift budget allocation paradigm when buying new hardware.
H_R4: EaaS separates economic turbulence independently when buying new hardware.

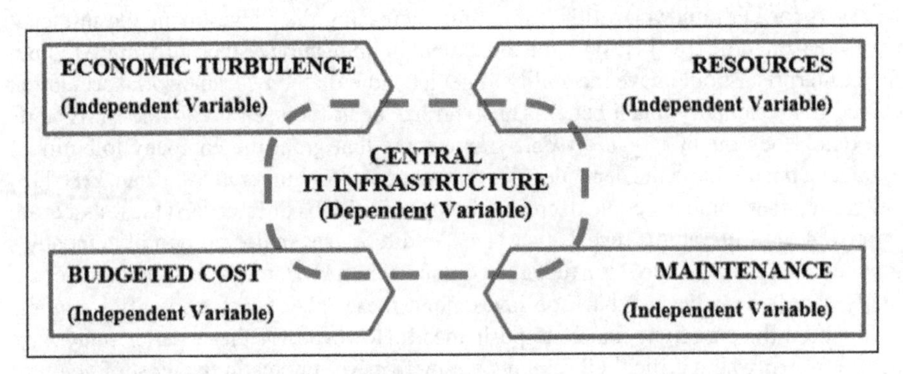

Fig. 1. Research model

Figure 1 summaries the association among the four exploration theories towards the focal technology framework. It additionally presents the component for Dependent Variable with the connected Independent Variable [33, 34, 40]. The speculation was tried utilizing Regression Analysis with a Model Fitting critical worth plus Goodness of Fit positive Chi-Square separation which characterize most of the respondents.

Utilizing the development and stability of IT foundation, MSEs can have the option to set out toward the more noteworthy perplexing business climate that was impractical already. While MSEs starts to advance, they should "put resources into data innovation that allows the endeavour to produce additional information without overburdening the infrastructure boundaries" [40]. The development and adjustment in component which has throughout the stage mixed should be expected business work process like automated creation have squeezed MSEs to be substantially more productive, consequently opening up their overall appeal.

The investigation draws an elective different methodology point of view that may be basic to determine the difficult Medium Size Enterprises. MSE confronted challenges on monetary requirements, extraordinary powerful business condition, testing scant assets and enormous cost-cognizant continuous upkeep to begin with. The investigation starts focusing on the new revelations from the three information gathering tools ("survey, meeting, and Delphi"). The reconsidered arrangement recommended for MSE framework improvement can likewise be deployed from Multi-National Corporation as a correlation idea paradigm. Eventually, it present investigation's commitment will create a viable marketing prudence advancement observing recommendation for the Medium Size Enterprise to amplify its current contributed foundation for natural strategic improvement.

2 Medium Size Enterprise ICT Venture Positioning

Company executive regularly accentuations their vital estimations for innovation separation and where undertakings they can use IT to grow every extraordinary component. Various enterprise have even intended to digitize' their strategies by using IT. Conventional accentuations such thinking relies upon a speculation that IT framework has incredible powers and uniqueness that can impact the enterprise key bearings. However, remaining an upper hand shouldn't be actually special highlights in every case except all the more significantly, being scant is the key factor. For an immense section of ventures, fundamental IT limits (data getting ready, data keeping, and data change) have ended up being immediately available and sensibly estimated. IT has advanced toward getting essentially an expense of digitized climate that should be paid by everything except give differentiation to none where the high speculation doesn't ensure critical upper hands.

"Entire venture framework and the tenacious appearance of big business are reliant on the uniformity, openness, adaptability of IT foundation" that MSE can send it as an instrument to do an essential reverse responsive guard within each business [6]. On the other hand, staying aware of development can be a costly instrument in the point of view for MSEs money related shortage. In this manner, MSEs ought to make a few trade offs in the midst of situating the most extreme front line advancement with coordinating monetary speculation to catch intensity.

"Enterprise can accomplish sensible separation by introducing new highlights that profoundly diminish costs while skilfully offering the benefit and highlights clients necessity" at the reasonable item lifecycle [7]. Situating the new highlights rashly includes more excellent venture to tap the chances while situating developed highlights doesn't create significant uniqueness to progress appeal. "The acknowledgment of the company be liable to on its ability to the upgraded advancement of the technology development direction to change in their venture concepts and at last to remain competitive particularly" to be acknowledge [7].

While fiscally and information imperatives are recognized issue for the MSE, in any case, it may not be their basic elements to keep away from situating feasible highlights as confining IT foundation turns out to be more developed. Furthermore, consolidating IT foundation into MSE digitize work process will just create obvious result after a completely fledged execution later, past customary administration primary concern skewed assumption. Consequently, IT framework is a "long yet on-going capital endeavors" all through the benefitting time frame and conceivably hauled towards the unstable business time frame.

From the challenges featured, it will affect the MSEs technology foundation design impending direction. Despite the fact that web engineering has developed with irrelevant uniqueness, the present monetary vulnerability gave expanded pressing factor and unpredictability to the MSEs. Technology business intergration performance period has contracted towards more modest period, compelling judgment decisions substantially more convoluted. With the heightening business peer compelled, numerous ventures, while hesitantly, have spent unnecessarily in IT foundation, influencing their undertaking money related outcomes (JosephNg, 2019/8/6). These unsatisfactory development hole, while hypothetically brings ponders, were unconventional to their activity were cleared under rug. The ware of IT framework in JosephNg's (2019/8/6) study has made innovation include an exceptional arrangement dependent on different undertaking condition. While moving towards shared services appears to be empowering, in any case, the more appropriate focuses ought not zero in on limiting shared assistance by internet based technology but on advancing in-house shared technology as the elective decision.

3 Methodology

This investigation utilized a DUAL blend mode approach whereby a double quantitative and double qualitative assortment is being utilized [33, 34]. The test-retest measures guarantee more steady examinations were performed from different essential information gathered. From on the exploration questions referenced beforehand, the investigation had built up some significant inquiries for the inquiry inside the review. These inquiries spotlight to emphasize and accentuation on the infrastructure arrangement, positioning basic conceptual, company directions and making do with the on-demand request. These activities tested are used to clarify the investigation conditions and aid the essential EaaS variables directions [21–32]. Primary information assortment by means of overview was sent to comprehend the remarkable advancement of the examination. Although the statistical information recognizes anomaly within these situation, Table 1 shows the investigation tracks up with interview and Delphi information for attribute translation.

A good data has an Alpha value nearer to 1.0 while a lower value of Alpha indicates lower chances of consistency. While the Cronbach Alpha value of 0.73 in Table 2 is still considered as good (Asad, M., Sharif, M.N.M. & Alekam, J.M. [35]; Veiga, A.D. [36]),

Table 1. Research methodology

Dimension	Sequential design
Methodology	Mix mode
Data collection	Mass survey, personal interview & focus group

this study focused more on the data reasoning from both the management interview and expert focus group. While the Cronbach Alpha is within the good range, this quantitative data served as outliner to the next stage of qualitative data reasoning.

Table 2. Cronbach alpha

Cronbach alpha	Cronbach alpha based on standardized items
0.730	0.731

Based on Table 3, a significant value of 0.028 is lesser than 0.05 shows higher acceptance of the data for a Final model and therefore the Independent Value can predict the Dependant Value (Kante, M., Chepken, C. & Oboko, R., [37], Flowers, E., Freeman, P., Flowers, E. P., Freeman, P., & Gladwell, V. F. [38]). It shows the 2 Log Likelihood's Final Model value of 476.753 is lower than the baseline Intercept Only value of 505.241 that support the acceptance of the Revised hypothesis. Furthermore, the Chi-Square value of 28.488 is a double-digit positive value shows the Final model (as a Revised hypothesis) has a significant difference from using Intercept Only model (as Null hypothesis). As a result, the Initial hypothesis has been overruled and the Revised hypothesis been established.

Table 3. Model fitting information

Model	−2 Log likelihood	Chi-square	df	Sig
Intercept only	505.241	28.488	16	0.028
Final	476.753			

Based on Table 4, the significant value starts from 0 (poor fit) to 1 (perfect fit) and the larger the significant value, the better the goodness-of-fit of the model (Kante, M., Chepken, C. & Oboko, R. [37], Flowers, E., Freeman, P., Flowers, E. P., Freeman, P., & Gladwell, V. F. [38]). Furthermore, Udo, E.N. & Akwukwuma, V.V.N. [39] and Ho, K.L.P. et al. (2018) emphasis that a good model generates higher significant value. Here, it shows 0.786 where the sample can fit 78.6% of the population with both the observed and expected value to be the same at 484. This result also shows the relevance of expanding this study to be a practical model for industry contribution.

Table 4. Goodness of fit

	Chi-square	df	Sig
Pearson	459.068	484	0.786
Deviance	385.682	484	1.000

4 Findings

From the problems featured, this has affected the MSEs framework engineering impending plans. Despite the fact that web engineering has developed with inconsequential uniqueness, the present financial vulnerability gave expanded pressing factor and multifaceted nature of each enterprise. Technology business performance has contracted to more modest situation, compelling judgment decisions considerably more confounded. With the heightening business peer constrained, numerous enterprise, while hesitantly, have spent unreasonably in IT framework, influencing their venture financial outcomes (JosephNg, 2019/8/6/5/2/1). These unsatisfactory advancement hole, while hypothetically brings ponders, were strange to their activity were cleared under floor covering. The ware of IT framework in JosephNg's (2019/8/6/5/2/1) study has made innovation highlight a special arrangement dependent on different enterprise condition. While relocating towards shared service appears to be empowering, in any case, the more relevant focuses ought not zero in on limiting shared services through internet base however on upgrading in-house shared service as the elective decision. The connection between Dependent Variable with Independent Variables for the four Revised Hypothesis are appeared in Table 5.

Economic Uncertainty is considered as Dependent Variable while Budget Cost, Maintenance, Product Knowledge and Complex Infrastructure are considered as Independent Variables. While Budget Cost is considered the Most Critical item at 45.0% in this study due to the traditional limitation associated with the smaller enterprise, Economic Uncertainty at 61.9% (34.7 + 27) should not be underestimated due to emerging environmental challenges. While having a complex infrastructure at 85.2% is critical to tap new technology innovation, it seems to be non-critical at 61.9% (33.6 + 28.3) for Product Knowledge. Furthermore, Medium Size Enterprise are very concerned at the ability to maintain these Complex Infrastructures due to the criticalness at 89.2% (43.7 + 45.5).

In Table 6, the overall Correlation Coefficient r values are below 0.25 showing "Very Weak Relationship" between all the variables. The overall significant p values are also below 0.05, signalling that the original null hypothesis can be rejected that leads to the acceptance of the revised hypothesis.

While it has been traditionally linked that with a limited budget in the Medium Size Enterprise, it harder to implement much more complex infrastructure that comes with more advance features. However, with a coefficient value of 0.243, it shows that on the ground, Medium Size Enterprise is not facing any major difficulties. Data from the management interviews shows that enterprises are utilizing freeware solution to achieved digital transforming without having to invest heavily in own infrastructure.

Table 5. Ordinal regression – case processing summary

		%
Eco_Uncertainty (Dependent variable)	Least critical	0.5%
	Slightly less critical	12.2%
	Neutral	28.4%
	Slight critical	27.0%
	Most critical	34.7%
Budgeted Cost – Q6a (Independent variable)	Least critical	1.8%
	Slightly less critical	2.3%
	Neutral	17.1%
	Slight critical	33.8%
	Most critical	45.0%
Maintenance – Q10b (Independent variable)	Least critical	1.4%
	Slightly less critical	1.4%
	Neutral	8.1%
	Slight critical	45.5%
	Most critical	43.7%
Prod Knowledge – Q8c (Independent variable)	Least critical	33.6%
	Slightly less critical	28.3%
	Neutral	22.1%
	Slight critical	10.2%
	Most critical	1.8%
Complex Infrastructure – Q4c (Independent variable)	Least critical	0.9%
	Slightly less critical	5.0%
	Neutral	9.0%
	Slight critical	25.7%
	Most critical	59.5%

The technical expert from the Delphi focus group has also reaffirmed that enterprise has already matured in the use of new technology into their daily standard operating procedures. These mixed-mode data from the quantitative survey, qualitative interview and qualitative focus group iteration have validated the consistency of findings. While this relationship is very weak, it shows positive implications to the enterprise from a technology acceptance and optimization level. The scarce resources traditionally associated with MSE does not hinder the enterprise to skim the benefits of deploying EaaS grid computing features.

The very weak value of 0.221 in Table 6 shows a lesser dependency for large maintenance cost associated with utilizing a complex infrastructure for the MSE and therefore

Table 6. Independent variable bivariate spearman correlations

			Budgeted cost	Complex infrastructure	Maintenance	Prod knowledge
Spearman rho	Budgeted cost (Q6a)	Correlation coefficient	1.0	0.243	0.838	0.218
		Sig (2-tailed)	–	−0.001	0.001	−0.002
	Complex infrastructure (Q10b)	Correlation coefficient	0.243	1.000	0.221	0.209
		Sig (2-tailed)	−0.001	–	0.001	0.001
	Maintenance (Q8c)	Correlation coefficient	0.838	0.221	1.000	−0.243
		Sig (2-tailed)	0.001	0.001	–	0.001
	Prod knowledge (Q4c	Correlation coefficient	0.218	0.209	0.243	1.000
		Sig (2-tailed)	−0.002	0.001	0.001	–

reducing the budgeted cost. As EaaS grid computing reutilized existing available desktop in the enterprise, it does not require new or additional investment for the sophisticated physical server and thus having a lesser footprint inside the already cramped MSE. Without the need for a dedicated data centre with its complicated electrical wiring and air-conditioning, more available resources (financial) can be utilized for other priority services. Furthermore, EaaS is available as a freeware software solution without expensive yearly software license cost nor annual software maintenance renewal cost. Even when there is a desktop breakdown is inevitable, the cost of repair is insignificantly lower than a full-fledged traditional server. Moreover, replacement spare parts for the desktop component is much easier to be sourced and expertise available in more cost-competitive options.

Besides saving upfront capital investment, Medium Size Enterprise can also reduce the need for costly operation expenses in skilled product knowledge manpower to implement the complex infrastructure as shown by the very weak relationship of 0.209 in Table 6. EaaS computing application is installed inside the desktop and can be configured to auto startup, thus reducing any complex maintenance for the user. Most desktop support is easily available at very competitive service charges. Furthermore, should a desktop faces any hardware issues, the replacement component is easily sourced and also a very low price. With the advancement of Enterprise Resource Management software system that is supported by outsourced expertise, the enterprise today only requires local processing and storage facilities on site for real-time local manufacturing automation.

For the maintenance versus product knowledge, the very weak value of 0.243 in Table 6 highlighted the minimum dependence of hiring expensive expertise purely for ICT operations. Being a desktop, most end-user can be upgraded to become onsite ICT support personnel with basic training provided, in line with the development of online training like YouTube. Remote support, if needed can be provided by external expertise to connect into the local server without the need for troublesome travelling time and parking expenses. EaaS grid computing application has been used intensively by many enterprises, especially the cost-sensitive environment to operate with bare minimum local expertise.

As the objective of the digital transformation is to improve efficiencies, EaaS has also helped to reduce the operation cost as shown in Table 6 under Budgeted Cost versus Maintenance at 0.838. This strong relationship value reflects the lower maintenance cost arise despite having complex infrastructure but at a lower budget cost with the lesser skilled staff needed. With the cost of technology lowering down, open-source software gaining more stable and outsourcing readily available from local and global remotely, this is the right moment to implement EaaS grid computing. This promotes much better new technology features adoption and thus steer the digital business for the Medium Size Enterprise, reducing the need for processes duplication and bottleneck operations.

5 Exostructure as a Service (EaaS)

While the company exposure to different economic storm, every wave may involve diverse technology framework prerequisites [40–42]. During the positive development of the company, the extension of necessity might be handily supported yet when the company faces monetary vulnerability, the undertaking might be compelled to bear the high weight cost of activities as appeared in Fig. 2.

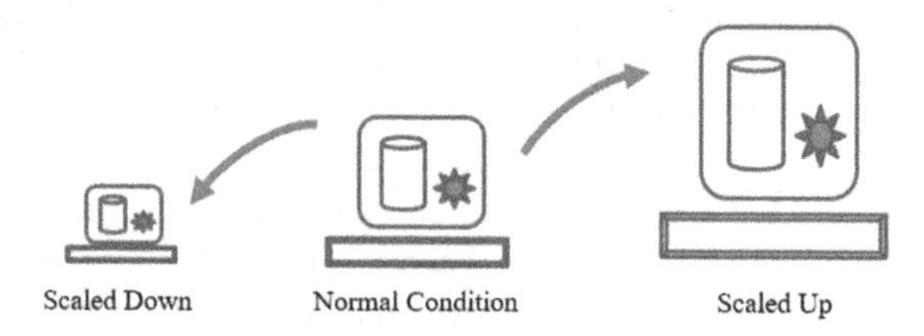

Fig. 2. Flexible organic BOINC infrastructure size

The innovation arrangement shows how each current work area's unused stockpiling and unused preparing force can be aggregated and upgrade as a greater virtual worker in a lattice climate as appeared in Fig. 3 [43]. This made more reserve funds on capital speculation and bringing down pointless activity costly for an undesirable new actual worker. Indeed, even the virtual BOINC regulator can be made utilizing the current actual PC inside the undertaking, accordingly, eliminating the requirement for a committed actual regulator, another cost-saving activity.

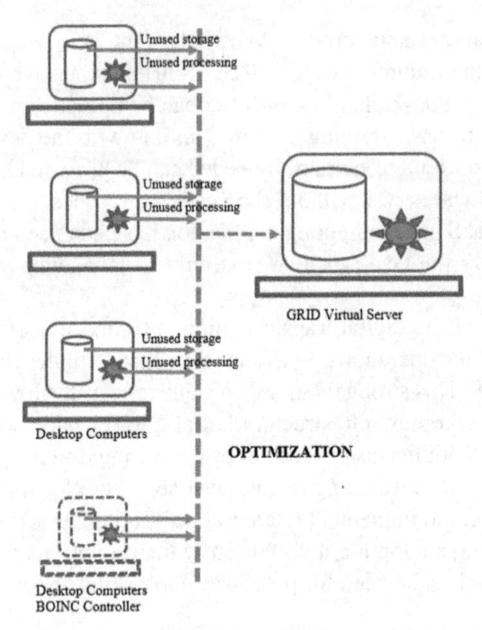

Fig. 3. Optimized BOINC grid connectivity

6 Conclusion

The discoveries of this investigation featured that legacy IT framework organization has become a basic factor of enterprise with undesirable revenue on venture, giving huge impact to undue competitive pressure and wastage spending plan. While most unpredictability, vulnerability, multifaceted nature and uncertainty issues keep on frequenting the undertaking, a significantly more adaptable IT framework model is requested. The investigation prompts the proposed natural IT foundation frameworks which can develop or lessen in accordance with the environment performance prompting the formation of Exostructure as a Service framework towards infrastructure development manual. BOINC interconnected system is envisioned towards a fundamental framework to make another simulated platform. Such deliverable arrangement contributed toward: (1) improve profitable ventures. Besides, it (2) advances proficiency within the company through giving incorporated preparing and capacity abilities and limit. This empowers between division work coordination and accordingly incredibly improve client fulfilment. Having supported monetary advantages that emerge from much better work measure stream will prompt (3) more noteworthy venture marking and notoriety, in accordance with improving venture's riches. All the more critically, this examination (4) incite a more prominent arousing of the aspiring developed GRID arrangement beyond TAM concepts in resource strapped yet high potential average size companies.

References

1. SMEcorp (2020). http://www.smecorp.gov.my/images/SMEAR/SMEAR2018_2019/final/english/SME%20AR%20-%20All%20Chapter%20Final%2024Jan2020.pdf. Accessed 18 Feb 2020
2. Guo, Z., Li, J., Ramesh, R.: Optimal management of virtual infrastructure under flexible cloud service agreement. Inf. Syst. Res. **30**(4), 1424–1446 (2019)
3. Rao J.J., Kumar V.: Technology adoption in the SME sector for promoting agile manufacturing practices. In: Satapathy, S., Bhateja, V., Das, S. (eds.) Smart Intelligent Computing and Applications. Smart Innovation, Systems and Technologies, vol. 105. Springer, Singapore (2019)
4. Jung, H., Hwang, J.T., Kim, B.K.: Does R&D investment increase SME survival during a recession? Technol. Forecast. Soc. Chang. **137**(December), 190–198 (2018)
5. Zhang, K., Alasmari, T.: Mobile learning technology acceptance in Saudi Arabian high education: an extended framework and a mixed-method study. Educ. Inf. Technol. **24**(3), 2127–2144 (2019)
6. Benitez, J., Ray, G., Henseler, J.: Impact of information technology infrastructure flexibility on mergers and acquisitions. MIS Q. **42**, 25–43 (2018)
7. Valacich, J.S., Wang, X., Jessup, L.M.: Did i buy the wrong gadget? How the evaluability of technology features influences technology feature preferences and subsequent product choice. MIS Q. **42**, 633–644 (2018)
8. EdwinCheong, L.T., Lim, L.J., JosephNg, P.S., MayKang, C.M., Phan, K.Y., Wong, S.W.: JomNetwork: reaffirming resource allocation through network monitoring. J. Inf. Syst. Res. Innov. **11**(2), 16–22 (2017)
9. Chris Foo, X.J., Siew, J.X., JosephNg, P.S., MayKang, C.M., Phan, K.Y., Lim, J.T.: JomNetwork: medium size enterprise, don't they also need network redundancy. J. Inf. Syst. Res. Innov. **11**(2), 1–5 (2017)
10. Siew, J.X., JosephNg, P.S., Lim, J.T.: JomNetwork: GLBP in medium size enterprise. Int. J. Trend Sci. Res. Dev. **3**(2), 725–728 (2019)
11. Lim, L.J., Sambas, H., MarcusGoh, N.C., Kawada, T., JosephNg, P.S.: ScareDuino: smart-farming with IoT. Int. J. Sci. Eng. Technol. **6**(6), 207–210 (2017)
12. Bañares, J.Á., Altmann, J.: Economics behind ICT infrastructure management. Electron. Mark. **28**(1), 7–9 (2018)
13. Khan, Z.H.: Exploring Strategies that IT Leaders Use to Adopt Cloud Computing, Doctor of Business Administration thesis. Walden University (2016)
14. Anwar, N., Masrek, M.N., Sani, M.K.J., Mohamad, A.N.: The proof of concept on the determinants of strategic utilization of information systems. Int. Inf. Inst. **19**(7A), 27–55 (2016)
15. Kang, C.M., JosephNg, P.S., Issa, K.: A study on intergrating penetration testing into the information security framework for Malaysian higher education institution. In: International Symposium on Mathematical Science and Computing Research, Ipoh, Malaysia, pp. 156–161 (2015)
16. Mesgari, M., Okoli, C.: Critical review of organization-technology sensemaking: towards technology materiality, discovery and action. Eur. J. Inf. Syst. **28**(2), 205–232 (2019)
17. da Silveira, B., et al.: Technology road mapping: a methodological proposition to refine Delphi results. Technol. Forecast. Soc. Changes **126**(2018), 194–206 (2018)
18. Urbach, N., Ahlemann, F.: Infrastructure as commodity: IT infrastructure services are traded on free markets and purchased as required. In: IT Management in the Digital Age, pp. 75–84. Springer (2016)
19. Narayanan, V.K.: Radical cost innovation. Strategy Leadersh. **47**(5), 53–54 (2019)

20. JosephNg, P.S., Loh, Y.F., Eaw, H.C.: Grid computing for MSE during volatile economy. In: International Conference on Control, Automation and Systems, IEEE Explore, Busan, Korea, pp. 709–714 (2020)
21. JosephNg, P.S., Kang, C.M., Ahmad, K.M., Wong, S.W., Phan, K.Y., Saw, S.H., Lim, J.T.: EaaS: available yet hidden infrastructure inside MSE. In: 5th International Conference on Network, Communication and Computing, Kyoto, Japan, pp. 17–20. ACM International Conference Proceeding Series (2016)
22. JosephNg, P.S., Kang, C.M., Choo, P.Y., Wong, S.W., Phan, K.Y., Lim, E.H.: Beyond cloud infrastructure services in medium size manufacturing. In: International Symposium on Mathematical Sciences & Computing Research, Ipoh, Malaysia, pp. 150–155 (2015)
23. Joseph Ng, P.S., Choo, P.Y., Wong, S.W., Phan, K.Y., Lim, E.H.: Malaysia SME ICT during economic turbulence. In: International Conference on Information & Computer Network, Singapore, pp. 67–71 (2012)
24. Joseph Ng, P.S., Yin, C.P., Wan, W.S., Nazmudeen, M.S.H.: Energizing ICT infrastructure for Malaysia SME during economic turbulence. In: Student Conference on Research and Development, Cyberjaya, Malaysia, pp. 328–322. IEEE Explore (2011)
25. JosephNg, P.S.: EaaS infrastructure disruptor for MSE. Int. J. Bus. Inf. Syst. **30**(3), 373–385 (2019)
26. JosephNg, P.S.: EaaS optimization: available yet hidden information technology infrastructure inside medium size enterprises. J. Technol. Forecast. Soc. Change **132**(July), 165–173 (2018)
27. JosephNg, P.S., Kang, C.M.: Beyond barebone cloud infrastructure services: stumbling competitiveness during economic turbulence. J. Sci. Technol. **24**(1), 101–121 (2016)
28. Joseph, N.P.S., Mahmood, A.K., Choo, P.Y., Wong, S.W., Phan, K.Y., Lim, E.H.: Barebone cloud IaaS: revitalization disruptive technology. Int. J. Bus. Inf. Syst. **18**(1), 107–126 (2015)
29. Joseph, N.P.S., Mahmood, A.K., Choo, P.Y., Wong, S.W., Phan, K.Y., Lim, E.H.: IaaS cloud optimization during economic turbulence for malaysia small and medium enterprise. Int. J. Bus. Inf. Syst. **16**(2), 196–208 (2014)
30. Joseph, N.P.S., Mahmood, A.K., Choo, P.Y., Wong, S.W., Phan, K.Y., Lim, E.H.: Battles in volatile information and communication technology landscape: the Malaysia small and medium enterprise case. Int. J. Bus. Inf. Syst. **13**(2), 217–234 (2013)
31. Joseph Ng, P.S., Kang, C.M., Mahmood, A.K., Choo, P.Y., Wong, S.W., Phan, K.Y., Lim, E.H.: Exostructure services for infrastructure resources optimization. J. Telecommun. Electron. Comput. Eng. **8**(4), 65–69 (2016)
32. Joseph Ng, P.S., Choo, P.Y., Wong, S.W., Phan, K.Y., Lim, E.H.: Hibernating ICT infrastructure during rainy days. J. Emerg. Trends Comput. Inf. Sci. **3**(1), 112–116 (2012)
33. Lin, T.T.C.: Multiscreen social TV system: a mixed-method understanding of users; attitude and adoption intention. Int. J. Hum.-Comput. Interact. **35**(2), 99–108 (2018)
34. Creswell, J.W.: Research Designs: Qualitative, Quantitative and Mixed Methods Approach, 5E. Sage publication, Thousand Oaks (2018)
35. Asad, M., Sharif, M.N.M., Alekam, J.M.: Moderating effect of entrepreneurial networking on the relationship between access to finance and performance of micro and small enterprises. Paradigms: Res. J. Commer. Econ. Soc. Sci. **10**(1), 1–13 (2016)
36. Veiga, A.D.: A cybersecurity culture research philosophy and approach to de-veloping a valid and reliable measuring instrument. In: SAI Computing Conference 2016, 13–15 July, London (2016)
37. Kante, M., Chepken, C., Oboko, R.: Effects of farmers' peer influence on the use of ICT-based farm, input information in developing countries: a case in Sikasso, Mali. J. Digit. Media Interac. **1**(1), 99–116 (2018)
38. Flowers, E., Freeman, P., Flowers, E.P., Freeman, P., Gladwell, V. F.: A cross-sectional study examining predictors of visit frequency to local green space and the im-pact this has on physical activity levels. BMC Public Health **16,** 420 (2016)

39. Udo, E.N., Akwukwuma, V.V.N.: Software adaptability metrics model using ordinary logistic regression. J. Softw. **14**(3), 116–128 (2019)
40. Zhang, D., Nault, B.R., Wei, X.: The strategic value of information technology in setting productive capacity. Inf. Syst. Res. **30**(4), 1124–1144 (2019)
41. Furstenau, D., Baiyere, A., Kliewer, N.: A dynamic model of embeddedness in digital infrastructure. Inf. Syst. Res. **30**(4), 1319–1342 (2019)
42. Ho, K.L.P., Nguyen, C.N., Adhikari, R., Miles, M.P., Bonney, L.: Exploring market orientation, innovation and financial performance in agricultural value chains in emerging economies. J. Innov. Knowl. **3**(3), 154–163 (2018)
43. Anderson, D.P.: BOINC: A Platform for Volunteer Computing, Computer Science, pp. 1–37. Cornell University Press, Ithaca (2018)

Development of a Software System for Simulation and Calculation of the Optimal Time of Upgrading Equipment

Askar Boranbayev[1(✉)], Seilkhan Boranbayev[2], Yuri Yatsenko[3], Yersultan Tulebayev[2], and Askar Nurbekov[2]

[1] Nazarbayev University, Nur-Sultan, Kazakhstan
aboranbayev@nu.edu.kz
[2] L.N. Gumilyov Eurasian National University, Nur-Sultan, Kazakhstan
[3] Houston Baptist University, Houston, USA
yyatsenko@hbu.edu

Abstract. The article is devoted to modeling the rational renewal of capital assets of an enterprise. The developed software system makes it possible to solve the problem of rational replacement of an asset in conditions when the enterprise does not have complete data on future technological changes. Replacing an asset on time can help reduce capital and operating costs. The system allows to solve the problem of asset replacement in cases where there is no complete information about the possible costs of future assets, when only a few certain data on technological changes are known.

Keywords: Software system · Model · Modeling · Optimization · Improving technology · Asset · Renovation

1 Introduction

This article is devoted to the development of a software system for modeling the rational replacement of enterprise assets in the context of an incomplete and uncertain forecast of future technological changes. The problem of asset replacement has been studied in many works, in particular in [1–10]. It is known from the scientific literature that the Infinite-Horizon (IH) method with the best technological prediction is the ideal benchmark for asset replacement.

There are different methods for replacing assets. However, many of them are not applied in engineering practice due to lack of data and other limitations. Therefore, in the literature on economics, for practical application, they recommend a simple and reliable method Economic Life (EL) [1–3]. This method allows you to find the optimal solution when replacing one asset.

On the other hand, it is known that if the cost of servicing assets grows, then the Economic Life method gives results different from the Infinite-Horizon method. Therefore, with the improvement of technology, the solution obtained by the Economic Life

K. Arai (Ed.): FICC 2021, AISC 1363, pp. 990–1002, 2021.
https://doi.org/10.1007/978-3-030-73100-7_68

method may not be optimal. To eliminate this drawback, the authors of [4] modified the classic Economic Life method - they introduced a corrected capital return ratio. In [4], it is proved that if exponential technological changes have the same effect on operating costs and the cost of new assets, then the modified Economic Life method and the replacement of Infinite-Horizon produce the same policy of replacement of assets with equal service life.

Further, in [5], this result was extended to the problem of asset replacement at stochastic costs.

2 Mathematical Models of Rational Renewal of Assets

We will use the notation, as well as some results and data from [4–10]:

$i = 1, 2,..$ - discrete time ($i = 0$ is the current year).
$P(i)$ - purchase (installation) cost of a new asset at year i.
$A(i, j)$ - operating and maintenance (O&M) cost of the asset of age j installed at year i.
$S(i, j)$ - salvage value of the asset of age j installed at year i.
M - maximal possible physical service life of assets.
$A_a(j)$ - age-dependent profile of O&M cost.
$S_a(j)$ - age-dependent profile of salvage value.
a - annual change factor of purchase cost P of a new asset.
q - annual change factor of O&M cost A of a new asset.
$L_1 = L$ - unknown lifetime of current asset.
L_k - unknown lifetime of the k-th asset, $k = 1, 2,...$
$\tau_{k\text{-}1}$ - unknown installation instant of the k-th asset, $k = 2, 3,...$
τ_0 - given installation instant of the current (first) asset.
d - given annual discount rate.
i_0 and i_1 - two years of measurements of new asset costs $P(i)$, $S(i, 0)$, and $A(i, 0)$.
$EEAC$ - effective equivalent annual cost of the present value PV_1 of total cost of current asset.

Consider the case with two dimensions of the value of new assets P(i), S(i, 0) and A(i, 0) in different years i_0 and i_1. Assuming geometric technological changes, these costs are reduced by constant factors after each time period:

$$P(i) = P(0)a^i, \quad A(i,j) = A_a(j)q^i, \quad 0 < \mathrm{a} < 1, \quad 0 < \mathrm{q} < 1, \tag{1}$$

Then we get the following formulas:

$$a = [P(i_1)/P(i_0)]^{1/(i_1 - i_0)}, \quad q = [A(i_1, 0)/A(i_0, 0)]^{1/(i_1 - i_0)}, \tag{2}$$

for the annual cost factors, a and q. With the improvement of technology, we will have. $0 < \mathrm{a} < 1$ и $0 < \mathrm{q} < 1$.

If there are no technological changes, then $\mathrm{a} = \mathrm{q} = 1$, and we only need one measurement of costs P (i), S (i, 0) and A (i, 0), usually in the current year $\mathrm{i} = 0$.

A common practical situation is when a firm has two cost dimensions P (i) and A (i, i) at the moment $\tau_0 < 0$ of the current asset installation, and currently i = 0. Then by (2),

$$a = [P(0)/P(\tau_0)]^{1/\tau_0}, \quad q = [A(0,0)/A(\tau_0,0)]^{1/\tau_0}. \tag{3}$$

Modified Economic Life method determines the optimal L1 of the first asset:

$$L_1 = \underset{1 \le L \le M}{\arg\min}\, EEAC(L), \tag{4}$$

where the function

$$EEAC(L) = \frac{d(1+d)^L}{(1+d)^L - q^L}\left[\frac{P(\tau_0)a^{L-\tau_0}}{(1+d)^{L-\tau_0}} - \frac{S(\tau_0, L)}{(1+d)^{L-\tau_0}} + \sum_{j=1}^{L}\frac{A(\tau_0, j)}{(1+d)^{j-\tau_0}}\right]. \tag{5}$$

describes the effective equivalent annual cost.

The modified EL method is identical to the original EL at a = q = 1, that is, without technological changes.

$$EAC(L) = \frac{d(1+d)^L}{(1+d)^L - 1^L}\left[\frac{P(\tau_0)}{(1+d)^{L-\tau_0}} - \frac{S(\tau_0, L)}{(1+d)^{L-\tau_0}} + \sum_{j=1}^{L}\frac{A(\tau_0, j)}{(1+d)^{j-\tau_0}}\right]. \tag{6}$$

The EL method becomes simpler if we assume that the disposal cost is negligible (i.e. S (,) = 0):

$$EAC(L) = \frac{d(1+d)^L}{(1+d)^L - 1^L}\left[\frac{P(\tau_0)}{(1+d)^{L-\tau_0}} + \sum_{j=1}^{L}\frac{A(\tau_0, j)}{(1+d)^{j-\tau_0}}\right]. \tag{7}$$

The above formulas are used to carry out calculations in the software system. The modified EL (2)–(5) algorithm requires the following input data:

1. Costs: P (0), S (0,0), A (0,0);
2. Age profiles: Sa(j) and Aa(j);
3. If the technology changes, then two additional values: $P(\tau_0)$ and $A(\tau_0, \tau_0)$.

It is very important for practice to resolve the issue of replacing an asset in the current year. For this, with the improvement of technologies, formulas (3) and (5) can be used. In this case, there is no need to solve the optimization problem. Since the current asset was established at time $\tau_0 < 0$, its lifetime at the current time $i = 0$ equals $-\tau_0 > 0$. Using (5), we can calculate two EEAC values: $EEAC(-\tau_0)$ for the current year and $EEAC(-\tau_0 + 1)$ for the coming year. Further, you can decide on the possibility of replacing the asset during the current year by comparing these two EEAC values:

If $EEAC(-\tau_0 + 1) > EEAC(-\tau_0)$, then the asset must be replaced in the current year.

If $EEAC(-\tau_0 + 1) \le EEAC(-\tau_0)$, then the asset continues to be used in the current year.

3 Software System for Simulating Rational Renewal of Assets

The developed software system calculates the effective equivalent annual cost for each asset life. The system uses EL method and modified EL method algorithms and calculates the optimal asset replacement year. The software system makes it possible to solve the problem of replacing an asset in cases where there is no complete information about the possible costs of future assets, when only a few certain data on technological changes are known. A stochastic modification of the Economic Life method is used to account for uncertain operating costs. The data for the calculations are taken from the respective databases.

In the development of software systems, the level of its reliability and safety plays an important role. It is impossible to achieve uninterrupted and trouble-free operation of the control system without ensuring the reliability and safety of automated information systems [11–37]. When developing a software system for modeling the rational renewal of assets, technologies and methods were used that ensure an increase in the level of their reliability and safety.

In the development of this program the systems were used: Delphi, ADO technology, MS SQL Server database management system.

3.1 Description of the Program and Interface.

First, you need to create a database (Capital) in the SQL Server DBMS (Fig. 1). To do this, click on the Databases tab in the Object Explorer window and select New Database.

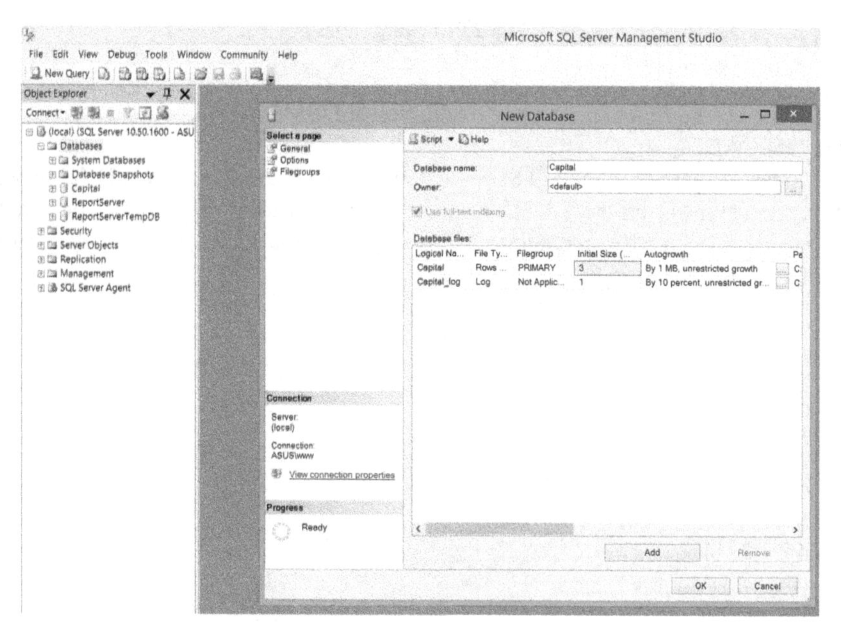

Fig. 1. Creating a database

To create a table, select the Tables tab, right-click, select New Table. The Capital database consists of three tables depicted in the diagram (Fig. 2).

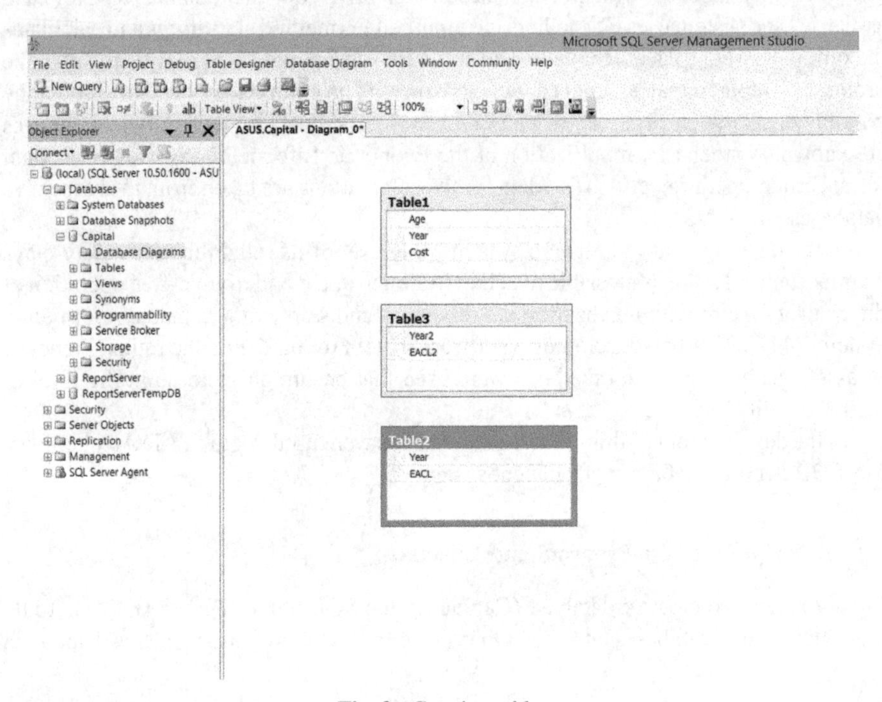

Fig. 2. Creating tables

To connect the Delphi program and SQL Server DBMS, ADO components are used: ADOTable, ADOQuery.

The tables that are in the Sql Server database can be extracted into a Delphi application. To do this, insert the AdoTable component onto the form in Delphi. Click on the component, select the ConnectionString property in the Object Inspector window. Next, click on the Build button (Fig. 3).

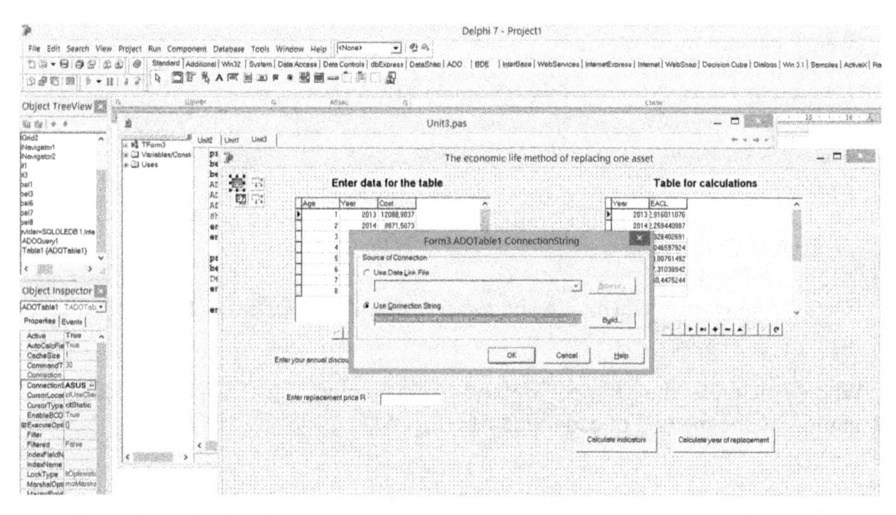

Fig. 3. Establishing Communication between Delphi Application and SQL Server

Figure 4 shows the main window of the program, where the choice of a method for replacing one asset is provided.

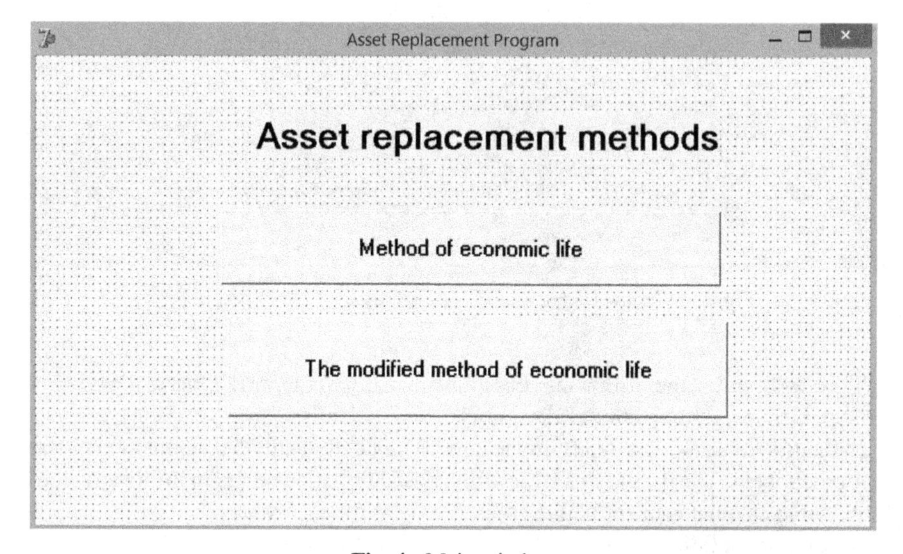

Fig. 4. Main window

This form uses 3 components: Label1, Button1, Button2. When you press the "Method of Economic Life EL" button, the program switches to the corresponding window for calculations by the method of economic life. Similarly, when you click the "Modified method of economic life MEL" button, the program goes to the window for calculations by the modified method of economic life.

Events and actions in this form:

procedure TForm2.Button2Click(Sender: TObject);
begin
Form1.Showmodal; - when the button is pressed, it switches to the first form
end;

procedure TForm2.Button1Click(Sender: TObject);
begin
Form3.Showmodal; - when the button is pressed, it switches to the third form
end;

When you press the button “Method of Economic Life EL”, we switch to the calculation by the method of economic life (Fig. 5).

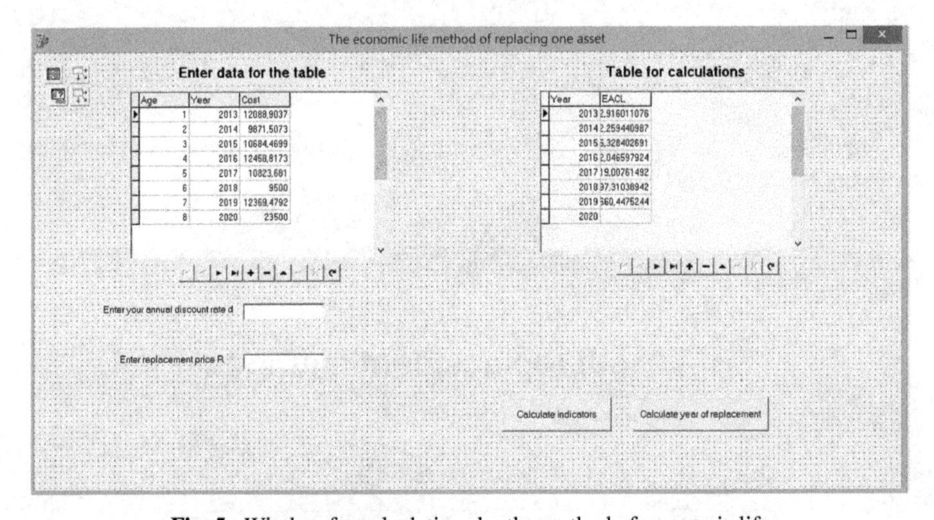

Fig. 5. Window for calculations by the method of economic life

This form uses the following components: AdoTable, AdoQuery, DataSource, DBGrid, DBNavigator, Button, Edit, Label.

In the first table, we use ready-made data or manually enter data using the DBNavigator component. The data in this table will be used for calculations in the program. The second table calculates the effective equivalent cost for each year.

In component Edit1, the value of the annual discount rate d is entered, in component Edit2, the value of the cost of replacing new equipment R is entered. Parameters a and q are not taken into account, since the calculation is carried out according to the method of economic life.

The calculation was made according to the following formula:

$$EAC(L) = \frac{d(1+d)^L}{(1+d)^L - 1^L}\left[\frac{P(\tau_0)}{(1+d)^{L-\tau_0}} + \sum_{j=1}^{L}\frac{A(\tau_0, j)}{(1+d)^{j-\tau_0}}\right].$$

Clicking the Calculate Metrics button displays the effective equivalent cost for each period.

Event and action that is responsible for the calculation:

```
procedure TForm3.Button2Click(Sender: TObject);
begin
LAbel6.Caption:= FloatToStr(     (exp(AdoTable1.FieldByName('Age').AsInteger )*
StrToInt(Edit3.Text)+(exp(AdoTable1.FieldByName('Age').AsInteger
)*AdoTable1.FieldByName('Cost').AsFloat))/
    (exp(AdoTable1.FieldByName('Age').AsInteger * ln(1+StrToFloat(Edit1.Text)))-
1)*exp(AdoTable1.FieldByName('Age').AsInteger                                  *
ln(1+StrToFloat(Edit1.Text)))*StrToFloat(Edit1.Text)) ;

AdoQuery1.Edit;
AdoQuery1.FieldByName('EAC(L)').AsFloat:=                                      (
(exp(AdoTable1.FieldByName('Age').AsInteger                                   )*
StrToInt(Edit3.Text)+(exp(AdoTable1.FieldByName('Age').AsInteger
)*AdoTable1.FieldByName('Cost').AsFloat))/
    (exp(AdoTable1.FieldByName('Age').AsInteger * ln(1+StrToFloat(Edit1.Text)))-
1)*exp(AdoTable1.FieldByName('Age').AsInteger                                  *
ln(1+StrToFloat(Edit1.Text)))*StrToFloat(Edit1.Text)) ;

end;
```

By pressing the button "Modified method of economic life MEL", we go to the next window (Fig. 6).

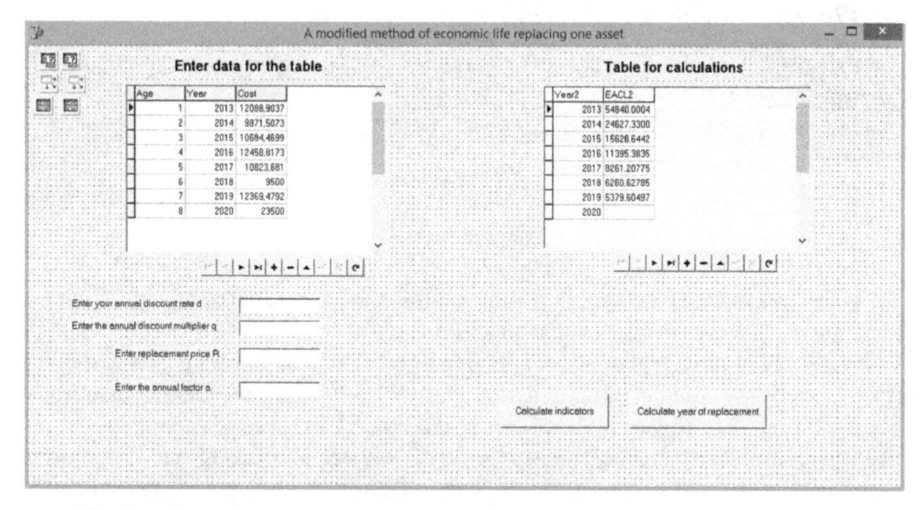

Fig. 6. Window for calculations according to the modified method of economic life

In the first table, we use ready-made data or manually enter data using the DBNavigator component. The data in this table will be used for calculations in the program. The second table calculates the effective equivalent cost for each year.

In component Edit1, the value of the annual discount rate d is entered, in component Edit3, the value of the cost of replacing new equipment R is entered. Parameters a and q are not taken into account, they are entered in the components Edit2 and Edit4.

The calculation was made according to the following formula:

$$EEAC(L) = \frac{d(1+d)^L}{(1+d)^L - q^L}\left[\frac{P(\tau_0)a^{L-\tau_0}}{(1+d)^{L-\tau_0}} - \frac{S(\tau_0, L)}{(1+d)^{L-\tau_0}} + \sum_{j=1}^{L}\frac{A(\tau_0, j)}{(1+d)^{j-\tau_0}}\right].$$

Event and action that is responsible for the calculation:

```
procedure TForm1.Button2Click(Sender: TObject);
begin
LAbel6.Caption:= FloatToStr( (exp(AdoTable1.FieldByName('Age').AsInteger *
ln(StrToFloat(Edit2.Text)))*
StrToInt(Edit3.Text)*(exp(AdoTable1.FieldByName('Age').AsInteger*ln(StrToFloat(E
dit4.Text))))+(exp(AdoTable1.FieldByName('Age').AsInteger *
ln(StrToFloat(Edit2.Text)))*AdoTable1.FieldByName('Cost').AsFloat))/
    (exp(AdoTable1.FieldByName('Age').AsInteger * ln(1+StrToFloat(Edit1.Text)))-
1)*exp(AdoTable1.FieldByName('Age').AsInteger *
ln(1+StrToFloat(Edit1.Text)))*StrToFloat(Edit1.Text)) ;

AdoQuery2.Edit;
AdoQuery2.FieldByName('EACL2').AsFloat:= (
(exp(AdoTable1.FieldByName('Age').AsInteger * ln(StrToFloat(Edit2.Text)))*
StrToInt(Edit3.Text)*(exp(AdoTable1.FieldByName('Age').AsInteger*ln(StrToFloat(E
dit4.Text))))+(exp(AdoTable1.FieldByName('Age').AsInteger *
ln(StrToFloat(Edit2.Text)))*AdoTable1.FieldByName('Cost').AsFloat))/
    (exp(AdoTable1.FieldByName('Age').AsInteger * ln(1+StrToFloat(Edit1.Text)))-
1)*exp(AdoTable1.FieldByName('Age').AsInteger *
ln(1+StrToFloat(Edit1.Text)))*StrToFloat(Edit1.Text)) ;

end;
```

Obtained results, parameters and their values:

d = 0.05; q = 0.9524; R = 45000; a = 0.95.

Figure 7 shows the calculations for the modified method of economic life. Figure 8 shows the calculations using the economic life method.

Figure 7 shows that in the case of applying the modified model of economic life, the effective equivalent annual cost decreased until 2020. This means that the asset must be replaced in 2020. Figure 8 shows that in the case of a model of economic life, the effective equivalent annual cost increases every year without taking into account the parameters a and q. Accordingly, it is difficult to figure out when to replace an asset.

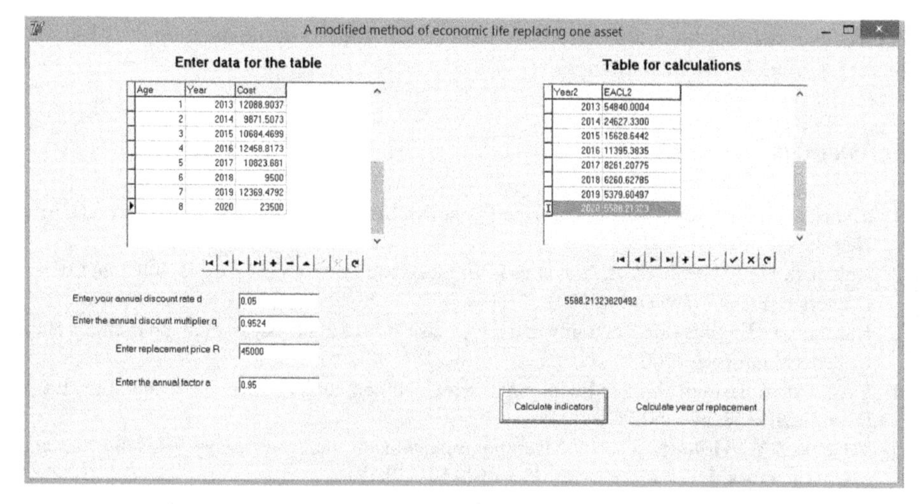

Fig. 7. Calculations by the modified method of economic life

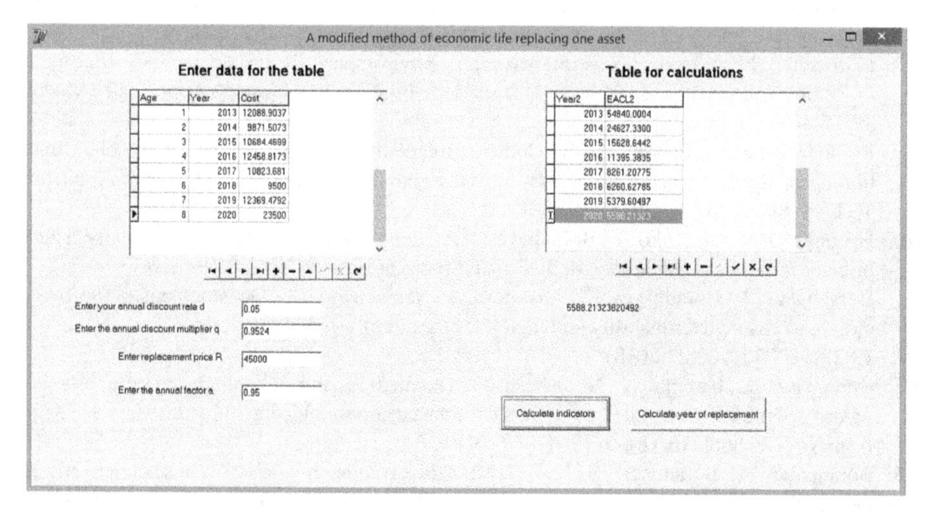

Fig. 8. Calculations by the method of economic life

4 Conclusion

An important challenge is the development and research of asset replacement methods when the future trend in maintenance costs is not exponential or not fully known even over a limited horizon. Some cost allocation methods are convenient for theoretical analysis, but they are difficult to apply and are rarely used in practice.

Usually, when solving practical problems, the time is taken discrete, there are measurement errors and insufficient information. Therefore, it is important that the proposed asset replacement models are consistent with practice, are clear and logically simple, so that the input data requirements for the model are reasonable. In practice, the choice of an effective asset replacement algorithm depends on the available data, on the observed dynamics of technological change.

References

1. Thuesen, G., Fabrycky, W.: Engineering Economy, 8th edn. Prentice Hall, Englewood Cliffs, New Jersey (1993)
2. Newman, D., Eschenbach, T., Lavelle, J.: Engineering Economic Analysis, 9th edn. Oxford University Press, New York (2004)
3. Hartman, J.: Engineering Economy and the Decision-Making Process. Pearson Prentice Hall, Upper Saddle River (2007)
4. Yatsenko, Y., Hritonenko, N.: Economic life replacement under improving technology. Int. J. Prod. Econ. **133**, 596–602 (2011)
5. Yatsenko, Y., Hritonenko, N.: Machine replacement under evolving deterministic and stochastic costs. Int. J. Prod. Econ. **193**, 491–501 (2017)
6. Yatsenko, Y., Hritonenko, N.: Asset replacement under improving operating and capital costs: a practical approach. Int. J. Prod. Res. **54**, 2922–2933 (2016)
7. Yatsenko, Y., Hritonenko, N., Boranbayev, S.: Non-equal-life asset replacement under evolving technology: a multi-cycle approach. Eng. Econ. **65**(4), 1–24 (2020)
8. Hritonenko, N., Yatsenko, Y., Boranbayev, S.: Environmentally sustainable industrial modernization and resource consumption: is the Hotelling's rule too steep? Appl. Math. Model. **39**(15), 4365–4377 (2015)
9. Boranbayev, S.N., Nurbekov, A.B.: Construction of an optimal mathematical model of functioning of the manufacturing industry of the Republic of Kazakhstan. J. Theor. Appl. Inf. Technol. **80**(1), 61–74 (2015)
10. Hritonenko, N., Yatsenko, Y., Boranbayev, A.: Generalized functions in the qualitative study of heterogeneous populations. Math. Popul. Stud. **26**(3), 146–162 (2019)
11. Boranbayev, A., Boranbayev, S., Nurusheva, A., Yersakhanov, K.: Development of a software system to ensure the reliability and fault tolerance in information systems. J. Eng. Appl. Sci. **13**(23), 10080–10085 (2018)
12. Boranbayev, S., Goranin, N., Nurusheva, A.: The methods and technologies of reliability and security of information systems and information and communication infrastructures. J. Theor. Appl. Inf. Technol. **96**(18), 6172–6188 (2018)
13. Boranbayev, A., Boranbayev, S., Nurusheva, A.: Development of a software system to ensure the reliability and fault tolerance in information systems based on expert estimates. In: Advances in Intelligent Systems and Computing, vol. 869, pp. 924–935 (2018)
14. Boranbayev, A., Boranbayev, S., Yersakhanov, K., Nurusheva, A., Taberkhan, R.: Methods of ensuring the reliability and fault tolerance of information systems. In: Advances in Intelligent Systems and Computing, vol. 738, pp. 729–730 (2018)
15. Boranbayev, S., Altayev, S., Boranbayev, A.: Applying the method of diverse redundancy in cloud based systems for increasing reliability. In: The 12th International Conference on Information Technology: New Generations (ITNG 2015), Las Vegas, Nevada, USA, 13–15 April 2015, pp. 796–799 (2015)
16. Turskis, Z., Goranin, N., Nurusheva, A., Boranbayev, S.: A fuzzy WASPAS-based approach to determine critical information infrastructures of EU sustainable development. Sustain. (Switz.) **11**(2), 424 (2019)
17. Turskis, Z., Goranin, N., Nurusheva, A., Boranbayev, S.: Information security risk assessment in critical infrastructure: a hybrid MCDM approach. Informatica (Neth.) **30**(1), 187–211 (2019)

18. Boranbayev, A.S., Boranbayev, S.N., Nurusheva, A.M., Yersakhanov, K.B., Seitkulov, Y.N.: Development of web application for detection and mitigation of risks of information and automated systems. Eurasian J. Math. Comput. Appl. **7**(1), 4–22 (2019)
19. Boranbayev, A.S., Boranbayev, S.N., Nurusheva, A.M., Seitkulov, Y.N., Sissenov, N.M.: A method to determine the level of the information system fault-tolerance. Eurasian J. Math. Comput. Appl. **7**(3), 13–32 (2019)
20. Boranbayev, A., Boranbayev, S., Nurbekov, A., Taberkhan, R.: The development of a software system for solving the problem of data classification and data processing. In: 16th International Conference on Information Technology - New Generations (ITNG 2019), vol. 800, pp. 621–623 (2019)
21. Boranbayev, A., Boranbayev, S., Nurusheva, A., Yersakhanov, K., Seitkulov, Y.: A software system for risk management of information systems. In: Proceedings of the 2018 IEEE 12th International Conference on Application of Information and Communication Technologies (AICT 2018), Almaty, Kazakhstan, 17–19 October 2018, pp. 284–289 (2018)
22. Boranbayev, S., Altayev, S., Boranbayev, A. and Nurbekov, A.: Mathematical model for optimal designing of reliable information systems. In: Proceedings of the 2014 IEEE 8th International Conference on Application of Information and Communication Technologies (AICT 2014), Astana, Kazakhstan, 15–17 October 2014, pp. 123–127 (2014)
23. Boranbayev, S., Boranbayev, A., Altayev, S., Seitkulov, Y.: Application of diversity method for reliability of cloud computing. In: Proceedings of the 2014 IEEE 8th International Conference on Application of Information and Communication Technologies (AICT 2014), Astana, Kazakhstan, 15–17 October 2014, pp. 244–248 (2014)
24. Boranbayev, S.: Mathematical model for the development and performance of sustainable economic programs. Int. J. Ecol. Dev. **6**(1), 15–20 (2007)
25. Boranbayev A., Boranbayev S., Nurusheva A.: Analyzing methods of recognition, classification and development of a software system. In: Advances in Intelligent Systems and Computing, vol. 869, pp. 690–702 (2018)
26. Boranbayev, A.S., Boranbayev, S.N.: Development and optimization of information systems for health insurance billing. In: ITNG2010 - 7th International Conference on Information Technology: New Generations, pp. 1282–1284 (2010)
27. Akhmetova, Z., Zhuzbayev, S., Boranbayev, S., Sarsenov, B.: Development of the system with component for the numerical calculation and visualization of non-stationary waves propagation in solids. Front. Artif. Intell. Appl. **293**, 353–359 (2016)
28. Boranbayev, S.N., Nurbekov, A.B.: Development of the methods and technologies for the information system designing and implementation. J. Theor. Appl. Inf. Technol. **82**(2), 212–220 (2015)
29. Boranbayev, A., Shuitenov, G., Boranbayev, S.: The method of data analysis from social networks using apache hadoop. In: Advances in Intelligent Systems and Computing, vol. 558, pp. 281–288 (2018)
30. Boranbayev, S., Nurkas, A., Tulebayev, Y., Tashtai, B.: Method of processing big data. In: Advances in Intelligent Systems and Computing, vol. 738, pp. 757–758 (2018)
31. Boranbayev, A., Boranbayev, S., Nurbekov, A.: Estimation of the degree of reliability and safety of software systems. In: Advances in Intelligent Systems and Computing, AISC , vol. 1129, pp. 743–755 (2020)
32. Boranbayev, A., Boranbayev, S., Nurbekov, A.: Development of the technique for the identification, assessment and neutralization of risks in information systems. In: Advances in Intelligent Systems and Computing, AISC, vol. 1129, pp. 733–742 (2020)
33. Boranbayev, A., Boranbayev, S., Nurusheva, A., Seitkulov, Y., Nurbekov, A.: Multi criteria method for determining the failure resistance of information system components. In: Advances in Intelligent Systems and Computing, vol. 1070, pp. 324–337 (2020)

34. Boranbayev, A., Boranbayev, S., Nurbekov, A., Taberkhan, R.: The software system for solving the problem of recognition and classification. In: Advances in Intelligent Systems and Computing, vol. 997, pp. 1063–1074 (2019)
35. Akhmetova, Z., Boranbayev, S., Zhuzbayev, S.: The visual representation of numerical solution for a non-stationary deformation in a solid body. In: Advances in Intelligent Systems and Computing, vol. 448, pp. 473–482 (2016)
36. Boranbayev, A., Boranbayev, S., Nurbekov, A.: Evaluating and applying risk remission strategy approaches to prevent prospective failures in information systems. In: Advances in Intelligent Systems and Computing, vol. 1134, pp. 647–651 (2020)
37. Boranbayev, A., Boranbayev, S., Nurbekov, A.: A proposed software developed for identifying and reducing risks at early stages of implementation to improve the dependability of information systems. In: Advances in Intelligent Systems and Computing, vol. 1134, pp. 163–168 (2020)

Comparative Evaluation of Several Classification Algorithms on News Posts Using Reddit Social Network Dataset

Ahmad Rawashdeh[1](✉), Mohammad Rawashdeh[1], and Omar Rawashdeh[2]

[1] University of Central Missouri, Lee's Summit, MO, USA
{arawashdeh,rawashdeh}@ucmo.edu
[2] University of Cincinnati, Cincinnati, OH, USA
rawashoy@mail.uc.edu

Abstract. This research investigates a comparative evaluation of the results of different classification algorithms applied on the Reddit News Social Network Dataset. A program, which is well described was written in C# Dot Net that could run the classification using different parameters. These parameters which were input to the program, include the data size to read, training and testing size, class to predict, vector representation and the classification algorithm including Logistic Regression (LR), Naïve Bayes (NB), Decision Tree (DT), and N-Nearest Neighbor (NN). The classification was performed multiple times (various parameters). Also, the training and testing data were selected at random. That, to get a better judgment on the results. The performance metrics: precision, recall, accuracy, f-measure, sensitivity, and specificity were calculated for every time any classification algorithm is executed. Results showed that the precision for the classification Algorithms ranged from a value greater than 50 to 100%. Regarding the average precision results, in most of the results, DT had the highest average precision (4 times as the highest with one time together with LR but for different data size). Also, it was found that regarding the time required for a classification algorithm to produce results, for a data of size 200, all algorithms were fast (less than 5 s) except for LR (2–3 min). For a data of size 500, LR was the slowest with a time of 32 min. Then NN was the second slowest with a time of 2 min. All others including NB, NN and DT were the fastest (time in less than 5 s). For a data of size 1000, LR took 1 day and it was not even enough for the LR classifier to finish executing and produce any results. All remaining algorithms also did not complete the calculation of results in 1 day.

Keywords: Machine learning · Classification · Comparative research · News · Reddit · Social network dataset · Experiment

1 Introduction

The influence of news stories on many lives is undeniable to anyone who constantly read and rely on news channel or websites to stay informed and updated about most of the important events that happens globally or locally. With the widespread of social

K. Arai (Ed.): FICC 2021, AISC 1363, pp. 1003–1026, 2021.
https://doi.org/10.1007/978-3-030-73100-7_69

networks, many have used the technology, intensively, to spread the word about their stories. On one hand, social networks such as Facebook, Instagram, Reddit, and twitter have provided the platform for different types of professional users to use their services. These professional users include but are not limited to: channels and news reporters, brands advocates, celebrities, public figures, companies and their products, institution (universities, schools etc.) and people who belong to or are affiliated with any of them. This broad set of audiences have used social networking websites to create profile/pages and started writing about what matter to them and started communicating to/and with the readers/viewers/fans/followers. On the other hand, bloggers (ranging from those who write about personal stories to those writing about global and major events) has become even popular and widespread nowadays.

With that in mind, the need for classifying the posts on social networks to their class types has become apparent. To achieve that goal, several algorithms of Machine Learning can assist, but first, the reader need to be introduced to what is meant by Machine Learning classification algorithms (supervised learning).

Supervised learning or what is called classification is one of the core concepts and topic of machine learning. Machine learning refer to those algorithms or concepts which attempts to solve problems for which no well-defined step-by-step algorithm exists to solve them. Machine learning, classification in particular, relies on having a labeled data (input mapped to output) that could be used to train a model (Logistic Regression, K nearest Neighbors, Neural Network, Decision tree, Naïve Bayse, C4.5, and others) [1, 2] that given a new instance, can predict the class/output based on the trained data. Caution must be taken not to have the model be biased or overfit the data. Unsupervised learning or Clustering is another type of machine learning algorithm which is closely related to classification. In clustering, the goal is trying to minimize intra-cluster similarities and maximize inter-clusters similarities. In clustering, the data is not labeled as in classification. But the goal is to group the data into clusters of similar data instances. In clustering and classification algorithm, the dataset is divided into training and testing (two-fold with usually 80% data size for training and 20% for testing). It is also worth mentioning that there are different measures for clustering quality including the silhouette which was investigated in [3]. Usually machine learning algorithms are evaluated and judged, as other algorithms/system (such as information retrieval, database, sorting algorithm, etc.) using the following metrics: precision, recall, F-measure, accuracy, sensitivity, specificity, accuracy, and Receive Operating Characteristics (AUC and ROC curves), in addition to time and space complexities.

To emphasize more, Machine learning algorithms became viral and interests on them grew drastically as new ideas and applications came into existence. Example of such applications includes: Facial recognition, Image processing and object recognition [4], autonomous driving vehicles [5], recommender systems in social networks, such as: Facebook[1] friends' recommendation (link prediction) [6–8], Netflix[2] movies recommendation, Youtube[3] suggested related videos, and Amazon retails recommendation of similar items of interest, sentiment analysis [9], are all example of applications of

[1] http://www.Facebook.com.

[2] http://www.Netflix.com.

[3] http://www.Youtube.com.

machine learning algorithms. Even a magical cloak application which has become a trend on social networking sites such as LinkedIn[4] in which a cloak worn by the person is made invisible in real time video while it is being captured [10]. In addition to applications in biology [11], drugs, security [12], and network and others. Moreover, classification has many applications which require the prediction of a class label given a training data. One of these applications is predicting the news type of a user's post which is the topic of this paper.

This work involved the following: choosing the appropriate dataset, data preprocessing, deciding on the idea of the research, coding and implementing the different classification algorithms (using the namespace "Accord.MachineLearning") and coding the classes for reading the dataset, producing results for different parameters by running the program many times, evaluating the algorithms, finalizing the program and writing and reporting the results.

The paper is organized as follows: next is literate review (Sect. 2), then the dataset (Sect. 3). After that, classification (Sect. 4). Then the classifier program (Sect. 5). Follows is the output (Sect. 6). Then details of the results and discussion (Sect. 7) followed by evaluation of the results (Sect. 8). Finally, the paper ends with conclusion and future work (Sect. 9) and last is the references.

2 Literature Review

Several papers surveyed classification algorithms. Some provided theoretical and conceptual based comparisons between such algorithms, while others have conducted an empirical experiment to compare and evaluate the performance of such methods.

As an example of the former (theoretical survey), is the work in [13] in which a survey of classification techniques and a good comparison in terms of learning speed, accuracy, speed of classification, tolerance to missing values, etc. was detailed. However, no empirical test was conducted. Just background information, definition, comparison between the different classification methods (theorical). Also, the work in [13] provided a review of classification method which is another example of a theoretical comparison between the classification algorithms.

While an example of the later (empirical evaluation) is the work in [14–16]. More details in the following paragraphs.

In [14] a Survey of five machine learning methods: support vector machine (SVM), Naïve Bayes (NB), Neural Network (NN), Decision Tree (DT), Radial basis function neural network (RBFNN), K-nearest-neighbor (KNN) was given. The paper also documented an experiment on three different datasets. And all three datasets are about sentiments of texts [14].

In [15] an empirical comparison of four different classification methods: k-nearest neighbor, Naïve Bayes, C4.5 Decision Trees and Support Vector Machine when applied on four different datasets was conducted. The results showed potential for all except for C4.5. Support vector achieved best results with two datasets, while the C4.5 decision tree showed potential for one dataset.

[4] http://www.linkedin.com.

In [16], four different classifications were examined: Naive Bayesian (NB) approach, Rocchio-style classifier, k-NN method and SVM system. Two benchmark collections were chosen as the testbeds: Reuters-21578 and small portion of Reuters Corpus Version 1 (RCV1), making the new results comparable to published results.

One might also consider an ensemble classifier. The ensemble classifier is based on combining the strength of many classifiers. However, according to [13] the application of ensemble methods is only recommended if we are interested in the best possible classification accuracy.

Another example of an application of classification algorithms is in social network. The work about using a classifier to detect cyberbullying in social network was researched in [17]. The work describes running the algorithms on two datasets: formspring.me and myspace.com; two algorithms: Lavenshtein and Naïve Bayes were considered.

The following is how this research addresses the shortcoming of the work mentioned above: the current work selects data to train and test randomly from the total dataset (unbiased). Second, the experiment was run many times and repeated for several values of the parameters (it is easy to repeat the experiment for any new values as the parameters were made as input to the program). Third, this work considers some classification algorithms (Logistic Regression and Naïve Bayes) which were not considered in all cited works. And finally, not only the confusion matrix or accuracy was used in the assessment, but all measures including precision, recall, f-measure, accuracy, sensitivity, and specificity were calculated.

More specifically, this work is different from the works in [13] in that it is an experimental evaluation which involves running classification methods on a dataset and evaluating the results. Whereas the works in the cited references were merely theoretical.

Also, in this work, the running times of the classification algorithms were measured and used in the evaluation of different algorithms, while in [14] it was not measured. Additionally, none of the works in [14–16] considered Logistic Regression in their evaluation, and the work in [14] didn't study Naïve Bayes. Another difference is that the work in [15] used confusion matrix and AUC but this work used many more measures. Also, the work in [16] includes a cited theoretical comparisons between the classification algorithms but not an experimental comparison made by the authors by running the algorithms on a dataset. Finally, none of the works cited previously used the same dataset as this work (Reddit Dataset).

This work aims at adding additional knowledge about the classification algorithm and their results in terms of performance metric (precision, recall, f-measure, accuracy, sensitivity, and specificity) and time by considering running the algorithms on Reddit dataset with several various parameters including: classification algorithm, data size to read, training and testing data sizes, vector representation, and definitely last but not least the class to predict.

3 Dataset

The name of the used dataset file is **reddit_data.csv**. The link to download all three dataset files (two twitter and one reddit can be found at [18]) and the related paper can be found at [19]. The original Reddit data set consists of 40,000 Reddit parent comments from May 2015 belonging to five subreddit pages, which are used as topic labels. The first 405 rows are well formatted. Starting from row 406 the data is not in proper format. Then from row 1084 it is in proper format again (four columns filled). The data is in five columns: parent_id, text, topic, length, size_range. Table 1 is snippet (first 3 rows) of the dataset:

Table 1. The first three rows of the reddit dataset

Parent_id	Text	Topic	Length	Size_range
t1_crojgfu	Thanks! Not sure if those links were up there before and now I'm confused whether to feel dumb or not	Pcmasterrace	103	101 to 200
t1_cquq97y	I think its unlikely someone would kill them selves by blunt force truma to the head and breaking their own spine. Unless they were on PCP or bathsalts then maybe	News	163	101 to 200
t1_cr92xnl	Hoult is another one that's important. But most of all, Fassbender and McAvoy, yes, I agree this	Movies	99	0 to 100

3.1 Dataset Preprocessing

The dataset was preprocessed: empty posts and empty class posts were removed, also unstructured paragraphs were removed (paragraphs which did not fit the format of the dataset). From 44285 rows, the csv file was reduced to 40236 rows which are well formatted and has all valid nonempty cell content.

The punctuations and special characters were removed. For punctuations removal, the following were removed from the posts ('[', ']', '(', ')', '!', '~', '$', '%', '^', '&', '*', '?', '<', '>', '"', '/', '\']. The characters '@' and '#' were preserved (# was preserved for its hashtag special meaning and '@' for preserving the email address if any was listed in the post for future use). Other punctuations such as ';', ':', ',', '.' Were used as delimiter characters in tokenizing the posts content (to get all words as a string array).

Stop words which were removed from the posts include [20] and [21]. The removed stop words were either pronouns or others. The removed pronouns were: he, she, they, it, him, her, their, its, you, I. Other stop words which were removed (mostly auxiliaries) are: is, am, was, are, were, be, we, being, been, can, could, do, does, did, done, have, has, having, Had, may, might, must, shall, should, will, would, of, off, in, on, the, a, non, not, and, if, to, by, that, at.

4 Classification (Supervised Learning)

Classification is predicting the class label of the data. Classifiers need to be trained and their performance is tested on a testing data. Usually the dataset is divided into training and testing where size of the training data is 80% of the overall size of the data and the size of the testing data is 20% of the overall data size. In this work, data of size 50, 100, 200, 500, and 1000 were used.

The following are the different classification methods, considered in this paper as well as implemented and tested, which are described with coding examples in [22]:

1. Logistic Regression: more in [2, 23].
2. KNN: using this algorithm, to classify a new example, the distance from that example to every training example is measured, and the k smallest distances are selected [1, 24].
3. Naïve Bayes: is based on the maximum a posterior probability which is given by the following formula [1, 25]:

$$\mathrm{P(C = c \| X_1 \ldots X_n)} = \frac{P(C = c)P(X_1, \ldots, X_N || C = c)}{P(X_1, \ldots, X_N)}$$

   Where $\frac{P(C=c)P(x1,...XN||C=c)}{P(X1,...,XN)} = P(C = c)P(X_1||C = c)...P(X_N||C = c)$,
4. Decision Tree: The tree infers a split of the training data based on available feature to produce a good generalization. Information gain is used to decide on which feature to use for splitting the data. A new example is classified using Decision tree by following a path from the root to a leaf node where a test is performed at each node (chosen based on information gain) on some feature. Two algorithms for building decision trees (ID3 and C4.5) [1].

An attempt to implement the three different methods of Support vector machine [26] was made. However, due to some technical problem, the algorithm did not produce any true positives at all, precision was 0 for all data sizes, and for different classes to predict [22]. The reader may also refer to[27].

4.1 Feature Selection

Selecting important features in text to represent each class instance as a vector.

Several methods for Feature selection including frequency, TF-IDF (document Frequency and Inverse Term Frequency), CHI, Information Gain (IG), Gain Ratio (GR) [22]. Only the basic frequency (0 or 1, regular) and TF-IDF were used in this work.

4.2 Vector Representation

Using frequency (bag of words) representation; one could represent each post as a binary vector (0 for word non occurring and 1 the word occurs at least once) which is called bag of words representation. Using this method, each post is represented as a vector of frequencies of selected words. For instance, if the selected words are: "play", "scene", "book" and the first post contain only "play" then its vector is (1, 0, 0). And if the posts contain the words "play" and "book" then its vector representation is (1, 0, 1). The posts were represented as vectors of length 10. The vector may be normalized using min and max to [0, 1.0], but in this work, that was not done [27].

Alternatively, one could represent the post as an integer vector based on the frequency of the terms in the training data and number of training vectors that contain the term, which is called Term Frequency- Inverse Document Frequency (TF-IDF). The formula for calculating the TF-IDF is in [28, 29].

In the implementation of the classification algorithm, we used the built-in methods:

```
var codebook = new BagOfWords() { MaximumOccurance = 1 }; //represetn the vector as binary 0 or 1
codebook.Learn(ex.TrainingWords);
var vocabulary = new TFIDF() { Tf = TermFrequency.Default, Idf = InverseDocumentFrequency.Default };
vocabulary.Learn(ex.TrainingWords);
// vector[i] is used to store the ith numerical vector representation of training post
if (VectRep.Equals("regular"))
      vectors[i] = codebook.Transform(ex.TrainingWords[i]);  // use BagOfWord representation
else
      vectors[i] = vocabulary.Transform(ex.TrainingWords[i]); // use TF-IDF representation
```

5 The Classifier Program

We wrote a program in C# Dot Net which trains and test the classification (Bag of word or TF-IDF) on a reddit social network posts dataset (described in the dataset section). It is a console application, that automate training and running the classifier, the researcher (this is how we used it) only need to input the class to be predicted (one of the following classes: news, movies, pcmasterrace, nfl, and relationships) and then input the training size (how many posts used for training to read which will be represented as vectors) and also input the testing size (how many posts used for testing the classifier). The program then train the models using the randomly selected posts vectors, then display the numbers of True positive, False Positives, True Negative, False Negative and the evaluation metric: precision, recall, F-measure, accuracy, sensitivity, specificity (may add ROC and AUC later). This runs in a loop, so the researcher has the chance to run the classifier multiple times using the dataset after providing the inputs and training the classifier. The results are expected to vary between the runs due to choosing different parameters: size of data to read from the entire dataset, training and testing sizes, the class to predict, the classification algorithm (Logistic Regression, Naïve Bayes, Decision Tree, or K-Nearest Neighbor) and different vector representation: regular binary of 0 s and 1 s (bag of words) or as a decimal of [0–9]s digits (TF-IDF). The results may also be different because even when choosing the same parameters, the training and testing data is chosen at random. So after all that, the program eventually outputs the results to text files (consisting of all vectors classified as either True positive or False positive along with the metric for every run) and to excel spreadsheets (with all metrics) one excel file for every algorithm and each with a sheet for every class. We had to create the charts manually afterwards.

In our evaluation of the classification algorithms, we chose the dataset sizes 50, 100, 200, 500, 1000, and 4000 for different runs such that the training data size for every case is 80% of the chosen data size to read and testing data size is 20% (both chosen at random). That means the training size and testing size for all previously mentioned data sizes are: [40, 10], [80, 20], [160, 40], [400, 100], [800, 200], and [3200, 800] respectively. The question that still has not been answered yet, how well the algorithms perform and scale with overall data size of 40236 rows. For data of size 1000, and 4000, the program was not able to finish and produce results even after 1 day of running.

The data has more than one class. So, it should not be a binary classifier where only one class is predicted. Notes [22].

6 Output

The following posts' vectors were classified as a news posts by the LR classifier (when the class news was to be predicted). Note: these vectors where classified when a data of size 200 was used to train the LR classifier with regular vector representation:

wouldn't, recommend, coming, contact, with, man, ever, again, sounds, like...[True Tp]

maybe, falcons, wanted, trade, with, us, supposedly, for, probably, had ....[True Tp]

this, back, future, strangely, bare, bottom, through, long, johns, go ....[True Tp]

The following were classified as true positive instance by the LR classifier with TF-IDF vector representation (when the class news was to be predicted)

maybe, falcons, wanted, trade, with, us, supposedly, for, probably, had...[True Tp]

get, your, point, thing, for, my, first, build, two, options ....[True Tp]

this, back, future, strangely, bare, bottom, through, long, johns, go ....[True Tp]

The performance metrics for the above outputs were:

Precision: 66.67%, recall: 16.67%, F-measure: 26.67%, Accuracy: 72.50%, Sensitivity: 16.67%, Specifity: 96.43%, correct_Predcitions: 29, incorrect_Predictions: 11

The following posts were classified as true positive class (when movies class was to be predicted) for a data size of 200 using regular vector representation:

it's, about, punishment, or, even, about, actually, beating, kid, thing...[True Tp]

his, name, odin, father, thor, when, years, ago, close, enough ....[True Tp]

yeah, private, lobbies, but, don't, anyone, pc, play, with, so ....[True Tp]

Notice the existence of the word "Thor" which is one of the well-known movies.

The following posts where one of the posts truly classified as a class "nfl" with a precision of 100% for that run using regular vector representation with data size 200 listed by classification algorithms:

DT: many, pats, extended, brady, before, season, right, after, winning, sb...[True Tp]

NN: just, any, title, doesn't, use, gameworks, or, limits, use, gameworks...[True Tp]

NB: cheated, win, super, bowl, love, see, full, year, suspension, let...[True Tp]

LR: actually, curious, see, how, play, out, found-footage, setting, we've, watching...[True Tp]

Notice the existence of the words "season", "winning", 'gamework', 'bowl', 'super', and 'win'.

Figure 1 shows the content of one of the Excel output files for the DT algorithm for the news class. The sheet contains the following columns: Algorithm_Conif, Precision, Recall, F-measure, Accuracy, Sensitivity, Specifity, Correct_Predictions, and Incorrect_Predictions. The format of the content of the cell "Algorithm_config" is: [algo]_[vector representation]_[class:news]_d[size]_[training, testing]_[counter]

Algorithm_Config	Precision	Recall	F-measure	Accuracy	Sensitivity	Specifity	Correct_Predictions	Incorrect_Predictions
DT_regular_news_d200_160,40_C:2	50.00%	18.18%	26.67%	72.50%	18.18%	93.10%	29	11
DT_regular_news_d200_160,40_C:3	40.00%	16.67%	23.53%	67.50%	16.67%	89.29%	27	13
DT_regular_news_d200_160,40_C:4	25.00%	20.00%	22.22%	65.00%	20.00%	80.00%	26	14
DT_regular_news_d200_160,40_C:5	12.50%	20.00%	15.38%	72.50%	20.00%	80.00%	29	11
DT_regular_news_d200_160,40_C:6	57.14%	30.77%	40.00%	70.00%	30.77%	88.89%	28	12
DT_regular_news_d200_160,40_C:7	33.33%	28.57%	30.77%	77.50%	28.57%	87.88%	31	9
DT_regular_news_d200_160,40_C:8	11.11%	12.50%	11.76%	62.50%	12.50%	75.00%	25	15
DT_regular_news_d200_160,40_C:9	80.00%	26.67%	40.00%	70.00%	26.67%	96.00%	28	12
DT_regular_news_d200_160,40_C:10	40.00%	16.67%	23.53%	67.50%	16.67%	89.29%	27	13
DT_regular_news_d500_400,100_C:11	27.27%	11.54%	16.22%	69.00%	11.54%	89.19%	69	31
DT_regular_news_d500_400,100_C:12	29.41%	19.23%	23.26%	67.00%	19.23%	83.78%	67	33
DT_regular_news_d500_400,100_C:13	22.22%	8.70%	12.50%	72.00%	8.70%	90.91%	72	28
DT_regular_news_d500_400,100_C:14	36.36%	12.90%	19.05%	66.00%	12.90%	89.86%	66	34
DT_regular_news_d500_400,100_C:15	27.27%	13.64%	18.18%	73.00%	13.64%	89.74%	73	27
DT_regular_news_d500_400,100_C:16	55.56%	19.23%	28.57%	75.00%	19.23%	94.59%	75	25
DT_regular_news_d500_400,100_C:17	50.00%	17.86%	26.32%	72.00%	17.86%	93.06%	72	28
DT_regular_news_d500_400,100_C:18	30.00%	11.11%	16.22%	69.00%	11.11%	90.41%	69	31
DT_TF-IDF_news_d200_160,40_C:19	50.00%	22.22%	30.77%	77.50%	22.22%	93.55%	31	9
DT_TF-IDF_news_d200_160,40_C:20	50.00%	7.14%	12.50%	65.00%	7.14%	96.15%	26	14
DT_TF-IDF_news_d200_160,40_C:21	50.00%	7.69%	13.33%	67.50%	7.69%	96.30%	27	13
DT_TF-IDF_news_d200_160,40_C:22	0.00%	0.00%	0.00%	75.00%	0.00%	96.77%	30	10
DT_TF-IDF_news_d200_160,40_C:23	42.86%	27.27%	33.33%	70.00%	27.27%	86.21%	28	12
DT_TF-IDF_news_d200_160,40_C:24	20.00%	10.00%	13.33%	67.50%	10.00%	86.67%	27	13

Fig. 1. Results DT algorithms. Class:news, representation:regular, size: 200

7 Results and Discussions

The following are the results for classification for each class, algorithm, and vector representation. The first set includes the average precision (calculated on several runs) for each classification algorithm, class, and data sizes of 100,200, and 500 using either regular or TF-IDF vector representation.

In Fig. 2 and 3, which show the average precision on News class: LR achieved the highest precision (49.17% and 50%) for both the regular and TF-IDF vector representation respectively calculated on a data of size 200. For a data of size 500, DT achieved the highest average precision (34.76% and 36.95%). Results for data size of 100 was not calculated for all classes.

"News" class Average Precisions.

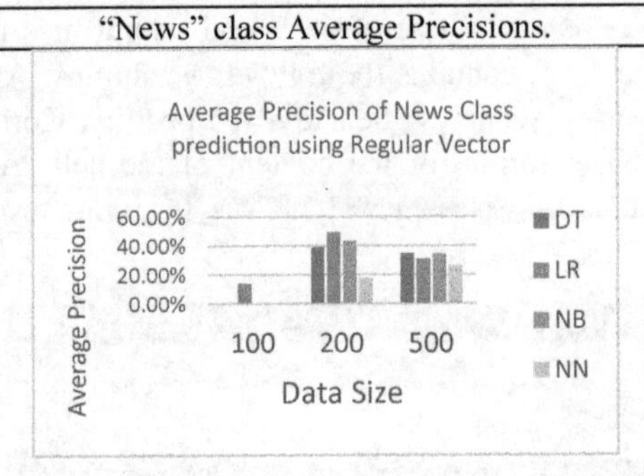

Fig. 2. Average precision of news class prediction using regular vector

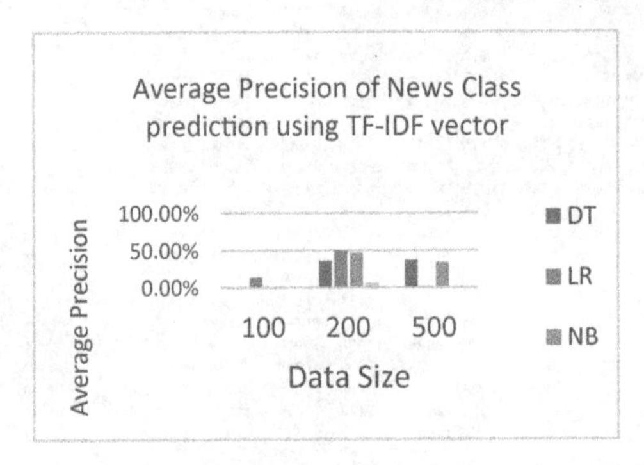

Fig. 3. Average precision of news class prediction using TF-IDF vector

Figures 4 and 5 show the average precision on Movies Data: the DT achieved the highest average precision (43.54% and 65.00%) for both regular and TF-IDF vector representation for a data of size 500.

"Movies" Average Precisions Charts

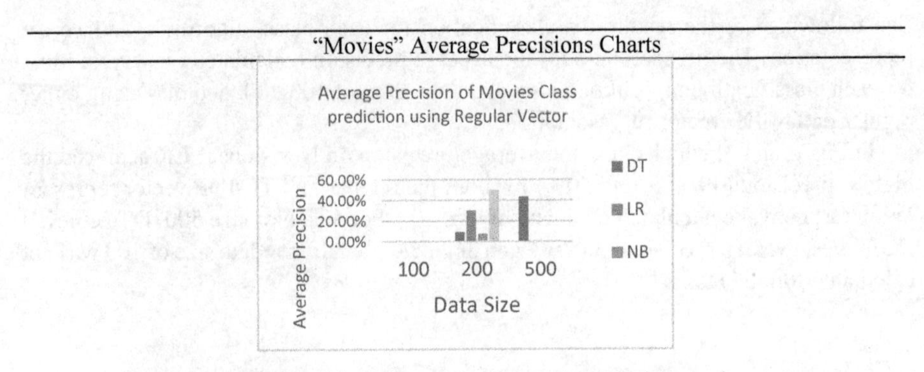

Fig. 4. Average precision of movies class prediction using regular vector

Figures 6 and 7 show the average precision on NFL class: all algorithms achieved almost the same average precision overall. However, to be more accurate, DT had the

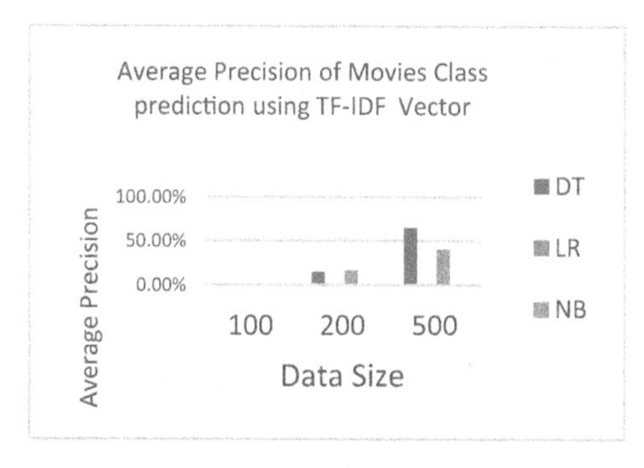

Fig. 5. Average precision of movies class prediction using TF-IDF vector

highest average precision (68.67%) for a data of size 200 using regular vector representation, but LR beat it using TF-IDF vector representation (with average precision of 45.07%). Also, LR had the highest average precision (52.91%) for a data of size 500 using regular vector representation.

"NFL" Average Precision Charts

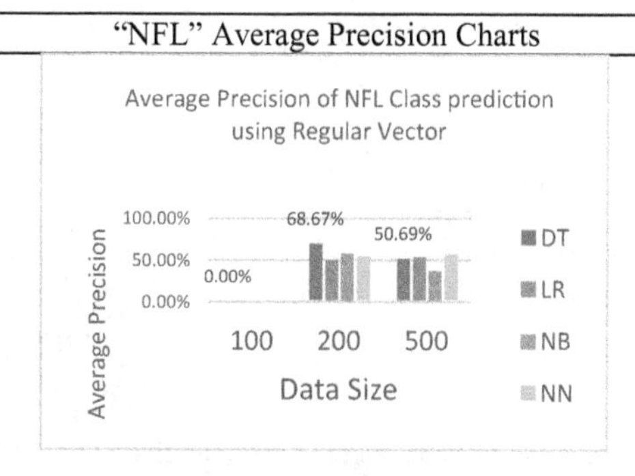

Fig. 6. Average precision of NFL class prediction using regular vector

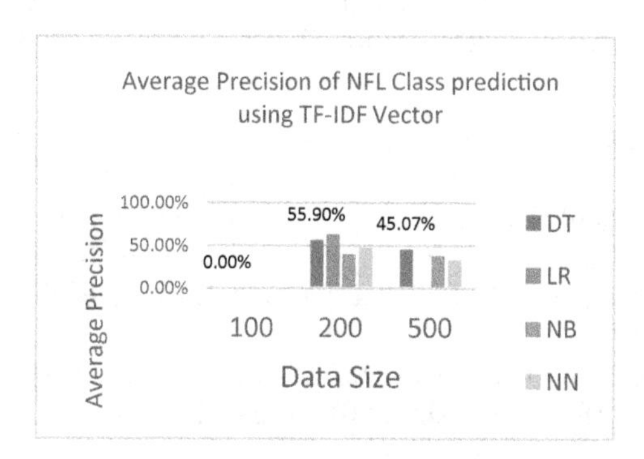

Fig. 7. Average precision of NFL class prediction using TF-IDF vector

Figures 8 and 9 show the average precision on Pcmasterrace class: DT had the highest average precision (57.57% and 52.91% for regular and TF-IDF vector representation respectively) for a data of size 500.

"Pcmasterrace" Average Precision Charts

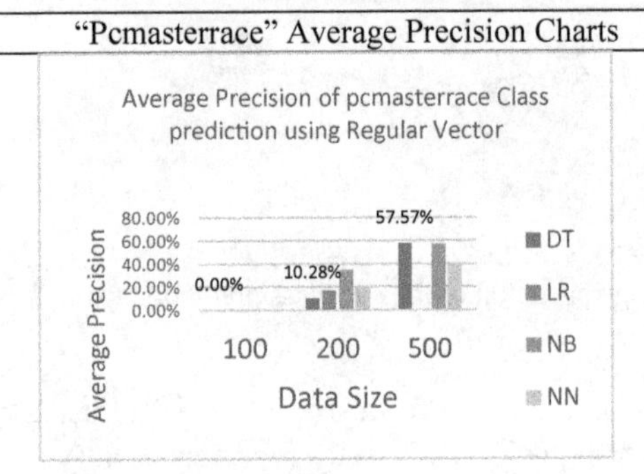

Fig. 8. Average precision of Pcmasterrace class prediction using regular vector

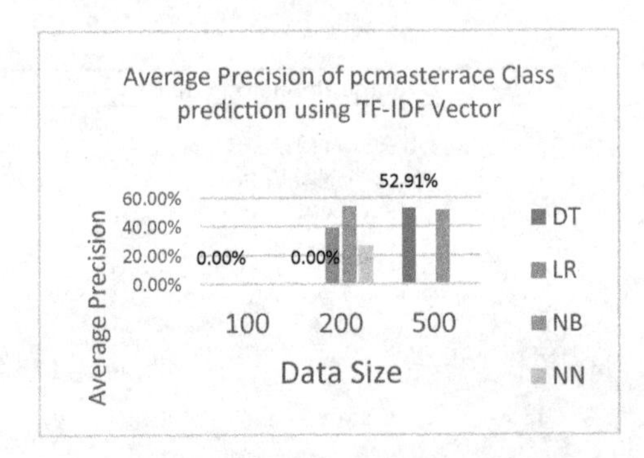

Fig. 9. Average precision of Pcmasterrace class prediction using TF-IDF vector

Figures 10 and 11 show the average precision on Relationships class: NB had the highest average precision using 500 data size (31.25% for regular and 34.00% for TF-IDF vector representations)

Summary (the highest average precision):

Class "news": (for data of size 200) LR. (for data of size 500) DT
Class "movies": (for data of size 500) DT for both representation
Class "NFL": (for data of size 200) DT has the highest using regular. LR has the highest for TF-IDF
Class "pcmasterrace": (for data of size 500) DT for both vector representation
Class "relationships": (for data of size 500) NB had the highest

The second set includes the max (highest) average precision for each classification algorithm, class, and data sizes of 100,200, and 500 using either regular or TF-IDF vector representation.

"Relationships" Average Precision Charts:

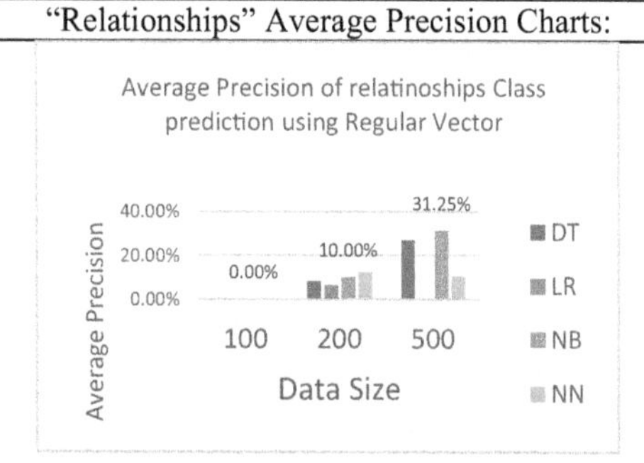

Fig. 10. Average precision of relationships class prediction using regular vector

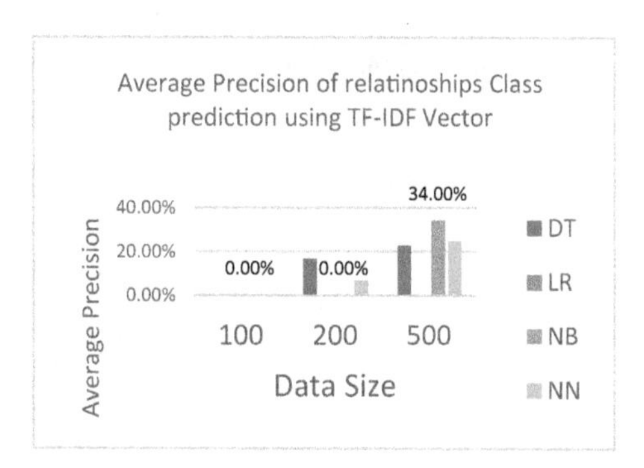

Fig. 11. Average precision of relationships class prediction using TF-IDF vector

Figures 12 and 13 show the maximum precision on News class: Almost all has the same maximum precision for a data of size 200 using regular vector representation with NN as the lowest (50.00%)

News Maximum Precision:

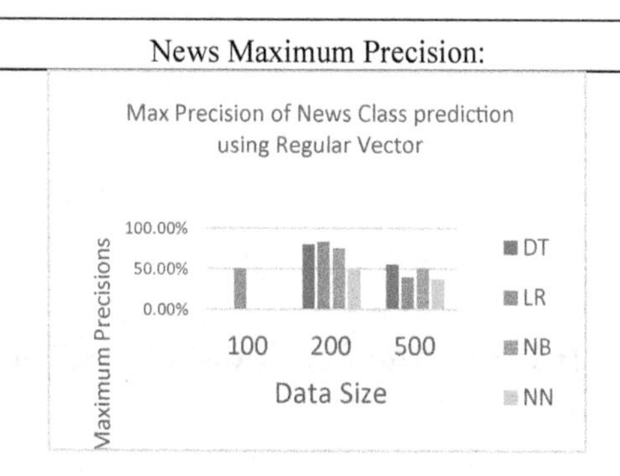

Fig. 12. Max precision of news class prediction using regular vector

Figures 14 and 15 show the maximum precision on Movies class: DT had the highest maximum for results using TF-IDF vector representation.

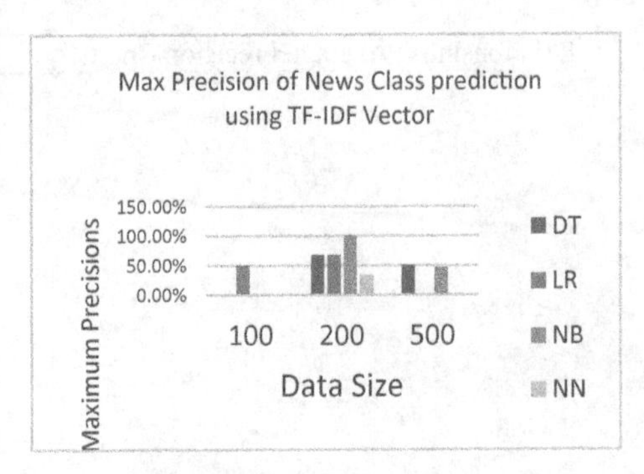

Fig. 13. Max precision of news class prediction using TF-IDF vector

Movies Maximum Precision

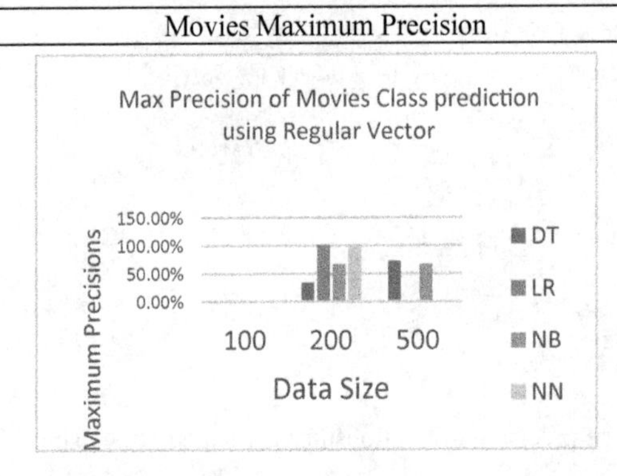

Fig. 14. Max precision of movies class prediction using regular vector

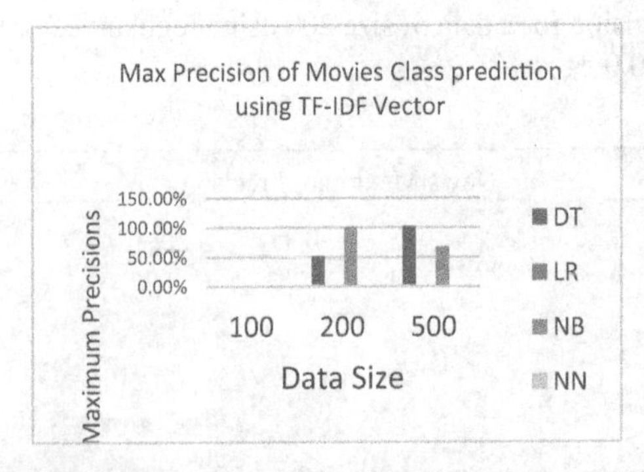

Fig. 15. Max precision of movies class prediction using TF-IDF vector

Figures 16 and 17 show the maximum precision on NFL class: for a data of size 200 all achieved a maximum value of 100% except for LR (71.43%). Moreover, for the same data size (200), using TF-IDF vector representation again all achieved 100% maximum value except for NB which had the minimum value of 60.00%.

NFL Maximum Precision

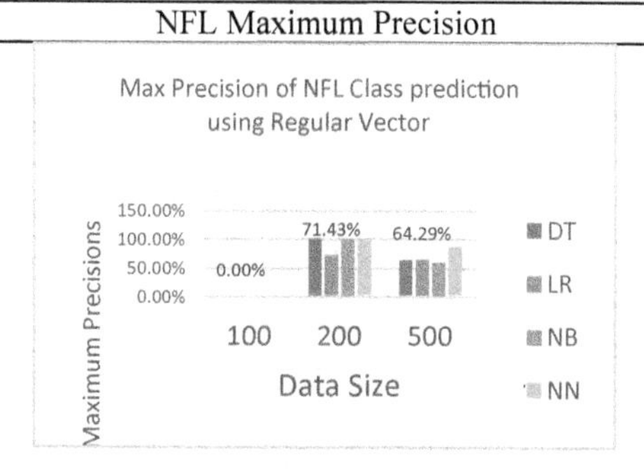

Fig. 16. Max precision of NFL class prediction using regular vector

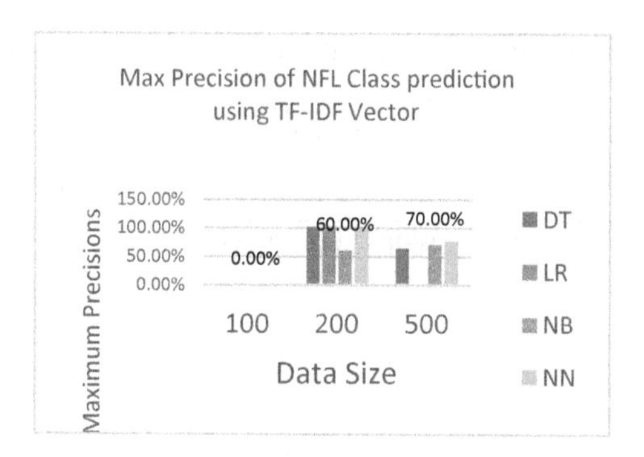

Fig. 17. Max precision of NFL class prediction using TF-IDF vector

Figures 18 and 19 show the maximum precision on Pcmasterrace class: DT had the highest maximum precision for a data of size 500 (for regular vector representation with max precision equals to 83.33% and TF-IDF vector representation with max precision of 85.71%).

Pcmasterrace Maximum Precision

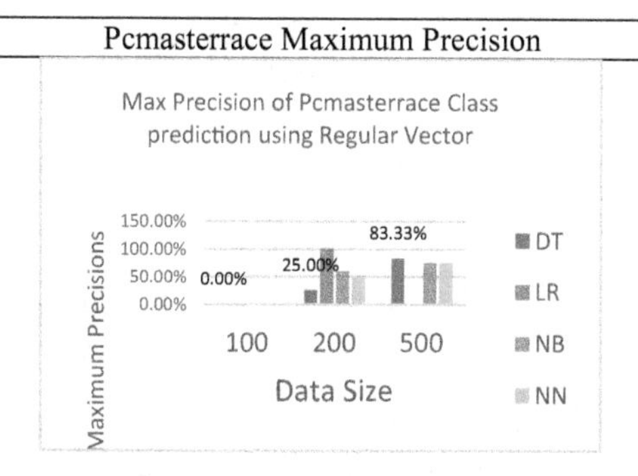

Fig. 18. Max precision of Pcmasterrace class prediction using regular vector

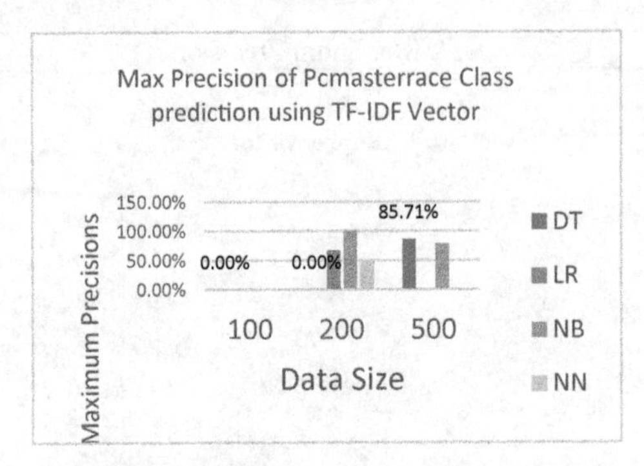

Fig. 19. Max precision of Pcmasterrace class prediction using TF-IDF vector

Figures 20 and 21 show the maximum precision on Relationships class: NN had the highest maximum value of 100% for data of size 200 using regular vector representation and a value of 66.67% for a data of size 500 using TF-IDF vector representation.

Relationships Maximum Precision

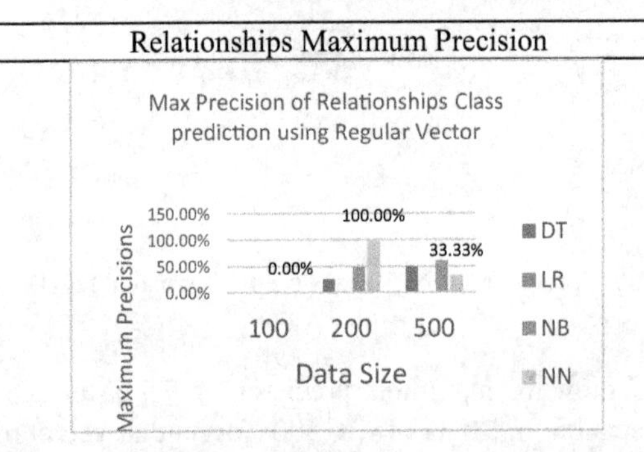

Fig. 20. Max precision of relationships class prediction using regular vector

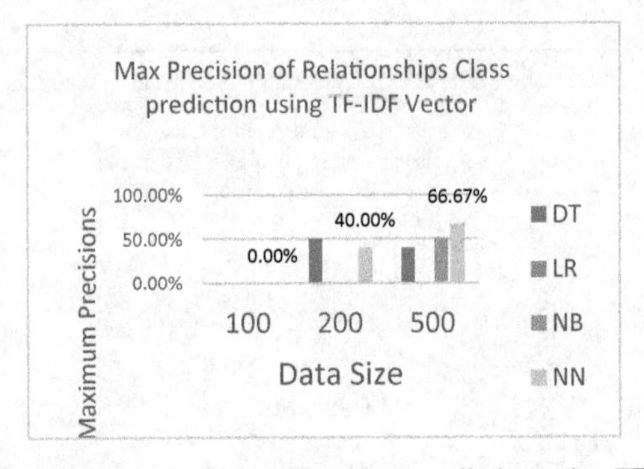

Fig. 21. Max precision of relationships class prediction using TF-IDF vector

Summary (maximum precision):

Class "news": All algorithms are quite equal with NN as the lowest
Class "movies": DT had the highest
Class "NFL": (for data of size 200) all achieved 100% except for LR using regular. using TF-IDF all except NB achieved 100%.
Class "pcmasterrace": (for data of size 500) DT had the highest
Class "relationships": (for data of size 200) NN had the highest

The following figures (Fig. 22 and 23) show the maximum average precision for all classification algorithms on the NFL dataset for every data size 100,200, and 500 (using regular and TF-IDF vector representation).

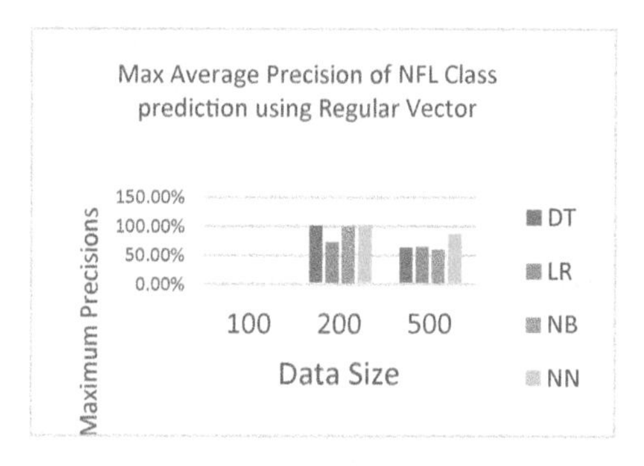

Fig. 22. Max average precision of NFL class prediction using regular vector

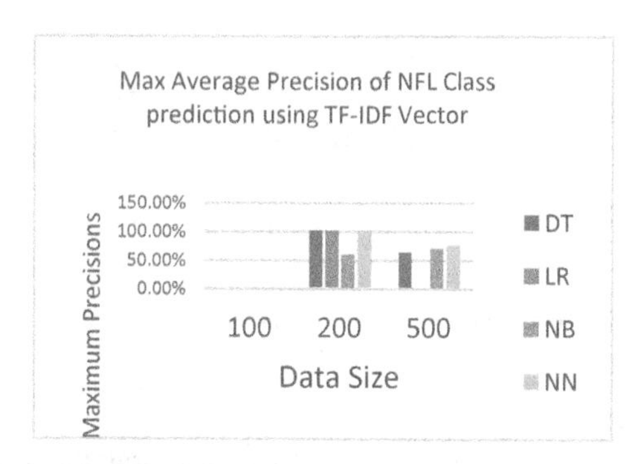

Fig. 23. Max average precision of NFL class prediction using TF-IDF vector

In summary (maximum average precision):
Class "NFL": (for data of size 200) using regular, all achieved 100% except for LR. For the same data size, using TF-IDF, all except NB achieved 100%.

The following figures (Fig. 24, 25, 26, 27 and 28) show the recall, f-measure, accuracy, sensitivity and specificity for the runs with the highest precision displayed by classification algorithm, class to predict, vector representation (regular or TF-IDF), and data size (100,200, and 500). This will help in getting a better understanding of all remaining metrics for when the algorithm achieves the highest precision. And when that occur (the highest precision) it may not be necessary for the other metrics including recall, f-measure, and the others to have the highest. The way these results were created is first the row in the result excel sheets with highest precision was selected from every criterion (regular vs TF-IDF and data size) then all values were sorted and the highest from every metric was included.

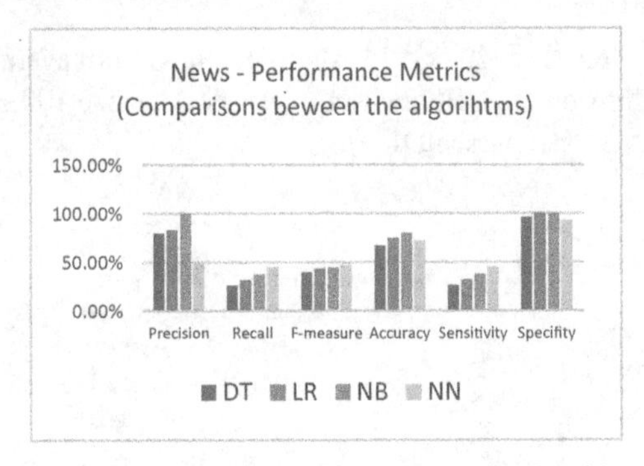

Fig. 24. News - performance metrics (comparisons between the algorithms) for the runs with the highest precision

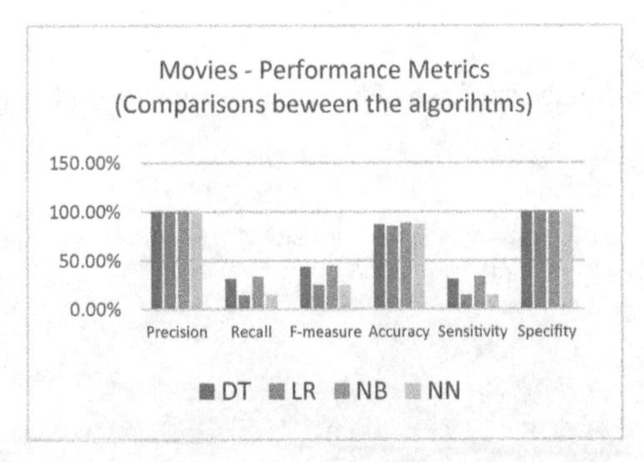

Fig. 25. Movies - performance metrics (comparisons between the algorithms) for the runs with the highest precision

The following figures (Fig. 29, 30, 31, 32 and 33) show the highest ever obtained values of precision, recall, f-measure, accuracy, sensitivity and specificity, unrelated and measured on different runs (for example, precision measured on the first run, recall measure on the second run) for every class and for regular or TF-IDF vector representation, grouped by classification algorithm, and data size (100, 200, and 500). The figures show the highest value for every metric (separately not from same run) by algorithm and class (category). So, precision of 100% for DT was not necessary from the same run that produced recall 60.00%. This is good to evaluate the algorithms using all different metrics independent of each other.

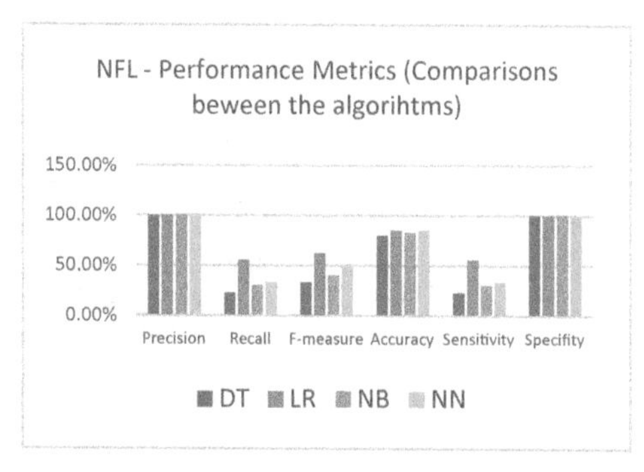

Fig. 26. NFL - performance metrics (comparisons between the algorithms) for the runs with the highest precision

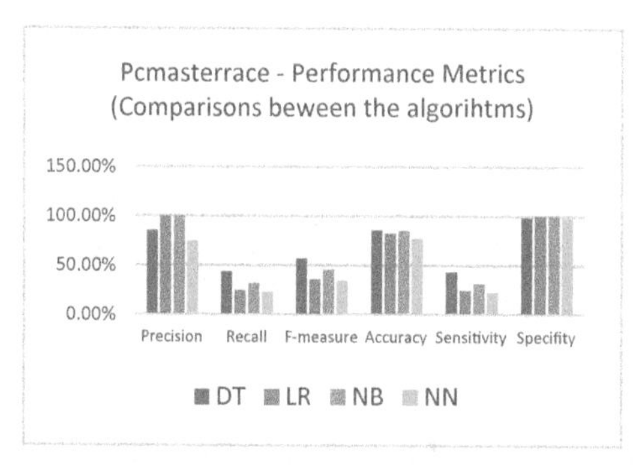

Fig. 27. Pcmasterrace - performance metrics (comparisons between the algorithms) for the runs with the highest precision

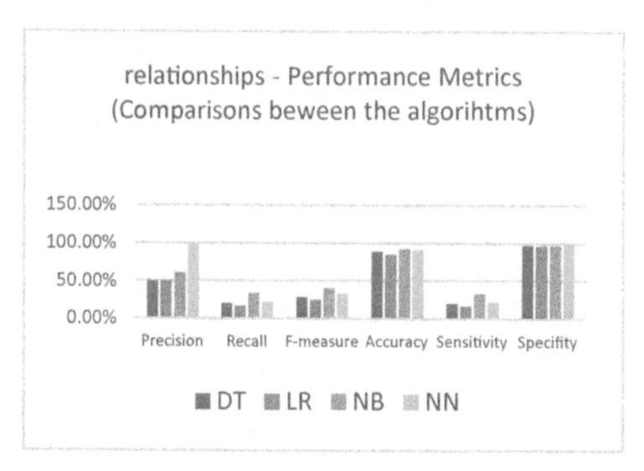

Fig. 28. Relationships - performance metrics (comparisons between the algorithms) for the runs with the highest precision

8 Evaluation of Results

Data was divided into training set comprised of 80% of size of the data to read and testing comprised of 20% of size of data to read.

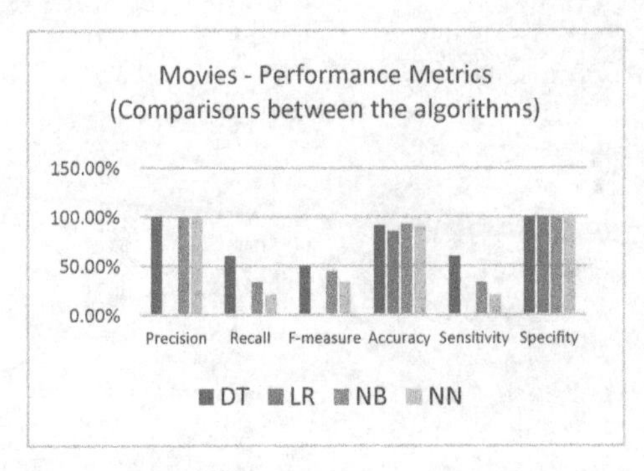

Fig. 29. Movies - The highest ever performance metrics (comparisons between the algorithms)

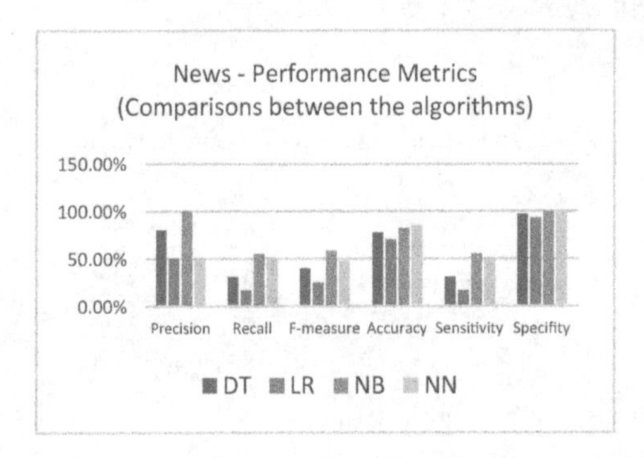

Fig. 30. News - The highest ever performance metrics (comparisons between the algorithms)

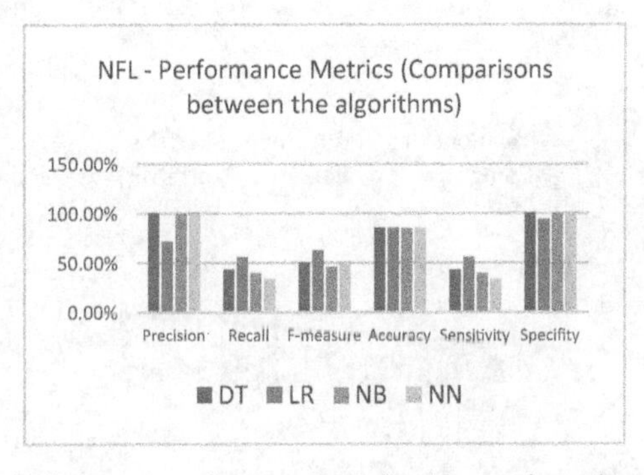

Fig. 31. NFL - The highest ever performance metrics (comparisons between the algorithms)

We calculated the precision, recall, F-measure (which combines precision and recall), accuracy, specificity, and sensitivity. To do that, we calculated the True Positive (Tp), False Positive (Fp), True Negative (Tn), and False Negative (Fn). All the four constitute the confusion matrix. Also, we calculated the number of correct predictions, and the number of incorrect predictions.

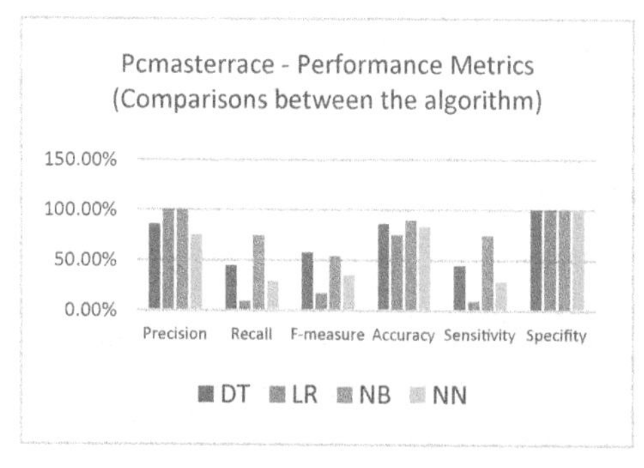

Fig. 32. Pcmasterrace - The highest ever performance metrics (comparisons between the algorithm)

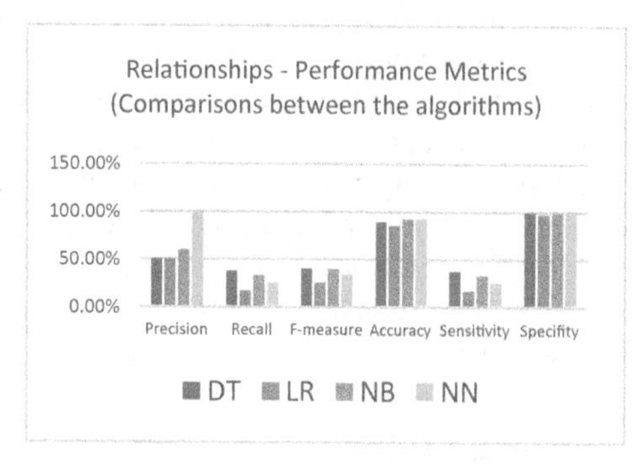

Fig. 33. Relationships - The highest ever performance metrics (comparisons between the algorithms)

The formula for precision, recall were used from the following website [30]. The formula for sensitivity and specificity were used from the following website [31]. The formula for accuracy was used from the following website [32]. Also, the reader may refer to the following [22].

When data to read was of size 50 (training 40 and testing 10) the precision was 0 for many running of the program (the accuracy and specificity were not 0, sometimes they were good enough such as 60% and 80%, respectively).

When data to read was increased to 100 (80 training and 20 testing), the precision was sometimes 100% and other times it was 0%. The accuracy and the specificity were not zero also in this time.

Hence, we ran the experiment multiple times with larger data size to read. No reason for justifying the low precision other than the data size to read was small (which is confirmed by the results). We ran the program with 50, 100, 200, and 500 as the size of the data to read. And repeated the classification about 5 to 7 times for every classification algorithm, class, data size and vector representation. For 50 as the size of data to read, no run achieved precision greater than 0% precision. But for accuracy and sensitivity it was ranging from 50 to 100%. When the program was run 100 times, only once the results achieved an accuracy of 50% or 30%.

For average precision results, the highest were (in most runs) for DL. Also, LR and NN were the highest for some runs (see Results and Discussion section).

8.1 Time Based Evaluation

For a data of small size (less than 50), the classifiers were fast. It took about 3-5 min to find the results once for given parameters. So, we were able to repeat the classification process multiple times in a very small amount of time (the program is efficient).

For a data of size 200, all classification algorithms were fast (less than 5 s) except for LR which was the slowest and it took about 2 to 3 min. For a data of size 500, the program took more time, and LR was the slowest with a time of 32 min. Then NN was the second slowest with a time of 2 min. SVM, NB, and NN were the fastest (time in less than 5 s). SVM was the fastest among all (less than 1 s) but it did not produce any true positive results. On the contrary, NB and NN produced true positives results. For a data size of 500, the program takes more time.

For a data of size 1000 (which is still even less than the entire dataset 40236), LR took 1 day and it was not even enough for the LR classifier to finish executing and producing any results. All remaining algorithms also did not complete the calculation of results in 1 day for a data of size 1000. And it became difficult to monitor and wait for the application to finish for more than that amount of time. Also, we attempted to read the entire dataset, and the classifier took about more than 1 day to train the classifier using the 1000 data size and it was not even enough to produce results.

9 Conclusion and Future Work

By running the program multiple times, the results vary since they were based on an element of randomness (randomly selection of training and testing vectors). Also, with the increase of the data size to read (from 100 to 200 to 500 vectors) the precision increased from 0 to even 100%. All classification algorithm, which were considered, produced results with precision greater than 0 and some reached 100% precision. In most cases, DT had the highest average precision. For a data of size greater than 1000, the classifier took more than one day and even did not produce results, so we took the decision to halt it. We attempted to run it for that same data size many times but with different classification algorithm, so that contributed to the conclusion that the data size affects the precision: the larger the higher.

This current work has the limitation of being applied only on one dataset and not including several algorithms even it contains intensive detailed comparisons though. Therefore, future work lies in applying the classification on more datasets, including: Twitter, Movies Review, and other datasets. In addition, to including more classification algorithms and running the algorithms on larger data size (if not the entire dataset).

Also, another aspect of future work lies in using semantic together with Natural Language Processing (stemmer, tagger, and parser) in classification. One can utilize a tagger to only consider words which belong to certain tags such as nominal subject (nsubj), direct object (dobj), etc., instead of considering all words. Also, the words which have the same stem (player, playing, plays, and played all have the same stem: play) should be treated equally by the classifier. This can be achieved with the assistance of a tagger and stemmer. Finally, words that are related by a semantic relationship (such as the hypernym: is-a relationship) should be treated equally as well. This can be obtained using ontologies such as wordnet, see [7]. All that, is expected, to improve the results of the classification algorithms.

One remaining question is how well the classification algorithms work with a very large dataset size.

References

1. Aly, M.: Survey on multiclass classification methods. Neural Netw. **19**, 1–9 (2005)
2. Dreiseitl, S., Lucila, O.M.: Logistic regression and artificial neural network classification models: a methodology review. J. Biomed. Inf. **35**(5–6), 352–359 (2020)
3. Rawashdeh, M., Ralescu, A.: Center-wise intra-inter silhouettes. In: International Conference on Scalable Uncertainty Management, Berlin, Heidelberg, pp. 406–419 (2012)
4. Uijlings, J.R., Van De Sande, K.E., Gevers, T., Smeulders, A.W.: Selective search for object recognition. Int. J. Comput. Vis. **104**(2), 154–171 (2013)
5. Fujiyoshi, H., Hirakawa, T., Yamashita, T.: Deep learning-based image recognition for autonomous driving. IATSS Res. **43**(4), 244–252 (2019)
6. Rawashdeh, A.: An experiment with link prediction in social network: two new link prediction methods. In: Arai, K., Bhatia, K. (eds.) Future Technologies Conference, 10 October 2019, pp. 563–581 (2019). https://link.springer.com/book/10.1007/978-3-030-32523-7
7. Rawashdeh, A., Rawashdeh, M., Díaz, I., Ralescu, A.: Measures of semantic similarity of nodes in a social network. In: Laurent, O.S., Bernadette, B.M., Ronald, R.Y. (eds.) International Conference on Information Processing and Management of Uncertainty in Knowledge-Based Systems, pp. 76–85 (2014). http://www.lirmm.fr/~lafourcade/pub/IPMU2014/papers/0443/04430076.pdf
8. Rawashdeh, A., Ralescu, A.: Similarity measure for social networks-a brief survey. In: MAICS, pp. 153–159, April 2015
9. Jain, A.P., Dandannavar, P.: Application of machine learning techniques to sentiment analysis. In: 2nd International Conference on Applied and Theoretical Computing and Communication Technology (iCATccT), pp. 628–632, July 2016
10. HaTri: How to make invisible cloak - magic trick you can do. https://www.youtube.com/watch?v=K8_1YxJch3Y. Accessed 9 Feb 2020
11. Libbrecht, M.W., Noble, W.S.: Machine learning applications in genetics and genomics. Nat. Rev. Genet. **16**(6), 321–332 (2015)
12. Maloof, M.A. (ed.): Machine Learning and Data Mining for Computer Security. Springer, Heidelberg (2006)
13. Kotsiantis, S.B., Zaharakis, I., Pintelas, P.: Supervised machine learning: a review of classification techniques. Emerg. Artif. Intell. Appl. Comput. Eng. **160**(1), 3–24 (2007)
14. Liu, Y., Bi, J.W., Fan, Z.P.: Multi-class sentiment classiÞcation: the experimental comparisons of feature selection and machine learning algorithms. Expert Syst. Appl. **20**(80), 323–339 (2017)
15. Podolsky, M.D., Barchuk, A.A., Kuznetcov, V.I., Gusarova, N.F., Gaidukov, V.S., Tarakanov, S.A.: Evaluation of machine learning algorithm utilization for lung cancer classification based on gene expression levels. Asian Pac. J. Cancer Prev. **17**(2), 835–838 (2016)
16. Khan, A., Baharudin, B., Lee, L.H., Khan, K.: Review of machine learning algorithms for text-documents classification. J. Adv. Inf. Technol. **1**(1), 4–20 (2010)

17. Nandhini, B.S., Sheeba, J.I.: Cyberbullying detection and classification using information retrieval algorithm. In: Proceedings of the 2015 International Conference on Advanced Research in Computer Science Engineering & Technology (ICARCSET 2015), pp. 1–5, March 2015
18. Curiskis, S., Kennedy, P., Osborn, T., Drake, B.: Data for: an evaluation of document clustering and topic modelling in two online social networks: Twitter and Reddit. https://data.mendeley.com/datasets/85njyhj45m/1. Accessed 2 Sept 2020
19. Curiskis, S.A., Drake, B., Osborn, T.R., Kennedy, P.J.: An evaluation of document clustering and topic modelling in two online social networks: Twitter and Reddit. Inf. Process. Manage. **57**(2), (2020)
20. CATHERINE. 23 auxiliary verbs. https://english109mercy.wordpress.com/2012/10/22/23-auxiliary-verbs/. Accessed 2 Sept 2020
21. Auxiliary verbs. https://www.myenglishpages.com/site_php_files/grammar-lesson-auxiliary-verbs.php. Accessed 2 Sept 2020
22. Hwang, Y.H.: C# Machine Learning Projects: Nine Real-World Projects to Build Robust and High-Performing Machine Learning Models with C#. Packt Publishing Ltd. (2018)
23. LogisticRegression Class. http://accord-framework.net/docs/html/T_Accord_Statistics_Models_Regression_LogisticRegression.htm. Accessed 9 Feb 2020
24. KNearestNeighbors <TInput> Class. http://accord-framework.net/docs/html/T_Accord_MachineLearning_KNearestNeighbors_1.htm. Accessed 9 Feb 2020
25. NaiveBayes Class. http://accord-framework.net/docs/html/T_Accord_MachineLearning_Bayes_NaiveBayes.htm. Accessed 2 Sept 2020
26. SupportVectorMachine Class. http://accord-framework.net/docs/html/T_Accord_MachineLearning_VectorMachines_SupportVectorMachine.htm. Accessed 2 Sept 2020
27. McCaffrey J. Support vector machines using accord.NET. https://visualstudiomagazine.com/articles/2019/02/01/support-vector-machines.aspx. Accessed 9 Feb 2020
28. Suite, R.A.: Getting Started in natural language processing: bag-of-words & TF-IDF, 21 June 2019. https://medium.com/datadriveninvestor/getting-started-in-natural-language-processing-bag-of-words-tf-idf-b62e9354eb7e. Accessed 2 Sept 2020
29. BagOfWords Class. http://accord-framework.net/docs/html/T_Accord_MachineLearning_BagOfWords.htm. Accessed 2 Sept 2020
30. Brownlee, J.: How to Calculate Precision, Recall, and F-Measure for Imbalanced Classification, January 2020. https://machinelearningmastery.com/precision-recall-and-f-measure-for-imbalanced-classification/. Accessed 2 Sept 2020
31. Part 1: Simple Definition and Calculation of Accuracy, Sensitivity and Specificity. https://www.ncbi.nlm.nih.gov/pmc/articles/PMC4614595/. Accessed 2 Sept 2020
32. Classification: Accuracy. https://developers.google.com/machine-learning/crash-course/classification/accuracy. Accessed 2 Sept 2020

Identifying Subject-Position and Power Relation Using Natural Language Processing with Fairclough Critical Discourse Analysis

Alcides Bernardo Tello[1,2,4](✉), Cayto Didi Miraval Tarazona[1], Anderson Torres Bernardo[3], Guillermo Arévalo Ríos[1], and Shi Jie[4]

[1] Universidad Nacional Hermilio Valdizán, Huánuco 100001, Perú
abernardo@udh.edu.pe
[2] Universidad de Huánuco, Huánuco 100001, Perú
[3] Institución Educativa Juana Moreno, Huánuco 100001, Perú
[4] University of Birmingham, Birmingham B15 2TT, UK

Abstract. Research in fields of social science and education is ready for the text exploration revolution with Artificial Intelligence educational solutions. This article uses the three-dimensional Fairclough's critical discourse analysis to explore essay number five, "The Religious Factor", by José Carlos Mariátegui in his work Seven Interpretative Essays on Peruvian Reality in relation with the current Peruvian education system. Once we gathered and filtered the essay, we applied Natural Language Processing on its complete text. We wielded the theoretical framework of critical discourse analysis to analyse the text that expounds the culturally constructed representation of reality, the power circulation in the population and the reproduction of power via education. The results of the analysis displayed a subject position related to the Peruvian government allying with the Catholic Church to enjoy power relationship. Further research might reveal what level of alliance persists at present.

Keywords: Critical discourse analysis · Natural Language Processing · State · Religion · Church

1 Introduction

Believing what we hear, see, read at face value as certain might direct us to the acceptance and reproduction of power. We might not be aware that such power subjects us to dominance. Social scientists search for ideologies or beliefs within a discourse or any utterance. Critical discourse analysis (CDA) is an essential tool in such endeavours. Computer scientists have developed frameworks, such as Natural Language Processing (NLP) and other Artificial Intelligence (AI) tools [1], which are now suitable for exploring discourses in any multimodal communicative event.

One widespread use of NLP and computational linguistic is opinion mining or sentiment analysis, which attempts to extract and quantify subjective information [2]. For example, classifying whether a given text is positive or negative, objective or subjective,

K. Arai (Ed.): FICC 2021, AISC 1363, pp. 1027–1035, 2021.
https://doi.org/10.1007/978-3-030-73100-7_70

in a binary class. Let us consider the likelihood of the word "w", given class "c" and vocabulary "V", with the training documents that contains "w" is classified as positive, the probability for this feature would be

$$P(w_i/c) = \frac{count(w_i, c) + 1}{\sum_{w \in V} count(w, c) + |V|} \quad (1)$$

However, advanced opinion mining classification dives further into emotional states, i.e., angry, gloomy, and cheerful. This task is a classification of a text into classes. Other approaches focus on estimating subjective sentences [2], implemented in programming languages, such as Python.

On the other hand, in Social Science, James Paul Gee makes distinctions in the concept of Discourse ("big D") and discourse ("small d"). They both refer to "everyday ways of talking", with the first relating to meaning-making and the second to the forms of talking about social reality, including medicine, food, customs, perspectives, discourse and other mixtures of language with social practices [3, 4]. Further, Foucault brought into focus the social power of discourse, i.e., its impact on social activities [5, 6].

Fairclough's approach [7] relates to the philosophy of critical realism in fathoming societal fabric. His approach has a connection to and complements Foucault's approach. Fairclough's approach has three dimensions. These dimensions involve analysis at three levels, viz., at the level of the words, at the level of the text and at the level of norms and social practices. Critical realism brings ontology and an epistemic relativism into the research of social phenomenon [9, 10].

Critical realism provides social researchers with an option in the middle of positivism and interpretivism, which are the two extreme paradigms [8]. The positivism paradigm uses law-like models whilst interpretivism uses hermeneutic forms "at the cost of causation". However, critical realists cannot rely on positivist rationale to understand the internalities of any social structure [10, 11]. Therefore, NLP, which can analyse content at the level of the text, can provide critical realists with a binocular to elicit in-depth interpretations of texts used in hermeneutics approaches, in conjunction with the two other dimensions of CDA. Fig. 1 depicts the communicative event in CDA that consists of three dimensions, viz., text, discursive practice and social practice [9, 10].

Consequently, we have used the power of both CDA and NLP to extract features of power relation and subjectivity in a country policy context under critical realism approach. We concluded with subject-positions in the discourse, as it befits CDA, from which we can make the most sense of the discourse. In concordance with Foucauldian concept of normalised power [3–5], the relation between the state and Catholicism has become a normalised power in Peruvian society, a power that makes its citizens want to live in whatever box the constructed social structure places them, which is invisible to ordinary perception. The structural relation tells people even what they should desire to smell.

In the next section, we justify our approach and our selection of data. In the third section, we explain our analysis, the fourth involves our finding and discussions, the fifth section explains our conclusion and the last section deals with further directions.

Fig. 1. The three dimensions of the communicative event

2 Research Approach and Data Collection

From the computer science perspective, as the first part of this study, concerning data cleansing for the detection and removal of unnecessary records or unwanted characters from a data set, such as whitespace, stop words, and punctuation that have no values in our analysis and study.

After cleansing, we checked the consistency of the data set. From there, the data set has been arranged into a format that the computer can explore via a programming language; we have used Python. Finally, we have processed the dataset and visualised the result (see Fig. 2). The next step corresponds to sense-making in connection with the CDA and the hermeneutics of the original text.

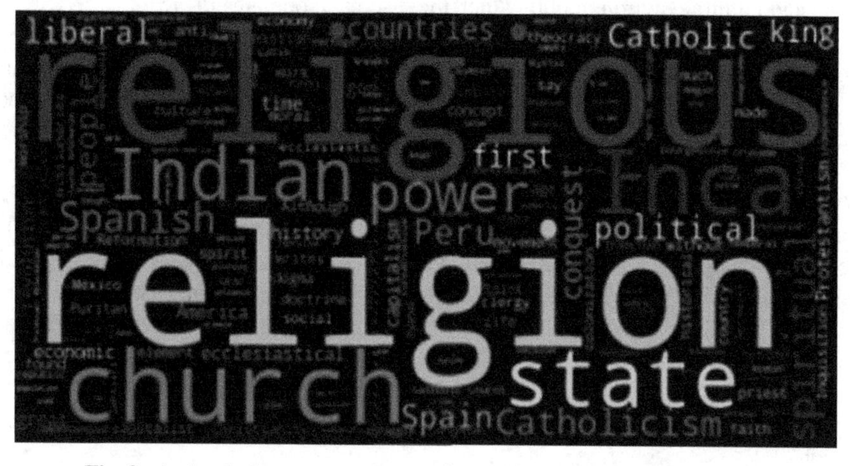

Fig. 2. A cloud of terms most frequently used in the fifth essay by JCMA.

The second part, from the social science perspective, this research takes a subjective posture due to the nature of CDA, which is both qualitative and interpretative. CDA assumes that not only does discourse construct the objects and the events that exist, but also constrains the possibility of understanding the objects [3, 4]. They also state that CDA research penetrates beyond perceived reality to investigate the actual reality.

Most scholars, such as those in [12], regard José Carlos Mariátegui Arellano (JCMA) as the greatest thinker and leading intellectual of his time, particularly for his work The Seven Interpretive Essays on Peruvian Reality [13]. Hegemony, ideology, state and church relationship discourse are most clearly exemplified and articulated in the material produced by JCMA. He has clearly articulated the ancient Peruvian beliefs and colonial heritage as well as the state goals and strategies in its fifth essay. Thus, essay number five represents the dataset of material in support of our state–religion study.

Therefore, essay number five was selected due to small-scale research. However small-scale, this study does offer valid insight by creating a framework for future social science research and more in-depth studies. This study shows the utility and power of NLP in CDA in social science studies in conjunction.

We have also collected the sentences related to religious education from the articles in the current constitution as part of the communicative event for our analysis.

3 Data Analysis

After processing the text (see Fig. 3 and Fig. 2), we interpreted how the discourse constructs meanings and identities and ultimately, communicates a view to the society. Our approach concerns exploring and explaining the relationship between the state, the church and the perceptions of society about the power relation. Power has been reproduced via education continuedly since colonialism. It connects both linguistic and social analyses [15]. Initially, we conducted a content analysis by categorising the basic information. Afterwards, we advanced to critical textual discourse analysis, all of which the method of CDA by adopting the examination of language patterns in the text discourse, social and cultural environment in which the discourse operates [5, 15].

By developing on the realm of critical linguistics aiming to uncover how discourse is used as an ideological tool, CDA exposes the relationship between language, power and ideology [13]. This method recognises that language is a social construct and reproduces the idea that discourse communicates 'constructed' views to society [4, 5].

From Fig. 2 and Fig. 3, with the help of NLP, we determined the words that were most used in the creation of the discourse in the essay to develop his communicative event. The analysis of textual discourse explores 'constructed' views by focusing on the way language is used rather than on the language itself and by taking a pragmatic stance in exploring what people intend their language rather than focusing on the literal meanings [15]. More broadly, the use of CDA within this research aimed to investigate the discourse in the text and inform readers about power, state and church as well as how these entities may be significant in not only shaping views of society, but in influencing society itself to accept and reinforce these views. These findings are explored in the next section.

Discourse can contain expressions which are objective or subjective. Objective expressions are facts, while subjective expressions are opinions describing feelings

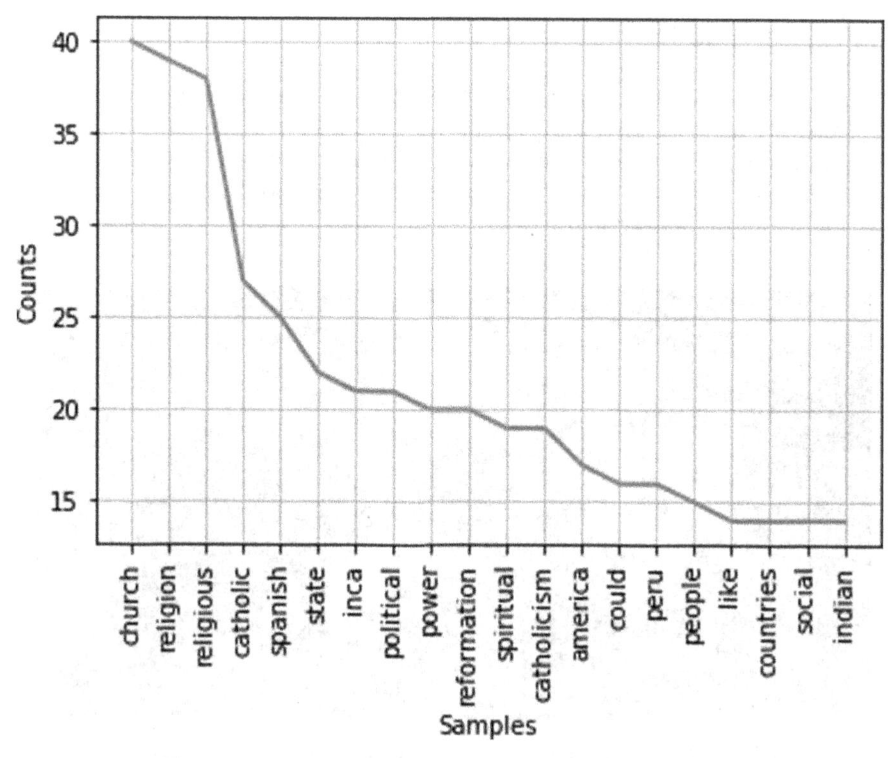

Fig. 3. The top-twenty ranking words in the essay by JCMA.

towards a particular area or idea. Table 1 shows the subjectivity that was found in the text related to religious education in the discourse of the current policy; the values cast doubt on the meaning of the policy's text.

Table 1. Subjectivity and Polarity for the Constitutional Discourse.

Sentence	Subjectivity
Everyone has the right to equality before the law	0.5357
No one should be discriminated against on any ground such as origin, race, sex, language, religion, opinion, economic condition or of any other nature	0.3583
There is no persecution because of ideas or beliefs	0.0000
Everyone has the right to keep their political, philosophical, religious or any other convictions confidential, as well as to maintain professional secrecy	0.2267
Religious education is taught with respect for the freedom of conscience	0.2500
The State coordinates educational policy	0.2500

4 Findings and Discussion

This study unearthed how the constructed entities, viz., the state, church and its people, are presented in the discourse and how NLP can help us analyse and interpret the relations among them in order to understand the interplay of those institutions and its representational strategies. Thus, language becomes a powerful tool that can have the power to shape people's worldviews.

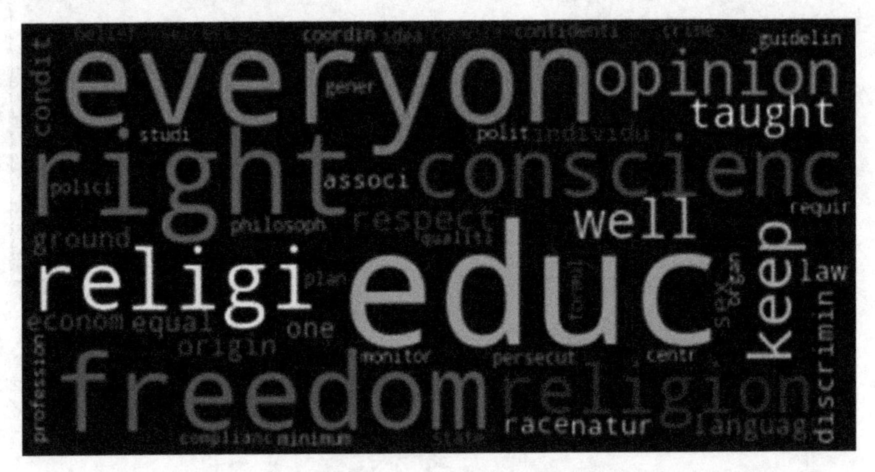

Fig. 4. The top-twenty ranking words in the state religious education policy.

CDA has approaches to answer questions related to the relationships between language, society and power [4]. Our findings include that NLP can reshape the boundaries of traditional CDA in the framework of hermeneutics. It can begin by examining how critical approaches to policy analysis, which examine power relationships. Without exploration, realities are taken-for-granted, and the assumptions in the structures of society have been used without questioning, including their identity, behaviour and belief.

The 300-year colonial Peruvian government has seen alliances between Catholicism and the Viceroyalty [13], which forced natives to convert into Christianity, including Santa Inquisition and wholesale slaughter, artefacts of which are currently displayed in the Museum. These actions have put the church at odds with their core values. However, it is paradoxical that catholic religious education is one of the subjects that are compulsory in primary and secondary schools for students of five and six years, respectively [16]. This is unlike other countries, such as England, which have implemented the study of different worldviews in educational institutions, which inclusive of several religious groups, rather than giving privilege to one religion [14].

CDA offers insights on how a particular discourse, taking Spanish rulers and Catholics as examples, becomes common sense and dominant. At the same time, it silences other interpretations of reality.

Nevertheless, according to the last Census that was conducted in 2017 in Peru by the National Institute of Statistics and Informatics (INEI), there is a dramatic decline in identification with Catholic denominations. In the census, there is also a substantial increase in non-religious affiliations as well as in those belonging to non-Christian faiths. Even though the national curriculum favours the teaching of purely Catholicism [16], the census is contradictory, showing that secularism is on the rise. The census report results regarding secularism is 94% higher than the figure reported in the 2007 Census.

According to Helmut Kessel, president of the Peruvian Secular Humanist Society, the leading causes for the decrease is the exposure of cases of sexual abuse and corruption in Catholic churches and its inconsistent profile history, the awakening of the citizens in the light of the abundance of information [17].

Notwithstanding the data from Table 1 and Fig. 4, the most apparent words across the policies are ‘education’, ‘religion’, ‘freedom’, ‘everyone’ and ‘rights’. This interplay of these words reveals that they possess the embedded nature of subjectivity and subjective experience with regards to the policy and the system of power. There is still a marriage between church and the state that instilled in children as normal and it is reproduced via education with subtle discourse of freedom, rights and education—providing children with perspectives, feelings, beliefs and desires for the biased agency. However, in CDA, it is considered true only from the perspective of the subjects making the policy.

5 Conclusions

In the light of CDA and NLP, this study has paid attention to the interplay between the church, the state and the public by exploring the fifth essay from Seven Interpretative Essays by JCMA. By analysing the text within the essay, it has become apparent that the relationship between the state, church and power existed in JCMA’s time. However, surprisingly, the alliance between state and church still operates at present, even instilling children in public schools. Historically maintained power circulates throughout Peruvian society. Although several elements centre upon general ideology themes, most of them use language that have connotations of power incorporated into it.

The usage of CDA and NLP in conjunction helps explain how those discourses are accepted as true despite face values are historically variable, showing unfair social relationships and power.

We drew the conclusion, with subject-positions in this discourse, in concordance with Foucault’s conception of normalised power, that the relation between the state and the Catholic church has become a normalised power in Peruvian society, a power that makes its citizens wish to live in whatever constructed social structure is outlaid to categorise them, something that is invisible in ordinary perception. The structural relation tells people even what they should desire to smell. We verified the results with respect to the current system, in which the power relations still operate, even in the educational system. The Catholic church has infused the educational system with its views, despite the fact that Peru is a multicultural society with worldviews from ancient times, such as Incan or Quechuan views, which are disregarded from formal implantation in school curricula as part of worldview’s studies. The mentality formed through these relations that are present in the discourse presents itself in regulations and policies, which are taken for granted without questioning the presence of colonialism legacy in the social structure, which is embedded in the experiences of policy and the system of power.

6 Future Work

CDA and NLP might help explore variables of uneven social relationships, technology and power that exist in social practices.

There is a need for mindful consideration of social sciences and ethical dilemmas in the field of AI. Every algorithm, as part of social practice, including artificial intelligence, is constructed by human beings and all human beings have their own biases and particular mindset. Their mindset is a product of their institutions [6]. Therefore, we need to ensure that algorithms benefit all of humanity while avoiding biased implementation. Codes should be created with diversity in mind and not with only one single mindset, for a diverse workforce. CDA might be a great companion for attaining those goals of a safe community code which is committed to addressing social and environmental injustice and challenging unequal power relations.

In machine learning, systems can learn from information, select patterns and make decisions without human intervention. Therefore, CDA can review the amount of power relations that are hidden in information before making decisions, either by humans or machines. It might be a challenge to ensure appropriate applications in machine learning.

References

1. Chowdhary, K.R.: Natural language processing. In: Fundamentals of Artificial Intelligence, pp. 603–649. Springer, New Delhi (2020)
2. Jurafsky, D., Martin, J.H.: Speech and language processing. Chapter A: Hidden Markov Models (Draft of September 11) (2018). Accessed 19 Aug 2020
3. Gee, J.P.: Discourse, small d, big D. In: The International Encyclopedia of Language and Social Interaction, pp. 1–5 (2015)
4. Derin, T., Nunung, S.P., Mutia, S.N., Budianto, H.: Discourse analysis (DA) in the context of English as a foreign language (EFL): a chronological review. ELSYA: J. English Lang. Stud. **2**(1), 1–8 (2020)
5. Mullet, D.R.: A general critical discourse analysis framework for educational research. J. Adv. Acad. **29**(2), 116–142 (2018)
6. Liu, K., Fang, G.: A review on critical discourse analysis. Theory Pract. Lang. Stud. **6**(5), 1076–1084 (2016)
7. Arribas-Ayllon, M., Valerie, W.: Foucauldian discourse analysis. In: The Sage Handbook of Qualitative Research in Psychology, pp. 91–108 (2008)
8. Archer, et al.: What Is Critical Realism? American Sociology Association. https://www.asatheory.org/current-newsletter-online/what-is-critical-realism. Accessed 11 Aug 2020
9. Kuhn, A., Guy, W.: A Dictionary of Film Studies. Oxford University Press, Oxford (2012)
10. Rogers, R: Critical discourse analysis and educational discourses. In: The Routledge Handbook of Critical Discourse Studies, p. 465 (2017)
11. Fairclough, N.: Discourse and Social Change, vol. 10. Polity Press, Cambridge (1992)
12. López Cruz, Y., Noblet Valverde,V.C.: Siete ensayos de la realidad peruana, ideología, emoción y voz de los pueblos americanos. pasado, presente y futuro. Caribeña de Ciencias Sociales (2019)
13. Arellano, J.C.M.: Siete ensayos de interpretación de la realidad peruana. Ediciones Era (1979)
14. Cush, D.: Should religious studies be part of the compulsory state school curriculum? Br. J. Religious Educ. **29**(3), 217–227 (2007)

15. Woods, L.A., Kroger, R.O.: Doing Discourse Analysis: Methods for Studying Action in Talk and Text. Sage, Thousand Oaks (2000)
16. Ministerio de Educación: Diseno Curricullar-Ministerio de Educacion (2016). https://www.minedu.gob.pe/curriculo/. Accessed 3 Aug 2020
17. Sociedad Secular y Humanista del Perú: Qué es el humanismo secular? (2018). https://ssh.org.pe/. Accessed 12 Aug 2020

Model-Driven Chats: Enabling Chatbot Development for Non-technical Domain Experts Through Chat Flow Visualization and Auto-generation

Amal Khalil[1(✉)], Fernando Hernandez Leiva[1], Akinkunmi Shonibare[1], Evan Marcel Arsenault[1], Laura Turner[1], Shadi khalifa[1], Linna Tam-Seto[2], Brooke Linden[2], Valerie Wood[2], Heather Stuart[2], Jennifer Nolan[3], and Colleen McDowell[3]

[1] Centre for Advanced Computing, Queen's University, Kingston, Canada
{khalil.a,f.hernandezleiva,a.shonibare,ema9,laura.turner, khalifa.s}@queensu.ca

[2] Centre for Health Services and Policy Research, Queen's University, Kingston, Canada
{linna.tam-seto,13bjd2,wood,hstuart}@queensu.ca

[3] IBM Canada Ltd., Markham, Canada
{jen.nolan,cmcdowel}@ca.ibm.com

Abstract. Chatbots can significantly reduce the workload on client-facing service agents by conducting a first screening and answering frequently asked questions. Reaping these benefits, various organizations now implement chatbots. As typical in an iterative project within multidisciplinary teams, information can be lost in translation. This paper presents the first proof-of-concept demonstration of creating an editable model for conversation flows development in the IBM Watson Assistant chatbot platform. Chat flows are auto-generating after changes are made to the model thus enabling our non-technical mental health domain expert partners to build a chatbot application by simply drawing the chat flow.

Keywords: IBM Watson Assistant · Conversational agents · Bidirectional model transformation · Software comprehension

1 Introduction

Chatbots, also known as conversational agents, are Artificial Intelligence (AI) and Natural Language Processing (NLP)-based computer programs that can engage in and guide conversations with humans about a particular topic, targeting a specific goal, in a way similar to conventional human-to-human conversations. Chatbots are emerging in various domains (e.g., healthcare, education, and customer support) to serve a diverse set of purposes including, but not limited

K. Arai (Ed.): FICC 2021, AISC 1363, pp. 1036–1050, 2021.
https://doi.org/10.1007/978-3-030-73100-7_71

to, counselling, enhancing client engagement, and providing smart and instant feedback [1]. Chatbot platforms are environments that facilitate the development and deployment of chatbots. Most of these platforms are cloud-hosted, such as Amazon's Lex [2], Google's Dialogflow (previously Api.ai) [3] and IBM's Watson Assistant (previously Watson Conversation) [4].

Building effective task-oriented chatbots is an iterative process that requires extensive manual work. A set of NLP-based trained models is used to process the user's phrases, identify the most fitted intents the system supports, and respond accordingly. A chatbot's dialog flow defines the steps for a multi-turn conversation to collect the input parameters necessary for the execution of detected intents.

The Here4U Military Version is an ongoing research project, funded by various Canadian organizations, aiming to create a conversational counseling mobile application to support mental health difficulties experienced by members of the military community including members, Veterans, and adult family member using IBMWatson Assistant (WA). The team working on this project consists of mental health researchers (referred to as the Research Team), data scientists, and software developers. The Here4U researchers are responsible for developing mental health-promoting dialogs guided by the well-known Cognitive Behavioral Therapy (CBT) theory. These dialogs are prepared in different formats (e.g., Word documents, Flowcharts, and Block diagrams), which are then mapped into WA dialog flows by the development team. The mapping process is not trivial and may involve design decisions that are not documented. Since these flows are subject to modification, having a one-to-one mapping system between the Research Team dialog flows and their corresponding WA ones is significant. Such a system can be used as the linking pin between the two representations of dialog flows. It can also be used to develop a broad spectrum of applications for visualizing, debugging, editing, and auto-generating WA dialog flows.

Model-Driven Engineering (MDE) is a well-known paradigm that has been adopted in building large-scale embedded systems such as avionics and automotive electronics [5]. MDE provides the core concepts of modeling, metamodeling, and model transformation, which can be applied in different disciplines for automating the conversion of one artifact representation to another. MDE also employs the use of domain-specific languages (DSLs), which are tailored to meet an application's unique needs. In the context of the Here4U project, we propose a Model-Driven Chat (MDC) approach based on MDE concepts for mapping the researchers' dialog flows into the ones that operate the chatbots, and vice versa.

The rest of this paper is structured as follows: Sect. 2 motivates the work with a running example and a discussion of the opportunities this work will bring to enhance the usability of the WA chatbot development platform, Sect. 3 overviews the details of our proposed visualization and auto-generation framework, Sect. 4 reflects on our work, Sect. 5 identifies related concepts, and Sect. 6 concludes the paper.

2 Illustrative Example

This section presents a walk-through of the process of developing and maintaining the Here4U chatbot. This process starts when the Research Team prepares conversation scenarios, in a simple textual or graphical format, for detecting mental health problems (e.g., depression, anxiety, PTSD, and sleep disorder) and providing the appropriate support. The Research Team then sends these scenarios to the development team to create their corresponding WA Dialog flow. The Research Team then tests the chatbot to verify the correctness of the translation. Quite often this process goes through multiple iterations for modification and refinement, which is time-consuming and has the potential for human-error.

Current WA tooling presents dialog nodes in a scalable tree-like structure that shows detail on one node at a time. This works well for most types of conversations, but the workspace can become large and complex when supporting mental-health conversations which are less straight forward than typical virtual assistant conversations. This makes it challenging for the Development Team to present and review these WA Dialog flows with the Research Team. Figure 1 shows a subset of the Here4U WA Dialog flow (currently consists of 5 nodes) as presented in the graphical user interface (GUI) tooling and its JSON format. Note the level of abstraction provided by the graphical nodes compared to their textual counterparts. Users need to click the "3-dots" menu of each node to set, change, or view the details of its features, such as the utterance phrases or actions of the chatbot in response to the user and any contextual information necessary for maintaining the Dialog global state.

It is evident that having a more comprehensive visual representation of the Dialog nodes and their structure with all necessary information will help both teams review, debug, and verify the chatbot control flow. Figure 2 shows a directed graph of the nodes presented in Fig. 1. It provides a more refined view of the current visual representation of the Dialog flow and a less granular view of its JSON representation. It highlights information about the node types, the relationships between the nodes, the execution order and conditions of the flow, the output responses, the non-default transitional behavior of the nodes (e.g., "Wait for user input", "Skip user input", or "Jump to another dialog node"), and the list of context variables defined or changed in each node.

If the graph in Fig. 2 can be modified and there is a mechanism to regenerate the JSON representation with the changes made (as seen in Fig. 3 and Fig. 4), this approach would enhance the chatbot development and maintenance processes.

In the rest of this section, we present examples of useful applications of our proposed representation.

Visualization. Understanding large WA Dialog flows is a difficult task. Using a standard graph representation makes it easier to use existing graph viewers and editors and benefit from their built-in functionalities (e.g., zooming, searching, navigating, auto-layout, editing, portability, etc.). In this work, we render our

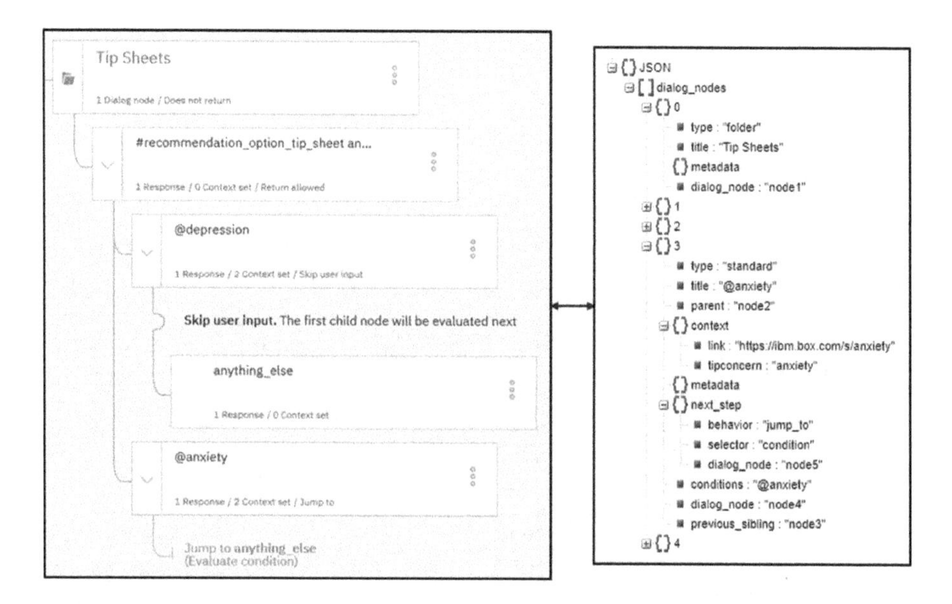

Fig. 1. A subset of an example WA Dialog (left) and its underlying JSON representation (right).

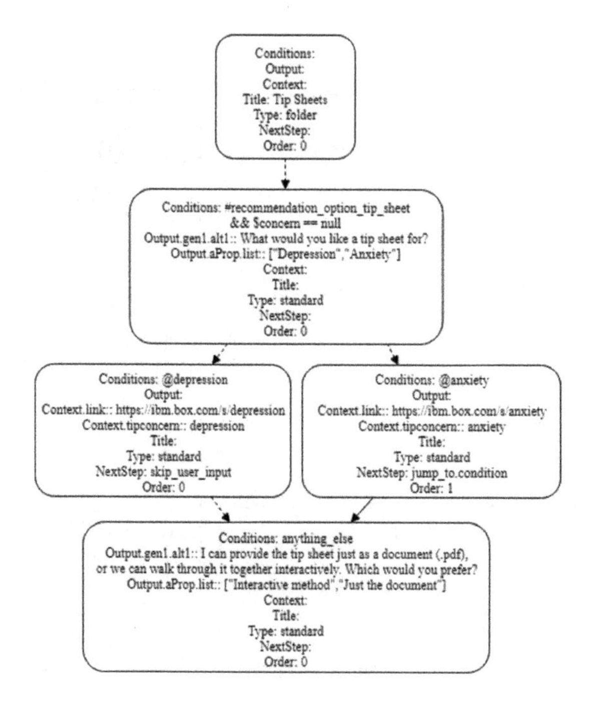

Fig. 2. A directed graph representation of the WA Dialog shown in Fig. 1

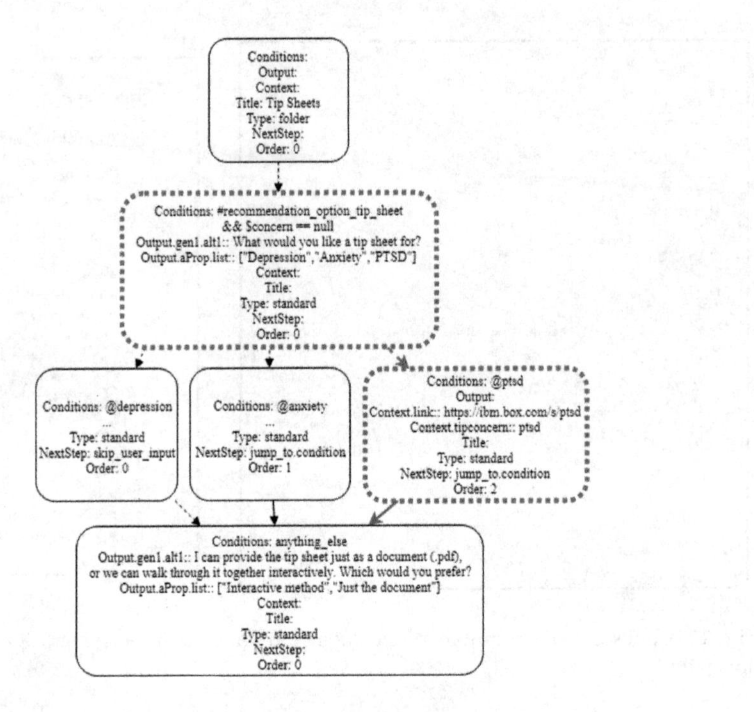

Fig. 3. An updated version of Fig. 2.

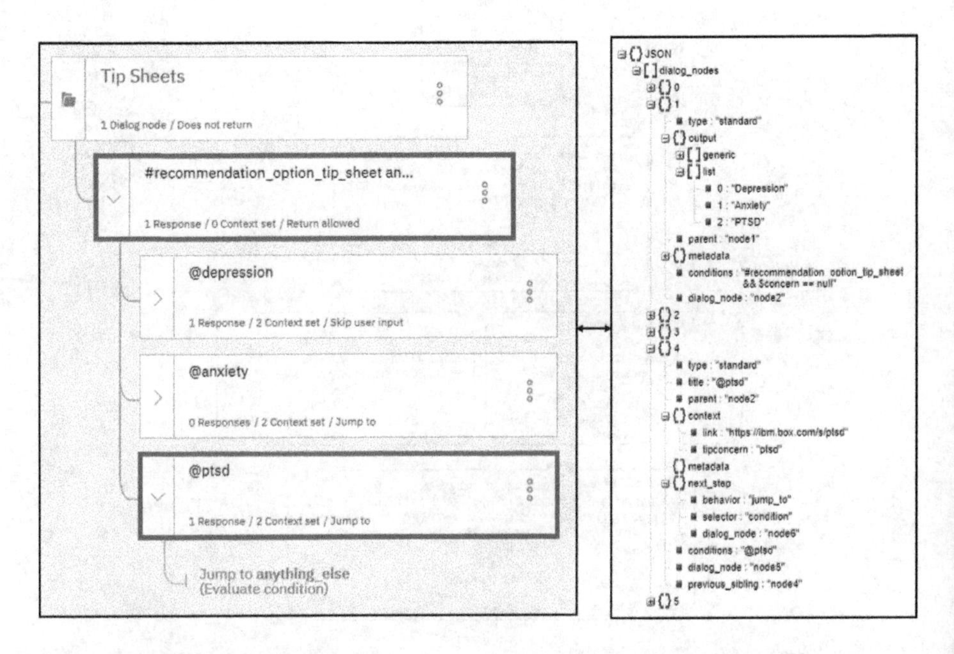

Fig. 4. An updated version of the WA Dialog shown in Fig. 3.

proposed directed graphs into two XML-based formats: 1) the SVG (Scalable Vector Graphics) file format and 2) the Draw.io native format. SVG is a standard graphic format that can be opened in every web browser application (e.g., Mozilla Firefox, Google Chrome, and Microsoft Edge) to be viewed, however Draw.io files can be easily imported in the Draw.io Online web application or the Draw.io Offline tool to be viewed and edited.

Debugging and Traceability. Keeping track of all conversation paths and tracing selected variables are key features of our proposed representation that provides the user with a global perspective of the entire Dialog flow, facilitating the detection of unexpected chatbot behavior.

Modification and Change Propagation. Building a bidirectional transformation framework contributes to maintaining consistency between the different representations. Such a mechanism would ease the modification and refactoring processes of the chatbot.

Dialog Documentation. The proposed representations display all the nodes' information, suggesting that they can be used as design documents for the developed conversation scenarios.

3 Methodology of Proposed Solution

The limitations discussed in Sect. 2 are not specific to the WA platform; they also exist in other chatbot development platforms. The motivating ideas presented here can be applied to all of them. In this section, we present our model transformation-based framework for WA Dialog visualization and auto-generation. The essentials of our approach are: 1) a formal definition that specifies the abstract syntax (known as a metamodel) of WA Dialog flows and the graph-based representation used for viewing and editing and 2) a transformation definition that specifies how to map elements from the WA Dialog representation to elements in the graph-based representation proposed, and vice versa. Additionally, we use existing tools to render SVG and Draw.io from the graphs described in our defined graph-based represenation. Our framework is shown in Fig. 5. Users can use directed graphs to visualize the chatbot flow in an SVG or Draw.io format, as highlighted by Option 1, and they can also use the draw.io tool to edit the flow and regenerate the WA representation, as indicated by Option 2.

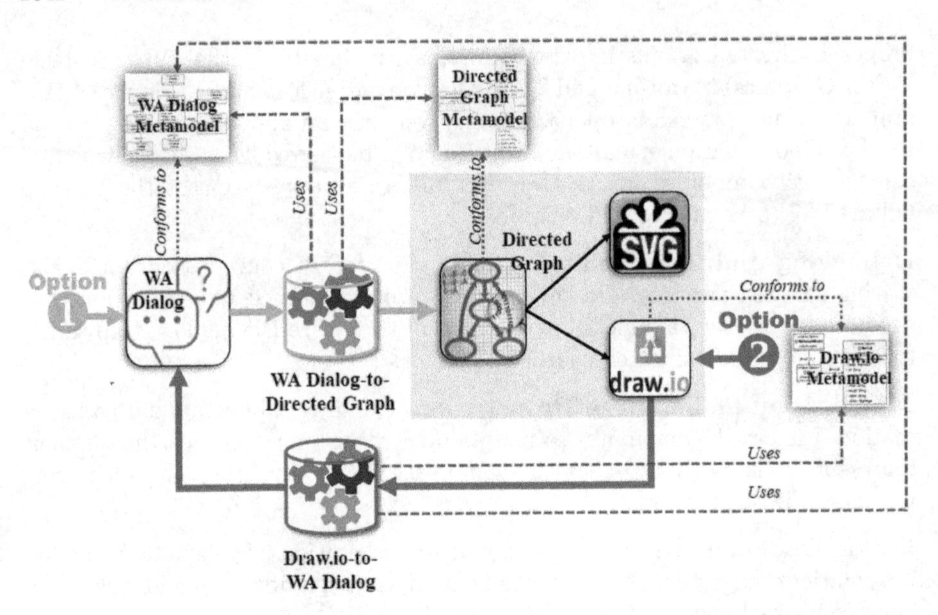

Fig. 5. Proposed framework

3.1 Defined Metamodels

Metamodels provide platform-independent descriptions, usually in an XML format, of how software artifacts are constructed. They are the key elements of model transformation. We developed the metamodels for the different artifacts used in the transformation processes as follows.

WA Dialog Metamodel. There are no formal specifications available for the WA Dialog notations. For this reason, we developed the part of the metamodel that defines the features used in the Here4U project. Figure 6 illustrates the meta-types of these features and their relationships. A `Dialog` meta-type specifies a dialog flow in WA. A `Dialog` is composed of multiple instances of the `DialogNode` meta-type, each of which has a unique dialog node identification, a type, a reference to the parent dialog node, one or more conditions, and a reference to the previous sibling dialog node, and can also have an output response of the `Output` meta-type, a list of context variables defined by the `Context` meta-type, and a next-step action represented by the `NextStep` meta-type. A typical dialog node is of type `standard`. Other node types include `folder`, `frame`, `slot`, `response_condition`, and `event_handler`. Only nodes of type `folder`, `standard`, and `frame` have a visual representation in the WA web tool. Other node types can be found in the JSON representation of the dialog nodes. In our transformation, we create visual representation for all node types.

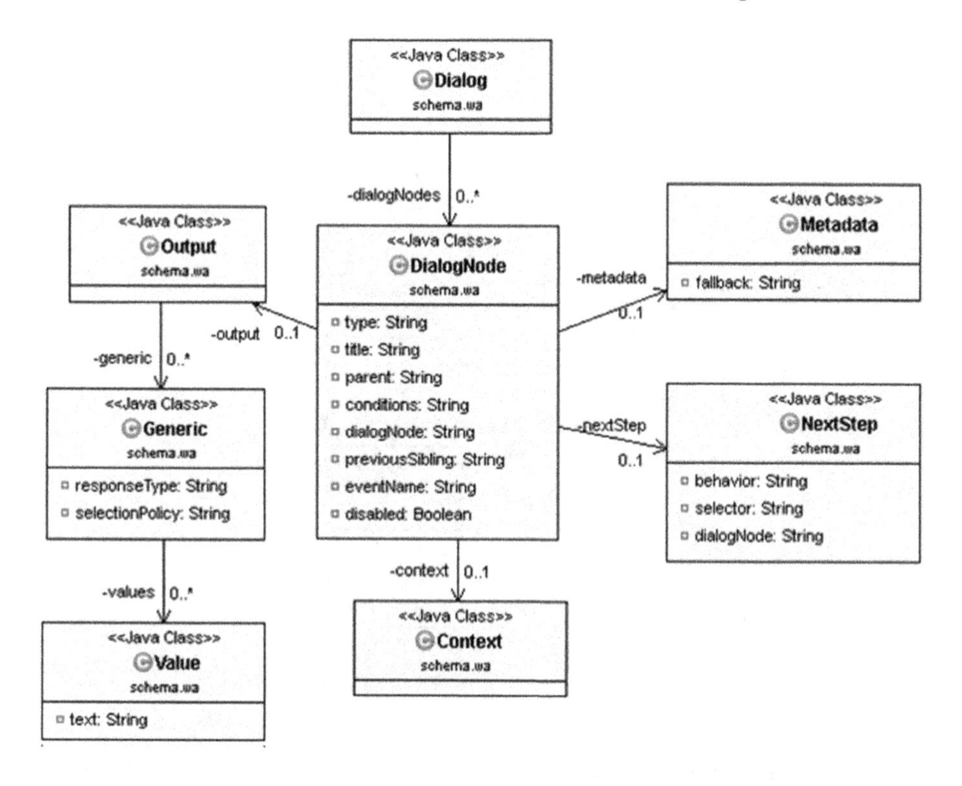

Fig. 6. The subset of the WA Dialog metamodel developed in our work

Directed Graph Metamodel. Directed Graphs are widely used for building common software visualizations (e.g., control flow graphs, dependency graphs, and system architecture diagrams). They are standard concepts which can be defined in terms of their nodes and edges as depicted in Fig. 7. We extended the standard metamodel to support the parent-child and the jump-to relationships present in the WA Dialog metamodel. In the Directed Graph metamodel, parent nodes are aware of their children, unlike in the WA metamodel. Each node in a Directed Graph model has a unique id and a label that combines the information presented in all other features.

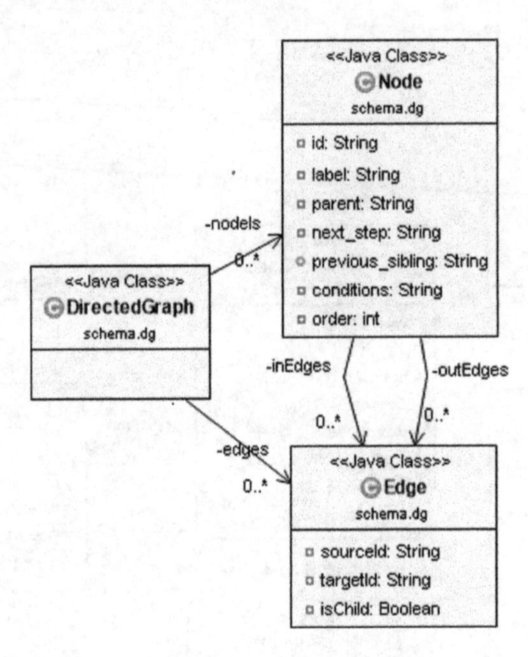

Fig. 7. Our customized directed graph metamodel

Draw.io Metamodel. Draw.io is the world's most used online diagramming tool for sketching various types of diagrams (e.g., flowcharts and UML diagrams) [6]. Like the WA Dialog metamodel, we created the subset of the Draw.io metamodel, shown in Fig. 8, from the XML representation of the diagram data exported from the tool. It shows the four basic meta-types used in our work. The `MxGraphModel` meta-type defines the graph model instances. The `MxCell` meta-type defines the cells (i.e., the elements) of the graph model. A graph cell can be either a layer for all other elements, a vertex, or an edge connecting the vertices. A graph model may have one or more layers, all of which share one root element, defined by the Root meta-type, and may contain multiple cells. The geometry of a graph cell is defined by the MxGeometry meta-type, however this is not part of our transformation.

3.2 Developed Transformations

A large variety of model transformation languages exists. These languages range from low-level imperative languages to high-level declarative ones [7]. For our proof-of-concept implementation, we decided not to use a dedicated model transformation language to develop our transformations. Instead, we used a general-purpose programming language following the direct manipulation approach [8],

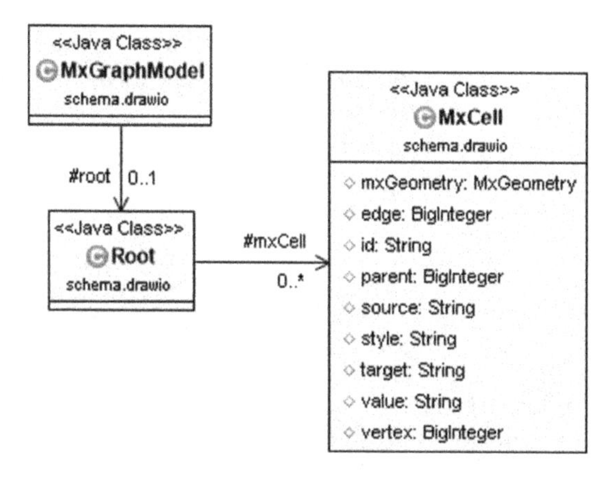

Fig. 8. The subset of the Draw.io metamodel developed in our work

which is harder to implement as it is done on the low-level but is easier to comprehend and maintain by the Here4U development team members who are not familiar with dedicated model transformation technologies.

WA Dialog-to-Directed Graph. The inputs of this transformation are the metamodel of the source artifact (WA Dialog), the metamodel of the target artifact (our customized Directed Graph), and a WA Dialog's JSON file. The transformation consists of rules that perform the required mapping between the source and the target metamodels. These rules are defined in collaboration[1] with the Here4U developers who were responsible for mapping the dialog flows prepared by the Research Team into their counterparts in WA. A summary of the top-level transformation rule and its associated sub-rules is listed in Algorithm 1. Some of these rules are simple rules for mapping primitive features such as a `DialogNode`'s type, label, and conditions, while others are more complex for mapping features such as a `DialogNode`'s response output, context variables, and next-step behavior and for creating edges between parent and children nodes and between a node and its successive one. Examples of simple rules include `getConditions()`, `getType()` and `getTitle()`. Examples of complex rules include `getOutput()`, `getContext()` and `getOrder()`.

Draw.io-to-WA Dialog. The inputs to this transformation are the metamodel of the source artifact (Draw.io Graph Model), the metamodel of the target artifact (WA Dialog), and a Draw.io GraphModel in XML format. We also use the metamodel of our customized Directed Graph as an intermediate representation in the mapping process. A summary of the top-level transformation rule

[1] There is no available document that describes the formal semantics for the WA Dialog flow. Users may need to navigate through many web pages to collect this information.

Algorithm 1. WADialog-to-WAGraph Transformation Pseudocode

Input: 1) *waDialog* – A WA *Dialog* instance, 2) the WA *Dialog* Metamodel, and 3) our customized *DirectedGraph* Metamodel
Output: *aDGraph* – A *DirectedGraph* instance

1: ▷ Constructing *waDirectedGraph*'s nodes
2: *dialogNodes* ← *waDialog.getDialogNodes*()
3: *graphNodes* ← ϕ
4: *graphEdges* ← ϕ
5: **for all** *aDialogNode* ∈ *dialogNodes* **do**
6: *aGraphNode* ← ϕ
7: *conditions* ← getConditions(*aDialogNode*)
8: *output* ← getOutput(*aDialogNode*)
9: *context* ← getContext(*aDialogNode*)
10: *title* ← getTitle(*aDialogNode*)
11: *type* ← getType(*aDialogNode*)
12: *nextStep* ← getNextStep(*aDialogNode*)
13: *previousSibling* ← getPreviousSibling(*aDialogNode*)
14: *order* ← getNodeOrder(*aDialogNode*, *dialogNodes*)
15: *id* ← getId(*aDialogNode*)
16: *parent* ← getParent(*aDialogNode*)
17: *label* ← concat(*conditions*, *output*, *context*, *title*, *type*, *nextStep*, *order*, *id*, *parent*)
18: ▷ Constructing *graphNode*'s edges
19: *parentEdge* ← createEdge(*parent*, *id*, *true*)
20: *children* ← getChildren(*aDialogNode*, *dialogNodes*)
21: *childrenEdges* ← createEdges(*id*, *children*, *true*)
22: *jumpToNodeId* ← getJumpToNodeID(*aDialogNode*)
23: *jumpToEdge* ← createEdge(*id*, *jumpToNodeId*, *false*)
24: *inEdges* ← {*parentEdge*}
25: *outEdges* ← {*childrenEdges*, *jumpToEdge*}
26: *aGraphNode* ← createGraphNode(*id*, *label*, *inEdges*, *outEdges*)
27: *graphNodes* ← *graphNodes* ∪ {*aGraphNode*}
28: *graphEdges* ← *graphEdges* ∪ {*inEdges*, *outEdges*}
29: **end for**
30: *aDGraph* ← createDirectedGraph(*graphNodes*, *graphEdges*)
31: **Return:** *aDGraph*

and its associated sub-rules is listed in Algorithm 2. Line 2 represents the step for extracting the graph nodes from a Draw.io *MxGraphModel* instance input. Lines 3–5 provide the rules for extracting and identifying the *parent*, *nextStep*, and *previousSibling* features of the graph nodes found in Line 2. Line 8 extracts the label of each graph node and lines 9–18 include the rules to extract all other features, such as the *type*, *title*, and *output*. Line 22 maps the graph nodes to WA dialog nodes.

Algorithm 2. Draw.io-to-WADialog Transformation Pseudocode

Inputs: 1) $waMxGraphModel$ – a Draw.io $MxGraphModel$ instance, 2) the $Draw.io$ Metamodel, and 3) WA $Dialog$ Metamodel
Output: $waDialog$ – A WA $Dialog$ instance
1: $graphNodes \leftarrow \phi$
2: $graphNodes \leftarrow$ getGraphNodes($waMxGraphModel$)
3: $graphNodes \leftarrow$ setParentRule($graphNodes$)
4: $graphNodes \leftarrow$ setNextStepRule($graphNodes$)
5: $graphNodes \leftarrow$ setPreviousSiblingRule($graphNodes$)
6: $dialogNodes \leftarrow \phi$
7: **for all** $aGraphNode \in graphNodes$ **do**
8: $label \leftarrow$ extractLable($aGraphNode$)
9: $type \leftarrow$ extractType($label$)
10: $title \leftarrow$ extractTitle($label$)
11: $output \leftarrow$ extractOutput($label$)
12: $parent \leftarrow$ extractParent($label$)
13: $context \leftarrow$ extractContext($label$)
14: $nextStep \leftarrow$ extractNextStep($label$)
15: $conditions \leftarrow$ extractConditions($label$)
16: $dialogNode \leftarrow$ extractDialogNodeId($label$)
17: $previousSibling \leftarrow$ extractPreviousSibling($label$)
18: $eventName \leftarrow$ extractEventName($label$)
19: $aDialogNode \leftarrow$ createDialogNode($type$, $title$, $output$, $parent$, $context$, $nextStep$, $conditions$, $dialogNode$, $previousSibling$, $eventName$)
20: $dialogNodes \leftarrow dialogNodes \cup \{aDialogNode\}$
21: **end for**
22: $waDialog \leftarrow$ createWADialog($dialogNodes$)
23: **Return:** $waDialog$

4 Discussion

The main goal of this work is to support the development of the Here4U chatbot application. The proof-of-concept tool was useful for the development team to review their work with the Research Team. The ability to slice parts of the dialog mitigates the complexity of reviewing the entire flow all at once. Using a graph database, such as Neo4J [9], as another visualization format would enable advanced querying, which is a useful feature to have. The current implementation relies heavily on the use of regular expressions to extract the different features of a dialog node from its corresponding graph node. A more robust alternative is to develop a DSL [10] that is tailored to the specific features required for describing WA graph nodes. Such a language can be used to create a more useable graph-editing environment or to extend the currently available diagrams in Draw.io.

The feedback collected from the development team on the benefits of the graph tool identifies the following key points: 1) Having the entire graph presented makes it easier to locate yourself in the dialog; being able to see the flow all at once in a visual way gives you contextual understanding; 2) It also provides additional information including all the connections to a given node; and 3) The graph tool makes it quite easy to see all the nodes that link to a given node simply by following the directed edges backwards.

5 Related Work

The work presented here employs two fundamental concepts that have been researched and applied for a long time; these are software visualization and model transformation. Software visualization is a powerful technique that helps in understanding and maintaining complex software systems [11]. These visualizations are being used for debugging, refactoring, reverse engineering [12], and computer science education [13]. MDE and model transformation are at the core of facilitating the construction of such visualizations [14,15]. They provide the modeling standards that help in developing a wide range of general-purpose and domain-specific modeling and sketching tools (e.g., Eclipse Modeling Framework, Eclipse Graphical Editing Framework, and Draw.io) [10,16]. Additionally, they provide the tools and mechanisms to elevate such visualizations from being static artifacts into dynamic ones that can be modified and used to propagate changes and auto-generate a more refined version of their source representations [17].

The use of MDE and textual DSLs to create chatbot-development environments is presented in [18–21]. Jarvis [18] and its updated version, Xatkit [19], were the first solutions to be proposed and they are platform-independent. [20] is a tool that uses the ReactiveML [22] programming language to define chatbot conversation flows, which can then be converted into JSON representations of WA chatbots. In [21], the authors present Conga (ChatbOt modelliNg lanGuAge), a new text-based DSL that is platform-independent and provides code generators for several commonly-used chatbot platforms such as Rasa [23] and Dialogflow [3]. The above-mentioned tools use textual representations to define chatbot flows, while we use visual representations. Similar to our tool, the one in [21] supports forward and reverse chatbot engineering, and similar to the implementation in [20], our environment is designed specifically for the WA platform.

6 Conclusion

Developing and maintaining sophisticated chatbot solutions is a complex and iterative task. Supporting current chatbot platforms with custom dynamic visualizations is significant for comprehending, debugging, and maintaining complex dialog flows like the ones developed for the Here4U project. This paper provides the details of our methodology to build such capabilities for the IBM Watson Assistant development platform. The feedback received from the Here4U Development Team highlights the usefulness of these custom visualizations for reviewing and inspecting conversation flow traces (execution paths) that generate undesirable response.

Acknowledgment. This work is funded and supported by IBM Canada, Mitacs, SOSCIP, the Canadian Institute for Military and Veteran Health Research (CIMVHR), and Canada Department of National Defense (Mental Health).

References

1. Klopfenstein, L.C., Delpriori, S., Malatini, S., Bogliolo, A.: The rise of bots: a survey of conversational interfaces, patterns, and paradigms. In: Proceedings of the 2017 Conference on Designing Interactive Systems, ser. DIS 2017, pp. 555–565. ACM, New York (2017). http://doi.acm.org/10.1145/3064663.3064672
2. Amazon: Amazon lex: Build conversation bots (2019). https://aws.amazon.com/lex/
3. Google: Google dialogflow (2019). https://dialogflow.com/
4. IBM: Watson assistant: IBM cloud (2019). https://www.ibm.com/cloud/watson-assistant/
5. Brambilla, M., Cabot, J., Wimmer, M.: Model-driven software engineering in practice. Synthesis Lect. Softw. Eng. **1**(1), 1–182 (2012)
6. Draw.io: Flowchart maker & online diagram software (2019). https://www.draw.io/
7. Czarnecki, K., Helsen, S.: Feature-based survey of model transformation approaches. IBM Syst. J. **45**(3), 621–645 (2006)
8. Akehurst, D.H., Bordbar, B., Evans, M.J., Howells, W.G.J., McDonald-Maier, K.D.: SiTra: simple transformations in java. In: International Conference on Model Driven Engineering Languages and Systems, pp. 351–364. Springer (2006)
9. Neo4J: Neo4j graph platform (2019). https://neo4j.com/
10. Van Deursen, A., Klint, P., Visser, J.: Domain-specific languages: an annotated bibliography. ACM Sigplan Not. **35**(6), 26–36 (2000)
11. Knight, C.: Visualisation for program comprehension: information and issues
12. Koschke, R.: Software visualization in software maintenance, reverse engineering, and re-engineering: a research survey. J. Softw. Maintenance Evol.: Res. Pract. **15**(2), 87–109 (2003)
13. Baecker, R.: Sorting out sorting: a case study of software visualization for teaching computer science. Softw. vis.: Program. Multimed. Exp. **1**, 369–381 (1998)
14. Bull, R.I., Favre, J.-M.: Visualization in the context of model driven engineering. In: MDDAUI, vol. 159 (2005)
15. Buckl, S., Ernst, A.M., Lankes, J., Matthes, F., Schweda, C.M., Wittenburg, A.: Generating visualizations of enterprise architectures using model transformations. Enterp. Model. Inf. Syst. Archit. Int. J. **2**(2) (2007)
16. Sprinkle, J., Karsai, G.: A domain-specific visual language for domain model evolution. J. Vis. Lang. Comput. **15**(3–4), 291–307 (2004)
17. Stevens, P.: A landscape of bidirectional model transformations. In:International Summer School on Generative and Transformational Techniques in Software Engineering. Springer, pp. 408–424 (2007)
18. Daniel, G., Cabot, J., Deruelle, L., Derras, M.: Multi-platform chatbot modeling and deployment with the jarvis framework. In: International Conference on Advanced Information Systems Engineering, pp. 177–193. Springer (2019)
19. Daniel, G., Cabot, J., Deruelle, L., Derras, M.: Xatkit: a multimodal low-code chatbot development framework. IEEE Access **8**, 15 332-15 346 (2020)

20. Baudart, G., Hirzel, M., Mandel, L., Shinnar, A., Siméon, J.: Reactive chatbot programming. In: Proceedings of the 5th ACM SIGPLAN International Workshop on Reactive and Event-Based Languages and Systems, pp. 21–30 (2018)
21. Pérez-Sole, S., Guerra, E., Deruelle, L., de Lara, J.: Model-driven chatbot development. In: 39th International Conference on Conceptual Modeling. Springer (2020)
22. ReactiveML: Reactiveml (2020). http://rml.lri.fr/index.html
23. R. T. Inc.: Rasa: Open source conversational AI - rasa (2020). https://rasa.com/

Job Recommendation Based on Curriculum Vitae Using Text Mining

Honorio Apaza[1], Américo Ariel Rubin de Celis Vidal[2](✉), and Josimar Edinson Chire Saire[3]

[1] Data Science Research Group, Universidad Nacional de Moquegua, Ilo, Moquegua, Peru
hapazaa@unam.edu.pe

[2] Universidad Nacional de Moquegua, Ilo, Moquegua, Peru
arubinv@unam.edu.pe

[3] Institute of Mathematics and Computer Science (ICMC), University of São Paulo (USP), São Carlos, SP, Brazil
jecs89@usp.br

Abstract. During the last years, the development in diverse areas related to computer science and internet, allowed to generate new alternatives for decision making in the selection of personnel for state and private companies. In order to optimize this selection process, the recommendation systems are the most suitable for working with explicit information related to the likes and dislikes of employers or end users, since this information allows to generate lists of recommendations based on collaboration or similarity of content. Therefore, this research takes as a basis these characteristics contained in the database of curricula and job offers, which correspond to the Peruvian ambit, which highlights the experience, knowledge and skills of each candidate, which are described in textual terms or words. This research focuses on the problem: how we can take advantage from the growth of unstructured information about job offers and curriculum vitae on different websites for CV recommendation. So, we use the techniques from Text Mining and Natural Language Processing. Then, as a relevant technique for the present study, we emphasize the technique frequency of the Term - Inverse Frequency of the documents (TF-IDF), which allows identifying the most relevant CVs in relation to a job offer of website through the average values (TF-IDF). So, the weighted value can be used as a qualification value of the relevant curriculum vitae for the recommendation.

Keywords: Natural language processing · Text mining · Data science · System recommender · Job offer · Curriculum vitae · Peru · South America

1 Introduction

In concordance with the report of Internet World Stats 2019 [1], the access to the Internet and web technology has grown fastly and let the users interact trough

K. Arai (Ed.): FICC 2021, AISC 1363, pp. 1051–1059, 2021.
https://doi.org/10.1007/978-3-030-73100-7_72

different platforms. Considering the ambit related to the working market, and the high demand and offer of job places, it is possible to observe that there is a set of unstructured data which contain important information related to the profile of candidates for a particular job, some of these websites as: Computrabajo, Bumeran, Info Jobs, etc. that usually has the profile of the positions and the Curriculum Vitae (CV) of the candidates.

Frequently, the users have specific preferences [2] related with some products or objects, these preferences are present in non-explicit text then the information about the likes or dislikes must be extracted from the text. The preferences of one user usually are represented using a matrix, and the content of the matrix has the level of preference for one specific product. Thus, the research on Text Mining (TM) [3] uses Natural Language Processing (NLP) to extract information from the text written by human beings. The automatization is necessary for the big number of sources (emails, documents, social networks, etc.).

Recommendation Systems (RS) are useful to suggest or recommend one item (product, object, etc.) to one user based on the information from other users. Usually, RS works with text then NLP algorithms [4–6] are used to extract information, relevant terms from the text source.

The present work explains the process step by step to build a Job Recommendation based on Web Scrapping, NLP an TM algorithms. Section 2 includes the review of the bibliography, Sect. 3 discusses the methodology to be used, Sect. 4 develops the work proposal, Sect. 5 discloses the results of the research and in Sect. 6 gives conclusions, last section presents future work.

2 Literature's Review

The study Employers expectations: A probabilistic text mining model [7], more than 20,000 job advertisements from various websites were processed, the method of text mining was applied to identify information skills derived from the web pages of the construction industry sector. In the research named Text Analysis for Job Matching Quality Improvement [5], in a context of data analysis that includes travel time, job location, job type, rates, candidate skill set, etc. And when applying keywords in a machine learning process using text mining tools, as a result, effective keywords are discovered for a job matching system. in the research entitled Natural Language Processing and Text Mining to Identify Knowledge Profiles for Software Engineering Positions [8], through the application of NLP and TM to analyze the unstructured text of the resumes and job offers, it manages to identify the knowledge profiles for software engineering positions. In the research entitled Data Mining Approach to Monitoring The Requirements of the Job Market: A Case Study [9], presents an approach based on data mining to identify the most demanded occupations in the modern labor market. To achieve this, have a latent semantic indexing model that is able to match the job announcement extracted from the 18web with the data of the occupation description in the database.

3 Methodology

In general terms the text mining is a process to turn the text unstructured in data structured for analysis, to achieve that, is necessary applied algorithms artificial intelligence and statistics techniques to text documents. That process can be combined in a workflow [10,11] as can see in Fig. 1.

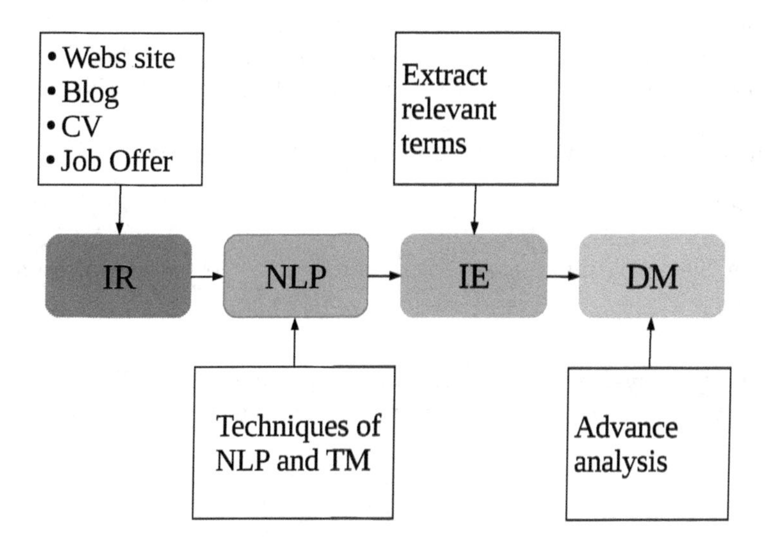

Fig. 1. Methodology flowchart.

- Information retrieval (RI), Collect data about one topic or many from the text sources (websites, emails, etc.)
- Natural Language Processing (NLP), Interpretation of the text considering linguistics supported by Machine Learning techniques.
- Information Extraction (IE), Identification of relevant terms for a summarized representation of documents [12].
- Data Mining (DM), Exploitation of structured data to discover knowledge about relationships.

4 Proposal

4.1 Information Retrieval

Initially was apply Scrapy to recover data in Spanish language about job offers and Curriculum Vitae (CVs) of the candidates from different websites as: computrabajo, Bumeran, etc. for the purpose of research. A collection a set of 10,000 job offers, with the next format: title, city, company's name, description, requirements, functions and previous knowledge for the job role. The set of 10,000 Curriculum Vitae was gathered in HTML code, with the next format: name, description, skills and experience, and that data was extracted thanks to BeautifulSoup library.

4.2 Natural Language Processing

Cleaning and Transformation of Data. Initially it was used the package NLTK (Natural Language Toolkit). Next, each of these data collected are processed by the functions:

- remove_non_ascii_(): For limitation of no-ASCII characters.
- to_lowercase_(): For data normalization to lowecase.
- remove_punctuation_(): To remove scores and accents of Spanish language.
- remove_stopwords_(): To remove empty words from the Spanish language (el, la, los, en, etc.).
- lemmatize_verbs_(): For the randomization of data verbs.

These function set allow obtain clean and standardized data, at this stage you are ready to apply TF-IDF (Term Frequency - Inverse Document Frequency) algorithm. The entry information is structured in two general lists, (resumes and job offers).

Term Frequency-Inverse Document Frequency (TF-IDF). TF-IDF represents a numerical relevance of one term for one document and set of documents (corpus).

$$tf\_idf_{t,d} = tf_{t,d} \cdot \log \frac{N}{df_t} \quad (1)$$

Where:
tf_idf = Term Frequency-Inverse Document. Frequency
$tf_{t,d}$ = Occurrences Frequency of term t in document d.
N = Total Number of Documents.
df_t = Number of documents with the term t.

Information Distribution. The data collected on CVs are divided into three groups: description (D1), experience (D2) and skills (D3), respectively.

The information collected on job offers are divided into five groups of terms, such as: Title, description, requirements, functions and knowledge. These groups are united in one only after applying NLTK, in addition the duplicate terms are removed to avoid redundancy when measuring the relevance of these terms in the CVs.

Application TF-IDF. To find the relevance value of each of the terms of the job offers in the curriculum vitae documents, the technique frequency of terms - inverse frequency of documents (TF-IDF) was applied.

Note that, the Eq. 1 of the TF-IDF algorithm consists of two important parts, the first is to find the frequency term value (tf) in a given document, the second is to find the inverse frequency value of N documents (idf).

The use of TF-IDF for filtering terms in documents for content-based recommendation system is used and recommended in the investigation of [13]. In this paper, the TF-IDF technique is applied to measure the relevance of the terms of job offers in resumes, the weighted value is an indicator of relevance of a CV with respect to job offers.

The relevance of the terms of the job offers are calculated in terms of CV, the terms are distributed in three parts, called documents (description, experience and skills).

Table 1 shows the example of the five terms, where: the fields TF-D1, TF-D2 and TF-D3 correspond to the calculation of frequency (tf) of terms of job offers in respective CV documents. The IDF field corresponds to calculation of inverse frequency of documents (idf), and in the TF-IDF-D1, TF-IDF-D2 and TF-IDF-D3 fields correspond to the $TF - IDF$ calculation for respective documents. Next, the tf-idf values are weighted for prescriptive documents, and finally the general average of all the documents is obtained.

Table 1. Term relevance calculation process

Terms	*TF-D1*	*TF-D2*	*TF-D3*	*IDF*	*TF-IDF-D1*	*TF-IDF-D2*	*TF-IDF-D3*
Minería	2/21	1/5	0/5	Log(3/2)	0.039	0.081	0
Scrapy	1/21	0/5	1/5	Log(3/2)	0.019	0	0.081
Aplicando	1/21	0/5	0/5	Log(3/1)	0.052	0	0
Sistema	2/21	1/5	0/5	Log(3/2)	0.039	0.081	0
Recomendación	2/21	1/5	0/5	Log(3/2)	0.039	0.081	0
n text	...	...	...	...	...	...	...
Average TF-IDF for documents					0.04	0.109	0.127
General average TF-IDF of the documents							0.092

4.3 Information Extraction

Table 2 show the averages of $TF - IDF$ detailed by document and last field show a average general, for this work could be useful for the recommendation of job offers and other data mining purposes. So that information was extracted to use in different contexts, on this case for CV recommendation.

4.4 Data Mining

On the results of the previous stage, we can analyze and visualize the usefulness of this data, the results can help different cases. In the Table 3, we can see a matrix of 5000 resumes for 9998 jobs, where the relevance value of each curriculum vitae is distributed on a matrix basis.

Table 2. Relevance of CV with respect to the job offer.

Idjob	*Idcv*	*Rating description*	*Rating experience*	*Rating skills*	*Averages general*
0	0	0.04	0.109	0.127	0.092
0	1	0	0	0	0
0	2	0.005	0.006	0	0.004
0	3	0.002	0.003	0	0.002
0	4	0	0	0.037	0.012

Table 3. Data mining about job offers and CV.

Id job	*Id Cv*						
	0	*1*	*2*	...	*4997*	*4998*	*4999*
0	0.092	0.000	0.004	...	0.015	0.001	0.000
1	0.000	0.008	0.013	...	0.011	0.012	0.004
2	0.000	0.023	0.013	...	0.012	0.001	0.002
3	0.000	0.023	0.013	...	0.017	0.001	0.002
4	0.000	0.025	0.013	...	0.015	0.001	0.002
5	0.000	0.022	0.013	...	0.014	0.001	0.003
...	...	...	...	...	...	...	...
9997	0.000	0.021	0.013	...	0.015	0.000	0.000

With this information can do analysis about more relevant CVs with respect a one job offer, as be see in the Fig. 2, where it can clearly be seen that some CV are more relevant than others for the first job offer (Job7). The CVs 2nd and 8th are evidently more relevant for the 7ht job offer.

5 Results

The present work provides a data model with averages of relevance of CV with respect to job offers, based on relevance of terms, that can be used to recommend CV or job offers, as shown in Fig. 3.

In the present case it is not necessary to test or validate results, because they are exact results based on relevance of terms, result of calculation of frequency of terms and inverse frequency of documents, therefore the results data are not probabilistic or approximations.

However, dataset is normalized base to max and min scaling, as can see in Eq. 2. when the value is approximate to 1 is most relevant and less relevant when it approximate to 0, in this dataset the highest value is 0.549 and the lowest 0.000.

$$X_{cs} = \frac{X - X_{min}}{X_{max} - X_{min}} \tag{2}$$

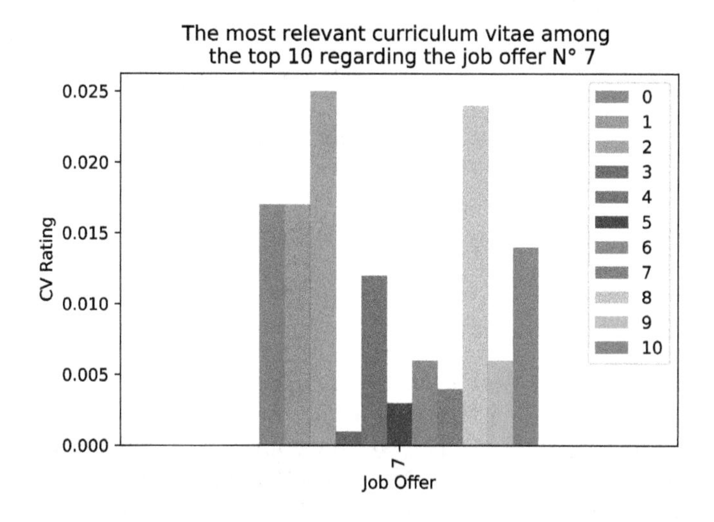

Fig. 2. Visual relevance analysis of CV.

Id_Job	Id_Cv	Rating_general	Titulo	Empresa	Ciudad
6193	4400	0.549	Chófer de Reparto //Todos las Zonas	Synergoz S.A.	Ate Lima
2080	4400	0.549	Chófer de Reparto//Cono Norte	Synergoz S.A.	Ate Lima
3807	4400	0.549	Chofer de Reparto// Cono Sur//Planilla desde e...	Synergoz S.A.	Ate Lima
5393	1560	0.366	Ayudante de cocina / Alimentación/ Linea de ca...	La Antojeria	Miraflores Lima
2370	583	0.366	Digitador	Grupo la Republica Publicaciones	Cusco Cusco

Fig. 3. Data model with averages of relevance of CV with respect to job offers.

The results obtained from the CVs analyzed, clearly establish a set of classifications related to the relevance of each candidate in relation to the employer's job offer.

Based on these results, the employer can determine if in the geographic scope of his interest it is possible to find a candidate with the profile that the company requires.

The Term frequency - reverse document frequency ($tf - idf$) proved to be effective in identifying more relevant information from a corpus of curriculum vitae data, the general averages of tf-idf could be considered as the rating value of the curriculum vitae for the recommendation of job offers.

6 Conclusions

Employability in the countries is one big issue, because people needs to get a job and afford their needs. Then, this work is focused to analyze Curriculum Vitaes and fit with job offers following Text Mining approach. Initial experiments show the affordability of the study. Classical TF-IDF algorithm to represent each profile is useful to find the match with job offers. The workflow used and every step explained related to it, can be expanded for similar studies.

7 Future Work

The recommendation is continue the research on Recommendation Systems to take advantage of the proposal dataset with peruvian information about jobs. First, the employability of graduates from public/private universities is necessary and helpful to tune their programs up. Second, understand how important is the content of the courses offered to the students in university classrooms. Third, build a web application to support job search and facilitate access of employment.

References

1. Internet World Stats "Internet Users in the World by Regions", March 2019. www.internetworldstats.com/stats.htm
2. Rajaraman, A., Ullman, J.D.: Recommendation Systems. Cambredge University Press, California (2014)
3. Torre, M.D.: Nuevas Técnicas de Minería de Textos: Aplicaciones. Departamento de ciencia de la computación e inteligencia artificial, Granada (2017)
4. Aggarwal, C.C.: Recommender Systems. Springer, Switzerland (2016)
5. Kinoa, Y., Kurokia, H., Machidab, T., Furuyab, N., Takanob, K.: Text analysis for job matching quality improvement. In: International Conference on Knowledge Based and Intelligent Information and Engineering Systems (2017)
6. Malherbe, E., Diaby, M., Cataldi, M., Viennet, E., Aufaure, M.: Selection for job categorization and recommendation to social network users. In: IEEE/ACM International Conference on Advances in Social Networks Analysis and Mining (ASONAM 2014) (2014)
7. Gao, L., Eldin, N.: Employers expectations: a probabilistic text mining model. In: Creative Construction Conference, CC 2014 (2014)

8. Almada, R.V., Elias, O.M., Gómez, C.E., Mendoza, M.D., López, S.G.: Natural language processing and text mining to identify knowledge profiles for software engineering positions. In: 5th 81International Conference in Software Engineering Research and Innovation (CONISOFT) (2017)
9. Karakatsanis, I., AlKhader, W., MacCrory, F., Alibasic, A., Omar, M.A., Aung, Z., Woon, W.L.: Data mining approach to monitoring the requirements of the job market: a case study. Electrical Engineering and Computer Science, Masdar Institute of Science and Technology, Abu Dhabi, United Arab Emirates (2016)
10. Zanini, N., Dhawan, V.: Text Mining: An introduction to theory and some applications. Research Matters (2015)
11. JISC: Text mining briefing paper. Joint Information Systems Committee (2008)
12. Brun, R.E., Senso, J.A.: Minería textual. El profesional de la información (2004). http://profesionaldelainformacion.com/contenidos/2004/enero/2.pdf
13. Okaka, R., Mwangi, W., Okeyo, G.: A hybrid approach for personalized recommender system using weighted TF-IDF on RSS contents. Int. J. Comput. Appl. Technol. Res

COVID-19's (Mis)Information Ecosystem on Twitter: How Partisanship Boosts the Spread of Conspiracy Narratives on German Speaking Twitter

Morteza Shahrezaye(✉), Miriam Meckel, Léa Steinacker, and Viktor Suter

Institute for Media and Communications Management, University of St. Gallen, Blumenbergpl. 9, 9000 St. Gallen, Switzerland
morteza.shahrezaye@unisg.ch

Abstract. In late 2019, the gravest pandemic in a century began spreading across the world. A state of uncertainty related to what has become known as SARS-CoV-2 has since fueled conspiracy narratives on social media about the origin, transmission and medical treatment of and vaccination against the resulting disease, COVID-19. Using social media intelligence to monitor and understand the proliferation of conspiracy narratives is one way to analyze the distribution of misinformation on the pandemic. We analyzed more than 9.5M German language tweets about COVID-19. The results show that only about 0.6% of all those tweets deal with conspiracy theory narratives. We also found that the political orientation of users correlates with the volume of content users contribute to the dissemination of conspiracy narratives, implying that partisan communicators have a higher motivation to take part in conspiratorial discussions on Twitter. Finally, we showed that contrary to other studies, automated accounts do not significantly influence the spread of misinformation in the German speaking Twitter sphere. They only represent about 1.31% of all conspiracy-related activities in our database.

Keywords: Social networks · Misinformation · Conspiracy theory · Political polarization

1 Introduction

In November 2019, a febrile respiratory illness caused by SARS-CoV-2 infected people in the city of Wuhan, China. On January 30th 2020, the World Health Organization (WHO) declared the spread of the virus a worldwide pandemic [22]. Shortly after, the WHO reported multiple COVID-19-related knowledge gaps relating to its origin, transmission, vaccinations, clinical considerations, and concerns regarding the safety of healthcare workers [29]. The organization warned of an "infodemic", defined by "an overabundance of information and the rapid spread of misleading or fabricated news, images, and videos" [28]. By

K. Arai (Ed.): FICC 2021, AISC 1363, pp. 1060–1073, 2021.
https://doi.org/10.1007/978-3-030-73100-7_73

August 2020, more than 22 million people worldwide had contracted the virus [27]. The Organization for Economic Co-operation and Development (OECD) put forward estimates of negative GDP growth for all member countries in 2020 due to the crisis [37].

COVID-19's indomitable dissemination around the globe combined with a lack of effective medical remedies [18,50] and its psychological and economic side effects [14,19,37,39] have left many people to uncertainty and fear of further developments. Previous research has shown that lack of certainty and control often results in the emergence and circulation of conspiracy theory narratives [48]. Popper defined conspiracy mentality as the "mistaken theory that, whatever happens in society – especially happenings such as war, unemployment, poverty, shortages, which people as a rule dislike – is the result of direct design by some powerful individuals and groups" [38]. This one-sided or even pathological method of reasoning regularly facilitates coping with uncertainty and fear by making the world more understandable and providing individuals with an illusion of control [34].

There are two main conditions conducive to the emergence of conspiracy narratives: individuals' psychological traits and socio-political factors. Regarding psychological traits, numerous laboratory studies demonstrate the correlation between conspiracy beliefs and psychological features like negative attitude toward authorities [30], self-esteem [1], paranoia and threat [35], powerlessness [1], education, gender and age [45], level of agreeableness [43], and death-related anxiety [36]. Another part of reasoning sees conspiracy mentality as a generalized political attitude [30] and correlates conspiracy beliefs to socio-political factors like political orientation. Enders et al. showed that conspiracy beliefs can be a product of partisanship [10]. Several other studies show a quadratic correlation between partisanship and the belief in certain conspiracy theories [46]. These insights imply that extremists on both sides of the political spectrum are more prone to believe in and to discuss conspiracy narratives.

We define conspiracy narratives as part of the overall phenomenon of misinformation on the internet. We use misinformation as the broader concept of fake or inaccurate information that is not necessarily intentionally produced (distinguished from disinformation which is regularly based on the intention to mislead the recipients). Among all the conspiracy narratives, we are interested in those propagated in times of pandemic crises. The spread of health-related conspiracy theories is not a new phenomenon [6,16,33] but seems to be even accelerated in world connected via social media.

The COVID-19 pandemic's unknown features, its psychological and economical side effects, the ubiquitous availability of Online Social Networks (OSNs) [25], and high levels of political polarization in many countries [13,51] make this pandemic a potential breeding ground for the spread of conspiracy narratives. From the outset of the crisis, "misleading rumors and conspiracy theories about the origin circulated the globe paired with fear-mongering, racism, and the mass purchase of face masks [...]. The social media panic travelled faster than the COVID-19 spread" [9]. Such conspiracy narratives can obstruct the efforts to properly inform the general public via medical and scientific findings

[17]. Therefore, investigating the origins and circulation of conspiracy narratives as well as the potential political motives supporting their spread on OSNs is of vital public relevance. With this objective, we analyzed more than 9.5M German language tweets about COVID-19 to answer the following research questions:

Research Question 1: What volume of German speaking Twitter activities comprises COVID-19 conspiracy discussions and how much of this content is removed from Twitter?

Research Question 2: Does the engagement with COVID-19 conspiracy narratives on Twitter correlate with political orientation of users?

Research Question 3: To what degree do automated accounts contribute to the circulation of conspiracy narratives in the German speaking Twitter sphere?

2 Data

We collected the data for this study during the early phase of the crisis, namely, between March 11th, the day on which the WHO declared the spread of the SARS-CoV-2 virus a pandemic [22] and May 31st, 2020. The data was downloaded using the Twitter's Streaming API by looking for the following keywords: "COVID", "COVID-19", "corona", and "coronavirus". Only Tweets posted by German speaking users or with German language were included. The final dataset comprises more than 9.5M tweets from which two categories of conspiracy narratives were selected: conspiracy narratives about the origin of the COVID-19 illness (Table 1) and those about its potential treatments (Table 2). The conspiracy narratives about the origin of the COVID-19 illness were selected based on Shahsavari et al., who automatically detected the significant circulation of the underlying conspiracy theories on Twitter using machine learning methods [41]. The second group of conspiracy narratives were chosen based on the fact that they were in the center of attention in German media [26] and thus a considerable number of tweets discussed them [20].

Table 1. Conspiracy narratives about the Origin of COVID-19

Case	Description
5G	Conspiracy narrative suggesting that the 5G network activates the virus
Bill Gates	Conspiracy narrative suggesting that Bill Gates aims to use COVID-19 to initiate a global surveillance regime
Wuhan laboratory	Conspiracy theory narrative suggesting that the virus originates from a laboratory in Wuhan, China

Table 3 indicates the number of tweets belonging to each conspiracy narrative and the keywords that are used to filter them out[1]. There were 68,466

[1] We used "homöopath" in order to match both German words "homöopathie" and "homöopathisch".

Table 2. Conspiracy narratives about potential treatments of COVID-19

Case	Description
Ibuprofen	Conspiracy narrative suggesting that Ibuprofen reduces COVID-19 symptoms
Homoeopathy	Conspiracy narrative suggesting that homeopathy medicines reduce COVID-19 symptoms
Malaria	Conspiracy narrative suggesting that a malaria drug is an antiviral against SARS-CoV-2 virus

tweets in total discussing the underlying conspiracy narratives. Figure 1 shows the timeline of the tweets.

Table 3. Number of tweets for each conspiracy narrative

Case	Number of tweets	Keywords
5G	5,762	5G, #5g
Bill Gates	24,653	Bill Gates, #billgates
Wuhan laboratory	9,366	#wuhanlab Wuhan Lab
Ibuprofen	7,016	Ibuprofen, #ibuprofen
Homeopathy	4,714	Homöopath, #Homöopath
Malaria	7,955	Malaria, #malaria
Control group	9,000	–

In addition to the six conspiracy narratives, 9000 tweets were randomly extracted from the dataset and served as a control group.

To answer research question 2, a list was extracted from official party websites; this list contains members of parliament (MPs) who are active on Twitter and belong to one of the six political parties in Germany's federal legislature. Each party runs several official Twitter pages that were added to the list of Twitter pages of each political party; for example, the official Twitter page of the Social Democratic Party (SPD) in the federal state of Bavaria, called "BayernSPD", was added to the SPD list. For each twitter account in the extracted list a maximum of 4000 tweet handles were downloaded from the Twitter API. Table 4 shows the relevant statistics on the political tweets.

In the next step, for each of the 68,466 users spreading conspiracy narratives (Table 3) the lists of their tweet handles were downloaded (Table 5). Finally, for each of them we counted the number of times they retweeted one of the political tweets in Table 4. Based on Boyd et al. retweets are mainly a form of endorsement [7]. Therefore, we assume if a user collects a discernible number

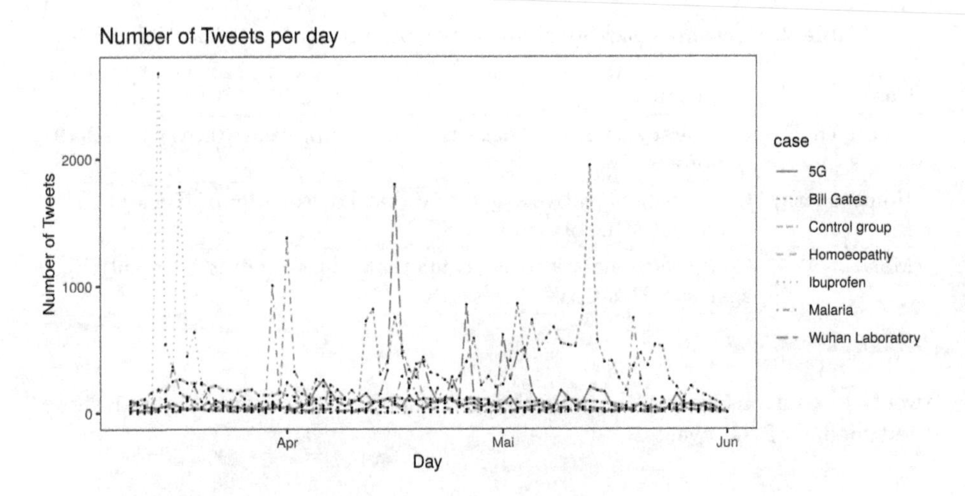

Fig. 1. Daily number of tweets for each conspiracy narrative

Table 4. Number of political tweets extracted from politicians' Twitter pages

Political party	Number of MPs on Twitter	Number of extra official Twitter pages	Total number of tweet handles
AfD	27	14	68,789
CDU/CSU	131	18	220,768
FDP	56	7	96,046
Bündnis 90/Die Grünen	56	11	169,864
Linke	50	12	155,794
SPD	110	17	221,029

Table 5. Number of tweets extracted from users spreading conspiracy narrative tweets

Case	Number of tweets	Number of downloaded tweets from the contributing users
5G	5762	10,967,158
Bill Gates	24653	35,144,536
Wuhan laboratory	9366	14,332,403
Ibuprofen	7016	14,855,267
Homoeopathy	4714	7,746,555
Malaria	7955	16,258,164
Control group	9000	12,217,082

of retweets from members of a certain political party, this user will most likely share the corresponding political orientation. This method of inference about the political orientation of users has been applied in similar studies [15].

3 Results

There are multiple studies showing that exposure to misinformation can lead to persistent negative effects on citizens. The respondents in a study adjusted their judgment proportional to their cognitive ability after they realized that their initial evaluation was based on inaccurate information. In other words, respondents with lower levels of cognitive ability tend to keep biased judgments even after exposure to the truth [32]. In another study, Tangherlini et al. found that conspiracy narratives stabilize based on the alignment of various narratives, domains, people, and places such that the removal of one or some of these entities would cause the conspiracy narrative to quickly fall apart [44]. Imhoff and Lamberty have shown that believing COVID-19 to be a hoax negatively correlated with compliance with self-reported, infection-reducing, containment-related behavior [31].

On that account, to assess a democratic information ecosystem that is balanced rather towards reliable information than misinformation we need to monitor and estimate if COVID-19 conspiracy theory narratives circulate significantly on Twitter. Based on a survey in mid-March 2020, about 48% of respondents stated that they have seen some pieces of likely misinformation about COVID-19 [24]. Shahsavari et al. used automated machine learning methods to automatically detect COVID-19 conspiracy narratives on Reddit, 4Chan, and news data [41]. Multiple other studies found evidence of COVID-19 misinformation spread on different OSNs [3,5,40].

To address the public concerns many of the service providers claimed that they will remove or tag this sort of content on their platforms. On March 16th 2020, Facebook, Microsoft, Google, Twitter and Reddit said they are teaming up to combat COVID-19 misinformation on their platforms [23]. On April 22nd, Twitter stated that they have removed over 2230 tweets containing misleading and potentially harmful COVID-19-related content [21]. On June 7th 2020, we examined how many of the German conspiracy-related tweets still exist on Twitter in order to understand if conspiracy-related tweets tend to exist on Twitter for a longer period of time compared to non conspiracy-related tweets. Table 6 shows the results.

3.1 Research Question 1

Based on Table 6, only about 0.61% of all COVID-19 German tweets are about one of the conspiracy narratives under consideration. These German tweets are posted by more than 36,000 unique Twitter users. While 0.61% is small in magnitude, it still comprises a relevant number of citizens. It is important to note though that this finding does not imply that only about 36,000 Twitter users believe in conspiracy theories. While our data shows the spread of conspiracy narratives, they do not reveal a user's stance towards the respective content. In terms of content moderation by Twitter, on average 7.3% of conspiracy narrative tweets are deleted after a certain period of time which is significantly higher than

Table 6. Share of conspiracy narratives among all COVID-19 tweets

Case	Number of tweets	Share (among all 9.5M COVID-19 tweets)	Share deleted on 7th June 2020
5G	5762	0.06%	6%
Bill Gates	24653	0.25%	7%
Wuhan laboratory	9366	0.098%	9%
Ibuprofen	7016	0.073%	14%
Homoeopathy	4714	0.049%	3%
Malaria	7955	0.083%	5%
Control group	9000	0.094%	6%

6% of tweets in the control group. We speculate that more of the conspiracy-related tweets are deleted because of Twitter's content moderation efforts that have been enforced due to recent public debates about misinformation on OSNs.

3.2 Research Question 2

There is a long list of laboratory studies that show a correlation between conspiracy mentality and extreme political orientation [10,46]. In this study we answer the slightly different question if the partisanship of Twitter users correlates with their contribution to conspiracy theory narrative discussions. Table 7 shows the distribution of the political orientations of users who discuss each of the underlying conspiracy narratives.

Table 7. Political orientation of users discussing conspiracy narratives

Case	AfD	CDU/ CSU	FDP	Bündnis 90/ Die Grünen	Linke	SPD	Unknown
5G	11%	3%	3%	8%	10%	14%	51%
Bill Gates	16%	3%	3%	8%	12%	14%	44%
Wuhan laboratory	27%	3%	15%	7%	5%	8%	35%
Ibuprofen	9%	3%	3%	8%	9%	16%	52%
Homoeopathy	5%	4%	6%	13%	15%	24%	33%
Malaria	10%	2%	2%	5%	6%	11%	64%
Control group	10%	2%	2%	4%	6%	10%	66%

Table 7 demonstrates that users who are likely to be supporters of AfD and SPD most actively discuss and spread COVID-related conspiracy narratives on Twitter. To check if contributions to conspiracy narratives are correlated with the political orientation of users, we ran a saturated Poisson log-linear model on the contingency Table 7. The model defines the counts as independent observations of a Poisson random variable and includes the linear combination and the interaction between conspiracy narratives and the political orientation of users [2].

$$log(\mu_{ij}) = \lambda + \lambda_i^N + \lambda_j^P + \lambda_{ij}^{NP} \quad (1)$$

where $\mu_{ij} = E(n_{ij})$ represents the expected counts, λs are parameters to be estimated and N and P stand for *Narrative* and *Political Orientation*. λ_{ij}^{NP}s corresponds to the interaction and association between conspiracy narratives and also reflects the departure from independence [2]. Since we suspect that beliefs in certain mutually contradictory conspiracy theories can be positively correlated [49], we aggregated the six conspiracy theory cases to two based on to which category they belong and formed Table 8 to remove any possible correlation.

Table 8. Political orientation of users who discuss two conspiracy narratives (absolute counts)

Case	AfD	CDU/ CSU	FDP	Bündnis 90/ Die Grünen	Linke	SPD	Unknown
Origins of COVID-19	4133	694	1396	1873	2449	3028	10296
Possible treatments of COVID-19	1263	432	497	1149	1347	2330	7841

Table 9 shows the ANOVA analysis of the underlying saturated Poisson log-linear model applied on Table 8. The last line of resulting p-values in Table 9 shows that in interaction parameter, $\mu_{ij} = E(n_{ij})$, is statistically significant. Therefore, we can reject the hypothesis that the contribution to conspiracy narratives is independent of the political orientation of users. The fact that there is evidence of a correlation between the contribution to conspiracy narratives and the political orientation of users, however, does not imply any causality.

Table 9. ANOVA of Poisson Log-Linear Model on the Contingency Table 7

	Df	Deviance	Resid. Df	Resid. Dev	Pr(>Chi)
NULL			13	31332.82	
Narrative	1	2115.49	12	29217.32	0.0000
Party	6	28294.31	6	923.02	0.0000
Narrative:party	6	923.02	0	0.00	0.0000

To further estimate the relative effect of political orientation on the contribution to conspiracy narratives on Twitter, we applied six Chi-Square goodness of fit tests on the control group and each of the other six conspiracy narratives. For all of the six tests the p-values were significantly less than 0.05, which suggests that the distributions of the contribution to the six different conspiracy narratives are statistically different compared to the control group. Figure 2 shows the distribution of the tests' residuals. The last column of Fig. 2 shows that the Twitter users without a certain political orientation contributed relatively less to conspiracy narratives in comparison to the control group. In other words, compared to the control group, users with certain political orientations contributed more to the circulation of conspiracy narratives.

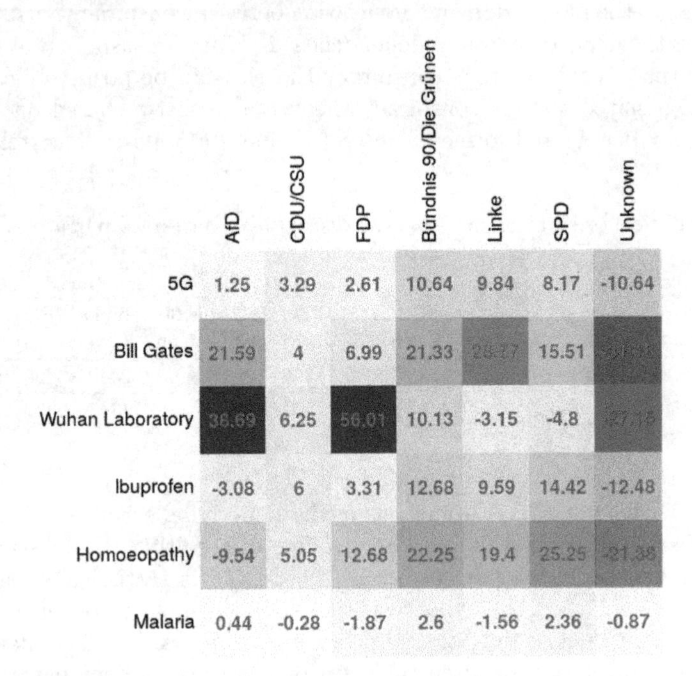

Fig. 2. Distribution of residuals of Chi-Square goodness of fit tests

3.3 Research Question 3

Automated accounts, or users who post programmatically, make up a significant amount of between 9% and 15% of Twitter users worldwide [8]. Multiple studies hold automated accounts responsible for political manipulation and undue influence on the political agenda [12,42]. However, more recent studies shed light on these previous results and showed that the influence of automated accounts is overestimated. Ferrara finds that automated accounts comprise less than 10% of users who post generally about COVID-19 [11].

There are multiple methods to automatically detect automated accounts on OSNs [4]. For this study, we used the method developed by Davis et al. [8]. They applied random forest classification trees on more than a thousand public metadata available using the Twitter API and on other human engineered features.

Table 10 displays the percentage of automated accounts (users with Complete Automation Probability higher than 0.5) and verified users who contribute to conspiracy narratives.

Table 10. Ratio of tweets posted by automated and verified users

Case	Share of tweets posted by verified users	Share of tweets posted by automated accounts	Ratio of automated accounts to verified users
5G	3.141%	1.578%	0.5
Bill Gates	1.85%	1.358%	0.73
Wuhan laboratory	9.065%	1.3%	0.14
Ibuprofen	3.349%	1.386%	0.41
Homoeopathy	1.039%	0.921%	0.89
Malaria	4.626%	1.343%	0.29
Control group	4.644%	0.89%	0.19

Based on this analysis, 1.31% of COVID-19 conspiracy narrative tweets are suspected to be posted by automated accounts. This number is significantly lower than many other studies on bot activities on Twitter. We speculate that this occurs due to three reasons. First, the importance of the topic might have captured a lot of public attention, so that significantly more users discuss COVID-19-related topics compared to usual Twitter discussions. Second, many service providers, including Twitter, have started to combat COVID-19 misinformation because of widespread warnings. Finally, we have concentrated on German tweets while the past estimates apply to tweets in English.

4 Discussions and Limitations

In this study we analyzed more than 9.5M German language tweets and showed that the volume of tweets that discuss one of the six considered conspiracy narratives represents about 0.6% of all COVID-19 tweets. This translates to more than 36,000 unique German speaking Twitter users. Imhoff and Lamberty found that "believing that COVID-19 was a hoax was a strong negative prediction of containment-related behaviors like hand washing and keeping physical distance". To provide the public with accurate information about the importance of such measures, social media intelligence can help elevate potential pitfalls of the Twitter information ecosystem.

Using more than 38,000 tweets and 36,000 unique Twitter users, we formed the contingency table of political orientation and of contribution to COVID-19 conspiracy narratives (Table 8). We then applied a saturated Poisson log-linear regression and showed that we cannot statistically reject independence among the underlying variables. This implies partisans have a higher motivation for taking part in COVID-19-related conspiracy discussions. This shows that politically polarized citizens increase the spread of health misinformation on Twitter.

Finally, we employed an automated accounts detection tool and showed that on average about 1.31% of the users who discuss COVID-19 conspiracy narratives are potentially automated accounts or bots. This number is much lower than estimations on general bot activity on Twitter, which is assumed to be up to 15% [8,47].

This study holds new insights as well as some limitations:

- Our results shed light on the problem of misinformation on Twitter in times of crises for a certain cultural and language context: Germany. We showed that the political orientation of politically polarized users translates to higher circulation of health-related conspiracy narratives on Twitter. Further research could compare the results of this study with other countries and language realms on Twitter.
- We also offer indications between political or ideological partisanship and engagement in the dissemination of misinformation on Twitter. In this study we examined if political partisanship motivates individuals to take part in conspiracy discussions. In other words, we did not distinguish between tweets promoting the conspiracy narratives and those rejecting them. One could extend the analysis and study the effect of partisanship on promoting conspiracy theories. Further research will also need to combine quantitative data analysis and qualitative content analysis to better understand the underlying motivations for engaging in conspiracy communication on OSNs.
- Finally, we offer a more nuanced view on the role of automated tweets regarding a highly emotionally-charged topic. There are numerous studies showing contradictory estimates of bot activity on OSNs. We found only about 1.31% of users who spread COVID-19 conspiracy tweets are potentially bots. This number is much lower than many of those put forward by other researchers. Further research could investigate this result in order to understand the reasons why this estimation is lower than other case studies.

References

1. Abalakina-Paap, M., Stephan, W.G., Craig, T., Larry Gregory, W.: Beliefs in conspiracies. Polit. Psychol. **20**(3), 637–647 (1999)
2. Agresti, A.: Categorical Data Analysis, vol. 482. Wiley, Hoboken (2003)
3. Ahmed, W., Vidal-Alaball, J., Downing, J., Seguí, F.L.: Covid-19 and the 5g conspiracy theory: social network analysis of Twitter data. J. Med. Internet Res.D **22**(5), e19458 (2020)
4. Alothali, E., Zaki, N., Mohamed, E.A., Alashwal, H.: Detecting social bots on Twitter: a literature review. In: 2018 International Conference on Innovations in Information Technology (IIT), pp. 175–180. IEEE (2018)

5. Boberg, S., Quandt, T., Schatto-Eckrodt, T., Frischlich, L.: Pandemic populism: Facebook pages of alternative news media and the corona crisis–a computational content analysis (2020). arXiv preprint arXiv:2004.02566
6. Bogart, L.M., Wagner, G., Galvan, F.H., Banks, D.: Conspiracy beliefs about HIV are related to antiretroviral treatment nonadherence among African American men with HIV. J. Acquir. Immune Defic. Syndr. **53**(5), 648 (2010)
7. Boyd, D., Golder, S., Lotan, G.: Tweet, tweet, retweet: conversational aspects of retweeting on Twitter. In: 2010 43rd Hawaii International Conference on System Sciences, pp. 1–10. IEEE (2010)
8. Davis, C.A., Varol, O., Ferrara, E., Flammini, A., Menczer, F.: BotOrNot: a system to evaluate social bots. In: Proceedings of the 25th International Conference Companion on World Wide Web, pp. 273–274 (2016)
9. Depoux, A., Martin, S., Karafillakis, E., Preet, R., Wilder-Smith, A., Larson, H.: The pandemic of social media panic travels faster than the COVID-19 outbreak. J. Travel Med. **27**(3), taaa031 (2020)
10. Enders, A.M., Smallpage, S.M., Lupton, R.N.: Are all 'birthers' conspiracy theorists? On the relationship between conspiratorial thinking and political orientations. Br. J. Polit. Sci. **50**(3), 849–866 (2020)
11. Ferrara, E.: What types of COVID-19 conspiracies are populated by Twitter bots? First Monday, May 2020
12. Ferrara, E., Varol, O., Davis, C., Menczer, F., Flammini, A.: The rise of social bots. Commun. ACM **59**(7), 96–104 (2016)
13. Fletcher, R., Cornia, A., Nielsen, R.K.: How polarized are online and offline news audiences? A comparative analysis of twelve countries. Int. J. Press/Polit. **25**(2), 169–195 (2020)
14. Frank, A., Hörmann, S., Krombach, J., Fatke, B., Holzhüter, F., Frank, W., Sondergeld, R., Förstl, H., Hölzle, P.: Covid-19 concerns and worries in patients with mental illness. Psychiatr. Prax. **47**(5), 267–272 (2020)
15. Garimella, K., De Francisc iMorales, G., Gionis, A., Mathioudakis, M.: Mary, mary, quite contrary: exposing Twitter users to contrarian news. In: Proceedings of the 26th International Conference on World Wide Web Companion, pp. 201–205 (2017)
16. Geissler, E., Sprinkle, H.S., Erhard Geissler and Robert Hunt Sprinkle: Disinformation squared: was the HIV-from-Fort-Detrick myth a Stasi success? Polit. Life Sci. **32**(2), 2–99 (2013)
17. Grimes, D.R.: On the viability of conspiratorial beliefs. PLOS ONE **11**(1), 1–17 (2016)
18. Guo, Y.-R., Cao, Q.-D., Hong, Z.-S., Tan, Y.-Y., Chen, S.-D., Jin, H.-J., Tan, K.-S., Wang, D.-Y., Yan, Y.: The origin, transmission and clinical therapies on coronavirus disease 2019 (covid-19) outbreak-an update on the status. Mil. Med. Res. **7**(1), 1–10 (2020)
19. Ho, C.S., Chee, C.Y., Ho, R.C.: Mental health strategies to combat the psychological impact of covid-19 beyond paranoia and panic. Ann. Acad. Med. Singapore **49**(1), 1–3 (2020)
20. Netzpolitik Homepage, 13 May 2020. https://netzpolitik.org/2020/wenn-die-eltern-ploetzlich-an-verschwoerungstheorien-glauben-coronapandemie/. Accessed 02 Sept 2020
21. Twitter, 22 March 2020. https://twitter.com/TwitterSafety/status/1253044734416711680. Accessed 2 Sept 2020
22. BBC Homepage, 11 March 2020. https://www.bbc.com/news/world51839944. Accessed 02 Sept 2020

23. Bloomberg Homepage, 17 March 2020. https://www.bloomberg.com/news/articles/2020-03-17/facebook-microsoft-google-team-up-against-virusmisinformation. Accessed 02 Sept 2020
24. Pew Research Center, 18 March 2020. https://www.journalism.org/2020/03/18/americans-immersed-in-covid-19-news-most-think-media-are-doing-fairly-well-coveringit/. Accessed 02 Sept 2020
25. Pew Research Center, 10 April 2019. https://www.pewresearch.org/fact-tank/2019/04/10/share-of-u-s-adults-using-social-media-including-facebook-is-mostly-unchanged-since2018/. Accessed 02 Sept 2020
26. Tagesschau Homepage, 24 March 2020. https://www.tagesschau.de/faktenfinder/corona-ibuprofen101.html. Accessed 02 Sept 2020
27. WHO Homepage, 01 June 2020. https://www.who.int/docs/default-source/coronaviruse/situation-reports/20200601-covid-19-sitrep133.pdf. Accessed 02 Sept 2020
28. WHO Homepage, 25 August 2020. https://www.who.int/news-room/feature-stories/detail/immunizing-the-public-againstmisinformation. Accessed 02 Sept 2020
29. WHO Homepage, 12 March 2020. https://www.who.int/who-documents-detail/a-coordinated-global-researchroadmap. Accessed 02 Sept 2020
30. Imhoff, R., Bruder, M.: Speaking (un-)truth to power: conspiracy mentality as a generalised political attitude. Eur. J. Pers. **28**(1), 25–43 (2014)
31. Imhoff, R., Lamberty, P.: A bioweapon or a hoax? The link between distinct conspiracy beliefs about the coronavirus disease (covid-19) outbreak and pandemic behavior. Soc. Psychol. Pers. Sci. **11**(8), 1110–1118 (2020)
32. De Keersmaecker, J., Roets, A.: 'Fake news': incorrect, but hard to correct. The role of cognitive ability on the impact of false information on social impressions. Intelligence **65**, 107–110 (2017)
33. Klofstad, C.A., Uscinski, J.E., Connolly, J.M., West, J.P.: What drives people to believe in zika conspiracy theories? Palgrave Commun. **5**(1), 1–8 (2019)
34. Kruglanski, A.W., Pierro, A., Mannetti, L., De Grada, E.: Groups as epistemic providers: need for closure and the unfolding of group-centrism. Psychol. Rev. **113**(1), 84 (2006)
35. Mancosu, M., Vassallo, S., Vezzoni, C.: Believing in conspiracy theories: evidence from an exploratory analysis of Italian survey data. South Eur. Soc. Polit. **22**(3), 327–344 (2017)
36. Newheiser, A.-K., Farias, M., Tausch, N.: The functional nature of conspiracy beliefs: examining the underpinnings of belief in the Da Vinci Code conspiracy. Pers. Individ. Differ. **51**(8), 1007–1011 (2011)
37. OECD: OECD Economic Outlook, Interim Report March 2020 (2020)
38. Popper, K.: The Open Society and its Enemies: Hegel and Marx. Routledge (2002)
39. Rajkumar, R.P.: Covid-19 and mental health: a review of the existing literature. Asian J. Psychiatry **52**, 102066 (2020)
40. Serrano, J.C.M., Papakyriakopoulos, O., Hegelich, S.: NLP-based feature extraction for the detection of covid-19 misinformation videos on outube (2020)
41. Shahsavari, S., Holur, P., Tangherlini, T.R., Roychowdhury, V.: Conspiracy in the time of corona: automatic detection of covid-19 conspiracy theories in social media and the news. arXiv preprint arXiv:2004.13783 (2020)
42. Shao, C., Ciampaglia, G.L., Varol, O., Flammini, A., Menczer, F.: The spread of fake news by social bots. arXiv preprint arXiv:1707.07592 (2017). vol. 96, p. 104

43. Swami, V., Coles, R., Stieger, S., Pietschnig, J., Furnham, A., Rehim, S., Voracek, M.: Conspiracist ideation in Britain and Austria: evidence of a monological belief system and associations between individual psychological differences and real-world and fictitious conspiracy theories. Br. J. Psychol. **102**(3), 443–463 (2011)
44. Tangherlini, T.R., Shahsavari, S., Shahbazi, B., Ebrahimzadeh, E., Roychowdhury, V.: An automated pipeline for the discovery of conspiracy and conspiracy theory narrative frameworks: bridgegate, pizzagate and storytelling on the web. PLOS ONE **15**(6), 1–39 (2020)
45. van Prooijen, J.-W.: Why education predicts decreased belief in conspiracy theories. Appl. Cogn. Psychol. **31**(1), 50–58 (2017)
46. van Prooijen, J.-W., Krouwel, A.P.M., Pollet, T.V.: Political extremism predicts belief in conspiracy theories. Soci. Psychol. Pers. Sci. **6**(5), 570–578 (2015)
47. Varol, O., Ferrara, E., Davis, C.A., Menczer, F., Flammini, A.: Online human-bot interactions: detection, estimation, and characterization. In: 11th International AAAI Conference on Web and Social Media (2017)
48. Whitson, J.A., Galinsky, A.D.: Lacking control increases illusory pattern perception. Science **322**(5898), 115–117 (2008)
49. Wood, M.J., Douglas, K.M., Sutton, R.M.: Dead and alive: beliefs in contradictory conspiracy theories. Soc. Psychol. Pers. Sci. **3**(6), 767–773 (2012)
50. Xie, P., Ma, W., Tang, H., Liu, D.: Severe covid-19: a review of recent progress with a look toward the future. Front. Pub. Health **8**, 189 (2020)
51. Yang, J., Rojas, H., Wojcieszak, M., Aalberg, T., Coen, S., Curran, J., Hayashi, K., Iyengar, S., Jones, P.K., Mazzoleni, G., Papathanassopoulos, S., Rhee, J.W., Rowe, D., Soroka, S., Tiffen, R.: Why are "others" so polarized? Perceived political polarization and media use in 10 countries. J. Comput. Mediated Commun. **21**(5), 349–367 (2016)

Sentiment Sentence Construction Algorithm of Newly-Coined Words and Emoticons Dictionary for Social Data Opinion Analysis

Jin Sol Yang[1], Jihun Kang[2], Kwang Sik Chung[3(✉)], and Kyoung-Il Yoon[1,2,3]

[1] Department of Computer Science, Graduate School of Korea National Open University, Dongsung-dong, Jongno-gu 110-791, Seoul, Korea
sanbbang@naver.com, K2j23h@korea.ac.kr
[2] BK 21 Computer Science Education Research Center, Korea University, 145 Anam-ro, Seongbuk-gu 02841, Seoul, Korea
[3] Department of Computer Science, Korea National Open University, Dongsung-dong, Jongno-gu 110-791, Seoul, Korea
kchung0825@knou.ac.kr

Abstract. With the development of social networks, Social Network Services (SNS) has been populated. And people create newly-coined words and emoticons. Accordingly, the newly-coined word and emoticon of SNS show the social phenomenon of modern society. For social opinion analysis with SNS newly-coined words and emoticons, additional and continued sentiment proactive manual measures are required with a lot of time and money. It is necessary to make an objective decision when classifying the sentiment dictionary. This paper proposes a method for automatically constructing a newly-coined word and emoticon sentiment dictionary by extracting newly-coined words and emoticons from SNS reviews. In addition, the sentiment sentence is determined from the collected SNS reviews to automatically determine the polarity and intensity of newly-coined words and emoticons. Sentiment sentences contain newly-coined words and emoticons with strong polarity intensity, and mean sentences with strong polarity. The automatically constructed newly-coined word and emoticon sentiment dictionary proposed in this paper showed similar analysis accuracy compared to the existing manual construction method. As a result, this study can extract newly-coined words and emoticons in real time from trend-sensitive SNS and apply them to the sentiment dictionary to improve analysis accuracy.

Keywords: Social media opinion · Social big data · Sentiment sentence · Emoticon · Newly-coined words · Sentiment sentence

1 Introduction

Social big data is a huge amount of unstructured data produced by SNS, and its amount is growing exponentially and spreading rapidly. Such social big data includes valuable information such as emotions, opinions, and opinions. Companies and public organizations use to analyze consumer trends and identify product assessments with social big

K. Arai (Ed.): FICC 2021, AISC 1363, pp. 1074–1085, 2021.
https://doi.org/10.1007/978-3-030-73100-7_74

data, as well as to plan for future strategies. Opinion mining technology based on social big data is used to collect and analyze reviews written by people to understand public opinion. The importance of the opinion mining technology that rapidly analyzes the information desired by the user from social big data and intelligently infers meaningful information [1–3]. However, SNS responds quickly to new issues and creates public opinion. As many newly-coined words and emoticons are quickly generated, the analysis accuracy of the opinion mining technology may be reduced. Since newly-coined words and emoticons implicitly contain social phenomena and trends in modern society, it should be an important factor in the analysis of opinion mining. Therefore, it is necessary to continuously update the newly-coined words and emoticons to be used as sentiment dictionary.

Most SNS users want their articles to be followed or recommended by other users. However, it is difficult to get recommendations from other users for reviews with a lot of content or uninteresting reviews. Therefore, SNS reviews imply their opinions in short sentences. Due to these characteristics, the frequency of use of new words and emoticons is rapidly increasing. The term coined as a term coined by the changes of the times, implicitly reflecting social phenomena such as the times and conflicts of the times, people's psychology, thoughts, and culture [4]. These new words are classified into a short form, a compound form, and a pure Korean form according to the form of the coin. First, the abbreviation means a new word that takes the form of an abbreviation, for example, selfie (self camera), spec (specification), ul-chan (face-chan), men-bung (mental-breaking), undo-don (movement) Women in the city), Solkamal (to be honest). Second, the compound word form means a new word created by combining words and words. For example, there are Chi-n-im (a compound word of chicken and God), childbirth refugees (a compound word of childbirth and refugees), a right-e-need (a compound word of doctors and God), and a giraffe (journalist waste). Thirdly, the pure Korean language means a new word intentionally made and spread by the National Institute of the Korean Language. Examples include netizens, chamsal, etc. [5]. Emoticons are electronic semi-languages made of letters and symbols, and express more emotionally than emotional text. Furthermore, there are various types of emoticons ranging from combinations of letters and symbols to 4BYTE characters and image tags. Some SNS companies develop their own emoticons and distribute them for free or for free. SNS users express their personality by downloading emoticon corpus provided by SNS companies for free or for free.

In remains of this paper, Sect. 2 discusses previous research related to opinion mining. Section 3 presents a method to automatic construction of newly-coined word and emoticon sentiment dictionary by extracting newly-coined words and emoticons from SNS reviews of twitters. In Sect. 4, the newly-coined word and emoticon sentiment dictionary construction process proposed in this paper is compared with manual construction method of the newly-coined word and emoticon sentiment dictionary. Finally, Sect. 5 concludes after discussing some improvements and future research plans.

2 Related Research

In the previous study [1], we improved the emotional analysis accuracy of social big data by constructing a newly-coined word and emoticon sentiment dictionary. However, the following problems were drawn in the process of constructing the sentiment dictionary of newly-coined words and emoticons by workers. First, classification workers are likely to judge the polarity and intensity of newly-coined words and emoticons by personal decision. For example, a large number of classification workers can determine the polarity as negative by judging the emoticon "TT, TT" as tears of sadness, or positively determining the polarity by judging as tears of joy. Since the polarity and strength can be determined differently according to the worker's personal feeling, an objective decision is necessary when construction of the sentiment dictionary. Second, since newly-coined words and emoticons are rapidly changed, continuous management of the dictionary is necessary. However, the speed and amount of newly-coined words and emoticons created on SNS is fast and huge. Sorting a large amount of newly-coined words and emoticons by workers is time consuming and expensive. In order to improve these problems, this paper introduces the process of automatically constructing a newly-coined word and emoticon sentiment dictionary rather than classification by workers. The online sentences have many spelling and spacing errors, and because the length of sentences is short, it is often impossible to grasp the exact meaning [6]. In order to solve this problem, a correction was made using a word pattern and a super/neutral dictionary, and a semantic selection method using the priority of parts of speech within a sentence was used. Song Eun-ji used SVM algorithm, which is an emotion analysis technique through machine learning, and is different from this paper based on sentiment dictionary [6]. The Korean Emotion Analysis Corpus (KOSAC) proposed a method to create a Korean emotion corpus, which is essential for emotional analysis and opinion analysis. KOSAC has a total of 17,582 emotion expressions annotated using 332 newspaper articles and 7,744 sentences as annotation objects to build a Korean emotion corpus essential for emotional analysis. In this paper, we propose a method for analyzing emotions using KOSAC as a basic sentiment dictionary and using a new word and emoticon sentiment dictionary [7, 8]. Yun-Fei Jia et al. [9] extracted sentiment words using similarity-based method after searching violent microblogs, and Hamidreza et al. [10] built an adaptive sentiment dictionary to improve the accuracy of polarity determination in microblogs. Sudhanshu Kumar et al. [11] investigated the influence of age and gender in sentiment analysis. Vasile-Daniel Păvăloaia et al. [12] analyzed customer reactions to photos and videos on six social networks (Facebook, Twitter, Instagram, Pinterest, Google+, Youtube). Jungkook An et al. [13] built a dictionary of emotional words using collective intelligence by voting for positive, neutral, and negative for each word targeting college students. Kyoungae Jang et al. [14] automatically built a positive/negative corpus by collecting product reviews through laptops, MP3s, and monitors in internet shopping malls. In [15, 16], polarity is classified between negative and positive. However, formal words and a formal dictionary are used to analyze SNS opinions.

3 Proposed Sentiment Sentence Construction Algorithm of Newly-Coined Words and Emoticons Dictionary

Sentences in social big data that contain newly-coined words and emoticons with strong polarity are likely to have the same polarity when they contain other newly-coined words and emoticons. In this paper, based on these assumptions, sentences containing newly-coined words and emoticons with strong polarity are determined as sentiment sentences, and polarity and sentiment weight are automatically assigned to newly extracted newly-coined words and emoticons using sentiment sentences. Sentiment sentences contain newly-coined words and emoticons with strong polarity, and are sentiment sentences. Figure 1 shows the process of automatically determining the polarity and sentiment weight of newly-coined words and emoticons extracted from social data, and consists of the social big data collection step, sentiment sentence decision step, newly-coined word and emoticon extraction step, newly-coined word and emoticon polarity and sentiment weight decision step.

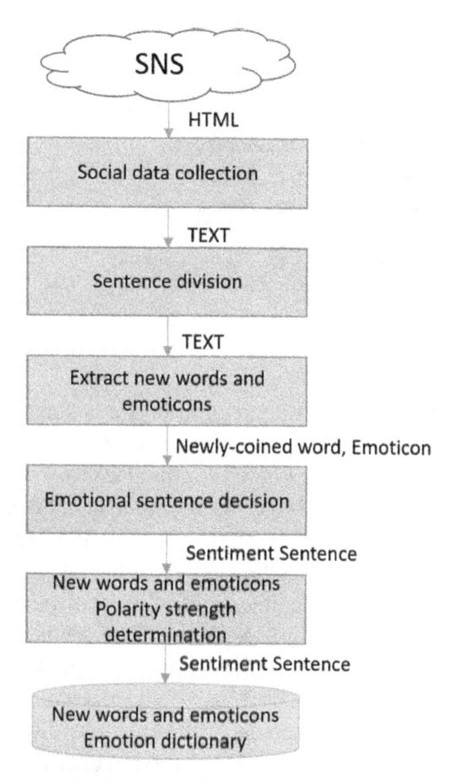

Fig. 1. The process of automatically determining the polarity and sentiment weight of newly-coined words and emoticons

3.1 Social Big Data Collection

We collect social big data to be used as base data for extracting newly-coined words and emoticons. As a method of collecting social data, it is collected using an open API provided by a social network service, or by developing a dedicated crawler that collects specific social data. The collected social big data includes some unnecessary information such as user name, URL, and hashtag. In order to minimize errors in newly-coined word and emoticon extraction, the user name, URL, and hashtag are removed [17]. Social big data that has been pre-processed is stored in the social big data DB. Table 1 is an example of stored social data.

Table 1. Social big data collected

Id	Content	Date
971240486651482114	@instiz Oh, BTS comes back April.... TT TT	2018-03-07 PM 1:31:47
971240579278540800	Comments are really spectacles…	2018-03-07 PM 1:32:09
971240579278475270	OMG !! I think it's been 14 years since I ate Shanghai Spice…	2018-03-07 PM 1:32:09
971240672488456193	My brother is.... Presbyopia....(Poooooor)	2018-03-07 PM 1:32:31
971240672362663937	The end time was completely overlooked.LOL	2018-03-07 PM 1:32:31
971240672270471169	.. Please sell yo sano…LoL	2018-03-07 PM 1:32:31

3.2 Sentence Division

Social big data may not have a clear distinction between sentences. Therefore, in this paper, sentences are classified as follows by extracting morphemes from social data. First, in social data, a sentence containing a newline character (\n), period (.), question mark (?), exclamation point (!), and ellipsis (…) is distinguished. Second, if the tag attribute of the morpheme extracted from the social big data is MAJ (connection adverb), EF (terminal ending), the sentence is classified. Third, if the tag attribute of the morpheme extracted from social big data is EC (connected ending) and the morpheme is '-silver/d', '-only', '-but', or '-just', the sentence is separated. The sentence is composed of short sentences and complex sentences. Complex sentences represent a variety of semantic relations between clauses and clauses or clauses and verb phrases through a connecting ending. The semantic functions presented in the previous discussion include 'listing, contrast, selection, background, good works, causes, conditions, results, concessions, simultaneous, purpose, reinforcement, analogy, comparison, and repetition' [2]. Among them, the contrast has a conjugation of '-silver/day', '-only', '-but', '-just', and '-only' and reverses the polarity between clauses in a sentence [2]. Figure 2 shows the process of segmentation of posts in which the linking ending is contrasted.

Usually, social media uploaders express their feelings by adding emoticons to the end of sentences. For example, the sentence in the SNS post 'Really Minhyun is so good… 💜Aaron also contributes too lol' is separated by an ellipsis (…). If the emoticon is included after the ellipsis (…) that separates the sentence boundary, the sentence boundary is separated even by the emoticon. Therefore, the first sentence is classified as 'I really like Minhyun… 💜' and the second sentence is 'Aron also contributes too haha'.

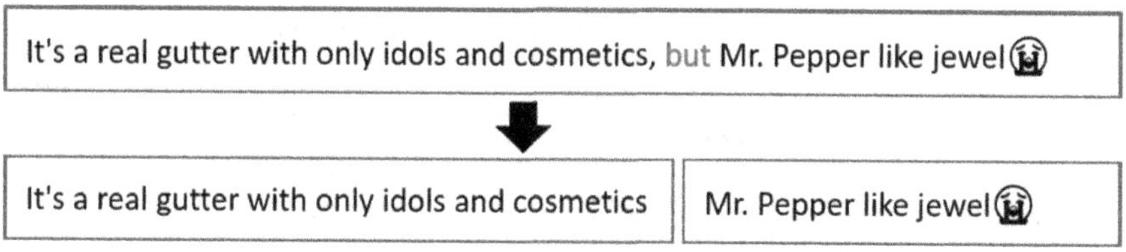

It's a real gutter with only idols and cosmetics, but Mr. Pepper like jewel 

It's a real gutter with only idols and cosmetics | Mr. Pepper like jewel

Fig. 2. Sentence segmentation by contrast

3.3 Extract Newly-Coined Words and Emoticons

In this section, newly-coined words, regular character emoticons, and 4byte character emoticons are extracted from the divided sentences. First, in the sentiment dictionary, newly-coined words are extracted using Naver's Open Dictionary. Naver's Open Dictionary is a user-participating dictionary, and newly-coined words in 32 languages including Korean, English, Chinese, and Japanese are registered on the website (open-dict.naver.com). Users of open dictionary can register newly-coined words directly, and can evaluate "Like" or "No" to newly-coined words registered by other users. In this paper, we create a dedicated crawler to collect open dictionaries with 10 or more "Like" languages on the website. Thereafter, the words in the open dictionary included in the sentimental sentence are used as a newly-coined word. Second, a general character emoticon is extracted from the sentiment sentence. Hangul expresses characters by combining vowels and consonants. Therefore, Hangul cannot express characters with a single element of vowels and consonants. However, most general character emoticons are not used by combining special characters and Korean vowels and consonants, such as "^^ㅋ" and "ㅠㅠ". Using these characteristics, a character, vowel, or consonant consecutively listed in two or more characters in a word phrase is defined as a general character emoticon. Third, 4Byte Unicode character emoticons are extracted from sentiment sentences. 4Byte Unicode characters have been expanded from the existing 2Byte Unicode characters to 4Byte, so that it is possible to express characters in the form of pictures. The emoticon using 4Byte Unicode characters is representative of Emoji developed by NTT Tokomosa of Japan. Emoji is being supported by Apple and Google, and is also supported by SNS company Facebook. The 4Byte Unicode character emoticon extraction process such as Emoji examines all the characters in the sentiment sentence and extracts the characters encoded in 4Byte Unicode. Table 2 shows the newly-coined words and emoticons extracted from the divided sentences.

3.4 Sentiment Sentence Decision

In this section, the algorithm proposed in this paper is applied to determine sentiment sentences from social big data divided into multiple short sentences. Sentiment sentences contain newly-coined words and emoticons with strong polarity intensity, and mean sentences with strong polarity. Sentiment sentence is composed of 'decision newly-coined word and emoticon' attribute, 'decision newly-coined word and emoticon' attribute, and 'polarity strength' attribute. The 'decision newly-coined word and emoticon' attribute means a newly-coined word and emoticon with strong polarity. The 'Dependent newly-coined words and emoticons' attribute means unregistered newly-coined words and

Table. 2. Extracting newly-coined words and emoticons

Sentence	Extract new words and emoticons
✪ It's like the hottest weather these days! Have a cofffeeee in the town to avoid heat♡	✪, cofffeeee, ♡
All of our favorite (and embarrassing) moments. ☺ □ •••	☺ □

emoticons whose polarity and intensity are determined by newly-coined words and emoticons with strong polarity. The 'polarity intensity' attribute is a polarity sentiment weight determined by a newly-coined word and emoticon with strong polarity. The newly-coined words and emoticons extracted from social big data divided into a plurality of short sentences are sentimental sentences through the newly-coined words and emoticon sentiment dictionary constructed in the previous study [1]. Figure 3 is a newly-coined word and emoticon sentiment dictionary constructed in the previous study [1]. The attributes of the newly-coined word and emoticon sentiment dictionary consist of 'No', 'Dictionary', 'Polarity', 'Sentiment weight', and 'Type'. The 'No' attribute means the unique number of the sentiment dictionary, and the 'dictionary' attribute means the newly-coined word or emoticon. The 'polarity' attribute has attribute values of 'positive', 'negative', and 'mix'. The 'sentiment weighted' attribute has attribute values from 1 to 5. In this paper, newly-coined words and emoticons whose 'polarity' attribute is 'positive' or 'negative' and whose 'sentiment weight' attribute is 4 or more are defined as newly-coined words and emoticons with strong polarity in the sentiment dictionary. If the extracted newly-coined words and emoticons are included in the existing newly-coined words and emoticon sentiment dictionary, it is determined as a sentiment sentence. If the newly-coined words and emoticons extracted from the sentiment sentence match words in the newly-coined words and emoticon sentiment dictionary, it is included in the determined newly-coined word emoticon attribute, and if not, it is included in the dependent newly-coined word and emoticon attribute. In the sentiment sentence, two or more decision newly-coined words and emoticons may be included. In this case, the polarity of the decision newly-coined words and emoticons must be the same, and if the polarities do not match, it cannot be a sentiment sentence. The 'polarity intensity' attribute of the sentiment sentence is calculated as (CNT * 2 * SENT). CNT refers to the number of newly-coined words and emoticons with strong polarity. SENT displays a value of 1 when the polarity of a newly-coined word or emoticon with a strong polarity is positive, and a value of −1 when it is negative. The 'polarity intensity' attribute of the sentiment sentence has an attribute value of −5 to 5 points, and an attribute value of 0 is not included. When the 'polarity intensity' attribute value of the sentiment sentence is positive, it indicates positive polarity, and the higher the absolute value, the stronger the positiveness. Conversely, when the 'polarity intensity' attribute value of the sentiment sentence is negative, it indicates negative polarity, and the higher the absolute value, the stronger the negative. Fig. 3 shows the process determined by sentiment sentences in social data [18, 19]. In Fig. 4, the social big data is divided into three short sentences by the sentence division described in Sect. 3.2. The divided second short sentence was determined as a sentimental sentence by decision newly-coined words and emoticons, and includes dependent newly-coined words and emoticons.

No	dictionary	polarity	weight	type
1	ㅋㅋ	positive	3	Character
2	^^;	positive	2	Character
3	Cuttee	positive	4	Character
4	 😍	positive	5	4Byte type
5	 💔	negative	4	4Byte type
6		negative	5	Image type
7		positive	3	Image type

Fig. 3. Sentiment dictionary of newly-coined words and emoticons

From V-App to Instagram photos today ㅠㅠ
Lovely, unique and infinitely special B2B, like to have an idol like this in the world♡😆
I was depressed before I watched V-App today, but I am relieved

Sentiment sentence	Decisional words and emoticons	Dependent new words and emoticons	Polarity strength
Lovely, unique and infinitely special B2B, like to have an idol like this in the world ♡😆	♡	😆	2

Fig. 4. Sentiment sentence determination in social data

3.5 Automatic Polarity Strength Determination of Newly-Coined Words and Emoticons

In the sentiment sentence, the dependent newly-coined words and emoticons are assigned to the polarity intensity attribute of the sentiment sentence and stored in the automatic polarity dictionary. As shown in Fig. 5, the properties of the automatic polarity dictionary are composed of 'dictionary' and 'polarity intensity'. The 'dictionary' attribute means a newly-coined word or emoticon. The 'polarity strength' attribute has an attribute value from −5 to 5. When the 'polarity intensity' attribute value is positive, it indicates positive polarity, and the higher the absolute value, the stronger the positive value. Conversely, when the 'polarity intensity' attribute value of the sentiment sentence is negative, it indicates negative polarity, and the higher the absolute value, the stronger the negative. When the 'polarity strength' attribute value is 0, it indicates neutral polarity. If the dependent newly-coined words and emoticons of the sentiment sentence already exist in the automatic sentiment dictionary, the polarity intensity attribute value of the automatic sentiment dictionary is recalculated. Figure 5 is the process of recalculating the polarity intensity of the automatic sentiment dictionary. In Fig. 5, the polarity strength of dependent newly-coined words and emoticons is added to the polarity strength history DB. Dependent newly-coined words and emoticons may have different polarity intensity attribute values according to newly-coined words and emoticons determined in the sentiment sentence. Therefore, the average value of the same dependent newly-coined word and emoticon stored in the polarity strength history DB is obtained. Then, the polarity intensity attribute value of the automatic polarity sensitivity dictionary is changed to an average value. Newly-coined words and emoticons are sensitive to trends, so they are used rapidly among people and are not used immediately. In addition, the polarity and intensity of newly-coined words and emoticons may change over time. Therefore, only the latest 1000 polarity attribute values stored in the polarity strength history DB are included in the average set.

4 Evaluation

In this section, new words and emoticon sentiment dictionaries are automatically built and compared with existing manual building methods. Twitter reviews are used as the basis for extracting new words and emoticons from social media. Twitter is one of the popular social media platforms for expressing opinions on all topics of people on social networks [11]. In order to collect Twitter reviews, we use the Twitter4j library, the Twitter search API. The new word and emoticon extraction and automatic polarity determination processes run on the Linux CentOS 6.5 OS and JAVA JDK 1.6 platform environments. And mysql 5.5 is used as a database for storing SNS reviews and new words and emoticons.

About 4 million reviews in Korean were collected on Twitter from March 7, 2018 to May 7, 2018 to extract new words and emoticons. From the collected Twitter reviews, 55,996 new words, 31,340 character emoticons, and 1,171 4Byte character emoticons were extracted. Image type emoticons were excluded from the emoticon extraction because they are not provided by Twitter. In this study, sample data of 410 new words and 201 emoticons were extracted from the extracted new words and emoticons to compare

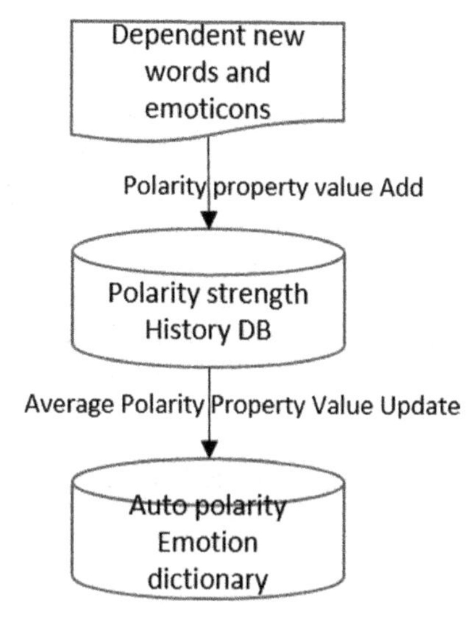

Fig. 5. Polarity intensity processing process

the existing manual emotional dictionary construction method with the automatic emotional dictionary construction method proposed in this paper. For the sample data, two emotional dictionaries were constructed using the existing manual construction method and the automatic construction method proposed in this paper. The evaluation method of this paper measures how much the polarity of the automatically constructed emotional dictionary matches the polarity of the manually constructed emotional dictionary.

5 Conclusion

As mobile devices have been developed and popularized, smartphones have become a necessity in everyday life. And the exchange of personal opinions on social issues using smartphones has been activated. By analyzing the emotions, opinions, and opinions of social big data and using them for corporate marketing and strategy establishment, companies can create emerging values in the online market. As a result, the importance of opinion mining technology in SNS is increasing. However, as SNS reacts quickly to new issues and creates public opinion, many newly-coined words and emoticons are created in this process. Accordingly, the prediction accuracy of the opinion mining technology may be reduced. In addition, the manual construction method of the sentiment dictionary by the worker takes a lot of time and money, and it is highly likely to weekly judge the polarity and intensity of newly-coined words and emoticons. Therefore, it is necessary to continuously update the sentiment dictionary, and since the polarity and strength can be determined differently depending on the operator, an objective decision is needed when classifying the sentiment dictionary. In order to improve these problems, this paper proposed a method for automatically constructing a newly-coined word and emoticon sentiment dictionary and continuously updating it, rather than a manual way to build a newly-coined word and emoticon sentiment dictionary. The proposed method showed similar analysis accuracy compared to the method of manually constructing a newly-coined word and emoticon sentiment dictionary. However, some problems were also discovered. First, it is difficult to use a general character emoticon such as "ㅠㅠ", which can have both positive and negative emotions in SNS, as sentiment dictionary. However, in the case of a 4Byte character emoticon, positive emotion (😂) and negative emotion (😱) are clearly distinguished. Second, SNS has a problem in that grammar or spelling is not clear, so the sentence is not properly divided during the sentence division process. In the future, the task of this study is to build a sentence division pattern dictionary to increase the accuracy of the automatic sentiment dictionary construction process, and to expand the research across the entire SNS, from research limited to texts to images, voice, and video.

Acknowledgment. This work was supported by the Korea Sanhak Foundation (KSF) in 2020.

References

1. Yang, J.S., Ko, M.-S., Chung, K.S.: Social emotional opinion decision with newly coined words and emoticon polarity of social networks services. Fut. Internet **11**(8), 165 (2019)
2. Nam, S., Chae, S.: A study on the use of connected language by Korean learners. **15**(1), 33–50 (2004)
3. Park, K., Park, H., Kim, H., Go, H.: Special name: social network service; research on opinion mining in SNS. J. Inf. Process. Soc. **18**(6), 68–78 (2011)
4. Hwan, K., Im, J.: Study on how to archive social phenomena using new words. Archival Research, vol. 52, pp. 315–342 (2017)
5. Lee, J.Y., Lee, J.S.: An exploratory study of consumption trends using neologism analysis. Consum. Policy Educ. Rev. **12**(4), 269–295 (2016)
6. Song, U.: Emotional analysis for customer feedback on social media. J. Korea Inf. Commun. Soc. (2015)
7. Kim, M.-H., Jang, H.-Y., Yoo, Y., Shin, H.-P.: Korean sentiment analysis corpus (KOSAC): corpus of Korean emotion and opinion analysis. J. Korea Inf. Commun. Soc. (2013)
8. Shin, H., Kim, M., Park, S.: A study on emotion analysis based on aspect information using Korean emotion analysis corpus. Linguistics **74**, 93–114 (2016)
9. Jia, Y.-F., Li, S., Renbiao, W.: Incorporating background checks with sentiment analysis to identify violence risky Chinese microblogs. Fut. Internet **11**(9), 200 (2019)
10. Keshavarz, H., Abadeh, M.S.: ALGA: adaptive lexicon learning using genetic algorithm for sentiment analysis of microblogs. Knowl. Based Syst. **122**, 1–16 (2017)
11. Kumar, S., Gahalawat, M., Roy, P.P., Dogra, D.P., Kim, B.-G.: Exploring impact of age and gender on sentiment analysis using machine learning. Electronics **9**(2), 374 (2020)
12. Păvăloaia, V.-D., Teodor, E.-M., Fotache, D., Danileţ, M.: Opinion mining on social media data: sentiment analysis of user preferences. Sustainability **11**(16), 4459 (2019)
13. An, J., Kim, H.-W.: Building a Korean sentiment lexicon using collective intelligence. J. Intell. Inf. Syst. **21**(2), 49–67 (2015)
14. Jang, K., Park, S., Kim, W.-J.: Automatic construction of a negative/positive corpus and emotional classification using the internet emotional sign. J. KIISE **42**(4), 512–521 (2015)
15. Kim, K., Lee, J.: Sentiment analysis of twitter using lexical functional information. In: Proceedings of the KISS 2014 Conference, Busan, South Korea (2014)

16. Hong, D., Jeong, H., Park, S., Han, E., Kim, H., Yun, I.: Study on the methodology for extracting information from SNS using a sentiment analysis. J. Korea Inst. Intell. Transp. Syst. **16**(6), 141–155 (2017)
17. Yu, Y., Wang, X.: World Cup 2014 in the Twitter world: a big data analysis of sentiments in US sports fans' tweets. Comput. Hum. Behav. **48**, 392–400 (2015)
18. Yang, J.S., Chung, K.S.: Newly-coined words and emoticon polarity for social emotional opinion decision. In: 2019 IEEE 2nd International Conference on Information and Computer Technologies (ICICT). IEEE (2019)
19. Yang, J.S., Chung, K.S.: Newly-coined words and emoticons dictionary construction for social data sentiment analysis. In: Proceedings of the 2019 11th International Conference on Future Computer and Communication, Rangoon, Myanmar, 27 February–1 March 2019, pp. 126–130 (2019)

Intelligent Chatbot Based on Emotional Model Including Newly-Coined Word and Emoticons

Kyoungil Yoon[1], Jihun Kang[2], and Kwang Sik Chung[3](✉)

[1] Graduate School, Department of Information Science, Korea National Open University, 86, Daehak-ro, Jongno-gu, Seoul, Republic of Korea
pofour@knou.ac.kr

[2] BK 21 Computer Science Education Research Center, Korea University, 145 Anam-ro, Seongbuk-gu, Seoul 02841, Korea
K2j23h@korea.ac.kr

[3] Department of Computer Science, Korea National Open University, Dongsung-dong, jongno-gu, Seoul 110-791, Korea
kchung0825@knou.ac.kr

Abstract. In this paper, we build a chatbot that identifies the user's emotions, including newly-coined words and emoticons, for office workers. Chatbot categorizes emotions that appear in user input sentences into eight emotions. Because emotions are a very subjective area, interpretations of the same word can vary depending on the situation and person. In addition, the contextual characteristics of dialogue, not general sentences, can be taken into account to classify the correct emotions. To increase the accuracy of the emotional analysis, the subject category is determined and proceeded at the beginning of the conversation. User input sentences identify the intention of utterance through analysis of morphemes and sentence. To classify emotions of standard language by building emotional dictionaries based on the Korean language dictionary. For the emotional classification of newly-coined words and emoticons, use the newly-coined words and emoticons sentiment dictionaries. Use a model that shows high accuracy by comparing feature-based classification models with artificial neural network models. The emotional decision model anticipates the user's emotional state by extracting mood values that represent the overall emotion of the conversation. Based on the expected user's mood value, chatbot is personified by adding newly-coined words and emoticons to answers.

Keywords: Chatbot · Emotion-aware · Newly-coined word analysis · Emoticon analysis

1 Introduction

Chatbot is a compound word of chat and bot, which is a computer program that provides answers and related information to user questions. It helps users to shop or provides relevant information. According to the definition of the Korea Internet Information Society Agency, a chatbot refers to "artificial intelligence-based communication software" that

K. Arai (Ed.): FICC 2021, AISC 1363, pp. 1086–1096, 2021.
https://doi.org/10.1007/978-3-030-73100-7_75

responds through text conversations with people and provides appropriate answers to questions or various related information (National Information Society Agency (NIA), 2018). In general, chatbots are based on artificial intelligence technology to quickly and accurately provide users' desired content through appropriate answers. Major messenger companies such as Facebook, WeChat, Line, and Kakao have developed and released chatbot platforms to encourage more companies to use their platforms. However, most chatbots are only used for functional purposes without emotional exchange with users. Emotion is a psychological state that a person feels, and it plays a major role in a person's natural conversation and social relationship [1].

People want to communicate their feelings to the other person and exchange their feelings while having a conversation. When the chatbot recognizes the user's emotion and reacts to exchange emotions, the user can feel intimacy through a human-like reaction. However, studies on analyzing and designing chatbots based on emotion models that respond to human emotions are still insufficient. The newly-coined word is one of the Internet terms and is used not only within the Internet or SNS but also in the real world. Most of the newly-coined word spread quickly and disappear depending on the fashion or SNS trends. Due to the characteristics of these newly-coined words, it is difficult to analyze the meaning of words or the emotions implied because there is no standardized corpus. Newly-coined words and emoticons that are mainly used when communicating via messenger are excluded from morpheme analysis and syntax analysis. This lowers the accuracy of the user's intention and emotion analysis. In this study, we design and implement a chatbot based on emotion model applying emotion analysis including newly-coined words and emoticons. The accuracy of emotion analysis is improved by including newly-coined words and emoticons in emotion analysis. The chatbot that recognizes and responds to the user's emotions is expected to enable users to feel intimacy and increase satisfaction with the use of chatbots.

This paper is composed of four sections. In the introduction, we wrote about chatbots and emotion classification, and the classification method and dictionary construction method used in related works were prepared. In the proposed Intelligent Chatbot System, Sect. 3, a method of responding by grasping the user's intention and emotion through the sentence entered by the user was written. In conclusions and future direction, the limitations of the study and future work were written.

2 Related Works

In 1950, Alan Turing said that machines were intelligent if they couldn't tell whether they were machines or humans when they texted them, and this idea is called the starting point of "smart chatbots" by many. In 1966, the MIT Artificial Intelligence Research Center developed the first chatbot called "ELIZA". "ELIZA" was the level of simply matching user input and answer. Later, in 1995, ALICE chatbot handled conversations using pattern matching [2].

In the past, chatbot services simply used a pattern matching method through a web browser based on HTML to recognize predefined keywords and output a designated response. Due to the advantage of being able to access the mobile Internet, interest has shifted to mobile platforms, and chatbots provide services in the form of apps. Currently, it is changing from mobile SNS to messenger platform. The main change factor is the inconvenience and hassle of using a new app. Chatbots are also changing to a messenger platform-oriented service. The reason for the messenger-based service is that users can receive the service they want by sending a message from the chat screen on the messenger without installing a new app. Recently, chatbots that have been advanced to the next level using machine learning and deep learning have been serviced. The chatbot is implemented to output a previously created answer through a specific word in the input sentence. Some other chatbots are being implemented using sophisticated natural language processing. Recently, mobile messenger companies are providing intelligent platforms optimized for users through artificial intelligence chatbots. Chatbots use natural language processing technology and pattern recognition technology to understand and analyze user input sentences. Chatbots have been studied since the early days of artificial intelligence research, but recently, with the development of machine learning and deep learning technologies, they began to develop rapidly together. Chatbots are divided into two types: Retrieval model [3] and Generative model [4], depending on the implementation method. The Retrieval model is a search-based model in which humans directly define and use sentences. This is a method of comparing a keyword in a user input sentence with a keyword in a predefined sentence and outputting a predetermined answer. The advantage of the Retrieval model is that a defined answer is output, so it outputs a grammatically completed sentence. However, if a predefined sentence does not exist, the answer cannot be output. Generative models create new answers from scratch, not predefined answers. Since you create a new answer every time, your answers will diversify and you can feel like you are talking to someone. However, the quality of the generated answers cannot be guaranteed, and a lot of data is required for general conversational learning.

Currently, there are not many studies on emotional classification for newly-coined words and emoticons. In a mobile environment, users communicate their feelings in short sentences using newly-coined words and emoticons. Therefore, more accurate results can be obtained by performing emotion classification including newly-coined words and emoticons for future emotion classification. A newly-coined word is an abbreviation for a long word, and there are also words that are composed of irregular shapes or are created by bringing the shape of a popular object or person. Most of the newly-coined words are words that are not registered in the dictionary. Therefore, if a morpheme analysis based on a Korean dictionary is performed for natural language processing, the newly-coined word does not have any form of part of speech. This is because the corpus to be provided is insufficient due to the characteristics of newly-coined words that are easily generated and then easily disappeared. Therefore, it builds a corpus for the meanings and emotions represented by newly-coined words and emoticons that are mainly used on the Internet as like in Fig. 1. Stores words and meanings with more than 10 likes in the Naver Open Dictionary. If the randomly collected data includes words from the Naver Open Dictionary, newly-coined words and emoticons are selected as candidates. In addition, character-type emoticons composed of initial consonant, image-type emoticons, and 4-byte Unicode emoticons are also selected as candidates for newly-coined words and emoticons. The candidates for newly-coined words and emoticons construct a dictionary for newly-coined words and emoticons by directly selecting valid newly-coined words and emoticons. The constructed newly-coined words and emoticons sentiment dictionary are used to identify newly-coined words afterwards.

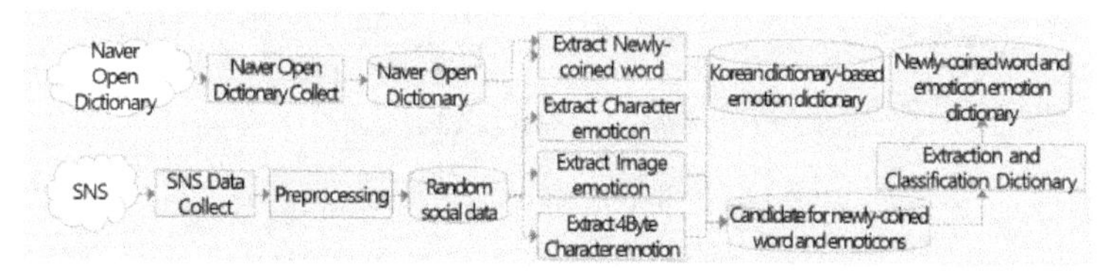

Fig. 1. Newly-coined word and emoticon emotion dictionary construction process.

3 Proposed Intelligent Chatbot System

Figure 2 is a structure of an emotion model-based chatbot including newly-coined words and emoticons, and has a structure of intention analysis and emotion analysis.

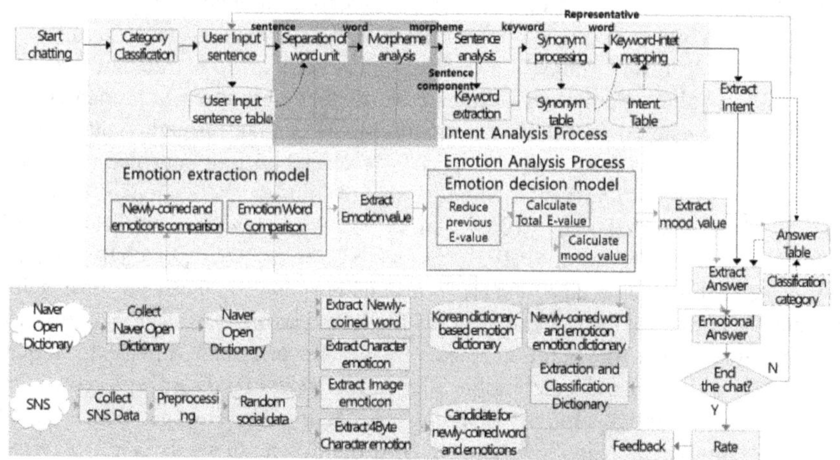

Fig. 2. Chatbot process.

The emotion model-based chatbot extracts emotions for user input sentences through the emotion extraction model. The classification method classifies the emotion of the sentence with the emotion value learned through the Word2vec algorithm. In addition, it is possible to emotionally respond to the user's emotional conversation through the emotion determination model.

Intent Analysis. When the chat starts, the topic of the chat is input and the category of the topic is determined in Fig. 3.

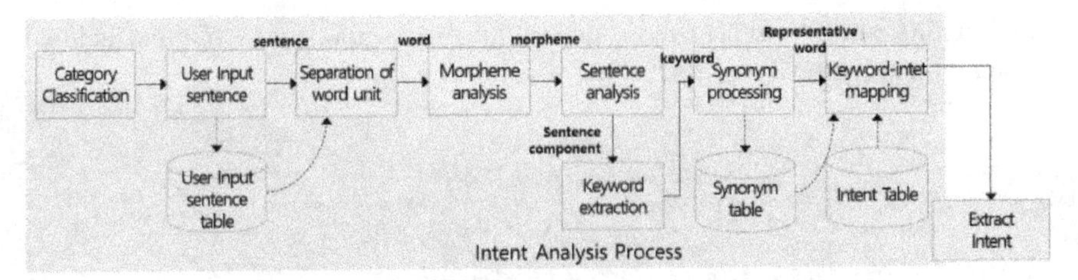

Fig. 3. Intent analysis process.

The subject of chat is determined by the user among the four types of interpersonal relationships, depression/anxiety, family, and sex. User input sentences are stored in the user input table. The elements of the user input table consist of token, sentence, and emotion. Users are classified according to tokens, user input sentences are stored in sentences, and sentiment values for sentences are extracted and updated in emotion. The user input sentence is separated into syntactic words and used to extract newly-coined words and emoticon emotion values. Syntactic words separated through morpheme analysis are separated into morphemes where each part of speech is tagged and used to extract sentimental words. For example, the user input sentence "I'm having a hard time because my boss talks so hard" is stored in the sentence in the user input table in the form of a sentence. Separated into syntactic word forms such as "I'm", "having", "a", "hard", "time", "because", "my", "boss", "talks", "so", "hard" they are used to extract new words and emoji emotion values. Sentiment words are words that can represent human emotions among words registered in the Korean dictionary. After that, through syntax analysis, the morphemes tagged with parts of speech are classified into sentence components with syntax tags according to the components constituting the sentence. When morpheme analysis is performed, "I/PRP", "m/VBP", "having/VBG", "a/DT", "hard/JJ", "time/NN", "because/IN", "my/PRP$", "boss/NN", "talks/NNS", "so/RB", "hard/JJ". The sentence component is tagged in the form. Chatbots use sentence analysis of subject, predicate, and object extracted as keywords for intention analysis. The intention value is a value representing the intention spoken by the user and is classified into 'request', 'command', 'warning', and 'celebration'. The extracted keywords are replaced with representative words through similarity/synonym processing. For example, data is stored in the pseudo/synonym table in the form of 'hard-difficult' and 'hard-demanding'. Among the user input sentences, the words 'demanding' and 'difficult' are replaced with the representative word 'hard' through similar/synonymous processing. The representative word is compared with an intention table composed of keywords and intention values. If a keyword is registered in the intention table, the representative word has the intention value, and the representative word not registered in the intention table does not have the intention value.

Emotion Extraction. The emotion value is a value obtained by numerically classifying emotions that a person feels in morphemes, words, newly-coined words, and emoticons as like in Fig. 4. The emotion values of newly-coined words and emoticons are extracted using the user input sentences separated by word units.

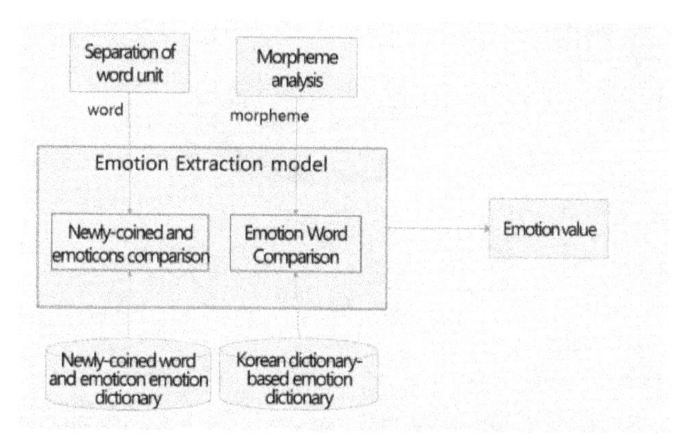

Fig. 4. Emotion extraction process.

The reason for separating and comparing by syntactic word unit is that in most cases, newly-coined words and emoticons are not listed in the Korean dictionary, so they are not recognized as morphemes during morpheme analysis. The chatbot separates the user input sentence into syntactic words and identifies newly-coined words and emoticons through a dictionary of newly-coined words and emoticons built in advance as like in Fig. 5. The chatbot extracts the emotion value when the separated syntactic word is registered in the newly-coined word and emoticon dictionary, and the newly-coined word and emoticon that are not registered in the dictionary cannot extract the emotion value. Newly-coined words and emoticons from which the emotion value has not been extracted are determined by the administrator and added to the newly-coined words and emoticons dictionary. All syntactic words are classified into morphemes with POS tags after morpheme analysis. The morphemes and words that can represent human emotions are used to construct an emotion dictionary based on a Korean dictionary and to classify emotions in the user's input sentences. The Korean language dictionary-based emotion dictionary refers to the sentiment polarity value of the Korean emotion analysis corpus (KOSAC), and the emotion classification is constructed by referring to Pluchick's emotion wheel. The morphemes with the POS tag are compared with the Korean dictionary-based emotion dictionary. The compared morphemes have emotion values based on the emotion words of the Korean dictionary-based emotion dictionary in Fig. 6.

The chatbot extracts the emotion value for the user input sentence by combining the emotion value of the newly-coined word and emoticon with the emotion value of the morpheme. The emotion value of the input sentence is stored by updating the corresponding emotion value to the sentence stored in the user input table. The emotion value of the input sentence is used as the input value of the emotion determination model. In the emotion determination model, the overall mood value of the conversation is calculated from the emotion values of sentences input by the user. The mood value is a predicted value of the emotion currently being felt by the user.

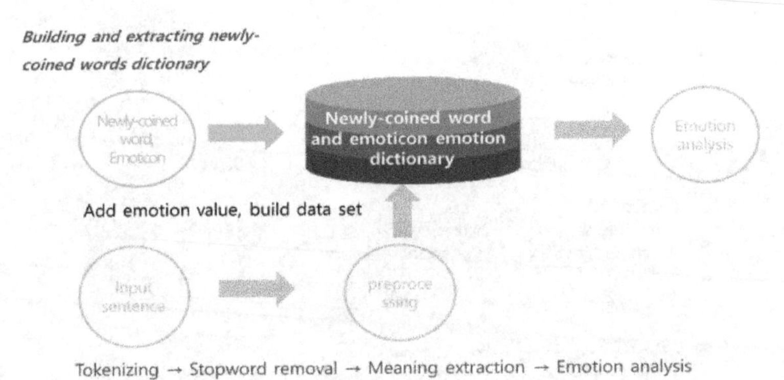

Fig. 5. Newly-coined word and emoticon emotion analysis process.

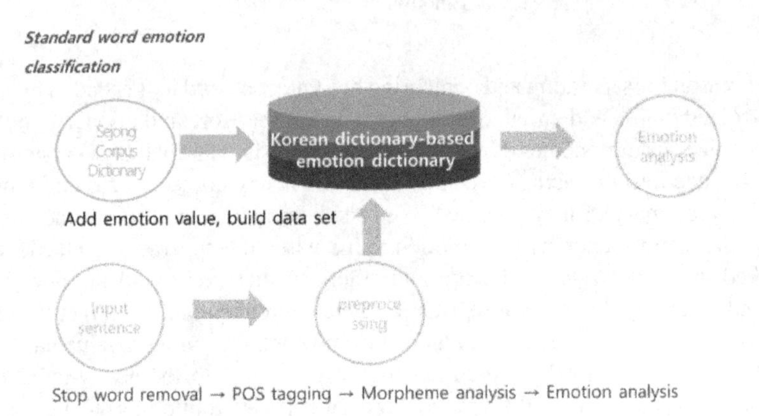

Fig. 6. Morpheme emotional analysis process.

Emotion Decision. The mood value of the emotion determination model is a value generated in the form of an array of emotions to be classified, and is expressed as each emotion and a corresponding value as like in Fig. 7.

The emotions within the mood value are expressed as emotion values, and the emotion values are expressed as emotion values as like in Fig. 8. When an input value comes into the emotion determination model, it is compared with the emotion value of the previous sentence in the user input table. If the same emotion is input continuously, the mood value is updated by increasing the corresponding emotion value by 2. If the same emotion is not continuously input, the corresponding emotion value is increased by 1. As the conversation progresses, the emotional value for the old input sentence is decreased by 0.1. When a continuous conversation is made, for example, if joy is used as an input value of the emotion determination model, the mood value has "joy-1.0, sadness-0, anger-0". Second, if sadness is used as an input value, it has a mood value of "joy-0.9, sadness-1.0, anger-0". If sadness is used as the third input value, since sadness has been input continuously, it increases by 2, and the emotion of the previous sentence decreases by 0.1 to have a mood value of "joy-0.8, sadness-2.9, anger-0". When the emotion value is 0.4 or higher, the corresponding emotion (weak) is output, and when the emotion value is 3.5 or higher, the corresponding emotion (strong) is output. When deciding emotions, strong emotions are prioritized to determine emotions. When there is more than one strong emotion, the emotion value is compared and the emotion value is determined as an emotion value having a larger emotion value. If there is more than one weak emotion without strong emotion, the same appraisal value is compared to determine the mood value. If the emotion value of all emotion values is less than 0.4, a "stable" emotion is output to express a neutral mood that is not biased against any emotion. The emotion determination model is updated every time the emotion value extracted from the user input sentence is used as an input value until the chat ends.

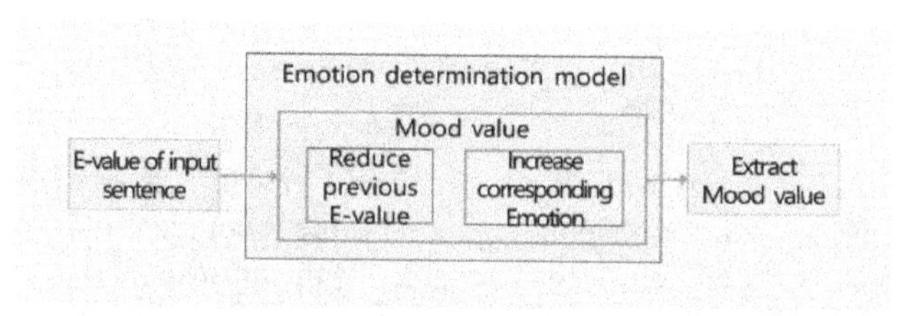

Fig. 7. Emotion determination model.

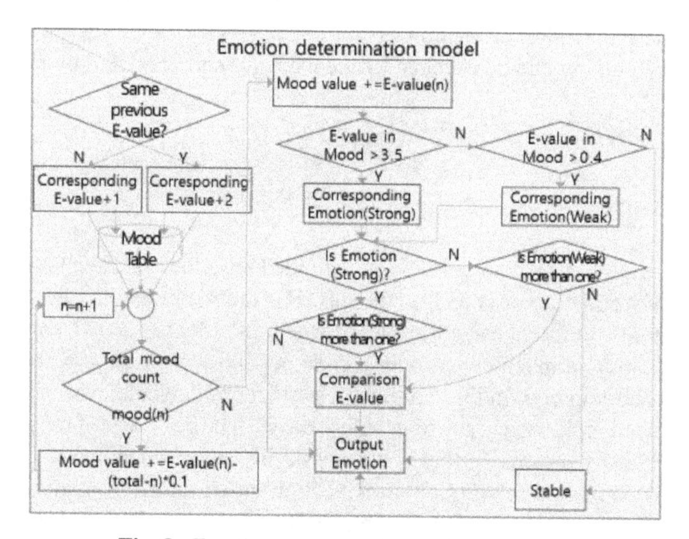

Fig. 8. Emotion determination model algorithm.

Answer Decision. Using the intent value extracted from the intention analysis process, the mood value extracted from the sentiment analysis process, and the category determined at the start of the chat, the chatbot extracts the appropriate answers to the user-entered statements from the answer table built for the answers to user input sentences as like in Fig. 9.

In the absence of an appropriate answer, "I don't know what you mean by it.", "Can you tell me exactly again?" The exception handling answer is displayed. When an

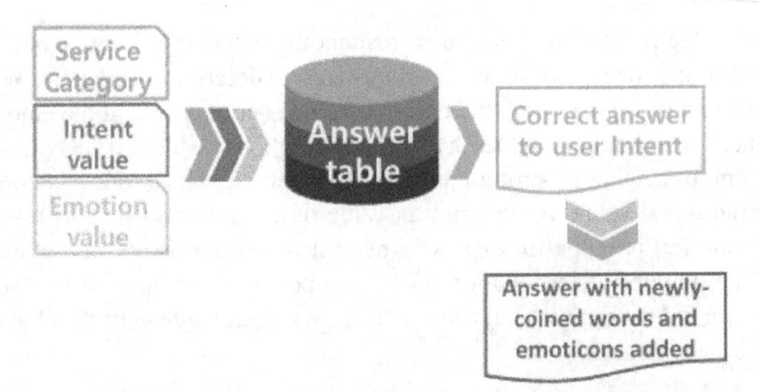

Fig. 9. Answer determination model process.

appropriate answer is determined, an emotional answer with a newly-coined word and emoticon appropriate to the corresponding mood value is output to the user. In addition, the newly-coined words and emoticons suitable for the corresponding mood value are extracted from the newly-coined words and emoticons emotion dictionary. The intention to end the chat is asked to the user through the number of input sentences of the user or a word indicating the end of the conversation. When the chatbot ends, the chatbot asks the user to rate the satisfaction of the conversation. Conversations with a satisfaction level below a certain number can be checked by the manager and corrected by making up for the deficiencies.

4 Conclusion

In this paper, a study was conducted for subdivided emotion classification of text data including newly-coined words and emoticons. For the detailed classification of emotions, a system of 8 emotion categories was defined. In order to classify emotions, naive Bayes classification, a machine learning technique, and word2vec, a word embedding technique, were compared. In the sentiment classification of sentences and paragraphs, word2vec showed higher accuracy than Naive Bayes. The accuracy of emotion classification is expected to improve further when classifying emotions in newly-coined words and emoticons compared to emotion classification based on the Korean dictionary. In order to improve the accuracy of emotion classification, a method with high accuracy was applied to the chatbot, including emotion analysis of newly-coined words and emoticons in both methods A newly-coined word and emoticon emotion analysis algorithm is applied to the chatbot to analyze the emotion of the user input sentence. The analyzed emotion was used as an emotion value as a temporary emotion when the user inputs a sentence. In order to measure the emotions of chatbot users, an emotion model was applied to the chatbot. The user's emotion measured in the emotion model is expressed as a mood value. The mood value is a value created in the form of an array of emotions to be classified, expressed in terms of each emotion and its corresponding numerical value, and is used to determine the answer of the chatbot. The emotion model used the emotion value of the user input text as an input value. The mood value increases the emotional value equal to the input emotion value. The emotion model expressed that emotion weakened over time by decreasing the emotion value of the assigned emotion value of the sentence with the old input sentence. In addition, when the same emotion value is continuously input, it expresses a strong emotion by increasing the increase in the emotion value. The emotion model was applied to the chatbot to measure the user's emotion in real time. Most chatbots are used only functionally without emotions as a function of order or guidance, making it difficult to exchange emotions with users. If the chatbot empathizes with the user's emotions and reacts with emotions that fit the situation, the user will be able to feel intimacy through anthropomorphic reactions. In this paper, we implemented a chatbot targeting office workers by applying an emotion analysis algorithm including newly-coined words and emoticons and an emotion model that predicts the user's mood. Chatbots provide user input sentences through morpheme analysis and sentence analysis, and extract subjects, predicates and objects as keywords. By comparing the extracted keywords with the intention table built, the intention indicated by the user input sentence is identified, and an answer appropriate to the intention is provided. The chatbot adds newly-coined words and emoticons that have the same emotion value as the emotion value measured by the emotion model to the answer. By designing and implementing an emotion model-based chatbot that empathizes with the user's emotions and reacts emotionally to the situation, using the emotion model, the user's intimacy was enhanced. The significance of the emotion classification chatbot including newly-coined words and emoticons proposed in this paper is as follows. First, machine learning and word embedding techniques were applied to Korean emotion classification and compared. Naive Bayes/word2vec offers a suitable method for conversations such as chat. Second, through the classification of emotions including newly-coined words and emoticons, the accuracy of emotion classification has been improved on SNS and chats where newly-coined words and emoticons are frequently used. It can be used as basic data for a study of a chatbot that analyzes and responds to the user's emotions in the trend of changing service types to SNS and messenger types.

But, there are limits that do not allow users to classify emotions other than the eight emotional categories that they set to classify in the study. And because each person has different levels of satisfaction and closeness, it is necessary to measure how much satisfaction and closeness a chatbot that recognizes and empathizes with the user's feelings through experiments.

In this paper, a method of classifying user's emotions into eight is proposed. However, the emotions a person feels are more complex, and various emotions can be felt at the same time. In addition, even in the same sentence, the level of emotion felt as soon as the user is different. Therefore, we will advance the emotion determination model in order to improve the classification method and classification accuracy that can classify more emotions. And we will conduct research on how different emotions can be mixed to express different emotions, such as 'anger + disgust = contempt'.

References

1. Yang, J.S., Ko, M.-S., Chung, K.S.: Social emotional opinion decision with newly coined words and emoticon polarity of social networks services. Future Internet **11**(8), 165 (2019)

2. Lee, S.: A study on the implementation of a chatbot based on an emotion model. Domestic Master's Thesis Soongsil University Graduate School of Software Specialization, Seoul (2018)
3. Kim, S., Jo, S.: An improved interactive helper agent using approximate pattern matching. J. Korean Inf. Sci. Soc. **28**(1B), 415–417 (2001)
4. Kim, M.-G., Cho, D., Kim, H.: Generative model chatbot technology analysis based on neural network. J. Korean Inst. Commun. Sci. 211–212 (2017)

Analyzing Performance of Classification Algorithms in Detection of Depression from Twitter

Aritra Bandyopadhyay(✉) and K. Manjula Shenoy

Manipal Institute of Technology, Manipal Academy of Higher Education, Manipal, India

Abstract. Depression is a mental disorder that is characterized by a general mood or feeling of low self-esteem, loss of interest towards daily activities and low energy within a particular person. It is a very serious mental condition and its automatic detection through online social media platforms like Twitter could help identifying depressed individuals remotely. This paper suggests a novel method to extract tweets indicating depression using word lists. Various classification algorithms like SVM, KNN, Naive Bayes and Random Forests have been used to classify the individual tweets as to whether they indicate depression in the subject or not. Metrics like F1-Score has been used the verify and compare the results of the models using an unseen test dataset.

Keywords: Depression · Natural Language Processing · Naive Bayes · Random forest · Support Vector Machine

1 Introduction

Depression is a type of mental disorder that is described as a constant and prolonged feeling of sadness, loss, or anger. It is a very serious mental condition and can adversely affect the mental health of a person. According to WHO more than 264 million individuals are adversely affected by depression. Approximately 800,000 people die from committing suicide every year. The main cause for committing suicide is depression in most of the cases.

With the recent surge is computational power and availability of data, a lot of research has been done in the field of sentiment analysis using machine learning algorithms [1–3]. Detecting depression is also a very popular area of research in the field of sentiment analysis [4–6]. Twitter is a 'microblogging' system that allows a user to send and receive short posts of up to 140 characters in length called tweets. Links to websites and other resources can also be included in a tweet. Twitter also allows a user to share or retweet tweets of other users. This feature allows tweets to be shared among users very quickly and efficiently.

While a lot of work has been done on detection of tweets containing hate speech and detection of tweets expressing negative emotions, very few works

K. Arai (Ed.): FICC 2021, AISC 1363, pp. 1097–1106, 2021.
https://doi.org/10.1007/978-3-030-73100-7_76

attempt to identify depression in users using only a single tweet that they have posted [7–9]. This is due to the lack of a proper dataset that contains tweets that indicate depression as most datasets contain a set of tweets posted by a user who is identified as a depressed user. We have tried a different approach in this paper as we try to identify whether a single given tweet indicates depression in the user without having any information about the user who posted the tweet or the other tweets the same user had posted. We have solved this issue by using word lists to extract tweets that may indicate depression in the author of the tweet. Tweets are a very informal method of posting information and thus they contain a lot of spelling errors and abbreviated words. Thus after a few steps involving data cleaning and preprocessing we train several popular machine learning classification algorithms on our data to predict whether a tweet indicates depression or not.

2 Related Works

Many researchers have worked on detecting depression using machine learning algorithms. A general discussion about the methodology researchers should follow while using Natural Language Processing or text mining methodologies to identify a particular sentiment or opinion is described in [10]. Some works attempt to detect symptoms of depression using text data obtained from social media platforms like internet forums and online chat rooms [11].

Depression can be detected from various types of data about the subject. De Melo et al. [12] detect depression using video of the subject as input into a deep distribution learning model. Some works use video, audio and text transcripts of interviews with the patient to determine whether a patient may be suffering from depression or not with the help of a bi-directional LSTM network [13]. While other researchers have treated speech data from subjects as a sequence of acoustic events to help them detect depression [14].

Our paper deals with detecting depression from twitter data. A few papers have been published in this field as well. Alexander et al. [15] discusses the challenges of mining data from microblogging websites like Twitter for performing any kind of sentiment analysis. Some researchers have used Deep Convolutional Neural Networks to conduct sentiment analysis based on data extracted from twitter [16].

3 Depression Dataset

3.1 Data Collection

Data collection for a problem such as detecting depression is quite challenging. There exist numerous open source datasets regarding tweets indicating negative sentiments. But only a small subset of tweets indicating negative sentiments are actually tweets indicating depression. Due to the unavailability of tweets indicating depression a lot of work has been done by extracting tweets which have the word "depression" in them. This method may be inefficient as there are many tweets that do not contain the word "depression" but are still extremely indicative of depression. An example of this could be the following tweet "I am feeling very hopeless and anxious. Plz want to die" which is surely indicative of depression but does not contain the exact word "depression" in it. Thus, we have used a slightly more intuitive method of collecting data regarding tweets indicating depression. We made a list of 27 words that depressed people are much more likely to use by referring articles and research papers that discuss this very issue. These words were divided into two categories which included important words and unimportant words. Important words were those words whose presence in a tweet was a very strong indicator of depression. While unimportant words were those words whose presence by themselves may or may not indicate depression. But if multiple unimportant words were found along with a few important words in a tweet then the tweet had a high chance of indication depression.

The Important Word List (list of four words whose presence in a tweet a tweet is a strong indicator of depression in the author of the tweet) include the words like:

- Depression
- Depressed
- Anxiety
- Suicide

The Unimportant Word List (list of 23 words whose presence in a tweet a tweet is a weak indicator of depression in the author of the tweet) includes words like:

- Anxious
- Sad
- Sorrow
- Despair
- Grief
- Hopelessness
- Helpless
- Abuse
- Alone
- Lonely
- Solitary
- Anger
- Anguish
- Fear
- Terrified
- Afraid
- Hopeless
- Panic

- Struggle
- Tragic
- Tragedy
- Troubled
- Worry

An advanced Twitter scraping tool called Twint was used to extract the depressive tweets. A tweet is considered to be indicating depression if it contains more than one important word or it contains 1 important word and at least 1 unimportant word or it contains at least three unimportant words. All tweets collected were posted between 1st January, 2015 and 31st December, 2016. Such tweets were added to the depressive tweets list. This made up our dataset of 915 depressive tweets. The non-depressive tweets were randomly sampled positive tweets from the Sentiment140 dataset [17]. We randomly sampled 2000 of the positive tweets to include non-depressive tweets in our dataset.

Fig. 1. Word Cloud of tweets that do not indicate depression

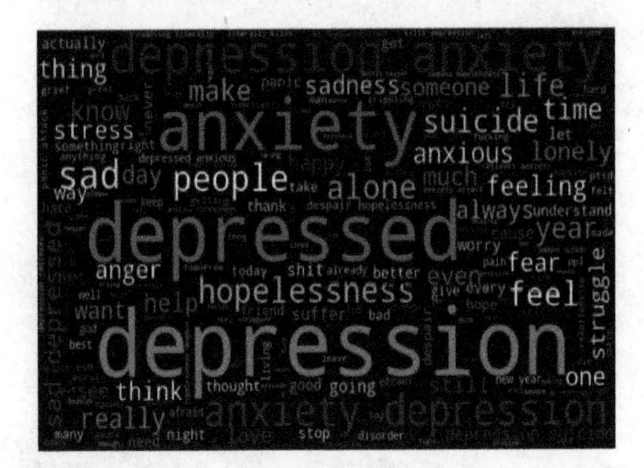

Fig. 2. Word Cloud of tweets that indicate depression

3.2 Data Preprocessing

Text data by itself cannot be used in any machine learning algorithm. They need to be converted to some numeric form by using a few NLP methods. Tweets contain certain hashtag words which is a word or phrase preceded by a hashtag (#), used within a message to identify a keyword or topic of interest and facilitate a search for it. All hashtag words in the tweets in our dataset has been considered as normal words appearing at the end of a sentence.

Text data contains stopwords, which are very common words used in any language that do not contain any relevant information about the context of the text. Such words are usually removed from the data in most Natural Language Processing projects. Here we have removed stopwords from the tweets that are contained in the stopword list of the nltk module. Another commonly used algorithm to preprocess textual data is stemming. Stemming reduces a word to its root word by removing the suffix. It ensures that words having the same meaning but having a different tense is represented as a single root word instead of two separate words. We have applied Porter Stemmer from the nltk library to implement stemming in our data.

As mentioned earlier, any text data must be converted and represented in some sort of numerical structure as computers cannot work with raw text data. We have used Count Vectorizer provided by the nltk library to achieve the same. The Count Vectorizer creates a vector for each tweet in which each element in the vector is the number of times a particular token occurs in the tweet. After preprocessing is completed we created wordcloud diagrams in Fig. 1 and Fig. 2 to confirm that our dataset approximately reflects the problem we are solving. The code for this paper which is in Python is publicly available in the github repository[1] given below.

4 Classification Techniques

Once the tweets have been preprocessed, we applied various popular machine learning classification algorithms to try and predict whether a given tweet is indicative of depression in the person who wrote that tweet or not. First we randomly sample 90% of tweets in the total dataset and we set this aside for training our various models. The remaining 10% used for testing the accuracy of our classification task. This is done so that the same data that is used to train the model is not used to test its accuracy as well. We trained and tested the accuracy of the following algorithms.

4.1 Naive Bayes Classifier

The Naive Bayes Classifier is one of the most simple yet effective algorithms to perform any kind of classification from textual data [18]. It assumes that each feature vector is conditionally independent of all other vectors within a given

[1] https://github.com/aritraban21/Depression-Twitter.

class. It predicts the conditional probability of a class given the set of evidences. The Naive Bayes algorithm is used to calculate the Posterior probability as shown in Eq. 1 where $P(Y|X)$ is the Posterior probability, $P(Y)$ is Prior probability and $P(X)$ is called evidence.

$$P(Y|X) = \frac{P(Y)\prod_i P(X_i|Y)}{P(X)} \tag{1}$$

In our paper we have used Multinomial Naive Bayes which is a slightly modified version of Naive Bayes Classifier and is shown to work well on text classification problems [19].

4.2 Random Forest

Decision trees form the basis of a random forest classification model. In a Decision Tree, each node tries to split the data such that the resulting subsets of the data are as different from each other as possible. It achieves this by splitting the dataset so that it causes the greatest reduction in the Gini Impurity which is the probability that a randomly chosen sample in a node would be incorrectly classified if it was labelled by the distribution of samples in the node. Equation 2 is used to compute the Gini impurity at a node n where J is the total number of classes and p_i is the ratio of the number of samples belonging to class i to the total number of samples present at node n. It repeats this process recursively until it reaches a node that contains samples from a single class.

$$I_G(n) = 1 - \sum_{i=1}^{J} (p_i)^2 \tag{2}$$

A single Decision Tree has a tendency to overfit the dataset as it tries to minimize the number of samples that are misclassified. Thus we use a Random Forest which is a collection of multiple Decision Trees. Each tree randomly samples data from the training dataset for learning. This is called bootstrapping and this helps in generating very different Decision Trees as Decision Trees are very sensitive to the data they are trained on. Only a subset of all features are considered for splitting each node in each Decision Tree which forces even more variation within the generated trees. The prediction of each decision tree in a random forest is averaged together to get the prediction of the random forest model.

4.3 Support Vector Machine

Support Vector Machines or SVM are one of the most popular classification algorithms used in machine learning. SVM tries to draw a hyperplane between groups of datapoints belonging to multiple classes so that it can classify any new datapoint using this hyperplane. It also ensures to find a plane that maximizes distance between the datapoints of several classes. It achieves this by using support vectors that are datapoints of each class that are closest to the hyperplane. Figure 3 shows exactly how a hyperplane works to classify data points.

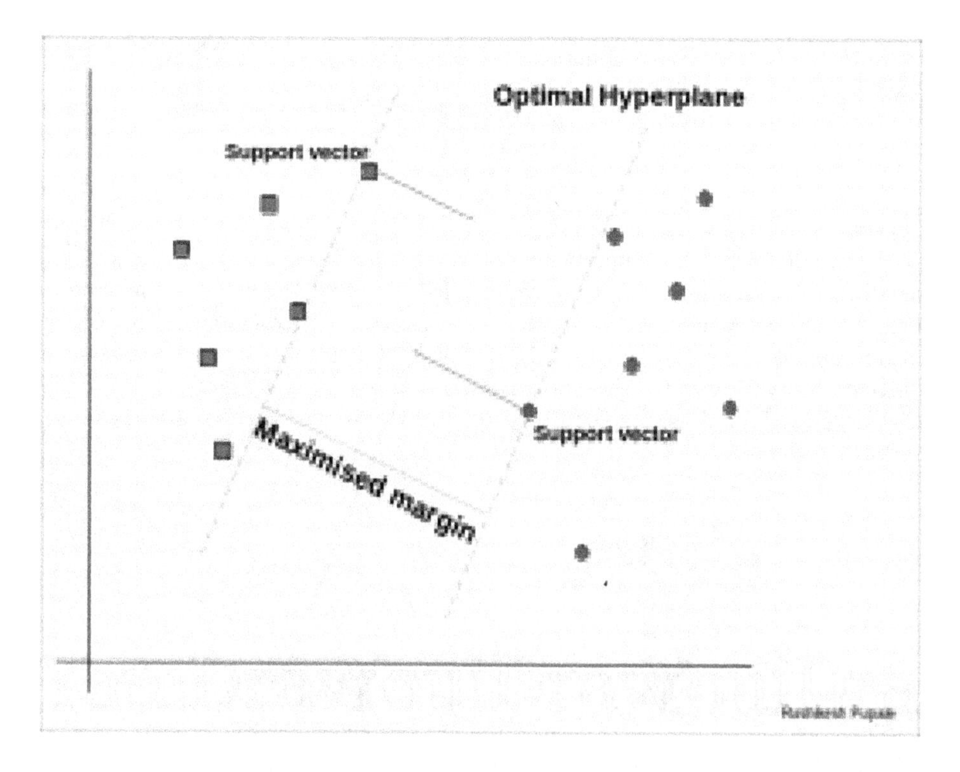

Fig. 3. SVM hyperplanes and support vectors

4.4 K Nearest Neighbours

The K Nearest Neighbour algorithm or KNN assumes that data points that are in close proximity to each other belong to the same class. Given a test data example we predict the class it belongs to by selecting the closest k neighbours to the test datapoint and predicting that class which occurs most frequently among the k closest data points. Here the closest k neighbours are determined by the Euclidean distances between the data points. The value of k is a hyper parameter in this algorithm and after testing a wide range of values we chose k to be 3 as this value gave our model the best accuracy.

5 Results

All the above models were trained on the same training data. The classification performance of these models were measured using the unseen test data. Metrics like Accuracy, Precision, Recall and F1-Score were used to compare the performance of the models on the test data. Accuracy is the ratio between correctly predicted observations to the total number of observations as shown in Eq. 3

where TP, TN, FP, FN are number of True Positives, True Negatives, False Positives and False Negatives in the test data, respectively.

$$Accuracy = \frac{TP + TN}{TP + FN + FP + TN} \quad (3)$$

However, Accuracy only works as a proper metric when the multiple classes have approximately same number of observations. Thus we use Precision and Recall to solve the problem of unsymmetrical datasets. Precision is the ratio of correctly predicted positive observations to the total predicted positive observations and is shown in Eq. 4 while Recall is the ratio of correctly predicted positive observations to all the positive observations and is shown in Eq. 5. F1 score is the weighted average of Precision and Recall and is calculated is shown in Eq. 6.

$$Precision = \frac{TP}{TP + FP} \quad (4)$$

$$Recall = \frac{TP}{TP + FN} \quad (5)$$

$$F1 - Score = \frac{2(Precision \times Recall)}{(Precision + Recall)} \quad (6)$$

The confusion matrices of all the models are shown in Fig. 4. Here, 0 label indicates "depression" while 1 label indicates "no depression". The comparison of the performance of the above models is shown in Table 1.

Table 1. Table Showing Performance Metrics of All Models

Models	Accuracy	Precision	Recall	F1-Score
Multinomial Naive Bayes	95.89	87.88	100.00	93.55
Random Forest	99.66	98.86	100.00	**99.43**
KNN	91.44	98.44	72.41	83.44
SVM	99.66	100.00	98.85	**99.42**

Both the Support Vector Classifier and the Random Forest Classifier outperformed the other two models significantly as shown by their high F1-Score values in Table 1. Thus, we can conclude that these two models should be used to correctly predict whether a single given tweet indicates depression in the author of the tweet or not.

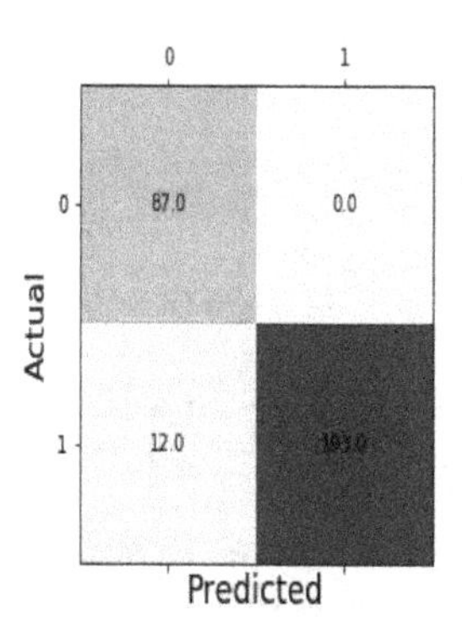

(a) Confusion Matrix for Naive Bayes model

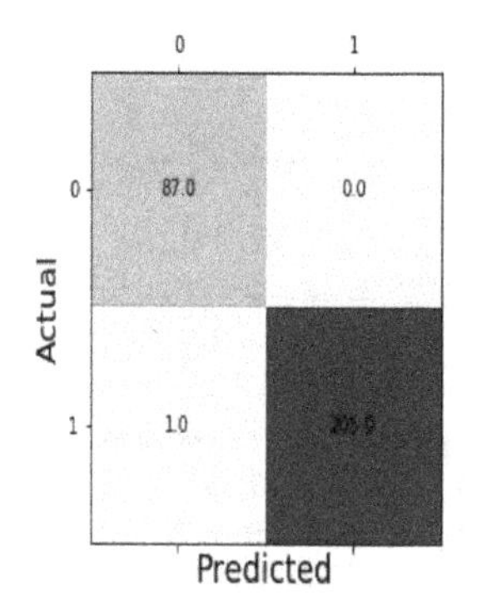

(b) Confusion Matrix for Random Forest model

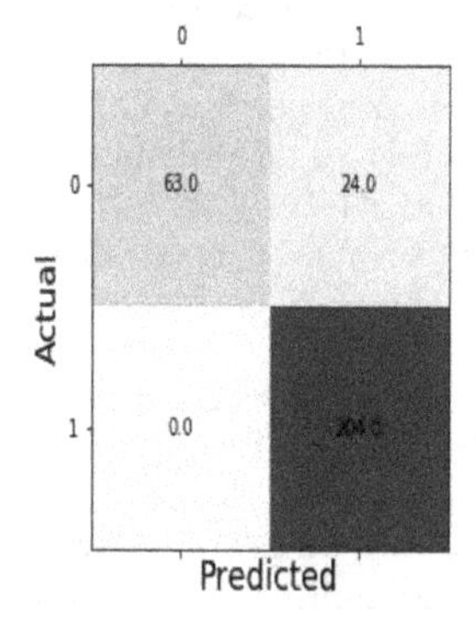

(c) Confusion Matrix for KNN model

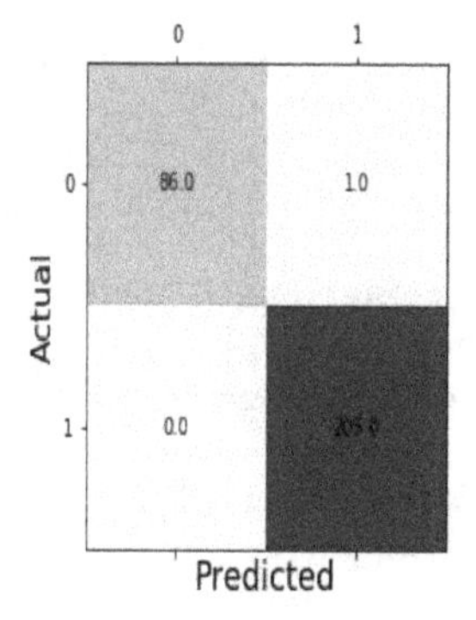

(d) Confusion Matrix for SVM model

Fig. 4. Confusion matrices for all models

6 Conclusion

There has been an exponential increase in awareness about mental health disorders over the last decade. Since a lot of people are very active in social media platforms, it is extremely important for us to be able to detect depression through the social media activities of an individual. In this paper we show how machine learning algorithms like Support Vector Machines or Random Forests can be used to detect depression using only a single tweet from any particular user. The proposed method can help in identifying depressed twitter users and providing them with the mental help and counselling they may require desperately, thus making the people of this world a lot more healthier mentally.

References

1. Nakov, P., Ritter, A., Rosenthal, S., Sebastiani, F., Stoyanov, V.: SemEval-2016 task 4: sentiment analysis in Twitter. arXiv preprint arXiv:1912.01973 (2019)

2. Cambria, E.: Affective computing and sentiment analysis. IEEE Intell. Syst. **31**(2), 102–107 (2016)
3. Hussein, D.M.E.D.M.: A survey on sentiment analysis challenges. J. King Saud Univ.-Eng. Sci. **30**(4), 330–338 (2018)
4. Tao, X., Zhou, X., Zhang, J., Yong, J.: Sentiment analysis for depression detection on social networks. In: International Conference on Advanced Data Mining and Applications, pp. 807–810. Springer, Cham (2016)
5. Jung, H., Park, H.A., Song, T.M.: Ontology-based approach to social data sentiment analysis: detection of adolescent depression signals. J. Med. Internet Res. **19**(7), e259 (2017)
6. Islam, M.R., Kamal, A.R.M., Sultana, N., Islam, R., Moni, M.A.: Detecting depression using K-nearest neighbors (KNN) classification technique. In: 2018 International Conference on Computer, Communication, Chemical, Material and Electronic Engineering (IC4ME2), pp. 1–4. IEEE, February 2018
7. Burnap, P., Williams, M.L.: Us and them: identifying cyber hate on Twitter across multiple protected characteristics. EPJ Data sci. **5**(1), 11 (2016)
8. Basile, V., Bosco, C., Fersini, E., Debora, N., Patti, V., Pardo, F.M.R., Sanguinetti, M.: Semeval-2019 task 5: multilingual detection of hate speech against immigrants and women in twitter. In: 13th International Workshop on Semantic Evaluation, pp. 54–63. Association for Computational Linguistics (2019)
9. Sreeja, I., Sunny, J.V., Jatian, L.: Twitter sentiment analysis on airline tweets in india using R language. In: Journal of Physics: Conference Series, vol. 1427, no. 1, p. 012003. IOP Publishing, January 2020
10. Zucco, C., Calabrese, B., Cannataro, M.: Sentiment analysis and affective computing for depression monitoring. In: 2017 IEEE International Conference on Bioinformatics and Biomedicine (BIBM), pp. 1988–1995. IEEE, November 2017
11. Karmen, C., Hsiung, R.C., Wetter, T.: Screening Internet forum participants for depression symptoms by assembling and enhancing multiple NLP methods. Comput. Methods Programs Biomed. **120**(1), 27–36 (2015)
12. De Melo, W.C., Granger, E., Hadid, A.: Depression detection based on deep distribution learning. In: 2019 IEEE International Conference on Image Processing (ICIP), pp. 4544–4548. IEEE, September 2019
13. Al Hanai, T., Ghassemi, M.M., Glass, J.R.: Detecting depression with audio/text sequence modeling of interviews. In: Interspeech, pp. 1716–1720, September 2018
14. Huang, Z., Epps, J., Joachim, D., Sethu, V.: Natural language processing methods for acoustic and landmark event-based features in speech-based depression detection. IEEE J. Sel. Top. Signal Process. **14**(2), 435–448 (2019)
15. Pak, A., Paroubek, P.: Twitter as a corpus for sentiment analysis and opinion mining. In: LREc, vol. 10, no. 2010, pp. 1320–1326, May 2010
16. Severyn, A., Moschitti, A.: Twitter sentiment analysis with deep convolutional neural networks. In: Proceedings of the 38th International ACM SIGIR Conference on Research and Development in Information Retrieval, pp. 959–962, August 2015
17. Alec, G., Bhayani, R., Huang, L.: Sentiment140. Site Functionality (2013c). http://help.sentiment140.com/site-functionality. Accessed 20 2016
18. Rish, I.: An empirical study of the naive bayes classifier. In: IJCAI 2001 Workshop on Empirical Methods in Artificial Intelligence, vol. 3, no. 22, pp. 41–46, August 2001
19. Singh, G., Kumar, B., Gaur, L., Tyagi, A.: Comparison between multinomial and bernoulli Naïve Bayes for text classification. In: 2019 International Conference on Automation, Computational and Technology Management (ICACTM), pp. 593–596. IEEE, April 2019

Author Index

K. Arai (Ed.): FICC 2021, AISC 1363, pp. 1107–1110, 2021.
https://doi.org/10.1007/978-3-030-73100-7

Z

CPSIA information can be obtained
at www.ICGtesting.com
Printed in the USA
LVHW010844190421
684889LV00001B/1